RF CIRCUIT DESIGN

WILEY SERIES ON INFORMATION AND COMMUNICATIONS TECHNOLOGIES

Series Editors: Russell Hsing and Vincent K. N. Lau

The Information and Communications Technologies (ICT) book series focuses on creating useful connections between advanced communication theories, practical designs, and end-user applications in various next generation networks and broadband access systems, including fiber, cable, satellite, and wireless. The ICT book series examines the difficulties of applying various advanced communication technologies to practical systems such as WiFi, WiMax, B3G, etc., and considers how technologies are designed in conjunction with standards, theories, and applications.

The ICT book series also addresses application-oriented topics such as service management and creation and end-user devices, as well as the coupling between end devices and infrastructure.

T. Russell Hsing, PhD, is the Executive Director of Emerging Technologies and Services Research at Telcordia Technologies. He manages and leads the applied research and development of information and wireless sensor networking solutions for numerous applications and systems. Email: thsing@telcordia.com

Vincent K.N. Lau, PhD, is Associate Professor in the Department of Electrical Engineering at the Hong Kong University of Science and Technology. His current research interest is on delay-sensitive cross-layer optimization with imperfect system state information. Email: eeknlau@ee.ust.hk

Wireless Internet and Mobile Computing: Interoperability and Performance
Yu-Kwong Ricky Kwok and Vincent K. N. Lau

RF Circuit Design
Richard C. Li

Digital Signal Processing Techniques and Applications in Radar Image Processing
Bu-Chin Wang

RF CIRCUIT DESIGN

RICHARD CHI-HSI LI

WILEY

A JOHN WILEY & SONS, INC., PUBLICATION

Published by John Wiley & Sons, Inc., Hoboken, New Jersey.
Published simultaneously in Canada.

ISBN 978-0-470-16758-8

Printed in the United States of America.

10 9 8 7 6 5 4 3 2 1

CONTENTS

PREFACE

I have worked on *RF* circuit design for more than 20 years. My motivation in writing this book is to share my *RF* circuit design experience, both successes and failures, with other readers. This book is aimed at *RF* circuit designers, and is organized into three parts, as shown in the figure.

PART I: INDIVIDUAL *RF* BLOCKS (CHAPTERS 1 TO 7)

There are many good books about *RF* circuit design on the market. Their arrangement is usually longitudinal, emphasizing the operating principles of individual blocks. These individual blocks include the *LNA*, mixer, filter, *VCO, PA*, and so on.

I have followed the longitudinal pattern shown in the figure:

- *LNA* (Chapter 1);
- Mixer (Chapter 2);
- Differential pair (Chapter 3);
- Balun (Chapter 4);
- Tunable filter (Chapter 5);
- *VCO* (Chapter 6);
- *PA* (Chapter 7).

Rather than emphasizing the operating principles, I provide practical engineering design examples, most of which I designed myself.

In Chapter 1, a new design procedure is presented. During the 1990s, I found that the maximum gain and minimum noise figure can be achieved simultaneously in the *LNA* design. This has been applied in many design products in my engineering projects; however, the positive results have not been previously published. I did

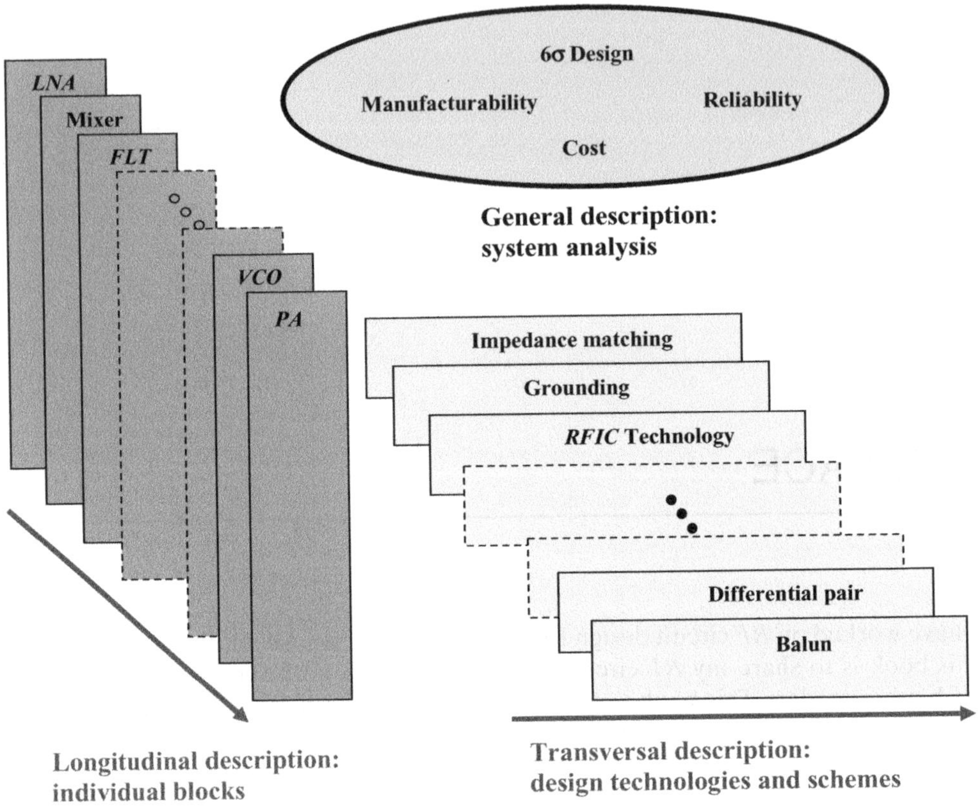

Three kinds of descriptions of *RF* circuit design.

introduce them in my lectures in recent years and received encouraging responses from the audiences.

In Chapter 4, the transformer balun and *LC* balun are emphasized. In past years, I have always used the special transformer balun with a ratio of $1{:}\sqrt{2}$ in my circuitry simulations. In this special transformer balun, I found an advantage. The simulation for a circuit with a differential pair configuration can be replaced and "interpreted" by the simulation for a circuit with a single-ended configuration. In the test lab, I prefer to apply the *LC* balun because of its simplicity, ease, and reliability. I developed the design equations in 1992, and they have been applied in practical engineering designs for many years, although no papers have been presented at conferences or published.

The content of Chapter 5 is abstracted from my U.S. patent. The bandwidth of a tunable filter can be kept unchanged over the entire expected frequency range only if the main coupling element is an inductor, not a capacitor or a combination of an inductor and a capacitor. In addition, a very deep imaginary rejection "zero" can be created by a small capacitor. This work proves that the performance of individual blocks can be greatly improved by means of simple schemes, even though the tunable filter was designed by engineers some 50 years ago.

PART II: DESIGN TECHNOLOGIES AND SCHEMES (CHAPTERS 8 TO 17)

As shown in the figure, the transversal description chapters contain four elements: impedance matching (Chapters 8 to 13), grounding and current coupling (Chapters 14 and 15), *RFIC* and *SOC* (Chapter 16), and 6σ design for manufacturability of product (Chapter 17).

As per my *RF* circuit design experience, understanding the operating principles of *RF* circuit blocks is much easier than developing a 6σ design for an *RF* module or *RFIC* chip; consequently, familiarity with the four basic technologies and schemes, including impedance matching, grounding, *RFIC* and *SOC*, and 6σ design, is essential. They are the basic requirements and "must" conditions for a qualified *RF* circuit designer.

Why is impedance matching technology so important? Because

- The main task of the *RF* circuit block is power transportation or manipulation, while the main task of the digital circuit block is status transportation or manipulation.
- Power transportation or manipulation is directly related to impedance matching. The necessary and sufficient condition for optimized power transportation or manipulation is the conjugate matching of input and output impedance between the *RF* blocks.
- Consequently, impedance matching must be done for almost all *RF* blocks. The only exceptions are:
  - The output of the oscillator or *VCO* and the input of the *VCO* buffer;
  - The *IF* digital input/output of the *RF* modulator/demodulator.

Impedance matching is a challenging task in a *UWB* system today. It is the core of *RF* circuit design technologies. That is why I have devoted one third of the book to this topic. My contribution to impedance matching is the division of the Smith chart into four regions so that the impedance matching network built by two parts can be directly designed in terms of couple equations. Chapter 4, which discusses impedance matching in the wide-band case, is abstracted from my recent research work on the *UWB* system. I think that my methodology for wide-band impedance matching is unique, and, again, it has not been previously published.

Why is grounding so important? Because

- At the *RF* frequency range, a metallic surface with good conductivity is very often not equipotential.
- At the *RF* frequency range, the ground points at two ends of a good *RF* cable are in most cases not equipotential.
- Very often, return current coupling on ground surface is ignored or de-emphasized.
- In today's *RFIC* design, the "zero" capacitor is a bottleneck.

Why are the *RFIC* and *SOC* so important? Because

- Compared with *RF* module design by discrete parts, *RFIC* design has the advantages of low cost, small size, and high reliability.

- The next step in circuit design is to reach *SOC* design. However, there are many barriers to overcome.

Why is 6σ design so important? Because

- The viability of a product on a mass production line depends on its approaching 6σ design or 100% yield rate.
- Prototype circuit design in the lab is not the same as 6σ design for a product in a mass production line. There is long way to go from a prototype circuit design level up to a 6σ design goal.
- 6σ design is the necessary and satisfactory criterion to measure the qualifications of an *RF* circuit designer.

PART III: *RF* SYSTEM ANALYSIS (CHAPTER 18)

As shown in the figure, the third part of this book provides a general description of the basic parameters and the necessary theoretical background of *RF* system analysis to control the individual *RF* circuit block design.

Most of this book is a summary of my design work and therefore may reflect my own imperfect understandings and prejudices. Comments from readers will be greatly appreciated. My email address is chihsili@yahoo.com.cn.

I have found the following books and articles very helpful in my engineering design work:

Paul R. Gray, Paul J. Hurst, Stephen H. Lewis, Robert G. Meyer, *Analysis and Design of Analog Integrated Circuits*, 4th ed., John Wiley & Sons, Inc., 2001; Thomas H. Lee, *The Design of CMOS Radio-Frequency Integrated Circuits*, Cambridge University Press, 1998; Donald R.J. White, *Electrical Filters, Synthesis, Design and Applications*, Don White Consultants, Inc., 1980; Barrie Gilbert, "The Multi-tanh Principle: A Tutorial Overview," *IEEE* Journal of Solid-State Circuits, Vol. 33, No. 1, January 1998, pp. 2–17; and H. A. Haus et al., "Representation of Noise in Linear Two Ports," Proceedings of the *IRE*, Vol. 48, January 1960, pp. 69–74.

Finally, I express my deepest appreciation to my lovely son, Bruno Sie Li, who checked and corrected my English and designed the front cover.

Richard Chi-Hsi Li

Fort Worth, Texas
March 2008

PART I

INDIVIDUAL *RF* BLOCKS

CHAPTER 1

LNA (LOW NOISE AMPLIFIER)

1.1 INTRODUCTION

In a wireless communication system, the *LNA* is the first circuit block in the receiver. It is one of most important blocks because:

- The sensitivity of the receiver is mainly determined by the *LNA* noise figure and power gain. The noise figure of the *LNA* significantly impacts the overall noise performance of the receiver. On the other hand, the power gain of the *LNA* significantly suppresses noise contributions from subsequent stages, so that it as well impacts the overall noise performance of the receiver.
- The *LNA* plays an important role in the linearity of the entire system. Its non-linearity must be reduced as much as possible.
- In a *CDMA* (Code Division Multiplex Access) wireless communication system, the *LNA* takes care of *AGC* (Automatic Gain Control) in the entire system as well.

This chapter covers

- Typical design procedures including selection of device size, raw device testing, input and output impedance matching, stability checking, and linearity examination and improvement. This has been important subject since the advent of more advanced wireless communication systems such as 64 *QAM*.
- Cascode *LNA*. As the wireless bandwidth is raised up to *GHz* or tens of *GHz*, the performance of the *LNA* is restricted by the input Miller capacitance. Increasing the isolation between the input and output in a *LNA* would be helpful to an advanced communication system. The cascode *LNA* would improve the performance from single-ended *LNA*.

- *AGC* (Automatic Gain Control). Without *AGC* capability, it is impossible for the wireless *CDMA* communication system to operate well.

In recent years, the differential *LNA* is specially required for the direct conversion or "zero *IF*" wireless communication system. This will be discussed in Chapter 3, where the differential pair discussed applies not only to the differential *LNA* but also to other *RF* circuit blocks.

The *LNA* has been developed over several decades. However, as the progress of electronic products moved forward, *LNA* design was required to reach higher and higher goals. For example, the voltage of *DC* power supply became lower and lower, from $3V$ to $1V$ in a cellular phone design. The current drain had to be reduced as much as possible so that the standby current of the overall receiver could be just a few mA to conserve battery consumption. It must be small and the cost must be low, and the performance must be maintained at a high level. *LNA* design becomes more complicated if trade-offs must be made between size, cost, and performance.

It is well known that the *LNA* must magnify the weak signal from the antenna and intensify it up to the power level required by subsequent stages. This implies that a *LNA* must have

- A low noise block so that the weak signal will not be "submerged" by noise;
- A high power gain block so that its output can drive the following stage well.

A *LNA* with maximum gain may not be in the state of minimum noise, or vice versa. A trade-off is usually made between maximizing gain and minimizing the noise figure. In past decades, much effort has been put into designing a *LNA* to reach both maximum gain and minimum noise figure simultaneously. This is a great challenge in *LNA* circuit design. This dilemma was solved more than 10 years ago in my designs but has not been previously published. Now I am going to share it with my readers.

1.2 SINGLE-ENDED SINGLE DEVICE *LNA*

In this section, the design procedures and schemes will be illustrated through a design example, in which a *MOSFET* transistor is selected as the single-ended device (it can of course be replaced by other types of devices).

A single-end *LNA* with a single device is the simplest low noise amplifier. Nevertheless, it is the essence or core in all other types of *LNA* designs, including cascode and differential designs. The design procedures and schemes described in this section are suitable to all types of *LNA* design.

The main goals for the design example are

- $V_{cc} = 3.0\,V$,
- $I_{cc} < 3.0\,mA$,
- frequency range $= 850$ to $940\,MHz$,
- $NF < 2.5\,dB$,
- gain $> 10\,dB$,

- $IP_3 > 0\,dBm$,
- $IP_2 > 40\,dBm$.

1.2.1 Size of Device

The first step in *LNA* circuit design is to decide the size of the device. Many tradeoffs must be taken into account between size, cost, performance, and so on. In this sub-section, only performance is counted in the selection of device size.

In digital *IC* circuit design, the *MOSFET* transistor has become dominant in recent years because the size of the device can be shrunk and the current drain can be reduced more than with other devices. Among *MOSFET* transistors, device length therefore becomes the key parameter in the selection of *IC* foundry and processing because it strongly impacts the total area of the *IC* die and therefore the cost, speed of performance, maximum data rate, current drain, and so on. The reason is simple: In digital *IC* circuit design, hundreds and even thousands of transistors are needed. The total area of the *IC* die, and therefore the cost, is significantly reduced as the device length decreases. *IC* scientists and engineers have worked very hard to shrink the size of transistors, which now approach unbelievably tiny sizes. In the 1990s, the length of a *MOSFET* device was in the order of μm; from 2000 to 2005 and the *IC* world entered the so-called "nanometers" era. Many foundries today have the capability to manufacture *MOSFET ICs* with lengths of 0.5, 0.35, 0.25, 0.18, $0.11\,\mu m$. In 2006, the length of a *MOSFET* device was further shrunk to 90, 45, $22.5\,nm$. The progress of *IC* processing is moving forward very fast, and, consequently, *IC* circuit design work becomes more and more challenging.

In the *RF* circuit design, bipolar transistors were applied to *RFIC* development in the 1990s. Meanwhile, the *MOSFET* device has been applied to the *RFIC* as well. The smaller size of *MOSFET* devices brings about the same advantages to *RF* circuit design as to digital circuit design, such as the reduction of cost and the increase in operating frequencies. It must be pointed out, however, that smaller size is not the main objective pursued in *RF* circuit design because the total number of devices applied in *RF* circuits is much smaller than the number of devices applied in digital circuits. Instead of pursuing smaller size, *RF* engineers prefer to select device lengths for which the technology of *IC* processing in the foundry is more advanced and the device model for simulation is more accurate. In addition, there are two important factors to be considered in the selection of the *MOSFET* device's size: the restriction of the device size due to the V_{gs} limitation and another due to the expectations of NF_{min}.

***1.2.1.1 Restrictions of* W/L *Due to Consideration of* V_{gs}** In *LNA* design, the *MOSFET* transistor is usually operated in its active region. Its *DC* characteristics can be expressed as:

$$I_d = \frac{\mu_n C_{ox}}{2}\frac{W}{L}(V_{gs} - V_{th})^2, \tag{1.1}$$

$$g_m = \frac{\partial I_d}{\partial V_{gs}} = \mu_n C_{ox}\frac{W}{L}(V_{gs} - V_{th}), \tag{1.2}$$

where

I_d = drain current,
g_m = transconductance of *MOSFET* transistor,
W = width of *MOSFET* transistor,
L = length of *MOSFET* transistor,
V_{gs} = gate-source voltage for n channel *MOSFET*,
V_{th} = threshold voltage for n channel *MOSFET*, the minimum gate-to-source voltage needed to produce an inversion layer beneath the gate,
V_{ds} = drain-source voltage for *n* channel *MOSFET*,
μ_n = channel mobility, typically 700 $cm^2/V\text{-}sec$,
C_{ox} = capacitance per unit area of the gate oxide,

and

$$C_{ox} = \frac{\varepsilon_{ox}}{t_{ox}}, \tag{1.3}$$

where

t_{ox} = thickness of the gate oxide.

From (1.1) and (1.2) we have

$$g_m = \sqrt{2\mu_n C_{ox}\frac{W}{L}I_d}. \tag{1.4}$$

$$V_{gs} = 2\frac{I_d}{g_m} + V_{th}. \tag{1.5}$$

Equation (1.4) shows that g_m is related to the ratio W/L. The increase of the ratio W/L is equal to the increase of g_m. On the other hand, from equation (1.5), it can be seen that there are two ways to make I_d reach a certain amount, either by increasing g_m through the increase of the ratio W/L for a given V_{gs} or by increasing V_{gs} through the factor of (V_{gs}-V_{th}). Should the selected value of the ratio W/L be too small, V_{gs} must be increased to an unacceptable value for a given I_d.

In order to illustrate the relationships between g_m, I_d, W/L and the corresponding values of V_{gs}, Table 1.1 lists the calculated V_{gs} values when I_d and the ratio W/L are selected in different levels or amounts and when the basic parameters applied in the calculations are assumed as follows:

$$\varepsilon_{ox} = 3.45\times10^{-13}\ F/cm, \tag{1.6}$$

$$t_{ox} = 23.3\ A^{\circ} = 23.3\times10^{-8}\ cm, \tag{1.7}$$

$$\mu_n = 170\ cm^2/V\text{-}sec, \tag{1.8}$$

$$C_{ox} = 14.81\ fF/\mu^2, \tag{1.9}$$

$$V_{tn} = 0.49\,V, \tag{1.10}$$

TABLE 1.1 V_{gs} limitation in the selection of device size

I_d (mA)	W (μm)	L (μm)	W/L	g_m (mA/V)	V_{gs} (V)
1.00	0.9	0.09	10.00	2.24	1.38
1.00	9	0.09	100.00	7.10	0.77
1.00	90	0.09	1000.00	22.44	0.58
1.00	180	0.09	2000.00	31.73	0.55
1.00	450	0.09	5000.00	50.17	0.53
1.00	900	0.09	10000.00	70.95	0.52
1.00	1800	0.09	20000.00	100.34	0.51
2.00	0.9	0.09	10.00	3.17	1.75
2.00	9	0.09	100.00	10.03	0.89
2.00	90	0.09	1000.00	31.73	0.62
2.00	180	0.09	2000.00	47.79	0.58
2.00	450	0.09	5000.00	70.95	0.55
2.00	900	0.09	10000.00	100.34	0.53
2.00	1800	0.09	20000.00	141.91	0.52
5.00	0.9	0.09	10.00	5.02	2.48
5.00	9	0.09	100.00	15.87	1.12
5.00	90	0.09	1000.00	50.17	0.69
5.00	180	0.09	2000.00	75.56	0.63
5.00	450	0.09	5000.00	11.19	0.58
5.00	900	0.09	10000.00	158.66	0.55
5.00	1800	0.09	20000.00	224.37	0.53
10.00	0.9	0.09	10.00	7.10	3.31
10.00	9	0.09	100.00	22.44	1.38
10.00	90	0.09	1000.00	70.95	0.77
10.00	180	0.09	2000.00	106.86	0.69
10.00	450	0.09	5000.00	158.66	0.62
10.00	900	0.09	10000.00	224.37	0.58
10.00	1800	0.09	20000.00	317.31	0.55
20.00	0.9	0.09	10.00	10.03	4.48
20.00	9	0.09	100.00	31.73	1.75
20.00	90	0.09	1000.00	100.34	0.89
20.00	180	0.09	2000.00	151.13	0.77
20.00	450	0.09	5000.00	224.37	0.67
20.00	900	0.09	10000.00	317.31	0.62
20.00	1800	0.09	20000.00	448.75	0.58
50.00	0.9	0.09	10.00	15.87	6.79
50.00	9	0.09	100.00	50.17	2.48
50.00	90	0.09	1000.00	158.66	1.12
50.00	180	0.09	2000.00	238.95	0.94
50.00	450	0.09	5000.00	354.77	0.77
50.00	900	0.09	10000.00	501.71	0.69
50.00	1800	0.09	20000.00	709.53	0.63

then

$$\mu_n C_{ox} = 251.72 \, \mu A/V^2. \tag{1.11}$$

In Table 1.1, the calculations are conducted for the cases of $I_d = 1, 2, 5, 10, 20$, and $50\,mA$ with the different levels of $W/L = 10, 100, 1000, 2000, 5000, 10000$, and 20000.

The underlined values of V_{gs} in the rightmost column in Table 1.1 are unacceptable because they are higher than $0.7V$, which is considered the highest acceptable value of V_{gs} when the *DC* power supply is low, say, 1.0 to $1.8V$. Therefore, the rows containing underlined values of V_{gs} in Table 1.1 must be abandoned in the selection of the ratio W/L. Hence, the values of the ratio W/L are restricted for the given values of I_d and g_m due to the constraint on V_{gs}. All other rows and their candidates in Table 1.1 are acceptable. They will be further narrowed down in consideration of the so-called "power-constrained noise optimization."

It should be noted that Table 1.1 is an example only. The selection of the ratio W/L for the device must be conducted by designers based on the basic parameters, ε_{ox}, t_{ox}, μ_n, C_{ox}, and V_{tn}, which actually apply to the device.

1.2.1.2 *Optimum Width* W_{opt} *of Device* Based on the theoretical derivation (Lee, 1998, pp. 230–232), the size selected for the device in *LNA* design is more reasonably considered from the expectation of a minimum of noise. By explicitly taking power consumption into account, the optimum width of a device W_{opt} for the minimum noise figure NF_{min} can be expressed as

$$W_{opt} = \frac{1}{3\omega L C_{ox} R_S}, \tag{1.12}$$

where

W_{opt} = optimum width of device (*MOSFET* transistor),
ω = operation angular frequency,
L = length of device (*MOSFET* transistor),
C_{ox} = capacitance per unit area of the gate oxide,
R_S = source resistance.

This results from the power-constrained noise optimization.

The value of the optimized width of the device is inversely proportional to the operating frequency, ω, the source resistance, R_S, the capacitance of the gate oxide area, C_{ox}, and the length of the device, L. The designer knows the first two parameters, ω and R_S. The other two, C_{ox} and L, are provided by the *IC* foundry, which may have a couple choices. For instance, device lengths of 0.25, 0.18, 0.13, 0.11, and $0.09\,\mu m$, are available in most *MOS IC* foundries at present. Based on the data that the *IC* foundry provides, the corresponding values of W_{opt} can be calculated from equation (1.12). Then, these W_{opt} and L values can be examined for a reasonable value of V_{gs} as in Table 1.1 and the best set of W_{opt} and L can be determined. Then, the final decision of *IC* processing can be made.

1.2.2 Raw Device Setup and Testing

Raw device testing is the second step in the block circuit design. It should be noted that it is a key step in a good LNA circuit design.

In the circuit design, a "device" is a general name for a transistor. The transistor can be bipolar, or a *MOSFET*, or GaAs, or some other type. The purpose of raw

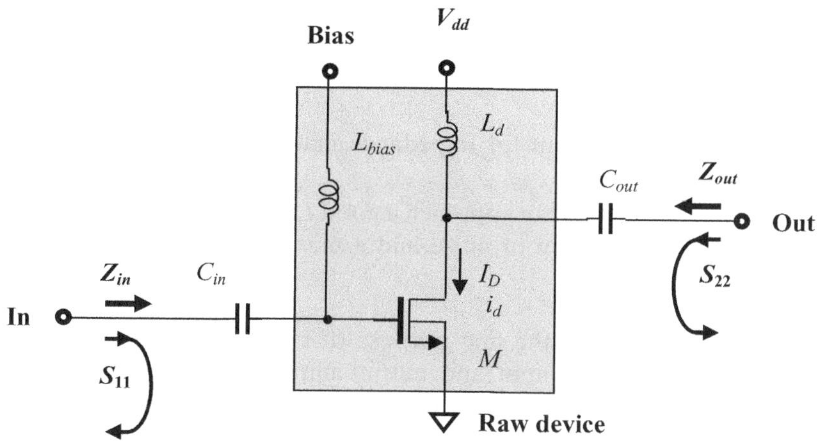

Figure 1.1 Setup for raw device testing, $f = 850$ to $940\,MHz$, $I_D = 2.6\,mA$ C_{in}, C_{out}: "zero" capacitor; L_{bias}, L_c, L_d: "infinite" inductor.

device testing is to determine the operating characteristics only of the device, and nothing else. However, an operating transistor must be provided with the *DC* power supply and bias, and therefore, some additional parts such as the *RF* choke and *DC* blocking or *AC* by-pass capacitors must be connected. The impedance of the additional parts must therefore approach either zero when they are connected in series, or infinity when they are connected in parallel. If so, the tested characteristics of the transistor are not disturbed by the addition of those parts.

Figure 1.1 shows the setup for raw device testing. The capacitors C_{in} and C_{out} are "zero" capacitors, while the inductors L_{bias} and L_d are "infinite" inductors. They are discussed in Chapter 14 where the "zero" capacitor and "infinite" inductor are selected from discrete chip parts. In the actual simulation for *IC* circuitry, the capacitors C_{in} and C_{out} can be large capacitors with a high value of capacitance so that their impedance approaches zero at operating frequencies, while the inductors L_{bias} and L_d can be a large inductors with a high value of inductance so that their impedance approaches infinity at operating frequencies. The desired current drain of the transistor, $I_D + i_d$, can be adjusted by the bias voltage, where I_D is the *DC* current drain portion and i_d is the *AC* current drain portion of the *MOSFET* transistor.

In Figure 1.1, the device is a *MOSFET* transistor with *CMOS IC* processing. Its size has been selected based on the considerations of V_{gs} and NF_{min} as discussed in the previous section. As mentioned above, the *DC* power supply is $3V$ and the drain current, adjusted by the bias, is $2.6\,mA$.

The operating frequency range is from 850 to $940\,MHz$. Its relative bandwidth is

$$\frac{\Delta f}{f_o} = \frac{940-850}{(940+850)/2} = 10.05\%. \tag{1.13}$$

This is a narrow-band block. Usually, a block or a system with a relative bandwidth greater than 15% is considered a wide-band block or system. A block or a system with a relative bandwidth less than 15% is considered a narrow-band block or system.

The purpose of raw device testing is twofold:

1) To create a starting point for impedance matching in order to continue the next design step.
2) To see if the raw device can approach a good *LNA* design. A good *LNA* design suggests that a minimum of noise and a maximum of gain can be obtained simultaneously.

It is easy to understand that the first purpose of raw device testing is to create a starting point for matching input and output impedance. The input and output impedances, Z_{in} and Z_{out}, are approximately related to the S parameters by the following equations:

$$Z_{in} = \frac{1+S_{11}}{1-S_{11}}, \tag{1.14}$$

$$Z_{out} = \frac{1+S_{22}}{1-S_{22}}. \tag{1.15}$$

This approximation is usually correct if the transistor's isolation between input and output is good and the testing calibration is well done. Through the testing of S_{11} and S_{22}, the input and output impedances, Z_{in} and Z_{out}, can be read directly from the Smith chart at the same locations of S_{11} and S_{22}, respectively.

Figure 1.2(a) shows the test results of S_{11} and S_{22}, and hence Z_{in} and Z_{out} on the Smith chart. The input and output impedances of a *MOSFET* transistor are usually capacitive and are located in the bottom half of the Smith chart, while the input and output impedances of a bipolar transistor can be either capacitive or inductive, depending on the device size, current drain, and operating frequency. Another difference between the *MOSFET* and the bipolar transistor is that the input and output impedances of a *MOSFET* transistor are usually located in the relatively higher impedance area, while the input and output impedances of a bipolar transistor are usually located in the relatively lower impedance area. This difference implies that impedance matching is more difficult for the *MOSFET* than for the bipolar transistor, and that the isolation between input and output in the *MOSFET* is better than in the bipolar transistor.

Figure 1.2(a) also shows that the location of S_{22} is much farther from $50\,\Omega$ and is in the very high impedance area, while S_{11} is located somewhat closer to $50\,\Omega$ than S_{22}. Correspondingly, Figure 1.2(b) shows that the magnitude of S_{22} is almost close to $-1\,dB$, of S_{11} around $-3\,dB$, and of S_{21} around $3.5\,dB$, which is much lower than expected. We do not mind S_{21} too much because it is tested under an unmatched case. The magnitude of S_{12} is usually around -20 to $-30\,dB$ and therefore disappears from the plot. Likewise, we do not mind S_{12} too much because the isolation in today's devices is usually sufficient unless a feedback circuit is added. A remarkable feature shown in Figure 1.2 is that the frequency response for all the S parameters is flattened, so that one does not need to worry about bandwidth at this point.

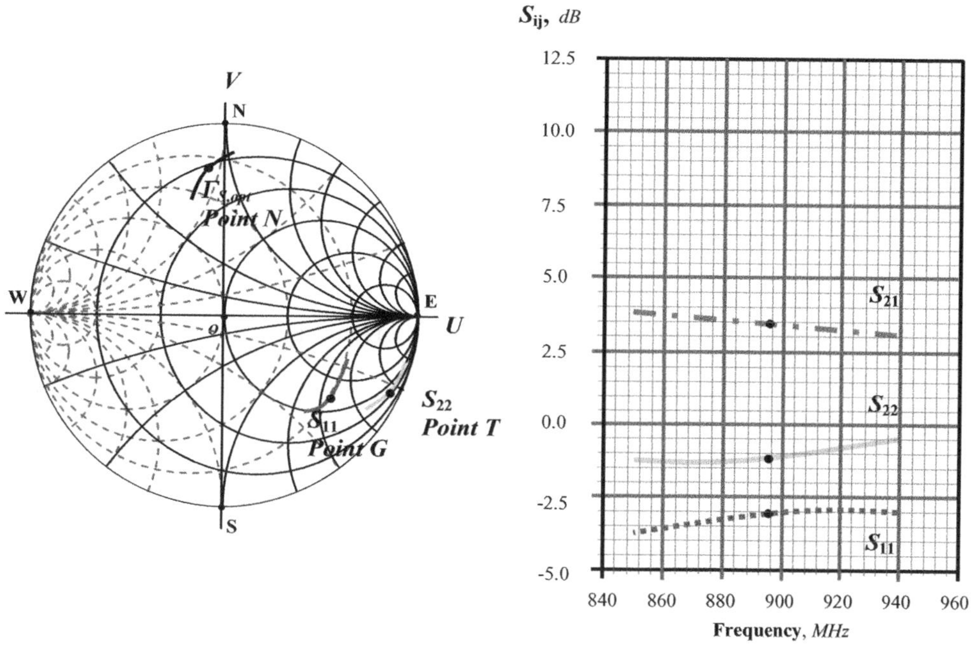

(a) S_{11}, S_{22} and $\Gamma_{S,opt}$ on Smith chart (b) Magnitude of S_{ij}, dB

Figure 1.2 S parameters from raw device testing, f = 850 to 940 MHz, I_D = 2.6 mA. (The intermediate frequency 895 MHz is marked with a dot on each trace.)

The second purpose of raw device testing is to examine the performance of the noise figure.

Based on Haus's theory (1960), the noise figure of a noisy block can be expressed by

$$NF = NF_{\min} + \frac{R_n}{G_s}\left[(G_S - G_{S,opt})^2 + (B_S - B_{S,opt})^2\right]. \tag{1.16}$$

where

NF = noise figure of noisy block,
NF_{min} = minimum of noise figure of noisy block,
R_n = equivalent noise resistance,
Y_S = admittance of input source,
G_S = conductance of input source,
B_S = subceptance of input source,
$Y_{S,opt}$ = optimum admitttance of input source,
$G_{S,opt}$ = optimum conductance of input source,
$B_{S,opt}$ = optimum subceptance of input source.

The noisy two-port block can reach a minimum of noise figure

$$NF = NF_{\min}, \tag{1.17}$$

when

$$G_S = G_{S,opt}, \tag{1.18}$$

$$B_S = B_{S,opt}. \tag{1.19}$$

Equations (1.18) and (1.19) can be written together, that is

$$Y_S = Y_{S,opt}, \tag{1.20}$$

where

$$Y_s = G_s + jB_s \tag{1.21}$$

$$Y_{S,opt} = G_{S,opt} + jB_{S,opt} \tag{1.22}$$

On the Smith chart, the optimum condition (1.20) is usually labeled by the corresponding reflection coefficient, $\Gamma_{S,opt}$, corresponding to $Y_{S,opt} = G_{S,opt} + j\ B_{S,opt}$,

$$\Gamma_S = \Gamma_{S,opt}. \tag{1.23}$$

Of course, $\Gamma_{S,opt}$ is a complicated function mainly determined by the type, size, and trans-conductance of the raw device. It has been formularized in some technical books. Usually, an optimum source reflection coefficient, $\Gamma_{S,opt}$, is computed by the computer simulation program and can be displayed on the Smith chart as shown in Figure 1.2(a). On the Smith chart, its corresponding parameters, G_{opt}, B_{opt}, R_{opt}, X_{opt}, can be read from the same point.

Noise figure would be expected to be at a minimum if the input impedance were at point N where its input impedance corresponds to $\Gamma_{S,opt}$. Instead, in raw device testing, the noise figure is tested at point G where its impedance corresponds to its S_{11}. Therefore its value is much higher than the expected minimum. Figure 1.3 shows the tested noise figure. In the entire frequency range, it is around 8.7 dB.

At this point, a question may be raised: Through impedance matching, the trace S_{11} can be pulled to 50 Ω, the center of the Smith chart. What would happen to the trace of $\Gamma_{S,opt}$ then? We expect that the trace of $\Gamma_{S,opt}$ would also be pulled to 50 Ω, the center of the Smith chart. Can we control the change of the trace of $\Gamma_{S,opt}$ when the impedance matching network is implemented?

Let's take a look at the performance of the noise figure (NF) in the entire frequency range from the raw device testing. Figure 1.3 shows that the noise figure in the operating frequency range is

$$NF = 8.52 \text{ to } 8.77\,dB, \quad \text{when } 850\,MHz < f < 940\,MHz, \tag{1.24}$$

$$NF = 8.7\,dB, \quad \text{when } f = 895\,MHz. \tag{1.25}$$

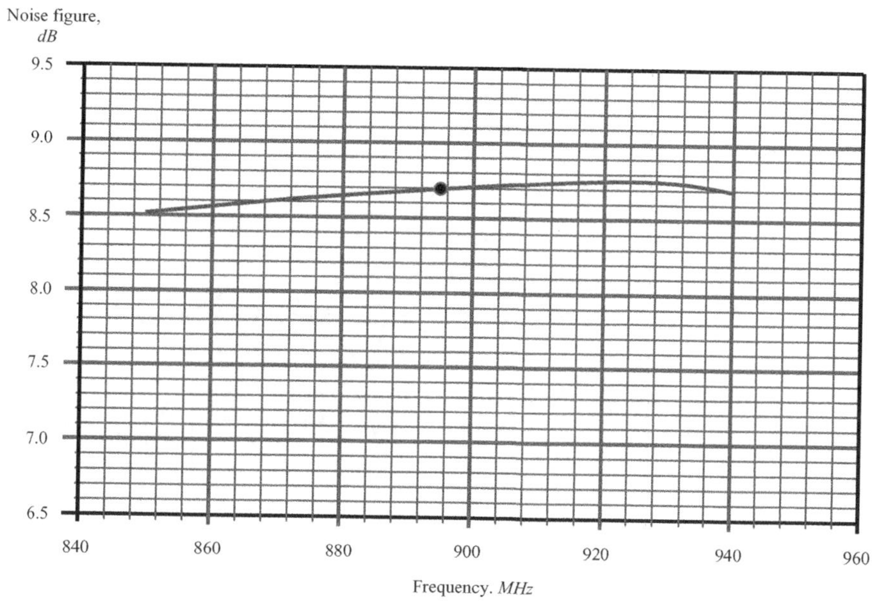

Figure 1.3 Noise figure from 850 to 940 MHz. $I_D = 2.6 mA$, $NF = 8.7 dB$ when $f = 895 MHz$.

The goal is

$$NF < 2.5 dB, \quad \text{when } 850 MHz < f < 940 MHz. \tag{1.26}$$

It can be seen that in the entire operating frequency range the noise figure is unacceptable.

Now let's examine the gain circles and noise figure circles at one frequency, say, $f = 895 MHz$, on the input reflection coefficient plane. Figure 1.4 plots both the gain circles and noise figure circles together.

The maximum of gain, $G = G_{max}$, is located at point G, which is 3.0 dB, as shown in Figure 1.2(b). However, its noise figure does not reach its minimum $NF_{min} = 5 dB$ as shown at point N, that is,

At point G,

$$G = G_{max} = 3.0 dB, \quad \text{and} \quad NF = 8.7 dB. \tag{1.27}$$

The minimum of the noise figure, $NF = NF_{min}$, is located at point N, which is quite far from point G. Its gain is, of course, much lower than the maximum gain of 3 dB at point G, that is,

At point N,

$$G = -4.8 dB, \quad \text{and} \quad NF = NF_{min} = 5 dB. \tag{1.28}$$

The raw device would be operating at point O, the center of the Smith chart, instead of at point G or N, if the internal impedance of the signal source is 50 Ω. Its gain would be higher than $-4.8 dB$ but lower than 3.0 dB, while its noise figure would be

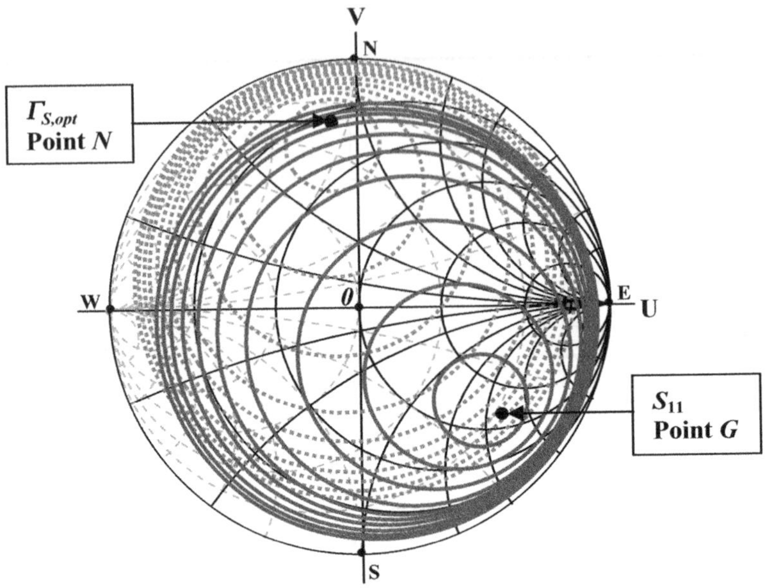

Figure 1.4 Constant gain circles and constant noise figure circles when $f = 895\,MHz$.
Gain circles: $G_{max} = 3.0\,dB$ at point G, step $= -1.0\,dB$.
Noise figure circles: $NF_{min} = 5\,dB$ at point N, step $= 0.5\,dB$.

lower than $8.7\,dB$ but higher than $5\,dB$. As a matter of fact, the raw device can be operated with any source impedance. Therefore its impedance can be correspondingly adjusted to any point on the Smith chart. The actual values of gain and noise figure can be read from Figure 1.4. These gains will not be higher than $3.0\,dB$, and the noise figures will not be lower than $5\,dB$. The ideal case is to find a raw device in which the maximum of gain, G_{max}, and the minimum of noise figure, NF_{min}, come together at one point on the Smith chart. This seems almost impossible without a special scheme being involved. However, we should ask the following questions: Might this be a temporary outcome because the design work is in the preliminary stage? The next step is to build the input and output impedance matching networks based on the raw device testing. Is it possible to pull points G and N together after the input impedance matching network is built?

We will temporarily submit the noise figure to the will of heaven, take care of the gain only, and move on to the task of input and output impedance matching as shown in Figure 1.5.

Impedance matching is a special scheme and the key technology in RF circuit design. It is discussed in some detail in this book. Because there are many ways to do impedance matching, many different results can be found. Figure 1.6 shows one such result. The point G of maximum gain, along with its gain circles, is moved from its original high impedance location as shown in Figure 1.4 to a location near the center of the Smith chart, $50\,\Omega$. The maximum of gain is increased from the $3.0\,dB$ of Figure 1.4 to $13\,dB$ as a result of the impedance matching. However, its

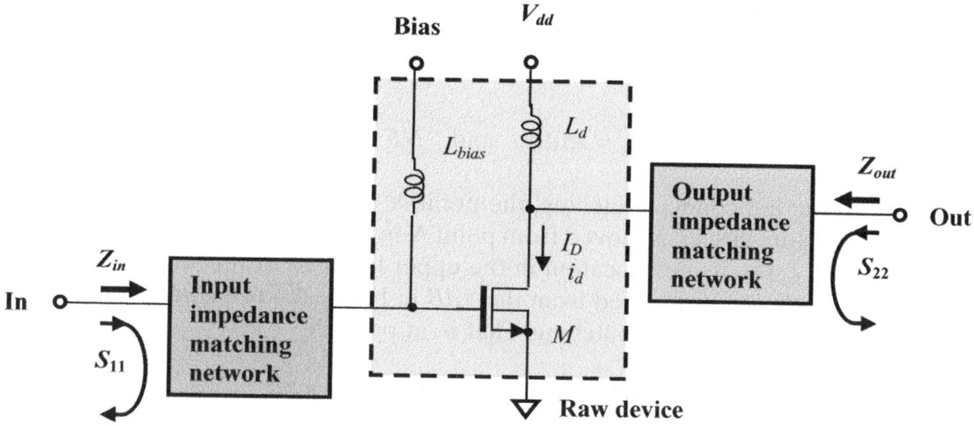

Figure 1.5 Input and output impedance matching networks are added to the raw device. $f = 850$ to $940\,MHz$, $I_D = 2.6\,mA$.

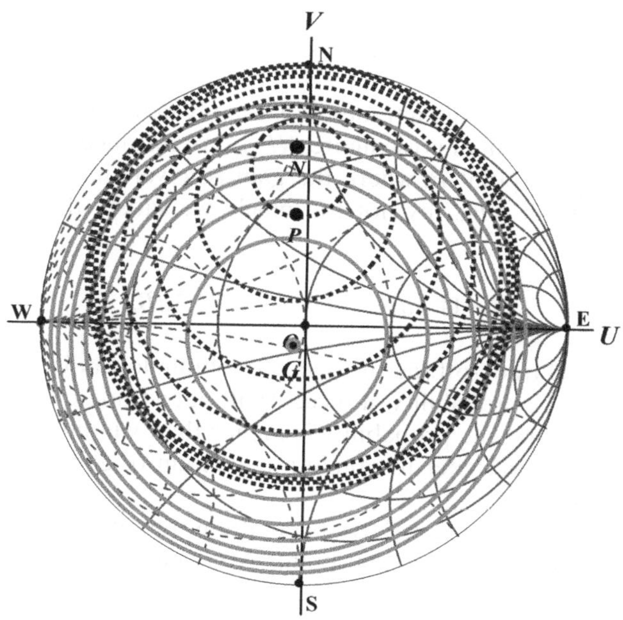

Input reflection coefficient Γ_S plane

Figure 1.6 Constant gain circles and constant noise figure circles when $f = 895\,MHz$.
Gain circles: $G_{max} = 13\,dB$ at point G, step $= 1.0\,dB$.
Noise figure circles: $NF_{min} = 1.8\,dB$ at point N, step $= 0.5\,dB$.

noise figure has not reached its minimum, $NF_{min} = 2\,dB$ as shown at point N. in Figure 1.6.

At point G,

$$G = G_{max} = 13\,dB, \quad \text{and} \quad NF = 3.5\,dB, \tag{1.29}$$

Also, due to the impedance matching, the point N of minimum noise figure, along with its noise figure circles, is moved from point N in the upper left area of the Smith chart in Figure 1.4 to its new location in the upper left area of the Smith chart. The minimum noise figure is dropped from the $5\,dB$ in Figure 1.4 to $1.8\,dB$ in Figure 1.6 as a result of the impedance matching. That is, at point N,

$$G = 8.2\,dB, \quad \text{and} \quad NF = NF_{min} = 1.8\,dB. \tag{1.30}$$

The goals as mentioned previously were:

$$G > 10\,dB, \qquad \text{when } 850\,MHz < f < 940\,MHz. \tag{1.31}$$

$$NF < 2.5\,dB, \qquad \text{when } 850\,MHz < f < 940\,MHz. \tag{1.32}$$

It can be seen that the input impedances at either point G or point N are still insufficient!

In order to satisfy these goals, a trade-off is usually made between the maximum gain, G_{max}, and the minimum of noise figure, NF_{min}. For instance, if we force the input impedance matching to point P, where the gain is lower than G_{max} but the noise figure is increased from NF_{min} by not too much, we satisfy the goals because at point P,

$$G = 11.3\,dB, \quad \text{and} \quad NF = 2.3\,dB. \tag{1.33}$$

It can be seen that the trade-off sacrifices gain to obtain low noise. This is, of course, adopted reluctantly. The probability of success without a trade-off just relies on luck, although with hard work, the probability is higher than winning a lottery. There is very little chance for a designer to obtain a maximum of gain and a minimum of noise figure at the same time.

However, this is what designers have been attempting in the past decades.

The ideal case to be pursued is that points N G, and hence the gain circles and noise figure circles, exactly overlap together at the center of the Smith chart when the input and output impedances are matched to $50\,\Omega$.

This is a very challenging but exciting project.

1.2.3 Challenge of a Good *LNA* Design

In order to solve the dilemma of overlapping the maximum gain with the minimum of noise figure at $50\,\Omega$ in the *LNA* circuit design, I suggest a repeated study of Haus's papers to thoroughly understand the meaning of the "optimum of source reflection coefficient, $\Gamma_{S,opt}$." In the history of the sciences of developing technology, every forward step requires a correct concept and diligent practice.

1.2.3.1 *Optimum of Source Voltage Reflection Coefficient,* Γ_{opt} Let's return to the raw device testing.

To clarify and distinguish the various reflection coefficients, Figure 1.7 emphasizes the directions of $\Gamma_{S,opt}$, Γ_S, S_{11}, Γ_{in}, and S_{11}^* immediately at the input and output of the raw device ($\Gamma_{S,opt}$ should not be misunderstood as the input reflection coefficient of the raw device, Γ_{in}, which is almost equal to S_{11} if S_{12} is negligible.) The following concepts are essential to solving the problem:

- In order to enable the raw device to approach the minimum of noise figure, $\Gamma_{S,opt}$ is a required optimum of source reflection coefficient, Γ_S, looking from the raw device toward the source.
- On the other hand, in order to enable the raw device to approach to the maximum gain, G_{max}, S_{11}^* is a required value of source reflection coefficient, Γ_S, looking from the raw device toward the source since the actual reflection coefficient looked from the source toward the raw device is S_{11}.

Based on Haus's theory, the minimum noise figure of the raw device will be simultaneously approached with the maximum gain if

$$\boxed{\Gamma_{S,opt} = S_{11}^*.} \tag{1.34}$$

Conversely, it is impossible to approach the minimum noise figure of the raw device with the maximum gain if

$$\Gamma_{S,opt} \neq S_{11}^*. \tag{1.35}$$

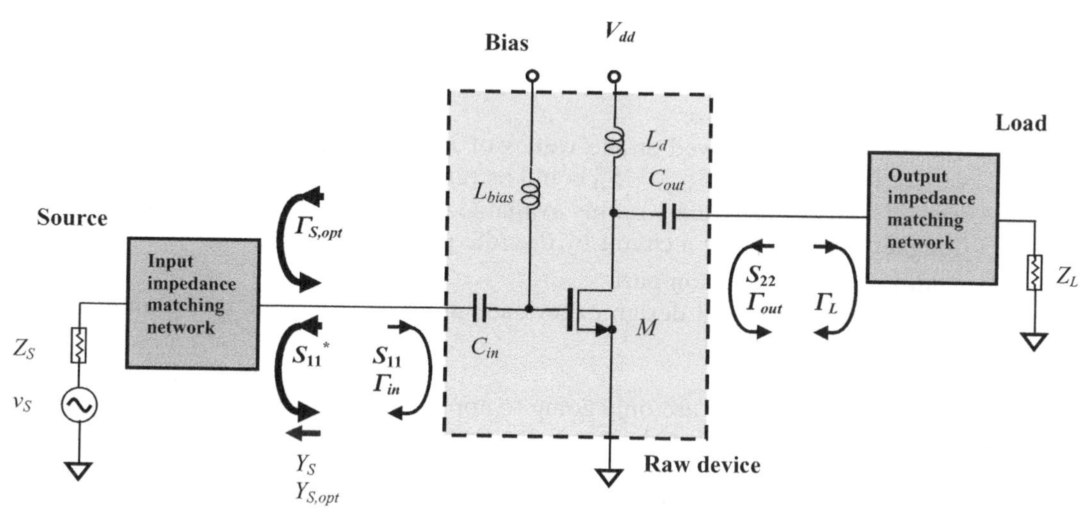

Figure 1.7 Directions of $\Gamma_{S,opt}$, Γ_{in} and Γ_{out}, S_{11} and S_{11}^* in the schematic for raw device testing.

I developed the condition (1.34) in the 1990s and have been applying it in many successful design projects, although it has not been published. We will now pursue equation (1.34). This equation represents the condition by which a perfect *LNA* with **a** minimum noise figure and maximum gain can be **simultaneously** approached after the input and output impedance matching networks are implemented.

Let's return to Figures 1.2 and 1.4 for the raw device testing. It can be found that the $\Gamma_{S,opt}$ and S_{11}^* do not satisfy condition (1.34), but are in the condition of (1.35). This is the reason for the problem in Figure 1.6, where the points G of maximum gain and N of minimum noise figure were very far apart and not in conjugate locations to each other; therefore the gain and noise figure circles did not overlap after the implementation of the input and output impedance matching networks. Consequently, in Figure 1.6, it was impossible to simultaneously achieve a maximum gain and a minimum noise figure.

It should be re-iterated that the implementation of the input impedance matching network cannot change the deviation status between $\Gamma_{S,opt}$ and S_{11}^*, since $\Gamma_{S,opt}$ and S_{11}^* are determined by the raw device only and are basically independent of the impedance matching network.

1.2.3.2 Simultaneous Approach of Both NF$_{min}$ *and* G$_{max}$

On the surface, it seems as though the purpose of raw device testing is the starting point for impedance matching. Actually, the more important purpose of this testing is to judge whether a raw device could build a good *LNA* or not, that is, to check whether condition (1.34) is satisfied or not. Should condition (1.34) not be satisfied, impedance matching design should be halted and possible means to satisfy or approach condition (1.34) should be sought.

Three major schemes are used to satisfy condition (1.34) in the step of raw device testing:

1) Increasing or decreasing the current drain, I_D;
   The S parameters as well as the values of $\Gamma_{S,opt}$ change as the current drain is varied. The condition $\boldsymbol{\Gamma_{S,opt}} = \boldsymbol{S_{11}^*}$ could be reached with an appropriate amount of current drain.
2) Changing the device size
   The S parameters as well as the values of $\Gamma_{S,opt}$ change as the device size is varied. The condition $\boldsymbol{\Gamma_{S,opt}} = \boldsymbol{S_{11}^*}$ could be reached with an appropriate device size. Of course, this scheme is only available to the *IC* designer and not to the designer implementing a circuit by discrete parts.
3) Addition of degeneration part
   According to empirical design experience in practical design, this is an easy way to achieve success.

Due to space limitations, we are only going to apply the third scheme to the design example. Readers are encouraged to apply the first and second in their designs since both are very effective as well. Usually, it is easier to approach the condition $\boldsymbol{\Gamma_{S,opt}} = \boldsymbol{S_{11}^*}$ by the use of multiple combined schemes rather than only one.

Figure 1.8 shows a degeneration inductor, L_{degen}, applied to the source of the *MOSFET* transistor.

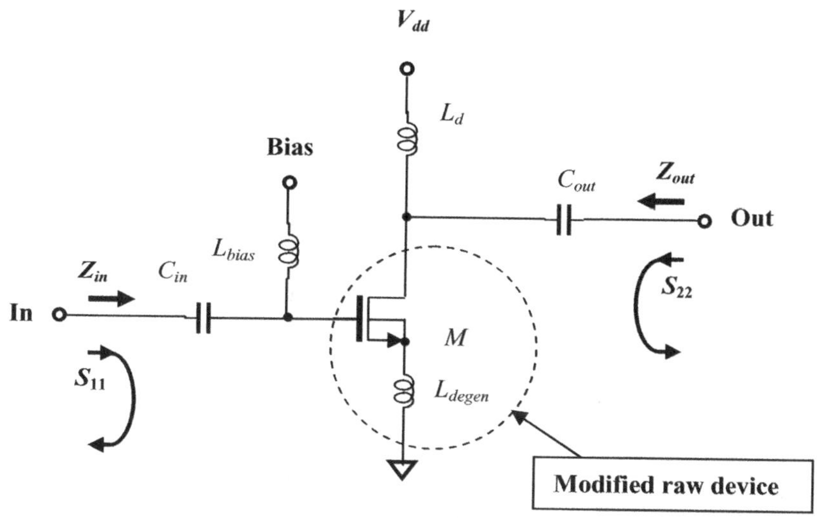

Figure 1.8 Setup for modified raw device testing by the addition of degeneration inductor. L_{degen}: degeneration inductor; C_{in}, C_{out}: "zero" capacitor; L_{bias}, L_c, L_d: "infinite" inductor.

The degeneration inductor, L_{degen}, is a simple and tiny part. For example, in the *RFIC* circuit design for *GHz* of operating frequency range, it is about 0.5 to 2 turns. The actual value of the degeneration inductor applied to this design sample is $10nH$.

In the simulation, it is connected to the source of the device. The device, combined with the degeneration inductor, is a modified raw device. The input impedance, of course, is different from the original impedance without the degeneration inductor.

By adjusting the value of the degeneration inductor to change its input impedance, condition (1.34) could be satisfied. As the degeneration inductor is added, the trace of S_{11} is moved from its original location as shown in Figure 1.2(a) to a new location as shown in Figure 1.9(a). Consequently, the traces of S_{11}^* and $\Gamma_{S,opt}$ on the input reflection coefficient plane are close to each other as shown in Figure 1.9. The corresponding variation of impedance can be determined by the equation derived in the following sub-section 1.2.3.3, in which both resistance and reactance are increased.

Figure 1.9(a) shows that the location of S_{11} is somewhat closer to $50\,\Omega$ than in Figure 1.2(a). Correspondingly, Figure 1.9(b) shows that the magnitude of S_{11} is now around $-3.75\,dB$, which is $0.75\,dB$ improved from that of Figure 1.2(b). The magnitude of S_{21} is about $4.0\,dB$, which is $0.5\,dB$ higher than the original shown in 1.2(b). Again, Figure 1.9 shows that the frequency response for all the S parameters is flat; we need not worry about the bandwidth for now. As a matter of fact, it does not matter too much in the variation of magnitude of S parameters because the input and output impedances are not matched yet.

From Figure 1.9(a) it is easy to imagine that S_{11} and $\Gamma_{S,opt}$ could be pulled close together near the center of the Smith chart after impedance matching is done, although S_{11}^* and $\Gamma_{S,opt}$ are still not at the same point.

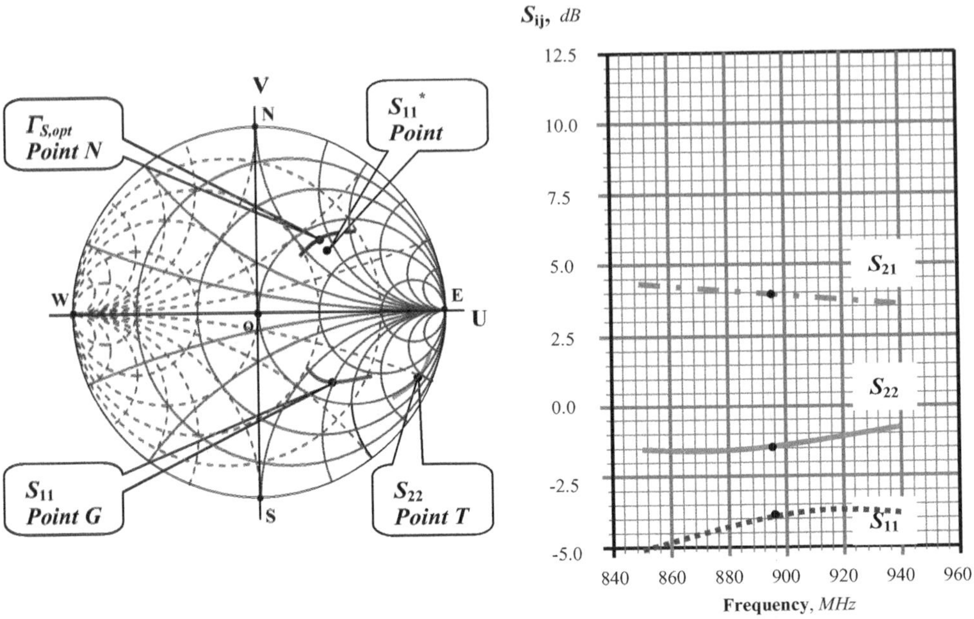

(a) S_{11}, S_{22} and $\Gamma_{S,opt}$ on Smith chart (b) Magnitude of S_{ij}, dB

Figure 1.9 S parameters from modified raw device testing, $f = 850\text{–}940\,MHz$, $I_D = 2.6\,mA$. (The intermediate frequency $895\,MHz$ is marked with a dot on each trace.)

Before discussing input and output impedance matching, we are going to examine the variation of the input impedance due to the addition of the degeneration inductor.

1.2.3.3 Variation of Input Impedance due to Degeneration Inductor Figure 1.10 plots the input stage of a *MOSFET* transistor and its equivalent. From Figure 1.10(b), we have

$$v_{in} = i_{in}\left(jL_g\omega + \frac{1}{jC_{gs}\omega}\right) + (i_{in} + i_d)\,jL_{degen}\omega = i_{in}j\left(L_g\omega - \frac{1}{C_{gs}\omega}\right) + (i_{in} + g_m V_{gs})\,jL_{degen}\omega, \tag{1.36}$$

$$v_{in} = i_{in}j\left(L_g\omega - \frac{1}{C_{gs}\omega}\right) + \left(i_{in} + g_m i_{in}\frac{1}{jC_{gs}\omega}\right)jL_{degen}\omega, \tag{1.37}$$

$$Z_{in} = \frac{v_{in}}{i_{in}} = R_{in} + jX_{in}, \tag{1.38}$$

$$Z_{in} = j(L_g + L_{degen})\omega + \frac{1}{jC_{gs}\omega} + \frac{g_m}{C_{gs}}L_{degen} = j\left[(L_g + L_{degen})\omega - \frac{1}{C_{gs}\omega}\right] + L_{degen}\omega_T, \tag{1.39}$$

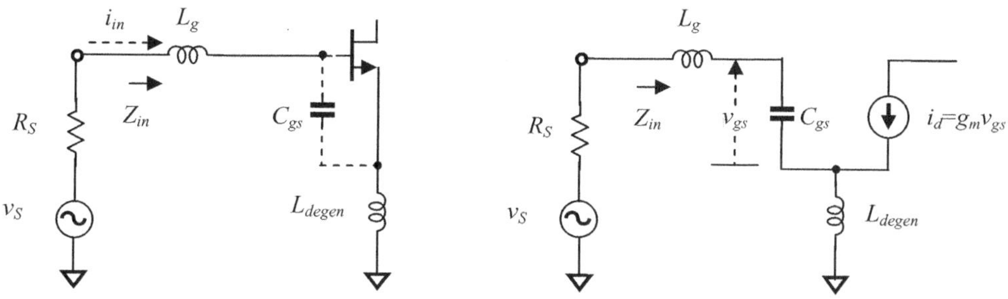

(a) Input stage (b) Equivalent of input stage

Figure 1.10 The input stage of a *MOSFET* transistor with common source configuration.

where

$$\omega_T = \frac{g_m}{C_{gs}}, \tag{1.40}$$

which is called the cut-off frequency.

$$R_{in} = L_{degen}\omega_T, \tag{1.41}$$

$$X_{in} = (L_g + L_{degen})\omega - \frac{1}{C_{gs}\omega}. \tag{1.42}$$

On the other hand,

$$G_{m,eff} = g_m Q_{in} = \frac{g_m}{\omega C_{gs}(R_S + L_{degen}\omega_T)} = \frac{\omega_T}{\omega R_S\left(1 + \frac{L_{degen}\omega_T}{R_S}\right)} = \frac{\omega_T}{2\omega R_S}, \tag{1.43}$$

when the input impedance is matched, that is,

$$R_s = L_{degen}\omega_T, \tag{1.44}$$

where Q_{in} is the effective input Q of the circuit. Note that R_g has been neglected relative to R_s. This equation is valid at resonance where the signal voltage across C_{gs} is equal to Q_{in} times the input signal. It should be noted that the overall trans-conductance is apparently independent of g_m, the intrinsic device trans-conductance.

From equations (1.41) and (1.42) it can be seen that the input impedance is increased

$$\Delta Z_{in} = L_{degen}\omega_T + jL_{degen}\omega, \tag{1.45}$$

due to the presence of L_{degen}.

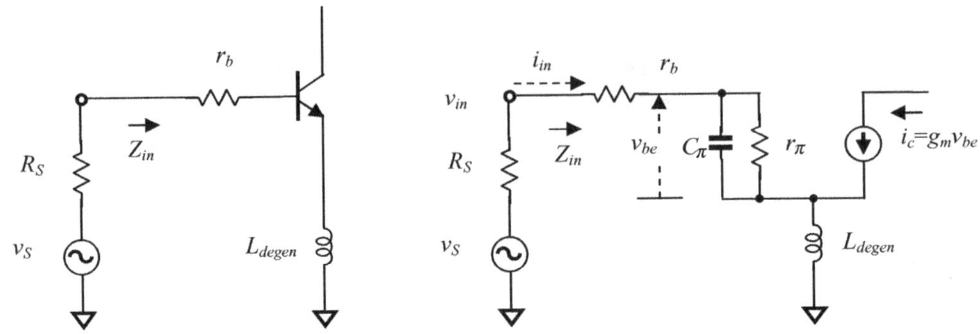

(a) Schematic of input stage (b) Equivalent of input stage

Figure 1.11 The input stage of a bipolar transistor with a degeneration inductor.

If the device is a bipolar transistor, the input portion can be drawn as shown in Figure 1.11. This figure shows the degeneration inductor connected from emitter to ground. The input impedance can be calculated as follows:

$$r_\pi // C_\pi = \frac{r_\pi}{1 + jC_\pi \omega r_\pi} = \frac{r_\pi}{1 + (C_\pi \omega r_\pi)^2} - j\frac{C_\pi \omega r_\pi^2}{1 + (C_\pi \omega r_\pi)^2}, \tag{1.46}$$

$$v_{in} = i_{in}(r_b + r_\pi // C_\pi) + (i_{in} + i_c) L_{degen}\omega, \tag{1.47}$$

$$i_c = g_m v_{be} = g_m i_{in}(r_\pi // C_\pi), \tag{1.48}$$

$$Z_{in} = \frac{v_{in}}{i_{in}} = r_b + r_\pi // C_\pi + j[1 + (r_\pi // C_\pi) g_m] L_{degen}\omega, \tag{1.49}$$

$$Z_{in} = r_b + r_\pi + \frac{L_{degen} C_\pi \omega^2 g_m r_\pi^2}{1 + (C_\pi \omega r_\pi)^2} + j\left[1 + \frac{\left(g_m - \frac{C_\pi}{L_{degen}} r_\pi\right) r_\pi}{1 + (C_\pi \omega r_\pi)^2}\right] L_{degen}\omega. \tag{1.50}$$

$$R_{in} = r_b + r_\pi + \frac{L_{degen} C_\pi \omega^2 g_m r_\pi^2}{1 + (C_\pi \omega r_\pi)^2}, \tag{1.51}$$

$$X_{in} = \left[1 + \frac{\left(g_m - \frac{C_\pi}{L_{degen}} r_\pi\right) r_\pi}{1 + (C_\pi \omega r_\pi)^2}\right] L_{degen}\omega. \tag{1.52}$$

It should be noted that when

$$g_m = \frac{C_\pi}{L_{degen}} r_\pi, \tag{1.53}$$

then

$$X_{in} = L_{degen}\omega. \tag{1.54}$$

In cases with low frequency, the capacitor C_π can be neglected. Then (1.50), (1.51), and (1.52) become

$$Z_{in} = r_b + r_\pi + j(1+g_m r_\pi) L_{degen}\omega. \tag{1.55}$$

$$R_{in} = r_b + r_\pi, \tag{1.56}$$

$$X_{in} = (1 + r_\pi g_m) L_{degen}\omega. \tag{1.57}$$

In cases with high frequency, R_{in} is increased while X_{in} is decreased in cases with low frequency.

From equations (1.51) and (1.52) it can be seen that the input impedance is increased

$$\Delta Z_{in} = \frac{L_{degen} C_\pi \omega^2 g_m r_\pi^2}{1+(C_\pi \omega r_\pi)^2} + j\left[1 + \frac{g_m r_\pi}{1+(C_\pi \omega r_\pi)^2}\right] L_{degen}\omega, \tag{1.58}$$

due to the presence of L_{degeh}.

In comparing (1.45) with (1.58), the increased resistance is almost the same while the increased reactance of the bipolar transistor has a higher slope than that of the *MOSFET* if C_π in the bipolar transistor is equivalent to C_{gs} in the *MOSFET* and $C_\pi \omega\, r_\pi >> 1$.

Based on these two equations, the designer is able to estimate how far the trace of S_{11} must be moved on the Smith chart, so as to satisfy condition (1.34), by which the maximum gain and minimum noise figure circles can be almost entirely overlapped near the center of the Smith chart.

1.2.4 Input and Output Impedance Matching

Impedance matching is a core technology in *RF* circuit design and is therefore discussed in detail in this book. The process of impedance matching for the design sample shown above will be omitted. Instead, the final input and output impedance matching networks and their performance are shown in Figures 1.12 and 1.13, respectively.

Figure 1.12 plots the impedance matching networks, which consist of parts

- In the input impedance network: $C_{S1,in} = 1pF$, $L_{P,in} = 20nH$, $C_{S2,in} = 39pF$, and
- In the output impedance network: $L_{P,out} = 15nH$ and $R_{P,out} = 1500\,\Omega$, $C_{S,out} = 1.6pF$.

The input impedance matching network is built with a *T* type circuitry of *C-L-C*, by which the trace of S_{11} is pulled to a location near the center of the Smith chart which corresponds to the reference impedance point of $50\,\Omega$. The capacitor $C_{S2,in} = 39pF$ is actually a "zero" capacitor and functions as the *DC* blocking part. By means

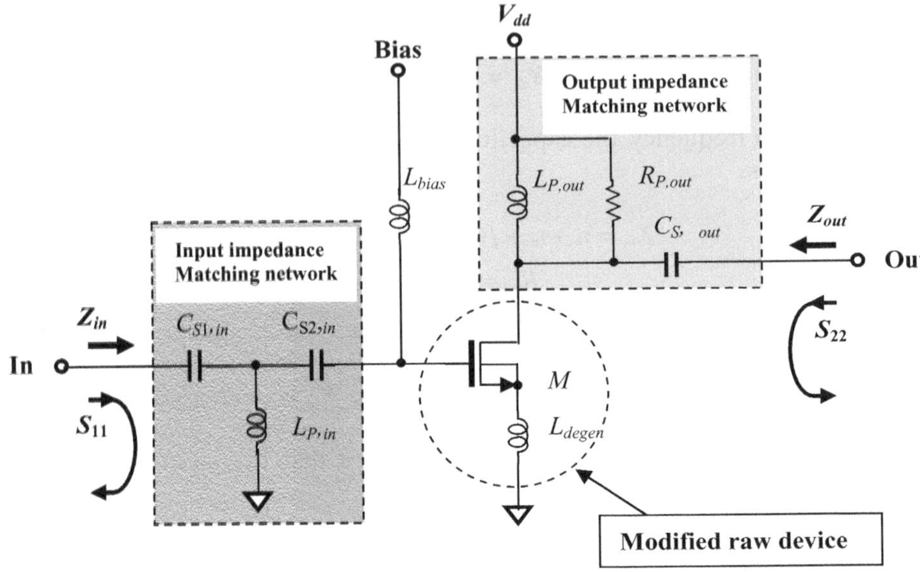

Figure 1.12 Impedance matching of modified raw device by parts. At input: $C_{S1,in} = 1pF$, $L_{P,in} = 20nH$, $C_{S2,in} = 39pF$. At output: $L_{P,out} = 15nH$, $R_{P,out} = 1500\,\Omega$, and $C_{S,out} = 1.6pF$.

of $L_{P,in}$ in parallel, the trace of S_{11} is pulled counter-clockwise to the resistance circle of $50\,\Omega$ along the constant conductance circle, and then, by means of $C_{S1,in}$ in series, the trace of S_{11} is pulled counter-clockwise to the area near the reference impedance point $50\,\Omega$ along the constant resistance circle.

The output impedance matching network consists of an inductor, a resistor, and a capacitor. The inductor, $L_{P,out} = 15nH$, is the main part of the output impedance matching network, by which the trace of S_{22} is pulled counter-clockwise to a location near the $50\,\Omega$ resistance circle along the constant conductance circle. The capacitor $C_{S,out} = 1.6pF$ is the next part, by which the S_{22} is drawn counter-clockwise to a location near the center of the Smith chart corresponding to the reference impedance point $50\,\Omega$ along the constant resistance circle. The resistor $R_{P,out}$ is a de-Q part, by which the bandwidth is widened and the *LNA* is changed from an unstable to a stable state.

From Figure 1.13(a) it can be seen that S_{11} and $\Gamma_{S,opt}$ are not completely superimposed on each other. The deviation arises from the fact that the S_{11}^* and $\Gamma_{S,opt}$ of the modified raw device shown in Figure 1.8 are not exactly overlapped. However, this is a normal phenomenon in the actual design because all parts have tolerance and in addition, absolutely ideal cases never happen in actual engineering design. The absolute accuracy of simulation processing is never realistic. As long as the two points S_{11}^* and $\Gamma_{S,opt}$ are pulled close enough to each other, it will be sufficient to satisfy the goals of the design.

Figure 1.13(b) shows that

- Input return loss, S_{11}

$$-13.8dB < S_{11} < -11.0dB, \quad \text{when } 850MHz < f < 940MHz, \tag{1.59}$$

$$S_{11} = -13.7dB, \quad \text{when } f = 895MHz, \tag{1.60}$$

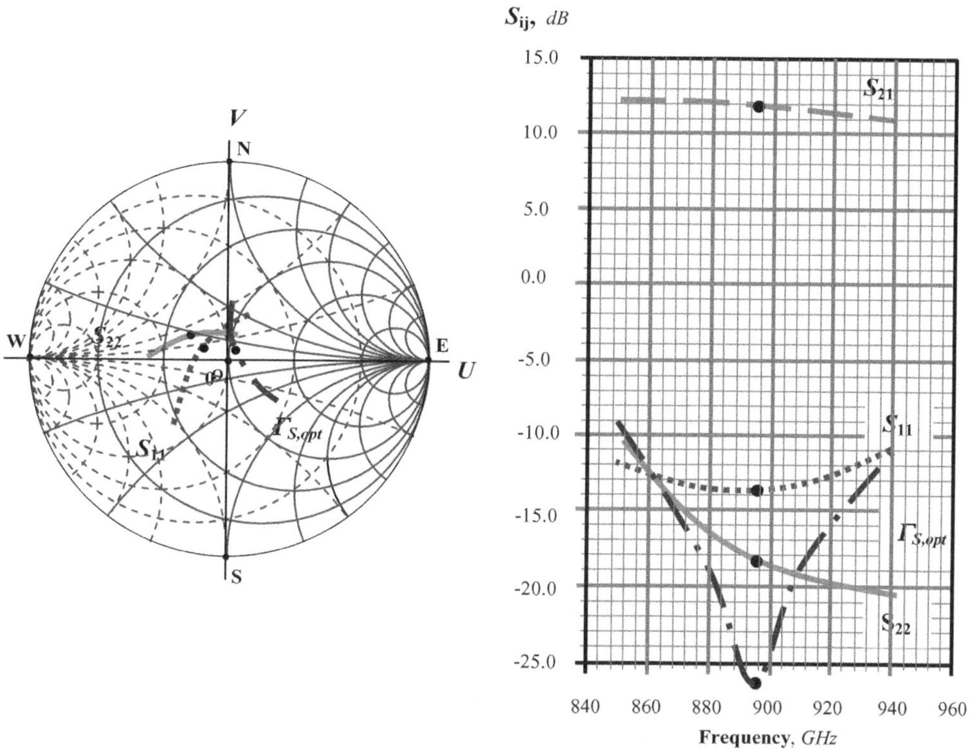

(a) S_{11}, S_{22} and $\Gamma_{S,opt}$ on Smith chart

(b) Magnitude of S_{ij}, dB

Figure 1.13 S parameters of the design example, $f = 850$–$940\,MHz$, $I_D = 2.6\,mA$. (The intermediate frequency $895\,MHz$ is marked with a dot on each trace.) The input and output impedances are matched by parts. At input: $C_{S1,in} = 1\,pF$, $L_{P,in} = 20\,nH$, $C_{S2,in} = 39\,pF$. At output: $L_{P,out} = 15\,nH$ and $R_{P,out} = 1500\,\Omega$, $C_{S,out} = 1.6\,pF$.

- Output return loss, S_{22}

$$-20.5\,dB < S_{22} < -10.5\,dB, \qquad \text{when } 850\,MHz < f < 940\,MHz, \tag{1.61}$$

$$S_{22} = -18.5\,dB, \qquad \text{when } f = 895\,MHz, \tag{1.62}$$

- Gain, S_{21}

$$11.0\,dB < S_{21} < 12.2\,dB, \qquad \text{when } 850\,MHz < f < 940\,MHz, \tag{1.63}$$

$$S_{21} = 12.0\,dB, \qquad \text{when } f = 895\,MHz, \tag{1.64}$$

- Optimum of source reflection coefficient, $\Gamma_{S,opt}$

$$-26.5\,dB < \boldsymbol{\Gamma}_{S,opt} < -9.0\,dB, \qquad \text{when } 850\,MHz < f < 940\,MHz, \tag{1.65}$$

$$\boldsymbol{\Gamma}_{S,opt} = -26.5\,dB, \qquad \text{when } f = 895\,MHz, \tag{1.66}$$

The isolation S_{12} is not shown in Figure 1.13(b) because it is lower than $25\,dB$ in the entire frequency range.

It can be seen that the gain is reasonable and the bandwidth is wide enough! In addition, the values of the parts are appropriate. Now let's check its noise performance.

1.2.5 Gain Circles and Noise Figure Circles

The noise performance of the design sample is good after impedance matching is done. This is expected because the modified raw device is designed for simultaneously approaching maximum gain and minimum noise figure. Figure 1.14 shows the performance of the noise figure in the entire frequency range, that is

$$1.37\,dB < NF < 1.97\,dB, \quad \text{when } 850\,MHz < f < 940\,MHz, \tag{1.67}$$

$$NF = 1.5\,dB, \quad \text{when } f = 895\,MHz, \tag{1.68}$$

The values of the noise figure shown in Figure 1.14 are denoted with two ordinates, one for simulated values and one for actual tested results. It can be seen that the actual tested values of the noise figure are $0.85\,dB$ higher than the simulated values. The good news is that the noise figure in the entire frequency range is better than the previously stated goals.

Exciting results are found from the plot of the gain and noise figure circles as shown in Figure 1.15. This is a plot for the operating frequency of $f = 895\,MHz$.

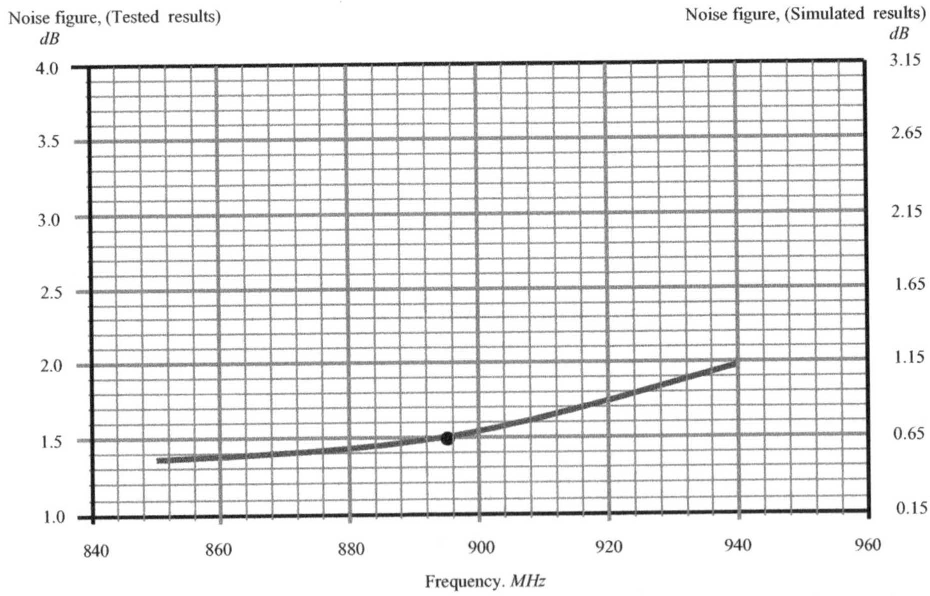

Figure 1.14 Noise figure of the design example, $f = 850$ to $940\,MHz$, $I_D = 2.6\,mA$. (The intermediate frequency 895 MHz is marked with a dot on the trace.)
$NF = 1.5\,dB$ when $f = 895\,GHz$.

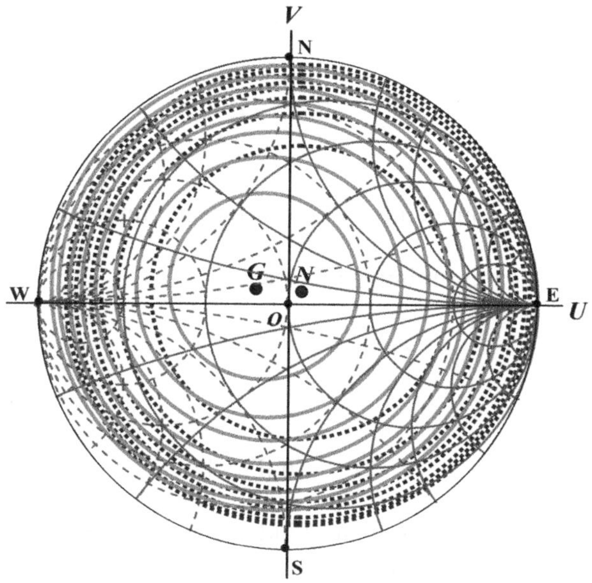

Input reflection coefficient Γ_S plane

Figure 1.15 Constant gain circles and constant noise figure circles when $f = 895\,MHz$.
Gain circles: $G_{max} = 12\,dB$ at point G, step = $1.0\,dB$.
Noise figure circles: $NF_{min} = 1.5\,dB$ at point N, step = $0.5\,dB$.

After the input and output impedances are matched, the center of the gain circles is point G, which is quite close to the center of the Smith chart. The maximum gain at this operating frequency is $12\,dB$ which is consistent with the results shown in Figure 1.13(b). On the other hand, the center of the noise figure circles is point N, which is also very closed to the center of Smith chart. The minimum of noise figure at this operating frequency is $1.5\,dB$ which is consistent with the results shown in Figure 1.14. The wonderful feature of Figure 1.15 is that the gain and noise circles overlap almost perfectly. The two centers are very close to each other and near the center of the Smith chart.

As mentioned above, the deviation between point N and point O is due to the incomplete overlapping in the modified raw device so that condition (1.34), $\Gamma_{S,opt} = S_{11}^*$, is not perfectly satisfied. In reality, it is impossible to reach an ideal goal predicted by theory. Deviation between practical design and circuit theory has always existed.

1.2.6 Stability

One of the important specifications for a *LNA* design is stability. It is obvious that a *LNA* may become an oscillator if it is unstable in the circuit performance. Therefore, the designer must examine its stability after the topology of the circuit and the value of parts, such as the design sample shown in Figure 1.12, are confirmed.

It is well known that the Smith chart is a reflection coefficient plane, called a Γ plane, in which the impedance, z, is plotted by means of the following relationship,

$$z = \frac{Z}{Z_o} = \frac{1+\Gamma}{1-\Gamma}, \tag{1.69}$$

where z is a normalized impedance by the reference impedance, which is usually $50\,\Omega$.

Equation (1.69) indicates an important fact, that is,

$$z > 0, \qquad \text{if } |\Gamma| < 1, \tag{1.70}$$

Otherwise,

$$z < 0, \qquad \text{if } |\Gamma| > 1, \tag{1.71}$$

This is a universal relationship for any kind of impedance and its corresponding reflection coefficient. If an impedance looking into a two-port block or terminal is positive, then the block or terminal is stable; if an impedance looking into a two-port block or a terminal is negative, then the block or terminal is unstable. As shown in Figure 1.16, there are 4 reflection coefficients, $\boldsymbol{\Gamma_S}$, $\boldsymbol{\Gamma_L}$, $\boldsymbol{\Gamma_{in}}$ and $\boldsymbol{\Gamma_{out}}$, and 4 S parameters, $\boldsymbol{S_{11}}$, $\boldsymbol{S_{22}}$, $\boldsymbol{S_{21}}$ and $\boldsymbol{S_{22}}$.

The conditions for unconditional stability for all of the reflection coefficients shown in Figure 1.16 are

$$|\Gamma_S| < 1, \tag{1.72}$$

$$|\Gamma_L| < 1, \tag{1.73}$$

$$|\Gamma_{in}| < 1, \tag{1.74}$$

$$|\Gamma_{out}| < 1. \tag{1.75}$$

The two-port block is unstable if the magnitude of any reflection coefficient above is greater than 1. A critical case is that in which the input and output reflection coefficients are equal to 1, that is,

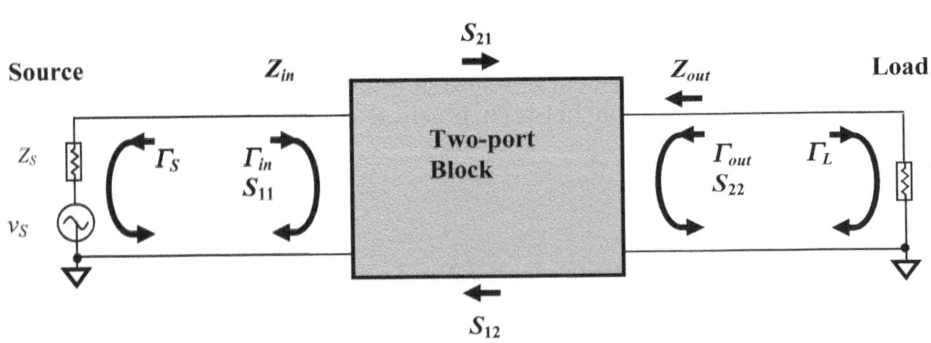

Figure 1.16 Different reflection coefficients and S parameters in a two-port block.

$$|\Gamma_{in}| = 1. \tag{1.76}$$

$$|\Gamma_{out}| = 1. \tag{1.77}$$

It is well known that

$$\Gamma_{in} = S_{11} + \frac{S_{12}S_{21}\Gamma_L}{1 - S_{22}\Gamma_L}, \tag{1.78}$$

$$\Gamma_{out} = S_{22} + \frac{S_{12}S_{21}\Gamma_S}{1 - S_{11}\Gamma_S}. \tag{1.79}$$

The conditions in the critical case therefore are

$$\left| S_{11} + \frac{S_{12}S_{21}\Gamma_L}{1 - S_{22}\Gamma_L} \right| = 1, \tag{1.80}$$

$$\left| S_{22} + \frac{S_{12}S_{21}\Gamma_S}{1 - S_{11}\Gamma_S} \right| = 1. \tag{1.81}$$

The critical values of Γ_S and Γ_L can be solved from equations (1.80) and (1.81), they are

$$\left| \Gamma_L - \frac{(S_{22} - \Delta S_{11}^*)^*}{|S_{22}|^2 - |\Delta|^2} \right| = \left| \frac{S_{12}S_{21}}{|S_{22}|^2 - |\Delta|^2} \right|, \tag{1.82}$$

$$\left| \Gamma_S - \frac{(S_{11} - \Delta S_{22}^*)^*}{|S_{11}|^2 - |\Delta|^2} \right| = \left| \frac{S_{12}S_{21}}{|S_{11}|^2 - |\Delta|^2} \right|. \tag{1.83}$$

where

$$|\Delta| = |S_{11}S_{22} - S_{12}S_{21}|. \tag{1.84}$$

On the Smith chart, they appear as two circles and are called the output and input stability circles. These two stability circles are very useful in the analysis of potentially unstable blocks. However, we are not going to repeat this analysis since it has already been discussed in great detail in many textbooks What we are interested in is how to quickly judge the stability of a designed block.

Historically, a K factor, which is a function of S parameters, has been defined as,

$$K = 1 - |S_{11}|^2 - |S_{22}|^2 + |\Delta|^2. \tag{1.85}$$

In terms of mathematical processing, the necessary and sufficient conditions for a two-port block to be unconditionally stable are

$$K > 1, \tag{1.86}$$

and

$$|\Delta| < 1. \tag{1.87}$$

For many years, the stability of a two-port block has been judged by the K factor and the intermediate parameter Δ in terms of equations (1.86) and (1.87).

In recent years, however, they have been replaced by the μ factor, which is defined as

$$\mu = \frac{1-[mag(S_{11})]^2}{mag[S_{22}-\Delta conj(S_{11})]+mag(S_{21}S_{12})}. \tag{1.88}$$

In equation (1.88), the symbols "mag(...)" and "mag[...]" denote the magnitude of the parameters in the parenthesis or brace, and the symbol "conj(...)" denotes the conjugated value of the parameters in the parenthesis.

The block is unconditionally stable if

$$\mu > 1. \tag{1.89}$$

Otherwise, it is potentially unstable.

Rather than the two conditions of the K and Δ factors, condition (1.89), using the factor of μ is simplest and most convenient criterion. From (1.88) it can be seen that the value of μ is dependent on the tested S parameters only. In today's simulation by *ADS* and Cadence simulation tools, the values of μ for the entire frequency range can be promptly and easily displayed on screen.

The stability of the raw device shown in Figure 1.8 is unstable because the simulation shows its μ value as

$$\mu = 0.95 < 1. \tag{1.90}$$

After the steps of raw device testing, the stability of the *LNA* block must be taken care of. One should perform analysis on the sources of instability.

In the schematic shown in Figure 1.8, the instability may come from the drain inductor, L_d, and the degenerator inductor, L_{degen}.

In order to retain as much voltage drop across the device as possible, so as to reduce intermodulation, it is preferred to use a drain inductor, L_d, rather than a resistor, although an inductor is much more expensive than a resistor. In order to satisfy the overlapping condition (1.34) for NF_{min} without a significant increase of the noise figure, it is preferred to use a degeneration inductor, L_{degen}, rather than a degeneration resistor, R_{degen}.

The drain inductor L_d in Figure 1.8 is adjusted and re-named $L_{P,out}$ as shown in Figure 1.12. The resistor, $R_{P,out}$, is used to de-Q the load inductor $L_{P,out}$, in order to prevent oscillation. On the other hand, the value of L_{degen}, should not be too high, otherwise the gain and stability both suffer.

In our design sample, $L_{degen} = 10nH$ is chosen. The resistor $R_{P,out}$ is the key part to keep the *LNA* block stable, while the lower value of the degeneration inductor L_{degen} provides good assistance.

The values of all the parts shown in the schematic as shown in Figure 1.12 are simultaneously optimized to obtain stability with a reasonable gain and noise figure level, that is,

$$\mu = 1.1 > 1, \tag{1.91}$$

when

$$G = 12\,dB, \tag{1.92}$$

and

$$NF = 1.5\,dB, \tag{1.93}$$

at $f = 895\,MHz$.

The μ value shown in (1.91) indicates that the designed *LNA* is in an unconditionally stable state.

1.2.7 Non-Linearity

1.2.7.1 Spectrum at* LNA *Output An overview of the spectrum at the *LNA* output can provide an intuitive feeling about the linearity of the design. Figure 1.17 shows that

$$P_{out} = -38.0\,dB_m, \quad \text{at } f = 895\,MHz, \tag{1.94}$$

$$P_{out} = -80.7\,dB_m, \quad \text{at } f = 1790\,MHz, \tag{1.95}$$

$$P_{out} = -119.1\,dB_m, \quad \text{at } f = 2685\,MHz, \tag{1.96}$$

$$\ldots$$

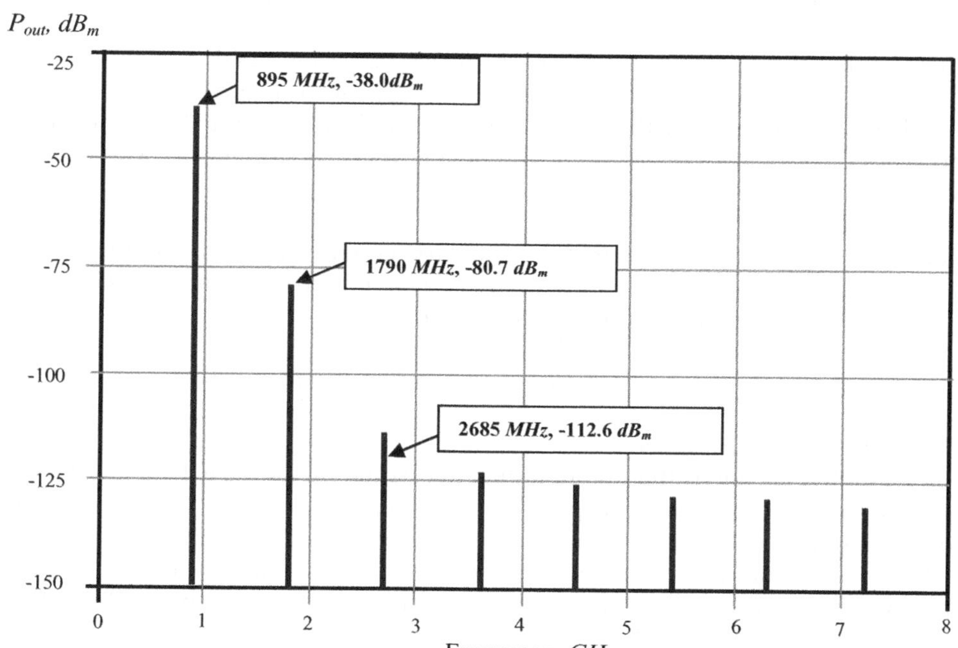

Figure 1.17 Spectrum at *LNA* output, $f_o = 895\,MHz$, $P_{in} = -50\,dB_m$.

when

$$P_{in} = -50\,dB_m. \tag{1.97}$$

Then the gain of LNA is

$$P_{out} - P_{in} = -38\,dB_m - (-50\,dB_m) = 12\,dB, \tag{1.98}$$

which is consistent with the gain testing shown in Figure 1.13(b).

The second harmonic is lower than the output power at the operating frequency by

$$\Delta P_{out,2} = P_{out}|_{1790\,MHz} - P_{out}|_{895\,MHz} = -80.7\,dB_m - (-38.0\,dB_m) = -42.7\,dB, \tag{1.99}$$

which implies that the spurious products at the half IF frequency, or the Able-Baker spurious products, would be low, and that the second order intercept point would be high. According to design experience, if the LNA block is applied to a popular cellular phone communication system, $\Delta P_{out,2}$ should be suppressed by $30\,dB$ at least at the operating frequency so that the distortion of the useful signal will not have conceivable distortion due to the second order spurious products. The second order non-linearity is of priority in a direct conversion or zero IF communication system because it produces DC-offset to disturb the desired signal.

The third harmonic is lower than the output power at the operating frequency by

$$\Delta P_{out,3} = P_{out}|_{1790\,MHz} - P_{out}|_{895\,MHz} = -116.0\,dB_m - (-38.0\,dB_m) = -78.0\,dB, \tag{1.100}$$

which implies that the third order non-linearity is very small, so that the probability of interference from adjacent channels will be very low or negligible.

Other harmonics higher than third order harmonic are at least $80\,dB$ lower than the output power at the operating frequency. There is nothing to worry about here.

1.2.7.2 1 dB Compression Point The $1\,dB$ compression point is the second easy method to overview the non-linearity of a circuit block. Figure 1.18 shows the $1\,dB$ compression point of our design sample, that is,

$$P_{1dB} = -1.25\,dB_m \qquad \text{when } f = 895\,MHz, \tag{1.101}$$

which satisfies the desired goal.

An intuitive feeling is that the non-linearity would be seriously hampered if the LNA is operated with an input power higher than $-5\,dB_m$. The $1\,dB$ compressed point looks like the demarcation point of input power. The LNA performs as a linear unit when its input power is below the $1\,dB$ compression point. On the other hand, the LNA performs as a non-linear unit when its input power is beyond the $1\,dB$ compression point.

The third order intercept point can be estimated to be 5 to $10\,dB$ higher than the $1\,dB$ compression point, which might be a value between 5 to $10\,dB_m$.

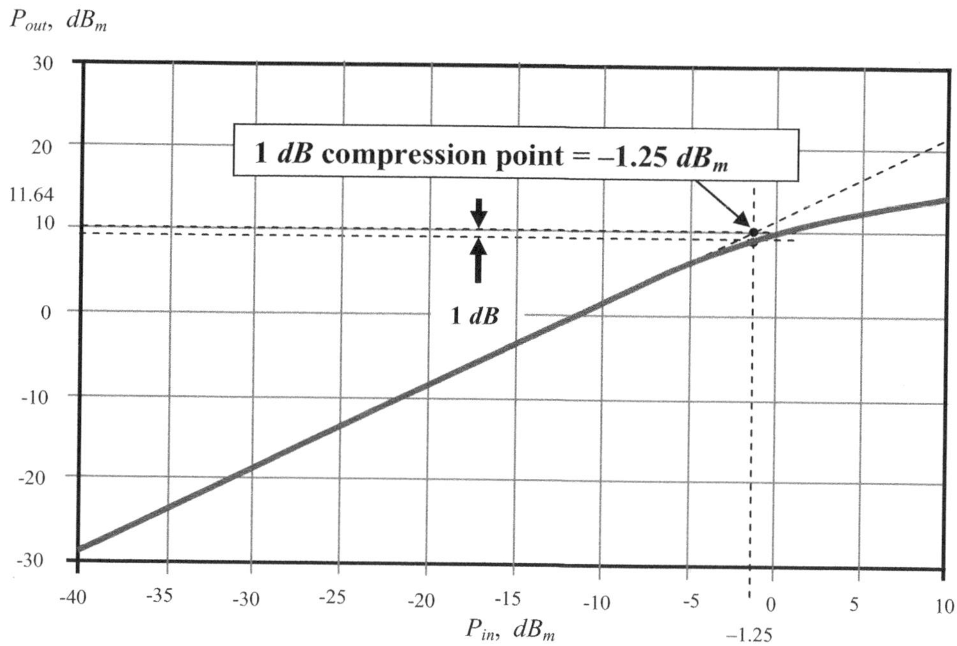

Figure 1.18 1 dB compression point when $f_o = 895\,MHz$.

1.2.7.3 ***Third and Second Order Intercept Points* IP_3 *and* IP_2** Figure 1.19 shows the third order intercept point, IP_3, that is,

$$IIP_3 = 8.6\,dB_m, \quad \text{and} \quad OIP_3 = IIP_3 + G = 20.6\,dB_m, \tag{1.102}$$

In Figure 1.19, the tested values of input and output powers are plotted with bold lines. The asymptote with the expected slope is plotted with dash lines. Point ***P**** is the actual intercept point. However, point P is the expected intercept point, which is the intercept point of the lines with slopes $n = 1$ and $n = 3$. The bent portion of the dash lines indicates another non-linearity of the designed block.

Figure 1.20 shows the second intercept point, IP_2, that is,

$$IIP_2 = 38.1\,dB_m, \quad \text{and} \quad OIP_2 = IIP_2 + G = 50.1\,dB_m. \tag{1.103}$$

Testing is conducted in the input power range below $0\,dB_m$ and the intercept point is obtained by the extension of the two lines with the slopes $n = 1$ and $n = 2$.

These results satisfy the aforementioned goals. Should the tested results be unsatisfactory, extra linearization work for the *LNA* must be conducted and the simulation may have to start from the beginning.

1.2.8 Design Procedures

Based on design experience, a flow chart of *LNA* design procedures is shown in Figure 1.21. It should be noted that

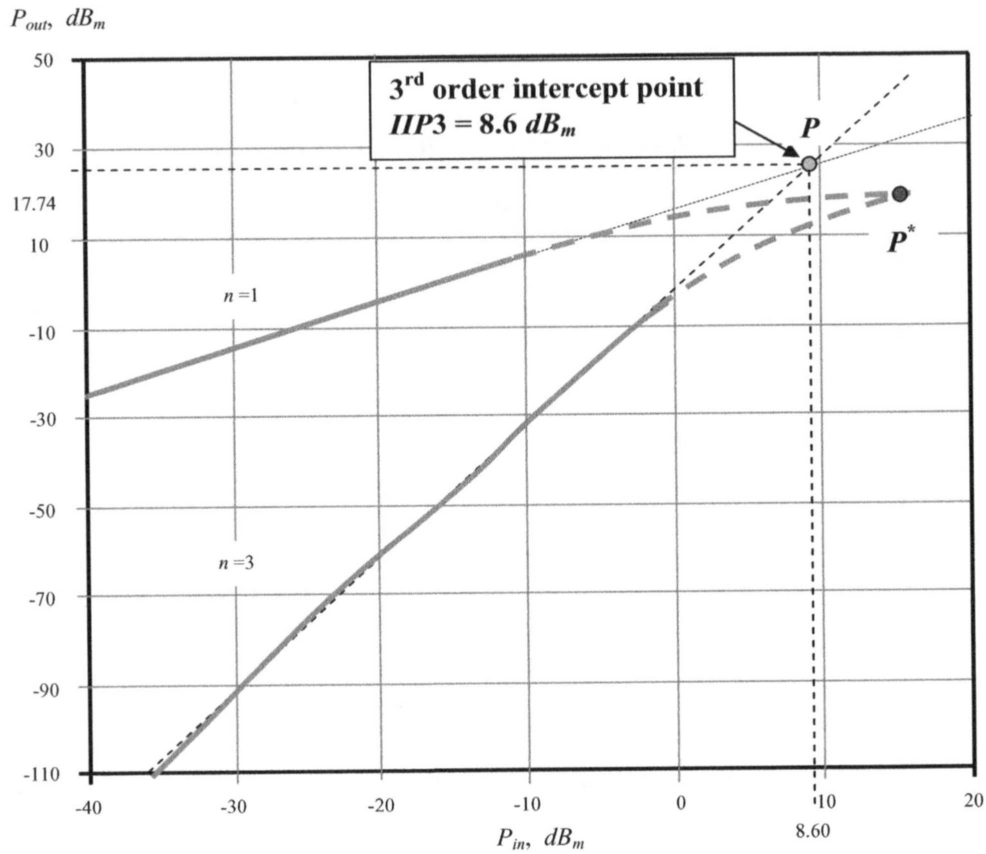

Figure 1.19 Third-order input intercept point when $f_o = 895\,MHz$.

- Raw device testing is the key step on which success or failure largely depends.
- The first iteration loop related to the raw device is judged by the condition of $S_{11}^* = \boldsymbol{\Gamma}_{\boldsymbol{S,opt}}$. There are three options to choose from to satisfy this condition. If the design work in this step is executed well, a good *LNA* design should be able to obtain maximum gain and minimum noise figure simultaneously. Consequently the gain and noise figure circles should almost overlap entirely near the center of the Smith chart.
- The second iteration loop related to the raw device is to judge the stability by the condition of $\boldsymbol{\mu > 1}$.
- The third iteration loop related to the raw device is to reach or exceed all the performance parameters.
- There is another iteration loop to examine the bandwidth in the process of input and output impedance matching.

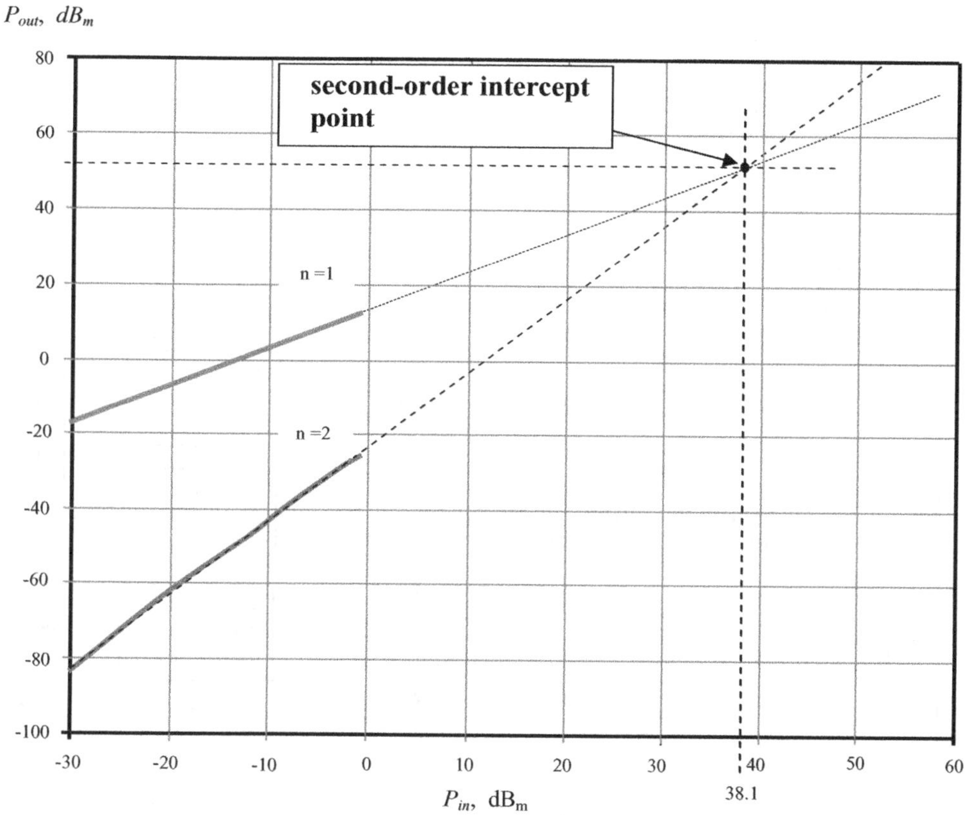

Figure 1.20 Second-order input intercept point when $f_o = 895\,MHz$.

1.2.9 Other Examples

The following three *LNA*s are actually implemented by discrete chip parts for different frequency bands. They are good examples to verify and illustrate the technology of the simultaneous approach to both NF_{min} and G_{max} as discussed in Section 1.2.3.2, even though they were designed by the author many years ago.

1.2.9.1* LNA *Design for* VHF *Frequency Band The *LNA* designed for *VHF* frequency band are outlined in the following three items:

- Electrical features (Table 1.2),
- The schematic (Figure 1.22),
- The gain and noise circles at the Γ_{in} (input voltage reflection coefficient) plane).

It should be noted that

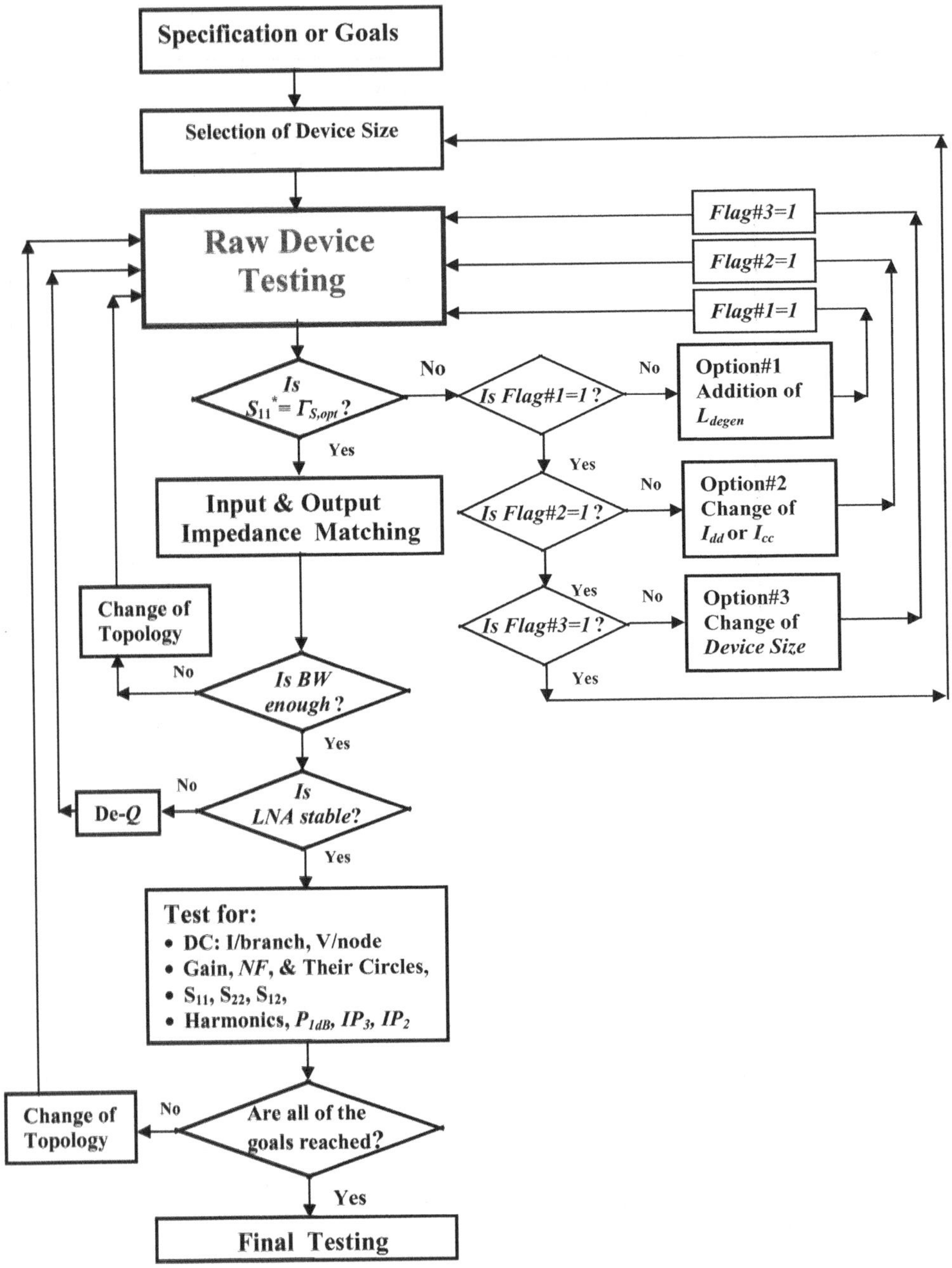

Figure 1.21 Flow chart for *LNA* design procedures.

- The inductor $L_3 = 33nH$ is the degeneration inductor mentioned above. The resistor $R_6 = 100\,\Omega$ is the "de-Q" resistor. The modified raw device consists of Q_1, L_3, and R_6.
- Capacitors, C_4, C_2, and C_6 are "zero" capacitors. Their specified values should be $1300pF$ and actual values are $1500pF$.

TABLE 1.2 Electrical features of *LNA* for *VHF* frequency band

	Specification		Final tested result	
Frequency range		130 MHz to 180 MHz		
Device		*BFQ67* (Manufacturer: Siemens)		
DC power supply	3	V		
Current drain	<4	mA	**3.47**	mA
Gain	>10	dB	**15.0**	dB
Noise figure	<2	dB	**1.75**	dB
IIP_3	>0	dB_m	**2.5**	dB_m
Input return	<−10	dB	**−40.0**	dB
Output return	<−10	dB	**−12**	dB

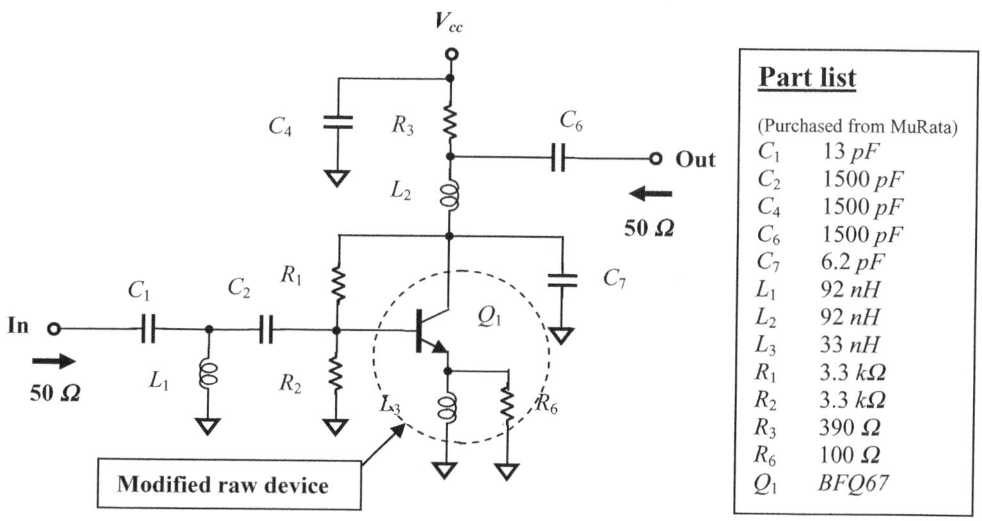

Figure 1.22 *LNA* designed for *VHF* frequency band.

- The resistors R_3 and R_6 are 390 Ω and 100 Ω respectively; they are both "de-Q" resistors.
- The resistors R_1 and R_2 are the combination of a voltage divider. They take the *DC* bias for the transistor from V_{cc}; they function as a feedback branch as well.
- The output impedance network consisting of R_3, L_2, and C_6 confers two advantages, lower part count and wider bandwidth.
- All parts in Figure 1.22 are chip parts and are purchased from MuRata.

Gain Circles and Noise Circles Figure 1.23 shows that for this circuit the noise figure circles overlap the gain circles well enough, although not perfectly. This verifies the assertion of simultaneously approaching both NF_{min} and G_{max}, because Figure 1.23 is obtained due to the satisfaction of condition (1.34).

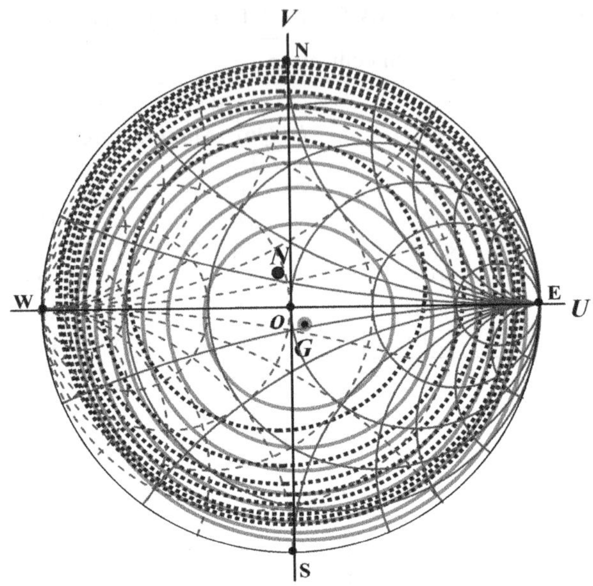

Input reflection coefficient Γ_S plane

Figure 1.23 Constant gain circles and constant noise figure circles when $f = 150\,MHz$.
Gain circles: $G_{max} = 15\,dB$ at point G, step = $1.0\,dB$.
Noise figure circles: $NF_{min} = 1.75\,dB$ at point N, step = $0.25\,dB$.

1.2.9.2 LNA Design for UHF Frequency Band The *LNA* designed for *UHF* frequency band can be outlined in the following three items:

- Electrical features (Table 1.3),
- The schematic (Figure 1.24),
- The gain and noise circles at the Γ_{in} (input reflection coefficient) plane.

It should be noted that

- The inductor $L_3 = 5.6nH$ is the degeneration inductor mentioned above. The resistor $R_6 = 390\,\Omega$ is the "de-Q" resistor. The modified raw device consists of Q_1, L_3, and R_6.
- The capacitors C_4, C_2, and C_6 are "zero" capacitors. Their value is $150pF$.
- The resistors $R_3 = R_6 = 390\,\Omega$ are "de-Q" resistors.
- The resistors, R_1 and R_2, are the combination of a voltage divider. They take the DC bias for the transistor from V_{cc}; they function as a feedback branch as well.
- The output impedance network consisting of R_3, L_2, and C_6 confers two advantages, lower part count and wider bandwidth.
- All parts in Figure 1.24 are chip parts and are purchased from MuRata.

TABLE 1.3 Electrical features of *LNA* for *UHF* frequency band

	Specification		Final tested result	
Frequency range	400 *MHz* to 470 *MHz*			
Device	*BFQ67* (Manufacturer: Siemens)			
DC power supply	3	*V*		
Current drain	<4	*mA*	**3.43**	*mA*
Gain	>10	*dB*	**12.0**	*dB*
Noise figure	<2	*dB*	**1.5**	*dB*
IIP_3	>0	dB_m	**5.0**	dB_m
Input return	<–10	*dB*	**–17.8**	*dB*
Output return	<–10	*dB*	**–16.0**	*dB*

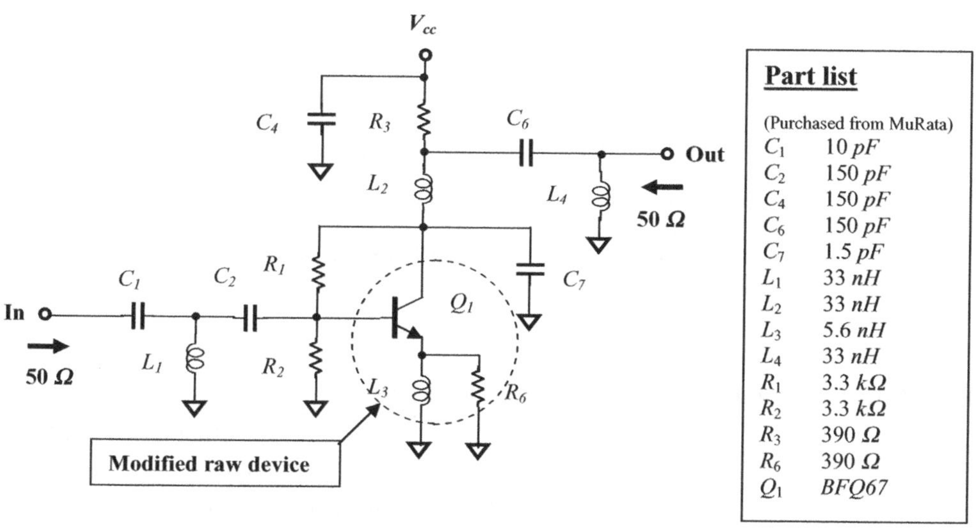

Figure 1.24 *LNA* designed for *UHF* frequency band.

Gain Circles and Noise Circles Figure 1.25 shows that for this circuit the noise figure circles overlap with the gain circles well enough, although not perfectly. This verifies the assertion of simultaneously approaching both NF_{min} and G_{max}, because Figure 1.25 is obtained due to the satisfaction of condition (1.34).

1.2.9.3* LNA *Design for 800/900* MHz *Frequency Band The *LNA* designed for 800/900 *MHz* frequency band is outlined in the following three items:

- Electrical features (Table 1.4),
- The schematic (Figure 1.26),
- The gain and noise circles at the Γ_{in} (input reflection coefficient) plane.

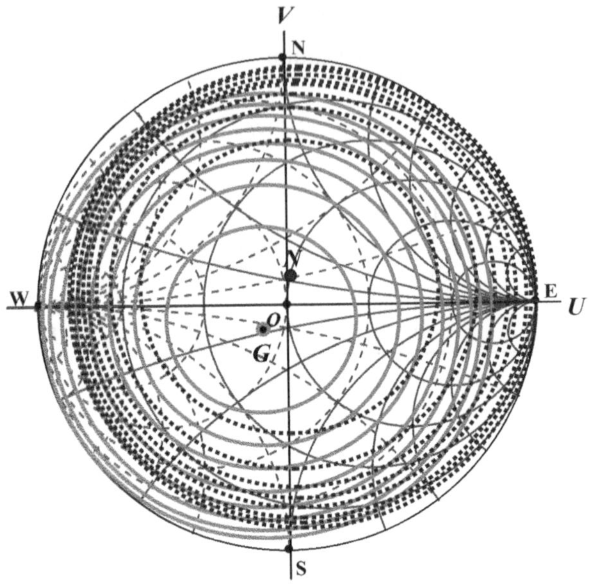

Input reflection coefficient Γ_S plane

Figure 1.25 Constant gain circles and constant noise figure circles when $f = 450\,MHz$.
Gain circles: $G_{max} = 12\,dB$ at point G, step $= 1.0\,dB$.
Noise figure circles: $NF_{min} = 1.5\,dB$ at point N, step $= 0.25\,dB$.

TABLE 1.4 Electrical features of *LNA* for 800/900 *MHz*

	Specification		Final tested result	
Frequency range	850 MHz to 940 MHz			
Device	$BFQ67$ (Manufactured by Siemens)			
DC power supply	3	V		
Current drain	<4	mA	**3.48**	mA
Gain	>10	dB	**11.0**	dB
Noise figure	<2	dB	**1.8**	dB
IIP_3	>0	dB_m	**8.6**	dB_m
Input return	<−10	dB	**−11.7**	dB
Output return	<−10	dB	**−17.0**	dB

It should be noted that

- The inductor $L_3 = 4.7\,nH$ is the degeneration inductor mentioned above. The resistor $R_6 = 390\,\Omega$ is the "de-Q" resistor. The modified raw device consists of Q_1, $L3$, and R_6.
- The capacitors, C_4, C_2, and C_6 are "zero" capacitors. Their value is $39\,pF$.
- The resistors $R_3 = R_6 = 390\,\Omega$ are "de-Q" resistors.

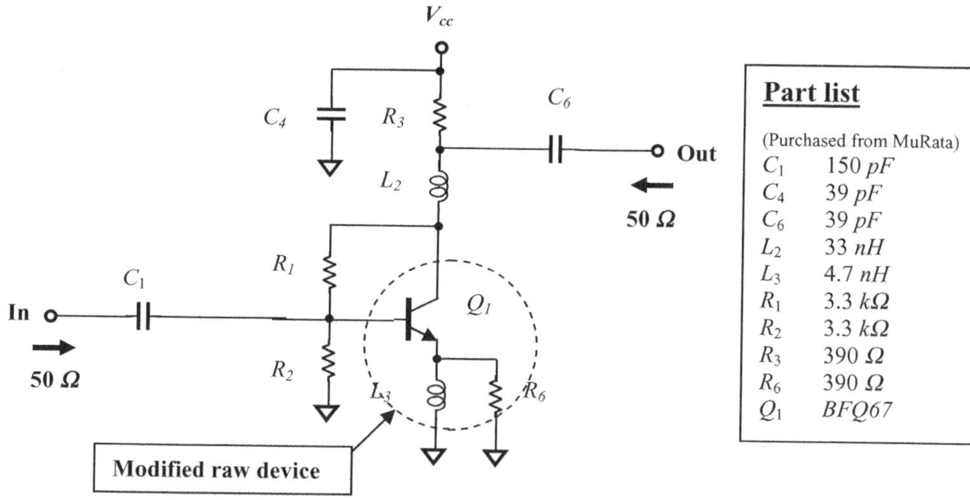

Figure 1.26 *LNA* designed for 800/900 *MHz*.

- The resistors, R_1 and R_2, are the combination of a voltage divider. They take the *DC* bias for the transistor from V_{cc}; they function as a feedback branch as well.
- The output impedance network consisting of R_3, L_2, and C_6 confers two advantages, lower part count and wider bandwidth.
- All parts in Figure 1.26 are chip parts and are purchased from MuRata.

Gain Circles and Noise Circles Figure 1.27 shows that for this circuit the noise figure circles are adequate, though not perfect, and overlap the gain circles. It verifies that the correctness of the assertion of simultaneously approach to both NF_{min} and G_{max}, because Figure 1.27 is obtained due to the satisfaction of the condition (1.34).

1.3 SINGLE-ENDED CASCODE *LNA*

1.3.1 Bipolar *CE-CB* Cascode Voltage Amplifier

Figure 1.28 shows a bipolar *CE-CB* cascode amplifier. The first stage is a voltage amplifier with a *CE* (Common Emitter) configuration because its emitter is the common *AC* and *DC* grounded terminal of the input and output. The second stage is a voltage amplifier with a *CB* (Common Base) configuration because its base is the common *AC* grounded terminal of the input and output. The collector of the first *CE* stage is connected to the emitter of the second *CB* stage. The input terminal is the base of the first *CE* stage, the output terminal is the collector of the second *CB* stage. *RF* chokes and a "zero" capacitor, C_{zero}, are connected to the devices for *DC* bias. In the operating frequency range, the impedance of the *RF* choke is assumed to approach infinity and the impedance of the capacitor C_{zero} is assumed to approach zero.

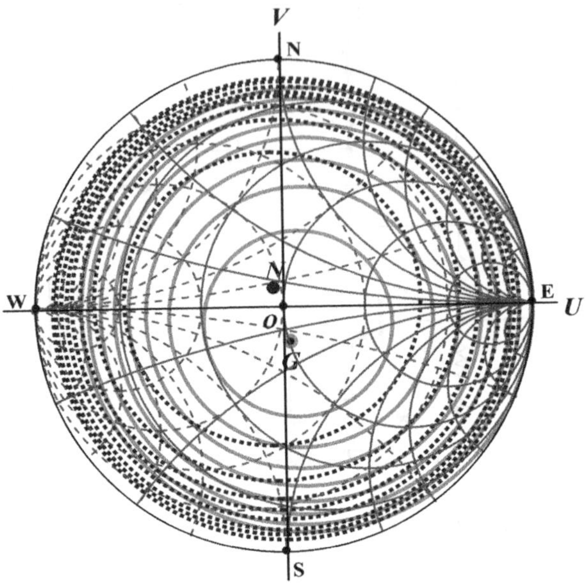

Figure 1.27 Constant gain circles and constant noise figure circles when $f = 850\,MHz$.
Gain circles: $G_{max} = 11\,dB$ at point G, step $= 1.0\,dB$.
Noise figure circles: $NF_{min} = 1.8\,dB$ at point N, step $= 0.25\,dB$.

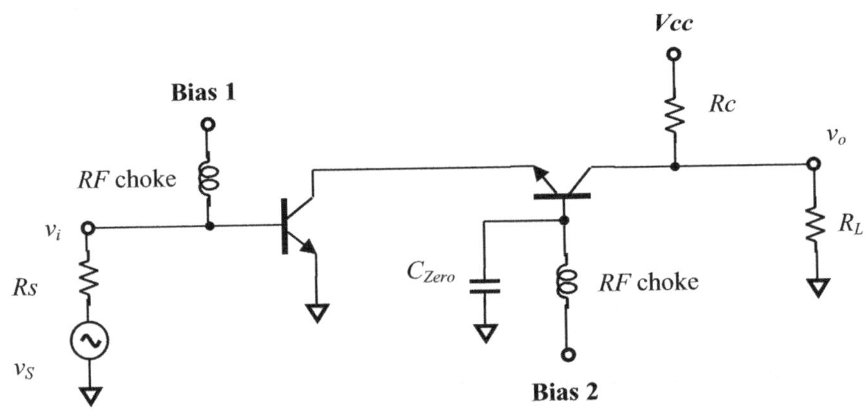

Figure 1.28 A bipolar cascode amplifier.

The equivalent of the bipolar cascode amplifier is shown in Figure 1.29(a). In order to simplify the analysis, let's consider the cases of low frequencies so as to neglect all the capacitors in the transistors, which is shown in Figure 1.29(b). Also, the resistors r_b, r_c and r_μ are neglected. In this section, the subscript "1" denotes the parameter in the first stage, the subscript "2" denotes the parameter in the second stage, the subscript "i" denotes the input parameter, the subscript "o" denotes the output parameter.

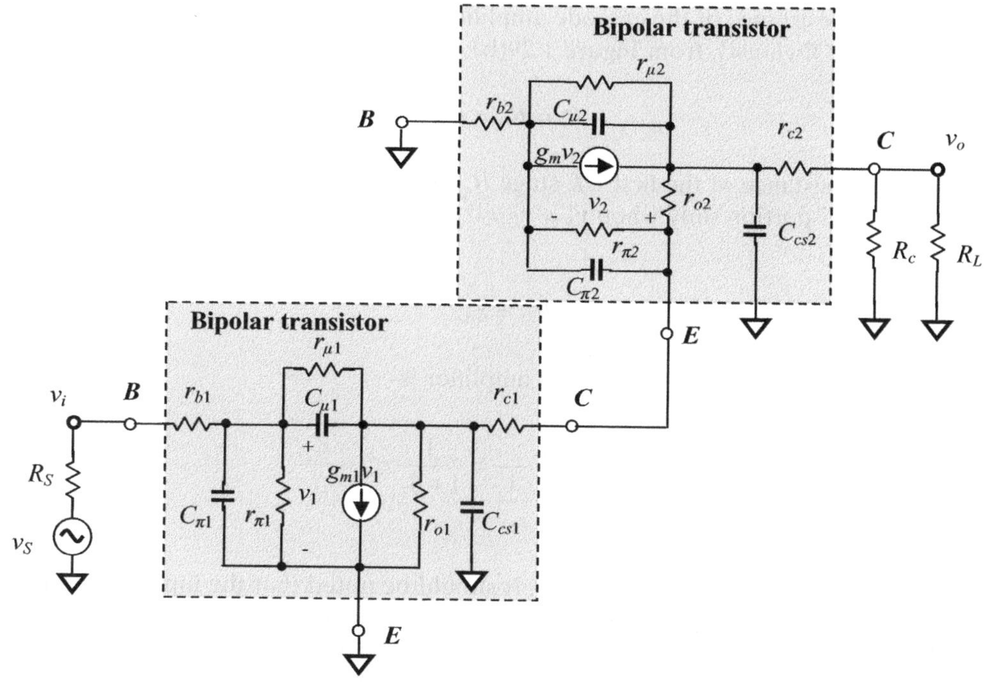

Figure 1.29(a) Equivalent circuit of bipolar cascode amplifier with *CE-CB* configuration.

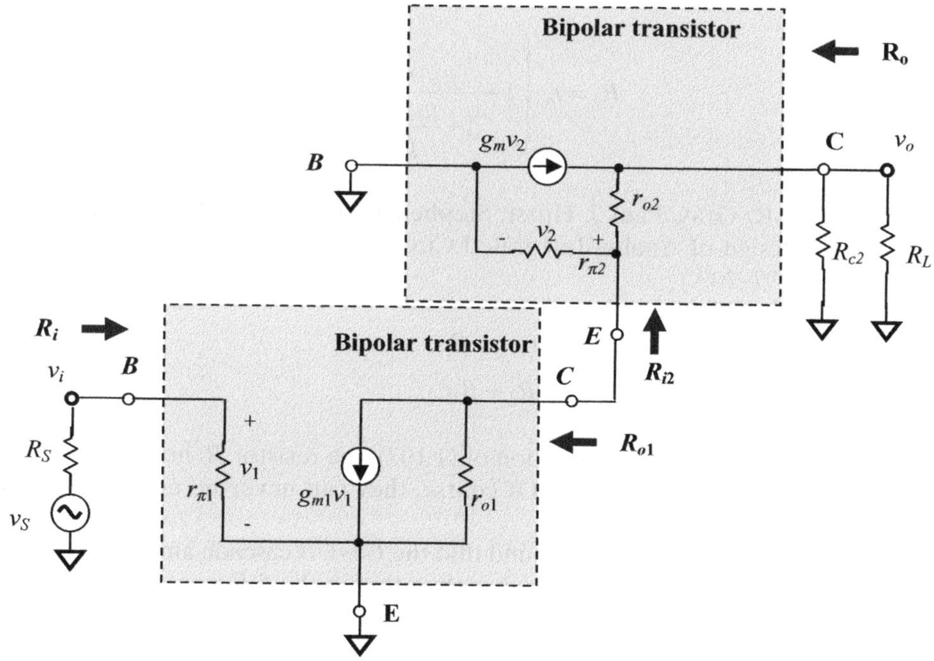

Figure 1.29(b) Equivalent circuit of bipolar cascode amplifier with *CE-CB* configuration at low frequencies, neglecting the resistors, r_b, r_μ, and r_c.

The input resistance of the cascode amplifier R_i is the input resistance of the first *CE* stage R_{i1}. Obviously, from Figure 1.29(b), it can be seen that

$$R_i = R_{i1} = r_{\pi 1}. \tag{1.104}$$

The output resistance of the first *CE* stage R_{o1} is the output impedance contributed by the first *CE* portion only when $v_S = v_1 = 0$ so that the generator $g_{m1}v_1$ is inactive. Then,

$$R_{o1} = r_{o1}. \tag{1.105}$$

The input resistance of the second *CB* amplifier is

$$R_{i2} = r_{e2} = \frac{1}{g_{m2} + \frac{1}{r_{\pi 2}}} = \frac{1}{1+\beta_2} r_{\pi 2} \approx \frac{r_{\pi 2}}{\beta_2}, \tag{1.106}$$

(discussed in Section 12.5, Chapter 12). It should be noted that the input resistance of the second *CB* stage is dropped by about β times from r_π in *CE* stage alone.

The output resistance is contributed by the upper *CB* portion only when $v_S = v_1 = 0$, so that the generator $g_{m1}v_1$ is inactive. The bottom portion looks just like a single resistor, r_{o1}. The entire *CE-CB* amplifier is equivalent to a *CE* stage with a degeneration resistor r_{o1} and a *AC* grounded input terminal. The output resistance has been formularized as

$$R_o = r_{o2}\left(1+\frac{g_{m2}r_{o1}}{1+\frac{g_{m2}r_{o1}}{\beta_2}}\right). \tag{1.107}$$

(Refer to: Paul R. Gray, Paul J. Hurst, Stephen H. Lewis, and Robert G. Meyer, Analysis and Design of Analog Integrated Circuits, John Wiley & Sons, Inc., 2004, 4th edition, pp. 197–207.)

If
$$g_{m2}r_{o1} >> \beta_2 >> 1, \tag{1.108}$$
then
$$R_o \approx \beta_2 r_{o2}. \tag{1.109}$$

It should be noted that in the derivation of (1.107), the resistor R_c in the second *CB* stage and the load R_L are ignored. Of course, they can never be neglected in *RF* circuit design.

From equation (1.109) it can be found that the *CE-CB* cascade amplifier displays an output resistance about β times higher than that in the *CE* stage alone.

Now let us consider the voltage and current gain.

The voltage gain of the first *CE* stage A_{v1} is not that expected from a normal *CE* stage, $A_{v1} = -g_{m1}r_{o1}$. The output resistor r_{o1} of the first *CE* stage is connected with the input resistor R_{i2} of the second *CB* stage in parallel. As shown in equation (1.106), the input resistance of the second *CB* stage is dropped by about β times from the r_π in a normal *CE* stage; consequently,

$$A_{v1} = -g_{m1} r_{o1} \,//\, R_{i2} \approx -g_{m1} \frac{r_{\pi 2}}{\beta_2}, \tag{1.110}$$

if the input voltage v_i but not v_S is considered as the reference for the voltage gain.

Should the main parameters of the *CE* and *CB* transistor be equal, that is,

$$g_{m1} = g_{m2}, \tag{1.111}$$

$$r_{\pi 1} = r_{\pi 2}, \tag{1.112}$$

and

$$\beta_1 = \beta_2, \tag{1.113}$$

then equation (1.110) becomes

$$A_{v1} = -1. \tag{1.114}$$

The current gain of first *CE* stage A_{i1} is

$$A_{i1} = \beta. \tag{1.115}$$

The current gain of second stage A_{i2} is

$$A_{i2} \approx 1, \tag{1.116}$$

because the input current is almost equal to its output current in the second *CB* stage.

$$I_{i2} \approx I_{o2} \quad \text{and} \quad i_{i2} \approx i_{o2}, \tag{1.117}$$

where the upper case letter 'I' denotes the *DC* current and the lower case letter '*i*' denotes the *AC* current.

From (1.115) and (1.116), the total current gain is

$$A_i \approx A_{i1} A_{i2} \approx \beta. \tag{1.118}$$

The trans-conductance from the input to output is

$$G_m = g_{m1}, \tag{1.119}$$

since $A_{i2} \approx 1$ as shown in (1.116).

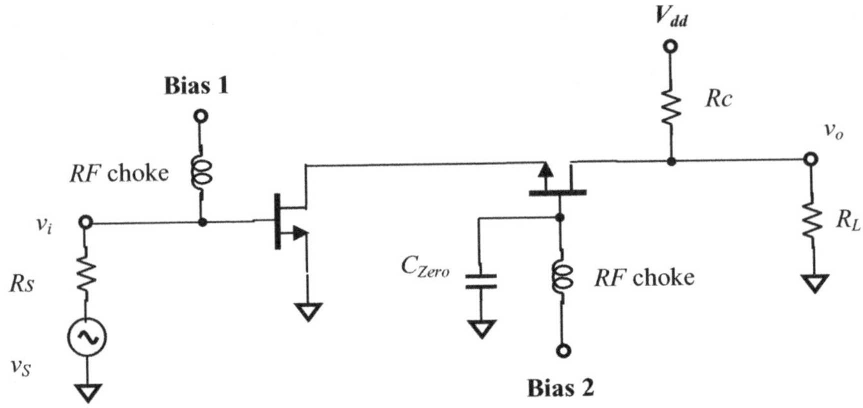

Figure 1.30 A *MOSFET* cascode amplifier.

The total voltage gain is

$$A_v = \frac{v_o}{v_i} \approx -G_m R_o \approx -g_{m1} r_{o2} \beta_2. \tag{1.120}$$

1.3.2 *MOSFET CS-CG* Cascode Voltage Amplifier

Figure 1.30 shows a *MOSFET CS-CG* cascode amplifier. The first stage is a voltage amplifier with a *CS* (Common Source) configuration because its source is the common *AC* and *DC* grounded terminal of the input and output. The second stage is a voltage amplifier with a *CG* (Common Gate) configuration because its gate is the common AC grounded terminal of the input and output. The drain of the first *CS* stage is connected with the source of the second *CG* stage. The input terminal is the gate of the first *CS* stage and the output terminal is the drain of the second *CG* stage. *RF* chokes and a "zero" capacitor, C_{zero}, are connected to the devices for *DC* bias. In the operating frequency range, the impedance of the *RF* choke is assumed to approach infinity and the impedance of the capacitor C_{zero} is assumed to approach zero.

The equivalent of the *MOSFET* cascode amplifier is shown in Figure 1.31. In order to simplify the analysis, let's consider the cases of low frequencies only so as to neglect all the capacitors in the transistors, shown in Figure 1.32. Also, the resistors r_b, r_c and r_μ are neglected. Figure 1.31 shows the equivalent circuit of the *MOSFET* cascode amplifier with a *CS-CG* configuration at low frequencies.

The input resistance of the *CS-CG* cascode amplifier is the input resistance of the first *CS* stage. Obviously, from Figure 1.32, it can be seen that

$$R_i \rightarrow \infty, \tag{1.121}$$

The output resistance of the first *CS* stage is contributed by the first *CS* portion only when $v_S = v_1 = 0$ so that the generator $g_{m1}v_1$ is inactive, then

$$R_{o1} = r_{o1}. \tag{1.122}$$

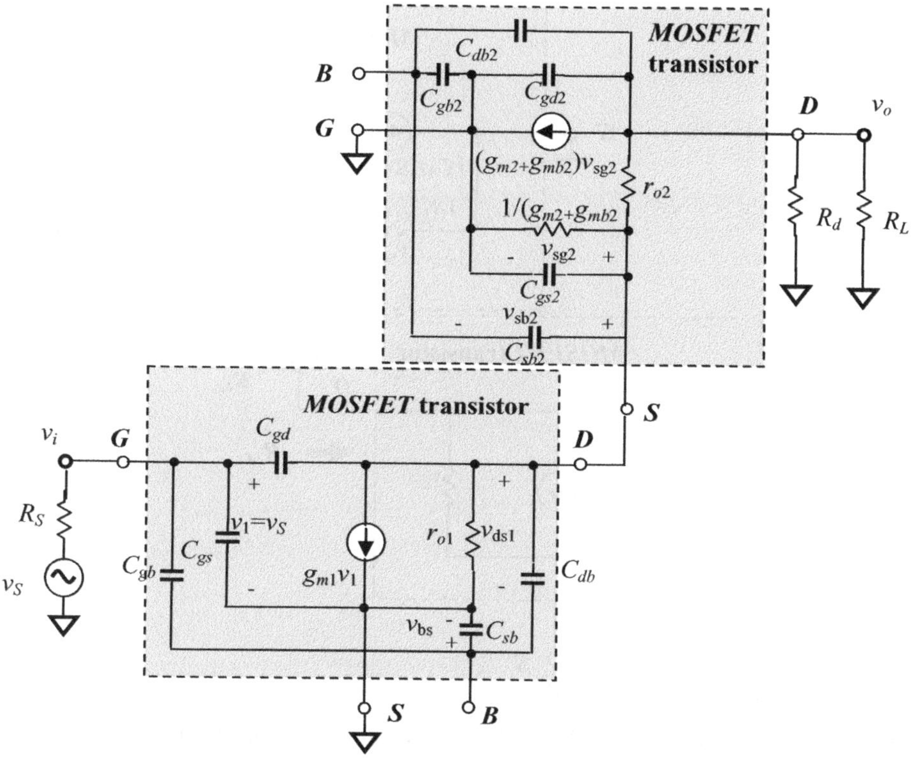

Figure 1.31 Equivalent circuit of a *MOSFET* cascode amplifier with *CS-CG* configuration.

In terms of *KCL* at the output and at the source of the second *CG* stage, the input resistance of the second *CG* amplifier is

$$R_{i2} = \frac{1}{(g_{m2}+g_{mb2})} + \frac{R'_L}{(g_{m2}+g_{mb2})r_{o2}}, \tag{1.123}$$

where

$$R'_L = R_d \,//\, R_L = \frac{R_d}{R_D + R_L} R_L, \tag{1.124}$$

and the trans-conductance is derived when $R'_L = 0$ so that $v_o = 0$, that is

$$G_m \approx g_{m1}, \tag{1.125}$$

The output resistance can be found when the g_{m1} generator is inactive, that is $v_i = 0$. Consequently the *CS* portion becomes a simple resistor r_{o1} and the entire cascode looks like a *CG* device with a source degeneration resistor. Then,

$$R_o = r_{o2} + r_{o1}[1+(g_{m2}+g_{mb2})]r_{o2} \approx (g_{m2}+g_{mb2})r_{o1}r_{o2}, \tag{1.126}$$

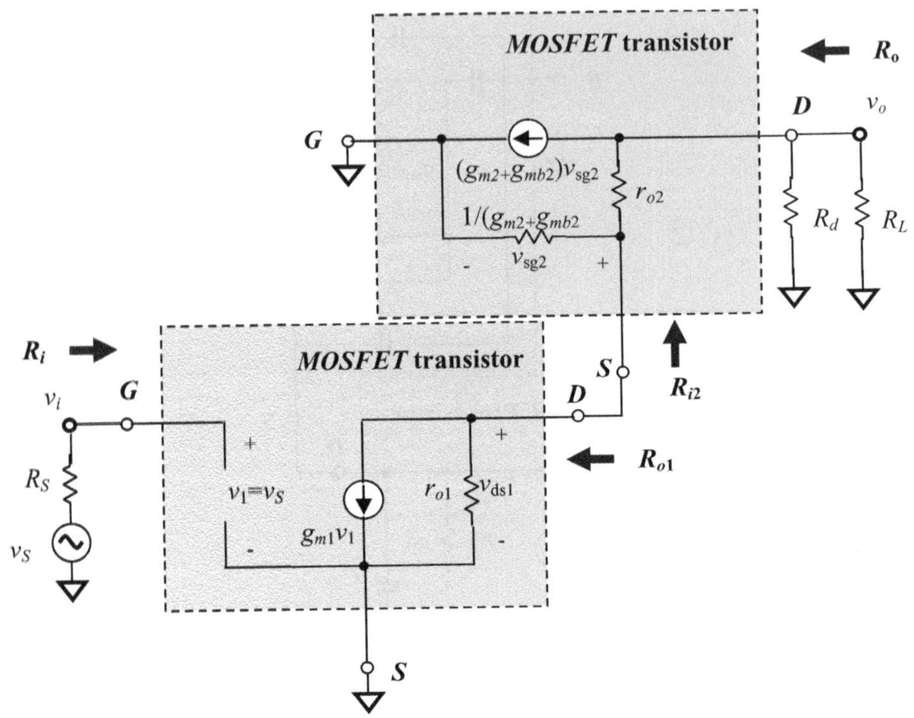

Figure 1.32 Equivalent circuit of *MOSFET* cascode amplifier with *CS-CG* configuration at low frequencies.

if

$$R_L' \to \infty, \tag{1.127}$$

(Refer to Paul R. Gray et al., *Analysis and Design of Analog Integrated Circuits*, pp. 208–209.)

Similarly to the *CE-CB* bipolar cascode amplifier, the input resistance of the second *CG* stage is quite low because the second term in equation (1.123) disappears when

$$R_L' = 0, \tag{1.128}$$

It becomes

$$R_{i2} \approx \frac{1}{(g_{m2} + g_{mb2})}. \tag{1.129}$$

which usually happens in cases where

$$R_{i2} << r_{o1}. \tag{1.130}$$

Now let us consider the voltage and current gain.

The voltage gain of the first *CS* stage is not that as expected from a normal *CS* stage, $A_{v1} = -g_{m1}r_{o1} // R_d$. The output resistor r_{o1} of the first *CS* stage is connected with the input resistor R_{i2} of the second *CG* stage in parallel. As shown in equation (1.129), the input resistance of the second *CG* stage is dropped by a good deal and is much lower than that of r_{o1} in a normal *CS* stage; consequently,

$$A_{v1} = -g_{m1}r_{o1} // R_{i2} \approx -g_{m1}R_{i2} \approx -\frac{g_{m1}}{g_{m2} + g_{mb2}}, \tag{1.131}$$

if the input voltage v_i but not v_S is considered as the reference for the voltage gain.

If the trans-conductances of the *CS* and *CG* transistors are the same, that is,

$$g_{m1} = g_{m2}, \tag{1.132}$$

then equation (1.131) becomes

$$A_{v1} \approx -1, \tag{1.133}$$

The current gain of first *CS* stage is

$$A_{i1} \to \infty. \tag{1.134}$$

The current gain of second *CG* stage is

$$A_{i2} \approx 1, \tag{1.135}$$

because the input current is almost equal to its output current in the second *CG* stage, that is

$$I_{i2} \approx I_{o2} \quad \text{and} \quad i_{i2} \approx i_{o2}, \tag{1.136}$$

where the upper case letter '*I*' denotes the *DC* current and the lower case letter '*i*' denotes the *AC* current.

Finally the total voltage gain can be evaluated from equations (1.124) and (1.126)

$$A_v = \frac{v_o}{v_i} \approx -G_m R_o \approx -g_{m1}(g_{m2} + g_{mb2}) r_{o1} r_{o2}. \tag{1.137}$$

1.3.3 Why Cascode?

Let's summarize the reasons why the cascode amplifier is superior to other amplifiers.

It increases the output impedance, which is particularly useful in desensitizing bias references from variations in power supply voltage and in achieving large amounts of voltage gain.

This is very beneficial to the digital circuit design. Digital circuits always operate under high impedances; the high output impedance of a cascode amplifier provides a powerful way to reach the goal of high voltage gain.

However, this remarkable advantage does not excite the *RF* circuit designers because RF circuits always operate under low impedances. *RF* designers pursue power gain but not voltage gain; here, high voltage gain does not make too much sense if its current gain is low.

However, both *RF* circuit designers and digital circuit designers are excited about these three advantages of the cascode amplifier.

1) It alleviates the Miller effect on a voltage amplifier.
Unwanted capacitive feedback always exists in a voltage amplifier with a *CE* or *CS* configuration. Alleviating the Miller effect therefore allows the amplifier operation at higher frequencies than would otherwise be possible.
The input impedance of the second *CG* stage is

$$R_{i2} = \frac{r_{\pi 2}}{\beta_2}, \tag{1.138}$$

which is β_2 times lower than the input impedance of a *CE* stage if the same device is applied to both stages. The voltage gain of the first stage is low due to its output impedance being pulled down by the low input impedance of the second stage, that is

$$A_{v1} = -g_{m1} R'_{L1} // R_{i2} = -g_{m1} R'_{L1} // \frac{r_{\pi 2}}{\beta_2} \approx -g_{m1} \frac{r_{\pi 2}}{\beta_2} \approx 1, \tag{1.139}$$

if

$$g_{m1} = g_{m2}. \tag{1.140}$$

Consequently, the input Miller capacitance in the first stage is kept the same as C_μ due to the low voltage gain.

$$C_{i,miller} = C_\mu A_{v1} \approx C_\mu, \tag{1.141}$$

Owing to the low Miller capacitance, the bandwidth in the first stage will be increased from

$$\omega_{T1} = \frac{g_{m1}}{C_{\pi 1} + C_{\mu 1} A_{v1}}, \tag{1.142}$$

to

$$\omega_{T1} = \frac{g_{m1}}{C_{\pi 1} + C_{\mu 1}}. \tag{1.143}$$

The bandwidth of the entire *CS-CB* cascode amplifier is then mainly determined by the second stage, that is,

$$\omega_{T2} = \frac{g_{m2}}{C_{\mu 2}}. \qquad (1.144)$$

It is therefore concluded that the bandwidth of the *CS-CB* cascode amplifier is approximately equivalent to the bandwidth of the single *CB* amplifier, which is much higher than that of a single *CE* amplifier.

This wide bandwidth is an outstanding advantage, which is why the cascode amplifier is widely applied in the amplifier design.

2) Better isolation
   The *CB* configuration of the second stage guarantees isolation between output and input. The base of the bipolar transistor is grounded and the capacitive feedback from the output to input is reduced to an insignificant level. In addition, the low input impedance of the second stage is also beneficial to the isolation between output and input.
3) It can magnify the signal: not only the voltage but also the power of the signal as well.
   In a cascade amplifier, the current is magnified in the first *CE* stage and the voltage is magnified in the second *CB* stage. Consequently, the voltage and power of signal can be magnified simultaneously, therefore satisfying both digital and *RF* circuit designers.

1.3.4 An Example

This is an *LNA RFIC* design example, which is designed for group #1, Band #2 of *UWB* (Ultra Wide Band) system. The main electrical features are

- *DC* power supply: $V_{dd} = 1.2\,V$,
- Current drain: $I_{total} < 5\,mA$,
- Operating frequency range: $f = 3.696$ to $4.4224\,GHz$,
- Gain: $G > 12\,\text{dB}$,
- Input return loss: $S_{11} < -10\,dB$,
- Output return loss: $S_{22} < -10\,dB$,
- Noise figure: $NF < 2.5\,\text{dB}$,
- Third order input intercept point: $IIP_3 > 5\,dBm$,
- Second order input intercept point: $IIP_2 > 35\,dBm$.

In consideration of the bandwidth, it is decided to use the cascode configuration, because its relative bandwidth is

$$BW = \frac{\Delta f}{f} = \frac{4224 - 3696}{(4224 + 3696)/2} = \frac{528}{3960} = 13.3\%. \qquad (1.145)$$

This can be categorized as neither a narrow- nor wide-band block. Usually, a block or a system with a relative bandwidth greater than 15% is considered a wide-band block or system; conversely, a block or a system with a relative bandwidth of less than 15% is considered a narrow-band block or system. We are hence quite hesitant to consider this design sample as a narrow-band block because 13.33% is close to 15%.

TSMC 90 *nm*, *CMOS* is selected to be the *IC* processing, and *n*-channel *MOSFET* transistors are chosen to be the *CS-CG* device. The size of the device is calculated based on the considerations described in Section 1.2.1.

1.3.4.1 Raw Device Testing Figure 1.33 shows the setup for raw device testing. The raw device is a combination of two transistors, M_1 and M_2. All the capacitors are "zero" capacitors and all the inductors are "infinite" inductors. At the operating frequencies, the impedance of a "zero" capacitor approaches zero and the impedance of an "infinite" inductor approaches infinity.

Figure 1.34(a) shows the tested *S* parameters and the optimum input reflection coefficient $\Gamma_{S,opt}$ on the Smith chart; Figure 1.34(b) shows the frequency response of the *S* parameters. The intermediate frequency, $f = 3.96\,GHz$, is marked with a dot on each trace.

The first impression is that the *CS-CG* configuration of the raw devices exhibits wide-band behavior as expected. In Figure 1.34(a), all of the traces corresponding to the entire frequency band, 3.696 to 4.224 GHz, are squeezed to a small trace segment on the Smith chart, and in Figure 1.34(b) the trace is quite flat over the entire frequency band.

The second impression is that, unfortunately, the optimum input reflection coefficient $\Gamma_{S,opt}$ is not at the expected location, which is supposed to be conjugate to S_{11}. The raw devices must be modified so that the overlapping condition (1.34) of maximum gain and minimum noise figure can be satisfied.

The magnitudes of the *S* parameters shown in Figure 1.34(b) can be outlined as follows:

$$2.45\,dB < S_{21} < 2.58\,dB, \quad \text{when } 3.696\,GHz < f < 4.224\,GHz, \tag{1.146}$$

$$S_{21} = 2.50\,dB, \quad \text{when } f = 3.960\,GHz, \tag{1.147}$$

$$-2.75\,dB < S_{11} < -2.20\,dB, \quad \text{when } 3.696\,GHz < f < 4.224\,GHz, \tag{1.148}$$

$$S_{11} = -2.5\,dB, \quad \text{when } f = 3.960\,GHz, \tag{1.149}$$

$$-0.19\,dB < S_{22} < -0.17\,dB, \quad \text{when } 3.696\,GHz < f < 4.224\,GHz, \tag{1.150}$$

$$S_{22} = -0.18\,dB, \quad \text{when } f = 3.960\,GHz, \tag{1.151}$$

For now, we will not worry about the low gain, poor return losses, stability, or other issues, because the input and output impedances are not matched yet. Let's go ahead to modify the raw devices. This modification is usually conducted many times, depending on how much experience the designer has. Figure 1.35 shows the final attempt, which is believed the best one, in which a degeneration inductor is connected between the source of the *CS* device and ground:

$$L_{degen} = 0.45\,nH, \tag{1.152}$$

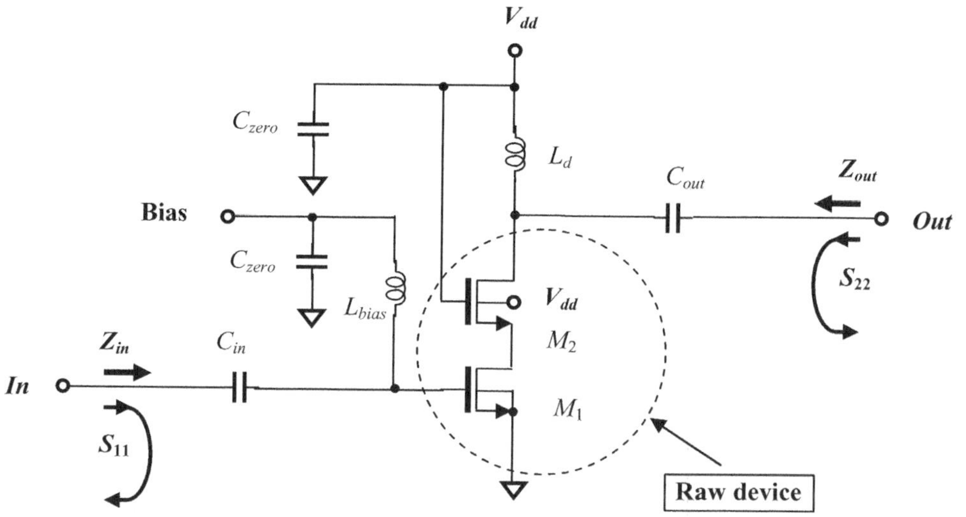

Figure 1.33 Schematic of a cascode *LNA* for raw device testing. C_{in}, C_{out}, C_{zero}: "zero" capacitor; L_{bias}, L_d: "infinite" inductor.

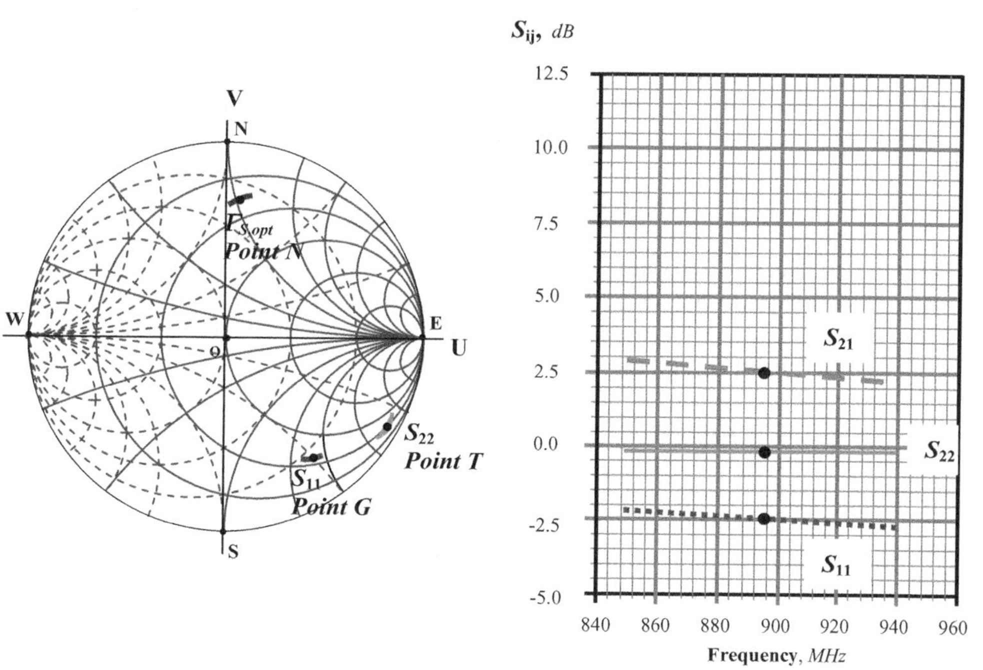

(a) S_{11}, S_{22} and $\Gamma_{S,opt}$ on Smith chart

(b) Magnitude of S_{ij}, *dB*

Figure 1.34 *S* parameters from raw device testing. (The intermediate frequency $f = 3.96\,GHz$ is marked by a dot on each trace.)

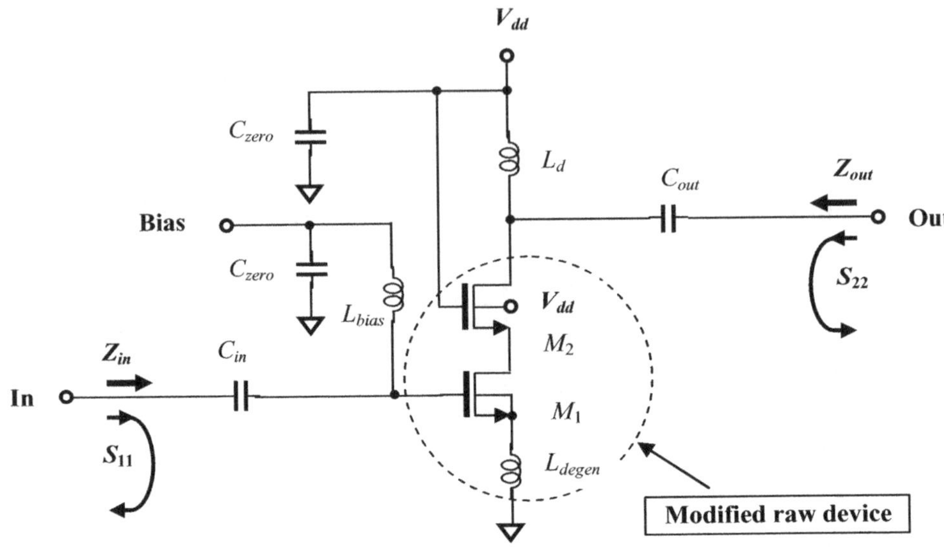

Figure 1.35 Schematic of setup for modified raw device testing. C_{in}, C_{out}, C_{zero}: "zero" capacitor; L_{bias}, L_d: "infinite" inductor.

The degeneration inductor, L_{degen}, is a simple and tiny part. In the *RFIC* circuit design for the operating frequency range of *GHz*, it is about 0.5 to 2 turns, with approximately a tenth to a couple *nH* of inductance.

Figure 1.36(a) shows the tested *S* parameters and the optimum input reflection coefficient $\Gamma_{S,opt}$ on the Smith chart from the modified raw device testing. The optimum input reflection coefficient $\Gamma_{S,opt}$ is at the expected location, which is very close to S_{11}^*. This means that the raw devices are quite close to fitting the overlapping condition of maximum gain and minimum of noise, (1.34), although $\Gamma_{S,opt}$ does not exactly overlap S_{11}^*. A little deviation is inevitable, mirroring a philosophy of life: "Nobody is perfect and nothing is perfect!"

Again, at this point, we are not worrying too much about the magnitude of the *S* parameters shown in Figure 1.36(b) which is far from the stated goals, because the input and output impedances are still not matched.

1.3.4.2 Gain and Bandwidth Impedance matching is one of the key issues in *RF* circuit design. A new designer without *RF* circuit design experience might need to spend a lot of time on it. Let's spend some time demonstrating the process of impedance matching in this example.

The first try to implement the impedance matching network is shown in Figure 1.37: it consists of parts

- In the input impedance network: $L_{S,in} = 4nH$, $C_{S,in} = 10pF$;
- In the output impedance network: $L_{P,out} = 10nH$ and $R_{P,out} = 500\,\Omega$, $C_{S,out} = 1pF$.

The inductor $L_{S,in} = 4nH$ is the main part in the input impedance matching network, by which the trace of S_{11} is supposed to be pulled to a location near the center of

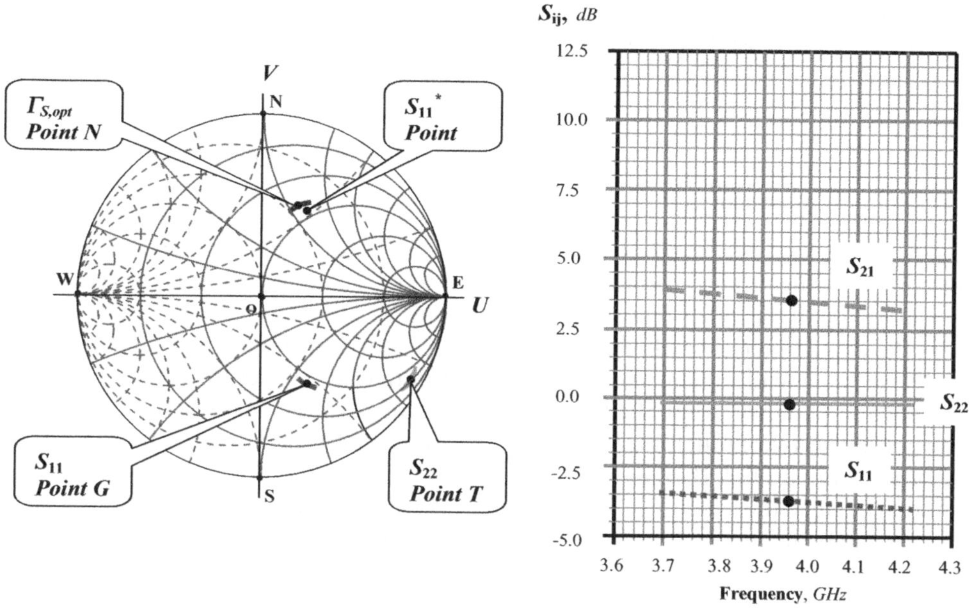

(a) S_{11}, S_{22} and $\Gamma_{S,opt}$ on Smith chart (b) Magnitude of S_{ij}, dB

Figure 1.36 S parameters from modified raw device testing. (The intermediate frequency $f = 3.96\,GHz$ is marked by a dot on each trace.)

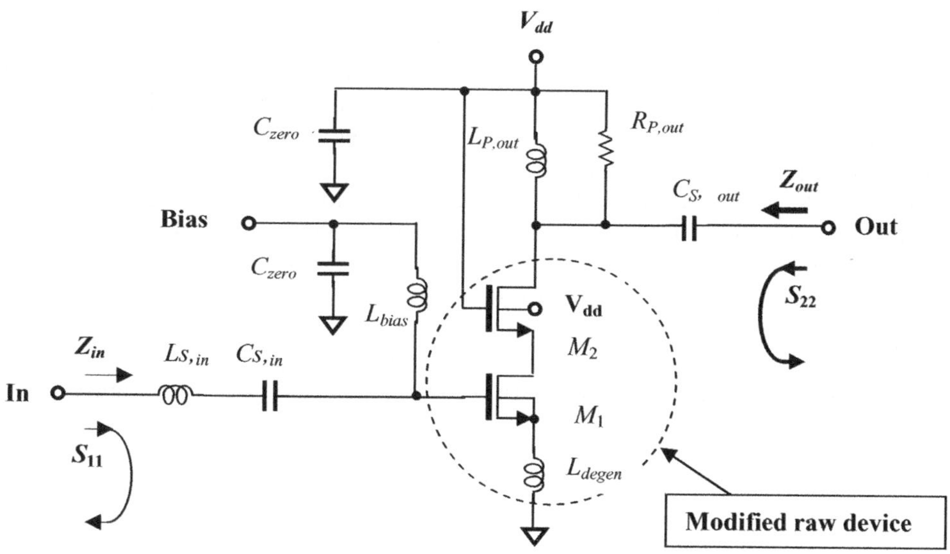

Figure 1.37 Impedance matching by parts. At input: $L_{S,in} = 4\,nH$, $C_{S,in} = 10\,pF$. At output: $L_{P,out} = 10\,nH$, $R_{P,out} = 500\,\Omega$, and $C_{S,out} = 1\,pF$.

the Smith chart, which corresponds to the reference impedance point $50\,\Omega$. The capacitor $C_{S,in} = 10pF$ functions as the DC blocking part because at the operating frequency range its impedance is very low, such that

$$Z_{10\,pF} = \frac{1}{j\omega C} = -j\frac{1}{2\pi fC} = 4.02\,\Omega, \qquad \text{when } f = 3960\,MHz. \tag{1.153}$$

The inductor $L_{P,out} = 10nH$ is the main part in the output impedance matching network, by which the trace of S_{22} is supposed to be pulled counter-clockwise up to a location near the $50\,\Omega$ resistance circle along the constant conductance circle; then, the capacitor $C_{S,out} = 1pF$ is the next part, by which the S_{22} is drawn counter-clockwise to a location near the center of the Smith chart, which corresponds to the reference impedance point $50\,\Omega$ along the constant resistance circle. The difference between these two inductors, $L_{S,in}$ and $L_{P,out}$, should be noted. One is in series and the other is in parallel. By means of $L_{S,in}$ in series, the trace of S_{11} is pulled clockwise to the area near the reference impedance point $50\,\Omega$ along the constant resistance circle, while by means of $L_{P,out}$ in parallel, the trace of S_{22} is pulled counter-clockwise to the area near the reference impedance point $50\,\Omega$ along the constant conductance circle.

The raw device shown in Figure 1.35 is unstable because the simulation shows its μ value as

$$\mu = 0.95 < 1. \tag{1.154}$$

Looking at the schematic shown in Figure 1.35, the instability may come from the load inductor, $L_{P,out}$, and the degenerator inductor, L_{degen}. However, in order to retain as much voltage drop across the device as possible, so as to reduce the intermodulation, I prefer to use a load inductor, $L_{P,out}$, rather than a resistor, even though an inductor is much more expensive than a resistor. On the other hand, in order to satisfy the overlapping condition (1.34) for NF_{min} without significant increase of the noise figure, I prefer to use a degeneration inductor, L_{degen}, rather than a degeneration resistor, R_{degen}. The value of L_{degen}, however, should not be too high, otherwise the gain and stability would both suffer. In design sample, $L_{degen} = 0.45\,nH$, the resistor, $R_{P,out}$, is used to de-Q the load inductor $L_{P,out}$ so as to prevent oscillation. It is the key part in keeping the LNA block stable, while the low value of the degeneration inductor L_{degen} provides good assistance.

The values of all the parts shown in the schematic in Figure 1.37 are simultaneously optimized to obtain stability with a reasonable gain and noise figure level, that is,

$$\mu = 1.23 > 1, \tag{1.155}$$

when

$$G = 8.0\,dB, \quad \text{and} \quad f = 3.960\,GHz. \tag{1.156}$$

The μ value shown in (1.155) indicates that the unconditionally stable condition is satisfied.

In summary, the input impedance matching network in this example basically consists of only one part. It is a simple impedance matching network indeed. The output impedance matching network consists of three parts, which is a typical part count in an *LNA* design. In general, impedance matching is difficult and it is one of the main tasks of an *RF* circuit designer. In the chapters about impedance matching, some useful schemes and technologies are presented for more complicated impedance matching networks.

Figure 1.38 displays the tested *S* parameters after the input and output impedance matching networks are implemented.

In the entire operating frequency range, the results are:

$$7.2\,dB < S_{21} < 8.8\,dB, \quad \text{when } 3.696 < f < 4.224\,GHz, \tag{1.157}$$

$$S_{21} = 8.0\,dB, \quad \text{when } f = 3.960\,GHz, \tag{1.158}$$

$$-27.1\,dB < S_{11} < -16.9\,dB, \quad \text{when } 3.696 < f < 4.224\,GHz, \tag{1.159}$$

$$S_{11} = -25.0\,dB, \quad \text{when } f = 3.960\,GHz, \tag{1.160}$$

$$-25.0\,dB < S_{22} < -23.7\,dB, \quad \text{when } 3.696 < f < 4.224\,GHz, \tag{1.161}$$

$$S_{22} = -23.9\,dB, \quad \text{when } f = 3.960\,GHz. \tag{1.162}$$

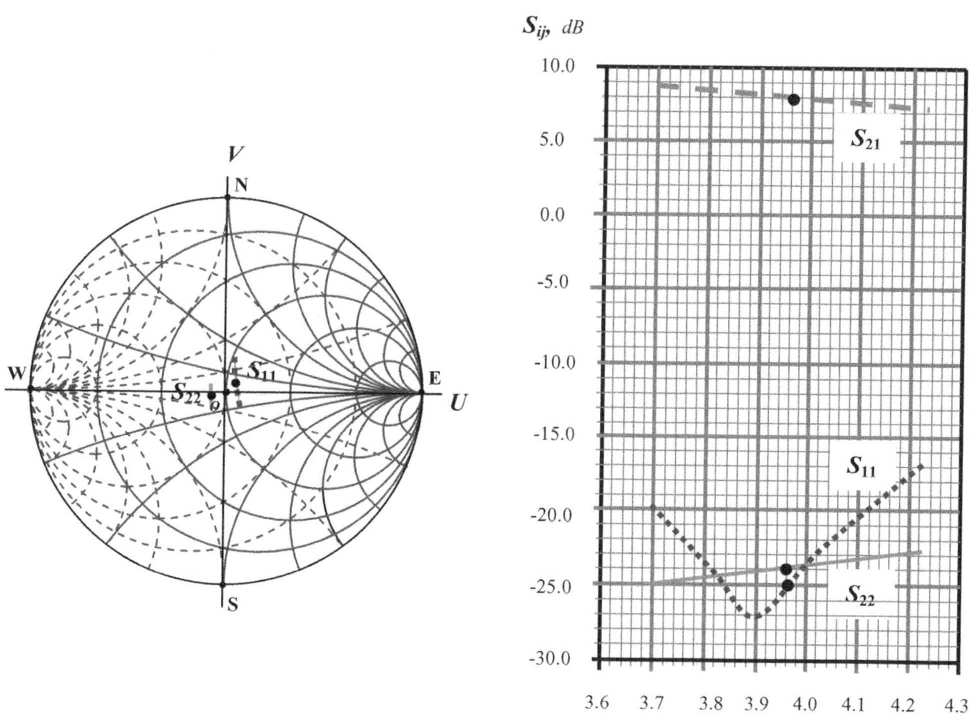

(a) S_{11} and S_{22} on Smith chart

(b) Magnitude of S_{ij}, *dB*

Figure 1.38 *S* parameters after input and output impedances are matched by parts. (The intermediate frequency $f = 3.96\,GHz$ is marked by a dot on each trace.) At input: $L_{S,in} = 4\,nH$, $C_{S,in} = 10\,pF$. At output: $L_{P,out} = 10\,nH$, $R_{P,out} = 500\,\Omega$, and $C_{S,out} = 1\,pF$.

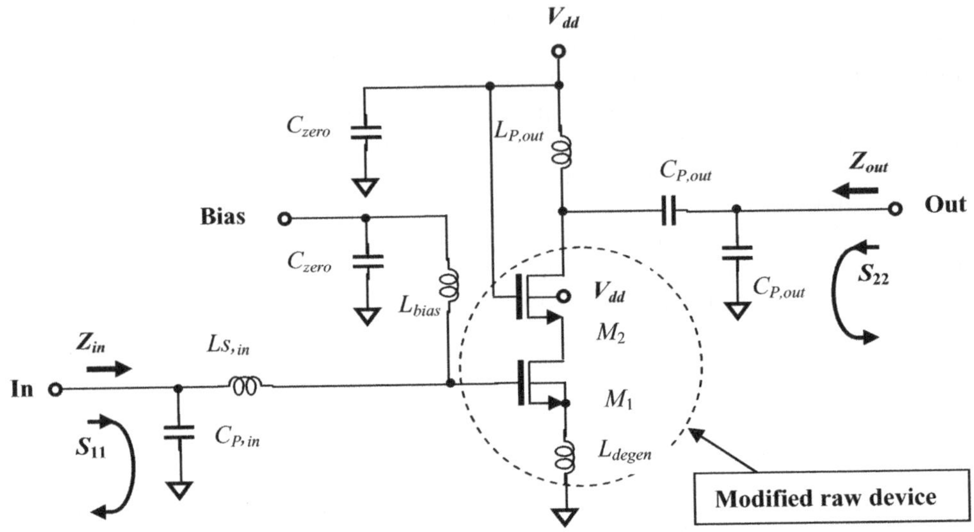

Figure 1.39 Modified impedance matching by parts. At input: $C_{P,in} = 0.3pF$, $L_{S,in} = 3.3nH$. At output: $L_{P,out} = 5nH$, $C_{P,out} = 0.13pF$, and $C_{S,out} = 0.3pF$.

By comparing the locations of S_{11} and S_{22} shown in Figures 1.36 and 1.38, it can be found that either the S_{11} or S_{22} traces are not moved exactly along the resistance or conductance circles. Both traces are moved to the location with higher resistance. In other words, the corresponding resistances of S_{11} and S_{22} as shown in Figure 1.38 are increased from those in Figure 1.36. This is due to the inductors $L_{S,in}$ and $L_{P,out}$, which are low Q parts and bring about additional resistance. The Q value of inductors applied in this section is assumed to be 10, which is a reasonable value in today's *RFIC* design.

The input and output impedances are matched well and the *LNA* is in a stable state. **However, the gain, S_{21} = 8.0 to 9.0 dB, is too low!**

One solution might be to change the topologies of the input and output impedance networks, even though they are quite simple, as shown in Figure 1.37, and have the feature of wide-band performance, as shown in Figure 1.38.

Figure 1.39 plots the modified impedance matching networks, which consists of parts

- In the input impedance network: $C_{P,in} = 0.3pF$, $L_{S,in} = 3.3nH$;
- In the output impedance network: $L_{P,out} = 5nH$ and $C_{S,out} = 0.13pF$, $C_{P,out} = 0.3pF$.

The topologies of the input and output impedance matching networks shown in 1.39 are different from those shown in Figure 1.37. In the input impedance network of Figure 1.39, the inductor $L_{S,in} = 3.3nH$ plays the same role as the inductor $L_{S,in} = 4nH$ in the input impedance network of Figure 1.37. However, after the inductor $L_{S,in} = 3.3nH$ is attached to the gate of the device, the location of S_{11} still does not exactly overlap with the center of the Smith chart. This implies that further improvements could be made if the trace S_{11} is made to exactly overlap the center of the Smith chart. This is the our clue or motivation to apply a capacitor, $C_{P,in} = 0.3pF$, to

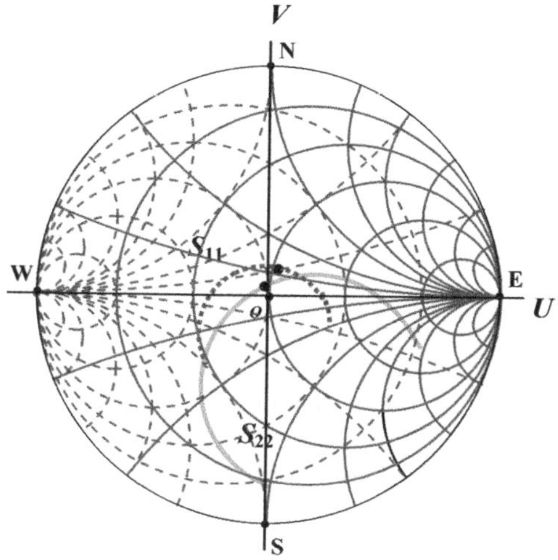

(a) S_{11} and S_{22} on Smith Chart

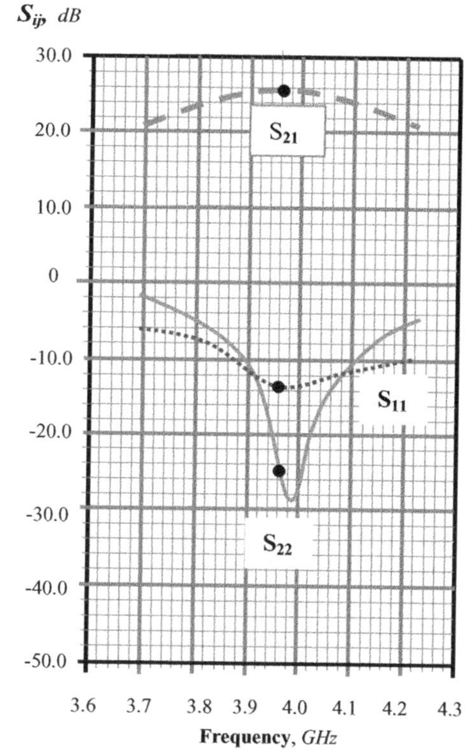

(b) Magnitude of S_{ij}, dB

Figure 1.40 S parameters after input and output impedances are matched by parts. (The intermediate frequency $f = 3.96\,GHz$ is marked by a dot on each trace.) At input: $C_{P,in} = 0.3\,pF$, $L_{S,in} = 3.3\,nH$. At output: $L_{P,out} = 5\,nH$, $C_{S,out} = 0.13\,pF$, and $C_{P,out} = 0.3\,pF$.

the input impedance network, so that the trace S_{11} can be pulled clockwise toward the center of the Smith chart along the constant conductance circle. This will definitively help enhance the power gain of *LNA*.

In the output impedance network shown in Figure 1.37, the de-Q resistor, $R_{P,out} = 500\,\Omega$, is removed. The DC blocking capacitor in Figure 1.37, $C_{S,out} = 1\,pF$, is replaced by a smaller capacitor, $C_{S,out} = 0.13\,pF$, and an additional capacitor, $C_{P,out} = 0.3\,pF$, is added at the output terminal as shown in Figure 1.39. By the addition of $L_{P,out} = 5\,nH$, the trace S_{22} is pulled counter-clockwise to somewhere in the upper portion of Smith chart along the constant conductance circle. Then, by the addition of $C_{S,out} = 0.13\,pF$, the trace S_{22} is drawn counter-clockwise to somewhere in the upper portion of the Smith chart along the constant resistance circle. Finally, by the addition of $C_{P,out} = 0.3\,pF$, the trace S_{22} is pulled clockwise toward the center of the Smith chart along the constant conductance circle.

The resulting S parameters are plotted in Figure 1.40:

$$20.8\,dB < S_{21} < 25.5\,dB, \quad \text{when } 3.696 < f < 4.224\,GHz, \tag{1.163}$$

$$S_{21} = 25.5\,dB, \quad \text{when } f = 3.960\,GHz, \tag{1.164}$$

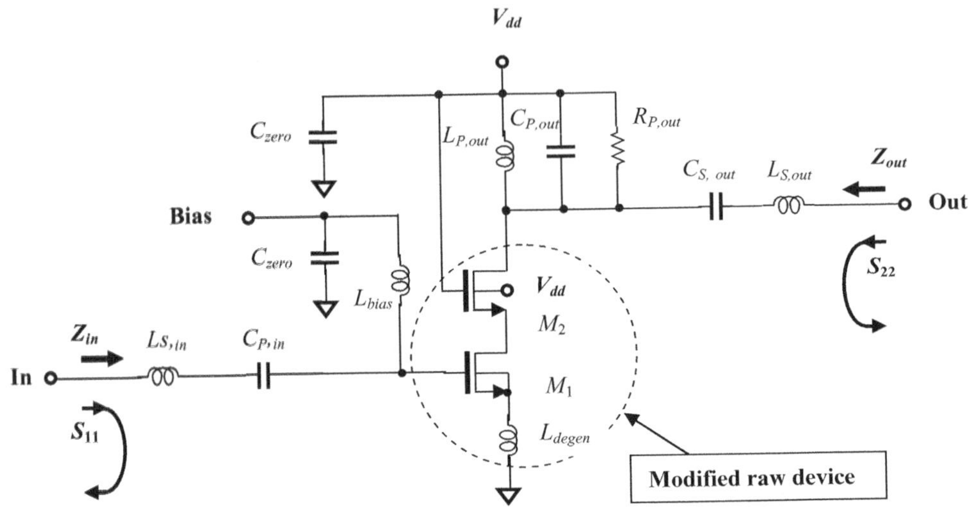

Figure 1.41 Re-modified impedance matching by parts. At input: $L_{S,in} = 3.8nH$, $C_{S,in} = 5pF$. At output: $R_{P,out} = 200\,\Omega$, $L_{P,out} = 2nH$, $C_{P,out} = 0.97pF$, and $C_{S,out} = 10pF$, $L_{S,out} = 3.3nH$.

$$-13.8dB < S_{11} < -6.2dB, \quad \text{when } 3.696 < f < 4.224GHz, \tag{1.165}$$

$$S_{11} = -13.6dB, \quad \text{when } f = 3.960GHz, \tag{1.166}$$

$$-28.8dB < S_{22} < -1.8dB, \quad \text{when } 3.696 < f < 4.224GHz, \tag{1.167}$$

$$S_{22} = -25.0dB, \quad \text{when } f = 3.960GHz. \tag{1.168}$$

In the entire frequency range, the gain, $S_{21} > 20.8$ dB, is satisfactory!

Unfortunately, from the frequency response of S_{11} and S_{22}, the bandwidth becomes narrow and does not cover the desired frequency range: 3.696 *GHz* < *f* < 4.224 *GHz*.

Another issue concerns the capacitors $C_{S,out} = 0.13pF$ and $C_{P,out} = 0.3pF$. Their capacitances are too small and therefore unrealistic! One cannot find any discrete capacitor with $0.13pF$ of capacitance on the market at all. A $0.3pF$ capacitor is in the same order as the parasitic or spray capacitance which exists in the circuit layout. These capacitors are therefore unacceptable parts in this circuit implementation.

So, more effort must be put into the impedance matching work! Let's try again!

In general, it is expected that input and output impedance networks can be constructed to satisfy the goals both of power gain and the bandwidth of *LNA*. Figure 1.41 shows a successful attempt, which consists of the parts:

- At input impedance matching network: $L_{S,in} = 3.8nH$, $C_{S,in} = 5pF$,
- At output impedance matching network: $R_{P,out} = 200\,\Omega$, $L_{P,out} = 2nH$, $C_{P,out} = 0.97pF$, and $C_{S,out} = 10pF$, $L_{S,out} = 3.3nH$.

Special topologies are applied to both input and output impedance matching networks.

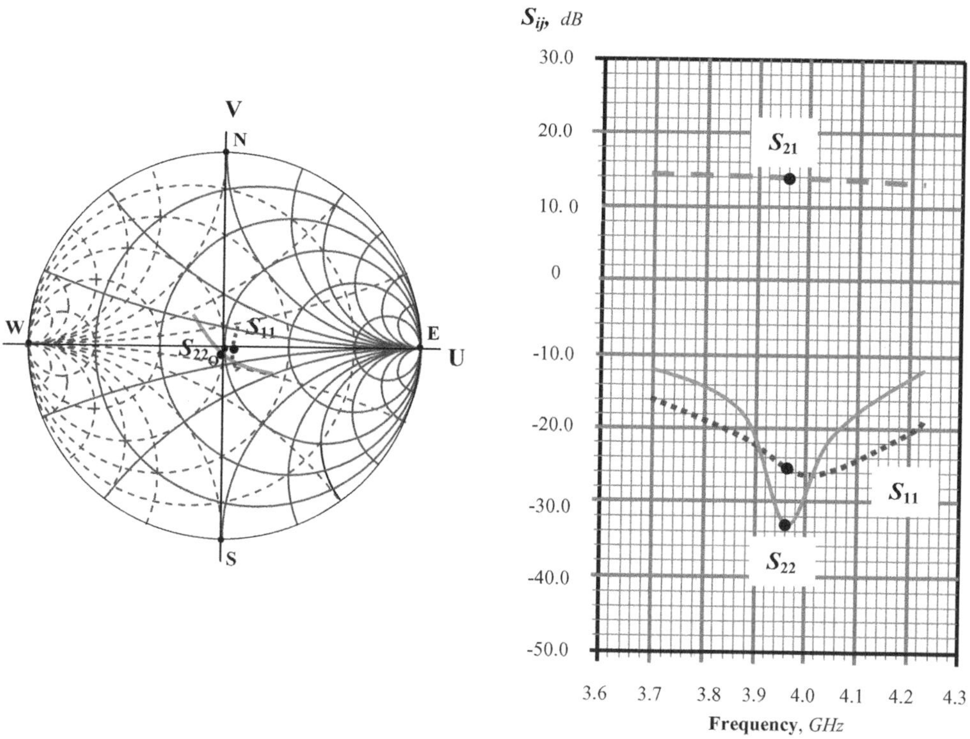

(a) S_{11} and S_{22} on Smith chart

(b) Magnitude of S_{ij}, dB

Figure 1.42 S parameters after input and output impedances are matched by parts. (The intermediate frequency $f = 3.96\,GHz$ is marked by a dot on each trace.) At input: $L_{S,in} = 3.8\,nH$, $C_{S,in} = 5\,pF$. At output: $R_{P,out} = 200\,\Omega$, $L_{P,out} = 2\,nH$, $C_{P,out} = 0.97\,pF$, $C_{S,out} = 10\,pF$, and $L_{S,out} = 3.3\,nH$.

The input impedance network is an LC "arm," in which two parts, $L_{S,in}$ and $C_{S,I}$, are connected in series. The output impedance matching network consists of one "branch" and one "arm". The "branch" is built by three parts, $L_{P,out}$, $C_{P,out}$, $R_{P,out}$, which are connected in parallel, while the "arm" is built by two parts, $C_{S,out}$ and $L_{S,out}$, which are connected in series. As discussed in Chapter 11, either the arm or branch is a special part, which is effectively applied in the wide-band impedance matching network. Readers are encouraged to read Chapter 11 about wide-band impedance matching so as to understand more schemes on wide-band impedance matching. Figure 1.42 shows its performance:

$$13.0\,dB < S_{21} < 14.3\,dB, \quad \text{when } 3.696 < f < 4.224\,GHz, \tag{1.169}$$

$$S_{21} = 14.0\,dB, \quad \text{when } f = 3.960\,GHz, \tag{1.170}$$

$$-26.2\,dB < S_{11} < -15.9\,dB, \quad \text{when } 3.696 < f < 4.224\,GHz, \tag{1.171}$$

$$S_{11} = -25.5\,dB, \quad \text{when } f = 3.960\,GHz, \tag{1.172}$$

$$-33.2\,dB < S_{22} < -11.9\,dB, \quad \text{when } 3.696 < f < 4.224\,GHz, \tag{1.173}$$

$$S_{22} = -33.1\,dB, \quad \text{when } f = 3.960\,GHz. \tag{1.174}$$

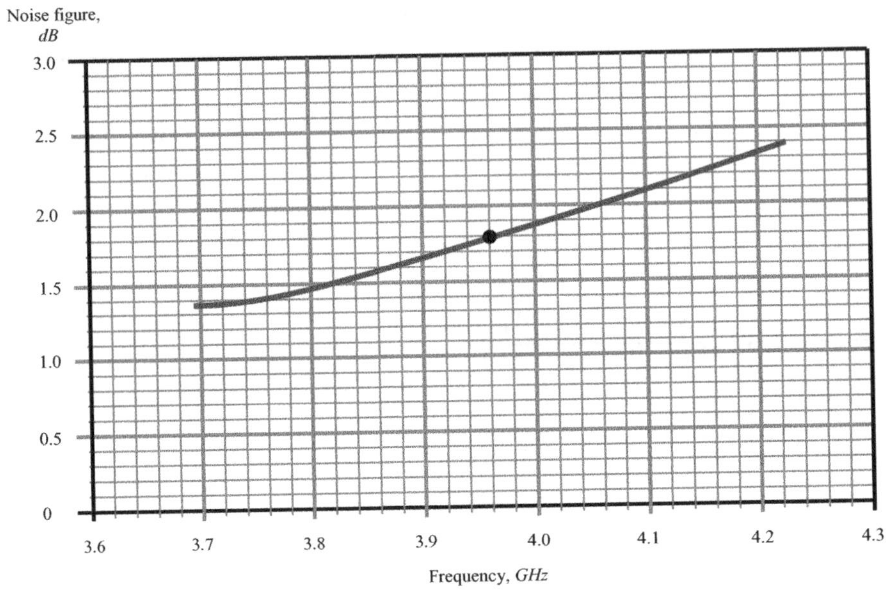

Figure 1.43 Noise figure from 3.696 to 4.224 GHz. $NF = 1.8\,dB$ when $f = 3.960\,GHz$.

Now the gain is reasonable and the bandwidth is wide enough! In addition, the values of the parts are appropriate!

From the demonstrations above, we recognize that impedance matching is not an easy task. Not only the gain but also the bandwidth needs to be taken care of. Additionally, there are many solutions to the same problem. The designer must select one of the possible solutions with the least part count, smallest size, lowest cost and current drain, and best performance.

1.3.4.3 Noise Figure The noise performance of the design sample is good after impedance matching is completed. This is expected because the modified raw device is designed to satisfy the condition (1.34) to simultaneously approach maximum gain and minimum noise figure. Figure 1.43 shows the performance of the noise figure over the entire frequency range:

$$NF = 1.37\,dB, \quad \text{when } f = 3.696\,GHz, \tag{1.175}$$

$$NF = 1.80\,dB, \quad \text{when } f = 3.960\,GHz, \tag{1.176}$$

and

$$NF = 2.40\,dB, \quad \text{when } f = 4.224\,GHz. \tag{1.177}$$

The exciting results can be seen on the plot of the gain and noise figure circles as shown in Figure 1.44. This is a plot for the operating frequency $f = 3.960\,GHz$. After the input and output impedances are matched, the center of the gain circles is point G, which is very close to the center of the Smith chart. The maximum gain at this operating frequency is 14 dB, which is consistent with the results shown in Figure 1.42(b). On the other hand, the center of the noise figure circles is point N, which

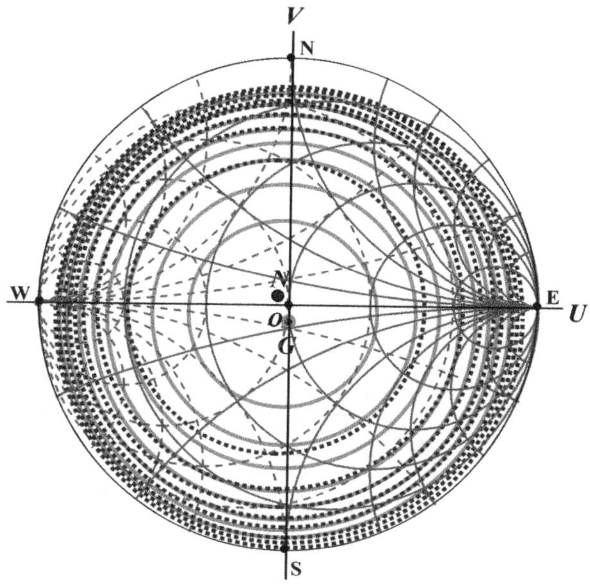

Input reflection coefficient Γ_S plane

Figure 1.44 Constant gain circles and constant noise figure circles when $f = 3.96\,GHz$.
Gain circles: $G_{max} = 14\,dB$ at point G, step = $1.0\,dB$.
Noise figure circles: $NF_{min} = 1.8\,dB$ at point N, step = $0.5\,dB$.

is also very close to the center of the Smith chart. The minimum noise figure at this operating frequency is $1.8\,dB$, which again is consistent with the results shown in Figure 1.43.

The wonderful feature in Figure 1.44 is that the circles and noise circles overlap each other almost completely. The two centers are very close to each other near the center of the Smith chart. The deviation between point N and point O is due to the tolerance existing in the condition (1.34), $\Gamma_{S,opt} = S_{11}^*$; the deviation between point G and point O is due to the tolerance of parts existing in the impedance matching networks. In practice, it is impossible to reach the perfect goal predicted by theory. Deviation between the practical design and the circuit theory always exists.

1.3.4.4 Non-linearity The frequency spectrum at the output of the *LNA* is shown in Figure 1.45. The first (the desired signal), second, and third harmonics are

$$P_{out} = -35.9\,dB_m, \qquad \text{at } f = 3.96\,GHz, \tag{1.178}$$

$$P_{out} = -91.7\,dB_m, \qquad \text{at } f = 7.92\,GHz, \tag{1.179}$$

$$P_{out} = -119.1\,dB_m, \qquad \text{at } f = 11.78\,GHz, \tag{1.180}$$

$$\ldots$$

when $P_{in} = -50\,dB_m$.

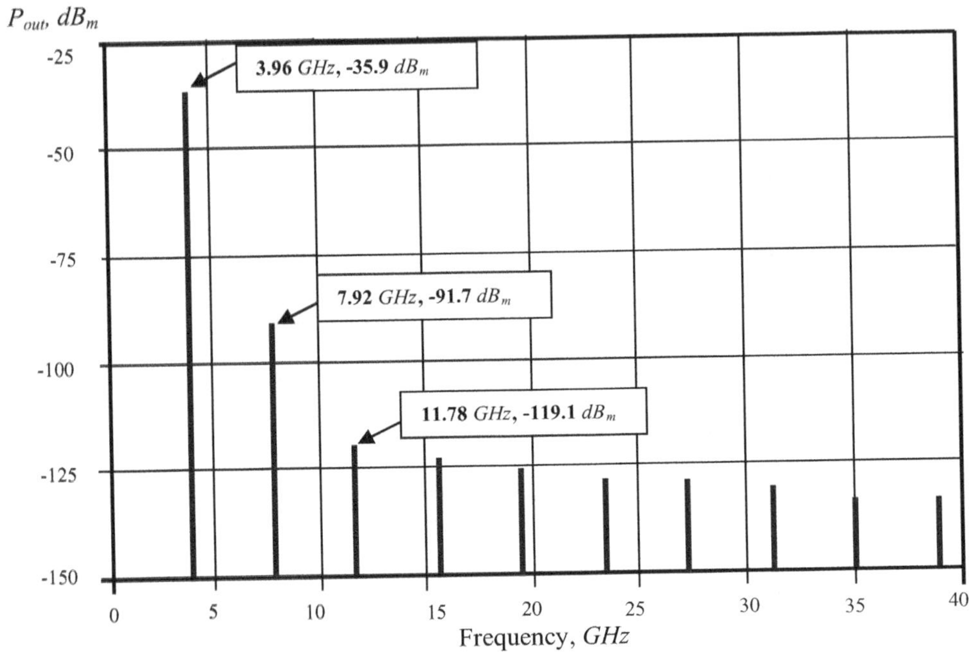

Figure 1.45 Spectrum at *LNA* output, $f_o = 3.96\,GHz$, $P_{in} = -50\,dB_m$.

The gain of the *LNA* is

$$P_{out} - P_{in} = -35.9\,dB_m - (-50\,dB_m) = 14.1\,dB, \tag{1.181}$$

which is approximately consistent with the gain testing of $S_{21} = 14.0\,dB$ as shown in Figure 1.42(b).

The second and third harmonics are lower than the output power at the operating frequency by

$$\Delta P_{out,2} = P_{out}|_{7.92\text{GHz}} - P_{out}|_{3.96GHz} = -91.7\,dB_m - (-35.9\,dB_m) = -55.8\,dB, \tag{1.182}$$

$$\Delta P_{out,3} = P_{out}|_{11.78\text{GHz}} - P_{out}|_{3.96GHz} = -119.1\,dB_m - (-35.9\,dB_m) = -83.2\,dB, \tag{1.183}$$

respectively, which implies that low value spurious products existed in this *LNA* block. The good news is that we need not worry about *DC* offset considerations and the interference of adjacent channels anymore.

Figures 1.46, 1.47, and 1.48 show the 1 *dB* compression point, third order intercept point, and second order intercept point respectively. The 1 *dB* compression point is

$$P_{1dB} = -3.64\,dB_m, \qquad \text{when } f = 3.96\,GHz, \tag{1.184}$$

The third order and second order intercept point are

$$IIP_3 = 3.74\,dB_m, \qquad \text{when } f = 3.96\,GHz, \tag{1.185}$$

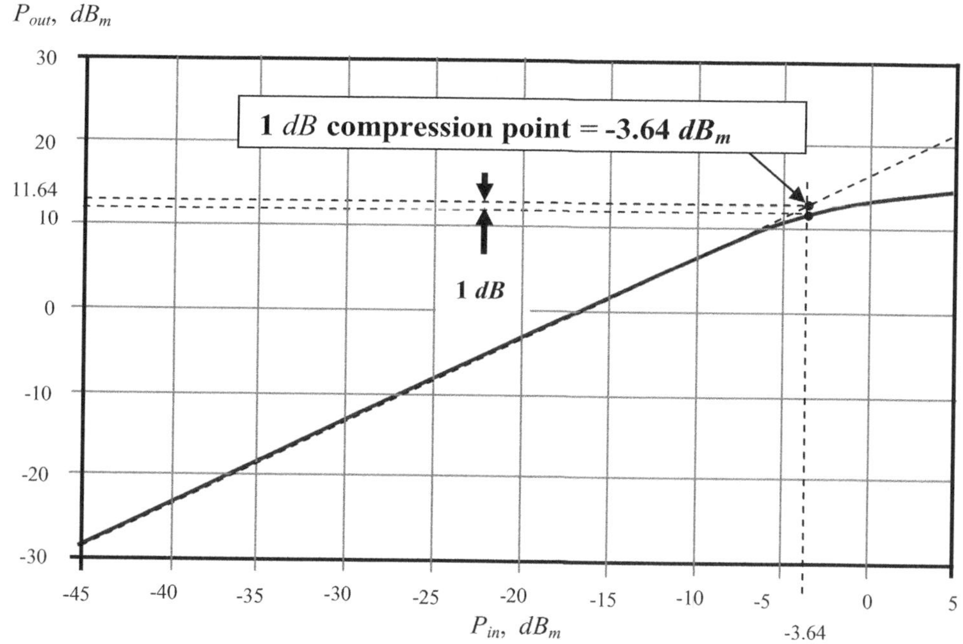

Figure 1.46 1 dB compression point when $f_o = 3.96\,GHz$.

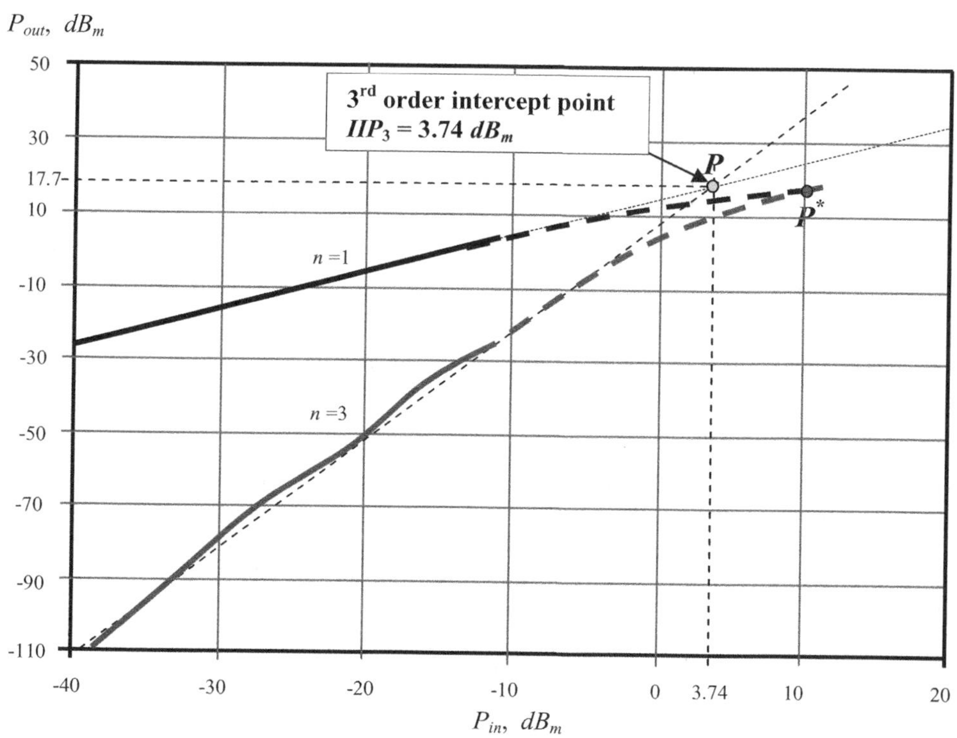

Figure 1.47 Third-order input intercept point when $f_o = 3.96\,GHz$.

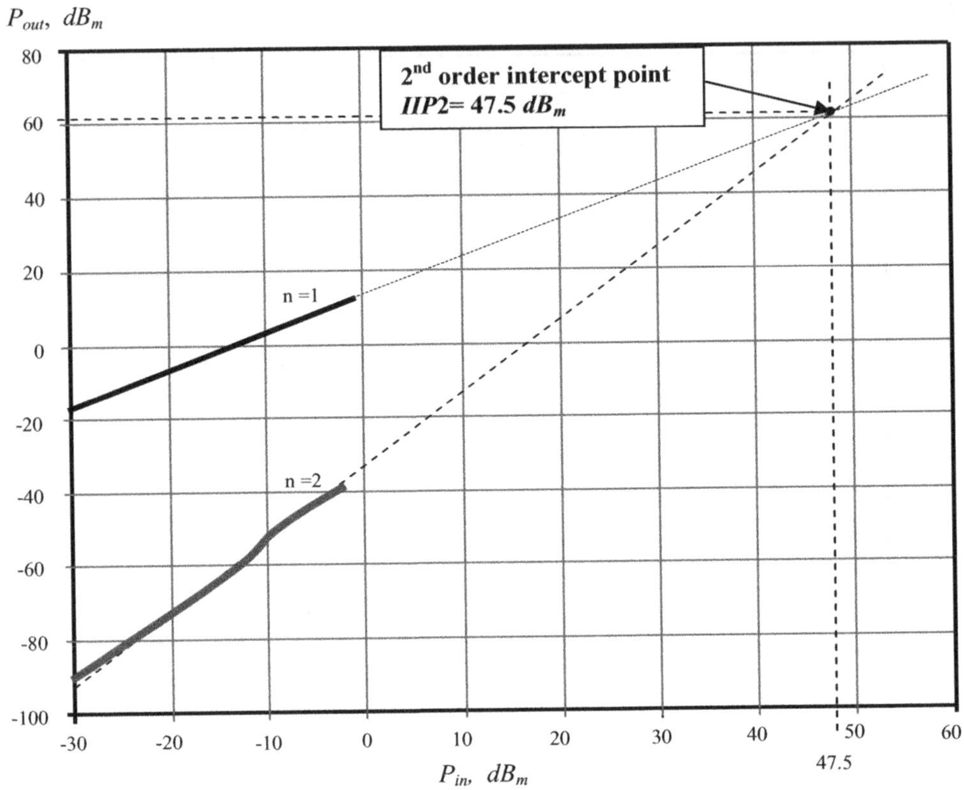

Figure 1.48 Second-order input intercept point when $f_o = 3.96\,GHz$.

$$IIP_2 = -47.5\,dB_m, \qquad \text{when } f = 3.96\,GHz, \tag{1.186}$$

respectively.

1.4 *LNA* WITH *AGC* (AUTOMATIC GAIN CONTROL)

1.4.1 *AGC* Operation

Received signals very often suffer from fading, so that fluctuation of the field strength of radio signals at the receiver antenna always exists. In order to maintain a relatively constant output to the sensor, such as the speaker, it is necessary to detect the strength of the received signal and then adjust the gain of the receiver automatically. This is the subject of *AGC* operation.

In the early development stages of portable radios and cellular phones, modulation technology was relatively simple and therefore it was not necessary to apply *AGC* technology. As advanced modulation technology develped, *AGC* technology became required and sometimes played a decisive role in the receiver.

In the early development stages of cellular phones, a $20\,dB$ dynamic range of *AGC* control in *LNA* design was occasionally required. However, in today's *CDMA* or *WCDMA* receivers, a *AGC* control range of up to $90\,dB$ is required. This is really an astronomic number! It is well-known that every $3\,dB$ variation corresponds to a 50% or double power variation. $90\,dB$ of variation implies that the received power or field strength around the antenna could vary by as many times as

$$2^{\frac{90}{3}} = 1{,}073{,}741{,}824!$$

The wide dynamic range of *AGC* control is indeed a challenge to circuit designers.

It is well known that the amount of power gain in a receiver is mostly contributed by the back-end after the de-modulator. In order to be able to set up $90\,dB$ of *AGC* in the back-end, the total power gain in the back-end must be higher than $90\,dB$. This will easily and inevitably bring about oscillation in those amplification blocks due to the high gain required. Eventually, the $90\,dB$ of power variation cannot rely on the *AGC* control in the back-end only. The front-end or *RF* designers must share the task of *AGC* control for a $90\,dB$ power variation. At present, the task of *AGC* control in a radio or a cellular phone is distributed to both front-end and back-end in an approximately 50 to 50% ratio. The *LNA* at the front end must then have an *AGC* control of 40 to $50\,dB$, as other blocks in front-end are difficult to work with for *AGC* control.

An *LNA* with *AGC* control is sometimes called a *VGA* (Variable Gain Amplifier).

An important concern is that the *LNA* with *AGC* control should not produce additional distortion on the modulated signal. At the present, *AGC* requirements for *AM* receivers are usually more stringent than those for *FM* or *PM* receivers.

The *AGC* control loop is usually implemented in the receiver of a wireless communication system, illustrated in Figure 1.49. There is a special block called the *RSSI* (Received Signal Strength Indicator) for *AGC* control in the receiver. It detects the received signal from the *IF* amplifier output and compares it with a reference signal, V_R, in a comparator. The difference between these two signals is the output of the *RSSI*, V_{agc}, which is used to control the gain of the *LNA*.

The analysis and design of the *AGC* control loop must be conducted at the system level. Here we will focus on the *LNA* with *AGC* block only. The characteristics of the *LNA* with an *AGC* block are shown in Figure 1.50. For low input received signals the *AGC* control is inactivated, so that the output of *LNA*, $V_{S,out}$ or $P_{S,out}$ is linearly related to the input received signal V_i. When the input received signal is increased above V_{i1}, the *AGC* loop is activated and will maintain the output level at a constant value or with a tiny increase until the input received signal increases above V_{i2}. The minimum and maximum output voltages from the *RSSI* block, $V_{agc,min}$ and $V_{agc,ma}$, correspond to V_{i1} and V_{i2} respectively. When the input received signal is stronger than V_{i2}, the *AGC* loses its control capability and the *LNA* output, $V_{S,out}$ or $P_{S,out}$, returns to the linear relationship with the input received signal V_i. Instead of a suddenly activating or inactivating of the *AGC* control, the actual performance of the *AGC* produces a somewhat smooth curve as shown in Figure 1.50.

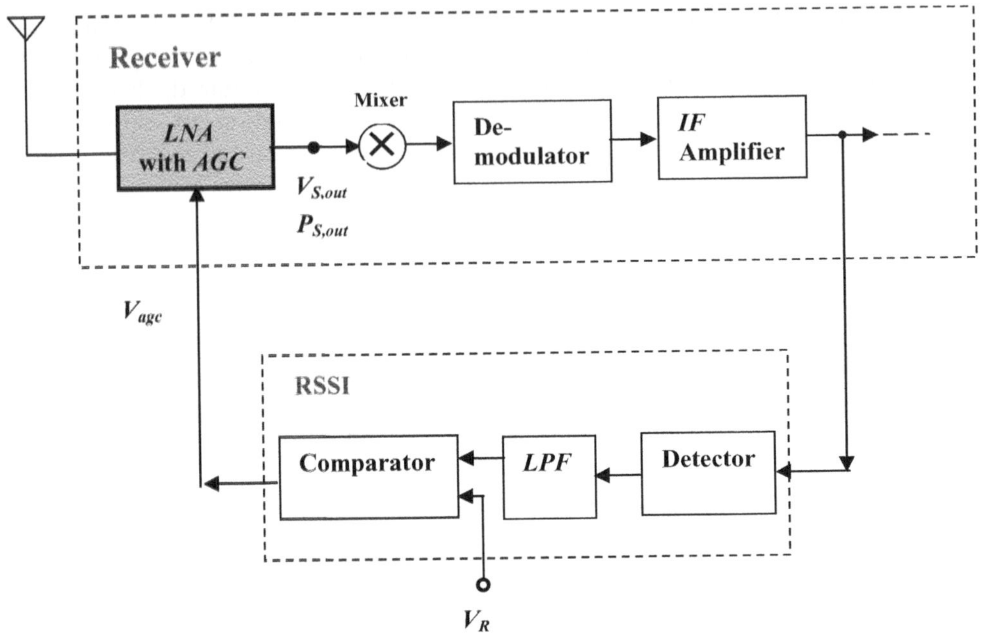

Figure 1.49 An *AGC* control loop in the receiver of a communication system.

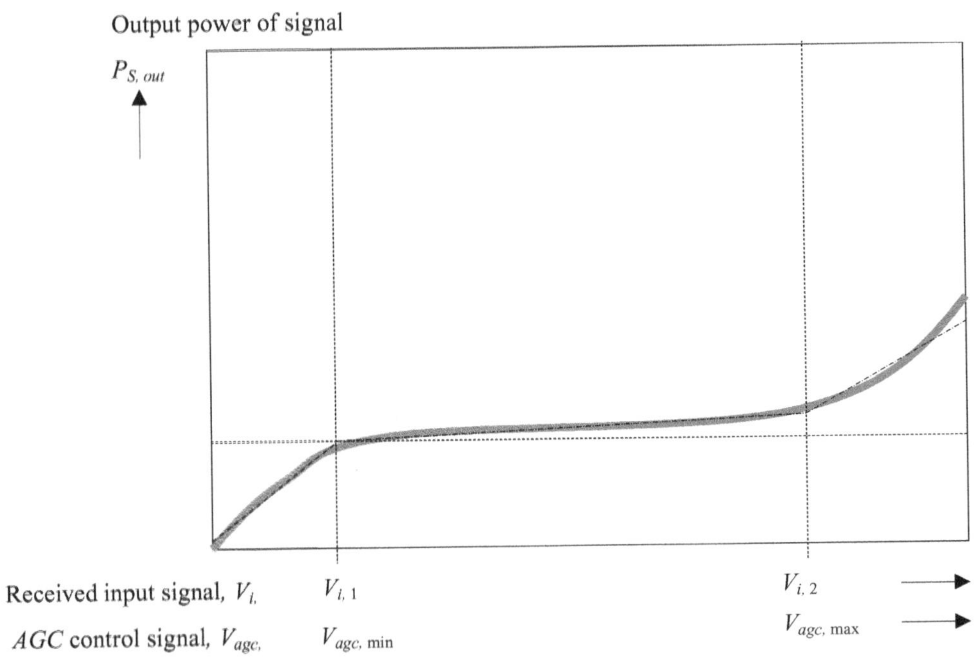

Figure 1.50 Expected characteristics of *AGC* control.
------------- Ideal
▬▬▬▬ Actual performance

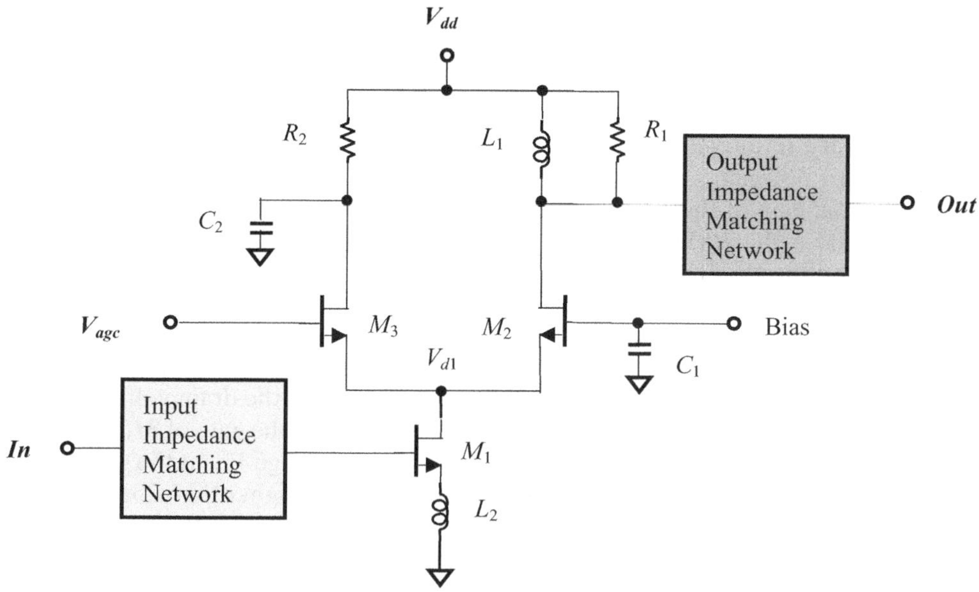

Figure 1.51 Traditional *LNA* with *AGC*.

The dynamic range of the *AGC* control voltage is

$$DR_{agc} = V_{i,2} - V_{i,1} = V_{agc,\max} - V_{agc,\min}, \tag{1.187}$$

in which the power variation of received signal at antenna, say, 40 to $50\,dB$, can be "swallowed."

1.4.2 Traditional *LNA* with *AGC*

Figure 1.51 shows a traditional *LNA* with *AGC*. Its configuration is a differential-type with a cascode *LNA* branch, which consists of M_1 and M_2, and an *AGC* control branch, which consists of M_1 and M_3. The function of each part is as follows:

- M_1, M_2: cascoded devices;
- C_1, C_2: *AC* short-circuited or "zero" capacitor;
- L_1: load inductor;
- R_1: stability resistor;
- L_2: degeneration inductor;
- M_3: *AGC* effective device;
- R_2: Current control resistor.

The input to the gate of the transistor M_3, V_{agc}, is a *DC* voltage from the *RSSI* output in the radio. When the *RSSI* voltage is lower than $V_{agc,min}$, the M_3 is turned off and *AGC* function is not activated. However, if the *RSSI* output voltage, V_{agc}, is increased above $V_{agc,min}$, the M_3 is turned on and shares the *AC* current from the M_1–M_2 cascode

LNA branch. As the *RSSI* voltage continuously rises, the M_3 will share more *AC* current. The *AC* current shared by M_3 is shorted to the ground by the capacitor C_2. Consequently, the gain of the cascode *LNA* branch M_1–M_2 will be reduced because the *AC* current is partially shared and shorted to the ground by the *AGC* control branch M_1–M_3. Intuitively, the higher the *RSSI* control voltage V_{agc}, the more the reduction of *LNA* gain.

The normal gain of the *LNA* is defined as the gain when the *AGC* branch is not activated, that is the *RSSI* voltage V_{agc} is lower than $V_{agc,min}$. The *LNA* gain can be reduced when the *AGC* branch becomes effective.

As mentioned above, the *AGC* dynamic range is required to be at least $40\,dB$. However, with the traditional *LNA* with *AGC* control design shown in Figure 1.51, the *AGC* dynamic range only approaches $20\,dB$. This limitation is due to the restrictions of the topology. Let's take a look at Figure 1.51. Should the drain voltage of M_1, V_{d1}, always trace the variation of *RSSI* output voltage V_{agc} at the gate of M_3, the *LNA* gain might be continuously reduced as the *RSSI* output voltage V_{agc} at the gate of M_3 is increased. Then the *AGC* dynamic range could be much higher than $20\,dB$.

Unfortunately, owing to the existence of the M_1–M_2 branch, the variation of voltage V_{d1} is not as expected. At the beginning, V_{d1} increases as the *RSSI* voltage increases, so that the *AC* current is shared by the *AGC* control branch and the *LNA* gain is lowered. However, as the increased *RSSI* output voltage V_{agc} reaches a certain value, V_{d1} stops increasing, or even inversely decreases. This leads to the increase of *LNA* gain back in the direction toward its normal gain without *AGC* control. That is why the *AGC* dynamic range is limited.

1.4.3 Increasing of *AGC* Dynamic Range

In order to increase the *AGC* dynamic range, there are three improvements to be adapted:

1) **Diode clamping**
   In order to force the voltage V_{d1} to trace the variation of the *RSSI* voltage, a device M_4 with a diode-connection is added to the gate and source of M_3 as shown in Figure 1.52.
2) **Avoiding *RSSI* operation as a current sink or source**
   Without the additional resistor R_3 and the additional device M_4 as shown in Figure 1.52, the gate of M_3 is the load of the *RSSI* output, which is extremely light since the input impedance of M_3 is quite high. Now, the addition of the device M_4 brings about a problem for the *RSSI* output. The device M_4 is *DC* connected from *RSSI* output to the source of M_3 and the rest of the circuit. This makes the *RSSI* output port a *DC* source or sink. The current from or into the *RSSI* output could be in the order of *mA* or more, which the *RSSI* output terminal cannot usually afford.

   The solution is to add a resistor R_3 between the V_{dd} and the gate of M_3 as shown in Figure 1.53. An appropriate adjustment of the value of R_3 can make the current either flowing from or into the *RSSI* output less than $0.5\,mA$.
3) **Selection of big device M_3**
   In the design of *LNA* with *AGC*, the selection of the device M_3 is very important. From a system design viewpoint, it is desirable to reduce the current drain

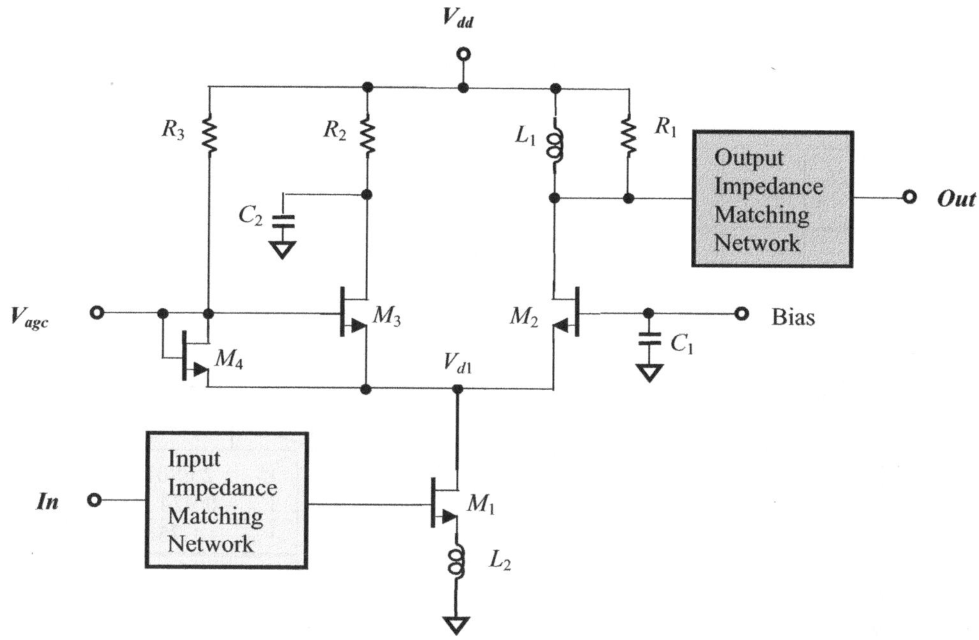

Figure 1.52 *LNA* with large dynamic range *AGC*.

as much as possible. Therefore, the I_{ds} current of M_3 should not be too high when it is turned on. On the other hand, its I_{ds} current should be very low when it is turned off. In order to satisfy these two requirements simultaneously, one must select a big device M_3 with a large number of "fingers.".

1.4.4 An Example

The schematic as shown in Figure 1.52 has been simulated with the *CMOS* process and the corresponding work on the bench has been done.

In this design, the number of fingers is 8 and 16 for M_1 and M_2 respectively, whereas the number of fingers for M_3 is selected as 100. (In the *CMOS* processing, the width of one finger is 16 microns.) The *DC* characteristics of M_3 are shown in Figure 1.53 and can be outlined as follows.

$$I_{ds} \approx 0.02\,mA \text{ or } 20\,\mu A, \qquad \text{when } V_{agc} = 1.2\,\text{V}; \tag{1.188}$$

$$I_{ds} = 1.25\,mA, \qquad \text{when } V_{agc} = 1.2\,V. \tag{1.189}$$

Table 1.5 lists its parts and Table 1.6 lists its performance.

This *LNA* with a large dynamic range of *AGC* design has been integrated with a 1 *watt PA* in an *IC* chip, which has been successfully applied to a radio production line.

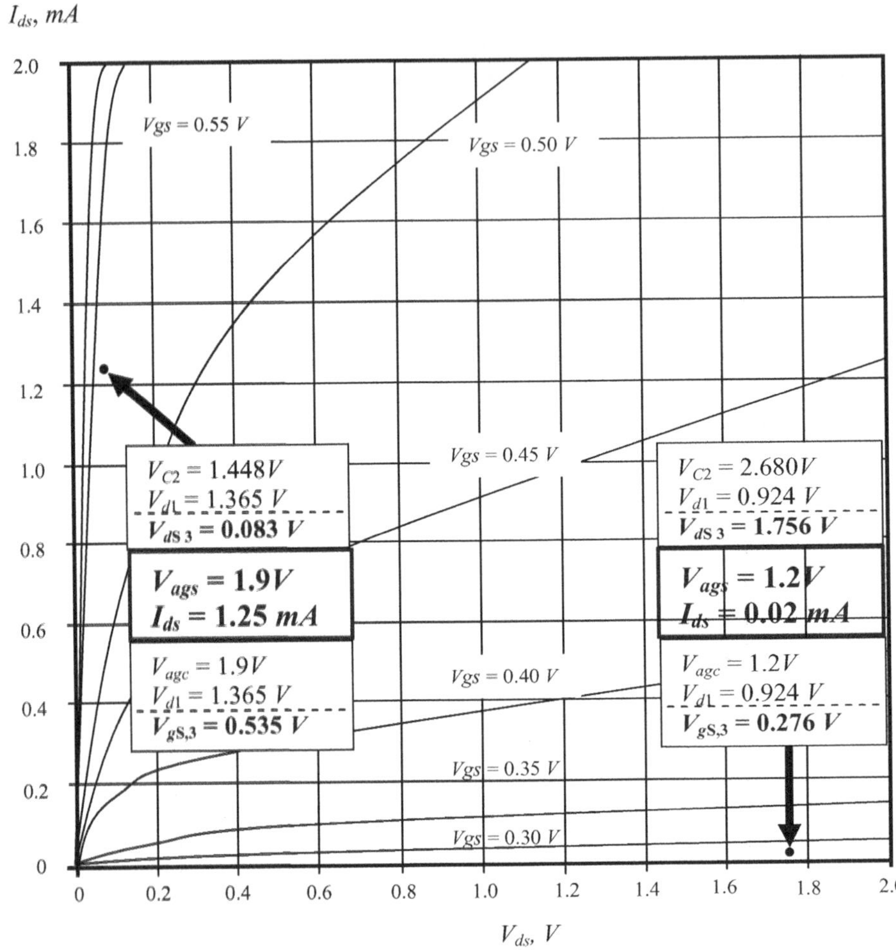

Figure 1.53 *DC* characteristics of transistor M_3.

TABLE 1.5 Part list of LNA with AGC

Part		Value.
M_1	Fingers = 8	$W_{total} = 128\,\mu m$
M_2	Fingers = 16	$W_{total} = 256\,\mu m$
M_3	Fingers = 100	$W_{total} = 1600\,\mu m$
M_4	Fingers = 16	$W_{total} = 1256\,\mu m$
R_1	$700\,\Omega$	
R_2	$1\,k\Omega$	
R_3	$10\,k\Omega$	
C_1	$20\,pF$	
C_2	$40\,pF$	
L_1	$50\,nH$	
L_2	$30\,nH$	

TABLE 1.6 Goals, simulation, and performance of an *LNA* with *AGC*

	Goal		Simulation		Performance		
DC power supply, *V*	3.0		3.0		3.0		*V*
Frequency, *f*	460 to 470		460 to 470		460 to 470		*MHz*
RSSI control voltage, V_{agc}	1.2	1.9	1.2	1.9	1.2	1.9	*V*
***AGC* dynamic range, DR_{agc}**	**0.00**	**–40**	**0.00**	**–43.7**	**0.0**	**–42**	***dB***
Current drain, I_{ds}	2.0	*	1.87	1.48	2.0	1.4	*mA*
Gain, *G*	1.0	–28	1.6	–31.1	15.0	–27.0	*dB*
Noise figure, *NF*	2.0	*	1.26	29.4	2.7	*	*dB*
3rd order intercept point, IIP_3	–15.0	*	–10.2	–7.5	–10.5	–8.0	dB_m
Input Return Loss, S_{11}	*	*	–15.3	–10.4	*	*	*dB*
Output Return Loss, S_{22}	*	*	–23.2	–11.7	*	*	*dB*

REFERENCES

[1] H. A. Haus et al., "Representation of Noise in Linear Two Ports," *Proceedings of the IRE*, Vol. 48, January, 1960, pp. 69–74.

[2] Guillermo Gonzalez, *Microwave Transistor Amplifiers*, Prentice-Hall, Inc., 1984.

[3] J. D. Sifri, "Matching Technique Yields Optimum LNA Performance," *Microwaves and RF*, February, 1986, pp. 87–90.

[4] Andrew N. Karanicolas, "A 2.7V 900 MHz CMOS LNA and Mixer," IEEE International Solid State Circuits Conference, 1996, pp. 50–53.

[5] Thomas H. Lee, *The Design of CMOS Radio-Frequency Integrated* Circuits, Cambridge University Press, 1998.

[6] G. Girlando and G. Palmisano, "Noise Figure and Impedance Matching in RF Cascode Amplifiers," Circuits and Systems II: Analog and Digital Signal Processing, *IEEE Transactions* on (see also IEEE Transactions on Circuits and Systems II: Express Briefs], Vol. 46, No. 11, November 1999, pp. 1388–1396.

[7] F. Svelto, G. Montagna, S. Deantoni, G. Braschi, and R. Castello, "Solution for Image Rejection CMOS LNA," ISCAS 2000 IEEE International Symposium on Circuits and Systems, May 28–31, Geneva, Swizerland, pp. III49–III52.

[8] "SPICE Simulation and Tradeoffs of CMOS LNA Performance with Source Degeneration Inductor," IEEE Transaction on Circuits and Systems-II: Analog and Digital Signal Processing, Vol. 47, No. 1, January 2000, pp. 62–65.

[9] G. Gramegna, A. Magazzu, C. Sclafani, and M. Paparo, "Ultra-Wide Dynamic Range 1.75 dB Noise Figure, 900 MHz CMOS LNA," ISSCC 2000, 20000 IEEE International Solid-State Circuit Conference, pp. 380–381.

[10] C. S. Kim, M. Park, C.-H. Kim, M.-Y. Park, S.-D. Kim, Y.-S. Youn, J.-W. Park, S.-H. Han, H. K. Yu, and H. Cho, "Design Guide of Coupling Between Inductors and Its Effect on Reverse Isolation of a CMOS LNA," 2000 IEEE MTT_S Digest, pp. 225–228.

[11] Ali Karimi-Sanjaani, Henrik Sjoland, and A. Abidi Asad, "A 2 GHz Merged CMOS LNA and Mixer for WCDMA," 2001 Symposium on VLSI Circuit Digest of Technical Papers, pp. 19–22.

[12] Francesco Gatta, Enrico Sacchi, Francesco Svelto, Paolo Vilmercati, and Rinaldo Castello, "A 2 dB Noise Figure 900 MHz Differential CMOS LNA," *IEEE Journal of Solid State Circuits*, Vol. 36, No. 10, October 2001, pp. 1444–1452.

[13] Paul R. Gray, Paul J. Hurst, Stephen H. Lewis, and Robert G. Meyer, *Analysis and Design of Analog Integrated Circuits*, 4th ed., John Wiley & Sons, Inc., 2001, pp. 206–212, 522–532.

[14] Frencesco Svelto, Stefano Deantoni, Giampiero Montagna, and Rinaldo Castello, "Implementation of a CMOS LNA Plus Mixer for GPS Applications with No External Components," *IEEE Transactions on Very Large Scale Integration (VLSI) Systems*, Vol. 9, No. 1, February 2001, pp. 100–104.

[15] Yi Lin, Michael Obrecht, and Tajinder Manku, "RF Noise Characterization of MOS Devices for LNA Design Using a Physical-Based Quasi-3-D Approach," *IEEE Transactions on Circuits and Systems-II: Analog and Digital Signal Processing*, Vol. 48, No. 10, October 2001, pp. 972–984.

[16] David J. Cassan and John R. Long, "A 1V 0.9dB NF Low Noise Amplifier for 5–6GHz WLAN in 0.18μm CMOS," IEEE 2002 Custom Integrated Circuits Conference, pp. 419–422.

[17] Jung-Suk Goo, Hee-Tae Ahn, Donald J. Ladwig Yu, Thomas H. Lee, and Robert W. Dutton, "A Noise Optimization Technique for Integrated Low Noise Amplifier," *IEEE Microwave and Guided Wave Letters*, Vol. 37, No. 8, August 2002, pp. 994–1002.

[18] Jiwei Chen and Bingxue Shi, "Impact of Intrinsic Channel Resistance on Noise Performance of CMPS LNA," *IEEE Electron Device Letters*, Vol. 23, No. 1, January 2002, pp. 34–36.

[19] David J. Cassan and John R. Long, "A1-V Transformer Feedback Low Noise Amplifier for 5GHz Wireless LAN in 0.18μm CMOS," *IEEE Journal of Solid-State Circuits*, Vol. 38, No. 3, March 2003, pp. 427–435.

[20] Richard Chi-Hsi Li, *Key Issues in RF/RFIC Circuit Design*, Higher Education Press, Beijing, 2005.

CHAPTER 2

MIXERS

2.1 INTRODUCTION

There are two kinds of mixers used to convert a signal from an *RF* to an *IF* frequency in a communication system: the active mixer and the passive mixer. Both have been developed and applied for several decades, and either one can be used in a communication system in the frequency range of *VHF*, *UHF*, and *GHz*. However, in the microwave frequency range, it is more realistic to apply the passive mixer.

Figures 2.1 and 2.2 show two typical active and passive mixers, respectively. The main device applied to an active mixer is the bipolar or *MOSFET* transistor, while the main device applied to a passive mixer is a diode or a matched quad diode.

The active mixer shown in Figure 2.1 is quite simple. It contains only one *MOSFET* transistor. The *RF* signal is fed to a self-coupling transformer and then goes to the gate of the *MOSFET* transistor. The self-coupling transformer or, say, an inductor with central tape, functions as an impedance matching part because the input impedance at the gate of the *MOSFET* transistor usually is quite high. The *LO* injection is fed to the source of the *MOSFET* transistor. At the drain of the *MOSFET* transistor, an inductor and three capacitors form an output impedance matching network.

The core of the passive mixer shown in Figure 2.2 is the quad diode. The four diodes are usually packaged together. They are, of course, expected to be as identical as possible, because the performance of the mixer largely depends on the uniformity of these four diodes. Both the *RF* signal and *LO* injection are fed to a transformer first and then go to the quad diode. This is due to the differential requirement from the quad diode: the input *RF* signal and *LO* injection into the quad diode must be as perfectly differential as possible. Two transformers function as a balun to transfer

RF Circuit Design, by Richard Chi-Hsi Li

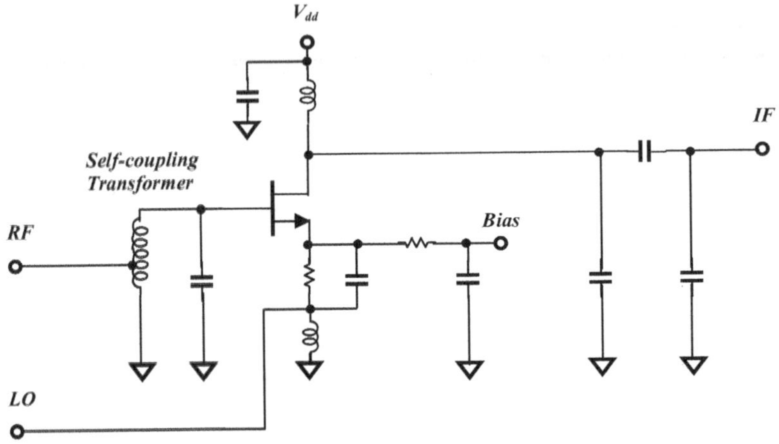

Figure 2.1 A typical *MOSFET* active mixer.

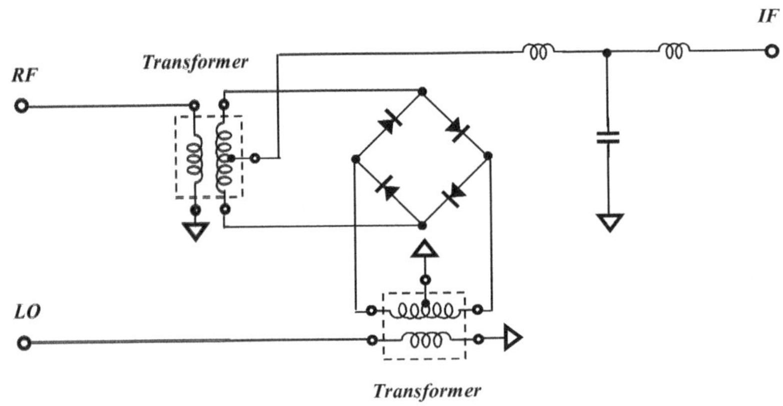

Figure 2.2 A typical quad diode passive mixer.

the single-ended input into a differential pair output. In addition, they function as an input impedance matching network. The central tap of the *RF* transformer is the *IF* output. The *T* type output impedance matching network consists of two inductors and one capacitor.

In a dual conversion portable radio or communication system, the *BPF* (Band Pass Filter) is always set in the path of the input *RF* signal and in the path of the *LO* injection before they get into the mixer, so that the spurious products can be considerably reduced. The helical filter has been applied for such a purpose. Figures 2.3 and 2.4 show the active and passive mixers with *BPF* helical filters, respectively.

The helical filters shown in Figures 2.3 and 2.4 are the adjustable *BPF* (Band Pass Filter) filters. They are filters with a cavity-type configuration, constructed as a casting aluminum box. There are a couple of small closets in the box. The helical coil is circled around a cylindrical plastic frame and then put into each small closet in the box. A ferrite core is inserted into each cylindrical frame and can be adjusted up and down along the cylindrical axis. A four-pole helical filter indicates that there

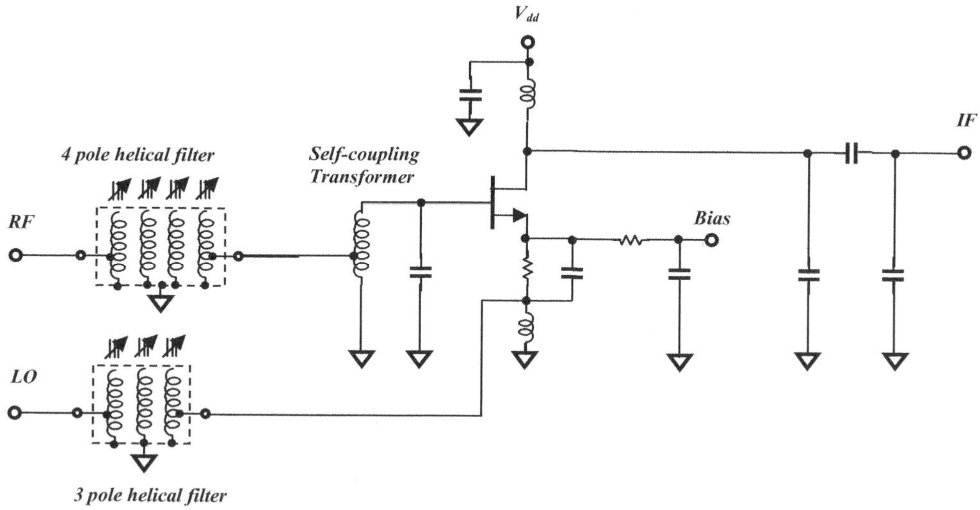

Figure 2.3 A typical *MOSFET* active mixer with *BPF* helical filter.

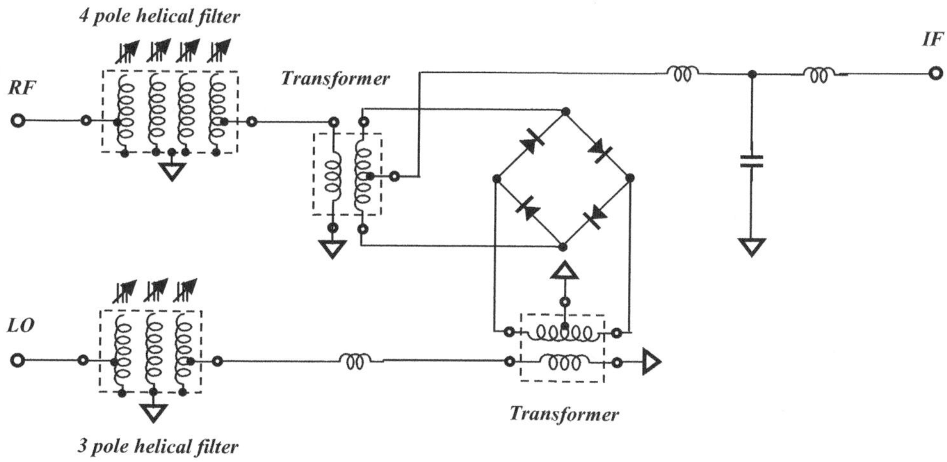

Figure 2.4 A typical quad diode passive mixer.

are four small closets with four helical coils contained in the metallic box; a three-pole helical filter indicates three small closets with three helical coils contained in a metallic box, and so on. In the *UHF* frequency range, the size of a helical filter is about $L \times W \times H = 2.0 \times 0.5 \times 1.5\,cm$ for a four-pole helical filter and about $L \times W \times H = 1.5 \times 0.5 \times 1.5\,cm$ for a three-pole helical filter.

At present, for considerations of cost and size, most helical filters have been replaced by other filters, such as the *SAW* filter. However, the performance of helical filters is superior to other filters in terms of a high Q value, low insertion loss, and wide bandwidth.

It is interesting, of course, to compare the respective advantages and disadvantages of mixers in order to decide which type of mixer to use. Table 2.1 lists the main

TABLE 2.1 Comparison between active and passive mixer

Item	Active mixer	Passive mixer	Unit
Current drain	~2 to 5	~0	mA
LO Injection	~−10 to 0	~5 to 10	dB_m
Conversion gain	~5 to 10	~−5 to −3	dB
Noise figure	~10 to 15	~3 to 5	dB
Bandwidth	Narrower	Wider	
Part count	~13	~6	
Reliability	Lower	Higher	
Cost	Lower	Higher	

Note: The values listed in this table are approximately correct.

items for comparison on the basis of Figures 2.1 and 2.2. In respect to the active mixer, the advantages of a passive mixer are no current drain, low noise figure, wide bandwidth, and lower part count (and hence high reliability); the disadvantages are a higher required *LO* injection, negative conversion gain, and high cost. It should be noted that in a passive mixer, two transformers with a central tap must be provided. Together with a matched quad diode device, symmetry becomes a key factor in their performance.

At present, by means of *RFIC* technology, the resistive mixer is being developed as a new configuration of a passive mixer, while the Gilbert cell is being developed as the core of the active mixer. We discuss them in the following sections.

2.2 PASSIVE MIXERS

A passive mixer implies that the current drain in the mixer circuitry is zero.

2.2.1 Simplest Passive Mixer

The simplest passive mixer can be implemented by a diode and a compound transformer as shown in Figure 2.5. The operating principle of this mixer is based on the even order non-linearity of the diode, that is,

$$i = a_o + a_1(k_r v_{RF} + k_l v_{LO}) + a_2(k_r v_{RF} + k_l v_{LO})^2 + a_3(k_r v_{RF} + k_l v_{LO})^3 + a_4(k_r v_{RF} + k_l v_{LO})^4 + \ldots \tag{2.1}$$

where

i = *RF* current flowing through the diode,
a_i = i^{th} order non-linearity coefficient,
k_r = coupling coefficient between *RF* and *IF* port of transformer,
k_l = coupling coefficient between *LO* and *IF* port of transformer,
v_{RF} = input voltage of RF signal,
v_{LO} = voltage of *LO* injection.

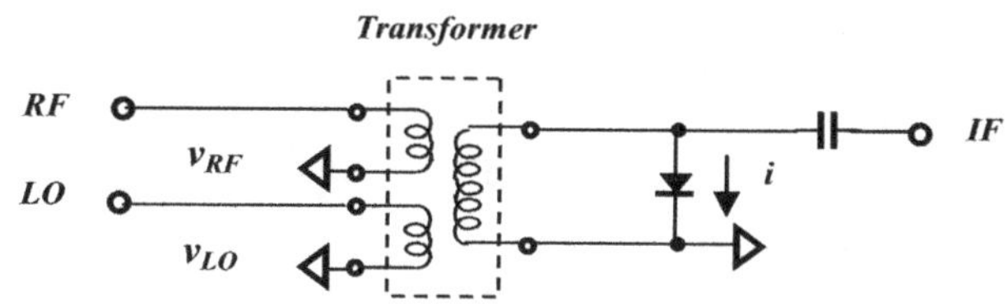

Figure 2.5 A simple passive mixer.

If

$$v_{RF} = v_{RFo} \cos(\omega_{RF}t + \varphi), \tag{2.2}$$

$$v_{LO} = v_{LOo} \cos \omega_{LO}t, \tag{2.3}$$

where

ω_{RF} = angular frequency of *RF* signal,
ω_{LO} = angular frequency of *LO* injection,

then the equation (2.1) becomes

$$\begin{aligned} i &= a_o + a_1[k_r v_{RFo} \cos(\omega_{RF}t + \varphi) + k_l v_{LOo} \cos \omega_{LO}t] + \\ &a_2\left[(k_r v_{RFo})^2 \frac{1 + \cos 2(\omega_{RF}t + \varphi)}{2} + (k_l v_{LOo})^2 \frac{1 + \cos 2\omega_{LO}t}{2}\right] + \\ &a_2[+ k_r k_l v_{RFo} v_{LOo} (\cos\{(\omega_{RF} + \omega_{LO})t + \varphi\} + \cos\{(\omega_{RF} - \omega_{LO})t + \varphi\})] + \\ &a_3[k_r v_{RFo} \cos(\omega_{RF}t + \varphi) + k_l v_{LOo} \cos \omega_{LO}t]^3 + \\ &a_4[k_r v_{RFo} \cos(\omega_{RF}t + \varphi) + k_l v_{LOo} \cos \omega_{LO}t]^4 + \ldots \end{aligned} \tag{2.4}$$

It should be noted that in equation (2.4), one of the terms with 2nd order non-linearity in the low-frequency range is

$$k_r k_l v_{RFo} v_{LOo} \cos\{(\omega_{RF} - \omega_{LO})t + \varphi\} = k_r k_l v_{RFo} v_{LOo} \cos\{\omega_{IF}t + \varphi\}, \tag{2.5}$$

where ω_{IF} = angular frequency of *IF* signal and is

$$\omega_{RF} - \omega_{LO} = \omega_{IF}, \tag{2.6}$$

Equations (2.4) to (2.6) indicate that the term with second-order non-linearity of the diode produces the *IF* components. This is the operational principle of a mixer.

2.2.2 Double Balanced Quad Diode Mixer

Figure 2.6 shows a typical double balanced quad diode mixer, in which the main part is a quad diode or ring diode. The transformer at the *RF* port couples the *RF* signal to the ring diodes while the transformer at the *LO* port couples the *LO* injection to the ring diodes.

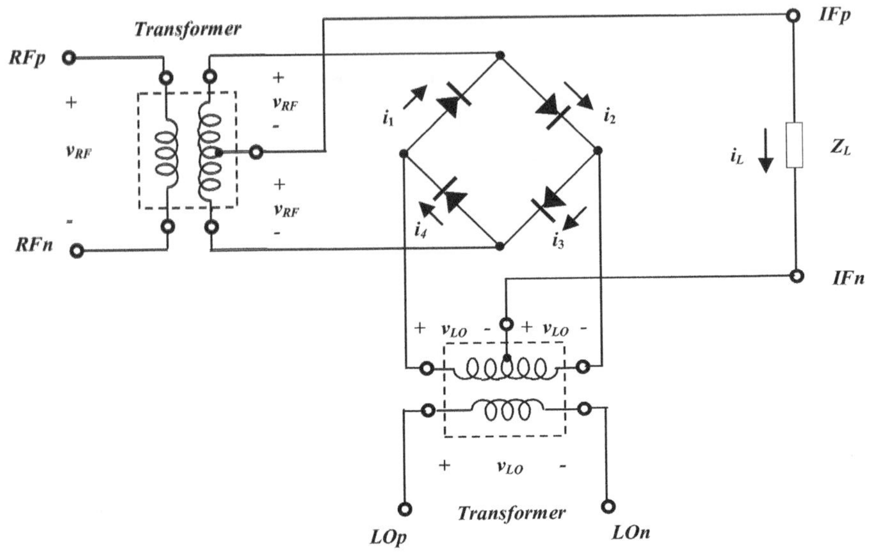

Figure 2.6 A typical double balanced quad diode mixer with ring configuration.

When the *RF* signal voltage and *LO* injection are coupled to the quad diodes, the corresponding currents go through all the *PN* junctions with different paths in the ring configuration. Owing to the non-linearity of the diodes, these currents interact with each other.

There are four sub-currents, i_1, i_2, i_3, and i_4, flowing through the diodes and the load, Z_L, with different respective paths. They are determined by the voltages from the *RF* signal and *LO* injection, v_{RF} and v_{LO}, on the individual loop containing the load Z_L. They are:

$$i_1 = a_0 + a_1(-v_{RF} + v_{LO}) + a_2(-v_{RF} + v_{LO})^2 + a_3(-v_{RF} + v_{LO})^3 + a_4(-v_{RF} + v_{LO})^4 + \ldots \tag{2.7}$$

$$i_2 = a_0 + a_1(v_{RF} + v_{LO}) + a_2(v_{RF} + v_{LO})^2 + a_3(v_{RF} + v_{LO})^3 + a_4(v_{RF} + v_{LO})^4 + \ldots \tag{2.8}$$

$$i_3 = a_0 + a_1(v_{RF} - v_{LO}) + a_2(v_{RF} - v_{LO})^2 + a_3(v_{RF} - v_{LO})^3 + a_4(v_{RF} - v_{LO})^4 + \ldots \tag{2.9}$$

$$i_4 = a_0 + a_1(-v_{RF} - v_{LO}) + a_2(-v_{RF} - v_{LO})^2 + a_3(-v_{RF} - v_{LO})^3 + a_4(-v_{RF} - v_{LO})^4 + \ldots \tag{2.10}$$

where $a_k = k^{\text{th}}$ order non-linearity coefficient of diode.

Finally the current flow through the load, i_L, consists of four sub-currents, i_1, i_2, i_3, and i_4, that is,

$$i_L = (i_1 - i_2) + (i_3 - i_4) = -a_2 8 v_{RF} v_{LO} - a_4 16(v_{RF}^3 v_{LO} + v_{RF} v_{LO}^3) + \ldots, \tag{2.11}$$

Equation (2.11) implies that the odd order non-linearity terms are cancelled with each other. The spurious products can be produced only by the even order non-linearity terms.

Assuming that

$$v_{RF} = v_{RFo}\cos(\omega_{RF}t + \varphi), \tag{2.12}$$

$$v_{LO} = v_{LOo}\cos\omega_{LO}t, \tag{2.13}$$

then,

$$\begin{aligned} i_L = &-a_2 4 v_{RFo} v_{LOo}\left[\cos\{(\omega_{RF}+\omega_{LO})t+\varphi\} + \cos\{(\omega_{RF}-\omega_{LO})t+\varphi\}\right] - \\ & a_4 8 v_{RFo} v_{LOo}\left[\cos\{(\omega_{RF}+\omega_{LO})t+\varphi\} + \cos\{(\omega_{RF}-\omega_{LO})t+\varphi\}\right]\cdot \\ & \left[v_{RFo}^2\{1+\cos 2(\omega_{RF}t+\varphi)\} + v_{LOo}^2\{1+\cos 2\omega_{LO}t\}\right] - \ldots \end{aligned} \tag{2.14}$$

or

$$\begin{aligned} i_L = &-\left[a_2 4 v_{RFo} v_{LOo} + a_4 8 v_{RFo} v_{LOo}\left(v_{RFo}^2 + v_{LOo}^2\right) + \ldots\right]\cos\{\omega_{IF}t+\varphi\} - \\ & \left[a_2 4 v_{RFo} v_{LOo} + a_4 8 v_{RFo} v_{LOo}\left(v_{RFo}^2 + v_{LOo}^2\right) + \ldots\right]\cos\{(\omega_{RF}+\omega_{LO})t+\varphi\} - \ldots \end{aligned} \tag{2.15}$$

In the right side of the above equation, the first term is the desired *IF* signal and will be preserved, while the other terms on the right side are high-frequency components and will be filtered out. Consequently, from equation (2.15), the *IF* current, i_{IF}, on the load, Z_L, is

$$i_{IF} = -\left[a_2 4 v_{RFo} v_{LOo} + a_4 8 v_{RFo} v_{LOo}\left(v_{RFo}^2 + v_{LOo}^2\right) + \ldots\right]\cos\{\omega_{IF}t+\varphi\}. \tag{2.16}$$

It can be seen that the second order non-linearity of the diode is the main part or key contribution in producing the *IF* signal. Other even order non-linearities contribute just a few percent to the *IF* signal but produce a lot of spurious products.

Also, from equation (2.16) it can be seen that the performance of a double balanced mixer with ring quad diodes depends on

- The non-linearity of diodes, such as $a_2, a_4, a_6, \ldots$
- *LO* injection, v_{LO},
- *RF* signal itself, v_{RFo},
- Frequency, ω.

Its typical performance, for example, is

- *LO* injection: $5\,dB_m$
- Conversion gain: –4.0 to –4.5 dB,
- Noise figure: 4.0 to 4.5 dB,
- IIP_3: –5 to –2 dB_m,
- Relative bandwidth: 10%.

In the quad diode mixer, instead of conversion gain, there is always conversion loss. The conversion loss depends on many factors, such as the *LO* injection, the non-linearity of device, the input and output impedance matching status, and the Q values of the parts in the mixer. It should be especially noted that the conversion loss is dependent on the *LO* injection level. Figure 2.7 shows an example of this relationship.

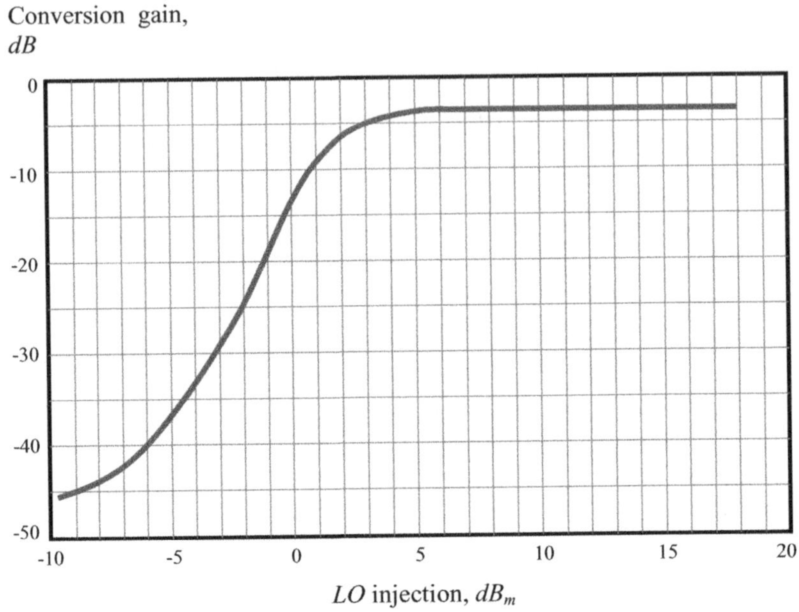

Figure 2.7 An example of the relationship between conversion gain and *LO* injection in a passive mixer.

The conversion loss is high when the *LO* injection is low. It is decreased when the *LO* injection is increased. Usually, a passive mixer begins to reach its normal operation state only after the *LO* injection is increased to over $3\,dB_m$.

For a passive *RF* block, the noise figure is reasonably equivalent to the conversion loss. However, in the quad ring mixer, a remarkable feature of a passive mixer is that its noise figure depends on the *LO* injection. Figure 2.8 shows that there is a minimum of the noise figure as the *LO* injection is varied. As shown in Figure 2.8, the minimum noise figure is $4.8\,dB$ when the *LO* injection is $7.8\,dB_m$. The non-linearity is mainly determined by the expressions of even orders of non-linearity of quad diodes except the second order of non-linearity.

It should be noted that in equations (2.7) to (2.10), the non-linearity characteristics of each diode in the quad diode are assumed to be perfectly identical, so that their non-linearity coefficients are the same. Obviously, imperfect uniformity among the quad diodes brings about more spurious problems, in which the odd order non-linearity of the diodes do not cancel each other. Then, the linearity of the mixer would be degraded.

In addition, it is also assumed that the central tap of transformers at either the *RF* or *LO* port is at a perfect symmetrical position so that the voltage at the secondary winding of the transformer is exactly divided by two identical voltages, $v_{RF} + v_{RF}$ or $v_{LO} + v_{LO}$. Should the perfectly symmetrical condition of the transformers be unsatisfied, the final expressions of the current flowing on the load or *IF* current, (2.15) and (2.16), would be much complicated, depending on how much asymmetry is present.

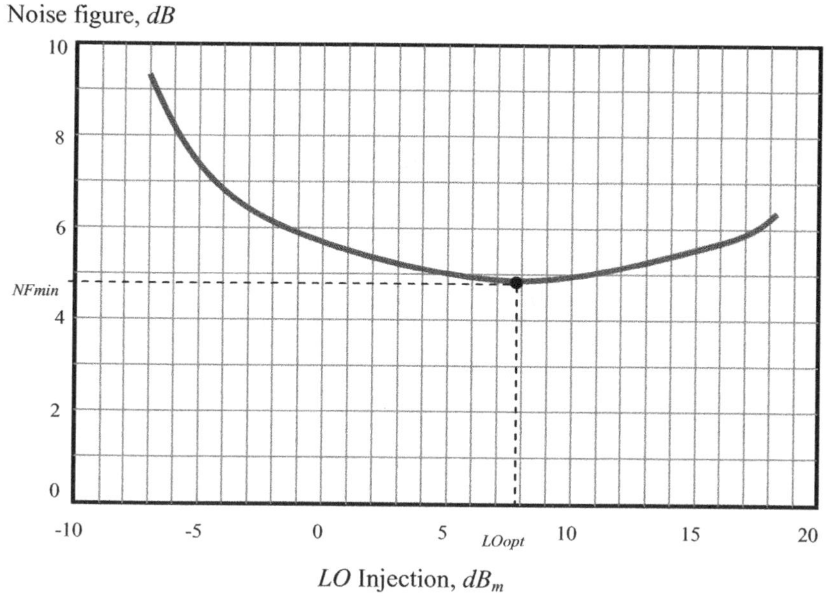

Figure 2.8 An example of the relationship between conversion gain and noise figure in a passive mixer.

2.2.3 Double Balanced Resistive Mixer

Figure 2.9 shows the core of a double balanced resistive mixer. Figure 2.10 is exactly the same as Figure 2.9, but is drawn in a different style.

The capacitors C_1 to C_6 block all the *DC* currents so that they are "zero" capacitors in the operating frequency range. The special feature of a resistive mixer is that there is no *DC* current flow through it. At first glance, the resistive mixer looks like a ring diode mixer. However, this is not true because it is not a real ring but a circuit with a double balanced configuration. Essentially, it is a Gilbert cell without *DC* power supply and bias.

This mixer's basic operating principle is that the *RF* signal current flows through the resistive channels under the gate of the double balanced transistors and is "modulated" by, or interacts with, the *LO* injection voltage at the gates of the double balanced transistors.

The differences between the quad diode and the resistive mixer can be found in their operating principles:

- In the quad diode mixer, the current of *RF* signal flows through the *PN* junction of the diode, while in the resistive mixer, the current of *RF* signal flows through the resistive channel of the *MOSFET* transistor under the gate.
- In the quad diode mixer, the *LO* injection current flows through the *PN* junction of the diode, just like the *RF* signal current, while in the resistive mixer, the *LO* injection appears as a control voltage to control the resistive channel of the *MOSFET* transistor.

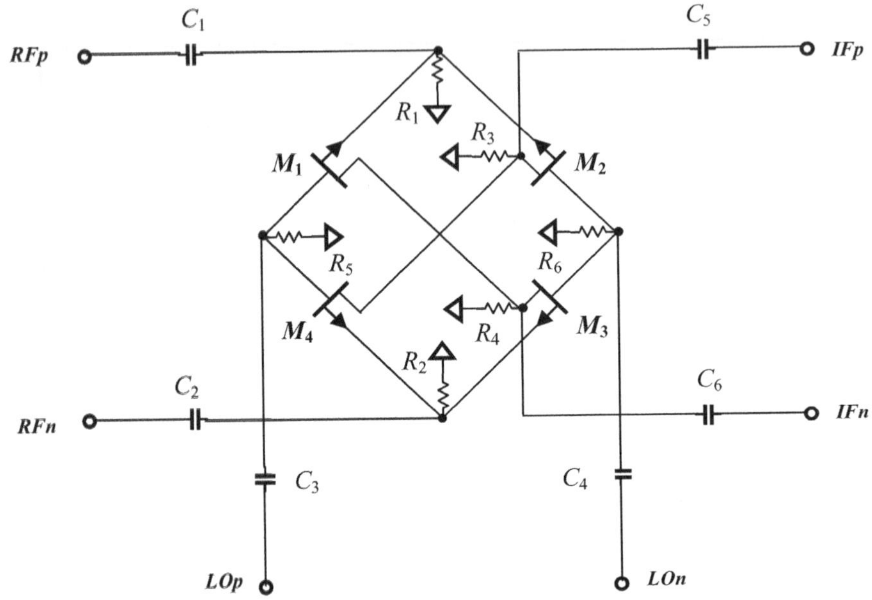

Figure 2.9 Core of a passive resistive mixer.

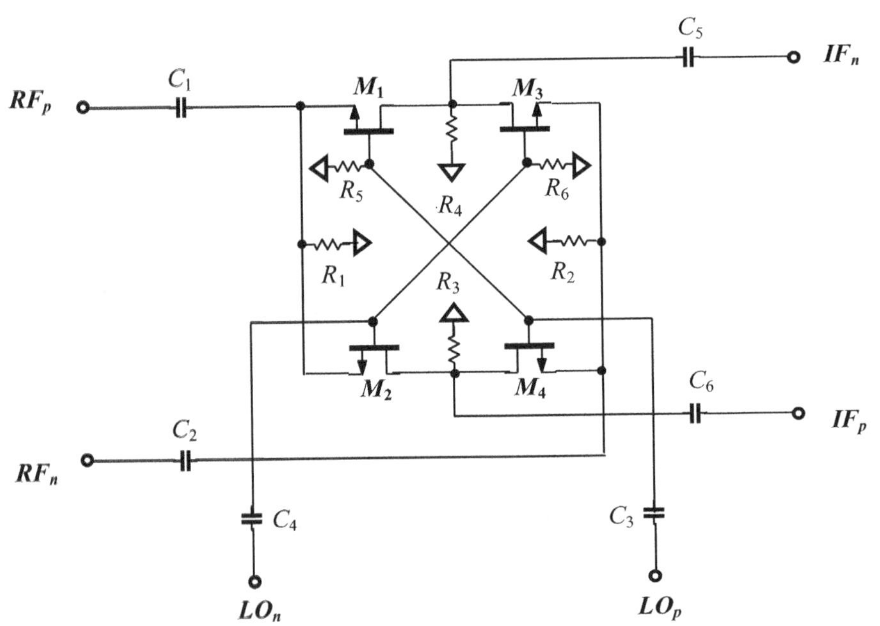

Figure 2.10 Alternative drawing of a passive resistive mixer core.

- In the quad diode mixer, the *IF* signal is a product resulting from the interaction of the *RF* signal current and *LO* injection current flowing through the *PN* junction together, while in the resistive mixer, the *IF* signal is a product resulting from the *RF* signal current flowing through the resistive channel and modulated by the *LO* injection control voltage at the gate of the *MOSFET* transistor.

In summary, the same *IF* signal is produced by both, but the mechanism of production is different.

The following is a design example for 800–900 *MHz* radios. Figure 2.11 is the schematic of the resistive mixer with three simple *LC* baluns for the *RF* input signal, *LO* injection, and *IF* output signal.

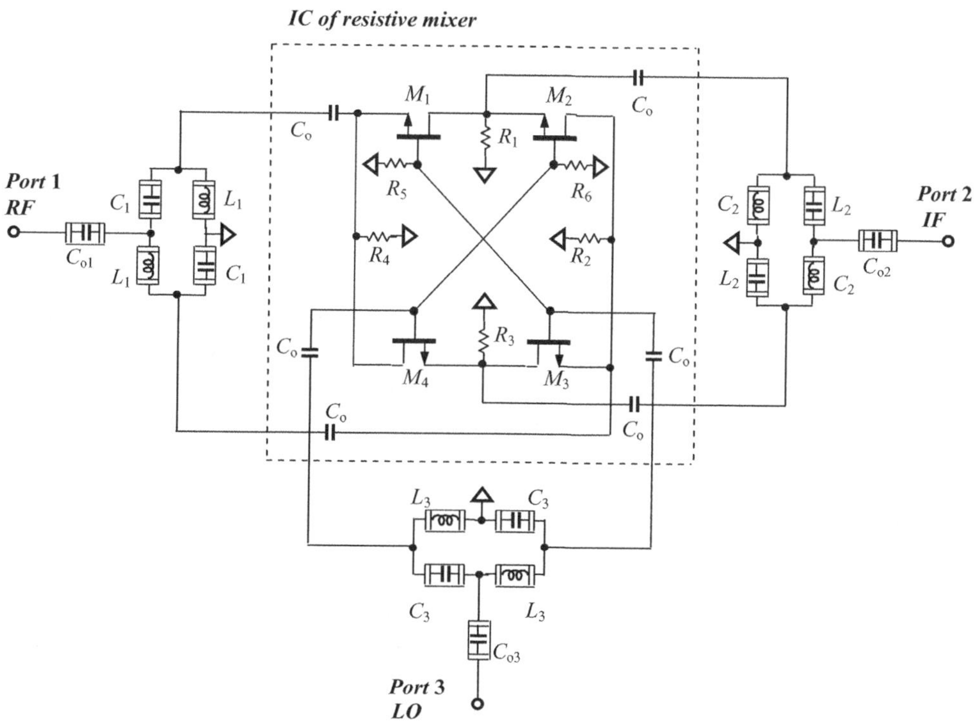

RF	IF	LO
f_{RF}=938 MHz	f_{IF}=73.35MHz	f_{LO}=864.65MHz
C_{o1} = 5.5 pF,	C_{o2} = 155 pF,	C_{o3} = 0.52 pF,
C_1 = 1.5 pF,	C_2 = 18 pF,	C_3 = 0.4 pF,
L_1 = 16.6 nH,	L_2 = 250 nH,	L_3 = 82 nH,
C_o = 20 pF,	C_o = 20 pF,	C_o = 20 pF,
R_1 = 1000 Ω,	R_2 = 1000 Ω,	R_3 = 1000 Ω,

W_{total} = 200 μm (M_1, M_2, M_3, M_4)

$R_5 = R_6 = 5000\ \Omega$.

Figure 2.11 A 800–900 *MHz MOSFET* resistive mixer.
chip capacitor outside *RFIC*
chip inductor outside *RFIC*

The core of the resistive mixer is designed for the *RFIC* chip. The three baluns for the *RF*, *LO*, and *IF* ports are designed with discrete parts, chip capacitors, and chip inductors. The design of the baluns will be described in Chapter 4; the values of their parts are listed in Figure 2.11.

The main features of this resistive mixer are

- $P_{LO} = -5\,dB_m$,
- $CG = -7.7\,dB$,
- $IIP3 = 2.4\,dB_m$,
- $NF = 4.1\,dB$.

By comparing its performance with that of the quad diode mixer introduced above, the apparent differences are:

- The required *LO* injection in resistive mixer is approximately $-5\,dB_m$, while it is $+5\,dB_m$ in the quad diode mixer. This is a big difference. The quad diode mixer needs a more powerful *LO* injection because it directly promotes the *LO* injection current flowing through the diodes. The resistive mixer needs a low *LO* injection because the *LO* injection is only applied for modulating the *RF* signal. In the resistive mixer, the *LO* injection is a control voltage appearing on the gate of the *MOSFET* transistor. So far, the resistive mixer is much superior to the quad diode mixer.
- The resistive mixer has better linearity than the quad diode mixer: the IIP_3 in the resistive mixer is $2.4\,dB_m$, while it is -5 to $-2\,dB_m$ in the quad diode mixer. The reason for this is that in a resistive mixer, the resistive channel under the gate of transistor is much more linear than a diode's character. This is another advantage to the resistive mixer.
- The main drawback of the resistive mixer is its conversion gain. Its conversion gain is $-7.7\,dB$, while the quad diode's conversion gain is -4.0 to $-4.5\,dB$.

Is it necessary to take a trade-off between the quad diode mixer and resistive mixer due to the advantages and disadvantages of the above comparison? The best solution, of course, is to preserve all the advantages and remove all the disadvantages. Figure 2.12 shows an alternative way to overcome the drawbacks of the resistive mixer. The new mixer consists of two blocks: the first block is a resistive mixer and the second is an *IF* amplifier. Let's call it a compound resistive mixer.

Figure 2.13 is the desired *IF* amplifier. It consists of a *MOSFET* transistor, chip capacitors, and chip inductors. It has a T type input impedance matching network built by two capacitors, C_1 and C_2, and one inductor, L_1. The output impedance matching network consists of only the inductor L_2 and a capacitor C_3. Their values are listed in Figure 2.13.

The performance of this *IF* amplifier is

- $G = 16.6\,dB$,
- $RL_{in} = -27\,dB$,
- $RL_{out} = -28\,dB$,
- $\mu = 1.16$,

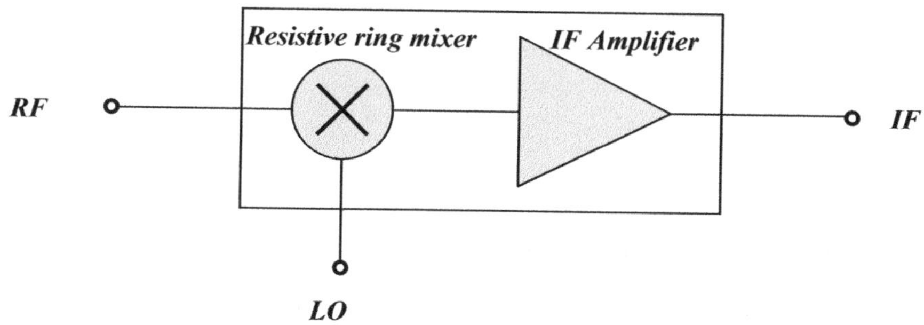

Figure 2.12 A compound resistive mixer built by resistive mixer and an *IF* amplifier connected together in series.

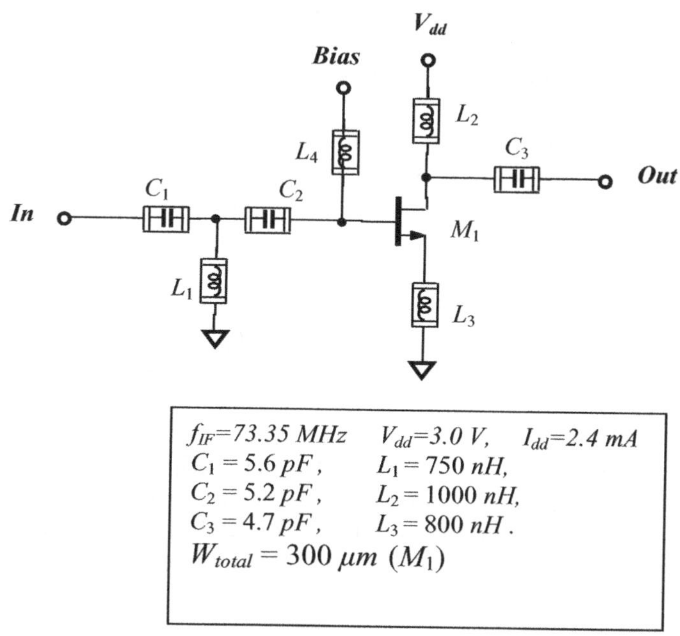

Figure 2.13 *IF* amplifier after 800-900 *MHz MOSFET* resistive mixer.
chip capacitor outside *RFIC*
chip inductor outside *RFIC*

- $IIP_3 = 4.4\,dB_m$,
- $NF = 0.76\,dB$.

The performance of the full compound mixer is

- $V_{dd} = 3.0\,V$, $I_{dd} = 2.4\,mA$,
- $f_{RF} = 938\,MHz$, $P_{RF} = -40\,dB_m$,
- $f_{LO} = 864.65\,MHz$, $P_{LO} = -5\,dB_m$,
- $f_{IF} = 73.35\,MHz$,

- $CG = 8.9\,dB$,
- $IIP3 = 7.1\,dB_m$,
- $NF = 5.8\,dB$.

2.3 ACTIVE MIXERS

2.3.1 Single-end Single Device Active Mixer

Figure 2.14 shows a simple active mixer built by a single *MOSFET* device. The *RF* signal is matched to the gate of the *MOSFET* transistor and the *LO* injection goes to the source of the *MOSFET* transistor through a transformer. The output *IF* signal is coupled from the drain of the *MOSFET* transistor through a Π type impedance matching network. The expensive part of this mixer, of course, is the transformer in either the *RFIC* or module configuration with discrete parts.

Figure 2.15 shows a simple active mixer built by a dual gate *MOSFET* device. It is still a mixer built by a single device, but it is a special device. The *RF* signal is matched to the first gate and the *LO* injection goes to the second gate of the *MOSFET* transistor. The output *IF* signal is coupled from the drain of the *MOSFET* transistor through a Π type impedance matching network. Its operating frequency range is around $1.5\,GHz$.

- $V_{dd} = 5.0\,V$, $I_{dd} = 2.9\,mA$,
- $f_{RF} = 1513 - 1525\,MHz$, $P_{RF} = -40\,dB_m$,
- $f_{LO} = 1459.45 - 1471.45\,MHz$, $P_{LO} = -4.5\,dB_m$,
- $f_{IF} = 53.55\,MHz$.

In order to lower the cost, it is best to apply micro strip lines as much as possible in this design. Furthermore, in order to shrink the size of the module, the substrate is not a plastic *PCB*, but a ceramic aluminum *PCB* (Printed Circuit Board) with a high permittivity, $\varepsilon_r = 10.7$. On the other hand, all the micro strip lines in the layout

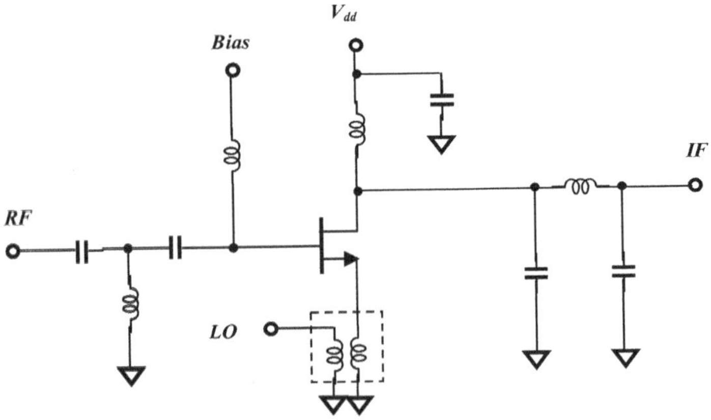

Figure 2.14 A simple active mixer built by a single device.

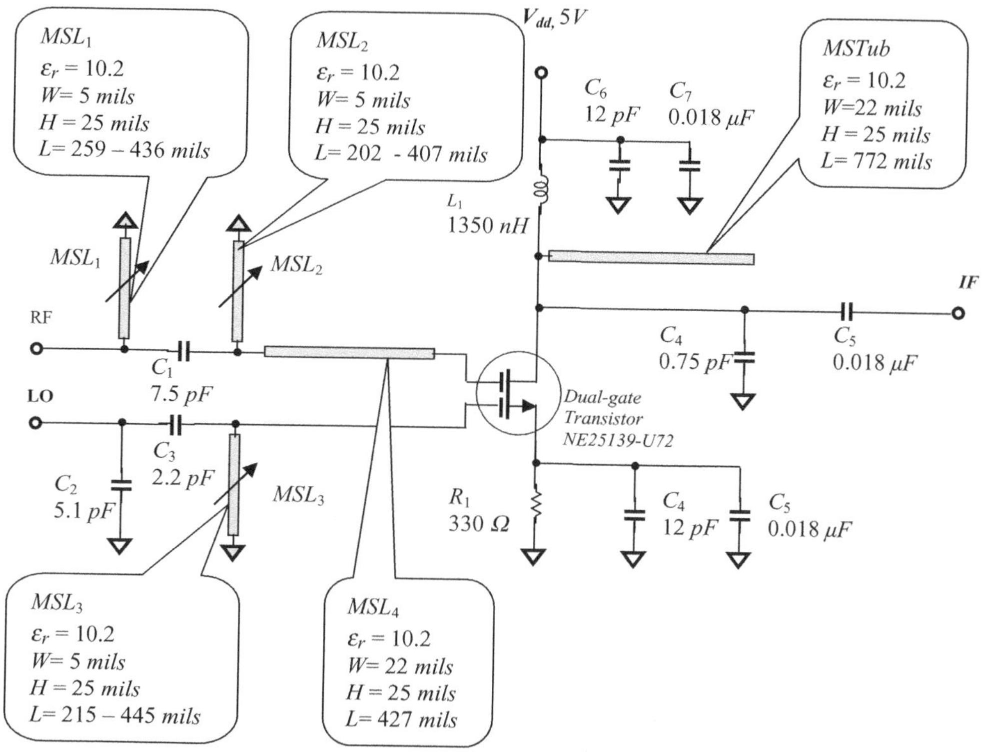

Figure 2.15 A simple active mixer built by a dual gate *GaAs* device. (Except for the dual gate device and micro strip lines, all discrete parts in the schematic are chip parts.)

have a "worm" shape. The adjustable micro strip lines, MSL_1, MSL_2, and MSL_3, are trimmed by a laser trimmer so that the input impedances at the *RF* and *LO* port can be well-matched from 50 Ω to the dual gate device.

There is some trouble in the isolation between the input *LO* and input *RF* signal and between the input *LO* and output *IF* signal. The first isolation problem between the input *LO* and *IF* signal is imaginable since the two gates are crowded in a common device. In the layout, the two input impedance matching networks built by micro strip lines are kept separated as far as possible. This does not cause any big problems. However, the second isolation problem between the input *LO* and output *IF* signal stubbornly persists. The *LO* signal "penetrates" the device and appears at the drain of the device without being attenuated enough from the original *LO* injection level. Therefore a micro strip line stub is applied at the drain of the device. The electrical length of the micro strip stub is a quarter wavelength of the *LO* injection frequency. One end is connected to the drain of the device and the other end is open-circuited. Consequently, this quarter wavelength open-circuited micro strip line stub forces the drain of device to be grounded at the *LO* injection frequency so that it is impossible for the *LO* injection to establish its voltage there. The isolation problem is thus solved in a very satisfactory way. The *LO* injection signal at the drain of the device is lowered by 30 *dB* before and after the open-circuited micro strip line stub with a quarter wavelength is applied.

The final performance from the actual testing is

- $CG = 8.0\,dB$,
- $RL_{RF} = -11\,dB$,
- $RL_{LO} = -8\,dB$,
- $IIP_2 = 3.0\,dB_m$,
- $IIP_3 = -8.0\,dB_m$,
- $NF = 8.0\,dB$.

2.3.2 Gilbert Cell

Figure 2.16 shows a typical *MOSFET* Gilbert cell. As shown in Figure 2.16, the basic element of a Gilbert cell is two differential pairs on the top and one differential pair on the bottom. The two differential pairs on the top form a double balanced configuration. The outputs are a cross combination of the drains of the two differential pairs on the top. The current drain is controlled by the tail current I_{EE}.

Figure 2.17 is exactly the same as Figure 2.16, but it is drawn like a passive mixer built by a ring or quad diode as shown in Figure 2.6. However, the Gilbert cell is

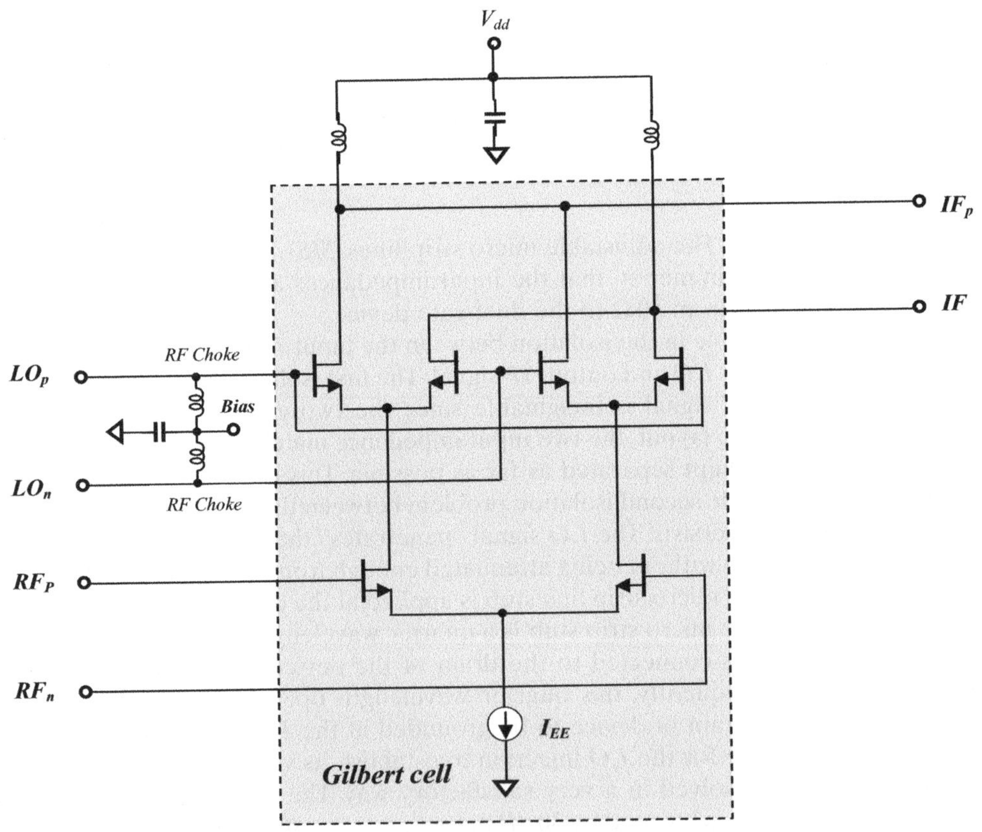

Figure 2.16 An active mixer with a *MOSFET* Gilbert cell.

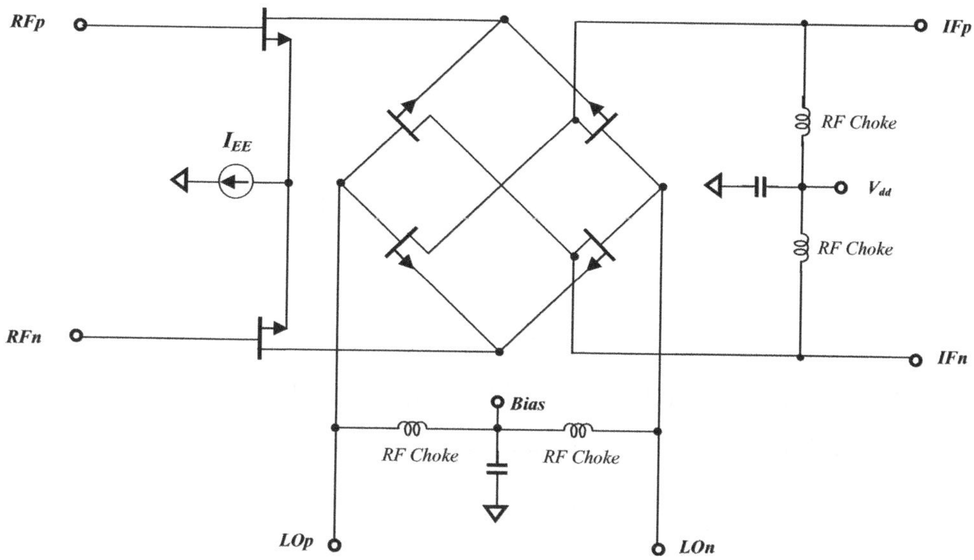

Figure 2.17 Alternative drawing of an active mixer with *MOSFET* Gilbert cell.

not a cell with a "ring" configuration, but is a cell with a dual balanced configuration. As a matter of fact, the Gilbert cell is a cell that is not only "double balanced" but also "double differential."

Theoretically, the apparent advantages of an ideal differential pair are

- The odd orders of non-linearity of the devices cancel each other.
- Additionally, an ideal differential pair has the capability to reject the common mode components of the incoming signal.

The Gilbert cell has all of these features since its configuration is "double differential."

From another viewpoint, that is, in comparison to a passive mixer,

- One of the outstanding features in an active mixer built by a Gilbert cell is the positive conversion gain. This could alleviate the problem of the gain required in the *LNA* block of a receiver.
- The *LO* required injection level in an active mixer is much lower than that in a passive mixer.

The disadvantages of an active mixer, as mentioned earlier, are:

- Its noise figure is higher than that of a passive mixer.
- An active mixer needs a *DC* power supply.

If the *MOSFET* transistors are simply replaced by bipolar transistors, a *MOSFET* Gilbert cell becomes a bipolar Gilbert cell, and the active mixer with a *MOSFET*

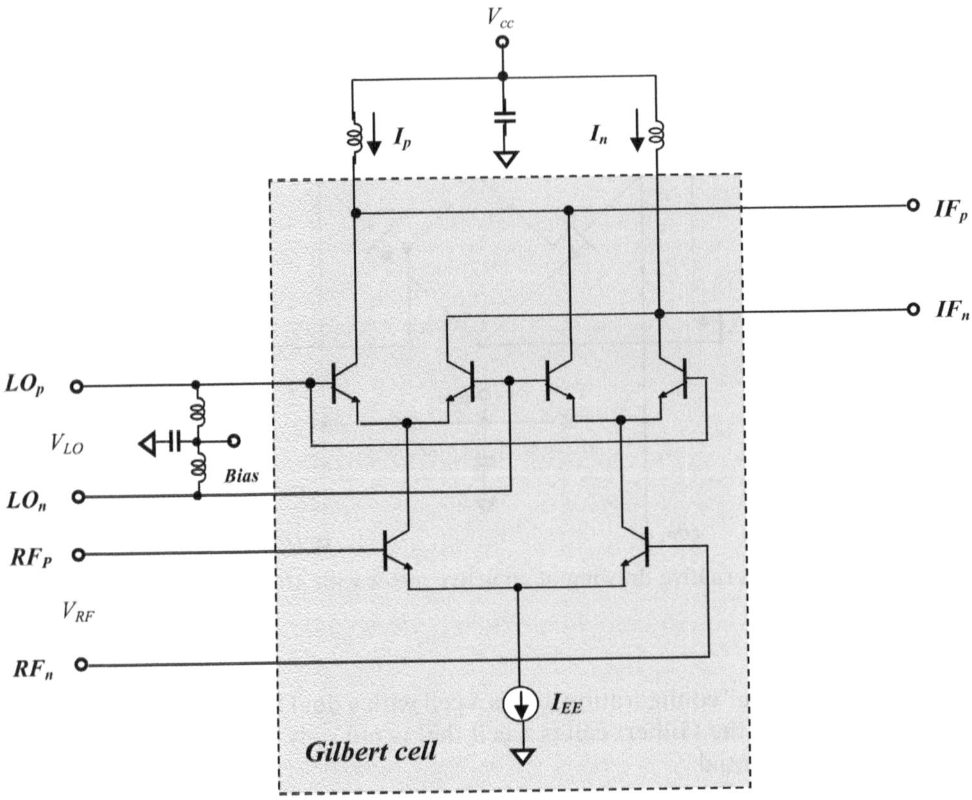

Figure 2.18 An active mixer with a bipolar Gilbert cell.

Gilbert cell as shown in Figure 2.16 becomes an active mixer with a bipolar Gilbert cell as shown in Figure 2.18.

An active mixer with bipolar Gilbert cell has been well analyzed (see Paul R. Gray et al., *Analysis and Design of Analog Integrated Circuits*). In a bipolar Gilbert cell, the differential output currents can be expressed as

$$\Delta I = I_P - I_n = I_{EE}\tanh\left(\frac{v_{RF}}{2V_T}\right)\tanh\left(\frac{v_{LO}}{2V_T}\right), \tag{2.17}$$

where v_{LO} and v_{RF} are the differential voltages at the *RF* and *LO* differential ports,

$$V_T = kT/q \approx 26mV \text{ at } 300K^\circ.$$

Note that

$$\tanh x = x - \frac{1}{3}x^3 + \frac{2}{15}x^5 - \frac{17}{315}x^7 + \frac{62}{2835}x^9 - \ldots \tag{2.18}$$

if

$$x < 1. \tag{2.19}$$

Usually v_{RF} is a small signal, that is,

$$v_{RF} << V_T, \tag{2.20}$$

so equation (2.17) can be approximated as

$$\Delta I = I_{EE}\frac{v_{RF}}{2V_T}\tanh\left(\frac{v_{LO}}{2V_T}\right), \tag{2.21}$$

However, v_{LO} is usually not a small signal, but comparable with or larger than V_T, that is,

$$v_{LO} \approx V_T, or > V_T. \tag{2.22}$$

In order to linearize the output differential current ΔI, the input differential voltage at the *LO* port must be pre-distorted so as to compensate for the hyperbolic tangent transfer characteristic. In other words, a "$\tanh^{-1}$" block must be added to the *LO* port before the *LO* injection v_{LO} enters the Gilbert cell. This $\tanh^{-1}$ block can be implemented by two diode-connected transistors based on the relation between $\tanh^{-1}$ and the natural logarithm:

$$\tanh^{-1} x = \frac{1}{2}\ln\left(\frac{1+x}{1-x}\right). \tag{2.23}$$

The addition of the $\tanh^{-1}$ block could be called the linearization processing of the Gilbert cell. Implementation of the $\tanh^{-1}$ block is introduced in Appendix 2.A.2.

Consequently, equation (16.21) becomes

$$\Delta I = I_{EE}\frac{v_{RF}}{2V_T}\frac{v_{LO}}{2V_T}. \tag{2.24}$$

If the *RF* signal and *LO* injection are sinusoidal as expressed in equations (2.2) and (2.3), that is,

$$v_{RF} = v_{RFo}\cos(\omega_{RF}t + \varphi), \tag{2.2}$$

$$v_{LO} = v_{LOo}\cos\omega_{LO}t. \tag{2.3}$$

And note that if the LO injection is a low-side LO injection,

$$\omega_{RF} - \omega_{LO} = \omega_{IF}, \tag{2.6}$$

Then (2.24) becomes

$$\left.\begin{aligned} \Delta I &= I_{EE}\frac{v_{RF_o}v_{LO_o}}{2V_T^2}\{\cos[(\omega_{RF}-\omega_{LO})t+\varphi]+\cos[(\omega_{RF}+\omega_{LO})t+\varphi]\},\\ \Delta I &= I_{EE}\frac{v_{RF_o}v_{LO_o}}{2V_T^2}\{\cos(\omega_{IF}t+\varphi)+\cos[(\omega_{RF}+\omega_{LO})t+\varphi]\}. \end{aligned}\right\} \tag{2.25}$$

If the high RF component, which is the second term in equation (2.25), is filtered out, then the remaining component is the desired IF signal, that is,

$$\Delta I = I_{EE} \frac{v_{RF_o} v_{LO_o}}{2V_T^2} \cos(\omega_{IF} t + \varphi). \tag{2.26}$$

This is the basic principle by which the Gilbert cell operates as a mixer.

2.3.3 Active Mixer with Bipolar Gilbert Cell

Let's present an example of an active mixer with a bipolar Gilbert cell. Figure 2.19 is a plan to design an active mixer for an 800 MHz radio. The main body is the active mixer with a bipolar Gilbert cell. The RF input and LO injection are single-ended and the IF output is to be single-ended. Therefore, three baluns for the RF, LO and IF ports must be included. Ideally, everything might be contained in an $RFIC$ chip. However, the IF balun is hard to design as an on-chip block, because the size of the inductors and capacitors required for the low IF frequency are unacceptable on an IC chip. It therefore is designed as an off-chip block. In addition, for considerations of cost-saving, the LC type balun is selected to implement the IF balun.

The LO and RF baluns can be put on an $RFIC$ chip. An active balun has been developed for such a purpose. Figure 2.20 shows the schematic of an active balun. In this figure, the collector and the base of Q_4 are connected. They function as a current mirror between the left and right branches of the circuitry. More importantly, they function to reverse the phase for the single-ended input signal, SN, between the left and right branches: at point B, the phase is basically the same as that in the input point P, whereas at point A, the phase is reversed 180° from the phase at point P. Consequently, the phase between points A and B is 180°. It ensures the output is differential.

As the single-ended input point of a balun, the impedance at point P must satisfy two conditions:

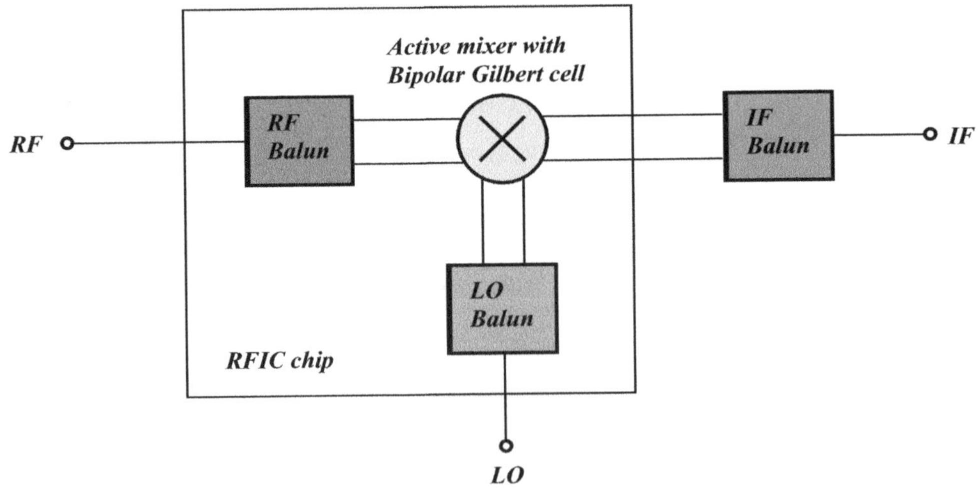

Figure 2.19 A plan to design an active mixer with a bipolar Gilbert cell.

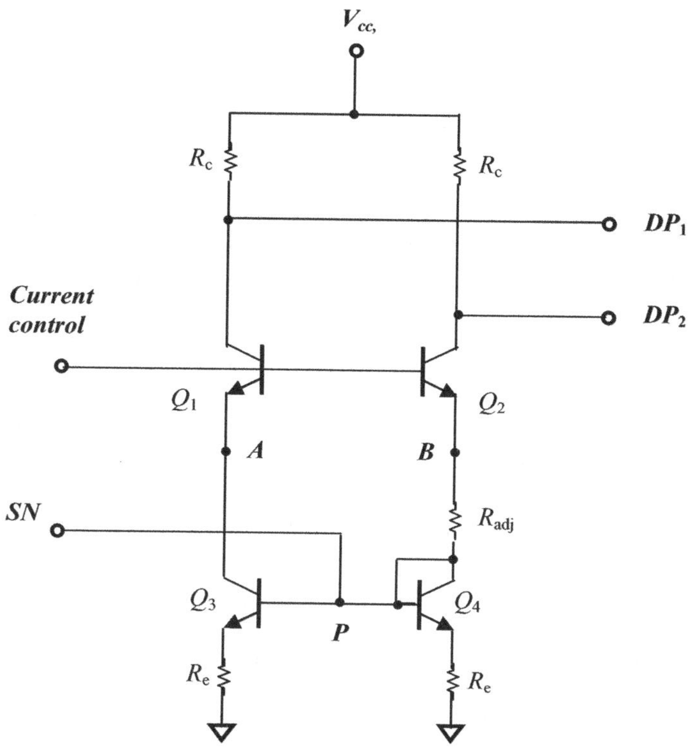

Figure 2.20 Schematic of an active balun.

- At point P, the impedance must be $50\,\Omega$.
- From point P, the impedances looking toward the left hand side and the right hand side must be equal to each other.

Without the adjusting resistor, R_{adj}, the second condition is impossible to satisfy. On the contrary, by adjusting the resistors R_{adj} and R_e, the two conditions above can be reached.

The transistors Q_1 and Q_2 are identical. They are applied for current control; they also play the role of isolating the differential output from the single-ended input. Usually, Q_3 and Q_4 are also identical. However, they could be different as long as the above two conditions are satisfied.

Figure 2.21 shows the active mixer with a bipolar Gilbert cell and active *LO* and *RF* baluns. The left hand side of the figure is the *LO* balun and the right hand side is the Gilbert cell and *RF* balun. The *RF* balun is based on same operating principle as the *LO* balun. It can be seen that the *RF* balun is also part of the Gilbert cell. The current flowing through the *LO* balun is controlled by the current source $I_{bias,1}$, and the current flowing through the Gilbert cell and *RF* balun is controlled by the current source $I_{bias,2}$. The transistors Q_9 to Q_{16} are the parts of the current mirror and solely serve for current control.

The outstanding achievement in this design is that no capacitors are needed and the current re-use is applied to the Gilbert cell and *RF* balun. Since there are no capacitors, the *RFIC* chip size is reduced.

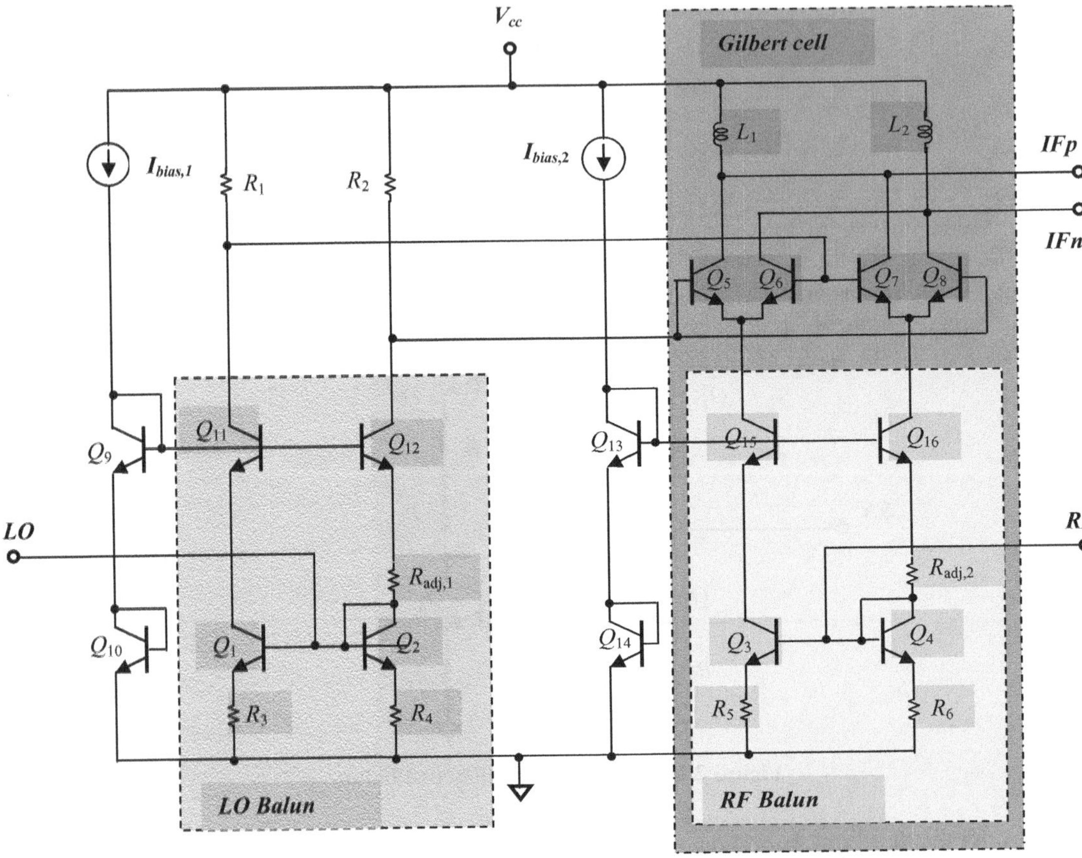

Figure 2.21 Active mixer with bipolar Gilbert cell and active *LO* and *RF* balun.

Now let's examine the off-chip *IF* balun. Figure 2.22 shows its schematic: it is constructed by chip inductors and chip capacitors. The *LC* balun design is based on the input and output impedances and the operating frequencies, which are introduced and discussed in Chapter 4, Section 4.3. The operating frequency in this design is $45\,MHz$ and the input and output impedances are shown in the figure.

The active mixer with a bipolar Gilbert cell and active baluns introduced in this section has been successfully packaged into actual products.

The operating frequencies and *DC* power supply are:

- $f_{RF} = 860\,MHz$,
- $f_{LO} = 905\,MHz$,
- $f_{IF} = 45\,MHz$,
- $V_{cc} = 3.0\,V$,
- $I_{cc} = 3.14\,mA$.

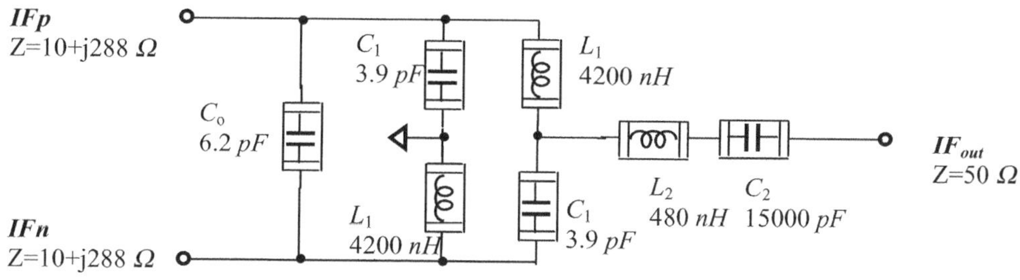

Figure 2.22 *LC* balun for *IF* differential terminals in the active mixer with bipolar Gilbert cell.

chip capacitor outside *RFIC*
chip inductor outside *RFIC*

Its performance is:

- $P_{LO} = -10\,dB_m$,
- $CG = 3.0\,dB$,
- $RL_{\mathrm{RF}} = -12.8\,dB$,
- $RL_{\mathrm{LO}} = -2.0\,dB$,
- $NF = 11.18\,dB$,
- $IIP3 = 3.98\,dB_m$.

2.3.4 Active Mixer with *MOSFET* Gilbert Cell

The *MOSFET* and bipolar transistors mainly differ in input impedance. The impedance at the gate of the *MOSFET* transistor is much higher than the impedance at the base of the bipolar transistor.

Owing to the high impedance at the gate of a *MOSFET* transistor, building an active balun by the *MOSFET* device seems to be difficult compared with building one by bipolar transistors. Passive baluns might be good candidates for the active mixer with a *MOSFET* Gilbert cell.

The difficult part of this design task is the implementation of the input impedance matching network. In narrow-band cases, a reasonable topology as shown in Figure 2.23 is recommended.

Both the *LO* and *RF* impedance matching networks are connected from 50 Ω to the gates of *MOSFET* transistors. As mentioned above, the impedance at the gate of the *MOSFET* transistor is quite high and usually is located in region 4 on the Smith chart. It is reasonable to insert an inductor, L_{S3} or L_{S1}, so as to move the original impedance at the gates of the transistors to region 1 on the Smith chart. Then, using the two capacitors, C_{P3} and C_{S3}, or C_{P1} and C_{S1}, the impedance is pulled to approach to 50 Ω. Two capacitors in both of the *LO* and *RF* impedance matching networks play the role of *DC* blocking as well since *DC* bias must be provided to the gate of the transistors. Thus the topology of the impedance matching network for *LO* or *RF* ports is the most economic, with a minimum part count.

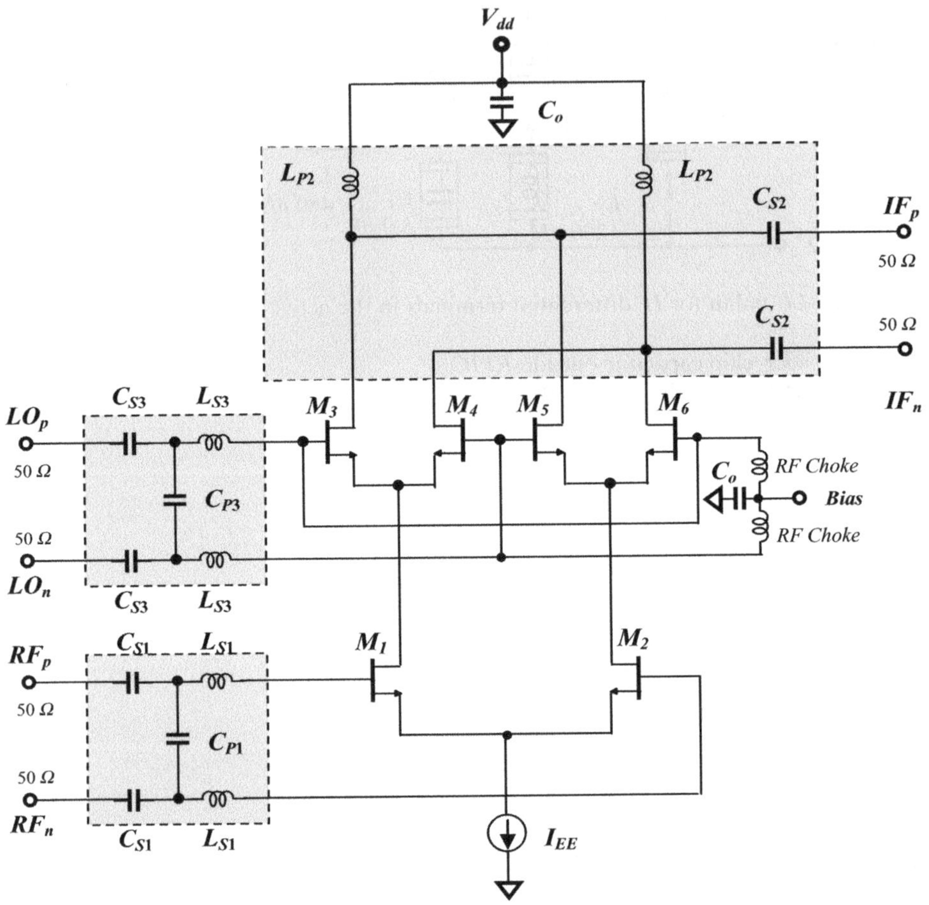

Figure 2.23 Recommended impedance matching networks for an active mixer with *MOSFET* Gilbert cell in narrow-band cases. C_o: "zero" capacitor.

At the *IF* output, the *IF* impedance matching network is inserted from the drain of transistor to 50 Ω. The first part is an inductor L_{P2} in parallel. This can definitely not be replaced by a capacitor because this part is expected to feed the *DC* current from the *DC* power supply as well. However, it may be replaced by a resistor. Indeed, using a resistor to provide *DC* current to the drain of the transistor should not be a problem. However, this resistor would occupy quite a high percentage of the *DC* voltage drop and thus reduce the *DC* voltage drop from the drain of the transistor to the ground, eventually degrading the linearity of the transistor. Therefore, the first part in the *IF* impedance matching network must be an inductor. There seems to be no other choice. After this part is inserted between the drain and V_{dd}, the original impedance is expected to move to the vicinity of the circle $r = 1$. Then, in terms of the capacitor C_{S2} in series, the impedance is pulled down to around 50 Ω. It is concluded that such an *IF* impedance matching network is the most reasonable and economic.

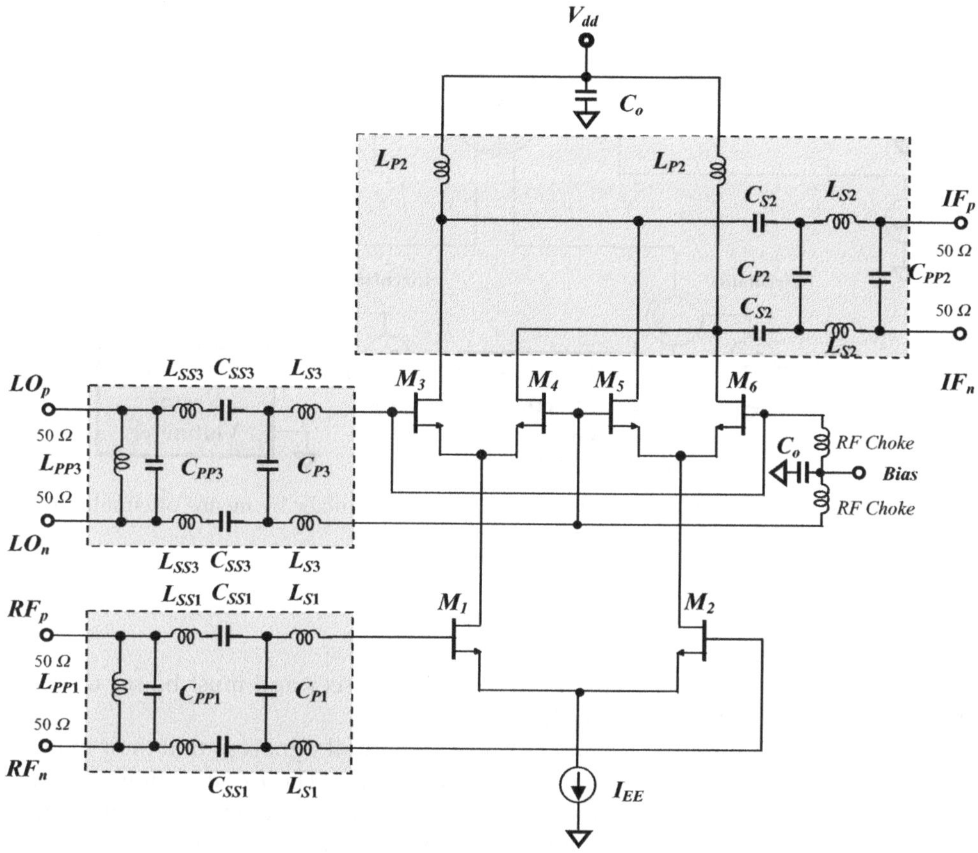

Figure 2.24 Recommended impedance matching networks for an active mixer with *MOSFET* Gilbert cell in wide-band cases. C_o: "zero" capacitor.

In wide-band or *UWB* cases a reasonable topology is recommended as shown in Figure 2.24, which will be discussed in Chapter 11. The unsatisfied point here is that there are five inductors appearing in the impedance matching network. This is not good for either cost saving or noise performance. Our excuse is that it is a design for a wide-band circuit, and therefore is allowed a greater cost and higher part count. However, the potential for improvement exists and further efforts should be made.

2.4 DESIGN SCHEMES

2.4.1 Impedance Measuring and Matching

Among *RF* blocks, one of the more complicated blocks is the mixer, because it has three ports with three different frequencies: the *RF*, *LO*, and *IF* ports. This complexity is enhanced if all three ports are differential, since then the number of ports increases from three to six as shown in Figure 2.25. The power sources v_{RF1} and v_{RF2},

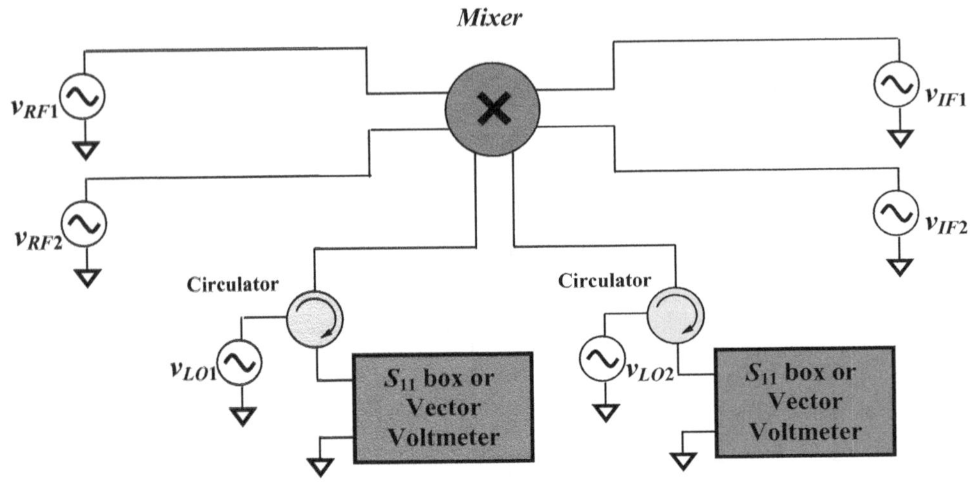

Figure 2.25 Impedance measurement for a differential mixer by means of circulator.

v_{LO1} and v_{LO2}, and v_{IF1} and v_{IF2} are all differential sources and must be set up with phase differences of 180° between the source pairs.

Among these six ports, the *RF* and *IF* ports have small signals so that their impedances can be measured directly by the network analyzer. The *LO* injection is a large signal, so it is necessary to apply the circulator for an accurate impedance measurement. In the simulation stage, a S_{11} box can be used to measure the voltage returned from the port in the mixer. In actual testing, a vector voltmeter would be a good assistant to measure the voltage returned from the port in the mixer.

As a matter of fact, as shown in Figure 2.26, the differential configuration of a mixer can be simplified from six differential ports to three single ports by means of the balun, either in the simulation phase or in the testing stage. This greatly reduces the complexity of measurement. Of course, the impedances of the baluns from the single-ended to the differential must be well known before they are applied to the test setup.

In the impedance testing or measuring, special attention must be paid to the *IF* portions. One can calculate the impedance from the ratio of the reflected and incident power, or the reflection coefficient. However, there is another unexpected *IF* power, the product of the *RF* input and *LO* injection, which also flows into the v_{IF} port as another incident power. In order to separate these powers, there are two ways to measure the *IF* impedance:

- The *IF* frequency at the *IF* power sources can be shifted a little bit so as to avoid the overlap with the product of *RF* input and *LO* injection. The measured impedance at the *IF* ports, of course, corresponds to the shifted frequency but not to the expected *IF* frequency. We can approximately treat the measured impedance at the shifted frequency as the *IF* impedance because the frequency shift is very small.

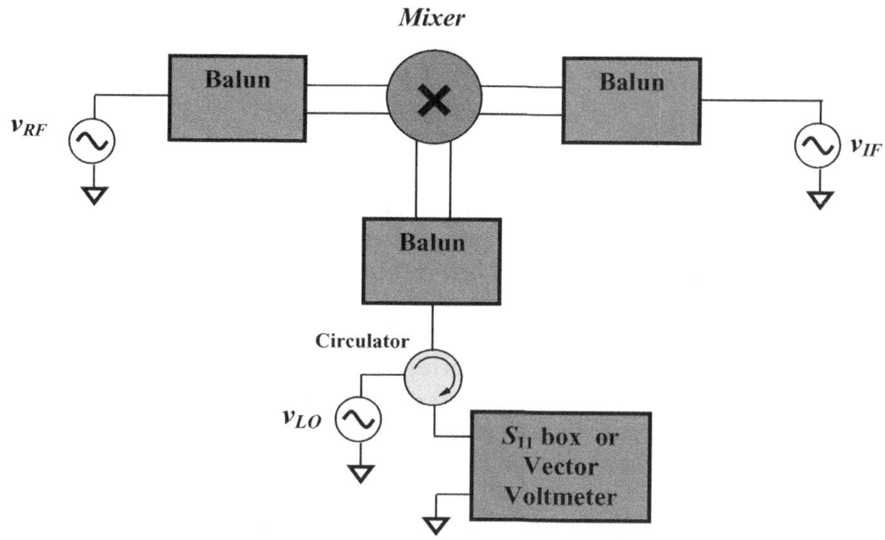

Figure 2.26 Change of impedance measurement for a differential mixer by means of a balun.

- Instead of shifting the *IF* frequency, the *IF* product due to the *RF* input and *LO* injection can be prohibited if the *RF* input is turned off.

Another possible way to prohibit the *IF* product due to the *RF* input and *LO* injection is to turn off the *LO* injection. However, this is not allowed do so because the real operating status of a mixer is determined and maintained by the *LO* injection.

Impedance matching is one of the key areas in *RF* circuit design. It is not an easy task sometimes. One of the examples is to impedance-match a differential mixer. Owing to the imperfect isolation between the *RF*, *LO*, and *IF* ports, impedance matching must be conducted with many iterations back and forth between three ports, especially between the *LO* and the *RF* ports, because the isolation between them is poor. The matching procedures should be:

1) Matching of the *LO* port first:
   The *LO* injection is the main source controlling the operating status of a mixer. The operating status of a mixer never approaches a normal state until the *LO* port is well matched.
2) Matching of the *RF* port second;
3) Repeating these two steps until the variation between steps is negligible:
   Owing to the imperfect isolation between the *LO* and *RF* ports, the variation resulting from the matching at the *RF* port would change the matching status at the *LO* port, and vice versa.
4) Finally, matching the *IF* port:
   Usually, the isolation between the *LO* and *IF* ports or between the *RF* and *IF* ports is good enough so that the variation resulting from the matching at

the *IF* port will not impact the matching status in either the *LO* or *RF* ports.

2.4.2 Current Bleeding

S. G. Lee and J. K. Choi (2000) proposed a new mixer topology with current bleeding as shown in Figure 2.27. It is a single-balanced mixer, in which a p-channel transistor M_4 functions as a bleeding current source as well as a part of the driver amplifier. This technology leads to better performance of conversion gain, linearity, noise figure, and *LO* isolation.

Conversion gain and IP_3 are proportional to the square root of the current flowing into M_1. Without the bleeding current flowing through M_4, the increase of the current flowing into M_1 must be traded off with the reduction of the load resistors R_{L1} and R_{L2}. This results in the reduction of conversion gain and the value of IP_3. With the bleeding current flowing through M_4, the increase of current flowing into M_1 does not have to be traded off with the reduction of the load resistors R_{L1} and R_{L2} since the increased current is flowing through M_4 but not M_2 and M_3. This results an increase of conversion gain and the value of IP_3.

Second, M_4 is a part of the driver amplifier together with M_1; the higher overall transconductance reduces the noise figure.

Finally, the switching transistors M_2 and M_3 can be operated at a lower gate-source voltage or smaller size due to the bleeding current introduced, so that the lower *LO* injection is good enough to have the same performance as that in the case without bleeding current introduced.

The experiment indicated that the performance of the mixer topology with bleeding current demonstrates an approximately $4dB$ higher conversion gain, $0.9dB$ lower noise figure, $2.4dB$ higher IP_3 at the output, and $3.3dB$ lower *LO* injection than that of the conventional mixer.

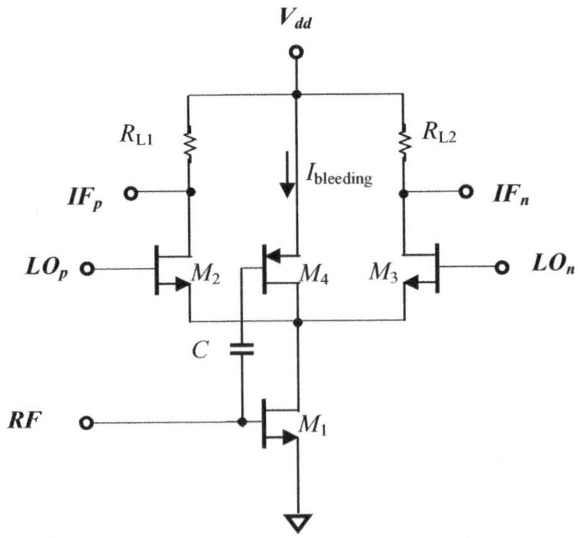

Figure 2.27 Schematic diagram of single-balanced mixer with current bleeding.

2.4.3 Multi-tanh Technique

In order to enhance the linearity of mixer with Gilbert cell, the multi-tanh technique was developed by Barrie Gilbert (1998). The key idea of the multi-tanh technique is to connect multiple differential pairs together in parallel, in which the trans-conductance function of each individual pair is arranged to cover a different range of input voltage. A much more linear overall function is approached as all of trans-conductance functions from all the individual pairs are summed together. In other words, the individually non-linear trans-conductance functions are separated along the input-voltage axis so that the trans-conductance can be kept constant over a wide range of the input voltage.

For an individual differential pair, let's recall equation (2.21)

$$\Delta I = I_{EE}\frac{v_{RF}}{2V_T}\tanh\left(\frac{v_{LO}}{2V_T}\right). \tag{2.21}$$

where v_{RF} is assumed to be small signal and condition (2.20) is satisfied.

Now let's focus on the linear relationship between ΔI and v_{LO}.

For simplicity, let's re-denote

$$I_{out} = \Delta I, \tag{2.27}$$

$$I_E = I_{EE}\frac{v_{RF}}{2V_T}, \tag{2.28}$$

$$v_{in} = v_{LO}, \tag{2.29}$$

Then, equation (2.21) becomes

$$I_{out} = I_E \tanh\frac{v_{in}}{2V_T}, \tag{2.30}$$

its trans-conductance is

$$g_m = \frac{\partial I_{out}}{\partial V_{in}} = \frac{I_E}{2V_T}\sec h^2\frac{v_{in}}{2V_T}, \tag{2.31}$$

The hyperbolic tangent relationship between g_m and v_{in} is plotted in Figure 2.28. The value of g_m changes considerably as v_{in} is varied. This indicates that it is difficult to keep the trans-conductance constant over a wide range of input voltage, resulting in the non-linearity between the input and output signal.

Using the multi-tanh technique, in which the linearity between the input and output can be improved significantly, the Gilbert cell is no longer a simple differential pair but is a combination of n differential pairs so that the resultant trans-conductance is

$$g_m = \sum_{j=1}^{n}\frac{I_j}{2V_T}\sec h^2\left(\frac{v_{in}+v_j}{2V_T}\right), \tag{2.32}$$

Figure 2.29 shows the schematic of n multi-tanh differential pairs. The outputs of all the differential pairs are connected together as the terminal I_{out} in parallel. The resultant trans-conductance, g_m, is a simple sum of all the individual trans-conductances, g_{mj}. The individual input voltage $v_{in} + v_j$ must and can be produced by

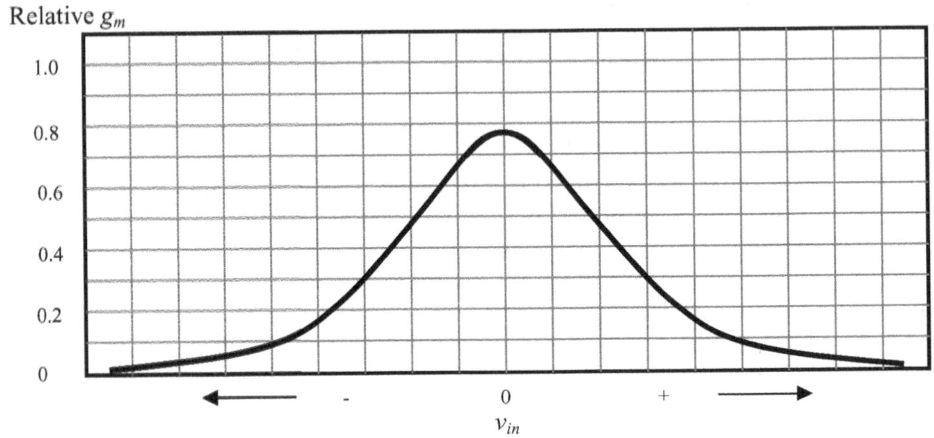

Figure 2.28 Typical transconductance g_m of a Gilbert cell.

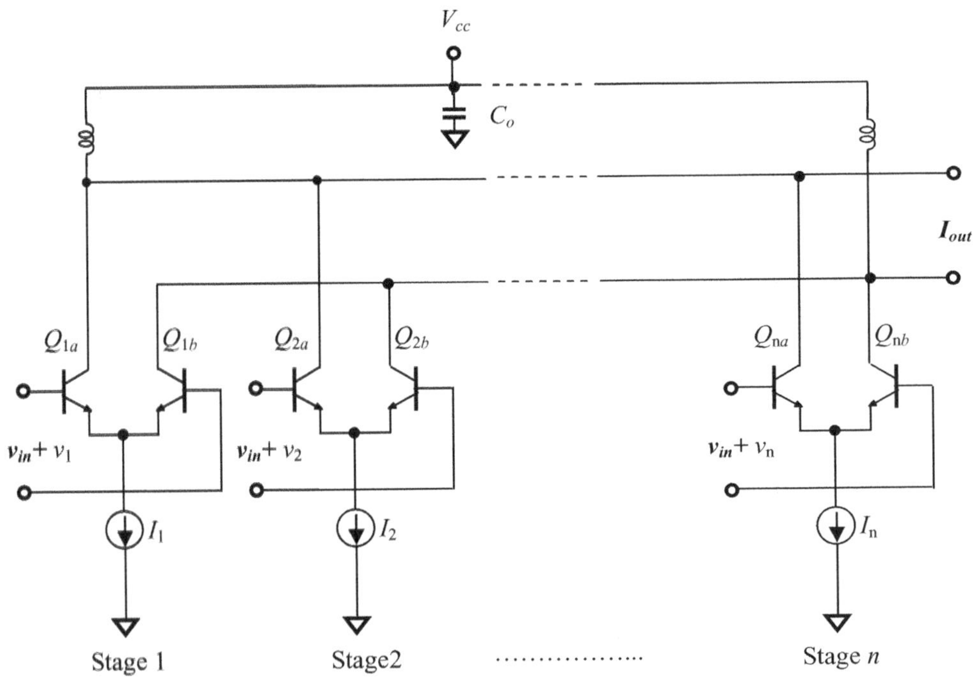

Figure 2.29 n multi-tanh differential pairs.

V_{in} through a voltage amplifier/follower or an transformer. By adjusting all the individual input voltages $v_{in} + v_j$, the individual trans-conductance g_{mj} can be reasonably arranged over a different range of input voltages so that the sum of the trans-conductances g_m is kept constant over a wide range of input voltages, as shown in Figure 2.30.

Double-tanh differential pairs are the simplest multi-tanh differential pairs; one is shown in Figure 2.31. Instead of different individual input voltages $v_{in} + v_j$, the input voltage for both differential pairs is kept the same, but the emitter area of the transistors is different in each differential pair. As shown in Figure 2.31, the emitter area of Q_{1b} and Q_{2a} is kept the same and is equal to a while the emitter area of Q_{1a} and Q_{2b} is kept the same and is equal to Aa, which is A times larger than the emitter area of Q_{1b} and Q_{2a}.

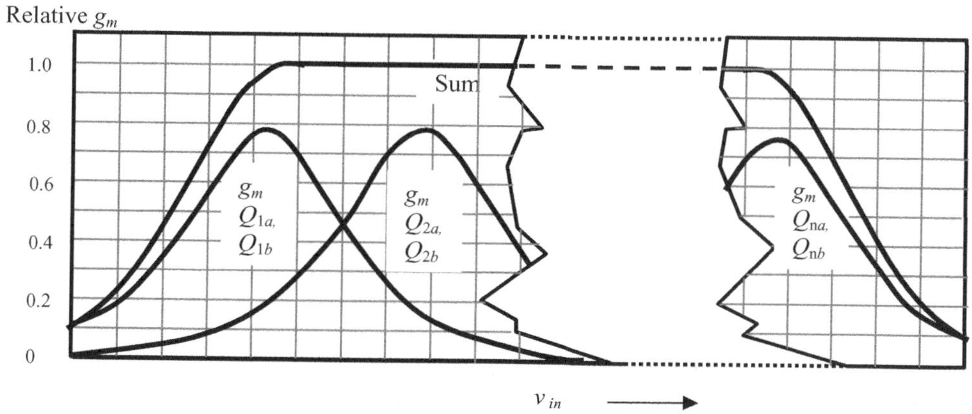

Figure 2.30 Multi-g_m components of n tanh differential pairs.

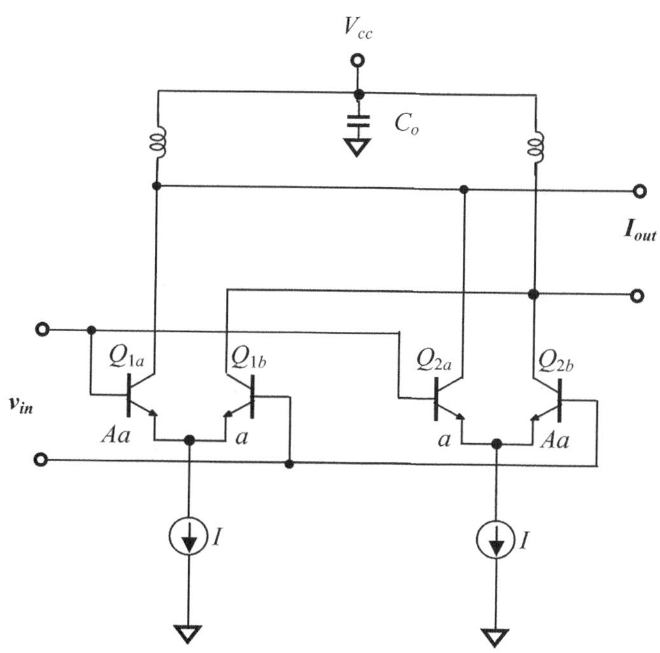

Figure 2.31 Doublet-tanh differential pairs. Aa = emitter area of Q_{1a} and Q_{2b}. a = emitter area of Q_{1b} and Q_{2a}.

As a matter of fact, the difference of the emitter area is equivalent to the difference of the different individual input voltages $v_{in} + v_j$. The first differential pair, marked "Stage 1" in Figure 2.31, results in the g_{m1} curve in Figure 2.32, which is shifted to the left-hand or negative direction of the input voltage v_{in} from the case when $v_j = 0$. The second differential pair, marked "Stage 2" in Figure 2.31, results in the g_{m2} curve in Figure 2.32, which is shifted to the right-hand or positive direction of the input voltage v_{in} from the case when $v_j = 0$. The sum of g_m is therefore kept constant over a wider range of input voltages.

2.4.4 Input Types

There are two *RF* input types of mixers built by Gilbert cells: the *CS* (Common Source) and *CG* (Common Gate). Figure 2.33 shows a *CS* type *RF* input, while Figure 2.34 shows a *CG* type *RF* input. The impedance of the mixer at the *RF* ports with a *CG* type *RF* input is much lower that that of the mixer with a *CS* type *RF* input since the input impedance for a *MOSFET* transistor with a *CG* configuration is much lower than the input impedance of that with a *CS* configuration. Low impedance is easier for *RF* circuit design than high impedance. As shown in Figure 2.33, there is a transformer applied in the *RF* ports in the mixer with a *CG* type *RF* input. In addition, a mixer with a *CG* type *RF* input provides a wider bandwidth of frequency response than a *CS* type *RF* input. Consequently, in most practical mixer designs, the *CG* type *RF* input is adapted.

Similarly, the *CB* (Common Base) type of *RF* input in a mixer built by bipolar transistors with a Gilbert cell has some advantages over that of the *CE* (Common Emitter) type *RF* input.

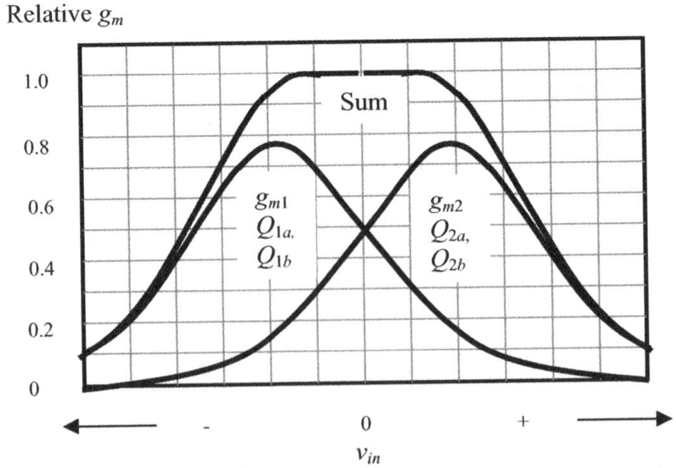

Figure 2.32 Dual g_m components of the doublet-tanh differential pairs.

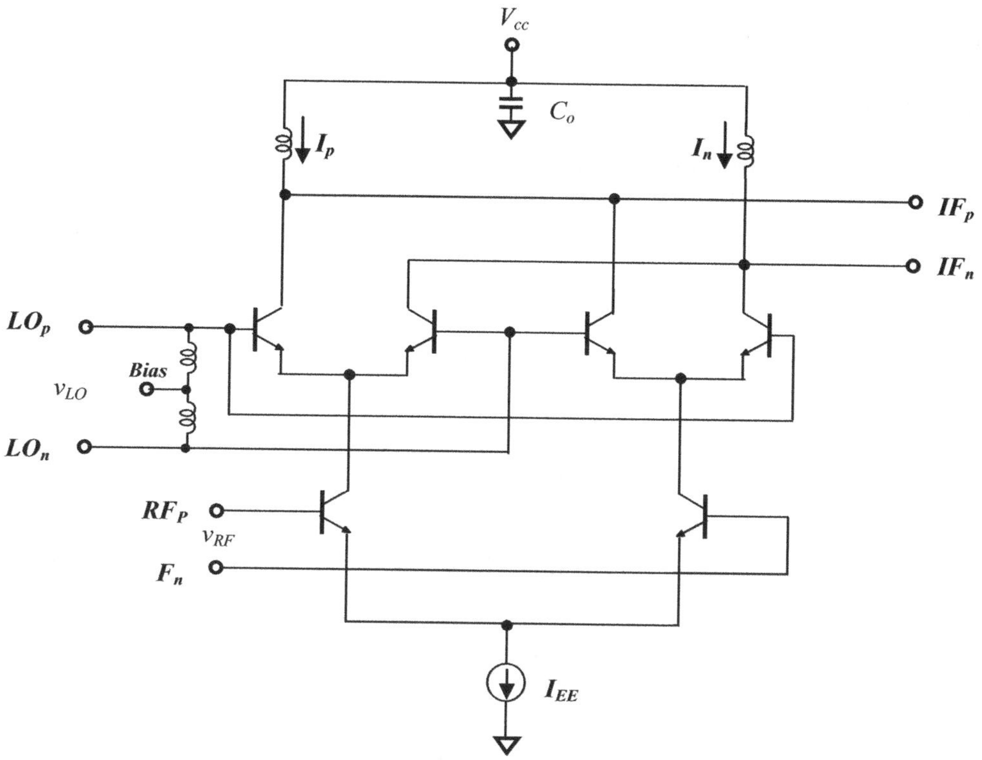

Figure 2.33 An active mixer with bipolar Gilbert cell and with *CS* input for *RF* signal. C_o: "zero" capacitor.

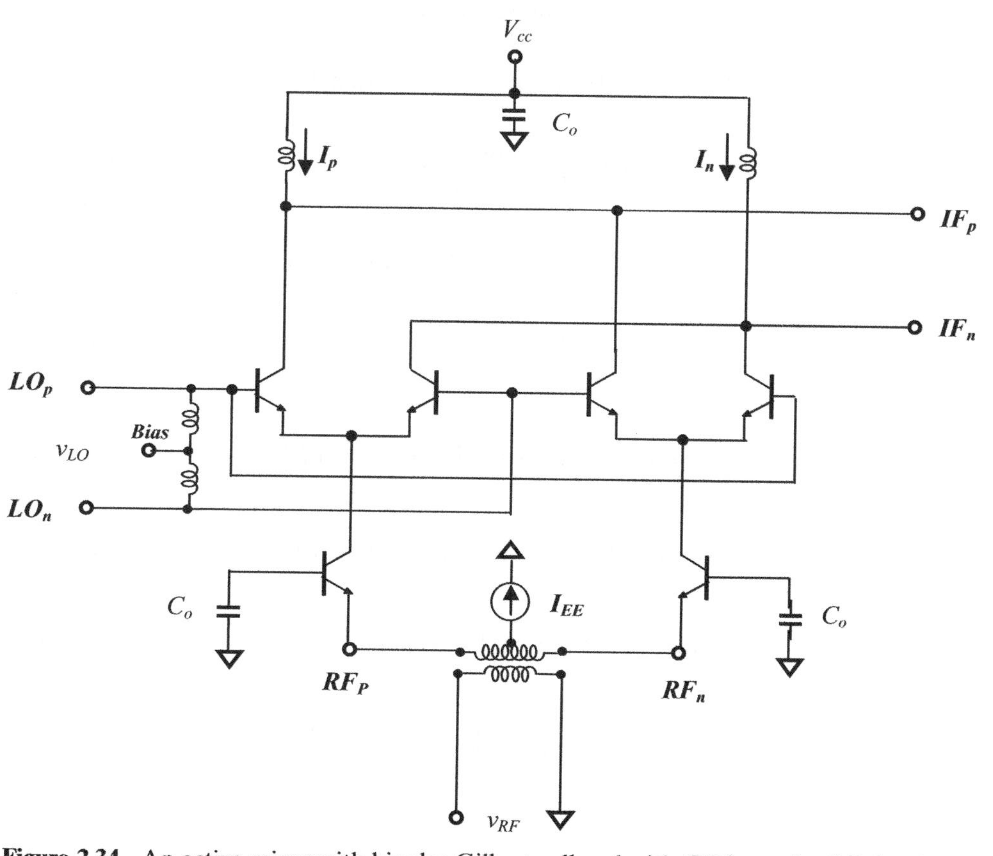

Figure 2.34 An active mixer with bipolar Gilbert cell and with *CG* input for *RF* signal. C_o: "zero" capacitor.

APPENDICES

2.A.1 Trigonometric and Hyperbolic Functions

$$e^{j\theta} = \cos\theta + j\sin\theta, \tag{2.A.1}$$

$$e^{-j\theta} = \cos\theta - j\sin\theta, \tag{2.A.2}$$

$$\cos\theta = \frac{e^{j\theta} + e^{-j\theta}}{2}, \tag{2.A.3}$$

$$\sin\theta = \frac{e^{j\theta} - e^{-j\theta}}{2j}, \tag{2.A.4}$$

$$\cosh\theta = \frac{e^{\theta} + e^{-\theta}}{2}, \tag{2.A.5}$$

$$\tanh\theta = \frac{\sinh\theta}{\cosh\theta} = \frac{e^{\theta} - e^{-\theta}}{e^{\theta} + e^{-\theta}} = \frac{e^{2\theta} - 1}{e^{2\theta} + 1}, \tag{2.A.6}$$

$$\sinh\theta = \frac{e^{\theta} - e^{-\theta}}{2}, \tag{2.A.7}$$

$$\sin(\alpha \pm \beta) = \sin\alpha\cos\beta \pm \cos\alpha\sin\beta, \tag{2.A.8}$$

$$\cos(\alpha \pm \beta) = \cos\alpha\cos\beta \mp \sin\alpha\sin\beta, \tag{2.A.9}$$

$$\tan(\alpha \pm \beta) = \frac{\tan\alpha \pm \tan\beta}{1 \mp \tan\alpha\tan\beta}, \tag{2.A.10}$$

$$\sin\alpha\cos\beta = \frac{1}{2}\sin(\alpha - \beta) + \frac{1}{2}\sin(\alpha + \beta), \tag{2.A.11}$$

$$\cos\alpha\sin\beta = -\frac{1}{2}\sin(\alpha - \beta) + \frac{1}{2}\sin(\alpha + \beta), \tag{2.A.12}$$

$$\sin\alpha\sin\beta = \frac{1}{2}\cos(\alpha - \beta) - \frac{1}{2}\cos(\alpha + \beta), \tag{2.A.13}$$

$$\cos\alpha\cos\beta = \frac{1}{2}\cos(\alpha - \beta) + \frac{1}{2}\cos(\alpha + \beta), \tag{2.A.14}$$

$$\sin\alpha + \sin\beta = 2\sin\left(\frac{\alpha + \beta}{2}\right)\cos\left(\frac{\alpha - \beta}{2}\right), \tag{2.A.15}$$

$$\sin\alpha - \sin\beta = 2\cos\left(\frac{\alpha + \beta}{2}\right)\sin\left(\frac{\alpha - \beta}{2}\right), \tag{2.A.16}$$

$$\cos\alpha + \cos\beta = 2\cos\left(\frac{\alpha+\beta}{2}\right)\cos\left(\frac{\alpha-\beta}{2}\right), \tag{2.A.17}$$

$$\cos\alpha - \cos\beta = -2\sin\left(\frac{\alpha+\beta}{2}\right)\sin\left(\frac{\alpha-\beta}{2}\right), \tag{2.A.18}$$

$$\tanh x = x - \frac{1}{3}x^3 + \frac{2}{15}x^5 - \frac{17}{315}x^7 + \frac{62}{2835}x^9 - \ldots, \tag{2.A.19}$$

$$\tanh^{-1} x = x + \frac{1}{3}x^3 + \frac{1}{5}x^5 + \frac{1}{7}x^7 + \frac{1}{9}x^9 - \ldots, \tag{2.A.20}$$

$$\tanh^{-1} x = \frac{1}{2}\ln\left(\frac{1+x}{1-x}\right). \tag{2.A.21}$$

2.A.2 Implementation of the "tanh^{-1}" Block

The "tanh^{-1}" block can be implemented by two diode-connected transistors, which are shown in Figure 2.A.1.

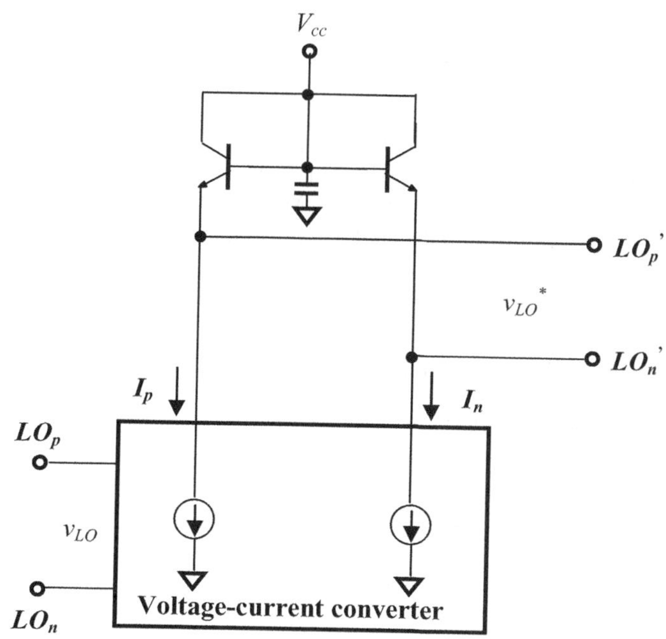

Figure 2.A.1 An inverse hyperbolic tangent circuit is implemented by two diode-connected transistors.

The function of the voltage-current converter is

$$I_p = I_o + Kv_{LO}, \tag{2.A.22}$$

$$I_n = I_o - Kv_{LO}, \tag{2.A.23}$$

Where

I_o = DC component of I_p or I_n,
K = trans-conductance of the voltage to current converter.

The differential voltage developed across the two diode-connected transistor is

$$v_{LO}^* = V_T \ln\frac{I_o + Kv_{LO}}{I_S} - V_T \ln\frac{I_o - Kv_{LO}}{I_S} = V_T \ln\frac{I_o + Kv_{LO}}{I_o - Kv_{LO}}, \tag{2.A.24}$$

$$v_{LO}^* = 2V_T \tanh^{-1}\left(\frac{Kv_{LO}}{I_o}\right). \tag{2.A.25}$$

By replacing v_{LO} in equation (2.A.21) with v_{LO}^* as shown in equation (2.A.25), we have

$$\Delta I = I_{EE}\frac{v_{RF}}{2V_T}\frac{Kv_{LO}}{I_o}, \tag{2.A.26}$$

which is the desired differential current.

REFERENCES

[1] T. H. Chen, K. W. Chang, S. B. T. Bui, L. C. T. Liu, and G. S. Pak, "Broadband Single- and Double-Balanced Resistive HEMT Monolithic Mixers," *IEEE Transactions on Microwave Theory and Techniques*, Vol. 43, No. 3, March 1995, pp. 477–484.

[2] Barrie Gilbert, "The Multi-tanh Principle: A Tutorial Overview," *IEEE Journal of Solid-State Circuits*, Vol. 33, No. 1, January 1998, pp. 2–17.

[3] S.-G. Lee and J.-K. Choi, "Current-reuse Bleeding Mixer," *Electronics Letters*, April 13, 2000, Vol. 36, No. 8, pp. 696–697.

[4] Edwin E. Bautista, Babak Bastini, and Joseph Heck, "A High IIP2 Down-conversion Mixer Using Dynamic Matching," *IEEE Journal of Solid-state Circuits*, Vol. 35, No. 12, December 2000, pp. 1934–1941.

[5] Kalle Kivekäs, Aarno Pärssinen, and Kari A. I. Halonen, "Characterization of IIP2 and DC-Offsets in Transconductance Mixers," *IEEE Transactions on Circuits and Systems—II: Analog and Digital Signal Processing*, Vol. 48, No. 11, November 2001, pp. 1028–1038.

[6] Paul R. Gray, Paul J. Hurst, Stephen H. Lewis, and Robert G. Meyer, *Analysis and Design of Analog Integrated Circuits*, 4th ed., John Wiley & Sons, Inc., 2001, pp. 206–212, 522–532.

[7] Faulkner, M., "DC Offset and IM2 Removal in Direct Conversion Receivers," *Communications, IEE Proceedings*, Vol. 149, No. 3, June 2002, pp. 179–184.

[8] Liwei Sheng and Lawrence E. Larson, "An Si–SiGe BiCMOS Direct-Conversion Mixer with Second-Order and Third-Order Nonlinearity Cancellation for WCDMA Applications," *IEEE Transactions on Microwave Theory and Techniques*, Vol. 51, No. 11, November 2003, pp. 2211–2220.

[9] Arif A. Siddiqi and Tad Kwasniewski, "2.4 GHz RF Down-conversion Mixers in Standard CMOS Technology," Department of Electronics, Carleton University, Ottawa, Canada, 2004, pp. 321–324.

[10] Hung-Che Wei, Ro-Min Weng, and Kun-Yi Lin, "A 1.5 V High-Linearity CMOS Mixer for 2.4 GHz Applications," IEEE, ISCAS 2004, pp. I-561–I-564.

[11] Shoji Otaka, Mitsuyuki Ashida, Masato Ishii, and Tetsuro Itakura, "A +10-dBm IIP3 SiGe Mixer with IM3 Cancellation Technique," *IEEE Journal of Solid-State Circuits*, Vol. 39, No. 12, December 2004, pp. 2333–2339.

[12] Parag Upadh yaya, Mallesh Rajashekharaiah, Deukhyoun Heo, and Yi-Jan Emery Chen, "A High IIP2 Doubly Balanced Sub-Harmonic Mixer in 0.25-pm CMOS for 5-GHz ISM Band Direct Conversion Receiver," *IEEE Radio Frequency Integrated Circuits Symposium*, 2005, pp. 175–178.

[13] Stedano Cipriani, Guglielmo Sirna, Paolo Cusinato, Lorenzo Carpineto, Francis Monchal, Christian Sorace, and Eric Duvivier, "Low-IF 90nm CMOS Receiver for 2.5G Application," *IEEE MELECON*, May 12-15, 2004, Dubrovnik, Croatia, pp. 151–154.

[14] Richard Svitek and Sanjay Raman, "5–6 GHz SiGe Active I/Q Subharmonic Mixers with Power Supply Noise Effect Characterization," *IEEE Microwave and Wireless Components Letters*, Vol. 14, No. 7, July 2004, pp. 319–321.

[15] Young-Kyun Jang, Changjae Kim, and HyungJoun Yoo, "Wide-band CMOS Mixer Design Using Flexible Input Matching Method," *Fourth International Conference on Microwave and Millimeter Wave Technology Proceedings*, 2004 IEEE, pp. 590–593.

[16] Michael Margraf and Georg Boeck, "Analysis and Modeling of Low-Frequency Noise in Resistive FET Mixers," *IEEE Transactions on Microwave Theory and Techniques*, Vol. 52, No. 7, July 2004, pp. 1709–1718.

[17] Hooman Darabi and Janice Chiu, "A Noise Cancellation Technique in Active RF-CMOS Mixers," *IEEE Journal of Solid-State Circuits*, Vol. 40, No. 12, December 2005, pp. 2628–2632.

[18] Karim W. Hamed, Alois P. Freundorfer, and Yahia M. M. Antar, "A Monolithic Double-Balanced Direct Conversion Mixer with an Integrated Wideband Passive Balun," *IEEE Journal of Solid-State Circuits*, Vol. 40, No. 3, March 2005, pp. 622–629.

[19] Amin Q. Safarian, Ahmad Yazdi, and Payam Heydari, "Design and Analysis of an Ultrawide-Band Distributed CMOS Mixer," *IEEE Transactions on Very Large Scale Integration (VLSI) Systems*, Vol. 13, No. 5, May 2005, pp. 618–629.

[20] David D. Wentzloff and Anantha P. Chandrakasan, "A 3.1-10.6 GHz Ultra-Wideband Pulse-Shaping Mixer," *IEEE Radio Frequency Integrated Circuits Symposium*, 2005, pp. 83–86.

[21] Mario Valla, Giampiero Montagna, Rinaldo Castello, Riccardo Tonietto, and Ivan Bietti, "A 72-mW CMOS 802.11a Direct Conversion Front-End with 3.5-dB NF and 200-kHz 1/f Noise Corner," *IEEE Journal of Solid-State Circuits*, Vol. 40, No. 4, April 2005, pp. 970–976.

[22] Sining Zhou and Mau-Chung Frank Chang, "A CMOS Passive Mixer with Low Flicker Noise for Low-Power Direct-Conversion Receiver," *IEEE Journal of Solid-State Circuits*, Vol. 40, No. 5, May 2005, pp. 1084–1092.

[23] Tero Tikka, Jussi Ryynanen, Mikko Hotti, and Kari Halonen, "Design of a High Linearity Mixer for Direct-Conversion Base-Station Receiver," IEEE, ISCAS 2006, pp. 2321–2324.

[24] Woonyun Kim, Dongjin Keum, Sanghoon Kang, and Byeong-Ha Park, "A Mixer with Third-Order Nonlinearity Cancellation Technique for CDMA Applications," *IEEE Microwave and Wireless Components Letters*, Vol. 17, No. 1, January 2007, pp. 76–78.

CHAPTER 3

DIFFERENTIAL PAIRS

3.1 WHY DIFFERENTIAL PAIRS?

3.1.1 Superficial Differences Between the Single-ended and Differential Pair

In the early years of electronic circuit development, only single-ended stages were studied and implemented. Gradually, it was found that some special types of performance could only be obtained by a differential pair instead of a single-ended stage, such as a push-pull amplifier or a double-balanced mixer. At present, both single-ended and differential pairs have been developed for most *RF* blocks, such as the *LNA*, mixer, *VCO*, filter, and *PA*. In recent years increasing numbers of differential circuits have appeared in *RF* or *RFIC* designs, especially in the "zero" *IF* or direct conversion communication systems.

Without input and output impedance matching networks, typical single-ended blocks are shown in Figure 3.1, typical differential pairs in Figure 3.2. The differential pair appears as two single-ended blocks joined together with a common emitter or source resistor, R_e or R_s, in a bipolar or a *MOSFET* differential pair. There is a current source, I_e or I_s, connected to this common resistor in parallel. The pair of two devices should have the the same size and configuration, as well as identical collector or drain resistors.

For the bipolar differential pair,

$$R_{c1} = R_{c2}, \tag{3.1}$$

$$Q_1 = Q_2, \tag{3.2}$$

RF Circuit Design, by Richard Chi-Hsi Li

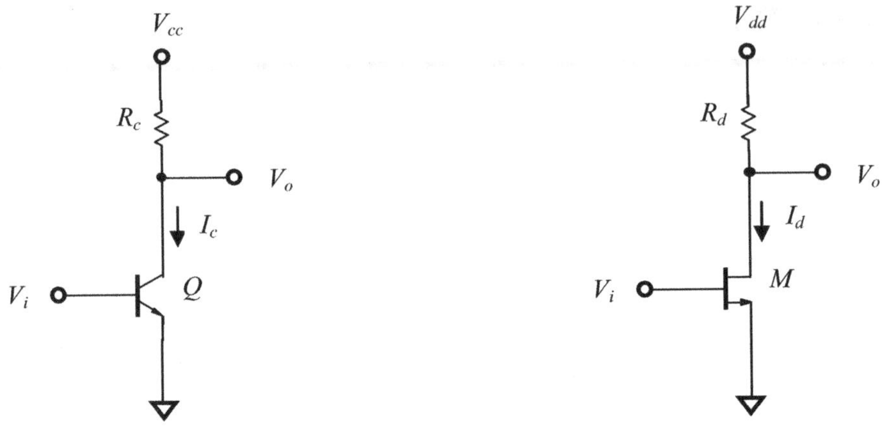

(a) A bipolar single-ended block (b) A *MOSFET* single-ended block

Figure 3.1 Typical single-ended blocks.

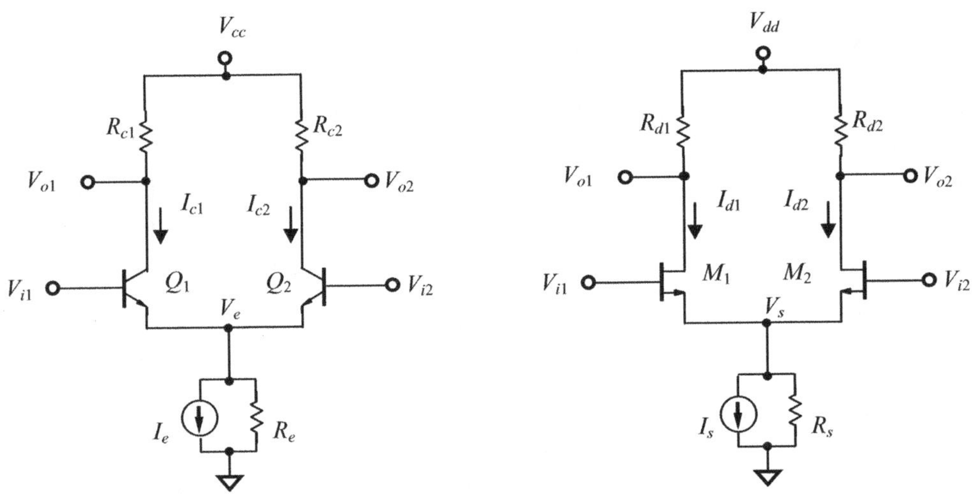

(a) A bipolar differential pair (b) A *MOSFET* differential pair

Figure 3.2 Typical differential pairs.

For the *MOSFET* differential pair,

$$R_{d1} = R_{d2}, \tag{3.3}$$

$$M_1 = M_2, \tag{3.4}$$

In short, both differential branches should be symmetrical, including their values, size, configuration, and layout.

The apparent differences are shown in Figures 3.1 and 3.2. The number of parts in a differential pair is about double that of a single-ended stage. For instance, as shown in Figure 3.1, the part count of each branch in a differential pair is equal to 2, which is the part count of a single-ended block. Actually, there are two more parts in the differential pair. These are the current sources I_e or I_s and the tail resistors R_e or R_s.

Consequently, either in the *RF* circuit with discrete parts or in the *RFIC* design, the layout area for the differential pair is about double that of the single-ended stage. The cost in an *RFIC* design is proportional to the layout area on the *IC* chip; this means that the cost of a differential pair is double that of a single-ended stage. In practice, the part count and layout area would actually be more than doubled, as one of the parts in the differential pair is a current source, which is usually not a single part and is not present in a single-ended stage.

If a differential pair is built by two single-ended stages, its current drain is usually double that of each single-ended stage if each differential output has approximately the same voltage gain as that in the output of the single-ended stage.

In general, the noise figure of a differential pair is higher than that of a single-ended stage. One of the reasons is that if the current flow through the device remains the same, shot noise is doubled from that of a single-ended stage. Another reason is that thermal noise is also approximately doubled from that of a single-ended stage because there are twice as many resistors. At this point, different opinions do exist and some people disagree with the assertion that the noise figure is increased in the differential pair. On the contrary, they believe that "common mode" noise is eliminated or reduced substantially in a differential pair compared with a single-ended stage.

A differential input signal between two input ports is expected to be faithfully magnified or transported by a differential pair and presented to its differential output signal between two output ports. Assuming that the input of the differential pair is a pure differential signal, the output of the differential pair should thus be a pure differential signal as well. To approach this ideal differential performance, the symmetry of the layout must be maintained as much as possible. Symmetry is meaningless for a single-ended stage.

Finally, a special part, called the balun, is needed for the differential pair. A balun splits a single-ended signal to a pair of differential signals or combines a pair of differential signals into a single-ended signal. Chapter 4 introduces and discusses baluns.

From the comparison above, it seems apparent that the differential pair is devoid of any merit compared to a single-ended stage. If so, what is use of the differential pair? The answer can be found in Section 3.1.2.

Many negative comments have been leveled at the differential pair in comparing it with the single-ended stage. The "apparent differences between the single-ended stage and the differential pair" have been deduced solely from their schematics. Their essential differences, however, are discovered from their performance and other aspects. As a matter of fact, the remarkable difference between the single-ended stage and the differential pair is their performance in non-linearity. Let's find this difference.

3.1.2 Non-linearity in the Single-ended Stage

In reality, all transistors, bipolar or *MOSFET*, have non-linear parts. Any circuit block containing a non-linear part is called a non-linear block. Therefore, a single-

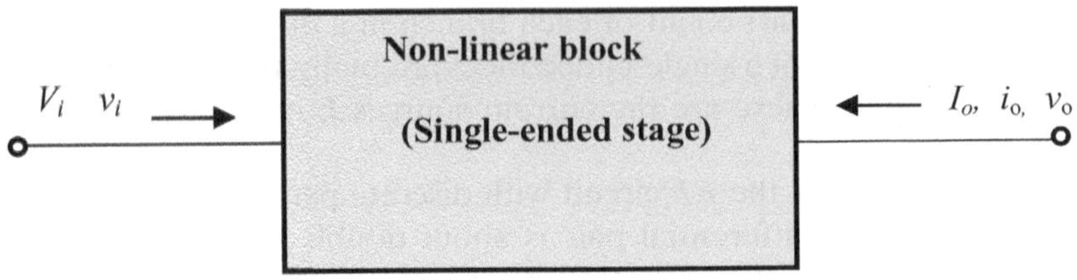

Figure 3.3 Non-linear block containing a single-end stage.

ended stage is a non-linear block as shown in Figure 3.3. The transfer function of a single-ended stage can be expressed in a general form:

$$I_o = I_{DC} + i_o = a_o + a_1 V_i + a_2 V_i^2 + a_3 V_i^3 + a_4 V_i^4 + \cdots, \tag{3.5}$$

where

I_o = resultant output current,
I_{DC} = output DC current,
i_o = output AC current,
V_i = input voltage,
v_i = input AC voltage,
a_i = i^{th} order of trans-conductance coefficient.

In general, the non-linearity of the single-ended stage includes both odd and even orders from 0, 1, 2, 3, to infinity.

If the input signal is an AC sinusoidal voltage, that is,

$$V_i = v_i = v_{io} \cos \omega t, \tag{3.6}$$

where

v_i = input AC sinusoidal signal,
v_{io} = amplitude of the input signal,
ω = angular frequency of the sinusoidal signal

then the transfer function (3.5) becomes

$$I_o = \sum_{n=0}^{\infty} a_n v_i^n = \sum_{n=0}^{\infty} a_n v_{io}^n \cos^n \omega t, \tag{3.7}$$

$$I_o = a_o + a_1 v_{io} \cos \omega t + a_2 v_{io}^2 \cos^2 \omega t + a_3 v_{io}^3 \cos^3 \omega t + a_4 v_{io}^4 \cos^4 \omega t + \cdots$$

$$I_o = a_o + a_1 v_{io} \cos \omega t + a_2 v_{io}^2 [1/2 + (1/2)\cos 2\omega t] + a_3 v_{io}^3 [(3/4)\cos \omega t + (1/4)\cos 3\omega t + a_4 v_{io}^4 [3/8 + (4/8)\cos 2\omega t + (1/8)\cos 4\omega t] + \cdots$$

$$I_o = a_o + [(1/2) a_2 v_{io}^2 + (3/8) a_4 v_{io}^4 + \cdots] + [a_1 v_{io} + (3/4) a_3 v_{io}^3 + \cdots] \cos \omega t + [(1/2) a_2 v_{io}^2 + (4/8) a_4 v_{io}^4 + \cdots] \cos 2\omega t + [(1/4) a_3 v_{io}^3 + \cdots] \cos 3\omega t + [(1/8) a_4 v_{io}^4 + \cdots] \cos 4\omega t + \cdots$$

It can be seen that when the input is a sinusoidal signal, the output current of a single-ended stage contains both DC and AC currents, and can be rewritten as

$$I_o = I_{DC} + i_o, \tag{3.8}$$

$$I_{DC} = a_o + v_{io}^2\left[(1/2)a_2 + (3/8)a_4 v_{io}^2 + \cdots\right], \tag{3.9}$$

$$i_o = \left[a_1 v_{io} + (3/4)a_3 v_{io}^3 + \cdots\right]\cos\omega t + \left[(1/2)a_2 v_{io}^2 + (4/8)a_4 v_{io}^4 + \cdots\right]\cos 2\omega t + \left[(1/4)a_3 v_{io}^3 + \cdots\right]\cos 3\omega t + \left[(1/8)a_4 v_{io}^4 + \cdots\right]\cos 4\omega t + \cdots \tag{3.10}$$

From equation (3.9) it can be seen that the output DC current consists of two parts. The first part is the first term a_o, which is irrelevant to v_i. The second part is the second term containing the factor v_{io}^2, which is dependent on the input AC sinusoidal voltage v_i and is called the DC offset, that is:

$$DC\ offset = v_{io}^2\left[(1/2)a_2 + (3/8)a_4 v_{io}^2 + \cdots\right] \tag{3.11}$$

Obviously, the DC offset relies directly on the input AC signal, so it is unstable and always fluctuates since the received signal varies from time to time.

A remarkable feature of the DC offset is that it is contributed by only the even orders, not the odd orders of its non-linearity.

From equation (3.10) it can be seen that the output AC current consists of all the harmonics. The odd order harmonics come from the odd order non-linearity terms, the even order harmonics from the even order non-linearity terms.

3.1.3 Non-linearity in the Differential Pair

The derivation of the non-linearity of a non-linear block containing a differential pair is similar to that for a non-linear block containing a single-ended stage. Keep in mind the following relation:

$$v_{i1} = -v_{i2}. \tag{3.12}$$

If

$$v_{i1} = v_{io}\cos\omega t, \tag{3.13}$$

then,

$$v_{i2} = -v_{io}\cos\omega t. \tag{3.14}$$

where

v_{i1} = input AC sinusoidal signal at differential branch 1,
v_{i2} = input AC sinusoidal signal at differential branch 2,
v_{io} = amplitude of each differential input signal.

Refer to Figure 3.4. The transfer function of each differential branch can be expressed in a general form:

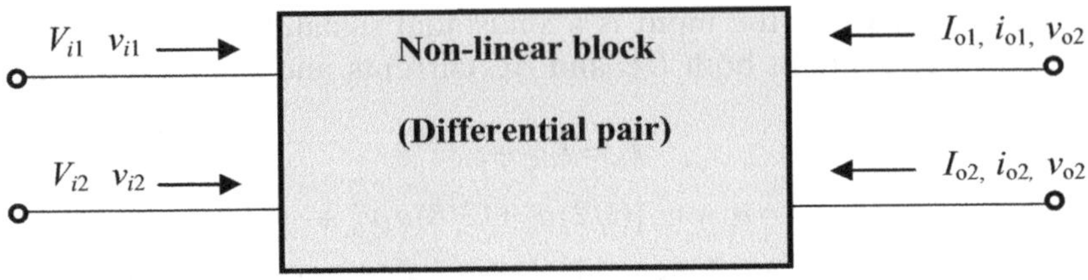

Figure 3.4 Non-linear block containing a differential pair.

$$I_{o1} = I_{DC1} + i_{o1} = a_{o1} + a_{11}V_{i1} + a_{21}V_{i1}^2 + a_{31}V_{i1}^3 + a_{41}V_{i1}^4 + \cdots, \tag{3.15}$$

$$I_{o2} = I_{DC2} + i_{o2} = a_{o2} + a_{12}V_{i2} + a_{22}V_{i2}^2 + a_{32}V_{i2}^3 + a_{42}V_{i2}^4 + \cdots, \tag{3.16}$$

where the second subscript denotes the branch number of the differential pair. And, if it is an ideal differential pair, we have

$$a_{i1} = a_{i2} = a_i. \tag{3.17}$$

If the input signal is an *AC* differential sinusoidal voltage, that is:

$$V_{i1} = v_{i1} = v_{io}\cos\omega t, \tag{3.18}$$

$$V_{i2} = v_{i2} = -v_{io}\cos\omega t, \tag{3.19}$$

then equations (3.15) and (3.16) become

$$\begin{aligned} I_{o1} &= a_o + a_1 v_{io}\cos\omega t + a_2 v_{io}^2\cos^2\omega t + a_3 v_{io}^3\cos^3\omega t + a_4 v_{io}^4\cos^4\omega t + \cdots \\ &= a_o + a_1 v_{io}\cos\omega t + a_2 v_{io}^2[1/2 + (1/2)\cos 2\omega t] + a_3 v_{io}^3[(3/4)\cos\omega t + \\ &\quad (1/4)\cos 3\omega t + a_4 v_{io}^4[3/8 + (4/8)\cos 2\omega t + (1/8)\cos 4\omega t] + \cdots \\ &= [a_o + (1/2)a_2 v_{io}^2 + (3/8)a_4 v_{io}^4 + \cdots] + [a_1 v_{io} + (3/4)a_3 v_{io}^3 + \cdots]\cos\omega t + \\ &\quad [(1/2)a_2 v_{io}^2 + (4/8)a_4 v_{io}^4 + \cdots]\cos 2\omega t + [(1/4)a_3 v_{io}^3 + \cdots]\cos 3\omega t + \\ &\quad [(1/8)a_4 v_{io}^4 + \cdots]\cos 4\omega t + \cdots \end{aligned} \tag{3.20}$$

$$\begin{aligned} I_{o2} &= a_o - a_1 v_{io}\cos\omega t + a_2 v_{io}^2\cos^2\omega t - a_3 v_{io}^3\cos^3\omega t + a_4 v_{io}^4\cos^4\omega t + \cdots \\ &= a_o - a_1 v_{io}\cos\omega t + a_2 v_{io}^2[1/2 + (1/2)\cos 2\omega t] - a_3 v_{io}^3[(3/4)\cos\omega t + \\ &\quad (1/4)\cos 3\omega t + a_4 v_{io}^4[3/8 + (4/8)\cos 2\omega t + (1/8)\cos 4\omega t] + \cdots \\ &= [a_o + (1/2)a_2 v_{io}^2 + (3/8)a_4 v_{io}^4 + \cdots] - [a_1 v_{io} + (3/4)a_3 v_{io}^3 + \cdots]\cos\omega t + \\ &\quad [(1/2)a_2 v_{io}^2 + (4/8)a_4 v_{io}^4 + \cdots]\cos 2\omega t - [(1/4)a_3 v_{io}^3 + \cdots]\cos 3\omega t + \\ &\quad [(1/8)a_4 v_{io}^4 + \cdots]\cos 4\omega t + \cdots \end{aligned} \tag{3.21}$$

The transfer function of the differential pair is

$$\begin{aligned} \Delta I &= I_{o1} - I_{o2} = i_o \\ &= 2[a_1 v_{io} + (3/4)a_3 v_{io}^3 + \cdots]\cos\omega t + 2[(1/4)a_3 v_{io}^3 + \cdots]\cos 3\omega t + \cdots \end{aligned} \tag{3.22}$$

The output differential current contains only the input differential signal and its odd order of harmonics. Thus the *DC* offset is obviously zero, that is,

$$DC\ offset = 0. \tag{3.23}$$

To summarize, in an ideal differential pair,

- The even orders of non-linearity are cancelled by each other;
- The *DC* offset does not exist.

This is the answer to the question: What is the differential pair for?

If an *RF* block is built with an ideal differential configuration and if the input is perfectly differential, the *DC* offset and the even orders of non-linearity disappear.

The next question is, of course, to ask why we pursue the goal of zero *DC* offset and canceling of the even orders of non-linearity. For the answer, let's move to the next section.

3.1.4 Importance of Differential Configuration in a Direct Conversion or "Zero *IF*" Communication System

In a dual conversion communication system, the spectrum of the de-modulated signals is allocated around the central frequency f_c as shown in Figure 3.5(a). The central frequency is not zero and its spectrum is spread over the bandwidth around the central frequency. The problem of *DC* offset does not exist because the spectrum of de-modulated signals does not contain the *DC* component.

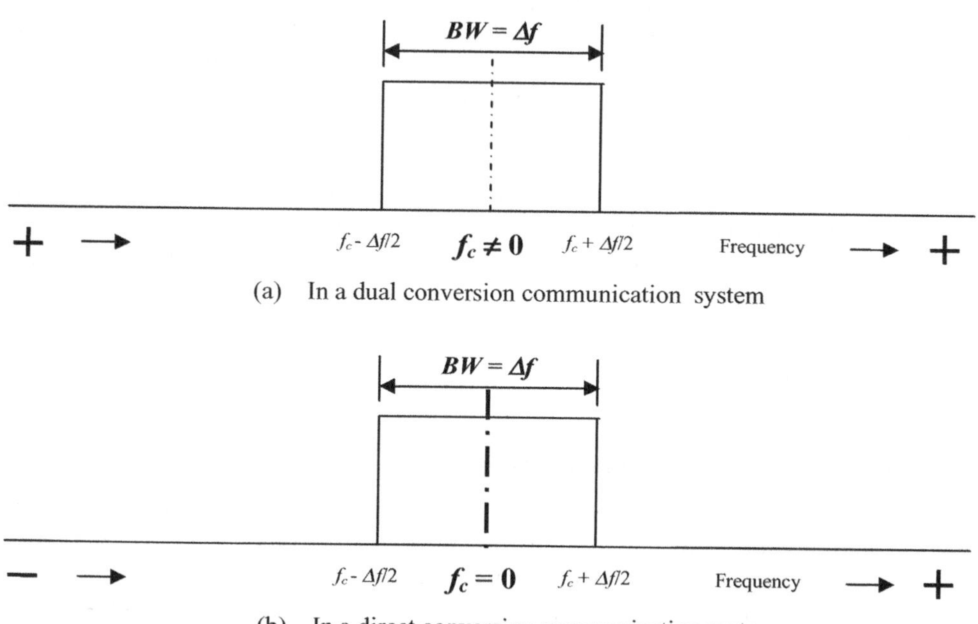

Figure 3.5 Spectrum of demodulated signal in a communication system.

In a direct conversion communication system, also called the "zero *IF*" communication system, the spectrum of de-modulated signals is allocated around the frequency $f = f_c = 0$ as shown in Figure 3.5(b). The central frequency is zero and its spectrum is spread across the zero frequency point and covers both negative and positive frequencies. This implies that the *DC* or zero frequency is the main component of the signals. Any conceivable interference at the *DC* or zero frequency might disturb the received signal and even put the communication system out of work completely.

The interference on the *DC* component in the spectrum can come from either outside or inside the circuit block. The interference from the outside depends on the environment, circuit configuration, and system design, which are outside the scope of this book.

The interference from inside the circuit block is also a complicated procedure in which many factors are involved. The main issue, however, is that the *DC* offset is inevitably produced from the non-linear devices of the circuitry. This is the main source of the *DC* offset problem in the "zero" *IF* communication system. Hence it becomes the key issue in success or failure in the design of a communication system.

Now, let's return to the discussion in Sections 3.1.2 and 3.1.3.

In a circuit block with single-ended configuration, *DC* offset is inevitably produced by the non-linearity of the device. However, in a circuit block with an ideal differential pair configuration, the produced *DC* offsets cancel each other. Furthermore, in a circuit block with an ideal differential pair configuration (discussed in Section 3.1.3), all the even orders of non-linearity cancel each other. The differential configuration of the circuit block is obviously a powerful means to remove or reduce *DC* offset. This is why in a "zero" *IF* radio, most *RF* blocks, including the *LNA*, mixer, and modulator, are differential blocks.

However, it must be pointed out that this outstanding behavior of a differential pair is so far only a theoretical prediction. In order to realize such an ideal objective, the engineering of the circuit design must be strict and accurate. There are four essential tasks:

- To guarantee the symmetry of the circuit layout; although perfect symmetry is impossible to achieve, it must be as close as possible.
- To strictly control the accuracy or tolerance of parts in either the discrete circuit or *IC* chip design.
- To avoid any interference.
- To have good *RF* grounding.

3.1.5 Why Direct Conversion or "Zero *IF*"?

Recalling all the disadvantages of the differential pair mentioned above, the advantage of building differential circuits for the "zero" *IF* communication system seems doubtful, unless the "zero" *IF* system brings about a huge benefit.

From an engineering design viewpoint, designing a dual conversion communication system is easier than designing a direct conversion or "zero *IF*" communication system.

From the viewpoint of system performance, a dual conversion system is better than a direct conversion or "zero *IF*" communication system. At present, dual conversion systems are used only in the military and the most advanced communication systems.

In spite of these various viewpoints, for a popular product such as a cell phone or *WLAN*, cost is the first priority.

A direct conversion or "zero" *IF* communication system is much cheaper than a dual conversion communication system. The price reduction is two-fold:

- First, all the *RF* filters applied in the dual conversion communication system can be removed in a direct conversion communication system.
- It is possible to put a full direct conversion communication system on one *IC* chip since all the *RF* filters are gone.

Cost is the first priority. This is the final answer to the question "Why differential pair?".

3.2 CAN *DC* OFFSET BE BLOCKED BY A CAPACITOR?

In order to clarify this question, let's examine how a capacitor blocks *DC* voltage from the source to the load in the following three cases (Figures 3.6 to 3.8). A capacitor *C* is inserted between the source and the load. The waveform at the source v_S is drawn in the upper left corner of these figures. The pulses are positive, with an average *DC* voltage V_o and a repetition period *T*. The waveform at the load v_L is drawn in the upper right corner and is different in three cases.

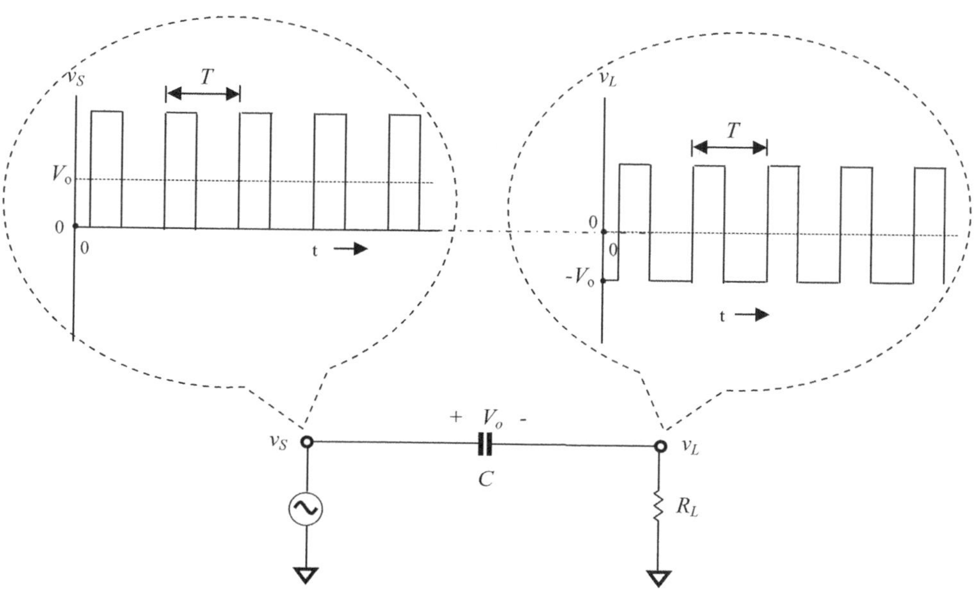

Figure 3.6 A capacitor blocks *DC* voltage V_o from the source v_S to v_L when $T >> R_LC$.

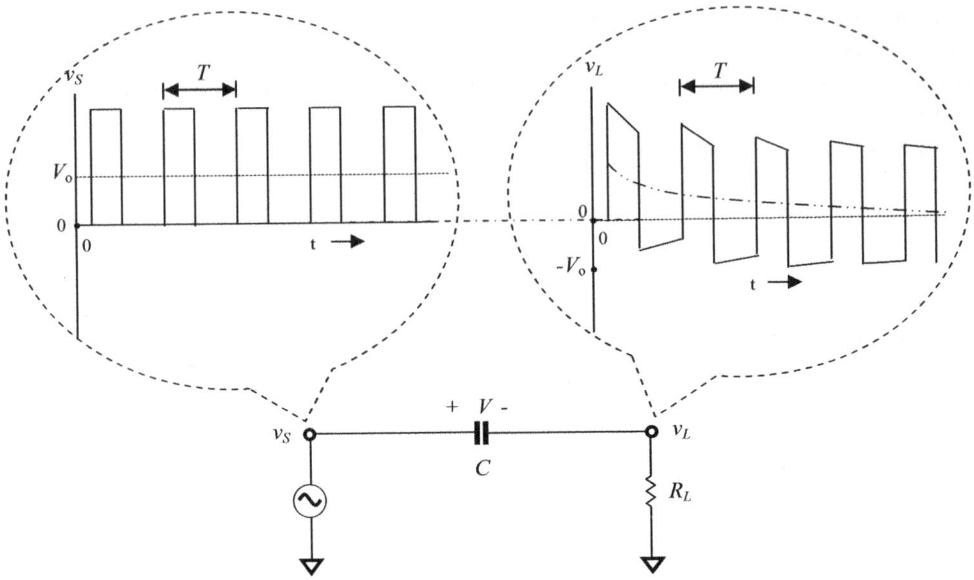

Figure 3.7 A capacitor blocks DC voltage V_o from the source v_S to v_L when $T \sim R_LC$.

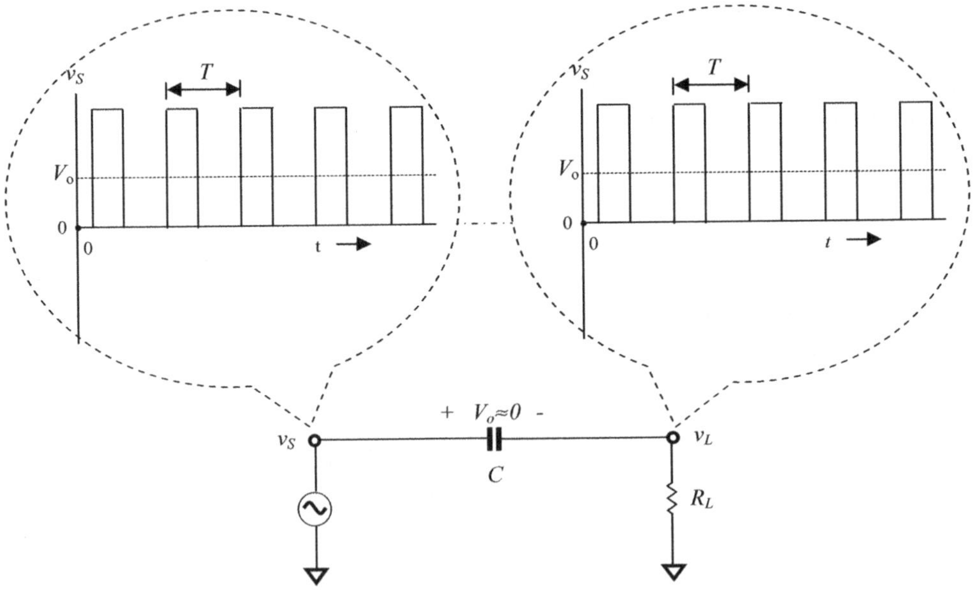

Figure 3.8 A capacitor blocks DC voltage V_o from the source v_S to v_L when $T << R_LC$.

The first case is when

$$T >> R_L C. \tag{3.24}$$

Figure 3.6 shows the waveforms at the source v_S and at the load v_L when $T >> R_L C$.

In this case, the charge and discharge period of the capacitor is much shorter than the repetition period. This implies that the capacitor is able to follow the variation of the source voltage v_S. Consequently, the average *DC* voltage V_o is built across the capacitor. The waveform at the load v_L as shown in the upper right corner is kept unchanged, but its average *DC* voltage is dropped from V_o to 0, the actual ground point.

Therefore, in this case the average *DC* voltage V_o existing in the source is blocked by the capacitor so that at the load the average *DC* voltage is 0.

The second case is when

$$T \sim R_L C. \tag{3.25}$$

Figure 3.7 shows the waveforms at the source v_S and at the load v_L when $T \sim R_L C$. Here, the charge and discharge period of the capacitor is around the same order as the repetition period. Hence, the capacitor is not able to follow the variation of the source voltage v_S well. Consequently, the average *DC* voltage built across the capacitor is gradually increased from 0 to V_o. At the load R_L, the waveform is changed and experiences the charged and discharged processes alternatively. The voltage across the capacitor is as follows:

When the capacitor is charged,

$$V = V_{oi}\left(1 - e^{-R_L C}\right), \tag{3.26}$$

where V_{oi} = voltage to be charged to.

When the capacitor is discharged,

$$V = V_{oj} e^{-R_L C}, \tag{3.27}$$

where V_{oj} = voltage to be discharged from.

Correspondingly, from the waveform at the load v_L as shown in the upper right corner of Figure 3.7, it can be seen that its average *DC* voltage has dropped from V_o toward 0 with a time constant $\tau = R_L C$. In this case, the average *DC* voltage V_o existing in the source is blocked from the source to the load by the capacitor after a time slot of about $\tau = R_L C$.

The third case is when

$$T << R_L C. \tag{3.28}$$

Figure 3.8 shows the waveforms at the source v_S and at the load v_L when $T << R_L C$. The charge and discharge time of the capacitor is much longer than the repetition period. This implies that the capacitor is not able to follow the variation of the source voltage v_S well. Consequently, the time for the average DC voltage built across the capacitor from 0 to V_o is very long and is almost zero during a limited time period. At the load R_L, the waveform is correspondingly shown in the upper right corner. In this case, the average DC voltage V_o at the load is still maintained at the same level as at the source.

From these three cases it can be seen that a capacitor can function as a DC blocking capacitor only when condition (3.24) is satisfied. Let's convert this inequality (3.24) to an approximate equation, that is,

$$T = 10R_L C_{max}, \tag{3.29}$$

where C has been replaced by C_{max}, which implies that the condition is satisfied even in the worst case.

Now let's return to our case. Owing to the non-linearity of the device, the output current of a single-ended stage contains both DC current, DC offset, and AC currents, with all the harmonics as shown in equations (3.8) to (3.11). Similarly, in the discussion about the DC blocking capacitor above, the pulse signal from the source can be expressed by the Fourier expansions, in which the DC and AC components with all the harmonics are also contained. The capability of a capacitor for DC blocking can be judged in the same way qualitatively, though not quantitatively.

The repetition frequency of the pulses f corresponding to T in Figures 3.6, 3.7, and 3.8 should be in the order of $100\,MHz$ to $10\,GHz$. In terms of (3.29), the value of the DC blocking capacitor can be evaluated. In Table 3.1, the values of R_L are selected from $10\,\Omega$ to $10\,k\Omega$. In actual circuits, the values of R_L are usually much higher than $50\,\Omega$ before impedance matching.

Table 3.1 lists the maximum values of a capacitor functioning as a DC blocking capacitor. They are much lower than the values of the "zero" chip capacitor if the seventh column when $R_L = 1\,k\Omega$ in Table 3.1 is compared with Table 3.2. For instance,

- If the operating frequency is $800\,MHz$,
  - From Table 3.1 it can be found that the capacitance of the DC blocking capacitor must be less than $0.125\,pF$ if $R_L = 1\,k\Omega$, while from Table 3.2 it can be found that the capacitance of a "zero" chip capacitor is $46\,pF$. Or,
  - From Table 3.1 it can be found that the capacitance of the DC blocking capacitor must be less than $2.5\,pF$ if $R_L = 50\,\Omega$, while from Table 3.2 it can be found that the capacitance of a "zero" chip capacitor is $46\,pF$.
- If the operating frequency is $2400\,MHz$,
  - From Table 3.1 it can be found that the capacitance of DC blocking capacitor must be less than $0.042\,pF$ if $R_L = 1\,k\Omega$, while from Table 3.2 it can be found that the capacitance of a "zero" chip capacitor is $5.1\,pF$.
  - From Table 3.1 it can be found that the capacitance of DC blocking capacitor must be less than $0.833\,pF$ if $R_L = 50\,\Omega$, while from Table 3.2 it can be found that the capacitance of a "zero" chip capacitor is $5.1\,pF$.

It is well known that the signal is not attenuated if a "zero" capacitor is inserted at the input or output in series. However, it is impossible for the "zero" capacitor to

TABLE 3.1 Maximum capacitance of a *DC* blocking capacitor, *pF*

f,MHz	$R_L = 10\,\Omega$	$R_L = 30\,\Omega$	$R_L = 50\,\Omega$	$R_L = 100\,\Omega$	$R_L = 500\,\Omega$	$R_L = 1\,k\Omega$	$R_L = 2\,k\Omega$	$R_L = 3\,k\Omega$	$R_L = 5\,k\Omega$	$R_L = 10\,k\Omega$
100	100.000	33.33	20.000	10.000	2.000	1.000	0.500	0.333	0.200	0.100
200	50.000	16.67	10.00	5.000	1.000	0.500	0.250	0.167	0.100	0.050
300	33.333	11.11	6.667	3.333	0.667	0.333	0.167	0.111	0.067	0.033
400	25.000	8.333	5.000	2.500	0.500	0.250	0.125	0.083	0.050	0.025
500	20.000	6.667	4.000	2.000	0.400	0.200	0.100	0.067	0.040	0.020
600	16.667	5.556	3.333	1.667	0.333	0.167	0.083	0.056	0.033	0.017
700	3.286	4.762	2.857	1.429	0.286	0.143	0.071	0.048	0.029	0.014
800	12.500	4.167	2.500	1.250	0.250	0.125	0.063	0.042	0.025	0.013
900	11.111	3.704	2.222	1.111	0.222	0.111	0.056	0.037	0.022	0.011
1,000	10.000	3.333	2.000	1.000	0.200	0.100	0.050	0.033	0.020	0.010
1,500	6.667	2.222	1.333	0.667	0.133	0.067	0.033	0.022	0.013	0.007
2,400	4.167	1.389	0.833	0.417	0.083	0.042	0.021	0.014	0.008	0.004
5,800	1.724	0.575	0.345	0.172	0.034	0.017	0.009	0.006	0.003	0.002
10,000	1.000	0.333	0.200	0.100	0.020	0.010	0.005	0.003	0.002	0.001

TABLE 3.2 Some SRF_C (Self-Resonant Frequency) values of chip capacitors

SRF_C (MHz)	Value of chip capacitor (pF)
40	18,000
50	11,664
100	2,916
150	1,296
450	144
500	117
800	46
900	36
1,000	29
1,500	13
2,400	5.1
5,400	1.0
6,235	0.75

function as a DC blocking capacitor because its value is too high. Should the "zero" capacitor be replaced by the capacitor listed in Table 3.1, the low value of the capacitor would bring about a significant attenuation of the signal. More importantly, capacitors with values of less than $0.1pF$ or $0.2pF$ are meaningless in a practical circuit design, no matter whether in discrete parts or on the $RFIC$ chip.

Consequently, one cannot expect the DC offset current to be blocked by a capacitor unless a special design scheme is developed. Theoretically DC offset absolutely exists in a single-ended stage, but disappears in an ideal differential pair. In reality, an ideal differential pair does not exist; DC offset persists in an imperfect differential pair.

The cancellation of DC offset thus becomes a hot topic in the development of direct conversion or "zero IF" communication systems.

3.3 FUNDAMENTALS OF DIFFERENTIAL PAIRS

3.3.1 Topology and Definition of a Differential Pair

In Section 3.1.1, the topology of a differential pair was defined and described by equations (3.1) to (3.4). In order to clarify the DC and AC characteristics of the differential pair, it is better to denote the voltages and currents with DC and AC components, such as

$$V_{i1} = V_{I1} + v_{i1}, \tag{3.30}$$

$$V_{i2} = V_{I2} + v_{i2}, \tag{3.31}$$

$$V_{o1} = V_{O1} + v_{o1}, \tag{3.32}$$

$$V_{o2} = V_{O2} + v_{o2}, \tag{3.33}$$

$$I_{c1} = I_{C1} + i_{c1}, \tag{3.34}$$

$$I_{c2} = I_{C2} + i_{c2}, \tag{3.35}$$

$$I_{d1} = I_{D1} + i_{d1}, \tag{3.36}$$

$$I_{d2} = I_{d2} + i_{d2}, \tag{3.37}$$

where

V_{i1} = sum of DC and AC voltage at input port 1,
V_{i2} = sum of DC and AC voltage at input port 2,
V_{I1} = DC voltage at input port 1,
V_{I2} = DC voltage at input port 2,
v_{i1} = AC voltage at input port 1,
v_{i2} = AC voltage at input port 2,
V_{o1} = sum of DC and AC voltage at output port 1,
V_{o2} = sum of DC and AC voltage at output port 2,
V_{O1} = DC voltage at output port 1,
V_{O2} = DC voltage at output port 2,
v_{o1} = AC voltage at output port 1,
v_{o2} = AC voltage at output port 2,
I_{c1} = sum of DC and AC current into collector of Q_1,
I_{c2} = sum of DC and AC current into collector of Q_2,
I_{C1} = DC current into collector of Q_1,
I_{C2} = DC current into collector of Q_2,
i_{c1} = AC current into collector of Q_1,
i_{c2} = AC current into collector of Q_2,
I_{d1} = sum of DC and AC current into drain of M_1,
I_{d2} = sum of DC and AC current into drain of M_2,
I_{D1} = DC current into drain of M_1,
I_{D2} = DC current into drain of M_2,
i_{d1} = AC current into drain of M_1,
i_{d2} = AC current into drain of M_2,

For a differential pair, the input and output signals can be expressed by voltage as shown below:

$$V_{I1} = V_{I2}, \tag{3.38}$$

$$v_{i1} = -v_{i2}, \tag{3.39}$$

or

$$|v_{i1}| = |v_{i2}|, \quad \angle v_{i1} - \angle v_{i2} = 180°, \tag{3.40}$$

and,

$$V_{O1} = V_{O2}, \tag{3.41}$$

$$v_{o1} = -v_{o2}, \tag{3.42}$$

or

$$|v_{o1}| = |v_{o2}|, \quad \angle v_{o1} - \angle v_{o2} = 180° \tag{3.43}$$

and the collector currents of a bipolar differential pair or the drain currents of a *MOSFET* differential pair are in a perfectly symmetrical state, that is,

$$I_{c1} = I_{c2}, \tag{3.44}$$

or

$$I_{d1} = I_{d2}, \tag{3.45}$$

Equations (3.30) to (3.45) are tedious. However, they are important in distinguishing the *DC* and *AC* components in the expression of voltages and currents. Some books and articles discuss the "*DC* transfer characteristics" of a device and then treat them as "*AC* transfer characteristics." As matter of fact, the transfer characteristics of a device cover both *DC* and *AC* components.

3.3.2 Transfer Characteristics of a Bipolar Differential Pair

Figure 3.9 shows a differential pair with bipolar devices. The emitter resistor R_e is renamed as the tail resistor, R_{TL}. The tail current I_{TL} can be provided by a current mirror and controlled by a control voltage V_{cv} as shown in Figure 3.10.

The transfer characteristics can be derived from the exponential relationship between the base voltage and the collector current.

$$I_{c1} = I_{S1}\left(1+\frac{V_{ce1}}{V_{A1}}\right)\exp\left(\frac{V_{be1}}{V_T}\right), \tag{3.46}$$

$$I_{c2} = I_{S2}\left(1+\frac{V_{ce2}}{V_{A2}}\right)\exp\left(\frac{V_{be2}}{V_T}\right), \tag{3.47}$$

where

I_{c1} = collector current of Q_1,
I_{c2} = collector current of Q_2,
I_{S1} = saturated current of Q_1,
I_{S2} = saturated current of Q_2,
V_{ce1} = collector-emitter voltage drop of Q_1,
V_{ce2} = collector-emitter voltage drop of Q_2,

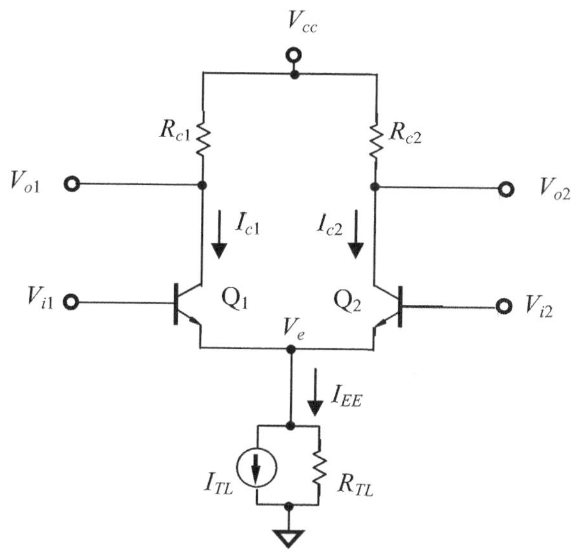

Figure 3.9 Plot of bipolar differential pair for discussion of its transfer characteristics.

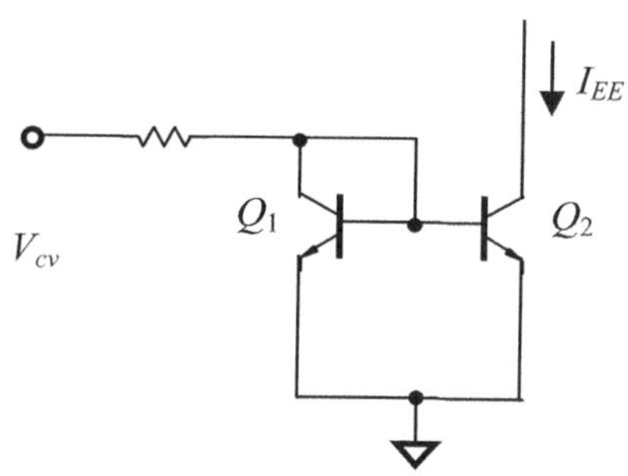

Figure 3.10 Tail current is provided by a current mirror and controlled by V_{cv}.

V_{be1} = base-emitter voltage drop of Q_1,
V_{be2} = base-emitter voltage drop of Q_2,
V_{A1} = early voltage of Q_1,
V_{A2} = early voltage of Q_2,
V_T = thermal voltage = kT/q = $26mV$ at $300\,K^o$,
k = Boltzmann constant = $8.617 \cdot 10^{-5} eV{\cdot}deg^{-1}$,
T = environment temperature, K^o,
q = charge of an electron = $1.6 \cdot 10^{-19}$ *Coulomb*.

Assuming that the transistors and collector resistors are identical, that is

$$I_{S1} = I_{S2} = I_S, \tag{3.48}$$

$$V_{ce1} = V_{ce2} = V_{ce}, \tag{3.49}$$

$$V_{A1} = V_{A2} = V_A, \tag{3.50}$$

$$\alpha_{F1} = \alpha_{F2} = \alpha_F, \tag{3.51}$$

$$R_{c1} = R_{c2} = R_c. \tag{3.52}$$

where

α_{F1} = ratio of collector to emitter current of Q_1,
α_{F2} = ratio of collector to emitter current of Q_2,
R_{c1} = collector resistor of Q_1,
R_{c2} = collector resistor of Q_2.

Then, from equations (3.46) to (3.47), we have

$$V_{be1} = V_T \ln \frac{I_{c1}}{I_S\left(1+\frac{V_{ce}}{V_A}\right)}, \tag{3.53}$$

$$V_{be2} = V_T \ln \frac{I_{c2}}{I_S\left(1+\frac{V_{ce}}{V_A}\right)}. \tag{3.54}$$

To define the input and output differential voltages, V_{id} and V_{od},

$$V_{id} = V_{i1} - V_{i2}, \tag{3.55}$$

$$V_{od} = V_{o1} - V_{o2}. \tag{3.56}$$

and in the loop of $V_{i1} \rightarrow$ emitter of Q_1 and $Q_2 \rightarrow V_{i2}$,

$$V_{be1} - V_{be2} = V_{i1} - V_{i2}, \tag{3.57}$$

from equation (3.46) to (3.57), we have

$$\frac{I_{c1}}{I_{c2}} = \exp\left(\frac{V_{be1} - V_{be2}}{V_T}\right) = \exp\left(\frac{V_{i1} - V_{i2}}{V_T}\right) = \exp\left(\frac{V_{id}}{V_T}\right), \tag{3.58}$$

By introducing the tail current I_{TL}, we have

$$I_{TL} = \frac{I_{c1} + I_{c2}}{\alpha_F}, \tag{3.59}$$

from equations (3.58) and (3.59), we have

$$I_{c1} = \frac{\alpha_F I_{TL}}{1+\exp\left(-\frac{V_{id}}{V_T}\right)}, \tag{3.60}$$

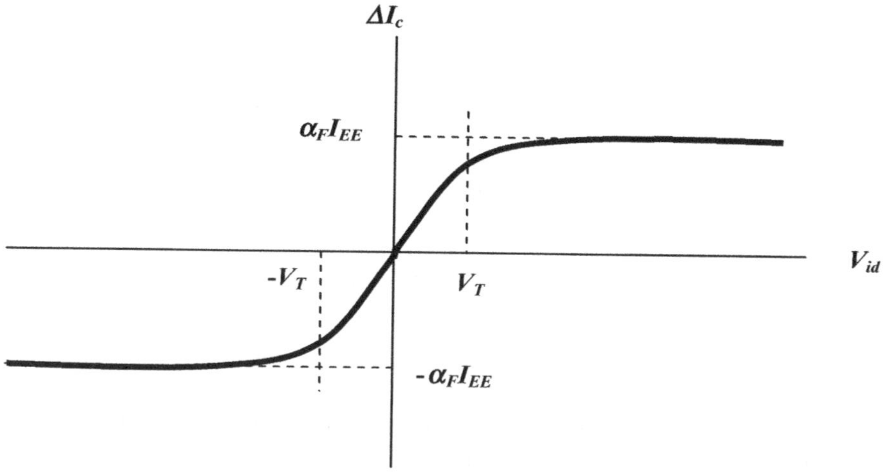

Figure 3.11 Transfer characteristic of a bipolar differential pair with emitter coupling.

$$I_{c2} = \frac{\alpha_F I_{TL}}{1 + \exp\left(\frac{V_{id}}{V_T}\right)}. \tag{3.61}$$

$$\Delta I_c = I_{c1} - I_{c2} = \alpha_F I_{TL} \tanh\left(\frac{V_{id}}{2V_T}\right). \tag{3.62}$$

This is the transfer characteristic of the bipolar differential pair and is plotted in Figure 3.11.

The output differential current in the transfer characteristic equation (3.62) can be converted to the output differential voltage.

Note that

$$V_{o1} = V_{cc} - I_{c1}R_c, \tag{3.63}$$

$$V_{o2} = V_{cc} - I_{c2}R_c, \tag{3.64}$$

Finally,

$$V_{od} = V_{o1} - V_{o2} = \alpha_F I_{TL} R_c \tanh\left(\frac{-V_{id}}{2V_T}\right). \tag{3.65}$$

Equations (3.62) and (3.65) present a relationship between the output differential current or voltage and the input differential voltage of a bipolar differential pair. Obviously, both of them are proportional to the tail current I_{TL}. On the other hand, the output differential current or voltage does not have a linear relationship with the input differential voltage. Instead, they are tangent hyperbolic functions of the input differential voltage. However, from Figure 3.11 it can be seen that they can be approximated as a linear relationship if

$$|V_{id}| < V_T. \tag{3.66}$$

3.3.3 Small Signal Approximation of a Bipolar Differential Pair

It should be noted that all the current and voltages in equation (3.65) contain both DC and AC components. From (3.30) to (3.33), (3.38) to (3.43), and (3.55) and (3.56), we have

$$V_{od} = V_{o1} - V_{o2} = (V_{o1} + v_{o1}) - (V_{o2} + v_{o2}) = (V_{o1} - V_{o2}) - (v_{o1} - v_{o2}) = v_{o1} - v_{o2} = v_{od}, \tag{3.67}$$

$$V_{id} = V_{i1} - V_{i2} = (V_{I1} + v_{i1}) - (V_{I2} + v_{i2}) = (V_{I1} - V_{I2}) - (v_{i1} - v_{i2}) = v_{i1} - v_{i2} = v_{id}. \tag{3.68}$$

Then, equation (3.65) can be changed into a relationship between the AC output differential voltage v_{od} and the AC input differential voltage v_{id}:

$$v_{od} = V_{o1} - V_{o2} = \alpha_F I_{TL} R_c \tanh\left(\frac{-v_{id}}{2V_T}\right). \tag{3.69}$$

if the AC input differential voltage v_{id} is a small signal, that is,

$$v_{id} << 2V_T. \tag{3.70}$$

It is well known that a tangent hyperbolic function in the equation can be extended to a Maclaurin series, such as, if

$$|x| < \pi/2, \tag{3.71}$$

then

$$\tanh x = x - \frac{1}{3}x^3 + \frac{2}{15}x^5 - \frac{17}{315}x^7 + \frac{62}{2835}x^9 - \cdots \tag{3.72}$$

Now, in our case, if

$$x = -\frac{v_{id}}{2V_T}, \tag{3.73}$$

then equation (3.69) becomes

$$v_{od} = \alpha_F I_{TL} R_c \left[\left(\frac{-v_{id}}{2V_T}\right) - \frac{1}{3}\left(\frac{-v_{id}}{2V_T}\right)^3 + \frac{2}{15}\left(\frac{-v_{id}}{2V_T}\right)^5 - \frac{17}{315}\left(\frac{-v_{id}}{2V_T}\right)^7 + \frac{62}{2835}\left(\frac{-v_{id}}{2V_T}\right)^9 - \cdots\right]$$

$$v_{od} = \alpha_F I_{TL} R_c \left[-\frac{1}{2}\left(\frac{v_{id}}{V_T}\right) + \frac{1}{24}\left(\frac{v_{id}}{V_T}\right)^3 - \frac{1}{240}\left(\frac{v_{id}}{V_T}\right)^5 + \frac{17}{40320}\left(\frac{v_{id}}{V_T}\right)^7 - \frac{62}{1451520}\left(\frac{-v_{id}}{2V_T}\right)^9 - \cdots\right], \tag{3.74}$$

in which the terms with orders higher than 2 could be neglected. Then, approximately, we obtain

$$v_{od} \approx -\frac{1}{2}\alpha_F I_{TL} R_c \frac{v_{id}}{V_T}. \tag{3.75}$$

In actual *RF* circuit design, equations (3.69) and (3.75) correspond to the cases of a modulator and a mixer design, respectively. In present designs, the Gilbert cell is the core of the designed block in both modulators and mixers. The big difference is that if the input is not a small signal in the modulator design, an arc-tangent-hyperbolic block must be added prior to the Gilbert cell so as to "neutralize" the tangent-hyperbolic function as shown in (3.69) and results in a linear modulation.

3.3.4 Transfer Characteristics of a *MOSFET* Differential Pair

Figure 3.12 shows a *MOSFET* differential pair. The emitter resistor R_s is renamed as the tail resistor, R_{TL}. The transfer characteristics can be derived from the square relationship between the gate voltage and the drain current.

$$I_{d1} = \frac{\mu_n C_{ox}}{2}\frac{W_1}{L_1}(V_{gs1} - V_{th})^2(1 + \lambda V_{ds1}), \tag{3.76}$$

$$I_{d2} = \frac{\mu_n C_{ox}}{2}\frac{W_2}{L_2}(V_{gs2} - V_{th})^2(1 + \lambda V_{ds2}), \tag{3.77}$$

where

I_{d1} = drain current of n channel *MOSFET* M_1,
I_{d2} = drain current of n channel *MOSFET* M_2,

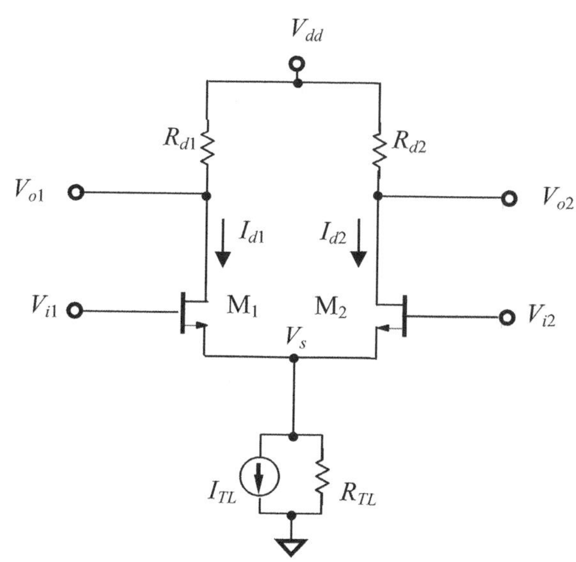

Figure 3.12 Plot of *MOSFET* differential pair for discussion of its transfer characteristics.

W_1 = width of n channel *MOSFET* M_1,
W_2 = width of n channel *MOSFET* M_2,
L_1 = length of n channel *MOSFET* M_1,
L_2 = length of n channel *MOSFET* M_2,
V_{gs1} = gate-source voltage for n channel *MOSFET* M_1,
V_{gs2} = gate-source voltage for n channel *MOSFET* M_2,
V_{th} = threshold voltage for n channel *MOSFET*, the minimum gate-to channel voltage needed for n carriers in the channel to exist,
V_{ds1} = drain-source voltage for n channel *MOSFET* M_1,
V_{ds2} = drain-source voltage for n channel *MOSFET* M_2,

and

$$C_{ox} = \frac{\varepsilon_{ox}}{t_{ox}}, \tag{3.78}$$

where

t_{ox} = thickness of the gate oxide.
ε_{ox} = permittivity of oxide.

From equations (3.76) and (3.77), we have

$$V_{gs1} - V_t = \sqrt{\frac{2I_{d1}}{\mu_n C_{ox} \frac{W_1}{L_1}(1+\lambda V_{ds1})}}, \tag{3.79}$$

$$V_{gs2} - V_t = \sqrt{\frac{2I_{d2}}{\mu_n C_{ox} \frac{W_2}{L_2}(1+\lambda V_{ds2})}}. \tag{3.80}$$

Assuming that the transistors are identical, that is,

$$W_1 = W_2 = W, \tag{3.81}$$

$$L_1 = L_2 = L, \tag{3.82}$$

$$V_{ds1} = V_{ds2} = V_{ds}, \tag{3.83}$$

then, from (3.79) to (3.83), we have

$$V_{id} = V_{i1} - V_{i2} = V_{gs1} - V_{gs2} = \frac{\sqrt{I_{d1}} - \sqrt{I_{d2}}}{\sqrt{\frac{\mu_n C_{ox}}{2} \frac{W}{L}(1+\lambda V_{ds})}}. \tag{3.84}$$

On the other hand,

$$I_{d1} + I_{d2} = I_{TL}, \tag{3.85}$$

from (3.84) and (3.85),

$$I_{d1} = \frac{I_{TL}}{2} + \frac{\mu_n C_{ox}}{4}\frac{W}{L}(1+\lambda V_{ds})V_{id}\sqrt{\frac{4I_{TL}}{\frac{\mu_n C_{ox}}{2}\frac{W}{L}(1+\lambda V_{ds})} - V_{id}^2}, \tag{3.86}$$

$$I_{d2} = \frac{I_{TL}}{2} - \frac{\mu_n C_{ox}}{4}\frac{W}{L}(1+\lambda V_{ds})V_{id}\sqrt{\frac{4I_{TL}}{\frac{\mu_n C_{ox}}{2}\frac{W}{L}(1+\lambda V_{ds})} - V_{id}^2}, \tag{3.87}$$

consequently,

$$\Delta I_d = I_{d1} - I_{d2} = \frac{\mu_n C_{ox}}{2}\frac{W}{L}(1+\lambda V_{ds})V_{id}\sqrt{\frac{4I_{TL}}{\frac{\mu_n C_{ox}}{2}\frac{W}{L}(1+\lambda V_{ds})} - V_{id}^2}, \tag{3.88}$$

$$V_{od} = V_{o1} - V_{o2} = (V_{dd} - I_{d1}R_d) - (V_{dd} - I_{d2}R_d) = -\Delta I_d R_d, \tag{3.89}$$

$$V_{od} = -R_d\frac{\mu_n C_{ox}}{2}\frac{W}{L}(1+\lambda V_{ds})V_{id}\sqrt{\frac{4I_{TL}}{\frac{\mu_n C_{ox}}{2}\frac{W}{L}(1+\lambda V_{ds})} - V_{id}^2}. \tag{3.90}$$

Equation (3.90) presents a relationship between the input and output differential voltages of a bipolar differential pair. This is called the *V-V* transfer characteristic of a *MOSFET* differential pair with source coupling. Obviously, the transfer characteristic of the input and output differential voltages in a *MOSFET* differential pair as shown in equation (3.90) is not a linear relationship. The output differential voltage is a complicated function of the input differential voltage.

3.3.5 Small Signal Approximation of a *MOSFET* Differential Pair

It should be noted that all of the current and voltages in equation (3.90) contain both *DC* and *AC* components. From (3.30) to (3.33), (3.38) to (3.43), and (3.55) and (3.56), it is found that equation (3.90) can be changed to a relationship between the *AC* output differential voltage v_{od} and the *AC* input differential voltage v_{id}:

$$v_{od} = -R_d\frac{\mu_n C_{ox}}{2}\frac{W}{L}(1+\lambda V_{ds})v_{id}\sqrt{\frac{4I_{TL}}{\frac{\mu_n C_{ox}}{2}\frac{W}{L}(1+\lambda V_{ds})} - v_{id}^2}. \tag{3.91}$$

The square root term in equation (3.91) can be extended to a Maclaurin series form: if

$$x^2 < 1, \tag{3.92}$$

$$\sqrt{a^2 - x^2} = a\left[1 - \frac{1}{2}\left(\frac{x}{a}\right)^2 - \frac{1}{8}\left(\frac{x}{a}\right)^4 - \frac{3}{48}\left(\frac{x}{a}\right)^6 - \cdots\right], \tag{3.93}$$

If the input differential voltage v_{id} is a small signal, equation (3.93) can be applied to equation (3.91).

Then equation (3.91) becomes

$$v_{od} = -R_d\sqrt{\frac{\mu_n C_{ox}}{2}\frac{W}{L}(1+\lambda V_{ds})I_{TL}}\left[v_{id} - \frac{\mu_n C_{ox}\frac{W}{L}(1+\lambda V_{ds})}{4I_{TL}}v_{id}^3 - \frac{1}{32}\left\{\frac{\mu_n C_{ox}\frac{W}{L}(1+\lambda V_{ds})}{I_{TL}}\right\}^2 v_{id}^5 - \cdots\right] \tag{3.94}$$

in which we can neglect the terms with orders higher than 2, approximately obtaining

$$v_{od} \approx -\left(\sqrt{\frac{\mu_n C_{ox}}{2}\frac{W}{L}(1+\lambda V_{ds})I_{TL}}\right)R_d v_{id}. \tag{3.95}$$

In actual *RF* circuit design, equations (3.91) and (3.95) correspond to the cases of a modulator and a mixer design, respectively. In present designs, the Gilbert cell is the core of the designed block in both modulator and mixer. The big difference is that if the input is not a small signal in the modulator design, an additional block must be added prior to the Gilbert cell so that the square-root term as shown in (3.91) can be "neutralized" to result in a linear modulation.

3.3.6 What Happens If the Input Signal Is Imperfectly Differential?

In Sections 3.3.2 to 3.3.5, the transfer characteristic of a bipolar and *MOSFET* differential pair was discussed. Through the study of the transfer characteristic, a remarkable special feature of the differential pair is found, that is, that there is zero *DC* offset in an ideal differential pair. However, it must be noted that this conclusion was obtained under two ideal conditions:

1) The differential pair is ideal, with a perfectly symmetrical configuration, perfectly identical parts, perfectly symmetrical layout, perfectly symmetrical grounding, and provision of a *DC* power supply.
2) The input signal must be perfectly differential.

The first ideal condition is easily understood via the derivation of equations (3.17) to (3.22). However, the second ideal condition is relatively easy to overlook.

The *DC* offset cannot be zero and the even orders non-linearity cannot be cancelled by each other if the input signal is not perfectly differential. For example, if

$$v_{i1} = v_{ic} + \frac{v_{id}}{2}, \tag{3.96}$$

$$v_{i2} = v_{ic} - \frac{v_{id}}{2}, \tag{3.97}$$

where

v_{id} = differential portion of input signal,
v_{ic} = common portion of input signal.

This is not a perfect differential signal because

$$v_{i1} \neq -v_{i2}. \tag{3.98}$$

Consequently, the output is not a perfectly differential signal and can be expressed in the same form as the input, that is,

$$v_{o1} = v_{oc} + \frac{v_{od}}{2}, \tag{3.99}$$

$$v_{o2} = v_{oc} - \frac{v_{od}}{2}, \tag{3.100}$$

where

v_{od} = differential portion of output signal,
v_{oc} = common portion of output signal.

By replacing equations (3.18) and (3.19) of differential inputs with equations (3.99) and (3.100) of imperfect differential inputs in the derivation of the output differential current or voltage from (3.20) to (3.22), it can be found that the *DC* offset cannot be zero and the even orders of non-linearity of the device do not cancel each other.

In reality, the input signal is always imperfect. Since this is the case, the merits of the differential pair, zero *DC* offset, and cancellation of even orders of non-linearity all disappear. It seems meaningless to apply the differential pair in *RF* circuits.

However, we need not be without hope when facing adverse situations. It is now necessary to understand how well the differential pair can promote the differential portion of the input signal and, on the other hand, how strongly the differential pair can reject the common portion of the input signal. If it can greatly magnify the differential portion and effectively suppress the common portion of the input signal, the differential pair might become a powerful circuit element. If this capability exists, zero *DC* offset can be approached and the even orders of non-linearity of the device can be reduced to an insignificant level. In short, we now need to understand, in respect to the desired differential portion, what the capability of the rejection of the common portion is in a differential pair.

In a concrete way, we are going to study the *CMRR* (Common Mode Rejection Ratio) of the differential pair. The general form of the transfer equations for differential mode and common mode operation are

$$v_{od} = A_{dm} v_{id} + A_{cm-dm} v_{ic}, \tag{3.101}$$

$$v_{oc} = A_{dm-cm} v_{id} + A_{cm} v_{ic}, \tag{3.102}$$

where

A_{dm} = differential mode voltage gain, that is, v_{od}/v_{id} when $v_{ic} = 0$,
A_{cm-dm} = common mode to differential mode voltage gain, that is, v_{od}/v_{ic} when $v_{id} = 0$,
A_{dm-cm} = differential mode to common mode voltage gain, that is, v_{oc}/v_{id} when $v_{ic} = 0$,
A_{cm} = common mode voltage gain, that is, v_{oc}/v_{ic} when $v_{id} = 0$.

The calculation of A_{dm-cm} and A_{cm-dm} for the bipolar transistor and *MOSFET* is somewhat tedious. They are usually a couple of orders smaller than A_{dm} and A_{cm}, and can be neglected. Let's confine our discussion only to A_{dm} and A_{cm}, since we are interested in the *CMRR*, which is defined as

$$CMRR = \frac{A_{dm}}{A_{cm}}. \tag{3.103}$$

This is the ratio of differential mode gain to common mode gain. There are three cases:

1) If $CMRR < 1$, or $CMRR < 0\,dB$, it signifies that the differential pair magnifies the input common portion more than it magnifies the input differential portion. The common mode operation rejects the differential mode operation.
2) If $CMRR = 1$, or $CMRR = 0\,dB$, it signifies that the differential pair magnifies the differential portion at the same order as it magnifies the common portion. The differential pair operates like a single-ended stage.
3) If $CMRR > 1$, or $CMRR > 0\,dB$, it signifies that the differential pair magnifies the input differential portion more than it magnifies the input common portion. The differential mode operation rejects the common mode operation.

The first two cases are not good. A differential pair must operate in the third case, that is, $CMRR > 1$ or $CMRR > 0\,dB$. Otherwise, the application of the differential pair is meaningless. As a matter of fact, the value of *CMRR* is expected to be as high as possible. The higher the *CMRR*, the lower the *DC* offset, and the greater the cancellation of even orders of non-linearity of the device. Generally speaking, a good design of a differential pair must ensure a *CMRR* of over $20\,dB$.

3.4 *CMRR* (COMMON MODE REJECTION RATIO)

3.4.1 Expression of *CMRR*

It is not difficult to imagine that the *CMRR* depends on the configuration of the differential pair, current drain, characteristics of transistors, and so on. In order to determine the relationship between the *CMRR* and other circuit parameters in the differential pair, some basic assumptions are made about the differential pair:

1) The differential pair is ideal with a perfectly symmetrical configuration, perfectly identical parts, perfectly symmetrical layout, perfectly symmetrical grounding, and *DC* power supply.
2) The input is an imperfect signal. It contains both differential and common mode portions.

First, let's derive the expression of the voltage gain when the differential pair operates in the differential mode, A_{dm}. By replacing V_{i1}, V_{i2}, V_{o1}, V_{o2} with $v_{id}/2$, $-v_{id}/2$, $v_{od}/2$, $-v_{od}/2$, respectively, Figure 3.2 is modified as Figure 3.13. In this figure, the tail current, either i_{eg} or i_{sg}, which flows from the node v_e or v_s to the ground, is zero, because the currents contributing to i_{eg} from both devices Q_1 and Q_2 or i_{sg} from both devices M_1 and M_2 are the same in magnitude but 180° different in phase.

This implies that the emitters or sources of devices in Figure 3.13 are virtually grounded and that the tail resistor R_{TL} is short-circuited. In other words, in Figure 3.13(a) and (b), the left and right branches of the differential pair are two identical and independent half circuits. Figure 3.14 shows the two identical and independent half circuits after the left and right branches of the differential pair in Figure 3.13 are completely separated. In this case the tail resistor R_{TL} becomes a dummy part.

The equivalents of the differential mode half circuit in the differential pair are sketched in Figure 3.15, which represents a low-frequency approximation. All the capacitors in the transistor's model are neglected approximately. In the bipolar device model, only r_π and R_c are reserved and in the *MOSFET* device model only r_{gs} and R_d are reserved.

For the bipolar device, Figure 3.15(a), we have

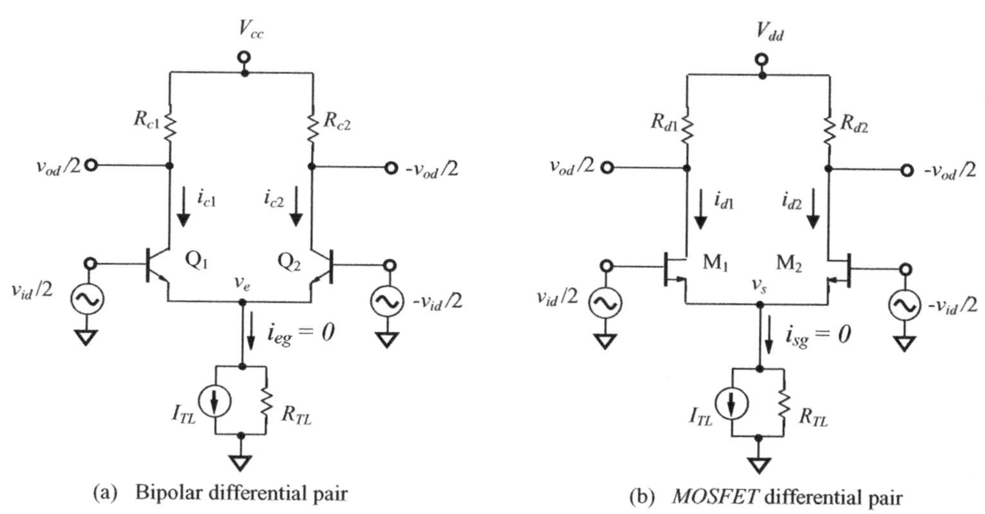

(a) Bipolar differential pair (b) *MOSFET* differential pair

Figure 3.13 Differential mode portions of differential pairs.

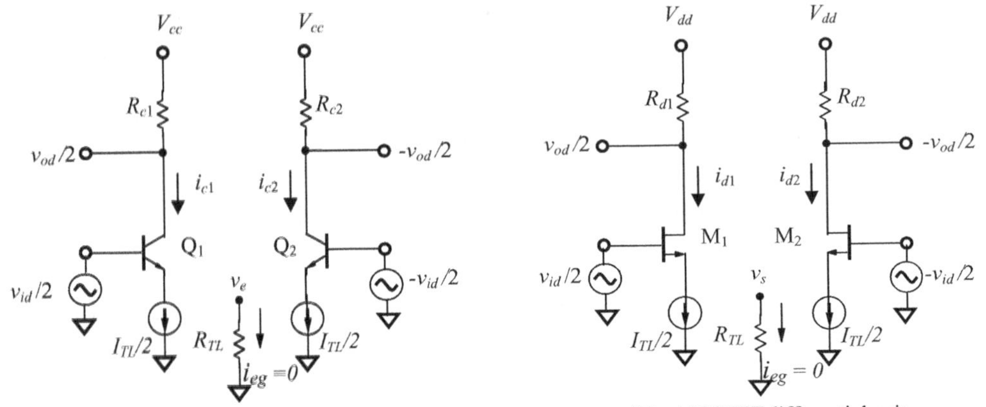

(a) Bipolar differential pair (b) *MOSFET* differential pair

Figure 3.14 Differential mode portions of differential pairs.

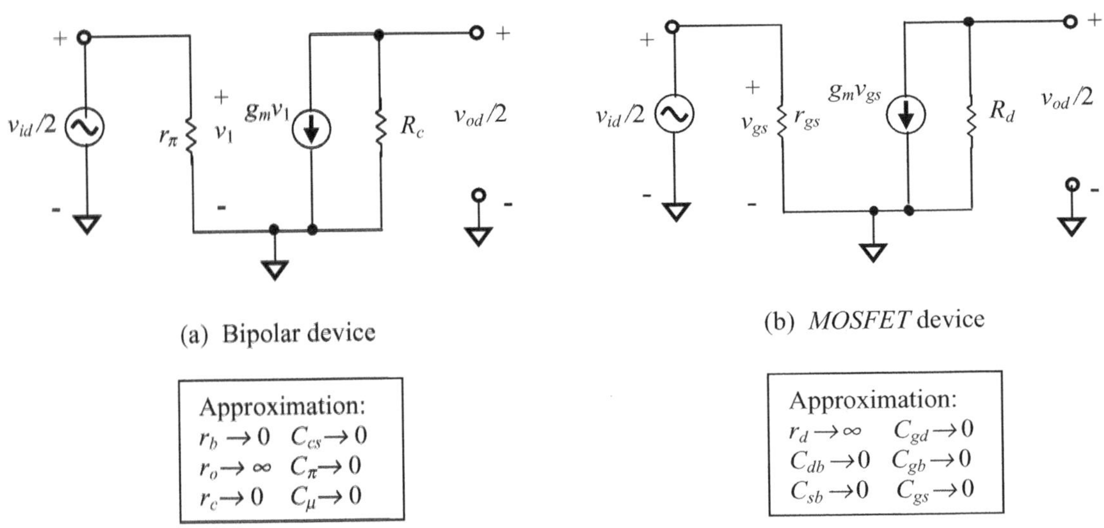

Figure 3.15 Equivalent of differential mode half circuit with low-frequency approximation.

$$\frac{v_{od}}{2} = -g_m v_1 R_c = -g_m \frac{v_{id}}{2} R_c. \tag{3.104}$$

For the *MOSFET* device, Figure 3.15(b), we have

$$\frac{v_{od}}{2} = -g_m v_{gs} R_d = -g_m \frac{v_{id}}{2} R_d, \tag{3.105}$$

where

$$R_c = R_{c1} = R_{c2}, \tag{3.106}$$

$$R_d = R_{d1} = R_{d2}. \tag{3.107}$$

Let's merge the two equations (3.104) and (3.105) into one:

$$\frac{v_{od}}{2} = -g_m \frac{v_{id}}{2} R, \tag{3.108}$$

if R_c and R_d are commonly denoted by R, then

$$A_{dm} = \frac{v_{od}}{v_{id}} = -g_m R. \tag{3.109}$$

The differential mode voltage gain A_{dm} is proportional to g_m and R. The A_{dm} can be enhanced by increasing the value of R (R_c or R_d). However, it might be detrimental to obtain a higher differential output power gain. Eventually, the effective way to enhance the A_{dm} is to increase the g_m of the devices.

Next, let's derive the expression of the voltage gain when the differential pair operates in the common mode, A_{cm}. By replacing V_{i1}, V_{i2}, V_{o1}, V_{o2} with v_{ic}, v_{ic}, v_{oc}, v_{oc}, respectively, Figure 3.2 is modified to Figure 3.16.

The left and right branches of the differential pairs in both Figure 3.16(a) and (b) are perfectly identical to each other. Since they are two identical, independent half-circuits, Figure 3.16 can be redrawn as Figure 3.17, in which the current between the emitters i_{ee} and the current between the sources i_{ss}, is zero because each half circuit is driven by the same input voltage, v_{ic}. The connection between the emitters or sources of the device can be removed so that Figure 3.17 becomes Figure 3.18.

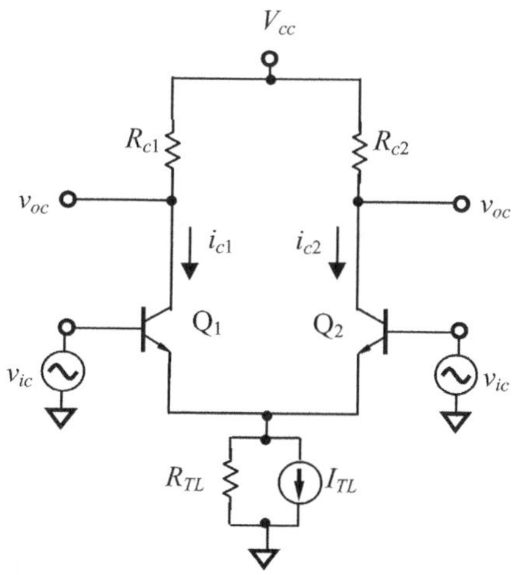

(a) Bipolar differential pair

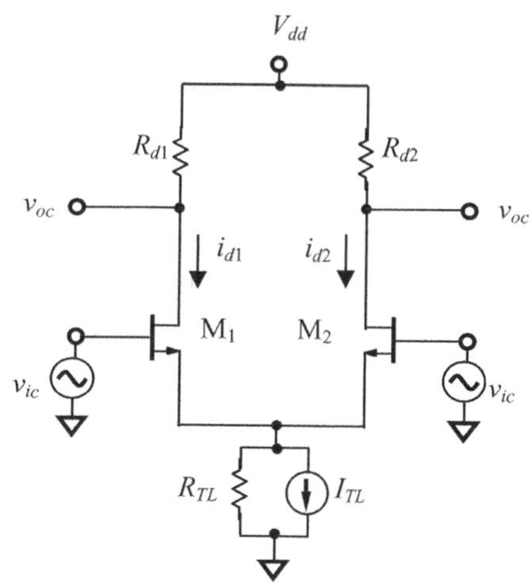

(b) *MOSFET* differential pair

Figure 3.16 Common mode portion of differential pairs.

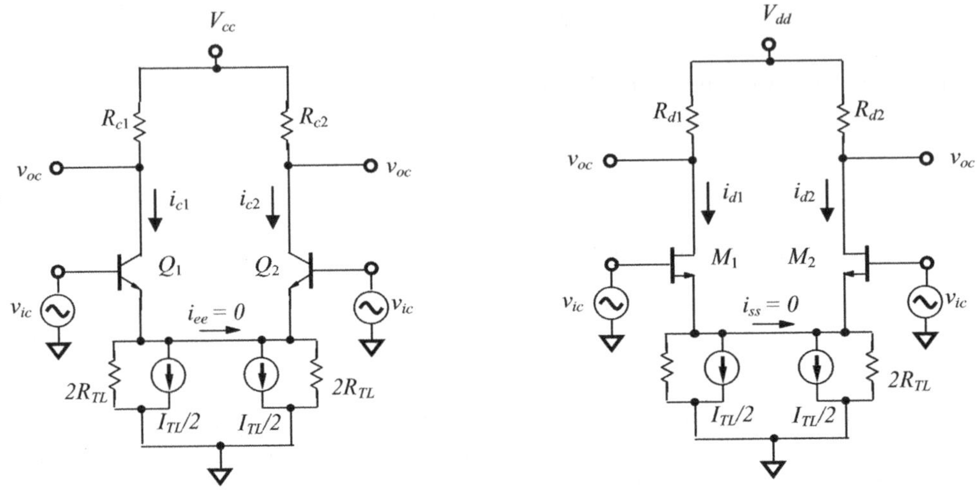

(a) Bipolar differential pair (b) *MOSFET* differential pair

Figure 3.17 Common mode portion of differential pairs connected together.

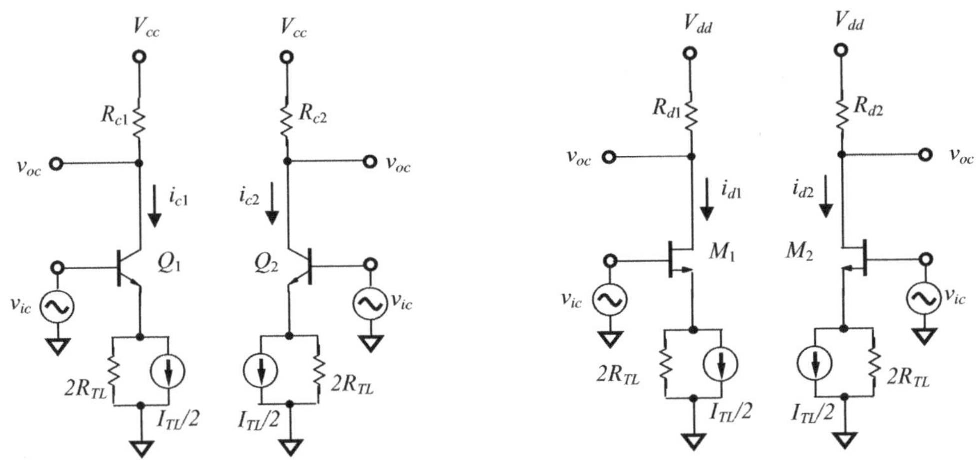

(a) Bipolar differential pair (b) *MOSFET* differential pair

Figure 3.18 Common mode portion of differential pairs disintegrated.

The equivalents of the common mode half circuit in the differential pair are sketched in Figure 3.19, which represents a low-frequency approximation. All the capacitors in the transistor's model are neglected approximately. In the bipolar device model, only r_π and R_c are reserved, and in the *MOSFET* device model, only r_{gs} and R_d are reserved.

As matter of fact, Figure 3.19 represents a common-emitter or common-source stage with an emitter or source degeneration. Its trans-conductance G_m can be calculated by means of simple *KCL* and *KVL*, although the calculation itself is some-

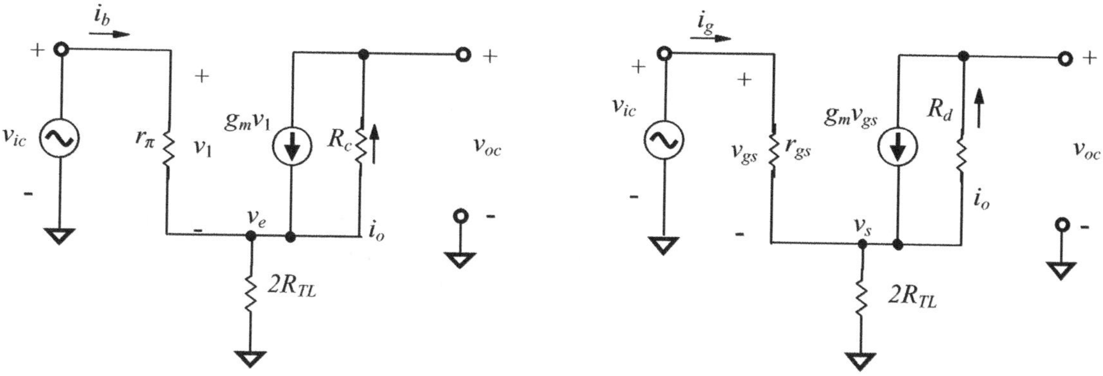

(a) Bipolar device

Approximation:
$r_b \to 0$ $C_{cs} \to 0$
$r_c \to 0$ $C_\pi \to 0$

(b) *MOSFET* device

Approximation:
$C_{gd} \to 0$ $C_{gs} \to 0$
$C_{db} \to 0$ $C_{gb} \to 0$
$C_{sb} \to 0$

Figure 3.19 Equivalent of common mode half circuit.

what tedious. For a detailed mathematical derivation, readers can refer to the book by Paul R. Gray et al., *Analysis and Design of Analog Integrated Circuits*.

For a differential pair built by bipolar devices,

$$G_m = \frac{i_o}{v_{ic}} = \frac{g_m}{1+2g_m R_{TL}(g_m r_\pi + 1)}, \qquad (3.110)$$

For a differential pair built by MOSFETs,

$$G_m = \frac{i_o}{v_{ic}} = \frac{g_m}{1+2g_m R_{TL}(g_m r_{gs} + 1)}, \qquad (3.111)$$

Again, let's merge the two equations (3.110) and (3.111) into one, obtaining

$$G_m = \frac{i_o}{v_{ic}} = \frac{g_m}{1+2g_m R_{TL}(g_m r_{\pi gs} + 1)}, \qquad (3.112)$$

where r_π and r_{gs} are commonly denoted by $r_{\pi gs}$

when

$$r_o >> R_{TL} \quad \text{or} \quad r_d >> R_{TL}. \qquad (3.113)$$

On the other hand,

$$v_{oc} = -i_o R = -G_m R v_{ic}, \qquad (3.114)$$

then

$$A_{cm} = \frac{v_{oc}}{v_{ic}} = -G_m R = -\frac{g_m R}{1 + 2g_m R_{TL}\left(1 + \frac{1}{g_m r_{\pi gs}}\right)}. \tag{3.115}$$

Common mode operation in a differential pair produces DC offset, which is very harmful to the RF signal in a direct conversion communication system. In a differential pair, the common mode voltage gain A_{cm} should be as low as possible. From equation (3.115) it can be seen that the tail resistor R_{TL} plays a key role in suppressing the common mode voltage gain.

If R_{TL} is zero, equation (3.115) becomes (3.109). Consequently, A_{cm} will be equal to A_{dm}, and $CMRR = 1$. In this case the differential pair functions as a single-ended stage in common mode rejection.

$CMRR$ is increased if the value of the tail resistor R_{TL} is higher. However, due to the limited DC power supply voltage, the value of R_{TL} must not be too high.

By equation (3.103), $CMRR$ is defined as the ratio of the differential mode voltage gain to the common mode voltage gain. From equations (3.109) and (3.115), we have

$$CMRR = \frac{A_{dm}}{A_{cm}} = 1 + 2g_m R_{TL}\left(1 + \frac{1}{g_m r_{\pi gs}}\right). \tag{3.116}$$

If the second term in the parenthesis is much smaller than 1 and can be overlooked, then equation (3.116) can be approximated as

$$CMRR = \frac{A_{dm}}{A_{cm}} = 1 + 2g_m R_{TL}. \tag{3.117}$$

This measures the rejection capacity opposing the common mode voltage and the magnifying capability to the differential mode voltage. A good differential pair must have a high A_{dm} but low A_{cm}. From equation (3.117) it can be seen that a higher g_m of the device and a higher tail resistance R_{TL} are preferred.

3.4.2 *CMRR* in a Single-ended Stage

It is meaningless to talk about $CMRR$ in a single-ended block because the "differential," or 180° phase difference, does not exist. However, as shown in Figure 3.20, if the single-ended input node and the ground are taken as an input differential pair, and the single-ended output node and the ground taken as an output differential pair, what would be the $CMRR$? In this figure, the voltage gain is obviously the same whether input is v_{id} or v_{ic}, that is,

$$A_{dm} = \frac{v_{od}}{v_{id}} = \frac{v_{oc}}{v_{ic}} = A_{cm}. \tag{3.118}$$

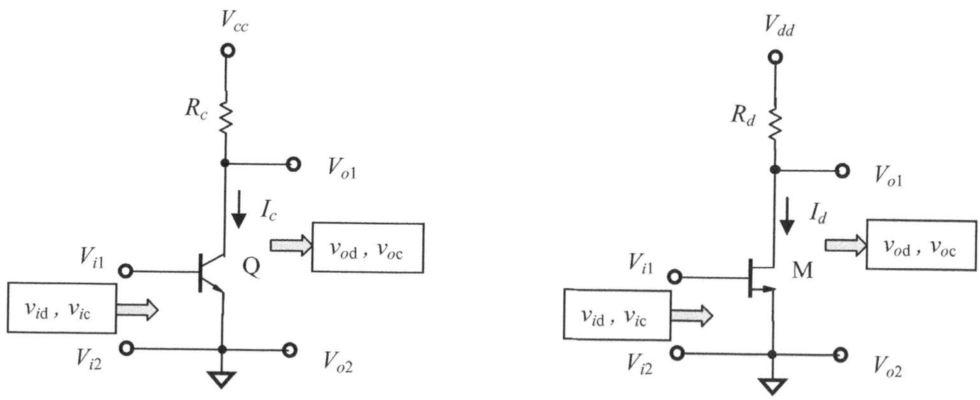

(a) Bipolar single-ended block (b) *MOSFET* single-ended block

Figure 3.20 Single-ended node and ground tentatively treated as differential pairs.

Then

$$CMRR = \frac{A_{dm}}{A_{cm}} = 1. \tag{3.119}$$

This means that there is no capability to reject common mode input in a single-ended stage. In other words, a single-ended stage cannot distinguish between differential mode input and common mode input.

3.4.3 *CMRR* in a Pseudo-differential Pair

In Figure 3.21(a) and (b), the "differential pair" is a simple combination of two identical and independent single-ended stages. If their inputs are perfect differential signals, their outputs will be differential signals.

The two single-ended stages are identical so that their voltage gains A_v are the same, that is,

$$A_v = \frac{v_{o1}}{v_{i1}} = \frac{v_{o2}}{v_{i2}}. \tag{3.120}$$

Mathematically,

$$A_v = \frac{v_{o1}}{v_{i1}} = \frac{v_{o2}}{v_{i2}} = \frac{v_{o1} - v_{o2}}{v_{i1} - v_{i2}}. \tag{3.121}$$

If

$$v_{i1} - v_{i2} = v_{id}, \tag{3.122}$$

$$v_{o1} - v_{o2} = v_{od}, \tag{3.123}$$

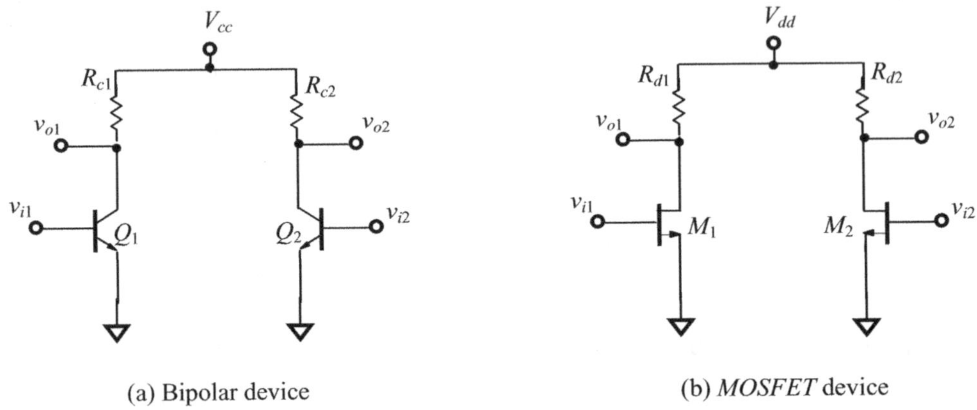

Figure 3.21 A "differential pair" is a simple combination of two identical and independent single-ended stages.

then from relation (3.121) we have

$$A_v = \frac{v_{od}}{v_{id}} = A_{dm}. \tag{3.124}$$

If

$$v_{i1} - v_{i2} = v_{ic}, \tag{3.125}$$

$$v_{o1} - v_{o2} = v_{oc}, \tag{3.126}$$

then from relation (3.121) we have

$$A_v = \frac{v_{oc}}{v_{ic}} = A_{cm}. \tag{3.127}$$

Finally, from equations (3.124) and (3.127), we have

$$CMRR = \frac{A_{dm}}{A_{cm}} = 1. \tag{3.128}$$

Just like the single-ended stage, the value of *CMRR* for a "differential pair" constructed by two identical and individual single-ended stages is the same, $CMRR = 1$ or $0\,dB$. The "differential pair" as shown in Figure 3.21 has no capability of common mode rejection, which is why it is called a pseudo-differential pair. In both the single-ended stage and pseudo-differential pair, the common point is that there is no common coupling part, the resistor R_{TL}, which combines the two individual single-ended blocks together. This is the reason why $CMRR = 1$ or $0\,dB$ in equations (3.119) and (3.128).

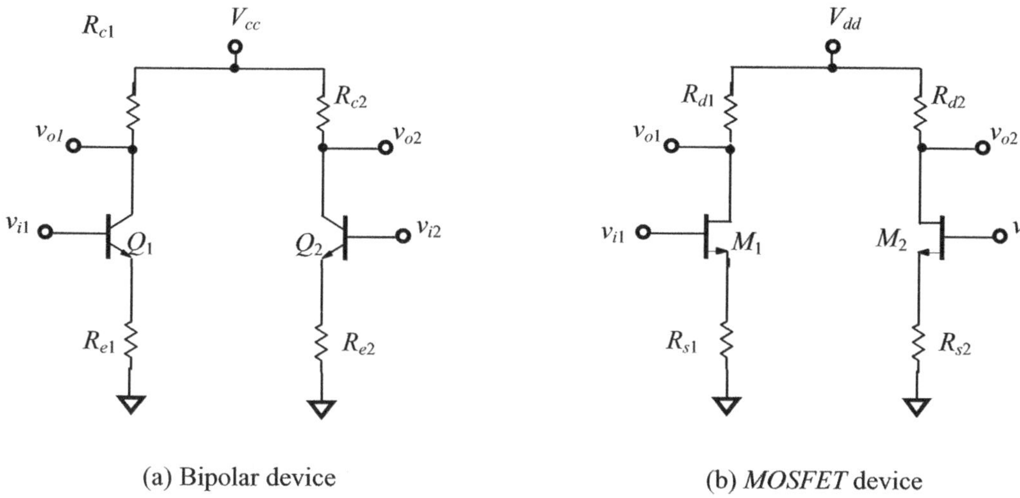

Figure 3.22 Another type of pseudo-differential pair.

Another type of pseudo-differential pair is shown in Figure 3.22. There are resistors, R_{e1} and R_{e2}, connected from the emitter of the bipolar transistor to the ground in Figure 3.22(a) and R_{s1} and R_{s2}, connected from the source of the *MOSFET* transistor to the ground in Figure 3.22(b). However, these resistors are not tail resistors. The tail resistor R_{TL} is the resistor by which the differential branches are coupled together. Without the tail resistor R_{TL}, the differential pair does not have the capability for common mode rejection, so that that $CMRR = 1$ or $0\,dB$.

3.4.4 Enhancement of *CMRR*

It may not be an exaggeration to say that the *CMRR* is the paramount goal in the design of a differential pair. Most of the effort in the design is placed on how to enhance the value of the *CMRR*.

As mentioned above, the existence of the tail resistance R_{TL} determines whether or not the differential pair has the capability of common mode rejection. The *CMRR* is proportional to the value of tail resistance R_{TL}; consequently, a direct way to enhance *CMRR* is to increase the value of R_{TL}. However, if the value of R_{TL} is too high, the *DC* voltage drop across the R_{TL} would be too great, and the *DC* voltage drop across the transistor would be too small. This would be harmful to the linearity of the transistor.

On the basis of equation (3.116) or (3.117), *CMRR* is proportional to the value of the trans-conductance of transistor, g_m. Therefore, another way to enhance the *CMRR* is to select the transistor and adjust its operating state so that the value of g_m is increased as much as possible.

CMRR can be improved through modifying topology. For example, we can insert a degeneration resistor R_{dg} between each emitter or source and R_{TL} as shown in Figure 3.23. It can be verified that equations (3.109), (3.115), and (3.116) are then modified to

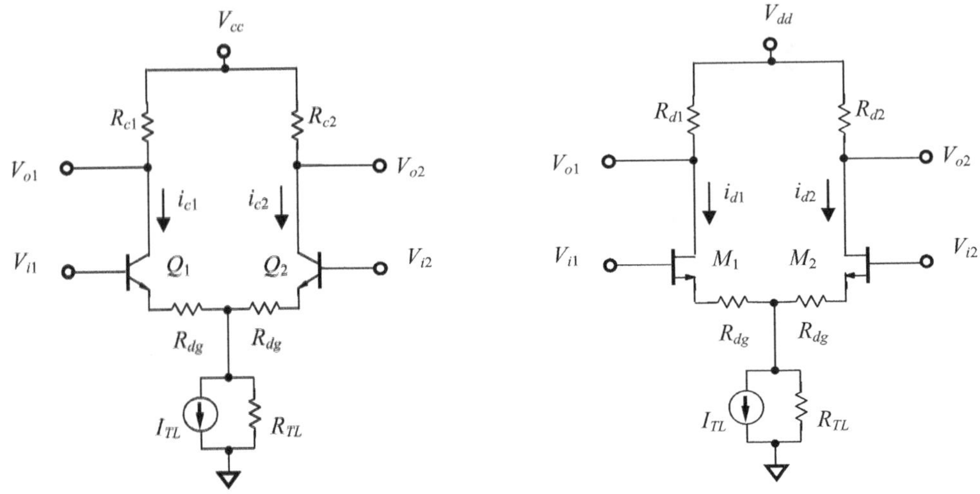

(a) Bipolar differential pair (b) *MOSFET* differential pair

Figure 3.23 Modified differential pairs by insertion of R_{dg}.

$$A_{dm} = \frac{-g_m R}{1 + g_m R_{dg}\left(1 + \frac{1}{g_m R_{\pi gs}}\right)}, \tag{3.129}$$

$$A_{cm} \approx \frac{-g_m R}{1 + 2g_m R_{TL}\left(1 + \frac{1}{g_m R_{\pi gs}}\right)\left(1 + \frac{R_{dg}}{2R_{TL}}\right)}, \tag{3.130}$$

$$CMRR \approx \frac{A_{dm}}{A_{cm}} = \frac{1 + 2g_m R_{TL}\left(1 + \frac{1}{g_m R_{\pi gs}}\right)\left(1 + \frac{R_{dg}}{2R_{TL}}\right)}{1 + g_m R_{dg}\left(1 + \frac{1}{g_m R_{\pi gs}}\right)}. \tag{3.131}$$

It should be noted that in all the expressions of *CMRR*, the low-frequency approach has been applied so that all the capacitors in the transistors' model are neglected approximately. In the actual engineering design, this approach must be abandoned if the operating frequency is high and the capacitors in the transistors' model cannot be neglected. However, all the expressions shown above still provide good references in *RF* circuit design.

APPENDICES

3.A.1 *DC* Offset Cancellation by Calibration

Theoretically, zero *DC* offset does not exist in an ideal differential pair; however, it definitely exists in an actual differential pair. In reality, an ideal differential pair does

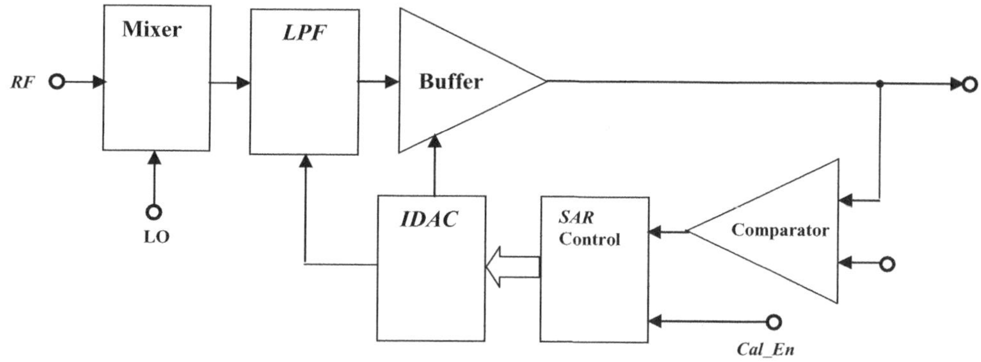

Figure 3.A.1 Simplified block diagram of *DC* offset calibration in Intersil's baseband *IC*, *HFA*3783.

not exist. In engineering design, asymmetry between the two branches of the differential pair (such as non-identical parts, inhomogeneous expansion or constriction of the substrate or *PCB*, and asymmetrical layout) is inevitable. Consequently, *DC* offset always exists in any circuit containing active devices, despite the differential configuration.

In current communication systems implementing direct conversion or "zero *IF*" modulation technology, *DC* offset cancellation is one of most important tasks to be undertaken, in addition to adapting the differential configuration in the circuitry. As long as the *DC* offset cancellation is well implemented, the *DC* offset will be kept below a threshold level and the system will operate in a normal state.

There are various ways to implement *DC* offset cancellation. Figure 3.A.1 shows *DC* offset cancellation executed in terms of *DC* offset calibration, which has been developed by Intersil in their baseband *IC*, *HFA*3783, for *WLAN* products.

To keep the *DC* offset constant, all the *DC* offsets produced by the mixer, *LPF* (Low-Pass Filter), and buffer are calibrated at an appropriate time during receiving, transmitting, or standby modes.

The comparator in Figure 3.A.1 senses the *DC* offset from the buffer output. Then, the output of the buffer is fed to a decision circuit called the *SAR* (Successive Approximation Register) state control. The eight digital outputs of the *SAR* control the *IDAC* (Current Output of Digital to Analog Conversion) which provides controls to the *LPF* and buffer so as to adjust and bring the *DC* offset down to a minimum. The *DC* offset calibration is initialized by the *Cal_En* (Calibration Enable) input either automatically or manually.

3.A.2 "Chopping" Mixer

In receiver, *DC* offset is mainly produced in the mixer block. A so-called "chopping mixer" is therefore developed for the *DC* offset cancellation.

Chopping technology was developed several decades ago when the super-regeneration radio was developed. By means of chopping the input and output signals of the mixer at the *RF* input port and *IF* output port respectively, the even orders of non-linearity of the mixer or the *DC* offset can be significantly reduced.

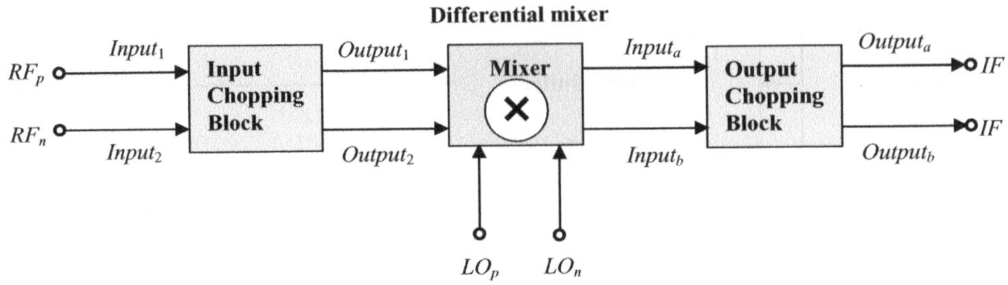

Figure 3.A.2 Block diagram of a "chopping" mixer.

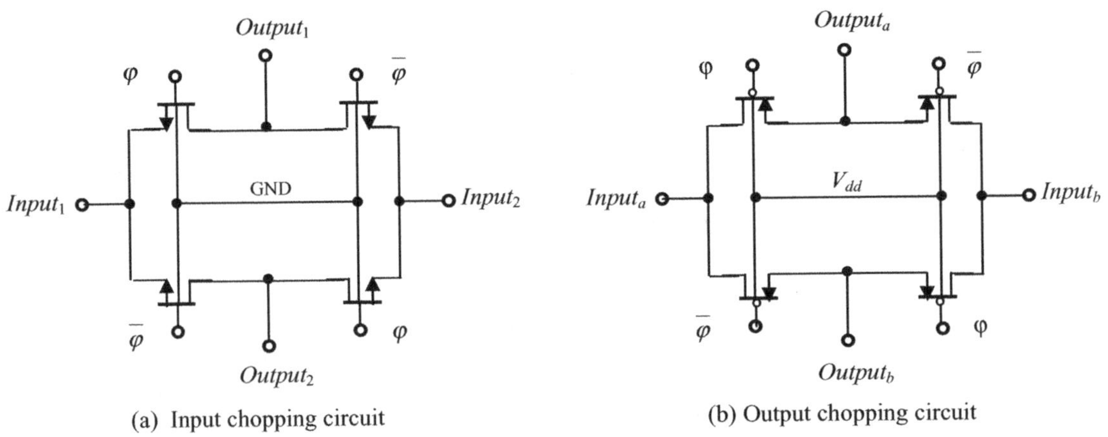

Figure 3.A.3 Input and output chopping circuits of a "chopping" mixer.

Figure 3.A.2 shows the block diagram of a chopping mixer. It has three parts: an input chopping block, a differential mixer core, and an output chopping block. The core of the differential mixer is a mixer with a Gilbert cell.

The input chopping block is inserted between the *RF* input port and the input of the differential mixer. It chops the *RF* signal with a repetition frequency of f_{CH}. The output chopping block is inserted between the *IF* output of the differential mixer and the *IF* output ports. It chops the *IF* signal with the same repetition frequency f_{CH} as that of the input chopping block.

Figure 3.A.3(a) and (b) depicts the input and output chopping blocks, respectively. Both of them are simply cross-toggled switches. The input chopping block consists of four *n*-channel *MOSFET*s and its substrate is grounded, while the output chopping block consists of four *p*-channel *MOSFET*s and its substrate is connected to the *DC* power supply V_{dd}. These differences are due to the different locations of the chopping circuits. The input chopping circuit is working in the low *DC* voltage portion, while the output chopping circuit is working in the high *DC* voltage portion.

Two differential inputs are alternatively connected to two differential outputs through two gate controls φ and $\overline{\varphi}$, which are fed with pulse signals at an appropriate chopping frequency f_{CH} and of the voltage amplitude with 0 and V_{dd} in an iterative manner. The phase of φ and $\overline{\varphi}$ is kept at a 180° difference: φ is at 0 when $\overline{\varphi}$ is

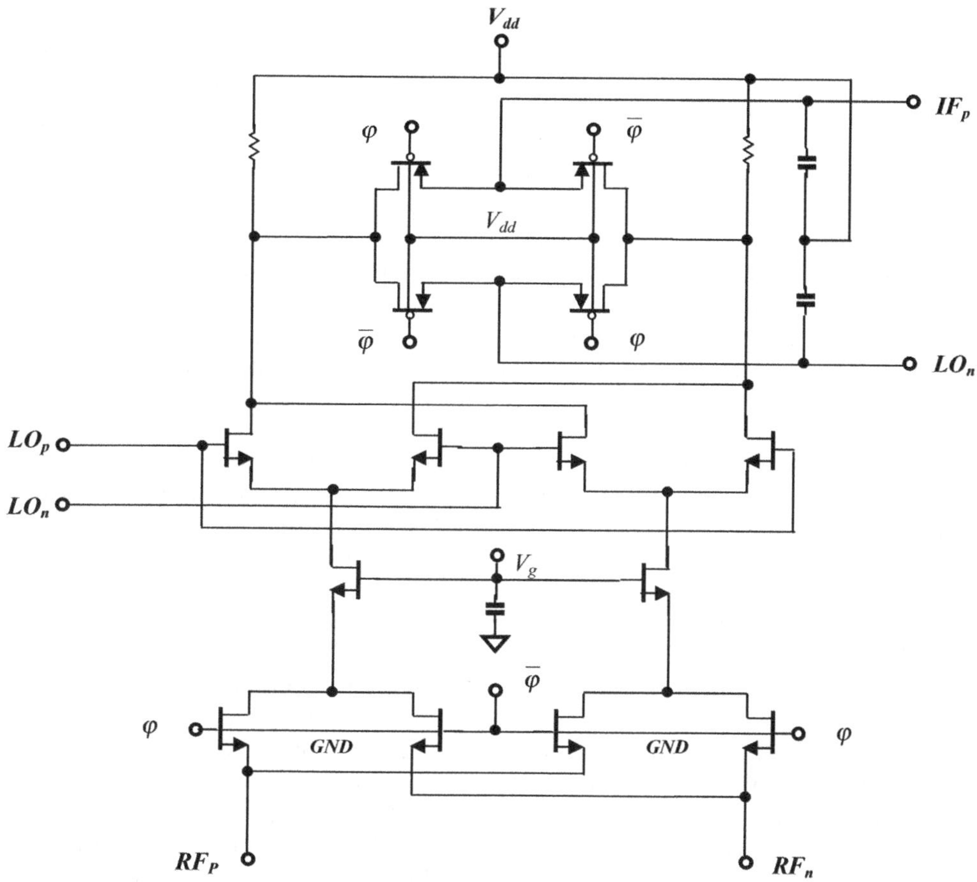

Figure 3.A.4 Topology of a "chopping" mixer.

at V_{dd}, and vice versa. The *RF* signal is therefore turned *ON/OFF* per the chopping frequency.

The chopping frequency is experimentally determined by simulation with benchwork. The criterion for determining this frequency is the maximization of IP_2 or minimization of *DC* offset for the entire chopping mixer. The IP_2 is a good measure of *DC* offset because it is dominant among the even order non-linearities which bring about *DC* offset. Like the differential mixer itself, the layouts of the input and output chopping blocks must be kept symmetrical as much as possible.

Figure 3.A.4 shows the schematic of the entire chopping mixer, including the input and output chopping circuits. The Gilbert cell and the input and output chopping blocks are built by *MOSFETs*. The input *RF* signals are fed to the inputs of the input chopping circuit; then the chopped *RF* signals from the outputs of the input chopping circuit are fed to the sources of the lower *MOSFET* pair. They are therefore called the *CG* configuration of the *RF* input. The special property of the *CG* configuration is that the *RF* input has a relatively wider bandwidth but lower conversion gain than other configurations, such as the *CS* or *CD*.

One possible way to connect the two *RF* input terminals, RF_p and RF_n, is to use an *RF* transformer. The *RF* input terminals, RF_p and RF_n, as shown in Figure 3.A.4, are connected to the second winding of the transformer with its center taped to the ground while the primary winding of the transformer can be connected as either a differential or single-ended *RF* input.

It is desirable to understand why the chopping mixer is able to cancel the *DC* offset. Figure 3.A.5 explains its operating principle, in which the variations of the frequency spectrums of various signals are depicted step by step. The frequency spectrum of a signal in Figure 3.A.5 is drawn within a rectangular frame.

Figure 3.A.5(a) and (b) shows the frequency spectrums of *RF* signal and *LO* injection at the input, respectively. The spectrums of the *RF* signal and *LO* injection are located in the same frequency range because they are designed for a direct conversion or "zero *IF*" system. Their central frequencies are marked with f_{RF} or f_{LO}.

The input chopping block chops the incoming *RF* signal *ON* and *OFF* with a repetition frequency of f_{CH}. Figure 3.A.5(c) shows the frequency spectrum after the input chopping block. The spectrum of the *RF* signal is split into two portions by the chopping: one is the spectrum shifted from the original central location f_{RF} to $f_{RF} - f_{CH}$; the other is the spectrum shifted from the original central location f_{RF} to $f_{RF} + f_{CH}$. The *LO* spectrum in Figure 3.A.5(c) is kept unchanged as shown in Figure 3.A.5(b).

Figure 3.A.5(d) shows the frequency spectrum after the Gilbert cell. The Gilbert cell mixes the *RF* signal and *LO* injection as shown in Figure 3.A.5(c) and produces two *IF* signals with spectrums located at $-f_{CH}$ and f_{CH}, respectively, as shown in Figure 3.A.5(d). Owing to imperfect isolation, a few percent of the power of the *RF* signal and *LO* injection is leaked from *RF* and *LO* input ports to the *IF* output ports; their spectrums can be seen in Figure 3.A.5(d). Meanwhile, *DC* offset is produced due to imperfect differential structure. This is marked with an upward arrow in Figure 3.A.5(d).

The output chopping block chops the *IF* signals *ON* and *OFF* with the same repetition frequency f_{CH} as in the input chopping block. Figure 3.A.5(e) shows the spectrum of the *IF* signals after the output chopping block. The spectrum of *IF* signal around $f = 0$ in Figure 3.A.5(d) is split into two spectrums: One shifts $-f_{CH}$ and the other shifts f_{CH} from its original location. The resultant spectrum is two split spectrums overlapping together as shown in Figure 3.A.5(e).

In the output impedance matching network (not shown in the circuit block diagram), the impedance is matched in the range of the "zero *IF*," so that the two *IF* signals with their central frequencies located at $-2f_{CH}$ and $+2f_{CH}$, (drawn with dashed rectangular frames in Figure 3.A.2(e)), and the spectrum around f_{LO} *and* f_{RF} are greatly reduced or removed.

3.A.3 Remarks on *DC* Offset Measurement

It is difficult to measure *DC* offset directly in the test laboratory because its absolute value is too low. *DC* offset can be indirectly measured by IP_2. There are two reasons: First, the *DC* offset is contributed by the even orders of the non-linearity of a device or block or by the even orders of the interference inside and outside the block. Second, the order of non-linearity or interference is, of course, the most effective

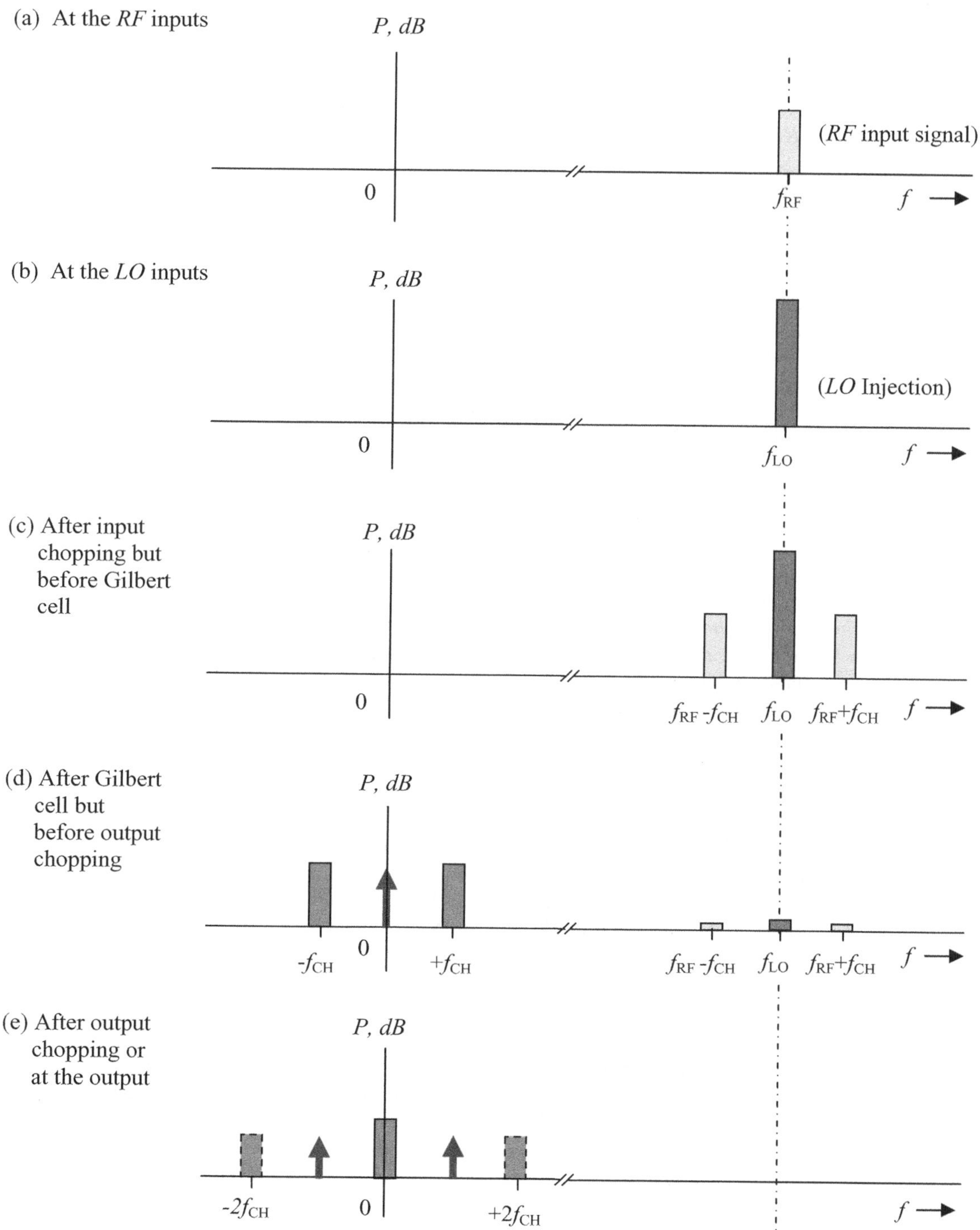

Figure 3.A.5 Principle of *DC* offset reduction in the "chopping" mixer.

contributor. It is therefore reasonable to take IP_2 as a measure of the DC offset of a device, a block, or a system. Further studies might be conducted on the equivalence between the DC offset and IP_2.

Empirically, in a direct conversion or "zero IF" communication system, an RF block or an RF system would not have any problems due to DC offset if its IIP_2 is greater than $80\,dB_m$.

REFERENCES

[1] A. Shoval, D. A. Johns, and W. M. Snelgrove, "DC Offset Performance of Four LMS Adaptive Algorithms," *IEEE International Symposium on Circuits and Systems*, 1994. *ISCAS*, Vol. 2, May 30–June 2, 1994, pp. 409–412.

[2] A. Shoval, D. A. Johns, and W. M. Snelgrove, "Comparison of DC Offset Effects in Four LMS Adaptive Algorithms," *IEEE Transactions on Circuits and Systems II: Analog and Digital Signal Processing* [see also *IEEE Transactions on Circuits and Systems II: Express Briefs*] Vol. 42, No. 3, March 1995, pp. 176–185.

[3] H. Yoshida, H. Tsurumi, and Y. Suzuki, "DC Offset Canceller in a Direct Conversion Receiver for QPSK Signal Reception," *IEEE International Symposium on Personal, Indoor and Mobile Radio Communications*, Vol. 3, Sept. 8–11, 1998, pp. 1314–1318.

[4] B. Matinpour, S. Chakraborty, and J. Laskar, "Novel DC-Offset Cancellation Techniques for Even-Harmonic Direct Conversion Receivers," *IEEE Transactions on Microwave Theory and Techniques*, Vol. 48, No. 12, Dec. 2000, pp. 2554–2559.

[5] B. Matinpour, S. Chakraborty, M. Hamai, C. Chun, and J. Laskar, "A Novel DC-Offset Cancellation Technique for Even-Harmonic Direct Conversion Receivers," *Microwave Symposium Digest*, IEEE MTT-S International, Vol. 2, June 11–16, 2000, pp. 631–634.

[6] Edwin E. Bautista, Babak Bastini, and Joseph Heck, "A High IIP2 Down-conversion Mixer Using Dynamic Matching," *IEEE Journal of Solid-state Circuits*, Vol. 35, No. 12, December 2000, pp. 1934–1941.

[7] Paul R. Gray, Paul J. Hurst, Stephen H. Lewis, and Robert G. Meyer, *Analysis and Design of Analog Integrated Circuits*, 4th ed., John Wiley & Sons, Inc., 2001.

[8] Po-Chiun Huang, Yi-Huei Chen, and Chorng-Kuang Wang, "A 2-V CMOS 455-kHz FM/FSK Demodulator Using Feedforward Offset Cancellation Limiting Amplifier," *IEEE Journal of Solid-state Circuits*, Vol. 36, No. 1, January 2001, pp. 135–138.

[9] M. Faulkner, "IM2 Removal in Direct Conversion Receivers," *IEEE VTS 53rd Vehicular Technology Conference*, Vol. 3, May 6–9, 2001, pp. 1897–1901.

[10] M. Faulkner, "DC Offset and IM2 Removal in Direct Conversion Receivers," *Communications, IEE Proceedings*, Vol. 149, No. 3, June 2002, pp. 179–184.

[11] O. V. Popov, "Dynamic DC Offset Impact on the 802.11a Receiver Performance," *Proceedings of ICCSC '02*. IEEE International Conference on Circuits and Systems for Communications, June 26–28, 2002, pp. 250–253.

[12] Huang Xinping and M. Caron, "Gain/phase Imbalance and DC Offset Compensation in Quadrature Modulators," *IEEE International Symposium on Circuits and Systems*, Vol. 4, May 26–29, 2002, pp. IV811–IV814.

[13] S. Shang, S. Mirabbasi, and R. Saleh, "A Technique for DC-Offset Removal and Carrier Phase Error Compensation in Integrated Wireless Receivers," *Proceedings of the 2003 International Symposium on Circuits and Systems*, Vol. 1, May 25–28, 2003, pp. I-173–I-176.

CHAPTER 4

RF BALUN

4.1 INTRODUCTION

As shown in Figure 4.1, a balun is a transformation block between a single-ended stage and a differential pair, either from the single-ended block to the differential pair as shown in Figure 4.1(a) or vice versa as shown in Figure 4.1(b). A balun transforming a signal from a single-ended to a differential pair splits a single-ended *RF* signal into a pair of differential signals with the same magnitude but 180° phase difference. A differential pair of signals has some special functions in the transportation and manipulation of signals. For instance, the capability of common mode rejection potentially exists in a differential pair but not in a single-ended signal. A balun transforming a signal from a differential pair to a single-ended signal combines a pair of differential signals into a single one, which simplifies both the simulation and testing of a block with a differential configuration when the special purpose of a differential pair signal is no longer needed.

The word "balun" is a portmanteau term formed from the words "balanced" and "unbalanced," in which "balanced" implies a differential configuration and "unbalanced" represents a single-ended configuration. It might seem reasonable that we rename the balun in Figure 4.1(b) as an "unbal." Sometimes, people refer to the balun type in Figure 4.1(a) as a "splitter" and the balun type in Figure 4.1(b) as a "combiner." The "splitter" is a reversed "combiner," and vice versa. From now on, we will only focus on the "splitter" unless otherwise noted.

The balun is usually implemented by passive parts and, if so, can be called a passive balun. However, active baluns are also available, in which the block contains both active and passive parts. In this chapter, our discussion is confined to passive baluns.

RF Circuit Design, by Richard Chi-Hsi Li

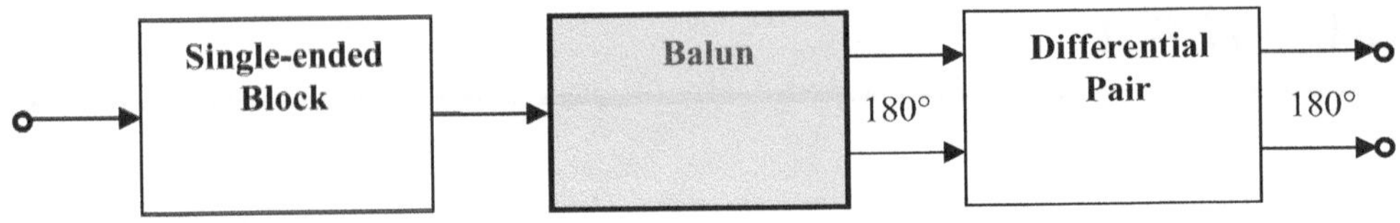

(a) From single-ended block to differential pair

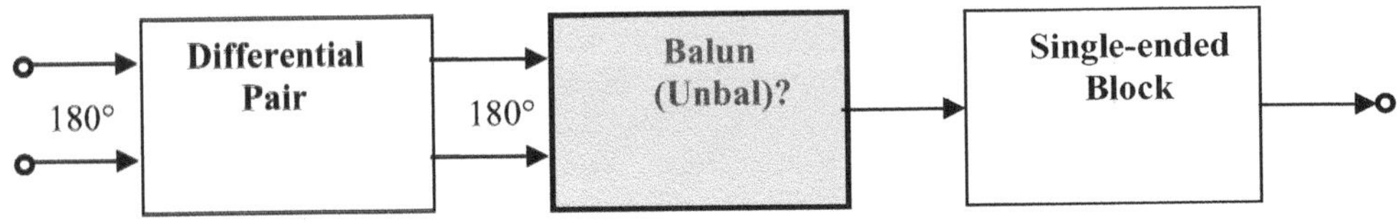

(b) From differential pair to single-ended block

Figure 4.1 The balun is a block between a single-ended block and a differential pair.

Figure 4.2 represents the characteristics of an ideal balun and can be outlined as follows:

1) There are three terminals: one single-ended and two differential. Typically, the impedance, looking into the terminal, is $50\,\Omega$ at all three terminals. However, other values are also possible, in which the impedances of the single-ended terminal and the two differential terminals are not the same. However, the two differential terminals' impedances must be kept equal.
2) At the two differential terminals, the magnitudes of the signal are equal but their phases are kept at a $180°$ difference. The absolute value of the phase at the three terminals depends on the type of balun and the reference ground point, *GND*, in the circuit block. Two phase ordinates are provided in Figure 4.2(b) for different phase scales.
3) Total insertion loss from the single-ended terminal to the differential terminals is zero. This implies that the power of the signal at each differential terminal is $3\,dB$ lower than that of the single-ended terminal because the signal power at the single-ended terminal is the sum of the signal powers at the two differential terminals. In reality, in a splitter the signal power at each differential terminal is more than $3\,dB$ lower than the signal power at the single-ended terminal because the total insertion loss is never zero in a practical balun.

A variety of *RF* baluns have been developed in past decades. In this chapter, three types of baluns are emphasized: the transformer, *LC*, and micro strip line.

In the simulation stage of *RF* circuit design, the ideal transformer balun is a good candidate for a transformation block between single-ended and differential pair blocks. The greatest advantage of the transformer balun is its frequency response with an almost infinite bandwidth. In addition, the insertion loss of the balun can be set to zero since it is an ideal balun.

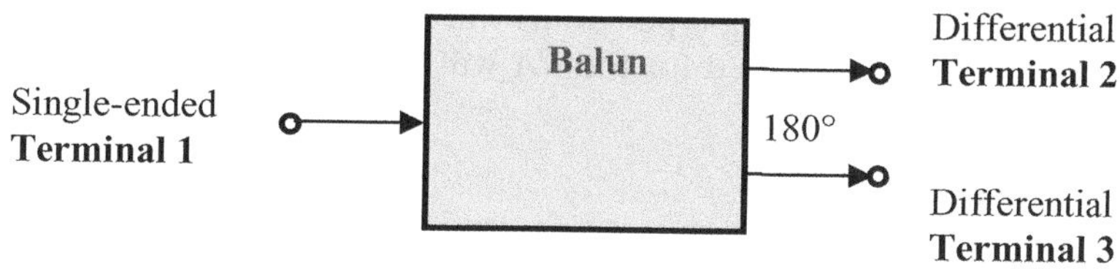

(a) Terminals of a balun

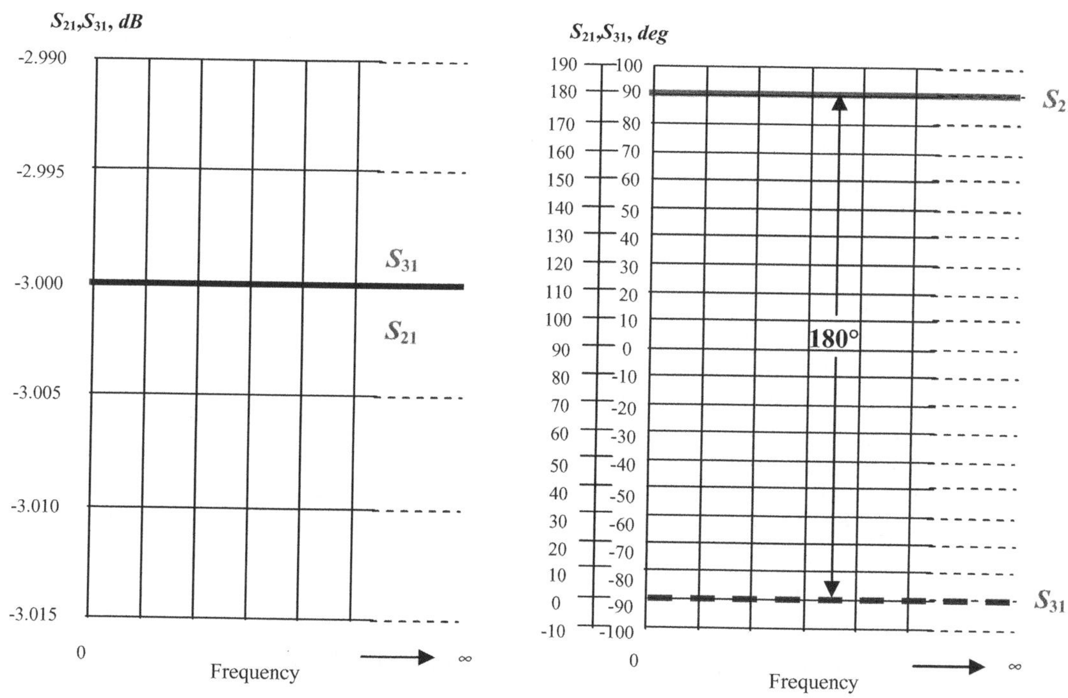

(b) Insertion loss and phase shift of an ideal balun

Figure 4.2 Characteristics of an ideal balun.

In the testing stage, the *LC* and micro strip line baluns are recommended. The greatest advantage of an *LC* balun is its simplicity: it can be implemented in a laboratory quickly and easily. Surprisingly, instead of a narrow-band response as might be expected, the *LC* balun behaves with a reasonably wide-band frequency response; this behavior will be explained in the corresponding sections.

Compared with the *LC* balun, the frequency response in a micro strip line balun is wider. In addition, its outstanding advantage is low cost, since its basic parts are micro strip lines.

As a matter of fact, these three types of baluns are the most popular in *RF* circuit design, while other types of baluns can also be found in practical applications. In the last section of this chapter, mixing type baluns will be introduced, while another two types of baluns specially designed for the *PA* will be presented in Chapter 7.

4.2 TRANSFORMER BALUNS

The distinguishing features of the transformer balun are

1) An ideal transformer balun has an infinitely wide frequency response and zero insertion loss.

   In reality, the ideal transformer balun is never realized; however, with respect to other types of baluns, a transformer balun has a wide frequency range and a low insertion loss if the operating frequency is lower than about 1 *GHz*.

2) The impedance matching can be adjusted by its turn ratio.

   If the impedances of the input source and output load are purely resistive without reactance portions, a transformer balun can function as an impedance matching network with a simple adjustment of its turn ratio. The task of impedance matching can therefore be avoided by the use of a transformer balun. This is a considerable benefit because in *RF* circuit design, the task of impedance matching is usually a "must," and can sometimes be complicated.

   If the impedance of input source and output load are not purely resistive, applying a transformer balun still eases the task of impedance matching. By adjusting its turn ratio, the real portion of the original impedance, at least, can be easily matched. The reactances of the impedances in either the single-ended or differential sides can be interpreted from each other. This makes the impedance matching work much easier. The interpretation of reactance between the single-ended and differential pair will be discussed in subsequent sections of this chapter.

4.2.1 Transformer Baluns in *RF* Circuit Design with Discrete Parts

In past years, the ferrite transformer balun has been widely applied in communication systems such as *VHF* and *UHF* radios, and even in radios with operating frequency ranges of around 800 to 900 *MHz*. Its cost is acceptable and its size has been reduced to the order of a couple of millimeters.

Unfortunately, the ferrite transformer balun is restricted by its upper frequency limit. If the operating frequency becomes higher than about 1 *GHz*, it simply goes out of order.

Instead, engineers have developed other types of baluns for application in higher *RF* frequency ranges above 1 *GHz* of communication and other systems.

4.2.2 Transformer Baluns in *RFIC* Circuit Design

In *RFIC* circuit design, one must seriously consider which kind of transformer balun is to be applied in the circuit implementation. The key point is to focus on whether an off-chip or on-chip transformer balun is chosen.

There are many types of off-chip and on-chip transformer baluns. The ferrite transformer is popular as an off-chip transformer, while the spiral-coil transformer is popular as an on-chip transformer.

The merit of a ferrite transformer as an off-chip discrete part is the higher Q value of the coil. However, its demerits are obvious: higher cost, larger size, and the upper limit of operating frequency, about 1 GHz. On the other hand, the merits of the spiral transformer on the *RFIC* chip are also obvious: lower cost, smaller size, and a much higher upper limit of operating frequency. Its fatal demerit is its extremely low Q value.

Among all the merits and demerits, cost is inevitably the first priority to be considered in a circuit design. Consequently, from the viewpoint of cost, it is preferable to build the transformer balun on the *RFIC* chip, despite its low Q value.

From the viewpoint of *IC* packaging, the off-chip transformer balun is likewise unwelcome, as one off-chip transformer balun requires at least three additional pins on the *IC* package. Neither is it welcome from the viewpoint of performance, because the bonding wires and pads for the balun inputs and outputs bring about uncertainty of performance and additional attenuation. These two viewpoints also support building the transformer on the *RFIC* chip despite its low Q value.

In summary, this is the decision that must be made:

- The on-chip transfornmer balun should be adapted in an *RFIC* circuit design if its low Q value is acceptable.
- Otherwise, the off-chip transformer balun should be adapted in an *RFIC* circuit design if a high Q value is required.

4.2.3 An Ideal Transformer Balun for Simulation

An ideal transformer balun for simulation is suggested as shown in Figure 4.3. The basic configuration of the suggested transformer balun is to combine two ideal, identical transformers stacked together as the core of the resultant transformer balun. At the single-ended portion, two primary windings of the two ideal transformers are connected in series; at the connected point, there is a resistor with a value of 1 $G\Omega$, which is meaningless in practical peformance but is merely a dummy part to satisfy computer requirements for a *DC* grounding path in the simulation. At the differential portion, two secondary windings of the two ideal transformers are also connected in series and their connected point is the central grounded point of the balun.

It should be specially noted that the turn ratio of each transformer is assigned as

$$Turn\_ratio = 1:n = 1:\sqrt{2}, \tag{4.1}$$

where n = Turn ratio of secondary winding of the transformer in respect to the primary winding.

It should also be noted that the turn ratio is kept unchanged before and after the two transformers are stacked together to form a transformer balun. In the resultant transformer balun, the total impedance of the differential pair is twice that of the single-ended side, that is,

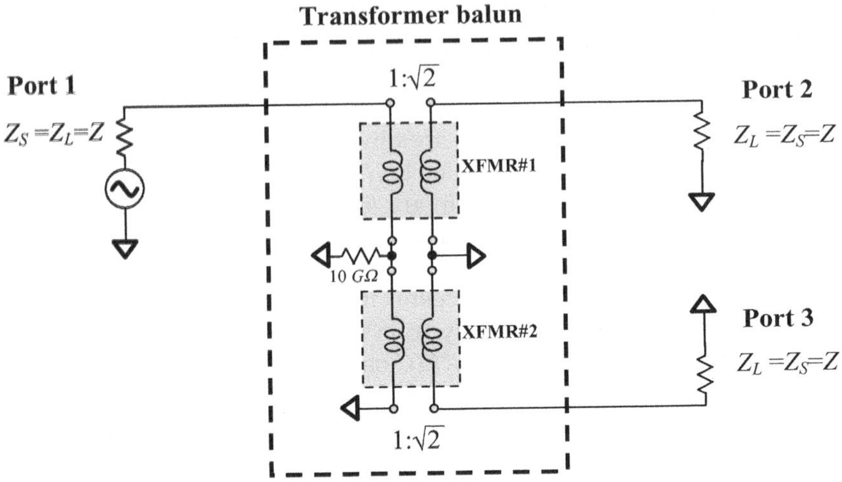

Figure 4.3 An transformer balun built by two ideal and identical transformers with turn ratio = $1:\sqrt{2}$ and $Z_S = Z_L = Z$.

$$Z_{L,T} = 2Z_L, \tag{4.2}$$

where

$Z_{L,T}$ = total load impedance from port 2 to port 3 in Figure 4.3,
$Z_{L,T}$ = individual load impedance at port 2 or port 3 in Figure 4.3.

It is well-known that the relationship between the turn ratio and the impedances of source and load in the resultant transformer balun is

$$n = \sqrt{\frac{Z_{L,T}}{Z_S}} = \sqrt{\frac{2Z_L}{Z_S}} = \sqrt{2}, \tag{4.3}$$

then, we have

$$Z_L = Z_S = Z. \tag{4.4}$$

We remove the subscript "*L*" or "*S*" since $Z_L = Z_S$. This is why the impedance at each port in Figure 4.3 is symbolized with "$Z_L = Z_S = Z$".

Now let's explore the features of this ideal transformer balun in terms of the following two simulations.

Simulation A: Simulation of the ideal transformer balun operating in the frequency range of group 1, *UWB* System: f = 3696–3960 to 4224 *MHz*, $Z_S = Z_L = 50\,\Omega$.

Figures 4.4 and 4.5 display the simulation results. Figure 4.4 shows the insertion loss and phase shift; Figure 4.5 shows the return loss of the ideal transformer balun

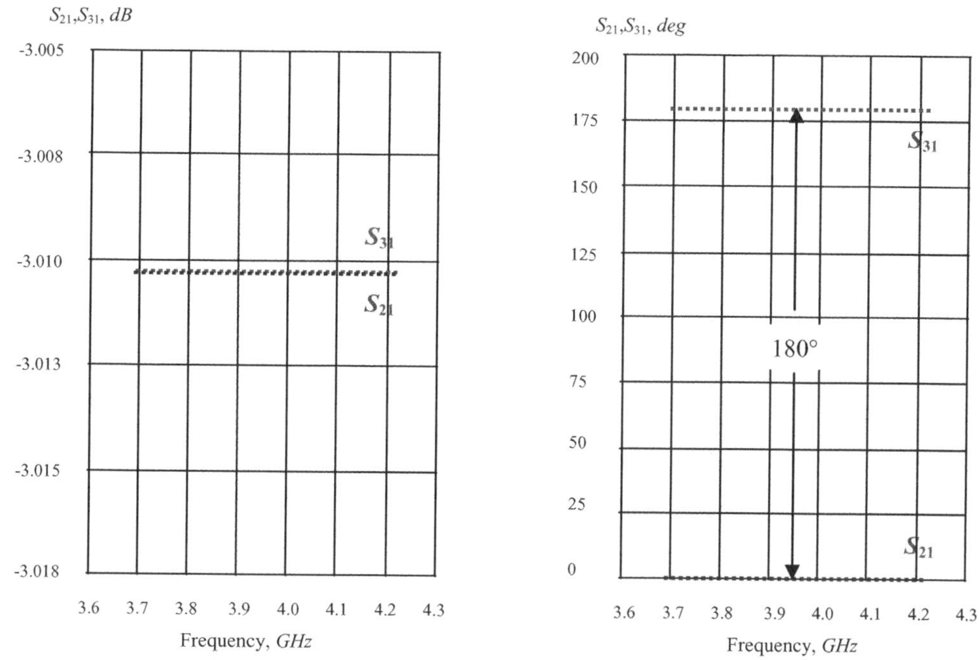

Figure 4.4 Insertion loss and phase shift of the ideal transformer balun operating in the frequency range of Group 1, *UWB* system: $f = 3696 - 3960 - 4224\,MHz$, $Z_S = Z_L = 50\,\Omega$.

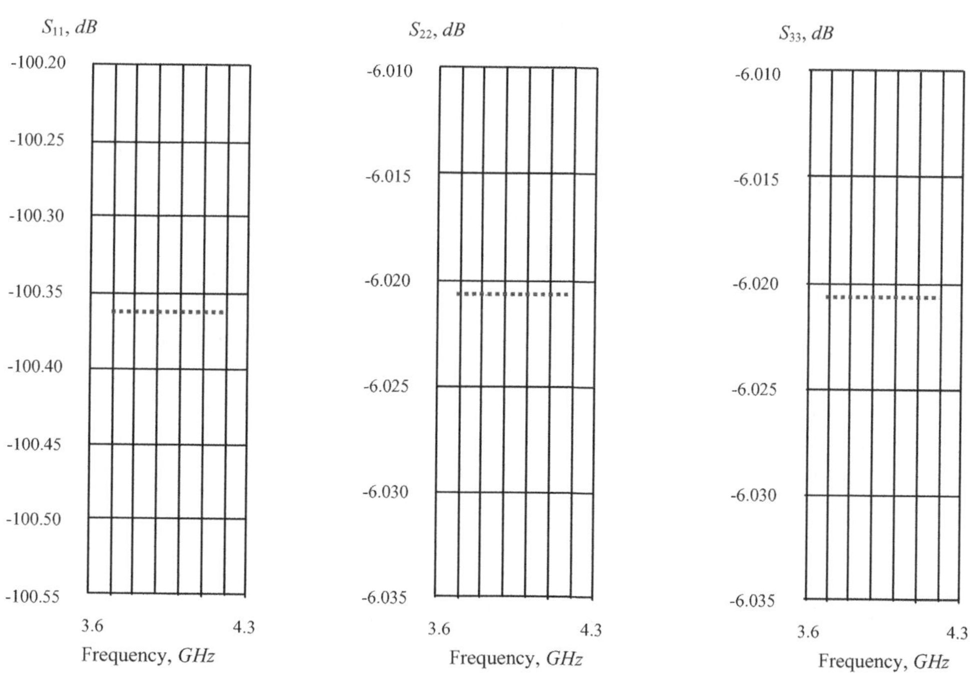

Figure 4.5 Return loss of the ideal transformer balun operating in the frequency range of group 1, *UWB* system: $f = 3696 - 3960 - 4224\,MHz$, $Z_S = Z_L = 50\,\Omega$.

TABLE 4.1 Values of relevant S parameters of the ideal transformer balun operating in the frequency range of group 1, UWB system: $f = 3696 - 3960 - 4224\,MHz$, $Z_S = Z_L = 50\,\Omega$

Frequencies	f		$3696 - 3960 - 4224$	MHz
Bandwidth	Δf		528	MHz
Insertion loss	S_{21}	mag.	-3.01 ± 0.01	dB
		phase	$0° \pm 0.1°$	
	S_{31}	mag.	-3.01 ± 0.01	dB
		phase	$+180° \pm 0.1°$	
Return loss	S_{11}	mag.	-100.36 ± 0.01	dB
	S_{22}	mag.	-6.02 ± 0.01	dB
	S_{33}	mag.	-6.02 ± 0.01	dB

operating in the frequency range of group 1, UWB system. Table 4.1 lists the values of the relevant S parameters of the performance.

As expected, the frequency responses of the S parameters are quite flat since the two stacked transformers are ideal. Unlike $S_{11} = -100.36\,dB$, the magnitudes of the return loss S_{22} and S_{33} are $-6.02\,dB$ due to the existence of three ports rather than two. However, this imperfection does not impact the simulation at all.

Simulation B: Simulation of the ideal transformer balun operated in the frequency range of group 3, UWB System: $f = 6864$–7128 to $7392\,MHz$, $Z_S = Z_L = 50\,\Omega$.

Figures 4.6 and 4.7 display the simulated results. Figure 4.6 shows the insertion loss and phase shift; Figure 4.7 shows the return loss of the ideal transformer balun operating in the frequency range of group 3, UWB system. Table 4.2 lists the values of the relevant S parameters of the performance.

Again, as expected, the frequency responses of the S parameters are quite flat since the two stacked transformers are ideal. The magnitudes of the return loss S_{22} and S_{33} are likewise not lowered like S_{11} due to the existence of three ports rather than two. However, this imperfection does not impact the simulation at all.

In summary, a balun with the same impedance value at all its terminals, including the one single-ended and two differential terminals, is preferred because the impedance matching for the differential pair can be replaced by the impedance matching for a single-ended stage. In other words, this circuitry with a differential configuration can be treated as a circuitry with a single-ended stage. This, of course, simplifies the simulation task significantly.

An ideal transformer balun is never found in reality. However, this is nevertheless a powerful tool at the simulation stage due to its advantage in simulation work. After the simulation is completed, the ideal transformer balun, of course, is replaced in the product by a practical balun.

4.2.4 Equivalency of Parts Between Single-ended and Differential Pairs with Respect to an Ideal Transformer Balun

The ideal transformer balun built by two ideal and identical transformers as shown in Figure 4.3 has very special features. One is that the same impedance appears in the single-ended port and two differential pair ports, that is, $Z_L = Z_S = Z$, if the turn

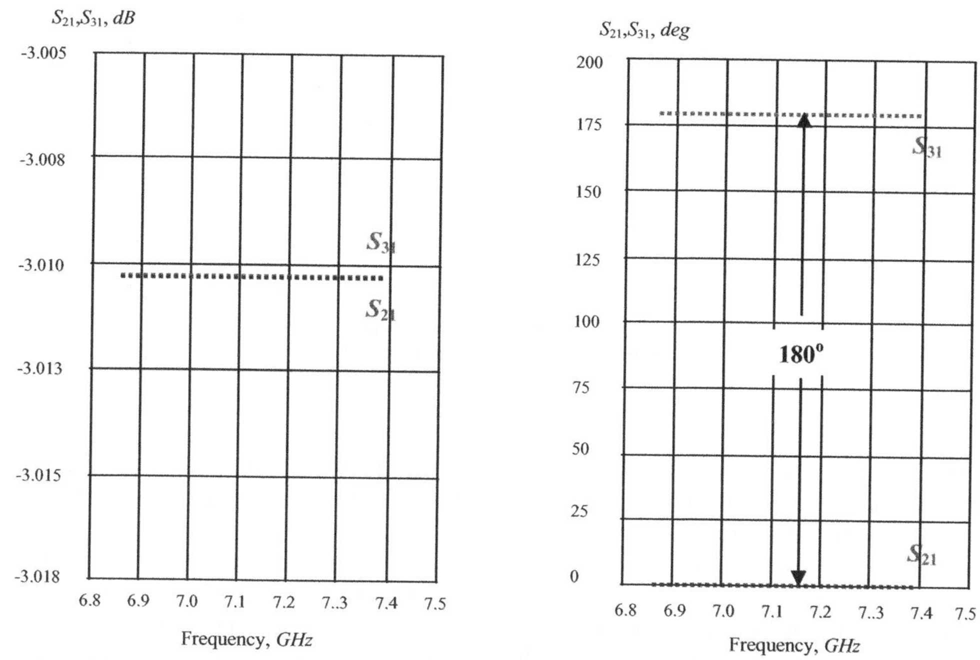

Figure 4.6 Insertion loss and phase shift of the ideal transformer balun operating in the frequency range of Group 3, *UWB* system: $f = 6864 - 7128 - 7392\,MHz$, $Z_S = Z_L = 50\,\Omega$.

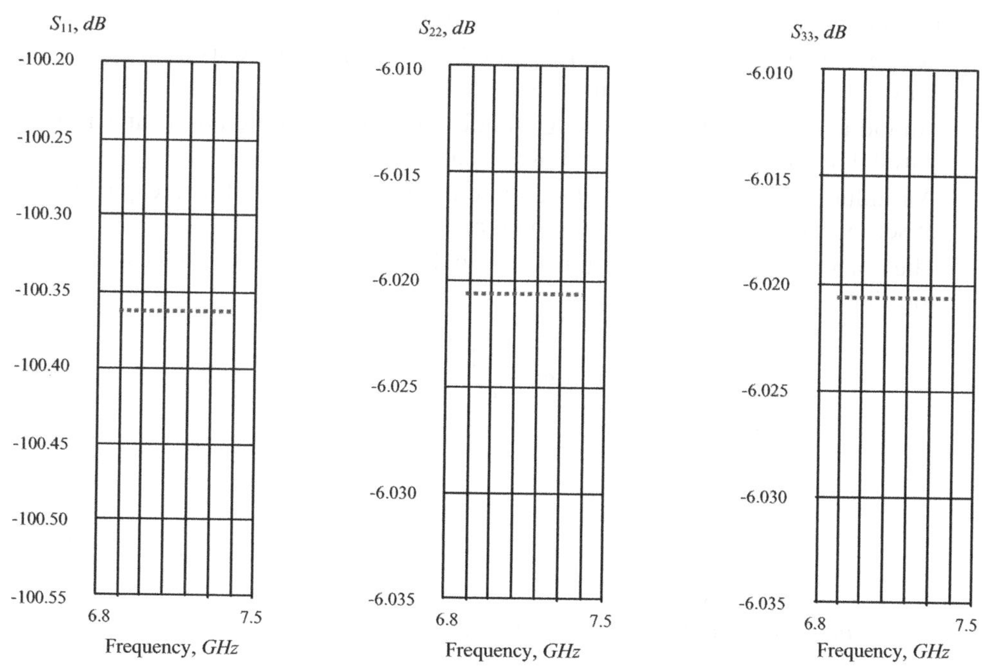

Figure 4.7 Return loss of the ideal transformer balun operating in the frequency range of Group 3, *UWB* system: $f = 6864 - 7128 - 7392\,MHz$, $Z_S = Z_L = 50\,\Omega$.

TABLE 4.2 Values of relevant *S* parameters of the ideal transformer balun operating in the frequency range of group 3, *UWB* system: f = 6864 – 7128 – 7392 *MHz*, $Z_S = Z_L = 50\,\Omega$

Frequencies	f		6864 – 7128 – 7392	*MHz*
Bandwidth	Δf		528	*MHz*
Insertion loss	S_{21}	mag.	–3.01 ± 0.01	*dB*
		phase	0° ± 0.1°	
	S_{31}	mag.	–3.01 ± 0.01	*dB*
		phase	+180° ± 0.1°	
Return loss	S_{11}	mag.	–100.36 ± 0.01	*dB*
	S_{22}	mag.	–6.02 ± 0.01	*dB*
	S_{33}	mag.	–6.02 ± 0.01	*dB*

ratio of each transformer is $1{:}\sqrt{2}$. Another special feature is that the equivalence of parts between the single-ended and differential pair enables these parts to be interpreted by each other.

What does "interpretation" of parts between the single-ended and the differential pair mean? If part *A* is added to the single-ended port and part *B* is added to each port of the differential pair simultaneously, then, at the operating frequency Ω_o or f_o, the *S* parameters or impedances at all three ports can be kept unchanged before and after these parts are added, as long as part *B* is "interpreted" from part *A* or part *A* is "interpreted" from part *B* based on the following special rules.

- A capacitor in parallel at the single-ended port is interpreted as an inductor in parallel at each port of differential pair, and vice versa,
- A capacitor in series at the single-ended port is interpreted as an inductor in series at each port of differential pair, and vice versa.
- An inductor in parallel at the single-ended port is interpreted as a capacitor in parallel at each port of differential pair, and vice versa.
- An inductor in series at the single-ended port is interpreted as a capacitor at in series each port of differential pair, and vice versa.
- The reactance of the inductor must be equal to the negative reactance of capacitor, that is

$$L\omega_o = \frac{1}{C\omega_o}, \tag{4.5}$$

or

$$LC = \frac{1}{\omega_o^2}, \tag{4.6}$$

where ω_o = operating angular frequency.

It should be noted that the inductor and capacitor mentioned in these special rules are added to different sides of the transformer balun. In other words, if the capacitor is added to the single-ended side, then the inductor is added to the differential pair side, and vice versa.

Figure 4.8 shows five cases where a part is added to the single-ended port and the interpreted part is added to each port of the differential pair respectively. Equation (4.5) is the relationship between the added parts, by which the added parts can be interpreted from each other. Simulations prove that the S parameters or impedances at all three ports are unchanged at the operating frequency Ω_o or f_o, and are approximately unchanged or slightly changed within certain bandwidths around the operating frequency, before and after the parts are added to the transformer balun.

This is a useful regulation! In terms of this regulation, simulation for a circuit with a differential pair configuration could be replaced by simulation for a circuit with a single-ended configuration!

To verify the interpretation regulation and to check the equivalence of impedances looking into and outward from the transformer balun, let's conduct seven more experimental simulations. The first five are conducted to verify the interpretation regulation. The added part number will be incremented from zero to four in steps so that they are not the same as shown in Figure 4.8, where only one part was added to either the single-ended port or each port of the differential pair. The last two simulations are conducted to check the equivalence of the impedances looking into and outward from the transformer balun. In all the simulations, the operating frequency is $f_o = 7128\,MHz$, which is the central frequency of group 3, UWB System. As matter of fact, we are also interested in seeing what would happen in the entire frequency bandwidth of group 3, UWB system, $f = 6864$ to $7392\,MHz$, with a central frequency of $f_o = 7128\,MHz$.

Simulation 1 is conducted for the original transformer balun without any parts added. The setup was shown in Figure 4.3; its simulation results were presented in Figures 4.6 and 4.7. However, for the integrity of description, it will be partially repeated in Figures 4.9 to 4.11. Figure 4.9 is the simulation setup with the transformer balun as shown in Figure 4.3; Figures 4.10 and 4.11 show the insertion loss, phase shift, and return loss.

It can be found that at the operating frequency, $f_o = 7128\,MHz$,

- Instead of the expected $-3\,dB$, the insertion losses S_{21} and S_{31} are both $-3.01\,dB$. This indicates that the attenuation of the transformer balun is $0.01\,dB$, which in an ideal transformer balun should be $0\,dB$. The difference of $0.01\,dB$ is due to the error of truncated decimal numbers in the digital computation.
- The phase of S_{21} is $0°$ and the phase of S_{31} is $180°$, so that the phase difference between S_{21} and S_{31} is $180°$. This indicates that ports 2 and 3 are a real differential pair.
- The magnitude of the return loss S_{11} at port 1 is dropped down to $-100\,dB$, which in an ideal transformer balun should be $-\infty\,dB$. Again, the difference is due to the error of truncated decimal numbers in the digital computation. Nevertheless, this indicates that port 1 is well matched to $50\,\Omega$. At ports 2 and 3, $S_{22} = S_{33} = -6\,dB$. Due to the existence of two rather than three ports, they are not dropped down as is S_{11}. Regardless, they can be ignored because they do not play any negative role in the simulated parameters.

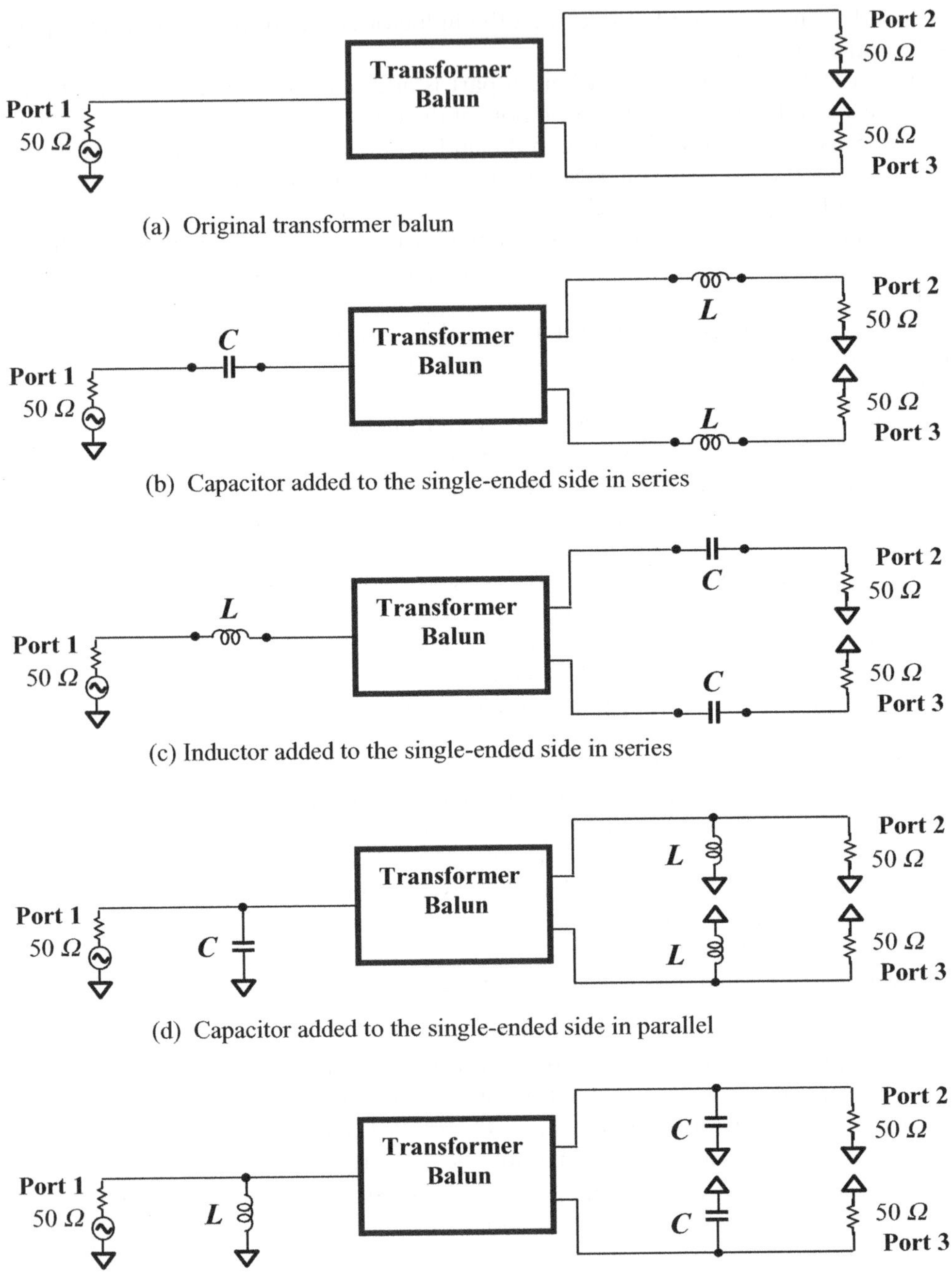

(a) Original transformer balun

(b) Capacitor added to the single-ended side in series

(c) Inductor added to the single-ended side in series

(d) Capacitor added to the single-ended side in parallel

(e) Inductor added to the single-ended side in parallel

Figure 4.8 Five ways to add parts to a transformer balun, in which the S parameters or impedances at all 3 ports as shown in this figure are kept unchanged before and after these parts are added. (Note that $LC = 1/\Omega_o$.)

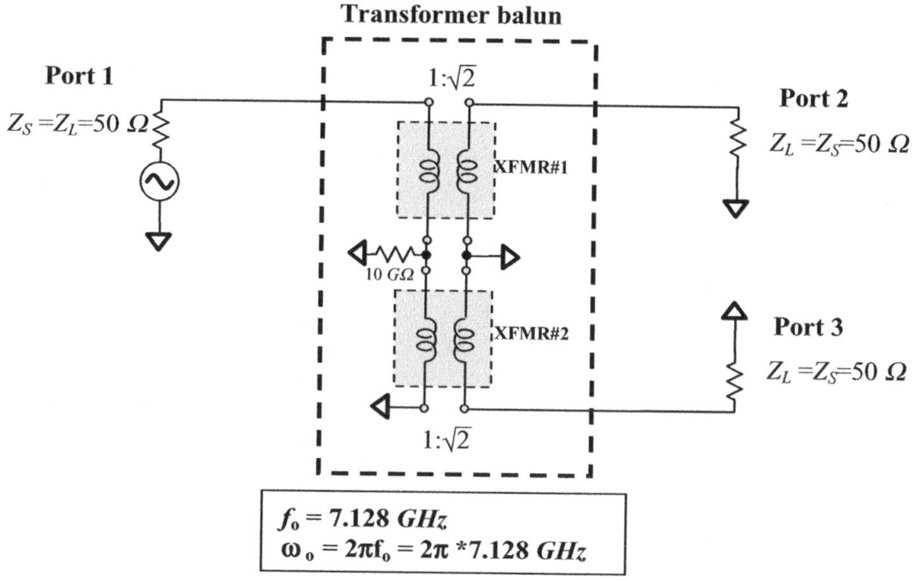

Figure 4.9 Simulation 1: transformer balun without any added parts.

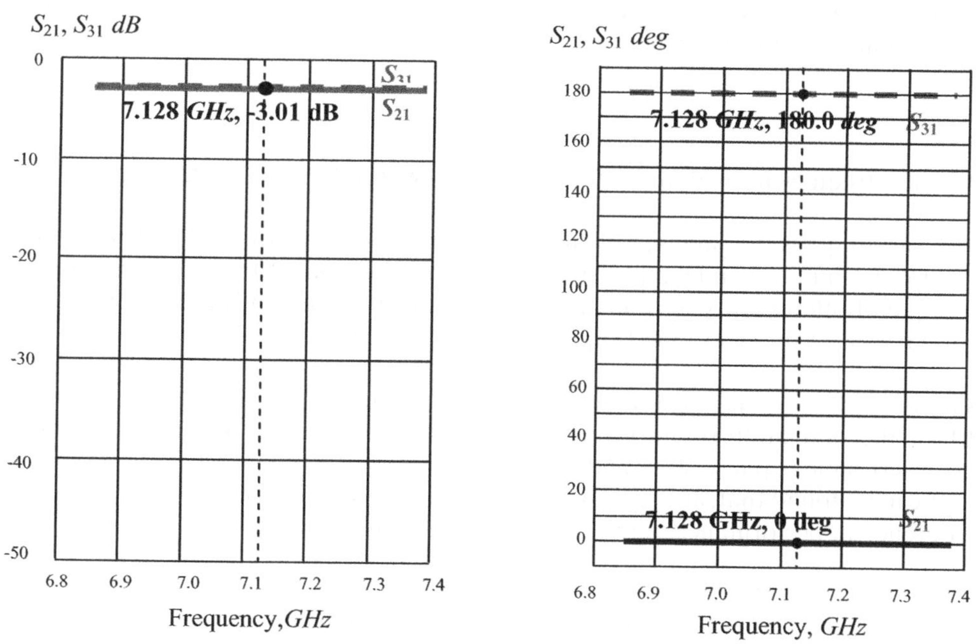

Figure 4.10 Insertion loss and phase shift of transformer balun for group 3 *UWB* system (S_{21}, S_{31}) in simulation 1: transformer balun without any added parts.

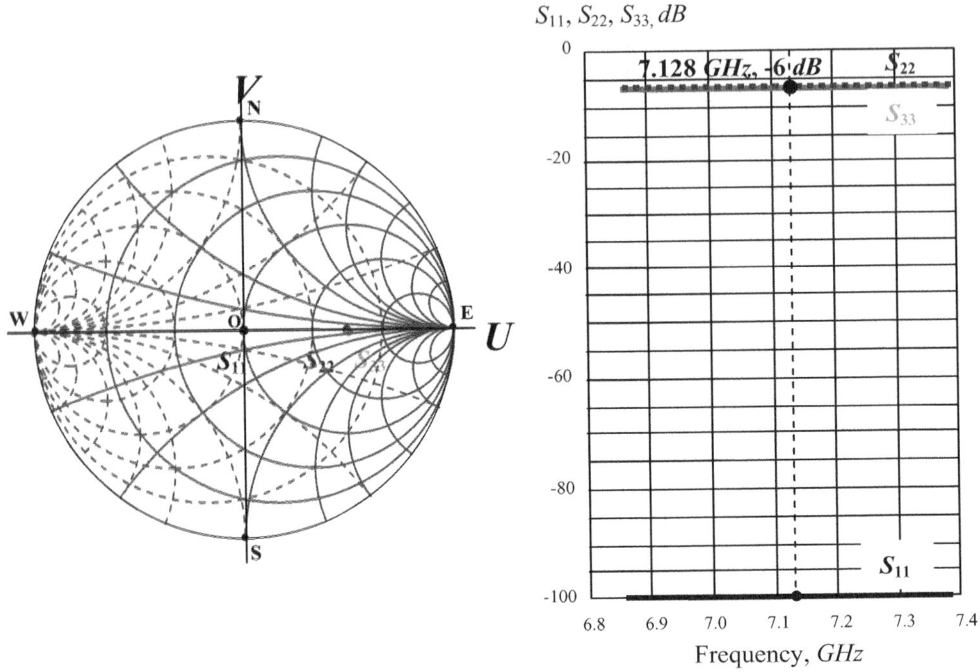

(a) On Smith chart (b) By rectangular coordination, *dB*

Figure 4.11 Return loss of transformer balun for group 3 *UWB* system (S_{11}, S_{22}, S_{33}) in simulation 1: transformer balun without any added parts.

At other frequencies within the frequency bandwidth, $f = 6864–7392\,MHz$,

- The frequency responses of all the S parameters are quite flat. This is due to the infinite bandwidth response resulting from ideal transformers. Consequently, the performance in all frequencies within the bandwidth is almost the same as at the operating frequency, $f_o = 7128\,MHz$.

Simulation 2 is conducted for the transformer balun with one part added. Figure 4.12 is the simulation setup of the transformer balun; Figures 4.13 and 4.14 show the insertion loss, phase shift, and return loss.

It can be found that, at the operating frequency, $f_o = 7128\,MHz$, there is not a big difference in performance between simulations 1 and 2, except that

- The phase of S_{21} is $-180°$ and the phase of S_{31} is $0°$, but the phase difference between S_{21} and S_{31} is still $180°$. This indicates that ports 2 and 3 are a real differential pair.
- The magnitude of the return loss S_{11} at port 1 is dropped down to $-96\,dB$, which should be $-\infty\ dB$ in an ideal transformer balun. Again, this difference is due to the error of truncated decimal numbers in the digital computation. Nevertheless, it indicates that port 1 is well matched to $50\,\Omega$.

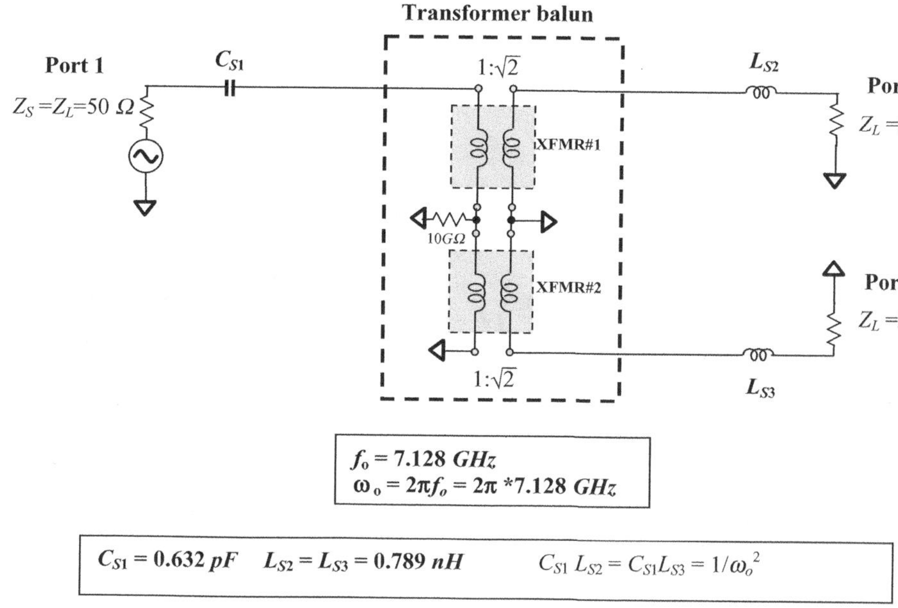

Figure 4.12 Simulation 2: transformer balun with one added part.

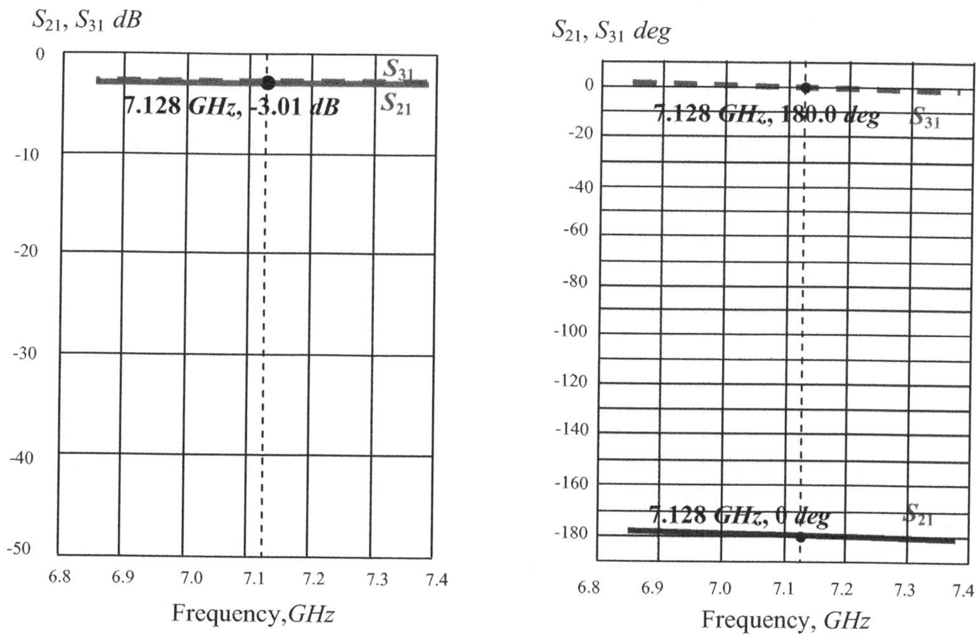

Figure 4.13 Insertion loss and phase shift of transformer balun for group 3 UWB system (S_{21}, S_{31}) in simulation 2: transformer balun with one added part.

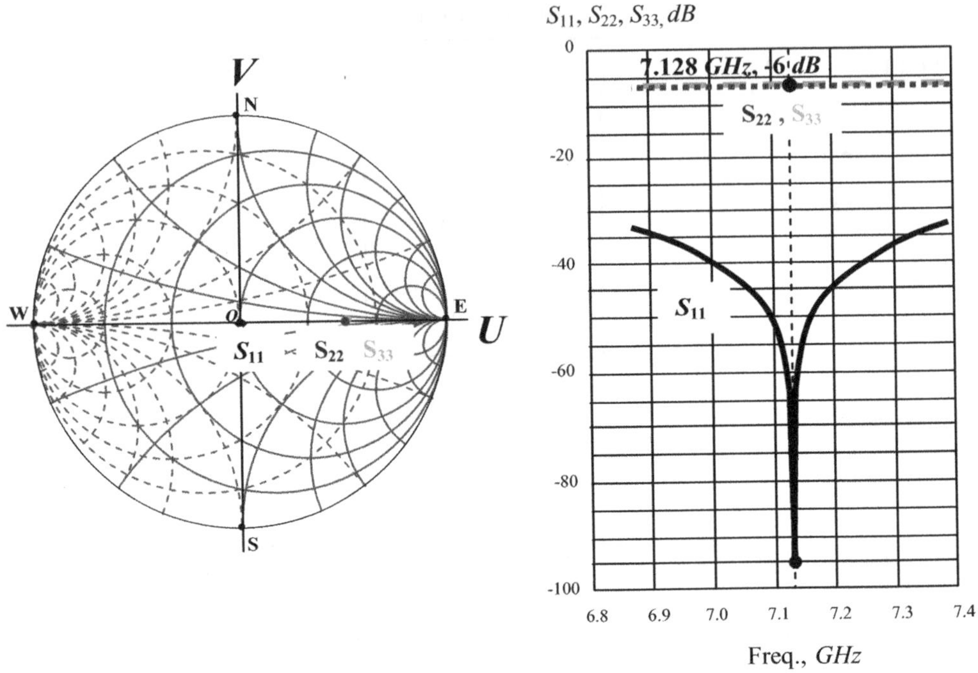

(a) On Smith chart (b) By rectangular coordination, *dB*

Figure 4.14 Return loss of transformer balun for group 3 *UWB* system (S_{11}, S_{22}, S_{33}) in simulation 2: transformer balun with one added part.

At other frequencies within the frequency bandwidth, $f = 6864–7392\,MHz$,

- The frequency responses of S_{21} and S_{31} are not entirely flat, but tilted a little bit, since the capacitor C_{S1} or the inductor L_{S2} are dependent on the frequency. The tilted magnitude is less than $0.5\,dB$ and the phase shift is less than 5°. Interestingly, the tilted directions of the phase of S_{21} and S_{31} are the same so that the phase difference of the two differential ports is kept at around 180°.
- The magnitude of the return loss at port 1, S_{11}, varies considerably within the frequency band. At low and high frequencies, the values of S_{11} are around $-40\,dB$, while at the central frequency, the value of S_{11} is dropped down to less than $-96\,dB$. The reason for this is the same as that for the previous variation: the impedance of the one additional part, either C_{S1} or L_{S2}, or L_{S3}, is dependent on the frequency. At ports 2 and 3, $S_{22} = S_{33} = -6\,dB$. These values are not dropped down as is S_{11} due to the existence of three ports rather than two ports. Regardless, they can be ignored because they do not play any negative role to the simulated parameters.

Simulation 3 is conducted for the transformer balun with two parts added. Figure 4.15 is the simulation setup with the transformer balun; Figures 4.16 and 4.17 show the insertion loss, phase shift, and return loss.

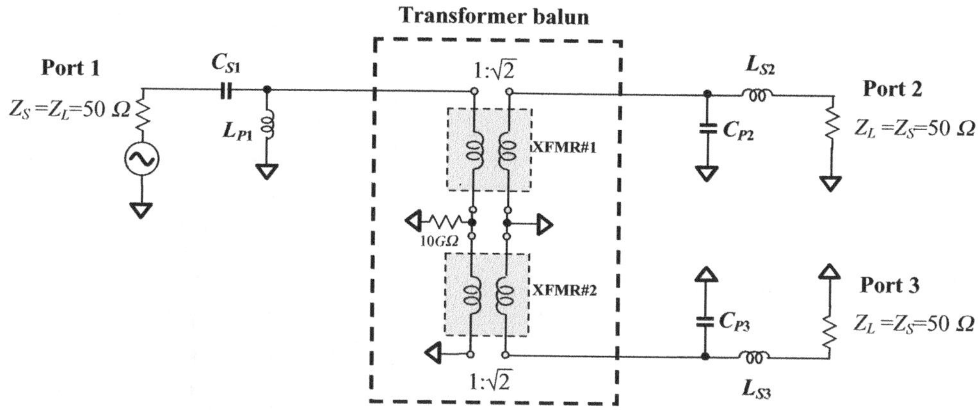

$f_o = 7.128\ GHz$
$\omega_o = 2\pi fo = 2\pi * 7.128\ GHz$

$C_{S1} = 0.631\ pF$	$L_{S2} = L_{S3} = 0.789\ nH$	$C_{S1} L_{S2} = C_{S1} L_{S3} = 1/\omega_o^2$
$L_{P1} = 0.395$ nH	$C_{P2} = C_{P3} = 1.263\ pF$	$L_{P1} C_{P2} = L_{P1} C_{P3} = 1/\omega_o^2$

Figure 4.15 Simulation 3: transformer balun with two added parts.

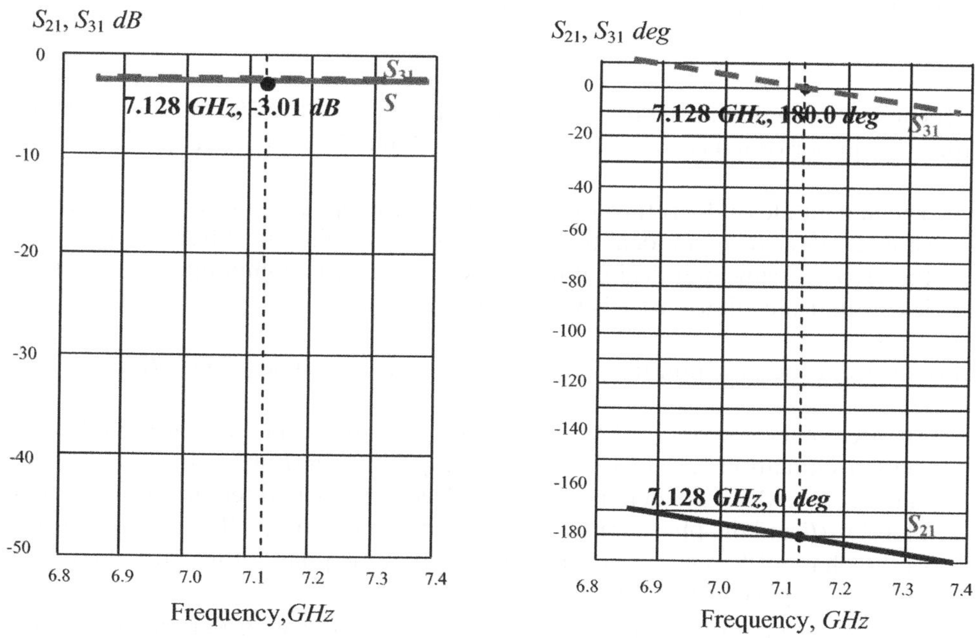

Figure 4.16 Insertion loss and phase shift of transformer balun for group 3 *UWB* system (S_{21}, S_{31}) in simulation 3: transformer balun with two added parts.

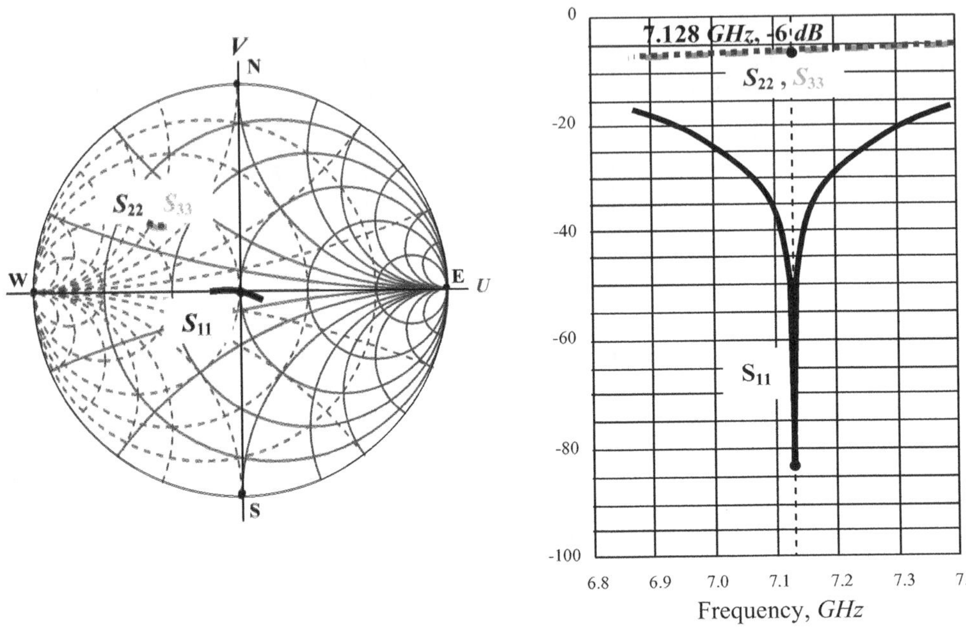

(a) On Smith chart (b) By rectangular coordination, *dB*

Figure 4.17 Return loss of transformer balun for group 3 *UWB* system (S_{11}, S_{22}, S_{33}) in simulation 3: transformer balun with two added parts.

It can be found that at the operating frequency, $f_o = 7128\,MHz$, there is not a big difference in performance between simulations 2 and 3, except that

- The magnitude of the return loss S_{11} at port 1 is dropped down to $-83\,dB$, which should be $-\infty\ dB$ in an ideal transformer balun. Again, this difference is due to the error of truncated decimal numbers in the digital computation. Nevertheless, it indicates that port 1 is well matched to $50\,\Omega$.

At other frequencies within the frequency bandwidth, $f = 6864\text{–}7392\,MHz$,

- The frequency responses of S_{21} and S_{31} are not entirely flat; they are tilted more than in simulation 2, since there are two capacitors and two inductors, C_{S1}, L_{P1}, and L_{S2}, C_{P2}, which are dependent on the frequency. The tilted magnitude is less than $1.0\,dB$ and the phase shift is less than $15\,°$. Interestingly, the tilted directions of the phase of S_{21} and S_{31} are the same, so that the phase difference of the two differential ports is kept at around $180\,°$.
- The magnitude of the return loss at port 1, S_{11}, varies considerably within the frequency band. At low and high frequencies the values of S_{11} are around -20 to $-40\,dB$ while at the central frequency the value of S_{11} is dropped down to less than $-83\,dB$. The reason for this is the same as that of the previous variation, that is, that the impedances of the additional parts, C_{S1}, L_{P1}, and L_{S2}, C_{P2}, are dependent on the frequency. At ports 2 and 3, S_{22} and S_{33} changed from -7

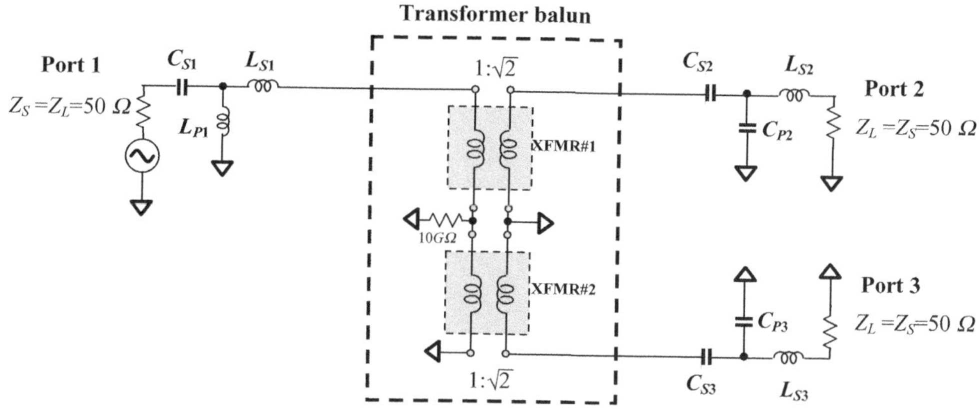

$f_o = 7.128\ GHz$
$\omega_o = 2\pi fo = 2\pi * 7.128\ GHz$

$C_{S1} = 0.632$ pF	$L_{S2} = L_{S3} = 0.789\ nH$	$C_{S1} L_{S2} = C_{S1} L_{S3} = 1/\omega_o^2$
$L_{P1} = 0.395$ nH	$C_{P2} = C_{P3} = 1.263\ pF$	$L_{P1} C_{P2} = L_{P1} C_{P3} = 1/\omega_o^2$
$L_{S1} = 1.000$ nH	$C_{S2} = C_{S3} = 0.499\ pF$	$L_{S1} C_{S2} = L_{S1} C_{S3} = 1/\omega_o^2$

Figure 4.18 Simulation 4: transformer balun with three added parts.

to $-5\,dB$. These values are not dropped as is S_{11} due to the existence of three rather than two ports. Regardless, they can be ignored because they do not play any negative role to the simulated parameters.

Simulation 4 is conducted for the transformer balun with three parts added. Figure 4.18 is the simulation setup with the transformer; Figures 4.19 and 4.20 show the insertion loss, phase shift, and return loss.

It can be found that at the operating frequency, $f_o = 7128\,MHz$, there is not a big difference in performance between simulations 3 and 4, except

- The magnitude of the return loss S_{11} at port 1 is dropped down to $-77\,dB$, which should be $-\infty\ dB$ in an ideal transformer balun. Again, this difference is due to the error of truncated decimal numbers in the digital computation. Nevertheless, it indicates that the port 1 is still well matched to $50\,\Omega$.

At other frequencies within the frequency bandwidth, $f = 6864$–$7392\,MHz$,

- The frequency responses of S_{21} and S_{31} are not entirely flat, but tilted more than in simulation 3, since there are three capacitors and three inductors, C_{S1}, L_{P1}, L_{S1}, and L_{S2}, C_{P2}, C_{S2}, which are dependent on the frequency. The tilted magnitude is less than $1.5\,dB$ and the phase shift is less than $25°$. Interestingly, the tilted directions of the phase of S_{21} and S_{31} are the same so that the phase difference of the two differential ports is kept at around $180°$.
- The magnitude of the return loss at port 1, S_{11}, varies considerably within the frequency band. At low and high frequencies the values of S_{11} are around -10

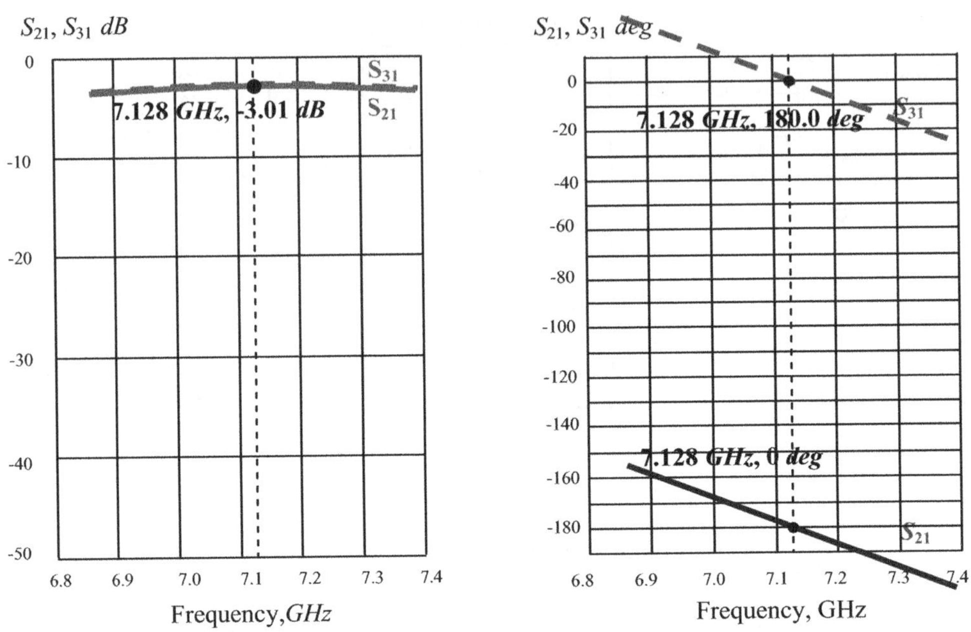

Figure 4.19 Insertion loss and phase shift of transformer balun for group 3 *UWB* system (S_{21}, S_{31}) in simulation 4: transformer balun with three added parts.

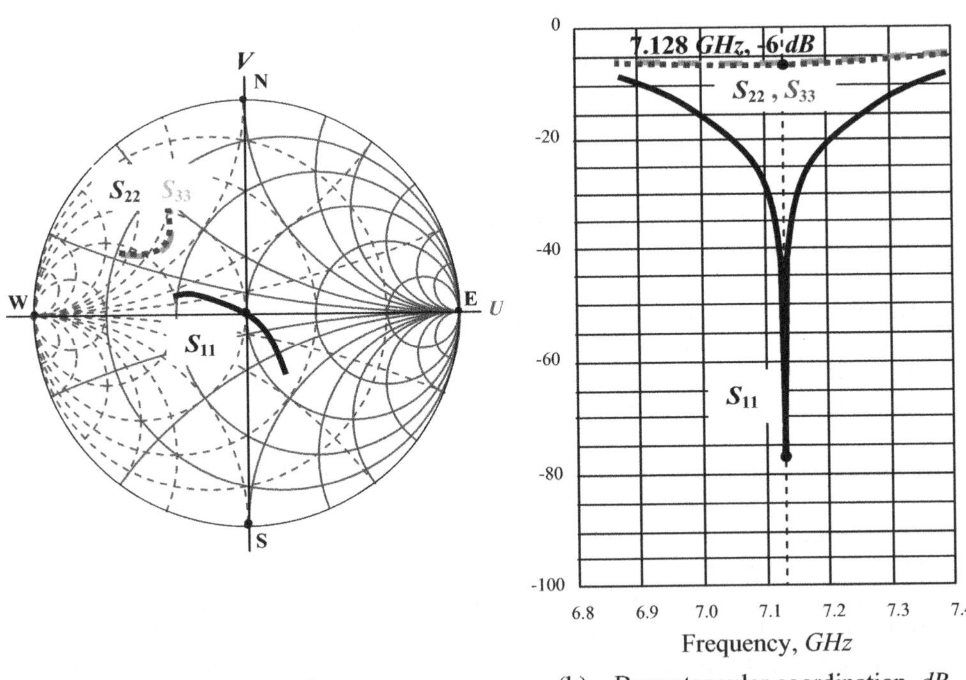

(a) On Smith chart (b) By rectangular coordination, *dB*

Figure 4.20 Return loss of transformer balun for group 3 *UWB* system (S_{11}, S_{22}, S_{33}) in simulation 4: transformer balun with three added parts.

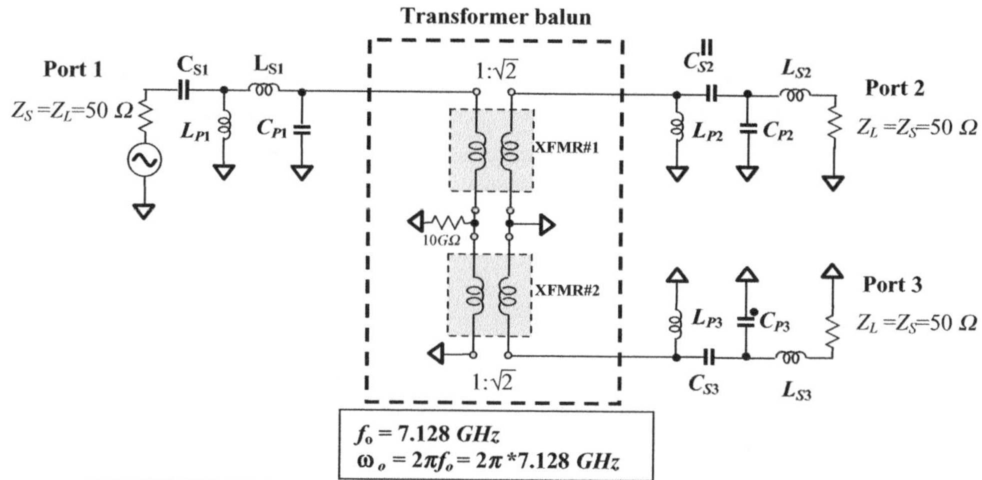

$C_{S1} = 0.632\ pF$	$L_{S2} = L_{S3} = 0.789\ nH$	$C_{S1} L_{S2} = C_{S1} L_{S3} = 1/\omega_o^2$
$L_{P1} = 0.395\ nH$	$C_{P2} = C_{P3} = 1.263\ pF$	$L_{P1} C_{P2} = L_{P1} C_{P3} = 1/\omega_o^2$
$L_{S1} = 1.000\ nH$	$C_{S2} = C_{S3} = 0.499\ pF$	$L_{S1} C_{S2} = L_{S1} C_{S3} = 1/\omega_o^2$
$C_{P1} = 0.249\ pF$	$L_{P2} = L_{P3} = 2.000\ nH$	$C_{P1} L_{P2} = C_{P1} L_{P3} = 1/\omega_o^2$

Figure 4.21 Simulation 5: transformer balun with four added parts.

to $-30\,dB$ while at the central frequency the value of S_{11} is dropped down to less than $-77\,dB$. The reason for this is the same as that of the previous variation, that is, that the impedance of the additional parts, C_{S1}, L_{P1}, L_{S1}, and L_{S2}, C_{P2}, C_{S2}, are dependent on the frequency. At ports 2 and 3, S_{22} and S_{33} are changed from -6 to $-4\,dB$. These values are not dropped down as is S_{11} due to the existence of three rather than two ports. Regardless, they can be ignored because they do not play any negative role to the simulated parameters.

Simulation 5 is conducted for the transformer balun with four parts added. Figure 4.21 is the simulation setup with the transformer balun; Figures 4.22 and 4.23 show the insertion loss, phase shift, and return loss.

It can be found that at the operating frequency, $f_o = 7128\,MHz$, there is not a big difference in performance between the simulation 4 and 5, except that

- The magnitude of the return loss S_{11} at port 1 drops down to $-81\,dB$, which should be $-\infty\ dB$ in an ideal transformer balun. Again, this difference is due to the error of truncated decimal numbers in the digital computation. Nevertheless, it indicates that port 1 is still well matched to $50\,\Omega$.
- At other frequencies within the frequency bandwidth, $f = 6864$–$7392\,MHz$,
- The frequency responses of S_{21} and S_{31} are not entirely flat; they are tilted more than in simulation 4, since there are four capacitors and four inductors, C_{S1}, L_{P1}, L_{S1}, C_{P1}, L_{S2}, C_{P2}, and C_{S2}, L_{P2}, which are dependent on the frequency. The tilted magnitude is less than $3.0\,dB$ and the phase shift is less than $30\,°$. Interestingly, the tilted directions of the phase of S_{21} and S_{31} are the same, so that the phase difference of the two differential ports is kept at around $180\,°$.

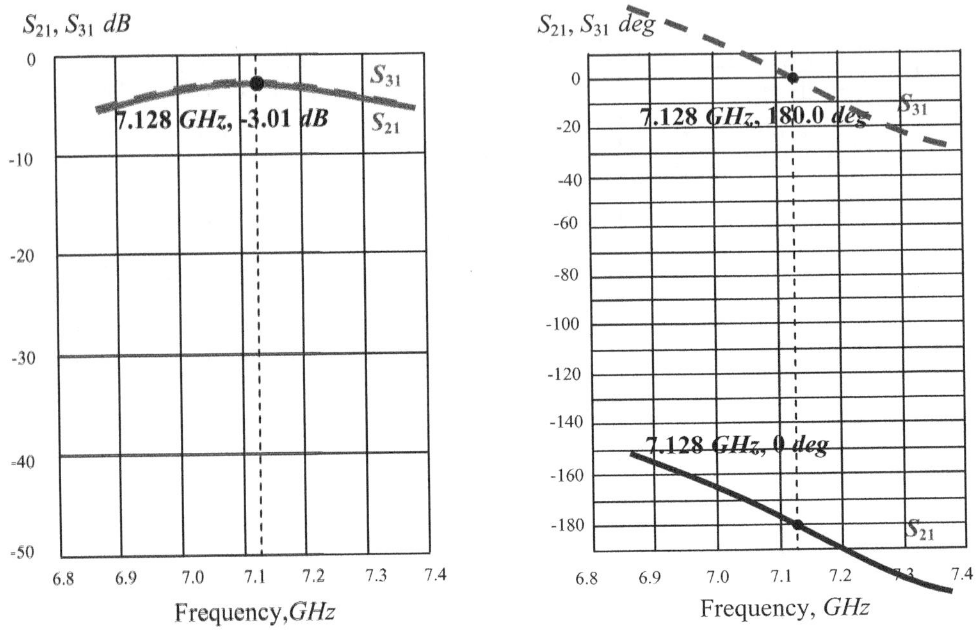

Figure 4.22 Insertion loss and phase shift of transformer balun for group 3 *UWB* system (S_{21}, S_{31}) in simulation 5: transformer balun with four added parts.

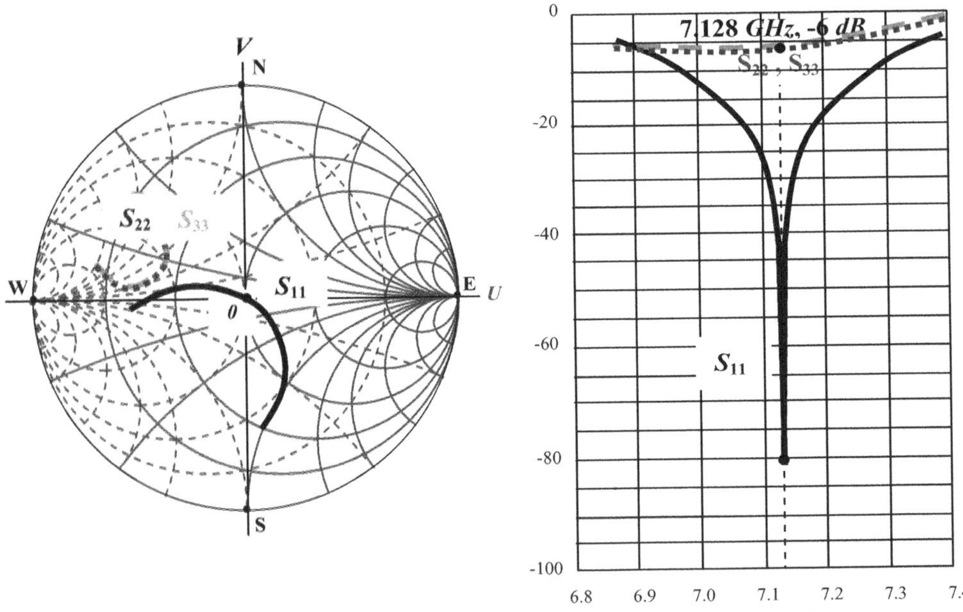

(a) On Smith chart (b) By rectangular coordination, *dB*

Figure 4.23 Return loss of transformer balun for group 3 *UWB* system (S_{11}, S_{22}, S_{33}) in simulation 5: transformer balun with four added parts.

- The magnitude of the return loss at port 1, S_{11}, varies considerably within the frequency band. At low and high frequencies the values of S_{11} are around −5 to $-20\,dB$ while at the central frequency the value of S_{11} drops down to less than −81 dB. The reason for this is the same as that of the previous variation, that is, that the impedances of the additional parts, C_{S1}, L_{P1}, L_{S1}, C_{P1}, and L_{S2}, C_{P2}, C_{S2}, L_{P2}, are dependent on the frequency. At ports 2 and 3, S_{22} and S_{33} are changed from −7 to $-2\,dB$. These values are not dropped down as is S_{11} due to the existence of three rather than two ports. They can be ignored because they do not play any negative role to the simulated parameters.

In summary, from these five simulations, it can be seen that the interpretation regulation in respect to the transformer balun is well verified because

1) At the assigned operating frequency, $f_o = 7128\,MHz$,
   - In all five simulations, S_{11} is always kept at very low levels, that is, $S_{11} < -70\,dB$. This means that at port 1, the impedance is always kept at a value very close to $50\,\Omega$.
   - In all five simulations, the magnitudes of S_{21} and S_{31} are kept at $-3.01\,dB$ and the phase between S_{21} and S_{31} is kept at $180°$. This means that ports 2 and 3 are a real differential pair with only $0.01\,dB$ attenuation from port 1 and that the impedance at both ports is kept at a value very close to $50\,\Omega$.
   - These results are due to the fact that in every simulation the added parts are interpreted from each other by equation (4.5). Consequently, it is concluded that the S parameters or impedances at all three ports are unchanged at the operating frequency, Ω_o or f_o, if the added parts at the single-ended side and at the differential pair side are interpreted from each other by equation (4.5).
2) At the frequency range around the operating frequency, $f = 6864\text{–}7392\,MHz$,
   - In all five simulations, S_{11} is always kept at low levels, although not as low as that at $f_o = 7128\,MHz$. This means that, at port 1, the impedance is always kept at a value around $50\,\Omega$.
   - In all five simulations, the magnitudes of S_{21} and S_{31} are less than $-3\,dB$ but not less than $-5\,dB$, and, interestingly, the phase between S_{21} and S_{31} is kept at around $180°$. This means that ports 2 and 3 can still be considered as a differential pair, but with an additional attenuation of 1 to $2\,dB$.
   - These results are due to the fact that in every simulation the added parts are interpreted from each other by equation (4.5). Consequently, it is concluded that the S parameters or impedances at all three ports are approximately unchanged in the frequency range of $f = 6864\text{–}7392\,MHz$ if the added parts at the single-ended side and at differential pair side are interpreted from each other by equation (4.5).

Now let's return to simulations 6 and 7.

Simulation 6 is conducted to check the impedance Z_1 looking into the transformer balun from the single-ended side if the added four parts, C_{S1}, L_{P1}, L_{S1}, and C_{P1}, at the single-ended side as shown in Figure 4.21 are removed. Figure 4.24 shows its simulation setup.

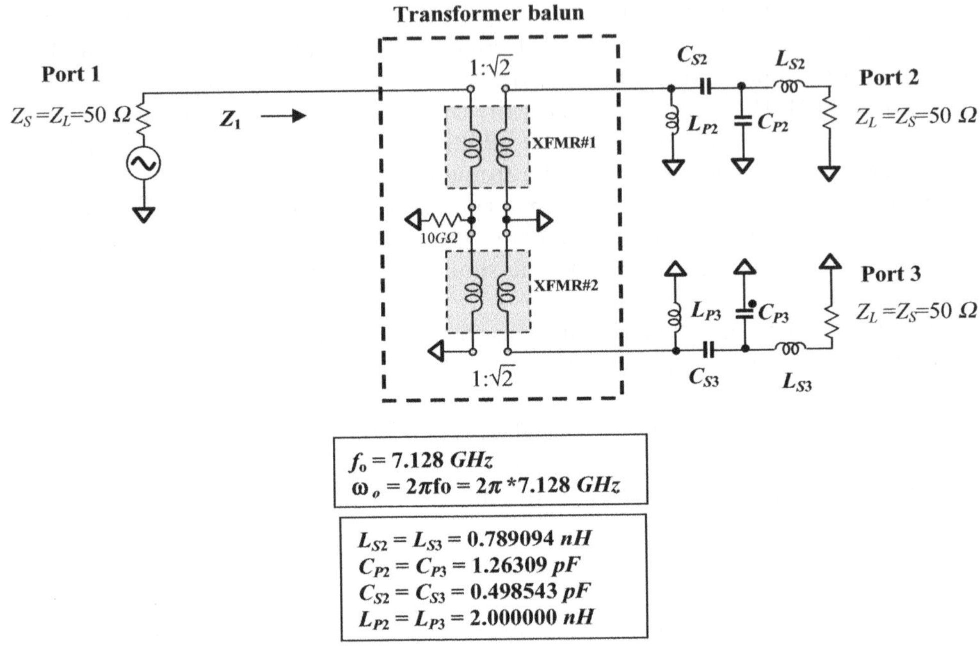

Figure 4.24 Simulation setup for testing impedance Z_1.

Simulation 7 is conducted to check the impedance Z_2 at one of the differential branches immediately after the transformer balun, looking into the differential port 2. Figure 4.25 shows its simulation setup.

Tested results from simulations 6 and 7 indicate that

$$Z_1 = Z_2 = Z, \tag{4.7}$$

and when $f = 7.1\,GHz$, $\quad Z_1 = Z_2 = Z = 10.16 + j\,56.7\,\Omega$;
when $f = 7.2\,GHz$, $\quad Z_1 = Z_2 = Z = 10.76 + j\,59.1\,\Omega$.

From simulations 6 and 7, it is concluded that the impedance looking from the single end before the transformer balun is the same as that looking into one of the differential branches after the transformer balun. In other words, the impedance of one differential branch Z_2 can be found at the single-ended port as Z_1. This transformer balun looks like a monster-revealing mirror. When you stand at the front of this magic mirror, you can discover the impedance behind it.

4.2.5 Impedance Matching for Differential Pairs by Means of a Transformer Balun

Now let's illustrate how to replace the impedance matching for a differential pair by impedance matching for a single-ended stage, so that a circuitry with a differential configuration can be treated as a circuitry with a single-ended stage.

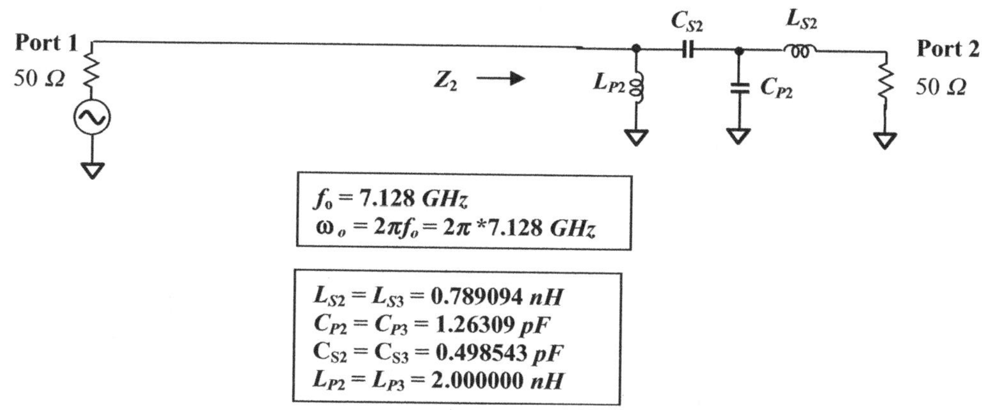

Figure 4.25 Simulation setup for testing impedance Z_2.

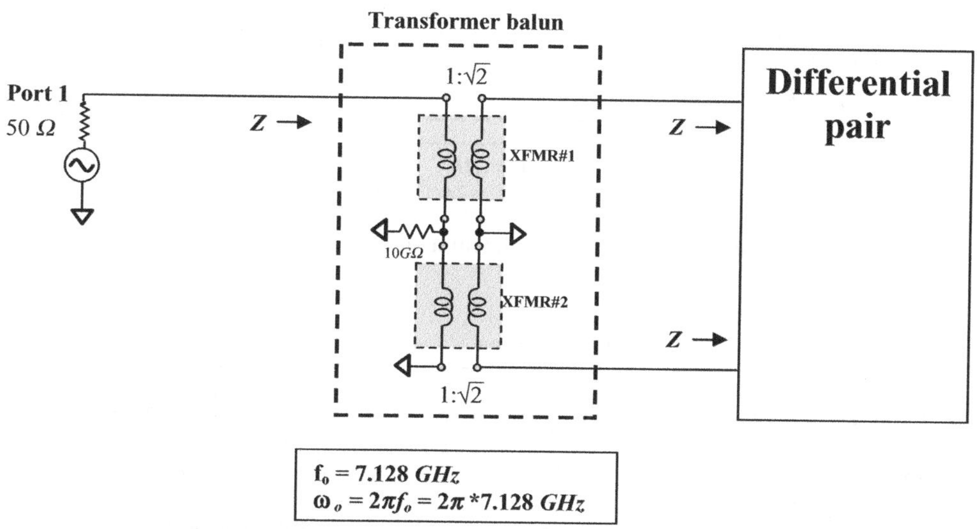

Figure 4.26 Differential pair is connected with a transformer balun built by two stacked transformers with a turn ratio = $1{:}\sqrt{2}$.

Assume that the impedance of each differential branch is Z. It must be impedance-matched to 50 Ω; the impedance matching can be conducted as follows:

1) As shown in Figure 4.26, we connect differential terminals of a transformer balun, which is built by two stacked transformers with their turn ratio = $1{:}\sqrt{2}$, to the terminals of the differential pair. As mentioned in simulations 6 and 7, the impedance looking from the single-ended stage before the transformer balun is the same as that looking into a differential branch after the transformer balun. All of them have same value Z as shown in Figure 4.26.
2) At the single-ended port, we do impedance matching between port 1 with 50 Ω source to the load Z and form the single-ended impedance matching network

as shown in Figure 4.27, in which the value of the parts are listed in bottom left corner.

3) We interpret the single-ended impedance matching network to the differential impedance matching network in terms of the rules described in Section 4.2.4. The results are shown in Figure 4.28.

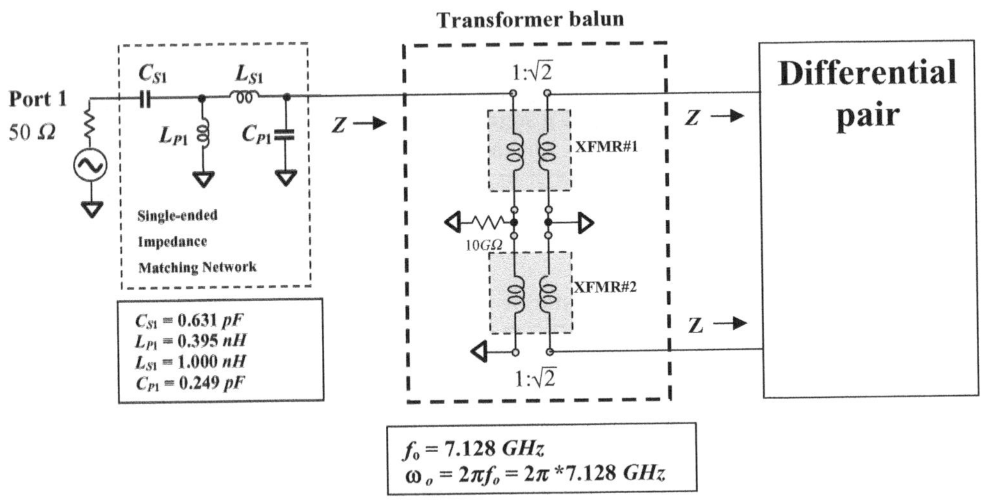

Figure 4.27 Impedance matching at single-ended side.

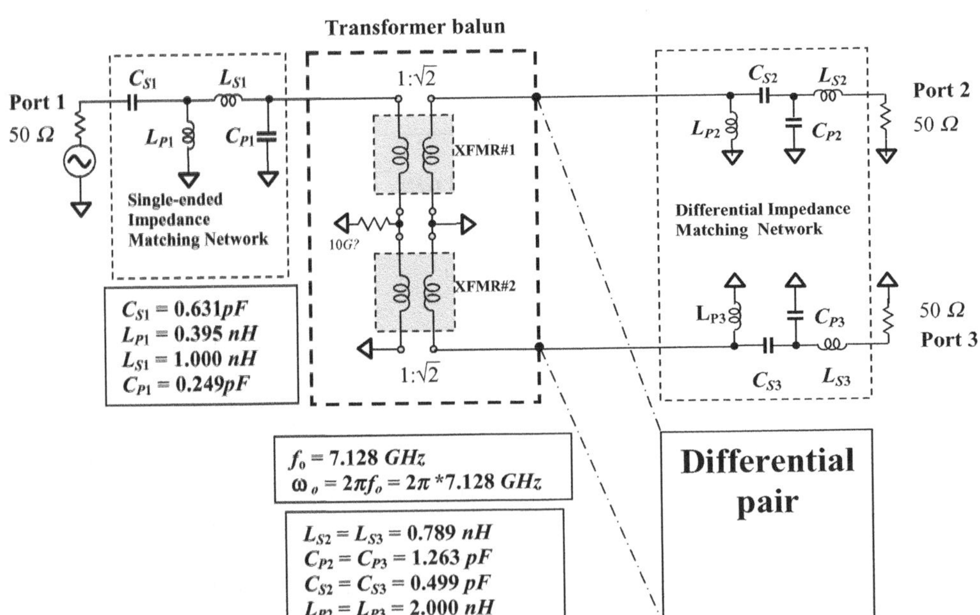

Figure 4.28 Interpreting the single-ended impedance matching network to differential impedance matching network (refer to Figure 4.21).

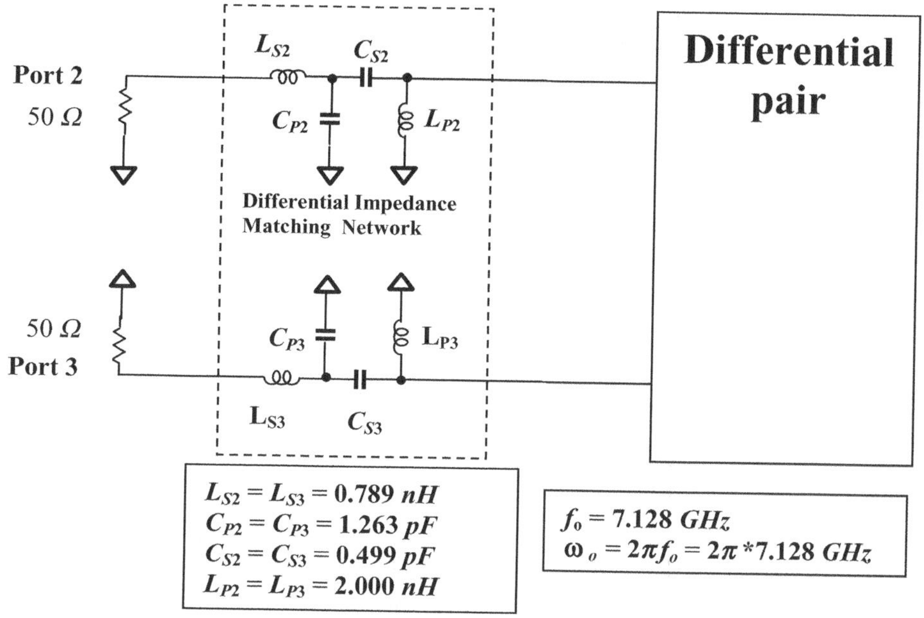

Figure 4.29 Final simulation results for the differential impedance matching network.

4) Removing port 1, the single-ended impedance matching network and the transformer balun, and re-aligning the differential impedance matching network and the differential pair, we have the result shown in Figure 4.29.

The four-step procedure demonstrates how to replace the impedance matching for a differential pair by impedance matching for a single-ended stage. It should be taken into account that this is exactly true only at the operating frequency, and is approximately true in the frequency range around the operating frequency within a certain frequency bandwidth. In cases of narrow bandwidth, with a relative bandwidth lower than 15%, the replacement of differential impedance matching by single-ended impedance matching should not be problematic in most cases. In cases of wide bandwidth, with a relative bandwidth higher than 15%, the replacement is not problematic at the assigned operating frequency but may show considerable discrepancies before and after replacement.

4.3 *LC* BALUNS

Figure 4.30 depicts a simple *LC* balun. It looks like a bridge and can serve as a splitter or a combiner. We will focus on the splitter since the combiner is simply a reversed splitter.

The distinguishing features of the *LC* balun are:

- Simplest configuration: The *LC* balun shown in Figure 4.30 is built by only two identical capacitors and two identical inductors. It is the simplest balun among all the baluns developed so far.

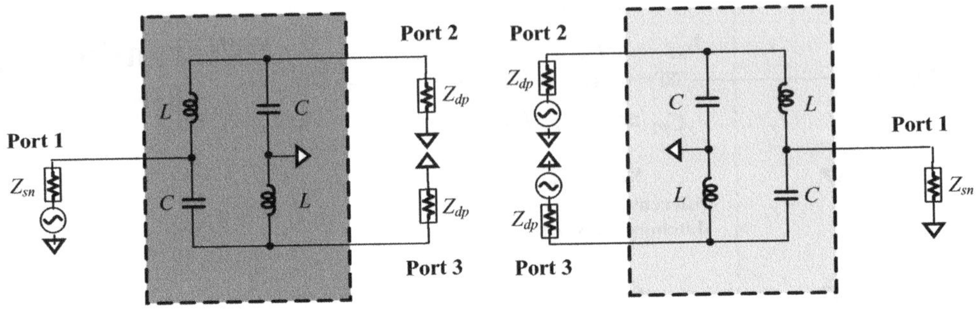

(a) *LC* Splitter (b) *LC* Combiner

Figure 4.30 *LC* balun functions as a splitter or combiner.

- Cost-effective: It is a balun with the lowest cost among all the baluns which have been developed so far, due to its aforementioned simplicity.
- Directly functions as an impedance matching network: The values of L and C in an LC balun are calculated on the basis of the impedances of the single-ended and differential pair. Therefore, the LC balun itself is the impedance matching network between the single-ended source and the differential pair.
- Existence of equivalence of parts between single-ended and differential pair enables these parts to be interpreted from each other. Just like the transformer balun discussed in the previous section, the equivalence of parts between the single-ended and differential pair allows the simulation of the differential pair circuitry to be replaced by a simulation of single-ended circuitry.
- The bandwidth of the LC balun is relatively narrow, compared with other baluns, because the body of the LC balun itself, two inductors and two capacitors, is dependent on frequency. This is the unique drawback of the LC balun.

4.3.1 Simplicity of *LC* Balun Design

An analysis of the simple balun shown in Figure 4.30(a) is found in Appendix 4.A.2. Designing a simple LC balun is tantamount to calculating the values of its inductors and capacitors. By the derivation in Appendix 4.A.2,

$$L\omega_o = \sqrt{2Z_{dp}Z_{sn}^*}, \tag{4.8}$$

and

$$C\omega_o = \frac{1}{\sqrt{2Z_{dp}Z_{sn}^*}}. \tag{4.9}$$

where

L = inductance of the inductor in the simple LC balun,
C = capacitance of the capacitor in the simple LC balun,

Z_{sn} = source impedance at single-ended port,
Z_{dp} = load impedance at each port of differential pair,
ω_o = operating angular frequency,

and

$$Z_{sn} = R_{sn} + jX_{sn}, \tag{4.10}$$

$$Z_{dp} = R_{dp} + jX_{dp}. \tag{4.11}$$

In terms of equations (4.8) and (4.9), a simple *LC* balun is easily designed.

4.3.2 Performance of a Simple *LC* Balun

Now let's explore the features of this simple *LC* balun in terms of the following two simulations.

Simulation A: Simulation of simple *LC* balun operated in the frequency range of group 1, *UWB* System: $f = 3696–3960$ to $4224\,MHz$, $f_o = 3960MHz$, $Z_S = Z_L = 50\,\Omega$.

Figure 4.31 shows the setup of the simulation for group 1, *UWB* system; Figures 4.32 and 4.33 display the simulated results. Figure 4.32 shows the insertion loss and phase shift while Figure 4.33 shows the return loss of the simple *LC* balun operating in the frequency range of group 1, *UWB* system. Table 4.3 lists the values of the relevant *S* parameters and their performance.

At the operating frequency, $f_o = 3960\,MHz$, $S_{21} = S_{31} = 3.01\,dB$, which means that at each port of the differential pair, the impedance is well matched with the *LC* balun. In addition, the phase at one port of the differential is 90° and the phase at the other is –90°, so that the phase difference between the two ports is 180°. The magnitude of the return loss at port 1, S_{11}, drops down to $-325\,dB$, so that it is not

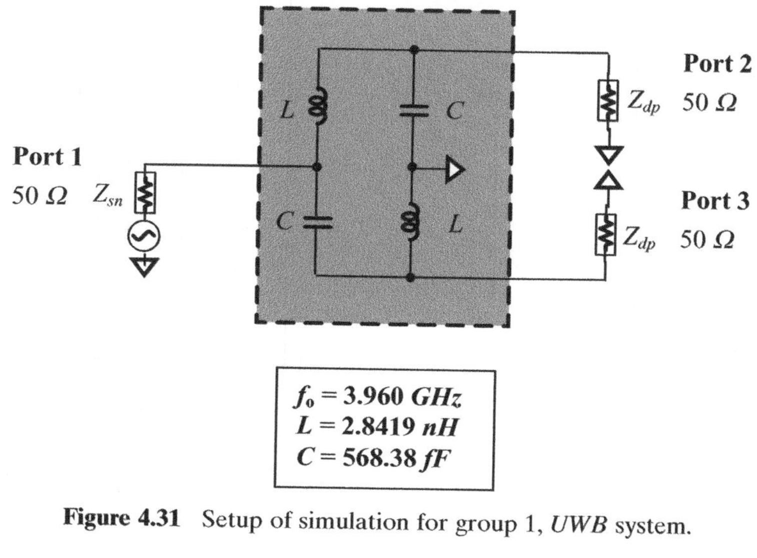

Figure 4.31 Setup of simulation for group 1, *UWB* system.

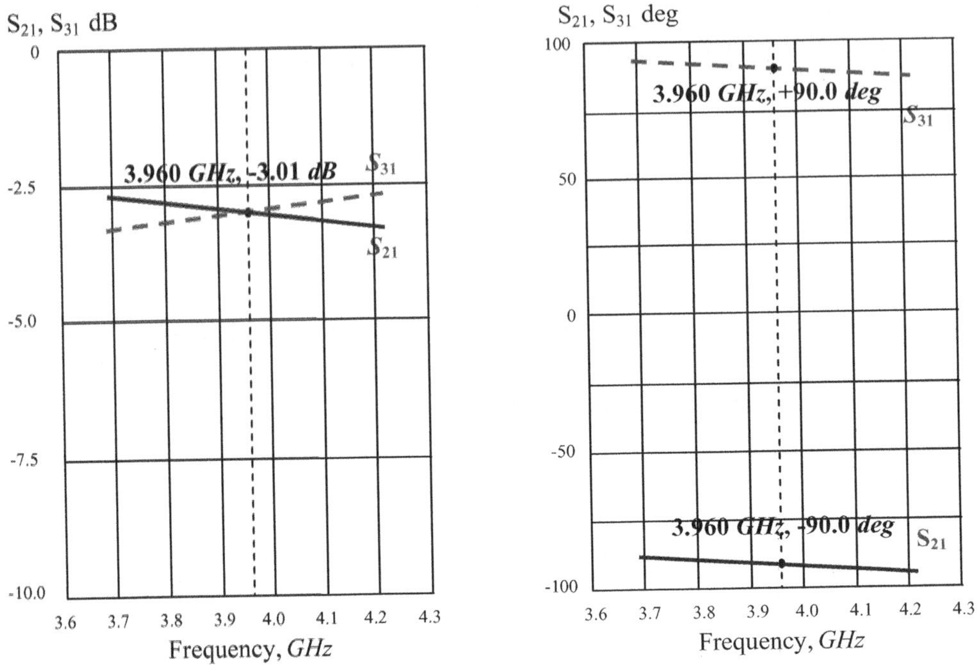

Figure 4.32 Insertion loss and phase shift of *LC* balun for group 1, *UWB* system.

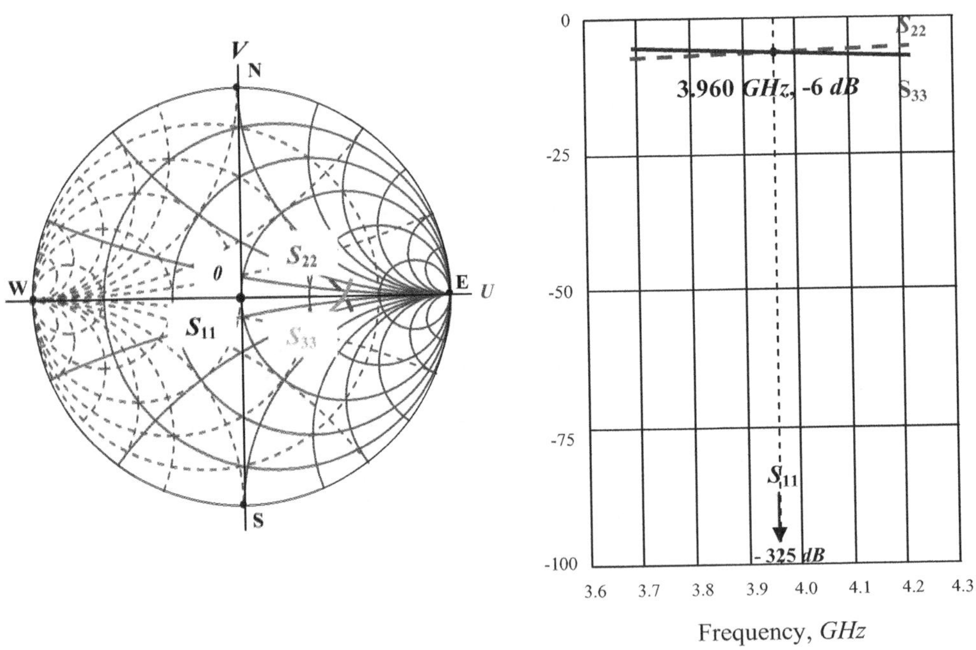

(a) On Smith chart (b) By rectangular coordination, *dB*

Figure 4.33 Return loss of *LC* balun for group 1 *UWB* system (S_{11}, S_{22}, S_{33}).

TABLE 4.3 Values of relevant S parameters of the ideal LC balun operating in the frequency range of group 1, UWB system: f = 3696 – 3960 – 4224 MHz, $Z_S = Z_L = 50\,\Omega$

Frequencies	f		3696 – 3960 – 4224	MHz
Bandwidth	Δf		528	MHz
Insertion loss	S_{21}	mag.	−3.01 ± 0.5 (max.)	dB
		phase	−90° ± 5° (max.)	
	S_{31}	mag.	−3.01 ± 0.5 (max.)	dB
		phase	+90° ± 5° (max.)	
Return loss	S_{11}	mag.	−325 ± …	dB
	S_{22}	mag.	−6.00 ± 0.75	dB
	S_{33}	mag.	−6.00 ± 0.75	dB

shown on the plot. At ports 2 and 3, $S_{22} = S_{33} = -6\,dB$. These values are not lowered as is S_{11}, due to the existence of three ports rather than two. However, this imperfection does not impact the simulation at all.

At other frequencies within the frequency bandwidth, f = 3696–4224 MHz, the frequency responses of S_{21} and S_{31} are not flat but tilted, since the capacitor C and inductor L are dependent on frequency. The tilted magnitude is less than 0.5 dB and the phase shift is less than 5°. Interestingly, the tilt directions of the phases of S_{21} and S_{31} are the same, so that the phase difference of the two differential ports is kept at around 180°. The magnitude of the return loss at port 1, S_{11}, drops down to about −325 dB so that it is not shown on the plot. At ports 2 and 3, S_{22} and S_{33} are approximately −6 dB, with a variation of less than 0.75 dB. These values are not lowered as is S_{11}, due to the existence of three ports rather than two. However, this imperfection, just like that at the operating frequency f_o = 3960 MHz, does not impact the simulation at all.

Simulation B: Simulation of the ideal LC balun operated in the frequency range of group 3, UWB System: f = 6864–7128 to 7392 MHz, f_o = 7128 MHz, $Z_S = Z_L$ = 50 Ω.

Figure 4.34 shows the setup of simulation for group 3, UWB system; Figures 4.35 and 4.36 display the simulated results. Figure 4.35 shows the insertion loss and phase shift while Figure 4.36 shows the return loss of the ideal transformer balun operating in the frequency range of group 3, UWB system. Table 4.4 lists the values of interested S parameters of their performance.

At the operating frequency, f_o = 7128 MHz, $S_{21} = S_{31}$ = 3.01 dB, which means that at each port of the differential pair, the impedance is well matched with the LC balun. In addition, the phase at one port of the differential is 90° and the phase at the other is −90°, so that the phase difference between the two ports is 180°. The magnitude of the return loss at port 1, S_{11}, drops down to −96 dB so that it is not shown on the plot. At ports 2 and 3, $S_{22} = S_{33} = -6\,dB$. These values are not lowered as is S_{11}, due to the existence of three ports rather than two. However, this imperfection does not impact the simulation at all.

At other frequencies within the frequency bandwidth, f = 6864–7392 MHz, the frequency responses of S_{21} and S_{31} are not flat, but tilted, since the capacitor C and the inductor L are dependent on frequency. The tilted magnitude is less than 0.5 dB and the phase shift is less than 5°. Interestingly, the tilt directions of the phases of

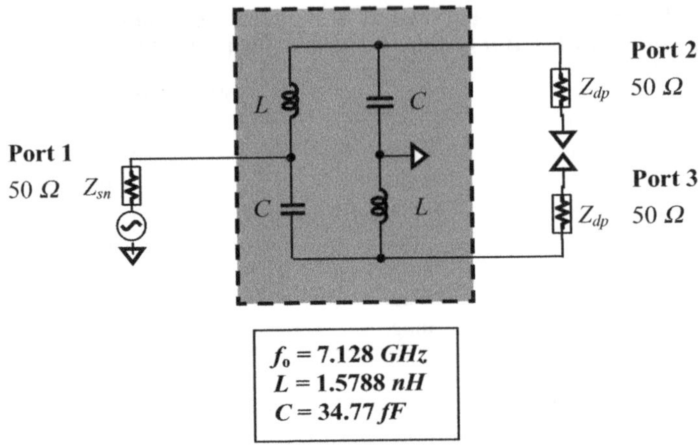

Figure 4.34 Setup of simulation for group 3, *UWB* system.

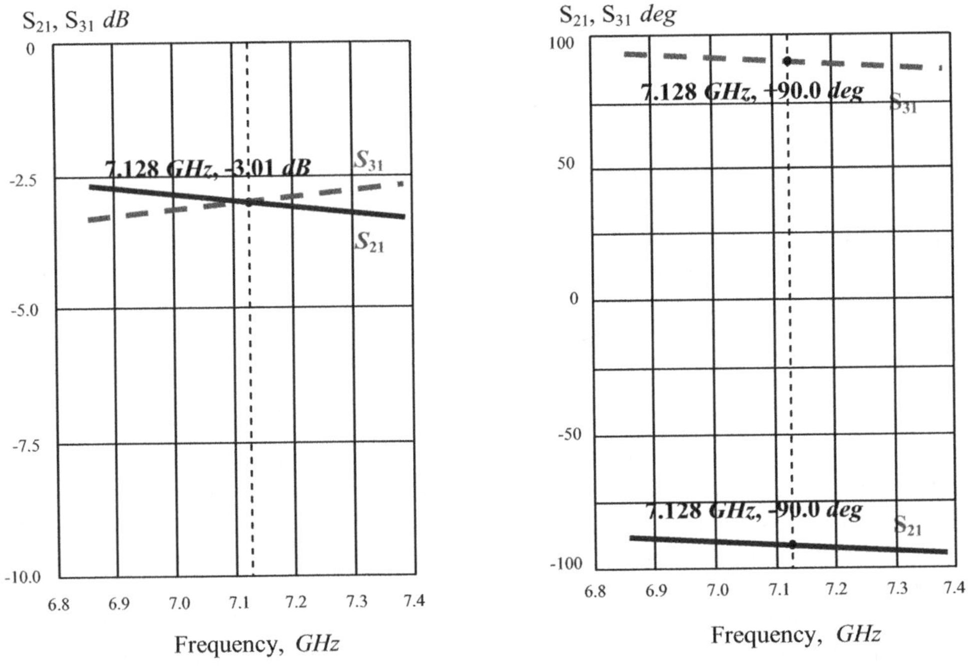

Figure 4.35 Insertion loss and phase shift of *LC* balun for group 3 *UWB* system (S_{21}, S_{31}).

S_{21} and S_{31} are the same, so that the phase difference of the two differential ports is kept at around 180°. The magnitude of the return loss at port 1, S_{11}, is dropped down to about $-96\,dB$ so that it is not shown on the plot. At ports 2 and 3, S_{22} and S_{33} are approximately $-6\,dB$, with a variation of less than $0.75\,dB$. These values are not lowered as is S_{11}, due to the existence of three ports rather than two. However, this imperfection, just like that at the operating frequency $f_o = 7128\,MHz$, does not impact the simulation at all.

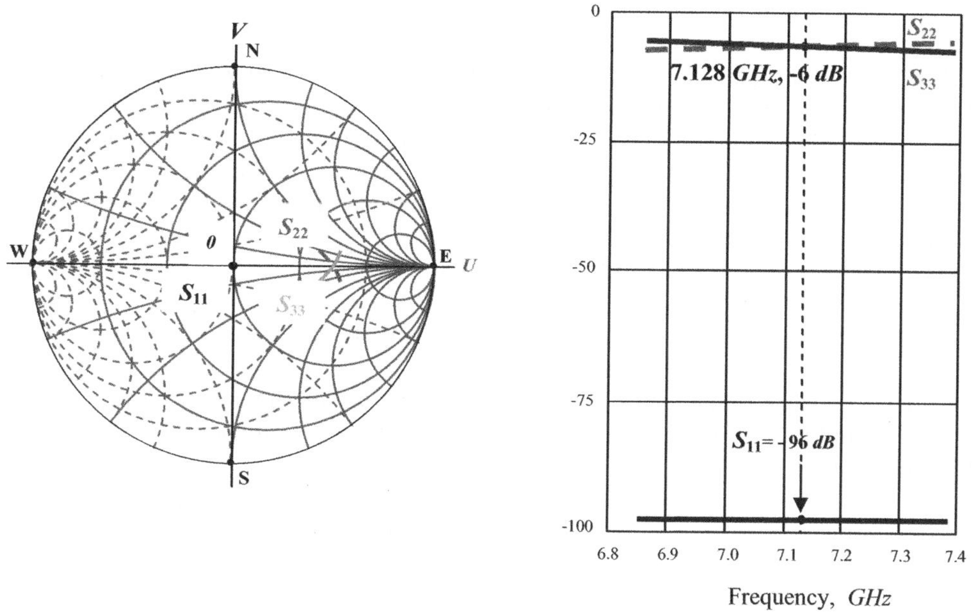

(a) On Smith chart (b) By rectangular coordination, *dB*

Figure 4.36 Return loss of *LC* balun for group 3 *UWB* system (S_{11}, S_{22}, S_{33}).

TABLE 4.4 Values of interested *S* parameters of the ideal *LC* balun operated in the frequency range of group 3, *UWB* system: f = 6864 – 7128 – 7392 *MHz*, $Z_S = Z_L = 50\,\Omega$

Frequencies	f		6864 – 7128 – 7392	*MHz*
Bandwidth	Δf		528	*MHz*
Insertion loss	S_{21}	mag.	−3.01 ± 0.5 (max.)	*dB*
		phase	−90° ± 5° (max.)	
	S_{31}	mag.	−3.01 ± 0.5 (max.)	*dB*
		phase	+90° ± 5° (max.)	
Return loss,	S_{11},	mag.	−96 ± …	*dB*
	S_{22},	mag.	−6.00 ± 0.75 (max.)	*dB*
	S_{33},	mag.	−6.00 ± 0.75 (max.)	*dB*

4.3.3 A Practical *LC* Balun

In Section 4.3.1 the simplicity of a *LC* balun was shown and in Section 4.3.2 its performance was demonstrated.

One might feel that the design of an *LC* balun is an easy task. It seems that all we need to do is calculate the values of *L* and *C* from equations (4.8) and (4.9) according to the impedance of the single-ended and differential pair ports, Z_{sn} and Z_{dp}. Unfortunately, it is generally not so simple.

First, it should be noted that in the demonstration of *LC* balun in Section 4.3.2, the impedances of all three ports were 50 Ω. This means that the *LC* balun applied in the demonstration was designed in a special case, in which Z_{sn} and Z_{dp} in the design equations (4.8) and (4.9) were both 50Ω. In most cases, Z_{sn} and Z_{dp} are complex values.

It follows that there is a problem with equations (4.8) and (4.9). In general cases, Z_{sn} and Z_{dp} are complex values, the product of Z_{sn}^* and Z_{dp} under the sign of square root is a complex number, by which it is impossible to get real values for the inductor L and the capacitor C. But, in reality, only inductors and capacitors with real values, not complex values, are available. As a matter of fact, this implies that there is a restriction in the design of the simple balun.

The product Z_{sn}^* and Z_{dp} can be expressed as

$$Z_{dp}Z_{sn}^* = (R_{dp} + jX_{dp})(R_{sn} - jX_{sn}) = (R_{dp}R_{sn} + X_{dp}X_{sn}) + j(X_{dp}R_{sn} - R_{dp}X_{sn}), \quad (4.12)$$

In order to ensure that the product of Z_{dp} and Z_{sn}^* under the square root is not a complex number, the imaginary portion of the product $Z_{dp}Z_{sn}^*$ must be a real number. This means that

$$X_{dp}R_{sn} - R_{dp}X_{sn} = 0, \quad (4.13)$$

or,

$$\frac{R_{sn}}{R_{dp}} = \frac{X_{sn}}{X_{dp}}. \quad (4.14)$$

The single or input impedance Z_{sn} must be proportional to Z_{dp} so that the ratio of their real parts is equal to the ratio of their imaginary parts as shown in equation (4.14).

Now, before the calculation for the values of L and C from equations (4.8) and (4.9), we must first check if condition (4.14) is satisfied or not. In special cases, if both the given Z_{sn} and the given Z_{dp} are pure resistances or reactances, this limitation will automatically disappear. However, for general cases, both the given Z_{sn} and Z_{dp} are neither pure resistance nor reactance, and the problem remains.

The unique solution to the problem is to modify either Z_{sn} or Z_{dp} by the addition of an extra part, either an inductor or a capacitor, so that the modified Z'_{sn} or Z'_{dp} satisfies condition (4.14). There are three considerations in the process of modification:

1) In order to avoid the increase of unnecessary insertion loss, only the imaginary part but not the real part of Z_{sn} or Z_{dp} is modified.
2) It is more economic if the modification is done at the single-ended side instead of the differential pair side.
3) It is more economic if a capacitor instead of an inductor is used for the modification.

In general cases, the original LC balun as shown in Figure 4.30 must be modified to that shown in Figures 4.37 and 4.38. Again, we are going to focus on the balun as a splitter not as a combiner.

In Figure 4.37, the capacitor C_{ms} is applied to modify the simple balun. It is inserted between port 1 and the input of the balun in series at the single-ended side.

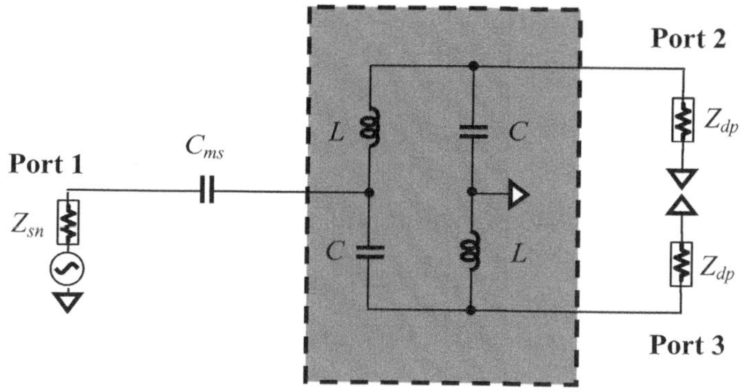

Figure 4.37 Simple *LC* balun is modified with insertion of a capacitor in series C_{ms} at the single-ended side.

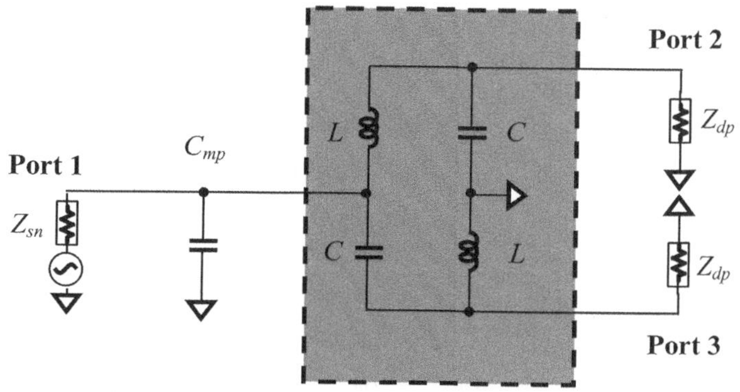

Figure 4.38 Simple *LC* balun is modified with insertion of a capacitor in parallel C_{mp} at the single-ended side.

In Figure 4.38, the capacitor C_{mp} is applied to modify the simple balun. It is connected between port 1 and the ground in parallel at the single-ended side.

Let's discuss the case as shown in Figure 4.37 where the capacitor C_{ms} is added in series to modify the simple *LC* balun. Another case as shown in Figure 4.38 where the capacitor C_{mp} is added *in parallel* to modify the simple *LC* balun will be left to the reader as an exercise.

Assuming that the original Z_{sn} and Z_{dp} are

$$Z_{sn} = r_{sn} + jx_{sn}, \tag{4.15}$$

$$Z_{dp} = r_{dp} - jx_{dp}, \tag{4.16}$$

where r_{sn}, x_{sn}, r_{dp}, and x_{dp} are assumed to be positive values.

At the single-ended port, the impedance will be changed from the original value of Z_{sn} as shown in (4.15) to the modified value Z'_{sn} by the insertion of the capacitaor C_{ms} in series, such that

$$Z'_{sn} = r_{sn} - jx'_{sn}, \tag{4.17}$$

where x'_{sn} is assumed to be a positive value as well, so that the values of Z'_{sn} and Z_{dp} satisfy limitation (4.14). This implies that

$$x'_{sn} = \frac{x_{dp}}{r_{dp}} r_{sn}. \tag{4.18}$$

The value of C_{ms} can be calculated from the difference ΔZ_{sn} between Z_{sn} and Z'_{sn}, that is

$$\Delta Z_{sn} = Z'_{sn} - Z_{sn} = (-jx'_{sn}) - jx_{sn} = -j(x'_{sn} + x_{sn}) = -j\left(\frac{x_{dp}}{r_{dp}} r_{sn} + x_{sn}\right). \tag{4.19}$$

$$C_{ms} = \frac{1}{\left(\frac{x_{dp}}{r_{dp}} r_{sn} + x_{sn}\right)\omega} \tag{4.20}$$

Where ω = operating angular frequency.

Here's an example with detailed values. Assuming that

$$Z_{sn} = r_{sn} + jx_{sn} = 80\,\Omega + j35\,\Omega,$$
$$Z_{dp} = r_{dp} - jx_{dp} = 100\,\Omega - j65\,\Omega,$$
$$f = 500\,MHz,$$

then

$$x'_{sn} = \frac{x_{dp}}{r_{dp}} r_{sn} = 52\,\Omega,$$

$$C_{ms} = \frac{1}{\left(\frac{x_{dp}}{r_{dp}} r_{sn} + x_{sn}\right)\omega} = 6.12\,pF,$$

$$Z'_{sn} = r_{sn} - jx'_{sn} = 80\,\Omega - j52\,\Omega,$$

$$Z_{dp} Z'^{*}_{sn} = r_{dp} r_{sn} + x_{dp} x'_{sn} = 11{,}380\,\Omega^2,$$

$$L = \frac{\sqrt{2 Z_{dp} Z'^{*}_{sn}}}{\omega} = \frac{\sqrt{2(r_{dp} r_{sn} + x_{dp} x'_{sn})}}{\omega} = \frac{\sqrt{22{,}760}}{3141.6 \cdot 10^6} = \frac{150.864}{3141.6} 10^{-6}\,H = 48.02\,nH$$

$$C = \frac{1}{\sqrt{2 Z_{dp} Z'^{*}_{sn}}\,\omega} = \frac{1}{\sqrt{2(r_{dp} r_{sn} + x_{dp} x'_{sn})}\,\omega} = \frac{10^{-6}}{150.864 \cdot 3141.6} = 2.11\,pF,$$

It should be noted that the frequency response of the modified *LC* balun shown in Figures 4.37 and 4.38 may be somewhat degraded from that of the simple balun shown in Figure 4.30(a). However, this modification is necessary because the product of Z_{sn} and Z_{dp} is not a real value in most cases.

Compared with the transformer balun, the drawback of the *LC* balun is its narrow-band frequency response. This is an inherent property of the *LC* balun because its basic parts, two inductors and two capacitors, are dependent on frequency. Nevertheless, the *LC* balun is widely applied in *RF* circuit design today, especially in the testing of differential circuitry. Besides the many merits mentioned above, the most attractive point of the *LC* balun is that it can be built by hand in minutes as long as the calculations and preparations for the two inductors and three capacitors, as shown in Figures 4.37 and 4.38, have been done correctly. Should the frequency bandwidth of the circuit block be wide, a *LC* balun with more than two inductors and three capacitors can be built to cover the entire bandwidth in the testing.

4.4 MICRO STRIP LINE BALUNS

In the low *RF* frequency range, the application of the micro strip line in the circuit design is restricted by its size. In high *RF* frequency range, however, the application of the micro strip line is receiving more attention year by year. A micro strip line could replace individual parts, such as inductors and capacitors, and even a simple network. For example, in many *RF* circuit blocks such as the *LNA* or *PA*, the recommendation is to implement all the input and output impedance matching networks by micro strip lines because

- In general, the simulation model of a micro strip line is more accurate than that of discrete parts.
- The tolerance of a micro strip line is more easily controlled than that of other parts.
- Most importantly, the application of the micro strip line is the most cost-effective, especially in *RF* module design or in *RF* block design by discrete parts.

In the last decades, many micro strip line type baluns have been developed. It would be impossible to describe all of them in this section. Instead, we are going to introduce the ring micro strip balun, which has been welcomed by *RF* designers.

4.4.1 Ring Baluns

The ring micro strip line balun is a simple but sophisticated balun applied in the frequency range from *UHF* to tens of *GHz*. It is an important part in *RF* circuit design and testing of *RF* blocks.

As shown in Figure 4.39, the main body of the ring micro strip line balun is a metallic ring on the *PCB* in *RF* module design or on the *IC* substrate on the *IC* chip. The ring circumference is equal to 1.5 wavelengths of the electric length and its width must be chosen so that its characterized impedance is equal to $2^{0.5}Z_o$. The characterized impedance of the four tapes must be equal to the source or load impedance $Z_o = 50\,\Omega$. The arc of the ring is one quarter-wavelength from port 1 to port 3, port 3 to port 2, and port 2 to port 4, but is three quarter-wavelengths from port 4 to port 1. If port 1 is chosen as the single-ended input port, then port 2 is a virtual grounding because it is distanced from port 1 by one-half wavelength. Ports 3 and 4 are assigned as the two differential ports.

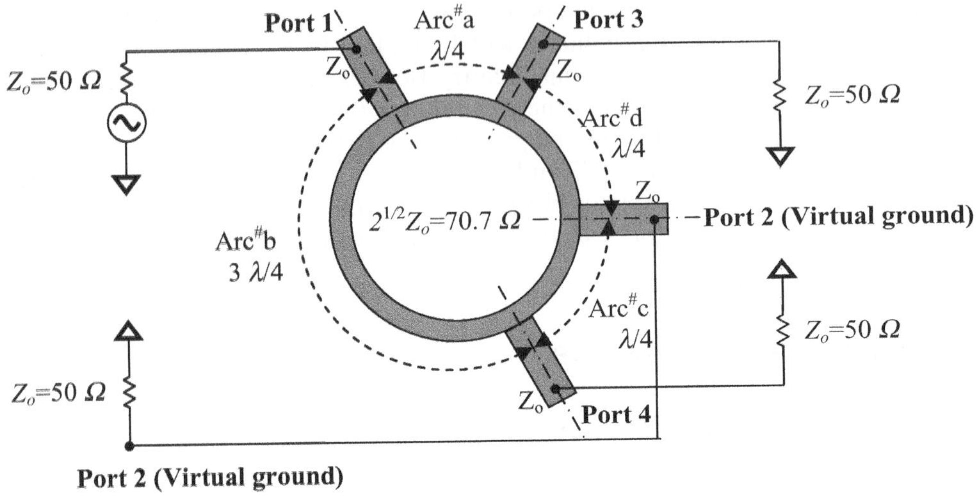

Figure 4.39 Layout of ring micro strip line balun.

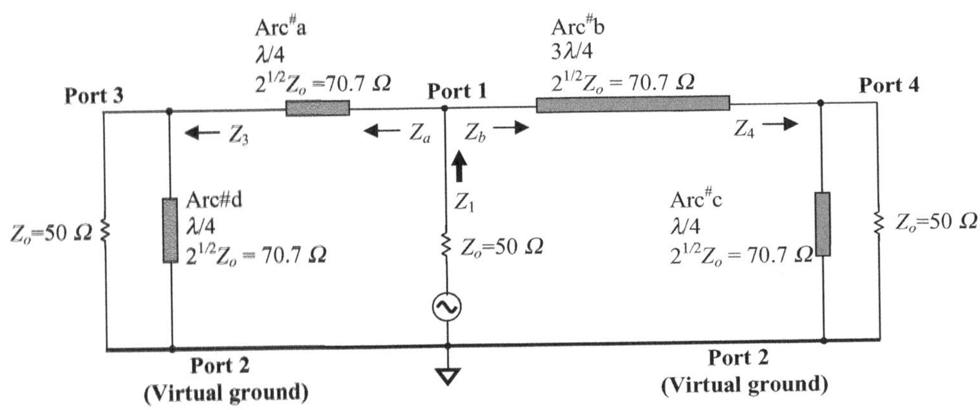

Figure 4.40 Equivalent circuit of ring micro strip balun.

Figure 4.40 shows its equivalent circuit.

- Port 2 is a virtual ground.
- The impedance looking from port 3 toward port 2 is

$$Z_3 = Z_o = 50\,\Omega, \tag{4.21}$$

where Z_3 = Impedance looking from port 3 toward port 2, because arc d is one quarter of the wavelength from the virtual grounded port 2, so that its impedance is infinite.

- The impedance looking from port 4 toward port 2 is

$$Z_4 = Z_o = 50\,\Omega, \tag{4.22}$$

where Z_4 = impedance looking from port 4 toward port 2,
because arc c is one quarter of the wavelength from the virtual grounded port 2, so that its impedance is infinite.

- The impedance looking from port 1 toward arc a or arc b is

$$Z_a = Z_b = 2Z_o = \frac{(2^{1/2} Z_o)^2}{Z_o}, \tag{4.23}$$

because arc a and arc b are odd multiples of quarter wavelength,
where

Z_a = impedance looking from port 1 toward arc a,

Z_b = impedance looking from port 1 toward arc b,

$2^{0.5}Z_o$ = characteristic impedance of the micro strip line.

- The impedance looking outward from port 1 is

$$Z_1 = Z_a // Z_b. \tag{4.24}$$

From expression (4.23), we have

$$Z_1 = Z_o. \tag{4.25}$$

It is therefore concluded that the impedance looking at port 1 into the ring is matched with the source impedance Z_o.

Relative to the virtual grounded port 2, the phase shift at port 1 is 180°. The phase shift at port 3 from the virtual grounded port 2 must be 90° because the length of arc a is one quarter-wavelength from port 1, while the phase shift at port 4 from the virtual grounded port 2 is –90° or 270° because the length of arc b is one quarter-wavelength from port 2 but three quarter-wavelengths from port 1.

The ring micro strip line balun is also the called "rat race" balun. It is a narrow-band balun, with a relative bandwidth usually less than 15%. It has been effectively applied in the development of cellular phones and other communication systems today.

4.4.2 Split Ring Baluns

For convenience in layout, the circular ring in a ring micro strip line balun can be cut into smaller pieces. Figure 4.41 is one design example, in which the ring is cut into ten pieces.

It is implemented on an aluminum ceramic substrate. Usually, the aluminum ceramic substrate has a higher permittivity, so that the size of ring balun can be reduced. In this example, the permittivity of the aluminum ceramic substrate, ε_r, is 10.5, while the permittivity of a plastic substrate is about 4.5. This means that the size of a module built with the ceramic substrate will be reduced by more than one-half compared with a module built with a plastic substrate. The thickness of the aluminum ceramic substrate is 25 *mils*.

The performance of this balun is shown in Figure 4.42. The magnitudes of S_{21} and S_{31} are between –3.58 to –3.32 dB in the frequency range from 800 to 900 MHz. Thus, the additional attenuation is only 0.3 to 0.6 dB throughout the entire frequency

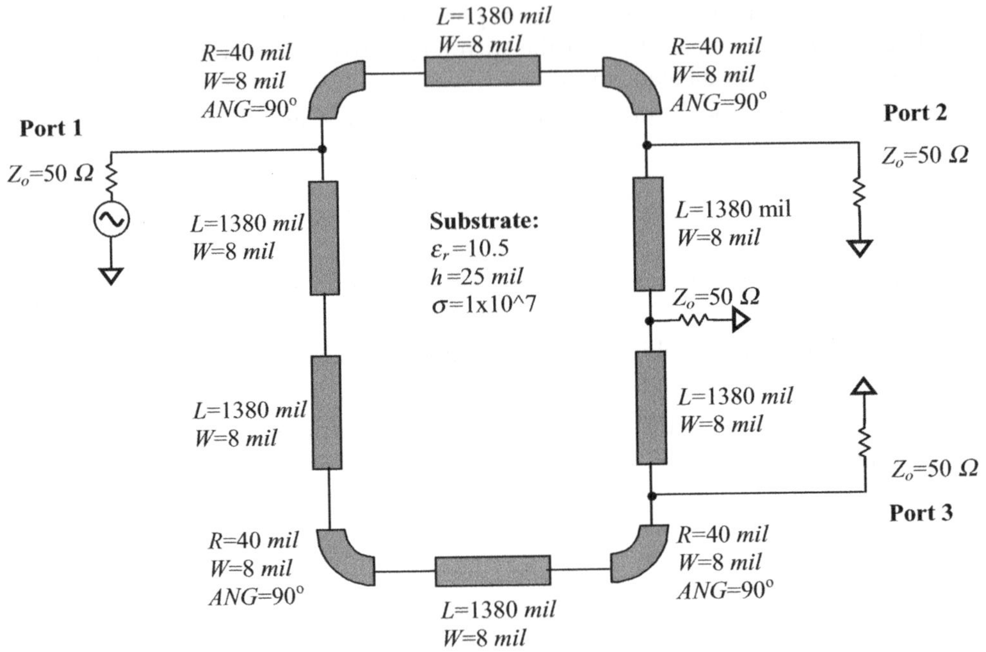

Figure 4.41 Split ring micro strip line balun. $f = 800$ to $900\,MHz$.

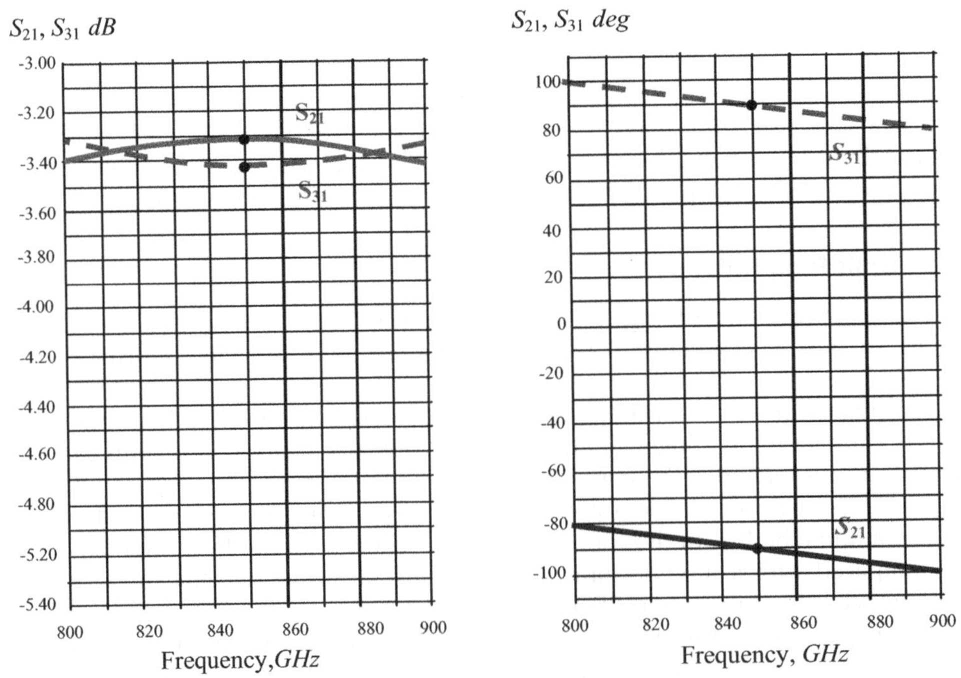

Figure 4.42 Insertion loss and phase shift of split ring balun. $f = 800$ to $900\,MHz$.

bandwidth. At the central frequency, $f = 850\,MHz$, the phases of S_{21} and S_{31} are exactly $90°$ and $-90°$ respectively, while at other frequencies, the frequency response of the phase is tilted over the bandwidth. Interestingly, the tilted directions of the phases of S_{21} and S_{31} are the same so that the phase difference is kept at around $180°$.

4.5 MIXED TYPES OF BALUNS

To reduce size, partial micro strip lines can be replaced by capacitors. In the first example, the micro strip line is replaced by a chip capacitor. Furthermore, if all the micro strip lines are replaced by inductors and capacitors, the size will be significantly reduced; this is demonstrated in the second example.

4.5.1 Baluns Built with Micro Strip Lines and Chip Capacitors

In the balun shown in Figure 4.41, the total length of the micro strip line is 8440 *mils*, or 8.44 inches. This is too long for some smaller products. It is therefore desirable to reduce this length so that the size of product can be shrunk.

The balun shown in Figure 4.43 is modified from the balun shown in Figure 4.41, in which partial micro strip lines are replaced by six chip capacitors. The total length of the micro strip line now is 5280 *mils*, or 5.28 inches, so that 37% of the total length has been saved. Its performance is shown in Figure 4.44. Comparing Figures 4.44

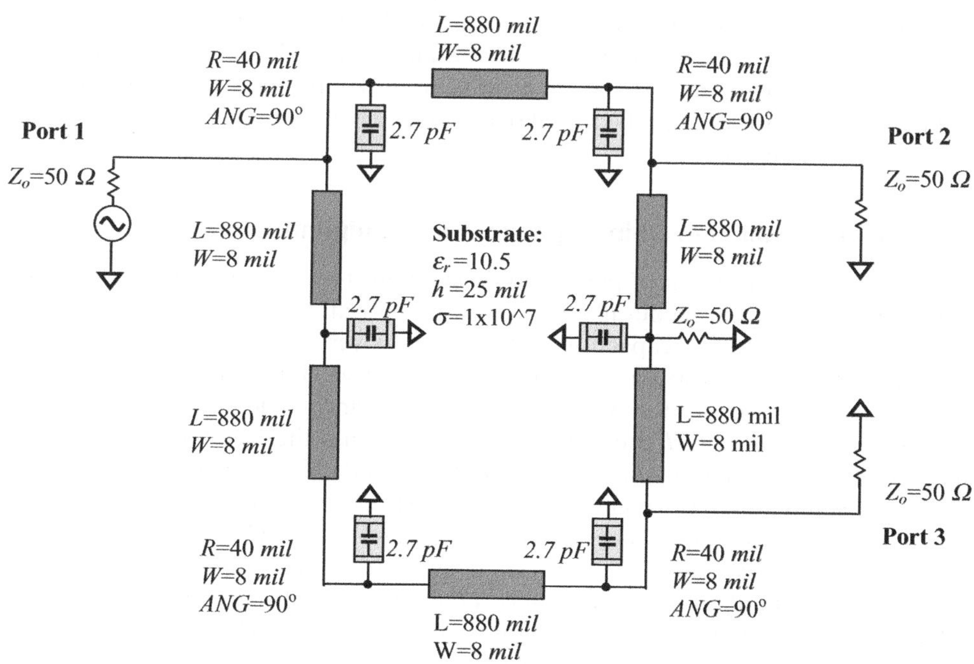

Figure 4.43 Modified split ring micro strip line balun. $f = 800$ to $900\,MHz$.
Chip capacitor

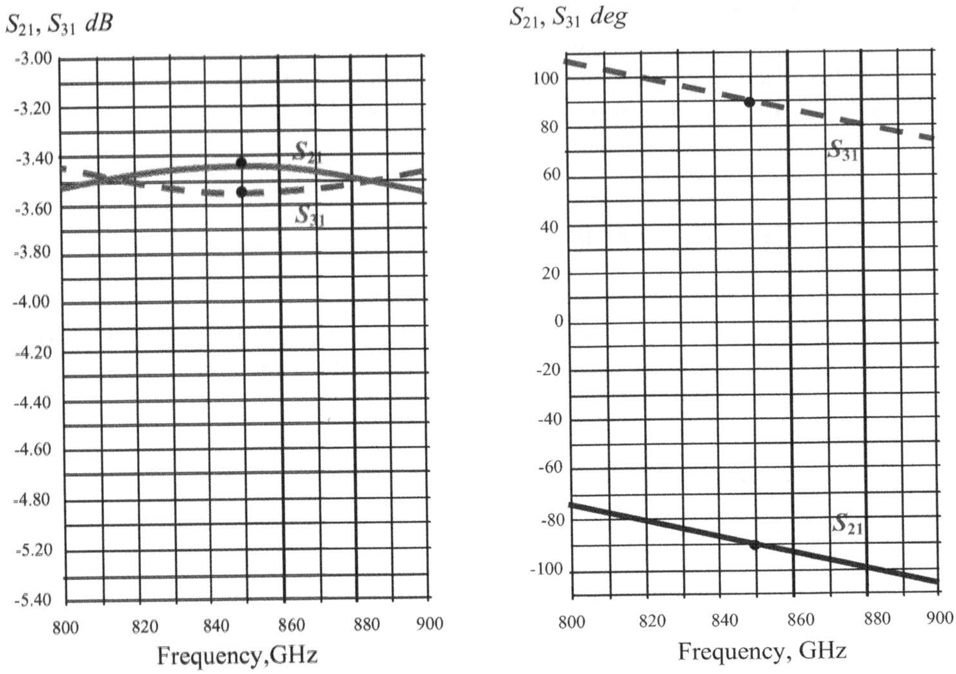

Figure 4.44 Insertion loss and phase shift of modified split ring balun. $f = 800$ to $900\,MHz$.

and 4.42, it is found they are similar. In the modified ring balun, the additional attenuation is about 0.1 to $0.2\,dB$ and the phase shift is tilted 10 to 20 degrees more, but the difference of the phase in the differential pair is still kept at around $180°$.

It should be noted that the actual model of chip capacitors is adapted in the simulation of circuitry so that the actual performance is close to the simulation result as shown in Figure 4.42.

4.5.2 Baluns Built with Chip Capacitors and Chip Inductors

The successful replacement of partial micro strip lines by chip capacitors shown in Figures 4.43 and 4.44 is encouraging. Is it possible for all the micro strip lines to be replaced by inductors and capacitors? The answer is yes. Figure 4.45 shows a balun built by only chip inductors and capacitors. There are six chip inductors with 7.3 nH of inductance and 6 capacitors with 7.5 pF of capacitance which form a "ring." The three capacitors with 39 pF are "zero" capacitors within the frequency bandwidth. In the simulation, actual models of chip inductors and chip capacitors are adapted so that the simulated performance and actual test results are closer together.

In Figure 4.45, the impedance the single-ended port is $50\,\Omega$, but the impedance at the differential pair ports is $20\,\Omega$. This is an actual design for a differential PA and it is implemented on the aluminum ceramic substrate. Its performance is shown in Figure 4.46. The additional attenuation is about 0.6 to $0.8\,dB$ and the phase shift is tilted more than the balun shown in Figure 4.44; however, the phase difference between differential pair ports is still kept around $180°$. This kind of balun should be quite useful in the products with RF module design.

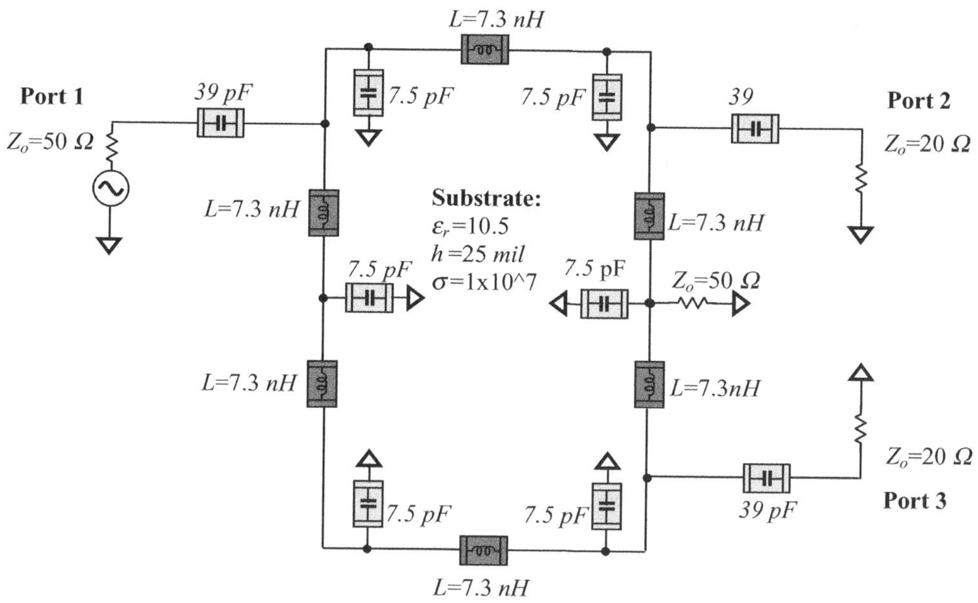

Figure 4.45 Ring balun is built by chip inductors and chip capacitors. $f = 800$ to $900\,MHz$. Chip capacitor Chip inductor

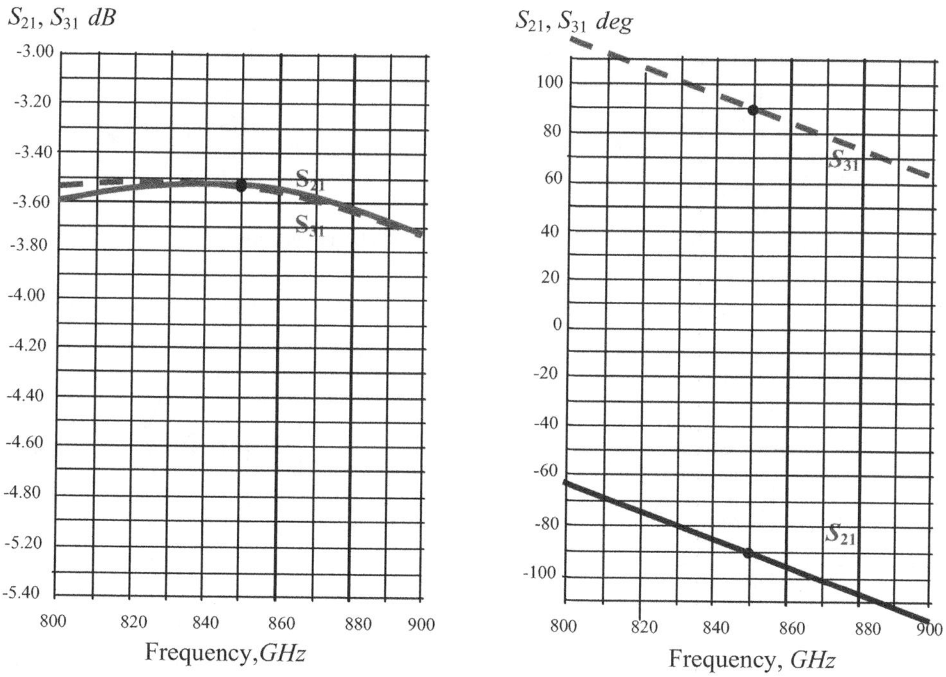

Figure 4.46 Insertion loss and phase shift of a balun built by chip inductors and chip capacitors. $f = 800$ to $900\,MHz$.

APPENDICES

4.A.1 Relationship Between Impedance and Turn Ratio of Transformer Baluns with Two Stacked Transformers

There are four ways to build a transformer balun by two stacked transformers with any turn ratio of 1: *n*.

a) ***SS-DS*** connection mode: "**S**ingle-ended, primary windings stacked in **S**eries, goes to **D**ifferential, secondary windings stacked in **S**eries."
b) ***SP-DS*** connection mode: "**S**ingle-ended, primary windings stacked in **P**arallel, goes to **D**ifferential, secondary windings stacked in **S**eries."
c) ***DS-SS*** connection mode: "**D**ifferential, primary windings stacked in **S**eries, goes to **S**ingle-ended, secondary windings stacked in **S**eries."
d) ***DS-SP*** connection mode: "**D**ifferential, primary windings stacked in **S**eries, goes to **S**ingle-ended, secondary windings stacked in **P**arallel."

Figure 4.A.1 shows four ways to build a transformer balun by two stacked transformers with their turn ratio, 1 : *n*. The 1 $G\Omega$ resistor is a dummy part to satisfy the requirement for a *DC* circuit-loop in either transformer 1 or transformer 2.

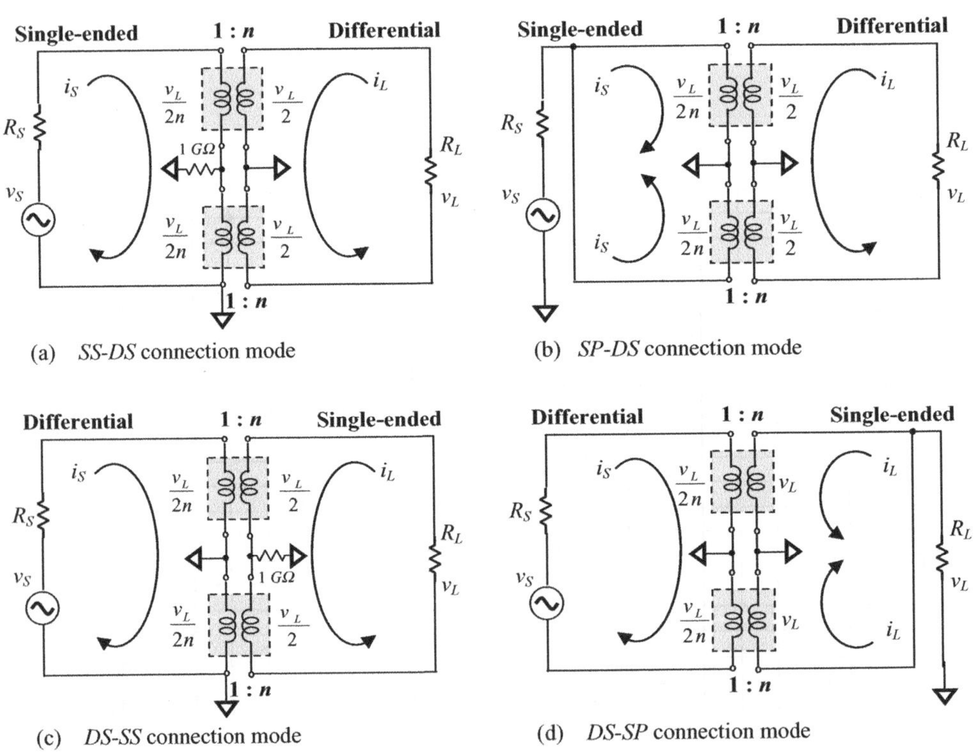

(a) *SS-DS* connection mode
(b) *SP-DS* connection mode
(c) *DS-SS* connection mode
(d) *DS-SP* connection mode

Figure 4.A.1 Four connection modes to build a transformer balun with two stacked transformers.

The relationship between *RL, RS* and the turn ratio *n* can be derived and expressed as follows:

a) ***SS-DS*** connection mode: "**S**ingle-ended, primary windings stacked in **S**eries, goes to **D**ifferential, secondary windings stacked in **S**eries."
From Figure 4.A.1(a), we have

$$v_S = \frac{2v_L}{2n} = \frac{v_L}{n} = \frac{i_L R_L}{n} = \frac{i_S}{n}\frac{R_L}{n} = \frac{i_S R_L}{n^2}, \tag{4.A.1}$$

then,

$$R_S = \frac{v_S}{i_S} = \frac{R_L}{n^2}. \tag{4.A.2}$$

b) ***SP-DS*** connection mode: "**S**ingle-ended, primary windings stacked in **P**arallel, goes to **D**ifferential, secondary windings stacked in **S**eries."
From Figure 4.A.1(b), we have

$$v_S = \frac{v_L}{2n} = \frac{i_L R_L}{2n} = \frac{i_S R_L}{2n^2}, \tag{4.A.3}$$

then,

$$R_S = \frac{v_S}{2i_S} = \frac{R_L}{4n^2}. \tag{4.A.4}$$

c) ***DS-SS*** connection mode: "**D**ifferential, primary windings stacked in **S**eries, goes to **S**ingle-ended, secondary windings stacked in **S**eries.
From Figure 4.A.1(c), we have

$$v_S = \frac{2v_L}{2n} = \frac{v_L}{n} = \frac{i_L R_L}{n} = \frac{i_S}{n}\frac{R_L}{n} = \frac{i_S R_L}{n^2}, \tag{4.A.5}$$

then

$$R_S = \frac{v_S}{i_S} = \frac{R_L}{n^2}. \tag{4.A.6}$$

d) ***DS-SP*** connection mode: "**D**ifferential, primary windings stacked in **S**eries, goes to **S**ingle-ended, secondary windings stacked in **P**arallel."
From Figure 4.A.1(d), we have

$$v_S = \frac{2v_L}{n} = 2\frac{2i_L R_L}{n} = 2\frac{2\frac{i_S}{n}R_L}{n} = 4\frac{i_S R_L}{n^2}, \tag{4.A.7}$$

then,

$$R_S = \frac{v_S}{i_S} = 4\frac{R_L}{n^2}. \tag{4.A.8}$$

4.A.2 Analysis of Simple *LC* Baluns

The following analysis will be aimed at the splitter as shown in Figure 4.A.2, because the combiner is simply a reversed splitter. (This analysis was developed by author in the 1990s and is abstracted from Richard Chi-Hsi Li, *Key Issues in RF/RFIC Circuit Design*, Higher Education Press, Beijing, 2005, pp. 220–225.)

There is a single port with its impedance Z_{sn} and two balanced ports with their impedance Z_{dp}. They are

$$Z_{sn} = R_{sn} + jX_{sn}, \tag{4.A.9}$$

$$Z_{dp} = R_{dp} + jX_{dp}, \tag{4.A.10}$$

where

Z_{sn} = iImpedance of single-ended port,
R_{sn} = real part of single-ended port impedance,
X_{sn} = imaginary part of single-ended port impedance,
Z_{dp} = impedance of one differential pair port,
R_{dp} = real part of impedance at one differential pair port,
X_{dp} = imaginary part of impedance at one differential pair port.

For the analysis, let's redraw Figure 4.A.2 and Figure 4.A.3. The impedances looking into the right-hand and left-hand sides from the single-ended port are

$$Z_a = Z_c + Z_L // Z_{dp}, \tag{4.A.11}$$

$$Z_b = Z_L + Z_c // Z_{dp}, \tag{4.A.12}$$

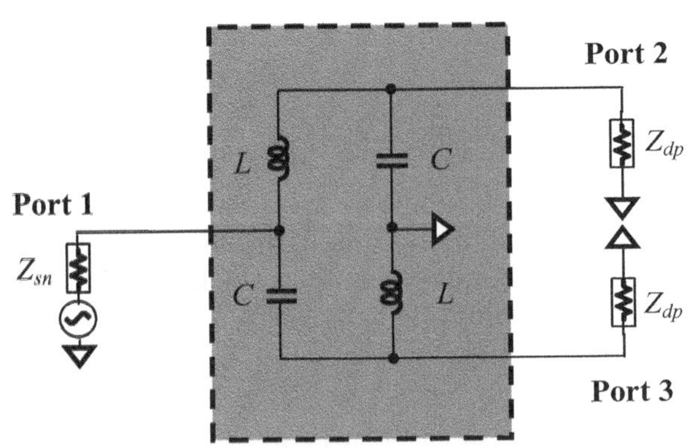

Figure 4.A.2 *LC* balun functions as a splitter.

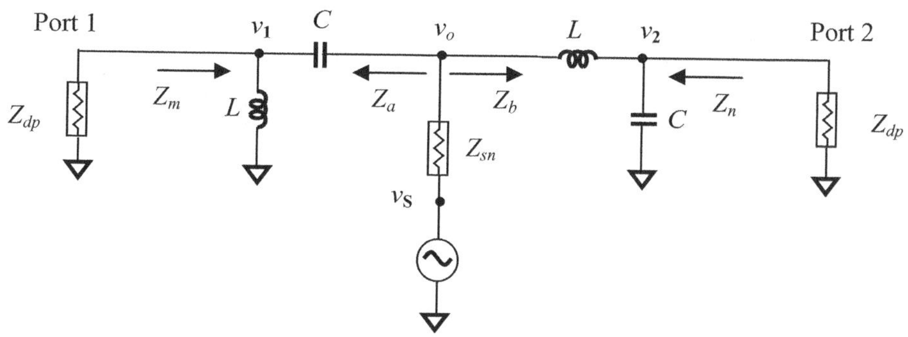

Figure 4.A.3 Redrawing of the splitter in Figure 4.A.2.

where

$$Z_c = \frac{1}{j\omega C}, \tag{4.A.13}$$

$$Z_L = j\omega L, \tag{4.A.14}$$

and ω = operating angular frequency.
Then,

$$Z_a = \frac{(1-\omega^2 LC)Z_{dp} + j\omega L}{-\omega^2 LC + j\omega C Z_{dp}}, \tag{4.A.15}$$

$$Z_b = \frac{(1-\omega^2 LC)Z_{dp} + j\omega L}{1 + j\omega C Z_{dp}}. \tag{4.A.16}$$

These two impedances are connected together in parallel from the single-ended port to the ground. The resulting impedance must be matched and is equal to the conjugated value of Z_{sn}, that is,

$$Z_{sn}^* = R_{sn} - jX_{sn} = \frac{Z_a Z_b}{Z_a + Z_b}. \tag{4.A.17}$$

By substituting (4.A.15) and (4.A.16) into (4.A.17), we have

$$Z_{sn}^* = \frac{(1-\omega^2 LC)Z_{dp} + j\omega L}{(1-\omega^2 LC) + j2\omega C Z_{dp}}. \tag{4.A.18}$$

$$Z_{sn}^* = \frac{(1-\omega^2 LC)R_{dp} + j[(1-\omega^2 LC)X_{dp} + \omega L]}{(1-\omega^2 LC - 2\omega C X_{dp}) + j2\omega C Z_{dp}}. \tag{4.A.19}$$

From (4.A.17) and (4.A.19),

$$R_{sn}=\frac{(1-\omega^2LC)R_{dp}(1-\omega^2LC-2\omega CX_{dp})+2\omega CR_{dp}[(1-\omega^2LC)X_{dp}+\omega L]}{(1-\omega^2LC-2\omega CX_{dp})^2+(2\omega CR_{dp})^2}. \quad (4.A.20)$$

$$X_{sn}=\frac{2\omega CR_{dp}(1-\omega^2LC)R_{dp}-(1-\omega^2LC-2\omega CX_{dp})[(1-\omega^2LC)X_{dp}+\omega L]}{(1-\omega^2LC-\omega CX_{dp})^2+(2\omega CR_{dp})^2}. \quad (4.A.21)$$

The equations (4.A.19), (4.A.20), and (4.A.21) can be simplified if

$$\omega^2LC=1. \quad (4.A.22)$$

This condition is expected. The power can be well transmitted from the single port to the differential ports or vice versa only when the balun core is a resonant-type-core. Under condition (4.A.22), equations (4.A.15), (4.A.16), (4.A.19), (4.A.20), and (4.A.21) can be simplified as

$$Z_a=\frac{j\omega L}{-1+j\omega CZ_{dp}}, \quad (4.A.23)$$

$$Z_b=\frac{j\omega L}{1+j\omega CZ_{dp}}, \quad (4.A.24)$$

$$Z_{sn}^*=\frac{L}{2CZ_{dp}}, \quad (4.A.25)$$

$$R_{sn}=\frac{L}{2C}\frac{R_{dp}}{(R_{dp}^2+X_{dp}^2)}, \quad (4.A.26)$$

$$X_{sn}=\frac{L}{2C}\frac{X_{dp}}{(R_{dp}^2+X_{dp}^2)}, \quad (4.A.27)$$

From equations (A.22) and (A.25), two conclusive equations listed below are obtained.

$$\omega L=\sqrt{2Z_{dp}Z_{sn}^*}, \quad (4.A.28)$$

$$\omega C=\frac{1}{\sqrt{2Z_{dp}Z_{sn}^*}}. \quad (4.A.29)$$

These two equations are very useful to the design because the values of L and C can be easily calculated from the given impedances Z_{dp} and Z_{sn}.

Now, let's examine the impedance looking from the balanced ports, Z_m and Z_n.

$$Z_m=(Z_b//Z_{sn}+Z_c)//Z_L, \quad (4.A.30)$$

$$Z_n=(Z_a//Z_n+Z_L)//Z_c. \quad (4.A.31)$$

From equations (4.A.11), (4.A.12), (4.A.28), and (4.A.29),

$$Z_m=\left(1+\frac{2Z_{sn}^*}{Z_{sn}}\right)Z_{dp}, \quad (4.A.32)$$

$$Z_n = \left(1 + \frac{2Z_{sn}^*}{Z_{sn}}\right) Z_{dp}. \tag{4.A.33}$$

Therefore,

$$Z_m = Z_n, \tag{4.A.34}$$

which proves that the two balanced ports are balanced.

It should be noted that from equation (4.A.32) or (4.A.33)

$$Z_m = Z_n = 3Z_{dp}, \tag{4.A.35}$$

if Z_{sn} is a pure resistor,

$$Z_{sn} = R_{sn}. \tag{4.A.36}$$

Expression (4.A.35) shows that Z_m and Z_n are not conjugate-matched to the balanced impedance of the two balanced ports respectively. This is not surprising because the LC balun is a network with three ports.

Furthermore, we are going to examine the phase relation between the single-ended port and the differential pair ports. Referring to Figure 4.A.3,

$$v_1 = \frac{Z_{dp} // Z_L}{Z_{dp} // Z_L + Z_c} v_o, \tag{4.A.37}$$

$$v_2 = \frac{Z_{dp} // Z_c}{Z_{dp} // Z_c + Z_L} v_o. \tag{4.A.38}$$

By means of equations (4.A.11) to (4.A.14), (4.A.28), and (4.A.29), and noting that

$$v_o = \frac{Z_{sn}^*}{Z_{sn} + Z_{sn}^*} v_s, \tag{4.A.39}$$

then

$$v_1 = j\frac{\sqrt{2(R_{dp}R_{sn} + X_{dp}X_{sn})}}{4R_{dp}} v_s, \tag{4.A.40}$$

$$v_2 = -j\frac{\sqrt{2(R_{dp}R_{sn} + X_{dp}Xs_{in})}}{4R_{dp}} v_s. \tag{4.A.41}$$

It means that the voltage at differential pair port 1 is 90° ahead of the single-ended port and the voltage at differential pair port 2 is 90° behind of the single-ended port. It is therefore concluded that the two differential pair ports are differential as long as condition (4.A.22) is satisfied.

TABLE 4.A.1 ***L* and *C* values of simple *LC* balun for group 1 and group 3 in *UWB* system**

Z_{sn}	Z_{dp}	(2*$Z_{sn}Z_{dp}$)^0.5	*f, Hz*	*ω, rad/s*	*L, nH*	*C, pF*
Group#1						
50	50	70.71067812	3168000000	19905177600	3.552376	0.710475
50	50	70.71067812	3432000000	21563942400	3.279116	0.655823
50	50	70.71067812	3696000000	23222707200	3.044894	0.608979
50	**50**	**70.71067812**	**3696000000**	**23222707200**	**3.044894**	**0.608979**
50	**50**	**70.71067812**	**3960000000**	**24881472000**	**2.841901**	**0.568380**
50	**50**	**70.71067812**	**4224000000**	**26540236800**	**2.664282**	**0.532856**
50	50	70.71067812	4224000000	26540236800	2.664282	0.532856
50	50	70.71067812	4488000000	28199001600	2.507560	0.501512
50	50	70.71067812	4752000000	29857766400	2.368251	0.473650
Group#3						
50	50	70.71067812	6336000000	39810355200	1.776188	0.355238
50	50	70.71067812	6600000000	41469120000	1.705141	0.341028
50	50	70.71067812	6864000000	43127884800	1.639558	0.327912
50	**50**	**70.71067812**	**6864000000**	**43127884800**	**1.639558**	**0.327912**
50	**50**	**70.71067812**	**7128000000**	**44786649600**	**1.578834**	**0.315767**
50	**50**	**70.71067812**	**7392000000**	**46445414400**	**1.522447**	**0.304489**
50	50	70.71067812	7392000000	46445414400	1.522447	0.304489
50	50	70.71067812	7656000000	48104179200	1.469949	0.293990
50	50	70.71067812	7920000000	49762944000	1.420950	0.284190

4.A.3 The Values of *L* and *C* of a Simple *LC* Balun for Group 1 and Group 3 in a *UWB* System

Table 4.A.1 presents the values of *L* and *C* of a simple *LC* balun for group 1 and group 3 in a *UWB* system.

4.A.4 Equivalency of Parts Between Single-ended and Differential Pairs with Respect to a Simple *LC* Balun

Just like the transformer balun, the equivalence of parts between the single-ended and differential pair also exists in a simple LC balun, so that these parts can be interpreted from each other in terms of the following rules:

- An inductor, $L/2$, in parallel at the single side, is interpreted as an inductor, L, in series at the differential branch, or vice versa.
- A capacitor, $2C$, in parallel at the single side is interpreted as a capacitor, C, in series at the differential branch, or vice versa.
- An inductor, $L/2$, in series at the single side is interpreted as an inductor, L, in parallel at the differential branch, or vice versa.
- A capacitor, $2C$, in series at the single side, is interpreted as a capacitor, C, in parallel at the differential branch, or vice versa.
- The value of the inductor or capacitor at the differential branch is determined by the impedances of the single-ended source and the differential pair load, such that,

$$L\omega_o = \sqrt{2Z_{dp}Z_{sn}^*}, \tag{4.A.28}$$

and

$$C\omega_o = \frac{1}{\sqrt{2Z_{dp}Z_{sn}^*}}, \tag{4.A.29}$$

where

L = inductance of the inductor in the simple LC balun,
C = capacitance of the capacitor in the simple LC balun,
Z_{sn} = source impedance at single-ended port,
Z_{dp} = load impedance at each port of differential pair,
ω_o = operating angular frequency,

It should be noted that, unlike the interpretation in transformer balun, if a capacitor is added to the single-ended side, then the corresponding part to be added to the differential pair side is also a capacitor; likewise, if a inductor is added to the single-ended side, then the corresponding part to be added to the differential pair side is also an inductor.

Figure 4.A.4 shows five cases in which a part is added to the single-ended port and the interpreted part is added to each port of differential pair respectively. Formulas (4.8) and (4.9) define the relationship between the added parts, by which the added parts can be interpreted from each other. Simulations prove that before and after the parts are added to the transformer balun, the S parameters or impedances at all three ports are unchanged at the operating frequency, ω_o or f_o, and are mostly unchanged or only changed slightly within a certain bandwidth around the operating frequency.

In order to verify the interpretation regulation and to check the equivalence of the impedances looking into and looking outwards from the transformer balun, let's conduct another five experimental simulations for the verification of the interpretation regulation. The number of parts added will be increased from zero to four in steps so that they are not the same as shown in Figure 4.A.4, where only one part is added to either the single-ended port or each port of the differential pair.

Figures 4.A.5 to 4.A.7 show simulation A, conducted for the original LC balun with no added parts. Figures 4.A.8 to 4.A.10 show simulation B, conducted for the original LC balun with one added part. Figures 4.A.11 to 4.A.13 show simulation C, conducted for the original LC balun with two added parts. Figures 4.A.14 to 4.A.16 show simulation D, conducted for the original LC balun with three added parts. Figures 4.A.17 to 4.A.19 show simulation E, conducted for the original LC balun with four added parts.

It should be noted that

- In the case of a simple LC balun, either L or C must be selected using formulas (4.A.28) and (4.A.29), which are dependent on the impedances of the single-ended and differential pair ports.

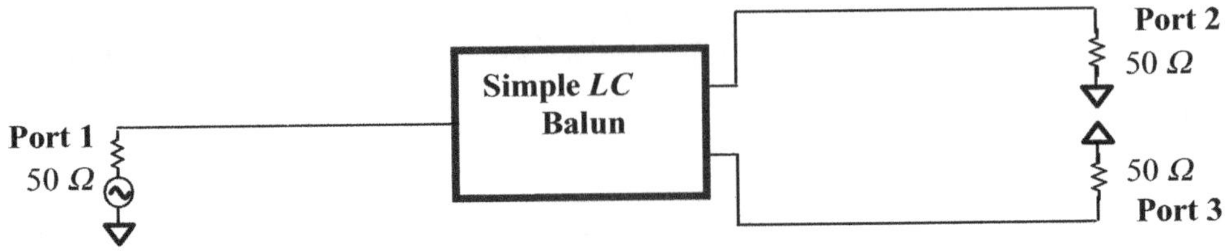

(a) Nothing added to the single-ended side

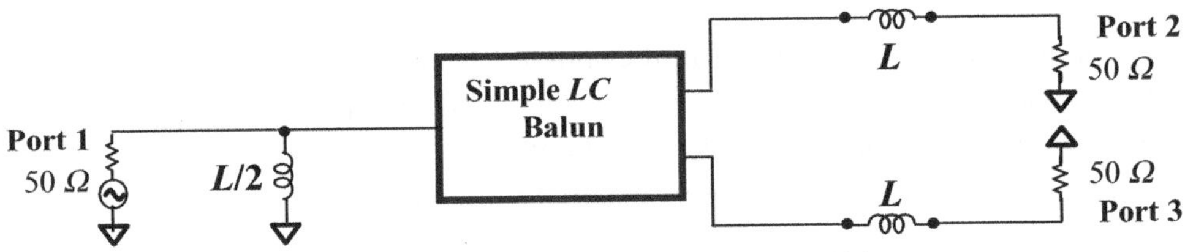

(b) Inductor added to the single-ended side in parallel

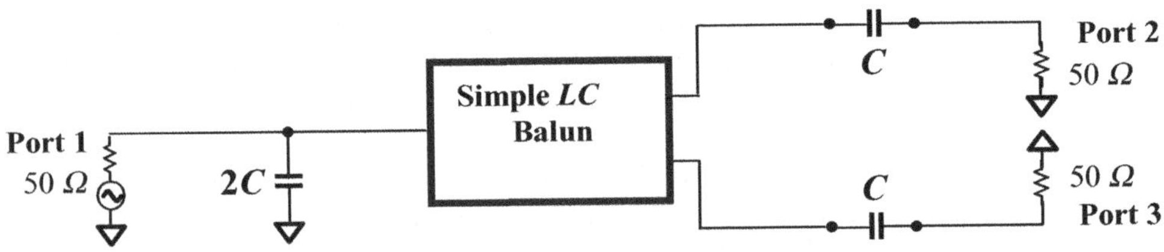

(c) Capacitor added to the single-ended side in parallel

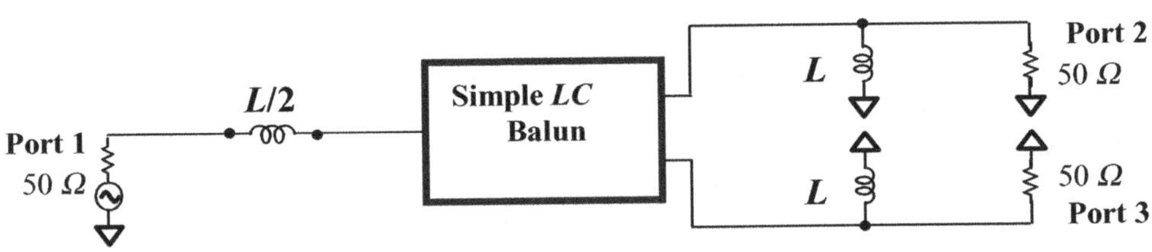

(d) Inductor added to the single-ended side in series

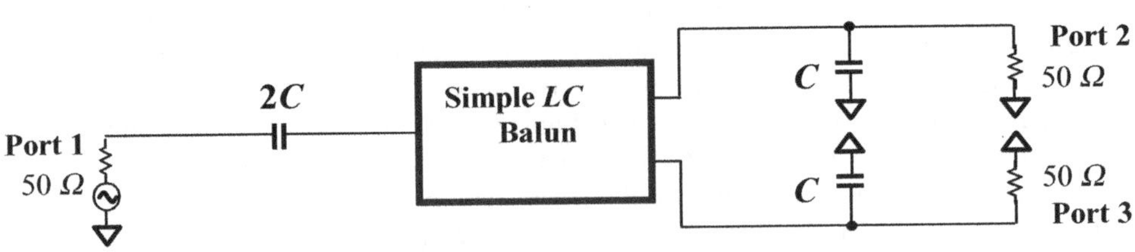

(e) Capacitor added to the single-ended side in series

Figure 4.A.4 Five ways to add parts to a simple LC balun, in which the S parameters or impedances at all three ports as shown in this figure are kept unchanged before and after these parts are added. (Note that $LC = 1/\omega_o^2$.)

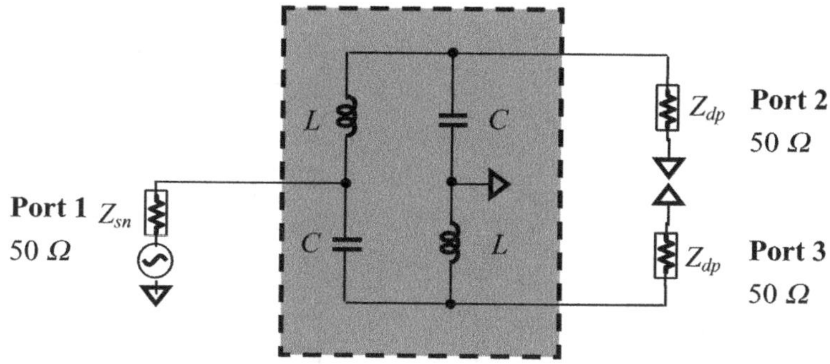

$$f_o = 7.128\ GHz$$
$$\omega_o = 2\pi f_o = 2\pi * 7.128\ GHz$$
$$L = 1.5788\ nH$$
$$C = 0.31577\ pF$$
$$L\omega_o = (2Z_{sn}Z_{dp})^{1/2},\ C\omega_o = (2Z_{sn}Z_{dp})^{-1/2}$$

Figure 4.A.5 Simulation A: *LC* balun with no added parts.

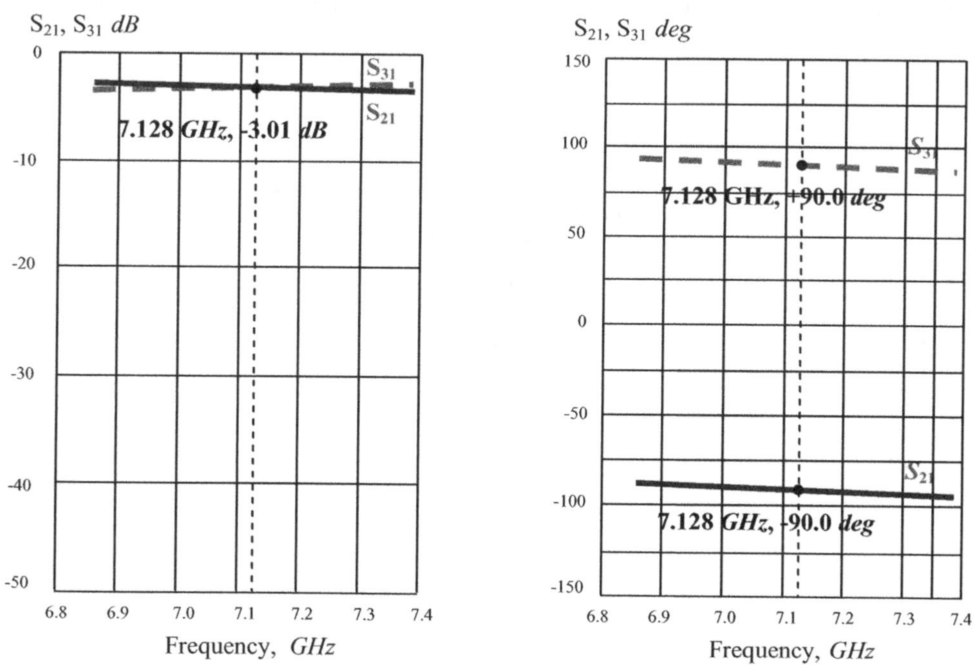

Figure 4.A.6 Insertion loss and phase shift of *LC* balun for group 3 *UWB* system (S_{21}, S_{31}) in simulation A: *LC* balun with no added parts.

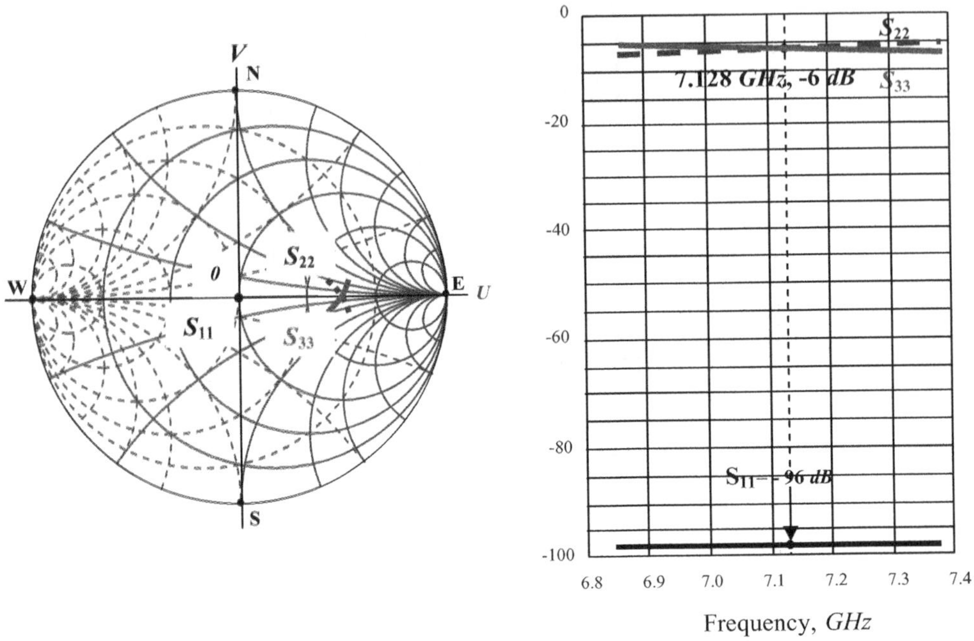

(a) On Smith chart

(b) By rectangular coordination, *dB*

Figure 4.A.7 Return loss of *LC* balun for group 3 *UWB* system (S_{11}, S_{22}, S_{33}) in simulation A: *LC* balun with no added parts.

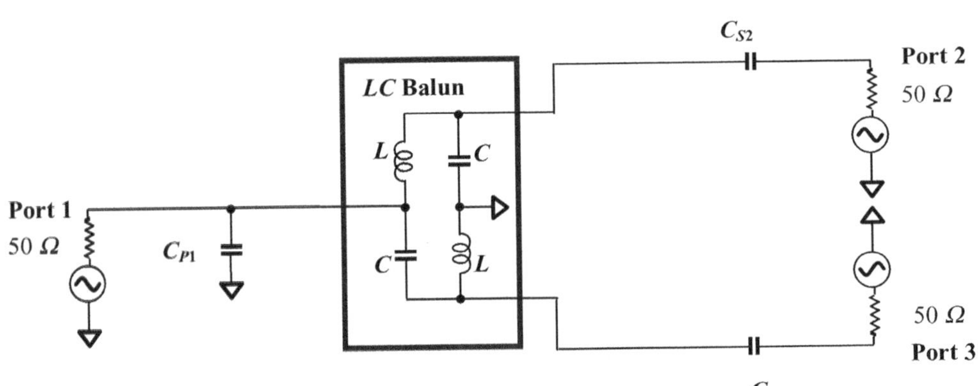

$f_o = 7.128$ GHz
$\omega_o = 2\pi f_o = 2\pi * 7.128\ GHz$
$L = 1.5788\ nH$
$C = 0.31577\ pF$

$C_{P1}/2 = C_{S2} = C_{S3} = 0.31577\ pF = (2Z_{sn}Z_{dp})^{-1/2}/\omega_o$

Figure 4.A.8 Simulation B: *LC* balun with one added part.

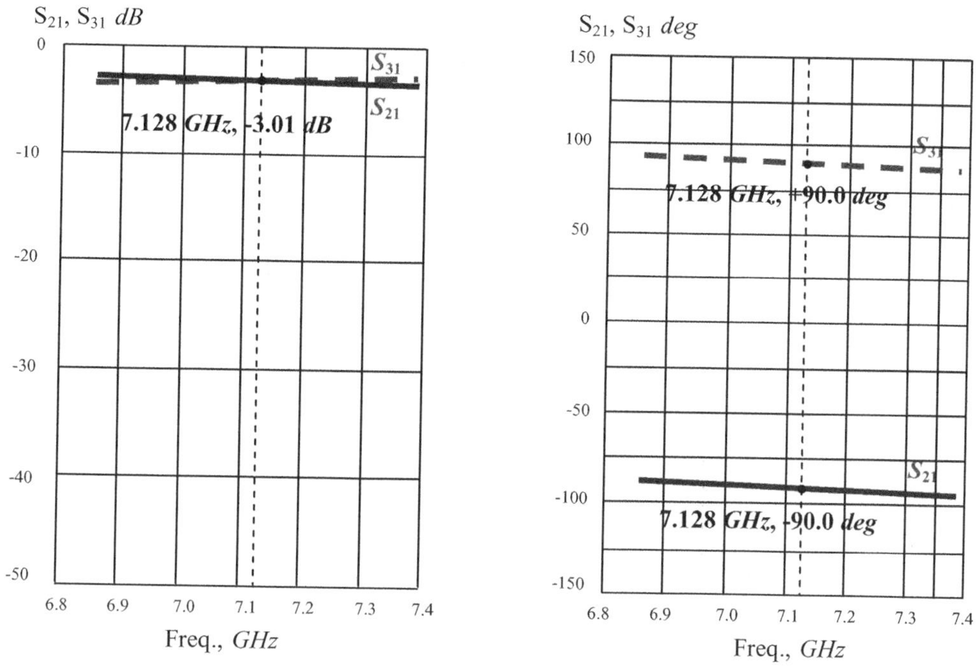

Figure 4.A.9 Insertion loss and phase shift of *LC* balun for group 3 *UWB* system (S_{21}, S_{31}) in simulation B: *LC* balun with one added part.

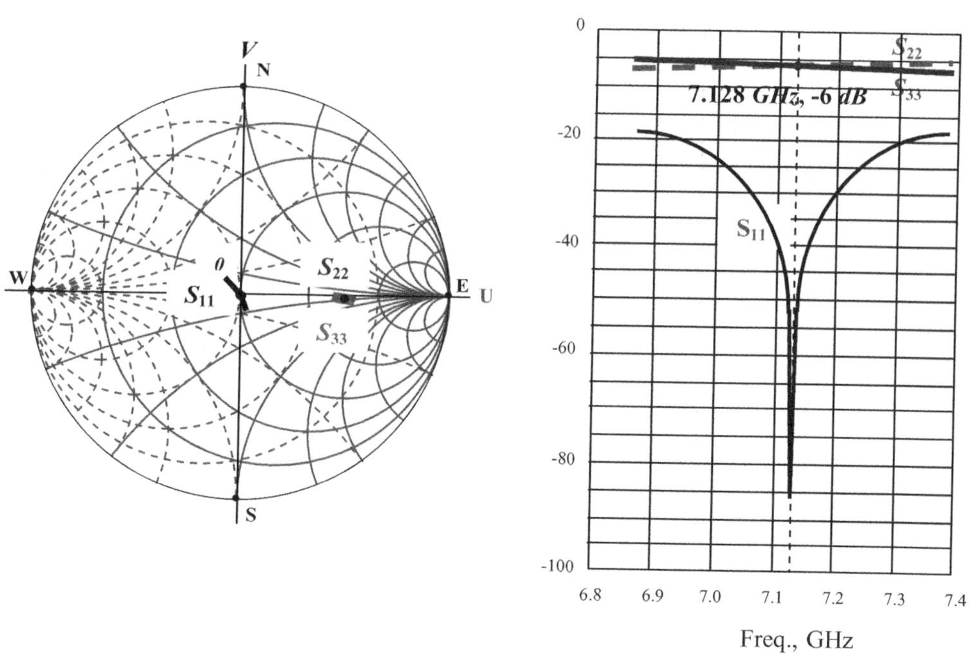

(a) On Smith chart

(b) By rectangular coordination, *dB*

Figure 4.A.10 Return loss of *LC* balun for group 3 *UWB* system (S_{11}, S_{22}, S_{33}) in simulation B: *LC* balun with one added part.

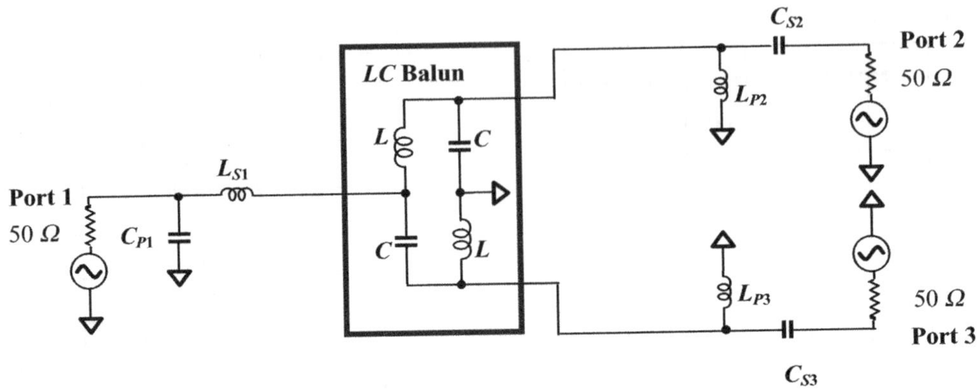

$f_o = 7.128\ GHz$
$\omega_o = 2\pi f_o = 2\pi * 7.128\ GHz$
$L = 1.5788\ nH$
$C = 0.31577\ pF$

$C_{P1}/2 = C_{S2} = C_{S3} = 0.31577\ pF = (2Z_{sn}Z_{dp})^{-1/2} / \omega_o$
$2L_{S1} = L_{P2} = L_{P3} = 1.5788\ nH = (2Z_{sn}Z_{dp})^{1/2} / \omega_o$

Figure 4.A.11 Simulation C: *LC* balun with two added parts. $C_{P1}/2 = C_{S2} = C_{S3}$, $2L_{S1} = L_{P2} = L_{P3}$.

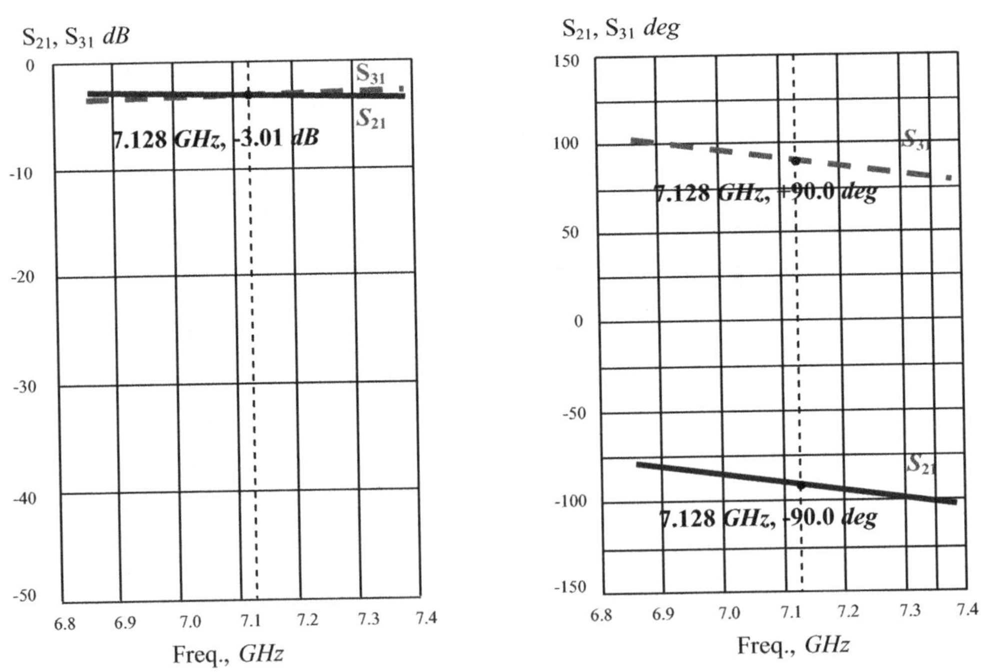

Figure 4.A.12 Insertion loss and phase shift of *LC* balun for group 3 *UWB* system (S_{21}, S_{31}) in simulation C: *LC* balun with two added parts. $C_{P1}/2 = C_{S2} = C_{S3}$, $2L_{S1} = L_{P2} = L_{P3}$.

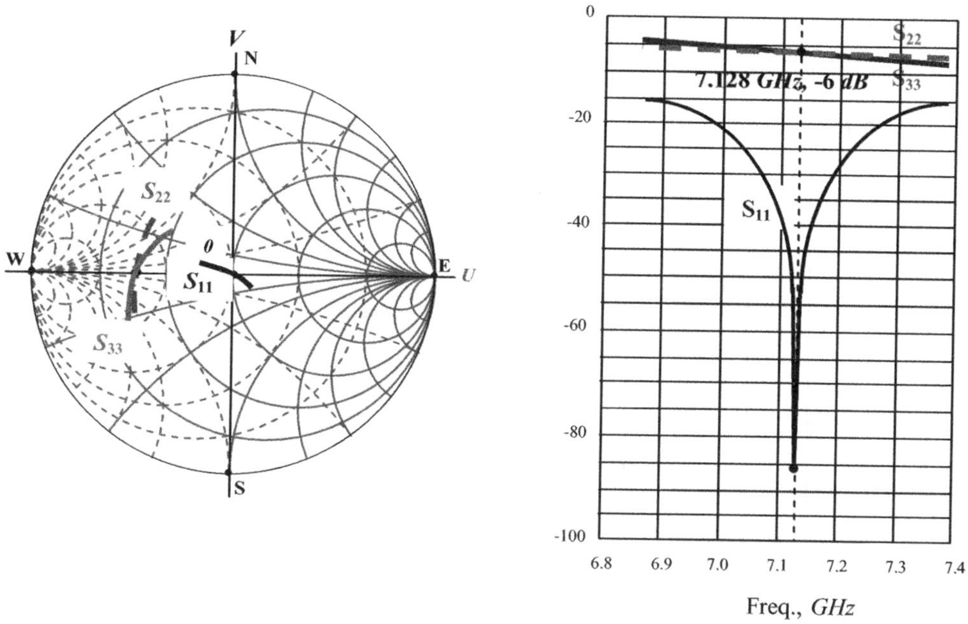

(a) On Smith chart (b) By rectangular coordination, *dB*

Figure 4.A.13 Return loss of *LC* balun for group 3 *UWB* system (S_{11}, S_{22}, S_{33}) in simulation C: *LC* balun with two added parts. $C_{P1}/2 = C_{S2} = C_{S3}$, $2L_{S1} = L_{P2} = L_{P3}$.

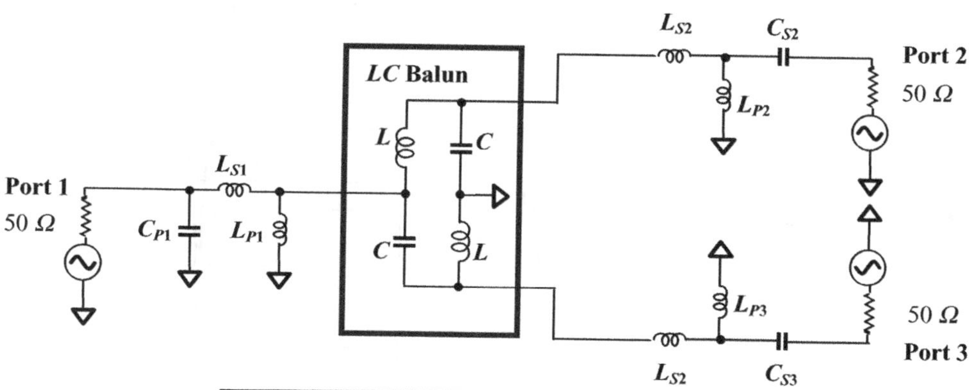

$f_o = 7.128\ GHz$
$\omega_o = 2\pi f_o = 2\pi * 7.128\ GHz$
$L = 1.5788\ nH$
$C = 0.31577\ pF$
$C_{P1}/2 = C_{S2} = C_{S3} = 0.31577\ pF = (2Z_{sn}Z_{dp})^{-1/2}/\ \omega_o$
$2L_{S1} = L_{P2} = L_{P3} = 1.5788\ nH = (2Z_{sn}Z_{dp})^{1/2}/\ \omega_o$
$2L_{P1} = L_{S2} = L_{S3} = 1.5788\ nH = (2Z_{sn}Z_{dp})^{1/2}/\ \omega_o$

Figure 4.A.14 Simulation D: *LC* balun with three added parts. $C_{P1}/2 = C_{S2} = C_{S3}$, $2L_{S1} = L_{P2} = L_{P3}$, $2L_{P1} = L_{S2} = L_{S3}$.

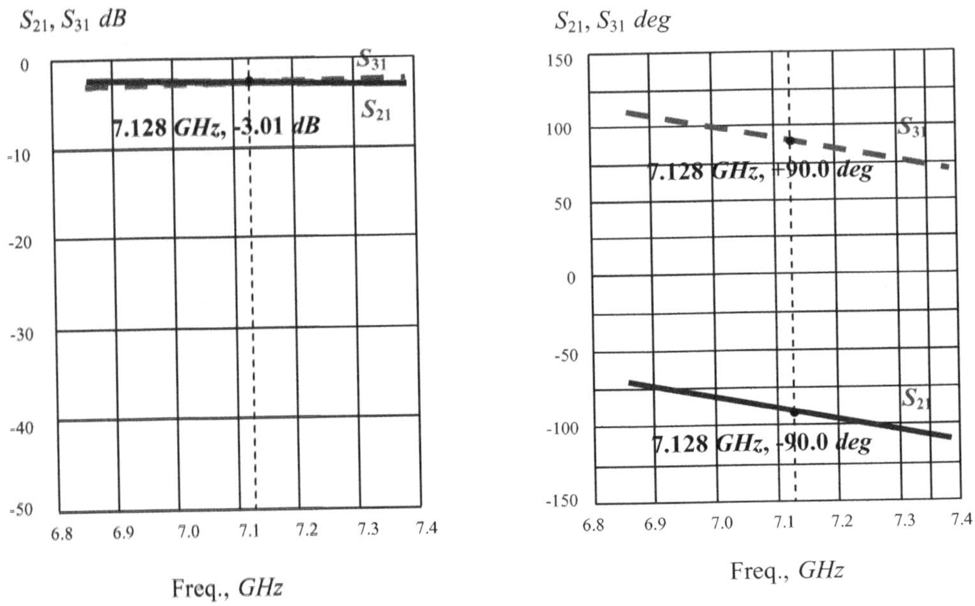

Figure 4.A.15 Insertion loss and phase shift of *LC* balun for group 3 *UWB* system (S_{21}, S_{31}) in simulation D: *LC* balun with three added parts. $C_{P1}/2 = C_{S2} = C_{S3}$, $2L_{S1} = L_{P2} = L_{P3}$, $2L_{P1} = L_{S2} = L_{S3}$.

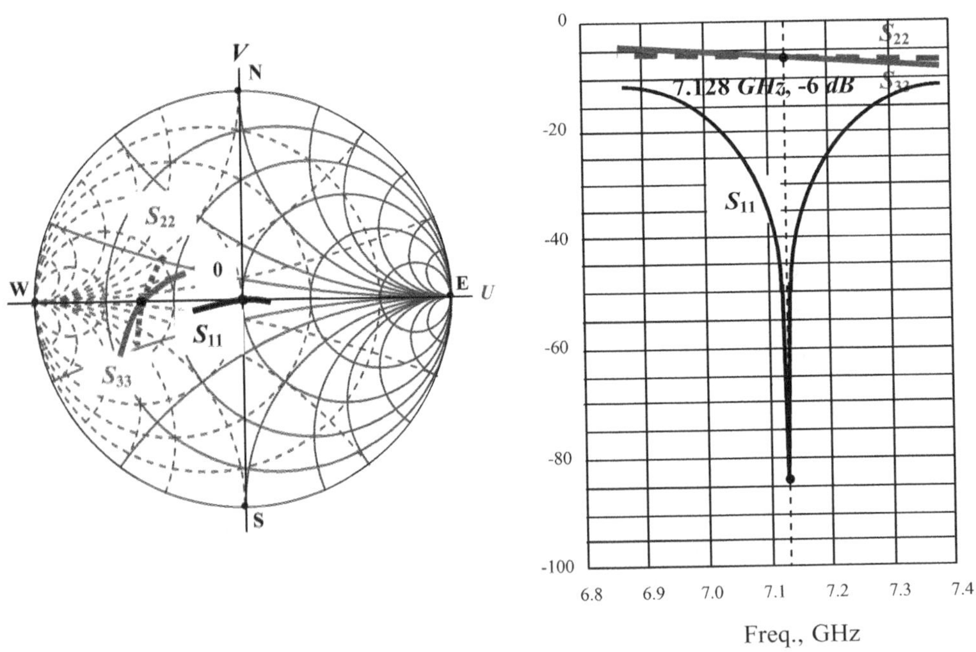

(a) On Smith chart (b) By rectangular coordination, dB

Figure 4.A.16 Return loss of *LC* balun for group 3 *UWB* system (S_{11}, S_{22}, S_{33}) in simulation D: *LC* balun with three added parts. $C_{P1}/2 = C_{S2} = C_{S3}$, $2L_{S1} = L_{P2} = L_{P3}$, $2L_{P1} = L_{S2} = L_{S3}$. $2L_{P1} = L_{S2} = L_{S3}$.

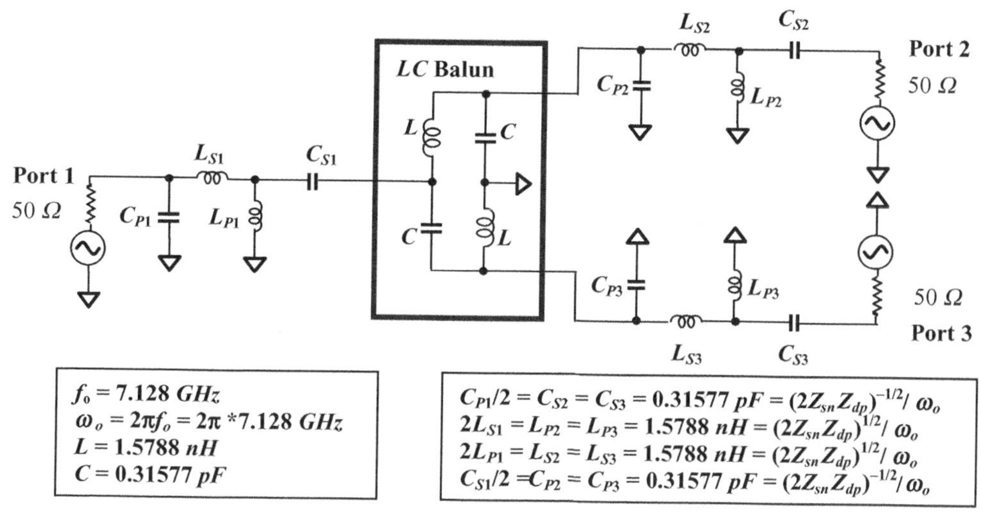

Figure 4.A.17 Simulation E: LC balun with four added parts. $C_{P1}/2 = C_{S2} = C_{S3}$, $2L_{S1} = L_{P2} = L_{P3}$, $2L_{P1} = L_{S2} = L_{S3}$, $C_{S1}/2 = C_{P2} = C_{P3}$.

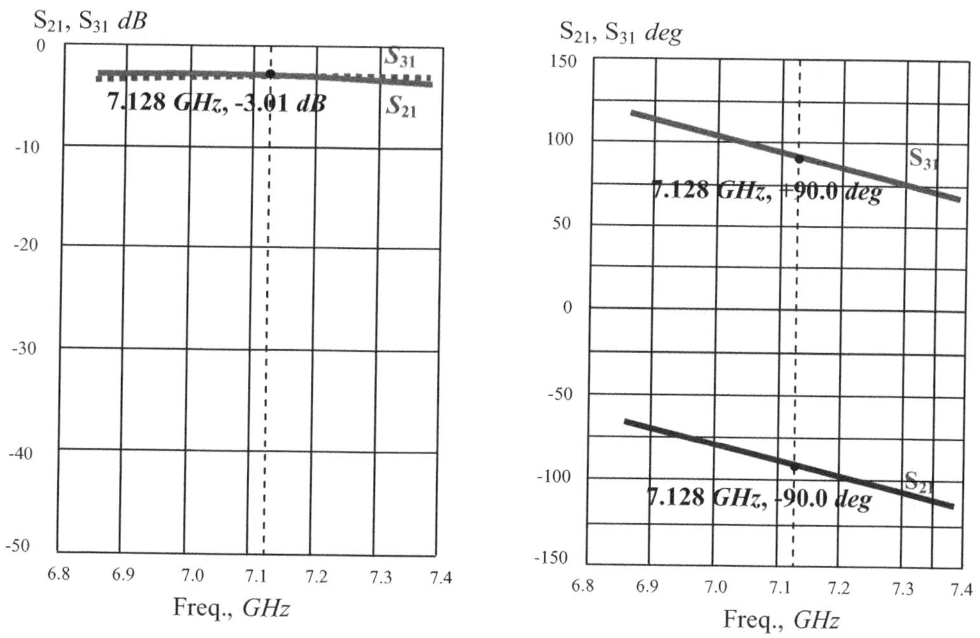

Figure 4.A.18 Insertion loss and phase shift of LC balun for group 3 UWB system (S_{21}, S_{31}) in simulation E: LC balun with four added parts. $C_{P1}/2 = C_{S2} = C_{S3}$, $2L_{S1} = L_{P2} = L_{P3}$, $2L_{P1} = L_{S2} = L_{S3}$, $C_{S1}/2 = C_{P2} = C_{P3}$.

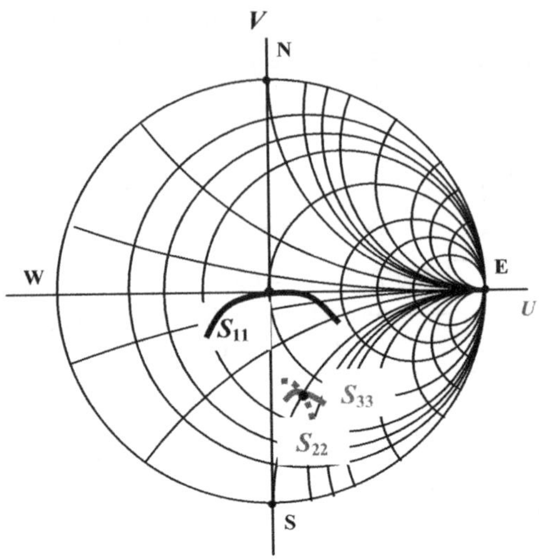

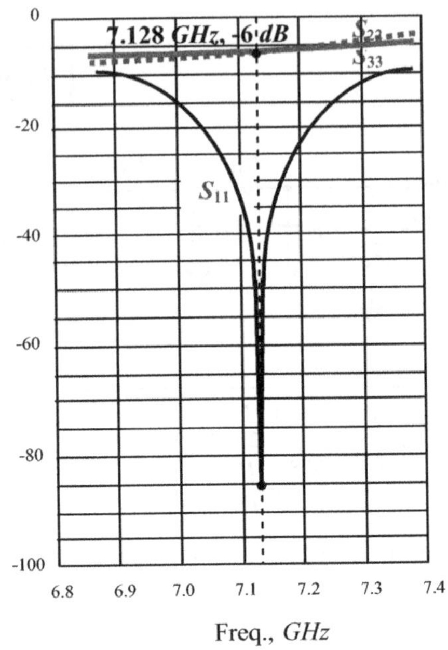

(a) On Smith chart (b) By rectangular coordination, dB

Figure 4.A.19 Return loss of LC balun for group 3 UWB system (S_{11}, S_{22}, S_{33}) in simulation E: LC balun with four added parts. $C_{P1}/2 = C_{S2} = C_{S3}$, $2L_{S1} = L_{P2} = L_{P3}$, $2L_{P1} = L_{S2} = L_{S3}$, $C_{S1}/2 = C_{P2} = C_{P3}$.

- In the case of the transformer balun, either L or C can be selected with any value as long as the following relation is satisfied, that is,

$$L\omega_o = \frac{1}{C\omega_o}, \tag{4.A.42}$$

or,

$$LC = \frac{1}{\omega_o^2}, \tag{4.A.43}$$

where ω_o = operating angular frequency,

This leads to the essential difference about "interpretation" in the transformer balun and in the simple LC balun.

- In the case of transformer balun, the "interpretation" can be made for any value of L or C as long as both of them obey the relation (4.A.42) or (4.A.43).

- In the case of a simple *LC* balun, the "interpretation" can be conducted only for values of *L* and *C* calculated from formulas (4.A.28) and (4.A.29), respectively.

Nevertheless, the "interpretation" feature of a *LC* balun does exist and might be applied for some other purpose. It is therefore kept as one of the Appendices of this chapter.

4.A.5 Some Useful Couplers

Wilkinson Coupler The Wilkinson coupler is a balun with a very simple configuration. As shown in Figure 4.A.20, it consists of only two micro strip lines of a quarter-wavelength length and a 100Ω resistor. Figures 4.A.20 (a) and (b) show a Wilkinson splitter and combiner, respectively.

The impedance Z_{dp} looking from one of the micro strip lines toward the adjacent differential port is

$$Z_{dp} = Z_o // (2Z_o + Z_o // Z_{dp}). \tag{4.A.44}$$

The impedance Z_{sn} looking from the micro strip line toward the single-ended port is

$$Z_{sn} Z_{dp} = \left(\sqrt{2} Z_o\right)^2. \tag{4.A.45}$$

From (4.A.44), we have

$$Z_{dp} = \frac{Z_o}{\sqrt{2}}, \tag{4.A.46}$$

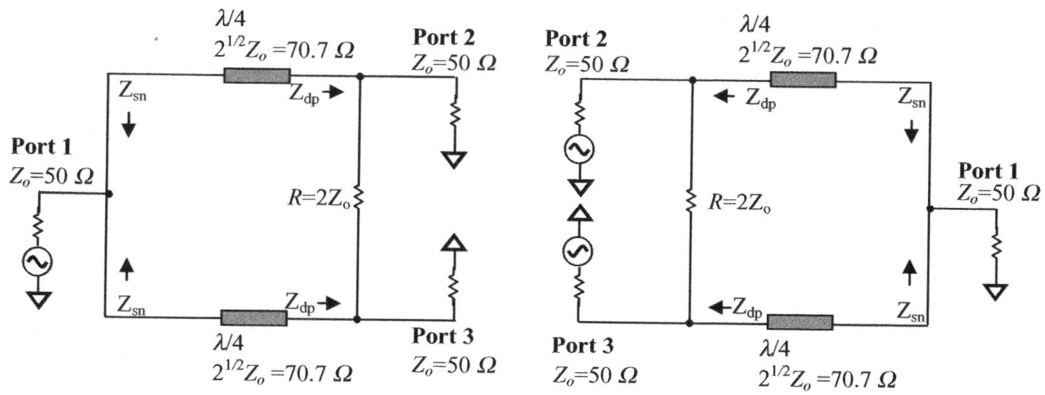

(a) Wilkinson splitter (b) Wilkinson combiner

Figure 4.A.20 Wilkinson coupler.

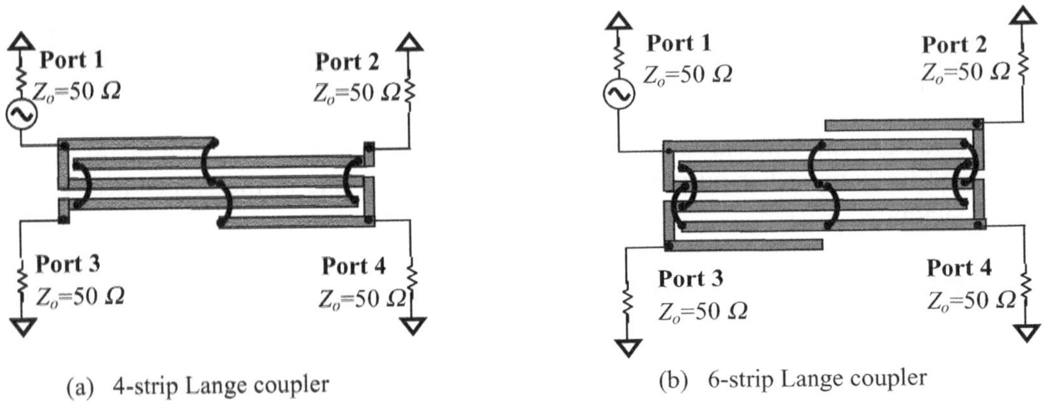

Figure 4.A.21 Micro Lange coupler.

From (4.A.45) and (4.A.46), we have

$$Z_{sn} = 2\sqrt{2}Z_o \tag{4.A.47}$$

Micro Strip Lange Coupler The micro strip Lange coupler is a quadrature hybrid balun. Figure 4.A.21 (a) and (b) shows its 4-strip and 6-strip configuration, respectively. The coupling between strips is very tight due to the interdigitated structure. Consequently, it achieves a very wide bandwidth performance. The relative bandwidth reached could be greater than 100%. For instance, if the operating frequency of a Lange coupler is available from 5 *GHz* to 20 GHz, then the relative bandwidth is $\Delta f/f_o = 15/12.5 = 120\%$.

Over a wide frequency range, the phase performance is very close to 90° and the variation of magnitude is usually less than 2 *dB*.

4.A.6 Cable Baluns

The cable balun was developed a couple of decades ago. At present, it is almost never applied in actual circuitry because of its large size.

However, it is still sometimes useful in the test laboratory where the size of a part is not a problem. Figure 4.A.22 shows the simplest cable balun. The length of the cable should be a quarter wavelength of the operating frequency. For

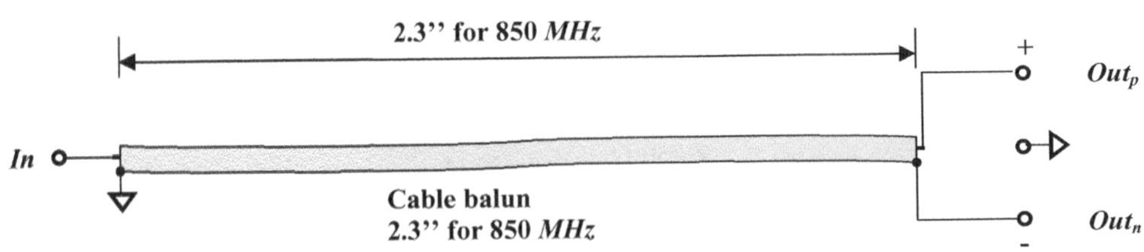

Figure 4.A.22 A simplest cable balun.

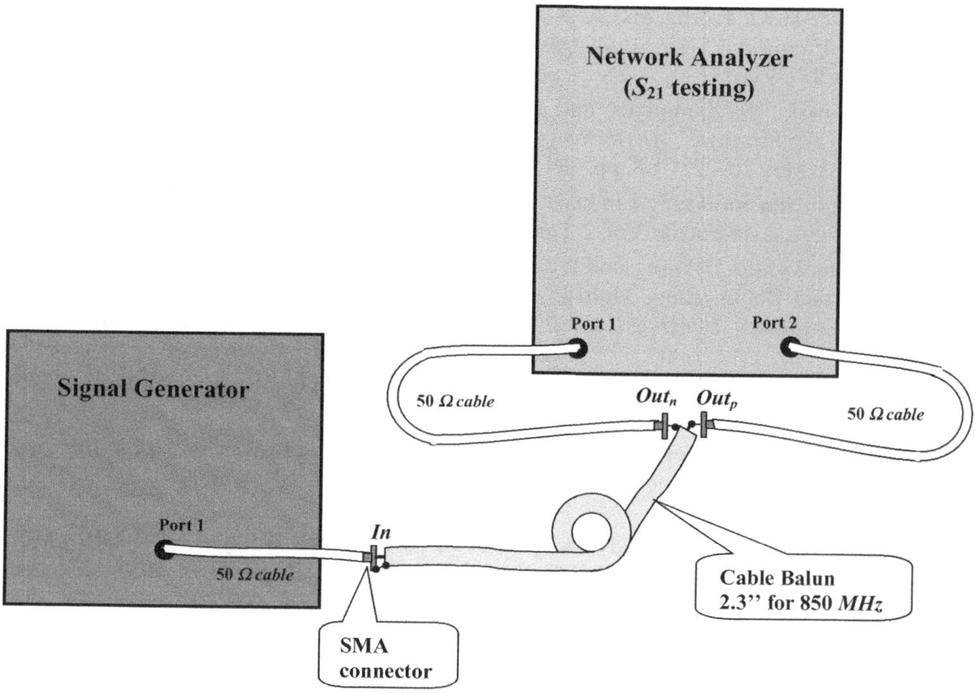

Figure 4.A.23 Verification of the simplest cable. balun.

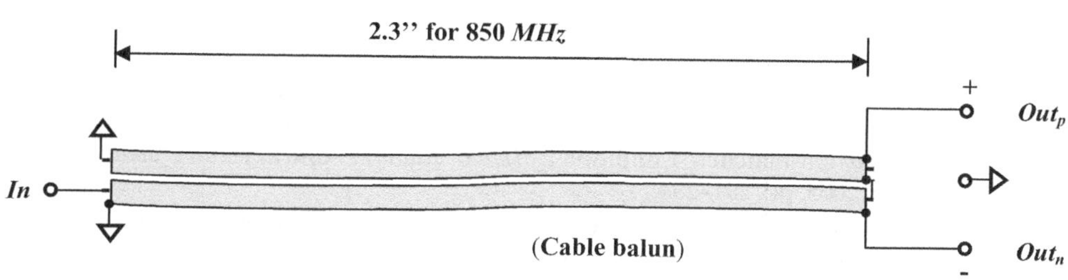

Figure 4.A.24 Cable balun with additional cable for compensation.

$f = 850\,MHz$, a regular RF cable of approximately 2.3 inches is required. This can be verified and tested by the setup as shown in Figure 4.A.23.

The simplest cable balun is not too balanced and can be improved as shown as Figure 4.A.24.

REFERENCES

[1] J. Rogers and R. Bhatia, "A 6 to 20 GHz Planar Balun Using a Wilkinson Divider and Lange Couplers," *Microwave Symposium Digest*, IEEE MTT-S International, Vol. 2, June 10–14, 1991, pp. 865–868.

[2] M. N. Tutt, H. Q. Tserng, and A. Ketterson, "A low-Loss, 5.5 GHz-20 GHz Monolithic Balun," *Microwave Symposium Digest*, IEEE MTT-S International, Vol. 2, June 8–13, 1997, pp. 933–936.

[3] T. Rutkowski, W. Zieniutycz, and K. Joachimowski, "Wideband Coaxial Balun for Antenna Application," *Microwaves and Radar*, International Conference on MIKON '98, Vol. 2, May 20–22, 1998, pp. 389–392.

[4] Jwo-Shium Sun and Tsung-Lin Lee, "Design of a Planar Balun," *Microwave Conference*, APMC 2001, Asia-Pacific, Vol. 2, December 3–6, 2001, pp. 535–538.

[5] Liang Tao, J. Gillis, D. Wang, and P. Cooper, "Design and Modeling of Compact On-chip Transformer/Balun Using Multi-Level Metal Windings for RF Integrated Circuits," *Radio Frequency Integrated Circuits (RFIC) Symposium* 2001, IEEE, May 20–22, 2001.

[6] Jyh-Wen Sheen and Ching-Wen Tang, "LTCC-MLC Balun for WLAN/Bluetooth," 2001 *IEEE MIT-S Digest*, pp. 315–318.

[7] Winfried Bakalski, Werner Simbirger, Herbert Knapp, Hans-Dieter Wohlmuth, and Arpad L. Scholtz, "Lumped and Distributed Lattice-type LC-Baluns," 2002 *IEEE MTT-S Digest*, pp. 209–212.

[8] H. Y. D. Yang, L. Zhang, and J. A. Castaneda, "Design and Analysis of a Multi-Layer Transformer Balun for Silicon RF Integrated Circuits," *Radio Frequency Integrated Circuits (RFIC) Symposium*, IEEE, June 2–4, 2002.

[9] Jong-Wook Lee and K. J. Webb, "Analysis and Design of Low-Loss Planar Microwave Baluns Having Three Symmetric Coupled Lines," *Microwave Symposium Digest*, IEEE MTT-S International, Vol. 1, June 2–7, 2002, pp. 117–120.

[10] Munenari Kawashima, Tadao Nakagawa, and Katsuhiko Araki, "A Novel Broadband Active Balun," *33rd European Microwave Conference, Munich 2003*, pp. 495–498.

[11] Kai-Ye Huang, Chia-Jen Hsu, and Len-Yi Leu, "Modeling Methodology of Integrated Five-Port Balun Using Two-Port RF Measurement," *IEEE Radio Frequency Integrated Circuits Symposium*, 2005, pp. 295–298.

[12] Francis Rotella, Gene Tkachenko, and Yuhua Cheng, "Characterization, Design, Modeling, and Model Validation of Silicon On-Wafer M:N Balun Components under Matched and Un-matched Conditions," *IEEE Radio Frequency Integrated Circuits Symposium*, 2005, pp. 291–294.

[13] Esa Tiiliharju and Kari A. I. Halonen, "An Active Differential Broad-Band Phase Splitter for Quadrature-Modulator Applications," *IEEE Transactions on Microwave Theory and Techniques*, Vol. 53, No. 2, February 2005, pp. 679–686.

[14] Marco A. Antoniades and George V. Eleftheriades, "A Broadband Wilkinson Balun Using Microstrip Metamaterial Lines," *IEEE Antennas and Wireless Propagation Letters*, 2005, Vol. 4, pp. 209–212.

[15] Alberto Costantini, Ben Lawrence, Simon Mahon, James Harvey, Gerry McCulloch, and Alexandre Bessemoulin, "Broadband Active and Passive Balun Circuits: Functional Blocks for Modern Millimeter-Wave Radio Architectures," *Proceedings of the 1st European Microwave Integrated Circuits Conference, September 2006, Manchester, UK*, pp. 421–424.

CHAPTER 5

TUNABLE FILTERS

5.1 TUNABLE FILTERS IN COMMUNICATION SYSTEMS

There are two kinds of filters in electrical circuit design: fixed and tunable. A fixed filter pass band is fixed and well-defined, while a tunable filter pass band can be tuned over a certain frequency range.

The theory for a fixed filter design was comprehensively developed many decades ago. The engineering developments of passive filters, including the *LPF* (Low-Pass Filter), *HPF* (High Pass Filter), *BPF* (Band Pass Filter), and *BRF* (Band Reject Filter), have been formulized and tabulated for the design with discrete parts. This simplifies the design procedures. The only problem remaining is how to convert the values of parts into reasonable ones through the so-called "Δ-*" or "π-T" transformation.

In past years, the frequency bands assigned for a communication system rose higher and higher. It became more and more difficult to implement filters by discrete parts because the required values of either capacitors or inductors became too small for the production line. Many types of filters with distributed parameters were developed, such as the crystal filter, ceramic block filter, micro strip line filter, and *SAW* filter.

The theory for a tunable filter design was developed many decades ago also. Its implementation, however, faces some hurdles. This is one of the reasons why it, and not another type of filter, was selected as the topic of this book. Today, the *SAW* filter is very often applied to a communication system, whether a dual conversion or a direct conversion system. As a matter of fact, a lot of effort is being put into the development of a *SAW* tunable filter at present.

RF Circuit Design, by Richard Chi-Hsi Li

In this chapter we are going to show how to remove the main obstacles in the tunable filter design. It might be helpful to provide some clues to assist in the design of a *SAW* tunable filter or other type of tunable filter, although in our example the tunable filter is implemented by discrete parts. This is the second reason to include this chapter. (The content of this chapter is abstracted from an US patent which was obtained by the author in 1992.)

5.1.1 Expected Constant Bandwidth of a Tunable Filter

In a communication system, a customer may occupy only one channel, although the number of available channels is usually huge. In other words, the bandwidth that a customer needs is much less than the entire bandwidth of the system. For example, in a *UHF* portable radio, the entire bandwidth is assigned to be from 403 to 520 *MHz*, while a customer occupying bandwidth for voice conversation needs no more than 25 *kHz*. To approach good performance, the ideal condition is to assign 25 *kHz* only to each customer by a tunable filter while the rest of the bandwidth is rejected. All of the hardware is the same for each customer: the difference lies only in the control voltage being provided to the tunable filter for each individual customer. Consequently, the goals of good performance and mass production can be realized simultaneously.

In practical *RF* tunable filter design, it is possible to request that the bandwidth be narrowed down to 10 *MHz* or so, but it is almost impossible to narrow the bandwidth down to the order of around 25 *kHz*. Nevertheless, narrowing the bandwidth is still very helpful in improving the selectivity of the receiver, reducing noise, and preventing a variety of spurious interference. As long as the bandwidth can be narrowed down somewhat from the entire bandwidth of the system, the performance of the portable radio can be appreciably improved. Without a tunable filter functioning in the front end of the receiver, spurious interference and noise may be significant.

Figure 5.1 shows how the tunable filter operates in a portable radio. When the control voltage, CV, is moved from its minimum, CV_L, to its maximum, CV_H, the capacitance of the varactor, C, is changed from its maximum, C_H, to its minimum, C_L, and the central frequency of the tunable filter, f_o, moves from its minimum, f_{oL}, to its maximum, f_{oH}, correspondingly.

To enable a tunable filter tuned over a wide frequency tuning range to be realistic, the bandwidth should be kept constant as shown in Figure 5.1 as much as possible when the central frequency, f_o, is tuned.

5.1.2 Variation of Bandwidth

Unfortunately, many tunable filter designs are far from ideal; the bandwidth varies significantly as shown in Figure 5.2 when the tuning frequency is tuned. In Figure 5.2(b), at the low end of the frequency range, the bandwidth of the tunable filter is narrow, as expected. As the control voltage is increased, the capacitance of the varactor is decreased and hence the central frequency of the tunable filter increases. On the other hand, its bandwidth also increases. Up to the high end of the frequency range, the bandwidth could be widened to an unacceptable amount, and the "tunable" in "tunable filter" becomes meaningless. In Figure 5.2(c), the

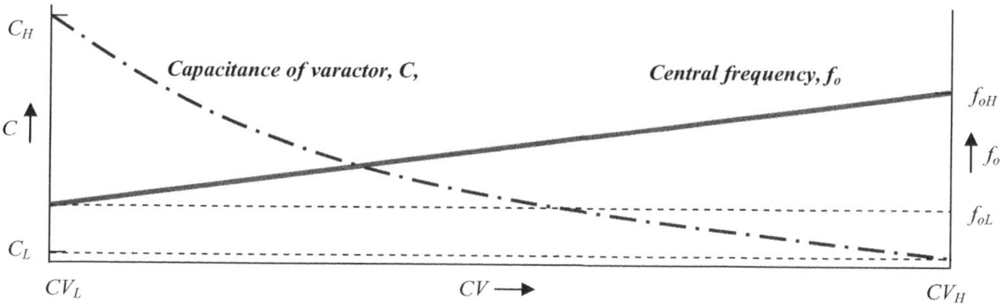

(a) Variation of the varactor capacitance and the central frequency as the control voltage is varied

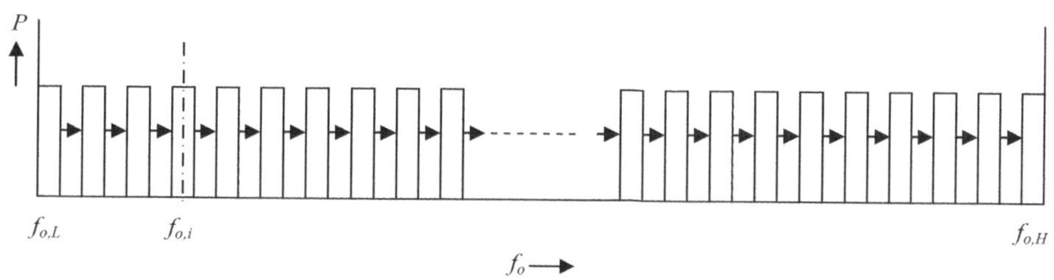

(b) Central frequency is moved to higher frequency with constant bandwidth as the control voltage is increased

Figure 5.1 Expected constant bandwidth as the control voltage is tuned from low to high voltage.

variation of bandwidth is just opposite from that in Figure 5.2(b). At the high end of the frequency range, the bandwidth of the tunable filter is narrow, as expected. As the control voltage is decreased, the capacitance of the varactor is increased and the central frequency of the tunable filter is lowered, but, on the other hand, its bandwidth is increased. At the low end of the frequency range, the bandwidth could be widened to an unacceptable amount, by which the "tunable" in "tunable filter" likewise becomes meaningless.

5.2 COUPLING BETWEEN TWO TANK CIRCUITS

A *RF* tunable filter is a *BPF* (Band Pass Filter) with a variable central frequency.

Typically, an *RF* tunable filter in a communication system consists of two tank circuits coupled by one or more coupling parts. Figure 5.3 shows its blocks.

The input and output tank circuits are implemented by an inductor and a capacitor in parallel. In order to tune the central frequency, the capacitor is a special one called a varactor, in which its capacitance is varied and tuned by a common control voltage, *CV*. Both the input and output tank circuits must be resonant at the same frequency and hence must be identical. Should the tunable filter be built by more than one tank circuit, its frequency response would be rolled up and down more sharply, and within the bandwidth of the pass band its frequency response would be

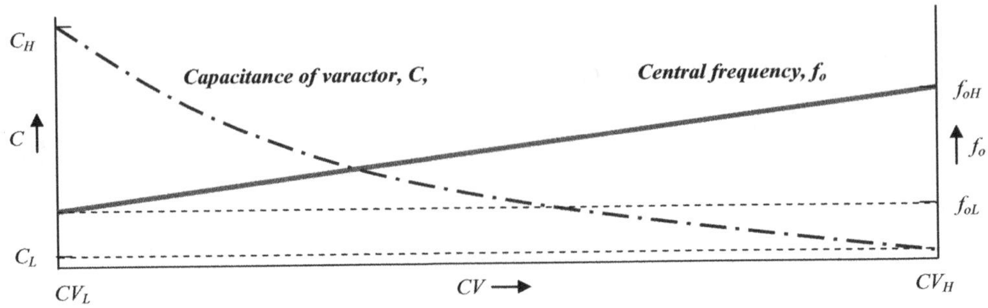

(a) Variation of the varactor capacitance and the central frequency as the control voltage is varied

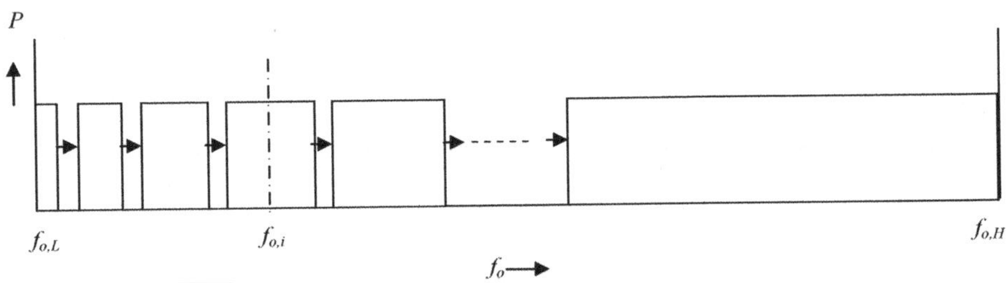

(b) Central frequency is moved to higher frequency with wider bandwidth as the control voltage is increased

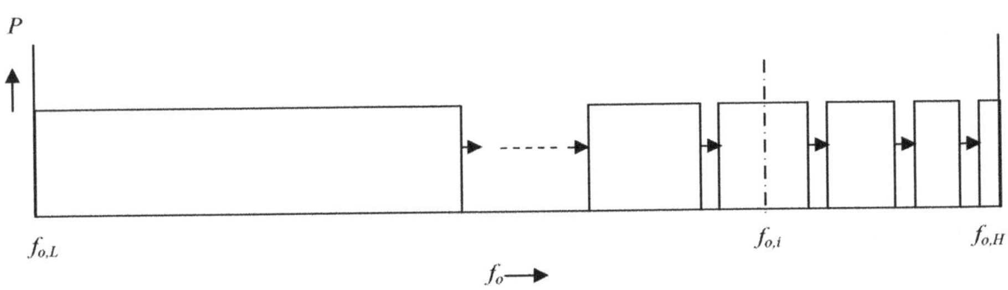

(c) Central frequency is moved to higher frequency with narrower bandwidth as the control voltage is increased

Figure 5.2 Variation of the bandwidth as the control voltage is changed.

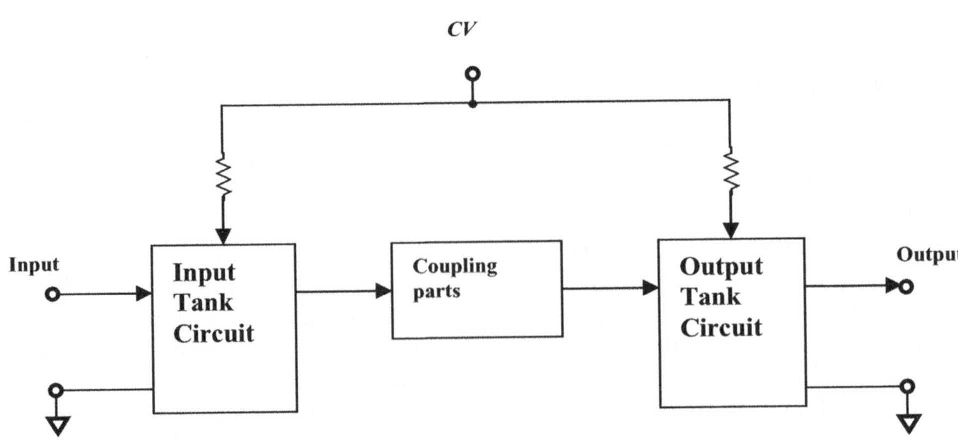

Figure 5.3 Blocks of a *RF* tunable filter.

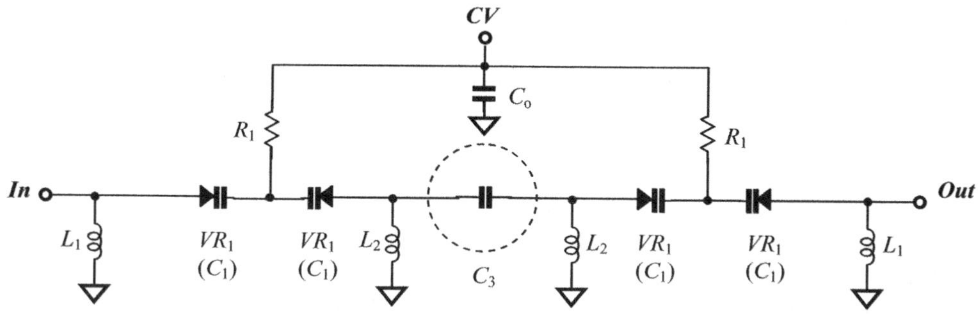

(a) Tunable filter implemented by two tank circuits and coupled by capacitor only

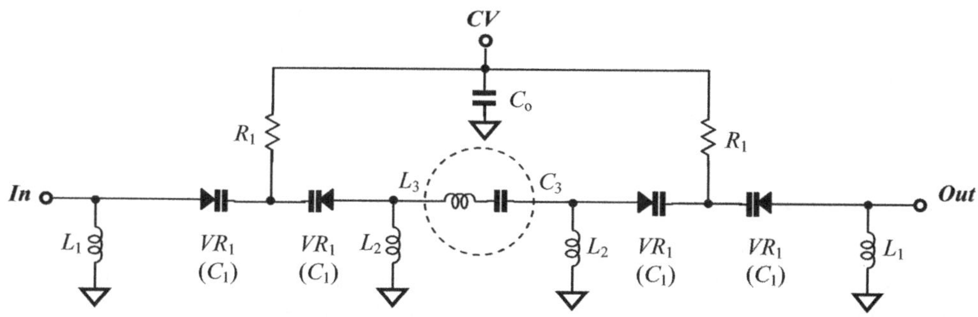

(b) Tunable filter implemented by two tank circuits and coupled by LC in series

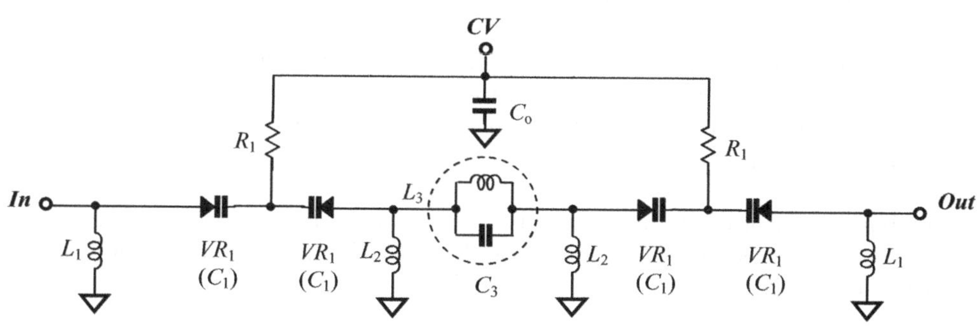

(c) Tunable filter implemented by two tank circuits and coupled by LC in parallel

Figure 5.4 Three coupling types of a tunable filter.

more flattened. However, the uniform requirement of the tank circuits restricts their number. The more the number of tank circuits, the more difficult it is to maintain their uniformity. This is why most *RF* tunable filters are implemented by only two tank circuits.

Figure 5.4 shows the schematics of an *RF* tunable filter with three different coupling styles. Instead of one inductor and one capacitor applied in one tank circuit, the tank circuit consists of two inductors and two varactors. This makes it convenient to apply the control voltage to the varactors and has the additional advantage of simplifying impedance matching to the input and output terminals. Figure 5.4(a) shows a tunable filter implemented by two tank circuits and coupled only by capacitor. Figure 5.4(b) and (c) shows tunable filters implemented by two tank circuits and coupled by an LC in series and in parallel respectively.

It should be noted that the tank circuits shown in Figure 5.4 have the same topology. The capacitors or varactors of the tank circuit are arranged as an arm in series with input or output terminals, while most of the inductors are connected as branches in parallel with the input or output terminals. Such a topology leads to a frequency response with a higher slope to the lower side of the pass band and a response with a lower slope to the higher side of pass band. Such a tunable filter is therefore more appropriate to be cooperated with a low-side injection mixer. On the contrary, if the capacitors or varactors of the tank circuit are arranged as a branch in parallel with the input or output while most of inductors are connected as arm in series with the input or output, it would result in a frequency response with a higher slope in the higher side of the pass band and a lower slope in the lower side of the pass band. Such a tunable filter is more appropriate to be cooperated with a high side injection mixer. In the following discussion, we will focus on the topology appropriate to the low-side injection mixer only.

5.2.1 Inappropriate Coupling

The tunable filter with three coupling types as shown in Figure 5.4 was designed in years past. One of their common problems is, as mentioned above, that the bandwidth changes significantly from the low end to high end of the frequency range.

Through a simple circuit analysis one can understand why this is so. As a matter of fact, the tank circuits are always tuned up to resonance at the central frequency. Therefore, the two tank circuits look like two small resistors, r_o, connected with the main coupling parts in series since the inductor and the capacitor are in resonant status so that their reactances are "neutralized" with each other. The tunable filters with three kinds of main coupling types as shown in Figure 5.4 can be replaced by their equivalents shown in Figure 5.5 when the tank circuits are resonant at the central frequency.

The bandwidth of the tunable filter is mainly determined by the Q value of the tank circuit. In cases where the main coupling part is the capacitor C only, as in Figures 5.4 (a) and 5.5(a),

$$Q = \frac{f_o}{BW} = \frac{1}{2r_o C\omega_o} = \frac{1}{4\pi r_o C f_o}, \tag{5.1}$$

where

Q = unloaded Q of the tunable filter;
f_o = central frequency of the tunable filter;
ω_o = central angular frequency of the tunable filter;
BW = frequency bandwidth of the tunable filter;
r_o = equivalent resistor when tank circuit is resonant at the central frequency.

From (5.1),

$$BW = 4\pi r_o C f_o^2, \tag{5.2}$$

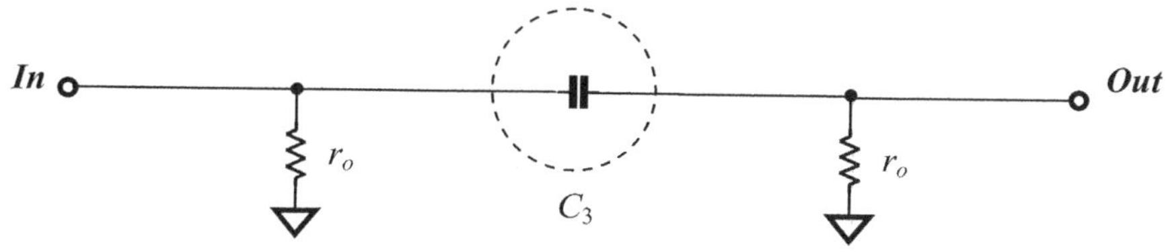

(a) At f_o, two tank circuits are in resonant state and coupled by C only

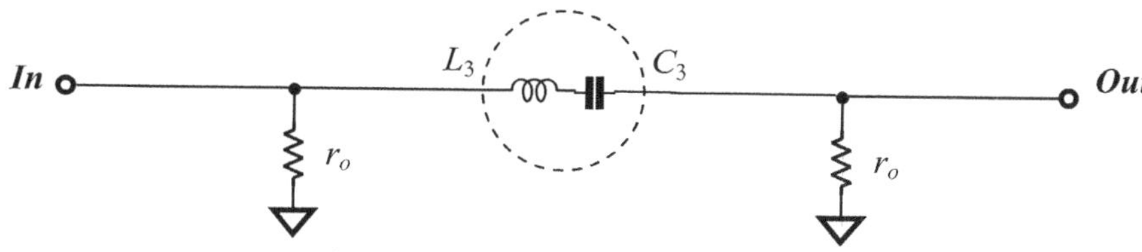

(b) At f_o, two tank circuits are in resonant state and coupled by *LC* in series

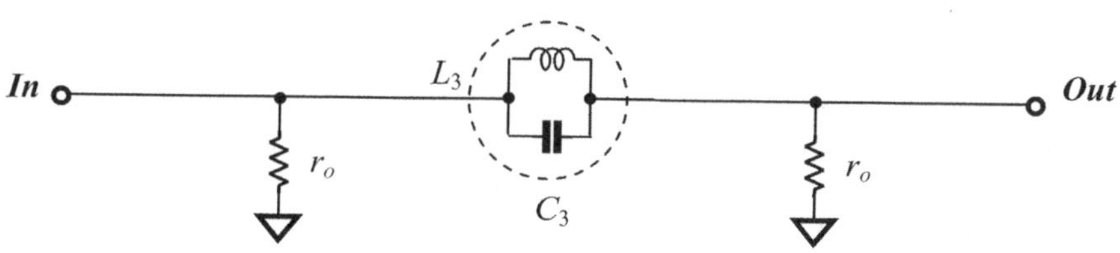

(c) At f_o, two tank circuits are in resonant state and coupled by *LC* in parallel

Figure 5.5 Equivalent circuit of a tunable filter with three coupling types in resonant state when $f = f_o$.

Then,

$$\frac{\partial(BW)}{\partial f_o} = 8\pi r_o C f_o. \tag{5.3}$$

In cases where the main coupling part is an *LC* in series as in Figures 5.4(b) and 5.5(b),

$$Q = \frac{f_o}{BW} = \frac{L\omega_o - \dfrac{1}{C\omega_o}}{2r_o} = \frac{4LC\pi^2 f_o^2 - 1}{4\pi r_o C f_o}, \tag{5.4}$$

From (5.4),

$$BW = \frac{4\pi r_o C f_o^2}{4\pi^2 LC f_o^2 - 1}, \tag{5.5}$$

Then,

$$\frac{\partial(BW)}{\partial f_o} = -\frac{8\pi r_o C f_o}{(4\pi^2 LC f_o^2 - 1)^2}. \tag{5.6}$$

In cases where the main coupling part is an LC in parallel, as in Figures 5.4(c) and 5.5(c),

$$Q = \frac{f_o}{BW} = \frac{\frac{L\omega_o}{1 - LC\omega_o^2}}{2r_o} = \frac{\pi L f_o}{r_o(1 - 4\pi^2 LC f_o^2)}. \tag{5.7}$$

From (5.4),

$$BW = \frac{(1 - 4\pi^2 LC f_o^2) r_o}{\pi L}, \tag{5.8}$$

then,

$$\frac{\partial(BW)}{\partial f_o} = -8\pi r_o C f_o. \tag{5.9}$$

As shown in equations (5.3), (5.6), and (5.9), in all three coupling types of tunable filters shown in Figure 5.4, the variation of the bandwidth is dependent on the central frequency, f_o.

Where the main coupling part is the capacitor C only, as shown in Figures 5.4(a) and 5.5(a), the variation of bandwidth is proportional to the central frequency f_o as shown in equation (5.3). The bandwidth becomes wider and wider as the central frequency increases. It may become unacceptably wide at the high-frequency end as shown in Figure 5.2(b). In the case where the main coupling part is an LC in series, as shown in Figures 5.4(b) and 5.5(b), the variation of bandwidth is dependent on the central frequency by a complicated function. In the case where the main coupling part is an LC in parallel as shown in Figure 5.5(c), the variation of bandwidth is negatively proportional to the central frequency. In both the cases shown in Figures 5.4(b) and 5.5 (b), and 5.4(c) and 5.5(c), the bandwidth becomes wider and wider as the central frequency is decreased. It may become unacceptably wide at the low-frequency end as shown in Figure 5.2(c).

It is therefore concluded that, generally speaking, all three coupling styles shown in Figure 5.4 are inappropriate in the tunable circuit design.

5.2.2 Reasonable Coupling

A reasonable coupling approach is shown in Figure 5.6, in which the coupling part between the two tank circuits consists of only an inductor L. When this filter is operating at its central frequency, its equivalent circuit operates as shown in Figure 5.7.

Similar to the derivation as in Section 5.2.1, we have

$$Q = \frac{f_o}{BW} = \frac{L\omega_o}{2r_o} = \frac{L\pi f_o}{r_o}. \tag{5.10}$$

From (5.10),

$$BW = \frac{r_o}{L\pi}, \tag{5.11}$$

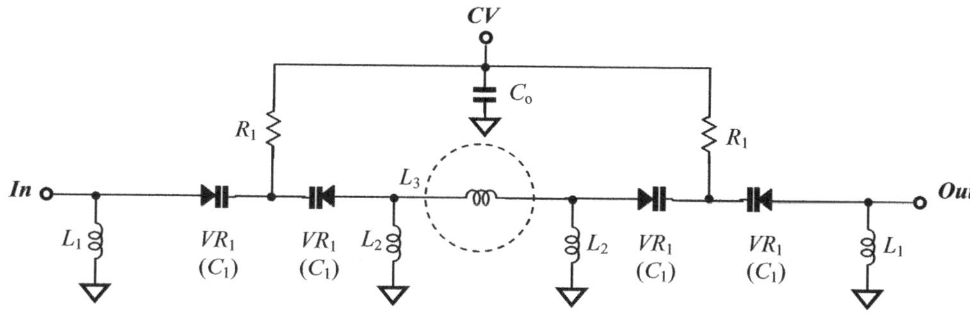

Figure 5.6 Tunable filter implemented by two tank circuits and coupled by L only.

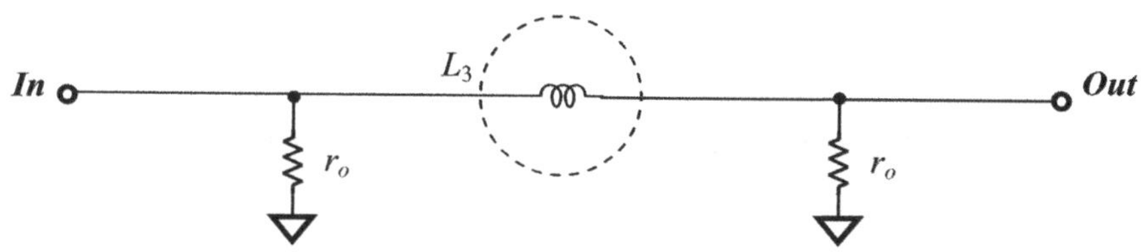

Figure 5.7 At f_o, two tank circuits are in resonant state and coupled by L only.

then,

$$\frac{\partial(BW)}{\partial f_o} = 0. \tag{5.12}$$

The expression (5.12) leads to an important conclusion: **The bandwidth of a tunable filter can be kept unchanged over the entire frequency range if the main coupling parts between the two tank circuits consist of only a pure inductor L.**

5.3 CIRCUIT DESCRIPTION

Figure 5.8 shows the schematic of a tunable filter which will be cooperated with a low-side injection mixer in a *UHF* portable radio. The main feature of the circuit is the pair of identical tank circuits coupled by an inductor, L_3. The simplest tank circuit consists of one inductor and one capacitor in parallel. As shown in Figure 5.8, the tank circuit consists of three inductors, two L_1 inductors, and one L_2 inductor, as well as two VR_1 varactors. The purpose of applying three inductors, instead of only one, is twofold: 1) To match the impedance of the input or output, $50\,\Omega$, without additional impedance matching parts; 2) To avoid too tight a coupling between the two tank circuits. In each tank circuit, two identical varactors are piggy-backed together. The tuning function is performed by these varactors. The capacitances of these varactors are controlled by the control voltage, CV, via the resistor $R_1 = 20\,k\Omega$, through which there is no current flowing.

The capacitor C_o is a "zero" capacitor in the *UHF* frequency range.

Another remarkable achievement in this tunable circuit design is that there is a second coupling capacitor, C_2, which creates two "zeros" in the skirt portion of the

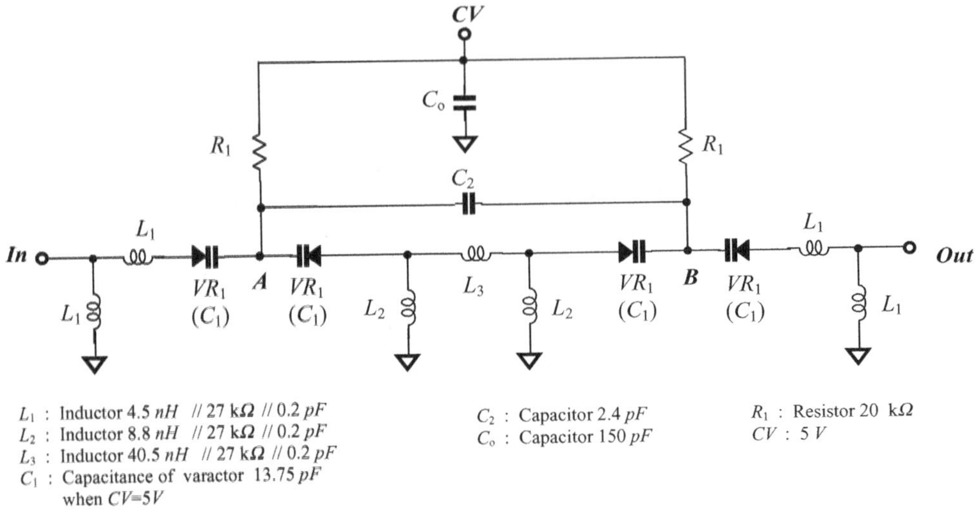

Figure 5.8 Schematic of a *UHF* tunable filter.

frequency response plot. One is located below the pass-band and another above. These two "zeros" can be adjusted by C_2, L_2, L_3, and VR_1, and can be applied to trace the imaginary spurious products in either the high side or low-side injections if the receiver is not operating in direct conversion mode. In the pass-band, the effect of the second coupling is negligible. Now, let's analyze how the second coupling works.

5.4 EFFECT OF SECOND COUPLING

Let's introduce the "π-T" or "Δ-*" transformation of the impedance.

Figure 5.9 shows the process of impedance transformation from the original network (a) to the final network (c).

The corresponding expressions between the impedances shown in Figure 5.9 are:

$$Z_2' = \frac{Z_2 Z_3}{Z_2 + Z_3 + Z_4}, \tag{5.13}$$

$$Z_3' = \frac{Z_3 Z_4}{Z_2 + Z_3 + Z_4}, \tag{5.14}$$

$$Z_4' = \frac{Z_4 Z_2}{Z_2 + Z_3 + Z_4}, \tag{5.15}$$

and

$$Z_a = Z_1 + Z_2' + Z_4' + Z_4' \frac{Z_1 + Z_2'}{Z_3' + Z_5}, \tag{5.16}$$

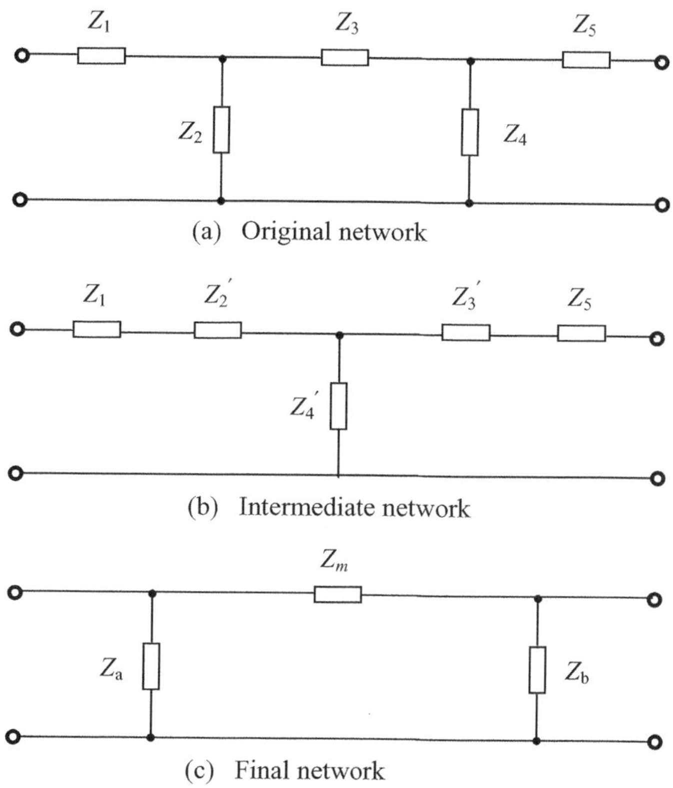

Figure 5.9 Z_2, Z_3, and Z_4 should have primes, not apostrophes.

$$Z_b = Z_5 + Z_3' + Z_4' + Z_4' \frac{Z_3' + Z_5}{Z_1 + Z_2'}, \tag{5.17}$$

$$Z_m = Z_1 + Z_5 + Z_2' + Z_3' + \frac{(Z_1 + Z_2')(Z_3' + Z_5)}{Z_4'}. \tag{5.18}$$

If

$$Z_1 = Z_5, \tag{5.19}$$

$$Z_2 = Z_4, \tag{5.20}$$

then

$$Z_2' = Z_3' = \frac{Z_2 \cdot Z_3}{2Z_2 + Z_3}, \tag{5.21}$$

$$Z_4' = \frac{Z_2^2}{2Z_2 + Z_3}, \tag{5.22}$$

$$Z_a = Z_b = Z_1 + Z_2, \tag{5.23}$$

$$Z_m = 2Z_1\left(1+\frac{Z_3}{Z_2}\right)+Z_3\left[1+\frac{Z_1^2}{Z_2^2}\left(1+2\frac{Z_2}{Z_3}\right)\right]. \tag{5.24}$$

Now let's apply the "π-T" or "Δ-*" transformation of impedance to the network between the nodes of A and B in Figure 5.8. The corresponding impedances are

$$Z_1 = Z_5 \Rightarrow \frac{1}{j\omega C_1}, \tag{5.25}$$

$$Z_2 = Z_4 \Rightarrow j\omega L_2, \tag{5.26}$$

$$Z_3 \Rightarrow j\omega L_3, \tag{5.27}$$

Substituting the relationships (5.25), (5.26), and (5.27) into (5.23) and (5.24), we have

$$Z_a = Z_b = \frac{1}{j\omega C_1} + j\omega L_3, \tag{5.28}$$

$$Z_m = \frac{1}{j\omega C_m} + j\omega L_m, \tag{5.29}$$

where

$$C_m = \frac{C_1}{2\left(1+\frac{L_3}{L_2}\right)}, \tag{5.30}$$

$$L_m = \left(1+\frac{1+2\frac{L_2}{L_3}}{L_2^2C_1^2\omega^4}\right)L_3. \tag{5.31}$$

The equivalent schematic of the tunable filter for *UHF* portable radio as shown in Figure 5.8 can be depicted as in Figure 5.10, which implies two possible "zeros" in the tunable filter. One of them is contributed by the series network, L_m, C_m, and C_2 and another exists in the parallel branch, $VR_1(C_1)$ and L_2. As a matter of fact, the original idea to have the second coupling is due to the recognition of these two "zeros".

The equivalent coupling network consists of L_m, C_m, and C_2. Its maximum impedance corresponds to a "zero" of the tunable filter at the frequency,

$$\omega_{Lo}^2 = \frac{C_2 + C_m}{L_m C_2 C_m}. \tag{5.32}$$

It is always lower than the central frequency of the tunable filter, ω_o, that is,

$$\omega_{Lo} < \omega_o, \tag{5.33}$$

ω_{Lo} is therefore called the low side "zero" frequency where a "zero" appears.

In a practical design, the difference between ω_o and ω_{Lo} is adjusted to about the value of $2\omega_{IF}$, so that ω_{Lo} could be treated as the imaginary frequency of the RF

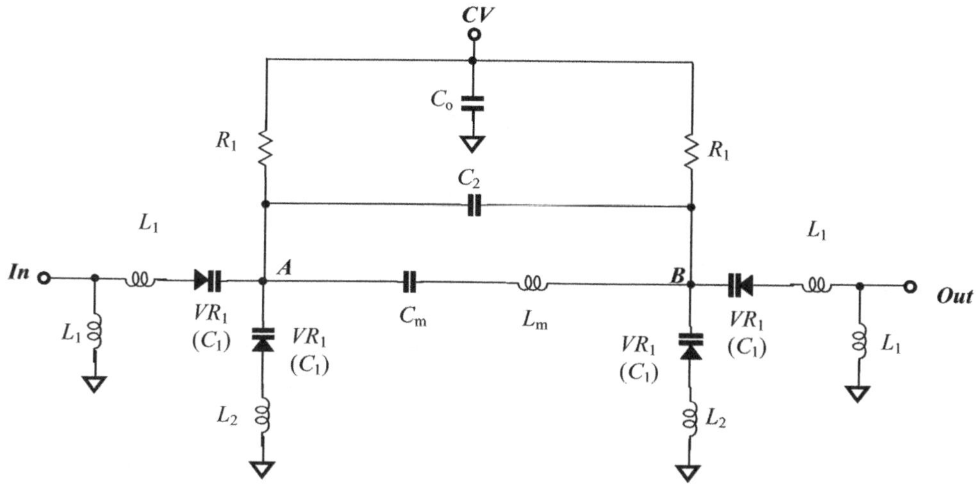

Figure 5.10 Equivalent schematic of a *UHF* tunable filter.

input, ω_{Lir} ,when a mixer is operated in a low-side injection mode. If the values of the parts in the tunable filter are selected and adjusted carefully, the difference between ω_o and ω_{Lo} is kept almost unchanged when the central frequency is tuned from the low-frequency end to the high-frequency end. This is a useful behavior of the imaginary rejection in co-operation with a mixer design.

Another "zero" of the tunable filter is created by the equivalent branch in parallel at either node A or node B as shown in Figure 5.10, which is formed by the parts C_1 and L_2 that is,

$$\omega_{Ho}^2 = \frac{1}{L_2 C_1}. \tag{5.34}$$

It should noted that the frequency ω_{Ho} is always higher than ω_o, that is,

$$\omega_{Ho} > \omega_o, \tag{5.35}$$

ω_{Ho} is therefore called the high side "zero" frequency where a "zero" appears.

When the tunable filter is cooperated with a mixer operating in the low-side injection mode, the "zero" at the frequency ω_{Ho} is not helpful to the imaginary rejection. However, it does help narrow the skirt of the entire frequency response curve.

Figure 5.11 compares the frequency responses between two cases with and without second coupling or, in other words, with or without the existence of the capacitor C_2. Both frequency response curves are tested when $f_o = 435.43\,MHz$.

Without second coupling, the skirt of the frequency response curve is wide open and the corresponding low-side imaginary rejection is poor, that is,

$$\text{Imag\_Rej}_{\text{at low side}} = -29.2\,dB, \tag{5.36}$$

down from the input signal level.

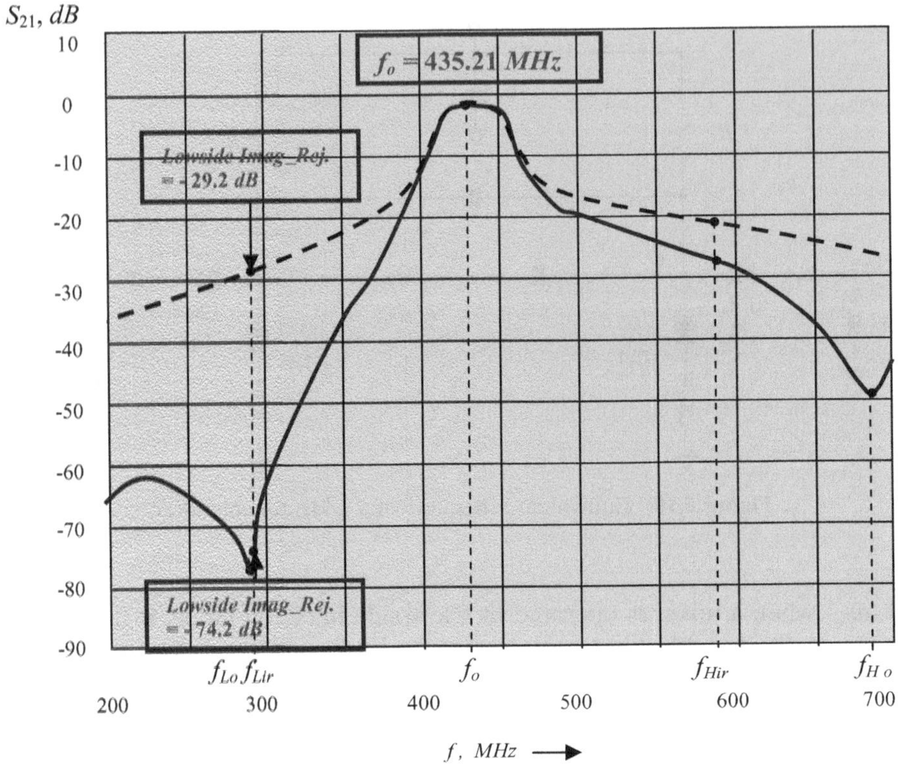

Figure 5.11 Comparison of frequency response of tunable filter.
——— with second coupling
– – – – without second coupling

With second coupling, the skirt of the frequency response curve is much narrowed since two "zeros" are created. The corresponding low-side imaginary rejection is much better, that is,

$$\text{Imag\_Rej}_{\text{at low side}} = -74.2\ dB, \tag{5.36}$$

down from the input signal level.

This is a 45 dB improvement! It is quite encouraging that such a small capacitor C_2 brings about such a huge advantage!

5.5 PERFORMANCE

The tested results of the tunable filter for a *UHF* portable radio are presented with four cases when the central frequency is tuned at

Low end of frequencies:	$f_{o1} = 403\,MHz$,
Intermediate frequency 1:	$f_{o2} = \sqrt{f_{o1} f_{o3}} = 435.21\,MHz$,

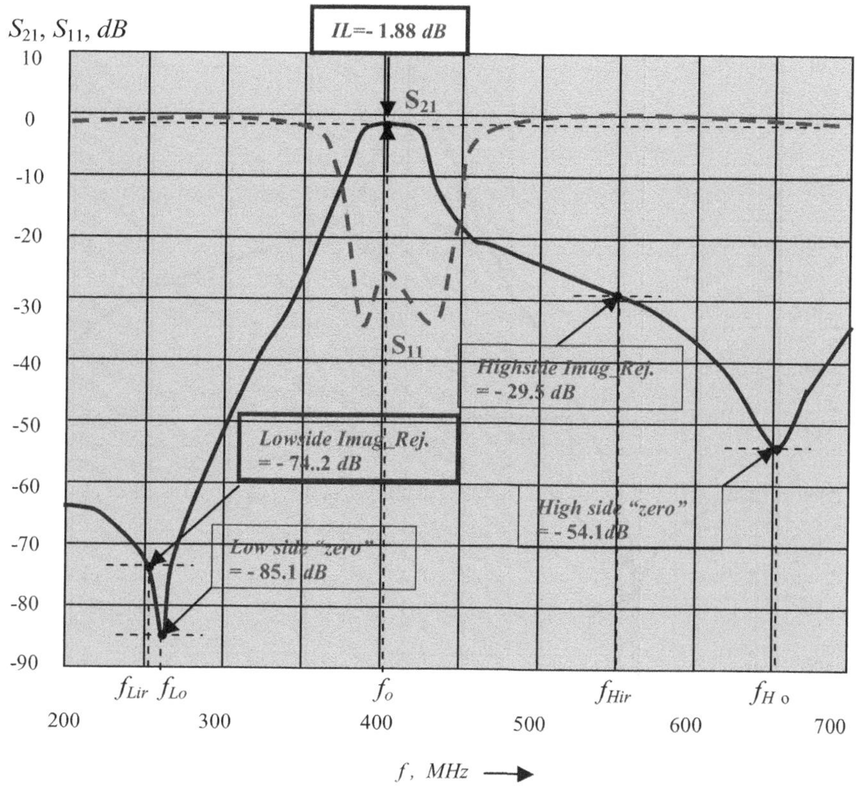

Figure 5.12 Frequency response of tunable filter when $f_o = 403.00\,MHz$.

Intermediate frequency 2:	$f_{o3} = 470\,MHz$,
High end of frequencies:	$f_{o4} = 512\,MHz$.

The frequency response of this tunable filter when $f_o = 403.00\,MHz$ is depicted in Figure 5.12. It shows that

- Central frequency, $f_o = 403.00\,MHz$,
- Bandwidth, $BW = 33.5\,MHz$,
- Insertion loss, $IL = -1.88\,dB$,
- Image rejection applied for low-side injection, when $f_{IF} = 73.35\,MHz$,

Low side	$Imag\_Rej = -74.2\,dB$,	at $f_{Lir} = 256.3\,MHz$,
	"zero" $= -85.1\,dB$,	at $f_{Lo} = 267.5\,MHz$,
High side	$Imag\_Rej = -29.5\,dB$,	at $f_{Hir} = 549.7\,MHz$.
	"zero" $= -54.1\,dB$,	at $f_{Ho} = 653.1\,MHz$.

The frequency response of this tunable filter when $f_o = 435.43\,MHz$ is depicted in Figure 5.13. It shows that

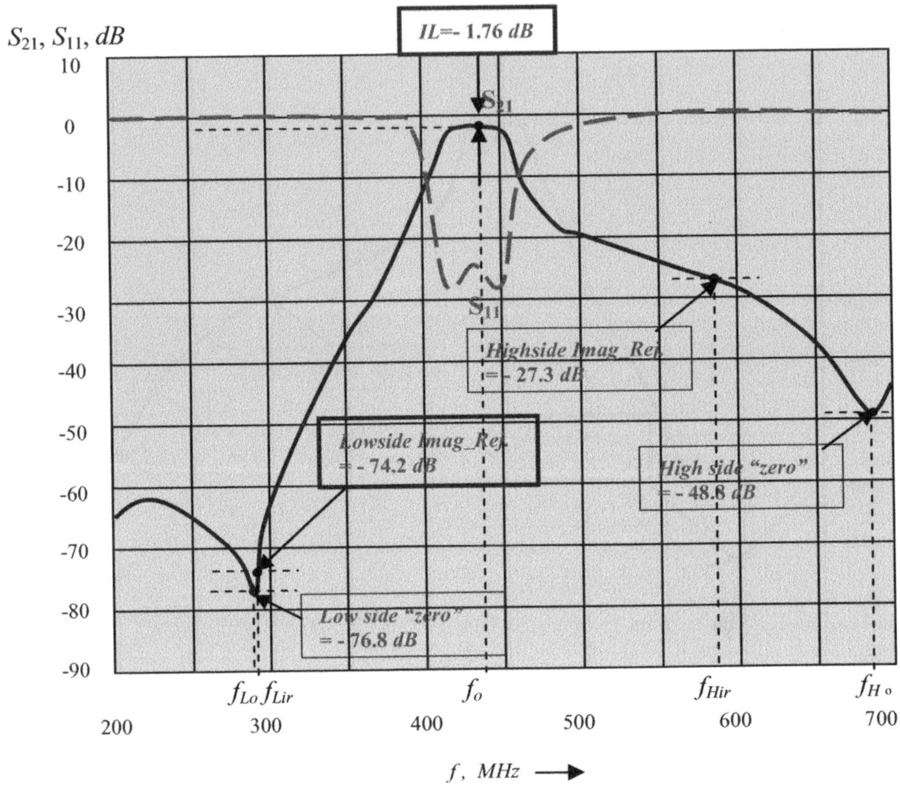

Figure 5.13 Frequency response of tunable filter when $f_o = 435.21\,MHz$.

- Central frequency, $f_o = 435.21\,MHz$,
- Bandwidth, $BW = 34.3\,MHz$,
- Insertion loss, $IL = -1.76\,dB$,
- Image rejection applied for low-side injection, when $f_{IF} = 73.35\,MHz$,
  Low side $Imag\_Rej = -74.2\,dB$, at $f_{Lir} = 288.73\,MHz$,
  "zero" $= -76.8\,dB$, at $f_{Lo} = 267.5\,MHz$,
  High side $Imag\_Rej = -27.3\,dB$, at $f_{Hir} = 582.13\,MHz$.
  "zero" $= -48.8\,dB$, at $f_{Ho} = 653.1\,MHz$.

The frequency response of this tunable filter when $f_o = 470.00\,MHz$ is depicted in Figure 5.14. It shows that

- Central frequency, $f_o = 470.00\,MHz$,
- Bandwidth, $BW = 35.1\,MHz$,
- Insertion loss, $IL = -1.56\,dB$,
- Image rejection applied for low-side injection, when $f_{IF} = 73.35\,MHz$,
  Low side $Imag\_Rej = -63.1\,dB$, at $f_{Lir} = 323.3\,MHz$,
  "zero" $= -72.5\,dB$, at $f_{Lo} = 267.5\,MHz$,

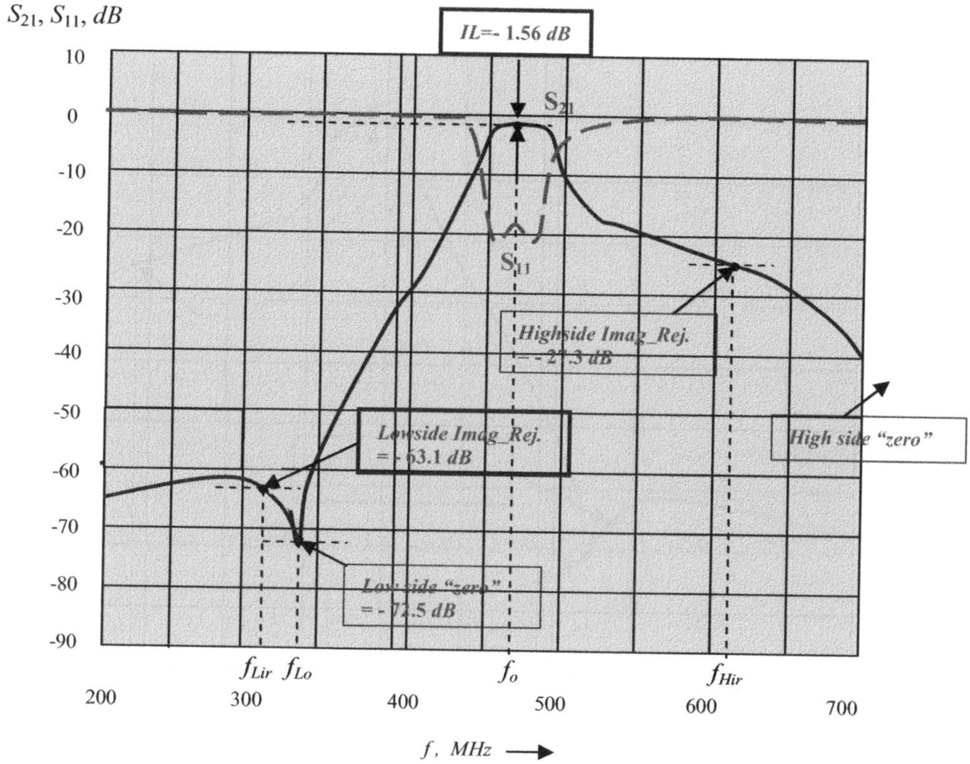

Figure 5.14 Frequency response of tunable filter when $f_o = 470.00\,MHz$.

High side	$Imag\_Rej = -24.8\,dB,$	at $f_{Hir} = 65.7\,MHz.$
	"$zero$" $= -?\ dB,$	at $f_{Ho} = ?\ MHz.$

The frequency response of this tunable filter when $f_o = 512.00\,MHz$ is depicted in Figure 5.15. It shows that

- Central frequency, $f_o = 512.00\,MHz,$
- Bandwidth, $BW = 35.8\,MHz,$
- Insertion loss, $IL = -1.42\,dB,$
- Image rejection applied for low-side injection, when $f_{IF} = 73.35\,MHz,$

Low side	$Imag\_Rej = -65.1\,dB,$	at $f_{Lir} = 365.3\,MHz,$
	"$zero$" $= -70.5\,dB,$	at $f_{Lo} = 377.5\,MHz,$
High side	$Imag\_Rej = -24.2\,dB,$	at $f_{Hir} = 658.7\,MHz.$
	"$zero$" $= -?\ dB,$	at $f_{Ho} = ?\ MHz.$

It can be concluded from Figures 5.12 to 5.15 that when $403 < f_o < 512\,MHz,$

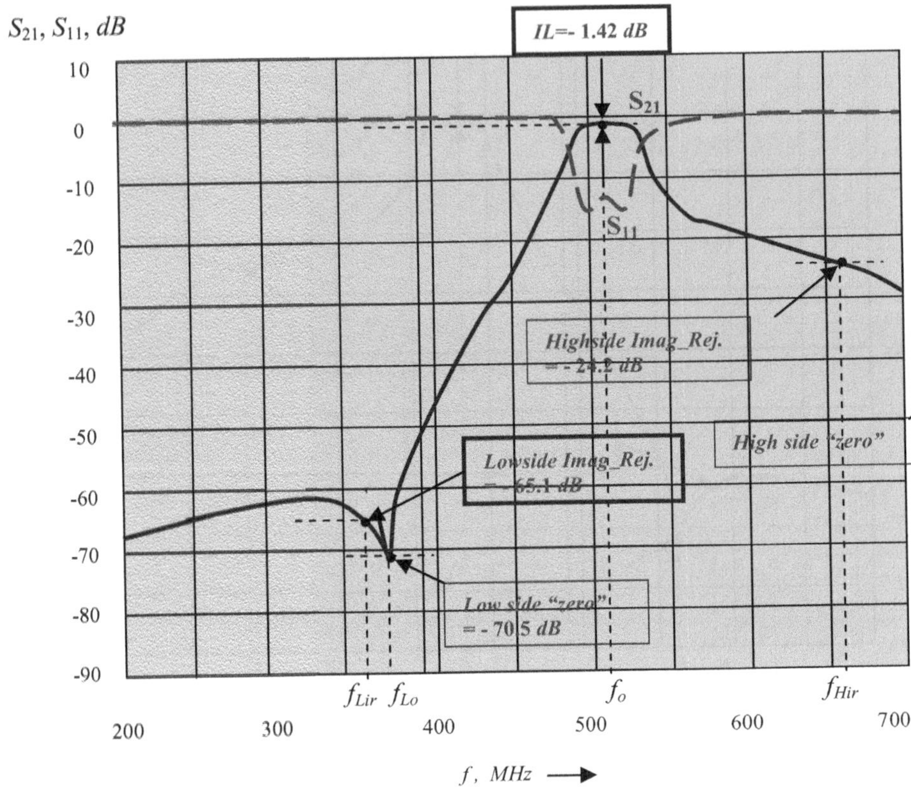

Figure 5.15 Frequency response of tunable filter when $f_o = 512.00\,MHz$.

1) The insertion loss is low;

$$\mathrm{IL} < 1.88\ dB, \tag{5.37}$$

2) The low side imaginary rejection is excellent if $f_{IF} = 73.35\,MHz$.

$$\mathrm{Imag\_Rej,}_{\ \mathrm{Lowside}} < -60\ dB. \tag{5.38}$$

3) The insertion loss and the imaginary rejection are tunable or traceable!

REFERENCES

[1] William H. Beyer, *Standard Mathematical Tables*, CRC Press, 24th ed., 1976.

[2] Donald R. J. White, *Electrical Filters*, Don White Consultants, Inc., 1980.

[3] Richard Chi-Hsi Li, "Tunable Filter Having Capacitive Coupled Tuning Elements," U.S. Patent 5,392,011, Motorola Inc., 1992.

CHAPTER 6

VCO (VOLTAGE-CONTROLLED OSCILLATOR)

6.1 "THREE-POINT" TYPE OSCILLATORS

There are many single-ended oscillators in *RF* circuit design. In this section, only the "three-point" type oscillators are discussed. Among the three-point type oscillators, the most popular one applied in communication systems is the Clapp oscillator, which will be discussed in Section 6.3 in some detail.

A three-point type oscillator contains only one device, either a bipolar or a *MOSFET* transistor. The *AC* equivalents of these three points are the base, collector, and emitter for a bipolar transistor, and the gate, drain, and source for a *MOSFET* transistor. Figure 6.1 shows the *AC* equivalent of a three-point oscillator built by a *MOSFET* transistor. The three nodes marked 1, 2, 3 are connected to the gate, drain, and the source of the *MOSFET* respectively. The three parts with impedances Z_1, Z_2, and Z_3 are connected between nodes 1 and 3, 3 and 2, and 2 and 1, respectively.

These three parts form a tank circuit loop. The *AC* equivalents plotted in Figure 6.1(a), (b), and (c) are exactly the same circuit. The only difference between Figures 6.1(a), (b), and (c) is that the tank circuit is at the input side in Figure 6.1(a), at the output side in Figure 6.1 (b), and at both the input and output sides in Figure 6.1(c).

Figure 6.2 shows that this oscillator looks like a common source (*CS*) amplifier with a feedback factor β, which is

$$\beta = \frac{Z_1}{Z_1 + Z_3}, \tag{6.1}$$

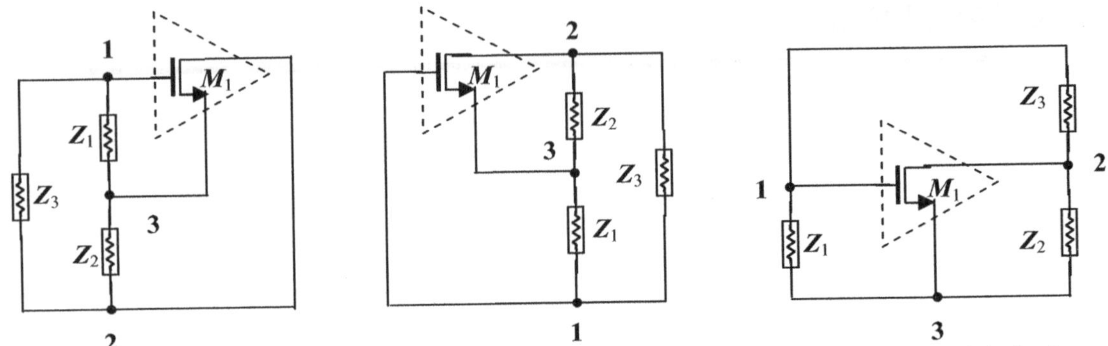

(a) Tank circuit in input side (b) Tank circuit in output side (c) Tank circuit in both input and output sides

Figure 6.1 *AC* equivalents of "three-point" type of oscillators.

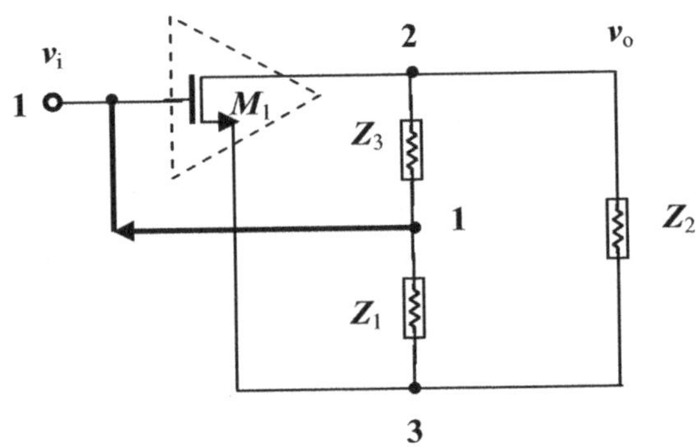

Figure 6.2 A "three-point" type oscillator looks like a common source (CS) amplifier with a feedback from output to input.

Figure 6.3 represents a voltage amplifier with feedback. The voltage gain without feedback A_v becomes a voltage gain with feedback A_f, that is,

$$A_v = \frac{v_o}{v_i}, \tag{6.2}$$

$$A_f = \frac{v_o}{v_i + \beta v_o} = \frac{A_v}{1 + \beta A_v}, \tag{6.3}$$

where

v_o = output voltage,
v_i = input voltage,
A_v = voltage gain without feedback,
A_f = voltage gain with feedback,
β = ratio of feedback voltage to output voltage

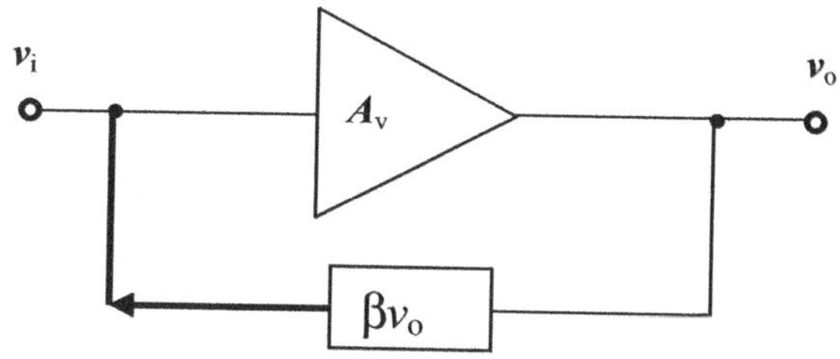

Figure 6.3 Voltage amplifier with feedback.

In the tank circuit, the impedance of a loop must be real or its phase shift must be zero at the operating frequency, that is,

$$Z_1 + Z_2 + Z_3 = 0, \tag{6.4}$$

then from expressions (6.1) and (6.4), we have

$$\beta A_v = A_v \frac{Z_1}{Z_1 + Z_3} = -A_v \frac{Z_1}{Z_2}. \tag{6.5}$$

From equations (6.3) and (6.5) it can be seen that in order to ensure that the amplifier with feedback is an oscillator, the factor βA_v must be negative so that it is an amplifier with positive feedback. Consequently from equation (6.5) it can be concluded that Z_1 and Z_2 must have the same sign since A_v is positive. In other words, Z_1 and Z_2 must be the same kind of reactance, either both inductive or both capacitive. On the other hand, Z_3 must have a reactance with an opposite sign of the reactance of Z_1 and Z_2. This was the basic clue in the development of the Hartley, Colpitts, and Pierce oscillators.

6.1.1 Hartley Oscillator

If Z_1 and Z_2 are inductors and Z_3 is a capacitor, the circuit is called a Hartley oscillator. Figure 6.4 shows the schematic and its *AC* equivalent circuit. The coupling between L_1 and L_2 can be described by their coupling coefficient, that is,

$$k_c = \frac{M}{\sqrt{L_1 L_2}}, \tag{6.6}$$

where

k_c = mutual coupling coefficient,
M = mutual inductance between L_1 and L_2.

In the case of strong coupling, that is,

$$k_c \approx 1. \tag{6.7}$$

The oscillation frequency can be approximated as

$$f = \frac{1}{2\pi\sqrt{C(L_1 + L_2 + 2M)}}. \tag{6.8}$$

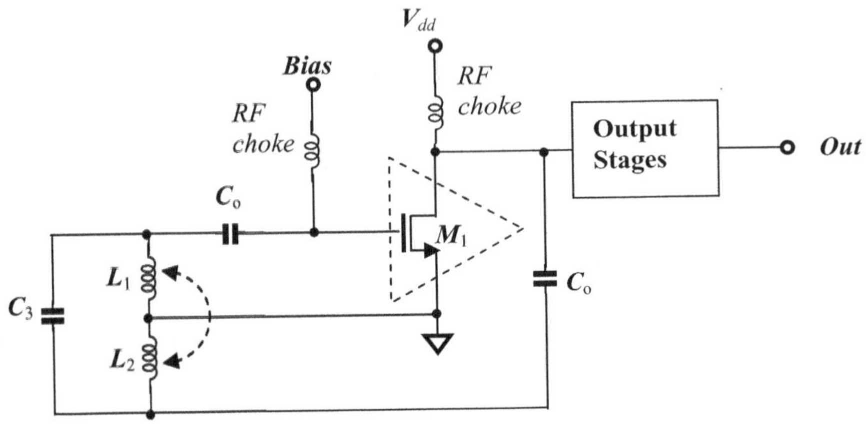

(a) Schematic

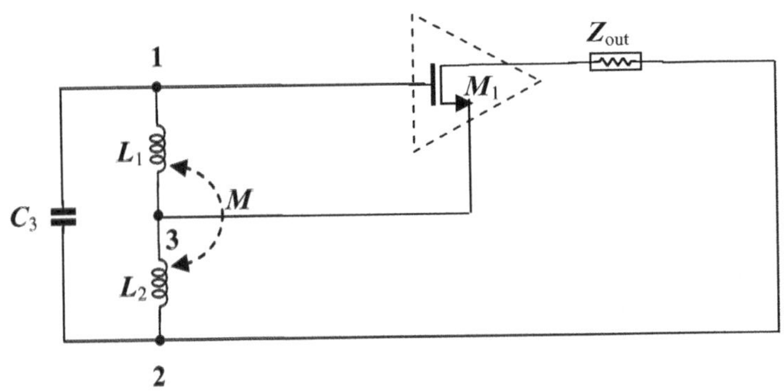

(b) AC equivalent circuit

Figure 6.4 Hartley oscillator. C_o: "zero" capacitor.

The condition of strong coupling (6.7) is fully satisfied if the two inductors in Figure 6.4 can be replaced by one inductor with an intermediate tap as shown in Figure 6.5. This replacement is beneficial to both cost and part count.

The Hartley oscillator has been applied to *RF* circuitry since the very early days of radio. One of its advantages is that the oscillation is easily agitated. In addition, the two inductors, L_1 and L_2, can be combined as one tapped inductor as shown in Figure 6.5. However, in this case the mutual inductance becomes an important factor in the oscillation so that the tapped point of the inductor must be selected carefully.

Today, the Hartley oscillator is much less common. One of the reasons for this is that two inductors or one tapped inductor must be applied. Another reason is that its phase noise is higher than that of other types of oscillators.

6.1.2 Colpitts Oscillator

If Z_1 and Z_2 are capacitors and Z_3 is an inductor, the circuit is called a Colpitts oscillator; one is plotted in Figure 6.6. The oscillation frequency can be approximated as

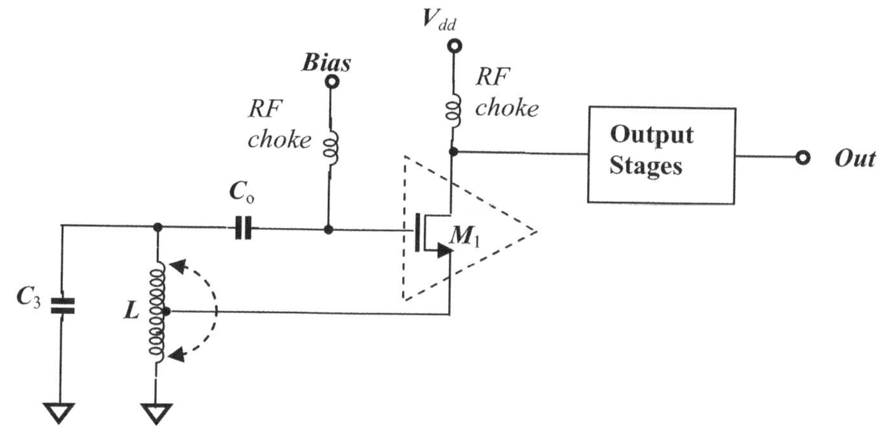

(a) Schematic

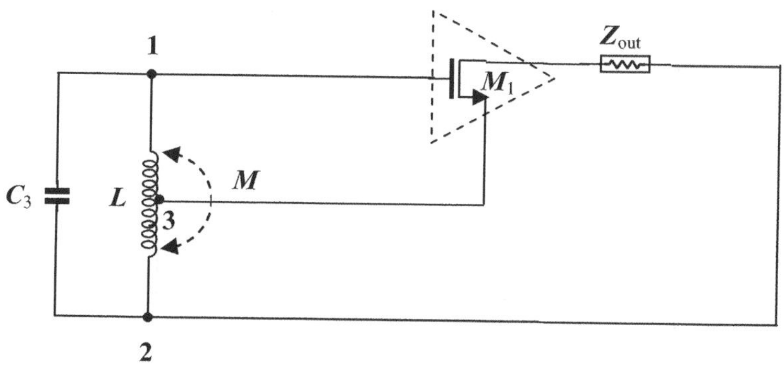

(b) *AC* equivalent circuit

Figure 6.5 Hartley oscillator with intermediate tapped inductor.

$$f = \frac{1}{2\pi\sqrt{L\dfrac{C_1C_2}{C_1+C_2}}}. \tag{6.9}$$

Compared with the Hartley oscillator, one of the Colpitts oscillator's advantages is that only one inductor is needed in the tank circuit, instead of two as in the Hartley. In addition, the phase noise in a Colpitts is much lower than that in a Hartley and other oscillators. This is why the Colpitts oscillator has become a popular type of *VCO* core today.

6.1.3 Clapp Oscillator

If Z_1 and Z_2 are capacitors and Z_3 is an inductor connected with two capacitors in series, the circuit is called a Clapp oscillator; one is plotted in Figure 6.7. As a matter of fact, the Clapp oscillator is a modified Colpitts oscillator. The improvement from

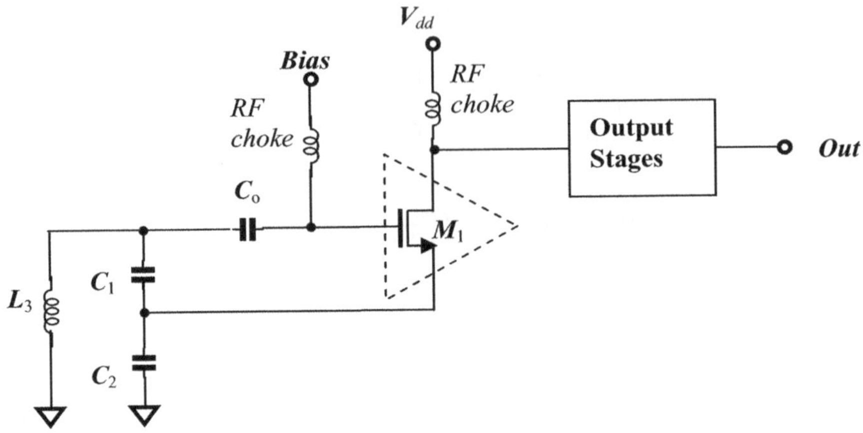

(a) Schematic

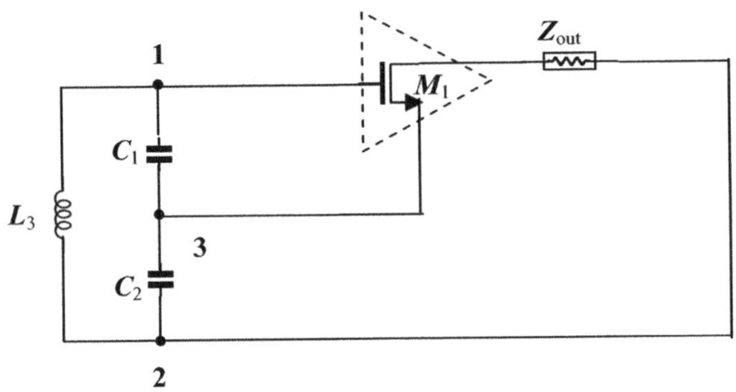

(b) *AC* equivalent circuit

Figure 6.6 Colpitts oscillator.

the Colpitts to the Clapp is that the frequency bandwidth is more easily covered by adjusting one of the two capacitors in series with the inductor. In Figure 6.7, C_R is the adjusted capacitor and is called a varactor, a capacitor in which the capacitance is adjusted or controlled by a control voltage, *C.V.*, so that the Clapp shown in Figure 6.7 is a *VCO*.

From Figure 6.7, we have

$$Z_3 = j\left(L_3\omega - \frac{C_3 + C_R}{C_3 C_R \omega}\right). \tag{6.10}$$

The oscillation frequency can be approximated as

$$f = \frac{1}{2\pi\sqrt{L\left(\frac{1}{C_1} + \frac{1}{C_2} + \frac{1}{C_3} + \frac{1}{C_R}\right)}}, \tag{6.11}$$

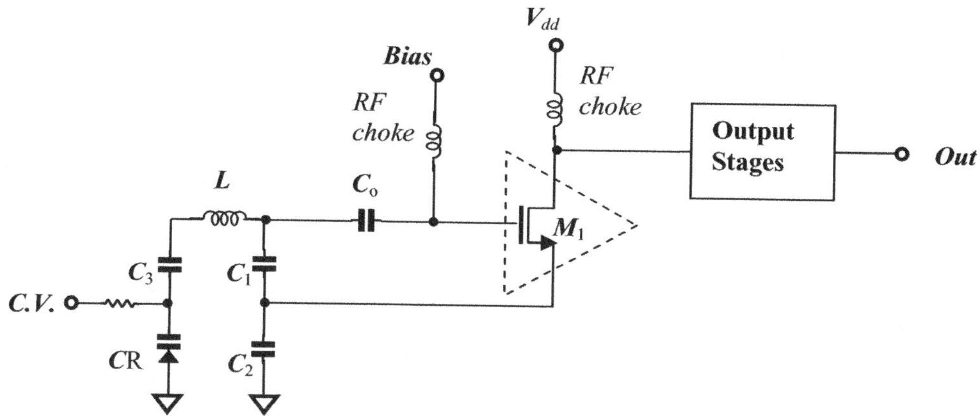

(a) Schematic

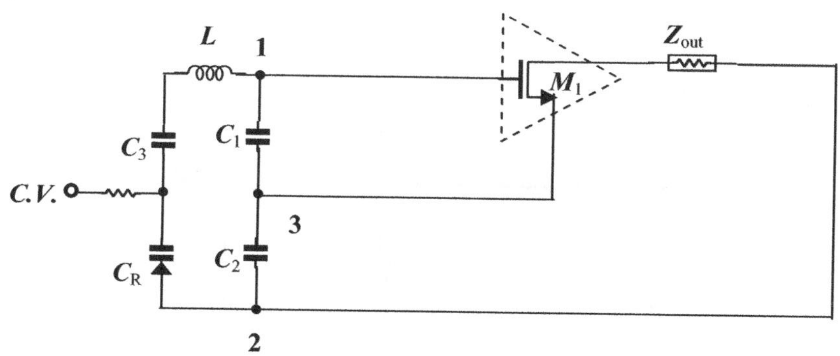

(b) *AC* equivalent circuit

Figure 6.7 Clapp oscillator.

where C_R = capacitance of varactor for a given value of control voltage *C.V.*

As a matter of fact, equations (6.10) and (6.11) both are only a rough approximation because the actual inductor *L* is a complicated part with spray capacitance.

It should be noted that in the practical design of the Hartley, Colpitts, and Clapp oscillators, as shown in Figures 6.4 to 6.7, the drain of the *MOSFET* transistor is not directly connected to the node 2 as shown in Figures 6.1 and 6.2. The *AC* equivalents of the three-point type oscillators shown in Figures 6.1 and 6.2 must be modified by adding the output impedance at the drain of *MOSFET* transistor; they become Figures 6.8 and 6.9 respectively. Figures 6.8 and 6.9 still behave as oscillators although their performances are somehow different from those shown in Figures 6.1 and 6.2. Additional attenuation and frequency shift may occur in the practical designs shown in Figures 6.8 and 6.9.

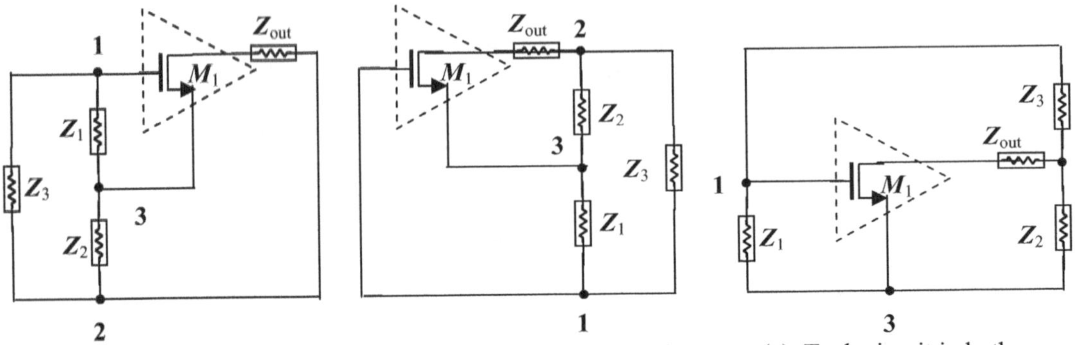

Figure 6.8 Modified *AC* equivalents of "three-point" type oscillators by the addition of output impedance Z_{out}.

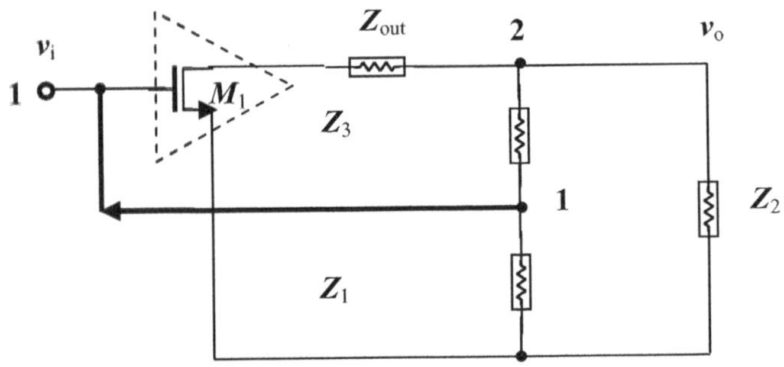

Figure 6.9 Modified Figure 6.2 by adding of output impedance Z_{out} between gate of *MOSFET* and node 2.

6.2 OTHER SINGLE-ENDED OSCILLATORS

6.2.1 Phase Shift Oscillator

The simplest *RC* phase shift oscillator is shown in Figure 6.10. Its frequency can be approximated as

$$f = \frac{4}{RC}. \tag{6.12}$$

In Figure 6.10, only one output is shown. Theoretically there could be a quad phase output from each node of the *RC* branches. However, due to the existence of M_1, *RF* chokes, and other additional parts shown in Figure 6.10, the expected quad phases are kept at a 90° difference between *RF* branches over the frequency bandwidth.

Instead of the RC shift, a simple ring phase-shift oscillator can be built by transistors only and is shown in Figure 6.11. There are four stages in series circularly. The

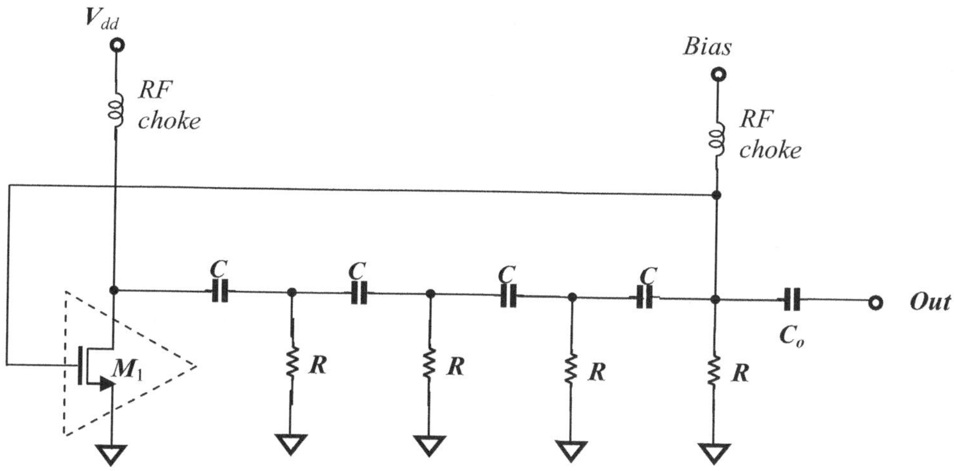

Figure 6.10 A simple *RC* phase-shift oscillator. C_o: "zero" capacitor.

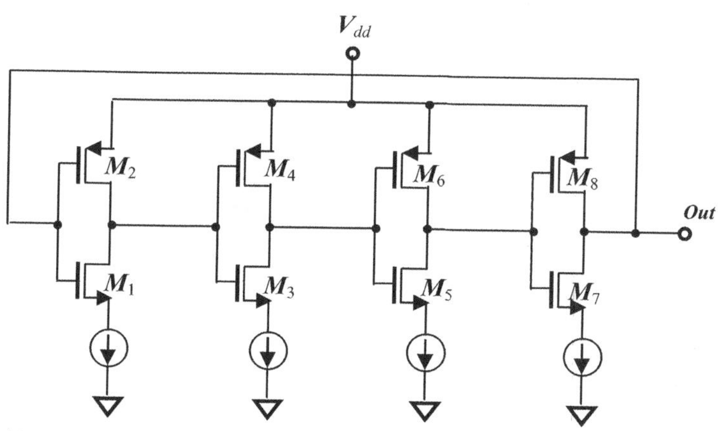

(Current source is 2 digits controlled by voltage V_c from *PLL*)

Figure 6.11 A simple phase-shift oscillator built by transistors only.

phase is shifted 90° in each stage. This oscillator can have quad phases outputs from each connecting node between stages.

Figure 6.12 shows a differential type of phase shift oscillator. The advantages of the phase shift oscillator are

1) Lower current drain consumption,
2) Considerably smaller size than that of a *LC VCO*,
3) Its quadrature signals are inherently produced from four stage ring *VCO*. Neither a *PPF* (Poly-Phase Filter) nor frequency divider to produce the quadrature signals is required.

Its disadvantage is that its phase noise is higher than that of the *LC VCO*.

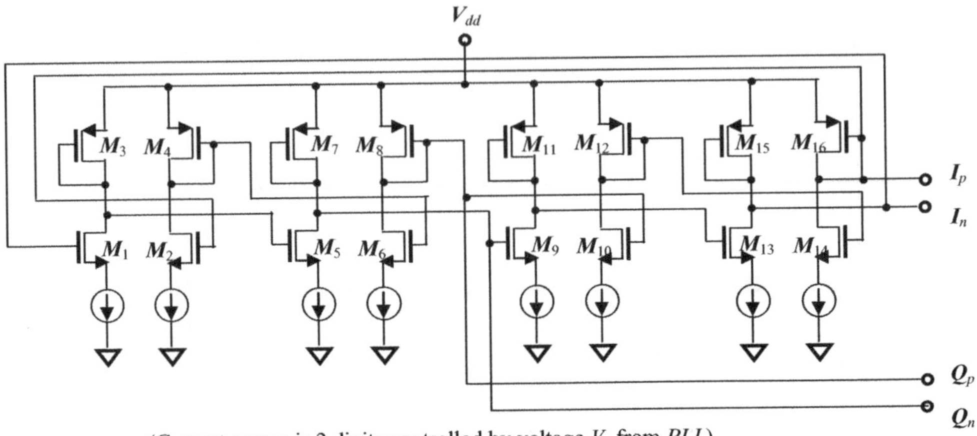

(Current source is 2 digits controlled by voltage V_c from *PLL*)

Figure 6.12 A simple differential phase-shift oscillator.

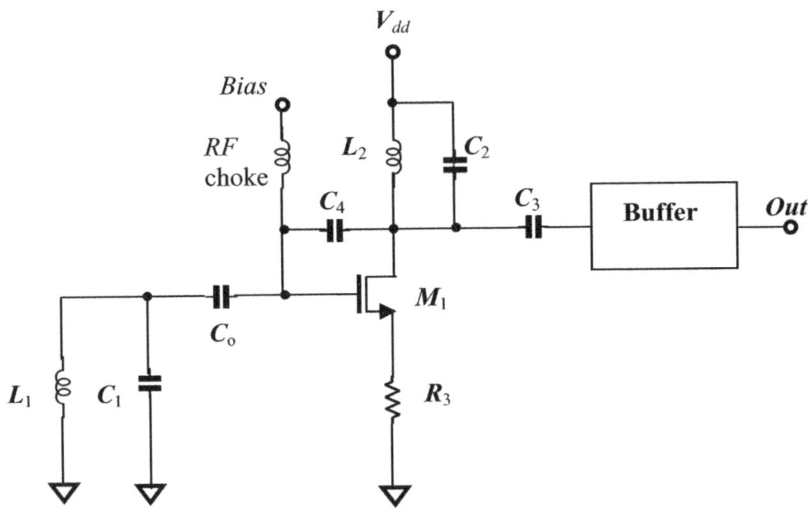

Figure 6.13 *TOTI* oscillator. C_o: "zero" capacitor.

6.2.2 *TITO* (Tuned Input and Tuned Output) Oscillator

Figure 6.13 shows a TITO oscillator, in which C_4 is a coupling capacitor. The oscillation frequency can be approximated as the resonant frequency of the input tank or output tank circuit, that is

$$f \approx \frac{1}{2\pi\sqrt{L_1C_1}} \approx \frac{1}{2\pi\sqrt{L_2C_2}}. \tag{6.13}$$

6.2.3 Resonant-Circuit Oscillator

Figure 6.14 shows a resonant oscillator. As a matter of fact, it is a positive feedback amplifier. There is a transformer connected between the drain and the gate. Some output power is coupled from the inductance of the secondary winding of the transformer L by the inductance of the primary winding of transformer L' through their mutual inductance M. The feedback power maintains the oscillation of the oscillator.

The oscillation frequency can be approximated as

$$f = \frac{1}{2\pi\sqrt{LC}}. \tag{6.14}$$

6.2.4 Crystal Oscillator

The key element in a crystal oscillator is a piezoelectric crystal. Figure 6.15 shows its characteristics. The Q value for a piezoelectric crystal is very high, usually in the

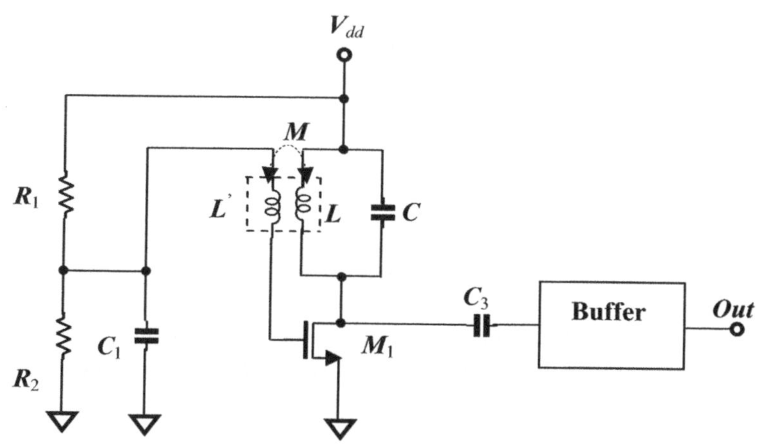

Figure 6.14 Resonant oscillator.

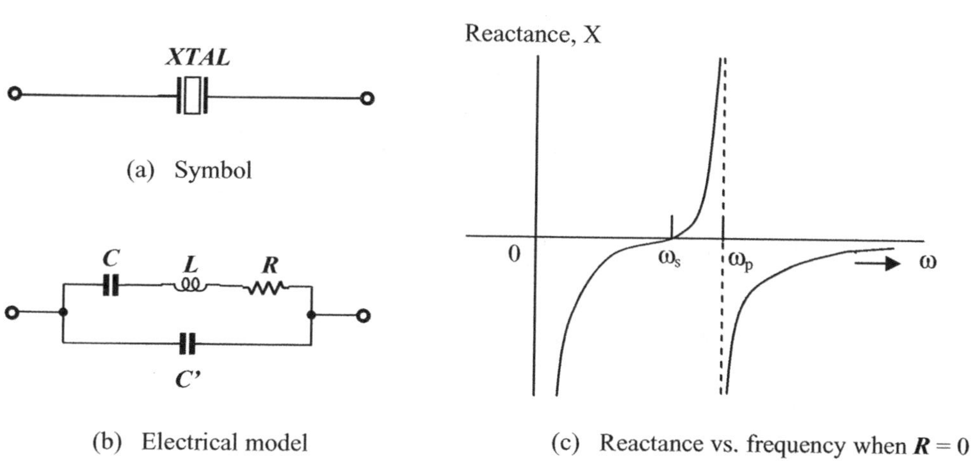

Figure 6.15 Characteristics of a piezoelectric crystal.

range of 5,000 to 10,000. For a 45 MHz crystal, the value of L is in the order of 100 mH, C is in the order of 0.0001 pF, and C' is in the order of 0.1 pF which is much higher than the value of C. In Figure 6.15(c),

$$jX = -j\frac{1}{\omega C'}\frac{\omega^2 - \omega_s^2}{\omega^2 - \omega_p^2}, \tag{6.15}$$

where

ω_s = series resonant angular frequency,
ω_p = parallel resonant angular frequency,
and

$$\omega_s^2 = \frac{1}{LC}. \tag{6.16}$$

$$\omega_p^2 = \frac{1}{L}\left(\frac{1}{C} + \frac{1}{C'}\right). \tag{6.17}$$

Usually

$$\omega_s^2 \approx \omega_p^2, \tag{6.18}$$

because

$$C' \gg C. \tag{6.19}$$

Figure 6.16 shows a crystal oscillator in which the crystal is used for Z_1 as shown in Figures 6.2 and 6.9, while the tuned LC combination $L_2//C_2$ is used for Z_2, and C_4 is used for Z_3 which is

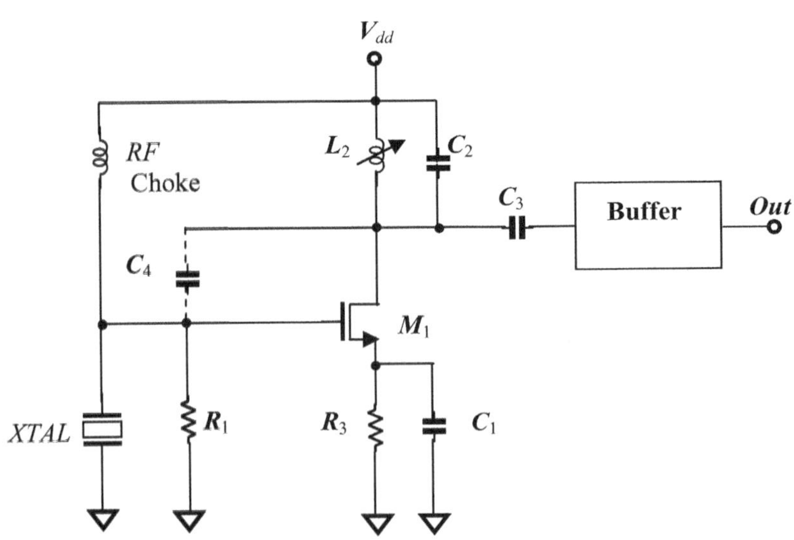

Figure 6.16 Crystal oscillator.

$$C_4 = C_{dg} + C_{stray}, \tag{6.20}$$

where

C_{dg} = capacitance between drain and gate of the *MOSFET*,
C_{stray} = stray capacitance existed in device and layout.

The oscillator frequency is essentially determined by the crystal, and not by the rest of the circuit.

6.3 *VCO* AND *PLL*

The main function of the *VCO* is to control the frequency of an oscillator by a control voltage, which is applied to a *PLL* (Phase Lock Loop), in most cases for frequency synthesizer, modulator, and demodulator, as well as other applications. To understand the *VCO* better it is necessary to introduce the fundamentals of the *PLL* and the relationship between the *VCO* and *PLL*.

6.3.1 Implications of *VCO*

A *VCO* is a frequency modulator (see Figure 6.17). The output voltage of a *VCO*, v_i , can be expressed by

$$v_{VCO}(t) = V_o \sin[\omega_c t + \theta(t)]. \tag{6.21}$$

The frequency deviation of the output, $d\theta / dt$, is proportional to the input voltage v_i.

$$\frac{d\theta(t)}{dt} = K_v v_c(t). \tag{6.22}$$

Its integral is the phase of *VCO*.

$$\theta(t) = K_v \int_0^t v_c(x)\,dx, \tag{6.23}$$

where

v_{VCO} = output voltage of *VCO*,
V_o = voltage amplitude of *VCO* output,
ω_c = angular frequency of *VCO*,

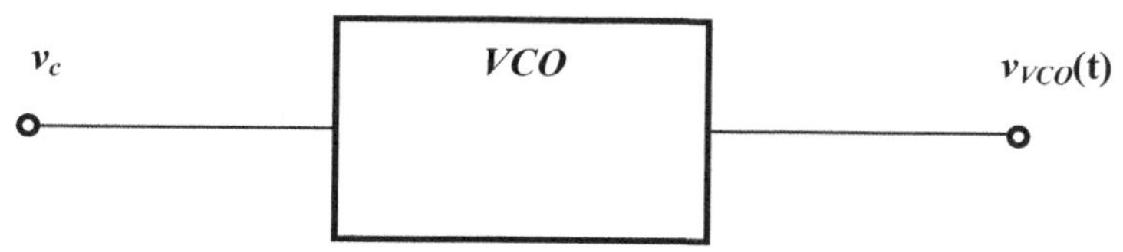

Figure 6.17 Function of a *VCO*.

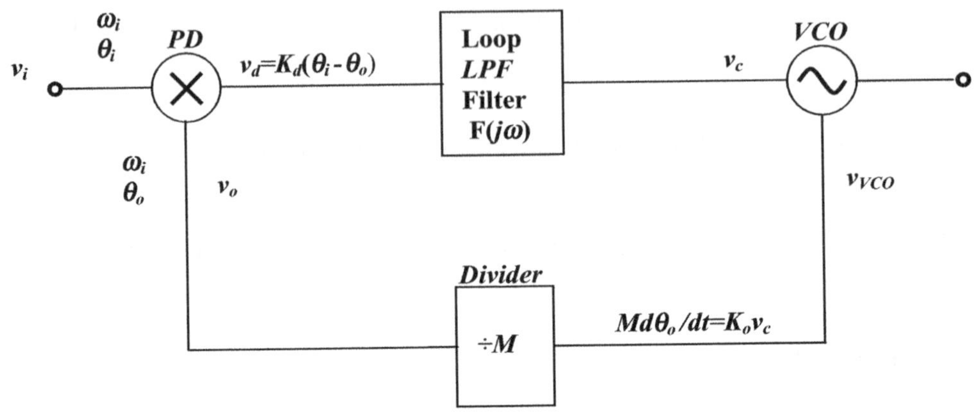

Figure 6.18 Basic block diagram of *PLL* (Phase Lock Loop).

θ = phase of *VCO* signal,
K_v = *VCO* constant, measured in radians per second per unit of input,
v_c = control voltage.

6.3.2 Transfer Function of *PLL*

Figure 6.18 shows a basic block diagram of a *PLL*, including a *PD* (Phase Detector), a loop filter, a *VCO* (Voltage Controlled Oscillator), and a divider. The phase detector is a multiplier when the two inputs are small signals. If these two inputs are assumed as

$$v_i(t) = V_s \sin(\omega_i t + \theta_i), \tag{6.24}$$

$$v_o(t) = V_o \cos(\omega_i t + \theta_o), \tag{6.25}$$

where

ω_i = input frequency of signal,
the basic loop equations are

$$v_d(t) = K_m v_i(t) \cdot v_o(t) = \frac{1}{2} K_m V_s V_o \sin(2\omega_i t + \theta_i + \theta_o) + \frac{1}{2} K_m V_s V_o \sin(\theta_i - \theta_o), \tag{6.26}$$

where

K_m = multiple factor of multiplier with dimensions of volts^{-1}.

Owing to the existence of the loop low-pass filter, the high-frequency term in (6.25) can be neglected so that

$$v_d(t) \approx v_i(t) \cdot v_o(t) = \frac{1}{2} K_m V_s V_o \sin(\theta_i - \theta_o) = K_d \sin(\theta_i - \theta_o), \tag{6.27}$$

where

K_d = phase detector gain factor of multiplier and is measured in unit of volts per radian.

The output of the loop *BPF* is

$$v_c(t) = v_d(t)F(j\omega). \tag{6.28}$$

The variation of the phase at the *VCO* output is linearly related to the control voltage at the *VCO* input, that is,

$$M\frac{d\theta_o}{dt} = K_o v_c, \tag{6.29}$$

where

K_o = *VCO* gain factor and is measured unit of *rad*/(*sec*·V),
M = divisor of the divider.

By using Laplace notation, the basic loop equations (6.27), (6.28), and (6.29) become

$$v_d(s) = K_d[\theta_i(s) - \theta_o(s)], \tag{6.30}$$

$$v_c(s) = v_d(s)F(s), \tag{6.31}$$

$$\theta_o = \frac{K_o v_c(s)}{sM}, \tag{6.32}$$

The open-loop transfer function or open-loop gain is

$$G(s) = \frac{\theta_o(s)}{\theta_e(s)} = \frac{\theta_o(s)}{\theta_i(s) - \theta_o(s)} = \frac{K_o K_d F(s)}{sM}. \tag{6.33}$$

The closed-loop transfer function or closed-loop gain is

$$H(s) = \frac{\theta_o(s)}{\theta_i(s)} = \frac{K_o K_d F(s)}{sM + K_o K_d F(s)} = \frac{G(s)}{1 + G(s)}, \tag{6.34}$$

$$1 - H(s) = \frac{\theta_e(s)}{\theta_i(s)} = \frac{\theta_i(s) - \theta_o(s)}{\theta_i(s)} = \frac{sM}{sM + K_o K_d F(s)}, \tag{6.35}$$

$$v_c(s) = v_d(s)F(s) = K_d\theta_e(s)F(s) = \frac{sMK_dF(s)\theta_i(s)}{sM + K_o K_d F(s)} = \frac{sM\theta_i(s)}{K_o}H(s), \tag{6.36}$$

where the phase difference between the two input signals into the phase detector is denoted by θ_e, that is

$$\theta_e(s) = \theta_i(s) - \theta_o(s). \tag{6.37}$$

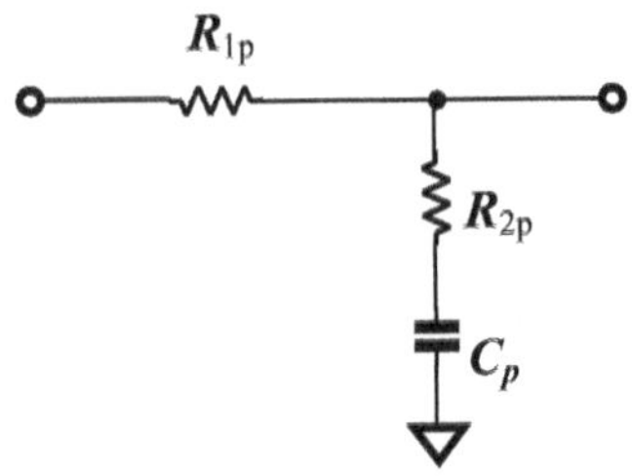

(a) Passive lag-lead filter

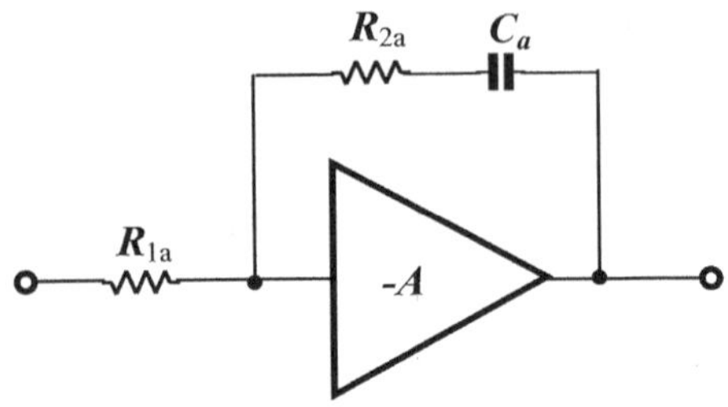

(b) Active filter

Figure 6.19 Filters applied in a second-order *PLL* loop.

The low-pass loop filter is very often a passive or an active second order loop filter. Figure 6.19(a) and (b) shows their respective schematics. The transfer function of the passive filter as shown in Figure 6.19(a) is

$$F_{psv}(s) = \frac{sC_pR_{2p}+1}{sC_p(R_{1p}+R_{2p})+1} = \frac{s\tau_{2p}+1}{s\tau_{1p}+1}, \tag{6.38}$$

where

$$\tau_{1p} = C_p(R_{1p}+R_{2p}), \tag{6.39}$$

$$\tau_{2p} = C_pR_{2p}. \tag{6.40}$$

The transfer function of the active filter as shown in Figure 6.17(b) is

$$F_{atv}(s) = \frac{-A(sC_aR_{2a}+1)}{sC_aR_{2a}+1+(1+A)sC_aR_{1a}}. \tag{6.41}$$

If

$$A \gg 1, \tag{6.42}$$

then

$$F_{atv}(s) = \frac{sC_aR_{2a}+1}{sC_aR_{1a}} = \frac{s\tau_{2a}+1}{s\tau_{1a}}, \tag{6.43}$$

where

$$\tau_{1a} = C_aR_{1a}, \tag{6.44}$$

$$\tau_{2a} = C_aR_{2a}. \tag{6.45}$$

The closed-loop transfer functions with passive and active second loop filters are
For the passive filter,

$$H_{psv}(s) = \frac{K_o K_d (s\tau_{2p} + 1)/\tau_{1p}}{s^2 + s(1 + K_o K_d \tau_{2p})/\tau_{1p} + K_o K_d/\tau_{1p}}. \tag{6.46}$$

For the active filter when condition (6.41) is satisfied,

$$H_{atv}(s) = \frac{K_o K_d (s\tau_{2a} + 1)/\tau_{1a}}{s^2 + s(K_o K_d \tau_{2a}/\tau_{1a}) + K_o K_d/\tau_{1a}}. \tag{6.47}$$

These two closed-loop transfer functions can be rewritten as

$$H_{psv}(s) = \frac{s(2\zeta_p \omega_{np} - \omega_{np}^2/K_o K_d) + \omega_{np}^2}{s^2 + 2\zeta_p \omega_{np} s + \omega_{np}^2}, \tag{6.48}$$

where

ω_{np} = natural frequency of loop,
ζ_p = damping factor,

$$\omega_{np} = \sqrt{\frac{K_o K_d}{\tau_{1p}}}, \tag{6.49}$$

$$\zeta_p = \frac{1}{2}\sqrt{\frac{K_o K_d}{\tau_{1p}}}\left(\tau_{2p} + \frac{1}{K_o K_d}\right), \tag{6.50}$$

and

$$H_{atv}(s) = \frac{2\zeta_a \omega_{na} s + \omega_{na}^2}{s^2 + 2\zeta_a \omega_{na} s + \omega_{na}^2}, \tag{6.51}$$

$$\omega_{na} = \sqrt{\frac{K_o K_d}{\tau_{1a}}}, \tag{6.52}$$

$$\zeta_a = \frac{\tau_{2a}}{2}\sqrt{\frac{K_o K_d}{\tau_{1a}}} = \frac{\tau_{2a}}{2}\omega_{na}. \tag{6.53}$$

6.3.3 White Noise from Input of *PLL*

Assuming that the input of *PLL* as shown in Figure 6.18, v_i, is a sinusoidal signal plus a stationary Gaussian bandpass noise, that is, expression (6.24) is modified as

$$v_i(t) = V_s \sin(\omega_i t + \theta_i) + n(t). \tag{6.54}$$

While the expression (6.25) is kept unchanged,

$$v_o(t) = V_o \cos(\omega_i t + \theta_o). \tag{6.55}$$

Expression (6.27) is modified to

$$v_d(t) = K_d[\sin(\theta_i - \theta_o) + n'(t)]. \tag{6.56}$$

If the noise power spectral density of $n(t)$ is

$$W_n(f) = N_o(V^2 \cdot Hz^{-1}), \tag{6.57}$$

it can be proved that the normalized spectrum density of $n'(t)$ is

$$\Phi_{n'}(f) = \frac{2N_o}{V_s^2}(Hz^{-1}). \tag{6.58}$$

The variance of the output phase is

$$\overline{\theta_{no}^2} = \int_0^\infty \Phi_{n'}(f)|H(j\omega)|^2 df, \tag{6.59}$$

$$\overline{\theta_{no}^2} = \frac{2N_o}{V_s^2}\int_0^\infty |H(j2\pi f)|^2 df = \frac{2N_o}{V_s^2} B_L, \tag{6.60}$$

where the noise bandwidth is defined as

$$B_L = \int_0^\infty |H(j\omega)|^2 df. \tag{6.61}$$

6.3.4 Phase Noise from *VCO*

The white noise from the input of the *PLL* or at the terminal v_i has just been considered and expression (6.24) has been modified to expression (6.54). The impact on the *PLL* loop is that there is a phase variance θ_{no}^2, as shown in expression (6.59) appearing at the output of the divider.

In the derivation of expression (6.60), it was assumed that there is no noise source inside the *PLL* loop, so that expression (6.25) is kept unchanged. This is, of course, not true in reality. As a matter of fact, another noise source exists in the *PLL* loop. In other words, all the blocks shown in Figure 6.18 can produce noise and bring about other phase variances. Among these noise sources, the phase noise from the *VCO* dominates.

Studies on the phase noise produced by the *VCO* have been conducted for many years.

The oscillation operation can be considered as a process of charging and discharging its tank circuit. The oscillator can be thought of as equivalent to two parts. One portion is the tank circuit, which consists of a resistor R, a capacitor, C, and an inductor L. The resistor R is an equivalent part which causes all the energy loss in the oscillation, so that the tank circuit is an energy consumer portion. To maintain oscillation, the other portion of the oscillator must provide the energy to the tank circuit, so that it is called the energy generator portion. For simplicity, this portion is assumed to be noiseless because all the possible resistance in this portion can be converted into an equivalent resistance and added to the value of the resistor R in the tank circuit. This is a simple model of an oscillator as shown in Figure 6.20.

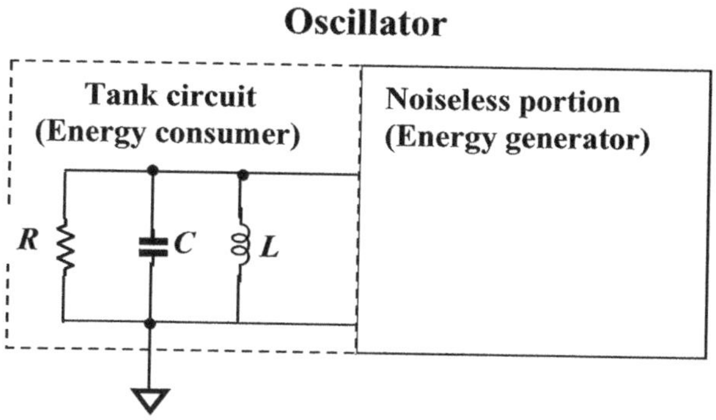

Figure 6.20 Oscillator can be equivalent to two portions: energy generator and energy consumer.

The noise produced by the R in the tank circuit is a thermal noise with a mean square spectral density of

$$\frac{\overline{i_n^2}}{\Delta f} = \frac{4kT}{R}. \tag{6.62}$$

where

$\overline{i_n^2}$ = mean square of thermal noise current of a resistor,
k = Boltzmann constant,
T = temperature,
R = resistance of R,
Δf = operating frequency bandwidth.

The impedance of a tank circuit is

$$Z(\omega_o + \Delta\omega) \approx j\frac{\omega_o L}{2\Delta\omega/\omega_o}, \tag{6.63}$$

$$\omega_o = \frac{1}{\sqrt{LC}}, \tag{6.64}$$

if $\Delta\omega$ is much smaller than ω_o, and
where

$Z(\omega_o+\Delta\omega)$ = impedance of tank circuit at $\omega = \omega_o+\Delta\omega$,
ω_o = operating angular frequency,
L = inductance in tank circuit,
C = capacitance in tank circuit.

Noting that the unloaded Q is

$$Q = \frac{R}{L\omega_o}. \tag{6.65}$$

Then the impedance of the tank circuit can be expressed in terms of Q, that is, from (6.63) and (6.65) we have

$$Z(\omega_o + \Delta\omega) = j\frac{\omega_o R}{2Q\Delta\omega}. \tag{6.66}$$

The spectral density of the mean-square noise voltage can be obtained by multiplying the spectral density of the mean-square noise current with the squared magnitude of the tank impedance, that is,

$$\frac{\overline{v_n^2}}{\Delta f} = \frac{\overline{i_n^2}}{\Delta f} Z^2 = 4kTR\left(\frac{\omega_o}{2Q\Delta\omega}\right)^2. \tag{6.67}$$

The normalized single-sideband noise spectral density is the ratio of noise power density divided by the power of signal,

$$L(\Delta\omega) = \frac{\dfrac{1}{2R}\dfrac{\overline{v_n^2}}{\Delta f}}{P_{sig}} = \frac{2kT}{P_{sig}}\left(\frac{\omega_o}{2Q\Delta\omega}\right)^2, \tag{6.68}$$

where P_{sig} = power of signal, or in the logarithmic scale with the unit of dB,

$$L(\Delta\omega) = 10 \cdot Log\left[\frac{2kT}{P_{sig}}\left(\frac{\omega_o}{2Q\Delta\omega}\right)^2\right]. \tag{6.69}$$

Figure 6.21 plots $L(\Delta\omega)$ vs. $\Delta\omega$ in the logarithmic scale based on equation (6.69), which is a straight line and when

$$\Delta\omega = \frac{\omega_o}{2Q}, \tag{6.70}$$

the normalized single-sideband noise spectral density $L(\Delta\omega)$ is

$$L(\Delta\omega)|_{\Delta\omega=\omega_0/(2Q)} = 10 \cdot Log\left(\frac{2kTR}{P_{sig}}\right). \tag{6.71}$$

The plot as shown in Figure 6.21 is usually called the phase noise plot because instead of the oscillation frequency ω_o, the abscissa is scaled by $\Delta\omega$, which represents the variation of oscillation.

As shown in Figure 6.21, the theoretical phase noise predicted by equation (6.69) is roughly close to the actual tested phase noise result in the region when $\Delta\omega$ is neither too low nor too high. This is a region where $L(\Delta\omega)$ is approximately proportional to $1/(\Delta\omega)^2$ or $1/f^2$. (Note: Traditionally, the symbol "f" is used in the phase noise plot. As a matter of fact, it is the offset frequency $\Delta\omega$). In the two extreme cases when $\Delta\omega$ is quite low or high, the theoretical phase noise predicted by equation (6.69) is far from the values of the actual tested phase noise result.

Leeson (1966) modified the phase noise from the simple model expressed by the equation (6.69). According to Leeson's model, the phase noise can be expressed by

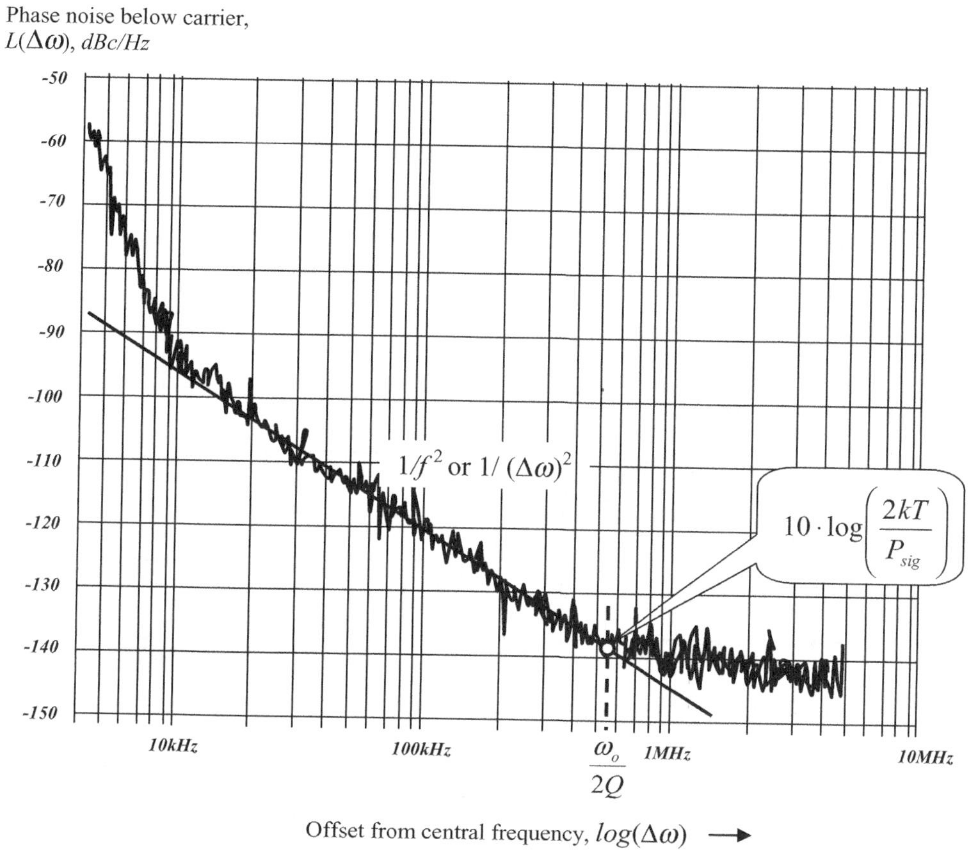

Figure 6.21 Spectrum of *VCO* phase noise.
Theoretical curve from equation (6.69)
Actually tested phase noise curve

$$L(\Delta\omega) = 10 \cdot Log\left[\frac{2FkT}{P_{sig}}\left(1 + \frac{\omega_o}{2Q\Delta\omega}\right)^2\left(1 + \frac{\Delta\omega_{1/f^3}}{|\Delta\omega|}\right)\right], \tag{6.72}$$

where

F = a factor to account for the increased noise in the $1/(\Delta\omega)^2$ or $1/f^2$ region,

$\Delta\omega_{1/f^3}$ = offset frequency turning point from $1/(\Delta\omega)^3$ or $1/f^3$ region to $1/(\Delta\omega)^2$ or $1/f^2$ region.

Comparing equation (6.72) with (6.68), there are three differences in equation (6.72):

1) A factor F is to account for the increased noise in the $1/(\Delta\omega)^2$ or $1/f^2$ region.
2) The additive factor of unity in the first parenthesis is to account for the noise floor.

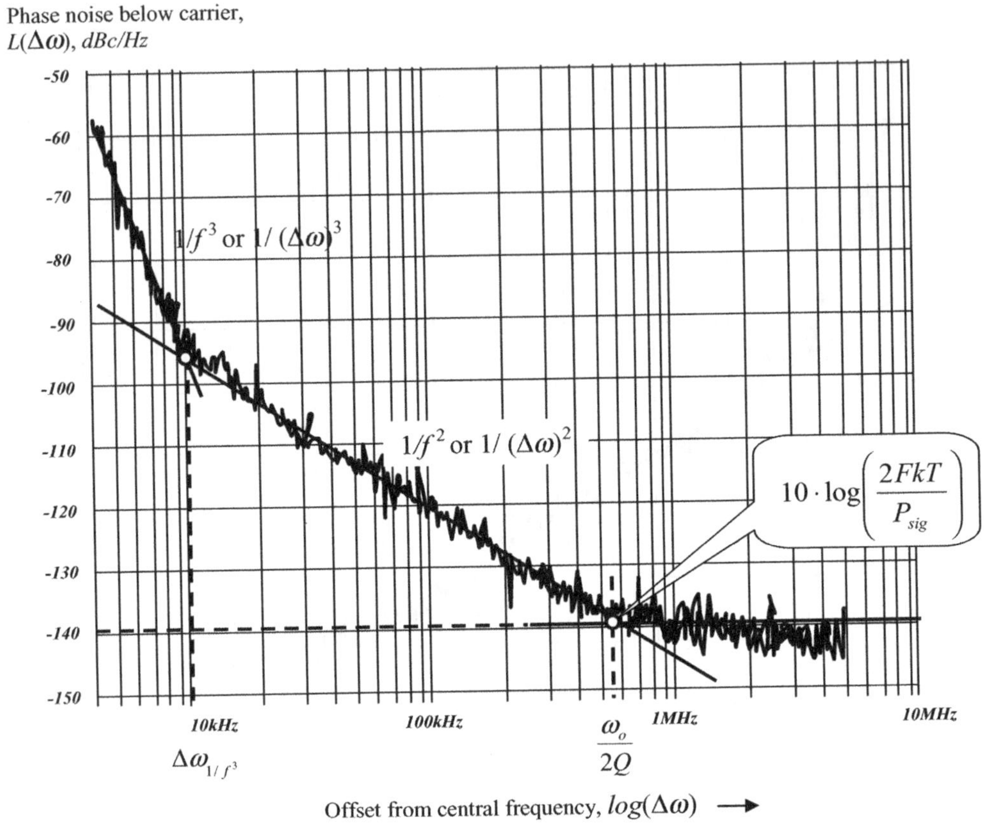

Figure 6.22 Spectrum of *VCO* phase noise.
——— Leeson's model, equation (6.72)
Actually tested phase noise curve

3) The multiplicative factor $1+\Delta\omega_{1/f^3}/|\Delta\omega|$ is to provide the $1/(\Delta\omega)^3$ or $1/f^3$ behavior in the $1/(\Delta\omega)^3$ or $1/f^3$ region.

Figure 6.22 plots $L(\Delta\omega)$ vs. $\Delta\omega$ in the logarithmic scale. The solid lines are asymptotes of equation (6.72).

The first line segment in Figure 6.22 is located in the $1/(\Delta\omega)^3$ or $1/f^3$ region. In this region, the flicker noise of the device dominates over other noises. It is well known that the flicker noise is proportional to the reciprocal of the frequency, that is,

$$\overline{i_{n,\,flick}{}^2} \propto \frac{I^a}{f}, \tag{6.73}$$

where

$i_{n,flick}$ = mean square of flicker noise current,
f = frequency,

I = DC current flowing through the device,

a = constant in the range from 0.5 to 2.

The flicker noise with $1/(\Delta\omega)$ or $1/f$ behavior and the noise with $1/(\Delta\omega)^2$ or $1/f^2$ behavior shown in equation (6.69) are converted to the noise with the $1/(\Delta\omega)^3$ or $1/f^3$ behavior. This is why the multiplicative factor $\left(1+\Delta\omega_{1/f^3}/|\Delta\omega|\right)$ is added in Leeson's equation (6.72).

The second line segment in Figure 6.22 is located in the $1/(\Delta\omega)^2$ or $1/f^2$ region. It is the same as that in the simple model described by the equation (6.69).

The third line segment in Figure 6.22 is located in a flat region. It represents the noise floor of the oscillator.

As shown in Figure 6.22, the actual tested phase noise curve can be approximated by Leeson's model. In recent years emphasis has been placed on the theoretical study of the factor F and the offset frequency turning point $\Delta\omega_{1/f^3}$ in equation (6.72).

More effort has been placed on the reduction of the phase noise of the oscillation because the phase noise directly impacts the performance of the phase detector and thus the entire PLL loop. Phase noise can cause PLL locked for an unacceptably long time, or the PLL being locked at unexpected frequencies, or in extreme cases put the PLL out of work altogether. Some special schemes have been reported, such as the reduction of phase noise by means of an on-chip filter or off-chip low-frequency inductor, and the modification of the oscillator topology.

6.4 DESIGN EXAMPLE OF A SINGLE-ENDED *VCO*

6.4.1 Single-Ended *VCO* with Clapp Configuration

This is a practical VCO design example with discrete parts which has been applied to a communication system in the UHF frequency range, from 440 to 470 MHz. Most of the design considerations and schemes are universally applicable to today's $RFIC$ design even though the VCO in this example is implemented by discrete parts.

Figure 6.23 shows the schematic of the VCO with Clapp configuration. The tank circuit consists of CR_1, C_1, L_2, C_2, and C_3. The varactor CR_1 is controlled by the control voltage CV. At the output portion, the Π type of Chebyshev filter consists of C_4, L_5, and C_5 and is inserted between the collector of the bipolar transistor and the buffer. The inductors L_3 and L_4 adjust the input and output impedances while the resistor R_4 partially plays the self-bias role. The primary goals are listed in Table 6.1.

In order to obtain a better coverage of the frequency bandwidth, we must be careful to measure, test the characteristics of the parts in the tank circuit, and then evaluate the frequency bandwidth through the adjustment of the parts' value.

6.4.2 Varactor

The capacitance of a varactor is changed as its control voltage is varied. As shown in Figure 6.24, the capacitance of a varactor is high when the control voltage is low and low when the control voltage is high. The operating frequency of the VCO is

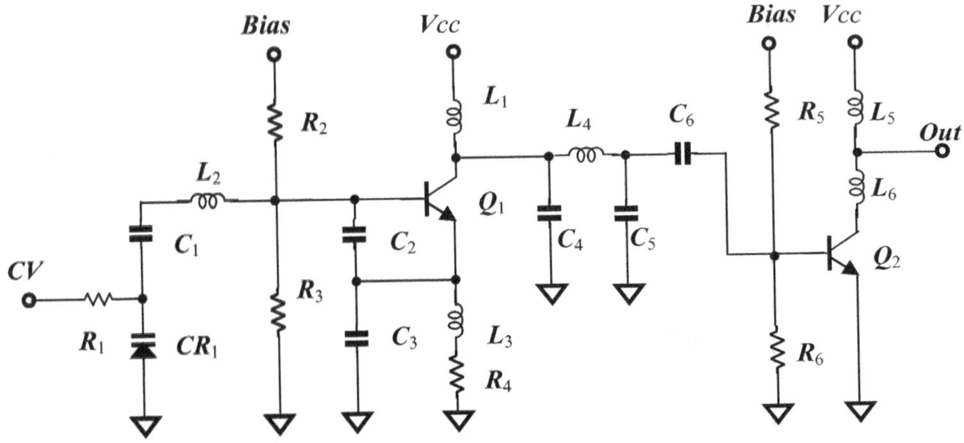

Figure 6.23 Single-end *VCO* with Clapp configuration.

TABLE 6.1 Primary goals of single-ended *VCO* with Clapp configuration

Frequency range:	440 to 470 MHz
DC power supply:	5 V
Current drain:	<15 mA
Output power:	>12 dB_m (16 mW)

therefore controlled by the control voltage through the variation of the varactor's capacitance.

Generally, the characteristics of a varactor are functions of frequency. Therefore, the variation of capacitance vs. control voltage must be carefully measured and evaluated as the frequency is varied. Fortunately, it has been found that such a frequency dependence for actual varactors is very small and within the range of experimental error. The values shown in Figure 6.24 are applicable for the frequency range to be covered. In other words, within experimental error, they are independent of frequency.

The test results for four varactors are shown in Figure 6.24. Varactor 1 might be applied for cases when the *VCO* is operated in the low-frequency range, since its capacitances on the range of control voltages are higher, while varactor 2 might be applied for cases when the *VCO* is operated in the high-frequency range since its capacitances on the range of control voltages are relatively lower. Varactor 3 seems to be the overall best candidate because its variation of capacitance vs. control voltage, $\Delta C/\Delta(CV)$, is higher than that of either varactor 1 or 2. Varactor 4 is a poor candidate because its variation of capacitance vs. control voltage, $\Delta C/\Delta(CV)$, changes too much over the range of the control voltage. The slope $\Delta C/\Delta(CV)$ is too steep when the control voltage is low whereas it is too flat when the control voltage is high.

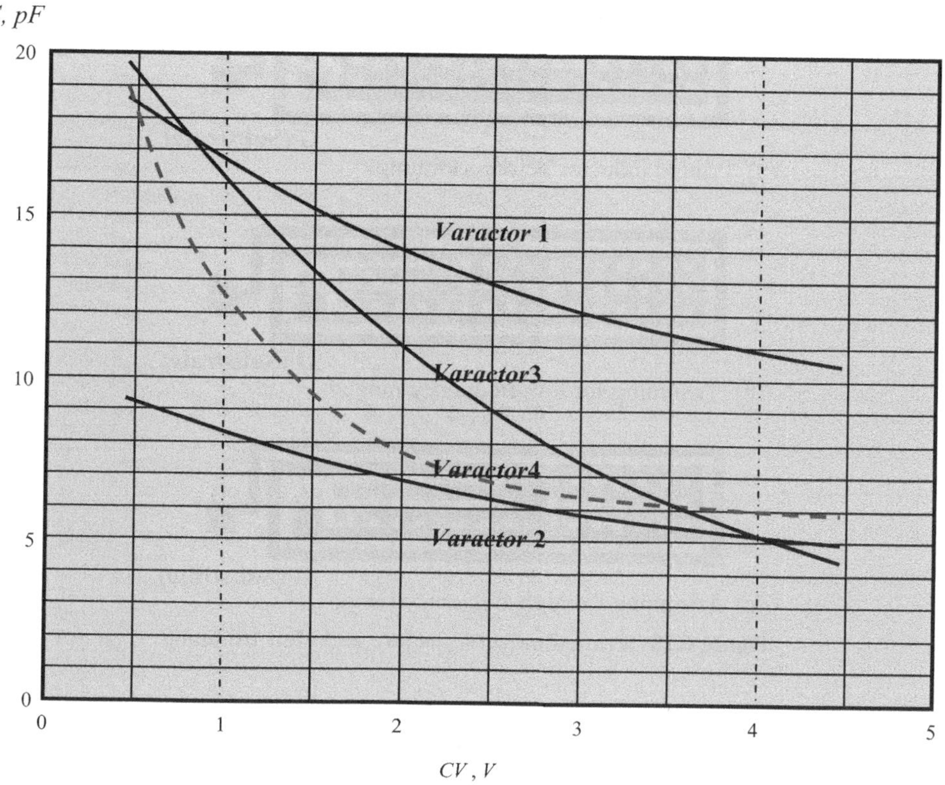

Figure 6.24 Capacitance vs. control voltage of a varactor.

6.4.3 Printed Inductor

For the sake of cost-saving, printed inductors are adapted for L_2 and L_5. Their configurations are shown in Figure 6.25. Their inductances can be adjusted by trimming the grill portion of the printed inductor. Figure 6.25(a) shows the primary printed inductor before trimming processing. Figures 6.25(b) and 6.25(c) show the printed inductor after it is trimmed for cases when the *VCO* operates in low- and high-frequency ranges. Its Q value is approximately 60 to 90, which is lower than that of an air coil but much higher than that of a spiral inductor in an *RFIC*.

Measuring the inductance of a printed inductor in the laboratory is a problem not only in theory, but also in practice. At the present, some calculation methods are available, but only for printed inductors of symmetrical or uniform geometry and not for irregular or un-symmetrically shaped printed inductors. It has been proven that any printed inductor can be modeled by the equivalent circuit as shown in Figure 6.26. Instead of distributed parameters, an equivalent circuit with lumped parameters is preferred.

The values of the parts in the equivalent circuit in Figure 6.26 can be obtained by measuring the impedance three times.

The measured impedance C'_A between the nodes A and G is contributed by two branches in parallel. One branch is the capacitor C_A. Another branch is C_{L+CB}, the inductor L connected with the capacitor C_B in series:

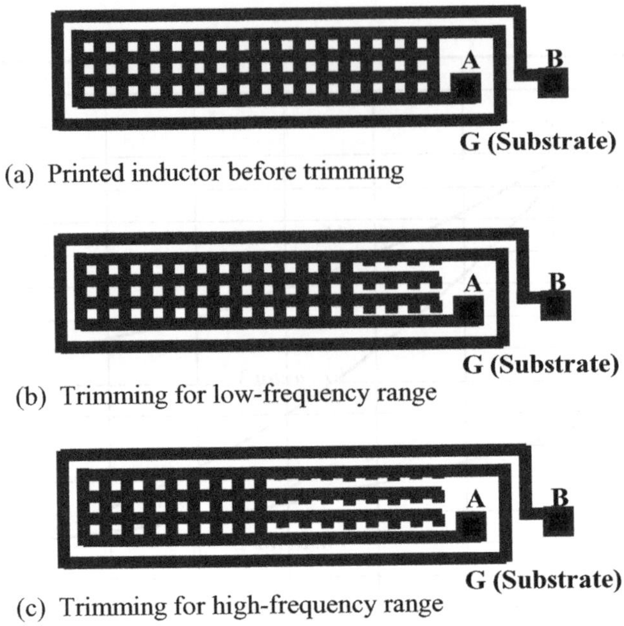

Figure 6.25 Printed inductors before and after trimming.

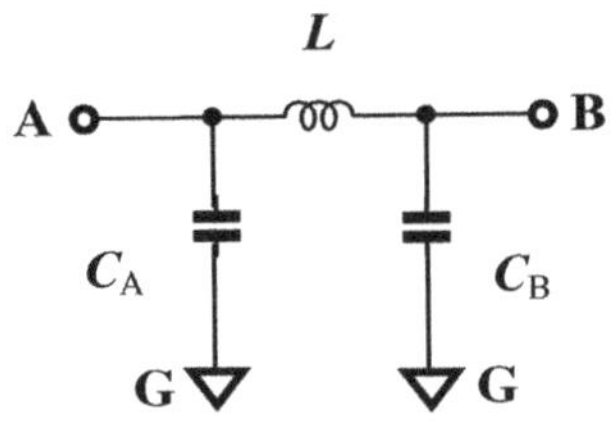

Figure 6.26 Equivalent circuit of printed inductor.

$$C_{L+C_B} = \frac{C_B}{1-\omega^2 LC_B}, \tag{6.74}$$

then, the measured C'_A should be

$$C'_A = C_A // C_{L+C_B} = C_A + \frac{C_B}{1-\omega^2 LC_B}, \tag{6.75}$$

Similarly, the measured capacitance, C'_B, between the nodes **B** and **G** is contributed by two branches in parallel. One branch is the capacitor C_B. Another branch is C_{L+CA}, the inductor L connected with the capacitor C_A in series:

$$C_{L+C_A} = \frac{C_A}{1-\omega^2 LC_A}, \tag{6.76}$$

then, the measured C'_B should be

$$C'_B = C_B // C_{L+C_A} = C_B + \frac{C_A}{1-\omega^2 LC_A}. \tag{6.77}$$

The measured inductance, L', between nodes $\boldsymbol{A}$ and $\boldsymbol{B}$, is contributed by two branches in parallel. One branch is the inductor L. The other branch is $C_A + C_B$, the capacitor C_A connected with the capacitor C_B in series:

$$C_{C_A+C_B} = \frac{C_A C_B}{C_A + C_B}, \tag{6.78}$$

then, the measured L' should be

$$L' = L // (C_A + C_B) = L\left(1 + \omega^2 L' \frac{C_A C_B}{C_A + C_B}\right). \tag{6.79}$$

By means of some simple but tedious algebraic manipulations, from expressions (6.74) to (6.79), we have

$$C_A = \frac{2C'_A C'_B(\omega^2 L' C'_A C'_B + C'_B - C'_A)}{(\omega^2 L' C'_A C'_B + C'_B - C'_A)^2 + 4\omega^2 L' C'_A C'_B}, \tag{6.80}$$

$$C_B = \frac{2C'_A C'_B(\omega^2 L' C'_A C'_B - C'_B + C_A)}{(\omega^2 L' C'_A C'_B + C'_B - C'_A)^2 + 4\omega^2 L' C'^2_A C'_B}, \tag{6.81}$$

$$L = \frac{(\omega^2 L' C'_A C'_B + C'_B - C'_A)^2 + 4\omega^2 L' C'^2_A C'_B}{2\omega^2 C'_A C'_B(\omega^2 L' C'_A C'_B + C'_B + C'_A)^2}. \tag{6.82}$$

The value of the parts in the equivalent circuit, C_A, L, and C_B, can be calculated from the measured values, C'_A, L', and C'_B by means of expressions (6.80) to (6.82).

Expressions (6.80) to (6.82) might be applicable to the spiral inductor implemented in the *RFIC*, as long as the spiral inductor can be measured or trimmed just like the printed inductor shown in Figure 6.26.

6.4.4 Simulation

Simulation for the *VCO* is somewhat different from the simulation for other individual blocks. Usually the entire *VCO* block can be simulated by *ADS* or cadence simulation system. In this example, instead of the simulation for the entire block, the *VCO* block is divided into four portions as shown in Figure 6.27, so that more accurate and detailed information can be obtained from the simulation. These four portions are

1) Half of tank circuit,
2) Another half of tank circuit plus amplifier,
3) *Π* type Chebyshev filter,
4) *VCO* buffer.

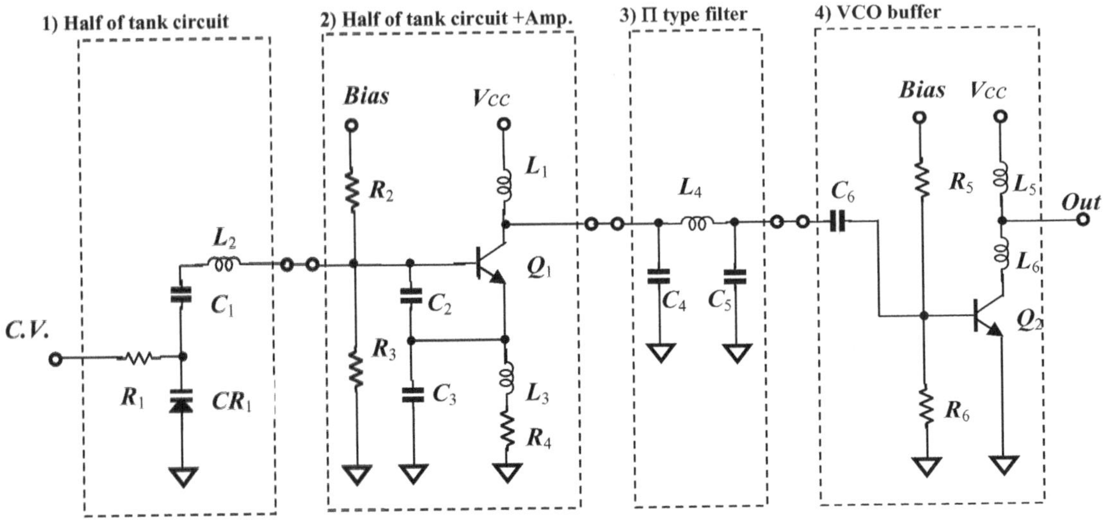

Figure 6.27 Sub-blocks of a single-end device *VCO* with Clapp configuration.

The main purpose of dividing the entire *VCO* block into sub-blocks is to examine the optimized oscillation condition between the tank circuit and the rest of the circuitry. The additional purpose is to optimize the performance of each sub-block more carefully and in detail.

The tank circuit consists mainly of CR_1, C_1, C_2, C_3, and L_2. However, in order to maintain the sub-blocks with only one input port and one output port so as to apply the *S* parameters testing for a two-port block, it was decided to separate the tank circuit into two halves. The first half consists of CR_1, C_1, and L_2. The second half consists of C_2 and C_3 and is combined with the bipolar transistor and other parts, which function as an amplifier element and form the second sub-block.

The first simulation run is to optimize the values of the parts in sub-block (2) as shown in Figure 6.28. The goals are:

1) $S_{11} = S_{11,\max}$,
2) To get (phase of S_{11}) when $S_{11} = S_{11,\max}$.

The second simulation run is to optimize the values of the parts in sub-block 1) as shown in Figure 6.29. The goals are:

1) (Phase of Γ) = −(Phase of $S_{11,\max}$),
2) $|\Gamma| > |1/S_{11,\max}|$.

where

$S_{11,\max}$ is obtained in the first simulation as shown in Figure 6.28, and
Γ is the voltage reflection coefficient as shown in Figure 6.29.

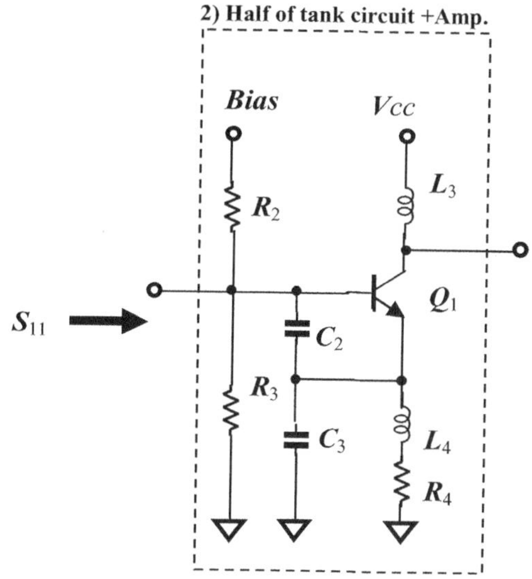

Figure 6.28 Optimization of sub-block 2 for $S_{11} = S_{11,\max}$.

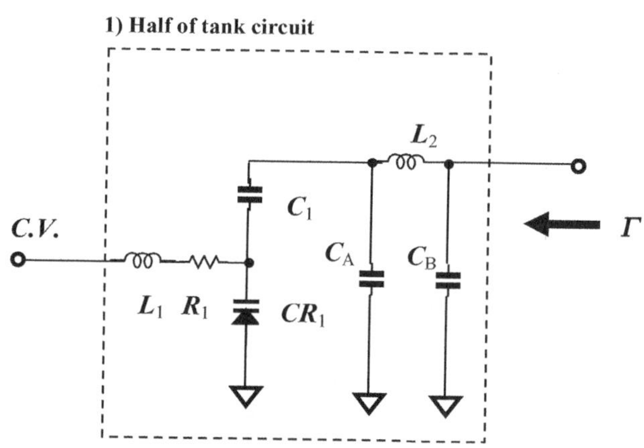

Figure 6.29 Optimization of sub-block 1 for $S_{11} = S_{11,\max}$ (where C_A and C_B are additional capacitors of the printed inductor L_2 as shown in Figure 6.26).

It should be noted that in Figure 6.29 the printed inductor L_2 is replaced by its equivalent circuit as shown in Figure 6.26, where the additional capacitors, C_A and C_B, are added.

Also, it should be noted that the range of the control voltage must be set up so that either the tank circuit or the amplifier stage can be operated over the desired frequency bandwidth. In other words, in the first and second simulation steps for sub-blocks 1) and 2), simulations must be conducted for three control voltages: CV_{max}, CV_o, and CV_{min}. These three control voltages correspond to three frequencies: f_{min}, f_o, and f_{max}, which cover the desired frequency bandwidth.

The third simulation run is to optimize the values of the parts in sub-block (3) as shown in Figure 6.30. The goals are:

1) To match its input to the output of sub-block (2),
2) To match its output to the input of sub-block (4).

The fourth simulation run is to optimize the values of the parts in sub-block (4) as shown in Figure 6.31. The goals are:

1) To match its output of sub-block (4) to 50Ω so that its power output reaches the maximum,
2) To see if its frequency response is flattened over the desired bandwidth.

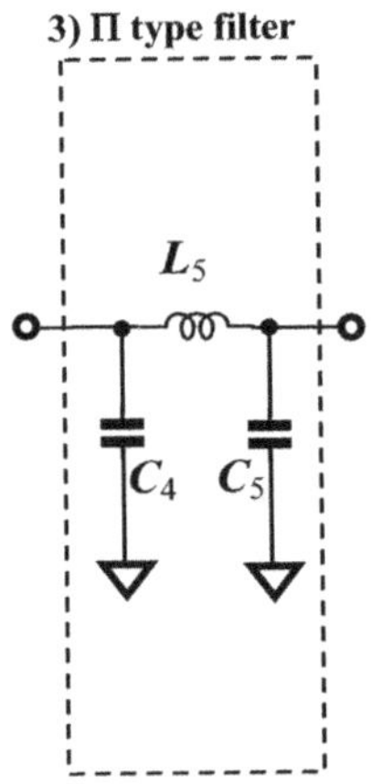

Figure 6.30 Optimization of sub-block 3 for impedance matching.

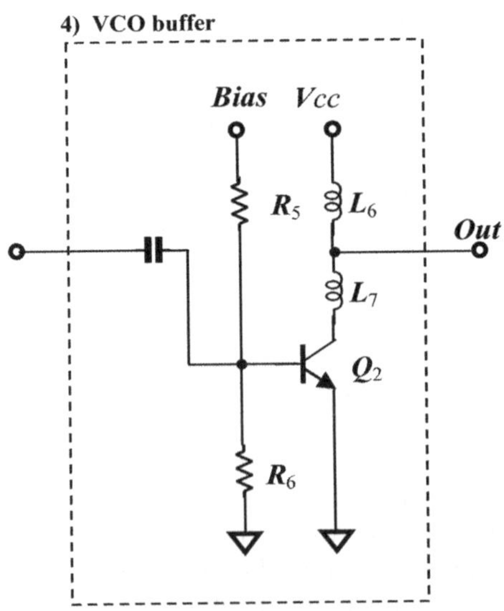

Figure 6.31 Optimization of sub-block 4 for maximum output.

The final simulation run is to check if all the blocks of *VCO* are in their normal operational states so that all the desired goals are reached or more than reached, including

1) Its power output reaches the maximum and its frequency response is flattened over the desired bandwidth,
2) The operating frequencies cover the entire desired frequency bandwidth,
3) Its phase noise is low enough so that the specification is satisfied over the desired frequency bandwidth,
4) The load pulling test is especially important to the stability of an oscillator.

Items 3 and 4 will be introduced and illustrated in the following sub-sections.

After repeating the above optimization simulations, the optimized part list is obtained as shown in Table 6.2.

6.4.5 Load Pulling Test and *VCO* Buffer

Another factor causing the instability of the oscillation frequency is the variation of the load. The load pulling test examines the variation of the oscillation frequency as the load impedance is changed. Figure 6.32 shows the *VCO* load pulling test and Table 6.3 lists an example of the tested results. From Table 6.3 it can be seen that when

$$R_L = R_o = 50\ \Omega, \qquad (6.83)$$

TABLE 6.2 Part list of the single-ended *VCO* with Clapp configuration shown in Figure 6.23

Q_1, Q_2	$2N3948$	$CR1$	$MV2103$	
R_1	$1\,k\Omega$	C_1	$62\,pF$	L_1 $300\,nH$
R_2	$1.5\,k\Omega$	C_2	$12\,pF$	L_2 **(Printed & Trimmed)**
R_3	$820\,\Omega$	C_3	$4.3\,pF$	L_3 $130\,nH$
R_4	$51\,\Omega$	C_4	$22\,pF$	L_4 **(Printed & Trimmed)**
R_5	$2.7\,k\Omega$	C_5	$30\,pF$	L_5 **(Printed & Trimmed**
R_6	$1.5\,k\Omega$	C_6	$0.018\,\mu F$	L_6 **(Printed & Trimmed)**

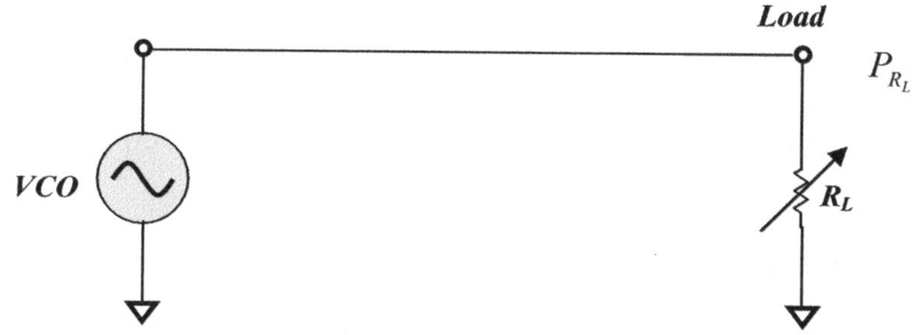

Figure 6.32 *VCO* load pulling test.

TABLE 6.3 An example of *VCO* load pulling test when $P_{RL} = 1$ watt or $30\,dB_m$ (R_o = Characteristic impedance = $50\,\Omega$)

R_o	R_L	Γ_L	VSWR	$VSWR_{dB}$	P_{RL}, dBm	f, MHz
50.00	**2500.00**	0.96	50.00	33.98	**0.3**	**463.846**
50.00	**1000.00**	0.90	20.00	26.02	**0.8**	**464.265**
50.00	**500.00**	0.82	10.00	20.00	**1.7**	**464.673**
50.00	**100.00**	0.33	2.00	6.02	**9.5**	**464.875**
50.00	**50.00**	0.00	1.00	0.00	**(16 mW) 12.0**	**465.000**
50.00	**25.00**	−0.33	2.00	6.02	**9.7**	**465.282**
50.00	**5.00**	−0.82	10.00	20.00	**1.8**	**465.453**

where

R_o = characteristic impedance,

and the output frequency,

$$f = 465\ MHz, \tag{6.84}$$

and when

$$25\,\Omega \le R_L \le 100\,\Omega, \tag{6.85}$$

or

$$VSWR = 2, \tag{6.86}$$

or

$$VSWR_{dB} = 6.02dB, \tag{6.87}$$

the maximum of frequency variation,

$$|\Delta f| \le 0.282\ MHz \tag{6.88}$$

or

$$\frac{|\Delta f|}{f} \le \frac{0.282}{465} \approx 0.061\%. \tag{6.89}$$

This is a good result in the load pulling test.

It must be noted that the impedance matching in most *RF* blocks is important and must be emphasized in *RF* circuit design. However, an exception exists in *VCO* buffer design.

In order to isolate the *VCO*, it is necessary to have a *VCO* buffer inserted between the *VCO* and the load R_L as shown in Figure 6.33. However, the *VCO* input should not be required to impedance-match with the *VCO* output. The reasons are:

- The output impedance of the *VCO* is negative because in an oscillator, the impedance at any node is negative except at the nodes for the *DC* power supply or grounded points. The oscillation is maintained by the negative resistance. Impedance matching to a node or port with negative impedance is meaningless.

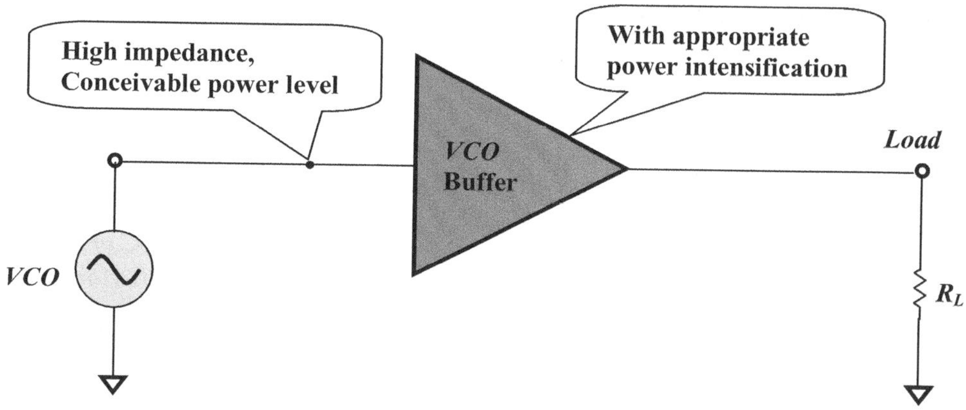

Figure 6.33 Unmatched *RF* block *VCO* coupled with a buffer.

- Rather than talking about impedance matching, let's say that the input of the *VCO* buffer is trying to couple some of the oscillating power from the *VCO*. There is expected to be as little coupling as possible because a strong coupling brings about disturbance to the *VCO* due to the variation of the load impedance and consequently, the degradation of the stability of the oscillation frequency. It is well-known that the stability of the oscillation frequency in a *VCO* or oscillator design is the most important goal.

Consequently, the input impedance of the *VCO* buffer will be kept at a high impedance so that the *VCO* output will not be disturbed. The *RF* power coupled from the *VCO* is kept at a minimum or a "conceivable" level. On the other hand, the sensed *RF* power must be intensified in the *VCO* buffer appropriately so as to provide enough *RF* power of the oscillating signal to other blocks. Figure 6.33 illustrates these basic ideas.

Now we can also explain why the simulation in this design example is unconventional. Instead of conducting a simulation of the entire block, the *VCO* block is divided into four portions as shown in Figure 6.27. By dividing it in this way, the function of each portion can be optimized with respect to its special target. If simulation is conducted for an entire block, the special targets, including the special load pulling concern for the *VCO* buffer, would be more or less neglected or ignored.

In addition to the load pulling test, other tests have been conducted for this *VCO* design example as shown in Figure 6.23. The performance is summarized in Table 6.4.

6.5 DIFFERENTIAL *VCO* AND QUAD PHASES *VCO*

Figure 6.34 shows the evolution of the differential *VCO*. Figure 6.34(a) shows two stages of the amplifier with positive feedback. In Figure 6.34(b), the two capacitors C_1 and C_2 in Figure 6.34(a) are combined into one capacitor C in series.

Figure 6.34(c) shows two stages of amplifier with positive feedback modified from Figure 6.34(b) by combining the two inductors L_1 and L_2 in Figure 6.34(b) into one

TABLE 6.4 Primary goals of single-ended *VCO* with Clapp configuration

Frequency range:	440 to 470 MHz
DC power supply:	5 V
Current drain:	13 mA
Output power:	12 dB_m (16 mW)
Frequency controlled range:	440 to 470 MHz
Corresponding control voltage:	1.0 V to 3.0 V
Control voltage sensitivity:	13 –15–17 MHz/V
Output power:	12 dB_m (16 mW)
Output impedance:	50 Ω
Pulling figure when ratio of standard wave:	0.282 MHz $VSWR = 2.0$
Phase noise when offset frequency:	$-95\,dB_c/Hz$ $f_{\text{offset}} = 10\,kHz$

inductor L in series in Figure 6.34(c), while the two resistors represent the loss of parts.

Figure 6.34(d) is a *VCO* with a pair of two stages of amplifiers stacked together. Instead of only one pair of n-channel *MOSFET* transistors as in Figures 6.34(a), (b), and (c), there is a pair of p-channel *MOSFET* transistors in the *VCO* block. Comparing the *VCO*s shown in Figures 6.34(c) and (d), it can be seen that the pair of p-channel *MOSFET* transistors look like the two resistors shown in Figure 6.34(c). On the other hand, the pair of n-channel *MOSFET* transistors looks like two resistors if the *VCO* is mainly built by the upper pair of p-channel *MOSFET* transistor. In other words, the *VCO* shown in Figure 6.34 consists of two sub-*VCO*s. One is mainly built by the bottom pair of n-channel *MOSFET* transistors while the upper pair of p-channel *MOSFET* transistors functions as its load. Another is mainly built by the upper pair of p-channel *MOSFET* transistors while the bottom pair of n-channel *MOSFET* transistors functions as its load. The remarkable advantage in such a stacked configuration is that the current drain can be re-used, so that the output voltage swing or output power is almost doubled from that of the *VCO*s without stacked configurations shown in Figure 6.34(a), (b), and (c).

A quad-phase *VCO* can be built by two differential *VCO*s. Figure 6.35 shows its block diagram. If the phases of input in the first *VCO* are 0° and 90°, its output phases are shifted 90° and become 90° and 180°, respectively. Then, if the outputs of the first *VCO* are connected to the inputs of second *VCO* inputs, its output phases are 180° and 0°, which are exactly the opposite phases of the inputs of the first *VCO*. To positively feedback to the first *VCO*, the outputs of the second *VCO* are exchanged and then fed to the inputs of the first *VCO*.

Figure 6.36 shows the schematic of the quadrature phase *VCO* built by two differential *VCO*s. The individual *VCO*s in Figure 6.36 are basically the same as the *VCO* shown in 6.34(b) except for two modifications:

- The two inductors L_1 and L_2 in Figure 6.34(b) are replaced by one inductor L with a central tap in Figure 6.36.
- There are four coupling *MOSFET* transistors added to the two *VCO*s for interconnection between I_p, I_n, Q_p, and Q_n.

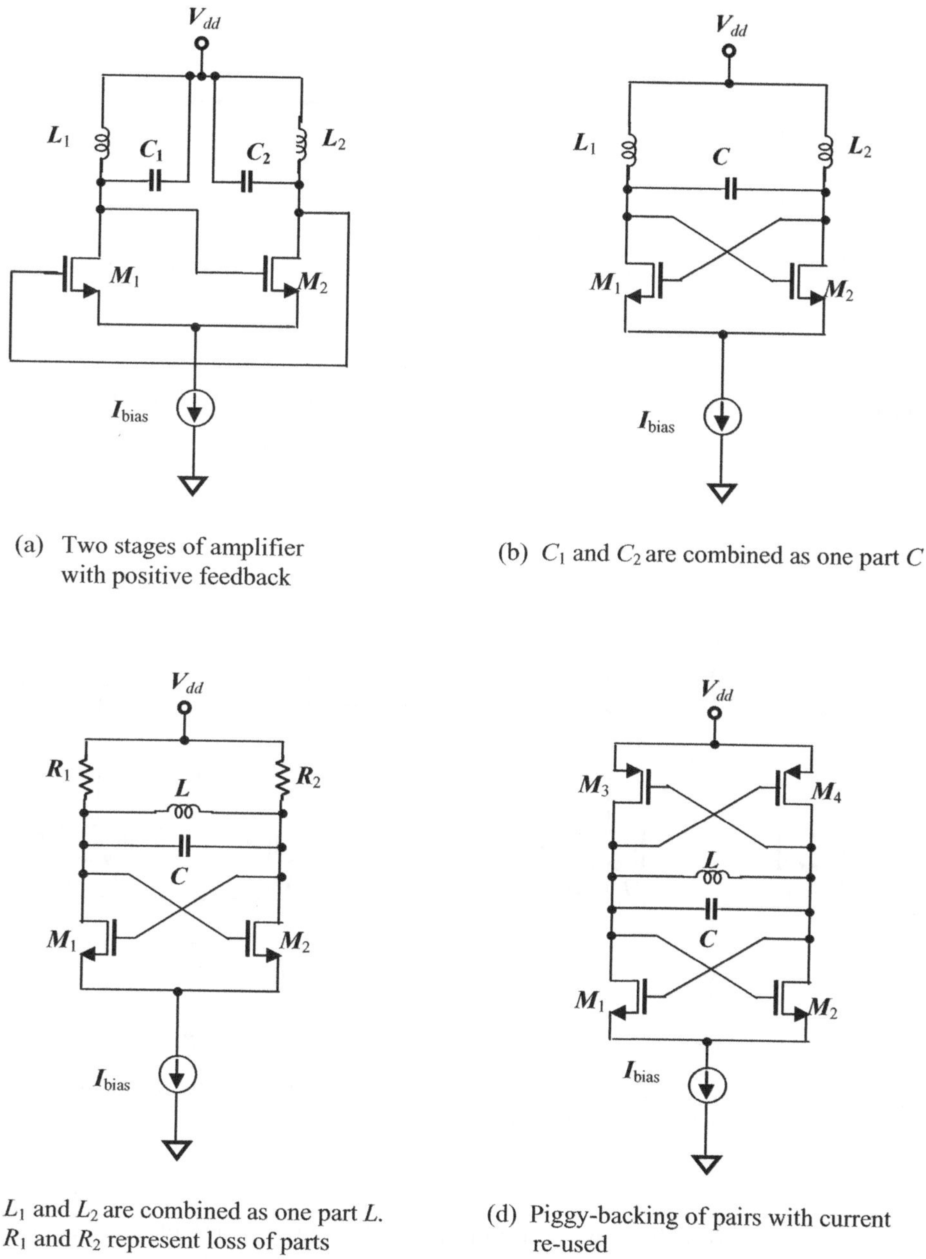

Figure 6.34 Evolution of differential *VCO*.

The quad phase *VCO* shown in Figure 6.37 is basically the same as the one shown in Figure 6.36. The difference is that the connection mode of the coupling transistors, M_3, M_4, M_7, and M_8, with M_1, M_2, M_5, and M_6 respectively in Figure 6.36 is in parallel, while the connection mode of the coupling transistors, M_3, M_4, M_7, and M_8 with M_1, M_2, M_5, and M_6, respectively, in Figure 6.37 is in series. The phase noise in the quad phase *VCO* shown in Figure 6.37 is much lower than that in Figure 6.36.

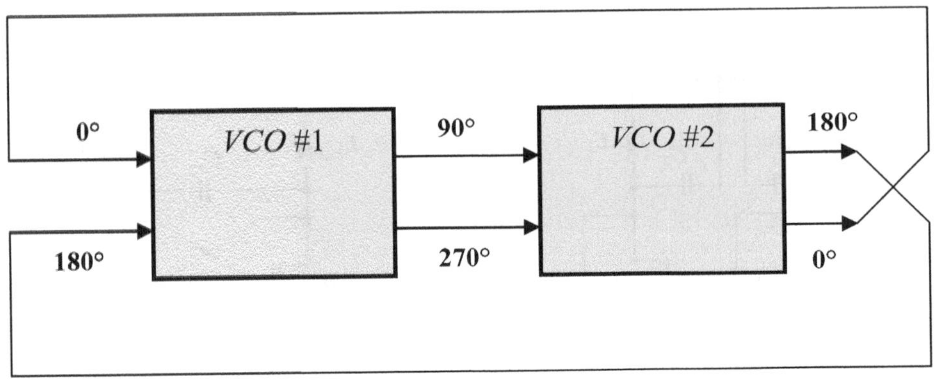

Figure 6.35 Block diagram of a quadrature phase *VCO* built by two differential *VCO*s.

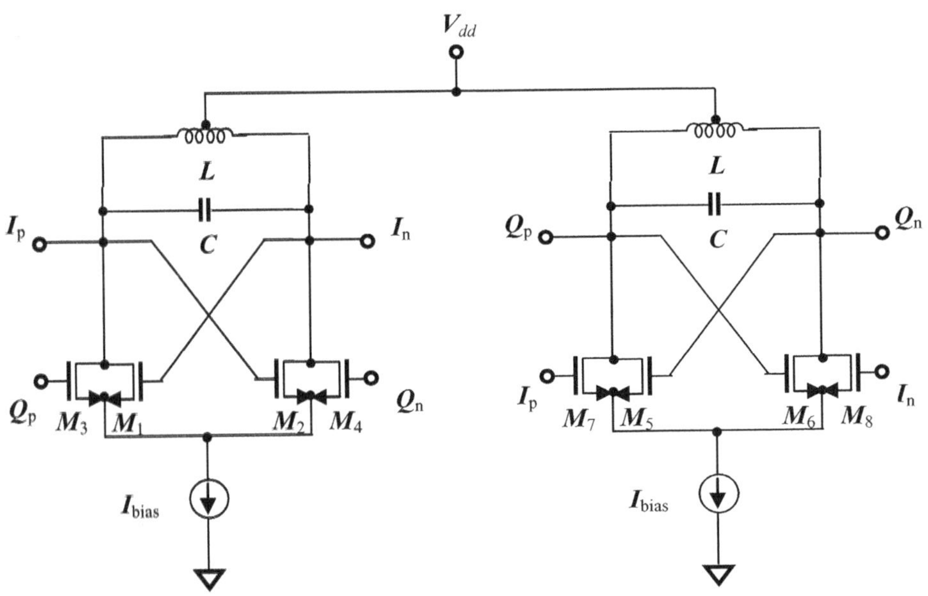

Figure 6.36 Schematic of quadrature phase *VCO* built by two differential *VCO*s. M_1 and M_3, M_2 and M_4, M_5 and M_7, M_6 and M_8 are connected in parallel.

Figures 6.38 and 6.39 show the quad phases *VCO* with current re-use. As mentioned in Section 6.4.5, a *VCO* buffer must be designed for the four terminals, I_p, I_n, Q_p, and Q_n, of the quad-phase *VCO* as shown in Figure 6.38. The coupling transistors in Figure 6.38 function as coupling parts only, and nothing else. The coupling transistors in Figure 6.39 function not only as coupling parts but also as buffer stages. Consequently, it may not be necessary to design a buffer for the quad phase *VCO*. It is therefore concluded that the quad-phase *VCO* as shown in Figure 6.39 is more economic than the quad phase *VCO* shown in Figure 6.38.

It should be noted that in both Figures 6.38 and 6.39, the capacitors are varactors with control voltages, V_{bias} and V_{tune}. The *VCO* operating frequency can be adjusted

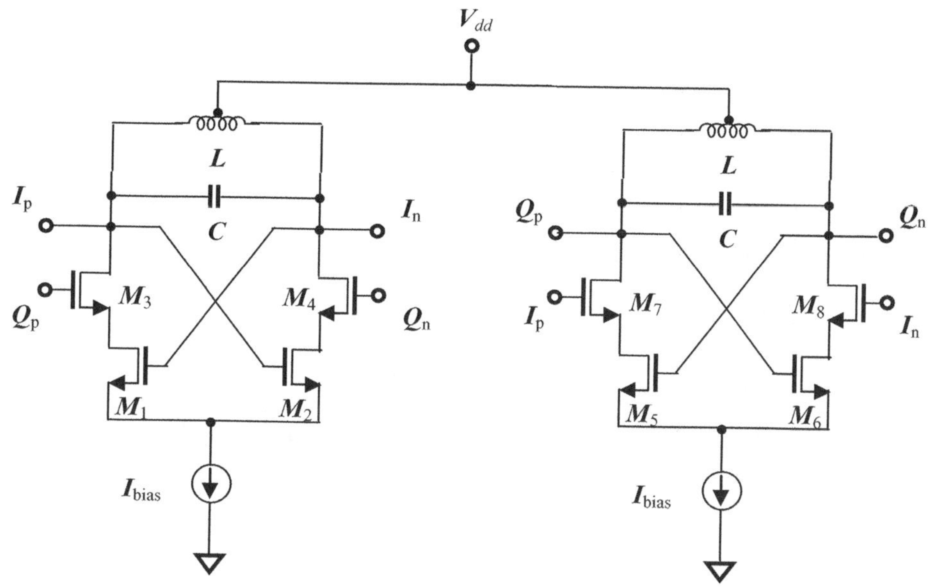

Figure 6.37 Schematic of quadrature phase *VCO* built by two differential *VCO*s. M_1 and M_3, M_2 and M_4, M_5 and M_7, M_6 and M_8 are connected in series.

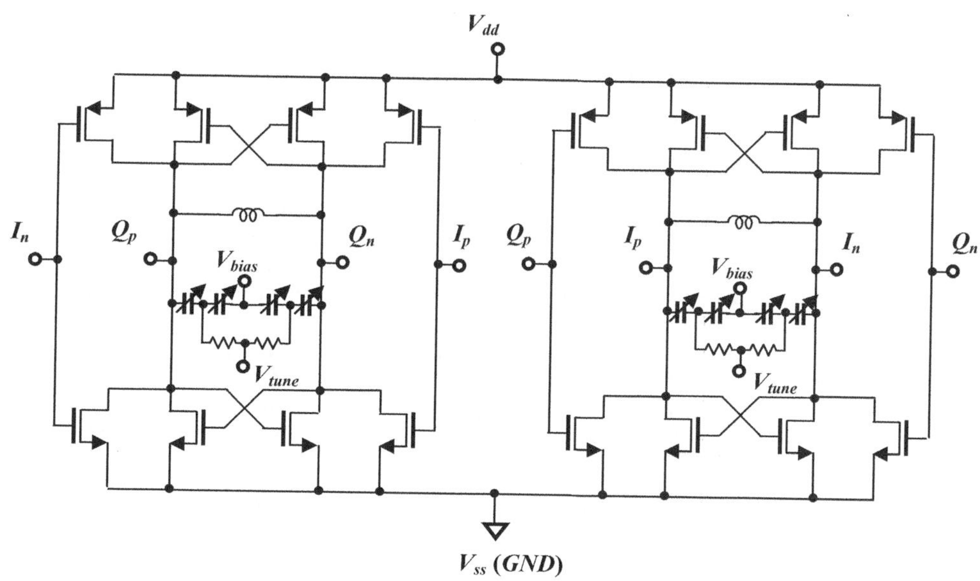

Figure 6.38 A quadrature phase *VCO* with coupling transistors.

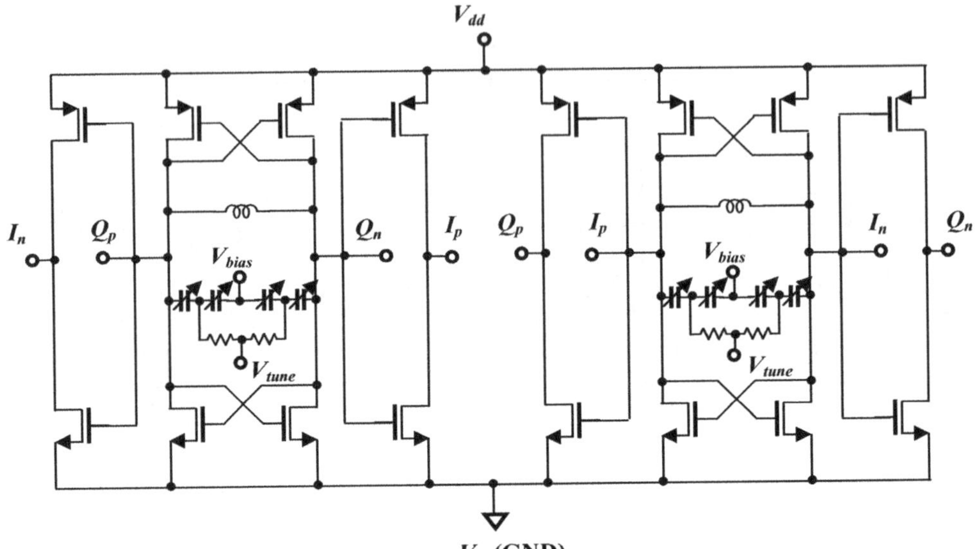

Figure 6.39 A quadrature phase *VCO* with coupling-buffer transistors.

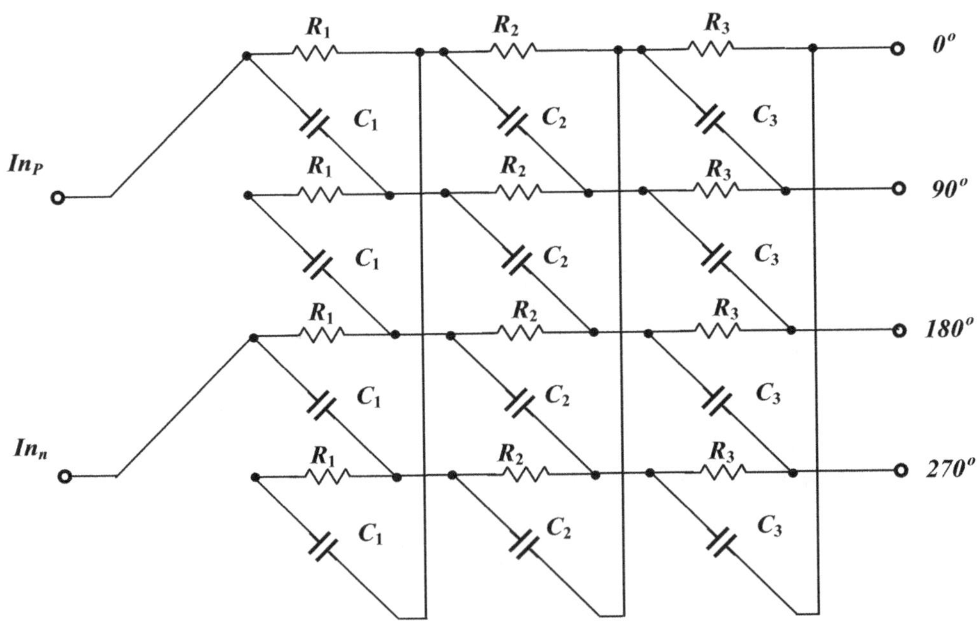

Figure 6.40 A quadrature *RC* phase split network.

by the control voltage through the variation of the varactors' capacitance. The varactors may be divided into two branches in parallel, one for coarse adjustment and one for fine adjustment. The varactors must be carefully selected so that the variation of *VCO* frequency can cover all the expected range.

From Figures 6.38 and 6.39 it can be seen that many transistors must be applied. If the performance of a quad-phase *VCO* is not expected to be paramount, it can be built by any differential *VCO* as shown in Figure 6.34 and then cooperated with a quadrature *RC* phase shift network such as the one shown in Figure 6.40.

The quadrature *RC* phase shift network shown in Figure 6.40 is a network with two differential inputs and four phase outputs. In each row there are three *RC* circuit sectors connected in series. The number of sectors can be chosen from one to any number by the designer, depending on the bandwidth required. The larger the number of sectors, the wider the bandwidth.

In general, the phase noise of the *VCO* with a quadrature *RC* phase shift network is much higher that the quad-phase *VCO* with *LC* tanks shown in Figures 6.38 and 6.39.

REFERENCES

[1] Sang-Woong Yoon, Stéphan Pinel, and Joy Laskar, "High-Performance 2-GHz CMOS LC VCO with High-Q Embedded Inductors Using Wiring Metal Layer in a Package," *IEEE Transactions on Advanced Packaging*, Vol. 29, No. 3, August 2006, pp. 639–646.

[2] Behzad Razavi, "A Study of Phase Noise in CMOS Oscillators," *IEEE Journal of Solid-State Circuits*, Vol. 31, No. 3, March 1996, pp. 331–343.

[3] Ali Hajimiri and Thomas H. Lee, "A General Theory of Phase Noise in Electrical Oscillators," *IEEE Journal of Solid-state Circuits*, Vol. 33, No. 2, February 1998, pp. 179–194.

[4] Ali Hajimiri and Thomas H. Lee, "Design Issues in CMOS Differential *LC* Oscillators," *IEEE Journal of Solid State Circuits*, Vol. 34, No. 5, May 1999, pp. 717–724.

[5] Thomas H. Lee and Ali Hajimiri, "Oscillator Phase Noise: A Tutorial," *IEEE Journal of Solid State Circuits*, Vol. 35, No. 3, March 2000, pp. 326–335.

[6] Bram De Muer, M. Borremans, M. Steyaert, and G. Li Puma, "A 2-GHz Low-Phase-Noise Integrated *LC*-VCO Set with Flicker-Noise Up-conversion Minimization," *IEEE Journal of Solid State Circuits*, Vol. 35, No. 7, July 2000, pp. 1034–1038.

[7] Donhee Ham and Ali Hajimiri, "Concepts and Methods in Optimization of Integrated *LC* VCOs," *IEEE Journal of Solid State Circuits*, Vol. 36, No. 6, June 2001, pp. 896–909.

[8] Pietro Andreani, Andrea Bonfanti, Luca Romanò, and Carlo Samori, "Analysis and Design of a 1.8-GHz CMOS *LC* Quadrature VCO," *IEEE Journal of Solid State Circuits*, Vol. 37, No. 12, December 2002, pp.1737–1747.

[9] Axel D. Berny, Ali M. Niknejad, and Robert G. Meyer, "A Wideband Low-Phase-Noise CMOS VCO," *IEEE 2003 Custom Integrated Circuits Conference*, pp. 555–558.

[10] KaChun Kwok and Howard C. Luong, "A 0.3-V 1.46-mW Low Phase Noise Oscillator with Transformer Feedback in Standard 0.18-μm CMOS Process,", *IEEE 2003 Custom Integrated Circuits Conference*, pp. 551–554.

[11] Reza Navid, Thomas H. Lee, and Robert W. Dutton, "Lumped Inductorless Oscillators: How Far Can They Go?" *IEEE 2003 Custom Integrated Circuits Conference*, pp. 543–546.

[12] Sander L. J. Gierkink, Salvatore Levantino, Robert C. Frye, Carlo Samori, and Vito Boccuzzi, "A Low-Phase-Noise 5-GHz CMOS Quadrature Using Superharmonic Coupling," *IEEE Journal of Solid State Circuits*, Vol. 38, No. 7, July 2003, pp. 1148–1154.

[13] Shenggao Li, Issy Kipnis, and Mohammed Ismail, "A 10-GHz CMOS Quadrature LC-VCO for Multirate Optical Applications," *IEEE Journal of Solid State Circuits*, Vol. 38, No. 10, October 2003, pp. 1626–1634.

[14] Pietro Andreani and Xiaoyan Wang, "On the Phase-Noise and Phase-Error Performances of Multiphase *LC* CMOS VCOs," *IEEE Journal of Solid State Circuits*, Vol. 39, No. 11, November 2004, pp. 1883–1893.

[15] Hyunwon Moon, Sungweon Kang, Youn Tae Kim, and Kwyro Lee, "A Fully Differential LC-VCO Using a New Varactor Control Structure," *IEEE Microwave and Wireless Components Letters*, Vol. 14, No. 9, September 2004, pp. 410–412.

[16] Sangsoo Ko and Songcheol Hong, "Noise Property of a Quadrature Balanced VCO," *IEEE Microwave and Wireless Components Letters*, Vol. 15, No. 10, October 2005, pp. 673–675.

[17] Seonghan Ryu, Yujin Chung, Huijung Kim, Jinsung Choi, and Bumman Kim, "Phase Noise Optimization of CMOS VCO through Harmonic Tuning," *IEEE Radio Frequency Integrated Circuits Symposium*, 2005, pp. 403–406.

[18] A. Koukab, Y. Lei, and M. Declercq, "Design and Optimization of a Linear Wide-band VCO for Multimode Applications," *IEEE Radio Frequency Integrated Circuits Symposium*, 2005, pp. 527–530.

[19] Babak Soltanian and Peter Kinget, "A Tail Current-Shaping Technique to Reduce Phase Noise in LC VCOs," *IEEE Custom Integrated Circuits Conference*, 2005, pp. 579–582.

[20] D. B. Leeson, "A Simple Model of Feedback Oscillator Noise Spectrum," *Proc. IEEE*, Vol. 54, February 1966, pp. 329–330.

CHAPTER 7

POWER AMPLIFIERS (*PA*)

7.1 CLASSIFICATIONS OF POWER AMPLIFIERS

The *PA* is an amplifier, shown in Figure 7.1, which intensifies the power of the input signal. Its output power is *G* times higher than its input power. The added portion of output power is *AC* power converted from the *DC* power supply. The ratio of the added portion of power to the input power is called its *PAE* (Power Added Efficiency), that is,

$$G_{watt} = \frac{P_{out,watt}}{P_{in,watt}}, \tag{7.1}$$

$$G_{dB} = P_{out,dB} - P_{in,dB}, \tag{7.2}$$

$$\Delta P = P_{out,watt} - P_{in,watt}, \tag{7.3}$$

$$PAE = \frac{\Delta P}{P_{DC}} = \frac{P_{out,watt} - P_{in,watt}}{P_{DC}}, \tag{7.4}$$

where

$P_{out,watt}$ = output power by watts,
$P_{in,watt}$ = input power by watts,
$P_{out,dB}$ = output power by dB,
$P_{in,dB}$ = input power by dB,
G_{watt} = power Gain by watts,
G_{dB} = power gain by dB,

RF Circuit Design, by Richard Chi-Hsi Li

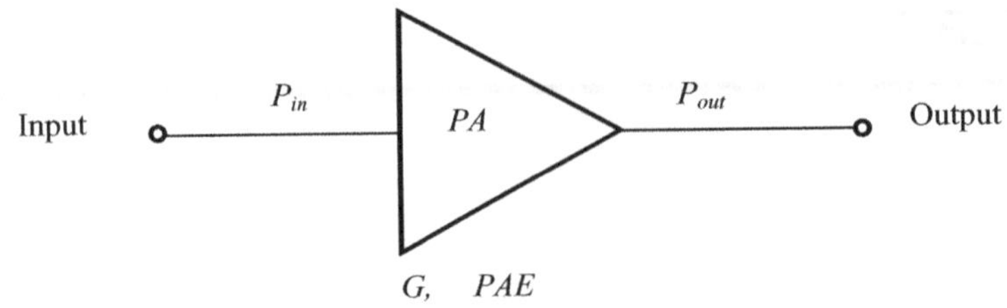

Figure 7.1 Representation of a power amplifier.

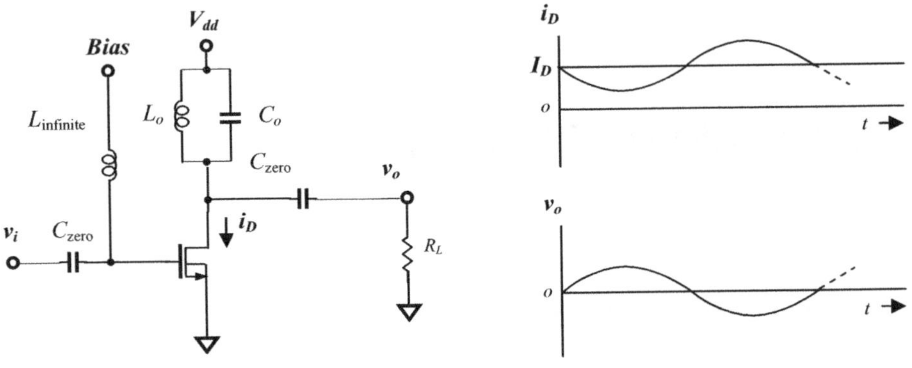

(a) Simple circuit of class *A* power amplifier
$L_{infinite}$ = *RF* choke,
C_{zero} = *DC* blocking capacitor.

(b) Waveforms of i_D and v_o

Figure 7.2 Class *A* power amplifier and its waveforms.

ΔP = difference between output and input power by watts,
P_{DC} = *DC* power from *DC* power supply,
PAE = power added efficiency by %.

In a *PA* design, the most important goal is the power converted efficiency from the *DC* power to the *AC* power, which combined with the input power forms the output power.

The classification of power amplifier is closely related to *PAE*.

7.1.1 Class A Power Amplifier

Figure 7.2 shows the operating principle of a class A power amplifier. The inductor L_o and capacitor C_o are resonant at the operating frequency. When the device is working as a class A power amplifier, the device is conducted over the entire cycle. The waveform of the current drain i_D is a normal sinusoidal wave and the waveform of the output voltage v_o is a normal sinusoidal wave as well. The *PAE* for a class A power amplifier is about 30% although theoretically it could approach a higher percentage.

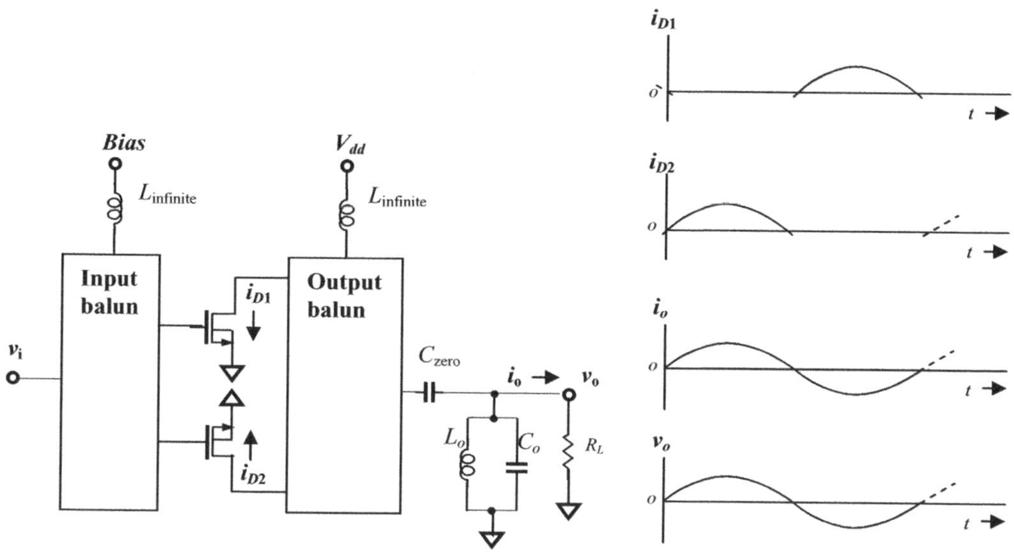

(a) Simple circuit of class *B* power amplifier
$L_{infinite}$ = *RF* choke
C_{zero}= *DC* blocking capacitor

(b) Waveforms of i_{D1}, and i_{D2}, i_o and v_o

Figure 7.3 Class *B* power amplifier and its waveforms.

7.1.2 Class B Power Amplifier

Figure 7.3 shows the operating principle of a class B power amplifier. The inductor L_o and capacitor C_o are resonant at the operating frequency. When the device is working as a class B power amplifier, the device is conducted for more than one half of but less than a whole cycle. The waveforms of the current drains i_{D1} and i_{D2} have a normal but "broken" sinusoidal wave lasting a little bit more than one half of a cycle. However, the waveform of the output current i_o and voltage v_o is a normal sinusoidal wave. The *PAE* for a class B power amplifier is about 60% although theoretically it could approach a higher percentage.

7.1.3 Class C Power Amplifier

Figure 7.4 shows the operating principle of a class C power amplifier. The inductor L_o and capacitor C_o are resonant at the operating frequency. When the device is working as a class C power amplifier, the device is conducted for less than half a cycle. The waveform of the current drain i_D is a piece of a sinusoidal wave, but the waveform of output voltage v_o is a normal sinusoidal wave. The *PAE* for a class C power amplifier is about 80% although theoretically it could approach a higher percentage.

7.1.4 Class D Power Amplifier

Figure 7.5 shows the operating principle of a class D power amplifier. The inductor L_o and capacitor C_o are resonant at the operating frequency. A class D power

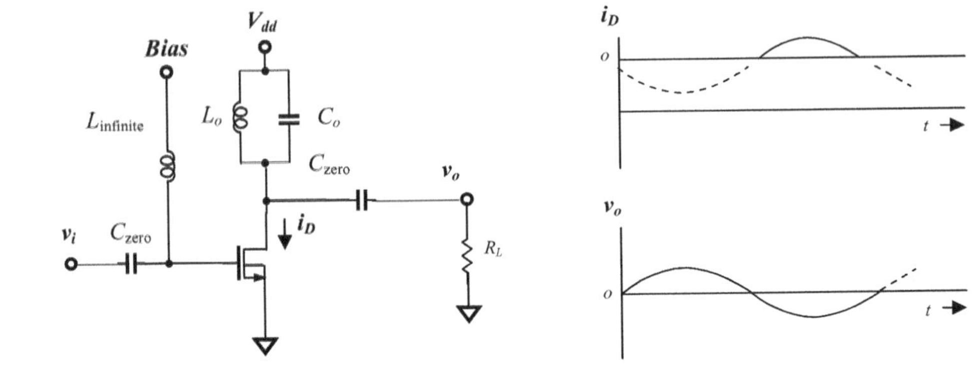

(a) Simple circuit of class C power amplifier
$L_{infinite}$ = RF choke
C_{zero}= DC blocking capacitor

(b) Waveforms of i_D and v_o

Figure 7.4 Class C power amplifier and its waveforms.

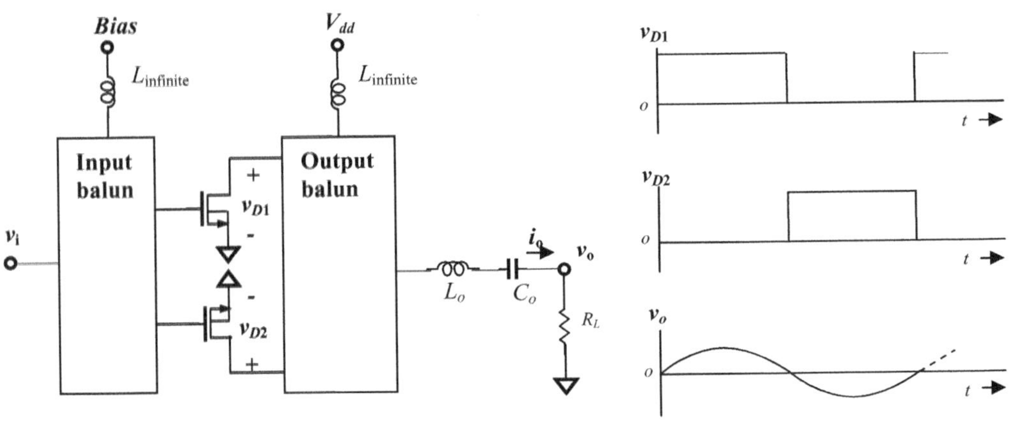

(a) Simple circuit of class D power amplifier
$L_{infinite}$ = RF choke
C_{zero}= DC blocking capacitor

(b) Waveforms of v_{D1}, v_{D2}, and v_o

Figure 7.5 Class D power amplifier and its waveforms.

amplifier is a pair of devices acting as a two pole switch, which defines a rectangular voltage or current.

The drain voltages, v_{D1} and v_{D2}, are rectangular waveforms. However, the waveform of the output voltage v_o is a normal sinusoidal wave. The PAE for a class D power amplifier is more than 80% although theoretically it could approach a higher percentage.

7.1.5 Class E Power Amplifier

Figure 7.6 shows the operating principle of a class E power amplifier. The inductor L_o and capacitor C_o are resonant at the operating frequency. When the device is

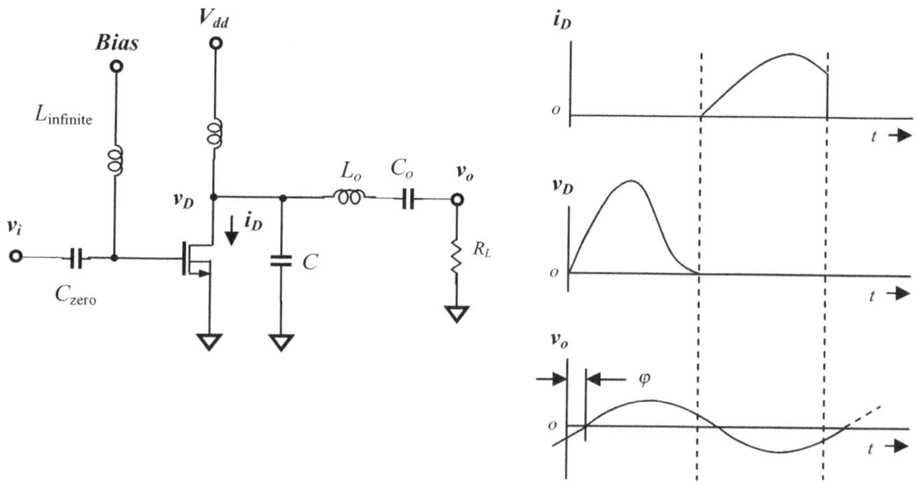

(a) Simple circuit of class E power amplifier
$L_{infinite} = RF$ choke
$C_{zero} = DC$ blocking capacitor

(b) Waveforms of i_D, v_D, and v_o

Figure 7.6 Class E power amplifier and its waveforms.

working as a class E power amplifier, the device is driven to act as a switch. The waveforms of the current i_D and voltage v_D are conjugated with each other but the waveform of the output voltage, v_o, is a normal sinusoidal wave with a phase delay φ. The PAE for a class E power amplifier is about 90% although theoretically it could approach a higher percentage.

7.1.6 Third Harmonic-peaking Class F Power Amplifier

Figure 7.7 shows the operating principle of a class F power amplifier. The inductor L_o and capacitor C_o are resonant at the operating frequency and the inductor L_3 and capacitor C_3 are resonant at the third harmonic frequency. When the device is working as a class F power amplifier, it is driven to act approximately as a switch. The waveform of the current $\boldsymbol{i_D}$ is about half that of the sinusoidal wave but the waveform of the output voltage $\boldsymbol{v_o}$ is a normal sinusoidal wave. The PAE for a class F power amplifier is about 80% although theoretically it could approach a higher percentage.

7.1.7 Class S Power Amplifier

Figure 7.8 shows the operating principle of a class S power amplifier. A LPF is built by the inductor L_o and the capacitor C_o. A class S power amplifier is a two position switch with a rectangular waveform applied to a low-pass filter, which allows the slowly varying DC or average component to appear on the load. The drain voltage $\boldsymbol{v_{D2}}$ is a rectangular waveform. However, the waveform of the output voltage v_o is a normal sinusoidal wave. The PAE for a class S power amplifier is more than 90% although theoretically it could approach a higher percentage.

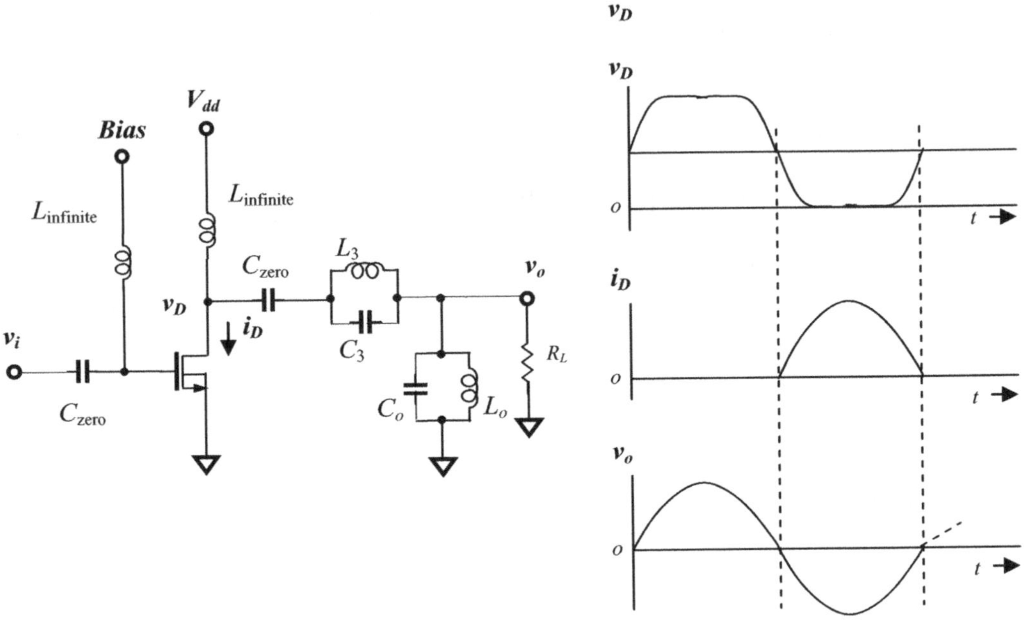

(a) Simple circuit of class F power amplifier
$L_{infinite} = RF$ choke
$C_{zero} = DC$ blocking capacitor

(b) Waveforms of i_D, v_D, and v_o

Figure 7.7 Third harmonic-peaking class F power amplifier and its waveforms.

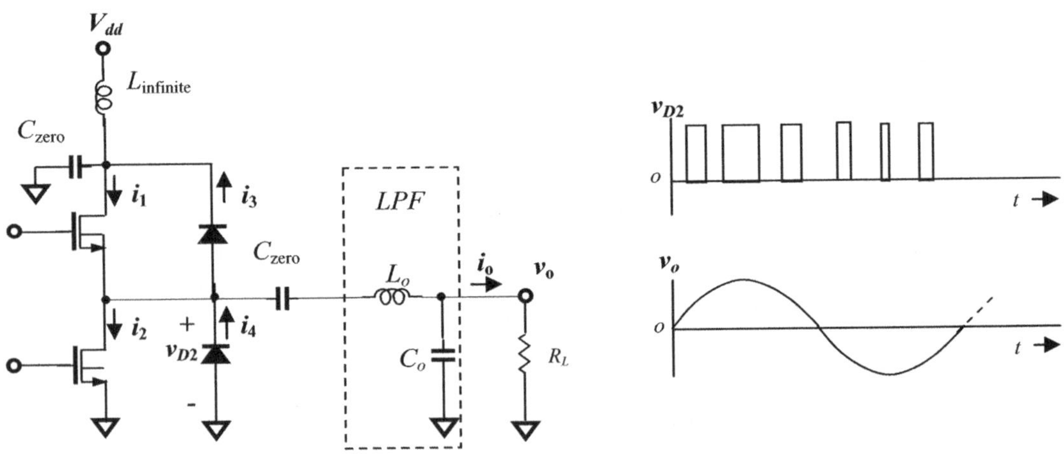

(a) Simple circuit of class S power amplifier
$L_{infinite} = RF$ choke
$C_{zero} = DC$ blocking capacitor

(b) Waveforms of v_{D1}, v_{D2}, and v_o

Figure 7.8 Class S power amplifier and its waveforms.

7.2 SINGLE-ENDED *PA* DESIGN

Usually the main design procedures for an *RF* block are

- Selection of device and obtaining the model of device from vendor or foundry,
- Simulation,
- Layout,
- Testing.

Very often, the first two procedures in a *PA* design are different. The power of a *PA* is much higher than that of a small signal. There is little chance that the model obtained from the vendor or foundry is accurate enough to get a reasonable simulation result. It therefore becomes necessary for the designer to test the device empirically. Even if the designer does provide a model of the device with an accuracy statement, testing it for verification is still necessary.

In the simulation stage, the main task is to work with the implementation of input and output impedance matching networks. The other procedures are basically the same as in the design for other *RF* circuit blocks. Therefore, in Section 7.2.1 we are going to focus on the practical procedures.

7.2.1 "Tuning on Bench"

The purpose of this test is to find out the input and output impedances of the device so that input and output impedance matching networks can be implemented in the next design step.

Figure 7.9 is a setup for the testing of the device by two *RF* tuners. The test fixture contains the tested device *M*, two "zero" capacitors, and two "infinite" inductors.

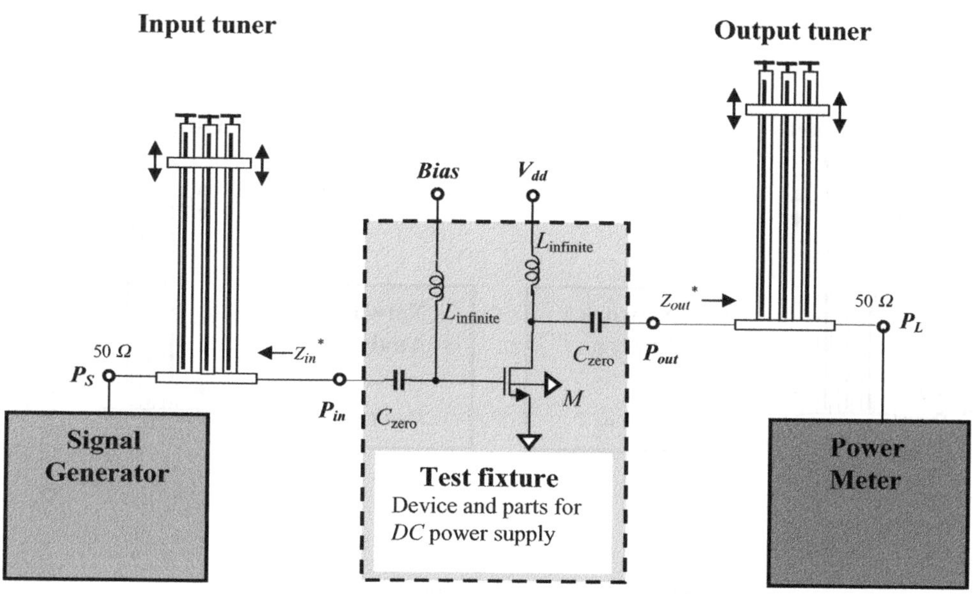

Figure 7.9 Device tuned by two tuners.

The device shown in Figure 7.9 is assumed to be a *MOSFET* transistor. All the parts are soldered on a small *PCB*. Two zero capacitors and two infinite inductors serve as the *DC* power supply to the device. Owing to the fact that their impedances are respectively zero and infinity, their existence does not disturb the *AC* or *RF* characteristics of the tested device. There are two *RF* tuners. The input tuner is inserted between the signal generator and the input of the test fixture. The output tuner is inserted between the output of the test fixture and the power meter.

For a specified *DC* power supply voltage, the *DC* current flowing through the device must be adjusted to the value specified by the *DC* bias. Then the work of "tuning on bench" can be started.

The input and output tuners can be slid up and down to adjust their own impedances. Therefore, the input and output impedances of the device can be regulated by the adjustment of the input and output tuners so that the output power displayed in the power meter approaches a maximum, upon which it is implied that either the input or output impedance of the device is matched well. The maximum must be verified carefully and patiently by several repeated adjustments of the tuners' positions.

When the power reading in the power meter reaches a maximum, the sliding positions of the input and output tuners must be fixed and then detached from the test setup. After the detachment is done, the two impedances, Z_{in}^* and Z_{out}^*, can be tested and obtained by means of the new test setup shown in Figure 7.10. The network analyzer can read the values of impedances Z_{in}^* and Z_{out}^*.

7.2.2 Simulation

Having the values of Z_{in}^* and Z_{out}^*, we can now design and implement the input and output matching networks. In the *ADS* (Advanced Design Systen) simulation system there is a "1 Port Opt" box, by which the corresponding impedance matching network can be optimized if the original impedance is known. For instance, if

$$z_{in}^* = r_{in}^* + jx_{in}^*, \tag{7.5}$$

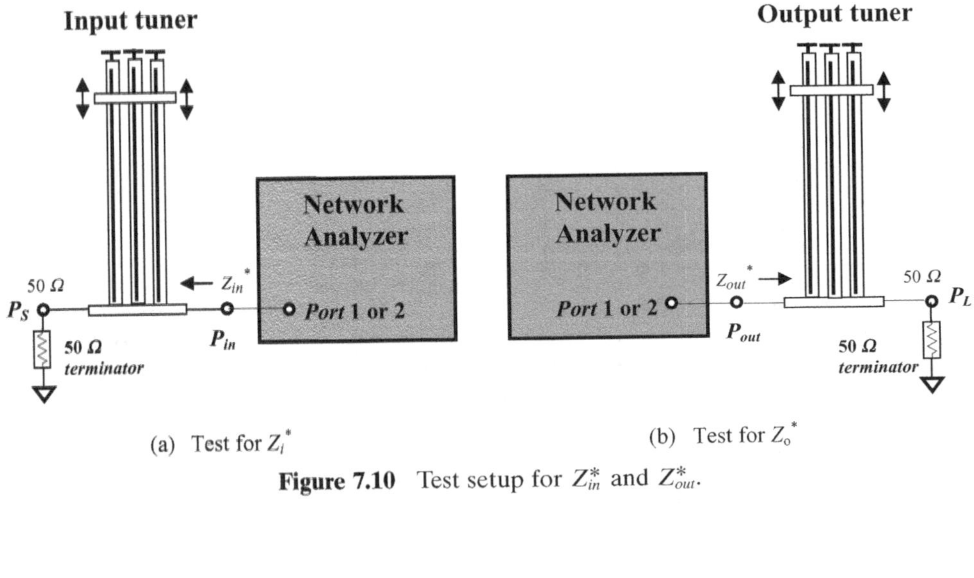

Figure 7.10 Test setup for Z_{in}^* and Z_{out}^*.

$$z_{out}^* = r_{out}^* + jx_{out}^*, \tag{7.6}$$

Where x_{in}^* is capacitive and x_{out}^* is inductive, so that

$$x_{in}^* < 0, \tag{7.7}$$

$$x_{out}^* > 0. \tag{7.8}$$

And then

x_{in} is inductive and x_{out} is capacitive.
x_{out} is capacitive and x_{out} is inductive.

Figures 7.11 and 7.12 show the simulation setup for optimization of the input and output impedance matching networks, respectively, by means of the 1 Port Opt box. As an example, the operating frequency is assumed to be

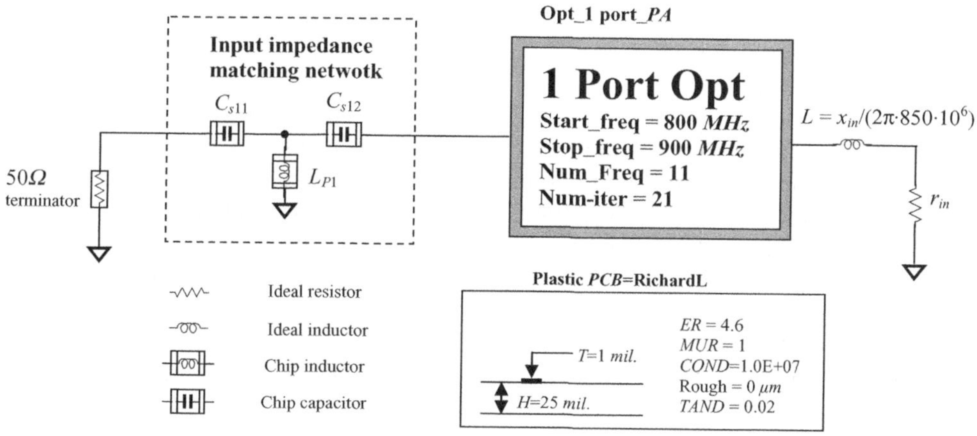

Figure 7.11 Simulation setup for optimization of input impedance matching network.

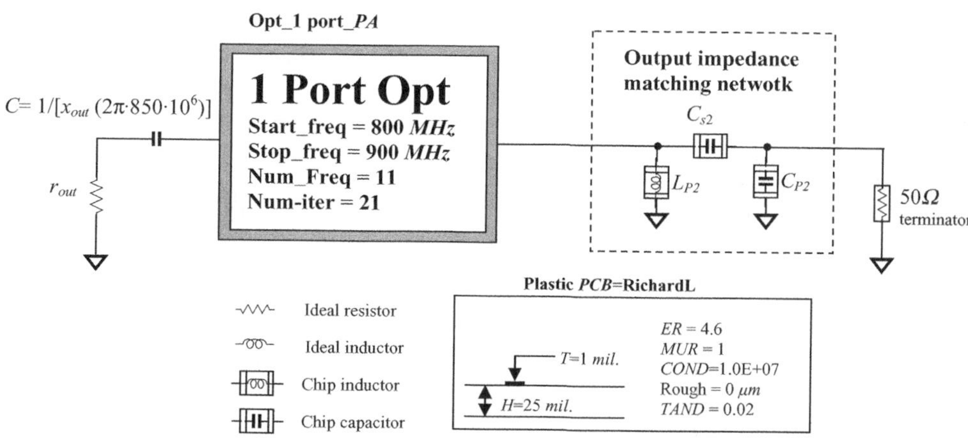

Figure 7.12 Simulation setup for optimization of output impedance matching network.

$$f = 800\text{–}900\,MHz \tag{7.9}$$

$$f_c = 850\,MHz. \tag{7.10}$$

All the parts of the input impedance matching network are assumed to be soldered on a plastic *PCB*. The main parameters of the plastic *PCB* are listed in both Figures 7.11 and 7.12. The actual models for the chip inductor and chip capacitor are applied in the simulation.

In order to reduce part count and cost, it is desirable that the zero capacitor at the input of the device be a part of the input impedance matching network. The capacitor C_{s12} in Figure 7.11 is intentionally arranged so that it can function both as an impedance matching part and as a *DC* blocking capacitor. On the other hand, it is desirable that the infinite inductor and zero capacitor at the output of the device become parts of the output impedance matching network. The inductor L_{P2} and capacitor C_{s2} in Figure 7.12 are intentionally arranged so that the inductor L_{P2} can function as an impedance matching part and as a *DC* voltage provider, and so that the capacitor C_{s2} can function as an impedance matching part and as a *DC* blocking capacitor as well.

Figure 7.13 shows the schematic after the impedance matching is done.

The remaining design procedures are layout and testing, which are basically the same as those in other *RF* circuit block designs.

Finally there is an important observation to be made about the "tuning on bench": special care must be paid to the "security". The test setup as shown in Figure 7.9, where the input and output impedances of the device are un-matched, is dangerous. The returned power from the load, the power meter in Figure 7.9, could burn and destroy the device. The solution is to insert a circulator between the output tuner and the power meter to to prevent the returned power from the power meter entering the device. An alternative solution is to turn on the signal from the signal generator carefully. It is best to turn it on starting from zero and then gradually increasing the signal step by step to the desired level. The increase of power in each step is kept to a minimum.

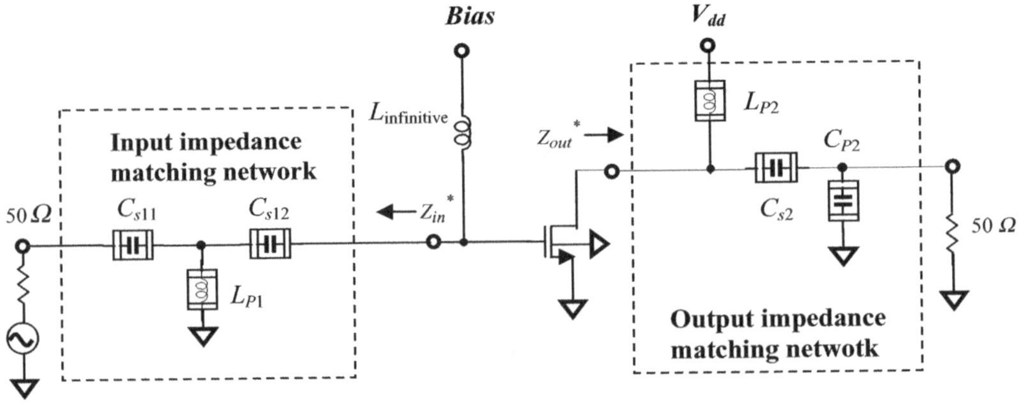

Figure 7.13 Two tuners replaced by input and output impedance matching networks.

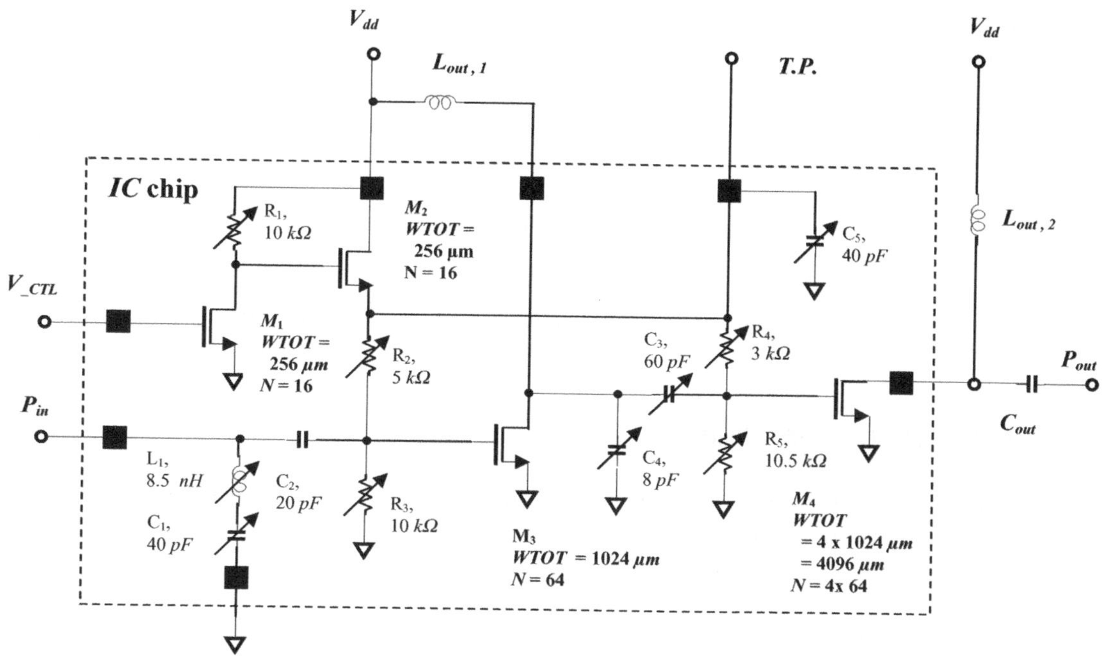

Figure 7.14 Schematic of 3 *V*, 1 *W*, *PA IC* (integrated circuit) with *CMOS* processing.

7.3 SINGLE-ENDED *PA-IC* DESIGN

The methodology of "tuning on bench" applied in *PA* design seems all right for a *PA* module or a *PA* built by discrete parts. However, it becomes very unreliable in *PA-IC* design. The bonding pads, bonding wires, and test fixture cause many non-uniformities and uncertainties.

How about forgetting the circuit simulation and directly jumping to the layout procedure? This is a simple but an "unorthodox" method. Surprisingly, it works! (The content of this section is abstracted from author's 1996 design work.)

Figure 7.14 shows the schematic of a 1–2 watt *PA* built on a *CMOS IC* chip. M_1 and M_2 is an electronic switch controlled by the control voltage $V_{\_CTL}$. When $V_{\_CTL}$ = "0", the *PA* is "*ON*" and when $V_{\_CTL}$ = "1", the *PA* is "*OFF*". M_3 is in the pre-amplifier stage and M_4 is in the final stage. M_4 is a huge transistor with 1024 fingers which is four times larger than M_3.

The main specifications and performance are:

- *DC* power supply = 3 *V*,
- frequency range = *UHF* with central frequency $f_c = 450\,MHz$,
- input power = $150\,mW$,
- output power = 1–2 *Watt*,
- current drain = 690 mA when output power = 1 *Watt*.
- $PAE > 50\%$

The key issues for success are:

- Selecting an expected topology,
- Hand-calculating the variable range for every part's value, including device size and the value of capacitors, inductors, and resistors,
- Designing all the parts in the circuit as variable parts. This requires a special layout scheme.
- Trimming every part to an expected value with a laser trimmer (also called a laser cutting system) on the *IC* die or on the wafer.

The outstanding achievement is that this *IC* chip has only one tape-out. It saves a lot of design time, at least a couple of months, and greatly reduces the design cost.

7.4 PUSH-PULL *PA* DESIGN

This is a push-pull *PA* design sample. (The content of this section is abstracted from author's 1992 design work.)

7.4.1 Main Specifications

- frequency range: 800–900 MHz, ($f_c = 850\,MHz$).
- output power: 2 watt and 5 watt.
- $PAE > 50\%$.
- *DC* power supply: 3 V.
- selected device: *GaAs FET*, *CLY*10, manufactured by Siemens.

7.4.2 Block Diagram

It is well known that a class A power amplifier has the best linearity but the lowest efficiency. Considering a trade-off between efficiency and linearity, I decided to design this *PA* with a push-pull style and with class B operation. Input and output baluns are needed for the push-pull configuration. Figure 7.15 shows the block diagram, which consists of five portions, including input and output impedance matching networks.

This is a push-pull *PA* but not a differential *PA*. The difference between the push-pull and the differential *PA*s is their *CMRR* capability. A *PA* with push-pull configuration does not have this capability, while a *PA* with differential configuration is able to reject the common mode input signal. The difference of topology is that a *PA* with a differential configuration has a common tail resistor between the two devices while a *PA* with push-pull configuration does not.

A question might be raised: Why is a push-pull *PA* preferred to a differential *PA*?

Usually, the power efficiency of a *PA* with a push-pull configuration is higher than a *PA* with a differential configuration. In a differential *PA*, the tail resistor is a negative feedback part. It lowers the output power for better performance in non-

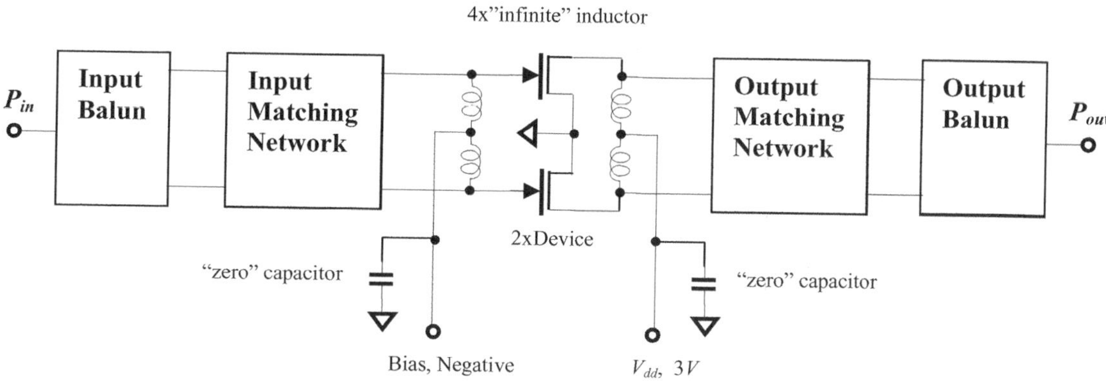

Figure 7.15 Block diagram of 3 V push-pull PA. Note: "zero" capacitor = bypass capacitor, "infinite" inductor = RF choke.

Definition:

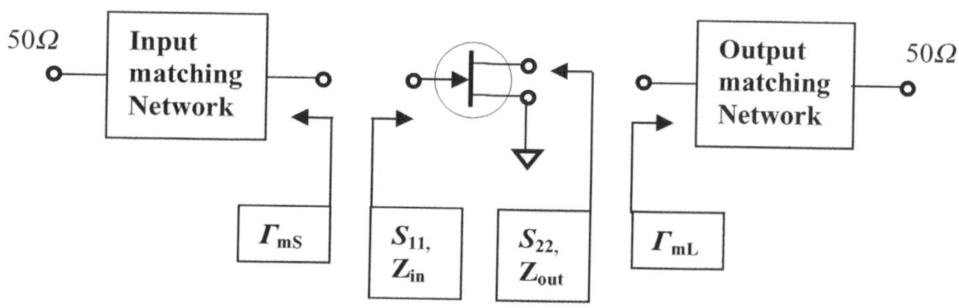

Measured data:

f, GHz	V_{dd}, V	I_d, mA	P_{1dB}, dBm	Gain, dB	$\Gamma_{mS,mag}$	$\Gamma_{mS,ang}$	$\Gamma_{mL,mag}$	$\Gamma_{mL,ang}$
0.9	3	700	26.7	15.3	0.58	169	0.68	-156
1.8	3	700	28.5	9.0	0.79	-110	0.75	-87
2.4	3	700	27.9	7.2	0.77	-86	0.73	-107

Figure 7.16 Siemens *CLY*10 power *GaAs-FET* matching conditions.

linearity and DC offset. In a push-pull PA, the tail resistor no longer exists so that its output power is higher than the differential one. Except in a design for a linear PA, efficiency is the most important factor for a PA design.

Of course, in the design of a linear PA, the priorities are reversed. The PA with a differential but not a push-pull configuration is preferred, because the first priority is now to have good performance in non-linearity and DC offset.

7.4.3 Impedance Matching

Siemens provided the primary data for a single device as shown in Figure 7.16.

The first row of measured data in Figure 7.16 is close to the expected frequency range, 800 to 900 MHz. From the values of $\Gamma_{mS,mag}$, $\Gamma_{mS,ang}$, $\Gamma_{mL,mag}$, and $\Gamma_{mL,ang}$, in this

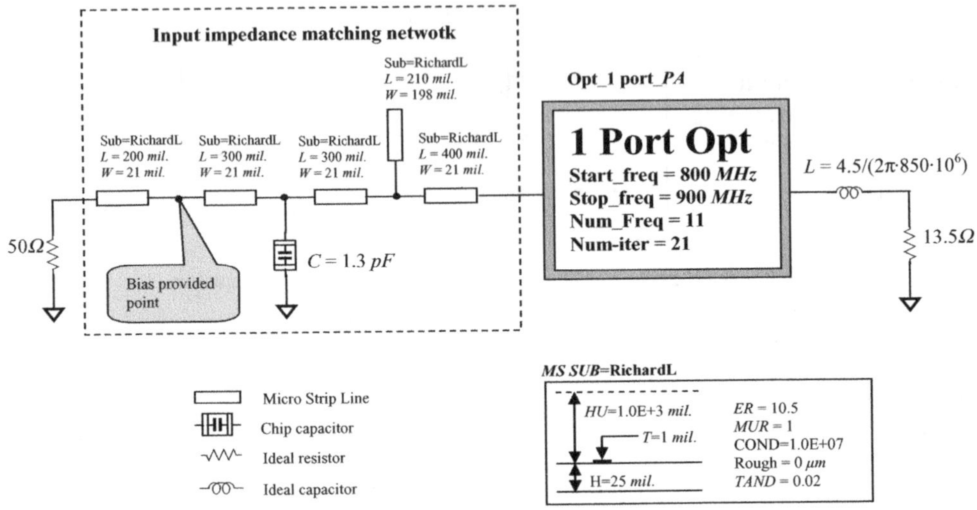

Figure 7.17 Optimization of *CLY* 10 input impedance matching.

row, the input and output impedances of the device *CLY*10 at the frequency $f = 850\,MHz$ can be approximately evaluated by Smith chart:

$$Z_{in} = 13.5 + j4.5\,\Omega, \tag{7.11}$$

$$Z_{out} = 10.0 - j10.0\,\Omega, \tag{7.12}$$

In order to reduce size and cost, the input and output impedance matching networks will be implemented only by capacitors and micro strip lines. The size of an inductor is much larger than that of a capacitor, and the cost of a chip inductor is about 10–20 times higher than that of a chip capacitor.

The input and output impedance networks can be optimized in terms of the "1 Port Opt" simulation. In this simulation we assume that all the parts, including the capacitor and micro strip lines, are put on a ceramic aluminum substrate. As shown in Figures 7.16 and 7.17, the relative permeability of the substrate is $\varepsilon_r = 10.5$ and its thickness is $25\,mils$.

Figure 7.17 shows the setup for optimizing the input impedance matching network. At the right hand side of the 1 Port Opt is $Z_{in} = 13.5 + j4.5\,\Omega$ as shown in expression (7.11). At the left hand side of the 1 Port Opt is the input impedance matching network.

Figure 7.18 shows the setup for optimizing the output impedance matching network. At the left hand side of the 1 Port Opt is $Z_{out} = 10 - j10.0\,\Omega$ as shown in expression (7.12). At the right hand side of the 1 Port Opt is the output impedance matching network.

The primary consideration for the topology of input and output impedance matching network is that they are simple *T* type matching networks. That is, one micro strip line segment is in series, one capacitor is in parallel, and one micro strip line segment is in series again. In the process of optimization, it is found that a micro strip opened-stub can improve the impedance matching and widen the bandwidth.

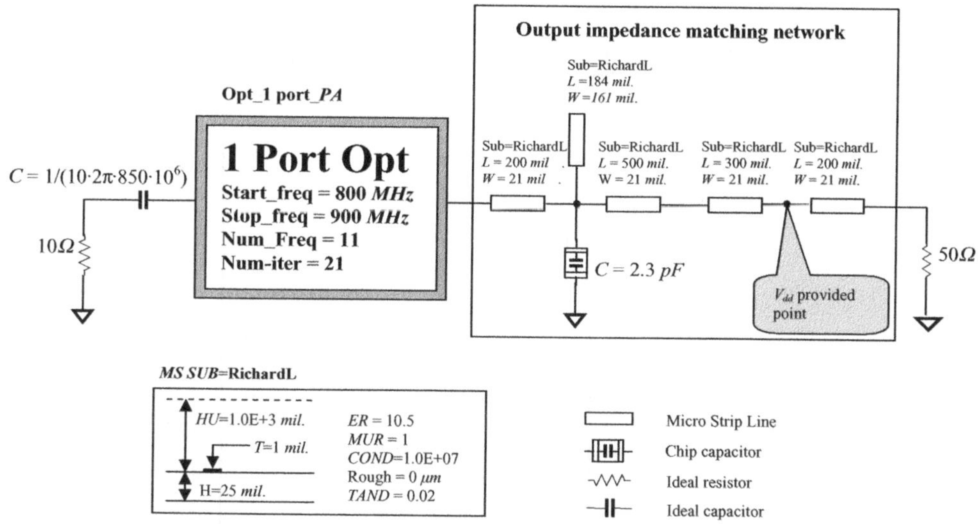

Figure 7.18 Optimization of *CLY* 10 output impedance matching.

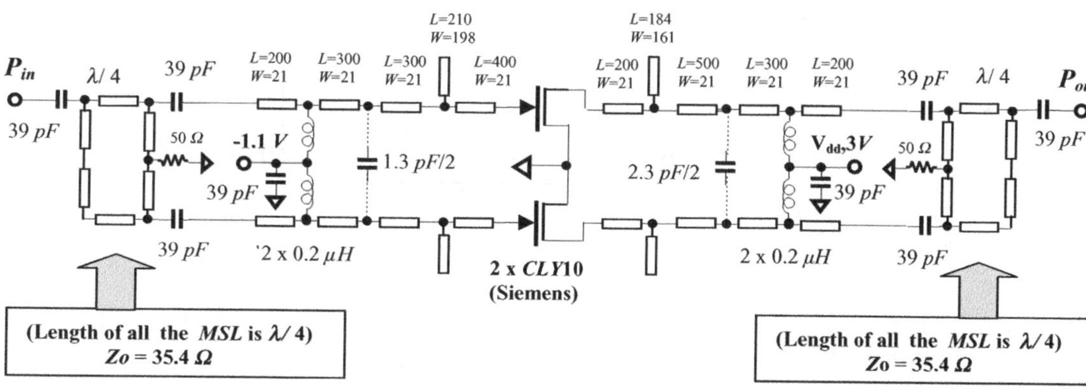

Figure 7.19 Schematic of 3 *V*, 2 *W*, push-pull *PA* (unit of length: *mil*). $f_o = 850\,MHz$, substrate: $H = 25\,mils$, $T = 1\,mil$, $\varepsilon_r = 10.5$.

In addition, a bias and V_{dd} providing point must be set up in the input and output impedance matching networks respectively. This is why there are five pieces of micro strip line segments total in both the input and output impedance matching networks.

The final optimized values of the capacitor and micro strip lines are shown in Figures 7.17 and 7.18. The impedance matching networks shown in Figures 7.17 and 7.18 are only half of the impedance matching network of the entire push-pull *PA*. Figure 7.19 shows the schematic of the entire push-pull 2 *watt PA*. The four inductors with values of $L = 0.2\,\mu H$ function as *RF* chokes.

The input and output matching networks are connected to the input and output baluns, respectively.

The impedance of all the terminals in both the input and output baluns is $50\,\Omega$. Both look like branch type baluns; however, they are micro strip baluns. The length of each micro strip line segment is a quarter wavelength corresponding to the central

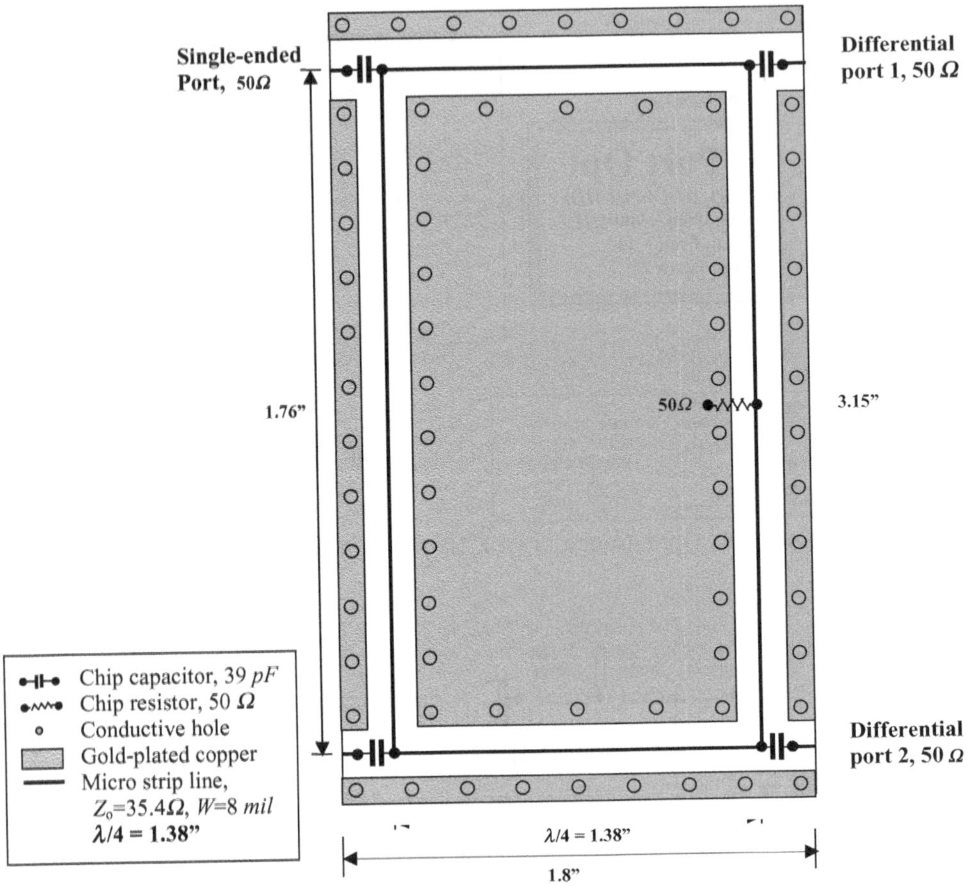

Figure 7.20 Layout of balun on a ceramic aluminate board. Thickness = 0.25 *mil*, $\varepsilon_r = 10.5$, $\mu_r = 1$, $f = 800$–$900\,MHz$.

frequency, $f = 850\,MHz$. The characteristic of the micro strip line segment is $35.4\,\Omega$. Figure 7.20 depicts the layout of the input and output baluns on the ceramic aluminate board. Figure 7.21 depicts the entire layout of the *PA* on the ceramic aluminate board.

The performance of this *PA* is outlined in Table 7.1. Figure 7.22 shows the curves of the output power and *PAE* vs. input power. It can be seen that when the input power is less than $20\,dB_m$, the output power and *PAE* are approximately proportional to the input power. The output power approach is saturated when the input power is about equal to or greater than $20\,dB_m$.

Figure 7.23 plots the curve the of output power vs. the *DC* power supply voltage V_{dd}. When V_{dd} is lower than $3\,V$, the output power is poor, but is proportional to the value of V_{dd}. When V_{dd} is higher than $3\,V$, the output power does not increase much as V_{dd} increases.

In parallel with the design for 2 *watt PA*, the design for a 5 *Watt PA* is conducted in a similar way. The devices *CS*-1 manufactured by Motorola are selected. In order to get more output power, the bias is changed from $-1.1\,V$ to $-1.58\,V$. Figure 7.24

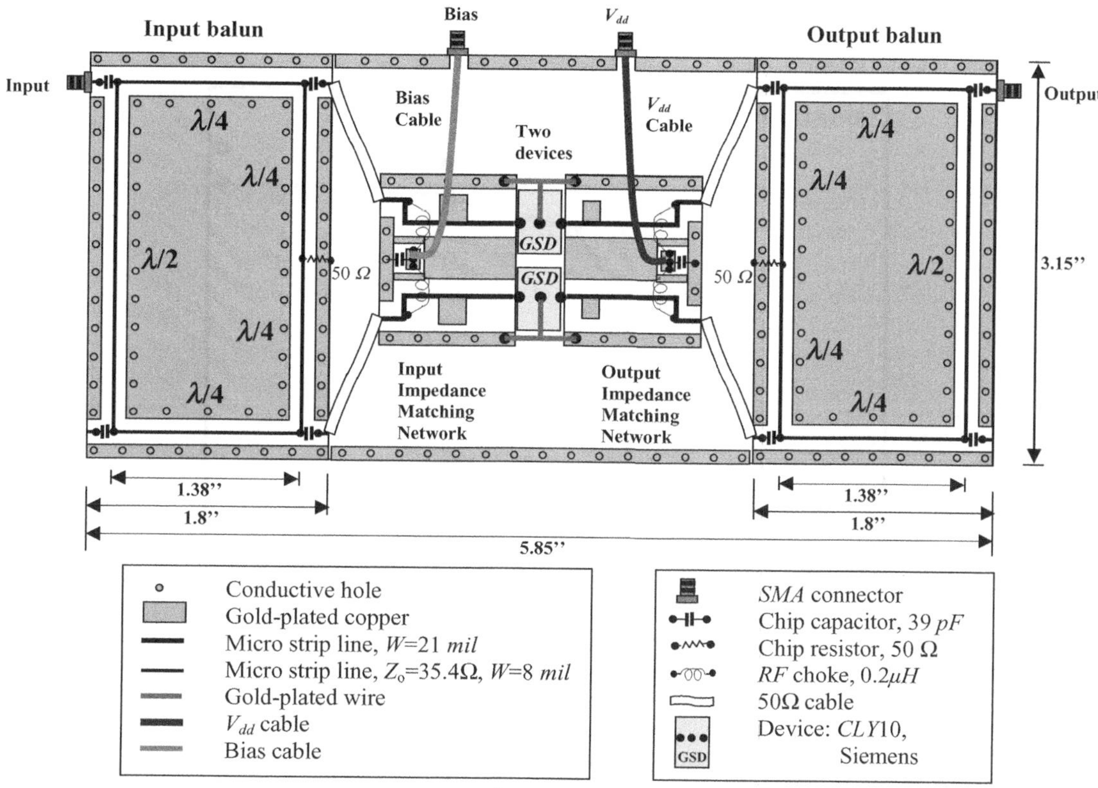

Figure 7.21 Layout of push-pull *PA* module on a ceramic aluminate board. Thickness = $0.25\,mil$, $\varepsilon_r = 10.5$, $\mu_r = 1$, $f = 800\text{–}900\,MHz$.

TABLE 7.1 Main performance of 3 *V*, 2 *W* push-pull *PA*

f_o	$= 850\,MHz$
P_{in}	$= 150\,mW$,
P_{out}	$= 2.2\,W$,
G	$= 12\,dB$
PAE	$= 56.8\%$,
I	$= 1682\,mA$

shows the schematic of the entire push-pull 5 watt *PA* after the input and output impedances of the devices are matched.

The performance of this *PA* is outlined in Table 7.2. Figure 7.25 shows the curves of the output power and *PAE* vs. input power. It can be seen that when the input power is less than $25\,dB_m$, the output power and *PAE* are approximately proportional to the input power. The output power approaches saturation when the input power is about equal to or greater than $25\,dB_m$.

Figure 7.26 plots the curve of the output power vs. the *DC* power supply voltage V_{dd}. When V_{dd} is lower than $3\,V$, the output power is poor, but is proportional to the value of V_{dd}. When V_{dd} is higher than $3\,V$, the output power does not increase much as V_{dd} increases.

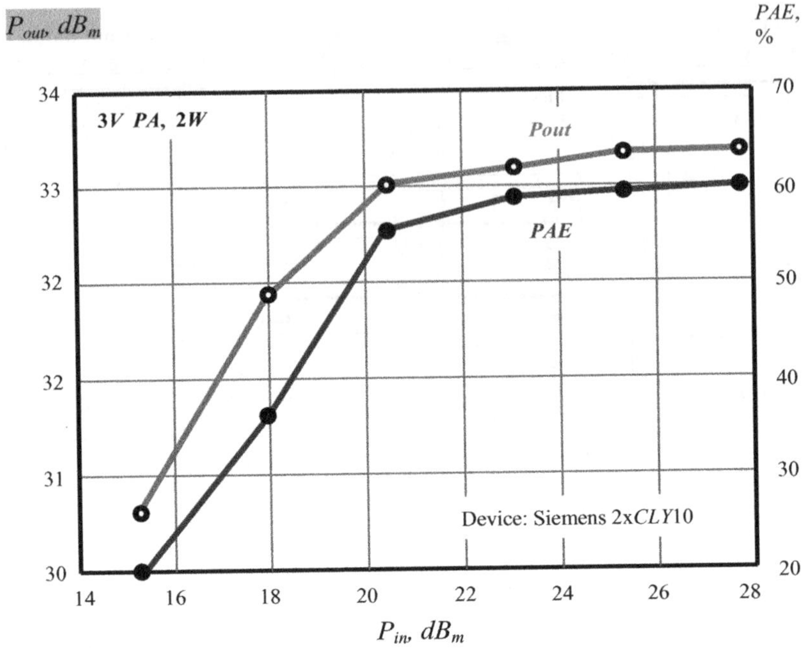

Figure 7.22 Plot of P_{out}, PAE vs. P_{in}, $f_o = 850\,MHz$, $2\,W$ push-pull PA.

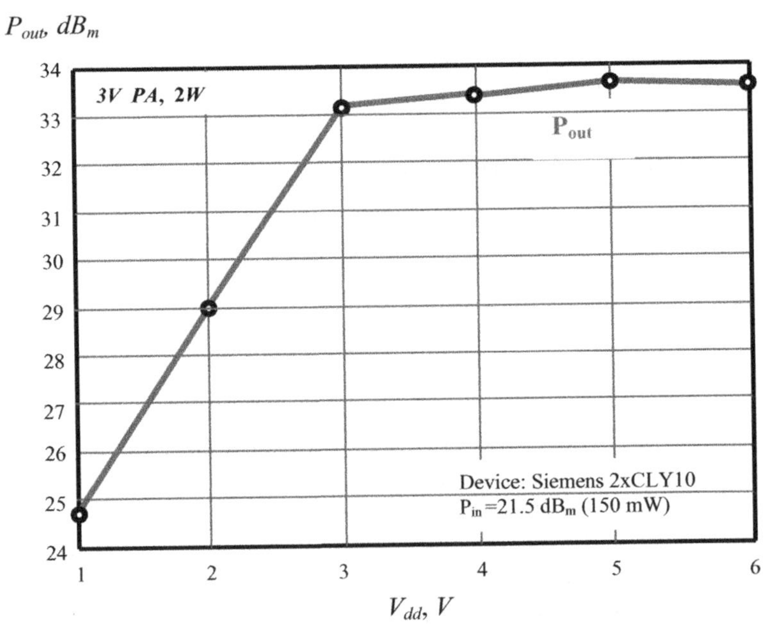

Figure 7.23 Plot of P_{out} vs. V_{dd}, $f_o = 850\,MHz$, $2\,W$ push-pull PA.

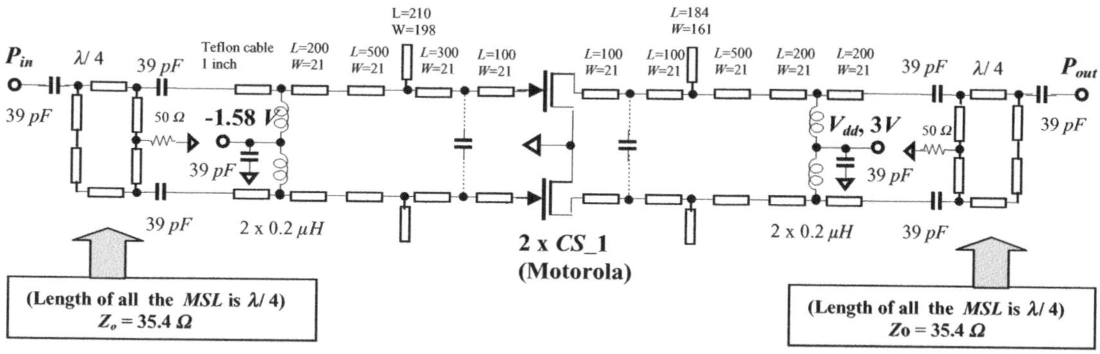

Figure 7.24 Schematic of 3 V, 5 W, push-pull PA (unit of length: mil). $f_o = 850\,MHz$, substrate: ceramic aluminate, $H = 25\,mils$, $T = 1\,mil$, $\varepsilon_r = 10.5$.

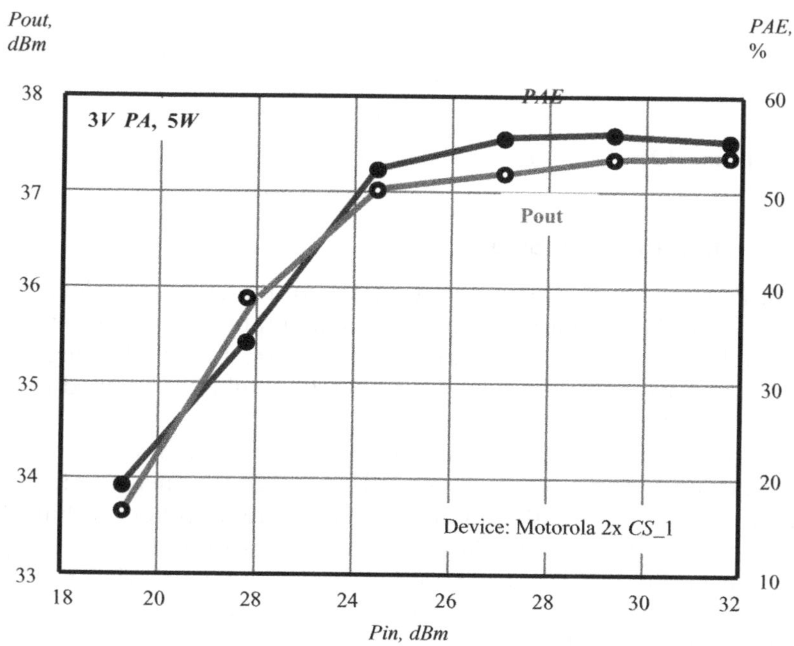

Figure 7.25 Plot of P_{out}, PAE vs. P_{in}, $f_o = 850\,MHz$, 5 W differential PA.

TABLE 7.2 Main performance of 3 V, 5 W push-pull PA

f_o	$= 850\,MHz$,
P_{in}	$= 640\,mW$,
P_{out}	$= 5.5\,W$,
G	$= 9.5\,dB$
PAE	$= 58\%$,
I	$= 2162\,mA$

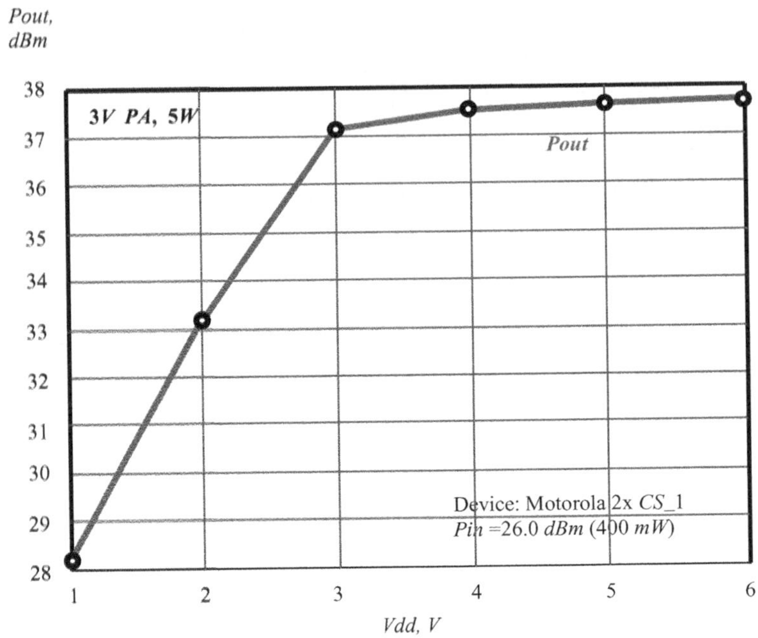

Figure 7.26 Plot of P_{out} vs. V_{dd}, $f_o = 850\,MHz$, $5\,W$ differential PA.

7.4.4 Reduction of Block Size

Looking at Figure 7.20, one can see that the size of the entire *PA* block is unsatisfactorily large. The largest parts are the input and output baluns as shown in Figure 7.20. We will therefore focus on shrinking the balun size. In the following discussion, only the 2 watt *PA* is considered since a 5 watt *PA* could be designed in the same way.

Figure 7.27 shows a modified balun built by a combination of micro strip lines and capacitors; its performance is shown in Figure 7.28. The magnitudes of S_{21} and S_{31} are kept within $-3.5 \pm 0.2\,\text{dB}$ and their phase difference is kept within $180°$ over the entire frequency bandwidth. The length of micro strip lines has been partially replaced by capacitors so that the total length of the micro strip line is reduced from 5520 mils to 3520 mils. Then, the layout of the push-pull PA is changed from Figure 7.21 to Figure 7.29. The size of the entire the PA module is reduced by about 40%.

Despite these changes, the size of the push-pull *PA* shown in Figure 7.29 is still not reduced to what is expected. Further effort is concentrated on three aspects:

1) Changing the input and output impedances of the device, Z_{in} and Z_{out}, to pure resistances, R'_{in} and R'_{out}, by connecting a capacitor or an inductor in parallel at the input or output respectively.

   Recalling the expression of Z_{in}, (7.11), at the gate G of the device, and the expression of Z_{out}, (7.12), at the drain D of device, that is,

$$Z_{in} = 13.5 + j4.5\,\Omega, \tag{7.11}$$

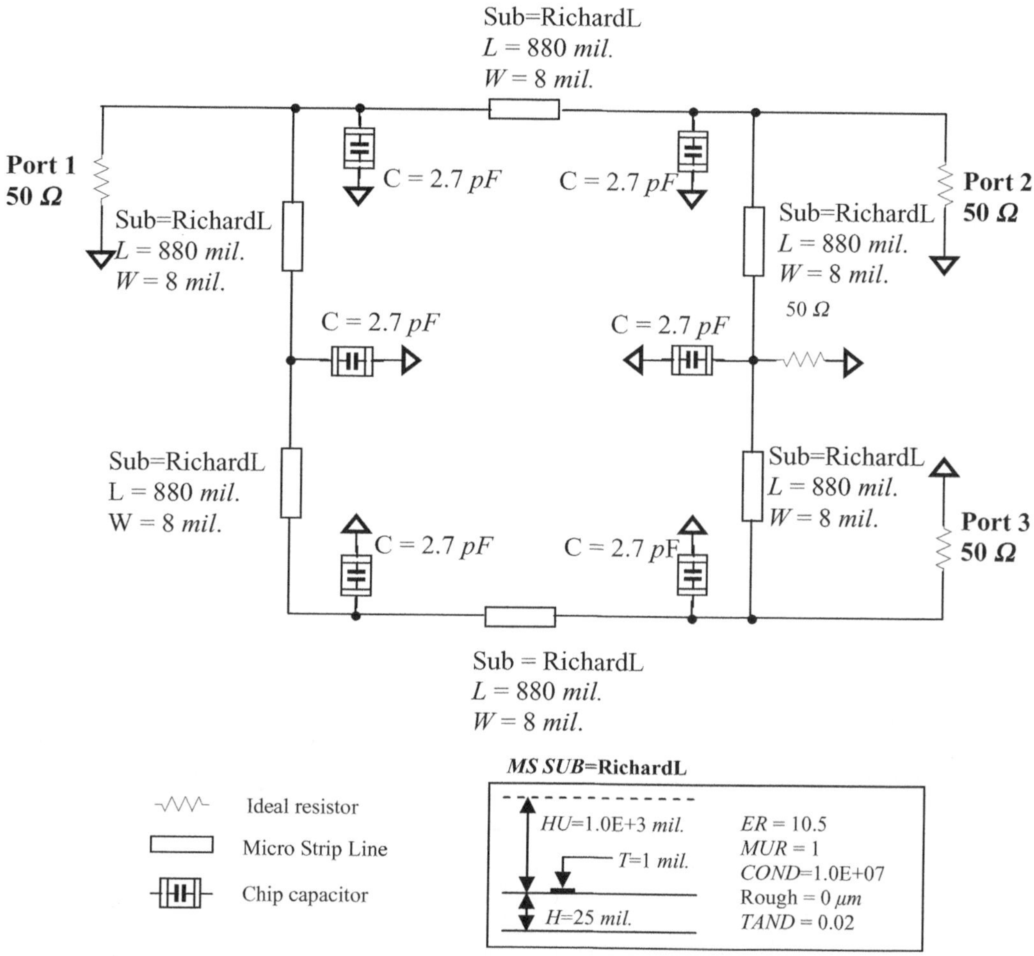

Figure 7.27 Modified balun with a combination of micro strip lines and capacitors. $f = 800$–$900\,MHz$, balun: $50\,\Omega$ to $2 \times 50\,\Omega$.

$$Z_{out} = 10.0 - j10.0\,\Omega, \tag{7.12}$$

From the Smith chart as shown in Figure 7.30, if Z_{in} at the gate G of the device is connected to a capacitor $C = 4.12\,pF$ in parallel, Z_{in} will be changed from $(13.5 + j4.5)\,\Omega$ to

$$Z'_{in} = R'_{in} = 15\,\Omega. \tag{7.13}$$

And, if Z_{out} at the drain D of the device is connected to an inductor $L = 3.75\,nH$ in parallel, Z_{out} will be changed from $(10 - j10)\,\Omega$ to

$$Z'_{out} = R'_{out} = 20\,\Omega. \tag{7.14}$$

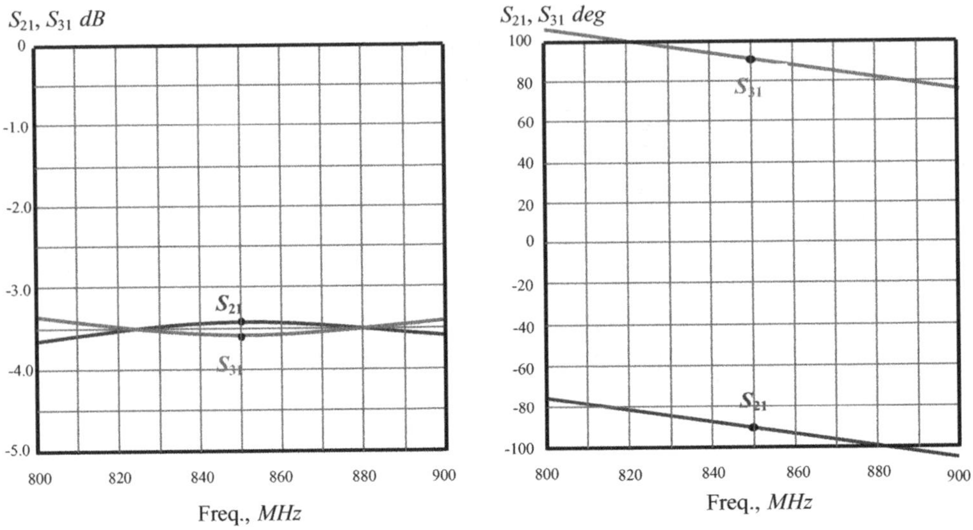

Figure 7.28 S_{21} and S_{31} of balun built by micro strip lines and capacitors. $f = 800\text{–}900\,MHz$, balun: $50\,\Omega$ to $2 \times 50\,\Omega$.

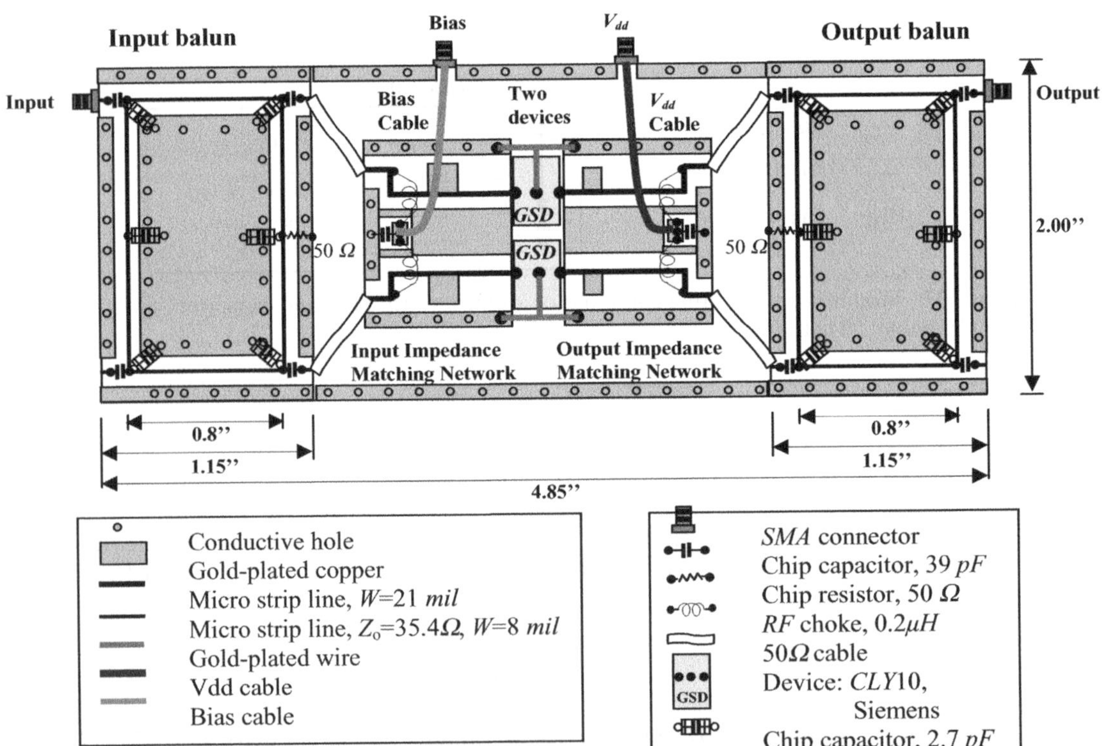

Figure 7.29 Modified layout of push-pull *PA* on a ceramic aluminate board. Thickness = $0.25\,mil$, $\varepsilon_r = 10.5$, $\mu_r = 1$, $f = 800\text{–}900\,MHz$.

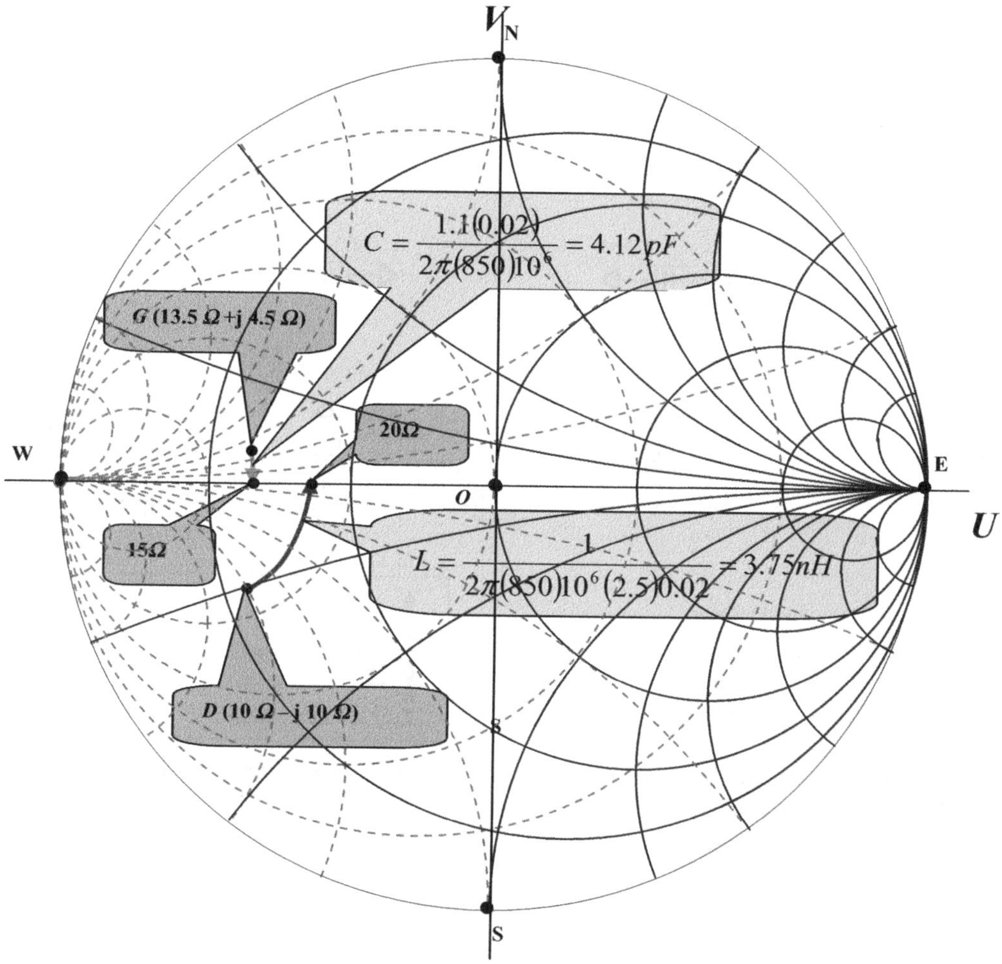

Figure 7.30 Drain D is impedance matched by an inductor $L = 3.75\,nH$ and gate G is impedance matched by a capacitor $C = 4.12\,pF$.

2) Removing the special impedance matching network as shown in Figures 7.21 and 7.29, and matching the new input and output impedances, R'_{in} and R'_{out}, directly by the input and output baluns respectively.
3) Modifying the input and output baluns as shown in Figure 7.19, again to further decrease the size. All the micro strip line segments will be replaced by chip capacitors and inductors.

Figures 7.31 and 7.32 show the input balun and its performance. Figures 7.33 and 7.34 show the output balun and its performance.

The magnitudes of S_{21} and S_{31} in Figure 7.28 are kept within $-3.75 \pm 0.5\,dB$ and their phase difference is kept within $180°$ over the entire frequency bandwidth. The magnitudes of S_{21} and S_{31} in Figure 7.34 are also kept in $-3.75 \pm 0.5\,dB$ and their phase difference is also kept within $180°$ over the entire frequency bandwidth. Within the frequency bandwidth, their performances are acceptable.

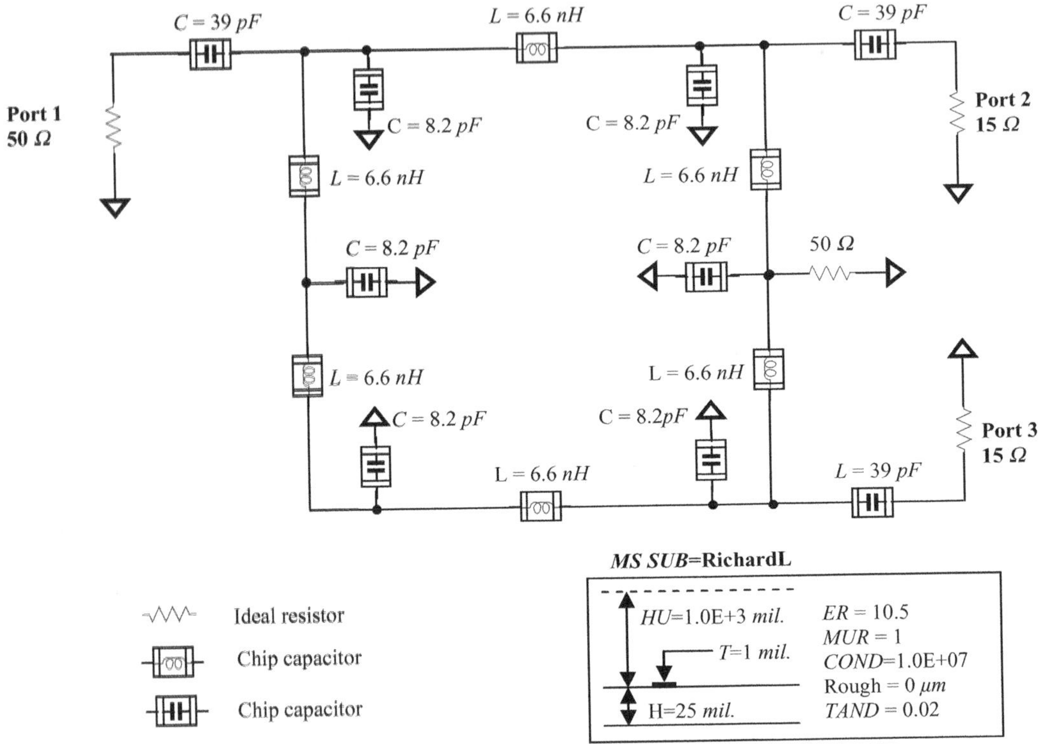

Figure 7.31 Input balun built by inductors and capacitors. $f = 800\text{–}900\,MHz$, balun: $50\,\Omega$ to $2 \times 15\,\Omega$, "zero" capacitor: $C = 39\,pF$.

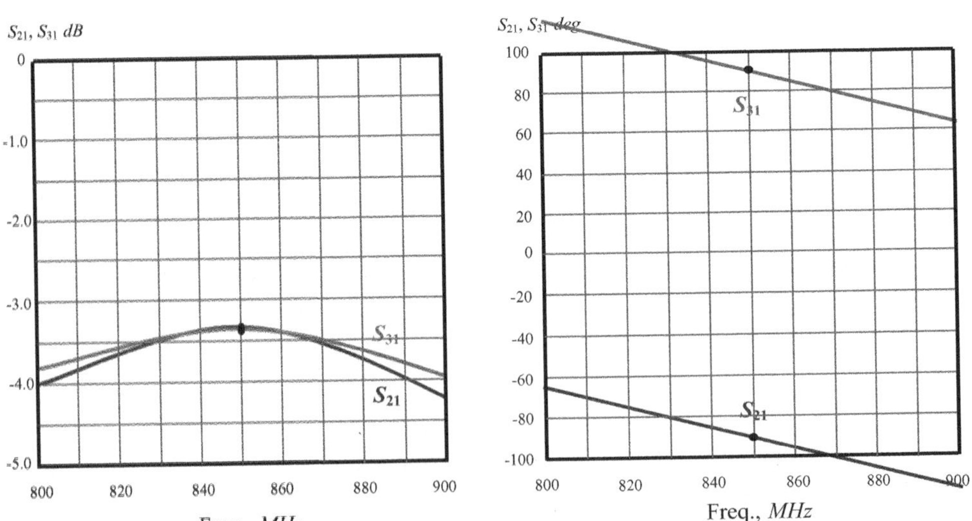

Figure 7.32 S_{21} and S_{31} of input balun built by inductors and capacitors. $f = 800\text{–}900\,MHz$, balun: $50\,\Omega$ to $2 \times 50\,\Omega$, "zero" capacitor: $C = 39\,pF$.

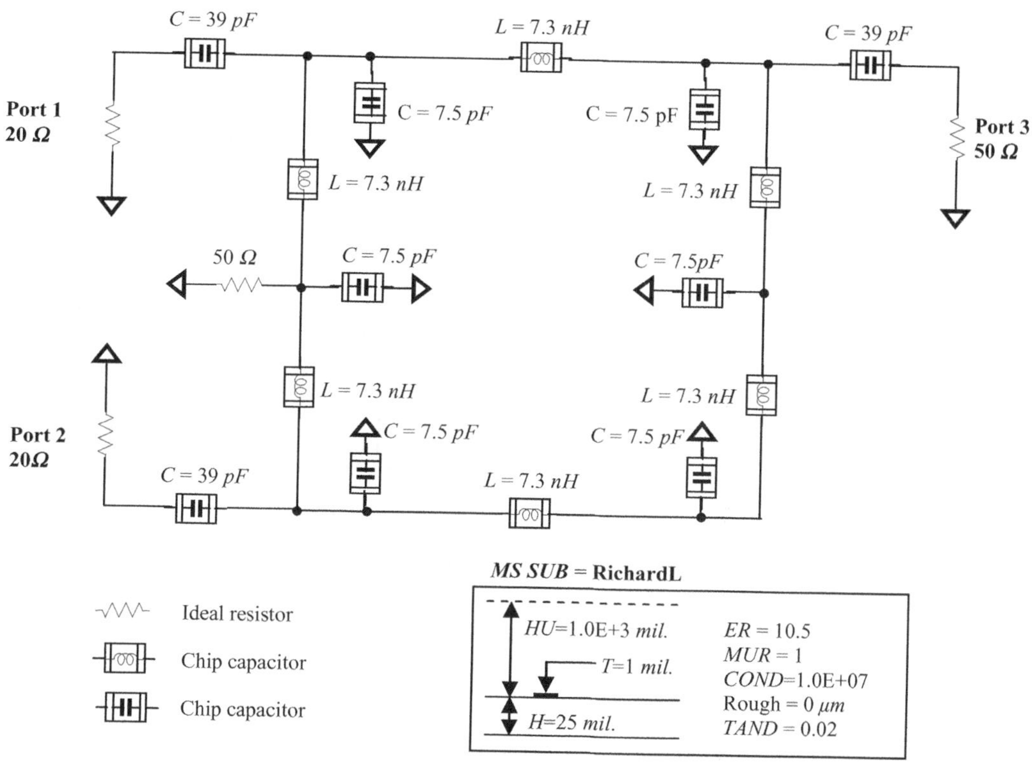

Figure 7.33 Output balun built by inductors and capacitors. f = 800–900 MHz, balun: $2 \times 20\,\Omega$ to $50\,\Omega$, "zero" capacitor: $C = 39\,pF$.

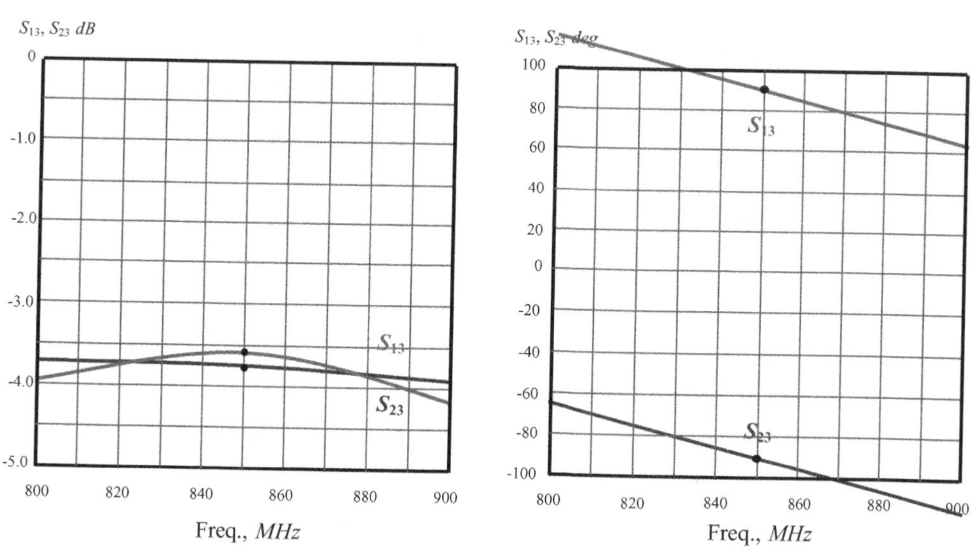

Figure 7.34 S_{13} and S_{23} of output balun built by inductors and capacitors. f = 800–900 MHz, balun: $2 \times 20\,\Omega$ to $50\,\Omega$, "zero" capacitor: $C = 39\,pF$.

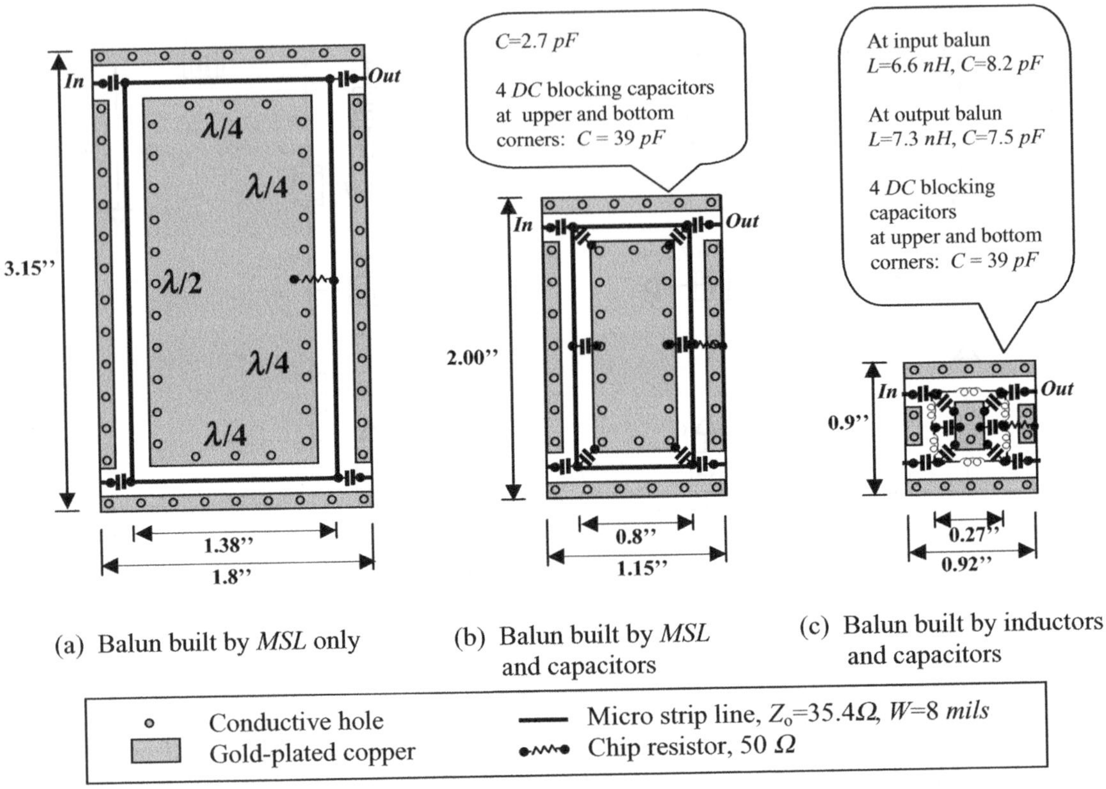

Figure 7.35 Evolution of balun applied to *PA* on a ceramic aluminate board. Thickness = $0.25\,mil$, $\varepsilon_r = 10.5$, $\mu_r = 1$, $f = 800$–$900\,MHz$.

Figure 7.35 shows that the size of the balun is reduced down from $3.15 \cdot 1.8\,in^2 = 5.67\,in^2$, to $0.9 \cdot 092\,in^2 = 0.83\,in^2$. Figure 7.36 shows the size of the entire *PA* with the minimized balun. The capacitor $C = 4.12\,pF$ changes the input impedance of the device Z_{in} to a pure resistance R'_{in}; the inductor $L = 3.75\,nH$ changes the output impedance of the device Z_{out} to a pure resistance R'_{out}. In addition, in order to avoid a crowded layout drawing, the parts in the input and output baluns are replaced by the parts' symbols.

As shown in Figure 7.21, the original size of the entire *PA* was $5.85 \cdot 3.15\,in^2 = 18.43\,in^2$. Now, from Figure 7.36, it can be seen that the size of entire *PA* is $2.88 \cdot 1.10\,in^2 = 3.17\,in^2$. The size of the entire *PA* has been reduced to approximately 5.8 times its original size.

7.4.5 Double Micro Strip Line Balun

As shown in Figure 7.36, the size of a push-pull *PA* can be shrunk to a reasonably small size if the baluns are completely built by chip capacitors and chip inductors.

However, this is not completely satisfactory because there are some drawbacks in the balun implementation. For instance, as shown in Figure 7.36, an input or an output balun will need at least six chip inductors and six capacitors, which leads to the following drawbacks:

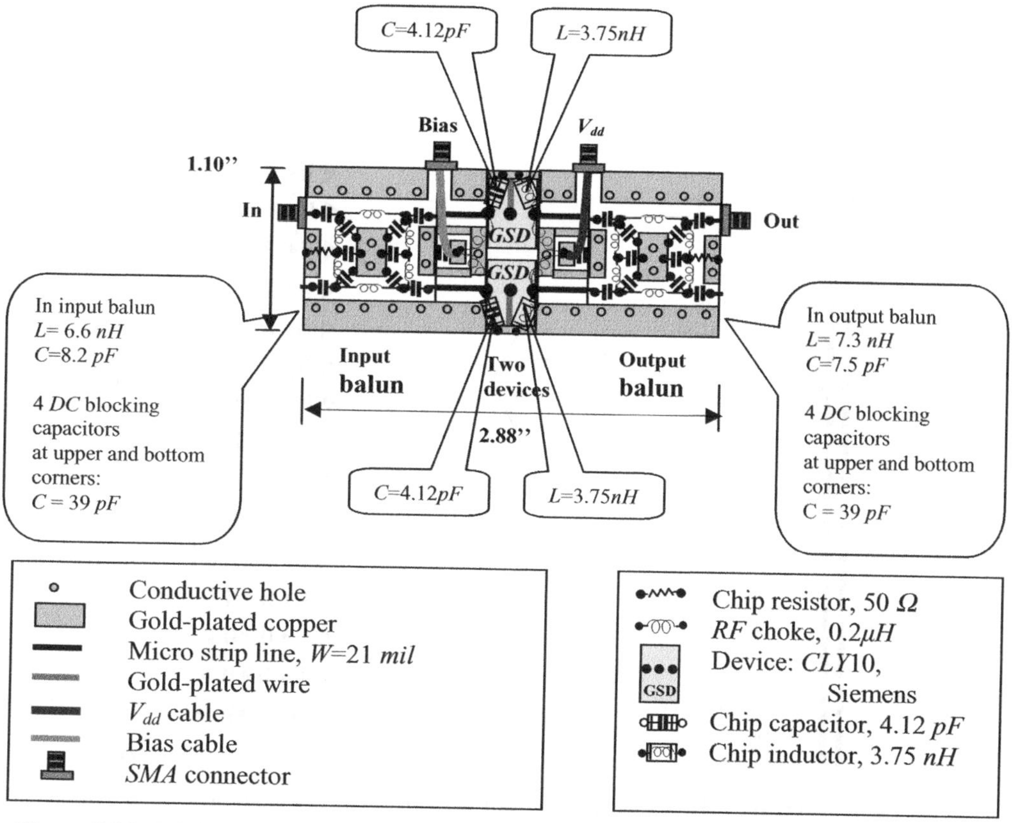

Figure 7.36 Minimized layout of push-pull *PA* on a ceramic aluminate board. Thickness = $0.25\,mil$, $\varepsilon_r = 10.5$, $\mu_r = 1$, $f = 800$–$900\,MHz$.

- The cost for a total of 12 chip inductors is quite high.
- Additional attenuation exists due to the 12 chip inductors.
- The parts' ununiformity degrades the performance of the balun, including inaccuracy or fluctuation of magnitude and phase.
- The chip inductor cannot bear such a high current on it.

Therefore, it is essential to develop a good balun for a push-pull *PA*. Such a balun must have the following features:

- Low cost,
- Very low insertion loss,
- Ability to operate in high current or high power status,
- Small size,
- Simple configuration.

The primary idea to develop a double micro strip line balun is illustrated by Figure 7.37. (The content of this section is abstracted from a US patent which was obtained by the author in 1993.)

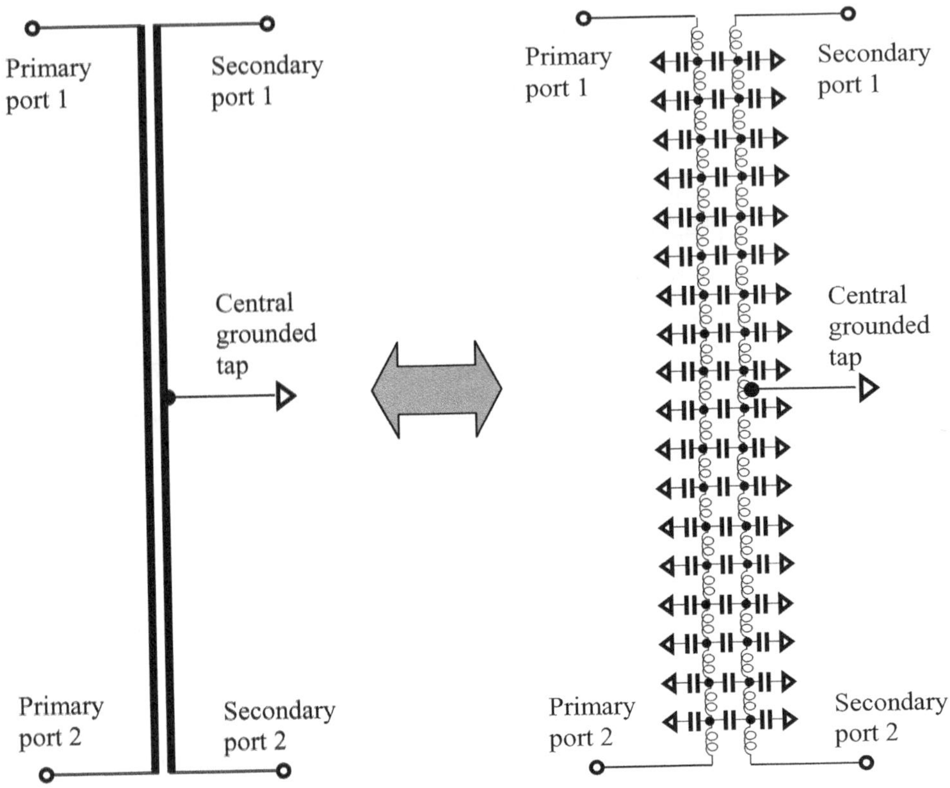

Figure 7.37 Conceptional drawing of double-line balun.

The model of the micro strip line can be represented by many "distributed" inductances in series and many "distributed" capacitances in parallel. The simple model of a double micro strip line in parallel can be represented by two individual micro strip line models connected with many "distributed" coupling capacitors as shown in Figure 7.37. It is imagined that if one of double micro strip lines is the primary path for the *RF* signal, its power can be coupled to the second micro strip line by those "distributed" coupling capacitors. These double lines could function like a transformer. If one end of the first micro strip line is grounded and the center of the second is also taped to the ground, then these double lines will look like a single-ended to differential pair balun.

If such an imagined model can be realized, its cost should be low. If the double micro strip lines are copper wire on a *PCB* or golden wire on an *IC* substrate, the insertion loss could be low because the resistance on the double lines is negligible. In addition, it can afford high current or high power. The most outstanding advantage would be its simple configuration. Should small size be an issue, this setup can be built on a *PCB* or *IC* substrate with higher permeability.

The next clue is to worry about radiation and hence interference with other parts in the circuitry. It would be helpful if these micro strip lines were enclosed by a grounded frame in order to prevent the magnetic flux from flying out to the sky. The double micro strip lines are bent into an "*S*" shape and seem like two loops connected together in series. Figure 7.38 shows such a configuration.

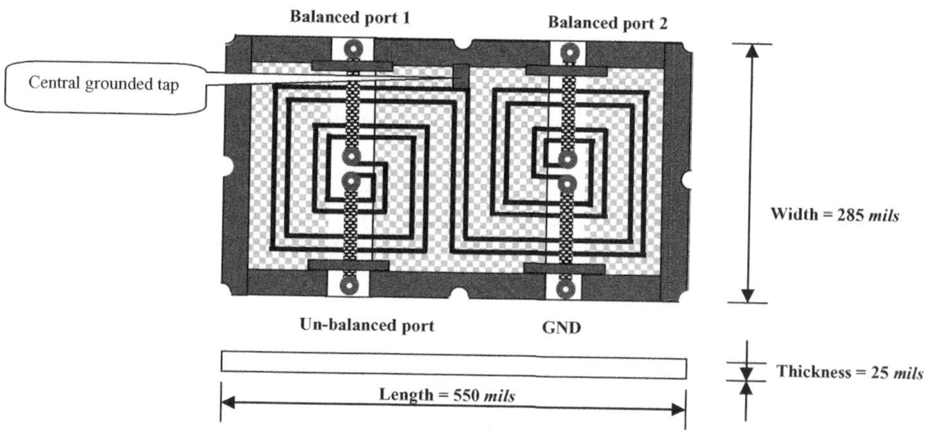

(a) Double micro strip line balun

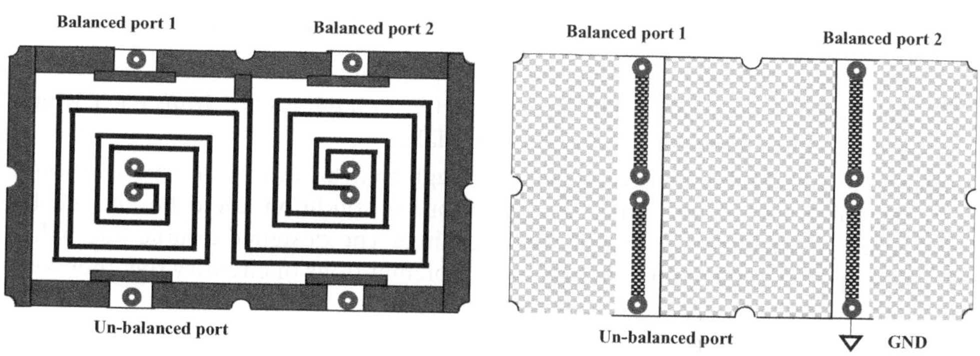

(b) Top side (c) Bottom side

Figure 7.38 Double-line balun build on a ceramic alumina substrate. Thickness of substrate: 25 *mils*, $\varepsilon_r = 10.5$, width of wire = 5 *mils*, spacing between double wire = 10 *mils*.
Top metal
Bottom metal
Bottom runner
MSL on top
Conductive via,

In Figure 7.38, the double micro strip lines in parallel are printed on a small piece of ceramic alumina with gold plating. The double micro strip line forms two windings with rectangular shapes of the same width and length. One of them functions as a single-ended or unbalanced port. The other with a central grounded tap forms two balanced ports. Obviously, the operating principle of this balun is based on the strong edge coupling between double lines.

In order to minimize insertion loss, two key issues are important to its performance:

1) One is its "*S*-type" configuration. Due to the "*S*" shape, the windings are clockwise on the left hand side and counter-clockwise on the right. Consequently, the magnetic flux generated by the loop on the left will be automati-

cally bent down into the loop on the right. The energy will not be scattered out of the balun block to lower the insertion loss and interfere with other parts. As a matter of fact, in the first design attempt, the loops on both left- and right-hand sides had the same clockwise winding and resulted in an increased insertion loss.

2) On the other hand, the total electrical length of the double line is about a half-wavelength of the central frequency. The advantage of the half-wavelength is that the energy of the standing wave is confined to the central area of the substrate. In other word, the leakage of energy, and hence, insertion loss, can be reduced to a lower level.

In order to minimize the size, four aspects of the design have been considered:

1) The material of the substrate is ceramic aluminate because its dielectric constant is high, in our case is $\varepsilon_r = 10.5$. Theoretically, the higher the dielectric constant, the smaller the size of the balun. However, a substrate with a higher dielectric constant brings about higher insertion loss and higher tolerance. As a matter of fact, there is a trade-off between size and insertion loss.
2) Instead of being circular, the loop is rectangular.
3) The width of the lines and the spacing between the lines are $5\,mils$ each, which is the limit of small size in the thin film lab.
4) The use of double lines only is not enough to achieve low insertion loss and minimize size at the same time. A resonant condition could be created by adding capacitors to assist the two windings. The designed double line balun is supposed to be applied in practical implementions of circuitry together with capacitors.

Much effort has been spent on the optimization of its size, including the length of the double lines, the width of the micro strip line, and the spacing between double the lines. In order to simplify the testing, one of the balanced ports is grounded while its central grounded tap is removed. The testing is then changed from testing for three ports into testing for two ports without loss of generality. In actual application, the test result corresponds to the condition when the balanced ports are treated as a terminal with 25 ohms of impedance, instead of $50\,\Omega$. The test setup is shown in Figure 7.39, where the capacitors, $2.7\,pF$, are the additional parts as mentioned above and are specially chosen for these specific dimensions of dual line balun: $L \times W \times T = 550 \times 285 \times 25\,mils$.

Figure 7.40 shows their typical frequency response and return loss. The successful modules are listed in Table 7.3.

The outstanding performance is the insertion loss,

$$IL < 0.6\,dB, \qquad \text{when } 485\,MHz < f < 940\,MHz. \tag{7.15}$$

As is well known, in the high RF frequency range, the insertion loss of a ferrite transformer balun is about 2–3 dB. Therefore, it is impossible to apply in a *PA* circuit block. Some other baluns, such as the ring or "Ratrace" baluns, might have the same advantage of low insertion loss. However, they are usually much larger than the

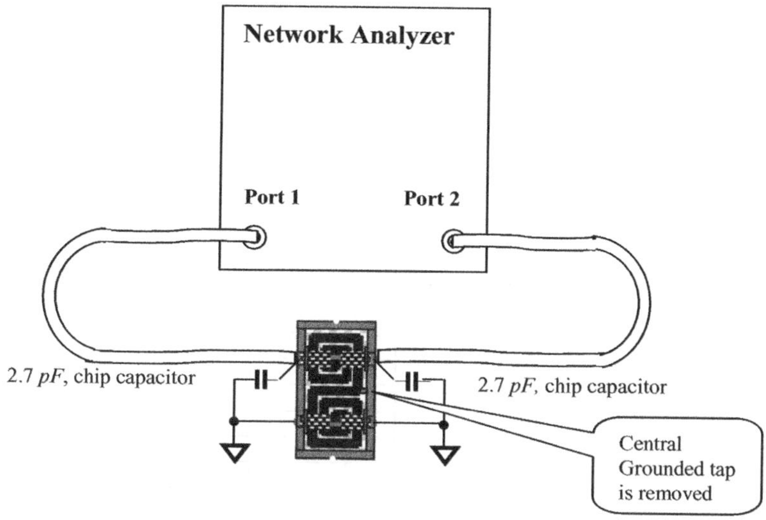

Figure 7.39 Setup for testing a double micro strip line balun.

double micro strip line balun. With the exception of these advantages in a double micro strip line balun, its simple configuration and low cost are superior to other types of baluns in the application in a *PA*.

As a matter of fact, the size of the double micro strip line balun can be further reduced if the dielectric constant of the ceramic alumina can be increased. A size reduction of more than 50% is possible.

7.4.6 Toroidal *RF* Transformer Balun

The idea of the toroidal *RF* transformer balun comes directly from that of the ferrite transformer. Figure 7.41 is a conceptual drawing in which the primary clue to implementing a transformer on a piece of ceramic alumina substrate is illustrated. (The content of this section is abstracted from author's 1993 design work.)

Figure 7.42 shows the designed configuration of a toroidal *RF* transformer balun. By using the thin film technology, a transformer is constructed on a ceramic alumina substrate with a high dielectric constant. Its basic components are the micro strip line segments and vias. There are two windings on the substrate, the single-ended and differential. The single-ended winding starts from the soldering port on the lateral side of the substrate (single-ended port 1 shown in Figure 7.42(a)) and runs between the top and bottom sides connected by the conductive vias. Finally, it terminates at another soldering port on the lateral side (*GND* port in Figure 7.42(a)), which is near the starting single-ended port. The differential windings start from the lateral side of the substrate (differential port 1 shown in Figure 7.42(a)) and also run between the top and bottom sides connected by conductive vias. Finally, it terminates at another soldering port on the lateral side (differential port 2 in Figure 7.42(a)), which is near the starting single-ended port. At the middle way there is a central tap by two lateral grounded ports (two *GND*s shown in Figure 7.42(a)). The two differential and single-ended windings are interlaced and strongly coupled. No additional parts are needed.

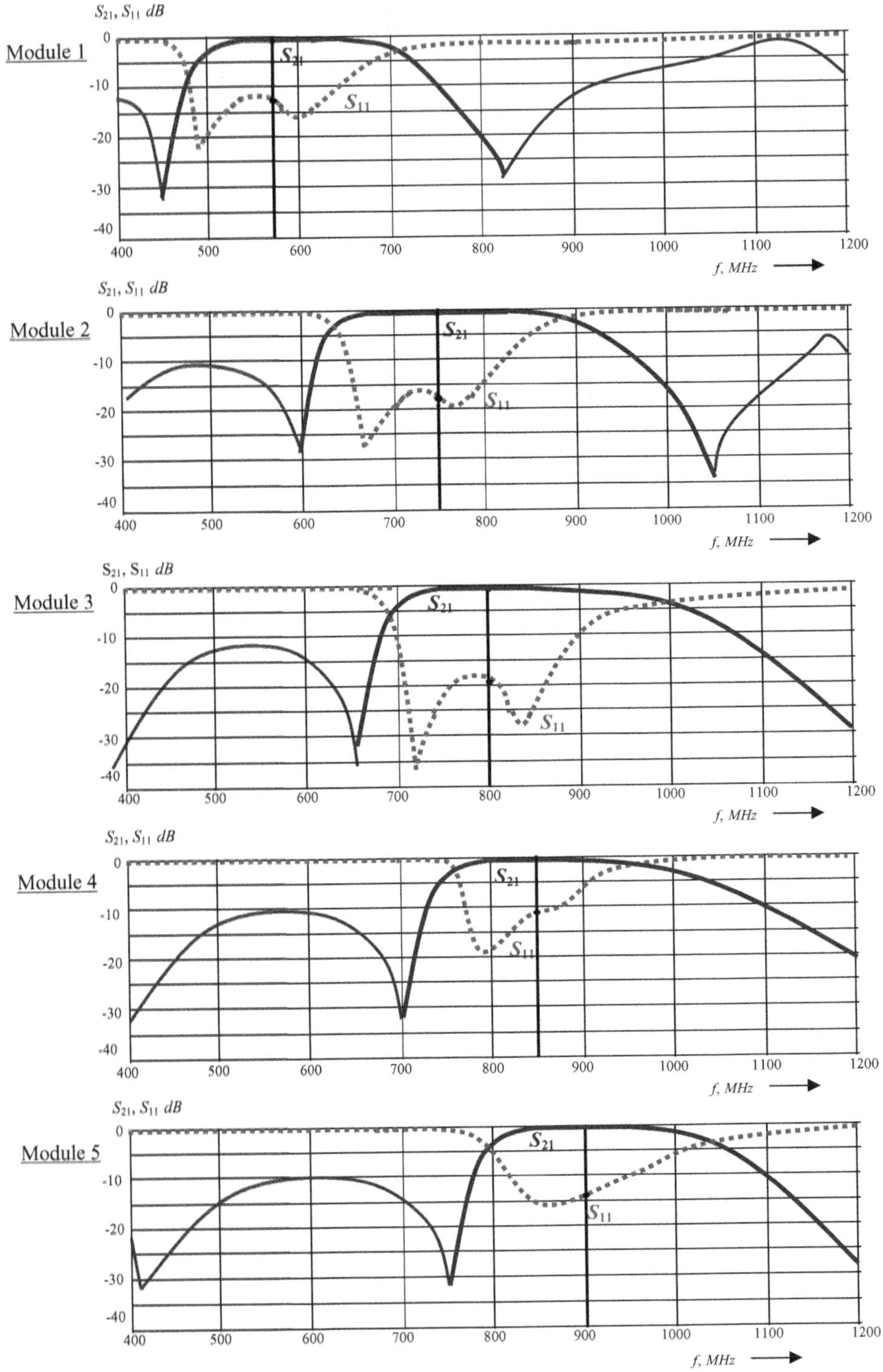

Figure 7.40 S_{21} and S_{11} of double micro strip line balun.

TABLE 7.3 Configuration and performance of double micro strip line balun

	Module 1	Module 2	Module 3	Module 4	Module 5
Substrate					
Material	Ceramic alumina				
Size, *mil*	$L \times W \times T = 550 \times 285 \times 25$				
Micro strip line					
Length, *mil*	5456	3463	3174	3034	2813
Width, *mil*	5	10	10	10	10
Spacing	5	5	5	5	5
Performance					
Freq., *MHz*	485–615	685–815	785–885	800–900	850–940
BW, *MHz*	130	130	100	100	90
Insertion loss, *dB*	**0.40**	**0.53**	**0.46**	**0.53**	**0.59**
Return loss, *dB*	<–11	<–15	<–18	<–12	<–11

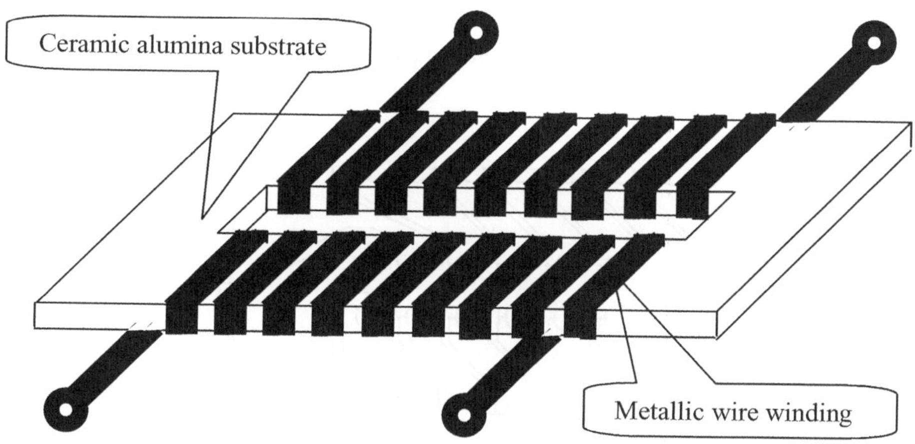

Figure 7.41 Conceptional drawing of a toroidal *RF* transformer balun.

The characteristics, of course, depend on many factors, such as the total length of the windings, the average width of the windings, the space between the windings, the diameter of the vias, the thickness of the substrate, and the dielectric constant of the substrate.

In order to minimize the insertion loss, two aspects have been considered:

1) The toroidal configuration of the transformer confines most of the *RF* energy within the toroidal windings. Consequently, the leakage of power, and hence, insertion loss, could be reduced to a very low level.
2) Therefore, the spacing between the two segments is set to 5 *mils* only, which is the limit of small size in the thin film lab.

In order to minimize the size, three aspects have been considered:

1) The material of the substrate is decided to be ceramic alumina substrate with a dielectric constant equal to 10.5. Theoretically, the higher the dielectric

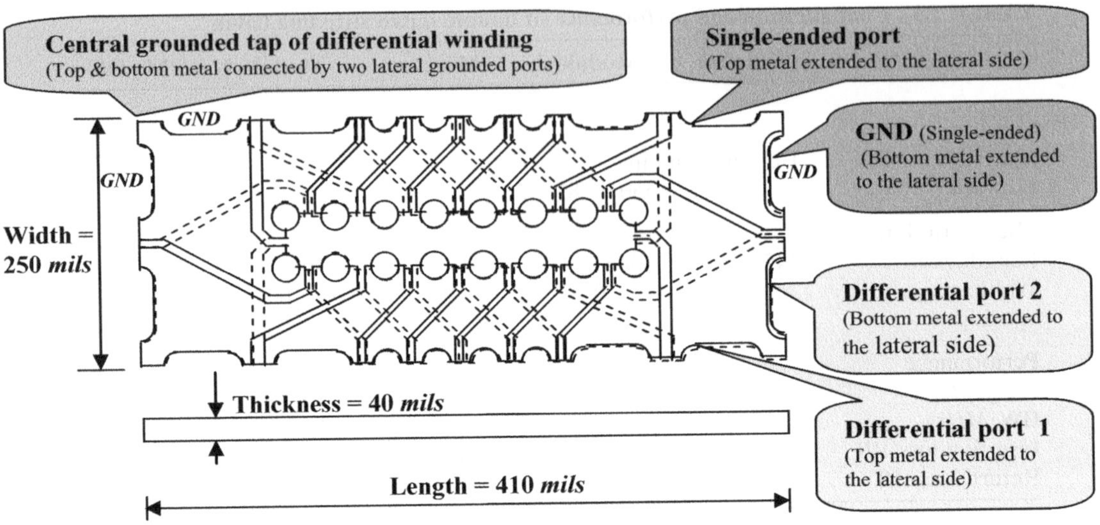

(a) Layout of toroidal balun. Substrate: ceramic aluminate, $\varepsilon_r = 10.5$ for $f = 800$–1650 *MHz*.

(b) Top view of toroidal balun layout

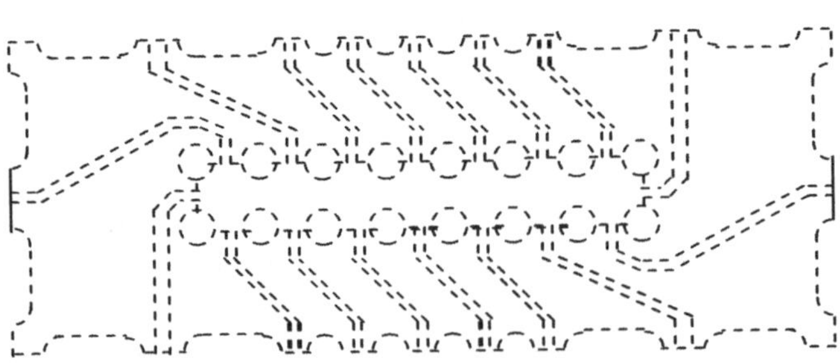

(c) Bottom view of toroidal balun layout

Figure 7.42 Toroidal balun built on ceramic alumina substrate. $\varepsilon_r = 10.5$, thickness = 40*mils*, spacing = 5*mils*, diameter of via = 20*mils*.

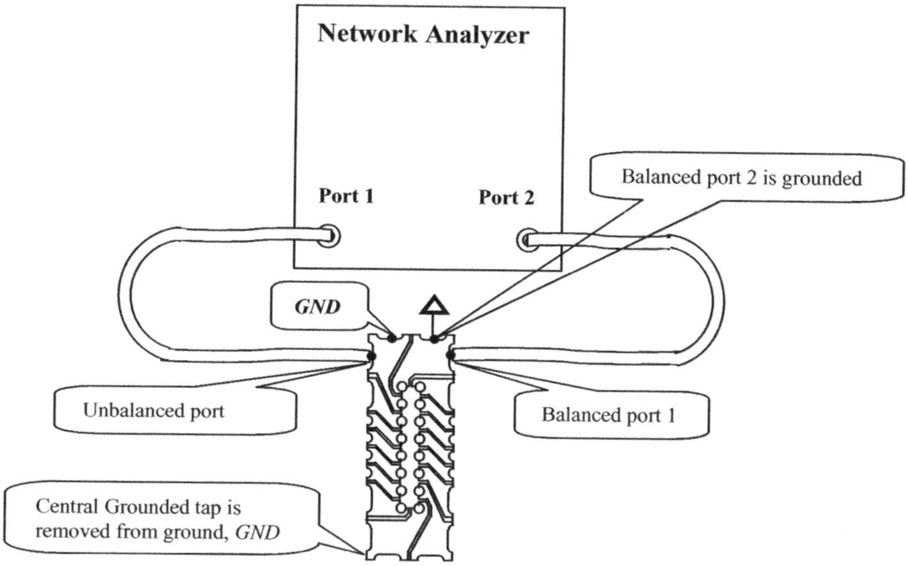

Figure 7.43 Setup for testing a toroidal *RF* transformer balun.

constant, the smaller the size of the balun. However, a substrate with a higher dielectric constant introduces a higher loss. It is a trade-off between size and insertion loss.

2) Instead of being circular, the the loop is rectangular.
3) The diameter of the vias is 20*mils* only, which is a critical limit in the thin film lab.

In order to simplify the testing work, the central grounded tap is removed from the ground, *GND*, and the test setup as shown in Figure 7.43. This is for the convenience of testing in the $50\,\Omega$ system only.

However, there is no loss of generality. The test results can be applied in actual circuitry when the impedance of a single-ended port is $50\,\Omega$ while the impedance of each differential port is $25\,\Omega$, with a $180°$ phase difference. Figure 7.44 shows one typical test result. The outstanding performance is

- Extremely low insertion loss,
- Extremely wide bandwidth.

From Figure 7.44 it can be seen that

$$0 > S_{21} > -0.35\,dB, \quad \text{when } f < 600\,MHz, \text{ and } f > 1650\,MHz, \tag{7.16}$$

$$S_{11} < -10\,dB, \quad \text{when } 600\,MHz < f < 1650\,MHz, \tag{7.17}$$

$$S_{11} = -35.3\,dB, \quad \text{when } f = 1000\,MHz, \tag{7.18}$$

The insertion loss in this module is less than $0.35\,dB$ over a relative bandwidth of 93%!

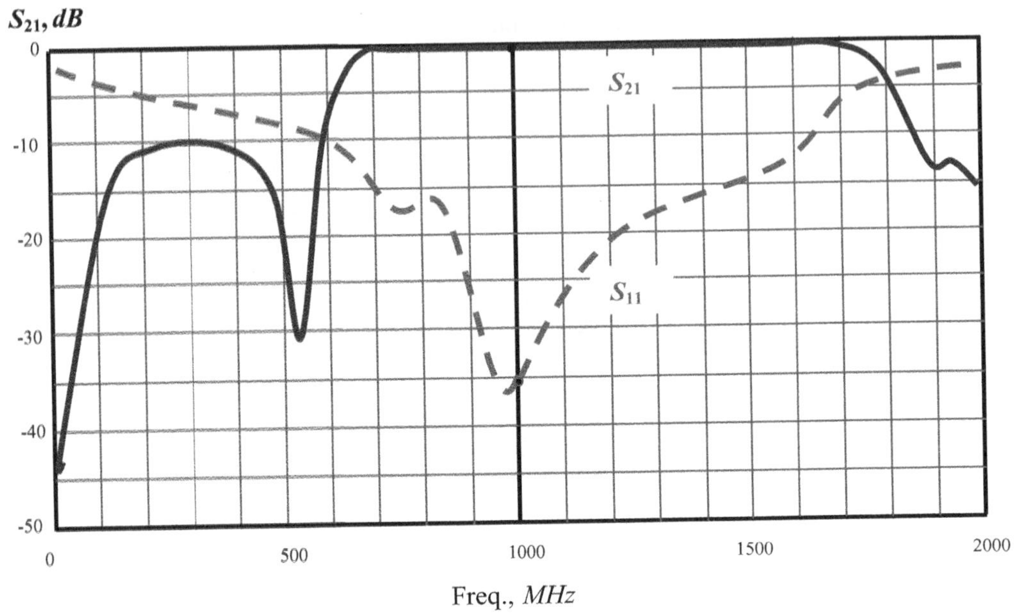

Figure 7.44 Frequency response of toroidal *RF* transformer balun. Balun size: $L \times W \times T = 410 \times 250 \times 40\,mils$. Substrate: ceramic aluminate, $\varepsilon_r = 10.5$. Micro strip line: width = $35\,mils$; spacing = $5\,mils$; via (dia.) = $20\,mils$.

Its performance is better than the double line balun mentioned above. Also, it is much better than the *RF* balun developed by *MLC* (Multi-Layer Ceramic) technology. Typically, the insertion loss of the *MLC* balun is about 1.5 to $2\,dB$ over a relative bandwidth of less than 50%. Other desirable features of the toroidal RF transformer balun are:

- Small size. Usually, *MLC* baluns are designed with the $\lambda/4$ or $\lambda/2$ micro strip lines. Their size is larger and their bandwidth much narrower than that of the toroidal RF transformer.
- Simple and reliable configuration. Indeed, it appears as merely a small piece of ceramic with some metallic pattern.
- Low cost. It contains only a top and bottom metallic layer while the *MLC* contains three or more metallic layers. Therefore, the former has a lower cost.

Successful modules have been implemented and their performances are similar to the module shown above. Table 7.4 lists their configuration and performance.

7.5 *PA* WITH TEMPERATURE COMPENSATION

There are many ways to compensate the change of *PA* power due to the variation of temperature. We are going to introduce a simple and easy way to assure the temperature compensation for *PA* (Yamauchi et al., 2001).

TABLE 7.4 Configuration and performance of toroidal *RF* transformer balun

	Module 1	Module 2	Module 3	Module 4	Module 5	Module 6	Module 7	Module 8
Substrate								
Material	Ceramic alumina							
ε_r	10.5	10.5	10.5	10.5	10.5	10.5	10.5	10.5
Length, *mils*	1070	1070	750	750	570	490	410	330
Width, *mils*	570	570	750	750	250	250	250	250
Thickness, *mils*	0	40	$2 \times 40^*$	$2 \times 40^*$	40	40	40	40
Micro strip line								
Width, *mils*	25	35	35	35	35	35	35	35
Spacing	5	5	5	5	5	5	5	5
Diameter of *via, mils*	20	20	20	20	20	20	20	20
Performance								
Freq., *MHz*	200–600	350–1150	160–500	300–1000	450–850	700–1250	850–1650	1200–2200
BW, MHz	400	800	340	700	400	550	850	1000
Insertion loss, *dB*	**0.31**	**0.24**	**0.15**	**0.15**	**0.20**	**0.26**	**0.25**	**0.32**
Return loss, *dB*	<–15	<–12	<–18	<–15	<–12	<–13	<–12	<–14

**Note*: These two modules consist of two identical substrates "piggy-backed" together.

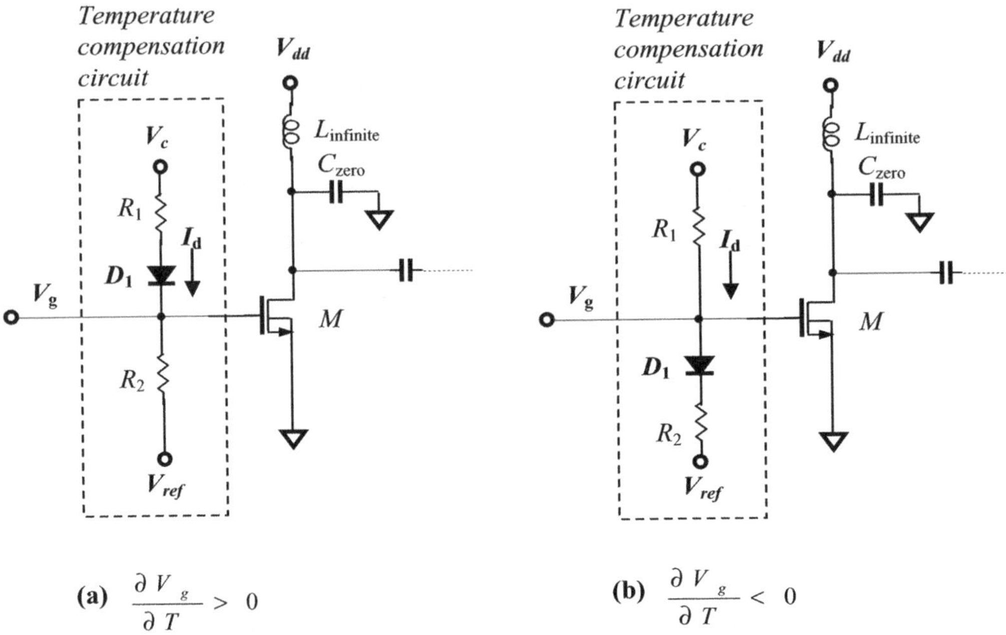

Figure 7.45 Temperature compensation circuit at gate of *MOSFET* transistor.

In general, the gain of a *MOSFET* amplifier increases when its gate voltage is increased.

On the other hand, the gain of a *MOSFET* amplifier can be increased or decreased when the temperature is increased. In order to compensate for the gain variation with temperature of an amplifier, changing the gate voltage of its *MOSFET* is the right approach.

Figure 7.45 shows the schematic diagram of the temperature-compensation circuit at the gate of the *MOSFET* transistor. The value of the gate voltage V_g is determined by the reference voltage V_{ref}, the gate control voltage V_c, and the current I_d flowing through the diode D_1. The current flowing through the diode is dependent on temperature. If the current flowing through the diode increases as the temperature increases, that is,

$$\frac{\partial I_d}{\partial T} > 0, \tag{7.19}$$

Then the temperature compensation circuit shown in Figure 7.45(a) is adapted. The gate voltage V_g is increased as the temperature is increased. Consequently, the decreased gain of the *MOSFET* due to temperature is compensated for by the increase of gain due to the increase of the gate voltage.

On the contrary, if the current flowing through the diode decreases as the temperature increases, that is,

$$\frac{\partial I_d}{\partial T} < 0, \tag{7.20}$$

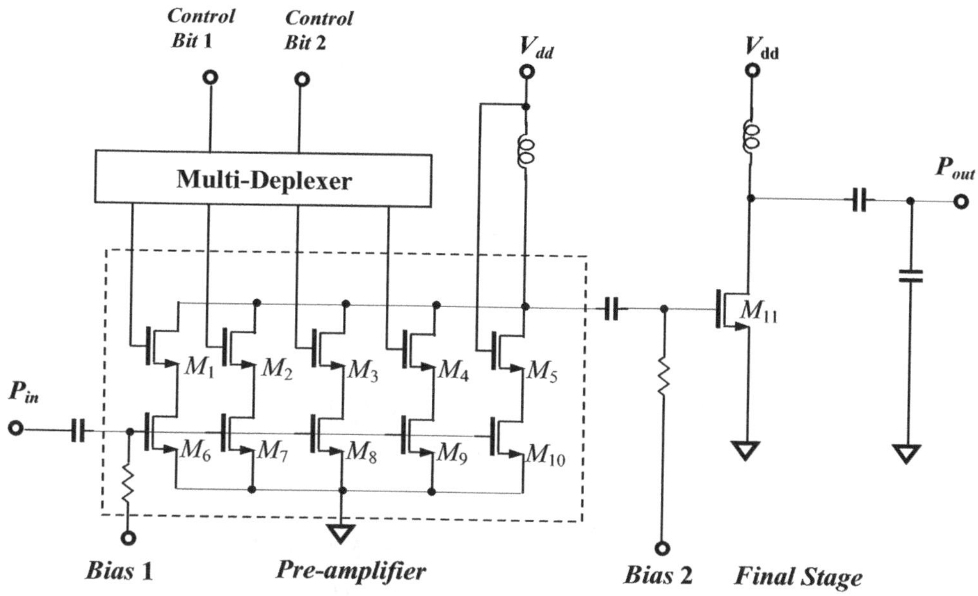

Figure 7.46 Digital power control for *MOSFET* power amplifier.

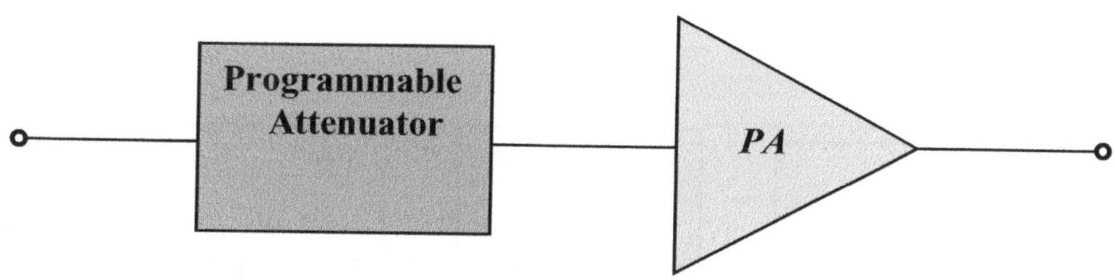

Figure 7.47 *PA* with programmable attenuator.

Then the temperature compensation circuit as shown in Figure 7.45(b) is adapted. The gate voltage V_g is also increased as the temperature is increases. Consequently the decreased gain of the *MOSFET* due to temperature is compensated for by the increase of gain due to the increase of the gate voltage.

7.6 *PA* WITH OUTPUT POWER CONTROL

It is desired to have output power control for power amplifiers in a communication system. Among the power controls, digital control is the most convenient and effective. One such digital control is shown in Figure 7.46.

The power control is set up in the pre-amplifier stage. The four control terminals can be controlled by the two bits of the multi-deplexer.

Another type of power control is to add a programmable attenuator before the *PA*. Figures 7.47 to 7.49 shows this technology (Khannur, 2003).

The programmable attenuator consists of 16 individual attenuators with attenuation from 0 to $7.5\,dB$ in 16 steps. The attenuation is increased by $0.5\,dB$ per step.

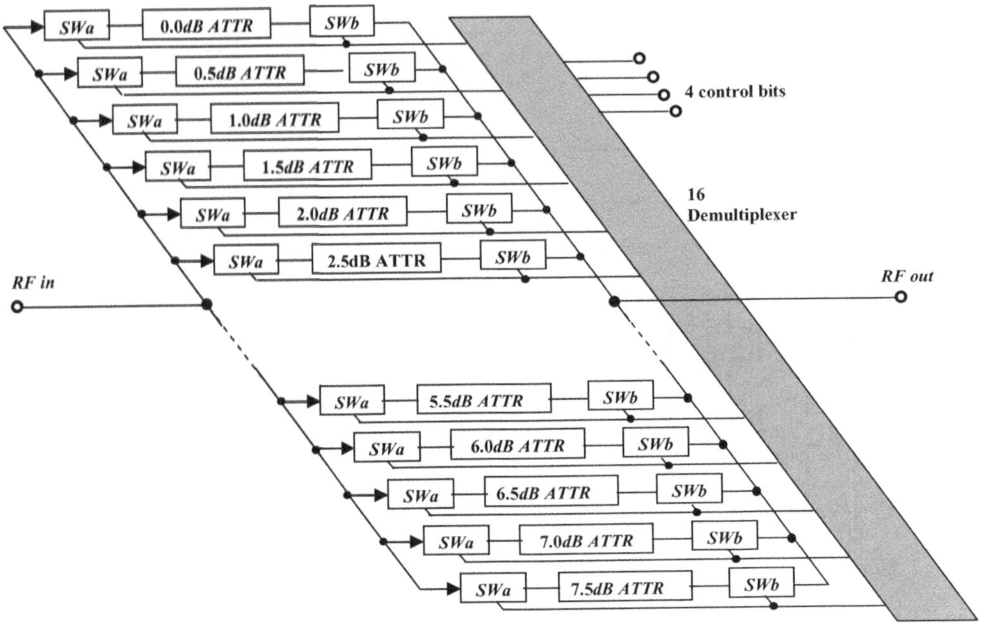

Figure 7.48 Full programmable attenuator.

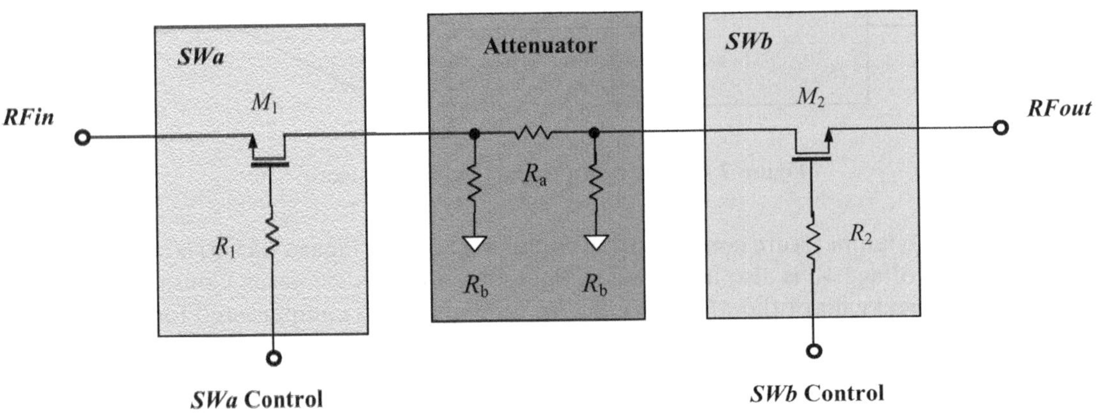

Figure 7.49 Individual attenuator.

These 16 individual attenuators are controlled by four bits of the demultiplexer. The total attenuation can be controlled from $0\,dB$ to $60\,dB$.

Each individual attenuator has two switched $MOSFET$ transistors. Their gates are called SWa and SWb. They are connected together and are controlled by the demultiplexer. The attenuator itself consists of three resistors with a Π type configuration. The attenuation depends on the values of resistors, which can be calculated through a simple mathematic derivation.

7.7 LINEAR *PA*

There are two kinds of non-linearities:

1) input-output power relationship, *AM-AM* conversion,
2) input-output phase relationship, *AM-PM* conversion.

They are uniquely related through the third order distortion phase. The non-linearity of a device can be represented by two components: the in phase non-linearity and the quadrature non-linearity. There are many different linearization techniques for the *PA*, such as,

- **Feedforward Error Cancellation**

  It compares the output or input to develop an error signal for summing with the output.

- **Continuous Feedback Error Cancellation**

  It compares the output or input to develop an error signal for predistorting the input.

- **Predistortion *(PD)***

  It uses a look-up table of premeasured corrections at the desired modulation values to predistort the input. This predistortion technology is sometimes called the complementary distortion technique. In order to achieve the optimal compensation condition, the *PD* circuit must have independently adjustable amplitude and phase for its third-order distortion output, which is referred to as a cuber predistortion linearizer. Theoretical investigation indicates that the IM_3 components of a power amplifier could be suppressed by a *CPL* (Cuber Predistortion Linearizer).

  The *PD* method is superior to the feedforward or continuous-feedback error correction techniques because of its simpler circuit configuration and low power consumption.

- **Cartesian Feedback**

  The power amplifier is linearized by using Cartesian feedback. It is called *Cartesian* feedback because the feedback is based on the Cartesian coordinates of the baseband symbols, *I* and *Q*, as opposed to polar coordinates. Basically, the concept behind this system is negative feedback. Figure 7.50 shows the block diagram of a typical Cartesian feedback system (Dawson and Lee, 2004).

- ***Transistor-level Linearized PA***

  The system-level methods to linearlize a power amplifier as mention above, such as predistortion, feedforward error-cancellation, and Cartesian feedback, require complicated hardware, and may be only suitable to the repeater or base stations.

  Now, let's introduce the transistor-level linearization for the *PA*, which may be more appropriate for power amplifiers in the handset (Tanaka et al., 1997). It is well known that a power amplifier with a differential configuration rejects the even orders of non-linearity very effectively. It is widely used in accurate continuous-time analog *MOSFET* circuits. However, there seems to be little

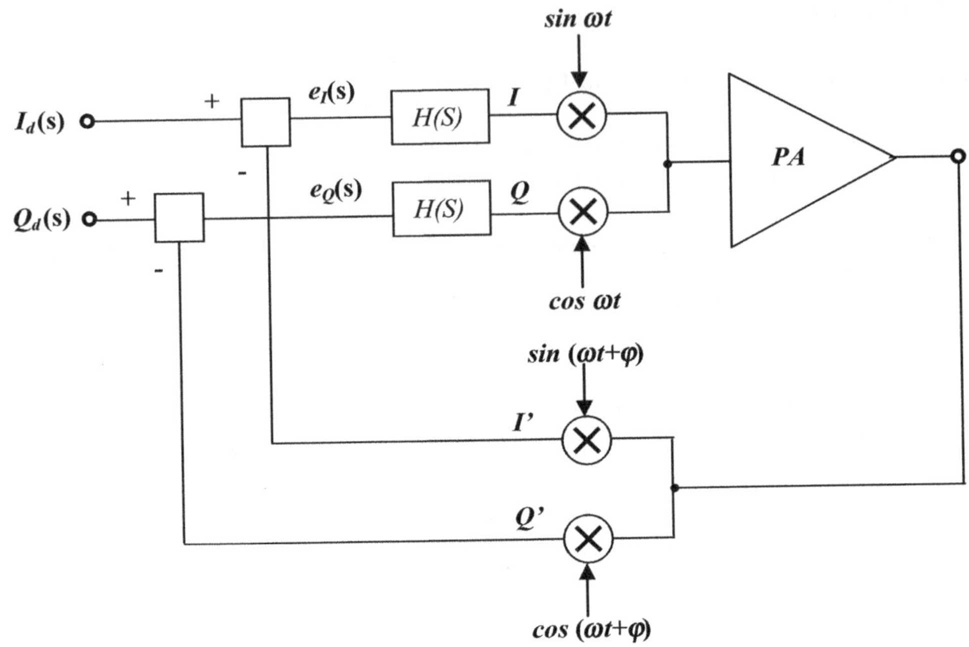

Figure 7.50 A typical Cartesian feedback system.

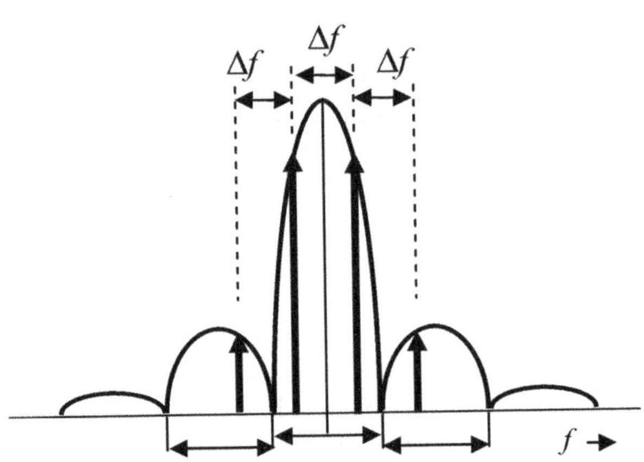

Figure 7.51 Two signals in main channel are the sources of interference due to third-order non-linearity.

special mention of the third-order and other odd orders of distortion and non-linearity. The odd order non-linearities become the main components of the non-linearity in a differential PA. The interference source of the odd non-linearity definitely exists because any pair of two signals within the operating bandwidth can be considered as interference sources to the adjacent channels. Figure 7.51 illustrates that the two signals are located within the main channels

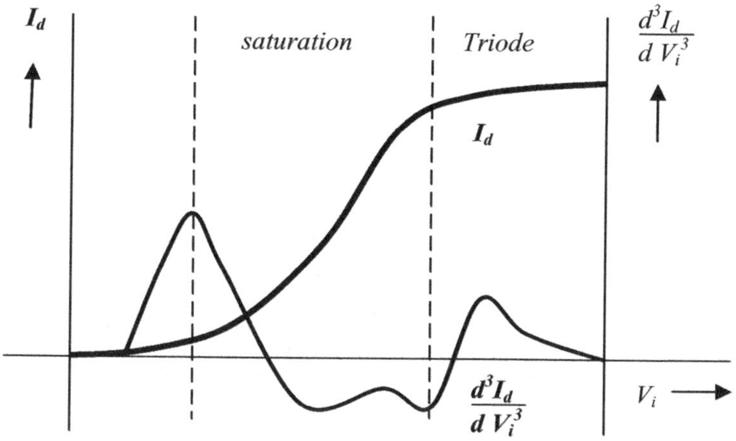

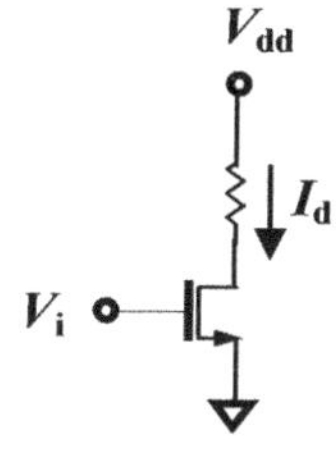

Figure 7.52 Characteristics of *MOSFET*. Curve of I_d vs. V_i and curve $\frac{d^3I_d}{dV_i^3}$ vs. V_i.

and their third-order non-linear products appear in the adjacent channels. Any pair of signals separated by Δf within the main channels are the interference source of the third-order non-linear products to the adjacent channels.

Instead of linearization of the *PA* by a complicated circuit loop, Tanaka et al. (1997) suggested the linearization of the *PA* at the transistor level. Similar to the multi-*tanh* devices applied to linearize a mixer, a compound *MOSFET* stage is developed and applied to linearize a *PA*.

Figure 7.52 shows the characteristics of a *MOSFET* transistor by two curves. It is well known that the curve of I_d vs. V_i can be divided into two regions. One is the saturated region and another the triode. The third-order non-linearity coefficient a_3 depends on V_i in a somewhat complicated relationship. As a matter of fact, the third-order differentiation of V_i is equal to $6a_3$, that is,

$$\frac{d^3I_d}{dV_i^3} = 6a_3, \tag{7.21}$$

if the transfer function of the *MOSFET* can be expressed as

$$I_d = a_o + a_1V_i + a_2V_i^2 + a_3V_i^3 + \cdots, \tag{7.22}$$

Figure 7.53(a) shows a new compound *MOSFET*, in which the bottom *MOSFET* M_1 is operated in the saturated region while M_2 and M_3 are operated in the triode region by the assistance of the control voltage V_c. It is found that the third-order non-linearity coefficient a_3 of compound *MOSFET* is very low, as shown in Figure 7.53(b) and is marked with a rectangular dashed frame, that is,

$$|a_3| < 0.01, \tag{7.23}$$

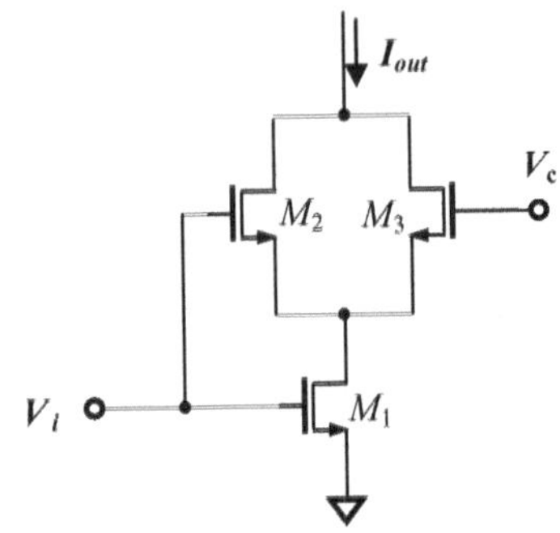

(a) A compound *MOSFET* stage

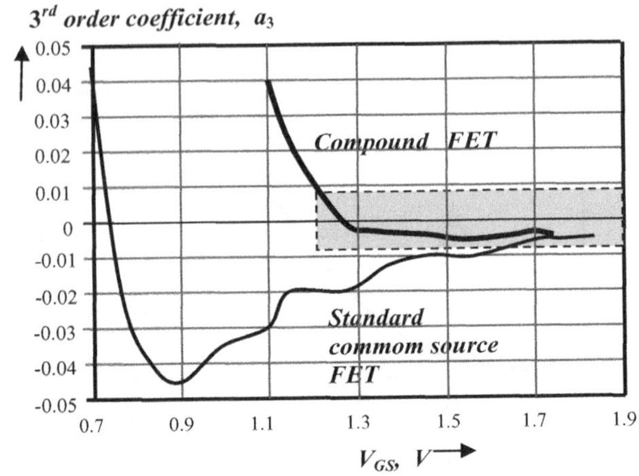

(b) The measured 3rd order coefficients of the standard common-source *FET* and the compound *FET*.

Figure 7.53 Comparison of the measured third-order coefficients between the standard common-source *FET* and the compound *FET*.

when

$$V_{GS} > 1.2\,V. \tag{7.24}$$

This is a comparison of the third-order non-linearity coefficient a_3 between the compound *MOSFET* and a standard common source *MOSFET*. The third-order non-linearity coefficient a_3 of a standard common source *MOSFET* has a deep negative notch around $V_{GS} = 0.9\,V$.

This is a very prospective way to improve the linearity of *PA*. A linear *PA* built by the compound *FET* is undoubtedly better than one built by the standard common-source *FET*. Of course, further work is anticipated so that condition (7.24) of the third-order non-linearity coefficient of the compound *FET* stage can be satisfied

when

$$V_{GS} > 0.8\,V, \tag{7.25}$$

or so.

REFERENCES

[1] H. Seidel, "A Microwave Feedforward Experiment," *Bell System Tech. J.*, vol. 50, 1971, pp. 2879–2916.

[2] T. Nojima and T. Konno, "Cuber Predistortion Linearizer for Relay Equipment in 800 MHz Band Land Mobile Telephone System," *IEEE Trans. on Vehicular Tech.*, Vol. VT-34, No. 4, 1985, pp. 169–177.

[3] Toshio Nojima and Tohru Konno, "Cuber Predistortion Linearizer for Relay Equipment in 800 MHz Band Land Mobile Telephone System," *IEEE Transactions on Vehicular Technology*, Vol. VT-34. No. 4, November 1985, pp. 169–177.

[4] M. Minowa, et al., "Backoff Improvement of an 800 MHz GaAs FET Amplifier for a QPSK Transmitter Using Adaptive Nonlinear Distortion Canceller," presented at *Vehicular Technology Conference*, 1990, pp. 542–546.

[5] Richard C. Li, "Double Line RF Balun Transformer," U.S. Patent 5,477,204, Motorola Inc., 1993.

[6] S. Tanaka, E. Behbahani, and A. A. Abidi, "A Linearization Technique for CMOS RF Power Amplifier," 1997, *Symposium on VLSI Circuits Digest of Technical Papers*, pp. 93–94.

[7] Ravi Gupta, Brian M. Ballweber, and David J. Allstot, "Design and Optimization of CMOS RF Power Amplifiers," *IEEE Journal of Solid-State Circuits*, Vol. 36, No. 2, February 2001, pp. 166–175.

[8] Kazuhisa Yamauchi, Yoshitada Iyama, Mamiko Yamaguchi, Yukio Ikeda, Shuji Urasaki, and Tadashi Takagi, "X-Band MMIC Power Amplifier with an On-Chip Temperature-Compensation Circuit," *IEEE Transactions on Microwave Theory and Techniques*, Vol. 49, No. 12, December 2001 pp. 2501–2506.

[9] Koen L. R. Mertens and Michiel S. J. Steyaert, "A 700-MHz 1-W Fully Differential CMOS Class-E Power Amplifier," *IEEE Journal of Solid-State Circuits*, Vol. 37, No. 2, February 2002, pp. 137–141.

[10] Pradeep B. Khannur, "A CMOS Power Amplifier with Power Control and T/R Switch for 2.45-GHz Bluetooth WISM Band Applications," *IEEE Radio Frequency Integrated Circuits Symposium*, 2003, pp. 145–148.

[11] Tirdad Sowlati and Domine M. W. Leenaerts, "A 2.4-GHz 0.18-μm CMOS Self-Biased Cascode Power Amplifier," *IEEE Journal of Solid-State Circuits*, Vol. 38, No. 8, August 2003, pp. 1318–1324.

[12] Lei Ding and G. Tong Zhou, "Effects of Even-Order Nonlinear Terms on Power Amplifier Modeling and Predistortion Linearization," *IEEE Transactions on Vehicular Technology*, Vol. 53, No. 1, January 2004, pp. 156–162.

[13] Biranchinath Sahu and Gabriel A. Rincón-Mora, "A High-Efficiency Linear RF Power Amplifier with a Power-Tracking Dynamically Adaptive Buck-Boost Supply," *IEEE Transactions of Microwave Theory and Techniques*, Vol. 52, No. 1, January 2004, pp. 112–120.

[14] Christian Fager, José Carlos Pedro, Nuno Borges de Carvalho, Herbert Zirath, Fernando Fortes, and Maria João Rosário, "A Comprehensive Analysis of IMD Behavior in RF CMOS Power Amplifiers," *IEEE Journal of Solid State Circuits*, Vol. 39, No. 1, January 2004, pp. 24–34.

[15] Joel L. Dawson and Thomas H. Lee, "Cartesian Feedback for RF Power Amplifier Linearization," *Proceeding of the 2004 American Control Conference*, Boston, 2004, pp. 361–366.

[16] Seung-Yup Lee, Yong-Sub Lee, Seung-Ho Hong, Hyun-Sik Choi, and Yoon-Ha Jeong, "An Adaptive Predistortion RF Power Amplifier with a Spectrum Monitor for Multicarrier WCDMA Applications," *IEEE Transactions on Microwave Theory and Techniques*, Vol. 53, No. 2, February 2005, pp. 786–793.

[17] Wangmyong Woo, Marvin D. Miller, and J. Stevenson Kenney, "A Hybrid Digital/RF Envelope Predistortion Linearization System for Power Amplifiers," *IEEE Transactions on Microwave Theory and Techniques*, Vol. 53, No. 1, January 2005, pp. 229–237.

[18] Fabien Lépine, Andreas Ådahl, and Herbert Zirath, "L-Band LDMOS Power Amplifiers Based on an Inverse Class-F Architecture," *IEEE Transactions on Microwave Theory and Techniques*, Vol. 53, No. 6, June 2005, pp. 2007–2012.

[19] Jongchan Kang, Daekyu Yu, Youngoo Yang, and Bumman Kim, "Highly Linear 0.18-µm CMOS Power Amplifier with Deep n-Well Structure," *IEEE Joural of Solid-State Circuits*, Vol. 41, No. 5, May 2006, pp. 1073–1080.

[20] Yu Zhao, Andre G. Metzger, Peter J. Zampardi, Masaya Iwamoto, and Peter M. Asbeck, "Linearity Improvement of HBT-Based Doherty Power Amplifiers Based on a Simple Analytical Model," *IEEE Transactions on Microwave Theory and Techniques*, Vol. 54, No. 12, December 2006, pp. 4479–4488.

[21] Magnus Isaksson, David Wisell, and Daniel Rönnow, "A Comparative Analysis of Behavioral Models for RF Power Amplifiers," *IEEE Transactions of Microwave Theory and Techniques*, Vol. 54, No. 1, January 2006, pp. 348–359.

PART II

DESIGN TECHNOLOGIES AND SCHEMES

CHAPTER 8

DIFFERENT METHODOLOGY BETWEEN *RF* AND DIGITAL CIRCUIT DESIGN

8.1 CONTROVERSY

For many years there has been an ongoing controversy between digital and *RF* circuit designers:

- *RF* circuit designers emphasize impedance matching. Digital circuit designers are indifferent.
- *RF* circuit designers are concerned with frequency response. Digital circuit designers are interested in the waveform, or "eye's diagram."
- *RF* circuit designers prefer to work in the frequency domain. Digital circuit designers like to work in the time domain.
- In the equipment budget, *RF* circuit designers opt for good network analyzers. Digital circuit designers prefer to buy the best oscilloscopes.
- *RF* circuit designers use the unit of dB_W. Digital circuit designers use the unit of dB_V.

Not only do the design methodologies differ; so do the technical terms.

- Digital circuit designers use the term "*AC* bypass" capacitors or "*DC* blocking" capacitors. *RF* circuit designers use the term "zero" capacitors.

It almost seems as if they were from two different planets. Even in some conferences and publications, these two "aliens" argue with each other. Each tries to prove that

RF Circuit Design, by Richard Chi-Hsi Li

his or her design methodology is superior to the other's. Eventually, nobody is the winner.

Let's outline their main controversies in the following sections.

8.1.1 Impedance Matching

The phrase "impedance matching" comes out of *RF* circuit designers' mouths almost every day. They were told by their supervisors that impedance matching is a "must" skill in circuit design. On the other hand, such terminology is never heard among the digital circuit designers. Their supervisors told them, "Ignore that 'foreign language.' Just focus on the 'eye diagram' or waveform."

It is not just digital circuit designers who ignore impedance matching. Even some *RF* circuit designers "discovered" something new in their "advanced" *RFIC* design. While it was necessary to take care of impedance matching in *RF* module design and in *RF* blocks built by discrete parts, where the incident and reflected power in the circuit really existed, they thought it meaningless to take care of "impedance matching" in an *RFIC* circuit design, because the size of an *IC* die is so small. Up to their assertions, in the *IC* realm the design methodology for the *RF* circuit should be more or less the same as that for the digital circuit. Since then, they have designed *RF* circuit blocks by the same method as that for digital circuit blocks. All of the individual *RF* blocks are simply crowded together since "impedance matching between the individual blocks is not necessary." Their design methodology for *RF* blocks is named the "combo" or "jumbo" design. Theoretically, they thought that all kinds of circuitry must obey Ohm's Law and follow *KCL* and *KCV* rules without exception. So, whence the difference in design methodology? From their viewpoint, it seems unnecessary to divide the circuit design team into an *RF* and a digital circuit design group.

RF circuit designers would be very happy if the impedance matching was unnecessary because it is the most difficult task in *RF* circuit design, especially in *RFIC* design for the *UWB* system. Unfortunately, the design experience indicates that the combo or jumbo design philosophy is absolutely wrong. For instance, without impedance matching, an *LNA* becomes a noisy attenuator or an oscillator in an *RFIC* chip. Without impedance matching, a mixer would become a "real" mixer indeed, blending all desired signals and undesired interference or noise together!

The key point to resolve in the controversy is whether the concept of voltage or power reflection is available in both *RF* and digital circuitry or not. Should the reflection of voltage or power not exist in a practical circuitry, the idea of a combo or jumbo design could be a correct design methodology. On the other hand, should the reflection of voltage or power exist in a practical circuitry, impedance matching would be important for power transportation or manipulation in a circuitry and then, the idea of combo or jumbo design would be an incorrect design methodology.

As a matter of fact, the existence of power or voltage reflection can be deduced from a rough analysis of an *RF* block. For example, without impedance matching, the insertion loss of a *LC* passive filter could be significant. However, if the *Q* values of the inductors or capacitors are high, the *LC* passive filter itself should not conceivably produce wear and tear on the power. This significant insertion loss demonstrates that quite a lot of power is reflected from the filter or load to the source. On the other hand, power or voltage reflection is not related to the size of the block but rather to the impedance matching status between source and load. A simple example illustrates the validity of such an assertion: Light is reflected from a mirror in the same way regardless of whether the light source is far from or close to the mirror.

8.1.2 The Key Parameter

This is a true story of a startup company researching and developing a wireless communication system. In spite of different opinions and various comments among his engineering teams, the engineering director asked both his *RF* and digital circuit design teams to work together for the design of a communication system. He ruled that **voltage** must be taken as the key parameter to measure the performance of every block, including both digital and *RF* blocks. In other words, the goal of input and output in every block, no matter *RF* or digital, must be specified with the voltage value. This engineering director hated the *RF* circuit designers' incessant gossip about power and impedance.

The engineers did try very hard to follow his instructions. There seemed to be no problem for the digital circuit blocks. However, the engineers were confused and didn't know how to specify the goals for *RF* blocks by voltage instead of power.

In the *RF* engineers' understanding, all of the parameters applied in *RF* circuit design, including gain, noise figure, IP_3, and IP_2, were expressed by power but not voltage. To follow the director's instructions, they spent a lot of time to convert all the parameters from power to voltage, since power was the traditional unit and showed in almost all the equipment readings. Sometimes the conversion was meaningless or uncertain. For instance, by the unit of voltage, *CNR* (Carrier to Noise Ratio) at the input of the demodulator was significantly dependent on the output impedance of the stage before the demodulator and the input impedance of the demodulator. Especially when the output impedance of the stage before demodulator and the input impedance of the demodulator were different, the conversion became impossible. Even more awkward, audiences at the presentation meeting held by this system design team couldn't understand why the values appearing in the system plan were surprisingly higher or lower than those values from other companies. Eventually, after learning of the extraordinary instructions given by the engineering director, those in the audience equipped with calculators could convert those values back from voltage to power.

In this system design team, the selection of a common key parameter for both *RF* and digital circuit designs became a hot topic. People argued without result, while the director still insisted on his original instructions. After a couple of weeks, the system design still hung in the air, and finally, for unknown reasons, the plan for the system design was wordlessly dropped. Some *RF* circuit designers were upset and left the company despite the director's exhortation: "Nothing is impossible!"

As a matter of fact, system design for a communication system must be divided into two parts: the digital and the *RF*. It is true that the key parameter in the digital circuit design is voltage or current. By means of voltage or current all the intermediate parameters can be characterized. However, the key parameter in the *RF* circuit design must be power or impedance. By means of power and impedance all the intermediate parameters in an *RF* circuit block can be characterized. Since impedance matching ensures the best performance of power transportation and manipulation in *RF* circuit blocks, impedance can be taken as the key parameter in an *RF* circuit design.

Why? The answer can be found in the following sections.

8.1.3 Circuit Testing and Main Test Equipment

In addition to the arguments about impedance matching and the key parameter, the difference between digital and *RF* circuit design can also be found in circuit testing and test equipment.

In a digital test lab, the test objective is always voltage and, occasionally, current. Of all the equipment available in a digital test lab, the oscilloscope is primary. It can sense the voltage at any node in the circuitry and display its "eye diagram," or waveform, on-screen, which characterizes the performance of a digital circuit intuitionally. In general, digital circuit designers prefer to analyze the circuitry in the time domain because responding speed is important to the performance of a digital circuit block.

In an *RF* test lab, the test objective is always power. Most *RF* test equipment, such as the spectrum analyzer, noise meter, and signal generator, measures the parameters of an *RF* circuit block in power but not in voltage. The main test equipment is the network analyzer. The performance of an *RF* circuit block can be characterized mainly by its frequency response on the network analyzer screen, expressed by power gain or loss in *dB*. The *RF* circuit designers prefer to analyze the circuitry in the frequency domain because coverage of bandwidth is important to the performance of an *RF* block.

In the test lab, testing a digital circuit block is somewhat easier than testing an *RF* circuit block. In testing for a digital circuit block, the probe of an oscilloscope is usually a sensor with high impedance. It does not disturb the circuit performance when it touches a node in the circuitry.

On the other hand, while using the network analyzer, the circuit designers may worry about the difference in circuit performance before and after the test equipment is connected to the desired test node, because the input and output impedance of the equipment is low, usually is $50\,\Omega$. In most cases it certainly will disturb the circuit performance.

Instead of voltage testing, *RF* circuit designers are concerned with power testing. Since all of the power testing must be conducted under a good impedance matching condition, the test equipment must be well calibrated. Unlike the testing for a digital circuit block by an oscilloscope, a buffer connected between the desired test node and the input of the network analyzer is not allowed because all the power tests for *RF* block must be conducted under the condition of impedance matching.

So far, the different methodology in *RF* and digital circuit design has been introduced only by the three main aspects noted above. More differences exist but will not be listed. We are going to focus on an explanation of where these differences come from.

8.2 DIFFERENCES BETWEEN *RF* AND DIGITAL BLOCKS IN A COMMUNICATION SYSTEM

8.2.1 Impedance

The input and output impedances of an *RF* circuitry are usually pretty low. In most cases, they are typically 50 Ω. On the other hand, the input and output impedances in a digital circuitry are usually pretty high. For example, the input and output impedances of an Op-Amp (Operating Amplifier) are mostly higher than $10\,k\Omega$.

The lower impedance in *RF* circuitry is beneficial to deliver power to a block or a part. It is well known that the power of a signal delivered to a block or a part with impedance Z can be expressed by

$$P = vi = \frac{v^2}{Z}, \tag{8.1}$$

where

P = power delivered to a block or a part,
v = voltage across the block or the part,
i = current flowing through the block or part,
Z = impedance of the block or the part.

For a given value of power, v^2 is proportional to Z. This implies that in order to deliver a given power to a block or a part, a higher voltage must be provided if its impedance is high. On the other hand, a lower voltage across the block is enough to deliver the same given power to a block or a part if its impedance is low. From the viewpoint of either cost or engineering design of the circuit, the application of lower voltage is much better than that of higher voltage. It is one of the reasons why the input and output impedance in the *RF* blocks is intentionally assigned to be low, because only a lower voltage is needed in order to deliver the same given power to a block or part with low impedance.

However, it is just the opposite for a digital signal. The higher impedance in digital circuitry is beneficial to the voltage swing in a digital block or part. For a given current, a higher impedance can have a higher voltage swing across a block or a part, and then the signal can ON/OFF the device more effectively, because

$$v = iZ. \tag{8.2}$$

The question is: Why is *RF* circuitry focused on power while digital circuitry is concerned about voltage?

8.2.2 Current Drain

In *RF* circuit blocks the current drains are usually in the order of mille-amperes while in digital circuit blocks they are usually in the order of micro-amperes. That is, the difference of the current drain magnitude between *RF* and digital circuit blocks is approximately 1,000.

In *RF* circuit blocks, it is desirable to increase the power of the *RF* signal as much as possible. This implies that higher current drains are preferred in *RF* circuit blocks because they are beneficial to deliver power to the block or the part for a given voltage.

In digital circuit blocks, it is desirable to reduce the power of the digital signal as much as possible. This implies that lower current drains are preferred in digital circuit blocks as long as the voltage swing is high enough.

Again, the question is: Why is *RF* circuitry focused on power while digital circuitry is concerned with voltage? The answer can be found in the next section.

8.2.3 Location

In a communication system, the demodulator is a remarkable demarcation in the receiver. As shown in Figure 8.1, before the demodulator, the blocks operate in the range of radio frequency so that they are called *RF* blocks. They are sometimes called the *RF* front end in the receiver, where *RF* circuit design is conducted. After demodulation, the blocks operate in the range of intermediate frequency or in the low digital data rate and are categorized as baseband blocks, or the digital/analog section. They are sometimes called the back end in the receiver, where digital/analog circuit design is conducted. The demodulator is a critical block, in which both digital and *RF* design technology are needed.

The order of blocks in the transmitter is just the opposite. Before the modulator, the blocks operate in the range of intermediate frequency or in the low digital data rate and are categorized as baseband blocks, or the digital/analog section. They are sometimes called the front end in the transmitter, where digital/analog circuit design is conducted. After the modulator, the blocks operate in the range of radio frequency so that they are called *RF* blocks and are sometimes called the *RF* back end in the transmitter, where *RF* circuit design is conducted. The modulator is also a critical block, in which both digital and *RF* design technology are needed.

A common feature can be seen in Figure 8.1, that is, in either the receiver or transmitter, the circuit portion closed to the antenna side contains *RF* blocks and another portion farther from the antenna side contains digital/analog circuit blocks.

In the receiver, the received modulated carrier is usually very weak. After it is power-magnified by the *LNA* and its frequency mixed down by the mixer, the modulated carrier can be demodulated only if its power is strong enough to suppress the noise power at the input of the de-modulator. Typically, the ratio of *RF* signal to noise power at the input of the demodulator is required to be more than $10\,dB$. It is therefore required that the *RF* signal be power-transported or power-operated before demodulation. After the de-modulator the digital-type message is demodulated from *RF* to base band. The digital signal is not required to be "power" transported but is only "status" or voltage transported between local blocks for digital signal processing. The voltage represents the status of signal. For the sake of power-saving, the power of the signal is reduced as much as possible. This is the answer to the questions: why voltage transportation or manipulation or the "status" transportation or manipulation is a unanimous task in the digital circuit design and why digital circuit designers are indifferent to the topic of impedance matching.

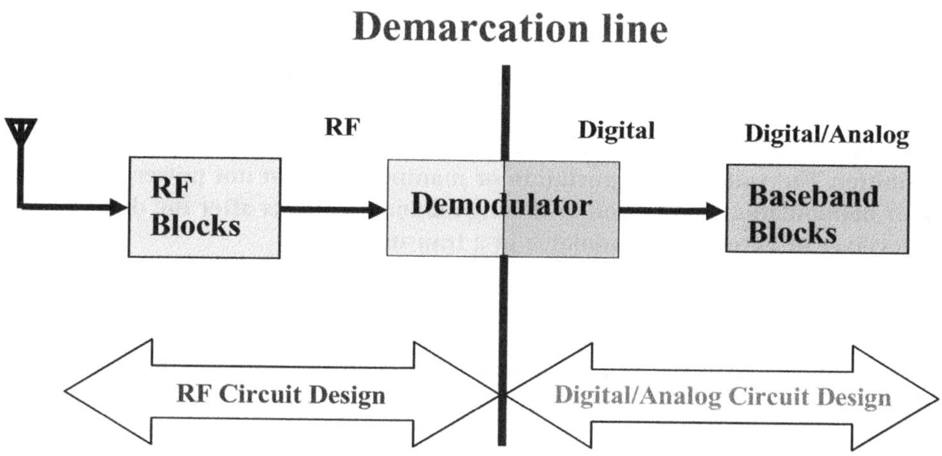

(a) Receiver

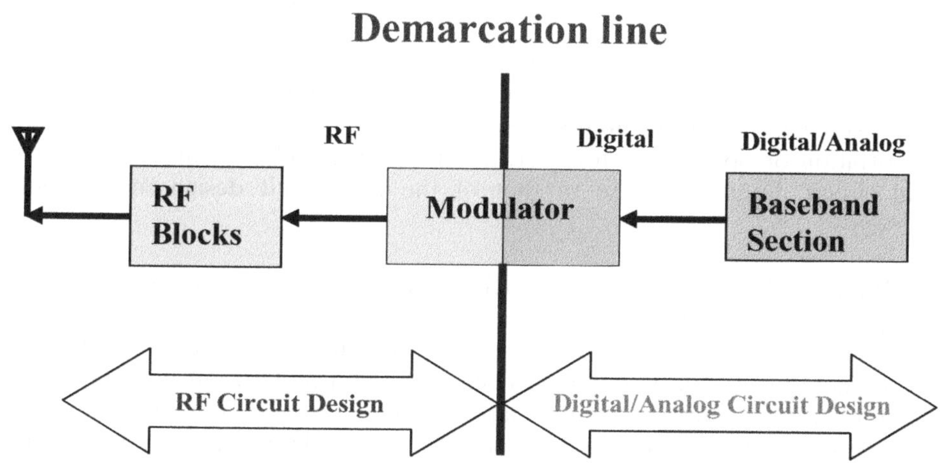

(b) Transmitter

Figure 8.1 Demarcation line in a communication system.

Similarly, in the transmitter, the digital signal is only required to reach the "modulation-effective status level" before the modulator. This signifies that the power or voltage of the input digital signal to the modulator can be as low as possible, as long as the input voltage or power reaches a level by which the carrier can be effectively modulated. In this case, the digital signal is transported or manipulated between the local circuit blocks and is not required to be power-transported but only voltage-transported. However, the modulated carrier after the modulator must be power-magnified and delivered to the antenna so that the modulated carrier is powerful enough to be propagated to a receiver located a long distance from the transmitter.

8.3 CONCLUSION

From the discussion in Section 8.2.3 it is concluded that: **The power transportation or manipulation but not the voltage transportation or manipulation is required in the *RF* blocks before the demodulator in a receiver and after the modulator in the transmitter. The voltage transportation or manipulation but not power transportation or manipulation is required in the digital/analog blocks after the demodulator in a receiver or before the modulator in a transmitter.**

This is the answer to the question in Sections 8.2.2 and 8.2.3: Why is *RF* circuitry focused on the power while the digital circuitry is concerned about voltage?

It can be seen that the controversy between digital and *RF* designers is unnecessary. **The difference in circuit design methodology arises from the different circuit design task.** The digital circuit is designed for voltage transportation or manipulation while the *RF* circuit is designed for power transportation or manipulation.

8.4 NOTES FOR HIGH-SPEED DIGITAL CIRCUIT DESIGN

In digital communication, the synonym of "high speed" is high data rate. In the case of high digital data rates, the tasks of both types of circuits are unchanged: the *RF* circuit is still working for the **power** while digital circuit is still working for the **status** transportation or operation. However, the design methodology in high data rate digital circuit designs is close to that of the *RF* circuit design methodology, because:

1) The remarkable variation of a digital circuit design from low speed to high speed is the change of the input and output impedance of a digital block. In the digital circuit design for low speeds, high input and output impedance of a digital block can be obtained from high input and output impedance of a transistor. The input and output capacitance of a transistor impacts the raising or dropping time of a pulse but does have a considerable impact on the input or output impedance. In the digital circuit design for high speeds, the input and output impedance of a transistor is reduced as its input or output capacitance impacts not only the raising or dropping time of a pulse but also has a significant impact on the input and output impedance of a digital block. The low input and output impedance leads to the necessity of impedance matching in the digital circuit design. Impedance matching is not only beneficial to power transportation but also to voltage transportation. Without impedance matching, reflected voltage will appear and interfere with the incident voltage. This brings about additional attenuation, additional jitter, an additional cross-talk, and eventually an additional bit error to the digital signal.
2) The traditional layout scheme is no longer reliable in digital circuits with high data rates. In the traditional layout for digital circuits with low data rates, the runners are always lined up in parallel so that the entire layout looks nice and neat. However, in the layout for digital circuits with high data rates, the runners lined up in parallel could cause appreciable co-planer capacitance and are coupled with each other; consequently, it may cause interference or

cross-talk. The layout for high-speed digital circuitry must be taken as seriously as for *RF* circuitry.

3) The traditional *AC* grounding scheme for a digital circuit with low data rate is no longer reliable for a digital circuit with high data rate. For high-speed digital circuitry the *AC* grounding must be taken as seriously as for *RF* circuitry.

4) In a digital circuitry with high data rate, isolation may become a serious problem. Usually a digital signal is a rectangular pulse while an *RF* signal is sinusoidal. The digital signal contains a wide-band spectrum while the *RF* contains a narrow-band spectrum. This implies that a digital circuit with a high data rate is easier to be interfaced with inside or outside interference sources because its frequency spectrum is much wider than that of a digital circuit with low data rate. Isolation between blocks becomes important and should be taken as seriously as for *RF* circuitry.

REFERENCES

[1] J. R. Turnbull, "Transistor Switching Circuits," February 21, 1963, University of Pennsylvania-University Museum, *International Solid State Circuits Conference, Digest of Technical Papers*, 1963, pp. 58–59.

[2] Enjun Xiao and J. S. Yuan, "RF Circuit Design In Reliability," 2003 *IEEE Radio Frequency Integrated Circuits Symposium*, 2003, pp. 575–578.

[3] Richard Chi-Hsi Li, *Key Issues in RF/RFIC Circuit Design*, Higher Education Press, Beijing, 2005.

CHAPTER 9

VOLTAGE AND POWER TRANSPORTATION

9.1 VOLTAGE DELIVERED FROM A SOURCE TO A LOAD

9.1.1 General Expression of Voltage Delivered from a Source to a Load

Figure 9.1 shows a voltage delivered from the source to the load and transported along a runner, either an *RF* cable, or a micro strip line with characteristic impedance Z_o. The impedances of the source and load are

$$Z_S = R_S + jX_S, \tag{9.1}$$

$$Z_L = R_L + jX_L, \tag{9.2}$$

where

Z_S = impedance of the source,
R_S = resistor of the source,
X_S = reactance of the source,
Z_L = impedance of the load,
R_L = resistor of the load,
X_L = reactance of the load,
Z_o = characteristic impedance of runner,
Γ_S = voltage reflection coefficients at source,
Γ_L = voltage reflection coefficients at load,
P_S = power available at source,
v_S = voltage available at source,

RF Circuit Design, by Richard Chi-Hsi Li

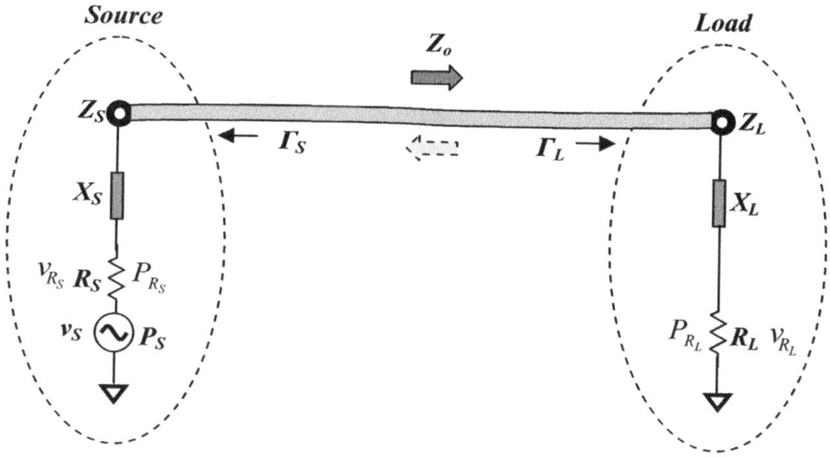

Figure 9.1 Voltage delivered from a source to a load.

P_{RS} = power on R_S,
v_{RS} = voltage across R_S,
P_{RL} = power on R_L,
v_{RL} = voltage across R_L.

In the expression (9.1) or (9.2), Xs is reactance of the source and X_L is reactance of the load.

Before the derivation of the general expression of the voltage delivered from a source to a load, it is first necessary to clarify the meaning of "delivered from a source to a load." As shown in Figure 9.1, the precise phrase must be "delivered from a source to the real part of a load," which does not include "the imaginary part of a load," because

(1) It is well known that the average power across X_S or X_L over one period of the *AC* signal is zero. In other words, an ideal capacitor or inductor experiences only a process of charging and discharging but never receives or consumes any net power.
(2) In practice, the attempt is to reduce the reactance of the source or load, X_S or X_L,and to have them neutralize each other as much as possible. In digital circuit design, the designers always try to select a device with less input capacitance, or try to neutralize it using an inductor or other method. In *RF* circuit design, the mutual neutralization of X_S or X_L is exactly one of the conditions of impedance conjugate matching.

Consequently, in the following discussion, the real meaning of "voltage delivered from a source to a load" is "voltage delivered from a source to the real part of a load."

According to transmission line theory, the voltage reflection coefficients at the source and load are

$$\Gamma_S = \frac{Z_S - Z_o}{Z_S + Z_o}, \tag{9.3}$$

$$\Gamma_L = \frac{Z_L - Z_o}{Z_L + Z_o}. \tag{9.4}$$

Usually Z_o is 50 Ω. And, in general,

$$Z_S \neq Z_o, \quad Z_L \neq Z_o, \tag{9.5}$$

the voltage suffers a reflection at the source or load terminal, that is,

$$\Gamma_S \neq 0, \quad \text{and} \quad \Gamma_L \neq 0. \tag{9.6}$$

The voltage delivered from source will be bounced back and forth between source and load. The resultant voltage on R_L is a sum of all the remaining voltages on R_L after every bounce back and forth of the voltage between source and load.

Let's assume that the source delivers a pure sinusoidal voltage,

$$v_S = v_{So}\, e^{j\omega t}, \tag{9.7}$$

where

v_{So} = amplitude of voltage,
ω = operating angular frequency,

the moment when the source delivers this voltage is

$$t = 0, \tag{9.8}$$

and the arrival time of the delivered voltage at the load is the delayed time, T_d.

Simultaneously, there are many other voltages arriving at the load, which were delivered at the time moments, $t = 0, -2T_d, -4T_d, -6T_d \ldots$ ago, where the negative sign indicates that it is past time. These voltages arrive at the load. Each voltage partially remains at the load and is partially bounced back to the source. The resultant voltage at the load when $t = T_d$, $v_{R_L}|_{t=T_d}$, is a sum of all the remaining voltages on R_L after the reflected voltage bounces back and forth between source and load, and can be expressed as follows:

$$v_{R_L}|_{t=T_d} = v_{So}\frac{R_L}{Z_S + Z_L}(1-\Gamma_L)\sum_{n=0}^{\infty} e^{j\omega[t-T_d+2nT_d]}(\Gamma_S\Gamma_L)^n. \tag{9.9}$$

or

$$v_{R_L}|_{t=T_d} = v_{So}\frac{R_L}{Z_S + Z_L}(1-\Gamma_L)e^{j\omega(t-T_d)} + v_{So}\frac{R_L}{Z_S + Z_L}(1-\Gamma_L)\sum_{n=1}^{\infty} e^{j\omega[t+(2n-1)T_d]}(\Gamma_S\Gamma_L)^n, \tag{9.10}$$

where

T_d = delayed time of the voltage traveling from source to load,
$v_{R_L}|_{t=T_d}$ = resultant remaining voltage on load.

The delayed time, $(t-T_d)$, is explicitly kept in every exponential factor $e^{j\omega(t-T_d+2nT_d)}$. The time interval, $2nT_d$, appearing in the exponential factor $e^{j\omega(t-T_d+2nT_d)}$, indicates the previous moment when the voltage term in expression was delivered from the source.

Expression (9.10) is exactly the same as expression (9.9). In expression (9.10) the first term in expression (9.9) is extracted from the general summation expression so as to intentionally distinguish it from others. The first term in expression (9.10) represents the the remaining voltage on R_L after the first reflection from the load. The second summation term in expression (9.10) represents the sum of the remaining voltage on R_L, including all the reflected voltages bounding back and forth between the source and the load in past time, which usually represents the interference due to the voltage reflection at load.

It should be noted that all the voltage terms in expression (9.9) and (9.10) arrive at the load at the same time $t = T_d$.

The first term on the right side of expression (9.9) and (9.10) represents the remaining voltage on R_L after the first reflection from the load with a delayed time T_d.

The second term on the right side of expression (9.9) and (9.10) represents the remaining voltage on R_L after the second reflection from the load with a delayed time $3T_d$. The additional factor, $(\Gamma_S\Gamma_L)$, represents reflected percentage after the voltage bounces back and forth once between source and load.

The third term on the right side of expression (9.9) and (9.10) represents the remaining voltage on R_L after the third reflection from the load with a delayed time $5T_d$. The additional factor, $(\Gamma_S\Gamma_L)^2$, represents reflected percentage after the voltage bounces back and forth twice between source and load.

And so on.

Figure 9.2 vividly depicts the voltages arriving at load simultaneously when $t = T_d$, but delivered from the source at different moments of t = ... 0, $-2T_d$, $-4T_d$, $-6T_d$, $-8T_d$ ... The number of voltages arriving at the load is infinite and it is therefore impossible to depict them totally, only eight are shown in Figure 9.2. They are sketched with

1. The source and load are located at the top-left and top-right respectively.
2. The distance from source to load is l, and the corresponding delay time for the delivered voltage moving from source to load is T_d,
3. The time axis is directed vertically downwards.
4. The voltages delivered from the source at different moments t = ... 0, $-2T_d$, $-4T_d$, $-6T_d$, $-8T_d$ ... are remarked in the left hand side of Figure 9.2. In each remark box the factor k_v denotes the portion of v_{So} which is delivered from source to R_L.
5. Finally, all of the delivered voltages from the source at different moments which arrive at the load simultaneously, are remarked with $v_{R_L}|_{t=T_d}$ in the bottom-right corner of Figure 9.2.

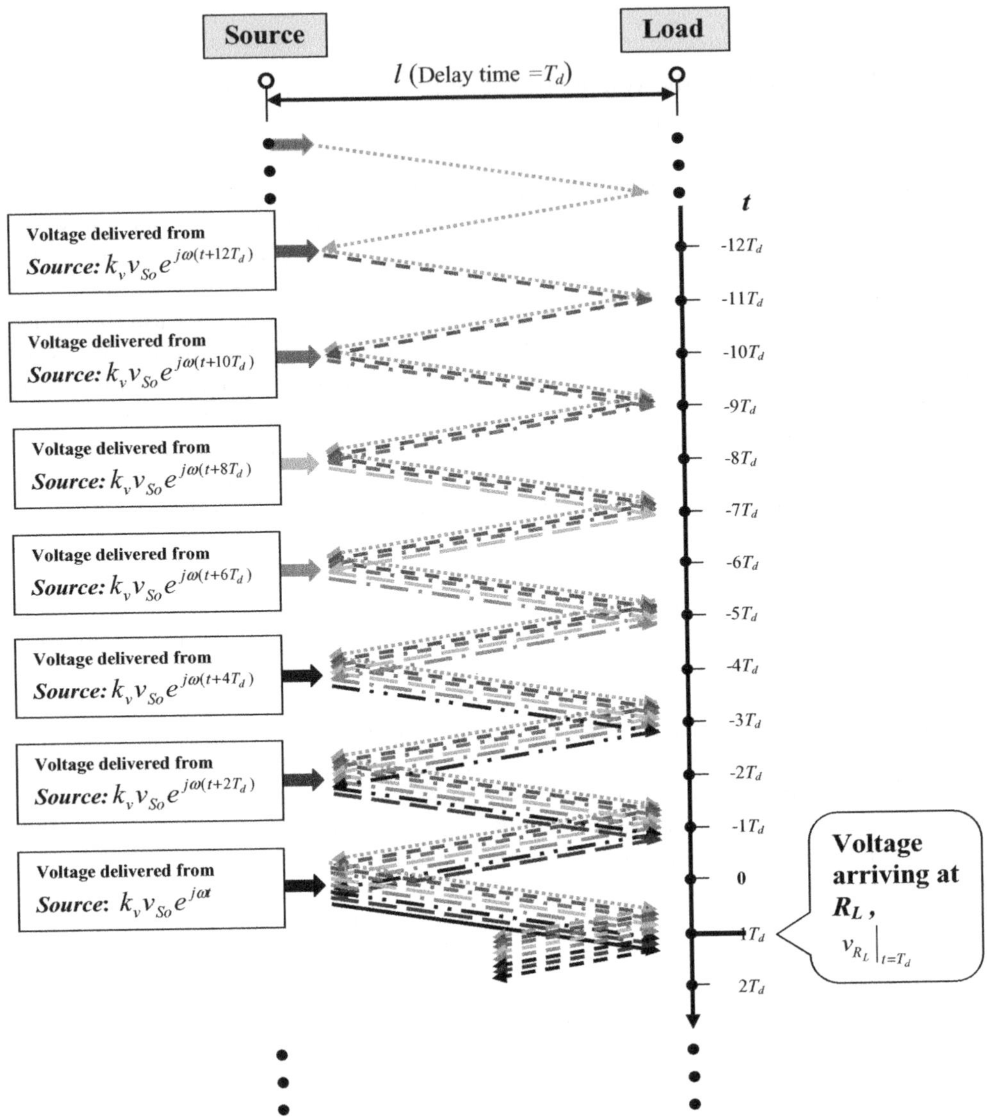

Figure 9.2 Voltages bouncing back and forth arrive at the load when $t = T_d$.
Note: $k_v = R_L/(Z_S + Z_L)$.

If

$$T_d \to 0, \tag{9.11}$$

Then the expression (9.9) becomes

$$v_{R_L}\big|_{t=T_d} = v_{So} e^{j\omega t} \frac{R_L}{Z_S + Z_L} (1 - \Gamma_L) \sum_{n=0}^{\infty} (\Gamma_S \Gamma_L)^n. \tag{9.12}$$

Note that

$$\sum_{n=0}^{\infty}(\Gamma_S\Gamma_L)^n = \frac{1}{1-\Gamma_S\Gamma_L}, \tag{9.13}$$

then

$$v_{R_L}\big|_{t=T_d} = v_{So}e^{j\omega t}\frac{R_L}{Z_S+Z_L}\frac{1-\Gamma_L}{1-\Gamma_S\Gamma_L}. \tag{9.14}$$

9.1.2 Additional Jitter or Distortion in a Digital Circuit Block

Now let's discuss the expression of the voltage transportation (9.12) in the cases when the approximation (9.11) is satisfied. Again, the expression (9.12) can be rewritten as

$$v_{R_L}\big|_{t=T_d} = v_{So}e^{j\omega t}\frac{R_L}{Z_S+Z_L}(1-\Gamma_L) + v_{So}e^{j\omega t}\frac{R_L}{Z_S+Z_L}(1-\Gamma_L)\sum_{n=1}^{\infty}(\Gamma_S\Gamma_L)^n. \tag{9.15}$$

As mentioned above, the first term on the right side of expression (9.15), represents the remaining voltage on R_L after the first reflection from the load with a delayed time T_d. It represents the desired voltage signal transported from the source to the load. The second summation term in expression (9.15) represents the sum of the remaining voltage on R_L, including all the reflected voltages bounding back and forth between the source and the load in the past time, which usually represents the interference due to the voltage reflection at load.

If the second summation term in expression (9.15) did not exist, the desired voltage signal would not be disturbed so that the distortion or jitter would not occur. In other words, the expression (9.15) corresponds to the following expression:

$$v_{R_L}\big|_{t=T_d} = v_{R_L}\big|_{t=T_d,1st} + v_{R_L}\big|_{t=T_d,srv}, \tag{9.16}$$

where

$v_{R_L}\big|_{t=T_d}$ = summed voltage across R_L,
$v_{R_L}\big|_{t=T_d,1st}$ = remaining voltage on R_L after the first reflection,
$v_{R_L}\big|_{t=T_d,srv}$ = sum of the reflected voltages remaining on R_L in the past time.

By comparing (9.15) with (9.16), we have

$$v_{R_L}\big|_{t=T_d,1st} = v_{So}e^{j\omega t}\frac{R_L}{Z_S+Z_L}(1-\Gamma_L), \tag{9.17}$$

$$v_{R_L}\big|_{t=T_d,sdv} = v_{So}e^{j\omega t}\frac{R_L}{Z_S+Z_L}(1-\Gamma_L)\frac{\Gamma_S\Gamma_L}{1-\Gamma_S\Gamma_L}, \tag{9.18}$$

The sum of the reflected voltages $v_{R_L}\big|_{t=T_d,srv}$ disturbs the incoming voltage $v_{R_L}\big|_{t=T_d,1st}$. The disturbance can be catalogued into two types: additional distortion or jitter, and additional interference. The additional distortion or jitter appears when the frequency of voltage is constant or when the sequential reflected voltage has the same frequency as that of the incoming voltage.

The additional distortion can be evaluated as the ratio of the sum of reflected voltages $v_{R_L}|_{t=T_d,srv}$ to the incoming voltage $v_{R_L}|_{t=T_d,1st}$, that is,

$$\Delta D_v|_{\%} = \frac{v_{R_L}|_{t=T_d,sdv}}{v_{R_L}|_{t=T_d,1st}} = \frac{\Gamma_S \Gamma_L}{1-\Gamma_S \Gamma_L}, \tag{9.19}$$

where $\Delta D_v|_{\%}$ = additional distortion in %.

Alternatively, assuming that the additional jitter has the same percentage as that of the additional distortion, that is,

$$\Delta(jitter)|_{\%} = \Delta D_v|_{\%} = \frac{\Gamma_S \Gamma_L}{1-\Gamma_S \Gamma_L}, \tag{9.20}$$

then the additional jitter can be evaluated as

$$\Delta(jitter)|_{\sec} = T \frac{v_{R_L}|_{t=T_d,sdv}}{v_{R_L}|_{t=T_d,1st}} = \frac{1}{f} \frac{\Gamma_S \Gamma_L}{1-\Gamma_S \Gamma_L}, \tag{9.21}$$

where

$\Delta(jitter)|_{\sec}$ = additional jitter in *sec.*,
$\Delta(jitter)|_{\%}$ = additional jitter in %,
$T = 1/f$ or $1/R$ = operating period,
f = operating frequency, or
R = digital data rate,
$\omega = 2\pi f$ = angular frequency.

Let's see how serious the additional distortion or jitter of a digital voltage would be, assuming that the data rate is 3.86 *Gbits/sec* or that the principal operating frequency is 3.86 *GHz*. From equations (9.19) to (9.21), the additional distortion and additional jitter can be calculated and is shown in Table 9.1.

It can be seen that when

$$\Gamma_L < 10\% \quad \text{and} \quad \Gamma_S < 10\%, \tag{9.22}$$

then

$$\Delta D|_{\%} = \Delta(jitter)_{\%} < 1.01\%, \tag{9.23}$$

$$\Delta(jitter)|_{\sec} < 2.6 ps, \tag{9.24}$$

In the cases of (9.22), the voltage reflection seems not too harmful to the digital voltage if impedance matching is ignored as usual. However, when

$$\Gamma_L > 10\%, \quad \text{and} \quad \Gamma_S > 10\%, \tag{9.25}$$

then

TABLE 9.1 Additional distortion and additional jitter in voltage transportation when $f = 3.86\,GHz$

Γ_S, %	Γ_L, %	ΔD, %	*f, GHz*	*T, ns*	*Jitter*, %	*Jitter, ps*
0	0	**0.00**	3.86	0.259	**0.00**	**0**
5	0	**0.00**	3.86	0.259	**0.00**	**0**
10	0	**0.00**	3.86	0.259	**0.00**	**0**
20	0	**0.00**	3.86	0.259	**0.00**	**0**
50	0	**0.00**	3.86	0.259	**0.00**	**0**
0	5	**0.00**	3.86	0.259	**0.00**	**0**
5	5	**0.25**	3.86	0.259	**0.25**	**0.6**
10	5	**0.50**	3.86	0.259	**0.50**	**1.3**
20	5	**1.01**	3.86	0.259	**1.01**	**2.6**
50	5	**2.56**	3.86	0.259	**2.56**	**6.6**
0	10	**0.00**	3.86	0.259	**0.00**	**0**
5	10	**0.50**	3.86	0.259	**0.50**	**1.3**
10	10	**1.01**	3.86	0.259	**1.01**	**2.6**
20	10	**2.04**	3.86	0.259	**2.04**	**5.3**
50	10	**5.26**	3.86	0.259	**5.26**	**13.6**
0	20	**0.00**	3.86	0.259	**0.00**	**0**
5	20	**1.01**	3.86	0.259	**1.01**	**2.6**
10	20	**2.04**	3.86	0.259	**2.04**	**5.3**
20	20	**4.17**	3.86	0.259	**4.17**	**10.8**
50	20	**11.1**	3.86	0.259	**11.1**	**28.8**
0	50	**0.00**	3.86	0.259	**0.00**	**0**
5	50	**2.56**	3.86	0.259	**2.56**	**6.64**
10	50	**5.26**	3.86	0.259	**5.26**	**13.6**
20	50	**11.1**	3.86	0.259	**11.1**	**28.8**
50	50	**33.3**	3.86	0.259	**33.3**	**86.4**

$$\Delta D|_{\%} = \Delta(jitter)_{\%} > 1.01\%, \tag{9.26}$$

$$\Delta(jitter)|_{\sec} > 2.6ps. \tag{9.27}$$

The voltage reflection becomes pernicious. When

$$\Gamma_L = 50\%, \quad \text{and} \quad \Gamma_S = 50\%, \tag{9.28}$$

then

$$\Delta D|_{\%} = \Delta(jitter)_{\%} = 33.3\%, \tag{9.29}$$

$$\Delta(jitter)|_{\sec} = 86.4ps. \tag{9.30}$$

The voltage reflection is horrible!

Similarly, the additional interference can also be evaluated on the basis of the expressions (9.17) and (9.18) when the frequency of incoming voltages is not the same as that of the reflected voltages in the past time. However, we are not going to do so. This analysis will remain as an exercise for the readers.

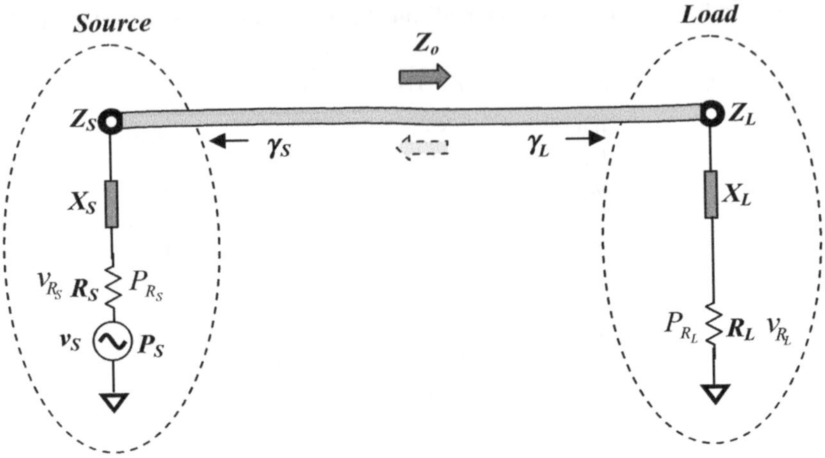

Figure 9.3 Power delivered from a source to a load.

9.2 POWER DELIVERED FROM A SOURCE TO A LOAD

9.2.1 General Expression of Power Delivered from a Source to a Load

Figure 9.3 shows the power delivered from a source to load. Similar to the discussion about "a voltage is delivered from a source to a load" above, as shown in Figure 9.3, the precise phrase must be "a power delivered from a source to the real part of a load," which does not include "the imaginary part of a load."

Let's assume that the delivered power suffers a reflection at the source or the load, that is,

$$\gamma_S \neq 0, \tag{9.31}$$

and

$$\gamma_L \neq 0. \tag{9.32}$$

where

γ_S = power reflection coefficient of source, and
γ_L = power reflection coefficient of load.

The relationships between the power reflection coefficients at source and load, γ_S and γ_L, and the voltage reflection coefficients at source and load, Γ_S and Γ_L, are

$$\gamma_S = \Gamma_S^2, \tag{9.33}$$

$$\gamma_L = \Gamma_L^2. \tag{9.34}$$

And the impedances of source and load are not equal to the characteristic impedance of the runner as shown in expression (9.5). The delivered power from the source will be bounced back and forth between source and load. The resultant power on R_L is a sum of all the remaining voltages on R_L after every bounce back and forth of the voltage between source and load.

Also, let's assume that the source delivers a pure sinusoidal voltage, $v_S = v_{So}e^{j\omega t}$, as shown in expression (9.7) and that the arrival time of delivered voltage at the load is the delayed time, T_d. The time reference point, $t = 0$, is defined as shown in expression (9.8) as the moment when the voltage is delivered from the source.

The general expression of power remaining on R_L is a sum of all the remaining powers on R_L after the reflected voltage bounces back and forth between source and load, and can be expressed as follows:

$$P_{R_L}|_{t=T_d} = v_{So}^2 \frac{R_L}{|Z_S+Z_L|^2}(1-\gamma_L)\sum_{n=0}^{\infty} e^{j2\omega[t-T_d+2nT_d]}(\gamma_S\gamma_L)^n, \tag{9.35}$$

or,

$$P_{R_L}|_{t=T_d} = v_{So}^2 \frac{R_L}{|Z_S+Z_L|^2}(1-\gamma_L)\sum_{n=0}^{\infty} e^{j2\omega[t+(2n-1)T_d]}(\gamma_S\gamma_L)^n, \tag{9.36}$$

where $P_{R_L}|_{t=T_d}$ is the resultant remaining power.

The delayed time, $(t - T_d)$, is explicitly kept in every exponential factor $e^{j2\omega(t-T_d+2nT_d)}$. The time interval, $2nT_d$, appearing in the exponential factor $e^{j2\omega(t-T_d+nT_d)}$, indicates the previous moment when the power term in expressions (9.35) and (9.36) was delivered from the source. It should be noted that all the power terms in expressions (9.35) and (9.36) arrive at the load at the same time, $t = T_d$.

The first term on the right side of expressions (9.35) and (9.36) represents the remaining power on R_L after the first reflection from the load with a delayed time T_d.

The second term on the right side of expressions (9.35) and (9.36) represents the remaining power on R_L after the second reflection from the load with a delayed time $3T_d$. The additional factor, $(\Gamma_S\Gamma_L)$, represents reflected percentage after the power bouncing back and forth once between source and load.

The third term on the right side of expressions (9.35) and (9.36) represents the remaining power on R_L after the third reflection from the load with a delayed time $5T_d$. The additional factor, $(\Gamma_S\Gamma_L)^2$, represents reflected percentage after the power bouncing back and forth twice between source and load.

And so on.

Similarly to Figure 9.2, Figure 9.4 vividly depicts the powers arrived at load simultaneously when $t = T_d$ but delivered from the source at different moment when $t = \ldots 0, -2T_d, -4T_d, -6T_d, -8T_d \ldots$ The number of powers arriving at load is infinite and it is therefore impossible to depict them totally, only eight are shown in Figure 9.4. They are sketched with

- The source and load are located at top-left and top-right respectively.
- The distance from source to load is l, and the corresponding delay time for the delivered power moving from source to load is T_d.
- The time axis is directed vertically downward.
- The powers delivered from the source at different moments when $t = \ldots 0, -2T_d, -4T_d, -6T_d, -8T_d \ldots$ are remarked in the left hand side of Figure 9.4. In each

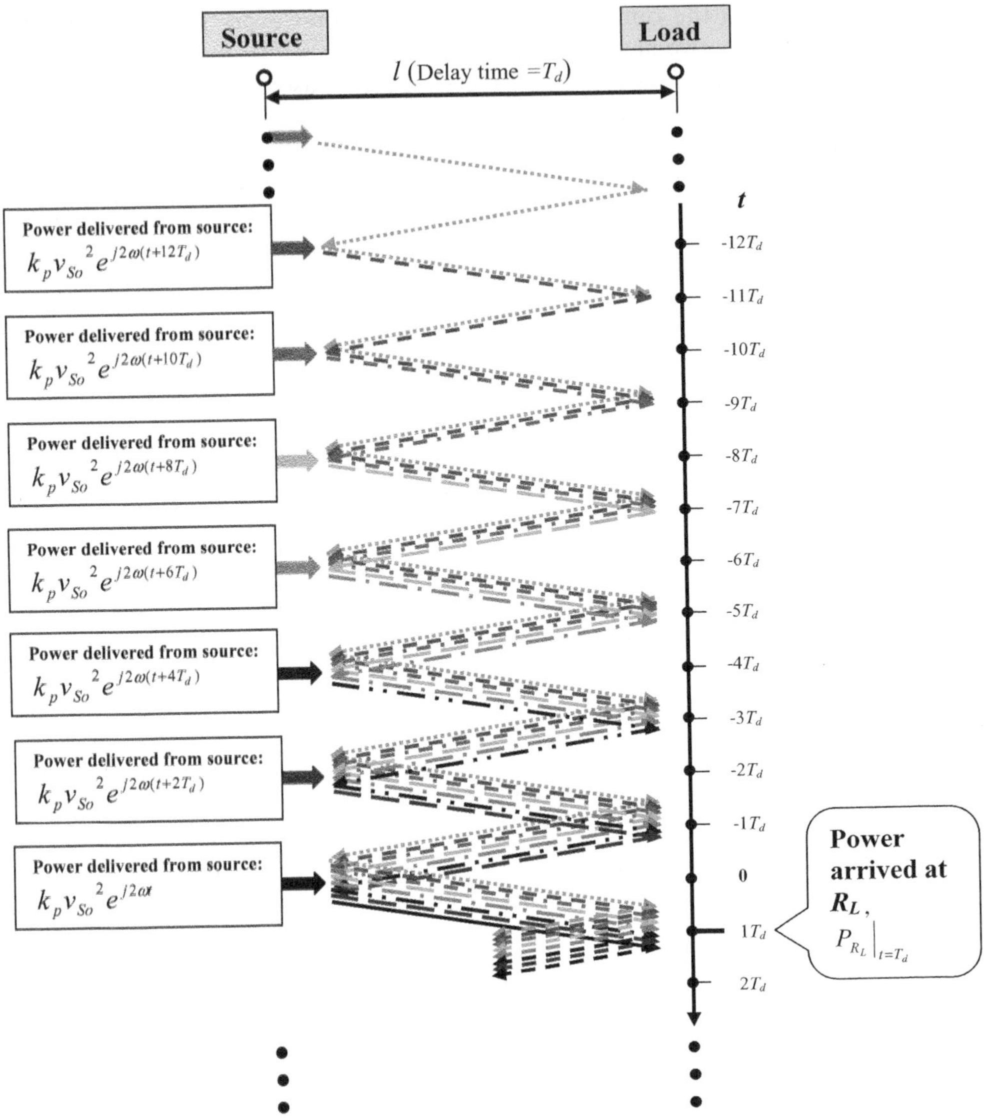

Figure 9.4 Powers bouncing back and forth arrive at the load when $t = T_d$.
Note: $k_p = R_L/|Z_S + Z_L|^2$.

remark box the factor k_p denotes the portion of power at the source which is delivered from source to R_L.

- Finally, all of the delivered powers from the source at different moments arrive at the load simultaneously and are remarked with $P_{R_L}|_{t=T_d}$ in the bottom-right corner of Figure 9.4.

Let's return to expression (9.35),

$$P_{R_L}|_{t=T_d} = v_{So}^2 \frac{R_L}{|Z_S + Z_L|^2}(1-\gamma_L)\sum_{n=0}^{\infty} e^{j2\omega[t-T_d+2nT_d]}(\gamma_S\gamma_L)^n, \tag{9.37}$$

If

$$T_d \to 0, \tag{9.38}$$

then

$$P_{R_L}|_{t=T_d} = v_{So}^2 e^{j2\omega t} \frac{R_L}{|Z_S + Z_L|^2} \frac{1-\gamma_L}{1-\gamma_S \gamma_L}, \tag{9.39}$$

or

$$P_{R_L}|_{t=T_d} = v_S^2 \frac{R_L}{|Z_S + Z_L|^2} \frac{1-\gamma_L}{1-\gamma_S \gamma_L}, \tag{9.40}$$

9.2.2 Power Instability

In the case when

$$\gamma_L \neq 0, \tag{9.41}$$

and

$$T_d \neq 0, \tag{9.42}$$

that is, when power reflection at the load exists and the delayed time is not negligible, the power delivered from the source to the load, P_{R_L}, is unstable. It can be found from expression (9.37) as long as T_d is not neglected. The power at the load is varied from time to time though the variation becomes smaller and smaller as the time elapses.

However, in most cases such a power instability can only be conceived in a very short period as long as T_d is a tiny time slot. This is especially true in integrated circuits where the length of the runner from the source to the load is usually much less than the corresponding quarter wavelength. Consequently, such a power instability could be ignored and thus expression (9.38) is a good approximation indeed.

9.2.3 Additional Power Loss

The power reflection at the load looks like the power delivered from the source to the load after experiencing an additional power loss. If there is no power reflection at the load, that is,

$$\gamma_L = 0, \tag{9.43}$$

then from expressions (9.39) and (9.40), we have

$$P_{R_L}|_{t=T_d, \gamma_L=0} = v_{So}^2 e^{j2\omega t} \frac{R_L}{|Z_S + Z_L|^2}, \tag{9.44}$$

or

$$P_{R_L}\big|_{t=T_d,\gamma_L=0} = v_S^2 \frac{R_L}{|Z_S+Z_L|^2}, \tag{9.45}$$

The difference between (9.39) or (9.40) and (9.44) or (9.45) is the additional power loss, that is,

$$\Delta P_{R_L}\big|_{t=T_d} = P_{R_L}\big|_{t=T_d} - P_{R_L}\big|_{t=T_d,\gamma_L=0} = -P_{R_L}\big|_{t=T_d,\gamma_L=0}\frac{(1-\gamma_S)\gamma_L}{1-\gamma_S\gamma_L}, \tag{9.46}$$

The relative additional power loss is

$$\frac{\Delta P_{R_L}\big|_{t=T_d}}{P_{R_L}\big|_{t=T_d,\gamma_L=0}} = -\frac{(1-\gamma_S)\gamma_L}{1-\gamma_S\gamma_L}. \tag{9.47}$$

The additional power attenuation factor, $(1-\gamma_s)\gamma_L/(1-\gamma_s\gamma_L)$, appearing in expression (9.47), is always less than 1 if $1 > \gamma_S > 0$ and $1 > \gamma_L > 0$.

As an example, Table 9.2 lists some calculated values of additional power loss due to the unmatched cases in terms of expressions (9.46) and (9.47). The special features of the additional power loss are:

TABLE 9.2 Additional power loss due to the unmatched case when $P_{R_L}\big|_{t=T_d,\gamma_L=0} = -30\,dBm$.

γ_S, %	γ_L, %	$\frac{(1-\gamma_S)\gamma_L}{(1-\gamma_S\gamma_L)}$	$P_{R_L}\big\|_{t=T_d,\gamma_L=0}$, dB_m	$\Delta P_{R_L}\big\|_{t=T_d}$, dB_m	$P_{R_L}\big\|_{t=T_d}$, dB_m
0	0	0.0000	**–30**	-infinitive	**–30**
5	0	0.0000	**–30**	-infinitive	**–30**
10	0	0.0000	**–30**	-infinitive	**–30**
20	0	0.0000	**–30**	-infinitive	**–30**
50	0	0.0000	**–30**	-infinitive	**–30**
0	5	0.0500	**–30**	–43.01	**–30.22**
5	5	0.0476	**–30**	–43.22	**–30.21**
10	5	0.0452	**–30**	–43.45	**–30.20**
20	5	0.0404	**–30**	–43.94	**–30.18**
50	5	0.0256	**–30**	–45.91	**–30.11**
0	10	0.1000	**–30**	–40.00	**–30.46**
5	10	0.0955	**–30**	–40.20	**–30.44**
10	10	0.0909	**–30**	–40.41	**–30.41**
20	10	0.0816	**–30**	–40.88	**–30.37**
50	10	0.0526	**–30**	–49.79	**–30.23**
0	20	0.2000	**–30**	–36.99	**–30.97**
5	20	0.1919	**–30**	–37.17	**–30.93**
10	20	0.1837	**–30**	–37.36	**–30.88**
20	20	0.1667	**–30**	–37.78	**–30.79**
50	20	0.1111	**–30**	–354	**–30.51**
0	50	0.5000	**–30**	–33.01	**–33.01**
5	50	0.4872	**–30**	–33.12	**–39.90**
10	50	0.4737	**–30**	–33.25	**–39.79**
20	50	0.4444	**–30**	–33.52	**–39.55**
50	50	0.3333	**–30**	–34.77	**–31.76**

- When $\gamma_L = 0$, there is no additional power loss.
- For a given value of γ_L, the additional power loss is somewhat reduced as the value of γ_S is increased.
- For a given value of γ_S, the additional power loss is somewhat increased as the value of γ_L is increased.
- The value of the additional power loss in Table 9.2 is $-3.01\,dB$ when $\gamma_S = 0$ and $\gamma_L = 50\%$.

The additional power loss due to the unmatched circuit design could seriously damage the performance of a communication or other system. For instance, a communication system with 64 QAM modulation would require a power accuracy between channels of less than 1 dB. The unmatched design for the RF circuit might be a serious killer in a dark corner.

9.2.4 Additional Distortion

Now let's discuss the general expression of the power transportation (9.37) in cases where the approximation (9.38) is satisfied. It is easy to understand that the sequentially reflected powers between the source and the load are overlapped over the incoming power and hence disturb the incoming power at the load. It is therefore desirable to distinguish the incoming power from all the sequentially reflected powers.

The summed powers arriving at the load consist of two parts. One is the remained power after first reflection. The second is the sequentially reflected powers between source and load which were delivered from the source at $t = -2T_d, -4T_d, -8T_d \ldots$ as shown in Figure 9.4. The negative time signifies that the reflected powers happened before the moment, $t = 0$. That is,

$$P_{R_L}\big|_{t=T_d} = P_{R_L}\big|_{t=T_d,1st} + P_{R_L}\big|_{t=T_d,srp}, \tag{9.44}$$

where

$P_{R_L}\big|_{t=T_d}$ = summed power on the load at the instant ($t = T_d$),

$P_{R_L}\big|_{t=T_d,1st}$ = remained power after the first reflection on the load at the instant ($t = T_d$),

$P_{R_L}\big|_{t=T_d,srp}$ = sum of reflected powers remained on the load at the instant ($t = T_d$).

Under the condition of (9.38), the expression (9.37) can be rewritten as

$$P_{R_L}\big|_{t=T_d} = v_{So}^2 e^{j2\omega t}\frac{R_L}{|Z_S+Z_L|^2}(1-\gamma_L) + v_{So}^2 e^{j2\omega t}\frac{R_L}{|Z_S+Z_L|^2}(1-\gamma_L)\sum_{n=1}^{\infty}(\gamma_S\gamma_L)^n. \tag{9.45}$$

The first term in the right hand side of expression (9.45) is the remained power on load after the first reflection. The second term in the right hand side of expression (9.46) is the sequentially reflected powers remained on R_L.

Note that

$$\sum_{n=1}^{\infty}(\gamma_S\gamma_L)^n = \sum_{n=0}^{\infty}(\gamma_S\gamma_L)^n - 1 = \frac{1}{1-\gamma_S\gamma_L} - 1 = \frac{\gamma_S\gamma_L}{1-\gamma_S\gamma_L}, \tag{9.46}$$

then, (9.45) becomes

$$P_{R_L}|_{t=T_d} = v_{So}^2 e^{j2\omega t}\frac{R_L}{|Z_S+Z_L|^2}(1-\gamma_L) + v_{So}^2 e^{j2\omega t}\frac{R_L}{|Z_S+Z_L|^2}(1-\gamma_L)\frac{\gamma_S\gamma_L}{1-\gamma_S\gamma_L}. \tag{9.47}$$

By comparing (9.44) with (9.47), we have

$$P_{R_L}|_{t=T_d,1st} = v_{So}^2 e^{j2\omega t}\frac{R_L}{|Z_S+Z_L|^2}(1-\gamma_L), \tag{9.48}$$

$$P_{R_L}|_{t=T_d,srp} = v_{So}^2 e^{j2\omega t}\frac{R_L}{|Z_S+Z_L|^2}(1-\gamma_L)\frac{\gamma_S\gamma_L}{1-\gamma_S\gamma_L}. \tag{9.49}$$

The sum of reflected powers $P_{R_L}|_{t=T_d,srp}$ disturbs the desired signal power $P_{R_L}|_{t=T_d,1st}$. The disturbance can be catalogued into two types: additional distortion and additional interference. The additional distortion appears when the frequency of power is constant or when the sum of the reflected powers in the past time has the same frequency as that of the incoming desired signal power. The additional interference appears when the frequency of power is not constant or when the sum of the reflected powers in the past time does not have the same frequency as that of the incoming desired signal power.

It should be noted that the "distortion of power" is meaningless. The distortion of voltage actually happened in reality. The additional distortion therefore is the square root of the ratio of $P_{R_L}|_{t=T_d,srp}$ to $P_{R_L}|_{t=T_d,1st}$, and can be calculated from (9.48) and (9.49), such as

$$\Delta D_p|_{\%} = \sqrt{\frac{P_{R_L}|_{t=T_d,srp}}{P_{R_L}|_{t=T_d,1st}}} = \sqrt{\frac{\gamma_S\gamma_L}{1-\gamma_S\gamma_L}}, \tag{9.50}$$

where $\Delta D_p|_{\%}$ = additional distortion in %.

Let's see how serious the additional distortion of a voltage would be. The additional distortion can be calculated and some values are shown in Table 9.3. From Table 9.3 it can be seen that

- In cases where $\gamma_L = 0$, there is no additional distortion. On the contrary, in the cases of $\gamma_L \neq 0$, the additional distortion is appreciable!
- The additional distortion is more sensitive to the value of γ_L than to the value of γ_S.
- For the given value of γ_L, the additional distortion is somewhat reduced as the value of γ_S is increased.
- For the given value of γ_S, the additional distortion is somewhat increased as the value of γ_L is increased.
- The highest value of the additional distortion in Table 9.3 is 70.71% when γ_S = 0 and γ_L = 50%.

TABLE 9.3 Additional distortion in power transportation

γ_S, %	γ_L, %	ΔD_p, %	γ_S, %	γ_L, %	ΔD_p, %
0	0	**0.00**	20	10	**28.57**
5	0	**0.00**	50	10	**29.94**
10	0	**0.00**	0	20	**44.72**
20	0	**0.00**	5	20	**43.81**
50	0	**0.00**	10	20	**49.86**
0	5	**29.36**	20	20	**40.82**
5	5	**21.82**	50	20	**33.33**
10	5	**21.27**	0	50	**70.71**
20	5	**20.10**	5	50	**68.82**
50	5	**16.01**	10	50	**68.00**
0	10	**31.62**	20	50	**66.67**
5	10	**30.90**	50	50	**57.74**
10	10	**30.15**			

9.2.5 Additional Interference

Now let's return to the expressions (9.48) and (9.49) in Section 9.2.4.

$$P_{R_L}|_{t=T_d,1st} = v_{So}^2 e^{j2\omega t}\frac{R_L}{|Z_S+Z_L|^2}(1-\gamma_L), \tag{9.48}$$

$$P_{R_L}|_{t=T_d,srp} = v_{So}^2 e^{j2\omega t}\frac{R_L}{|Z_S+Z_L|^2}(1-\gamma_L)\frac{\gamma_S\gamma_L}{1-\gamma_S\gamma_L}, \tag{9.49}$$

As mentioned above, the sum of the reflected powers in the past time disturb the desired signal power on the load. The additional interference appears when the frequency of power is not constant or when the sum of the reflected power in the past time does not have the same frequency as that of the desired signal power.

Consequently, the additional interference can be evaluated as the ratio of the sum of the reflected powers in the past time $P_{R_L}|_{t=T_d,srp}$ to the desired signal power $P_{R_L}|_{t=T_d,1st}$.

$$S_{R_L}|_{t=T_d} = P_{R_L}|_{t=T_d,1st} = v_{So}^2 e^{j2\omega t}\frac{R_L}{|Z_S+Z_L|^2}(1-\gamma_L), \tag{9.51}$$

$$\Delta I_{R_L}|_{t=T_d} = P_{R_L}|_{t=T_d,srp} = V_{So}^2 e^{j2\omega_2(t-T_d)}\frac{R_L}{|Z_S+Z_L|^2}(1-\gamma_L)\frac{\gamma_S\gamma_L}{1-\gamma_S\gamma_L}, \tag{9.52}$$

where

$S_{R_L}|_{t=T_d}$ = desired signal power arrived on the load at the instant ($t = T_d$),
$\Delta I_{R_L}|_{t=T_d}$ = additional interference power on the load at the instant ($t = T_d$),

From (9.51) and (9.52), we have

$$\frac{\Delta I_{R_L}|_{t=T_d}}{S_{R_L}|_{t=T_d}} = \frac{\gamma_S \gamma_L}{1-\gamma_S \gamma_L}. \tag{9.53}$$

The original ratio of signal to interference without additional interference, SIR_{R_Lo} is

$$SIR_{R_Lo}|_{t=T_d} = \frac{S_{R_L}|_{t=T_d}}{I_{R_L}|_{t=T_d}}. \tag{9.54}$$

From expressions (9.53) and (9.54), the ratio of signal to interference becomes

$$SIR_{R_L} = \frac{S_{R_L I}}{I_{R_L 1} + \Delta I_{R_L 1}} = \frac{SIR_{R_Lo}}{1 + \dfrac{\Delta I_{R_L 1}}{I_{R_L 1}}} = \frac{SIR_{R_Lo}}{1 + \dfrac{\gamma_S \gamma_L}{1-\gamma_S \gamma_L} SIR_{R_Lo}}, \tag{9.55}$$

Finally,

$$SIR_{R_L} = \frac{SIR_{R_Lo}}{1 + \dfrac{\gamma_S \gamma_L}{1-\gamma_S \gamma_L} SIR_{R_Lo}}, \tag{9.56}$$

where SIR_{R_L} = resulting ratio of signal to interference with additional interference due to unmatched impedance.

Table 9.4 lists some calculated ratios of signal to interference due to the unmatched impedance. From the table it can be seen that

- In cases where $\gamma_L = 0$, there is no additional interference, so the SIR remains unchanged.
- The additional interference is more sensitive to the value of γ_L than to the value of γ_S,
- For a given value of γ_L, the additional interference increases as the value of γ_S increases so that the SIR is reduced.
- For a given value of γ_S, the additional interference increases as the value of γ_L increases so that the SIR is reduced.
- The highest value of the additional interference in Table 9.4 is reached when $\gamma_S = 50\%$ and $\gamma_L = 50\%$. At this point the SIR drops from $15\,dB$ to $4.38\,dB$.

9.3 IMPEDANCE CONJUGATE MATCHING

From the discussions in Sections 9.1 and 9.2, it can be seen that the voltage or power reflection is very harmful to the performance of both digital and RF circuit blocks. Both voltage and power reflection are due to impedance unmatched condition existing in the source or the load. Impedance matching therefore becomes the key in all circuit designs, including RF, digital, and analog.

TABLE 9.4 Calculated ratio of signal to interference as the reflection coefficient, γ, is varied

γ_S, %	γ_L, %	$SIR_{R_{Lo}}$ dB	$SIR_{R_{Lo}}$ W	$\gamma_S\gamma_L$ %	$\gamma_S\gamma_L(1-\gamma_S\gamma_L)$ W	$1+\frac{\gamma_S\gamma_L}{1-\gamma_S\gamma_L}SIR_{R_{Lo}}$ W	SIR_{R_L} dB
0	**0**	**15**	31.62	0.0000	0.0000	1.00	**15.00**
5	**0**	**15**	31.62	0.0000	0.0000	1.00	**15.00**
10	**0**	**15**	31.62	0.0000	0.0000	1.00	**15.00**
20	**0**	**15**	31.62	0.0000	0.0000	1.00	**15.00**
50	**0**	**15**	31.62	0.0000	0.0000	1.00	**15.00**
0	**5**	**15**	31.62	0.0000	0.0000	1.00	**15.00**
5	**5**	**15**	31.62	0.0025	0.0025	1.08	**14.67**
10	**5**	**15**	31.62	0.0050	0.0050	1.16	**14.36**
20	**5**	**15**	31.62	0.0100	0.0101	1.32	**13.80**
50	**5**	**15**	31.62	0.0250	0.0256	1.81	**12.42**
0	**10**	**15**	31.62	0.0000	0.0000	1.00	**15.00**
5	**10**	**15**	31.62	0.0050	0.0050	1.16	**14.36**
10	**10**	**15**	31.62	0.0100	0.0101	1.32	**13.80**
20	**10**	**15**	31.62	0.0200	0.0204	1.65	**12.84**
50	**10**	**15**	31.62	0.0500	0.0526	2.66	**10.74**
0	**20**	**15**	31.62	0.0000	0.0000	1.00	**15.00**
5	**20**	**15**	31.62	0.0100	0.0101	1.32	**13.80**
10	**20**	**15**	31.62	0.0200	0.0204	1.65	**12.84**
20	**20**	**15**	31.62	0.0400	0.0417	2.32	**11.35**
50	**20**	**15**	31.62	0.1000	0.1111	4.51	**8.45**
0	**50**	**15**	31.62	0.0000	0.0000	1.00	**15.00**
5	**50**	**15**	31.62	0.0250	0.0256	1.81	**12.42**
10	**50**	**15**	31.62	0.0500	0.0526	2.66	**10.74**
20	**50**	**15**	31.62	0.1000	0.1111	4.51	**8.45**
50	**50**	**15**	31.62	0.2500	0.3333	11.54	**4.38**

9.3.1 Maximization of Power Transportation

First, let's discuss the case of power delivered from a source to a load without reflection, as shown in Figure 9.5,

$$\gamma_S = 0, \tag{9.57}$$

$$\gamma_L = 0. \tag{9.58}$$

the voltage and power delivered from source to the real part of the load, R_L, can be expressed by

$$v_{R_L}(t) = \frac{v_S}{Z_S + Z_L} R_L, \tag{9.59}$$

$$P_{R_L}(t) = v_S^2 \frac{R_L}{|Z_S + Z_L|^2} = v_S^2 \frac{R_L}{(R_S + R_L)^2 + (X_S + X_L)^2}. \tag{9.60}$$

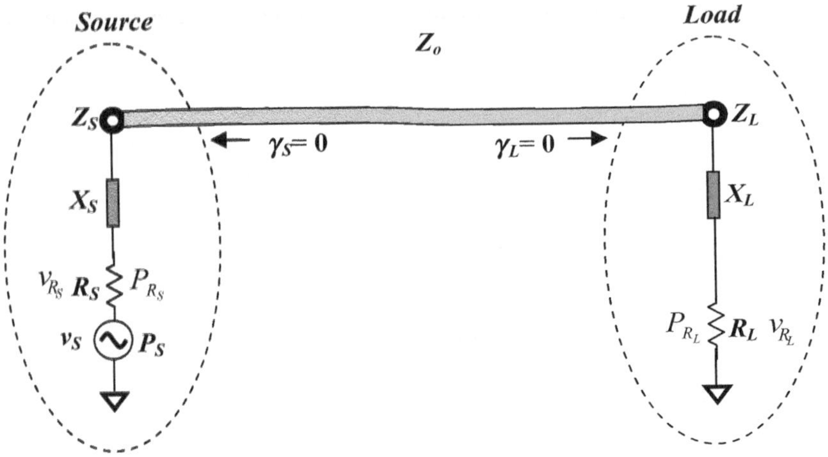

Figure 9.5 Power delivered from a source to a load without reflection.

It is desirable to achieve maximum power transportation from the source to the load resistor. From expression (9.60) it is easy to see that the maximum will be reached if the denominator in the expression (9.60) is at a minimum. The two terms in the denominator cannot be negative, but must be positive or zero, because they are the square term of real values. Theoretically, the smallest possible minimum is zero. The first term in the denominator is the resistors' value which cannot be zero. However, the second term in denominator can be artificially forced to equal zero, that is

$$X_S + X_L = 0,$$

or,

$$X_S = -X_L. \tag{9.61}$$

The relation (9.61) indicates that the reactance of the source and the load must have equal magnitude but opposite sign. It implies that the load reactance, X_L, must be inductive if the source reactance, X_S, is capacitive, and vice versa. Under the relation of (9.61), equation (9.60) becomes

$$P_{R_L} = v_S^2 \frac{R_L}{(R_S + R_L)^2}. \tag{9.62}$$

Now another relation to achieve maximum power transportation from the source to the load resistor can be found from the equation (9.62). Mathematically, by partially differentiating (9.62) in respect to R_L,

$$\frac{\partial P_{R_L}}{\partial R_L} = v_S^2 \left[\frac{1}{(R_S + R_L)^2} - 2\frac{R_L}{(R_S + R_L)^3} \right] = v_S^2 \frac{R_S - R_L}{(R_S + R_L)^3}. \tag{9.63}$$

The condition to maximize P_{R_L} is

$$\frac{\partial P_{R_L}}{\partial R_L} = 0. \tag{9.64}$$

From relations (9.62) and (9.63), we have

$$R_S = R_L \tag{9.65}$$

By combining the relations (9.61) and (9.65) we obtain

$$Z_S^* = Z_L, \quad \text{or} \quad Z_S = Z_L^*. \tag{9.66}$$

The relation (9.66) is called the condition of impedance conjugate matching or simply the condition of impedance matching.

9.3.2 Power Transportation without Phase Shift

The condition (9.61) indicates that when the reactance of source and load, X_S and X_L, are neutralized from each other, then the schematic shown in Figure 9.5 can be re-drawn as that shown in Figure 9.6. In the entire loop from source to load, there are only resistive parts but no parts with reactance. This implies that there is no phase shift when the voltage or power is transported from the source to the load, that is,

$$\angle v_L - \angle v_S = 0,$$

or,

$$\angle v_L = \angle v_S. \tag{9.67}$$

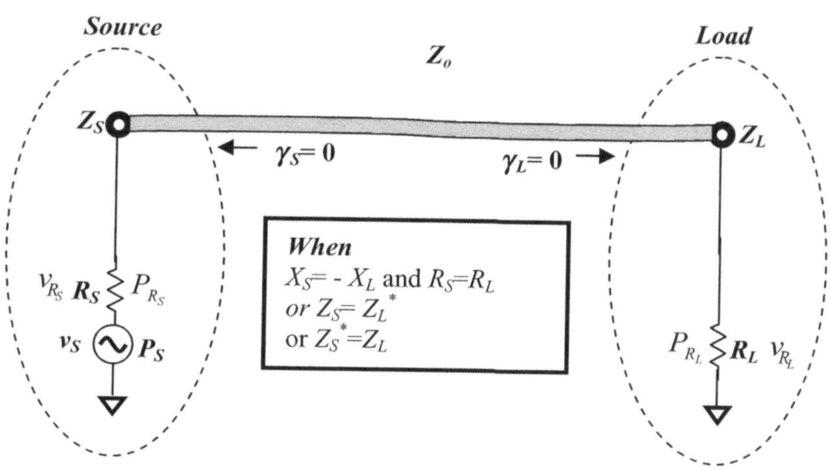

Figure 9.6 Power delivered from a source to a load without reflection when $X_S = -X_L$ and $R_S = R_L$.

where the symbol "$\angle x$" denotes angle of x.

Summarily, the significance of impedance conjugate matching is twofold:

1. **To maximize the power transportation from source to load.**
2. **To transport the voltage from source to load without phase shift.**

The second point, voltage or power transportation without phase shift, is another important feature of impedance conjugate matching. Unfortunately, it is very often ignored or mysteriously "swallowed" in the discussion about impedance conjugate matching.

In the past few decades modulation technology has been developing with great leaps, in which the phase modulation plays an important role in the effective utilization of the frequency bandwidth. Figure 9.7 illustrates the progress of modulation technology in a communication system. Historically, *PSK* (Phase Shift-Keying) technology has been developed from *BPSK* to *QPSK*, 8*PSK*, 16PSK etc. In a 16*PSK* system as shown in Figure 9.7(d), the minimum phase difference between two adjacent symbols is only 22.5°. This is somewhat difficult for the circuit designer. Instead of the *PSK* system, the *QAM* (Quadrant Amplitude Modulation) system enhances

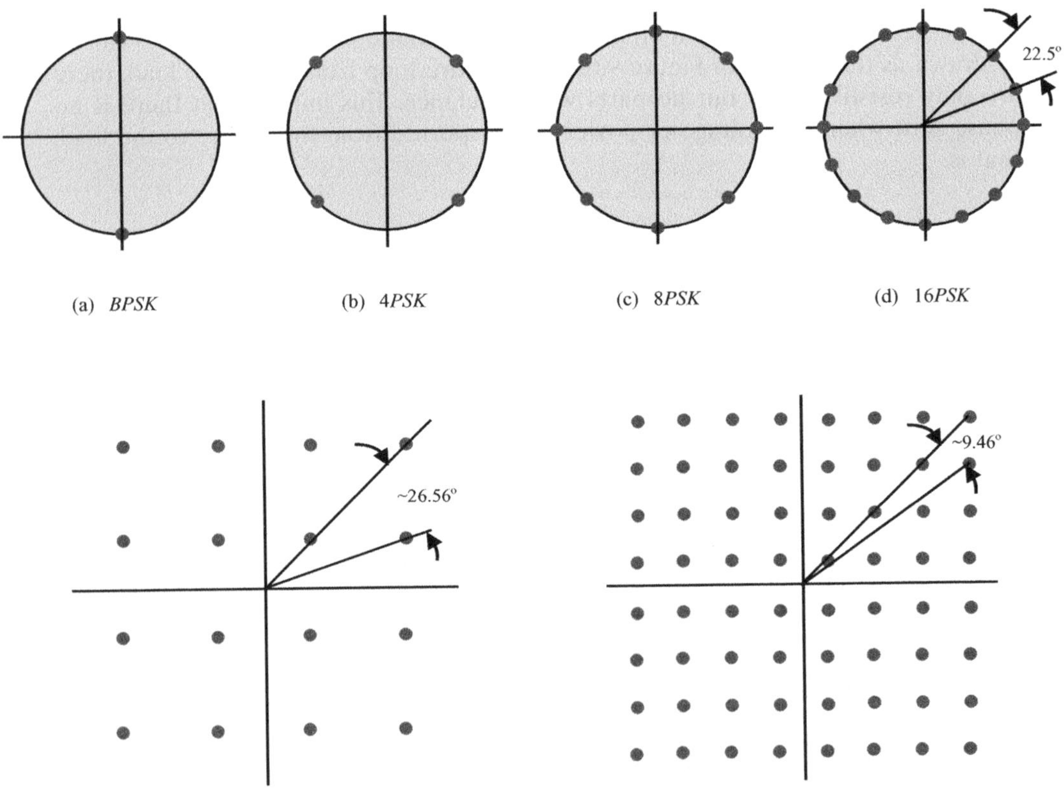

(a) *BPSK* (b) 4*PSK* (c) 8*PSK* (d) 16*PSK*

(e) 16*QAM* (f) 64*QAM*

Figure 9.7 Progress of modulation technology from *PSK* to *QAM*.

the resolution to recognize two adjacent symbols considerably. In a *QAM* system, both amplitude and phase are applied for allocation of a symbol, so that the system recognizes the symbol not only by phase but also by amplitude. It releases the phase difference between two adjacent symbols. In a 16*QAM* system, the minimum phase difference between two adjacent symbols is 26.56° as shown in Figure 9.7(e) which is greater than 22.5° in a 16*PSK* system as shown in Figure 9.7(d). In order to further enhance utilization of bandwidth, the 64*QAM* system was developed after the 16*QAM* system. It, of course, is a more sophisticated system but it brings about a 9.46° minimum of phase difference between two adjacent symbols, which is a big challenge to *RF* circuit designers!

Now, let's return to impedance conjugate matching.

It is clear from the discussion above that the phase shift exists in the loop from the source to load if the impedance of source and load does not match well. If the phase shift is unstable and varied, it puts the designed *RF* block or system inoperable. As mentioned above, the minimum phase difference in a 64*QAM* system is only 9.46°, which is the upper limit of phase shift tolerance for the entire system. For an individual *RF* block, the tolerance of the phase shift variation must be much less than 9.46°. It can therefore be concluded that impedance matching is absolutely necessary in *RF* circuit design.

Under the condition of impedance conjugate matching (9.66), the maximum power at the load can be obtained. By substituting equation (9.66) into (9.59) and (9.60), we have

$$v_{R_L}|_{Z_S=Z_L^*} = v_{R_L,\max} = \frac{v_S}{2}, \tag{9.68}$$

$$P_{R_L}|_{Z_S=Z_L^*} = P_{R_L,\max} = \frac{v_S^2}{4R_L}. \tag{9.69}$$

Similarly to expressions (9.59) and (9.60), the voltage, v_{R_S}, and its corresponding power, P_{R_S} on the source resistor, R_S, can be expressed as follows:

$$v_{R_S} = \frac{v_S}{Z_S + Z_L} R_S \tag{9.70}$$

$$P_{R_S}(t) = v_S^2 \frac{R_S}{|Z_S + Z_L|^2} = v_S^2 \frac{R_S}{(R_S + R_L)^2 + (X_S + X_L)^2}. \tag{9.71}$$

By substituting the impedance conjugate-matching condition (9.66) into (9.70) and (9.71), we have,

$$v_{R_S}|_{Z_S=Z_L^*} = v_{R_S,\max} = \frac{v_S}{2}, \tag{9.72}$$

$$P_{R_S}|_{Z_S=Z_L^*} = P_{R_S,\max} = \frac{v_S^2}{4R_S}. \tag{9.73}$$

From equations (9.68), (9.69), (9.72) and (9.73), it can be concluded that under the condition of impedance conjugate matching (9.66):

- The power delivered to the load reaches a maximum.
- The power delivered to the load is equal to the power remaining on the source.

The corresponding equations are:

$$v_{R_L,\max} = v_{R_S,\max} = \frac{v_S}{2}, \tag{9.74}$$

$$P_{R_L,\max} = P_{R_S,\max} = \frac{v_S^2}{4R_L} = \frac{v_S^2}{4R_S}. \tag{9.75}$$

The total available power P_S in the loop from source to load is the sum of the powers on R_S and R_L, that is

$$P_S = P_{R_S} + P_{R_L}, \tag{9.76}$$

then,

$$P_{R_L} = P_{R_S} = \frac{1}{2} P_S. \tag{9.77}$$

The total power is divided by two. One half remains on R_S and the other half is delivered to the load R_L.

9.3.3 Impedance Matching Network

Usually the impedance of a source does not conjugate match with the impedance of the load, that is,

$$Z_S \neq Z_L^*. \tag{9.78}$$

In order to maximize power transportation without phase shift from the source to the load, the impedance conjugate matching condition must be satisfied. Therefore, an impedance matching network must be inserted between the source and the load. As shown in Figure 9.8, the input impedance of the impedance matching network must be equal to Z_S^* and the output impedance of the impedance matching network must be equal to Z_L^*, that is,

$$Z_{in} = Z_S^*, \tag{9.79}$$

$$Z_{out} = Z_L^*. \tag{9.80}$$

As a matter of fact, the entire system shown in Figure 9.8 can be seen as three sub-impedance matching loops.

The first sub- impedance matching loop can be dawn in Figure 9.9.

The new source is the same as the old source shown in Figure 9.8. The new load consists of the impedance matching network and the old load shown in Figure 9.8. As shown in expression (9.79), the impedance of the new source is conjugate-

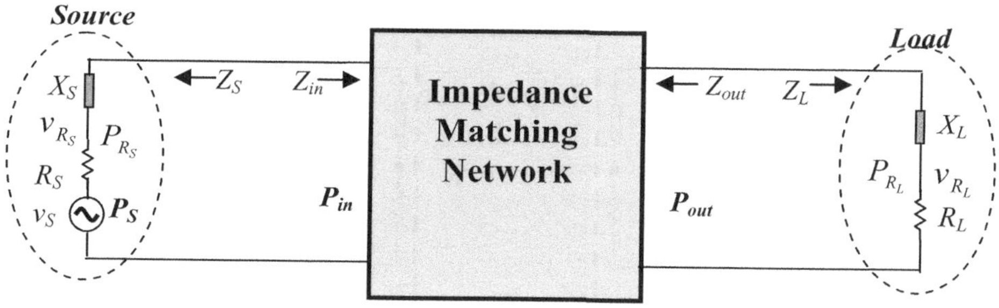

Figure 9.8 An impedance matching network is inserted between source and load when $Z_S \neq Z_L^*$.

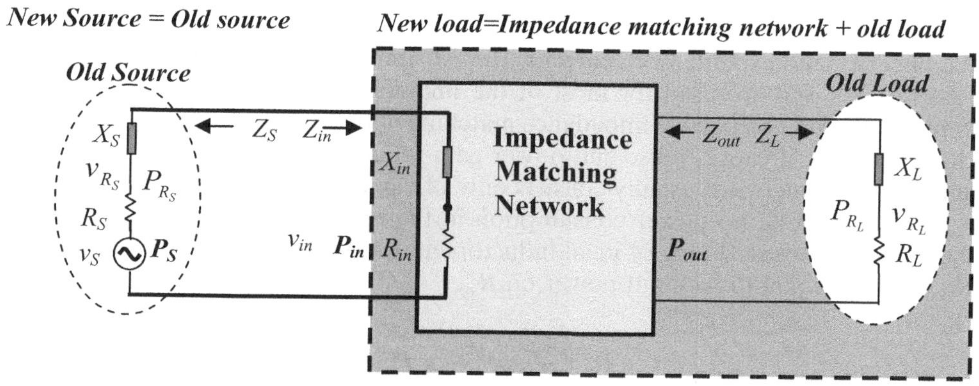

Figure 9.9 First sub-impedance matching loop: new source = old source, new load = impedance matching network + old load.

matched with that of the new load. Consequently, as shown in expression (9.75), the power delivered to R_{in} is equal to the power remaining on the R_S, that is,

$$P_{in} = \frac{v_{in}^2}{R_{in}} = \frac{v_S^2}{4R_{in}} = \frac{v_S^2}{4R_S} = \frac{v_{R_S}^2}{R_S} = P_{R_S}, \tag{9.81}$$

where

P_{in} = virtual or equivalent power resulting from the combination of the load and the impedance matching network,

R_{in} = virtual or equivalent resistance resulting from the combination of the load and the impedance matching network.

The second sub-impedance matching loop is the impedance matching network it self and is shown in Figure 9.10.

Theoretically, an impedance matching network could be constructed by either passive or active parts. The emitter follower, the source follower, and the buffer are examples of active impedance matching networks. However, implementation of an impedance matching network by active parts would increase noise and cost, degrade

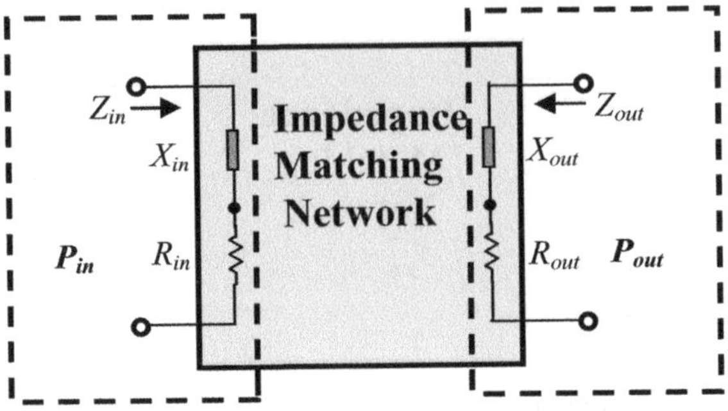

Figure 9.10 Second sub-impedance matching loop: the impedance matching network itself.

the linearity, and expend more current. The LC impedance matching networks are passive ones, and are used for most of the impedance matching networks. In the implementation of a passive impedance matching network, the resistor is not a good choice because it causes noise and power gain reduction. Consequently an impedance matching network usually consists only of capacitors and inductors.

There would be no power consumption in the impedance matching network if the network consisted only of ideal inductors and capacitors. The output power on R_{out} must be equal to its input power on R_{in}.

$$P_{in} = \frac{v_{in}^2}{R_{in}} = \frac{v_{out}^2}{R_{out}} = P_{out}, \tag{9.82}$$

where

P_{in} = virtual or equivalent power resulted from the combination of the load and the impedance matching network,

R_{in} = virtual or equivalent resistance resulted from the combination of the load and the impedance matching network.

P_{out} = virtual or equivalent power resulted from the combination of the source and the impedance matching network,

R_{out} = virtual or equivalent resistance resulted from the combination of the source and the impedance matching network.

The third sub-impedance matching loop is shown in Figure 9.11.

The new source consists of the old source, as shown in Figure 9.8, and the impedance matching network. The new load is the same as the old as shown in Figure 9.8. As shown in expression (9.80), the impedance of the new source is conjugate-matched with the impedance of the new load. As shown in the expressions (9.74) and (9.75), the power delivered to R_L is equal to the power remaining on the R_{out}, that is,

$$P_{R_L} = \frac{v_{R_L}^2}{R_L} = \frac{v_{out}^2}{R_{out}} = P_{out}. \tag{9.83}$$

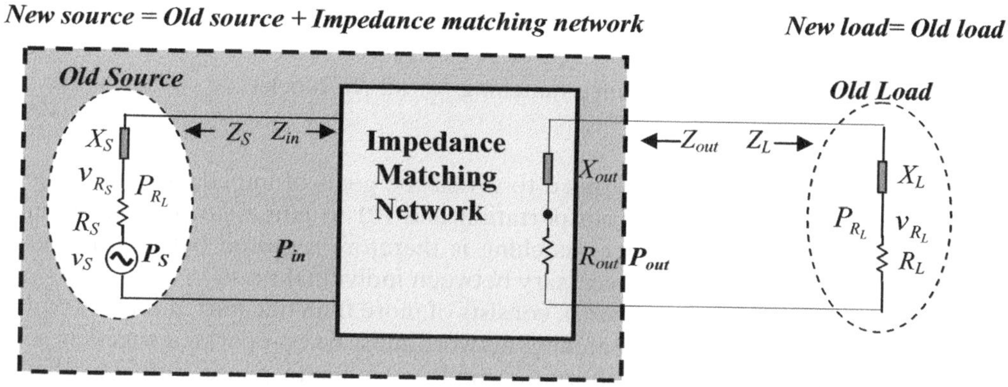

Figure 9.11 Third sub-impedance matching loop: new source = old source + impedance matching network, new load = old load.

Now, from equation (9.81) to (9.83) and (9.77) we have

$$P_{R_S} = P_{in} = P_{out} = P_{R_L} = \frac{1}{2} P_S. \tag{9.84}$$

It should be pointed out that the simple expression (9.84) implies two important concepts:

1) The powers, P_{R_S}, P_{in}, P_{out}, and P_{R_L}, are equal to each other. This is the basic theoretical background in practical power measurement, because the power, P_{R_S}, at the source can be measured as its equal value, P_{R_L}, at the load if an impedance matching network is inserted between the source and the load.
2) The insertion of an impedance matching network without power consumption will ensure maximum power transportation without phase shift of the voltage from the source to the load.

9.3.4 Necessity of Impedance Matching

So far we have emphasized the importance of impedance matching in *RF* circuit design. This is because the main task in *RF* circuit design is power transportation or power manipulation. To maximize power transportation without phase shift, impedance matching is an indispensable "must" technology. For most *RF* circuit blocks, input impedance must be matched with the output impedance of the preceding stage and its output impedance must be matched with the input impedance of the next stage.

Nobody would doubt the importance of impedance matching in an *RF* or *RFIC* design. However, there are paradoxes about impedance matching. For example,

- Is it necessary to do impedance matching from part to part?
- Within the impedance matching network, impedance between parts is generally not matched. Is it necessary to insert a "sub-impedance matching

network" between the two parts in the impedance-matching network itself?

- Is it necessary to do impedance matching for all *RF* blocks?

And so on.

To answer these questions, we need to revisit the goals of impedance matching: (1) to reach maximum power transportation and (2) to eliminate phase shift in power transportation. Impedance matching is therefore required between power transportation units, but is not necessary between individual parts.

If an impedance matching network consists of more than one part, it implies that all the parts in the impedance matching network must be co-operated together so as to maximize the power transportation and to eliminate phase shift. In other words, the matching network, but not its individual parts, is an indivisible basic entity or a "minimum cell." Their parts are indivisible in the power transportation. Impedance matching between the individual parts is meaningless.

In general, impedance must be matched between two basic power transportation units. The unique exceptional case in which we do not do so is when these two basic power transportation units are combined for special performance.

Let's illustrate this answer by examples.

Example 1: Figure 9.12(a) shows a raw source follower without impedance matching at its output; *DC* bias is neglected.

It consists of a *MOSFET* transistor and an output resistor, R_o. Looking at the emitter of the transistor in Figure 9.12 (a), it can be seen that its impedance, Z_x, is usually not conjugate-matched to the impedance of the emitter resistor, R_o. Is it beneficial to the power transportation if an impedance matching network is inserted

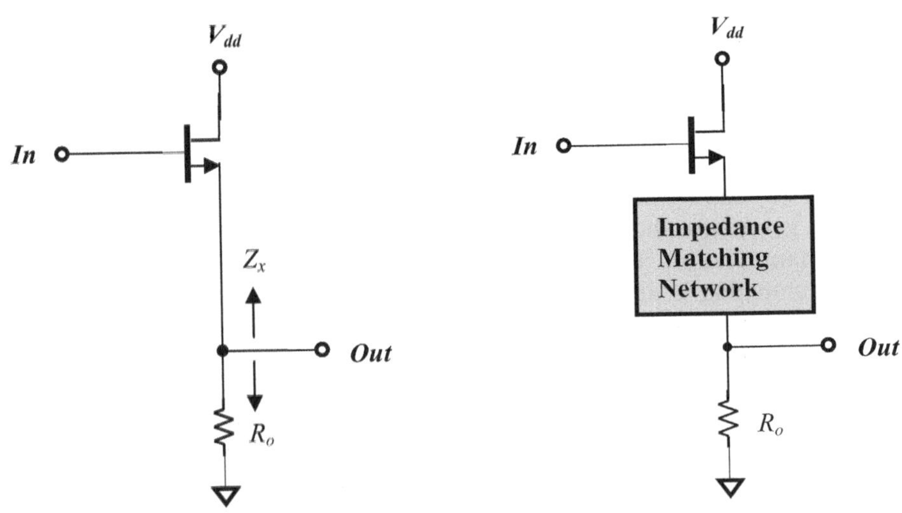

(a) Primary source follower (b) "Improved" source follower?

Figure 9.12 A source follower (*DC* bias neglected).

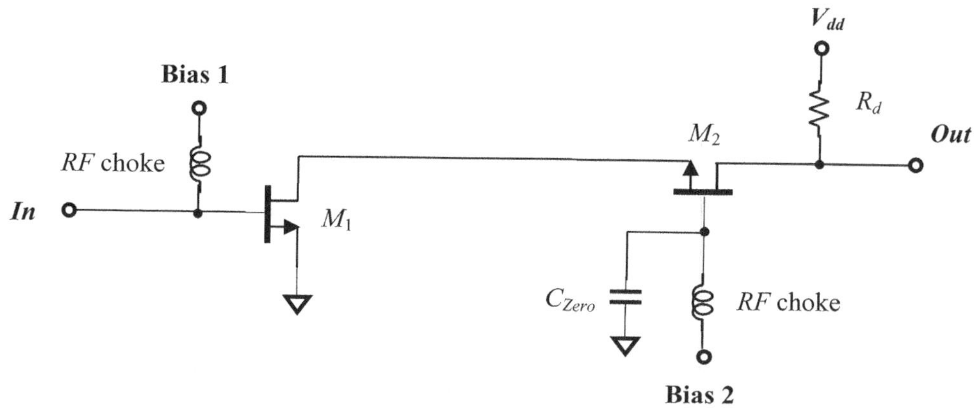

(a) A primary *MOSFET* cascode amplifier

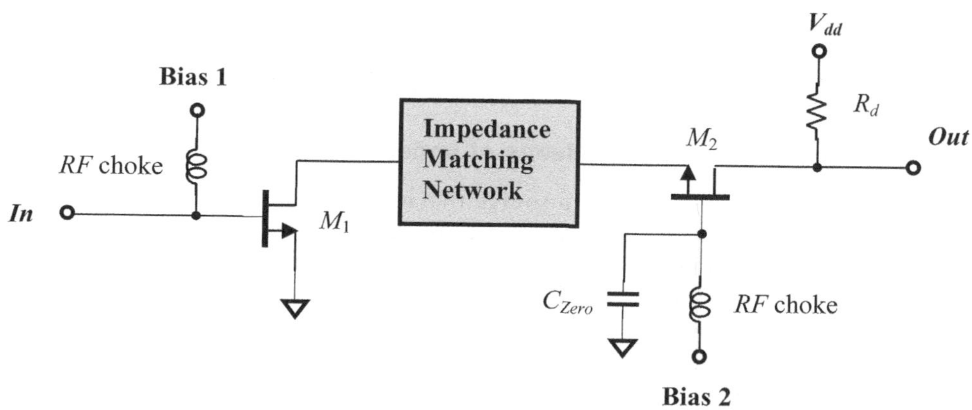

(b) An impedance matching network inserted between two stages

Figure 9.13 A *MOSFET* cascode amplifier.

between the source of the *MOSFET* transistor and the output resistor, R_o as shown in Figure 9.12(b)? The answer is no, because an individual *MOSFET* transistor or output resistor is not a basic or independent power transportation unit. Both combined form a basic power transportation unit and are indivisible.

Example 2: Figure 9.13 shows a *CS-CG* cascode *LNA*. Is it beneficial to the power transportation if a matching network is inserted between the collector of the *CS* transistor and the emitter of *CG* transistor?

In Figure 9.13(a), the first power transportation unit consists of the *CS* transistor and the parts connected to its source and gate. The second power transportation unit consists of the *CG* transistor and the parts connected to its drain. These are independent units in the power transportation. If an impedance matching network is inserted between the collector of the *CS* transistor and the emitter of the *CG* transistor as shown in Figure 9.13(b), it may be beneficial to maximize the power transportation. However, the *CS-CG* cascode configuration is seriously disturbed by the insertion of the impedance matching network, that is, its special feature, minimization of the input Miller capacitance, is removed away or greatly degraded. It is advisable not to insert such an impedance matching network between the collector of the *CS* transistor and the emitter of the *CG* transistor.

It should be noted that in a communication system some circuit blocks, such as the modulator or demodulator, are operated not only for *RF* but also for digital signal. In other words, these circuit blocks involve not only power but also digital transportation and manipulation. Consequently, only the terminals operated for power transportation or manipulation need to be impedance matched. At those terminals operated only for digital transportation or manipulation, impedance matching may be not necessary.

Finally it should be pointed out that *VCO* is a unique *RF* circuit block. Impedance matching is meaningless between *VCO* and other *RF* blocks. (See Chapter 6 for further discussion.)

9.4 ADDITIONAL EFFECTS OF IMPEDANCE MATCHING

9.4.1 Voltage Pumped Up by Means of Impedance Matching

Voltage and power delivered from a source to a load were discussed in Sections 9.1 and 9.2. In this sub-section, we discuss a cross-relationship between voltage and power transportation, that is, voltage delivered from a source to a load through an impedance matching network.

Figure 9.14 shows a voltage delivered from a source to a load with an impedance matching network. For simplicity, the reactance of our source or load is assumed to be zero. Based on the equation (9.84), the power delivered from source R_S to load R_L is equal to half of total power P_S in the loop while another half is retained at R_L, that is,

$$P_{R_L} = \frac{v_{R_L}^2}{R_L} = \frac{v_{R_S}^2}{R_S} = P_{R_S} = \frac{1}{2} P_S, \tag{9.85}$$

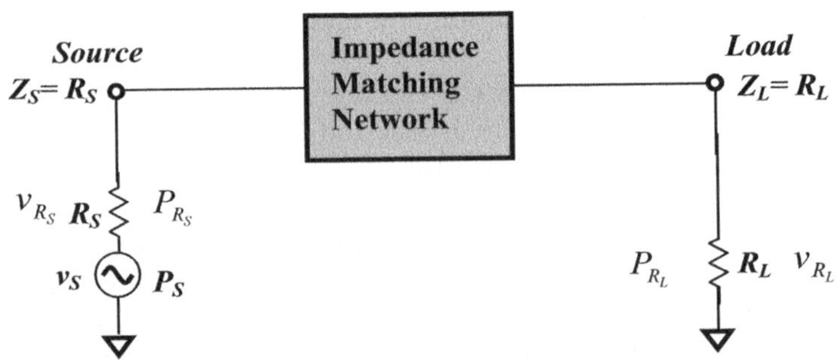

Figure 9.14 Voltage delivered from a source to a load through an impedance matching network.

TABLE 9.5 Relationship between v_{RS}, R_S and v_{R_L}, R_L for a given value of P_S = *0 dBm* in an impedance matching loop as shown in Figure 9.14

P_s dB_m	v_s V	P_{RS} dB_m	R_s, Ω	v_{RS} V	P_{RL} dB_m	R_L Ω	v_{RL} V
0	0.2236	−3	50	**0.1583**	−3	**10**	**0.0708**
0	0.2236	−3	50	**0.1583**	−3	**12**	**0.0776**
0	0.2236	−3	50	**0.1583**	−3	**16**	**0.0895**
0	0.2236	−3	50	**0.1583**	−3	**22**	**0.1050**
0	0.2236	−3	50	**0.1583**	−3	**30**	**0.1226**
0	**0.2236**	**−3**	**50**	**0.1583**	**−3**	**50**	**0.1583**
0	0.2236	−3	50	**0.1583**	−3	**100**	**0.2239**
0	0.2236	−3	50	**0.1583**	−3	**200**	**0.3166**
0	0.2236	−3	50	**0.1583**	−3	**500**	**0.5006**
0	0.2236	−3	50	**0.1583**	−3	**1k**	**0.7079**
0	0.2236	−3	50	**0.1583**	−3	**2k**	**1.0012**
0	0.2236	−3	50	**0.1583**	−3	**10k**	**2.2387**

Table 9.5 lists the correlated values of v_{RS} and R_S, v_{R_L} and R_L, calculated from equation (9.85) when impedance at source or load is well matched and when $P_S = 0\,dB_m$, with a corresponding voltage, $v_s = 0.2236\,V$ across a $50\,\Omega$ resistor. In Table 9.5, R_S is kept $50\,\Omega$ unchanged while R_L is varied from $10\,\Omega$ to $10\,k\Omega$.

Three cases are shown in Table 9.5:

1) Dropped-down case: v_{R_L} is dropped down from v_{R_S}:

$$v_{R_L} < v_{RS}, \tag{9.86}$$

if

$$R_L < R_S. \tag{9.87}$$

2) Unchanged case: v_{R_L} is equal to v_{R_S}:

$$v_{R_L} = v_{RS} = 0.1583\,V, \tag{9.88}$$

if

$$R_L = R_S = 50\,\Omega. \tag{9.89}$$

3) Pumped-up case: v_{R_L} is pumped up from v_{R_S}:

$$v_{R_L} > v_{RS}, \tag{9.90}$$

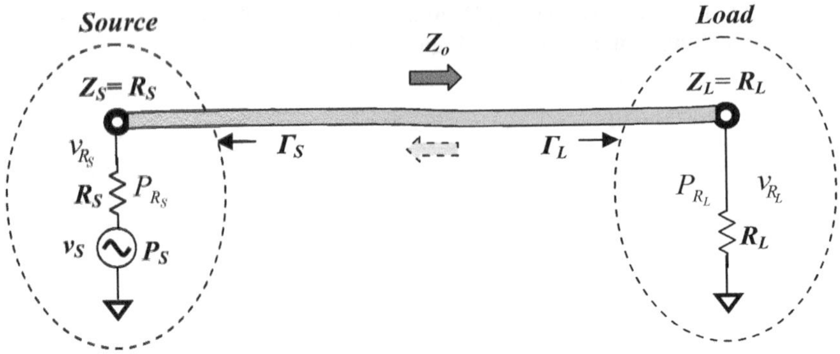

Figure 9.15 Voltage delivered from a source to a load directly without impedance matching network.

if

$$R_L > R_S. \tag{9.91}$$

As shown in the last row in Table 9.5, when $R_S = 50\,\Omega$ and $R_L = 10\,k\Omega$, the value of v_{R_L} is equal to $2.2387\,V$, which is pumped up more than 14 times from $v_{R_S} = 0.1583\,V$.

Now let's discuss another case when a voltage is directly delivered from source to load without impedance matching network.

Figure 9.15 depicts such a simple loop with $P_S = 0\,dB_m$, which corresponds to voltage, $v_{R_L} = 0.2236\,V$ across a $50\,\Omega$ resistor. The various voltages can be calculated using the following equations:

$$P_S = \frac{v_S^2}{50}, \quad \text{or} \quad v_S = \sqrt{P_S \cdot 50}, \tag{9.92}$$

$$v_{R_S} = \frac{R_S}{R_S + R_L} v_S, \tag{9.93}$$

$$v_{R_L} = \frac{R_L}{R_S + R_L} v_S, \tag{9.94}$$

The calculated values of v_{R_S} and v_{R_L} are listed in Table 9.6, where R_S is kept $50\,\Omega$ is unchanged while R_L is varied from $10\,\Omega$ to $10\,k\Omega$.

Three cases shown in Table 9.6:

1) Dropped-down case : v_{R_L} is dropped down from v_{R_S}.

$$v_{R_L} < v_{R_S}, \tag{9.95}$$

if

$$R_L < R_S. \tag{9.96}$$

2) Unchanged case: v_{R_L} or v_{R_S} is equal to v_{R_L}.

TABLE 9.6 Relationship between v_{RS}, R_S and v_{R_L}, R_L for a given value of P_S = *0 dBm* in an impedance matching loop as shown in Figure 9.15

P_S dB_m	v_s V	R_s Ω	v_{RS} V	R_L, Ω	v_{RL} V
0	0.2236	50	**0.1863**	**10**	**0.0373**
0	0.2236	50	**0.1803**	**12**	**0.0433**
0	0.2236	50	**0.1694**	**16**	**0.0542**
0	0.2236	50	**0.1553**	**20**	**0.0683**
0	0.2236	50	**0.1398**	**30**	**0.0839**
0	**0.2236**	**50**	**0.1118**	**50**	**0.1118**
0	0.2236	50	**0.0745**	**100**	**0.1491**
0	0.2236	50	**0.0447**	**200**	**0.1789**
0	0.2236	50	**0.0203**	**500**	**0.2033**
0	0.2236	50	**0.0106**	**1k**	**0.2130**
0	0.2236	50	**0.0055**	**2k**	**0.2182**
0	0.2236	50	**0.0011**	**10k**	**0.2225**

$$v_{R_L} = v_{R_S} = 0.1118\,V, \tag{9.97}$$

if

$$R_L = R_S = 50\,\Omega. \tag{9.98}$$

3) Pumped-up case : v_{R_L} is pumped up from v_{R_S}.

$$v_{R_L} > v_{R_S}, \tag{9.99}$$

if

$$R_L > R_S. \tag{9.100}$$

By comparing the two cases with and without impedance matching networks, as shown in Tables 9.5 and 9.6, when a voltage is delivered from a source to a load, it may be found that both cases are quite different.

The slight difference is that when $R_S = R_L = 50\,\Omega$, $v_{R_L} = v_{R_S} = 0.1583\,V$ as shown in expression (9.88) for the case with impedance matching network, but $v_{R_L} = v_{R_S} = 0.1118\,V$ as shown in the expression (9.92) for the case without impedance matching network.

The huge difference of the voltage at the load v_{R_L} appears when $R_L > R_S$, especially when $R_L >> R_S$. For example, when $R_S = 50\,\Omega$ and $R_L = 10\,k\Omega$,

- In the case with impedance matching network, the value of v_{R_L} is dramatically pumped up to $2.2387\,V$ from the source while $v_s = 0.2236\,V$, shown in the last row of Table 9.5.
- On the other hand, in the case without impedance matching network, the value of v_{R_L} is dropped down to $0.2225\,V$ from the source while $v_s = 0.2236\,V$, shown in the last row of Table 9.6.

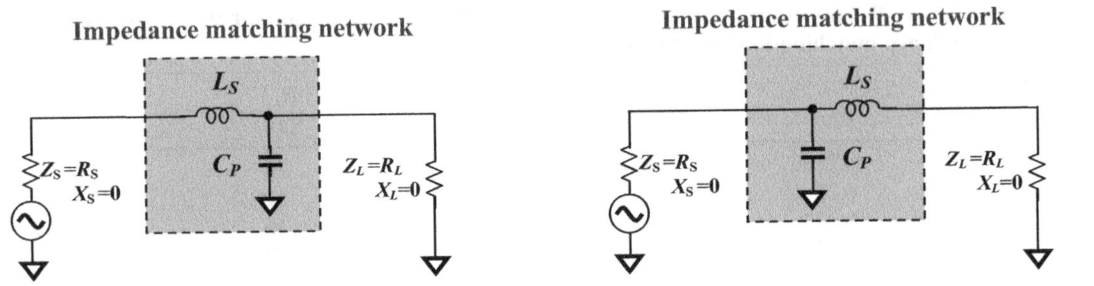

(a) Voltage pumped up from R_S to R_L by upward resistance transformer ($\boldsymbol{R_S < R_L}$)

(b) Voltage dropped down from R_S to R_L by downward resistance transformer ($\boldsymbol{R_S > R_L}$)

Figure 9.16 Voltage delivered from a source to a load through an impedance matching network, which is a $\boldsymbol{L_S - C_P}$ upward or downward resistance transformer.

The voltage can be pumped up or dropped down if it is delivered from a source to a load through an impedance matching network. Special attention must be paid to the significantly pumped-up case. This is very useful in *RF* circuit design when a pumped-up voltage is needed.

Pumped-up or dropped-down voltage can be realized by an upward or downward resistance transformer. An upward resistance transformer is applied for delivering a pumped-up voltage while a downward resistance transformer is applied for delivering a dropped-down voltage from source to load. Figures 9.16(a) and 9.16(b) show the impedance matching network in Figure 9.14 by replacing the upward and downward resistance transformers, respectively.

Both the source and load impedance, Z_S and Z_L, are purely resistive, that is,

$$Z_S = R_S, \quad X_S = 0, \tag{9.101}$$

$$Z_L = R_L, \quad X_L = 0. \tag{9.102}$$

In Figure 9.16(a),

$$R_S < R_L, \tag{9.103}$$

while in Figure 9.16(b),

$$R_S > R_L. \tag{9.104}$$

Figure 9.17 shows impedance matching of $L_S - C_P$ upward and downward resistance transformers in the Smith chart.

As a matter of fact, Figure 9.16(a) and (b) are imaginary to each other. They become the same if the position of the load and the source, R_L and R_S, are exchanged in either Figure 9.16(a) or (b), which corresponds to a change from an upward to a downward transformer.

Let's introduce the related formula from Thomas H. Lee (1998) about the downward impedance transformer as shown in Figure 9.16(b). The resistor R_L in series with L_S and the resistor R_S in parallel with C_P are impedance-matched if their Q values are equal to each other, such as

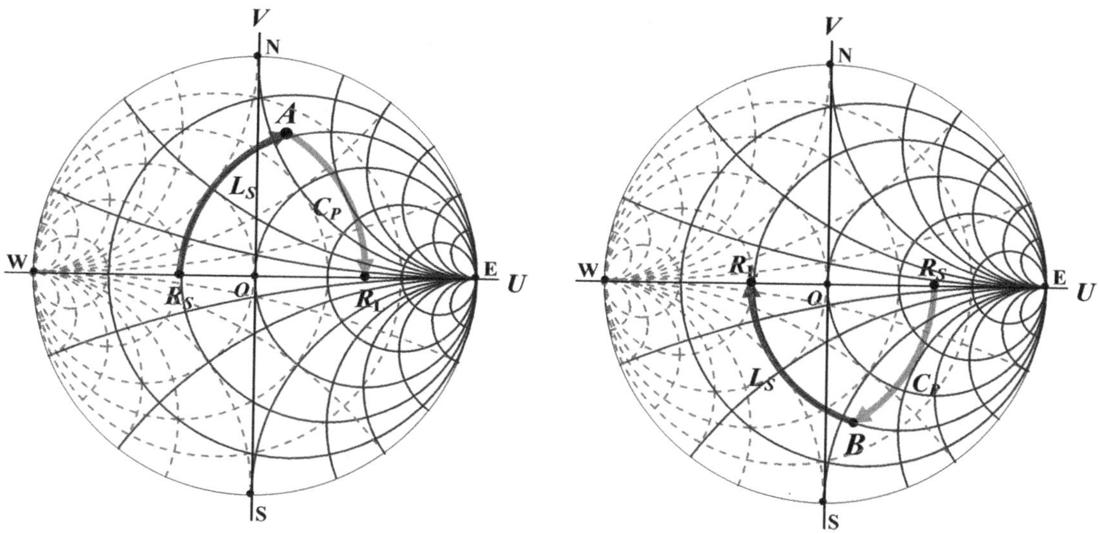

(a) Upward resistance transformer consists of L_S and C_P ($R_S < R_L$)

(b) Downward resistance transformer consists of C_P and L_S ($R_S > R_L$)

Figure 9.17 Impedance matching of a $L_S - C_P$ upward and downward resistance transformer shown in Smith chart.

$$Q = \omega_o C_P R_S = \frac{\omega_o L_S}{R_L}, \tag{9.105}$$

where Q is called the "quality factor", and ω_o is the operating angular frequency, and is

$$\omega_o^2 = \frac{1}{C_P L_S}, \tag{9.106}$$

$$R_S = R_L(1+Q^2) = R_L\left[1 + \frac{1}{(\omega_o R_L C_P)^2}\right] = R_L + \frac{1}{R_L}\frac{L_S}{C_P}, \tag{9.107}$$

then,

$$R_L(R_S - R_L) = \frac{L_S}{C_P}. \tag{9.108}$$

From equations (9.106) and (9.108), we have

$$C_P = \frac{1}{\omega_o \sqrt{R_L(R_S - R_L)}}, \tag{9.109}$$

$$L_S = \sqrt{\frac{R_L(R_S - R_L)}{\omega_o}}. \tag{9.110}$$

In these two equations, R_L, R_S, and ω_o, are given parameters. The values of C_P and L_S can be calculated from equations (9.109) and (9.110). They are real numbers because in a downward transformer, $R_S > R_L$.

Now let's shift to the upward impedance transformer with L_S and C_P as shown in Figure 9.16(a). As mentioned above, the expressions for L_S and C_P can be obtained by simply exchanging the positions of the load and the source, R_S and R_L, in the equations (9.109) and (9.110), that is

$$L_S = \sqrt{\frac{R_S(R_L - R_S)}{\omega_o}}, \tag{9.111}$$

$$C_P = \frac{1}{\omega_o\sqrt{R_S(R_L - R_S)}}. \tag{9.112}$$

In these two equations, R_L, R_S, and ω_o, are given parameters. The values of L_S and C_P can be calculated from equations (9.111) and (9.112). They are real numbers because in an upward transformer, $R_S < R_L$.

As a matter of fact, the upward and downward impedance transformer as shown in Figure 9.16(a) and (b) is a basic segment of a low-pass filter. We can have another form of the matching network between R_L and R_S. Figure 9.18(a) and (b) show an alternative $C_S - L_P$ upward and downward impedance transformer, or, in other words, a matching network with a basic segment of a high-pass filter. Figure 9.19 shows impedance matching of $C_S - L_P$ upward and downward resistance transformers in Smith chart respectively.

Again, as shown in Figure 9.18(a) and (b), these two types of upward and downward impedance transformers are imaginary to each other. That is, both of them become the same if the position of the load and the source, R_L and R_S, are exchanged in either Figure 9.18(a) or (b). The derivation of the two analytical expressions for C_S and L_P, which are similar to expressions, (9.109), (9.110), (9.111), and (9.112), is an exercise for the reader.

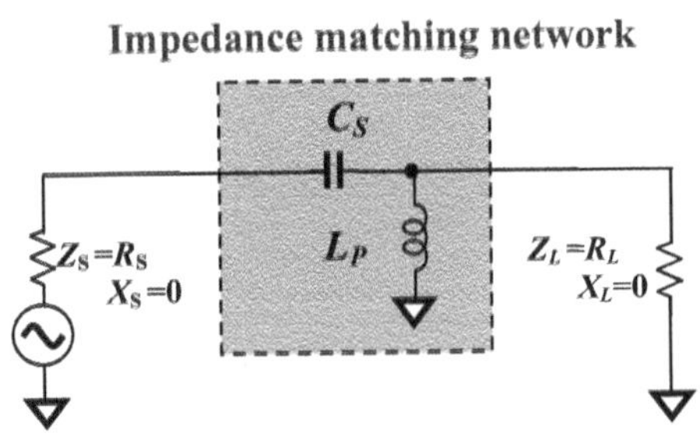

(a) Voltage pumped up from R_S to R_L by upward resistance transformer (**$R_S < R_L$**)

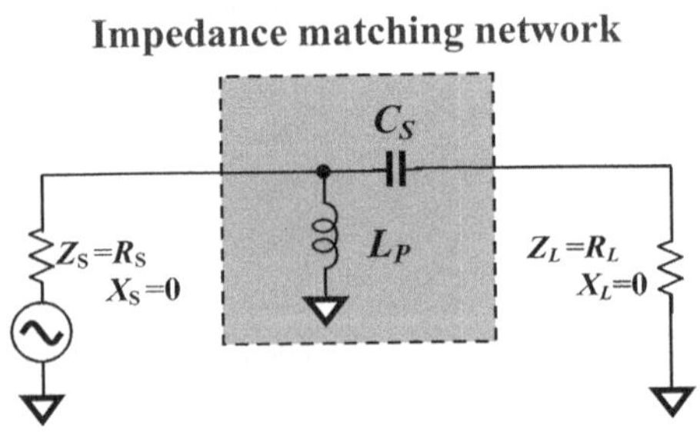

(b) Voltage dropped down from R_S to R_L by downward resistance transformer (**$R_S > R_L$**)

Figure 9.18 Voltage delivered from a source to a load through an impedance matching network, which is a $C_S - L_P$ upward or downward resistance transformer.

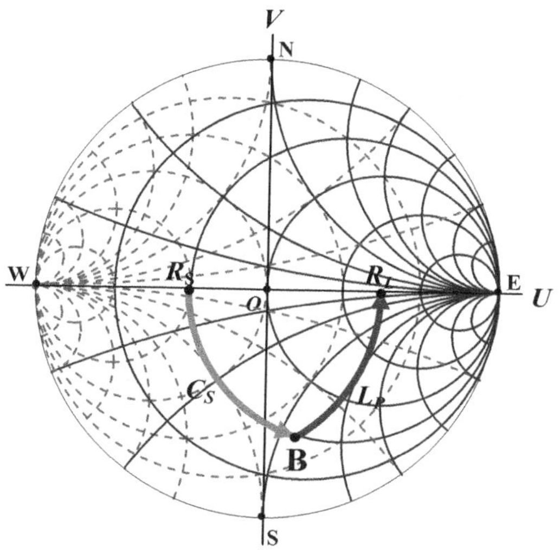

(a) Upward resistance transformer consists of C_S and L_P ($R_S < R_L$)

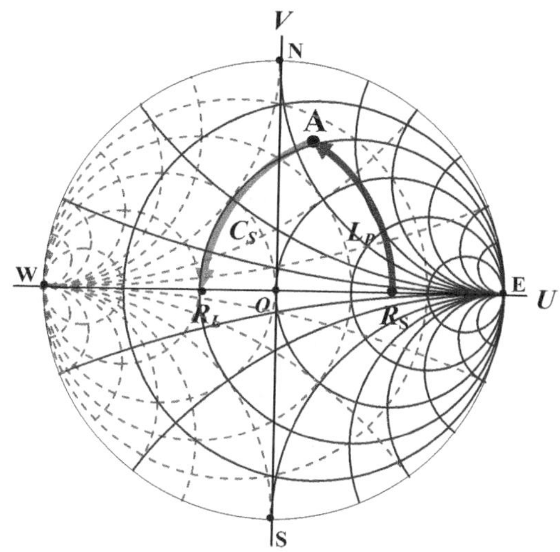

(b) Downward resistance transformer consists of L_P and C_S ($R_S > R_L$)

Figure 9.19 Upward and downward resistance transformer consists of C_S and L_P.

9.4.2 Power Measurement

Power measurement in an *RF* test laboratory is different from the testing of other parameters. It is important to understand whether the test is done under matched or unmatched conditions since the test outcomes are determined by the impedance matching status between the tested point and the spectrum analyzer. The reading of power measurement under unmatched cases is incorrect!

Usually power measurement is conducted by means of a power meter or a spectrum analyzer. The impedance of both the power meter or spectrum analyzer is typically $50\,\Omega$ and the unit of power reading is *dB*. Figure 9.20 shows two cases of power testing at point *P*. Assuming that the impedance and the voltage of the *DTU* (Desired Test Unit) is Z_S and v_S respectively, and that the input impedance of the network analyzer is $50\,\Omega$.

In the matched case (a), a matching network has been inserted between the tested point and the network analyzer. In Section 9.3.2 we showed that the power reading from the spectrum analyzer is the same as shown in the expression (9.69), such as

$$P_L = P_o = \frac{v_S^2}{4R_S}, \tag{9.113}$$

where P_o = maximum of matched power, which can be sensed by the spectrum analyzer.

In the unmatched case (b), the spectrum analyzer directly measures the power at point *P* without any assistance, and we have

$$P_L^* = v_S^2 \frac{50}{|50 + Z_S|^2}, \tag{9.114}$$

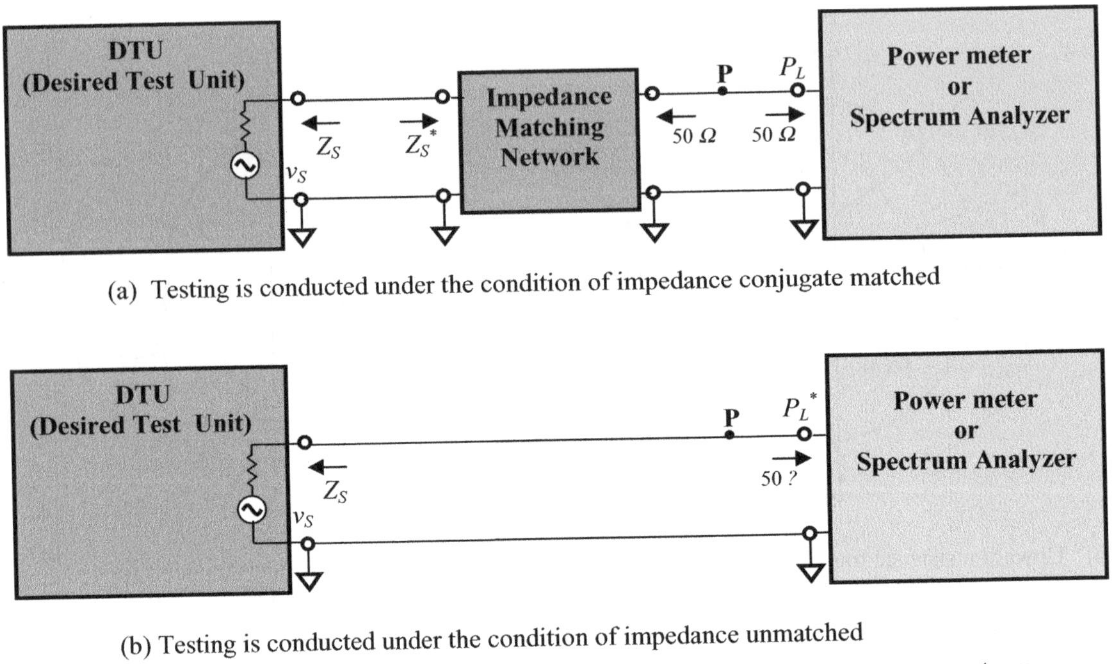

Figure 9.20 Output power of a tested block is measured by a power meter or a spectrum analyzer.

where P_L^* = measured power at point P.

The ratio of measured power under matched and unmatched cases is

$$\frac{P_L^*}{P_o} = 4R_S\frac{50}{|50+Z_S|^2} = \frac{200R_S}{(50+R_S)^2+X_S^2}. \tag{9.115}$$

The expression (9.115) represents the difference between measured powers under matched and unmatched cases. Table 9.7 lists the calculated results in dB. It can be seen that in the eleventh row of Table 9.7 where $R_S = 50\,\Omega$, $X_S = 0\,\Omega$, the impedance matching condition is satisfied so that the impedance of the tested point matches that of the spectrum analyzer. This results in the measured power, P_L, being equal to the expected power, P_o, as shown in matched case (a). In general, in the unmatched case (b), the measured power, P_L^*, will deviate from the expected power, P_o, by a certain amount. The deviation becomes significant when the impedance is far from a matched condition. The power reading in the unmatched case is always lower than the power reading under the impedance matched case. In some cases the difference is horrible! For instance, as shown in the tenth row of Table 9.7 where the impedance of DTU is $Z_S = 10\,\Omega + j10k\Omega$ while the input impedance of power meter or spectrum analyzer is $50\,\Omega$, the difference is $47\,dB$!

Figure 9.21 plots the calculated power ratio, P_L^*/P_o, vs. impedance of the DTU, Z_S. This "waterfall" diagram may give you an intuitive feeling for the inaccuracy of power measure due to the unmatched impedance.

TABLE 9.7 Calculated power ratio, P_L^*/P_o as the impedance of the *DTU* is varied

R_S, Ω	X_S, Ω	P_L^*/P_o	P_L^*/P_o, *dB*
10	0	0.555556	–2.6
10	20	0.500000	–3.0
10	50	0.327869	–4.8
10	100	0.147059	–8.3
10	200	0.500000	–13.4
10	500	0.327869	–21.0
10	1000	0.001993	–27.0
10	2000	0.500000	–33.0
10	5000	0.327869	–41.0
10	10000	0.000020	–47.0
50	0	1.000000	0.0
50	20	0.961538	–0.2
50	50	0.800000	–1.0
50	100	0.500000	–3.0
50	200	0.200000	–7.0
50	500	0038462	–14.1
50	1000	0.009901	–20.0
50	2000	0.002494	–26.0
50	5000	0.002494	–34.0
50	10000	0.000100	–40.0
100	0	0.888889	–0.5
100	20	0.873362	–0.6
100	50	0.800000	–1.0
100	100	0.615385	–2.1
100	200	0.320000	–4.9
100	500	0.073394	–11.3
100	1000	0.019560	–17.1
100	2000	0.004972	–23.0
100	500	0.073394	–31.0
100	10000	0.000200	–37.0
1000	0	0.181406	–7.4
1000	20	0.181340	–7.4
1000	50	0.180995	–7.4
1000	100	0.179775	–7.5
1000	200	0.175055	–7.6
1000	500	0.147874	–8.3
1000	1000	0.095125	–10.2
1000	2000	0.039196	–14.1
1000	5000	0.007662	–21.2
1000	10000	0.001978	–27.0
10000	0	0.019801	–17.0
10000	20	0.019801	–17.0
10000	50	0.019801	–17.0
10000	100	0.019800	–17.0
10000	200	0.019794	–17.0
10000	500	0.019753	–17.0
10000	1000	0.019607	–17.1
10000	2000	0.019047	–17.2
10000	5000	0.015873	–18.0
10000	10000	0.009950	–20.0

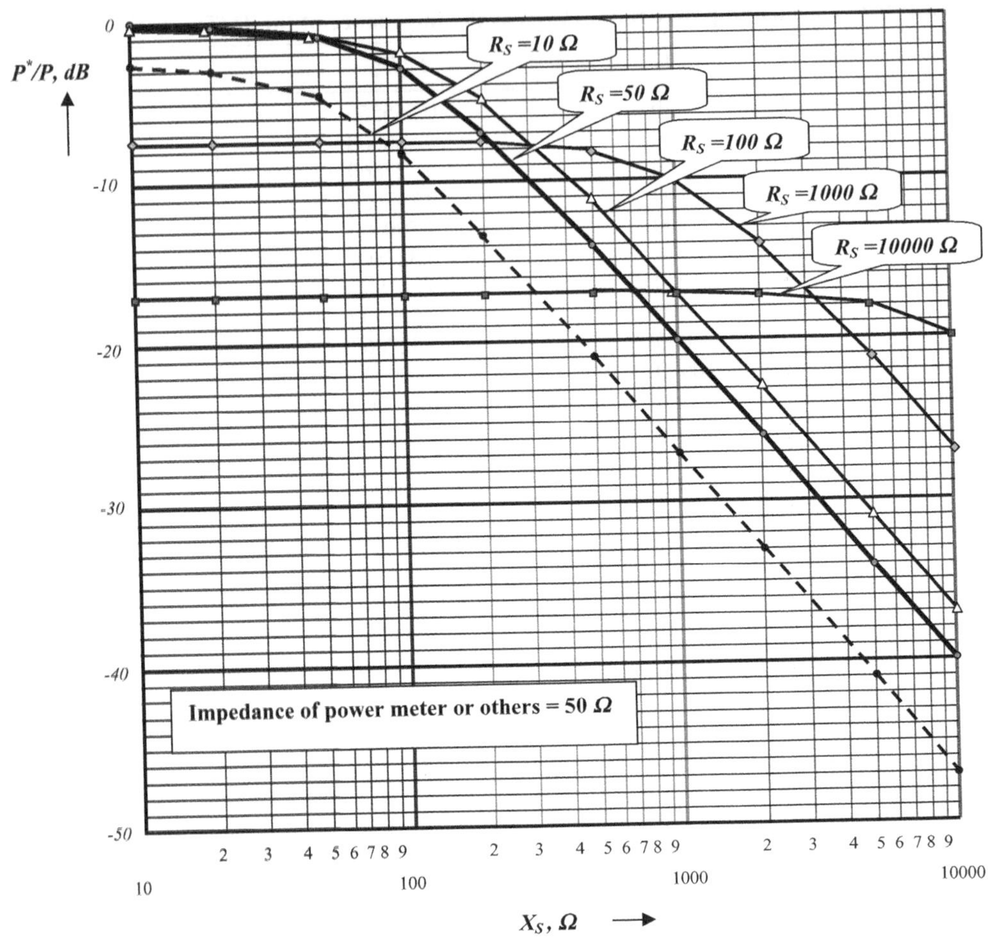

Figure 9.21 Plot of calculated power ratio, P_L^*/P_o, vs. impedance of *DTU*, Z_S.

APPENDICES

9.A.1 *VSWR* and Other Reflection and Transmission Coefficients

For the reader's convenience, Table 9.A.1 lists values calculated from the relationships between impedance, the reflection coefficient, transmission coefficient, and other parameters. The value of 50 Ω is taken as the reference impedance, R_o.

The equations applied to calculate the various parameters are:

- For voltage reflection coefficient,

$$\Gamma = \frac{R - R_o}{R + R_o}. \tag{9.A.1}$$

- For *VSWR* (Voltage Standing Wave Ratio) in Watt,

TABLE 9.A.1 Relationships between impedance, reflection coefficient, transmission coefficient, and other parameters along with the axis U ($V = 0$) in the complex plane of voltage reflection coefficientΓ

R, Ω	($R_o = 50\,\Omega$) $r = R/R_o$	$\|\Gamma\|$	$VSWR$	$VSWR_{dB}$	(S_{11} or S_{22}) RL_{dB}	TL_{dB}	γ	$P_{T,\%}$	$P_{R,\%}$
0.2878	0.0056	0.99	178.57	45.04	−0.100	−16.4280	0.98	2.22	97.78
1.4386	0.0290	0.94	34.48	30.75	−0.500	−9.6361	0.89	10.96	89.04
2.8750	0.0574	0.89	17.42	24.82	−1.000	−6.8683	0.79	20.53	79.47
5.7313	0.1146	0.79	8.73	18.82	−2.000	−4.3292	0.63	36.90	63.10
8.5500	0.1710	0.71	5.85	15.34	−3.000	−3.0206	0.50	49.88	50.12
11.3136	0.2263	0.63	4.42	12.91	−4.000	−2.2048	0.40	60.19	39.81
14.0066	0.2801	0.56	3.57	11.06	−5.000	−1.6509	0.32	68.36	31.64
16.6140	0.3323	0.50	3.01	9.57	−6.000	−1.2563	0.25	74.88	25.12
19.1236	0.3825	0.45	2.62	8.35	−7.000	−0.9665	0.20	80.04	19.96
21.5252	0.4305	0.40	2.32	7.32	−8.000	−0.7494	0.16	84.15	15.85
23.8110	0.4762	0.35	2.10	6.44	−9.000	−0.5843	0.13	87.42	12.58
25.9747	0.5195	0.32	1.93	5.69	−10.000	−0.4576	0.10	89.99	10.01
29.9240	0.5986	0.25	1.67	4.46	−12.000	−0.2830	0.06	93.70	6.30
34.9020	0.6980	0.18	1.43	3.12	−15.000	−0.1396	0.03	96.84	3.16
40.9091	0.8182	0.10	1.22	1.74	−20.000	−0.0436	0.01	99.00	1.00
44.6760	0.8935	0.06	1.12	0.98	−25.000	−0.0138	0.00	99.68	0.32
46.9347	0.9387	0.03	1.07	0.55	−30.000	−0.0043	0.00	99.90	0.10
48.2528	0.9651	0.02	1.04	0.31	−35.000	−0.0014	0.00	99.97	0.03
49.0099	0.9802	0.01	1.02	0.17	−40.000	−0.0004	0.00	99.99	0.01
49.4408	0.9888	0.01	1.01	0.10	−45.000	−0.0001	0.00	100.00	0.00
49.6848	0.9937	0.00	1.01	0.05	−50.000	0.0000	0.00	100.00	0.00
50.0000	1.0000	0.00	1.00	0.00	(-Infinite)	0.0000	0.00	100.00	0.00
50.3172	1.0063	0.00	1.01	0.05	−50.000	0.0000	0.00	100.00	0.00
50.5655	1.0113	0.01	1.01	0.10	−45.000	−0.0001	0.00	100.00	0.00
51.0101	1.0202	0.01	1.02	0.17	−40.000	−0.0004	0.00	99.99	0.01
51.8105	1.0362	0.02	1.04	0.31	−35.000	−0.0014	0.00	99.97	0.03
53.2656	1.0653	0.03	1.07	0.55	−30.000	−0.0043	0.00	99.90	0.10
55.9585	1.1192	0.06	1.12	0.98	−25.000	−0.0138	0.00	99.68	0.32
61.1111	1.2222	0.10	1.22	1.74	−20.000	−0.0436	0.01	99.00	1.00
71.6291	1.4326	0.18	1.43	3.12	−15.000	−0.1396	0.03	96.84	3.16
83.5450	1.6710	0.25	1.67	4.46	−12.000	−0.2830	0.06	93.69	6.31
96.2475	1.9244	0.32	1.93	5.69	−10.000	−0.4576	0.10	90.01	9.99
104.9942	2.0999	0.35	2.10	6.44	−9.00	−0.5844	0.13	87.41	12.59
116.1431	2.3229	0.40	2.32	7.32	−8.000	−0.7494	0.16	84.15	15.85
130.7280	2.6146	0.45	2.61	8.35	−7.000	−0.9665	0.20	80.05	19.95
150.4750	3.0095	0.50	3.01	9.57	−6.000	−1.2563	0.25	74.88	25.12
178.4900	3.5698	0.56	3.57	11.05	−5.000	−1.6509	0.32	68.38	31.62
220.9700	4.4194	0.63	4.42	12.91	−4.000	−2.2048	0.40	60.19	39.81
292.4050	5.8481	0.71	5.85	15.34	−3.000	−3.0207	0.50	49.88	50.12
436.2200	8.7244	0.79	8.72	18.81	−2.000	−4.3293	0.63	36.90	63.10
869.5500	17.3910	0.89	17.39	24.81	−1.000	−6.8683	0.79	20.57	79.43

$$VSWR = \frac{1+|\Gamma|}{1-|\Gamma|}. \quad (9.A.2)$$

- For $VSWR$ in dB,

$$VSWR_{dB} = 20Log(VSWR). \quad (9.A.3)$$

- For RL (Return Loss) in dB,

$$RL_{dB} = S_{11,dB} = 20Log|\Gamma|. \quad (9.A.4)$$

- For TL (Transmission Loss) in dB,

$$TL_{dB} = 10Log(1-\Gamma^2). \quad (9.A.5)$$

- For power reflection coefficient,

$$\gamma = \Gamma^2. \quad (9.A.6)$$

- For transmitted power in %,

$$P_{T,\%} = 100(1-\Gamma^2). \quad (9.A.7)$$

- For reflected power in %,

$$P_{R,\%} = 100\Gamma^2. \quad (9.A.8)$$

9.A.2 Relationships between Power (dB_m), Voltage (V), and Power ($Watt$)

The equations applied for calculation of voltage (mV) from power (dBm), and for power (mW) from voltage (mV) are as shown below. Values are shown in Table 9.A.2.

- Voltage (mV) calculated from power (dBm),

$$v_V = \sqrt{(P_{dBm} - 30)\, P_\Omega}, \quad (9.A.9)$$

$$v_{mV} = 10^3 \sqrt{(P_{dBm} - 30)\, R_\Omega}, \quad (9.A.10)$$

$$v_{\mu V} = 10^6 \sqrt{(P_{dBm} - 30)\, R_\Omega}. \quad (9.A.11)$$

- Power (mW) calculated from voltage (mV),

$$P_W = v_V^2 \frac{1}{R_\Omega}, \quad (9.A.12)$$

TABLE 9.A.2 Relationships between power (*dB*), voltage (*V*), and power (*W*)

Power, dB_m	Voltage, V	Power, W	Power, dB_m	Voltage, mV	Power, mW	Power, dB_m	Voltage, μV	Power nW
50	70.71	100.00	**0**	**223.61**	**1.0000**	−50	707.11	10.0000
49	63.02	79.43	−1	199.29	0.794328	−51	630.21	7.943282
48	56.17	63.10	−2	177.62	0.630957	−52	561.67	6.309573
47	50.06	50.12	−3	158.30	0.501187	−53	500.59	5.011872
46	44.62	39.81	−4	141.09	0.398107	−54	446.15	3.981072
45	39.76	31.62	−5	125.74	0.316228	−55	397.64	3.162278
44	35.44	25.12	−6	112.07	0.251189	−56	354.39	2.511886
43	31.59	19.95	−7	99.88	0.199526	−57	315.85	1.995262
42	28.15	15.85	−8	89.02	0.158489	−58	281.50	1.584893
41	25.09	12.59	−9	79.34	0.125893	−59	250.89	1.258925
40	22.36	10	−10	70.71	0.1000	−60	223.61	1.0000
39	19.93	7.94	−11	63.02	0.079433	−61	199.29	0.794328
38	17.76	6.31	−12	56.17	0.063096	−62	177.62	0.630957
37	15.83	5.01	−13	50.06	0.050119	−63	158.30	0.501187
36	14.11	3.98	−14	44.62	0.039811	−64	141.09	0.398107
35	12.57	3.16	−15	39.76	0.031623	−65	125.74	0.316228
34	11.21	2.51	−16	35.44	0.025119	−66	112.07	0.251189
33	9.99	2.00	−17	31.59	0.019953	−67	99.88	0.199526
32	8.90	1.58	−18	28.15	0.015849	−68	89.02	0.158489
31	7.93	1.26	−19	25.09	0.012589	−69	79.34	0.125893
30	7.07	1	−20	22.36	0.0100	−70	70.71	0.1000
29	6.30	0.79	−21	19.93	0.007943	−71	63.02	0.079433
28	5.62	0.63	−22	17.76	0.006310	−72	56.17	0.063096
27	5.01	0.50	−23	15.83	0.005012	−73	50.06	0.050119
26	4.46	0.40	−24	14.11	0.003981	−74	44.62	0.039811
25	3.98	0.32	−25	12.57	0.003162	−75	39.76	0.031623
24	3.54	0.25	−26	11.21	0.002512	−76	35.44	0.025119
23	3.16	0.20	−27	9.99	0.001995	−77	31.59	0.019953
22	2.82	0.16	−28	8.90	0.001585	−78	28.15	0.015849
21	2.51	0.13	−29	7.93	0.001259	−79	25.09	0.012589
20	2.24	0.1	−30	7.07	0.0010	−80	22.36	0.0100
19	1.99	0.0794	−31	6.30	0.000794	−81	19.93	0.007943
18	1.78	0.0631	−32	5.62	0.000631	−82	17.76	0.006310
17	1.58	0.0501	−33	5.01	0.000501	−83	15.83	0.005012
16	1.41	0.0398	−34	4.46	0.000398	−84	14.11	0.003981
15	1.26	0.0316	−35	3.98	0.000316	−85	12.57	0.003162
14	1.12	0.0251	−36	3.54	0.000251	−86	11.21	0.002512
13	1.00	0.0200	−37	3.16	0.000200	−87	9.99	0.001995
12	0.89	0.0158	−38	2.82	0.000158	−88	8.90	0.001585
11	0.79	0.0126	−39	2.51	0.000126	−89	7.93	0.001259
10	0.71	0.0100	−40	2.24	0.0001	−90	7.07	0.0010
9	0.63	0.0079	−41	1.99	0.000079	−91	6.30	0.000794
8	0.56	0.0063	−42	1.78	0.000063	−92	5.62	0.000631
7	0.50	0.0050	−43	1.58	0.000050	−93	5.01	0.000501
6	0.45	0.0040	−44	1.41	0.000040	−94	4.46	0.000398
5	0.40	0.0032	−45	1.26	0.000032	−95	3.98	0.000316
4	0.35	0.0025	−46	1.12	0.000025	−96	3.54	0.000251
3	0.32	0.0020	−47	1.00	0.000020	−97	3.16	0.000200
2	0.28	0.0016	−48	0.89	0.000016	−98	2.82	0.000158
1	0.25	0.0013	−49	0.79	0.000013	−99	2.51	0.000126

$$P_{mW} = 10^3\left(\frac{v_{mV}}{10^3}\right)^2 \frac{1}{R_\Omega}, \quad (9.A.13)$$

$$P_{nW} = 10^9\left(\frac{v_{\mu V}}{10^6}\right)^2 \frac{1}{R_\Omega}. \quad (9.A.14)$$

REFERENCES

[1] "S-Parameters, Circuit Analysis and Design," *Hewlett-Packard Application Note 95*, September, 1968.

[2] P. H. Smith, *Electronic Applications of the Smith Chart*, McGraw-Hill, New York, 1969.

[3] William H. Hayt Jr. and Jack E. Kemmerly, *Engineering Circuit Analysis*, McGraw-Hill, Inc., 1971.

[4] G. D. Vendelin, W. Alexander, and D. Mock, "Computer Analyzes RF Circuits with Generalized Smith Charts," *Electronics*, 1974, pp. 102–110.

[5] P. L. D. Abrie, *The Design of Impedance Matching Networks for Radio-Frequency and Microwave Amplifiers*, Artech House, Norwood, Mass., 1985.

[6] Richard Li, *Key Issues in RF/RFIC Circuit Design*, Higher Education Press, Beijing, 2005.

[7] Thomas H. Lee, *The Design of CMOS Radio-Frequency Integrated Circuits*, Cambridge University Press, 1998.

CHAPTER 10

IMPEDANCE MATCHING IN A NARROW-BAND CASE

10.1 INTRODUCTION

As emphasized in previous chapters, impedance matching is the key task in *RF* circuit design because the main function of the *RF* block is to ensure maximization of power transportation without phase shift. The task of impedance matching is to change the original impedance to the standard reference impedance, 50 Ω, or some other desired impedance through the construction of an impedance matching network.

There are some basic engineering fundamentals that circuit designers must bear in mind in the process of impedance matching:

First, in real-life *RF* circuit block designs, the input or output impedance of an *RF* block may be required to match a specific impedance which is not 50 Ω for cost-saving and performance-improving purposes. However, in order to enable the testing of these *RF* circuit blocks, the input and output impedances must nonetheless be matched to 50 Ω, the standard reference impedance that test equipment manufacturers adopt. Thus, a somewhat complicated design procedure must be used when the actual *RF* circuit design must match an impedance that is not 50 Ω. This will be discussed in detail in Section 10.6.

Second, the impedance matching network can be built with either active or passive parts. It is more desirable to build an impedance matching network using passive parts than it is to build one using active parts because the passsive are simpler, and more current-saving, and cost-effective than the active. The passive parts include inductors, capacitors, and resistors; however, the resistor is usually excluded because it attenuates the signal and introduces considerable noise. Even though its use is allowed, almost no one does so. Consequently, what we need to be familiar with is how to apply the capacitor and inductor to the matching network. To build a narrow-band impedance matching network, one, two, or three parts are needed. However, the number of parts may be increased, depending on the relative bandwidth.

Third, it is necessary to pay attention to the demarcation of "narrow" and "wide" bands. As a matter of fact, this demarcation is not derived strictly from theory, but rather from the engineering design experience. Generally speaking, an *RF* circuit block is categorized as a narrow-band block if its relative bandwidth of operation is less than 15%; an *RF* circuit block is categorized as a wide-band block if its relative bandwidth of operation is more than 15%. When the relative bandwidth of operation is around 15%, it is better to treat an *RF* circuit block as a wide-band block so as to have a reasonable performance for both wide-band and narrow-band cases.

In the return loss *RL* (S_{11} or S_{22}) testing, the trace of impedance *Z* or return loss *RL* (S_{11} or S_{22}) displayed in the Smith chart corresponds to the response for a frequency bandwidth. An example is shown in Figure 10.1, where the displayed trace shows the frequency response of impedance *Z* or return loss *RL* (S_{11} or S_{22}) from lowest frequency to highest frequency, that is,

$$\Delta f = f_{\max} - f_{\min}, \tag{10.1}$$

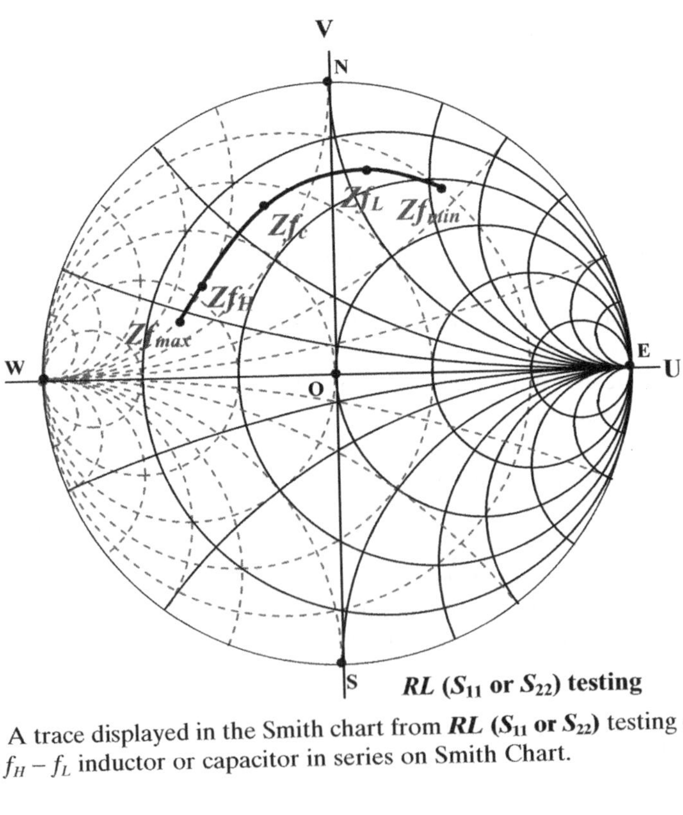

Figure 10.1 A trace displayed in the Smith chart from ***RL*** **(S_{11} or S_{22})** testing covers a bandwidth $BW = f_H - f_L$ inductor or capacitor in series on Smith Chart.

$$BW = f_H - f_L, \tag{10.2}$$

$$f_c = \frac{1}{2}(f_L + f_H). \tag{10.3}$$

where

Δf = frequency coverage of the trace,
f_{max} = highest frequency on the trace,
f_{min} = lowest frequency on the trace.
BW = frequency bandwidth,.
f_H = high-frequency end of desired bandwidth,
f_L = low-frequency end of desired bandwidth,
f_c = central frequency,

Each point on the displayed trace corresponds to a frequency. The values of impedance Z and return loss RL (S_{11} or S_{22}), can be read at each point simultaneously, such as,

Z_{fmin} = impedance at lowest frequency on the trace,
Z_{fmax} = impedance at highest frequency on the trace,
Zf_L = impedance at low-frequency end of desired bandwidth,
Zf_H = impedance at high-frequency end of desired bandwidth,
Zf_c = impedance at central frequency point.

In wide-band RF circuit design, the response for the whole frequency bandwidth must be taken care of. However, in the narrow-band RF circuit design, we can simply focus our attention on the central frequency because the frequency response at central frequency is a good approximation of the frequency response for the whole bandwidth. Consequently, in all the displays of the Smith chart in this chapter, only one point corresponding to the central frequency is shown since only narrow-band cases are discussed in this chapter.

Fourth, the Smith chart is a powerful tool in the impedance matching work. For a simple impedance matching work one can design the impedance matching network through hand-calculation by means of related equations. However, impedance matching tasks become much easier with the assistance of the Smith chart. The source or load impedance is no longer restricted to a pure resistance but can be a complex impedance, including both real and imaginary parts., though the standard reference impedance, say, 50 Ω, is always desired in the RF design.

There are many reference books about the applications of the Smith chart in engineering design. Here we are not attempting to repeat that information, but only to attach some important relations and equations between parameters in Appendix 10.A.1.

The passive impedance matching network for narrow-band cases will be discussed in this chapter and the passive impedance matching network for wide-band cases will be discussed in Chapter 11.

10.2 IMPEDANCE MATCHING BY MEANS OF RETURN LOSS ADJUSTMENT

10.2.1 Return Loss Circles on the Smith Chart

In the complex plane of voltage reflection coefficient, a family of power reflection coefficient, $\gamma = \Gamma^2$, and their corresponding return loss circles, either S_{11} or S_{22}, can be plotted using the Smith chart with impedance coordination since

$$\Gamma = U + jV, \tag{10.4}$$

then,

$$\gamma = \Gamma^2 = U^2 + V^2, \tag{10.5}$$

In the complex plane of voltage reflection coefficient, the scale in both the U and V axis is linear so that the equation (10.5) is a circle centered at reference impedance point, $50\,\Omega$, with a radius $|\Gamma|$. In other words, all the points for $|\Gamma|$ = constant are located on the same circle. Then, according to the definition of return loss,

$$RL(S_{11}.or.S_{22}) = 10Log|\Gamma|^2, dB \tag{10.6}$$

all the points for RL = constant are located on the same circle.

Figure 10.2 plots this family of circles. The center of these circles is the reference impedance point, $50\,\Omega$, that is, the center of the Smith chart. The constant value of each γ or RL circle can be determined, as in Table 9.A.1. For convenience, let's re-list the values of some relative columns from Table 9.A.1 on Table 10.1.

There are 39 rows in Table 10.1. In the last column, the values are symmetrical with respect to the twentieth row where S_{11} or S_{22} = –infinite. This indicates that there are two normalized resistances corresponding to a single value of S_{11} or S_{22} along the axis U ($V = 0$) in the complex plane of the voltage reflection coefficient. In other words, there are two points on the U axis ($V = 0$) for the same value of of S_{11} or S_{22}. One of them is located in the left-hand side of the circles' centers, $50\,\Omega$, while another is in the right-hand side. The distance between these two points is the diameter of the circle. For example,
when the return loss is

$$RL = S_{11} \quad \text{or} \quad S_{22} = -15\,dB, \tag{10.7}$$

the two corresponding normalized resistances or the two resistances are

$$r = 0.6980 \quad \text{and} \quad 1.4326, \tag{10.8}$$

$$R = 34.9020\,\Omega \quad \text{and} \quad 71.6291\,\Omega. \tag{10.9}$$

It should be noted that the value of the circles can be alternatively determined by the radially scaled ruler on a conventional Smith chart.

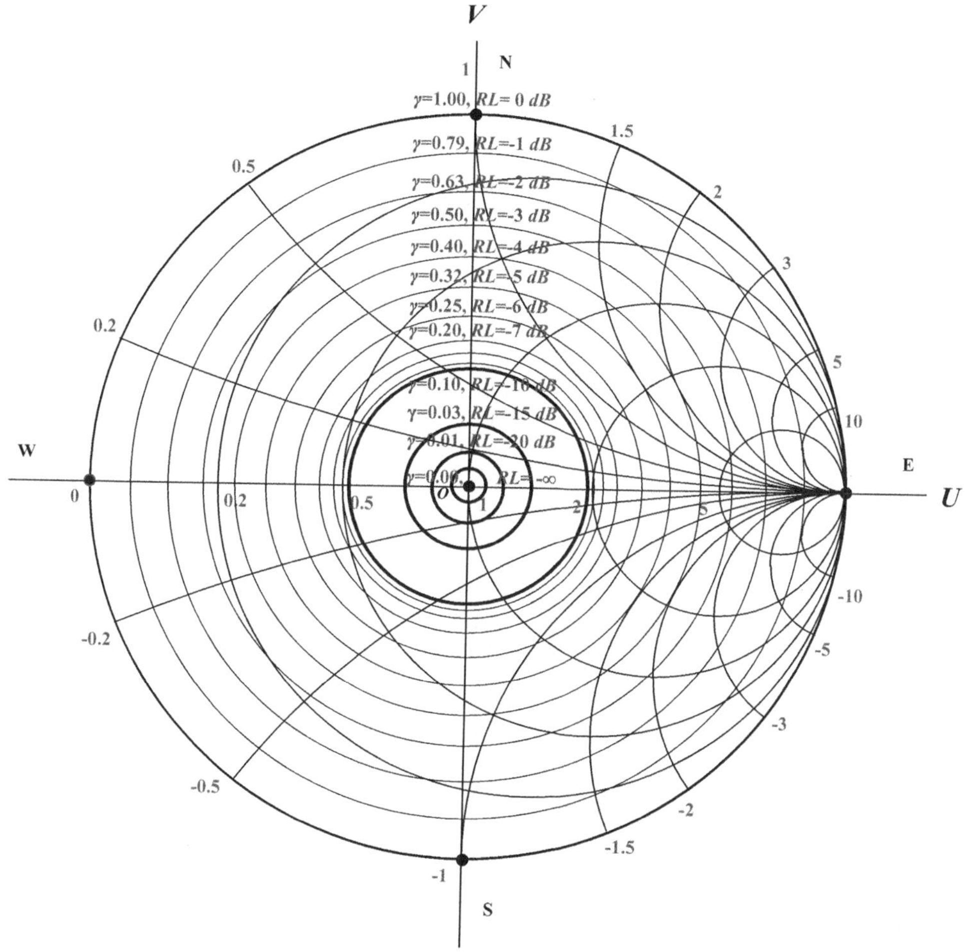

Note 1: Power reflection coefficient, γ, and return loss, *RL* (S_{11} or S_{22}) is set in boldface with values along the vertical axis, *V*, such as **γ =1, *RL*=0 *dB*, γ=0.79, *RL*=−1 *dB*, γ=0.63, *RL*=−2 *dB*.**

Note 2: Normalized resistance, *r*, is set in boldface with values around the biggest circle, such as **0, 0.1, 0.2, 0.5, 1, 1.5, 2, 3, 5, 10, ∞, −10, −5, −3, −2, −1.5, −1, −0.5, −0.2, −0.1.**

Note 3: Normalized reactance, *x*, is set in boldface with values along the horizontal axis *U*, such as **0, 0.2, 0.5, 1, 2, 5, ∞.**

Figure 10.2 Constant return loss $S_{11,\mathrm{dB}}$ or $S_{22,\mathrm{dB}}$ circles on the Smith chart.

10.2.2 Relationship between Return Loss and Impedance Matching

Figure 10.2 indicates on the Smith chart one of the important features about impedance. If an impedance is far from the center of the Smith chart, it is far from the impedance matching state, so that the return loss is high. In an extreme case, the impedance is located at the largest outer circle which implies a zero resistance. This leads to a maximum of return loss, $0\,dB$, in which the reflection voltage is equal to the incident voltage. A terminal with such an impedance is in a full reflection state.

TABLE 10.1 Variation of return loss *RL* along with *U* axis ($V = 0$)

Resistance, R, Ω	Reference resistance, R_o, Ω	Normalized resistance, $rr = R/R_o$	*Power Reflection coefficient* $\gamma = \Gamma^2$	Return loss, *RL* S_{11} or S_{22}, *dB*
2.8750	50	0.0575	0.79	−1
5.7313	50	0.1146	0.63	−2
8.5500	50	0.1710	0.50	−3
11.3136	50	0.2263	0.40	−4
14.0066	50	0.2801	0.32	−5
16.6140	50	0.3323	0.25	−6
19.1236	50	0.3825	0.20	−7
21.5252	50	0.4305	0.16	−8
23.8110	50	0.4762	0.13	−9
25.9747	**50**	**0.5195**	**0.10**	**−10**
29.9240	50	0.5986	0.06	−12
34.9020	50	0.6980	0.03	−15
40.9091	50	0.8182	0.01	−20
44.6760	50	0.8935	0.00	−25
46.9347	50	0.9387	0.00	−30
48.2528	50	0.9651	0.00	−35
49.0099	50	0.9802	0.00	−40
49.4408	50	0.9888	0.00	−45
49.6848	50	0.9937	0.00	−50
50.0000	50	1.0000	0.00	(−Infinite)
50.3172	50	1.0063	0.00	−50
50.5655	50	1.0113	0.00	−45
51.0101	50	1.0202	0.00	−40
51.8105	50	1.0362	0.00	−35
53.2656	50	1.0653	0.00	−30
55.9585	50	1.1192	0.00	−25
61.1111	50	1.2222	0.01	−20
71.6291	50	1.4326	0.03	−15
83.5450	50	1.6709	0.06	−12
96.2475	**50**	**1.9250**	**0.10**	**−10**
104.9942	50	2.0999	0.13	−9
116.1431	50	2.3229	0.16	−8
130.7280	50	2.6146	0.20	−7
150.4750	50	3.0095	0.25	−6
178.4900	50	3.5698	0.32	−5
220.9700	50	4.4194	0.40	−4
292.4050	50	5.8481	0.50	−3
436.2200	50	8.7244	0.63	−2
869.5500	50	17.3910	0.79	−1

On the other hand, another extreme case is when the impedance is equal to the reference impedance 50 Ω. It is a terminal with return loss = $-\infty$, an ideal impedance matched state without reflection. In other cases, as the impedance approaches the center of Smith chart from the largest circle, the power reflection coefficient γ and the return loss *RL* decrease simultaneously. Generally speaking, both the power reflection coefficient γ and the return loss *RL* are indicators of the impedance matching state. However, from the values of RL and γ as specified in Figure 10.2, it is easily shown that the return loss RL is more sensitive to the impedance matching state than the power reflection coefficient γ. This is why the return loss RL (S_{11} or S_{22}), instead of the power reflection coefficient γ, is always taken as an indicator of the impedance matching state.

In practical engineering design, the ideal impedance matching state is never reached. The RF circuit designers do always try their best to match an original impedance to the ideal matched state. Does any rule exist to measure the impedance matching state? No! There is no formal regulation or rule suggested by anyone or by any international organization. RF circuit designers could have their own rules to judge the impedance matching state. However, an approximate rule is very often adapted by RF circuit designers in their simulations, that is,

$$RL = S_{11} \quad \text{or} \quad S_{22} = -10\,dB, \tag{10.10}$$

from Table 10.1, the two corresponding normalized resistances and the two resistances are

$$r = 0.5195 \quad \text{and} \quad 1.9250, \tag{10.11}$$

$$R = 25.9747\,\Omega \quad \text{and} \quad 96.2475\,\Omega. \tag{10.12}$$

In Figure 10.2, the circle with $RL = S_{11}$ or $S_{22} = -10\,dB$ is plotted with a bold line and its radius is approximately one-third of the radius of the largest circle.

From now on, let's take expression (10.10) as an approximate demarcation criterion. This implies that

- The impedance matching state is acceptable if

$$RL = S_{11} \quad \text{or} \quad S_{22} < -10\,dB, \tag{10.13}$$

in which the impedance is close enough to the reference impedance $50\,\Omega$, or the center of Smith chart.

- On the other hand, the impedance matching state is unacceptable if

$$RL = S_{11} \quad \text{or} \quad S_{22} > -10\,dB, \tag{10.14}$$

in which the impedance is far from the reference impedance $50\,\Omega$, or the center of the Smith chart.

10.2.3 Implementation of an Impedance Matching Network

An original RF block usually has an impedance not equal to the reference impedance of $50\,\Omega$ and is expected to be matched. Therefore, an impedance matching network must be inserted between the original *RF* block and a network analyzer which provides 50 Ω of reference impedance.

Figure 10.3 shows a setup for implementation of the impedance matching network. It consists of three portions: a network analyzer to read return loss or impedance, the original *RF* block to be matched, and the impedance matching network to be built.

Impedance matching is a process of constructing an impedance network and is executed through the adjustment of return loss on the Smith chart.

The impedance matching network consists of arms and branches. The arm is a circuit segment in series and looks like a "horizontal" component while the branch is a circuit segment in parallel and looks like a "vertical" component. As shown in Figure 10.4, in the narrow-band case each arm or branch contains only one part, either a capacitor or an inductor.

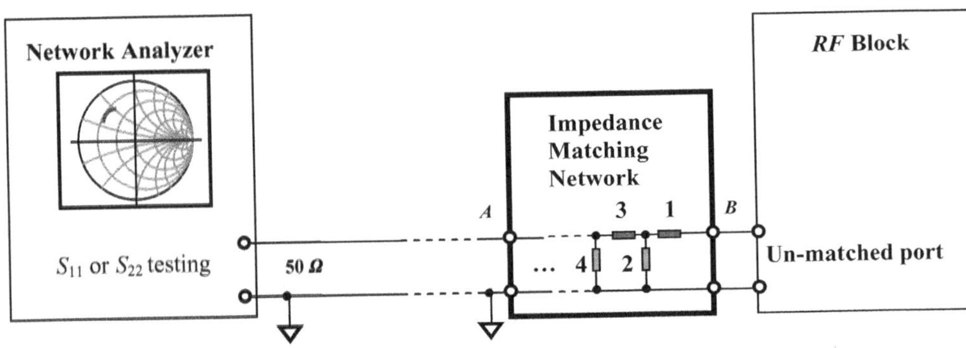

Figure 10.3 Setup for impedance matching from input impedance of an original *RF* block to 50 Ω.

Arm **Branch**

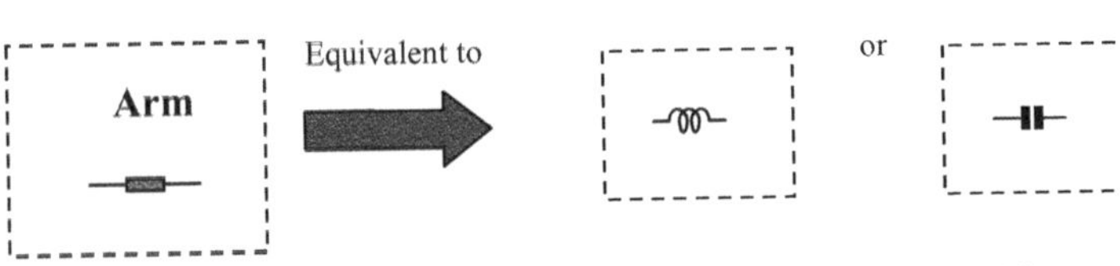

(a) An arm consists of one part in the impedance matching network. It is a connected component in "horizontal" direction (in series)

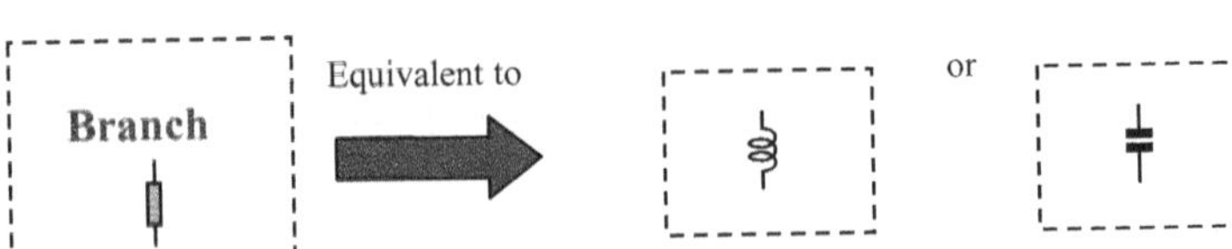

(b) A branch consists of one part in the impedance matching network. It is a connected component in "vertical" direction (in parallel)

Figure 10.4 Representation of arm or branch consisting of one part.

In Figure 10.3 the arms or branches, marked 1, 2, 3, 4, and so on, are inserted into the impedance matching network one by one sequentially. At the beginning, the impedance matching network is empty so that point A is directly connected to point B. The S_{11} or impedance trace is displayed on the screen of network analyzer. It represents the original impedance of the RF block. As the first arm marked with "1" is inserted between point A and B, the S_{11} or impedance trace will move to another location on the Smith chart. As further branches or arms are inserted into the impedance matching network, the S_{11} or impedance trace will move to other locations on the Smith chart accordingly. The S_{11} or impedance trace is expected to ultimately move to the center of the Smith chart, 50 Ω. If so, the process of impedance matching is complete and the resultant entity of arms and branches is the desired impedance matching network. Usually more than one arm or one branch must be added so that the final impedance, say, 50 Ω, at point A can be approached.

10.3 IMPEDANCE MATCHING NETWORK BUILT OF ONE PART

10.3.1 One Part Inserted into Impedance Matching Network in Series

When the impedance matching network is inserted with one part in series, it is preferable to use impedance, instead of admittance, to describe the variation of its electrical characteristics. We have

$$Z_o = R_o + jX_o, \tag{10.15}$$

$$Z = Z_o + \Delta Z = R + jX, \tag{10.16}$$

where

Z_o = original impedance before any part is inserted,
R_o = real part of Z_o,
X_o = imaginary part of Z_o,
Z = resultant impedance after the part is inserted,
ΔZ = variation of impedance,
R = resistance of resultant impedance, or real part of Z,
X = reactance of resultant impedance, or imaginary part of Z.

If the part inserted in the impedance matching network is either an ideal inductor or an ideal capacitor, then only the imaginary part of the resultant impedance is changed, that is,

$$\Delta R = R - R_o = 0, \tag{10.17}$$

$$\Delta X = X - X_o, \tag{10.18}$$

where

ΔR = variation of resistance,
ΔX = variation of reactance.

When the impedance matching network is inserted with an inductor in series,

$$\Delta X = \Delta X_L = +L\omega, \tag{10.19}$$

The resultant impedance is, of course, from (10.15) to (10.19),

$$Z = Z + \Delta Z_L = R + jX = R_o + j(X_o + \Delta X_L), \tag{10.20}$$

and when the impedance matching network is inserted with a capacitor in series,

$$\Delta X = \Delta X_C = -\frac{1}{C\omega}, \tag{10.21}$$

The resultant impedance is, of course, from (10.14) to (10.18),

$$Z = Z + \Delta Z_C = R + jX = R_o + j(X_o + \Delta X_C), \tag{10.22}$$

where

L = inductance of the part inserted to the impedance matching network,
C = capacitance of the part inserted to the impedance matching network,
ω = angular frequency,
ΔZ_L = inserted inductive impedance,
ΔZ_C = inserted capacitive impedance,
ΔX_L = inserted inductive reactance,
ΔX_C = inserted capacitive reactance,

Figure 10.5 shows the inserted reactance ΔX and Figure 10.6 shows the direction the impedance is pulled on the Smith chart when the impedance matching network is inserted with L or C in series. It can be seen that the variation of impedance in the Smith chart obeys the following rules of thumb:

When the impedance matching network is inserted with an inductor in series,

- The inserted inductive reactance, $\Delta X = \Delta X_L$, is positive and its magnitude increases as the operating frequency increases.
- The insertion of an inductor in series, L_S, results in the original impedance P moving clockwise along the r = constant impedance circle. The movement arc length depends on the value of the inductor.

When the impedance matching network is inserted with a capacitor in series

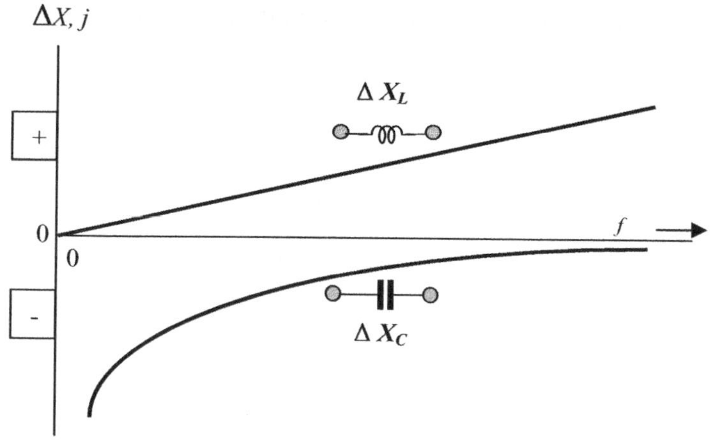

Figure 10.5 Variation of reactance when impedance matching network is inserted with one part, either inductor or capacitor, in series.

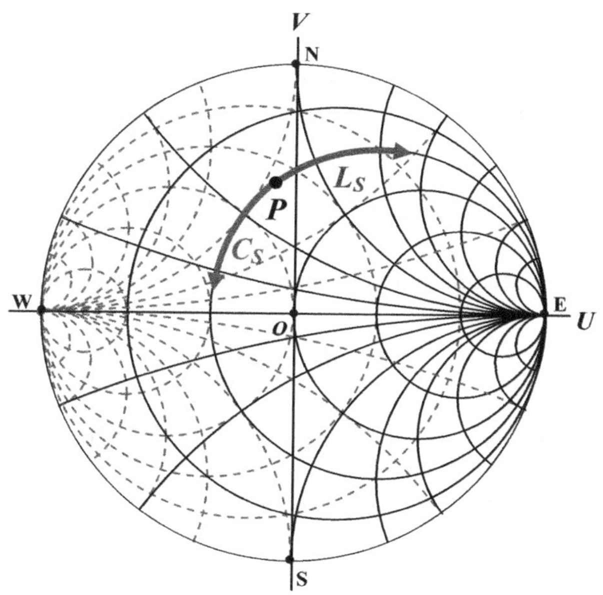

Figure 10.6 Pulled directions of impedance by the insertion of an inductor or a capacitor in series in a Smith chart.

- The inserted capacitive reactance, $\Delta X = \Delta X_C$, is negative and its magnitude decreases as the operating frequency increases.
- The insertion of a capacitor in series, C_S, results in the original impedance P moving counter-clockwise along the r = constant impedance circle. The movement arc length depends on the value of capacitor.

10.3.2 One Part Inserted into Impedance Matching Network in Parallel

When an impedance matching network is inserted with one part in parallel, it is preferable to use admittance, instead of impedance, to describe the variation of its electrical characteristics. We have

$$Y_o = G_o + jB_o \text{ i}, \tag{10.23}$$

$$Y = Y_o + \Delta Y = G + jB \text{ i}, \tag{10.24}$$

where

Y_o = original admittance before the part is inserted,
G_o = real part of Y_o,
G_o = imaginary part of Y_o,
Y = resultant admittance after the part is inserted,
ΔY = variation of admittance,
G = conductance of resultant admittance, or real part of Y,
B = susceptance of resultant admittance, Imaginary part of Y,

If the part inserted into the impedance matching network is either an ideal inductor or an ideal capacitor, then only the imaginary part of the resultant impedance is changed, that is,

$$\Delta G = G - G_o = 0, \quad \Delta B = B - B_o, \tag{10.25}$$

When an impedance matching network is inserted with an inductor in parallel,

$$\Delta B = \Delta B_L = -\frac{1}{L\omega}, \tag{10.26}$$

The resultant admittance is, of course, from (10.23) to (10.26),

$$Y = Y + \Delta Y_L = G + jB = G_o + j(B_o + \Delta B_L), \tag{10.27}$$

when the impedance matching network is inserted with a capacitor in parallel,

$$\Delta B = \Delta B_C = C\omega, \tag{10.28}$$

The resultant admittance is, of course, from (10.23) to (10.25),

$$Y = Y + \Delta Y_C = G + jB = G_o + j(B_o + \Delta B_C), \tag{10.29}$$

where

L = inductance of the part inserted to the block,
C = capacitance of the part inserted to the block,

ω = angular frequency,
ΔY_L = inserted inductive admittance,
ΔY_C = inserted capacitive admittance,
ΔB_L = inserted inductive susceptance,
ΔB_C = inserted capacitive susceptance.

Figure 10.7 shows the inserted susceptance ΔY and Figure 10.8 shows the pulled direction of impedance on the Smith chart when the impedance matching network

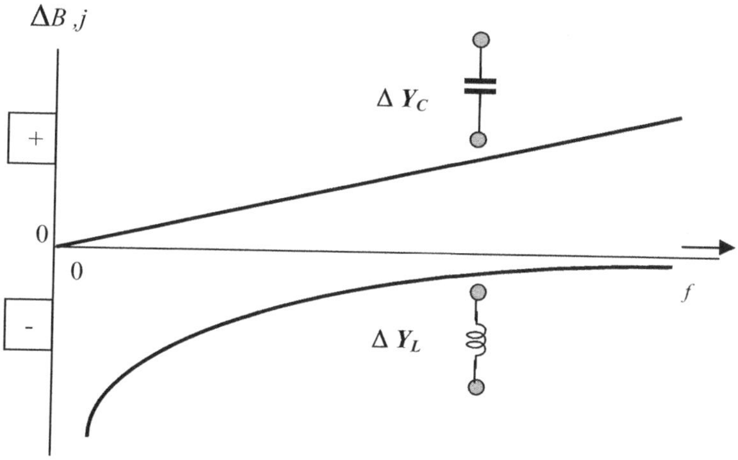

Figure 10.7 Variation of susceptance when impedance matching network is inserted with one part, either inductor or capacitor, in parallel.

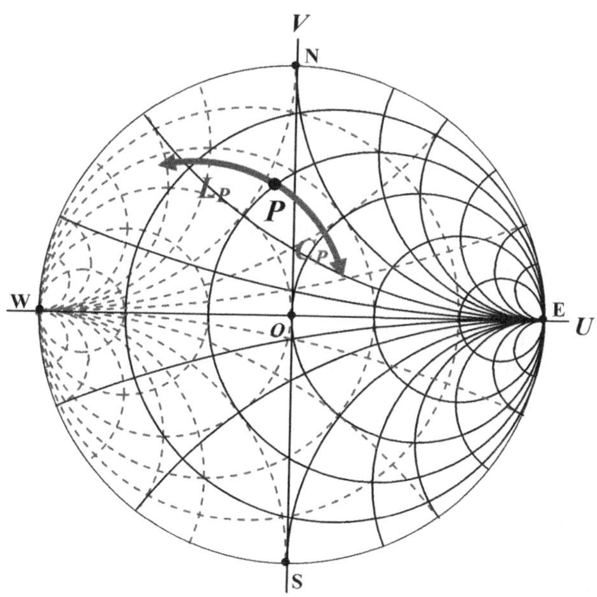

Figure 10.8 Pulled directions of impedance by the insertion of an inductor or a capacitor in parallel on a Smith chart.

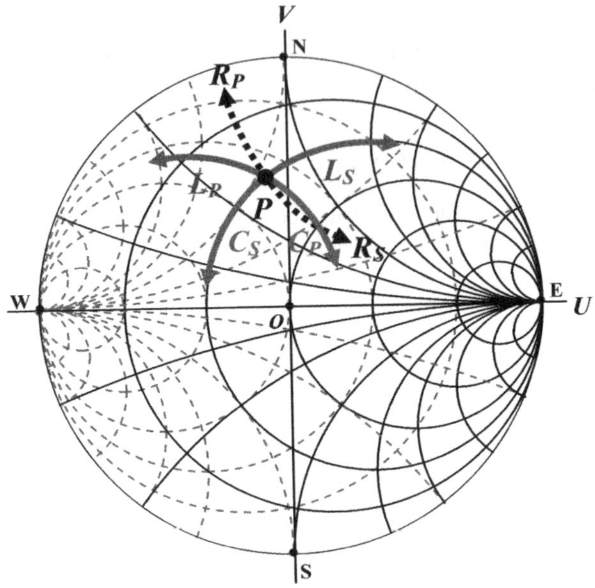

Figure 10.9 Pulled directions of impedance by the insertion of L, C, or R on a Smith chart.

is inserted with L or C in parallel. It can be seen that the variation of impedance on the Smith chart obeys the following rules of thumb:

When the impedance matching network is inserted with a capacitor in parallel,

- The inserted susceptance, $\Delta Y = \Delta Y_C$, is positive and its magnitude increases as the operating frequency increases.
- The insertion of a capacitor in parallel, C_P, results in the original impedance P moving clockwise along the g = constant admittance circle. The movement arc length depends on the value of capacitor.

When the impedance matching network is inserted with an inductor in parallel,

- The inserted susceptance, $\Delta Y = \Delta Y_L$, is negative and its magnitude decreases as the operating frequency increases.
- The insertion of an inductor in parallel, L_P, results in the original impedance P moving counter-clockwise along the g = constant admittance circle. The movement arc length depends on the value of inductor.

As a matter of fact, Figures 10.6 and 10.8 can be combined into Figure 10.9, where the pulled directions due to the insertion of a resistor in series and in parallel are also depicted, though it makes sense only theoretically, not in practical engineering design. Theoretically,

- The insertion of a resistor in series, R_S, results the original impedance P moving along the x = constant arc to a higher resistance circle. The moved distance depends on the value of the resistor.

- The insertion of a resistor in parallel, R_P, results in the original impedance P moving along the x = constant arc to a lower resistance circle. The moved distance depends on the value of the resistor.

Obviously, there is a very slim chance of being able to build an impedance matching network containing only one part because it is impossible in most cases to pull the original impedance to the reference impedance, $50\,\Omega$, by only one part, either inductor or capacitor, unless the original impedance is located on both circles with $r = 1$ and $g = 1$, or, in two narrow-ring areas adjacent to both the circles.

However, it is possible to use two parts to match the original impedance to a desired value if the block is operating for a narrow-band system. In most of *RF* circuit block design, it is quite conventional to implement an impedance matching network in two parts.

10.4 IMPEDANCE MATCHING NETWORK BUILT OF TWO PARTS

10.4.1 Regions in the Smith Chart

The topology of a matching network depends on the destination impedance to be matched to and the original impedance to be matched from. Usually, if the destination impedance is the standard reference $50\,\Omega$, then the topology of the impedance matching network depends only on the location of the original impedance to be matched on the Smith chart.

In order to be able to formulize the values of the impedance matching parts, it is suggested dividing the Smith chart into four regions. (Richard Chi-Hsi Li, "Key Issues in RF/RFIC Circuit Design", Higher Education Press, Beijing, 2005, pp. 45–46)

Figure 10.10 shows the demarcation of these four regions on the Smith chart and Table 10.2 lists the range of impedance in these four regions. It will be quite convenient to the analysis of impedance matching if the Smith chart is divided into four regions. Not only the topology but also the value of the parts in an impedance matching network can be easily determined and calculated.

10.4.2 Values of Parts

Let's denote the original impedance to be matched as

$$Z_m = R_m + jX_m, \tag{10.30}$$

where

Z_m = original impedance to be matched,
R_m = resistance of original impedance,
X_m = reactance of original impedance.

Figure 10.11 shows that in all regions there are two ways to pull the original impedance, at P_1, P_2, P_3, and P_4, to the center of the Smith chart, $\boldsymbol{O}$, by the addition of two

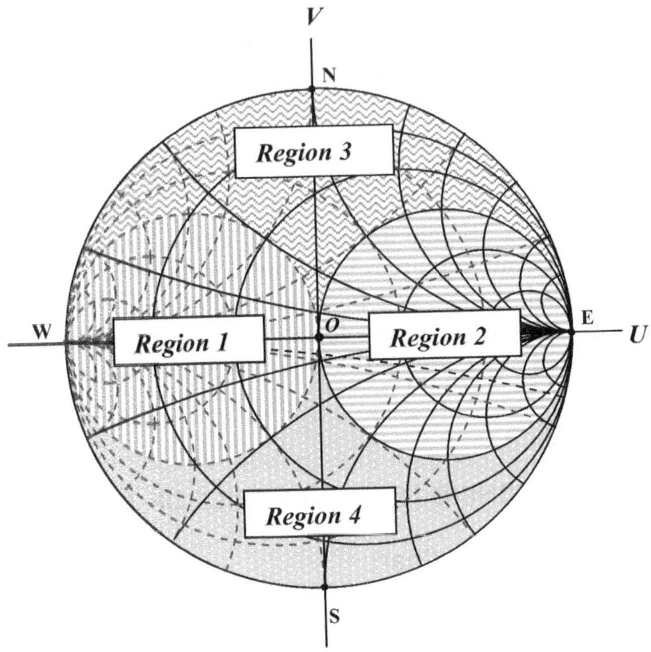

Figure 10.10 Demarcation of four regions on a Smith chart.

TABLE 10.2 Range of impedance in 4 regions on a Smith chart

Region 1 Low resistance or high conductance	Region 2 High resistance or low conductance	Region 3 Low resistance and low conductance	Region 4 Low resistance and low conductance
$r < 1$ $x < \|0.5\|$ $g > 1$ $-\infty < b < +\infty$	$r > 1$ $-\infty < x < +\infty$ $g < 1$ $b < \|0.5\|$	$r < 1$ $x > 0$ $g < 1$ $b < 0$	$r < 1$ $x < 0$ $g < 1$ $b > 0$

types of passive parts, the inductor and capacitor. The subscript of P represents the region where the impedance is located. From Figure 10.11 it can be seen that.

There are two ways to pull P_1 to the center of Smith chart, 50 Ω:

1) In Figure 10.11(a) P_1 is pulled to A by the addition of a capacitor C_S in series first, and then from A to O by the addition of an inductor L_P in parallel.
2) In Figure 10.11(a) P_1 is pulled to B by the addition of an inductor L_S in series first, and then to pull B to O by the addition of a capacitor C_P in parallel.

There are two ways to pull P_2 to the center of Smith chart, 50 Ω:

1) In Figure 10.11(b) P_2 is pulled to C by the addition of an inductor L_P in parallel first, and then from C to O by the addition of a capacitor C_S in series.

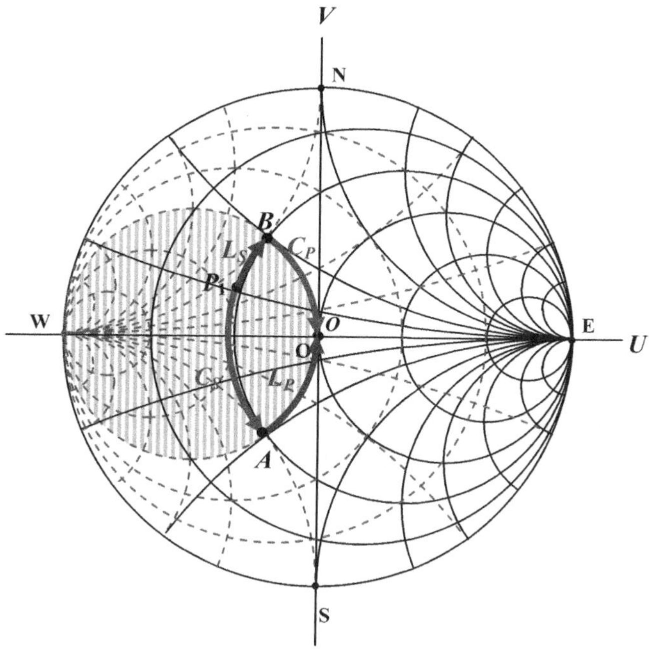

Figure 10.11(a) Two ways to pull the original impedance P_1 in region 1 to the center of a Smith chart, O, by addition of two passive parts.

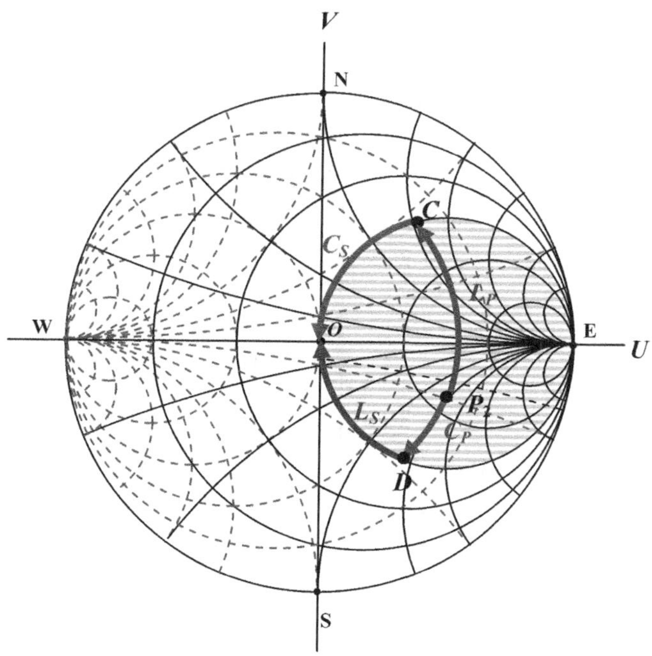

Figure 10.11(b) Two ways to pull the original impedance P_2 in region 2 to the center of a Smith chart, O, by addition of two passive parts.

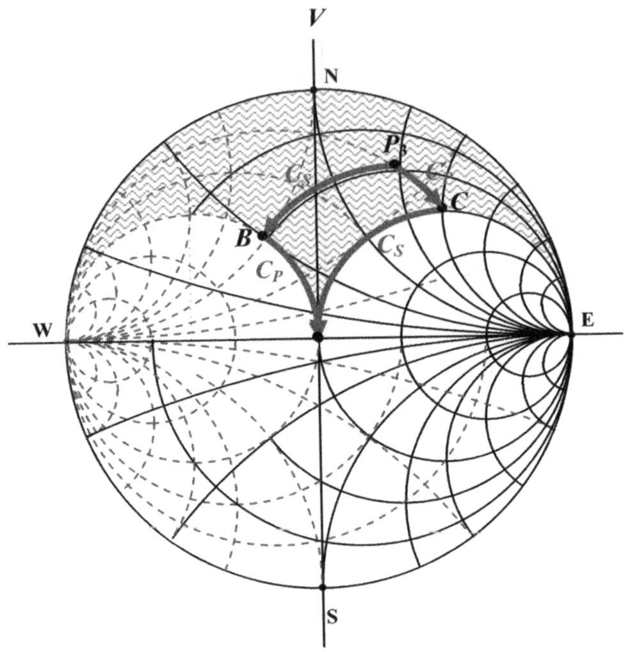

Figure 10.11(c) Two ways to pull the original impedance P_3 in region 3 to the center of a Smith chart, O, by addition of two passive parts.

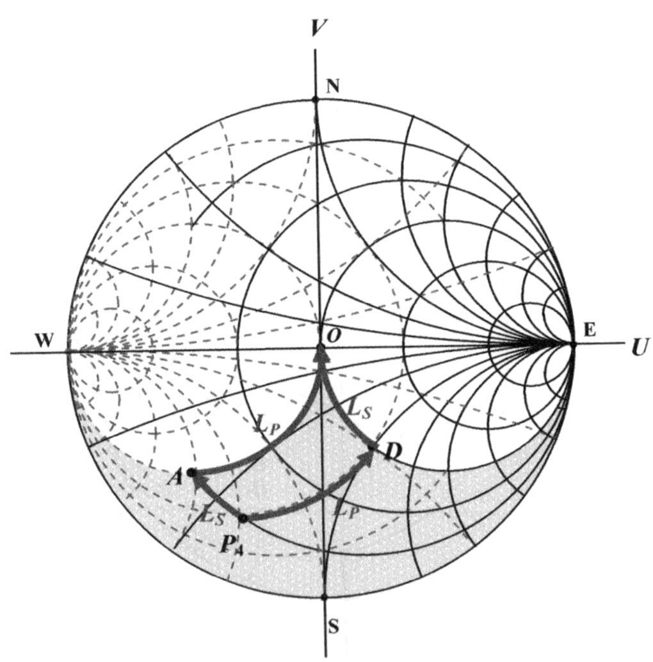

Figure 10.11(d) Two ways to pull the original impedance P_4 in region 4 to the center of a Smith chart, O, by addition of two passive parts.

2) In Figure 10.10(b) $\boldsymbol{P_2}$ is pulled to $\boldsymbol{D}$ by the addition of a capacitor C_P in parallel first, and then to pull $\boldsymbol{D}$ to $\boldsymbol{O}$ by the addition of an inductor L_S in series.

There are two ways to pull $\boldsymbol{P_3}$ to the center of the Smith chart, 50 Ω:

1) In Figure 10.11(c) $\boldsymbol{P_3}$ is pulled to $\boldsymbol{B}$ by the addition of a capacitor C_S in series first, and then from $\boldsymbol{B}$ to $\boldsymbol{O}$ by the addition of a capacitor C_P in parallel.
2) In Figure 10.11(c) $\boldsymbol{P_3}$ is pulled to $\boldsymbol{C}$ by the addition of a capacitor C_P in parallel first, and then to pull $\boldsymbol{C}$ to $\boldsymbol{O}$ by the addition of a capacitor C_S in series.

There are two ways to pull $\boldsymbol{P_4}$ to the center of Smith chart, 50 Ω:

1) In Figure 10.11(d) $\boldsymbol{P_4}$ is pulled to $\boldsymbol{D}$ by the addition of an inductor L_P in parallel first, and then from $\boldsymbol{D}$ to $\boldsymbol{O}$ by the addition of an inductor L_S in series.
2) In Figure 10.11(d) $\boldsymbol{P_4}$ is pulled to $\boldsymbol{A}$ by the addition of an inductor L_S in series first, and then to pull $\boldsymbol{A}$ to $\boldsymbol{O}$ by the addition of an inductor L_P in parallel.

Three common rules or features can be summarized from the description above:

1) The first part in a two part impedance matching network is to pull the original impedance to either one of the circles $g = 1$ or $r = 1$. Then,
2) The second part in a two part impedance matching network is to pull the impedance on the circle either $g = 1$ or $r = 1$, after it is pulled by the first part, to the standard reference impedance, 50 Ω.
3) One of two matching parts is in series and another one is in parallel. However, the designer can select the first part in series or in parallel because, as described above, for all possible original impedances there are two ways to be pulled to the center of the Smith chart. The first part in one way is in series while in another way it is in parallel, and vice versa.

On the basis of the common rules or features mentioned above, the two part type can be selected and the values of the matching parts, one in series and another one in parallel, can be formulized. This is a tedious mathematical process and will be introduced in Appendix 10.A.2. What we are going to do is to introduce the appropriate formulas from Appendix 10.A.2 so that the values of the two impedance matching parts can be calculated when the original impedance is located in any of the fjour regions. The following four cases correspond to four regions where the original impedance is located.

1. The original impedance to be matched is located in Region 1.

There are two ways to pull $\boldsymbol{P_1}$ to the center of the Smith chart, 50 Ω:

1) By the addition of a capacitor C_S in series first, and then by the addition of an inductor L_P in parallel.

$$C_S = \frac{C_m}{C_m \omega \sqrt{R_m(Z_o - R_m)} - 1}, \tag{10.31}$$

$$L_P = \frac{R_m Z_o}{\omega\sqrt{R_m (Z_o - R_m)}}, \tag{10.32}$$

where

C_S = impedance matching part, a capacitor in series,
L_P = impedance matching part, an inductor in parallel,
C_m = capacitance of original impedance,
ω = angular operating frequency,
Z_o = standard reference resistance, 50 Ω usually,
R_m = resistance of original impedance,

2) By the addition of an inductor L_S in series first, and then by the addition of a capacitor C_P in parallel.

$$L_S = \frac{\sqrt{R_m(Z_o - R_m)} - X_m}{\omega}, \tag{10.33}$$

$$C_P = \frac{\sqrt{R_m (Z_o - R_m)}}{R_m Z_o \omega}, \tag{10.34}$$

Where

L_S = impedance matching part, an inductor in series,
C_P = impedance matching part, a capacitor in parallel,
ω = ingular operating frequency,
Z_o = itandard reference resistance, 50 Ω usually,
R_m = resistance of original impedance,
X_m = reactance of original impedance.
$Z_m = R_m + X_m$ original impedance to be matched.

2. The original impedance to be matched is located in Region 2.

There are two ways to pull P_2 to the center of the Smith chart, 50 Ω:

1) By the addition of an inductor L_P first in parallel, and then by the addition of a capacitor C_S in series.

$$L_P = \frac{X_m Z_o + \sqrt{R_m Z_o (R_m^2 + X_m^2 - R_m Z_o)}}{(R_m - Z_o)\omega}, \tag{10.35}$$

$$C_S = \frac{R_m/\omega}{\sqrt{R_m Z_o (R_m^2 + X_m^2 - R_m Z_o)}}. \tag{10.36}$$

2) By the addition of a capacitor C_P in parallel first, and then by the addition of an inductor L_S in series.

$$C_P = \frac{(R_m - Z_o)/\omega}{\sqrt{R_m Z_o (R_m^2 + X_m^2 - R_m Z_o)} - X_m Z_o}, \tag{10.37}$$

$$L_S = \frac{\sqrt{R_m Z_o (R_m^2 + X_m^2 - R_m Z_o)}}{R_m \omega}. \tag{10.38}$$

3. The original impedance to be matched is located in Region 3.

There are two ways to pull $\boldsymbol{P_3}$ to the center of the Smith chart, $50\,\Omega$:

1) By the addition of a capacitor C_S in series first, and then by the addition of a capacitor C_P in parallel.

$$C_S = \frac{C_m}{C_m \omega \sqrt{R_m (Z_o - R_m)} - 1}, \tag{10.39}$$

$$C_P = \frac{\sqrt{R_m (Z_o - R_m)}}{R_m Z_o \omega}, \tag{10.40}$$

2) By the addition of a capacitor C_P in parallel first, and then by the addition of a capacitor C_S in series.

$$C_P = \frac{(R_m - Z_o)/\omega}{\sqrt{R_m Z_o (R_m^2 + X_m^2 - R_m Z_o)} - X_m Z_o}, \tag{10.41}$$

$$C_S = \frac{R_m/\omega}{\sqrt{R_m Z_o (R_m^2 + X_m^2 - R_m Z_o)}}. \tag{10.42}$$

4. The original impedance to be matched is located in Region 4.

There are two ways to pull $\boldsymbol{P_4}$ to the center of Smith chart, $50\,\Omega$:

1) By the addition of an inductor L_P in parallel first, and then by the addition of an inductor L_S in series.

$$L_S = \frac{\sqrt{R_m Z_o (R_m^2 + X_m^2 - R_m Z_o)}}{R_m \omega}, \tag{10.43}$$

$$L_P = \frac{X_m Z_o + \sqrt{R_m Z_o (R_m^2 + X_m^2 - R_m Z_o)}}{(R_m - Z_o)\omega}. \tag{10.44}$$

2) By the addition of an inductor L_S in series first, and then by the addition of an inductor L_P in parallel.

$$L_P = \frac{R_m Z_o}{\omega \sqrt{R_m (Z_o - R_m)}}, \tag{10.45}$$

$$L_S = \frac{\sqrt{R_m (Z_o - R_m)} - X_m}{\omega}. \tag{10.46}$$

Counting from expressions (10.31) to (10.46), there are 16 expressions in total. In fact there are only eight independent equations.

Before these equations were developed, the designer could only rely on the Smith chart and calculate the values of impedance matching parts by carefully estimating the location of the impedance and then measuring the arc length on the Smith chart. These calculations always bring about a value with significant inaccuracy, especially when the impedance to be measured is located in the high impedance area on the Smith chart. In addition, the calculations are quite tedious.

In terms of these eight equations, it is quite convenient to calculate the values of the impedance matching parts for all cases no matter in which region the original impedance Z_m to be matched is located, as long as the original impedance Z_m and the operating frequency ω are known. Armed with these eight equations, the designer does not need assistance from the Smith chart.

In summary, in each region there are two options corresponding to the two topologies of impedance matching networks as shown in Figure 10.11. It should be noted that the selection of whether to use an inductor or a capacitor for X_1 and X_2 depends on the region where the original impedance Z_m, or R_m and X_m is located. Figure 10.11 shows that

- If the original impedance is located in regions 1 or 2, the two matched parts consist of one capacitor and one inductor.
- If the original impedance is located in region 3, both impedance-matching parts are two capacitors.
- If the original impedance is located in region 4, both impedance-matching parts are two inductors.

10.4.3 Selection of Topology

In general, there are three factors that must be considered in the selection of the topology for an impedance matching network:

1) Consideration of the availability of topology

Table 10.3 lists eight possible topologies for an impedance matching network containing two parts.

TABLE 10.3 Eight possible topologies of an impedance matching network containing two passive parts

1)	$C_P - L_S$
2)	$L_S - C_P$
3)	$C_S - L_P$
4)	$L_P - C_S$
5)	$C_P - C_S$
6)	$C_S - C_P$
7)	$L_P - L_S$
8)	$L_S - L_P$

Notes: The first part is connected to the original impedance to be matched and the second part is connected to the standard reference impedance, 50 Ω. The subscript "*P*" stands for "in parallel" and the subscript "*S*" stands for "in series."

As discussed above, only two topologies are available to match a specific original impedance in the Smith chart to the center or the reference impedance of 50 Ω. In other words, not all of the possible topologies are available for the implementation of a two-part impedance matching network. The availability of topology depends on the location of the original impedance in the Smith chart. An impedance matching network with a specific topology is able to match an original impedance to 50 Ω only if this original impedance is located in some specific regions of Smith chart. These regions are called "applied regions" of the specific topology. On the other hand, an impedance matching network is unable to match an original impedance to 50 Ω if this original impedance is located in some specific regions of the Smith chart. These regions are called "prohibited regions" of the specific topology.

Appendix 10.A.3 presents the applied area and the prohibited area in a Smith chart for a specific topology of an impedance matching network containing two parts, also summarized in Table 10.4.

These restrictions of impedance matching networks containing two parts can be avoided as long as the designer selects a correct topology based on Table 10.4.

2) Consideration of cost

Obviously, from a cost-saving viewpoint, the last two topologies, 7) and 8), in Table 10.4 with two inductors are not good choices because they are too expensive. These two topologies should be abandoned.

Fortunately, some other topologies are available if the original impedance to be matched is located in region 4. Instead of two inductors, the impedance matching network can be built of one inductor and one capacitor as shown in Figure 10.12, where the original impedance $\boldsymbol{P_4}$ can be pulled to the center of

TABLE 10.4 Applied and prohibited regions of an impedance matching network with specific topology

Topology	Z_m is located in applied region	Z_m is located in prohibited region
1) $\boldsymbol{C_P - L_S}$	Regions 2, 3	Regions 1, 4
2) $\boldsymbol{L_S - C_P}$	**Regions 1, 4**	Regions 2, 3
3) $\boldsymbol{C_S - L_P}$	Regions 1, 3	Regions 2, 4
4) $\boldsymbol{L_P - C_S}$	**Regions 2, 4**	Regions 1, 3
5) $\boldsymbol{C_P - C_S}$	Region 3	Regions 1, 2, 4
6) $\boldsymbol{C_S - C_P}$	Region 3	Regions 1, 2, 4
7) $\boldsymbol{L_P - L_S}$	**Region 4**	Regions 1, 2, 3
8) $\boldsymbol{L_S - L_P}$	**Region 4**	Regions 1, 2, 3

Notes: Z_m = Original impedance to be matched. In the first column the first part is connected to the original impedance to be matched and the second part is connected to the standard reference impedance, 50 Ω. The subscript "P" stands for "in parallel" and the subscript "S" stands for "in series."

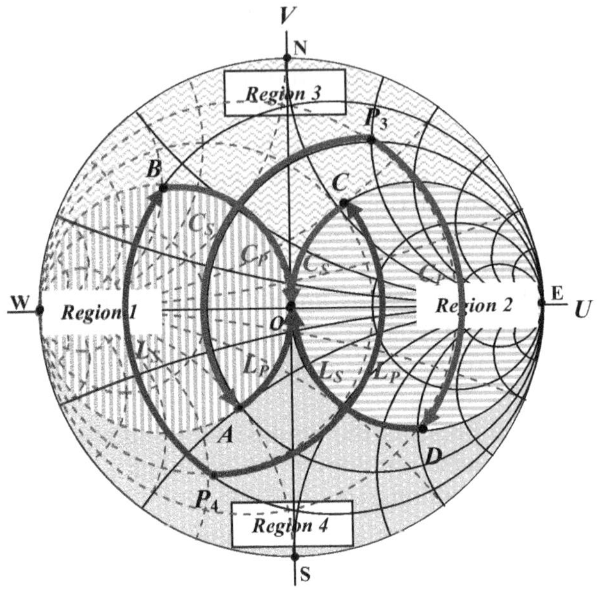

Figure 10.12 Alternative way to pull the original impedances P_3 and P_4 to the center of a Smith chart, O. One capacitor and one small inductor.

the Smith chart by one inductor and one capacitor by two options. This reduction of the number of inductors is a significant improvement in *RF* circuit design, especially in *RFIC* design. Figure 10.12 also shows that instead of two capacitors, one inductor and one capacitor can form an impedance matching network if the original impedance to be matched is located in region 3. It provides an alternative method but is not as good as that of implementing an impedance matching network by only two capacitors.

In real life engineering design, most impedance matching networks are built and inserted between a source and the input of a device, or between the output of a device and a load. Figure 10.13 shows the input and output impedance matching networks located before and after the device respectively.

The selection of topology for input and output impedance matching networks in such a case is somewhat different from that in general cases.

3) Consideration of *DC* blocking, *DC* feeding, and *DC* short circuiting

The input and output impedance matching network of a device must play the role of not only impedance matching but also that of *DC* blocking and *DC* feeding. An overview of all the topologies listed in Table 10.4, shows that only topologies (4) $L_P - C_S$ or (7) $L_P - L_S$ are available for *DC* feeding. However, topology (7) $L_P - L_S$, cannot serve for DC blocking, and, on the other hand, it consists of two inductors, so it is immediately discarded as an option in consideration of DC blocking and cost-saving.

It is therefore concluded that topology 4), $L_P - C_S$, is the exclusive choice for the input and output impedance matching network of the device. Figure

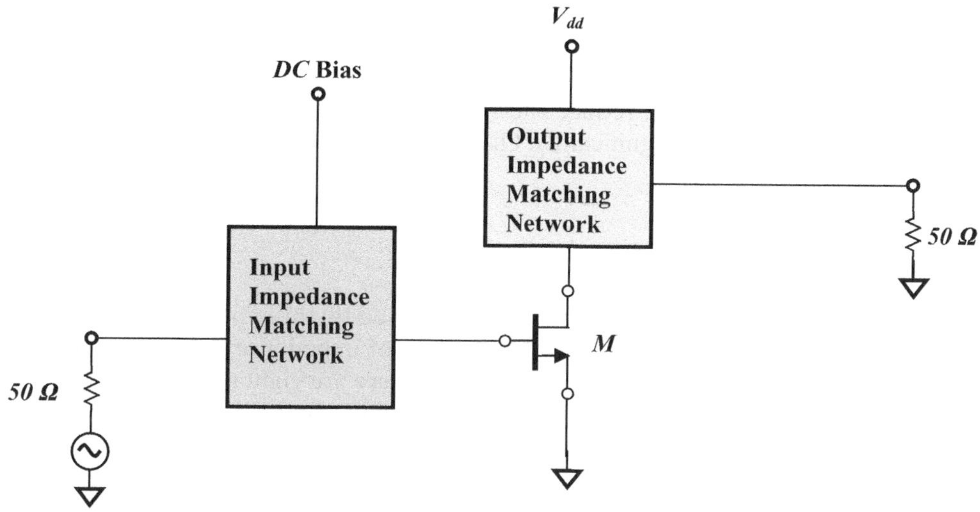

Figure 10.13 Location of input and output impedance matching networks for a device.

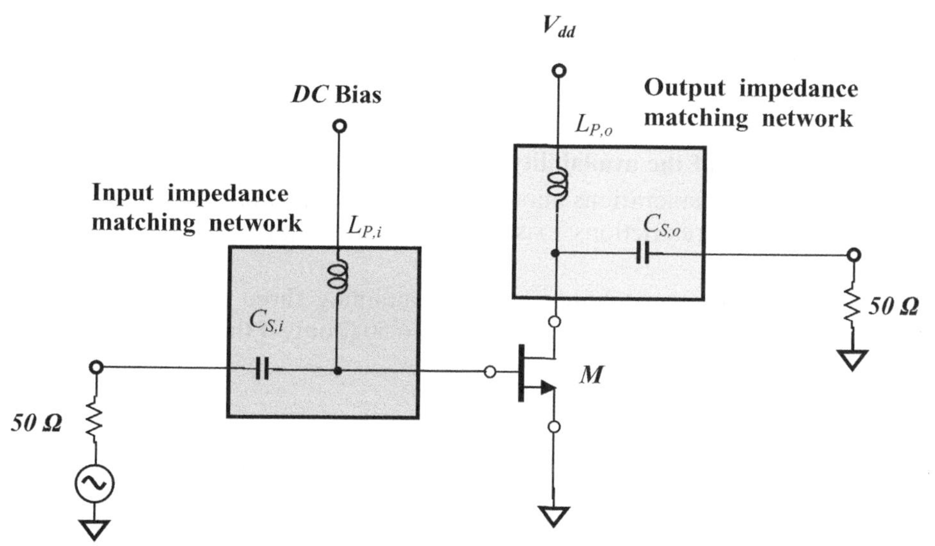

Figure 10.14 Conclusions on implementing input and output impedance matching networks by two parts.

10.14 illustrates the conclusions on the implementation of input and output impedance matching network by two parts.

Also, it should be noted that the applied regions of topology (4), $\boldsymbol{L_P - C_S}$, are regions 2 and 4. If the location of the original impedance is not located in region 2 or 4, it cannot but put an expectation on the impedance matching network containing three parts, discussed in Section 10.5.

10.5 IMPEDANCE MATCHING NETWORK BUILT OF THREE PARTS

An impedance matching network constructed by three parts, a combination of capacitors and inductors, can reduce the restrictions from an impedance matching network built of two parts significantly. Of course, the price to be paid is the cost of an extra part.

10.5.1 "Π" and "T" Types

All the topologies of a matching network constructed by three parts can be categorized into Π and T types. "Π" type and "T" types of impedance matching network contain three parts, either capacitor or inductor. There are eight possible alternative topologies in either the "Π" type or "T" type impedance matching networks, which are plotted in Appendix 10.A.6 and listed in Table 10.5.

10.5.2 Recommended Topologies

1) **Consideration of cost**

   From the viewpoint of cost-saving, those topologies listed in Table 10.5 containing two or three inductors will be ignored. In other words, only topologies (1), (2), (3), and (5) in Table 10.5 will be considered in the circuit design. They are listed in Table 10.6.

2) **Consideration of the availability of topology.**

   Just like the considerations about the two part impedance matching network, there are also restrictions existing in the three part impedance matching networks.

   An impedance matching network containing three parts with a specific topology is able to match an impedance to 50 Ω only if the original impedance

TABLE 10.5 Possible topologies of 3 parts impedance matching network

	"Π" type	"T" type
(1)	$C_{P1} - C_S - C_{P2}$,	$C_{S1} - C_P - C_{S2}$,
(2)	$L_{P1} - C_S - C_{P2}$,	$L_{S1} - C_P - C_{S2}$,
(3)	$C_{P1} - L_S - C_{P2}$,	$C_{S1} - L_P - C_{S2}$,
(4)	$L_{P1} - L_S - C_{P2}$,	$L_{S1} - L_P - C_{S2}$,
(5)	$C_{P1} - C_S - L_{P2}$,	$C_{S1} - C_P - iL_{S2}$,
(6)	$L_{P1} - C_S - L_{P2}$,	$L_{S1} - C_P - L_{S2}$,
(7)	$C_{P1} - L_S - L_{P2}$,	$C_{S1} - L_P - L_{S2}$,
(8)	$L_{P1} - L_S - L_{P2}$,	$L_{S1} - L_P - L_{S2}$.

Notes: In the "topology" list, the first part is connected to the original impedance to be matched, and the second part is connected to the standard reference impedance, 50 Ω. The subscript "P" stands for "in parallel" and the subscript "S" stands for "in series."

TABLE 10.6 Applied and prohibited regions of an impedance matching network with specific topology (from there the impedance Z_m is going to be matched to 50 Ω)

Topolgy	Applied regions	Prohibited regions
"*Π*" type		
(1) $C_{P1} - C_S - C_{P2}$	Region 3	Regions 1, 2, 4
(2) $\boldsymbol{L_{P1} - C_S - C_{P2}}$,	**Regions 2, 3, 4**	**Region 1**
(3) $\boldsymbol{C_{P1} - L_S - C_{P2}}$,	**Regions 1, 2, 3, 4**	**None**
(5) $\boldsymbol{C_{P1} - C_S - L_{P2}}$,	**Regions 1, 3**	**Regions 2,4**
"*T*" type		
(1) $C_{S1} - C_P - C_{S2}$,	Region 3	Regions 1, 2, 4
(2) $\boldsymbol{L_{S1} - C_P - C_{S2}}$,	**Regions 1, 3, 4**	**Region 2**
(3) $\boldsymbol{C_{S1} - L_P - C_{S2}}$,	**Regions 1, 2, 3, 4**	**None**
(5) $\boldsymbol{C_{S1} - C_P - L_{S2}}$,	**Regions 2, 3**	**Regions 1, 4**

Notes: In the column of "topology" the first part is connected to the original impedance to be matched and the second part is connected to the standard reference impedance, 50 Ω. The subscript "*P*" stands for "in parallel" and the subscript "*S*" stands for "in series."

to be matched is located in specific regions of the Smith chart. Appendix 10. A.4 presents the applied areas and the prohibited areas on a Smith chart for a specific topology of the impedance matching network containing three parts; this is also summarized in Table 10.6. Usually the original impedance Z_m is located in regions 1, 2, or 4. Consequently, the topologies that we are interested in are topologies (2), (3), and (5), for both the "*Π*" and "*T*" type of impedance matching networks. The topology (1) is immediately discarded because its prohibited regions are region 1, 2, and 4.

3) Consideration of *DC* blocking, *DC* feeding, and *DC* short circuiting.

Figures 10.15 to 10.20 plot the topologies (2), (3), and (5), for both the "*Π*" and "*T*" type of impedance matching networks. Table 10.7 lists the main performance of these plots on *DC* blocking, *DC* feeding, and *DC* short-circuited. From the table it can be seen that in a "*Π*" type of impedance matching network, only topology (2) $\boldsymbol{L_{P1} - C_S - C_{P2}}$ has no problem with *DC* blocking, *DC* feeding, and *DC* short-circuit. Topology (3) $\boldsymbol{C_{P1} - L_S - C_{P2}}$ has a problem with *DC* blocking, and, in addition, it needs two *RF* chokes, or two $L_{infinite}$ inductors. The topology (5) $\boldsymbol{C_{P1} - C_S - L_{P2}}$ needs two *RF* chokes, or two $L_{infinite}$ inductors.

In a "*T*" type of impedance matching network, topology (2) $\boldsymbol{C_{P1} - L_S - C_{P2}}$, topology (3) $\boldsymbol{C_{P1} - L_S - C_{P2}}$, and topology (5) $\boldsymbol{C_{P1} - L_S - C_{P2}}$ have no problems with *DC* blocking, *DC* feeding, and *DC* short-circuit. However, all of them need two RF chokes, or two $L_{infinitive}$ inductors so that they become a four part impedance matching networks.

Obviously the "*Π*" type of topology (2) $\boldsymbol{L_{P1} - C_S - C_{P2}}$ is the best choice among the three parts impedance matching network.

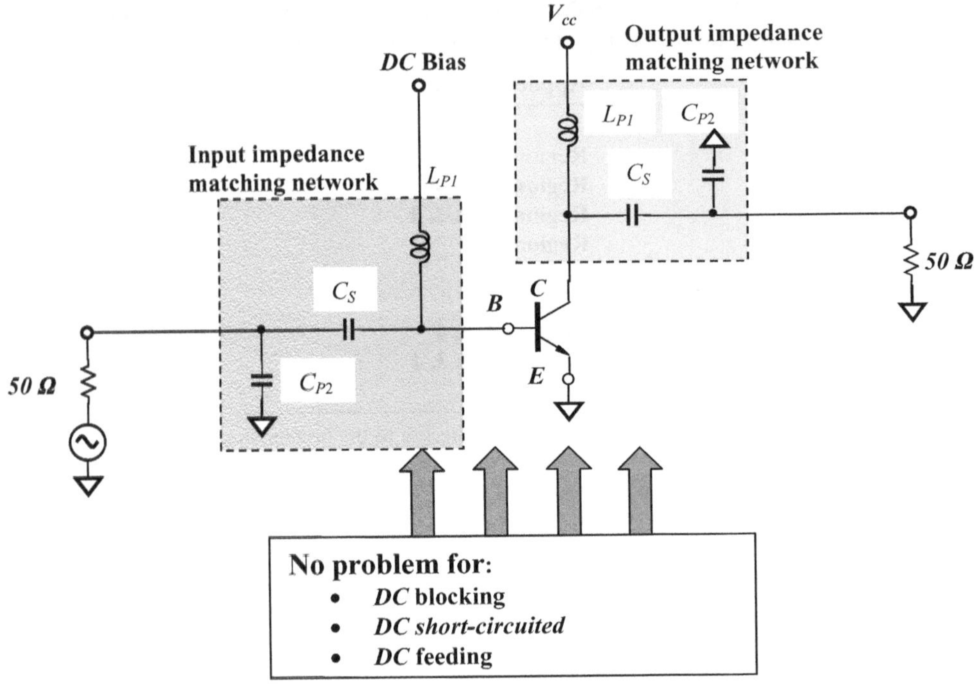

Figure 10.15 $L_{P1} - C_S - C_{P2}$ topology of impedance matching network available for input and output of a device.

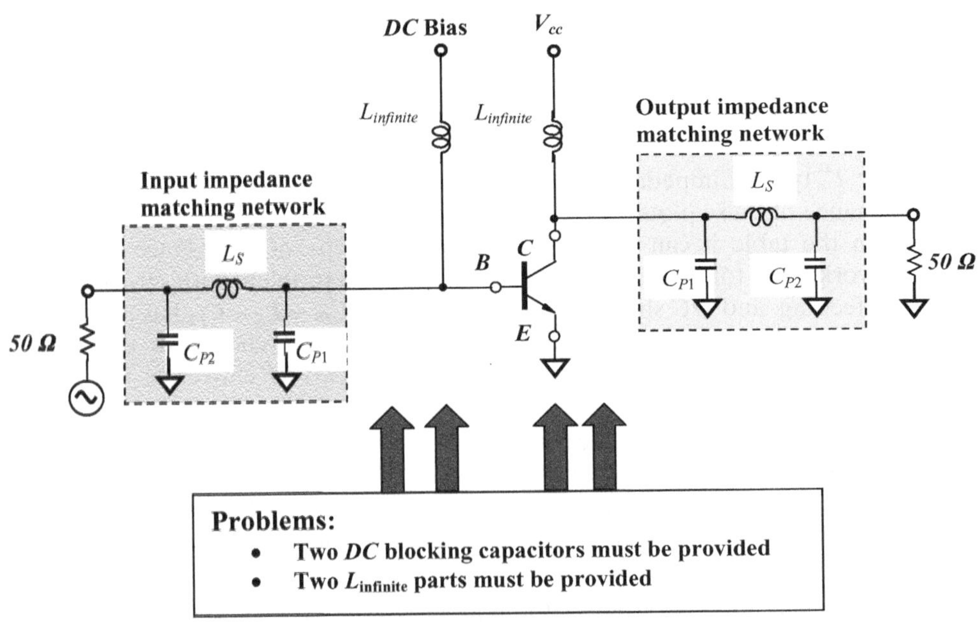

Figure 10.16 $C_{P1} - L_S - C_{P2}$ topology of impedance matching network problematical for input and output of a device.

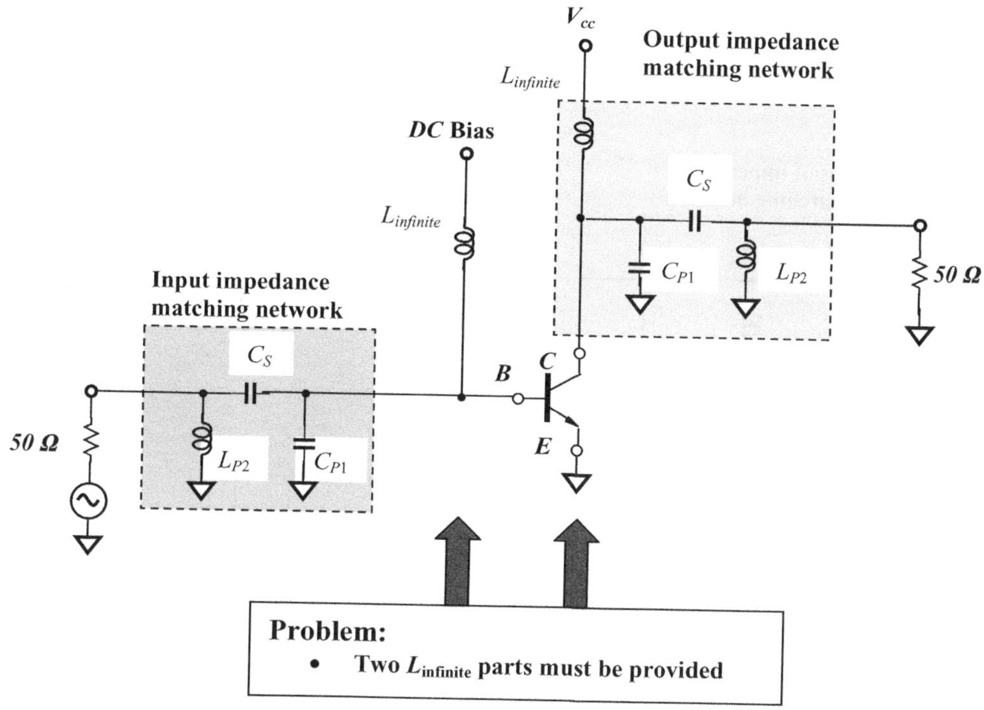

Figure 10.17 $C_{P1} - C_S - L_{P2}$ topology of impedance matching network problematical for input and output of a device.

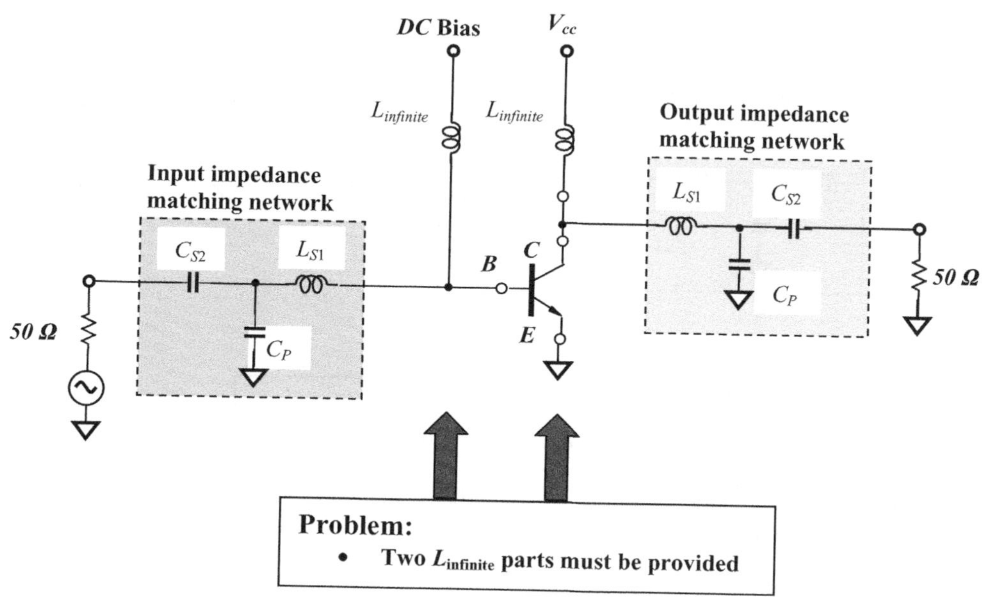

Figure 10.18 $L_{S1} - C_P - C_{S2}$ topology of impedance matching network problematical for input and output of a device.

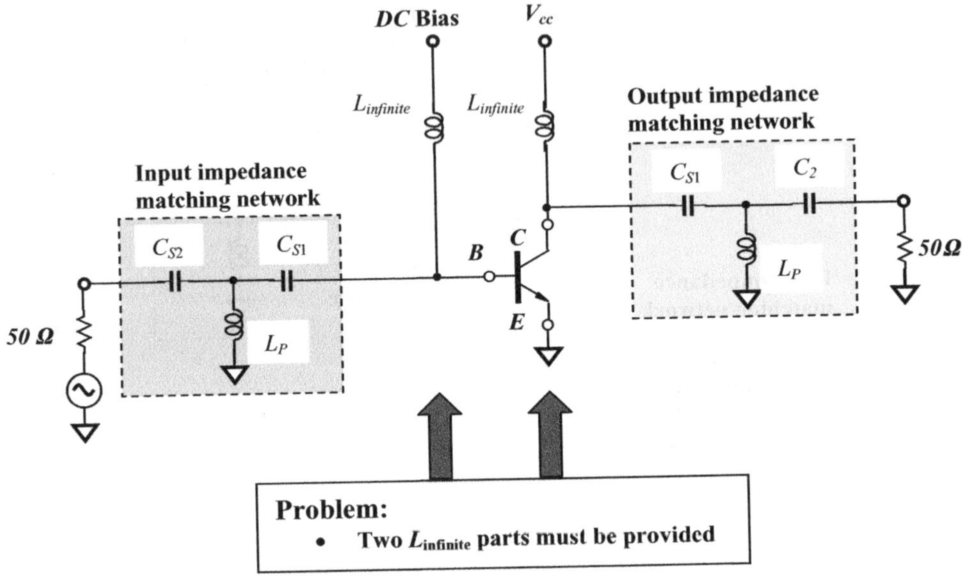

Figure 10.19 $C_{S1} - L_P - C_{S2}$ topology of impedance matching network problematical for input and output of a device.

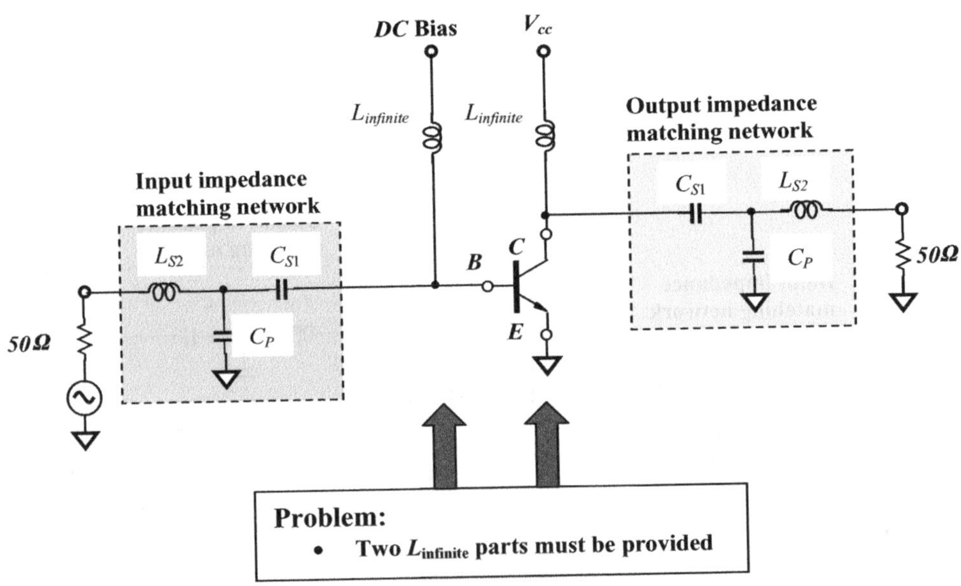

Figure 10.20 $C_{S1} - C_P - L_{S2}$ topology of impedance matching network problematical for input and output of a device.

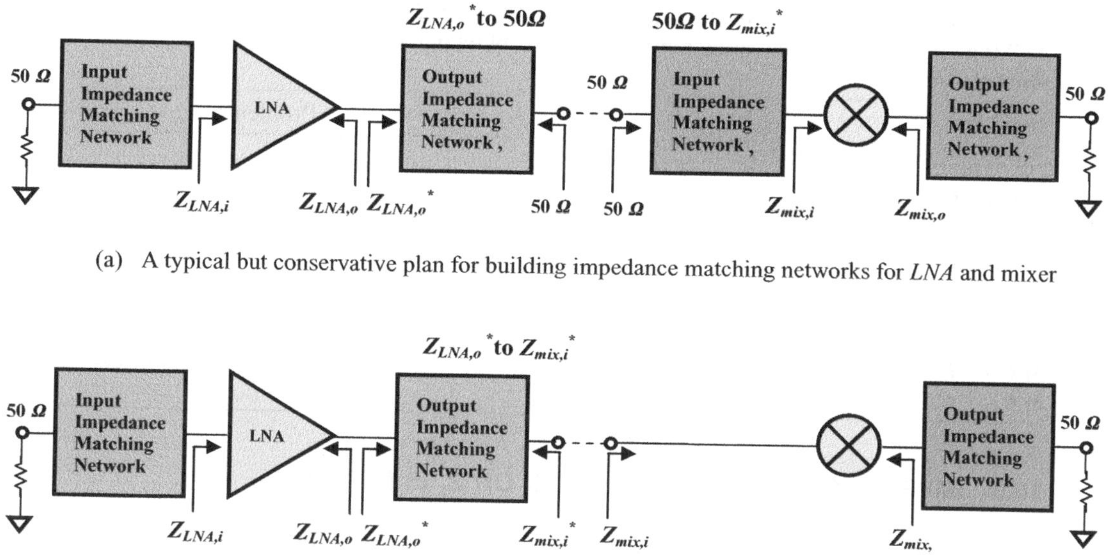

(a) A typical but conservative plan for building impedance matching networks for *LNA* and mixer

(b) A better plan for building impedance matching networks for *LNA* and mixer

Figure 10.21 Two plans in building impedance matching networks for *LNA* and mixer.

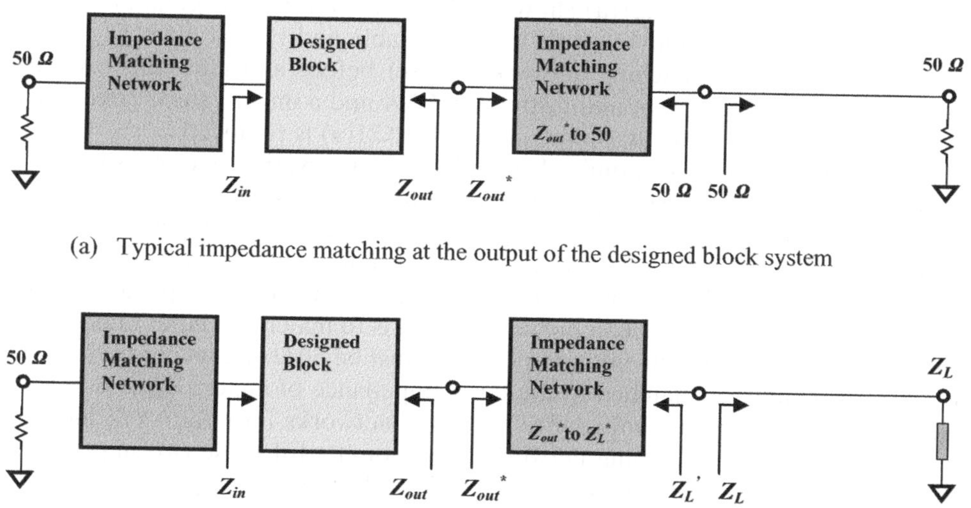

(a) Typical impedance matching at the output of the designed block system

(b) Output impedance, Z_{out}, of a designed block is assigned to match to Z_L but not 50 Ω.

Figure 10.22 Impedance matching when Z_S or Z_L is or is not 50 Ω.

TABLE 10.7 Performance of the topologies (2), (3), and (5), for both of the "*Π*" and "*T*" type of impedance matching networks on *DC* blocking, *DC* feeding, and *DC* short-circuited

Topolgy	Problems
"*Π*" type	
2) $L_{P1} - C_S - C_{P2}$	**None**
3) $C_{P1} - L_S - C_{P2}$	***DC* blocking**
	Two $L_{infinite}$ are needed
5) $C_{P1} - C_S - L_{P2}$	Two $L_{infinite}$ are needed
"*T*" type	
2) $L_{S1} - C_P - C_{S2}$	Two $L_{infinite}$ are needed
3) $C_{S1} - L_P - CS_2$	Two $L_{infinite}$ are needed
5) $C_{S1} - C_P - L_{S2}$	Two $L_{infinite}$ are needed

10.6 IMPEDANCE MATCHING WHEN Z_S OR Z_L IS NOT 50 Ω

In the actual *RF* circuit design, most of the input and output impedances of blocks are matched to 50 Ω. The main reason is that the standard reference impedance, 50 Ω, is adopted in almost all the equipment for *RF* block testing. In the testing, the input and output impedance of the desired test *RF* unit or block must be matched with the input and output impedance of the test equipment. Otherwise, the tested result will be incorrect since the test objective in the *RF* block testing is always power, not voltage. This is a special test feature in the testing of *RF* blocks.

As an example, Figure 10.21(a) shows a block diagram for the combination of *LNA* and a mixer block. Obviously, it is a reasonable and typical plan. The input and output impedance matching networks are set before and after the *LNA* or mixer, respectively. The design and testing for *LNA* and a mixer can be conducted independently as long as the dash line in Figure 10.21(a) is removed.

However, this is not an optimized plan, but a conservative one. As shown in Figure 10.21(a), there are two impedance matching networks between *LNA* output and the mixer input. The first impedance matching network is to match the output impedance of *LNA* down to 50 Ω so that its input impedance must be conjugate-matched to the output impedance of *LNA*, $Z^*_{LNA,o}$, and its output impedance must be 50 Ω. The second impedance matching network is to match the input impedance of mixer from 50 Ω so that its input impedance must be 50 Ω and its output impedance must be conjugate-matched to the input impedance of mixer, $Z^*_{mix,i}$. Is it possible to combine these two impedance matching networks into one? Yes, indeed. One whole impedance matching network can be omitted if these two impedance matching networks are combined into one. Obviously, the advantages are:

- It is more cost-effective,
- Insertion loss could be reduced,
- Less noise is expected.

Figure 10.21(b) shows the block diagram after combining the two impedance matching networks between the *LNA* and mixer into one. The combined impedance matching network is to conjugate-match the output impedance of *LNA* directly to

TABLE 10.8 Corresponding relationships between Figure 10.21 and Figure 10.22

In Figure 10.21		In Figure 10.22
LNA	⇔	Designed block
$Z_{LNA,i}$	⇔	Z_{in}
$Z_{LNA,o}$	⇔	Z_{out}
$Z_{LNA,o}{}^*$	⇔	$Z_{out}{}^*$
$Z_{mix,i}{}^*$	⇔	$Z_L{}'$
$Z_{mix,i}$	⇔	Z_L

the input impedance of the mixer. Therefore, its input impedance must be $Z^*_{LNA,o}$ and its output impedance must be $Z^*_{mix,i}$.

This is the special subjective of impedance matching for the cases when the Z_S or Z_L of an *RF* block is not 50 Ω.

Figure 10.22 interprets this subject into a general form. The interpretation is focused on the first three blocks, or the *LNA* portion as shown in Figure 10.21. The corresponding relationships between Figures 10.21 and 10.22 are listed in Table 10.8.

Now let's copy Figure 10.22(b) as Figure 10.23(a) and denote it as block *A*. Theoretically and practically, to match the output impedance of a designed block, $Z_{out,}$ to Z_L, but not to 50 Ω, is not a problem. It looks like a better plan. However, it brings about a new problem. The testing for block *A* is not available because its output is not 50Ω. The testing is only available in a 50 Ω system since the input and output impedance of most test equipments in an *RF* laboratory is 50 Ω.

In order to make testing and measuring possible, an additional impedance matching network from Z_L to 50 Ω must be added to block *A*. Figure 10.23(b) depicts the entire system of blocks. Let's denote the additional impedance matching block as block *B* and the entire block as a block *C*.

It must be noted that only block *A* will be applied and block *B* will be removed in the product. What we need to do is to get the performance of block A from testing. Unfortunately, one can test the performance for the entire block *C*, but not for any individual sub-block in block *C* because there is not any node with 50 Ω impedance between these individual sub-blocks. Thus, it is impossible to verify the performance of any individual sub-block in block *C*. In other words, it is still impossible to directly test for the performance of block A.

Alternatively, the performance of block *A* can be obtained by subtracting the performance contributed by block *B* from the performance of block *C*. So, let us focus on the testing of performance of block *B*.

In order to evaluate the performance contributed by block *B*, it must be tested or measured in the 50 Ω system as well. Figure 10.23(c) shows a simple but effective means: There are two identical blocks, *B*, piggy-backed together to form a new block *D*, so that they can be tested or measured exactly in the 50 Ω system. The performance of block *D* can be characterized by testing its power loss, noise figure, intercept points, and so on. The contribution of block *B* can be evaluated by testing block *D* in terms of cascaded equations.

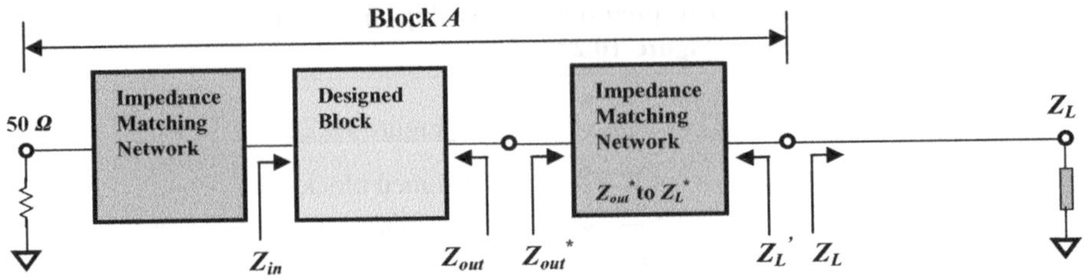

(a) Output impedance, Z_{out}, of a designed block is assigned to match to Z_L but not 50 Ω.

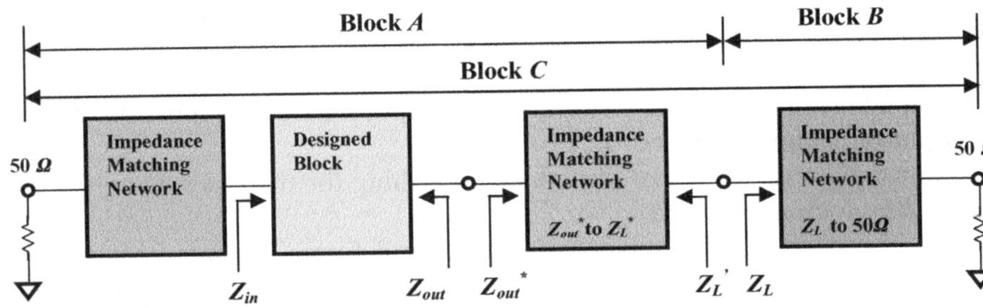

(b) Block B, an additional impedance matching network from Z_L to 50 Ω, is added to block A.

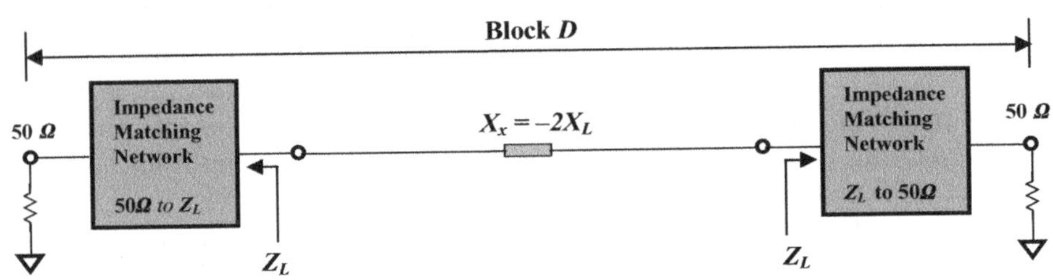

(c) Two piggy-backed blocks B form a new block D with 50 Ω. An additional reactance, $X_X = -2X_L$, must be added between two piggy-backed blocks so that their impedances are conjugate-matched with each other

Figure 10.23 Design and testing for a *RF* block when its output is asked to match an impedance Z_L but not 50 Ω.

It should be noted that in order to reach the impedance matching condition for these two piggy-backed blocks, a part with reactance, $X_x = -2X_L$, must be inserted between the piggy-backed sides so as to "neutralize" the reactance existing between the two piggy-backed blocks. X_L is the reactance of the load impedance Z_L which is a known parameter for the circuit design.

Testing these two piggy-backed blocks, or block D, will be conducted for all the required parameters to characterize their performance, such as power gain or loss, noise figure, IP_3 or IP_2, and so on. Then, the value of the required parameters for one of the piggy-backed blocks or block B can be calculated in terms of cascaded equations, rather than a plausible mathematical treatment to divide the tested value by two. The calculation is somewhat tedious and will be demonstrated in the following exercise.

TABLE 10.9 Calculated values for all the required parameters from tested values

	Tested values	**Tested values**	**Calculated values**	**Calculated values**
	For blocks shown in Figure 10.23(b)	For blocks shown in Figure 10.23(c)	For last block shown in Figure 10.23(b)	For blocks shown in Figure 10.23(a)
Column	1	2	3	4
Assigned	Block *C*	Block *D*	Block *B*	Block *A*
***G* (Gain)**	**12.5 *dB***	**−2.7 *dB***	**−1.35 *dB***	**11.15 *dB***
NF	**2.8 *dB***	**2.7 *dB***	**1.35 *dB***	**2.74 *dB***
IIP_3	**5 dB_m**	**100 dB_m**	**102.39 dB_m**	**5.0 dB_m**
IIP_2	**32 dB_m**	**150 dB_m**	**155.37 dB_m**	**40.0 dB_m**

The next testing is for block *C* as shown in Figure 10.23(b). The tested value for all the required parameters is contributed by blocks *A* and *B*.

Finally, the actual performance of block *A* can be obtained from the performance of blocks *C* and *B* if the contribution of block *B* is subtracted from the performance of block *C* through the cascaded equations.

Here's an exercise to see how to calculate the values of the required parameters for an assigned block *A* by the subtraction of the tested values for block *B* from the tested values for block *C*.

Assuming that the circuit block *C* as shown in 10.23(b) has been well designed, the tested results for block *C* as shown in Figure 10.23(b) and block *D* as shown in Figure 10.20(c) are listed in the first and second columns of Table 10.9 respectively, while the calculated values for the additional impedance matching block or block *B* and the assigned blocks or block *A* as shown in Figure 10.23(a) are listed in the third and fourth columns, respectively. From Figure 10.23, it can be found that

$$\boxed{\textbf{Block } B} + \boxed{\textbf{Block } B} \Rightarrow \boxed{\textbf{Block } D},$$

The cascaded equations are therefore

$$G_B + G_B = G_D\ (dB), \tag{10.47}$$

or,

$$G_B = \frac{G_D}{2}\ (dB), \tag{10.48}$$

$$NF_B + \frac{NF_B - 1}{G_B} = NF_D\ (Watt), \tag{10.49}$$

or,

$$NF_B = \frac{1 + G_B NF_D}{1 + G_B}\ (Watt), \tag{10.50}$$

$$\frac{1}{(IIP3)_B} + \frac{G_B}{(IIP3)_B} = \frac{1}{(IIP3)_D}\ (Watt), \tag{10.51}$$

or,

$$(IIP3)_B = (IIP3)_D(1+G_B)\,(Watt), \tag{10.52}$$

$$\left[\frac{1}{(IIP2)_B}\right]^{1/2} + \left[\frac{G_B}{(IIP2)_B}\right]^{1/2} = \left[\frac{1}{(IIP2)_D}\right]^{1/2} (Watt), \tag{10.53}$$

or,

$$(IIP2)_B = (IIP2)_D\left(1+G_B^{1/2}\right)^2 (Watt), \tag{10.54}$$

where the subscripts "B" and "D" denote the parameter of the corresponding block B and block D.

Also,

,

$$G_A + G_B = G_C\ (dB), \tag{10.55}$$

or,

$$G_A = G_C - G_B\ (dB), \tag{10.56}$$

$$NF_A + \frac{NF_B - 1}{G_A} = NF_C\ (Watt), \tag{10.57}$$

or,

$$NF_A = NF_C - \frac{NF_B - 1}{G_A}\,(Watt), \tag{10.58}$$

$$\frac{1}{(IIP3)_A} + \frac{G_A}{(IIP3)_B} = \frac{1}{(IIP3)_C}\,(Watt), \tag{10.59}$$

or,

$$(IIP3)_A = \frac{(IIP3)_C(IIP3)_B}{(IIP3)_B - G_A(IIP3)_C}\,(Watt), \tag{10.60}$$

$$\left[\frac{1}{(IIP2)_A}\right]^{1/2} + \left[\frac{G_A}{(IIP2)_B}\right]^{1/2} = \left[\frac{1}{(IIP2)_C}\right]^{1/2} (Watt), \tag{10.61}$$

or,

$$(IIP2)_A = \frac{1}{\left\{\left[\frac{1}{(IIP2)_C}\right]^{1/2} - \left[\frac{G_A}{(IIP2)_B}\right]^{1/2}\right\}^2}\,(Watt), \tag{10.62}$$

where the subscripts "A," "B," and "C" denote the parameter of the corresponding block A, block B, and block C, respectively.

10.7 PARTS IN AN IMPEDANCE MATCHING NETWORK

In Sections 10.3, 10.4, and 10.5, a passive impedance matching network built of one, two, and three parts have been discussed. It is rare to see a passive impedance matching network built of only one part. In most practical engineering designs of the narrow-band type, it is usually built of two or three parts. A passive impedance matching network built of two parts is quite popular, though there is a topology limitation. If a passive impedance matching network is built of three parts, the topology constraints are significantly decreased and infinitive choices of the parts' values are provided.

Empirically, in order to design a wide-band matching network, more than three parts may be necessary. It should be noted that, in reality, the passive parts are not ideal. They always attenuate the signal with their resistive components. Too many parts in an impedance matching network may bring about a serious attenuation of the signal.

Theoretically. resistors could be applied to an impedance matching network. They are, however, never used in a practical impedance matching network because they bring about serious attenuation and noise. Usually an impedance matching network is built of only capacitors and inductors, but not resistors; moreover, capacitors are preferable to inductors. In the circuit designed by discrete parts, the cost of the capacitor is much less than that of inductor. In the *IC* circuit design, the area of a capacitor is much less than that of an inductor. Therefore, the cost of a capacitor is also much cheaper than that of an inductor. In addition, the Q value of an inductor is much lower than that of a capacitor in the *IC* circuits.

APPENDICES

10.A.1 Fundamentals of the Smith Chart

The Smith chart is a powerful tool in *RF* circuit design, reflecting the relationship between voltage reflection coefficient and impedance. Based on transmission line theory, an *AC* voltage transported along a transmission line and incident on a load will be partially reflected as long as the characteristic impedance of the transmission line Z_o is different from the impedance of the load Z. The ratio of the reflected voltage to the incident voltage is defined as the voltage reflection coefficient, Γ, and can be expressed as

$$\Gamma = \frac{Z - Z_o}{Z + Z_o}, \tag{10.A.1}$$

where

Γ = voltage reflection coefficient,
Z = impedance of the load,
Z_o = characteristic impedance of transmission line.

In general, the voltage reflection coefficient Γ is a complex parameter, that is

$$\Gamma = U + jV, \tag{10.A.2}$$

where

Γ = voltage reflection coefficient;
U = real part of voltage reflection coefficient;
V = real part of voltage reflection coefficient.

The impedance Z is also a complex parameter, that is

$$Z = R + jX, \tag{10.A.3}$$

where

R = real part or resistance of the impedance;
X = imaginary part of the impedance.

By the introduction of the normalized impedance, that is,

$$z = \frac{Z}{Z_o}, \tag{10.A.4}$$

expression (10.A.4) becomes

$$z = r + jx, \tag{10.A.5}$$

$$r = \frac{R}{Z_o}, \tag{10.A.6}$$

$$x = \frac{X}{Z_o}, \tag{10.A.7}$$

where

r = normalized resistance,
x = normalized resistance.

And the equation (10.A.1) becomes

$$\Gamma = \frac{z-1}{z+1} = \frac{(r-1)+jx}{(r+1)+jx} = U + jV, \tag{10.A.8}$$

where

$$U = \frac{r^2 - 1 + x^2}{(r+1)^2 + x^2}, \tag{10.A.9}$$

$$V = \frac{2x}{(r+1)^2 + x^2}. \tag{10.A.10}$$

By eliminating x from (10.A.9) and (10.A.10), the result is that

$$\left(U-\frac{r}{r+1}\right)^2+V^2=\left(\frac{1}{r+1}\right)^2. \tag{10.A.11}$$

By eliminating r from (10.A.9) and (10.A.10), the result is that

$$(U-1)^2+\left(V-\frac{1}{x}\right)^2=\left(\frac{1}{x}\right)^2. \tag{10.A.12}$$

The equation (10.A.11) contains a family of circles centered at $U = r/(r + 1)$, $V = 0$ with radii $1/(r + 1)$, and the equation (10.A.12) contains a family of arcs on circles centered at $U = 1$, $V = 1/x$ with radii $1/x$.

Figure 10.A.1 shows these circles and arcs in the Smith chart. It presents the impedance coordination in the complex plane of the voltage reflection coefficient. The value of r for each circle is marked along the central horizontal line while the value of x for each arc is marked along the biggest circle. At the left-most point, denoted by W, $r = x = 0$, so that $z = Z_o = 0$. It is a zero impedance point or a short-circuited point. At the right-most point, denoted by E, $x = 0$ and $0 \le r < \infty$, so that $0 \le z < \infty$, or $0 \le Z < \infty$. It is an open-circuited point or a singular point mathematically. At the upper-most point, denoted by N, is the point where $r = 0$, $x = 1$. It is the point where the impedance is purely inductive and $X = Z_o$. At the bottom-most point, denoted by S, is the point where $r = 0$, $x = -1$. It is the point where the impedance is purely capacitive and $X = -Z_o$. At the center point, denoted by O, $r = 1$, $x = 0$, so that $z = 1$ or $Z = Z_o = 50\,\Omega$. It is a standard reference impedance point.

The coordination of impedance, z, in the Smith chart can be converted into admittance coordination, y.

$$y=\frac{1}{z}=g+jb, \tag{10.A.13}$$

where

y = normalized admittance,
g = normalized conductance,
b = normalized susceptance.

The relationship between the admittance, y, and the voltage reflection coefficient, Γ, can be found from (10.A.1) and (10.A.4), that is,

$$z=\frac{Z}{Z_o}=\frac{1+\Gamma}{1-\Gamma} \tag{10.A.14}$$

and then from the definition of y as shown in (10.A.13), we have

$$y=\frac{1}{z}=\frac{1-\Gamma}{1+\Gamma}=\frac{1+\Gamma e^{j\pi}}{1-\Gamma e^{j\pi}}. \tag{10.A.15}$$

In equation (10.A.15), we rewrite $-\Gamma$ as $\Gamma e^{j\pi}$ so that equations (10.A.14) and (10.A.15) both look similar in their mathematical expression. By comparing

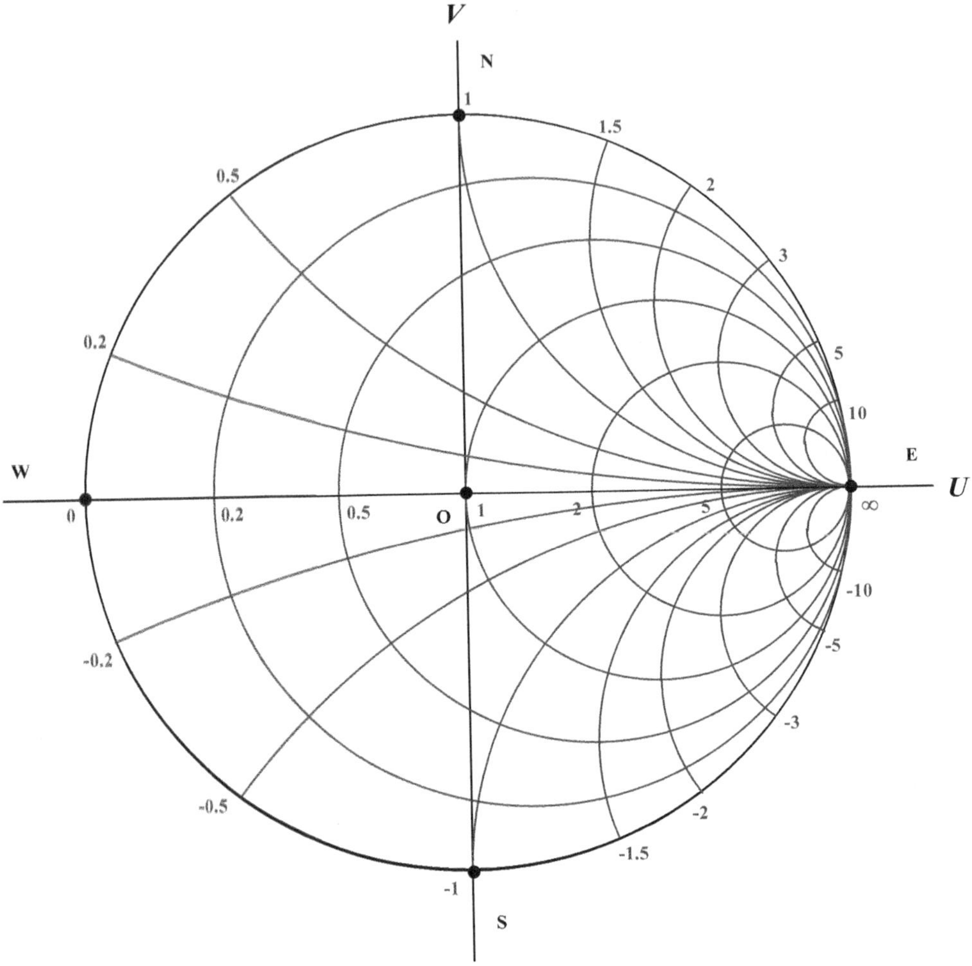

Note 1: Normalized resistance, r, is set in boldface with values around the biggest circle, **0, 0.2, 0.5, 1, 1.5, 2, 3, 5, 10, ∞, −10, −5, −3, −2, −1.5, −1, −0.5, −0.2.**
Note 2: Normalized reactance, x, is set in boldface with values along the horizontal axis U, **0, 0.2, 0.5, 1, 2, 5, ∞.**

Figure 10.A.1 Impedance coordination of the Smith chart with constant resistance or reactance curves.

equations (10.A.14) and (10.A.15), it is found that the conversion from z to y in the Smith chart can be done if the complex plane Γ is simply rotated 180°.

Figure 10.A.2 presents the admittance coordination of the Smith chart in the complex plane of the voltage reflection coefficient. The value of g for each circle is marked along the central horizontal line while the value of b for each arc is marked along the biggest circle. At the left-most point, denoted by W, $b = 0$ and $0 \le g < \infty$, so that $0 = y < \infty$, or $0 = Y < \infty$. This is a short-circuited point, or, mathematically,

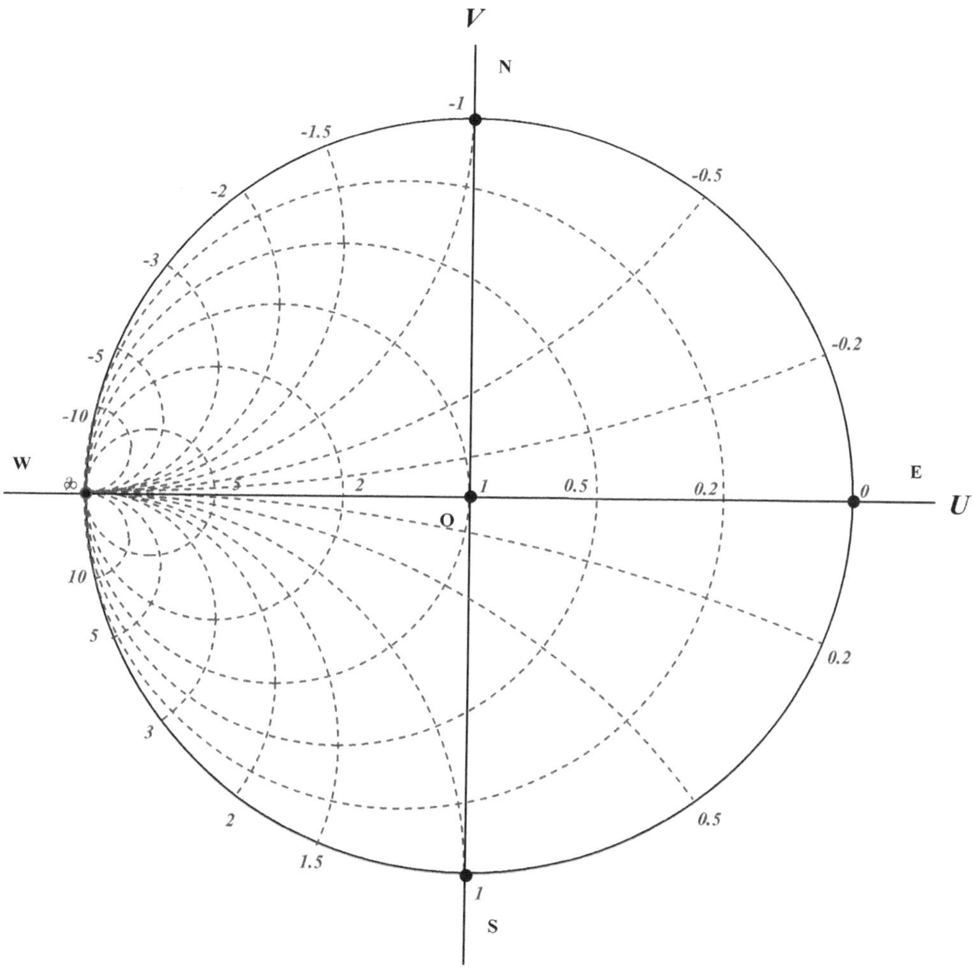

Note 1: Normalized conductance, *g*, is set in italics with values around the biggest circle, *0, 0.2, 0.5, 1, 1.5, 2, 3, 5, 10, ∞, -10, -5, -3, -2, -1.5, -1, -0.5, -0.2.*
Note 2: Normalized subceptance, *b,* is set in italics with values along the horizontal axis *U*, *0, 0.2, 0.5, 1, 2, 5, ∞.*

Figure 10.A.2 Admittance coordination of the Smith chart with constant conductance or subceptance curves.

a singular point. At the right-most point, denoted by E, $g = b = 0$, so that $y = Y_o = 0$. It is a zero admittance point or an open-circuited point. At the center point, denoted by O, $g = 1$ and $b = 0$, so that $y = 1$.

It is more convenient for an *RF* circuit designer if both impedance and admittance coordination can be marked in the Smith chart simultaneously. Figure 10.A.3 shows such a combination of Figures 10.A.1 and 10.A.2, that is, a Smith chart showing both impedance and admittance coordination. If a matching network

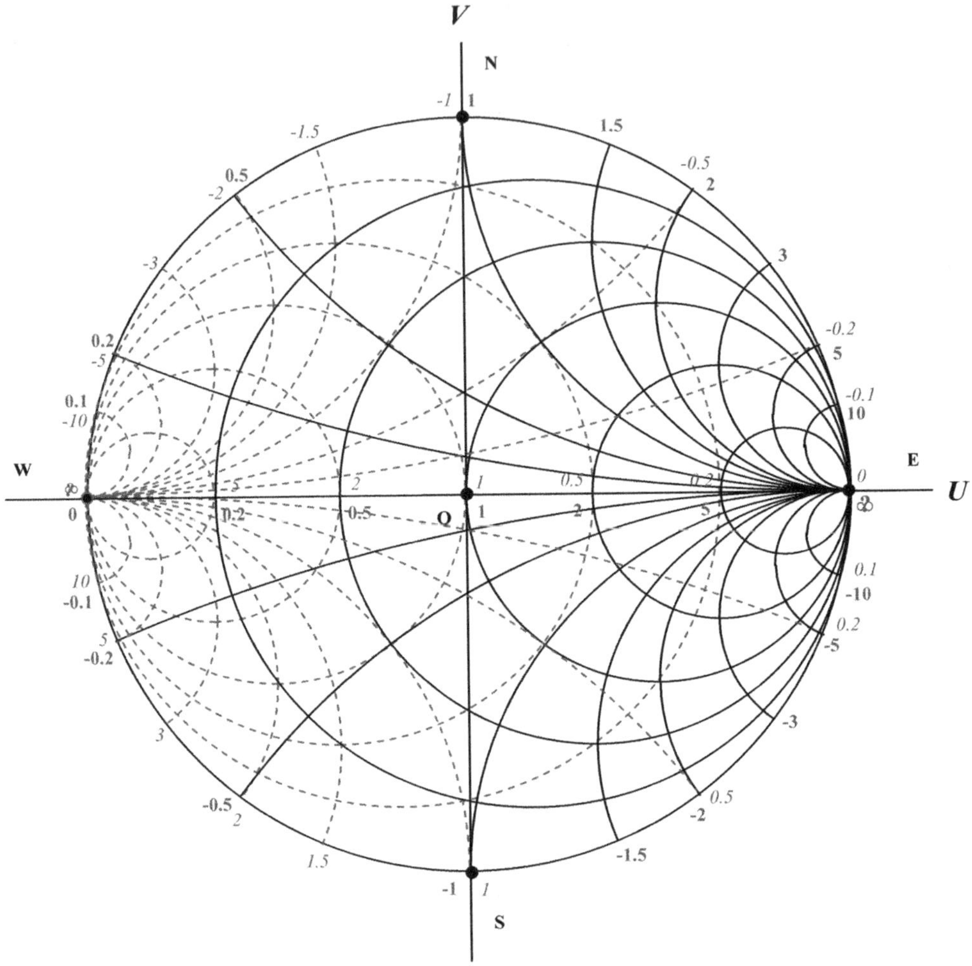

Note 1: Normalized resistance, *r*, is set in boldface with values around the biggest circle, **0, 0.1, 0.2, 0.5, 1, 1.5, 2, 3, 5, 10, ∞, -10, -5, -3, -2, -1.5, -1, -0.5, -0.2, -0.1.**

Note 2: Normalized reactance, *x*, is set in boldface with values along the horizontal axis *U*, **0, 0.2, 0.5, 1, 2, 5, ∞.**

Note 3: Normalized conductance, *g*, is set in italics with values around the biggest circle, *0, 0.1, 0.2, 0.5, 1, 1.5, 2, 3, 5, 10, ∞, -10, -5, -3, -2, -1.5, -1, -0.5, -0.2, -0.1*

Note 4: Normalized subceptance, *b*, is set in italics with values along the horizontal axis *U*, *0, 0.2, 0.5, 1, 2, 5, ∞.*

Figure 10.A.3 Impedance and admittance coordination of the Smith chart with constant resistance or reactance curves (solid curves) and with constant conductance and subceptance curves (dashed curves).

consists of only passive parts either in series or in parallel, then the impedance coordination is applied when the part is added in series, while the admittance coordination is applied when the part is added in parallel. Consequently, with the help of a Smith chart as shown in Figure 10.A.3, both variations of impedance and admittance for a matching network can be figured out simultaneously and without difficulty.

Usually, not only z, y, and Γ but also other related parameters are marked in a Smith chart. They are:

- Reflection coefficient of power, γ
- Return loss in dB,
- Reflection loss in dB,
- *VSWR* (Voltage Standing Wave Ratio),
- *VSWR* in dB, and
- Transmission loss coefficient,

The multiple coordinates make the Smith chart more convenient in the *RF* circuit design. Most parameters can be read from the same chart simultaneously.

10.A.2 Formula for a Two-Part Impedance Matching Network

The following formulas for two-part impedance matching network were developed and derived by the author. From the discussion of two-part impedance matching network in Section 10.4.2, and directly from Figure 10.11, three common rules or features are summarized:

1. **The first step in two-part impedance matching network is to pull the original impedance to the circle either $g = 1$ or $r = 1$.**
2. **The second step in two-part impedance matching network is to pull the impedance on the circle either $g = 1$ or $r = 1$, after it is pulled by the first part, to the standard reference impedance, 50 Ω.**
3. **One of two matching parts is in series and another is in parallel. However, the designer can select the first part in series or in parallel, because, as described above, in all possible original impedance there are two ways to be pulled to the center of the Smith chart. The first part in one way is in series, while in another way it is in parallel, and vice versa.**

On the basis of the common rules or features mentioned above, the type of two parts can be selected and the values of the matching parts, one in series and another in parallel, can be formalized.

For the purpose of the derivation of the formula, let's list the relations between impedance and admittance first. An impedance can be expressed by

$$Z = R + jX, \qquad (10.A.16)$$

its normalized impedance and admittance are

$$z = r + jx, \qquad (10.A.17)$$

where

$$r = \frac{R}{Z_o}, \qquad (10.A.18)$$

$$x = \frac{X}{Z_o}, \tag{10.A.19}$$

where Z_o = characteristic impedance.

$$y = g + jb = \frac{1}{z} = \frac{1}{r + jx} = \frac{r - jx}{r^2 + x^2}, \tag{10.A.20}$$

$$g = \frac{r}{r^2 + x^2}, \tag{10.A.21}$$

$$b = -\frac{x}{r^2 + x^2}. \tag{10.A.22}$$

The selection of the type and the calculation of the matching parts' value can be conducted for four cases:

Let's repeat Figure 10.11 as Figure 10.A.4.

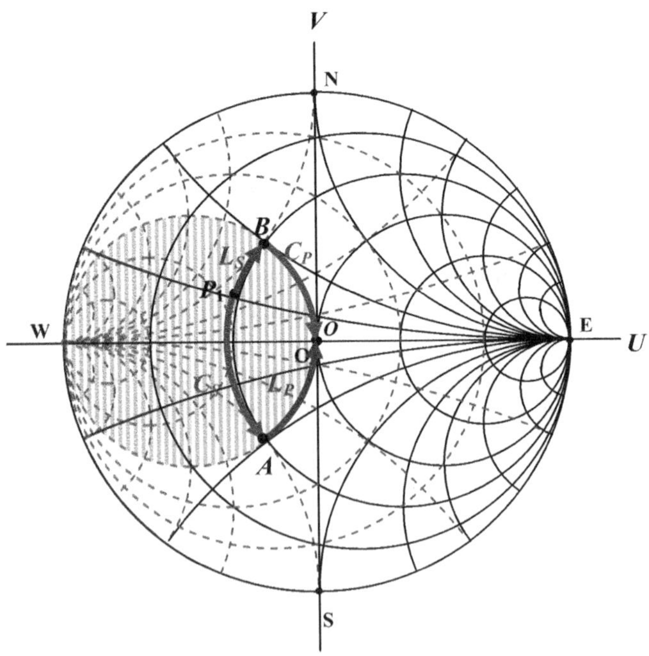

Figure 10.A.4(a) Two ways to pull the original impedance P_1 in region 1 to the center of a Smith chart, O, by addition of two passive parts.

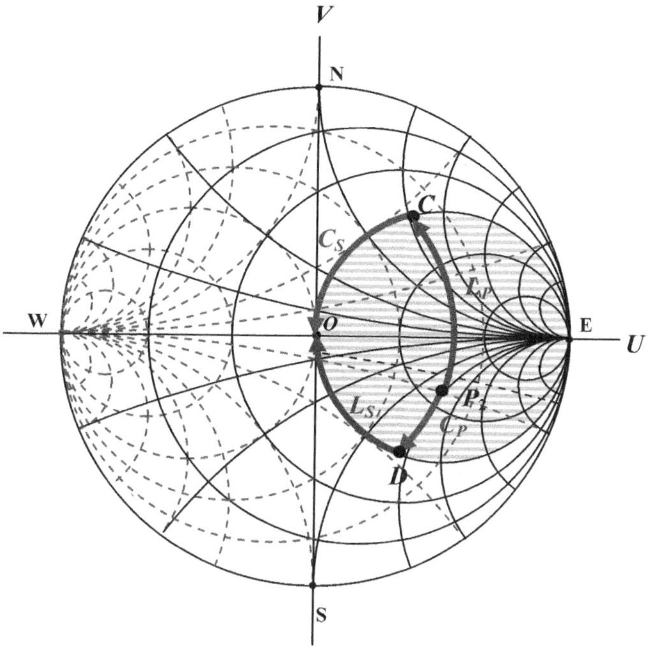

Figure 10.A.4(b) Two ways to pull the original impedance P_2 in region 2 to the center of a Smith chart, O, by addition of two passive parts.

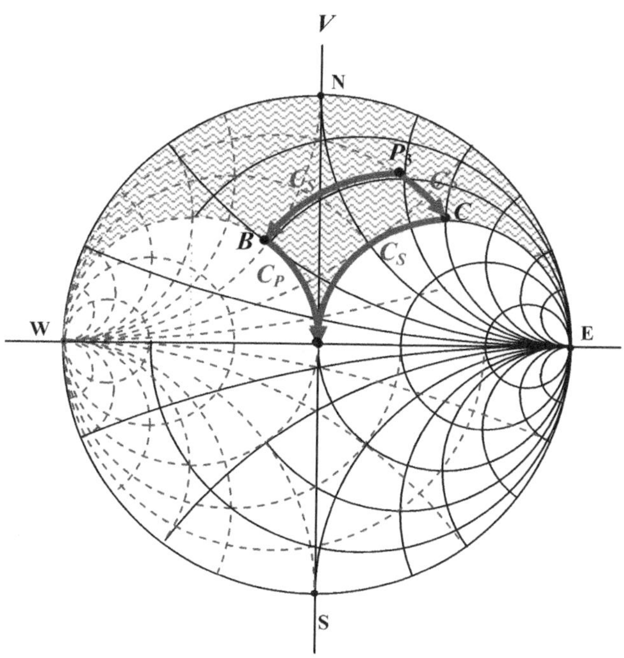

Figure 10.A.4(c) Two ways to pull the original impedance P_3 in region 3 to the center of a Smith chart, O, by addition of two passive parts.

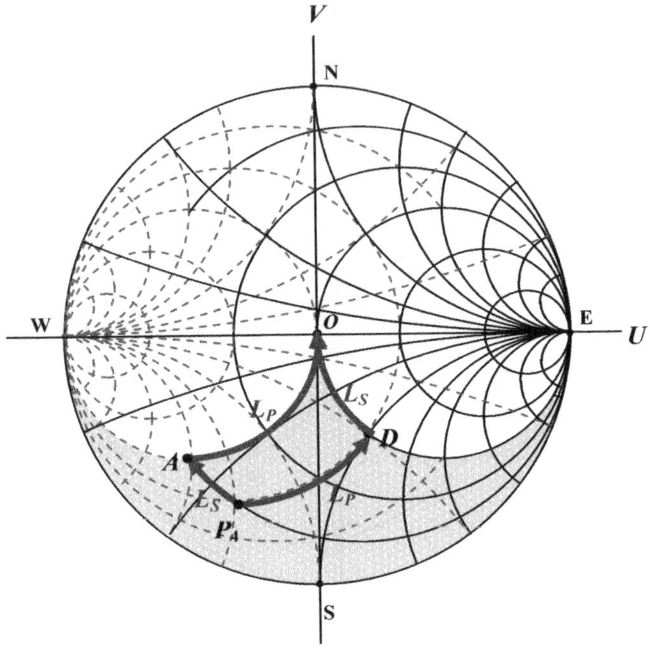

Figure 10.A.4(d) Two ways to pull the original impedance P_4 in region 4 to the center of a Smith chart, O, by addition of two passive parts.

Case 1: When the original impedance is located in region 1

The first impedance matching part, X_1, or x_1, denoted by subscript "1", is added with the original impedance Z_m in series so that the original impedance, Z_m, can be moved to the circle with $g = 1$.

Let's denote the original impedance to be matched as Z_m. It consists of real and imaginary part R_m and X_m, that is,

$$Z_m = R_m + jX_m, \tag{10.A.23}$$

and its normalized impedance is

$$z_m = r_m + jx_m, \tag{10.A.24}$$

where

$$r_m = \frac{R_m}{Z_o}, \tag{10.A.25}$$

$$x_m = \frac{X_m}{Z_o}, \tag{10.A.26}$$

After the first matching part X_1 is added, the original impedance Z_m is moved to the circle with $g = 1$ and becomes Z_{g1}, which is

$$Z_{g1} = R_{g1} + jX_{g1}, \tag{10.A.27}$$

or its normalized impedance

$$z_{g1} = r_{g1} + jx_{g1}, \tag{10.A.28}$$

Correspondingly,

$$Y_{g1} = G_{g1} + B_{g1}, \tag{10.A.29}$$

$$y_{g1} = g_{g1} + b_{g1}, \tag{10.A.30}$$

where

$$r_{g1} = \frac{R_{g1}}{Z_o}, \tag{10.A.31}$$

$$x_{g1} = \frac{X_{g1}}{Z_o}, \tag{10.A.32}$$

$$g_{g1} = \frac{G_{g1}}{Z_o} = \frac{r_{g1}}{r_{g1}^2 + x_{g1}^2}, \tag{10.A.33}$$

$$b_{g1} = \frac{B_{g1}}{Z_o} = -\frac{x_{g1}}{r_{g1}^2 + x_{g1}^2}, \tag{10.A.34}$$

The subscript "g1" denotes the parameter located at the circle with $g = 1$ in the Smith chart. The real part of Z_{g1} or z_{g1} is unchanged but the imaginary part of Z_{g1} or z_{g1} is changed due to the addition of X_1 or x_1, that is,

$$r_{g1} = r_m, \tag{10.A.35}$$

and

$$x_{g1} = x_1 + x_m, \tag{10.A.36}$$

Consequently, from expressions (10.A.33), (10.A.35), and (10.A.36), we have

$$g_{g1} = \frac{r_{g1}}{r_{g1}^2 + x_{g1}^2} = \frac{r_m}{r_m^2 + (x_1 + x_m)^2} = 1. \tag{10.A.37}$$

Then,

$$r_m^2 + (x_1 + x_m)^2 = r_m, \tag{10.A.38}$$

$$\left(\frac{R_m}{Z_o}\right)^2 + \left(\frac{X_1 + X_m}{Z_o}\right)^2 = \frac{R_m}{Z_o}, \tag{10.A.39}$$

$$X_1 + X_m = \pm\sqrt{R_m(Z_o - R_m)}, \tag{10.A.40}$$

$$X_1 = \pm\sqrt{R_m(Z_o - R_m)} - X_m. \tag{10.A.41}$$

The positive sign in equation (10.A.41) is for the selection of an inductive X_1 while the negative sign is for the selection of a capacitive X_1.

The second matching part, X_2, or x_2, denoted by subscript "2", in parallel is to pull Z_{g1} or z_{g1} on the circle $g = 1$ to the center of the Smith chart or to the reference impedance, 50 Ω.

In other word, the second matching part, X_2, or x_2, is to "neutralize" the imaginary part of Y_{g1} or y_{g1}, that is

$$\frac{1}{x_2} + (-b_{g1}) = 0, \tag{10.A.42}$$

From expressions (10.A.42) and (10.A.34) we have

$$\frac{1}{x_2} = b_{g1} = -\frac{x_1 + x_m}{r_m^2 + (x_1 + x_m)^2}. \tag{10.A.43}$$

From (10.A.41) and (10.A.43), we have

$$\frac{1}{X_2} = -\frac{X_1 + X_m}{R_m^2 + (X_1 + X_m)^2} = \frac{\mp\sqrt{R_m(Z_o - R_m)}}{R_m^2 + R_m(Z_o - R_m)}, \tag{10.A.44}$$

$$\frac{1}{X_2} = \frac{\mp\sqrt{R_m(Z_o - R_m)}}{R_m Z_o}, \tag{10.A.45}$$

The positive sign in equation (10.A.45) is for the selection of inductive X_2 while the negative sign is for the selection of capacitive X_2.

Also, a simple relationship between X_1 and X_2 can be found from equations (10.A.41) and (10.A.45):

$$-X_2(X_1 + X_m) = R_m Z_o. \tag{10.A.46}$$

In terms of equations, (10.A.41) and (10.A.45), the values of two matching parts, X_1 and X_2, can be calculated.

As shown in Figure 10.A.4, there are two options for the impedance matching from original value Z_m to the reference impedance 50Ω. However, if first impedance matching part is capacitive, the second must be inductive, and vice versa. Consequently, two possible combinations to build the impedance matching network are:

1) By adding a capacitor C_S in series first, and then by adding of an inductor L_P in parallel secondly.

$$X_1 = -\sqrt{R_m(Z_o - R_m)} - X_m. \tag{10.A.47}$$

$$C_S = \frac{C_m}{C_m\omega\sqrt{R_m(Z_o - R_m)} - 1}, \tag{10.A.48}$$

and

$$\frac{1}{X_2} = \frac{+\sqrt{R_m(Z_o - R_m)}}{R_m Z_o}, \tag{10.A.49}$$

$$L_P = \frac{R_m Z_o}{\omega\sqrt{R_m(Z_o - R_m)}}. \tag{10.A.50}$$

2) By adding an inductor L_S in series first, and then by adding a capacitor C_P in parallel.

$$X_1 = +\sqrt{R_m(Z_o - R_m)} - X_m. \tag{10.A.51}$$

$$L_S = \frac{\sqrt{R_m(Z_o - R_m)} - X_m}{\omega}, \tag{10.A.52}$$

and

$$\frac{1}{X_2} = \frac{-\sqrt{R_m(Z_o - R_m)}}{R_m Z_o}, \tag{10.A.53}$$

$$C_P = \frac{\sqrt{R_m(Z_o - R_m)}}{R_m Z_o \omega}, \tag{10.A.54}$$

These two options correspond to the two topologies of impedance matching networks as shown in Figure 10.A.4. It should be noted that the expressions are reasonable only

If

$$\sqrt{R_m(Z_o - R_m)} > X_m. \tag{10.A.55}$$

The inductor and capacitor, L_S and C_P, in the first option would become a capacitor and an inductor, C_S and L_P, and the capacitor and inductor, C_S and L_P, in the second option would become an inductor and a capacitor, L_S and C_P,
if

$$\sqrt{R_m(Z_o - R_m)} < X_m. \tag{10.A.56}$$

Case 2: when the original impedance is located in region 2

The first impedance matching part, X_1, or x_1, denoted by subscript "1", is added with the original impedance Z_m in parallel so that the original impedance, Z_m, can be moved to the circle with $r = 1$.

After the first matching part X_1, is connected with the original impedance, Z_m, in parallel, the resultant impedance is

$$Z_{r1} = \frac{(R_m + jX_m)jX_1}{(R_m + jX_m) + jX_1} = \frac{(-X_m X_1 + jX_1 R_m)[R_m - j(X_m + X_1)]}{R_m^2 + (X_m + X_1)^2}, \tag{10.A.57}$$

$$R_{r1} = \frac{R_m X_1^2}{R_m^2 + (X_m + X_1)^2}, \tag{10.A.58}$$

$$X_{r1} = \frac{R_m^2 + X_m(X_m + X_1)}{R_m^2 + (X_m + X_1)^2} X_1. \tag{10.A.59}$$

The subscript "$r1$" indicates that the denoted parameter is on the circle with $r = 1$ on the Smith chart.

The real part of the impedance Z_{r1} must be equal to Z_o because Z_{r1} is on the circle with $r = 1$, that is

$$R_{r1} = \frac{R_m X_1^2}{R_m^2 + (X_m + X_1)^2} = Z_o, \tag{10.A.60}$$

$$(R_m - Z_o)X_1^2 - 2X_m Z_o X_1 - (R_m^2 + X_m^2)Z_o = 0, \tag{10.A.61}$$

$$X_1 = \frac{X_m Z_o \pm \sqrt{R_m Z_o (R_m^2 + X_m^2 - R_m Z_o)}}{R_m - Z_o}, \tag{10.A.62}$$

The positive sign in equation (10.A.62) is for the selection of inductive X_1 while the negative sign is for the selection of capacitive X_1.

The second matching part, X_2, or x_2, denoted by subscript "2", in series is to pull Z_{r1} or z_{r1} on the circle $r = 1$ to the center of the Smith chart or to the reference impedance, 50 Ω, that is,

$$X_2 = -X_{r1} = -\frac{R_m^2 + X_m(X_m + X_1)}{R_m^2 + (X_m + X_1)^2} X_P = \left[\frac{(X_m + X_1)Z_o}{R_m X_1} - 1\right] X_1, \tag{10.A.63}$$

$$X_2 = \frac{Z_o}{R_m} X_m + \frac{Z_o - R_m}{R_m} X_1, \tag{10.A.64}$$

Substituting of X_1 from (10.A.62) into (10.A.64), we have

$$X_2 = \mp \frac{\sqrt{R_m Z_o (R_m^2 + X_m^2 - R_m Z_o)}}{R_m}. \tag{10.A.65}$$

There are also two options. The positive sign in equation (10.A.65) is for the selection of inductive X_2 while the negative sign is for the selection of capacitive X_2.

Also, a simple relationship between X_1 and X_2 can be found from equation (10.A.62) and (10.A.65):

$$[X_m Z_o - X_1(R_m - Z_o)]X_2 = Z_o(R_m^2 + X_m^2 - R_m Z_o). \tag{10.A.66}$$

In terms of equations, (10.A.62) and (10.A.65), the values of two matching parts, X_1 and X_2, can be calculated.

As shown in Figure 10.A.4, there are two options for the impedance matching from original value Z_m to the reference impedance 50 Ω in the expressions (10.A.62) and (10.A.65). However, if the first impedance matching part is capacitive, then the second must be inductive, and vice versa. Consequently, two possible combinations to build the impedance matching network are:

1) By adding an inductor L_P in parallel first, and then by adding a capacitor C_S in series second.

$$X_1 = \frac{X_m Z_o + \sqrt{R_m Z_o (R_m^2 + X_m^2 - R_m Z_o)}}{R_m - Z_o}, \tag{10.A.67}$$

$$L_P = \frac{X_m Z_o + \sqrt{R_m Z_o (R_m^2 + X_m^2 - R_m Z_o)}}{(R_m - Z_o)\omega}, \tag{10.A.68}$$

$$X_2 = -\frac{\sqrt{R_m Z_o (R_m^2 + X_m^2 - R_m Z_o)}}{R_m}. \tag{10.A.69}$$

$$C_S = \frac{R_m/\omega}{\sqrt{R_m Z_o (R_m^2 + X_m^2 - R_m Z_o)}}. \tag{10.A.70}$$

2) By adding a capacitor C_P in parallel first, and then by adding an inductor L_S in series second.

$$X_1 = \frac{X_m Z_o - \sqrt{R_m Z_o (R_m^2 + X_m^2 - R_m Z_o)}}{R_m - Z_o}, \tag{10.A.71}$$

$$C_P = \frac{(R_m - Z_o)/\omega}{\sqrt{R_m Z_o (R_m^2 + X_m^2 - R_m Z_o)} - X_m Z_o}, \tag{10.A.72}$$

$$X_2 = +\frac{\sqrt{R_m Z_o (R_m^2 + X_m^2 - R_m Z_o)}}{R_m}. \tag{10.A.73}$$

$$L_S = \frac{\sqrt{R_m Z_o (R_m^2 + X_m^2 - R_m Z_o)}}{R_m \omega}. \tag{10.A.74}$$

Case 3: when the original impedance is located in region 3

The first impedance matching part, X_1, or x_1, denoted by subscript "1", is added with the original impedance Z_m in series or in parallel so that the original impedance, Z_m, can be moved to the circle with $g = 1$ or $r = 1$. X_1 is a capacitor.

In series

$$X_1 = -\sqrt{R_m (Z_o - R_m)} - X_m, \tag{10.A.75}$$

In parallel

$$X_1 = \frac{X_m Z_o - \sqrt{R_m Z_o (R_m^2 + X_m^2 - R_m Z_o)}}{R_m - Z_o}, \tag{10.A.76}$$

Equations (10.A.75) and (10.A.76) correspond to equations (10.A.41) and (10.A.62), respectively, when the negative sign in equations (10.A.41) and (10.A.62) is selected for a capacitive X_1.

The second matching part, X_2, or x_2, denoted by subscript "2", in parallel or in series, is to pull Z_{g1} or z_{g1} on the circle $g = 1$ or to pull Z_{r1} or z_{r1} on the circle $r = 1$ to the center of the Smith chart or to the reference impedance, 50 Ω. X_2 is a capacitor as well.

In parallel

$$\frac{1}{X_2} = \frac{-\sqrt{R_m(Z_o - R_m)}}{R_m Z_o}, \tag{10.A.77}$$

In series

$$X_2 = -\frac{\sqrt{R_m Z_o(R_m^2 + X_m^2 - R_m Z_o)}}{R_m}. \tag{10.A.78}$$

Equations (10.A.77) and (10.A.78) correspond to equations (10.A.45) and (10.A.65), respectively, when the negative sign in equations (10.A.45) and (10.A.65) is selected for a capacitive X_2.

As shown in Figure 10.A.4, there are two options for the impedance matching from original value Z_m to the reference impedance 50Ω. Two impedance matching parts are capacitors.

Consequently, two possible combinations to build the impedance matching network are:

1) By adding a capacitor C_S in series first, and then by adding a capacitor C_P in parallel second.

$$X_1 = -\sqrt{R_m(Z_o - R_m)} - X_m. \tag{10.A.79}$$

$$C_S = \frac{C_m}{C_m \omega \sqrt{R_m(Z_o - R_m)} - 1}, \tag{10.A.80}$$

and

$$\frac{1}{X_2} = \frac{-\sqrt{R_m(Z_o - R_m)}}{R_m Z_o}, \tag{10.A.81}$$

$$C_P = \frac{\sqrt{R_m(Z_o - R_m)}}{R_m Z_o \omega}. \tag{10.A.82}$$

2) By adding a capacitor C_P in parallel first, and then by adding a capacitor C_S in series second.

$$X_1 = \frac{X_m Z_o - \sqrt{R_m Z_o(R_m^2 + X_m^2 - R_m Z_o)}}{R_m - Z_o}, \tag{10.A.83}$$

$$C_P = \frac{(R_m - Z_o)/\omega}{\sqrt{R_m Z_o(R_m^2 + X_m^2 - R_m Z_o)} - X_m Z_o}, \tag{10.A.84}$$

and

$$X_2 = -\frac{\sqrt{R_m Z_o(R_m^2 + X_m^2 - R_m Z_o)}}{R_m}. \tag{10.A.85}$$

$$C_S = \frac{R_m/\omega}{\sqrt{R_m Z_o(R_m^2 + X_m^2 - R_m Z_o)}}. \tag{10.A.86}$$

Case 4: when the original impedance is located in region 4

The first impedance matching part, X_1, or x_1, denoted by subscript "1", is added with the original impedance Z_m in series or in parallel so that the original impedance, Z_m, can be moved to the circle with $g = 1$ or $r = 1$. X_1 is an inductor.

In series

$$X_1 = +\sqrt{R_m(Z_o - R_m)} - X_m. \tag{10.A.87}$$

In parallel

$$X_1 = \frac{X_m Z_o + \sqrt{R_m Z_o(R_m^2 + X_m^2 - R_m Z_o)}}{R_m - Z_o}, \tag{10.A.88}$$

Equations (10.A.87) and (10.A.88) correspond to equations (10.A.41) and (10.A.62), respectively, when the positive sign in equations (10.A.41) and (10.A.62) is selected for an inductive X_1.

The second matching part, X2, or x2, denoted by subscript "2", in parallel or in series is to pull Z_{g1} or z_{g1} on the circle $g = 1$ or $r = 1$ to the center of the Smith chart or to the reference impedance, 50 Ω. X_2 is an inductor as well.

In parallel

$$\frac{1}{X_2} = \frac{+\sqrt{R_m(Z_o - R_m)}}{R_m Z_o}, \tag{10.A.89}$$

In series

$$X_2 = +\frac{\sqrt{R_m Z_o(R_m^2 + X_m^2 - R_m Z_o)}}{R_m}. \tag{10.A.90}$$

Equations (10.A.89) and (10.A.90) correspond to equations (10.A.45) and (10.A.65), respectively, when the positive sign in equations (10.A.45) and (10.A.65) is selected for an inductive X_2.

As shown in Figure 10.A.4, there are two options for the impedance matching from original value Z_m to the reference impedance 50Ω. Two impedance matching parts are inductors.

Consequently, two possible combinations to build the impedance matching network are:

1) By adding an inductor L_S in series first, and then by adding a capacitor L_P in parallel second.

$$X_1 = +\sqrt{R_m(Z_o - R_m)} - X_m. \tag{10.A.91}$$

$$L_S = \frac{\sqrt{R_m(Z_o - R_m)} - X_m}{\omega}, \tag{10.A.92}$$

$$\frac{1}{X_2} = \frac{+\sqrt{R_m(Z_o - R_m)}}{R_m Z_o}, \tag{10.A.93}$$

$$L_P = \frac{R_m Z_o}{\omega\sqrt{R_m(Z_o - R_m)}}. \tag{10.A.94}$$

2) By adding a capacitor L_P in parallel first, and then by adding of an inductor L_S in series secondly.

$$X_1 = \frac{X_m Z_o + \sqrt{R_m Z_o(R_m^2 + X_m^2 - R_m Z_o)}}{R_m - Z_o}, \tag{10.A.95}$$

$$L_P = \frac{X_m Z_o + \sqrt{R_m Z_o(R_m^2 + X_m^2 - R_m Z_o)}}{(R_m - Z_o)\omega}, \tag{10.A.96}$$

$$X_2 = + \frac{\sqrt{R_m Z_o(R_m^2 + X_m^2 - R_m Z_o)}}{R_m}, \tag{10.A.97}$$

$$L_S = \frac{\sqrt{R_m Z_o(R_m^2 + X_m^2 - R_m Z_o)}}{R_m \omega}, \tag{10.A.98}$$

10.A.3 Topology Restrictions of a Two-Part Impedance Matching Network

The reader is referred to Figures 10.A.5 to 10.A.8.

10.A.4 Topology Restrictions of a Three-Part Impedance Matching Network

The reader is referred to Figures 10.A.9 to 10.A.16.

10.A.5 Conversion between "*Π*" and "*T*" Type Matching Networks

If a matching network consists of three parts, it can be built with "*T*" or "*Π*" configuration as shown in Figure 10.A.17. It is difficult to say which one is better. Generally, the selection is based on the following considerations:

- Type "*T*" is better than type "*Π*" if the *DC* blocking to the next block is involved.
- Type "*Π*" is better than type "*T*" if the spray reactance in the input of next block is concerned.

As a matter of fact, "*T*" and "*Π*" type of block can be converted from each other. The conversion equations from *Π* to *T* are:

$$Z_{T1} = \frac{Z_{\Pi 1} Z_{\Pi 2}}{Z_{\Pi 1} + Z_{\Pi 2} + Z_{\Pi 3}}, \tag{10.A.99}$$

$$Z_{T2} = \frac{Z_{\Pi 2} Z_{\Pi 3}}{Z_{\Pi 1} + Z_{\Pi 2} + Z_{\Pi 3}}, \tag{10.A.100}$$

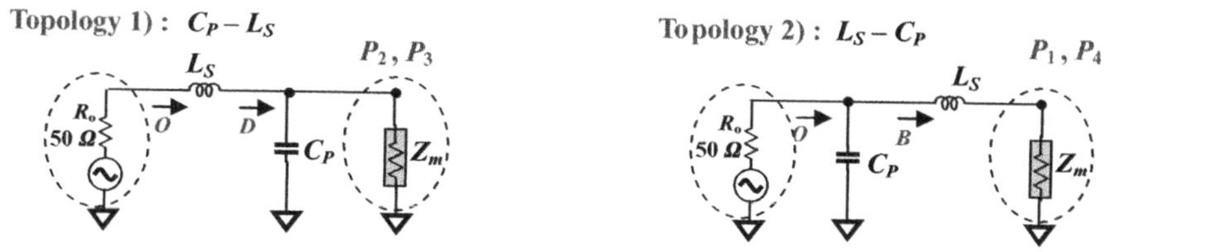

Note 1: Z_m is the original impedance to be matched.

Note 2: In the title of circuit topology "C_P-L_S" or "L_S-C_P," the first part is connected to original impedance to be matched and second part is connected to reference impedance, 50 Ω.

Note 3: Subscript "P" stands for "in parallel" and subscript "S" stands for "in series."

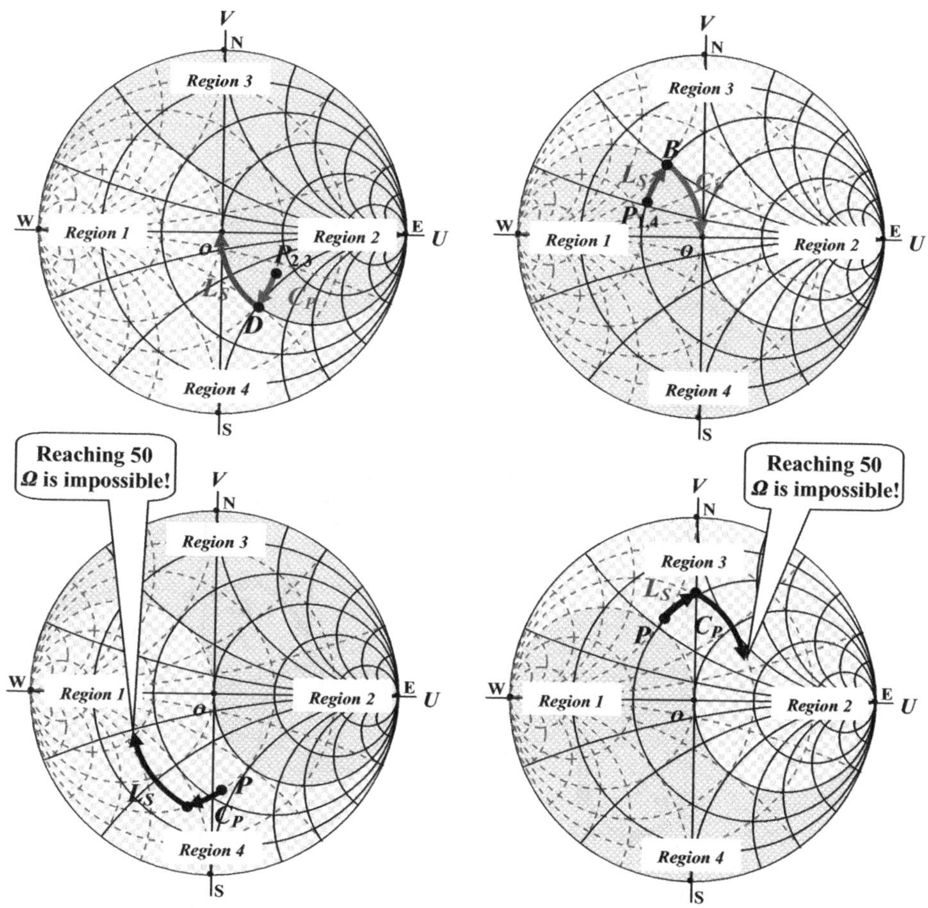

Applied regions:	Prohibited regions:
o Region 2 o Region 3	o Region 1 o Region 4

Applied regions:	Prohibited regions:
o Region 1 o Region 4	o Region 2 o Region 3

Figure 10.A.5 Applied and prohibited regions of the topology: $C_P - L_S$ or $L_S - C_P$.

Note 1: Z_m is the original impedance to be matched.
Note 2: In the title of circuit topology "C_S-L_P" or "L_P-C_S," the first part is connected to original impedance to be matched and second part is connected to reference impedance, 50 Ω.
Note 3: Subscript "P" stands for "in parallel" and subscript "S" stands for "in series."

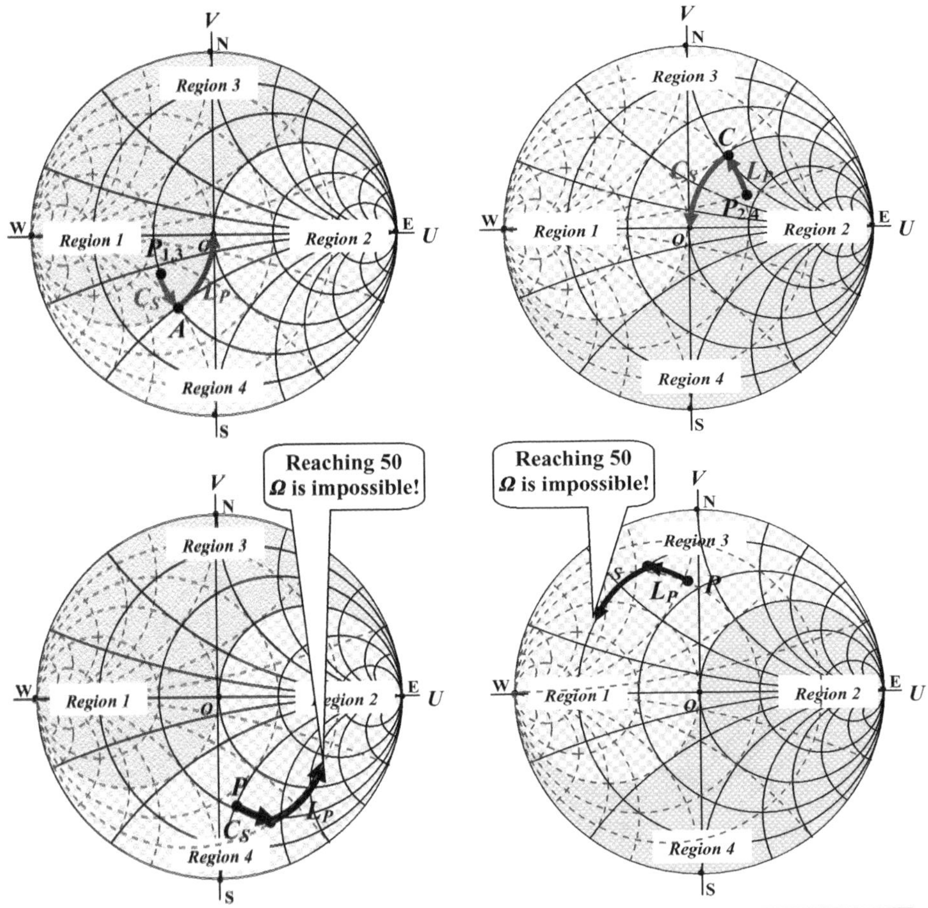

Applied regions:	Prohibited regions:
o Region 1	o Region 2
o Region 3	o Region 4

Applied regions:	Prohibited regions:
o Region 2	o Region 1
o Region 4	o Region 3

Figure 10.A.6 Applied and prohibited regions of the topology: $C_S - L_P$ or $L_P - C_S$.

Note 1: Z_m is the original impedance to be matched.
Note 2: In the title of circuit topology "C_S-C_P" or "C_P-C_S," the first part is connected to original impedance to be matched and second part is connected to reference impedance, 50 Ω.
Note 3: Subscript "P" stands for "in parallel" and subscript "S" stands for "in series."

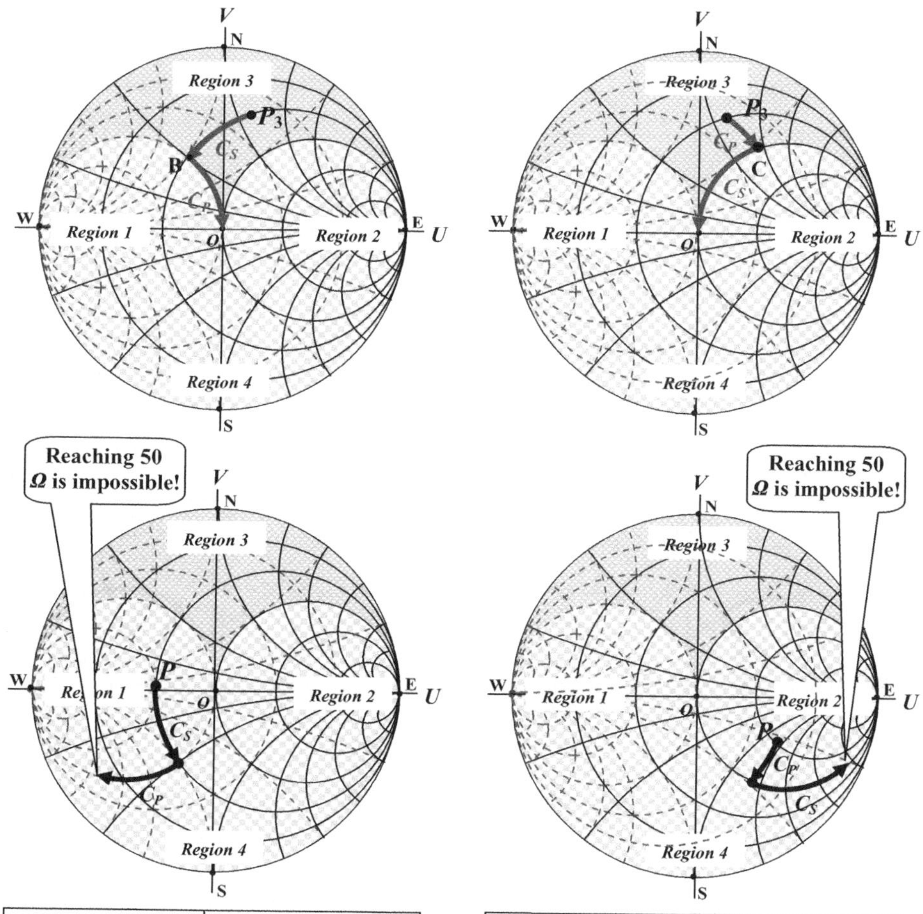

Applied regions:	Prohibited regions:
o Region 3	o Region 1 o Region 2 o Region 4

Applied regions:	Prohibited regions:
o Region 3	o Region 1 o Region 2 o Region 4

Figure 10.A.7 Applied and prohibited regions of the topology: $C_S - C_P$ or $C_P - C_S$.

Note 1: Z_m is the original impedance to be matched.
Note 2: In the title of circuit topology, "L_P-L_S" or "L_S-L_P," the first part is connected to original impedance to be matched and second part is connected to reference impedance, 50 Ω.
Note 3: Subscript "P" stands for "in parallel" and subscript "S" stands for "in series."

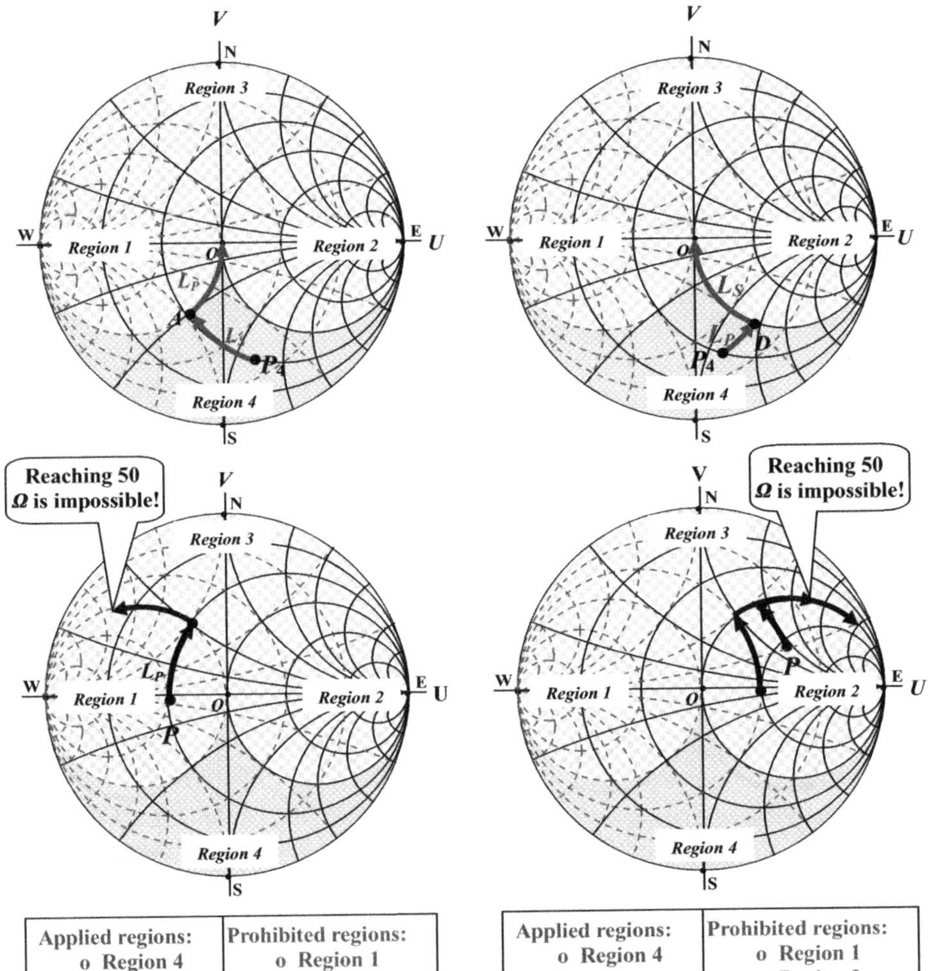

Applied regions:	Prohibited regions:
o Region 4	o Region 1 o Region 2 o Region 3

Applied regions:	Prohibited regions:
o Region 4	o Region 1 o Region 2 o Region 3

Figure 10.A.8 Applied and prohibited regions of the topology: $L_S - L_P$ or $L_P - L_S$.

Topology *Π* 1: $C_{P1} - C_S - C_{P2}$

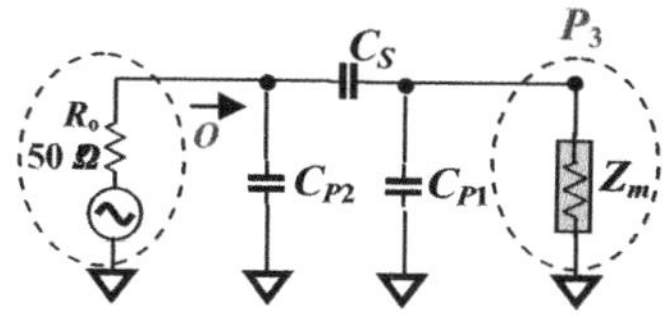

Applied regions:	Prohibited regions:
o Region 3	o Region 1 o Region 2 o Region 4

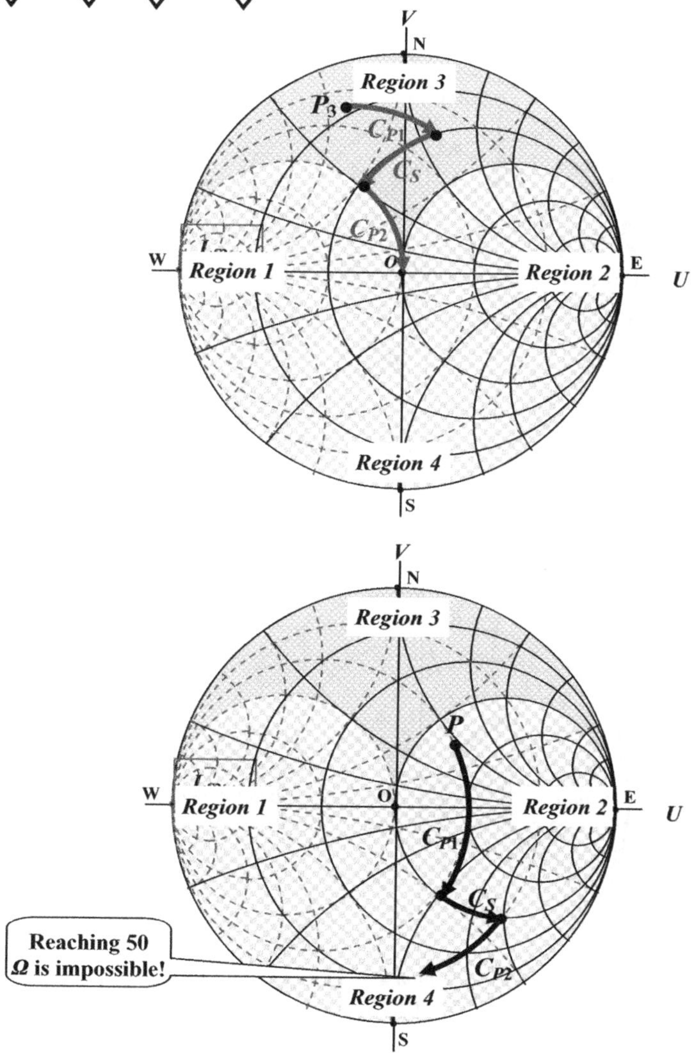

Note 1: Z_m is the original impedance to be matched.
Note 2: In the title of circuit topology, first part marked with subscript "1" is connected to original impedance to be matched and second part marked with subscript "2" is connected to reference impedance, 50 Ω.
Note 3: Subscript "*P*" stands for "in parallel" and subscript "*S*" stands for "in series."

Figure 10.A.9 Applied and prohibited areas of the topology: $C_{P1} - C_S - C_{P2}$.

Topology Π 2: L_{P1} - C_S - C_{P2}

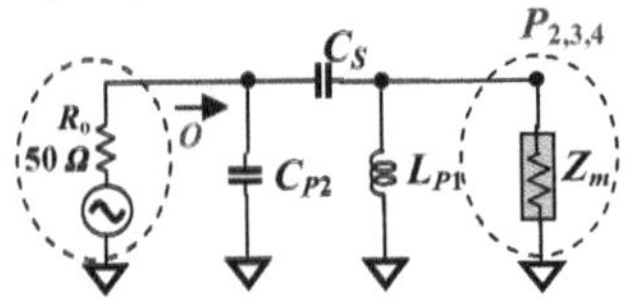

Applied regions:	Prohibited regions:
o Region 2 o Region 3 o Region 4	o Region 1

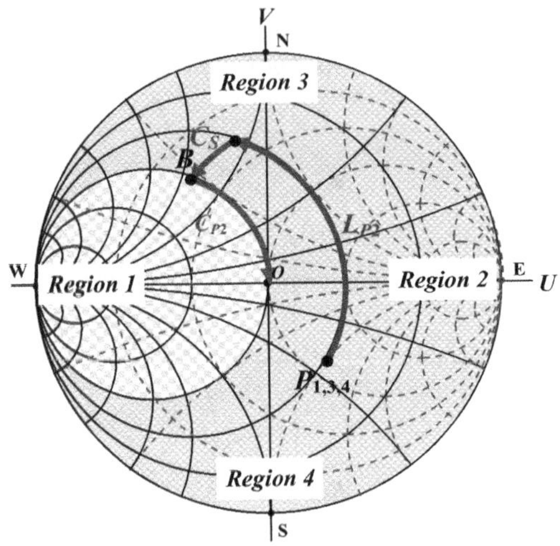

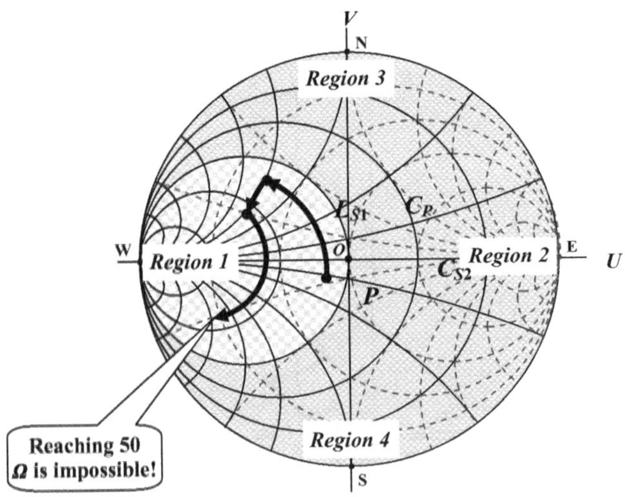

Note 1: Z_m is the original impedance to be matched.
Note 2: In the title of circuit topology, first part marked with subscript "1" is connected to original impedance to be matched and second part marked with subscript "2" is connected to reference impedance, 50 Ω.
Note 3: Subscript "P" stands for "in parallel" and subscript "S" stands for "in series."

Figure 10.A.10 Applied and prohibited areas of the topology: $L_{P1} - C_S - C_{P2}$.

Topology *Π* 3: C_{P1} - L_S - C_{P2}

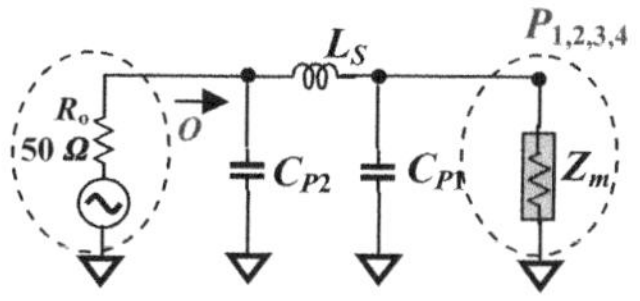

Applied regions:	Prohibited regions:
o Region 1 o Region 2 o Region 3 o Region 4	0 !

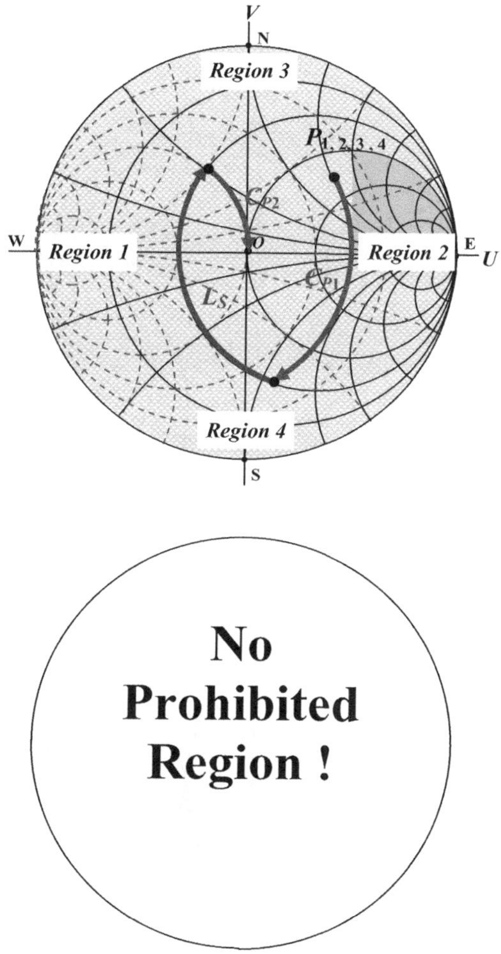

Note 1: Z_m is the original impedance to be matched.
Note 2: In the title of circuit topology, first part marked with subscript "1" is connected to original impedance to be matched and second part marked with subscript "2" is connected to reference impedance, 50 *Ω*.
Note 3: Subscript "*P*" stands for "in parallel" and subscript "*S*" stands for "in series."

Figure 10.A.11 Applied and prohibited areas of the topology: $C_{P1} - L_S - C_{P2}$.

Topology Π 5: C_{P1} - C_S - L_{P2}

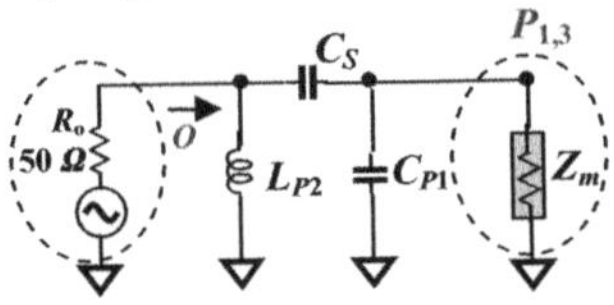

Applied regions:	Prohibited regions:
o Region 1	o Region 2
o Region 3	o Region 4

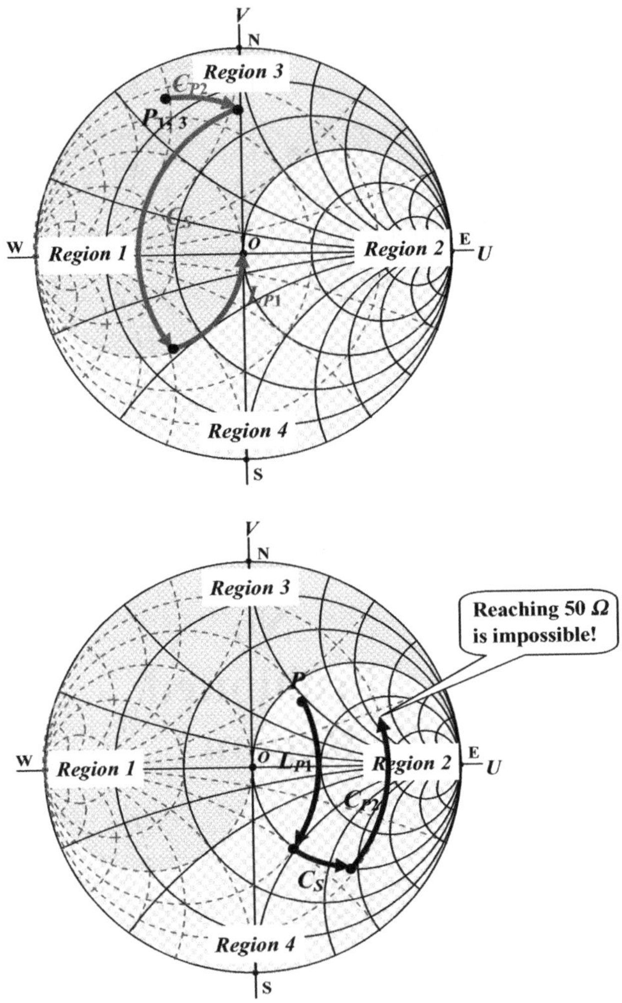

Note 1: Z_m is the original impedance to be matched.
Note 2: In the title of circuit topology, first part marked with subscript "1" is connected to original impedance to be matched and second part marked with subscript "2" is connected to reference impedance, 50 Ω.
Note 3: Subscript "P" stands for "in parallel" and subscript "S" stands for "in series."

Figure 10.A.12 Applied and prohibited areas of the topology: $L_{P1} - C_S - C_{P2}$.

Topology $T1$: C_{S1} - C_P – C_{S2}

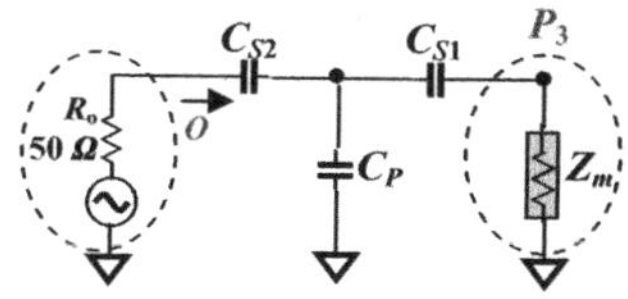

Applied regions:	Prohibited regions:
o Region 3	o Region 1 o Region 2 o Region 4

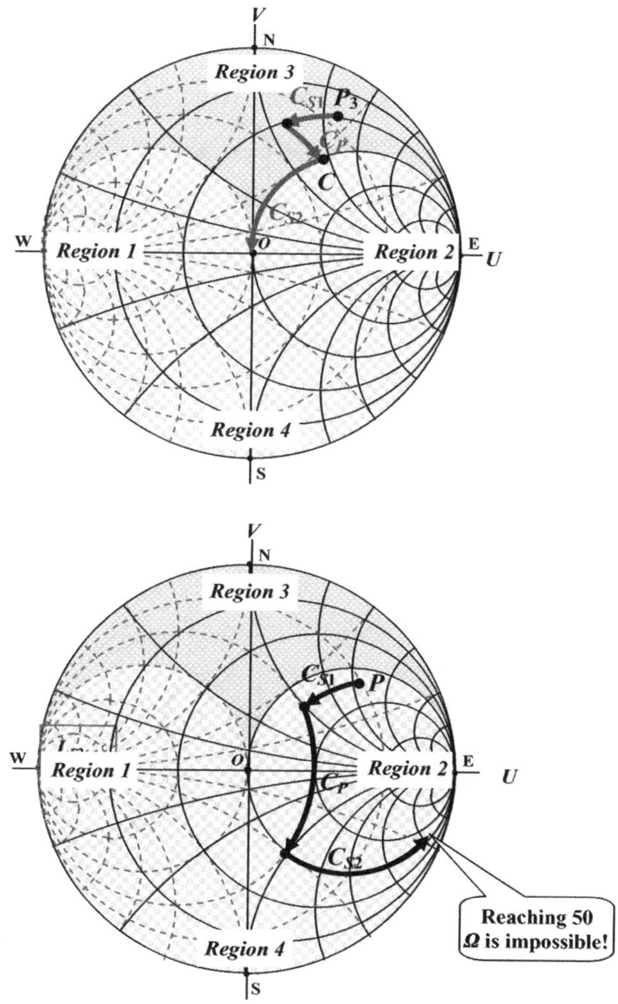

Note 1: Z_m is the original impedance to be matched.
Note 2: In the title of circuit topology, first part marked with subscript "1" is connected to original impedance to be matched and second part marked with subscript "2" is connected to reference impedance, 50 Ω.
Note 3: Subscript "P" stands for "in parallel" and subscript "S" stands for "in series."

Figure 10.A.13 Applied and prohibited areas of the topology: $C_{S1} - C_P - C_{S2}$.

Topology *T2*: L_{S1} - C_P − C_{S2}

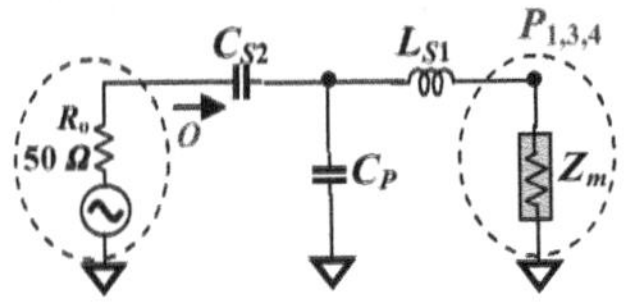

Applied regions:	Prohibited regions:
o Region 1 o Region 3 o Region 4	o Region 2

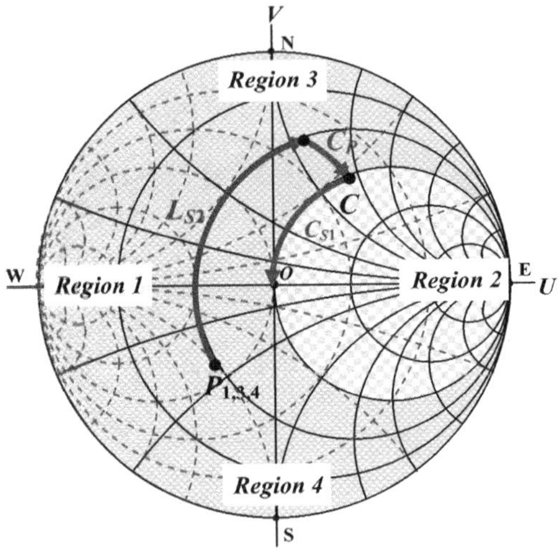

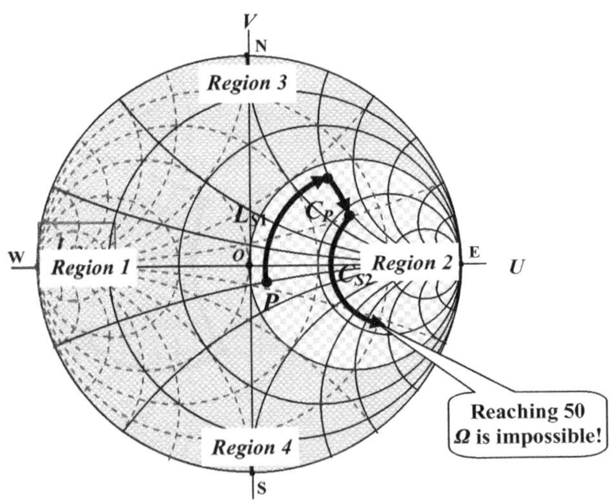

Note 1: Z_m is the original impedance to be matched.
Note 2: In the title of circuit topology, first part marked with subscript "1" is connected to original impedance to be matched and second part marked with subscript "2" is connected to reference impedance, 50 Ω.
Note 3: Subscript "*P*" stands for "in parallel" and subscript "*S*" stands for "in series."

Figure 10.A.14 Applied and prohibited areas of the topology: $L_{S1} - C_P - C_{S2}$.

Topology *T3*: C_{S1} - L_P - C_{S2}

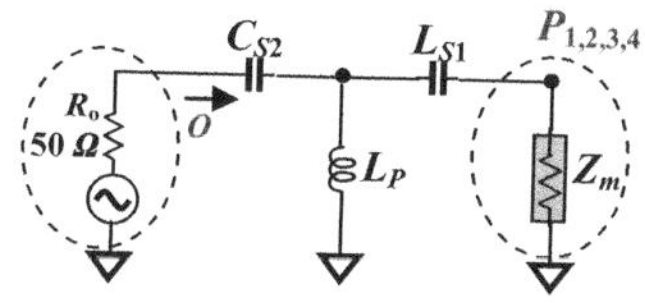

Applied regions:	Prohibited regions:
o Region 1 o Region 2 o Region 3 o Region 4	0 !

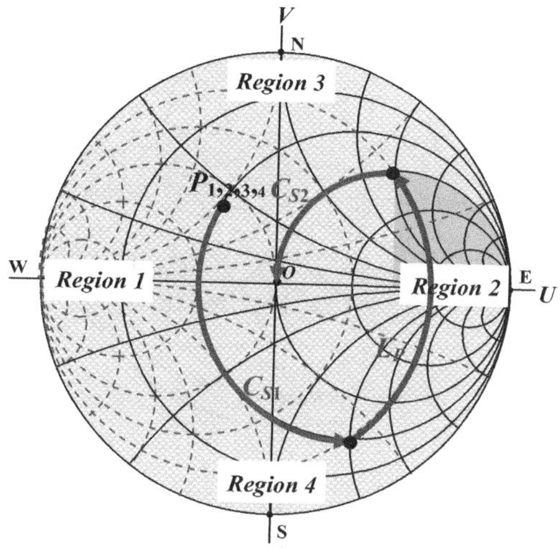

Note 1: Z_m is the original impedance to be matched.
Note 2: In the title of circuit topology, first part marked with subscript "1" is connected to original impedance to be matched and second part marked with subscript "2" is connected to reference impedance, 50 Ω.
Note 3: Subscript "*P*" stands for "in parallel" and subscript "*S*" stands for "in series."

Figure 10.A.15 Applied and prohibited areas of the topology: $C_{S1} - L_P - C_{S2}$.

Topology *T5*: C_{S1} - C_P–L_{S2}

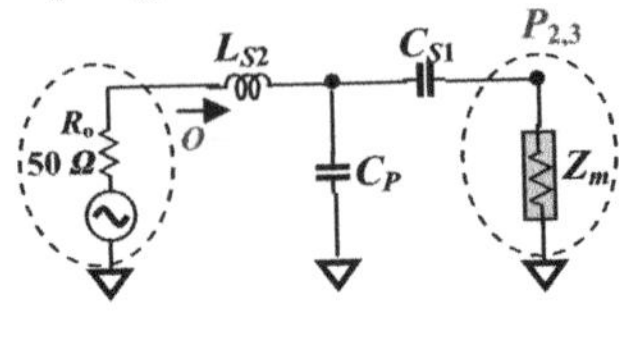

Applied regions:	Prohibited regions:
o Region 2 o Region 3	o Region 1 o Region 4

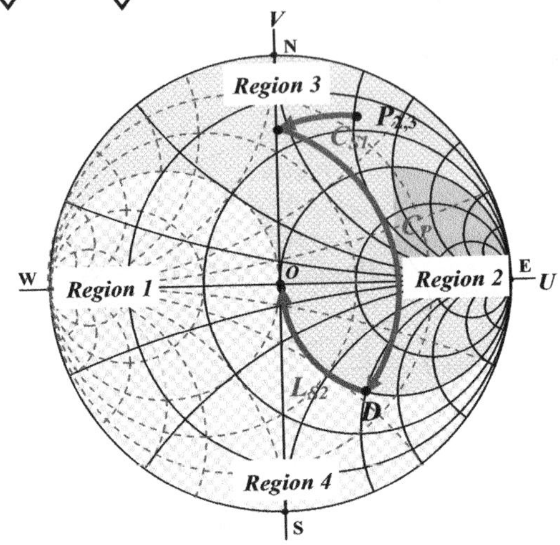

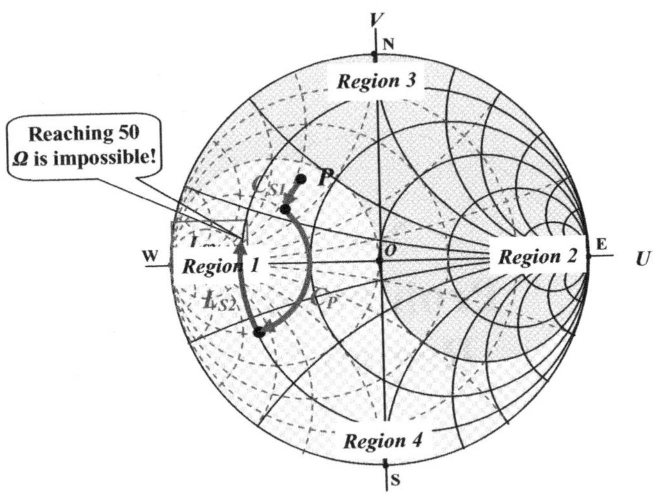

Note 1: Z_m is the original impedance to be matched.
Note 2: In the title of circuit topology, first part marked with subscript "1" is connected to original impedance to be matched and second part marked with subscript "2" is connected to reference impedance, 50 Ω.
Note 3: Subscript "*P*" stands for "in parallel" and subscript "*S*" stands for "in series."

Figure 10.A.16 Applied and prohibited areas of the topology: $C_{S1} - C_P - L_{S2}$.

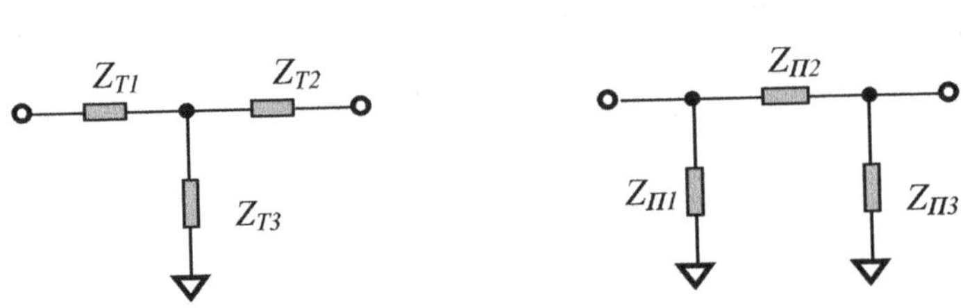

Figure 10.A.17 "*T*" and "*Π*" type matching networks with three parts.

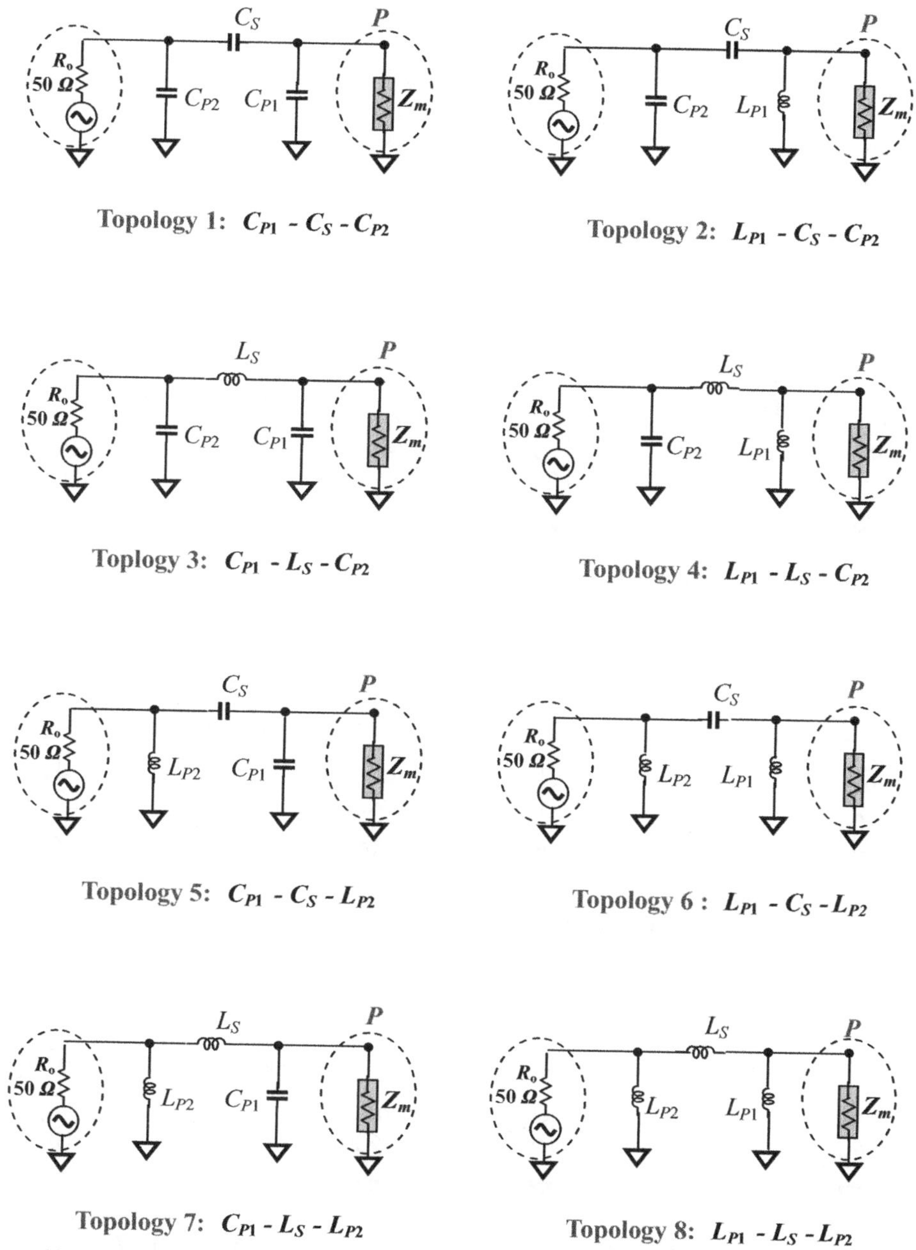

Figure 10.A.18 Topologies of Π type impedance matching networks.
Note 1: Z_m is the original impedance to be matched.
Note 2: In the title of circuit topology, first part marked with subscript "1" is connected to the original impedance to be matched, second part marked with subscript "2" is connected to the reference impedance, 50 Ω.
Note 3: Subscript "P" stands for "in parallel" and subscript "S" stands for "in series."

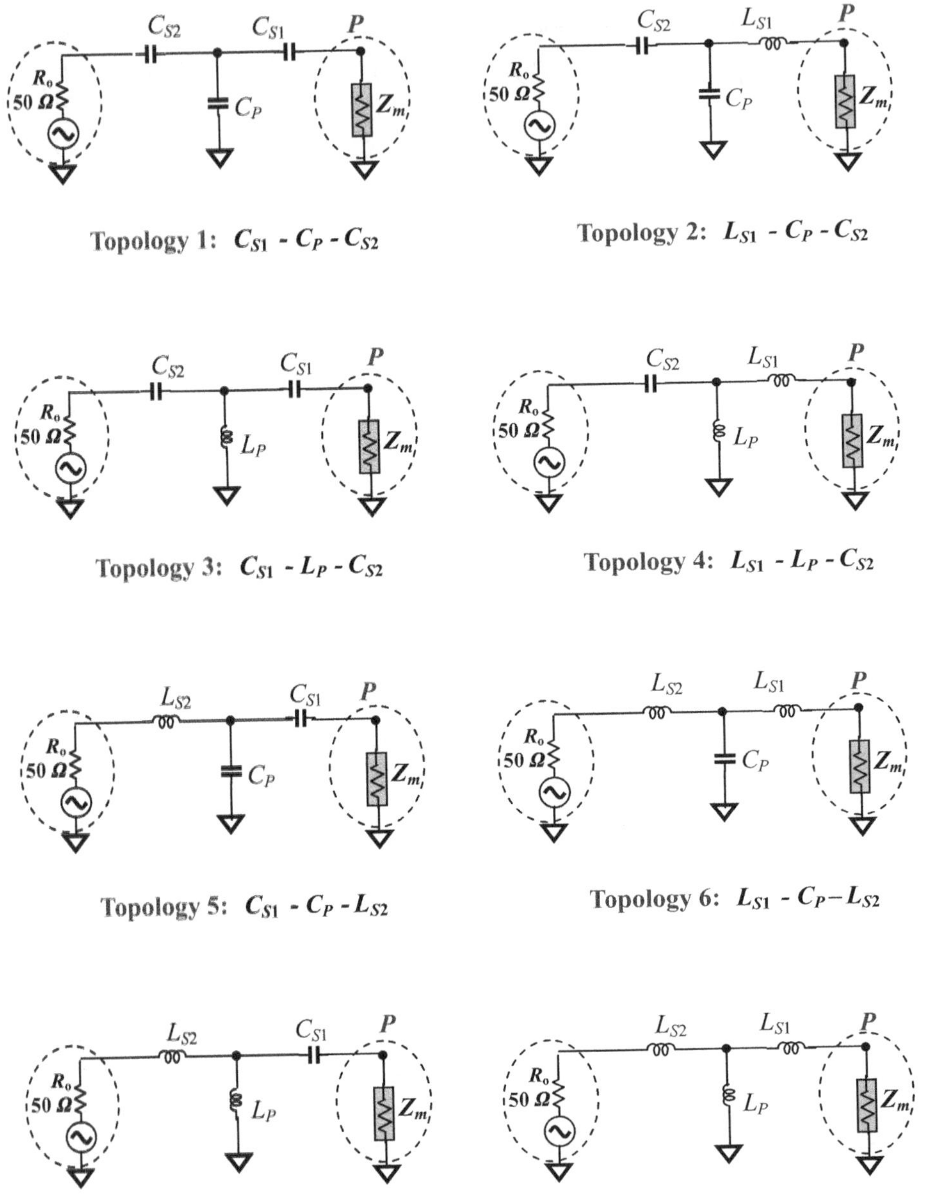

Topology 1: C_{S1} - C_P - C_{S2}

Topology 2: L_{S1} - C_P - C_{S2}

Topology 3: C_{S1} - L_P - C_{S2}

Topology 4: L_{S1} - L_P - C_{S2}

Topology 5: C_{S1} - C_P - L_{S2}

Topology 6: L_{S1} - C_P – L_{S2}

Topology 7: C_{S1} - L_P - L_{S2}

Topology 8: L_{S1} - L_P – L_{S2}

Figure 10.A.19 Topologies of T type impedance matching networks.
Note 1: Z_m is the original impedance to be matched.
Note 2: In the title of circuit topology, first part marked with subscript "1" is connected to the original impedance to be matched, second part marked with subscript "2" is connected to the reference impedance, 50 Ω.
Note 3: Subscript "P" stands for "in parallel" and subscript "S" stands for "in series."

$$Z_{T3} = \frac{Z_{\Pi 3} Z_{\Pi 1}}{Z_{\Pi 1} + Z_{\Pi 2} + Z_{\Pi 3}}. \quad (10.A.101)$$

The conversion equations from T to Π are:

$$Z_{\Pi 1} = Z_{T3} + Z_{T1} + \frac{Z_{T3} Z_{T1}}{Z_{T2}}, \quad (10.A.102)$$

$$Z_{\Pi 2} = Z_{T1} + Z_{T2} + \frac{Z_{T1} Z_{T2}}{Z_{T3}} \quad (10.A.103)$$

$$Z_{\Pi 3} = Z_{T2} + Z_{T3} + \frac{Z_{T2} Z_{T3}}{Z_{T1}}. \quad (10.A.104)$$

"*T-Π*" conversion is also called "* – Δ" conversion because "*T*" looks like "*" and "*Π*" looks like "Δ".

Figure 10.A.17 illustrates that the "*T* – *Π* " or "* – Δ" conversion provides the flexibility of the topology for a matching network, while its input and output impedance can be kept unchanged.

The "*T* – *Π* " or "* – Δ" conversion is important to a designer. The reasons are:

- Sometimes the part's value is too low and is comparable with the parasitic parameters. In this case the reliability of its performance is nothing to talk about. At another extreme end, its performance is also not reliable if the part's value is too high.
- Sometimes the designer cannot find the expected value of a part in the market. By means of the "*T* – *Π* " or "* – Δ" conversion, the value of a part can be changed.
- The total part number can be reduced or increased.

Sometimes the part's value is found to be unreasonably high or unacceptably low in the circuit design. In terms of "*T* – *Π* " or "* – Δ" conversion, the value of parts may become reasonable and appropriate to the implementation of the circuitry.

10.A.6 Possible "*Π*" and "*T*" Impedance Matching Networks

The reader is referred to Figures 10.A.18 and 10.A.19.

REFERENCES

[1] G. P. Young and S. O. Scanlan, "Matching Network Design Studies for Microwave Transistor Amplifiers," *IEEE Transactions on Microwave Theory and Techniques*, Vol. MTT–29, No. 10, October 1981, pp. 1027–1035.

[2] P. L. D. Abrie, *The Design of Impedance Matching Networks for Radio-Frequency and Microwave Amplifiers*, Artech House, Norwood, Mass. 1985.

[3] U. L. Rohde, "Designing a Matched Low Noise Amplifier Using CAD Tools," *Microwave Journal*, Vol. 29, No. 10, October, 1986, pp. 154–160.

[4] George D. Vendelin, Anthony M. Pavio, and Ulrich L. Rohde, *Microwave Circuit Design Using Linear and Nonlinear Techniques*, John Wiley & Sons, Inc., 1990.

[5] J. K. Fidler and Y. Sun, "Computer-aided Determination of Impedance Matching Domain," Digital and Analogue Filters and Filtering Systems, IEEE Twelfth Saraga Colloquium, November 6, 1992, pp. 1/1–1/6.

[6] Yichuang Sun and J. K. Fidler, "Design of Π Impedance Matching Networks," Circuits and Systems, 1994. ISCAS '94, *IEEE International Symposium*, Vol. 5, May 30–June 2, 1994, pp. 5–8.

[7] Y. Sun and J. K. Fidler, "Design Method for Impedance Matching Networks," *Circuits, Devices and Systems, IEEE Proceedings*, Vol. 143, No. 4, August 1996, pp. 186–194.

[8] M. Lapinoja and T. Rahkonen, "An Active Tuning and Impedance Matching Element," Circuits and Systems, 1998. ISCAS '98. *Proceedings of the 1998 IEEE International Symposium*, Vol. 1, No. 3 May–3 June 1998, pp. 559–562.

[9] G. Girlando and G. Palmisano, "Noise Figure and Impedance Matching in RF Cascode Amplifiers," *Circuits and Systems II: Analog and Digital Signal Processing, IEEE Transactions*, Vol. 46, No. 11, November 1999, pp. 1388–1396.

[10] J.-M. Collantes, R. D. Pollard, and M. Sayed, "Effects of DUT Mismatch on the Noise Figure Characterization: A Comparative Analysis of Two Y-factor Techniques," *Instrumentation and Measurement, IEEE Transactions*, Vol. 51, No. 6, December 2002, pp. 1150–1156.

[11] Richard Chi-Hsi Li, *Key Issues in RF/RFIC Circuit Design*, Higher Education Press, Beijing, 2005.

CHAPTER 11

IMPEDANCE MATCHING IN A WIDE-BAND CASE

11.1 APPEARANCE OF NARROW- AND WIDE-BAND RETURN LOSS ON A SMITH CHART

The methodology of impedance matching is basically the same between narrow- and wide-band cases. However, in the narrow-band case, only one frequency needs to be taken care of in the process of impedance matching because a narrow bandwidth can be represented approximately by the central frequency. In the wide-band case, instead of a single frequency, the entire wide bandwidth must be taken care of, which makes impedance matching somewhat difficult. In this chapter, we will apply the same process of impedance matching as that in the narrow-band case, but our main effort will be dealing with the expansion scheme from narrow to wide bandwidth. (The methodology of impedance matching in a wide-band case introduced and discussed in this chapter was developed in 2005 by the author specially for an ultrawide band (UWB) system and has not been made public until now.)

Figure 11.1 shows the difference of the return loss or impedance on the Smith chart between narrow- and wide-band cases. The dashed-line circle in each plot is the demarcation circle of return loss as mentioned in Chapter 10, where S_{11} or $S_{22} = -10\,dB$.

In the narrow-band cases as shown in Figure 11.1(a), the return loss or impedance is approximated by a central frequency, which is a single point on the Smith chart denoted by Z_{fc}. The impedance matching state is clear if the critical circle of return loss, S_{11} or $S_{22} = -10\,dB$, is taken as the unique criterion, that is,

- In the upper-left plot, Z_{fc} is located outside the demarcation circle so that the impedance is in an unmatched state.

RF Circuit Design, by Richard Chi-Hsi Li

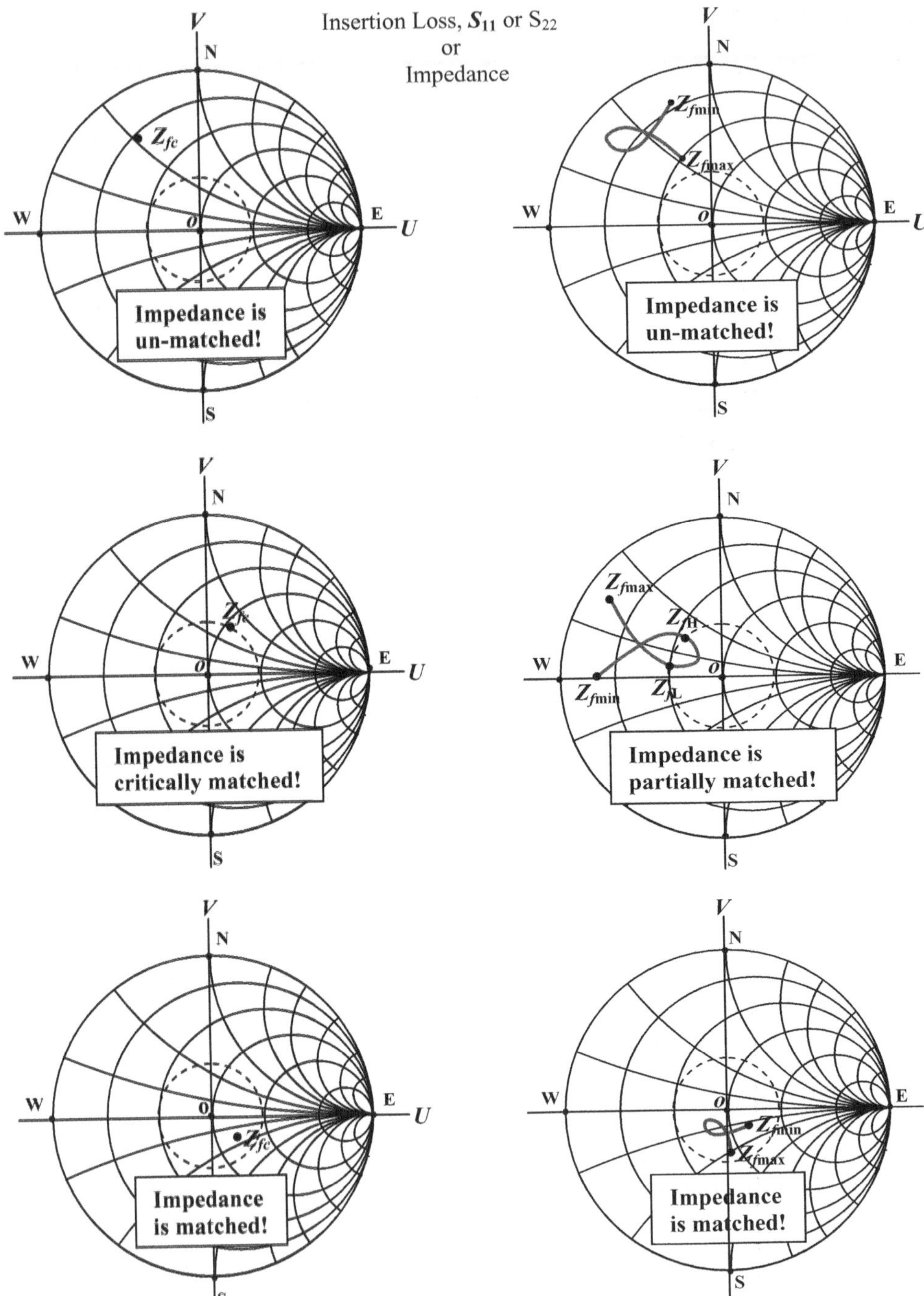

(a) In narrow band case (b) In wide band case

Figure 11.1 Appearance of return loss in narrow and wide band cases.

- In the middle-left plot, Z_{fc} is exactly located on the demarcation circle so that the impedance is in a critical matched state.
- In the bottom-left plot, Z_{fc} is located inside the demarcation circle so that the impedance is in a matched state.

In the wide-band cases as shown in Figure 11.1(b), the trace of S_{11}, S_{22} or impedance on the Smith chart is not a point, but a line segment, which is the collective return loss or impedance responding for all the frequencies over the bandwidth. Instead of one point, the impedance matching state must be examined over the entire line segment of S_{11}, S_{22}, or impedance. One end of the trace is marked with Z_{fmin} which denotes the impedance at the minimum frequency f_{min}. Another end of the trace is marked with Z_{fmax} which denotes the impedance at the maximum frequency f_{max}. The impedance matching state can still be judged if the demarcation circle of return loss, S_{11} or $S_{22} = -10\,dB$, is taken as the unique criterion again, that is, as shown in Figure 11.1(b).

- In the upper-right plot, the entire trace of return loss or impedance is located outside the demarcation circle so that the impedances are entirely in the unmatched state.
- In the middle-right plot, the trace of return loss or impedance is located partially outside and partially inside the demarcation circle so that the impedances are in a partially matched state.
- In the bottom-right plot, the trace of return loss or impedance is located entirely inside the demarcation circle so that the impedances over the entire frequency bandwidth are in the matched state.

In order to explore the bandwidth from the appearance of return loss on the Smith chart, let's focus on the middle-right plot in Figure 11.1(b). The return loss trace has two intercept points with the demarcation circle, Z_{fL} and Z_{fH}. These are the impedances at the frequencies f_L and f_H, respectively. The impedances are in a matched state in the frequency interval from f_L to f_H so that the bandwidth is found as

$$BW = f_H - f_L, \tag{11.1}$$

whereas the impedances are in the unmatched state in the frequency intervals from f_{min} to f_L and f_H to f_{max}.

Obviously, the frequency bandwidth (Figure 11.1) is specially defined for impedance matching but not for the power gain or noise figure, even though they are closely related. This definition is reasonable because those impedances within the bandwidth are closer to the center of the Smith chart, $50\,\Omega$, and are therefore in an acceptable impedance matching state. Although not a formal rule, engineering design experience has for many years proved that the demarcated return loss circle, S_{11} or $S_{22} = -10\,dB$, is quite a good criterion for acceptable impedance matching.

The return loss on the Smith chart can be converted into a Cartesian plot, which provides a more intuitive understanding of the bandwidth in respect to the impedance matching status. Figure 11.2 shows the corresponding relationship between readings from the Smith chart and from the curves of the return loss, S_{11} or S_{22}, vs.

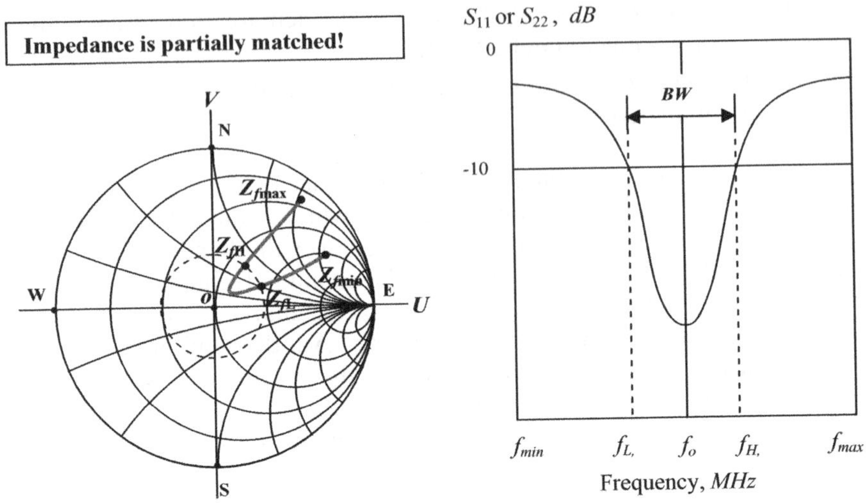

(a) Return loss on Smith chart (b) Return loss vs. frequency in Cartesian coordinates

Figure 11.2 Return loss on Smith chart and in Cartesian coordinates.

the frequency using Cartesian coordinates. The impedances are in a partially matched state.

The bandwidth for impedance matching expressed in Cartesian coordinates is much clearer than on the Smith chart. Then, why do engineers prefer to use the Smith chart rather than the Cartesian coordination in implementing impedance matching networks? The reason is that in the successive building processes of impedance matching networks, one can judge whether to insert an inductor or capacitor into the impedance matching network from the Smith chart, but not from the return loss vs. frequency plot in Cartesian coordinates. Also, one can calculate the desired value of the necessary inductor or capacitor from the Smith chart but not from the Cartesian plot. In other words, the implementation of an impedance matching network relies on the Smith chart but not on the return loss *vs.* frequency plot in the Cartesian coordination.

It is therefore desirable to study how to evaluate and control the bandwidth on the basis of the appearance of S_{11} or S_{22} on the Smith chart. What we are more interested in, additionally, is how to expand the bandwidth in the impedance matching process through the implementation of the impedance matching network.

Let's observe the appearance of the trace S_{11} or S_{22} on the Smith chart.

Figure 11.3 shows the dependence of the bandwidth on the location of the return loss on the Smith chart, or on the distance from the trace of S_{11} or S_{22} to the center of the Smith chart, $50\,\Omega$.

The return loss traces of S_{11} or S_{22} in Figure 11.3(a) and (b) have the same shape and occupy the same area, but the trace in Figure 11.3(a) is farther from the center of Smith chart than the trace in Figure 11.3(b). A large portion of the trace in Figure 11.3(b) is located inside the demarcation circle, $S_{11} = -10\,dB$, while most of the trace in Figure 11.3(a) is located outside the demarcation circle. Consequently, the trace

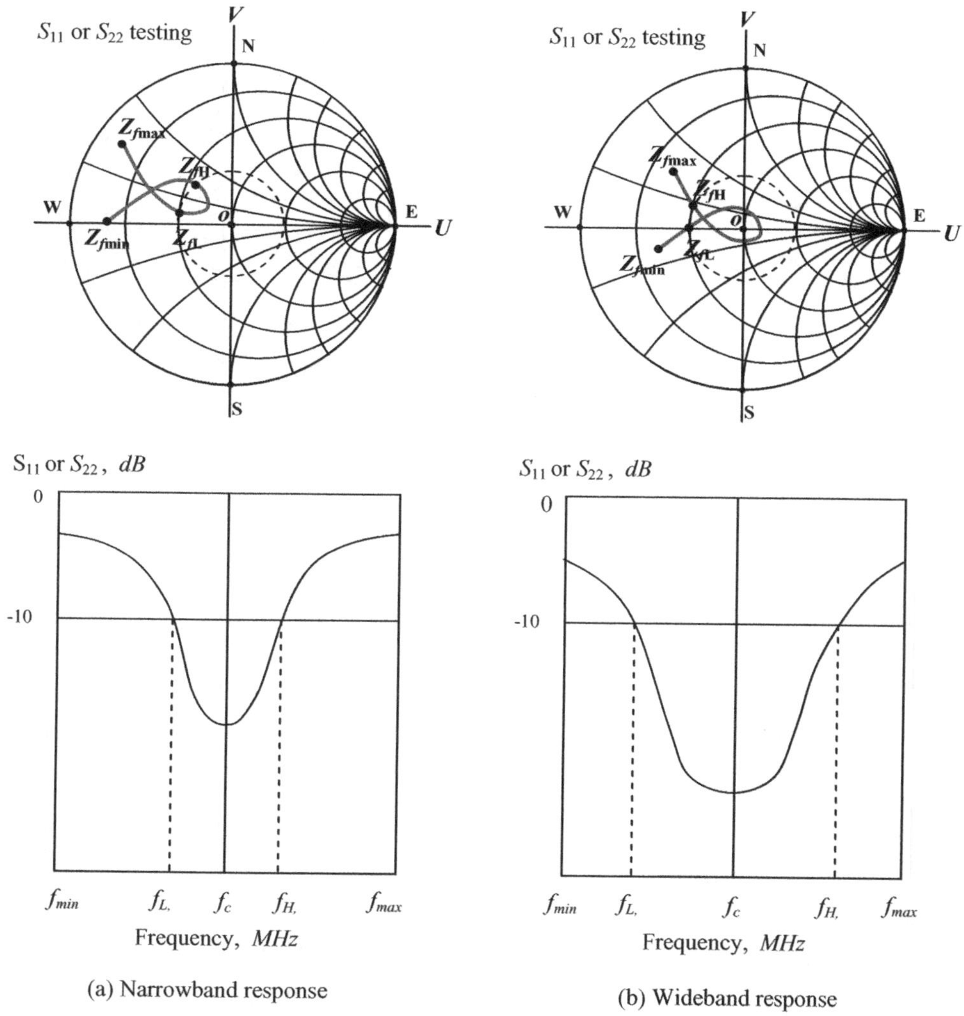

Figure 11.3 Bandwidth is dependent on the location of return loss trace on the Smith chart.

in Figure 11.3(b) corresponds with a wide-band case, while the trace in Figure 11.3(a) corresponds with a narrow-band case.

Figure 11.4 shows the dependence of the bandwidth on the density of the return loss on the Smith chart, or on the area occupied by the trace of S_{11} or S_{22} on the Smith chart.

The return loss traces of S_{11} or S_{22} in Figure 11.4(a) and (b) have the same shape but the occupied area is different. The area occupied by the trace in Figure 11.4(a) is much larger than the area occupied by the trace in Figure 11.4(b). The portion of the trace in Figure 11.4(b) located inside the demarcation circle, $S_{11} = -10dB$, is larger than that of the trace in Figure 11.4(a). Consequently, the trace in Figure 11.4(b) corresponds with a wide-band case, while the trace in Figure 11.4(a) corresponds with a narrow-band case.

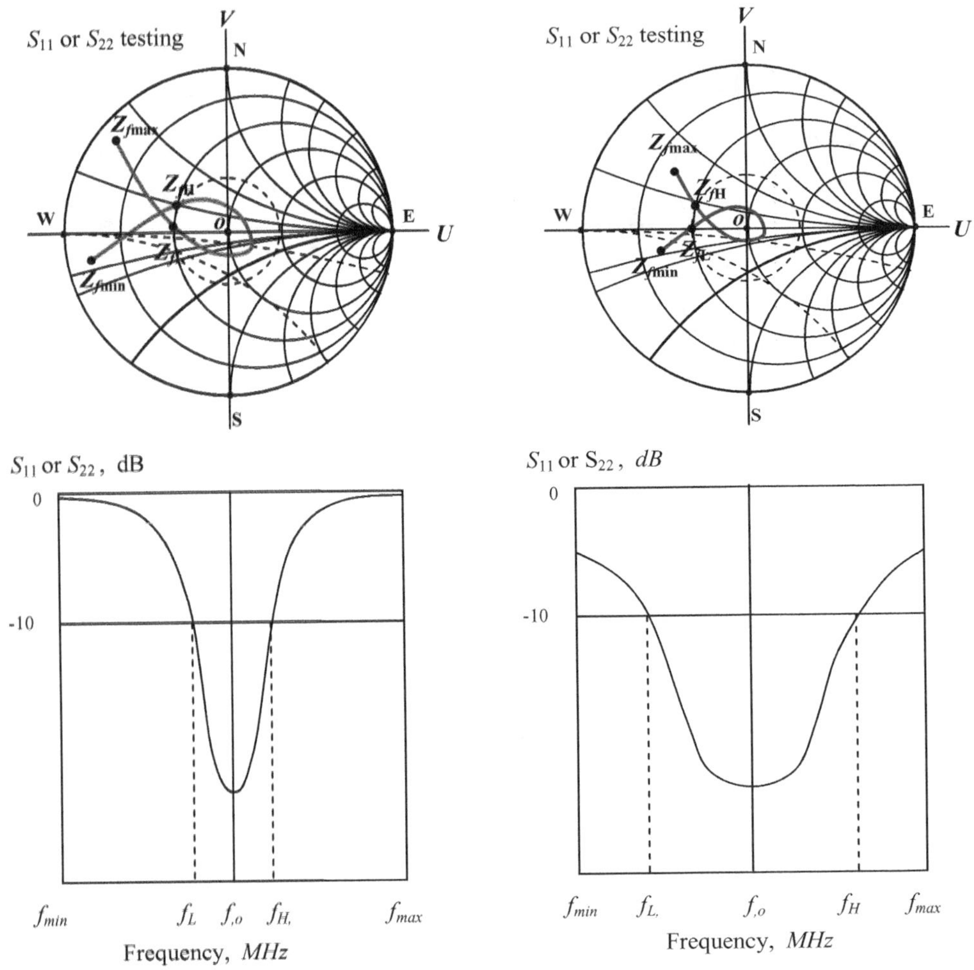

(a) Narrowband response (b) Wideband response

Figure 11.4 Bandwidth is dependent on the density of the return loss trace in Smith chart.

From the two observations made in Figures 11.3 and 11.4, it is concluded that

- In order to change the impedance matching at a terminal from a narrow-band to a wide-band response, the distance from the trace of the return loss, S_{11} or S_{22} (or impedances) to the center of the Smith chart must be as short as possible.
- In order to change the impedance matching at a terminal from a narrow-band to a wide-band response, the area the impedance trace or the return loss S_{11} or S_{22} occupies on the Smith chart must be shrunk to as small an area as possible.

In short, in order to change the impedance matching at a terminal from a narrow-band to wide-band response, squeeze the return loss trace and move it to the center of the Smith chart!

11.2 IMPEDANCE VARIATION DUE TO INSERTION OF ONE PART PER ARM OR PER BRANCH

In the passive impedance matching network discussed in Chapter 10, only one part was set forth on one branch (in parallel) or one arm (in series). In Chapter 10, we were only concerned with the one frequency which approximated the narrow-band case. In this chapter we are going to extend our discussion from the narrow-band to the wide-band case. The first thing to do is to expand our discussion of impedance matching in Chapter 10 from one original impedance, or only one point on the Smith chart, to a collectivity of impedances contained within a frequency bandwidth. We also resume the discussion begun in Chapter 10 about the variation of impedance due to the insertion of one arm and one branch into the impedance matching network.

In wide-band cases the original impedance appearing on the Smith chart is a segment of the impedance curve and is no longer a point. Each point on the segment of impedance curve represents an impedance value corresponding to a frequency within the bandwidth. Each time a part, either a capacitor or an inductor, is inserted into the impedance matching network, the impedance curve segment will be moved to a new location on the Smith chart and its shape, length, and orientation will usually change, depending on the part's value and the connection mode. As in the discussion in Chapter 10, there are four cases of impedance variation if only one part in one arm or one branch is inserted into the impedance matching network at a time.

11.2.1 An Inductor Inserted into an Impedance Matching Network in Series

When an inductor is inserted into the impedance matching network in series, the variation of the original impedance ΔZ is equal to the variation of inductive reactance ΔZ_L, that is,

$$\Delta Z = \Delta Z_L = j\Delta X_L = jL\omega. \tag{11.2}$$

Figure 11.5 plots the curve of ΔZ_L vs. f from equation (11.2). The variation of the impedance is an increase of reactance as the frequency is increased.

On the Smith chart, the impedance trace moves clockwise around the circle of r = constant, where r is the resistance value of the original impedance. The reactance increases more at the high-frequency end than at the low-frequency end because the increase of reactance is proportional to the frequency as shown in equation (11.2) and in Figure 11.5.

Figure 11.6 shows two examples, in which both the impedance changes from trace 1 to trace 2 after an inductor is inserted into the impedance matching network in series. Figure 11.6(a) shows an example in which the impedance trace has the same even shape before and after the inductor is inserted. Figure 11.6(b) shows an example in which the impedance trace changes its shape from even, before the inductor is inserted, to curly, after the inductor is inserted.

The symbols appearing in Figure 11.6, $Z_{fL,1}$, $Z_{fH,1}$, $Z_{fL,2}$, $Z_{fH,2}$, denote impedances at the low and high frequencies of trace 1 and trace 2, and are specified in Figure

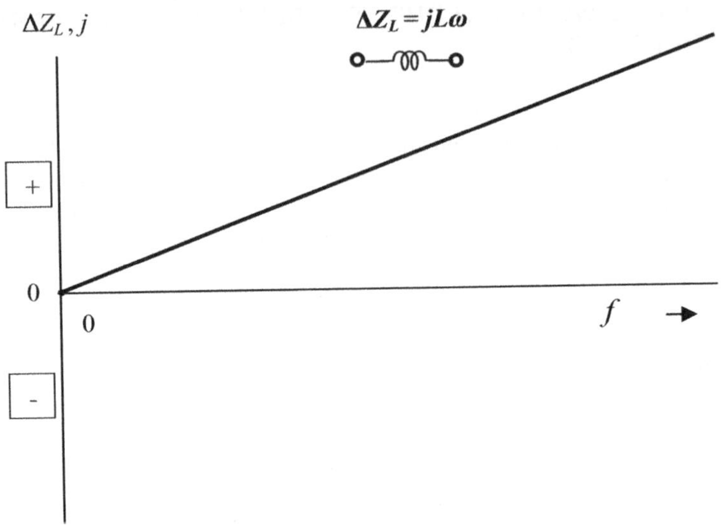

Figure 11.5 Variation of impedance when an inductor is inserted into impedance matching network in series.

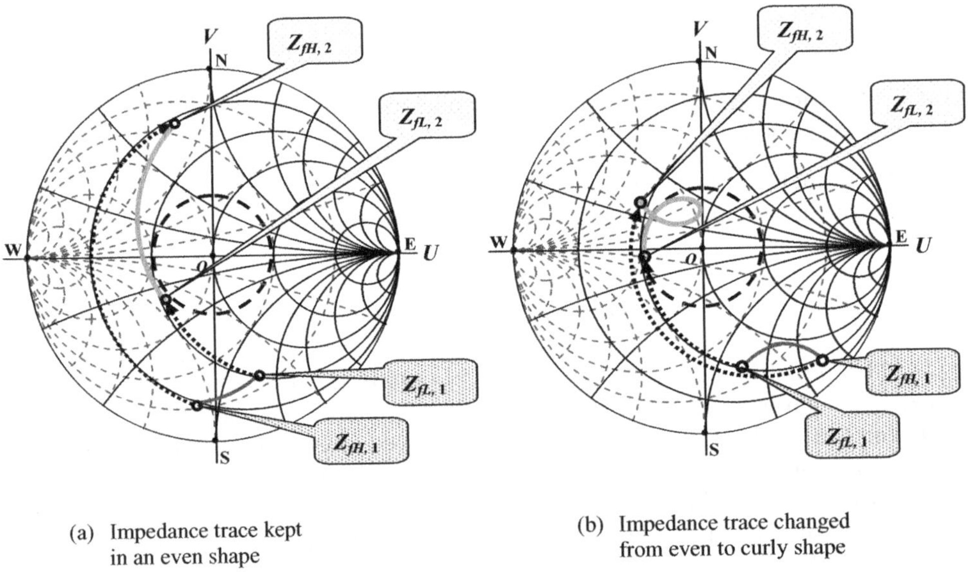

Figure 11.6 Variation of impedance from trace 1 to trace 2 if an inductor is inserted into impedance matching network in series.
— impedance trace 1, Z_{fL}: impedance at low-frequency end,
— impedance trace 2, Z_{fH}: impedance at high-frequency end,
◄···· moving direction of impedance subscript "1," "2" denotes number of impedance trace
demarcation circle, $RL = -10\,\text{dB}$.

11.6. The subscripts "f_L" and "f_H" denote the low and high-frequency ends within the bandwidth respectively; subscripts "1" and "2" denote trace 1 and trace 2, respectively.

The variation of the impedance trace from trace 1 to trace 2 can be summarized as follows:

1) All the impedances on 1 basically move clockwise around the resistance circle, although their resistances increase a little due to the equivalent resistance of the inductor in series.
2) The variation of impedance at the high-frequency end is greater than the variation of impedance at the low-frequency end due to the relation (11.2).
3) The general purpose of inserting an inductor into the impedance matching network in series is usually to move trace 1 to trace 2, so that trace 2 is closer to the reference impedance or the center of the Smith chart, as shown in Figure 11.6(a) and (b). In Figure 11.6, the circle drawn by the dashed line is the demarcation circle, $RL = -10dB$. Most of trace 2 in Figure 11.6(b) is located within the demarcation circle.
4) A remarkable event is the impedance trace's change from an even shape to a curly shape as shown in Figure 11.6(b) so that the return loss trace is squeezed or rolled up. As mentioned in Section 11.1, this change indicates that the bandwidth of the impedance trace for impedance matching has changed from narrow to wide.

The reason for the "rolling or squeezing of the trace" is twofold:

- In Figure 11.6(b), the reactance of the original impedance at the high-frequency end is higher than that at the low-frequency end, that is

$$X_{fL,1} > X_{fH,1}, \tag{11.3}$$

- The variation of impedance at the high-frequency end is much greater than the variation of impedance at the low-frequency end as shown in the relation (11.2) and Figure 11.5.

On the other hand, in Figure 11.6(a),

$$X_{fL,1} < X_{fH,1}. \tag{11.4}$$

The trace of the original impedance is neither rolled up nor squeezed, and thus it is kept in its even shape before and after the inductor is inserted into the impedance matching network, even though the variation of impedance at the high-frequency end is much greater than the variation of impedance at the low-frequency end due to the relation (11.2).

In order to expand the frequency bandwidth through the implementation of an impedance matching network, the designer always looks for or tries to create an original impedance trace satisfying the condition (11.3).

11.2.2 A Capacitor Inserted into an Impedance Matching Network in Series

When a capacitor is inserted into the impedance matching network in series, the variation of the original impedance ΔZ is equal to the variation of capacitive reactance ΔZ_C, that is,

$$\Delta Z = \Delta Z_C = j\Delta X_C = \frac{1}{jC\omega} = -j\frac{1}{C\omega}. \tag{11.5}$$

Figure 11.7 plots the curve of ΔZ_C vs. f from equation (11.5). The variation of the impedance is a decrease of reactance in magnitude as frequency is increased.

On the Smith chart, the impedance trace moves counter-clockwise along the circle of r = constant, where r is the resistance value of the original impedance. The reactance decreases more at the low-frequency end than at the high-frequency end because the decrease of reactance is inversely proportional to the frequency as shown in equation (11.5) and in Figure 11.7.

Figure 11.8 shows an example in which the impedance variation from trace 3, $Z_{fL,3}$ and $Z_{fH,3}$, to trace 4, $Z_{fL,4}$ and $Z_{fH,4}$, over the bandwidth f_L to f_H, before and after a capacitor is inserted into the impedance matching network in series.

The variation of the impedance trace from trace 3 to trace 4 can be summarized as follows:

1) All the impedances on trace 3 basically move counter-clockwise around their own resistance circles.
2) The variation of impedance at the low-frequency end is greater than the variation of impedance at the high-frequency end due to the relation (11.5).

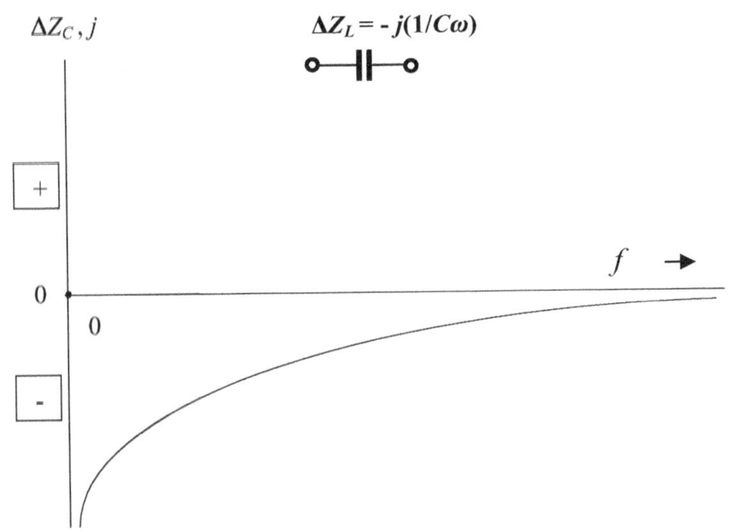

Figure 11.7 Variation of impedance when a capacitor is inserted into impedance matching network in series.

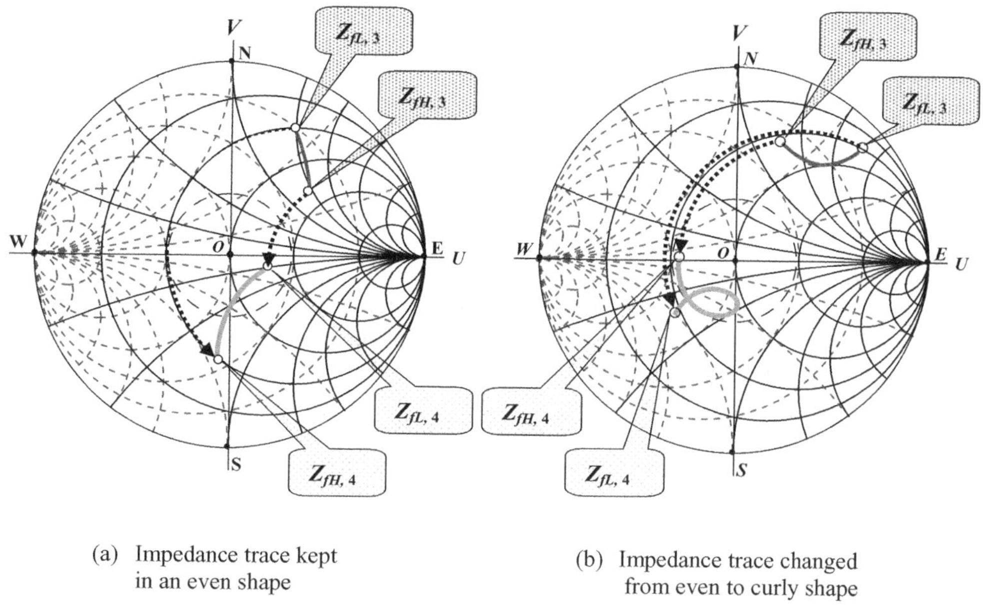

Figure 11.8 Variation of impedance from trace 3 to trace 4 if a capacitor is inserted into impedance matching network in series.
— impedance trace 3, Z_{fL}: impedance at low-frequency end,
— impedance trace 4, Z_{fH}: impedance at high-frequency end,
◄···· moving direction of impedance subscript "3," "4" denotes number of impedance trace
( ) demarcation circle, $RL = -10\,\text{dB}$.

3) The general purpose of inserting the capacitor, C_S, into the impedance matching network in series is usually to move trace 3 to trace 4, so that trace 4 is closer to the reference impedance, the center of the Smith chart as shown in both Figure 11.8(a) and (b). In Figure 11.8, the circle drawn by the dashed line is the demarcation circle, $RL = -10\,dB$. Most of trace 4 in Figure 11.8(b) is located within the demarcation circle.
4) A remarkable event is the impedance trace's change from an even shape to a curly shape as shown in Figure 11.8(b) so that the return loss trace is squeezed or rolled up. As mentioned in Section 11.1, this change indicates that the bandwidth of the impedance trace for impedance matching has changed from narrow band to wide band.

The reason for the "rolling up or squeezing of the trace" is twofold:

- In Figure 11.8(b), the reactance of the original impedance at low-frequency end is higher than that at high-frequency end, that is,

$$X_{fL,3} > X_{fH,3}, \tag{11.6}$$

- The variation of impedance at the low-frequency end is much greater than the variation of impedance at the high-frequency end as shown in the relation (11.5) and Figure 11.7.

On the other hand, in Figure 11.8(a),

$$X_{fL,3} < X_{fH,3}. \tag{11.7}$$

The trace of the original impedance is neither rolled up nor squeezed, and thus it is kept in its even shape before and after the capacitor is inserted into the impedance matching network, even though the variation of impedance at the high-frequency end is much greater than the variation of impedance at the low-frequency end due to the relation (11.5).

In order to expand the frequency bandwidth through the implementation of an impedance matching network, the designer always looks for or tries to create an original impedance trace satisfying the condition (11.6).

11.2.3 An Inductor Inserted into an Impedance Matching Network in Parallel

When an inductor is inserted into the impedance matching network in parallel, the variation of the original admittance ΔY is equal to the variation of inductive admittance ΔY_L, that is,

$$\Delta Y = \Delta Y_L = \frac{1}{\Delta Z_L} = j\Delta B_L = \frac{1}{jL\omega} = -j\frac{1}{L\omega}. \tag{11.8}$$

Figure 11.9 plots the curve of ΔY_L vs. f from equation (11.8). The variation of impedance is also a decrease of subceptance in magnitude as frequency is increased.

On the Smith chart, the impedance trace moves counter-clockwise around the conductance circle. The subceptance decreases more at the low-frequency end than at the high-frequency end because the decrease of subceptance is inversely proportional to the frequency as shown in equation (11.8) and in Figure 11.9.

Figure 11.10 shows an example in which the impedance variation from trace 5, Z_{fL5} and $Z_{fH,5}$, to trace 6, $Z_{fL,6}$ and $Z_{fH,6}$, over the bandwidth f_L to f_H, before and after an inductor is inserted into the impedance matching network in parallel.

The variation of the impedance trace from trace 5 to trace 6 can be summarized as follows:

1) All the impedances on the trace 5 basically move counter-clockwise around their own conductance circles.
2) The variation of impedance or admittance at the low-frequency end is greater than the variation of impedance at the high-frequency end due to the relation (11.8).
3) The general purpose of inserting the inductor, C_P, into the impedance matching network in parallel is usually to move trace 5 to trace 6, so that trace 6 is closer to the reference impedance, the center of the Smith chart as shown in Figure 11.10(a) and (b). In Figure 11.10, the circle drawn by the dashed line

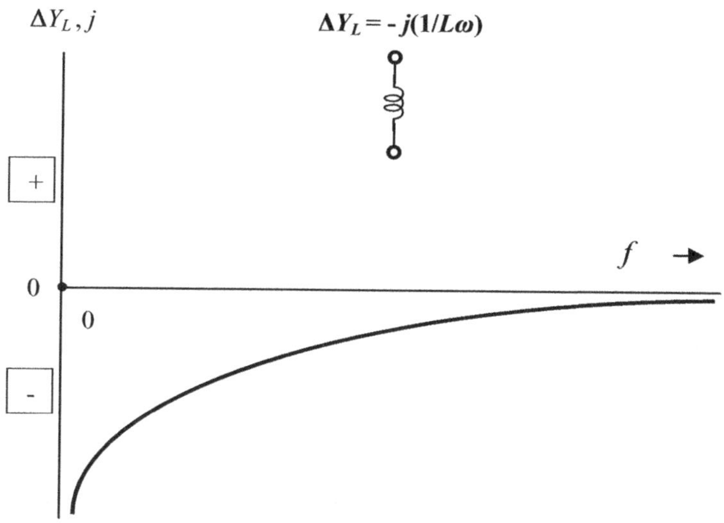

Figure 11.9 Variation of admittance when an inductor is inserted into impedance matching network in parallel.

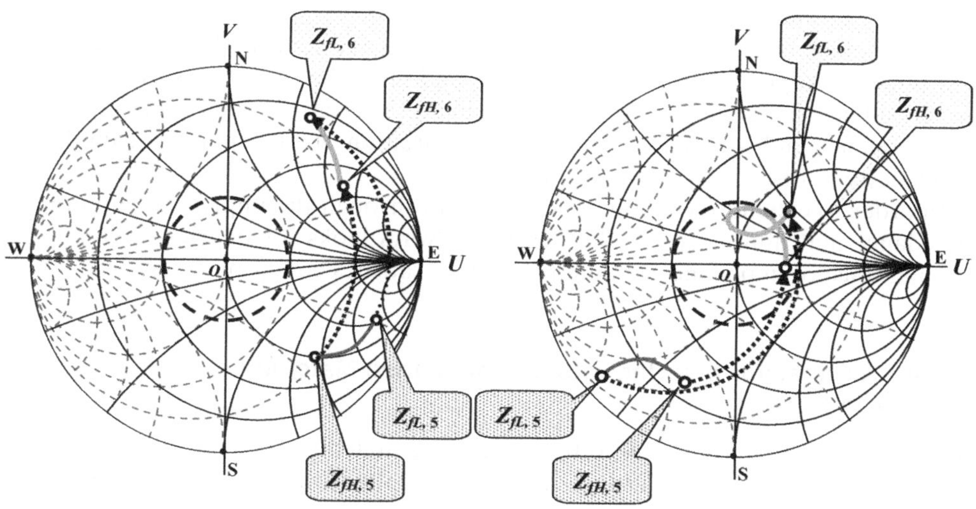

(a) Impedance trace kept in an even shape

(b) Impedance trace changed from even to curly shape

Figure 11.10 Variation of impedance from trace 5 to trace 6 if an inductor is inserted into impedance matching network in parallel.
— impedance trace 5, Z_{fL}: impedance at low-frequency end,
— impedance trace 6, Z_{fH}: impedance at high-frequency end,
◄···· moving direction of impedance subscript "5," "6" denotes number of impedance trace
( ) demarcation circle, $RL = -10\,dB$.

is the demarcation circle, $RL = -10\,dB$. Most of trace 6 in Figure 11.10(b) is located within the demarcation circle.

4) A remarkable event is the impedance trace's change from an even shape to a curly shape as shown in Figure 11.10(b) so that the return loss trace is squeezed or rolled up. As mentioned in Section 11.1, this change indicates that the bandwidth of the impedance trace for impedance matching has changed from narrow band to wide band.

The reason for the "rolling or squeezing of the trace" is twofold:

- In Figure 11.10(b), the subceptance of the original impedance at low-frequency end is higher than that at high-frequency end, that is

$$B_{fL,5} > B_{fH,5}, \tag{11.9}$$

- The variation of subceptance at the low-frequency end is much greater than the variation of impedance at the high-frequency end, as shown in equation (11.7) and Figure 11.9.

On the other hand, in Figure 11.10(a),

$$B_{fL,5} > B_{fH,5}. \tag{11.10}$$

The trace of the original impedance or admittance is neither rolled up nor squeezed, and thus it is kept in its even shape before and after the inductor is inserted into the impedance matching network, even though the variation of subceptance at the low-frequency end is much greater than the variation of subceptance at the high-frequency end due to equation (11.8).

In order to expand the frequency bandwidth through the implementation of an impedance matching network, the designer always looks for or tries to create an original trace of impedance or admittance that satisfies condition (11.9).

11.2.4 A Capacitor Inserted into an Impedance Matching Network in Parallel

When a capacitor is inserted into the impedance matching network in parallel, the variation of the original admittance ΔY is equal to the variation of capacitive admittance ΔY_C, that is,

$$\Delta Y = \Delta Y_C = \frac{1}{\Delta Z_C} = j\Delta B_C = jC\omega. \tag{11.11}$$

Figure 11.11 plots the curve of ΔY_C vs. f from equation (11.11). The variation of admittance is also an increase of subceptance ΔB_C as frequency is increased.

On the Smith chart, the impedance trace moves clockwise along the circle of g = constant, where g is the resistance value of the original impedance. The subceptance increases more at the high-frequency end than at the low-frequency end

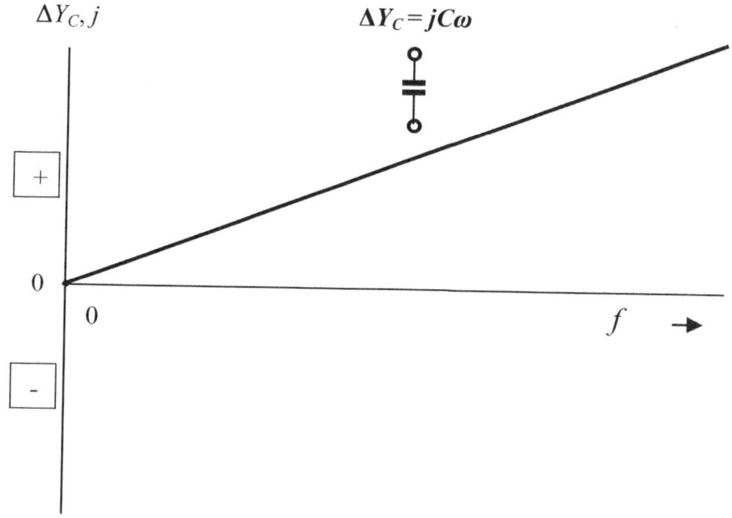

Figure 11.11 Variation of admittance when a capacitor is inserted into impedance matching network in parallel.

because the increase of subceptance is proportional to the frequency as shown in equation (11.11) and Figure 11.11.

Figure 11.12 shows an example in which the impedance variation from trace 7, Z_{fL7} and $Z_{fH,7}$, to trace 8, $Z_{fL,8}$ and $Z_{fH,8}$, covering the bandwidth f_L to f_H, before and after a capacitor is inserted into the impedance matching network in series.

The variation of the impedance trace from trace 7 to trace 8 can be summarized as follows:

1) All the impedances on the trace 7 basically move clockwise around their own conductance circles.
2) The variation of admittance at the high-frequency end is greater than the variation of admittance at the low-frequency end due to the relation (11.11).
3) The general purpose of inserting the capacitor in parallel is to move trace 7 to trace 8 so that trace 8 is closer to the reference impedance, the center of the Smith chart as shown in Figure 11.12(a) and (b). In Figure 11.12, the circle drawn by the dashed line is the demarcation circle, $RL = -10\,dB$. Most of trace 8 in Figure 11.12(b) is located within the demarcation circle.
4) A remarkable event is the impedance trace's change from an even shape to a curly shape as shown in Figure 11.12(b) so that the return loss trace is squeezed or rolled up. As mentioned in Section 11.1, this change indicates that the bandwidth of the impedance trace for impedance matching has changed from narrow band to wide band.

The reason for the "rolling up or squeezing of the trace" is twofold:

- In Figure 11.12(b), the subceptance of the original impedance at the low-frequency end is higher than that at the high-frequency end, that is,

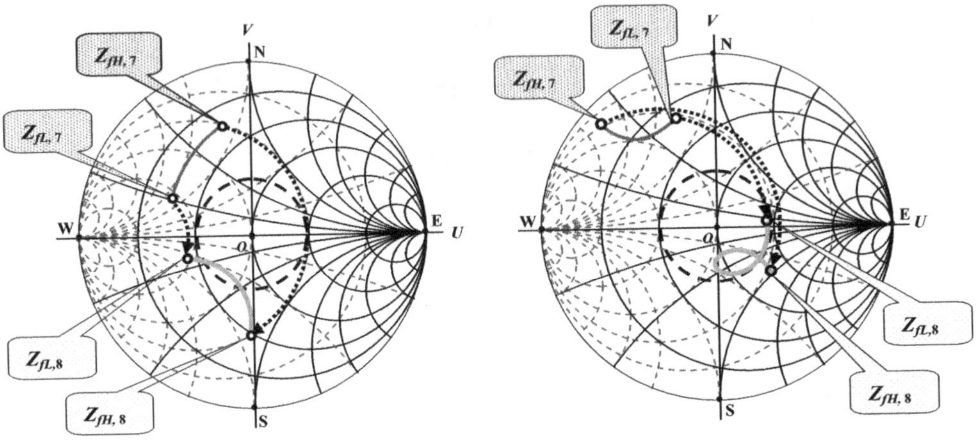

(a) Impedance trace kept in an even shape

(b) Impedance trace changed from even to curly shape

Figure 11.12 Variation of impedance from trace 7 to trace 8 if a capacitor is inserted into impedance matching network in parallel.
impedance trace 7, Z_{fL}: impedance at low-frequency end,
impedance trace 8, Z_{fH}: impedance at high-frequency end,
moving direction of impedance subscript "7," "8" denotes number of impedance trace
demarcation circle, $RL = -10\,\text{dB}$.

$$B_{fL,7} > B_{fH,7}, \tag{11.12}$$

- The variation of subceptance at the high-frequency end is much greater than the variation of subceptance at the low-frequency end as shown in equation (11.11) and Figure 11.11.

On the other hand, in Figure 11.12(a),

$$B_{fL,7} < B_{fH,7}. \tag{11.13}$$

The trace of the original impedance or admittance is neither rolled up nor squeezed, and thus it is kept in its even shape before and after the inductor is inserted into the impedance matching network, even though the variation of subceptance at the low-frequency end is much greater than the variation of subceptance at the high-frequency end due to the relation (11.11).

In order to expand the frequency bandwidth through the implement of an impedance matching network, the designer always looks for or tries to create an original trace of impedance or admittance satisfying the condition (11.12).

11.3 IMPEDANCE VARIATION DUE TO THE INSERTION OF TWO PARTS PER ARM OR PER BRANCH

In order to expand bandwidth in the process of impedance matching, "rolling up" or "squeezing" the trace of impedance on the Smith chart is the right direction.

In the examples shown in Figures 11.6(b), 11.8(b), 11.10(b), and 11.12(b), the trace of impedance is changed from an even to a curly shape. This indicates that it is possible to roll up or squeeze the trace of the impedance on the Smith chart by inserting only one part per arm or per branch. However, the orientation of the original impedance trace on the Smith chart is restricted by any one of the conditions (11.3), (11.6), (11.9), or (11.12). As a matter of fact, this restriction results from the fact that the low- and high-frequency ends of the original impedance trace move in the same directions no matter if the inserted part is an inductor or a capacitor, whether in series or in parallel. This can be easily verified from Figures 11.5, 11.7, 11.9, and 11.11. The variation of reactance or subceptance for all the frequencies has the same sign, either positive or negative, though the magnitude of variation over the frequencies is different.

Should the low- and high-frequency ends at the impedance trace move in opposite directions after one arm or one branch is inserted into the impedance matching network, the impedance trace could be easily squeezed, rolled up, or shrunk to a small area on the Smith chart and therefore the bandwidth could be expanded to what is expected. Instead of one part per arm or one part per branch, two parts per arm or two parts per branch can execute such a function. In other words, in order to expand the bandwidth in a more effective way, one new arm and one new branch of parts will be added to the list of arms and branches. Figure 10.4 now is modified to Figure 11.13. The new arm is a LC combination in series and the new branch is a LC combination in parallel, shown in the last column in Figure 11.13. We'll introduce their performances in Section 11.3.1.

11.3.1 Two Parts Connected in Series to Form One Arm

When a combination of LC in series is inserted into the impedance matching network as an arm, the variation of the original impedance ΔZ is equal to

$$\Delta Z = \Delta Z_S = jL_S\omega + \frac{1}{jC_S\omega} = jL_S\omega\left(1 - \frac{1}{jL_SC_S\omega^2}\right) = jL_S\omega\left(1 - \frac{\omega_S^2}{\omega^2}\right), \quad (11.14)$$

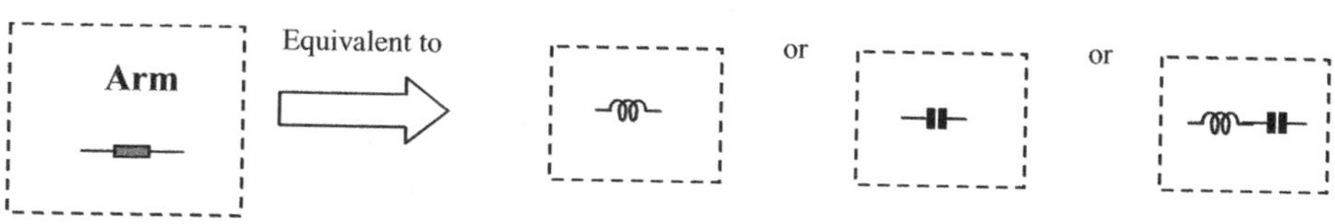

(a) Arm consists of one part or two parts and, in the impedance matching network; it is a connected component in "horizontal" direction (in series)

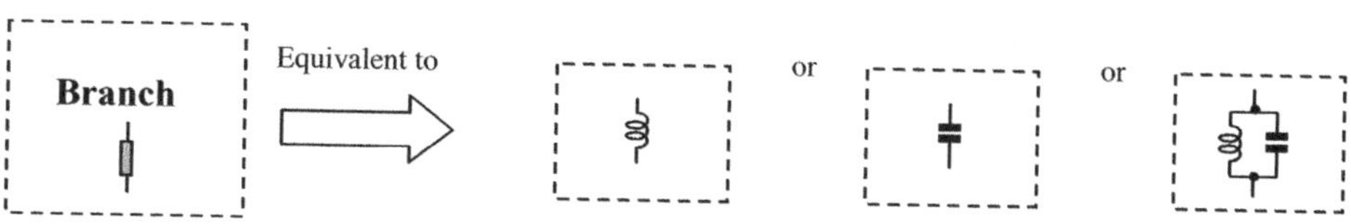

(b) Branch consists of one part or two parts and, in the impedance matching network; it is a connected component in "vertical" direction (in parallel)

Figure 11.13 Representation of arm or branch consists of one part or two parts.

where subscript "P" denotes that a part is connected "in parallel," ΔZ_S is the resultant impedance of LC combination in series, and

$$\omega_S^2 = \frac{1}{L_S C_S}, \tag{11.14a}$$

where

ω_S = angular SRF (Self-Resonant Frequency) of LC combination in series,
and f_S = SRF (Self-Resonant Frequency) of LC combination in series and it appears in Figure 11.14.

Figure 11.14 plots the curve of ΔZ_S vs. f from equation (11.13). It can be seen that

1) In the case of LC combination in series,
   - The variation of impedance is capacitive reactance when the operating frequency is lower than the SRF.
   - The variation of impedance is inductive reactance when frequency is the higher than the SRF.
   - The variation of impedance is zero at the SRF.
2) LC combination in series can be inserted into an impedance matching network as an arm in series only. It is prohibited to apply it as a branch in parallel because the network would be short-circuited at the SRF where its resultant impedance would be zero.

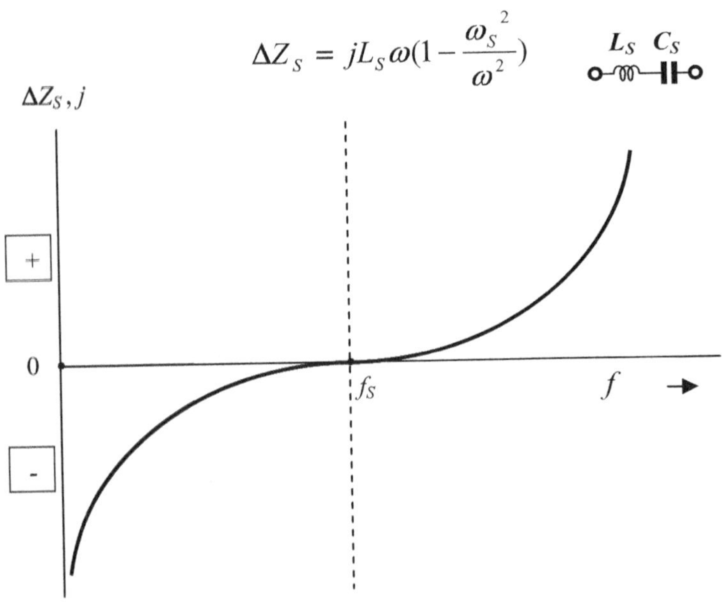

Figure 11.14 Impedance of LC combination in series.

Distinguished from the previous case where one part per arm or one part per branch is inserted into the impedance matching network, the variation of impedance at low frequencies is *different* from that at high frequencies if the values of L_S and C_S are adjusted so that the *SRF* of the *LC* combination is the central frequency of the desired bandwidth. Then, the variation of impedance at the low-frequency end is capacitive, or negative, and the variation of impedance at the high-frequency end is inductive, or positive. This means that the low-frequency end of the impedance trace will be moved counter-clockwise, whereas the high-frequency end will be moved clockwise on the Smith chart. They move in opposite directions from one another, so the impedance trace is expected to be rolled up into a smaller area. This is the goal we pursue.

Figure 11.15 shows an example before and after the *LC* combination in series is inserted into the impedance matching network as an arm. The impedance trace is moved from trace *A*, $Z_{fL,A}$ and $Z_{fH,A}$, to trace *B*, $Z_{fL,B}$ and $Z_{fH,B}$, over the bandwidth f_L to f_H.

The variation of the impedance trace from trace *A* to trace *B* can be summarized as follows:

1) The low-frequency end on the impedance trace *A*, $Z_{fL,A}$, basically moves counter-clockwise along the resistance circle to $Z_{fL,B}$, while

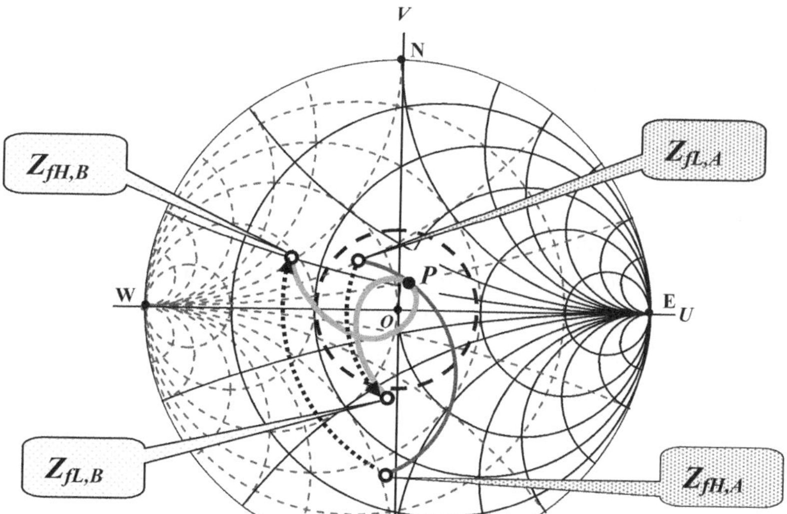

Figure 11.15 Variation of impedance from trace *A* to trace *B* before and after the *LC* combination in series is inserted into impedance matching network as an arm.
—— impedance trace *A*, Z_{fL}: impedance at low-frequency end,
—— impedance trace *B*, Z_{fH}: impedance at high-frequency end,
◄···· moving direction of impedance subscript "*A*," "*B*" denotes number of impedance trace
( ) demarcation circle, $RL = -10\,dB$.

2) The high-frequency end on the impedance trace A, $Z_{fH,A}$, moves clockwise along the resistance circle to $Z_{fH,B}$. A slight deviation of the trace from the resistance circle exists due to a little resistance in the imperfect inductor L_S.
3) The common point P on trace A and trace B in Figure 11.15 corresponds to the impedance at ω_S, or the SRF frequency, where the variation of impedance is zero on the basis of the equation (11.14).
4) Consequently, trace B is rolled around the reference impedance, the center of the Smith chart, and therefore increases the bandwidth from trace A.

To insert an LC combination in series into an impedance matching network as an arm is a powerful scheme to squeeze the trace and consequently to expand the bandwidth.

11.3.2 Two Parts Connected in Parallel to Form One Branch

When a combination of LC in parallel is inserted into impedance matching network as a branch, the variation of the original admittance ΔY is also equal to the variation of subceptance ΔB_P, that is,

$$\Delta Z = j\Delta X_P = \frac{jL_P\omega \dfrac{1}{jC_P\omega}}{jL_P\omega + \dfrac{1}{jC_P\omega}} = \frac{1}{jC_P\omega\left(1 - \dfrac{\omega_P^2}{\omega^2}\right)}, \tag{11.15}$$

$$\Delta Y = j\Delta B_P = \frac{1}{\Delta Z} = jC_P\omega\left(1 - \frac{\omega_P^2}{\omega^2}\right), \tag{11.16}$$

where

ΔZ = variation of impedance due to insertion of LC combination in parallel,
ΔX_P = variation of reactance due to insertion of LC combination in parallel,
ΔY = variation of admittance due to insertion of LC combination in parallel,
ΔY_P = variation of subceptance due to insertion of LC combination in parallel,
subscript "P" denotes that a part is connected "in parallel,"

and

$$\omega_P^2 = \frac{1}{L_P C_P}, \tag{11.17}$$

where

ω_P = angular SRF (Self-Resonant Frequency) of LC combination in parallel,

and f_P = SRF (Self-Resonant Frequency) of LC combination in parallel and is applied in Figure 11.16.

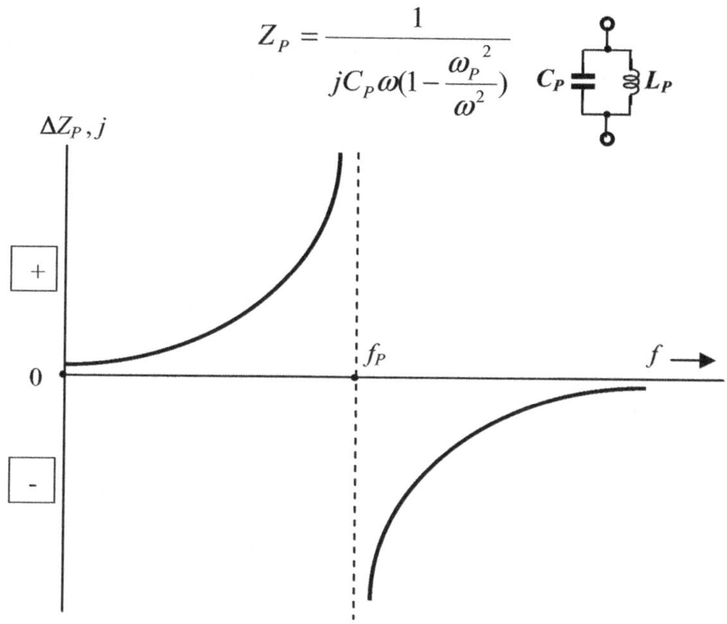

(a) Impedance of a LC combination in parallel

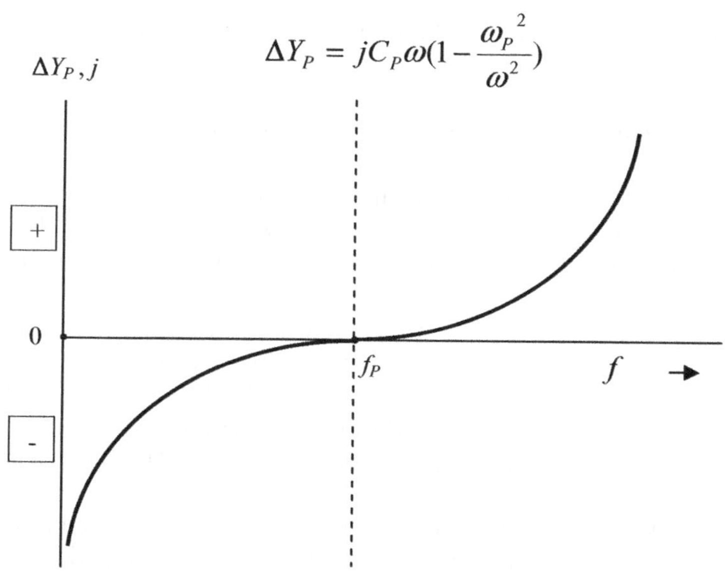

(b) Admittance of a LC combination in parallel

Figure 11.16 Impedance or admittance of a LC combination in parallel.

Based on equations (11.15) and (11.16), Figure 11.16(a) and (b) plots the curve of ΔZ_P vs. f and the curve ΔY_P vs. f, respectively. Figure 11.16(b) looks like Figure 11.14 but they are essentially different, Figure 11.16(b) draws the curve ΔY_P vs. f, whereas Figure 11.14 draws the curve ΔZ_S vs. f.

In the case of *LC* combination in parallel, admittance is a more convenient parameter than impedance. Therefore, only Figure 11.16(b) but not 11.16(a) is applied to our discussion. From Figure 11.16(b) it can be seen that

1) In the case of *LC* combination in parallel,
   - The variation of admittance is a decrease of subceptance when the frequency is lower than f_P or the *SRF*,
   - The variation of admittance is an increase of subceptance when the frequency is higher than f_P or the *SRF*. And,
   - The variation of subceptance is zero at the *SRF*.
2) *LC* combination in parallel can be inserted into an impedance matching network as a branch in parallel only. It is prohibited to apply it as an arm in series because the network would be open-circuited at f_P or *SRF* where its resultant impedance is infinite as shown in Figure 11.16(a).

Figure 11.17 shows an example before and after a *LC* combination in parallel is inserted into impedance matching network as a branch, in which the impedance variation from trace *C*, $Z_{fL,\mathrm{C}}$ and $Z_{fH,\mathrm{C}}$, to trace *D*, $Z_{fL,\mathrm{D}}$ and $Z_{fH,\mathrm{D}}$, over the bandwidth, f_L to f_H.

The variation of impedance or admittance trace from trace *C* to trace *D* can be summarized as follows:

1) The low-impedance end on trace *C*, $Z_{fL,C}$, moves counter-clockwise along the conductance circle to $Z_{fL,\mathrm{D}}$, while
2) The high-frequency end on the impedance trace *C*, $Z_{fH,C}$, moves clockwise along with the conductance circle to $Z_{fH,D}$. A slight deviation of the trace from the conductance circle exists due to a slight resistance existing in the imperfect inductor L_P.
3) The common point *P* on trace *C* and trace *D* in Figure 11.17 corresponds to the impedance or admittance at ω_P or the SRF frequency in parallel, where the variation of admittance is zero on the basis of the equation (11.16).
4) Consequently, trace *C* is squeezed to trace *D* and is now located around the reference impedance, the center of the Smith chart. The bandwidth is therefore increased.

   To insert an *LC* combination in parallel into the impedance matching network is another powerful scheme to squeeze the impedance trace and, consequently, to expand the bandwidth.

11.4 IMPEDANCE MATCHING IN *IQ* MODULATOR DESIGN FOR A *UWB* SYSTEM

So far we have been exploring how to expand the bandwidth of an *RF* circuit block through the implementation of an impedance matching network. This discussion has

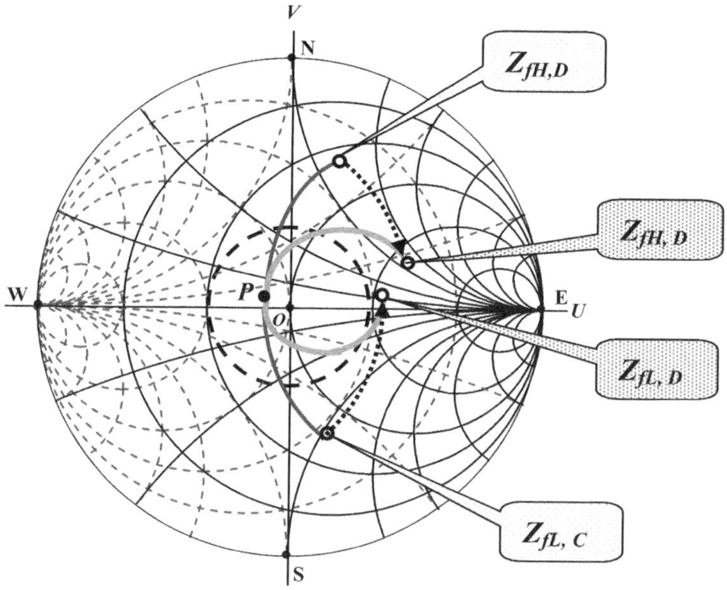

Figure 11.17 Variation of impedance from trace *C* to trace *D* before and after the *LC* combination in parallel is inserted into impedance matching network as a branch as a branch.
—— impedance trace *C*, Z_{fL}: impedance at low-frequency end,
—— impedance trace *D*, Z_{fH}: impedance at high-frequency end,
◄···· moving direction of impedance subscript "*C*," "*D*" denotes number of impedance trace
( ) demarcation circle, $RL = -10dB$.

been conducted qualitatively but not quantitatively. Now, as a quantitative example, let's introduce partial impedance matching work for an *IQ* modulator in a *UWB* system. This is enough to confirm the possibility of expanding the bandwidth of an *RF* block in a practical circuit design, although this example is only a partial and not a complete *IQ* modulator design.

11.4.1 Gilbert Cell

A Gilbert cell is selected as the core of *IQ* modulator. As shown in Figure 11.18, it consists of six devices, which in this example are *MOSFET* transistors. It has two pairs of differential inputs and one pair of differential outputs. For a modulator design, it is supposed that

- The *IF* signal goes to the gates of M_1 and M_2, which are a pair of differential inputs.
- The *LO* injection goes to the gates of M_3 to M_6, which are another pair of differential inputs.
- The *RF* modulated carrier comes from the drains of M_3 to M_6, which are a pair of differential outputs.

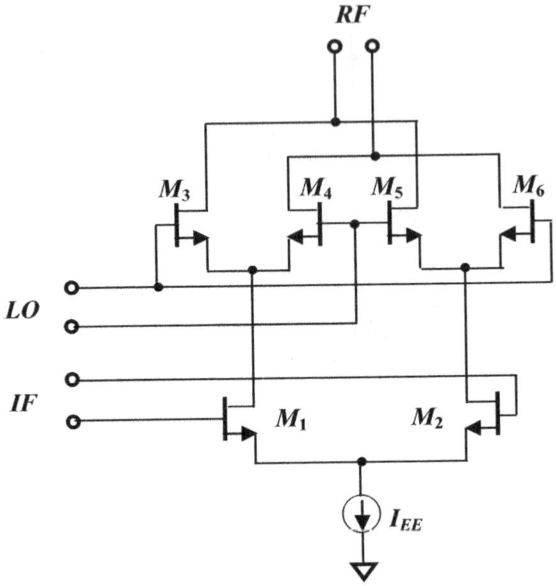

Figure 11.18 Gilbert cell.

To make the Gilbert cell operate, some necessary parts must be connected. Figure 11.19 shows the setup of an operative Gilbert cell for the purpose of circuitry simulation. The *DC* power supply with output voltage V_{dd} provides *DC* electricity to the Gilbert cell through inductors L_{P1} and L_{P2} while a bias voltage source also provides an appropriate voltage to the main switching transistors M_3 to M_6 through inductors L_{P1} and L_{P2}.

To avoid the disturbance of the Gilbert cell, the impedances of all the inductors, L_{P1}, L_{P2}, L_{P3}, and L_{P4} must approach infinity. The capacitor, C_o, is a "zero" capacitor, which approaches zero impedance at the operating frequency. In short, the combinations, (L_{P1}, L_{P2}, and C_o), (L_{P3}, L_{P4} and C_o), provide *DC* voltage or current but do not disturb the performance of the Gilbert cell.

On the other hand, for simplicity in the simulation stage, all three differential pairs are connected with an ideal transformer balun, respectively, so that the differential pairs become single-ended terminals. Each transformer balun consists of two identical transformers stacked together. The turn ratio the of transformer is $1:\sqrt{2}$ at the *IF* and *LO* inputs but is $\sqrt{2}:1$ at the *RF* output. Figure 11.20 depicts this special transformer balun, which is easily set up for the simulation of circuitry in a Cadence or *ADS* system. Its special feature is that the impedances, looking into the transformer either at the individual differential port or at the single-ended port, are the same. The parts attached to the differential ports can be interpreted as the parts at the single-ended port and vice versa, as discussed in Chapter 4.

The transformer balun provides a convenient means to convert a differential port to a single-ended port. It reduces complexity a lot in the simulation stage of design. However, it is only temporarily applied for the circuitry simulation since ideal parts are never applied in a product. It will be removed and replaced by a practical balun if needed.

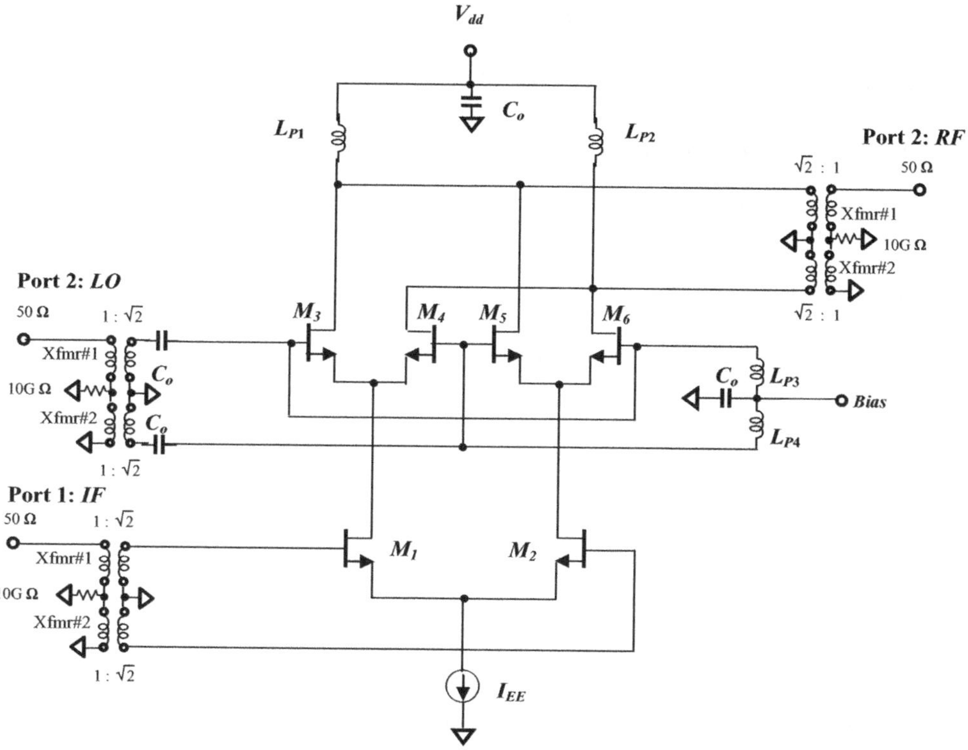

Figure 11.19 Setup of an operative Gilbert cell for simulation.

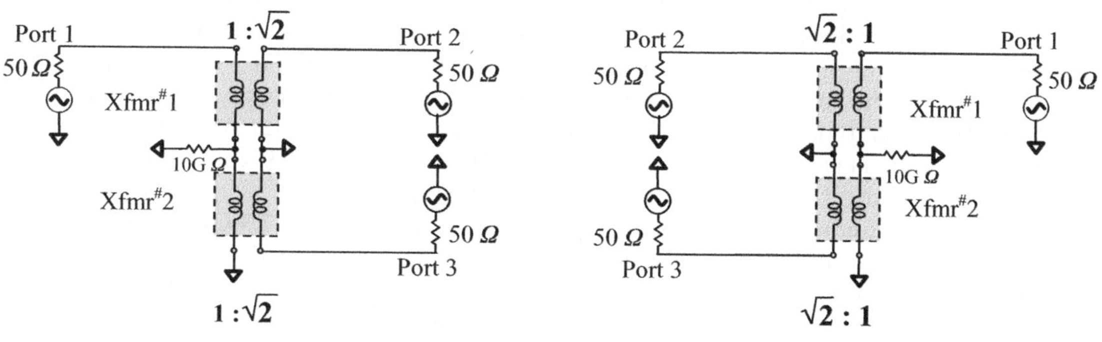

(a) For *LO* and *IF* ports, turn ratio $=1:\sqrt{2}$ (b) For *RF* ports , turn ratio $=\sqrt{2}:1$

Figure 11.20 Transformer balun built by two transformers stacked together.

In the setup for an operational Gilbert cell, inductors with infinite impedance or capacitors with zero impedance never exist in reality. As a reasonable approximation, the values of inductors and capacitors in Figure 11.19 are taken as

$$Lp_1 = Lp_2 = Lp_3 = Lp_4 = 1000\,nH,$$

$$C_o = 1000\,pF,$$

where the operating frequency is assumed to be in the order of GHz.

11.4.2 Impedances of a Gilbert Cell

The input and output impedances of a Gilbert cell can be obtained by return loss, S_{11} or S_{22}, testing. The testing is conducted under the conditions of

- Frequency range

$$f = 10\,MHz \text{ to } 4.3GHz,$$

- DC power supply and bias

$$V_{dd} = 3V,$$

$$\text{Bias} = 0.685V,$$

- Current drain

$$I_{EE} = 2 \times 3.7\,mA$$

Figures 11.21 and 11.22 show the S parameters, S_{11}(IF port), S_{22}(RF port), and S_{33}(LO port) of the Gilbert cell when the impedances at all three ports have not been matched.

It can be seen that the impedance matching state is very poor at all three ports. Figure 11.21 shows that the return losses, S_{11}, S_{22}, and S_{33}, are all higher than $-1\,dB$ in the frequency range from $10\,MHz$ to $4.3\,GHz$. Figure 11.22 shows that the traces of S_{11} and S_{33} at the IF and LO ports are located in regions 4 and 2 of the Smith chart, where the reactances are capacitive and their values are mostly higher than $50\,\Omega$. The magnitudes of their reactances are much higher than $50\,\Omega$. As a matter of fact, the impedances at IF (port 1), RF (port 2), and LO (port 3) can be directly read from Figure 11.22: they are partially listed in Table 11.1.

Within the frequency range as shown in Table 11.1, $f = 3.696$ to $7.392\,GHz$, which is the frequency range UWB systems are concerned with, the variation of reactance approximately doubles from the low-frequency end to the high-frequency end at LO (port 3) and at RF (port 2). Impedance matching at these two ports is a difficult task because the frequency range to be covered is so wide. Impedance matching for wide-band RF block is very much a challenge in the design of UWB systems.

11.4.3 Impedance Matching for *LO*, *RF*, and *IF* Ports Ignoring Bandwidth

Let's go ahead and design an IQ modulator for only band 2, group 1 in a UWB system, in which the frequencies and their bandwidth are

$$f_{max} = 4224\,MHz,$$

$$f_{min} = 3696\,MHz,$$

$$BW = f_{max} - f_{min} = 4224 - 3696\,MHz = 528\,MHz$$

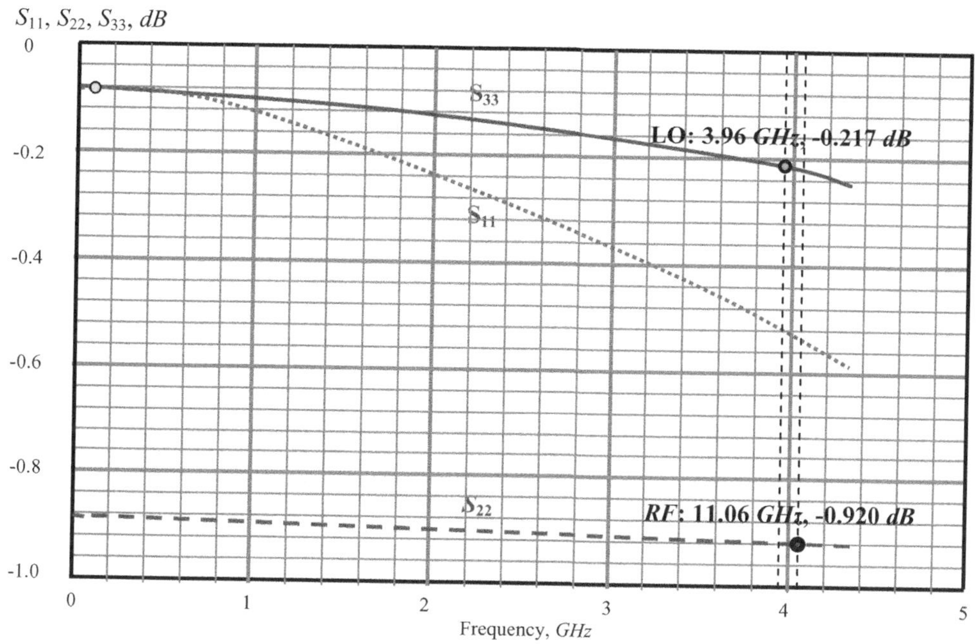

Figure 11.21 The magnitudes (*dB*) of S_{11}, S_{22}, S_{33} when the impedances at *IF* (port 1), *RF* (port 2), and *LO* (port 3) have not been matched.
○ S_{11} at 100 *MHz* ● S_{22} at 11.06 *GHz* ● S_{33} at 3.96 *GHz*

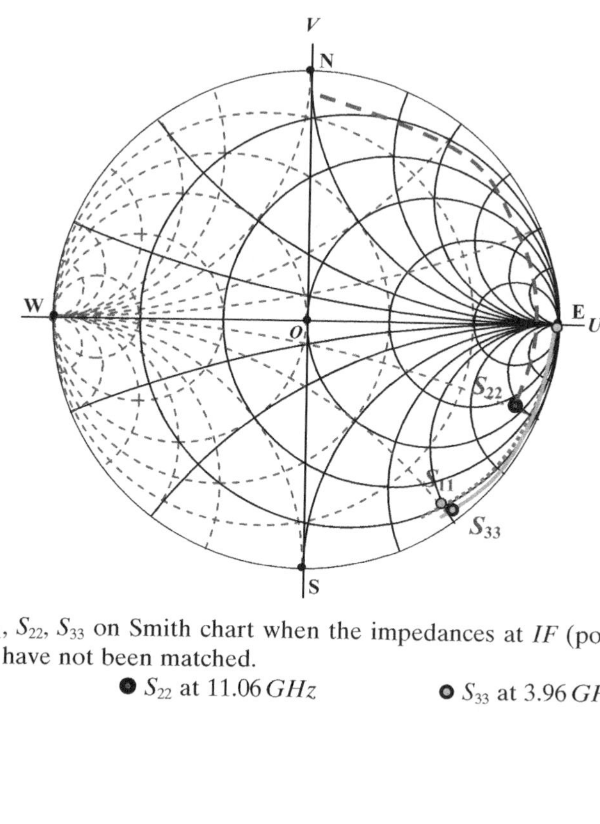

Figure 11.22 S_{11}, S_{22}, S_{33} on Smith chart when the impedances at *IF* (port 1), *RF* (port 2), and *LO* (port 3) have not been matched.
○ S_{11} at 100 *MHz* ● S_{22} at 11.06 *GHz* ● S_{33} at 3.96 *GHz*

TABLE 11.1 Impedances at 3 ports of Gilbert cell impedances are matched

Port	Frequency	Impedance
IF (port 1)	$100\,MHz$	$1144\,\Omega - j3171\,\Omega$
	$264\,MHz$	$186\,\Omega - j1331.8\,\Omega$
	$528\,MHz$	$52.1\,\Omega - j675.2\,\Omega$
RF (port 2)	$3.696\,GHz$	$78.6\,\Omega - j249.1\,\Omega$
	$3.960\,GHz$	$66.9\,\Omega - j2311.2\,\Omega$
	$4.060\,GHz$	$61.9\,\Omega - j225.8\,\Omega$
	$4.224\,GHz$	$60.8\,\Omega - j222.6\,\Omega$
	$6.864\,GHz$	$26.3\,\Omega - j138.6\,\Omega$
	$7.128\,GHz$	$25.2\,\Omega - j1311.0\,\Omega$
	$7.392\,GHz$	$24.0\,\Omega - j129.2\,\Omega$
LO (port 3)	$3.696\,GHz$	$4.1\,\Omega - j109.9\,\Omega$
	$3.960\,GHz$	$4.0\,\Omega - j102.2\,\Omega$
	$4.224\,GHz$	$3.6\,\Omega - j96.5\,\Omega$
	$6.864\,GHz$	$2.9\,\Omega - j59.8\,\Omega$
	$7.128\,GHz$	$2.7\,\Omega - j57.5\,\Omega$
	$7.392\,GHz$	$2.8\,\Omega - j55.1\,\Omega$

As our first attempt at the implementation of an input and output impedance matching network, let's ignore the bandwidth requirement and only take care of one frequency, the central frequency of Band 2 in the UWB system at each port, that is,

$$f_{LO} = 3960\,MHz,$$

$$f_{RF} = 4060\,MHz,$$

and

$$f_{IF} = 100\,MHz.$$

This is easy work because the bandwidth is ignored. As a matter of fact, this is simply an extreme narrow-band case. In terms of the design scheme discussed in Chapter 3, the impedance matching can be done in a short time period since either the input or output impedance matching network can basically be implemented by only two parts, which are

- At IF (port 1):

$$C_P = 2\,pF,$$

$$L_S = 1\,\mu F.$$

- At RF (port 2):

$$Lp_1 = Lp_2 = 3.8\,nH, (r = 9.45\,\Omega, \text{assumed } Q = 10),$$

$$C_S = 0.275\,pF.$$

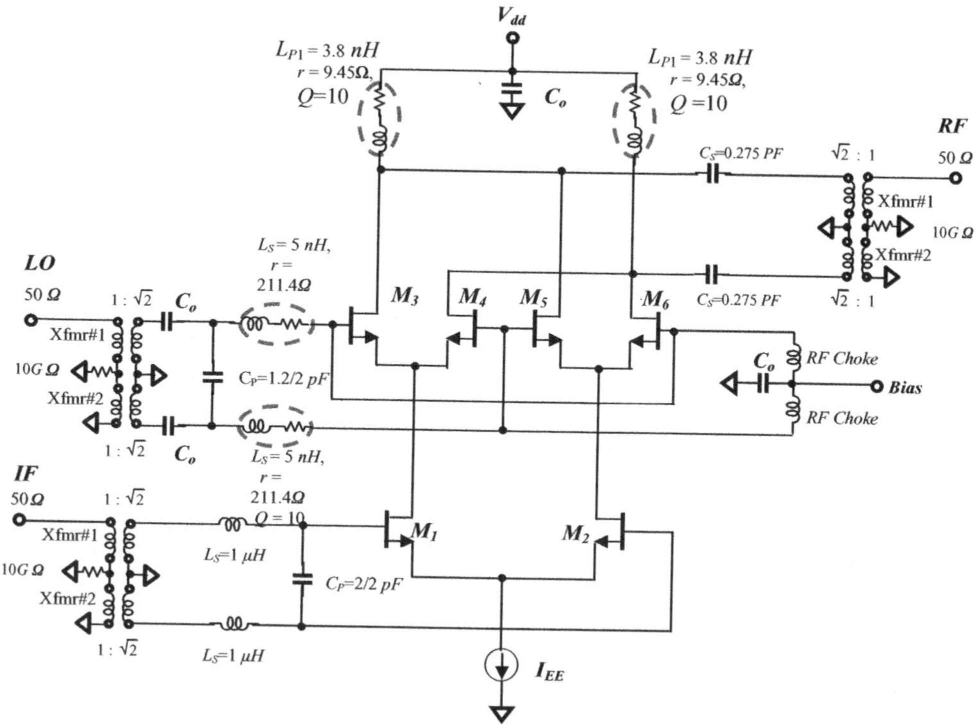

Figure 11.23 Input and output impedance matching networks of the Gilbert cell for the simple try: Only one frequency at each port is taken care of, that is, $f_{LO} = 3960\,MHz$, $f_{RF} = 4060\,MHz$, $f_{IF} = 100\,MHz$.

- At *LO* (port 3):

$$C_P = 1.2\,pF,$$

$$L_S = 5\,nH\ (r = 12.44\,\Omega,\ \text{assumed}\ Q = 10).$$

Figure 11.23 shows the parts' value from the simulated results. In the input and output impedance matching network, the inductor is connected with a small resistor in series, which is calculated based on $Q = 10$ and $f = 3960\,MHz$. The combination of the inductor and resistor approximates a practical spiral inductor on an *RFIC* chip. In Figure 11.23, it is circled by a dashed ellipse. In addition, the "zero" capacitor C_o does not count as a part in the impedance matching network.

Figures 11.24 and 11.25 show the parameters S_{11}, S_{22}, S_{33} after the impedances at the *IF* (port 1), *RF* (port 2), and *LO* (port 3) of the Gilbert cell have been matched.

Figure 11.24 shows that

$$BW_{RF} = 3.75\ \text{to}\ 4.33\,GHz; \quad (S_{22} < -10\,dB),$$

$$BW_{LO} = 3.82\ \text{to}\ 4.15\,GHz. \quad (S_{33} < -10\,dB),$$

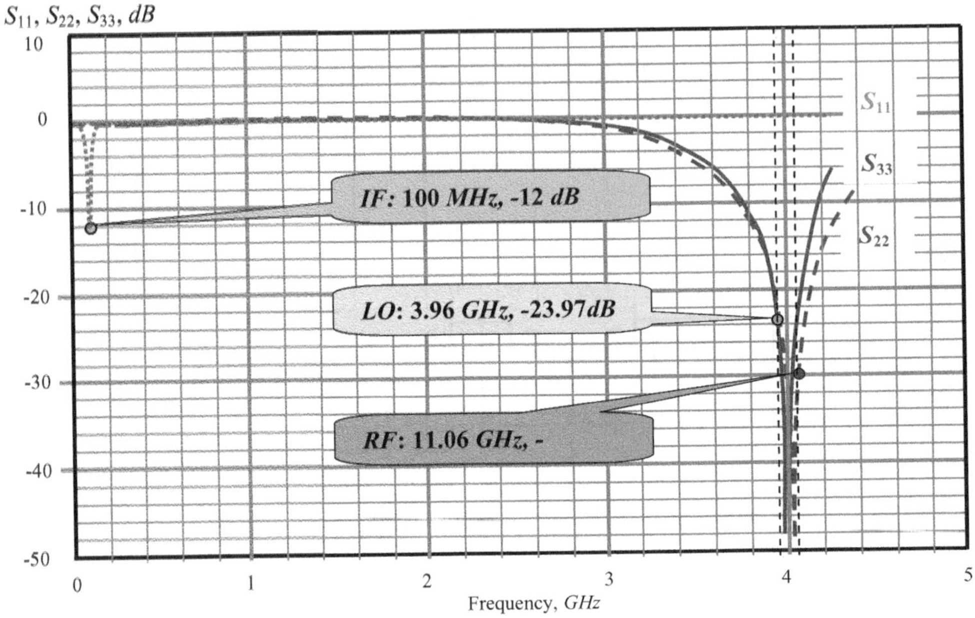

Figure 11.24 The magnitudes (dB) of S_{11}, S_{22}, S_{33} after the impedances at IF (port 1), RF (port 2), and LO (port 3) have been matched.

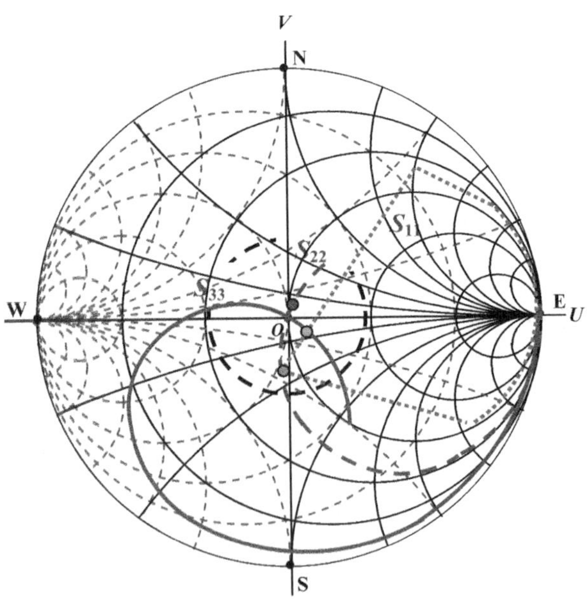

Figure 11.25 S_{11}, S_{22}, S_{33} on Smith chart after the impedances at IF (port 1), RF (port 2), and LO (port 3) have been matched.

○ S_{11} at 100 MHz ● S_{22} at 11.06 GHz ● S_{33} at 3.96 GHz

where

BW_{RF} = bandwidth at *RF* port,
BW_{LO} = bandwidth at *LO* port,

if the bandwidth is judged by the criterion of the return loss,

$$S_{11} = S_{22} = S_{33} = -10\,dB.$$

In a *UWB* system, the bandwidth of the *IQ* modulator designed for band 2, group 1 is expected to be

$$BW_{RF} = 3.796 \text{ to } 4.324\,GHz,$$

$$BW_{LO} = 3.696 \text{ to } 4.224\,GHz.$$

Obviously the bandwidth is not wide enough to cover band 2, group 1 in a *UWB* system.

Figure 11.25 also displays a very narrow bandwidth response since most portions of the traces of S_{11}, S_{22}, and S_{33} are outside the demarcation circle, $S_{11} = S_{22} = S_{33} = -10\,dB$. They are scattered over a large area on the Smith chart, although the three impedances corresponding to the specific frequencies, S_{11} at 100 MHz, S_{22} at 4.06 GHz, and S_{33} at 3.96 GHz, are inside the demarcation circle.

From now on we are going to ignore the *IF* port because

- The input *IF* signal of the *IQ* modulator comes from the digital block. The design work for the *IF* port must be discussed, negotiated, and cooperated with the digital circuit designer.
- The *IF* port is operated in *IF* but not *RF* frequency.

Instead, we are going to focus on the *LO* and *RF* ports because

- Usually the *LO* port is assigned to the gate and the *RF* port is assigned to the drain of the transistor if the modulator is built by *MOSFET* transistors. Impedance matching with the gate and drain of *MOSFET* transistor represents the two main technologies of impedance matching, which are widely demanded in the design of other *RF* blocks.
- The *LO* and *RF* ports operate in the *RF* frequency range.

11.4.4 Wide Bandwidth Required in a *UWB* System

Portable radios or cellular phones are narrow-band systems because their relative bandwidths are usually less than 15%. The emergence of the *UWB* system in recent years challenges the *RF* circuit designer to design *RF* circuits with a *UWB* response.

Table 11.2 lists the band group allocation in a *UWB* system. The frequency range in a *UWB* system is

$$f = 3.168 \text{ to } 10.560\,GHz.$$

TABLE 11.2 Frequency plan of *UWB* system

Primary Band Plan

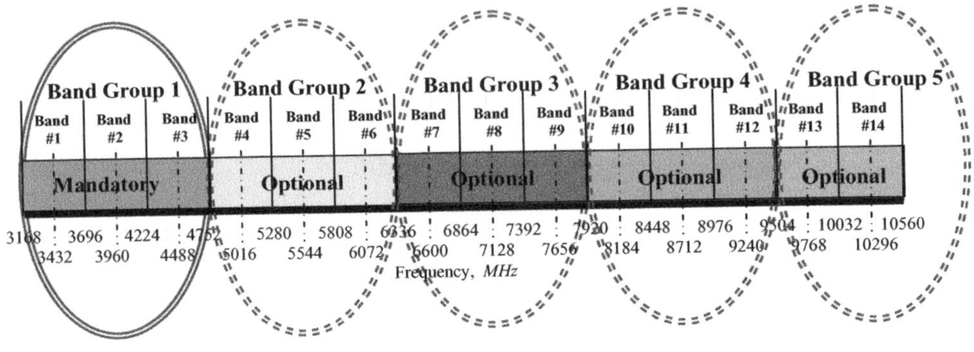

New World-wide Compliant Band Plan

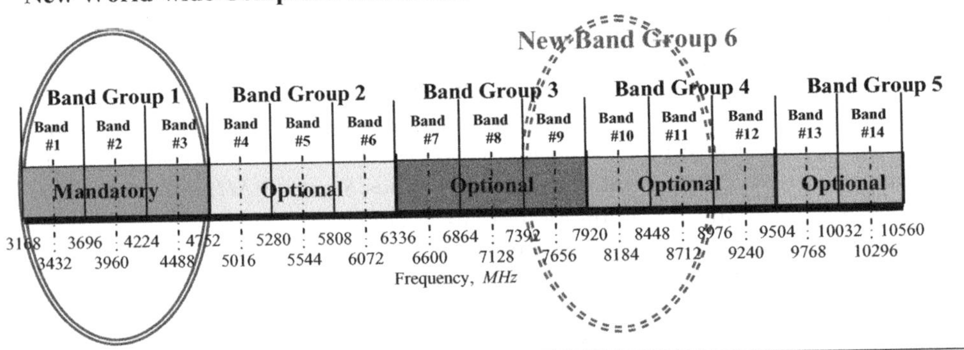

Which is divided into 14 sub-bands and the bandwidth of each sub-band is

$$BW_{\text{band}} = 528\,MHz.$$

These 14 sub-bands are assigned to five groups. Each of the first four groups contain three sub-bands, while the fifth group contains only two sub-bands. The first of these groups in the frequency range is a mandatory group to be covered in any *UWB* system, while the other groups are optional. The second group is always ignored because it is occupied by the *WLAN* (Wireless Local Area Network) communication system. Owing to restrictions made by the governments of various countries for military or other special purposes, the other sub-bands contained in the third, fourth, and fifth groups are selected for use by *UWB* companies or enterprises in different ways. A new "sixth group", containing sub-bands 9, 10, 11, is found to have fewer restrictions in the world wireless market. Currently, most *UWB* companies and enterprises are focusing on the first and sixth groups.

TABLE 11.3 Band Group allocation for *UWB*

	Group #	Band #	f_L, MHz	f_C, MHz	f_H, MHz	Δf, MHz	$\Delta f/f_C$	$\Delta f/f_C$
	1	1	**3168**	**3432**	**3696**	528	**15.38%**	
	1	2	**3696**	**3960**	**4224**	528	**13.33%**	40.00%
	1	3	**4224**	**4488**	**4752**	528	**11.76%**	
	2	4	**4752**	**5016**	**5280**	528	**10.53%**	
	2	5	**5280**	**5544**	**5808**	528	**9.52%**	
	2	6	**5808**	**6072**	**6336**	528	**8.70%**	
	3	7	**6336**	**6600**	**6864**	528	**8.00%**	
	3	8	**6864**	**7128**	**7392**	528	**7.41%**	
6	3	9	**7392**	**7656**	**7920**	528	**6.90%**	34.48%
6	4	10	**7920**	**8184**	**8448**	528	**6.45%**	
6	4	11	**8448**	**8712**	**8976**	528	**6.06%**	
	4	12	**8976**	**9240**	**9504**	528	**5.71%**	
	5	13	**9504**	**9768**	**10032**	528	**5.41%**	
	5	14	**10032**	**10296**	**10560**	528	**5.13%**	

As mentioned in Chapter 3, a system or a block is categorized as narrow-band if its relative bandwidth is lower than 15% while an *RF* block is categorized as a wide-band system or block if its relative bandwidth is higher than 15%. The 15% of relative bandwidth is a rough but reasonable value as a demarcation between narrow- and wide-band cases, though it is not a strict criterion. Table 11.3 lists the frequencies for each frequency band and calculates all of their absolute bandwidths, $BW = \Delta f$, and relative bandwidths, $\Delta f/f_C$. It can be seen that the wide-band design is required if the system is going to cover only group 1, in which the relative bandwidth is 40%. In order to lower the cost and simplify the circuit design, the relative bandwidth of a system should be expanded as much as possible. For example, it is desirable to combine group 3 and group 6 together, which results in a relative bandwidth of $\Delta f/f_C$ = 34.48% if the system is to cover the bandwidth containing Bands 7 to 11 or to cover frequencies from 6336 to 8976 MHz.

The key issue in a wide-band system or circuit design is, of course, the wide-band impedance matching.

11.4.5 Basic Idea to Expand the Bandwidth

As pointed out in Section 11.1, in order to change the *RF* block's impedance response from narrow band to wide band, the following approaches are used in the implementation process of the impedance matching network:

1) The distance between the trace of return loss, S_{11} or S_{22}, and the center of the Smith chart should be reduced to be as short as possible;
2) The area the impedance trace or the return loss S_{11} or S_{22} occupies on the Smith chart must be squeezed or shrunk to as small an area as possible.

A shorter distance implies that the *RF* block has a higher return loss and therefore a correspondingly wider bandwidth. Should the trace be squeezed to the center of the Smith chart, the *RF* block would definitely have wide-band performance.

There are many ways to shrink the return loss, S_{11} or S_{22} on the Smith chart. The *LC* combination, either in series or in parallel, is quite an effective scheme to increase the bandwidth. The single part per arm or per branch scheme can also effectively shrink the trace's covered area if the original trace is located in certain positions.

The part count should be kept as low as possible. Above all, inductors should be employed as little as possible because the inductor is much more expensive and has a lower Q value than the capacitor. However, in order to expand the impedance matching network from narrow to wide bandwidth, between one to five inductors are usually required.

Keeping these ideas in mind, it has been proven that impedance matching for a *UWB* system can be done well through the reasonable implementation of impedance matching networks. In the following sub-sections we present two examples of partial works in the *I* or *Q* modulator design for a *UWB* system. Instead of the entire design work for the *IQ* modulator, we show only the portion of how to expand the impedance response from narrow band to wide band at the *LO* and *RF* ports. In particular, we show the evolution of the bandwidth as the impedance matching parts are inserted into the impedance matching network step by step.

11.4.6 Example 1: Impedance Matching in *IQ* Modulator Design for Group 1 in a *UWB* System

Figure 11.26 shows an *I* or *Q* modulator for a *UWB* system with impedance matching networks. In this example the operating frequency range is from 3168 to 4752 *MHz* for group 1 in the *UWB* system so that the relative bandwidth is, therefore, 40%. The modulator is supposed to be implemented on an *RFIC* chip.

Figures 11.27 and 11.28 plot the input impedance matching network at the *LO* port and the output impedance matching network at the *RF* port, respectively.

The input impedance matching network at the *LO* port shown in Figure 11.27 consists of two arms and two branches. An inductor is always connected with a small resistor in series and is circled by a dashed-line ellipse. The small resistor represents the attenuation of the inductor and its value is calculated on the basis of the assumed quality factor, $Q = 10$, and the operating frequency, $f = 3960\,MHz$. Two resistance values appear in the r expression of L_{S1} and L_{P1}. The first value is the calculated resistance based on $Q = 10$ and $f = 3960\,MHz$ and the second is additional resistance for more bandwidth expansion. One of the two arms consists of one inductor, and the other consists of a *LC* combination in series. Similarly, one of the two branches consists of one capacitor and another branch consists of a *LC* combination in parallel. The symbols "2×", appearing in L_{P1}, and "/2", appearing in C_P and C_{P1}, indicate that there are two identical parts connected together in series as a branch bridging between two differential nodes.

The output impedance matching network at the *RF* port is shown in Figure 11.28. It consists of two arms and three branches. Again, an inductor is always connected with a small resistor in series and is circled by a dashed-line ellipse. The small resistor represents the attenuation of the inductor and its value is calculated from the

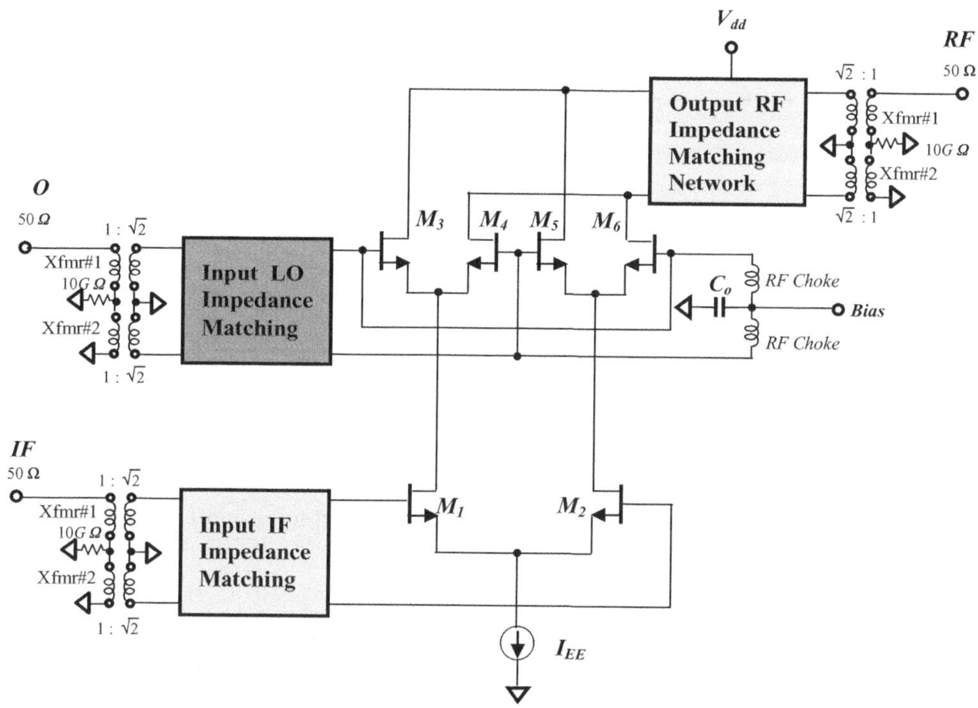

Figure 11.26 *I* or *Q* modulator for *UWB* system with impedance matching networks.

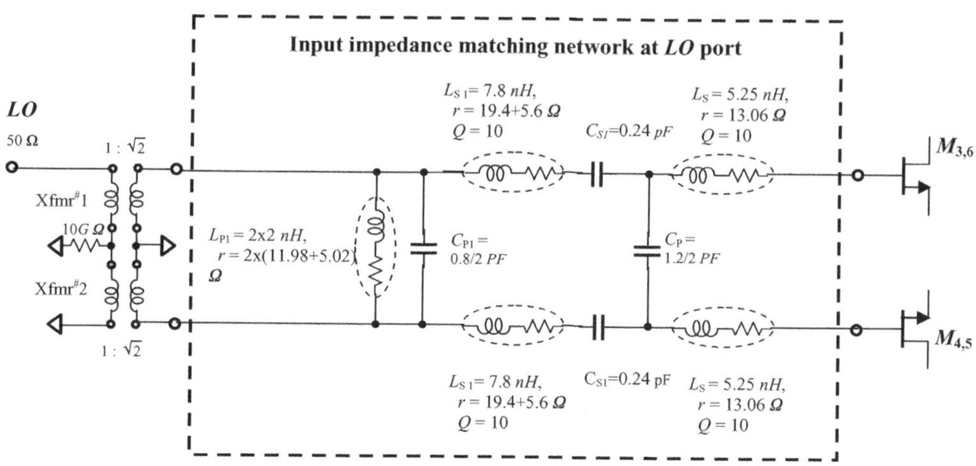

Figure 11.27 Input impedance matching network at *LO* ports for group 1 in *UWB* system, $f = 3168$ to $4752\,MHz$.

assumed quality factor, $Q = 10$ and the operating frequency, $f = 3960\,MHz$. There is only one part contained in an arm or a branch. The symbol "/2" appearing in C_P and C_{P1} indicates that there are two identical parts connected in series as a branch bridging two differential nodes. The *DC* power supply, V_{dd}, provides the current to the devices through two L_P inductors.

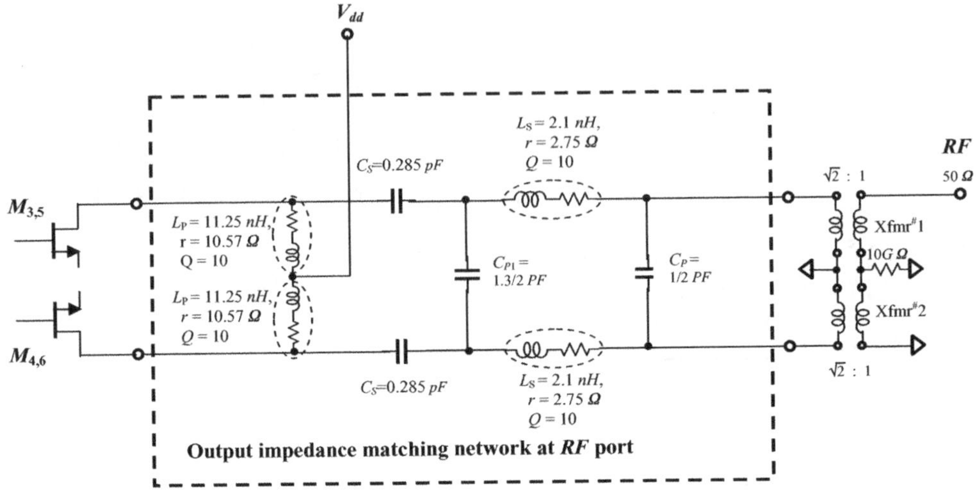

Figure 11.28 Output impedance matching network at *RF* port for group 1 in *UWB* system, $f = 3168$ to $4752\,MHz$.

The input impedance matching network at the *LO* port, as shown in Figure 11.27, and the output impedance matching network at the *RF* port, as shown in Figure 11.28, successfully expand the bandwidth from very narrow to very wide band. Figure 11.29 shows the expected value of the bandwidth being reached for an *IQ* modulator working in group 1 of a *UWB* system.

From Figure 11.29(a) it can be seen that most portions of the return loss trace, either S_{22} or S_{33}, are located within the demarcation circle. This indicates that the impedances at both port 2 and port 3 are matched to $50\,\Omega$ in the wide-band case. The bandwidth can be read from Figure 11.29(b) accordingly, that is,

At the *RF* port,

$$BW_{RF} = \Delta f = 4860 - 3100\,MHz = 1760\,MHz,$$

$$(BW_{RF})\text{relative} = \Delta f / f_C = 1760/[(4860 + 3100)/2] = 44.22\%.$$

At the *LO* port,

$$BW_{LO} = \Delta f = 4750 - 3150\,MHz = 1600\,MHz,$$

$$(BW_{LO})\text{relative} = \Delta f / f_C = 1600/[(4750 + 3150)/2] = 41.56\%.$$

On the other hand, the bandwidth required in the *IQ* modulator design for group 1 of *UWB* system is

$$BW_{RF\ or\ LO} = \Delta f = 4752 - 3168\,MHz = 1584\,MHz,$$

$$(BW_{RF\ or\ LO})\text{relative} = \Delta f / f_C = 1584/[(4752 + 3168)/2] = 40.00\%.$$

Obviously the design of the input impedance matching network at the *LO* port as shown in Figure 11.27 and the output impedance matching network at the *RF* port as shown in Figure 11.28 is successful!

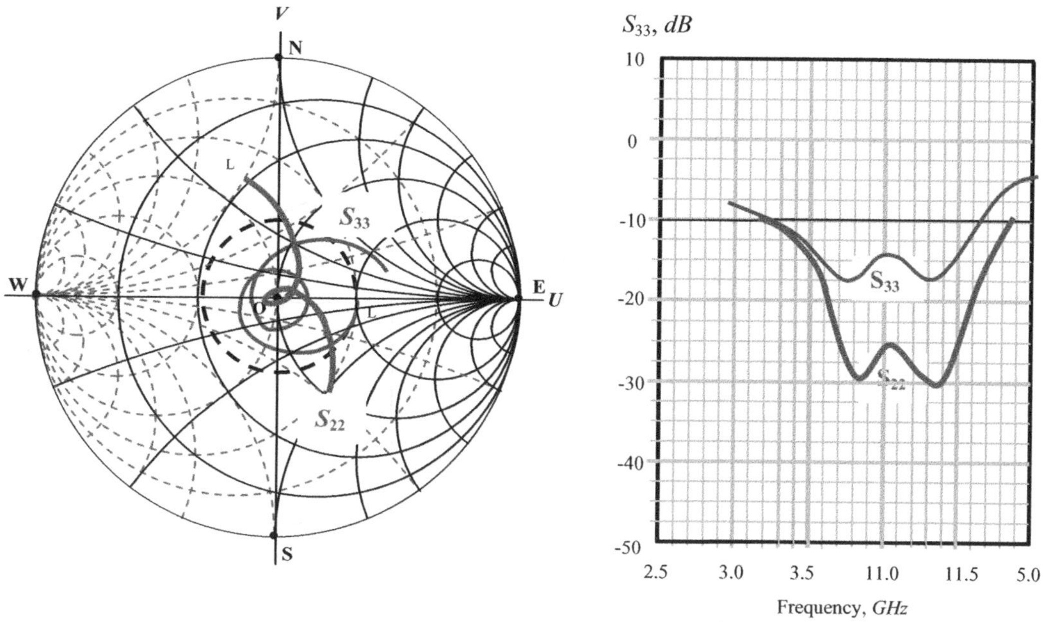

(a) Return loss on Smith chart after impedance matching is done at port *RF* (S_{22}) and *LO* (S_{33})

(b) S_{22} and S_{33} *vs. f* after impedance matching is done at port *RF* (S_{22}) and LO (S_{33})

Figure 11.29 At *LO* and *RF* port, relative bandwidth is expanded to the expected value for group 1, *UWB* system through impedance matching.
— S_{22}
— S_{33}
( ) demarcation circle, $RL = -10\,dB$

It is very desirable and of interest to study how to approach the goal of wide-band impedance matching. In the following paragraphs I'll demonstrate the evolution of bandwidth at the *LO* and *RF* ports as the impedance matching arms/branches are inserted one by one.

Let's demonstrate the evolution of bandwidth at an *LO* port first.

Figure 11.30(a) shows each impedance trace on the Smith chart after an arm or a branch of parts is inserted into the impedance matching network. A letter "*L*", attached to each trace, denotes the low-frequency end of the trace. Correspondingly, Figure 11.30(b) shows each curve of S_{33} vs. frequency after an arm or a branch of parts is inserted into the impedance matching network. In Figure 11.30(c), the LO_p and LO_n are the *LO* differential inputs of the *I* or *Q* modulator and are assigned as port 3. These are, in fact, the gates of *MOSFET* devices. At the outputs of LO_p and LO_n, the round square frame marked with "1" corresponds to trace 1 in both the Smith chart and the Cartesian coordinates with the magnitude of S_{33} *dB*. From this point, an impedance matching network is constructed by successive insertion of parts. Each time an arm or branch is added to the impedance matching network, the trace number of the return loss in the Cartesian coordinates or the trace number of the impedance trace in the Smith chart increases by 1. Figure 11.30 shows the evolution of impedance as the traces change from trace 1 all the way to trace 5 as individual arms or branches are inserted into the impedance matching network. At

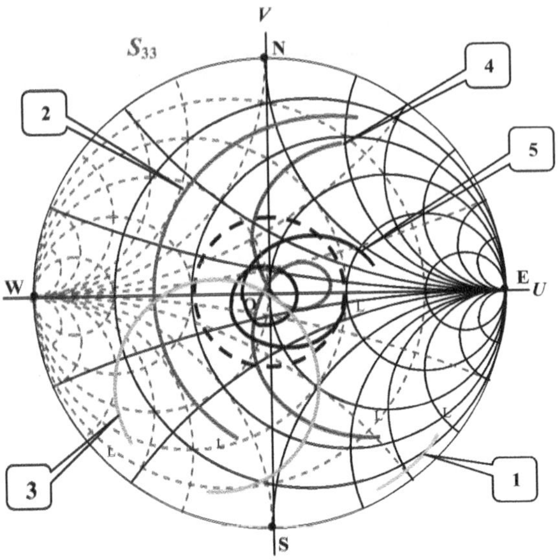

(a) S_{33} relocated on Smith chart after one arm or branch is inserted into the impedance matching network

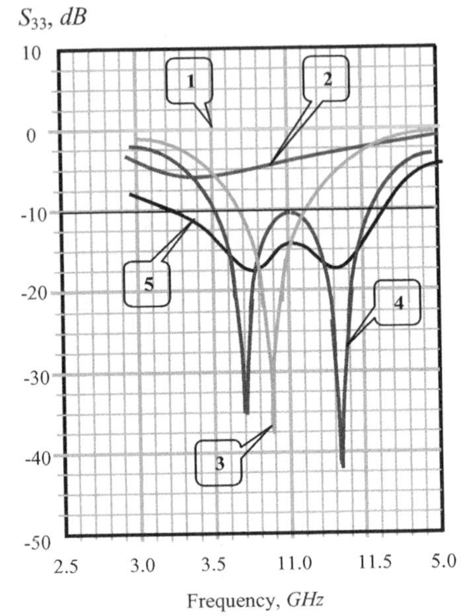

(b) S_{33} vs. f curves for each location of S_{33} on Smith chart in the adjacent left hand side

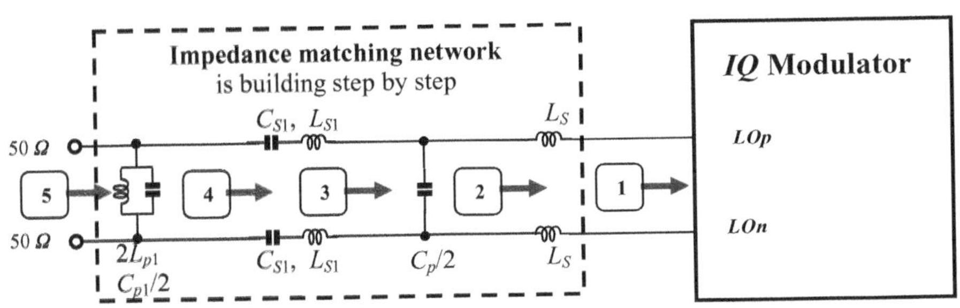

1. L_S= 5.25 pH (as if L_S does not exist).
2. L_S= 5.25 nH (R_S=13.06 Ω, Q=10).
3. L_S= 5.25 nH (R_S=13.06 Ω, Q=10),
   C_P= 1.2 pF.
4. L_S= 5.25 nH (R_S=13.06 Ω, Q=10),
   C_P= 1.2 pF,
   L_{S1}= 7.8 nH(Rs=19.4 Ω for Q=10, additional Rs=5.6 Ω), C_{S1}= 0.24 pF.
5. L_S= 5.25 nH (R_S=13.06 Ω, Q=10),
   C_P= 1.2 pF,
   L_{S1}= 7.8 nH(Rs=19.4 Ω for Q=10, additional Rs=5.6 Ω), C_{S1}= 0.24 pF,
   L_{p1}= 2 nH(R_S=11.98 Ω for Q=10, additional Rs=5.02 Ω), C_{P1}= 0.8 pF.

(c) Impedance matching in LO port

Figure 11.30 Evolution of bandwidth at LO port as the parts are inserted into impedance matching network (f = 3168 to 4752 MHz).

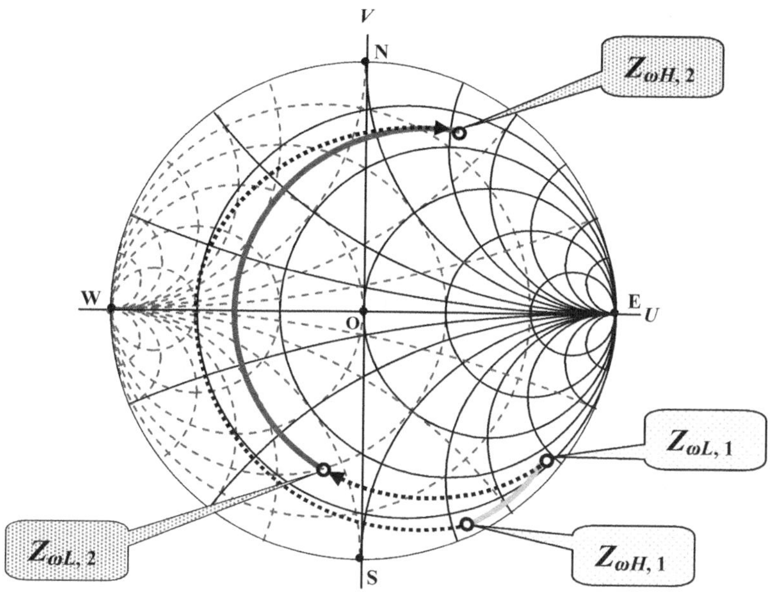

Figure 11.31 Variation of impedance from trace 1 to trace 2 at *LO* port as the part, L_S, is inserted into impedance matching network (f = 3168 to 4752 *MHz*).

the bottom of Figure 11.30(c), the parts for all the arms and branches are listed for each corresponding trace.

It can be found that the return loss, S_{33}, of trace 1 is very high. The magnitude of S_{33} is almost equal to zero *dB* in the entire frequency bandwidth of interest, from 3168 to 4752 *MHz*. Therefore, the first step to match the *LO* port is to insert an inductor in series, L_S, so that trace 1 at LO_p and LO_n, which is located in the high impedance area, can be changed to trace 2, located in the low impedance area. It should noted that, as shown in Figure 11.31, all the impedances on trace 1 more or less move clockwise around their own resistance circles, though their resistances are increased due to the equivalent resistance of the inductor in series. On the other hand, the variation of impedance at the high-frequency end is much greater than the variation of impedance at the low-frequency end.

Trace 2 on the Smith chart is closer to the reference impedance, 50 Ω, the center of the Smith chart, than trace 1. Therefore, the magnitude of S_{33} is improved from zero to a couple *dB* as shown in the Cartesian coordinates of Figure 11.30. However, a couple negative *dB* of S_{33} indicates too much return power and is not enough to reach a good impedance matching performance, although it improved from trace 1. Typically, the return loss must be greater than 10 *dB* (or, the return loss must be lower than −10 *dB*).

Consequently, the second step to match the *LO* port is to insert a capacitor in parallel, C_p, so that trace 2 is bent to trace 3 as shown in Figure 11.32. It is found that all the impedances on trace 2 basically move clockwise around their own subceptance circle. On the other hand, the variation of impedance at the high-frequency end is much greater than the variation of impedance at the low-frequency end. The impedance corresponding to an intermediate frequency crosses the reference

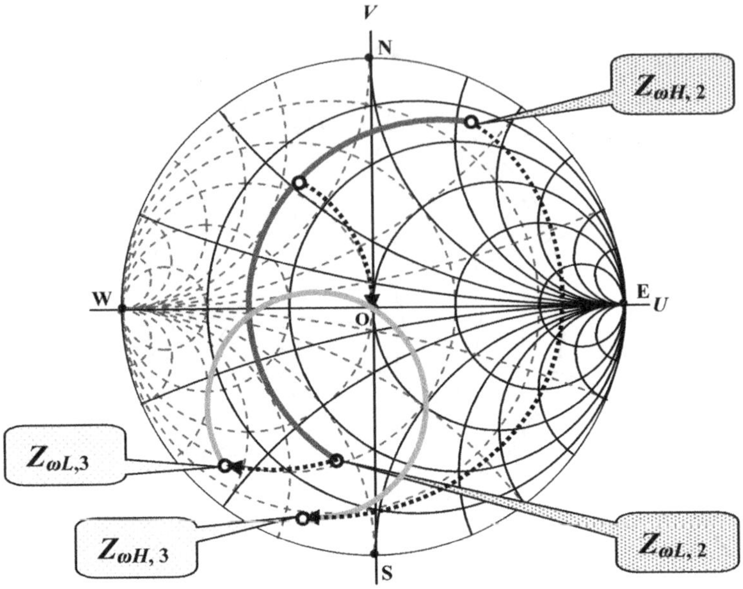

Figure 11.32 Variation of impedance from trace 2 to trace 3 at LO port as the part, $C_p = 1.2$ pF, is inserted into impedance matching network ($f_L = 3168\,MHz$, $f_H = 4752\,MHz$).

impedance, the center of the Smith chart. The return loss, S_{33}, of trace 3 as shown in Cartesian coordinates in Figure 11.30, is greatly improved from that of trace 2. Especially at the intermediate frequency, the magnitude of S_{33} reaches approximately $-35\,dB$. For narrow-band RF block designs, the bandwidth in which the $S_{33} < -10\ dB$ is easily satisfied. However, trace 3 is still not wide enough for wide-band performance.

In order to further improve the bandwidth, a combination of LC in series is inserted into the impedance matching network in series. This moves trace 3 to trace4 as shown in Figure 11.33. It should be noted that the high-frequency end of trace 3 moves clockwise while its low-frequency end moves counter-clockwise while the location of its intermediate frequency nearby the center of Smith chart, $50\,\Omega$, is kept unchanged. This special feature "twists" or "rolls" the trace and therefore widens the bandwidth since the twisting or rolling action shrinks the area covered by the trace on the Smith chart.

The rolling or twisting of the trace can be generated by a combination of LC in series. It could also be formed by a single inductor L in series, which is shown in Figure 11.6(b). The trace must be located on the Smith chart in such a way that:

- The low- and high-frequency end of the trace are located approximately on the same resistance circle.
- The high-frequency end is "more capacitive" or "less inductive" than the low-frequency end.

The difference is that in the case when a combination of LC in series is inserted, the entire trace does not move clockwise in such a simple way as in the case when a simple inductor L is inserted. It might instead be moved to a location where its

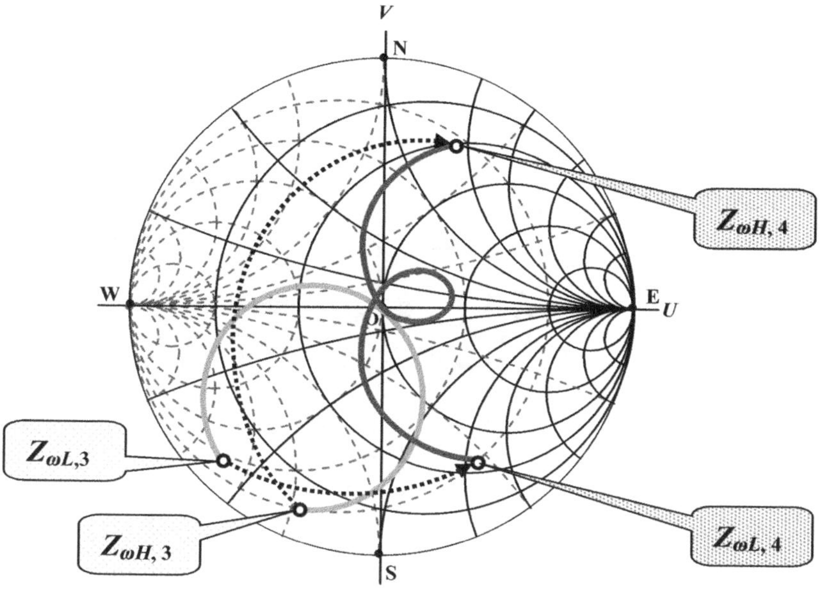

Figure 11.33 Variation of impedance from trace 3 to trace 4 at *LO* port as the part, L_{S1}, C_{S1}, is inserted into impedance matching network (f = 3168 to 4752 *MHz*).

resistance circles would be quite different from the original ones. It is therefore possible to move its intermediate portion to the area around the reference impedance, 50 Ω, in order to accomplish the widening of the bandwidth. In cases when a single inductor L is inserted, the entire trace would be moved clockwise to a new location, where its resistance circles and return loss would remain more or less unchanged, and consequently, its variation of bandwidth would be slight.

Rolling or twisting of the trace by the insertion of a combination LC in series is undoubtedly a powerful way to increase bandwidth.

Very often, the bandwidth of trace 4 is still not enough for a *UWB* system. Thus, we must widen the bandwidth even more. At this point we do not wish to move the intermediate portion of the trace out of the area around the reference impedance, 50 Ω. Instead, it is preferable to squeeze both trace portions corresponding to the low and high ends of the frequency toward the reference impedance, 50 Ω, the center of the Smith chart. Figure 11.34 shows this step, in which a combination of LC in parallel is inserted into the impedance matching network in parallel. The high-frequency end of trace 4 is moved clockwise along the subceptance circle while the low-frequency end of trace 4 is moved counter-clockwise along the subceptance circle. Trace 5 is now squeezed around the reference impedance most excellently, and the bandwidth is greatly increased so that it satisfies the desired goal.

It can be seen that the squeezing of the trace by inserting of a combination LC in parallel into the impedance matching network is also an undoubtedly powerful scheme to increase bandwidth. By now, the bandwidth is wide enough to satisfy application to a *UWB* system. As a matter of fact, Figures 11.30 to 11.34 show a typical wide-band impedance matching process for the *LO* port of a Gilbert cell when the *MOSFET* devices are applied, no matter whether the Gilbert cell is being constructed for a mixer or for a modulator.

Now, let's demonstrate the evolution of bandwidth at an *RF* port.

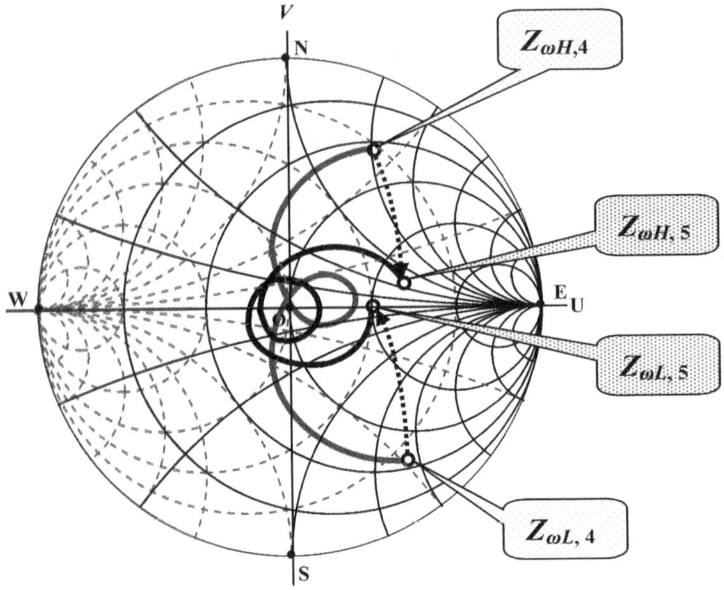

Figure 11.34 Variation of impedance from trace 4 to trace 5 at *LO* port as the part, L_{p1}, C_{p1}, is inserted into impedance matching network ($f = 3168$ to $4752\,MHz$).

Figure 11.35 shows the evolution of the bandwidth at the *RF* port as the parts are inserted into the impedance matching network.

Figure 11.35(a) shows each impedance trace on the Smith chart after an arm or a branch of parts is inserted into the impedance matching network. The letter "*L*", attached to each trace, denotes the low-frequency end of the trace. Correspondingly, Figure 11.35(b) shows each curve of S_{22} vs. frequency after an arm or a branch of parts is inserted into the impedance matching network. In Figure 11.35(c), the RF_p and RF_n are the *RF* differential outputs of the *I* or *Q* modulator and are assigned as port 2. These are, in fact, the drains of *MOSFET* devices. At the outputs of RF_p and RF_n, the round square frame marked with "1" corresponds to trace 1 in both the Smith Chart and the Cartesian coordinates with the magnitude of S_{22}, *dB*. From this point an impedance matching network is constructed by successive insertion of parts. Each time an arm or branch is inserted into the impedance matching network, the trace number of the return loss in the Cartesian coordinates or the trace number of the impedance on the Smith chart trace increases by 1. Figure 11.35 shows the evolution of impedance as the traces change from trace 1 to trace 6 as the individual arms and branches are inserted into the impedance matching network. At the bottom of Figure 11.35, the parts for all the arms and branches are listed for each corresponding trace.

It can be found that the return loss, S_{22}, of trace 1 is very high. The magnitude of S_{22} is almost equal to zero *dB* in the entire frequency bandwidth of interest. The first step to match the *RF* port is to insert an inductor in parallel, L_p, as shown in Figure 11.36.

The next step is to add a capacitor in series, C_S, so that trace 2 is moved to trace 3 as shown in Figure 11.37.

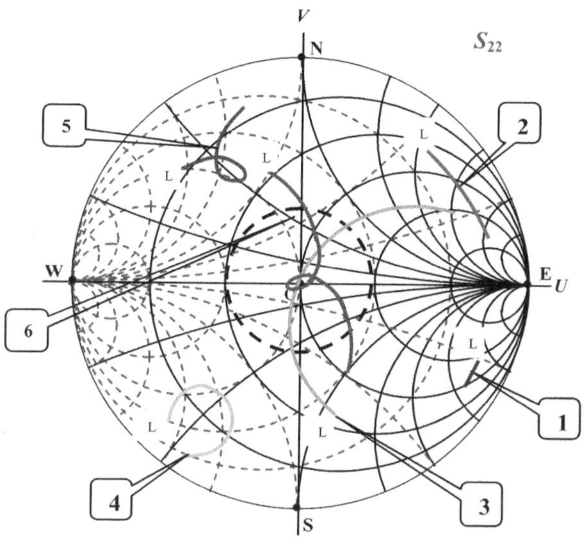

(a) S_{22} relocated on Smith chart after one arm or branch is inserted into the impedance matching network

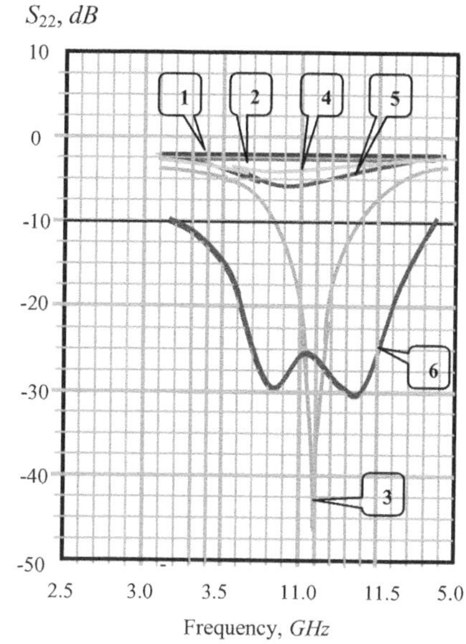

(b) S_{22} and S_{33} vs. f after impedance matching is done at port *RF* (S_{22}) and *LO* (S_{33})

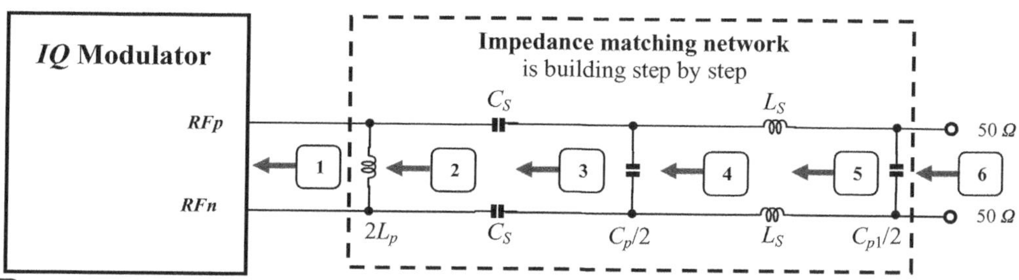

1 L_P= 38 *H* (as if L_P does not exist).

2 L_P= 3.8 nH (R_S=9.45 Ω, Q=10).

3 L_P= 3.8 *nH* (R_S=9.45 Ω, Q=10), C_S= 0.275 *pF*.

4 L_P= 3.8 *nH* (R_S=9.45 Ω, Q=10), C_S= 0.275 *pF*, C_P= 1.3*pF*.

5 L_P= 3.8 *nH* (R_S=9.45 Ω, Q=10), C_S= 0.275 *pF*, C_P= 1.3*pF*, L_S= **1.8 *nH*** (R_S=11.48 Ω, Q=10).

6 L_P= 3.8 *nH* (R_S=9.45 Ω, Q=10), C_S= 0.275 *pF*, C_P= 1.3*pF*. L_S= **1.8 *nH***(R_S=11.48 Ω, Q=10), C_{P1}= **0.85** *pF*

(c) Impedance matching in *RF* port

Figure 11.35 Evolution of bandwidth at *RF* port as the parts are added to impedance matching network.
Note: The parts, C_p, L_s, and C_{p1}, are added for the expansion of bandwidth! Resultant *BW* from simulation: 3100 to 4860 *MHz*; desired *BW* for group 1, band 2 in *UWB* system: 3596 to 4324 *MHz*; desired *BW* for entire group 1 in *UWB* system: 3168 to 4752 *MHz*.

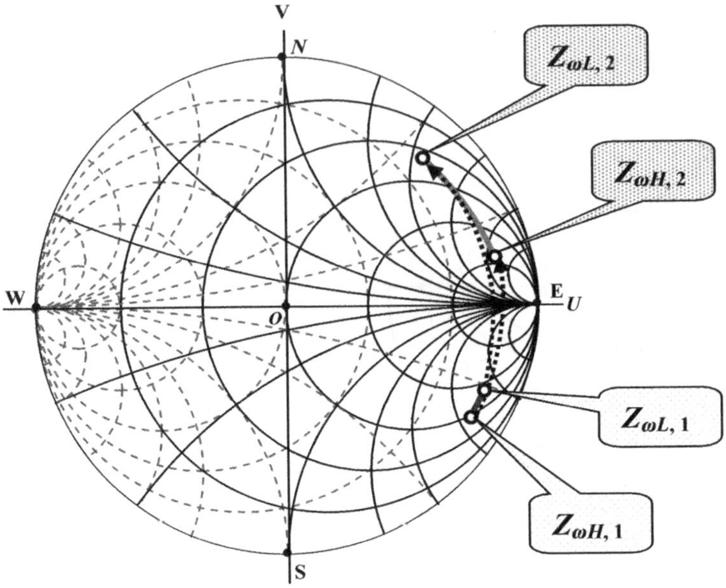

Figure 11.36 Variation of impedance from trace 1 to trace 2 at *RF* port as the part, L_p, is inserted into impedance matching network (f = 3168 to 4752 MHz).

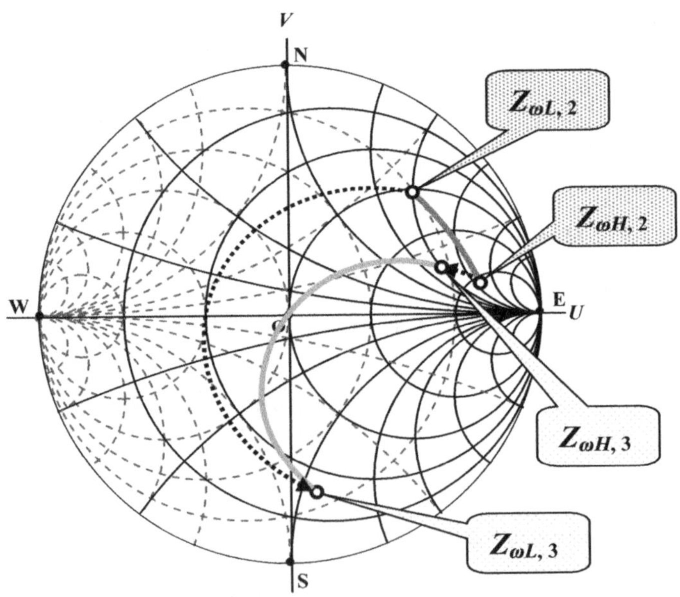

Figure 11.37 Variation of impedance from trace 2 to trace 3 at *RF* port as the part, C_S, is inserted into impedance matching network (f = 3168 to 4752 MHz).

It is found that all the impedances on trace 2 are basically moved counter-clockwise around their own resistance circles. On the other hand, the variation of impedance at the low-frequency end is much greater than the variation of impedance at the high-frequency end. The impedance corresponding to an intermediate frequency crosses the reference impedance, the center of the Smith chart. The return loss, S_{22}, of trace 3 as shown in Cartesian coordinates in Figure 11.35, is greatly improved from that of trace 2. Especially at the central frequency, the magnitude of S_{22} reaches approximately $-46\,dB$. For narrow-band *RF* block designs, the bandwidth in which the $S_{22} < -10\,dB$ is easily satisfied. However, trace 3 is still not wide enough for wide-band performance.

In order to further improve the bandwidth, we could insert a *LC* combination in parallel to the impedance matching network to squeeze the trace as shown in Figure 11.34.

An alternative method for the third step is shown in Figure 11.38, in which a single capacitor, *Cp*, is inserted in parallel. This makes trace 3 move away from the reference impedance, $50\,\Omega$, but it bends trace 3 to trace 4, which covers a much smaller area. Both the low- and high-frequency end are moved clockwise along their respective subceptance circles. The variation of impedance at the low-frequency end is much greater than that at the high-frequency end. It should be noted that in trace 4 the impedance at the high-frequency end is more "capacitive" than the impedance at the low-frequency end. As discussed above, this trace 4 could potentially be rolled up or twisted to further widen the bandwidth.

Figure 11.39 shows the movement from trace 4 to trace 5 after a single inductor, L_S, is inserted into the impedance matching network in series as the fourth step. As we have expected, trace 4 is rolled or twisted into trace5. However,

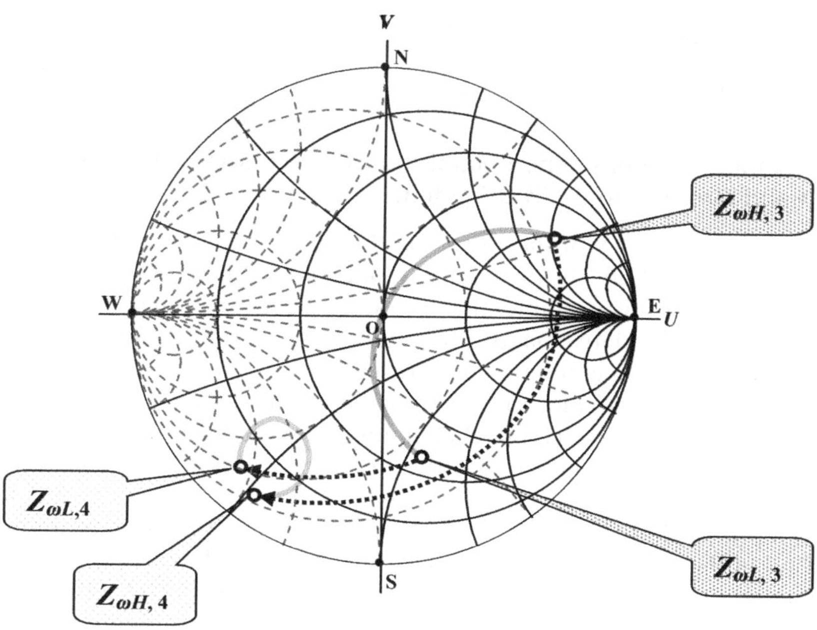

Figure 11.38 Variation of impedance from trace 3 to trace 4 at *RF* port as the part, C_p, is inserted into impedance matching network ($f = 3168$ to $4752\,MHz$).

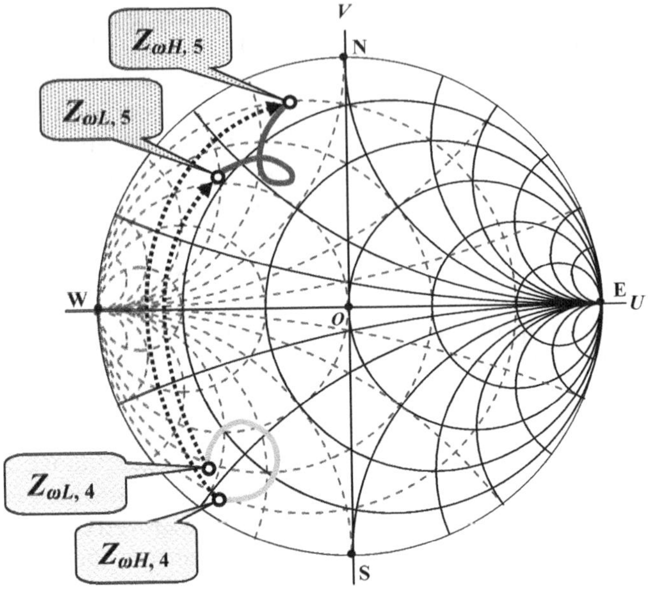

Figure 11.39 Variation of impedance from trace 4 to trace 5 at *RF* port as the part, L_S, is inserted into impedance matching network (f = 3168 to 4752 MHz).

as shown in the Cartesian coordinates of Figure 11.35, the magnitudes of S_{22} of the traces 4 and 5 are almost in the same order and are unsatisfactory.

The fifth step is to insert a single capacitor into the impedance matching network in parallel, so that trace 5 is pulled to trace 6, located around the area of the reference impedance, as shown in Figure 11.40. The bandwidth can be read from the Cartesian coordinates of the S_{22} magnitude in Figure 11.35.

11.4.7 Example 2: Impedance Matching in *IQ* Modulator Design for Group 3 and Group 6 in a *UWB* System

In example 2 the topology of the *I* or *Q* modulator is the same as that in example 1 as shown in Figure 11.26. The difference is that it is designed for bandwidth groups 3 and 6 together. In this example, the operating frequency range is from 6336 to 8976 MHz, so that the relative bandwidth is therefore 34.48%. Again, the modulator is supposed to be implemented on an *RFIC* chip.

Figures 11.41 and 11.42 plot the input impedance matching network at the *LO* port and the output impedance matching network at the *RF* port, respectively. The input impedance matching network at the *LO* port shown in Figure 11.41 consists of two arms and two branches. An inductor is always connected with a small resistor in series and is circled by a dashed-line ellipse. The small resistor represents the attenuation of the inductor and its value is calculated on the basis of the assumed quality factor, Q = 10, and the operating frequency, f = 7656 MHz. There are two resistance values appearing in the r expression of L_{S1} and L_{P1}. The first value is the calculated resistance based on Q = 10 and f = 7656 MHz and the second value is additional resistance for more bandwidth expansion. One of the two arms consists of one inductor, the other consists of a *LC* combination in series. Similarly, one of

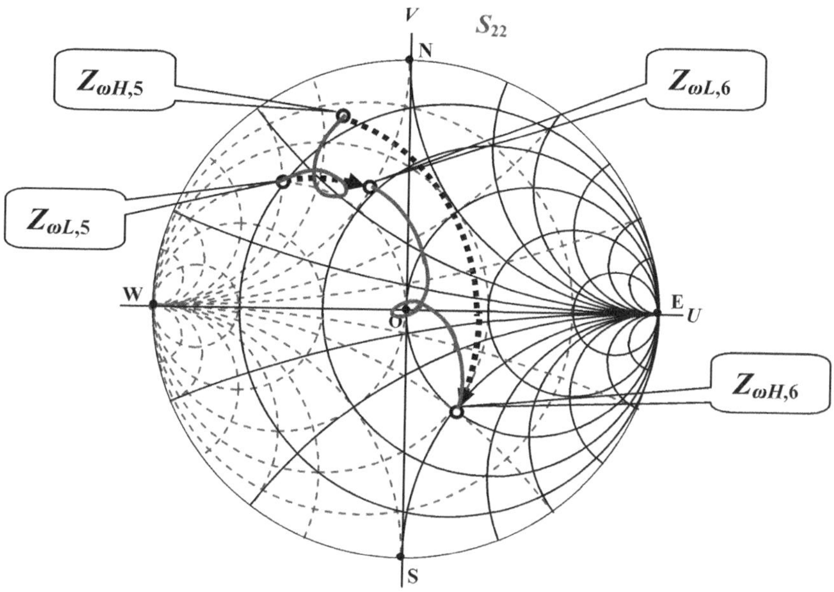

Figure 11.40 Variation of impedance from trace 5 to trace 6 at *RF* port as the part, C_{p1}, is inserted into impedance matching network (f = 3168 to 4752 *MHz*).

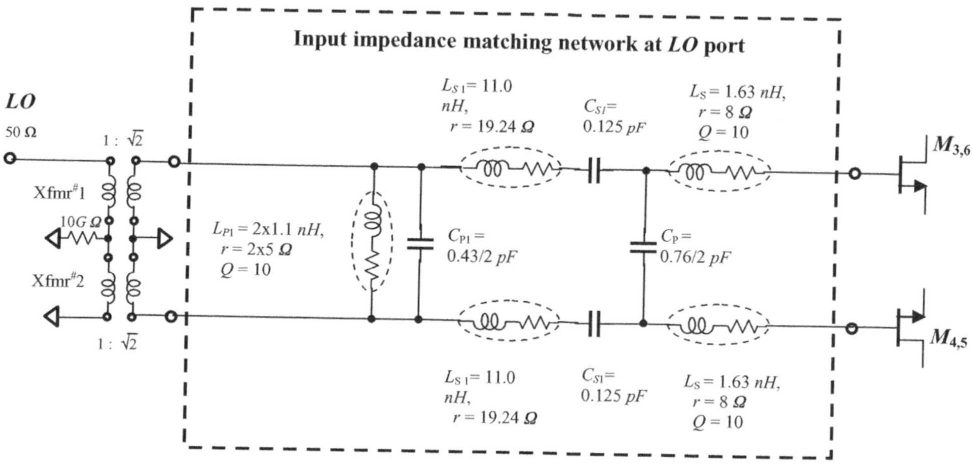

Figure 11.41 Input impedance matching network at *LO* ports for group 3 + group 6 in *UWB* system (f = 6336 to 8976 *MHz*).

the two branches consists of one capacitor and another branch consists of a *LC* combination in parallel. The symbols "2×", appearing in L_{P1}, and "/2", appearing in C_P and C_{P1}, indicate that there are two identical parts connected together in series as a branch bridging two differential nodes.

The output impedance matching network at the *RF* port is shown in Figure 11.42. It consists of two arms and three branches. Again, an inductor is always connected with a small resistor in series and is circled by a dashed-line ellipse. The small resistor represents the attenuation of the inductor and its value is calculated from the assumed quality factor, Q = 10 and the operating frequency, f = 7656 *MHz*. Only a

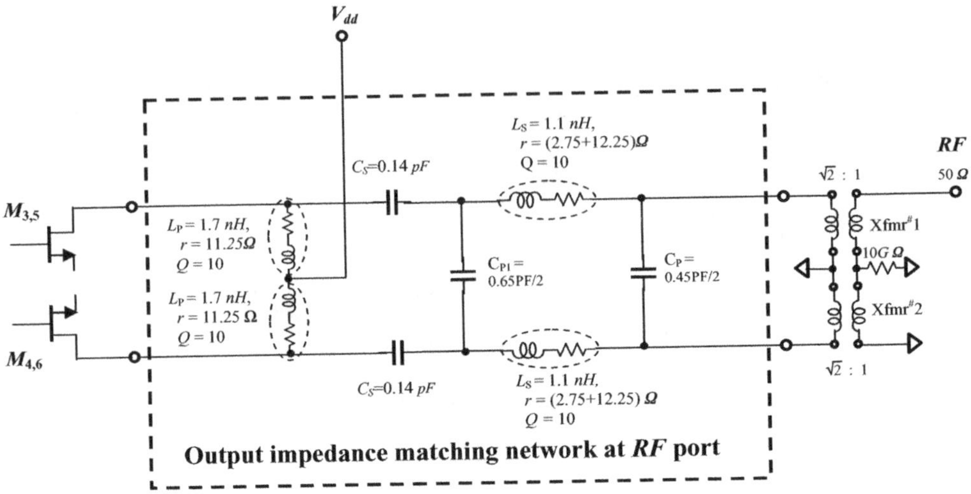

Figure 11.42 Output impedance matching network at *RF* port for group 3 + group 6 in *UWB* system (f = 6336 to 8976 MHz).

single part is contained in an arm or a branch. The symbol "/2" appearing in C_P and C_{P1} indicates that there are two identical parts connected together in series as a branch bridging two differential nodes. The DC power supply, V_{dd}, provides the current to the devices through two L_P inductors.

The input impedance matching network at the *LO* port as shown in Figure 11.41, and the output impedance matching network at the *RF* port, as shown in Figure 11.42, successfully expand the bandwidth from very narrow to very wide band. Figure 11.43 shows the expected value of the bandwidth being reached for an *IQ* modulator working in bandwidth groups 3 and 6 in a *UWB* system.

From Figure 11.43(a) it can be seen that most portions of the return loss trace, either S_{22} or S_{33}, are located within the demarcation circle. This indicates that the impedances at both port 2 and port 3 are matched to 50 Ω in the wide-band case. The bandwidth can be read from Figure 11.43(b) accordingly, that is,
at the *RF* port,

$$BW_{RF} = \Delta f = 9080 - 6100\,MHz = 2980\,MHz,$$

$$(BW_{RF})\text{relative} = \Delta f / f_C = 2980/[(9080 + 6100)/2] = 39.26\%.$$

at the *LO* port,

$$BW_{LO} = \Delta f = 9000 - 6330\,MHz = 2670\,MHz,$$

$$(BW_{LO})\text{relative} = \Delta f / f_C = 2670/[(9000 + 6330)/2] = 34.83\%.$$

On the other hand, the bandwidth required in the *IQ* modulator design for group 3 + group 6 of a *UWB* system is

$$BW_{RF\ or\ LO} = \Delta f = 8976 - 6336\,MHz = 2640\,MHz,$$

$$(BW_{RF\ or\ LO})\text{relative} = \Delta f / f_C = 2640/[(8976 + 6336)/2] = 34.48\%.$$

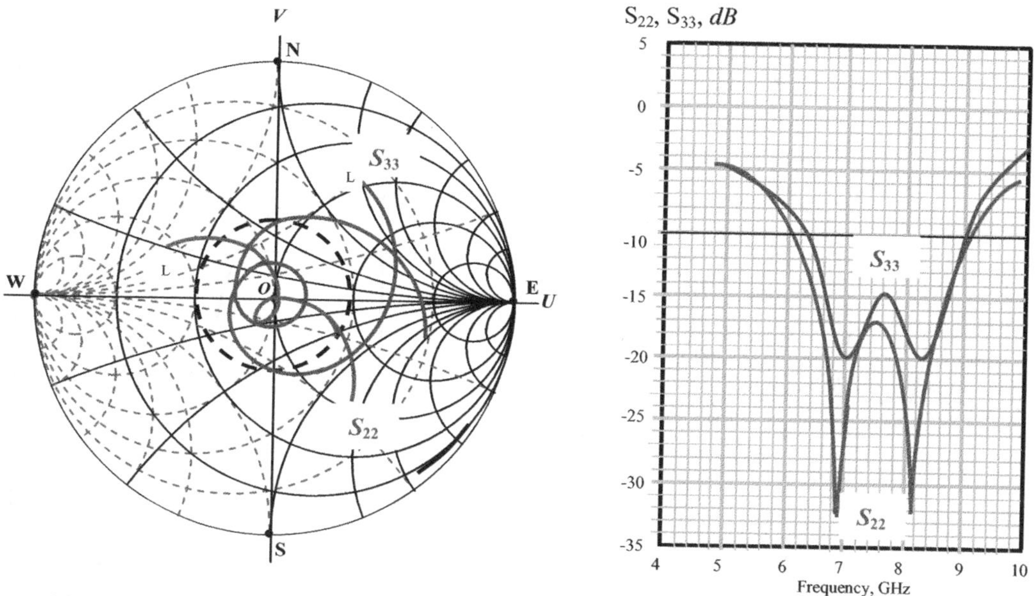

(a) Return loss on Smith chart after impedance matching is done at port *RF* (S_{22}) and *LO* (S_{33})

(b) S_{22} and S_{33} vs. f after impedance matching is done at port *RF* (S_{22}) and *LO* (S_{33})

Figure 11.43 At *LO* and *RF* ports, relative bandwidth is expanded to the expected value for group 3 + group 6, *UWB* system through impedance matching.
— S_{22}
— S_{33}
( ) demarcation circle, $RL = -10\,dB$

Obviously, the design of the input impedance matching network at the *LO* port as shown in Figure 11.41 and the output impedance matching network at the *RF* port as shown in Figure 11.42 is successful!

Figure 11.44 shows the evolution of S_{33} at the *LO* port as the impedance matching parts are inserted successively. Figure 11.45 shows the evolution of S_{22} at the *RF* port as the impedance matching parts are inserted successively. By the same method as described in example 1, readers can follow the evolution of the bandwidth accordingly.

11.5 DISCUSSION OF WIDE-BAND IMPEDANCE MATCHING NETWORKS

In Section 11.4, the designs of the *IQ* modulators with *MOSFET* transistors were taken as examples in the discussion of wide-band impedance matching. As a matter of fact, they represented the most interesting cases because *MOSFET* devices have become much more popular than other devices nowadays. On the other hand, all the schemes and ideas about wide-band impedance matching can be applied to circuit designs with other types of devices.

Based on the discussion in the previous sections, a passive wide-band impedance matching network for *MOSFET* devices can be formalized into two cases as follows.

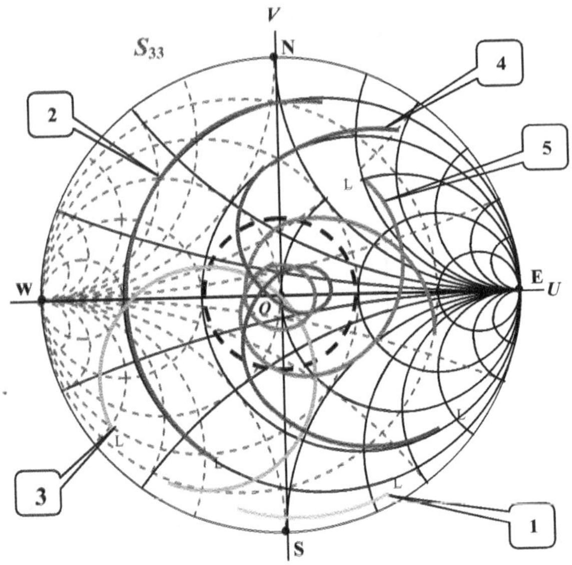

(a) S_{33} relocated on Smith chart after one arm or branch is inserted into the impedance matching network

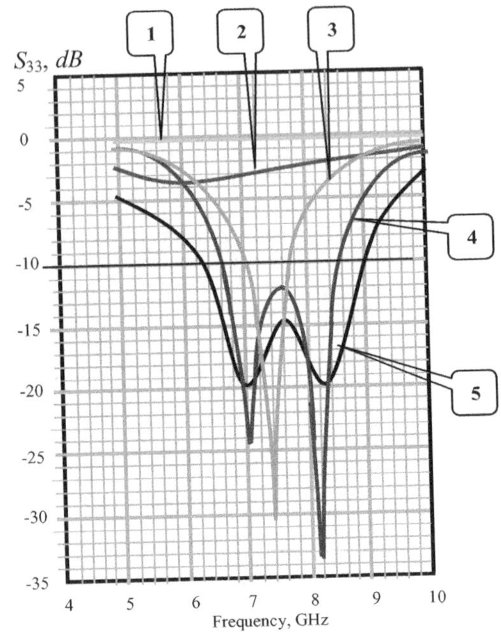

(b) S_{33} vs. f curves for each location of S_{33} on Smith chart in the adjacent left hand side

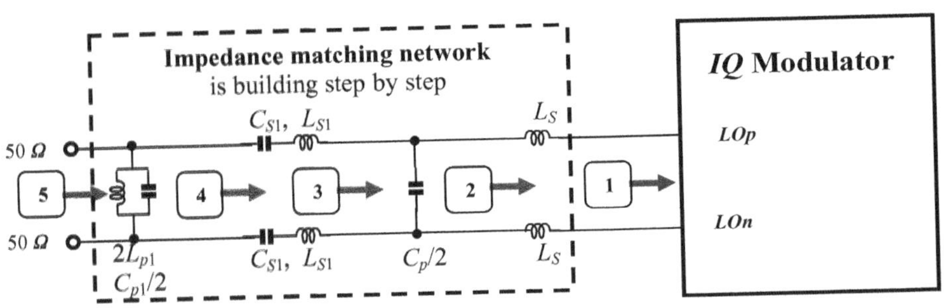

1. L_S= 1.63 pH (as if L_S did not exist).
2. L_S= 1.63 nH (R_S=8.0 Ω, Q=10).
3. L_S= 1.63 nH (R_S=8.0 Ω, Q=10), C_P= 0.8 pF.
4. L_S= 1.63 nH (R_S=8.0 Ω, Q=10), C_P= 0.76 pF, L_{S1}= 11.0 nH (R_S=19.24 Ω, Q=10), C_{S1}= 0.125 pF.
5. L_S= 1.63 nH (R_S=8.0 Ω, Q=10), C_P= 0.76 pF, L_{S1}= 11.0 nH (R_S=19.24 Ω, Q=10), C_{S1}= 0.125 pF, L_{P1}= 1.1 nH (R_S=5 Ω, Q=10), C_{P1}= 0.43 pF.

(c) Impedance matching at RF port

Figure 11.44 Evolution of bandwidth at LO port as the parts are inserted into impedance matching network (f = 6336 to 8976 MHz).

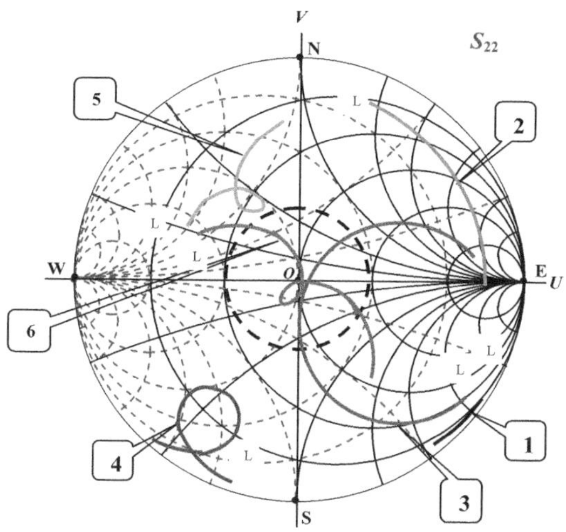

(a) S_{22} relocated on Smith chart after one arm or branch is inserted into the impedance matching network

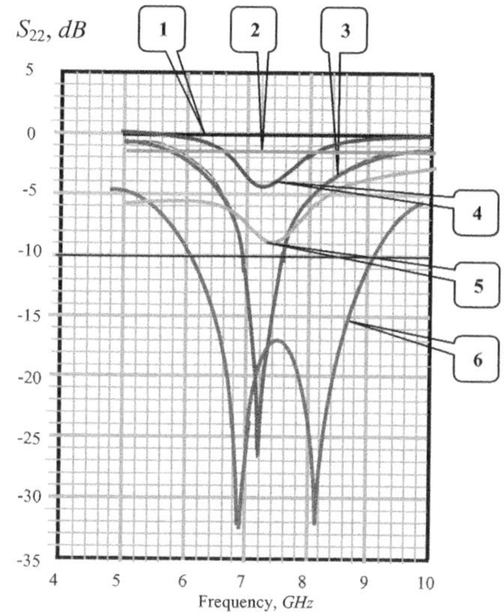

(b) S_{22} vs. f curves for each location of S_{22} on Smith chart in the adjacent left hand side

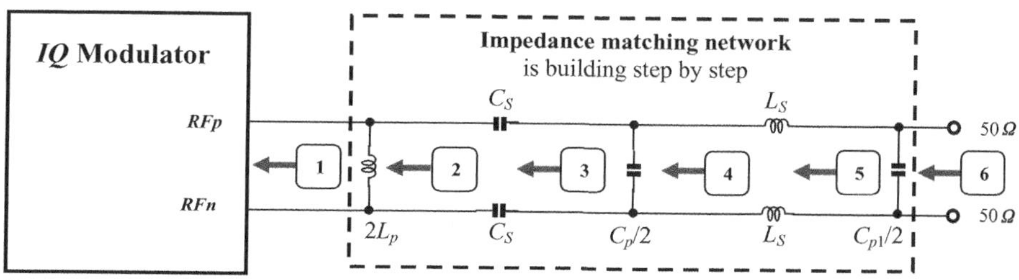

1 L_P= 17 H (as if L_P did not exist)

2 L_P= 1.7 nH (R_S=11.25 Ω, Q=10)

3 L_P= 1.7 nH (R_S=11.25 Ω, Q=10),
C_S= 0.14 pF

4 L_P= 1.7 nH (R_S=11.25 Ω, Q=10),
C_S= 0.14 pF,
C_p= 0.65 pF

5 L_P= 1.7 nH (R_S=11.25 Ω, Q=10),
C_S= 0.14 pF,
C_p= 0.65 pF,
Ls=1.1 nH (Rs=2.83 Ω, Q=10, additional Rs= 12.17 Ω)

6 L_P= 1.7 nH(R_S=11.25 Ω, Q=10),
C_S= 0.14 pF,
C_p= 0.65 pF,
Ls=1.1 nH (Rs=2.83 Ω, Q=10, additional Rs= 12.17 Ω),
C_{P1}= 0.45 pF

(c) Impedance matching at RF port

Figure 11.45 Evolution of bandwidth at RF port as the parts are inserted into impedance matching network (f = 6336 to 8976 MHz).

11.5.1 Impedance Matching for Gates of *MOSFET* Devices

A RF block built by $MOSFET$ devices, in which the gates of the $MOSFET$ devices are high impedance terminals to be matched, is quite popular. There are many examples available, such as the input of LNA, the inputs of the RF and LO ports of a Gilbert cell, and so on.

In general we have four choices for the first part applied to the impedance matching network:

- A capacitor in parallel:
  This is not welcome because it will short the incoming *RF* signal to the ground and bring about unexpected attenuation.
- A capacitor in series:
  This is not welcome because it will move the impedance at the gate of the *MOSFET* device to a higher impedance area since the impedance at the gate of the *MOSFET* device is originally capacitive high impedance.
- An inductor in parallel:
  This is, in fact, not qualified to be the first matching part because in such a case a *DC* blocking capacitor must be inserted between this inductor and the gate so as to avoid a short-circuited *DC* bias.
- An inductor in series:
  This is welcome because it can move the high capacitive impedance at the gate to a point of low impedance.

Consequently, the topology of a wide-band impedance matching network for the gates of MOS devices is suggested and is shown in Figure 11.46. The examples have been presented in Section 11.5 and the simulation results are shown in Figures 11.30 and 11.44, where the LO_p and LO_n are the LO differential inputs of the I or Q modulator, which are equivalent to *INp* (gate 1) and *INn* (gate 2) as shown in Figure 11.46. In Section 11.5, this topology worked successfully in the designs of both frequency bands: group 1 of the *UWB* system and the combination of group3 plus group 6 of the *UWB* system. This implies that this topology can expand the relative bandwidth by up to 40%!

The impedance matching network in Figure 11.46 contains five inductors and four capacitors. The five inductors will incur high costs and complicate the circuit design. However, this price must be paid for the wide-band requirement. As a matter of fact, if it were a narrow-band design, two inductors and two capacitors, C_{s1}, L_{s1}, $C_{p1}/2$ and $2L_{p1}$, could be omitted, and only two inductors, $2 \times L_s$, and one capacitor, $C_p/2$, would be required.

In order to lower the cost and simplify the circuit design, an alternative topology, shown in Figure 11.47, could be applied for the construction of a wide-band impedance matching network.

This topology contains five capacitors but only three inductors, so the part count, especially the part count of inductors, is reduced.

11.5.2 Impedance Matching for Drains of *MOSFET* Devices

A *RF* block built by *MOSFET* devices, in which the drain of the *MOSFET* device is a high impedance terminal to be matched, is a quite popular case. There are many examples available, such as, the output of *LNA*, the output of a Gilbert cell for an *IQ* modulator, and so on.

In general we only apply an inductor as the first part to the impedance matching network.

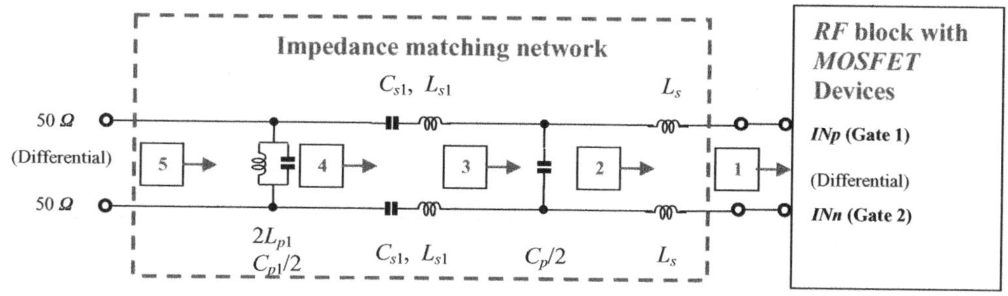

Figure 11.46 Topology of impedance matching network at gates of *MOSFET* devices with five inductors and four capacitors.

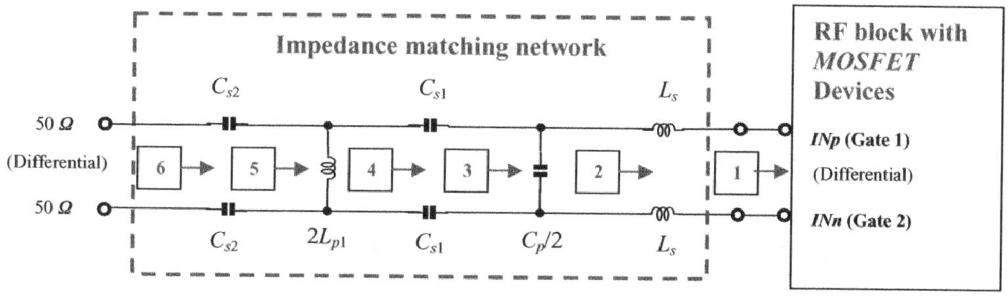

Figure 11.47 An alternative topology of impedance matching network at gates of *MOSFET* devices with three inductors and five capacitors.

- The first part must be connected between the drain and *DC* power supply. This implies that the first part must be connected in parallel.
- Obviously, a capacitor is not a viable candidate.
- A resistor is also not a valid candidate because it forces the *DC* power supply to drop a certain amount of *DC* voltage. Then the *DC* voltage drop across the device may be much lower than the *DC* power supply voltage. Consequently, it increases the non-linearity of the device and thus degrades its performance.

Next, the topology of a wide-band impedance matching network for the drain of *MOSFET* devices is suggested and is shown in Figure 11.48. The examples have been presented in Section 11.5 and the simulation results are shown in Figures 11.35 and 11.45, where the RF_p and RF_n are the *RF* differential outputs of the *I* or *Q* modulator, which are equivalent to OUT_p (Drain 1) and OUT_n (Drain 2) as shown in Figure 11.48.

The impedance matching network in Figure 11.48 contains three inductors and four capacitors. The three inductors will incur a somewhat high cost and also complicate the circuit design. However, this price must be paid for the wide-band requirement. As a matter of fact, if this were a narrow-band design, two inductors and two capacitors, $C_p/2$, $2 \times L_s$, and $C_{p1}/2$, could be omitted, and only one inductor, $2L_p$, and two capacitors, $2 \times C_s$, would be needed.

In Section 11.5, this topology worked successfully in the design for the combined frequency band of group 3 and group 6 in a *UWB* system. The alternative topology

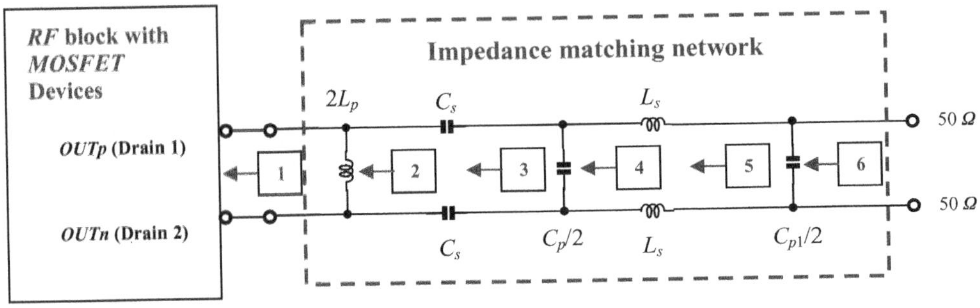

Figure 11.48 Topology of impedance matching network at drains of *MOSFET* devices with three inductors and four capacitors.

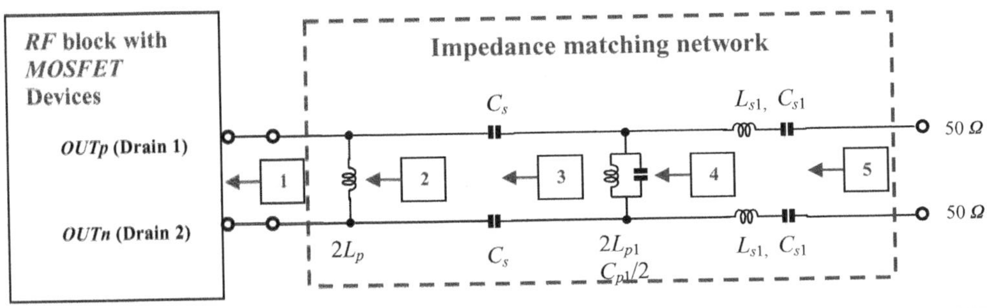

Figure 11.49 An alternative topology of impedance matching network at drains of *MOSFET* devices with four inductors and five capacitors.

is suggested in Figure 11.49. The price of this topology is the extra inductor and capacitor compared with the number of parts in the topology shown in Figure 11.48.

REFERENCES

[1] D. C. Youla, "A New Theory of Broad-Band Matching," *IEEE Transactions on Circuit Theory*, Vol. CT–11, March 1964, pp. 30–50.

[2] K. R. Cioffi, "Broad-band Distributed Amplifier Impedance-Matching Techniques," *Microwave Theory and Techniques, IEEE Transactions*, Vol. 37, No. 12, December 1989, pp. 1870–1876.

[3] T. O'Meara, "Very-Wide-Band Impedance-Matching Network," Component Parts, *IRE Transactions*, Vol. 9, No. 1, March 1962, pp. 38–411.

[4] G. J. Laughlin, "A New Impedance-Matched Wide-Band Balun and Magic Tee," *Microwave Theory and Techniques, IEEE Transactions*, Vol. 24, No. 3, March 1976, pp. 135–141.

[5] H. J. Carlin and P. Amstutz, "On Optimum Broad-Band Matching," *IEEE Transactions on Circuits and Systems*, Vol. CAS-28, No. 5, May 1981, pp. 401–405.

CHAPTER 12

IMPEDANCE AND GAIN OF A RAW DEVICE

12.1 INTRODUCTION

In electrical science and technology, the single-stage amplifier is almost indispensable. However, its implications in digital circuit design are different from those in *RF* circuit design. Figure 12.1 shows such a difference.

A single-stage amplifier in digital circuit design consists of a transistor and a couple of passive parts which provide *DC* power, *DC* blocking, and *DC* bias, and offer input and output. Its performance is characterized by the input and output voltages or current, V_{in}, i_{in}, V_{out}, i_{out}, in which the voltage and current gain, A_v and A_i (or β), are mainly concerned.

A single-stage amplifier in *RF* circuit design consists of three parts: the input impedance matching network, the raw device itself, and the output impedance matching network. A single-stage amplifier in the digital realm is, in fact, merely a raw device in the *RF* realm. A raw device without impedance matching cannot be an independently functioning block because the most important objective of an *RF* block is its power gain, $\boldsymbol{G}$, and not its voltage or current gain. Impedance matching takes priority in order to ensure superior power transportation or manipulation. The input impedance matching network must be matched between the source impedance z_s and the input impedance of the device z_{in}, while the output impedance matching network must be matched between the output impedance of the device z_{out} and the load impedance z_L. In *RF* circuit design the main concerns of the raw device are its input and output impedances, z_{in} and z_{out}.

It must be noted that the input and output impedance matching networks should not contain the source or the load impedances, z_S and z_L, because our examination

RF Circuit Design, by Richard Chi-Hsi Li

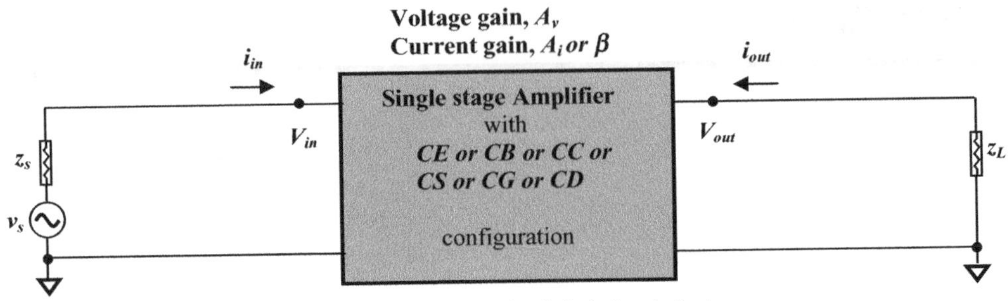

(a) Single stage amplifier in digital circuit design.

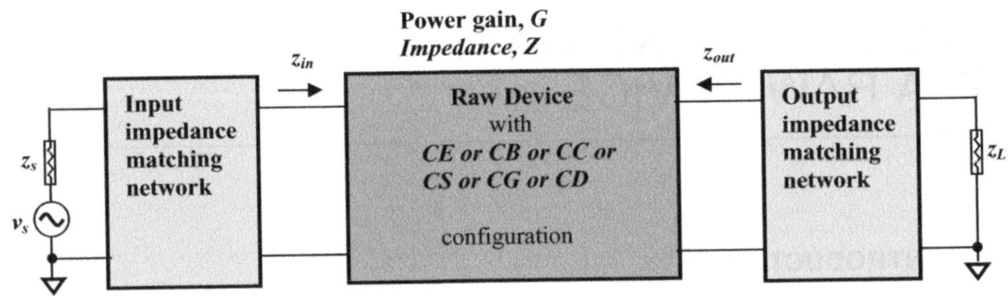

(b) Single stage amplifier in *RF* circuit design.

Figure 12.1 Different viewpoints between single stage amplifier and raw device.

is constrained to the single-stage amplifier as shown in Figure 12.1(a) or the raw device block as shown in the Figure 12.1(b), and is not concerned with source or load. In some technical books the equations of the input and output impedances do contain the source or the load impedances, z_S and z_L. This is not a mistake, but it is not suitable for *RF* circuit design.

For an *RF* circuit designer, the parameters of primary concern are the input and output impedances of the raw device, z_{in} and z_{out}, by which the input and output impedance matching networks can be built for power gain. Nevertheless, *RF* designers must also be familiar with voltage gain and current gain because, as a matter of fact, the power gain consists of voltage gain and current gain.

Consequently, this chapter will cover voltage gain and current gain of a single-stage amplifier, but, since this book is aimed at *RF* designers, will focus more on the input and output impedances. Study of the input and output impedances of a device with different configurations is a prerequisite in *RF* circuit design.

We are going to discuss six configurations of the raw device.

There are three configurations for the bipolar transistor:

- *CE* (Common Emitter), including *CE* with emitter degeneration;
- *CB* (Common Base);
- *CC* (Common Collector, or Emitter Follower);

And there are three configurations for the *MOSFET* transistor:

- *CS* (Common Source), including *CS* with source degeneration;
- *CG* (Common Gate);
- *CD* (Common Drain, or Source Follower).

The characterization of the device is usually presented through the following parameters:

- Open-circuited voltage gain: This tests the voltage magnification capability of the device. The testing condition is that the output terminal is opened or is not loaded, that is, $Z_L \to \infty$. "Open-circuited" implies that the testing represents the character of the device itself without other circuits connected to it, so that the testing indeed represents the character of the device itself;
- Short-circuited current drain: This tests the current magnification capability of the device. The test conditions are that the output terminal is shorted and the input terminal is provided with a small testing current source. Again, "short-circuited" implies that the testing represents the character of the device itself without other circuits to disturb it, so that the testing indeed represents the character of the device itself;
- Input and output impedance testing: This tests the power transportation capability of the device. From the testing one can observe how well the device can be matched with its source and load so that the power transportation or manipulation can be optimized. The tests of both input and output impedances are conducted inward to the device so that the results truly represent the character of the device itself.

12.2 MILLER EFFECT

Before discussing of the characterization of a device, let us introduce the Miller effect and its impact on a device.

A device operating at a high frequency has feedback of signal from output to input if its isolation between output and input is imperfect. Figure 12.2(a) shows a device block, in which a feedback element with impedance Z is connected between output and input. The open-circuited voltage gain of the device is

$$A_v = \frac{v_o}{v_i}. \tag{12.1}$$

where

A_v = open-circuited voltage gain;
v_o = output voltage at output node;
v_i = input voltage at input node.

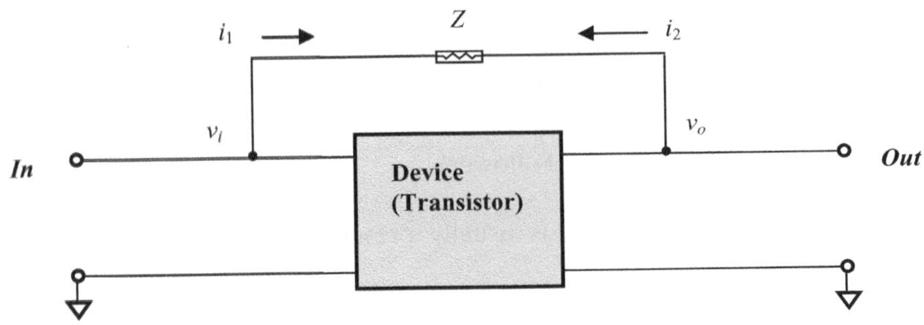

(a) A device with feedback element, impedance Z, connected between input and output

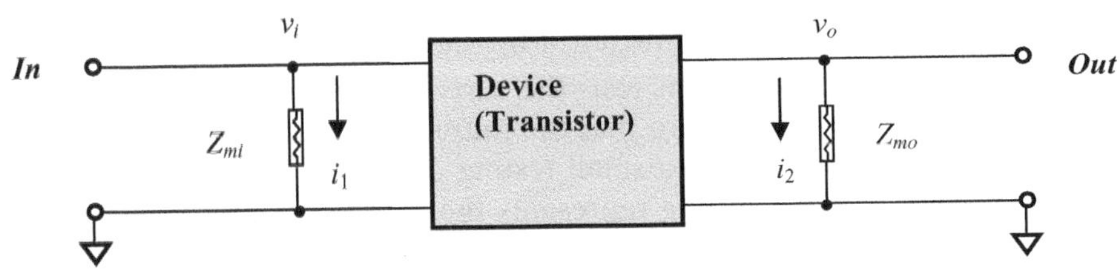

(b) Equivalent circuit of the device shown in (a) above, $A_v = v_o/v_i$

Figure 12.2 Miller effect on a device (transistor).

Miller's theorem states that

If a device has open-circuited voltage gain A_v between input and output nodes, then the feedback element with impedance Z connected between these two nodes can be replaced by an equivalent element connected to ground at each node. The impedance of the equivalent element is

At the input node,

$$Z_{mi} = \left|\frac{1}{A_v - 1}\right| Z, \tag{12.2}$$

and at the output node,

$$Z_{mo} = \left|\frac{A_v}{A_v - 1}\right| Z, \tag{12.3}$$

where

Z_{mi} = equivalent input Miller impedance at input node connected to the ground,

Z_{mo} = equivalent output Miller impedance at output node connected to the ground.

Let's verify the expressions (12.2) and (12.3) below. In Figure 12.2(a), a current i_1 is drawn from the input to output or a current i_2 is drawn from the output to input through the feedback element with impedance Z. The direction of the actual current flow depends on the magnitudes of v_i and v_o. In Figure 12.2(a), the direction of i_1 corresponds to cases when $v_i > v_o$ while the direction of i_2 corresponds to those cases when $v_i < v_o$. Let us set up a common rule to define the direction of either i_1 or i_2: **The current flows away from the node.** Consequently, the mathematical derivations must be distinguished with two cases, $v_i > v_o$ and $v_i < v_o$.

1) In cases when $v_i > v_o$ or $A_v < 1$.
At the input node in Figure 12.2(a),

$$i_1 = \frac{v_i - v_o}{Z} = \frac{v_i - A_v v_i}{Z} = \frac{v_i}{\dfrac{1}{(1-A_v)}Z}, \tag{12.4}$$

and at the input node in Figure 12.2(b),

$$i_1 = \frac{v_i}{Z_{mi}}. \tag{12.5}$$

By comparing (12.4) with (12.5),

$$Z_{mi} = \frac{1}{1-A_v}Z. \tag{12.6}$$

At the output node in Figure 12.2(a),

$$i_2 = -\frac{v_o - v_i}{Z} = -\frac{v_o - v_o/A_v}{Z} = \frac{v_o}{\dfrac{A_v}{(1-A_v)}Z}, \tag{12.7}$$

and at the output node in Figure 12.2(b),

$$i_2 = \frac{v_o}{Z_{mo}}. \tag{12.8}$$

By comparing (12.7) with (12.8),

$$Z_{mo} = \frac{A_v}{1-A_v}Z. \tag{12.9}$$

2) In cases when $v_i < v_o$ or $A_v > 1$.
At the input node in Figure 12.2(a),

$$i_1 = -\frac{v_i - v_o}{Z} = -\frac{v_i - A_v v_i}{Z} = \frac{v_i}{\dfrac{1}{(A_v-1)}Z}. \tag{12.10}$$

By comparing (12.10) with (12.5),

$$Z_{mi} = \frac{1}{A_v - 1} Z. \tag{12.11}$$

At the output node in Figure 12.2(a),

$$i_2 = \frac{v_o - v_i}{Z} = \frac{v_o - v_o/A_v}{Z} = \frac{v_o}{\frac{A_v}{(A_v - 1)} Z}. \tag{12.12}$$

By comparing (12.12) with (12.8),

$$Z_{mo} = \frac{A_v}{A_v - 1} Z. \tag{12.13}$$

As a matter of fact, equations (12.6) and (12.9) and equatoins (12.11) and (12.13) for the two cases can be combined into two expressions, which are

$$Z_{mi} = \left| \frac{1}{A_v - 1} \right| Z, \tag{12.2}$$

$$Z_{mo} = \left| \frac{A_v}{A_v - 1} \right| Z. \tag{12.3}$$

So, the derivation from (12.4) to (12.13) verifies Miller's theorem.

When the absolute value of the voltage gain A_v is much greater than 1, that is,

$$|A_v| >> 1, \tag{12.14}$$

then,

$$Z_{mi} \approx \frac{Z}{|A_v|}, \tag{12.15}$$

$$Z_{mo} \approx Z. \tag{12.16}$$

This illustrates that the equivalent input Miller impedance Z_{mi} is A_v times less than the feedback impedance Z, while the equivalent output Miller impedance Z_{mo} is almost the same as the feedback impedance Z. In other words, the variation of input impedance may be significant while the variation of output impedance is approximately equal to the feedback impedance Z. For example, if the feedback element is a capacitor C and its impedance is

$$Z = \frac{1}{j\omega C}, \tag{12.17}$$

then, from (12.15), the equivalent input, the Miller impedance is

$$Z_{mi} \approx \frac{Z}{|A_v|} = \frac{1}{j\omega C|A_v|}, \tag{12.18}$$

which corresponds to an equivalent input Miller capacitance C_{mi}, that is,

$$C_{mi} \approx |A_v|C. \tag{12.19}$$

And from (12.16) the equivalent input Miller impedance is

$$Z_{mo} \approx Z = \frac{1}{j\omega C}, \tag{12.20}$$

which corresponds to an equivalent output Miller capacitance C_{mo}, that is,

$$C_{mo} \approx C. \tag{12.21}$$

This implies that

The existence of a feedback capacitor *C* in a device is equivalent to the addition of a capacitor with A_v times the feedback capacitance at the input terminal in parallel plus a capacitor with feedback capacitance at the output terminal in parallel (A_v being the open-circuited voltage gain of the device). This equivalent relationship is called the Miller effect.

The Miller effect gives designers a serious warning that the feedback capacitor *C* may greatly degrade device performance because the input and output capacitances of the device could be significantly increased.

On the other hand, the Miller effect is very helpful to the circuit designer because the analysis of circuit performance is much easier if there is no feedback part. Through the Miller effect described above, the device model can be modified so as to approach an equivalent model with no feedback element, as expected.

However, it must be noted that the impedance *Z* of the feedback element is dependent on the device configuration. Therefore, the Miller effect and the equivalent variations of input and output impedance due to the existence of the feedback element are dependent on the device configuration as well. By examining all the configurations of the device model, the feedback element in the configurations of *CB*, *CG*, *CC*, and *CD* is negligible. The Miller effect mainly exists in *CE* and *CS* configurations of the device. Consequently in the following discussion, the analysis of the Miller effect will be conducted only for the configurations of *CE* and *CS*.

12.3 SMALL SIGNAL MODEL OF A BIPOLAR TRANSISTOR

Figure 12.3 shows the *DC* characteristics of a bipolar transistor and Figure 12.4 shows the *DC* characteristics of a bipolar transistor with early voltage.

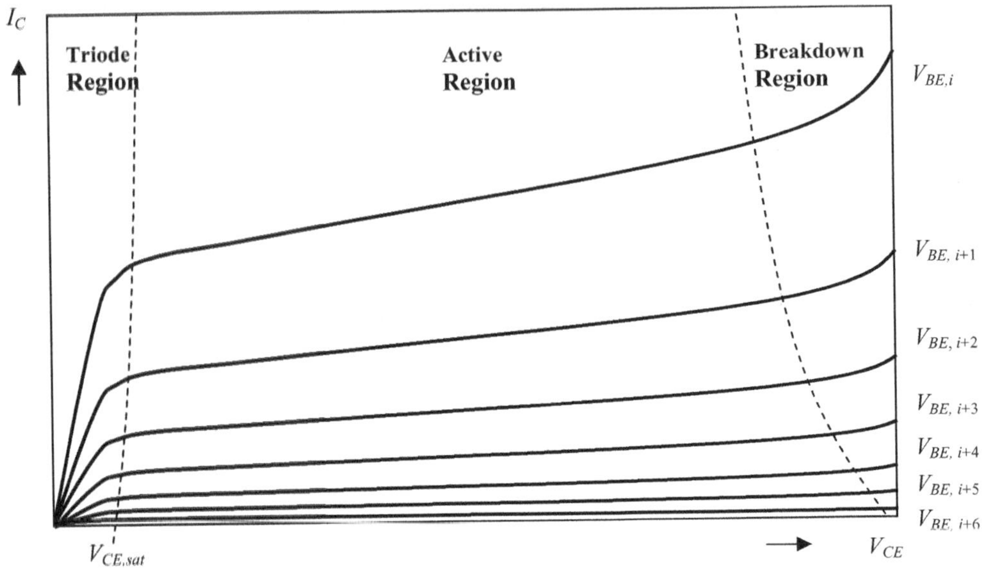

Figure 12.3 *DC* characteristics of a bipolar transistor (I_C vs. V_{CE}).

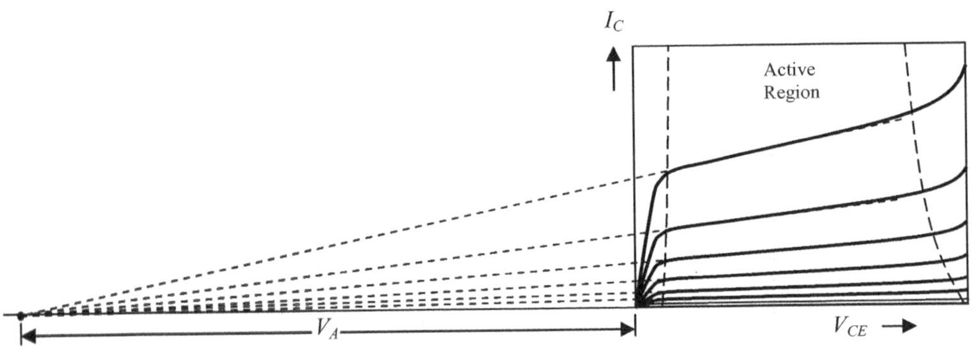

Figure 12.4 *DC* characteristics of a bipolar transistor (I_C vs. V_{CE}) and its early voltage.

The *DC* characteristics of a bipolar transistor can be expressed by

$$I_c = I_s\left(1+\frac{V_{ce}}{V_A}\right)\exp\left(\frac{V_{be}}{V_T}\right) = I_s\left(1+\frac{V_{ce}}{V_A}\right)\exp\left(\frac{qV_{be}}{kT}\right). \tag{12.22}$$

Its trans-conductance is the derivation of collector current I_c in respect to the voltage drop from base to emitter V_{be}, that is,

$$g_m = \frac{\partial I_c}{\partial V_{be}} = I_s\left(1+\frac{V_{ce}}{V_A}\right)\frac{q}{kT}\exp\left(\frac{qV_{be}}{kT}\right) = \frac{qI_c}{kT}, \tag{12.23}$$

$$C_\pi = C_b + C_{je}, \tag{12.24}$$

$$C_b = \tau_F g_m, \tag{12.25}$$

$$C_{je} = \frac{C_{jeo}}{\sqrt[3]{1 - \frac{V_D}{\psi_o}}}, \tag{12.26}$$

$$C_\mu = \frac{C_{\mu o}}{\left(1 - \frac{V_{cb}}{\psi_o}\right)^n}, \tag{12.27}$$

$$r_\pi = \frac{\beta_o}{g_m}, \tag{12.28}$$

$$r_o = \frac{V_A}{I_c}, \tag{12.29}$$

$$r_\mu = (1 \rightarrow 10)\beta_o r_o, \tag{12.30}$$

where

g_m = transconductance (= $38\,mA/V$ if $I_c = 1\,mA$ at room temperature),
I_c = total collector current,
I_s = collector current when $V_{be} = 0$,
q = charge of electron = 1.6×10^{-19} *Coulomb*,
k = Boltzmann constant = $8.617 \times 10^{-5} ev.deg^{-1}$,
T = absolute temperature = 300 °C (room temperature),
$kT/q \approx 26\,mV$ at room temperature,
V_{cb} = voltage drop from collector to base,
V_{be} = voltage drop from base to emitter,
V_{cb} = forward bias on the collector-base junction,
C_b = base-charging capacitance,
τ_F = base transit time in the forward direction,
C_{je} = emitter depletion-region capacitance,
C_{jeo} = emitter depletion-region capacitance when $V_D = 0$,
V_D = bias on the junction,
Ψ_o = built-in potential, a voltage across the junction when bias is zero,
C_μ = collector-base capacitance,
$C_{\mu o}$ = value of C_μ when $V_{cb} = 0$,
n = exponent between 0.1 to 0.5,
β_o = small-signal current gain,
V_A = early voltage,
r_π = input resistance,
r_o = output resistance,
r_μ = collector-base resistance.

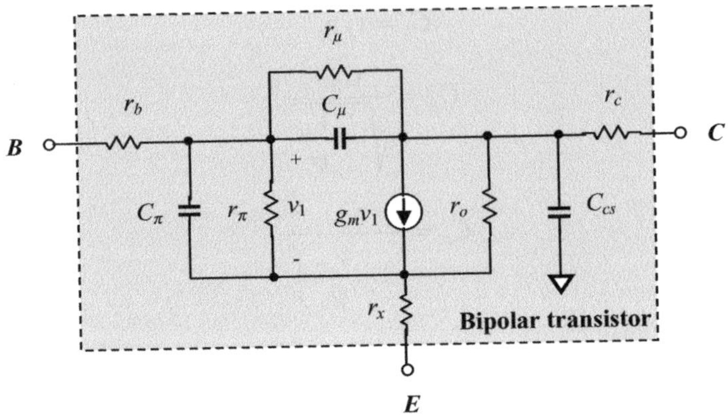

Figure 12.5 Small-signal equivalent model of a bipolar transistor.

Figure 12.5 shows a small-signal equivalent model of a bipolar transistor, also called the hybrid-π model, where

E = emitter of the bipolar transistor,
B = base of the bipolar transistor,
C = collector of the bipolar transistor,
r_b = contact resistance of base and the resistance between base and emitter (about a couple tens to a couple hundreds of Ω),
C_π = input capacitance, composed of the base-charging capacitance C_b and the emitter-base depletion layer capacitance C_{je} (about 0.1 to $2pF$),
r_π = input resistance (about a few tens of $k\Omega$),
C_μ = collector-base capacitance (very small and is usually only a few fF),
r_μ = collector-base resistance (has the highest value, usually in the range of tens of mega-Ω);
r_x = emitter lead resistance (only a few Ω and is neglected if the emitter current is not too high),
C_{cs} = collector-substrate capacitance (in the range of tens of fF),
r_o = output resistance (about tens of $k\Omega$),
r_c = collector resistance, composed of resistance from collector to buried layer, resistance with buried layer, and resistance from buried layer to base (about couple tens to couple hundreds of Ω).

In most circuit designs two resistors, r_μ and r_x, are always neglected because usually

- The value of resistor r_μ is much higher than for other resistors, that is,

$$r_\mu >> r_b, r_c, r_o, r_\pi, R_c. \tag{12.31}$$

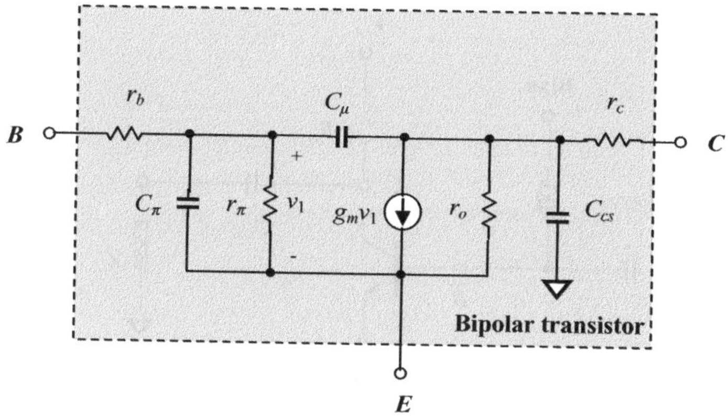

Figure 12.6 Simplified small-signal model of a bipolar transistor with negligence of r_μ and r_x.

- On the contrary, the value of resistor r_x is much lower than that of other resistors, that is,

$$r_x << r_b, r_c, r_o, r_\pi, R_c. \tag{12.32}$$

Then, the small signal model shown in Figure 12.5 is simplified to the one shown in Figure 12.6 and is applied to all the chapters in this book.

12.4 BIPOLAR TRANSISTOR WITH *CE* (COMMON EMITTER) CONFIGURATION

A bipolar transistor with *CE* configuration indicates that the emitter is a common reference point or *AC* grounded point of input and output terminals, which is plotted in Figure 12.7. In Figure 12.7(a), two "zero" capacitors, which approach zero impedance at the operating frequency, function as *DC* blocking capacitors and are symbolized by C_{zero}. The "infinite" inductor functions as an *RF* choke, which approaches infinite impedance at the operating frequencies and which is therefore symbolized as $L_{infinite}$. The collector resistor R_c leads the *DC* power to the collector.

Figure 12.7(b) is a small signal model of a *CE* device with an *AC* grounded emitter, in which all "zero" capacitors and the "infinite" inductor have been ignored.

12.4.1 Open-circuited Voltage Gain $A_{v,CE}$ of a *CE* Device

The open-circuited voltage gain, A_v, of a *CE* device can be analyzed on the basis of Figure 12.7(b), that is

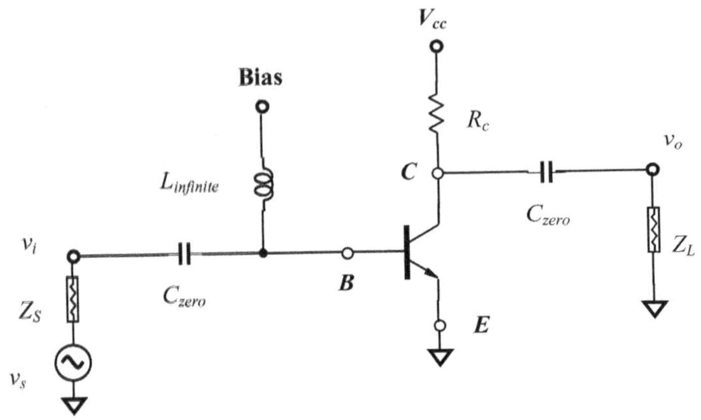

(a) Schematic of a bipolar transistor with *CE* configuration, C_{zero}: "zero" capacitor approach capacitance at operating frequencies, $L_{infinite}$: "infinite" inductor approaches infinite inductance at operating frequencies

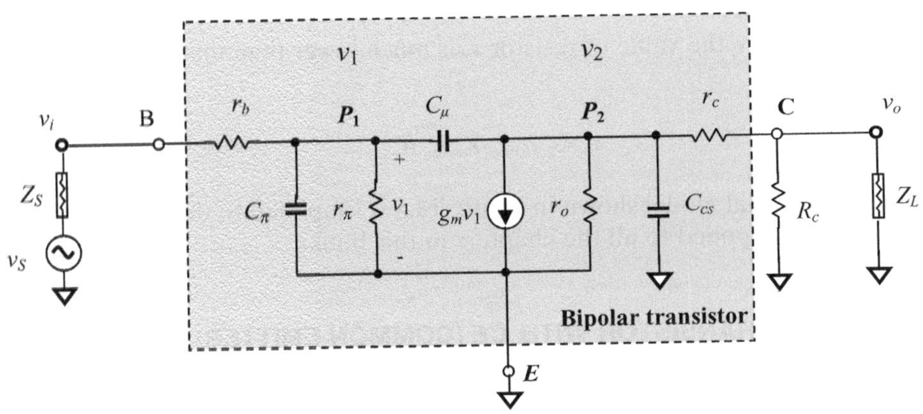

(b) Small signal model of a bipolar transistor with *CE* configuration

Figure 12.7 A bipolar transistor with *CE* configuration. Z_S: source impedance, Z_L: load impedance.

KCL at node *C*

$$\frac{v_2 - v_o}{r_c} - \frac{v_0}{R_c // Z_L} = 0, \tag{12.33}$$

KCL at node P_1

$$\frac{v_i - v_1}{r_b} = \frac{v_1 - v_2}{\dfrac{1}{j\omega C_\mu}} + \frac{v_1}{\dfrac{r_\pi}{1 + j\omega C_\pi r_\pi}}, \tag{12.34}$$

KCL at node P_2

$$g_m v_1 + \frac{v_2}{(r_c + R_c // Z_L) // r_o // C_{cs}} + \frac{v_2 - v_1}{\frac{1}{j\omega C_\mu}} = 0. \tag{12.35}$$

These three equations can be rewritten as

$$v_2 = \left(1 + \frac{r_c}{R_c // Z_L}\right) v_0, \tag{12.36}$$

$$v_i = \left[1 + \frac{r_b}{r_\pi} + j\omega(C_\mu + C_\pi) r_b\right] v_1 - j\omega C_\mu r_b v_2, \tag{12.37}$$

$$-(g_m - j\omega C_\mu) v_1 = \left[\frac{1}{(r_c + R_c // Z_L) // r_o // C_{cs}} + j\omega C_\mu\right] v_2. \tag{12.38}$$

Substituting v_2 in (12.36) into (12.38), we have

$$v_1 = \frac{1}{-(g_m - j\omega C_\mu)} \left[\frac{1}{(r_c + R_c // Z_L) // r_o // C_{cs}} + j\omega C_\mu\right] \left(1 + \frac{r_c}{R_c // Z_L}\right) v_0. \tag{12.39}$$

Replacing v_1 in (12.39) by (12.37) and (12.36),

$$v_i = \left\{ \frac{\left[1 + \frac{r_b}{r_\pi} + j\omega(C_\mu + C_\pi) r_b\right]}{-(g_m - j\omega C_\mu)} \left[\frac{1}{(r_c + R_c // Z_L) // r_o // C_{cs}} + j\omega C_\mu\right] - j\omega C_\mu r_b \right\} \left(1 + \frac{r_c}{R_c // Z_L}\right) v_o. \tag{12.40}$$

Finally, the open-circuited voltage gain of a *CE* device is

$$A_{v,CE} = \frac{v_o}{v_i} = -\frac{R_c // Z_L}{r_c + R_c // Z_L} \frac{(g_m - j\omega C_\mu)}{\left[1 + \frac{r_b}{r_\pi} + j\omega(C_\mu + C_\pi) r_b\right]\left[\frac{1}{(r_c + R_c // Z_L) // r_o // C_{cs}} + j\omega C_\mu\right] + j\omega C_\mu r_b (g_m - j\omega C_\mu)}, \tag{12.41}$$

$$A_{v,CE} = \frac{v_o}{v_i} = -\frac{R_c // Z_L}{r_c + R_c // Z_L} \frac{(g_m - j\omega C_\mu)}{\left[1 + \frac{r_b}{r_\pi} + j\omega(C_\mu + C_\pi) r_b\right]\left[\frac{1}{r_o} + j\omega(C_\mu + C_{cs}) + \frac{1}{r_c + R_c // Z_L}\right] + j\omega C_\mu r_b (g_m - j\omega C_\mu)}. \tag{12.42}$$

The open-circuited voltage gain $A_{v,CE}$ is in the case when

$$Z_L \to \infty, \tag{12.43}$$

then,

$$A_{v,CE} = -\frac{R_c}{r_c + R_c}\frac{(g_m - j\omega C_\mu)}{\left[1+\frac{r_b}{r_\pi}+j\omega(C_\mu + C_\pi)r_b\right]\left[\frac{1}{r_o}+j\omega(C_\mu + C_{cs})+\frac{1}{r_c+R_c}\right]+j\omega C_\mu r_b(g_m - j\omega C_\mu)}. \tag{12.44}$$

The expression (12.44) can be furthermore simplified by the following approximations:

- The resistor r_c will be neglected because

$$r_c << R_c. \tag{12.45}$$

- The resistor r_b is not neglected but its value is much smaller than the value of r_π, that is,

$$r_b << r_\pi. \tag{12.46}$$

- The value of capacitor C_π is much higher than the values of C_μ and C_{cs}, that is,

$$C_\pi >> C_\mu. \tag{12.47}$$

$$C_\pi >> C_{cs}. \tag{12.48}$$

Then, the open-circuited voltage gain becomes:

$$A_{v,CE} = \frac{v_o}{v_i} = -\frac{(g_m - j\omega C_\mu)}{(1+j\omega C_\pi r_b)\left[\frac{r_o + R_c}{r_o R_c}+j\omega(C_\mu + C_{cs})\right]+j\omega C_\mu r_b(g_m - j\omega C_\mu)}. \tag{12.49}$$

Mathematically, the denominator can be written with two poles, p_1 and p_2, that is

$$D(j\omega) = \left(1-\frac{j\omega}{p_1}\right)\left(1-\frac{j\omega}{p_2}\right) = 1 - j\omega\left(\frac{1}{p_1}+\frac{1}{p_2}\right) - \frac{\omega^2}{p_1 p_2} \approx 1 - \frac{j\omega}{p_1} - \frac{\omega^2}{p_1 p_2}, \tag{12.50}$$

if

$$|p_2| >> |p_1|, \tag{12.51}$$

which is usually the case. Then, it is easy to write the expressions for the two poles from (12.49), that is

$$p_1 \approx -\frac{1}{r_b}\frac{1}{C_\pi + C_\mu\left(1+g_m\dfrac{r_oR_c}{r_o+R_c}+\dfrac{1}{r_b}\dfrac{r_oR_c}{r_o+R_c}\right)+C_{cs}\dfrac{1}{r_b}\dfrac{r_oR_c}{r_o+R_c}}, \tag{12.52}$$

$$p_2 \approx -\frac{1}{(C_\mu+C_{cs})\dfrac{r_oR_c}{r_o+R_c}}-\frac{1}{\left(1+\dfrac{C_{cs}}{C_\mu}\right)C_\pi\dfrac{1}{g_m+\dfrac{1}{r_b}+\dfrac{r_o+R_c}{r_oR_c}}}-\frac{1}{\left(1+\dfrac{C_\mu}{C_{cs}}\right)C_\pi r_b}. \tag{12.53}$$

The first pole is the low-frequency pole and is dominant over the second pole. The magnitude of p_1 shows that the voltage gain is $3dB$ below its low-frequency value at a frequency

$$\omega_{-3dB} = |p_1|, \tag{12.54}$$

The expression (12.49) is a general expression of the open-circuited voltage gain for both low and high frequencies.

In low-frequency cases, the terms containing capacitors in the expression (12.49) could be omitted, in which the open-circuited voltage gain becomes

$$A_{v,CE}\big|_{\omega\to 0} = -g_m\frac{r_oR_c}{r_o+R_c}. \tag{12.55}$$

This is the low-frequency approximation of open-circuited voltage gain.

It should be noted that

- First, in some publications the voltage gain is defined as

$$A_{v,CE} = \frac{v_o}{v_S}, \tag{12.56}$$

  where v_S is the source voltage as shown in Figure 12.7(b). The expression (12.49) must be modified by multiplying by a factor of

$$\frac{(Z_S+r_b)//r_\pi}{Z_S+r_b} = \frac{r_\pi}{Z_S+r_b+r_\pi}, \tag{12.57}$$

  on its right side. In actual engineering circuit design, this factor is actually close to 1, so that the tolerance is acceptable.

  However, the input voltage to the device should be v_i and not v_S, as shown in Figure 12.7(b), because the definition of voltage gain examines only the gain contributed by the device itself and not the gain contributed by the source impedance.

- Second, in some publications, the terms containing C_{cs} in the derivations for $A_{v,CE}$ are neglected. This seems contradictory, since the value of C_{cs} is usually higher than the value of C_μ.

12.4.2 Short-circuited Current Gain β_{CE} and Frequency Response of a *CE* Device

The frequency response of a transistor is most often specified by the transition frequency, f_T, where the shorted-circuited current gain in a *CE* configuration falls to unity. The short-circuited current gain is the ratio of i_o to i_i when the output is short-circuited to the ground, so $v_o = 0$. The transition frequency, f_T, is a measure of the maximum useful frequency of the transistor when used as an amplifier.

Figure 12.8 is drawn for the calculation of short-circuited current gain and frequency response of a bipolar transistor with a *CE* configuration.

Note that the output node is short-circuited, $v_o = 0$. Then, we have

$$v_1 \approx \frac{r_\pi}{1 + j\omega r_\pi (C_\pi + C_\mu)} i_i. \tag{12.58}$$

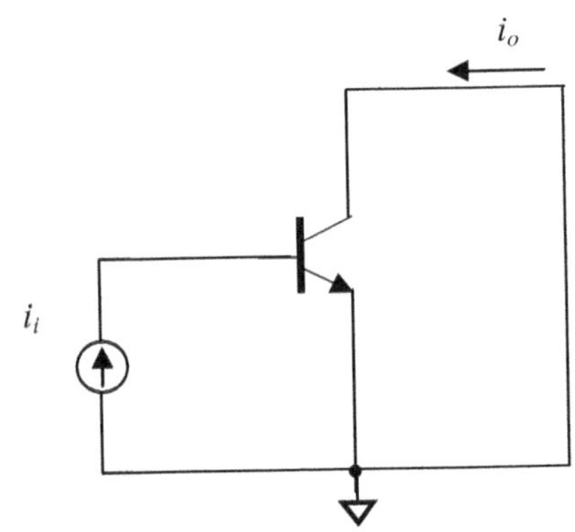

(a) Output is short-circuited for calculation of current gain β and frequency response

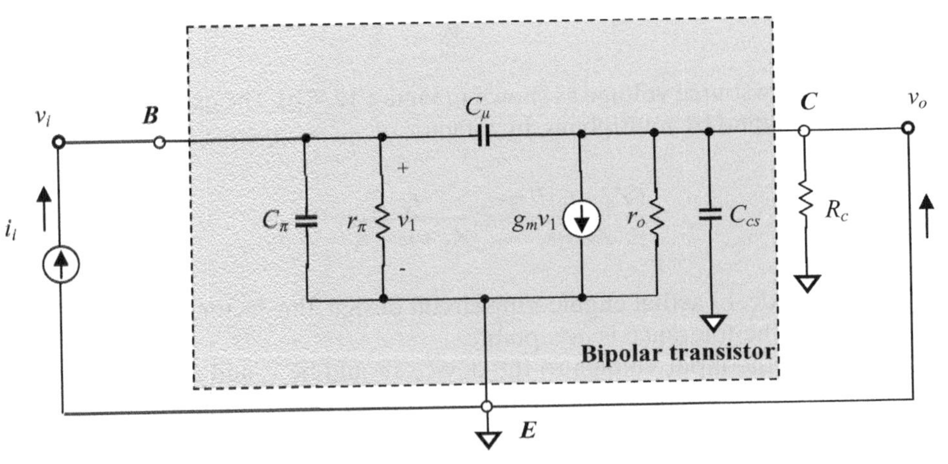

(b) Small signal model for calculation of current gain β_{CE} and frequency

Figure 12.8 Output short-circuited for calculation of current gain β_{CE} and frequency response of a bipolar with *CE* configuration.

And, if the current i_o flowing through C_μ is neglected,

$$i_o \approx g_m v_1 = \frac{g_m r_\pi}{1 + j\omega r_\pi (C_\pi + C_\mu)} i_i, \tag{12.59}$$

$$\beta_{CE}(j\omega) = \frac{i_o}{i_i} = \frac{\beta_o}{1 + j\beta_o \dfrac{(C_\pi + C_\mu)}{g_m}\omega}, \tag{12.60}$$

where $\beta_{CE}(j\omega)$ is the short-circuited current gain of *CE* device.

$\beta_o = g_m r_\pi$ is the current gain in low frequencies.

At high frequencies,

$$j\beta_o \frac{(C_\pi + C_\mu)}{g_m}\omega >> 1, \tag{12.61}$$

$$\beta_{CE}(j\omega) \approx \frac{g_m}{j(C_\pi + C_\mu)\omega}. \tag{12.62}$$

When the input current is equal to the output current in magnitude, that is,

$$|\beta_{CE}(j\omega)| = 1, \tag{12.63}$$

then from (12.62) it can be found that the corresponding frequency ω_T is

$$\omega_{T,CE} = \frac{g_m}{C_\pi + C_\mu}, \tag{12.64}$$

or,

$$f_{T,CE} = \frac{1}{2\pi}\frac{g_m}{C_\pi + C_\mu}. \tag{12.65}$$

$\omega_{T,CE}$ is called transition frequency of CE device.

On the other hand, when β_{CE} is $3\,dB$ less than the low-frequency value β_o, that is,

$$|\beta_{CE}(j\omega)| = \frac{\beta_o}{\sqrt{2}}, \tag{12.66}$$

then from (12.60) it can be found that the beta cutoff, or corner frequency, $\omega_{\beta,CE}$, is

$$\omega_{\beta,CE} = \frac{1}{\beta_o}\frac{g_m}{C_\pi + C_\mu} = \frac{\omega_{T,CE}}{\beta_o}, \tag{12.67}$$

or,

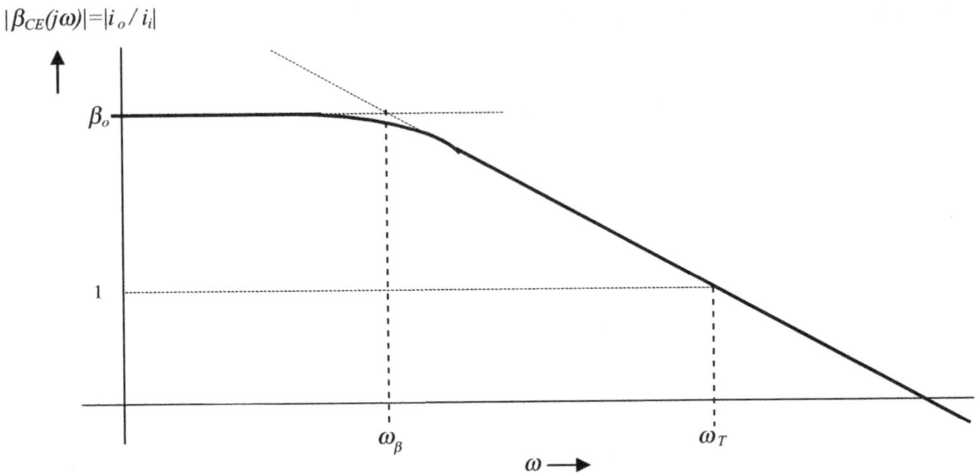

Figure 12.9 Frequency response of short-circuited current gain in a bipolar transistor with *CE* configuration.

$$f_{\beta,CE} = \frac{1}{2\pi\beta_o}\frac{g_m}{C_\pi + C_\mu} = \frac{f_{T,CE}}{\beta_o}. \tag{12.68}$$

At low frequencies, from (12.60)

$$\beta_{CE}(j\omega)|_{\omega\to 0} = \beta_o. \tag{12.69}$$

Figure 12.9 shows an example of the frequency response of a *CE* device.

12.4.3 Primary Input and Output Impedances of a *CE* Device

The primary input and output impedances of a *CE* device can be calculated based on Figure 12.10, that is,

The input impedance is

$$Z_{in} = r_b + c_\pi // r_\pi = r_b + \frac{r_\pi}{1 + j\omega C_\pi r_\pi} = \frac{r_b + r_\pi + j\omega C_\pi r_\pi r_b}{1 + j\omega C_\pi r_\pi}, \tag{12.70}$$

or

$$Z_{in} = \left[r_b + \frac{r_\pi}{(r_\pi C_\pi \omega)^2 + 1}\right] - j\frac{r_\pi^2 C_\pi \omega}{(r_\pi C_\pi \omega)^2 + 1}. \tag{12.71}$$

The output impedance is

$$Z_{out} = (r_c + r_o // C_{cs}) // R_c = \left(r_c + \frac{r_o}{1 + j\omega C_{cs} r_o}\right) // R_c, \tag{12.72}$$

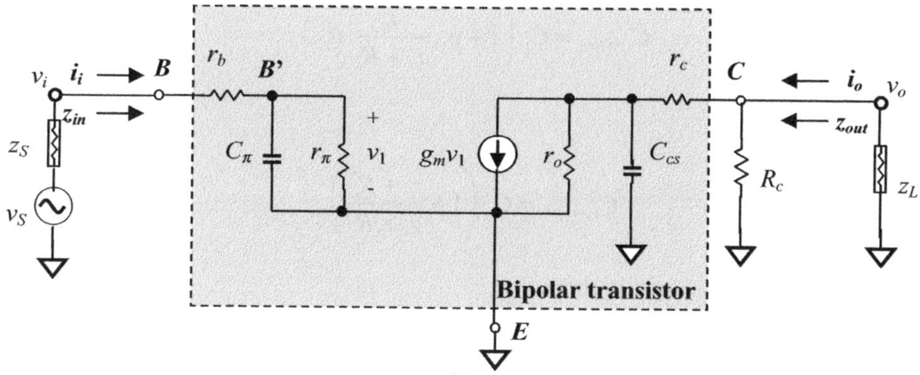

Figure 12.10 Simplified small signal model of a *CE* (common emitter) device for impedance calculation.

$$Z_{out} = \frac{r_c + r_o + jC_{cs}\omega r_c r_o}{R_c + r_c + r_o + jC_{cs}\omega r_o(r_c + R_c)} R_c. \tag{12.73}$$

It can be seen that the input impedance of a *CE* device is mainly determined by r_π and C_π and r_b is negligible because usually $r_b << r_\pi$, and that the output impedance of a *CE* device is mainly determined by r_o, R_c, and C_{cs} because usually $r_c << r_o$ and R_c.

In low frequencies, the capacitors C_π and C_{cs} are negligible. Therefore expressions (12.71) and (12.73) become

$$Z_{in}|_{\omega \to 0} = r_b + r_\pi, \tag{12.74}$$

and

$$Z_{out}|_{\omega \to 0} = (r_c + r_o) // R_c. \tag{12.75}$$

respectively.

12.4.4 Miller Effect on a *CE* Device

In order to simplify the discussion of the Miller effect, the value of the open-circuited voltage gain, $A_{v,CE}$, for the *CE* device is taken from the low-frequency cases, that is,

$$A_{v,CE} = \frac{v_o}{v_i} = -g_m \frac{r_o}{r_o + R_c} R_c. \tag{12.76}$$

This is, of course, a rough approximation.

Note that the feedback element in a *CE* device is the capacitor C_μ. Then, from expressions (12.19), (12.21), and (12.76), the input and output Miller capacitors can be found, such as

$$C_{i,miller} = C_\mu \left(1 + g_m \frac{r_o}{r_o + R_c} R_c\right), \tag{12.77}$$

and

$$C_{o,miller} = C_\mu \left(1 + \frac{r_o + R_c}{g_m r_o R_c}\right). \tag{12.78}$$

The total input capacitance is therefore

$$C_{in} = C_\pi + C_{i,miller} = C_\pi + C_\mu \left(1 + g_m \frac{r_o}{r_o + R_c} R_c\right), \tag{12.79}$$

And the total output capacitance is

$$C_{out} = C_{cs} + C_{o,miller} = C_{cs} + C_\mu \left(1 + \frac{r_o + R_c}{g_m r_o R_c}\right). \tag{12.80}$$

From now on, in the analysis of circuitry involving a *CE* device, it will be more convenient to replace the equivalent model of the *CE* device shown in Figure 12.7(b) by Figure 12.11, in which the feedback capacitor C_μ is contained in the input and output capacitors, C_{in} and C_{out}.

By means of the model modified with the Miller effect, shown in Figure 12.11, let's re-calculate the open-circuited voltage gain.

The output voltage is

$$v_o = -g_m v_1 R_c // (r_c + r_o // C_{out}). \tag{12.81}$$

Note that

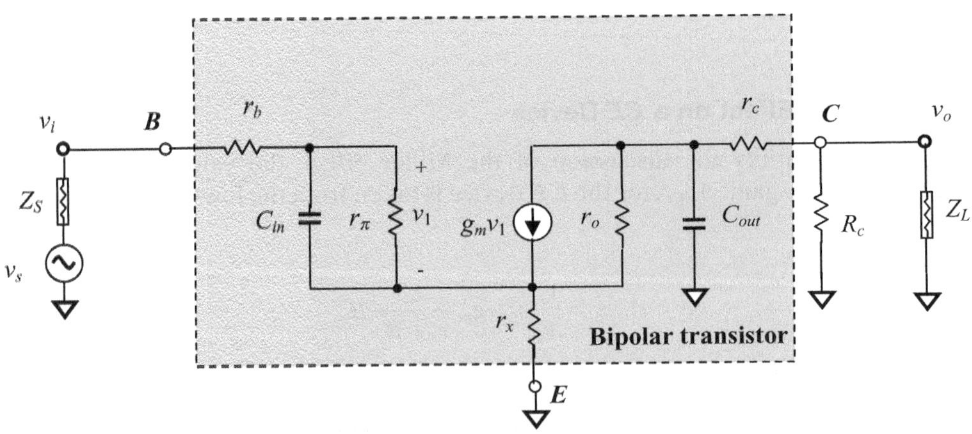

Figure 12.11 Model of bipolar transistor with *CE* configuration and with Miller capacitors included in C_{in} and C_{out}. Z_S: source impedance, Z_L: load impedance.

$$r_c + \frac{r_o}{1 + j\omega C_{out} r_o} = \frac{r_c + j\omega C_{out} r_c r_o + r_o}{1 + j\omega C_{out} r_o} \approx \frac{1 + j\omega C_{out} r_c}{1 + j\omega C_{out} r_o} r_o, \tag{12.82}$$

$$R_c // (r_c + r_o // C_{out}) = R_c // \left(\frac{1 + j\omega C_{out} r_c}{1 + j\omega C_{out} r_o} r_o \right) = \frac{R_c \dfrac{1 + j\omega C_{out} r_c}{1 + j\omega C_{out} r_o} r_o}{R_c + \dfrac{1 + j\omega C_{out} r_c}{1 + j\omega C_{out} r_o} r_o}, \tag{12.83}$$

$$R_c // (r_c + r_o // C_{out}) = \frac{R_c r_o (1 + j\omega C_{out} r_c)}{R_c (1 + j\omega C_{out} r_o) + r_o (1 + j\omega C_{out} r_c)} \approx \frac{R_c r_o}{R_c + r_o} \frac{1 + j\omega C_{out} r_c}{1 + j\omega C_{out} \dfrac{R_c r_o}{R_c + r_o}}. \tag{12.84}$$

If

$$r_c << R_c. \tag{12.85}$$

Substituting (12.84) into (12.81), we have

$$v_o = -g_m v_1 R_c // (r_c + r_o // C_{out}) \approx -g_m v_1 \frac{r_o R_c}{r_o + R_c} \frac{1 + j\omega C_{out} r_c}{1 + j\omega C_{out} \dfrac{r_o R_c}{r_o + R_c}}. \tag{12.86}$$

The voltage across the capacitor C_{in} is

$$v_1 = \frac{r_\pi // C_{in}}{r_b + r_\pi // C_{in}} v_i = \frac{1}{1 + j\omega C_{in} r_b} v_i, \tag{12.87}$$

if

$$r_b << r_\pi. \tag{12.88}$$

The open-circuited voltage gain of a *CE* device then becomes

$$A_{v,CE} = \frac{v_o}{v_i} = -g_m \frac{r_o R_c}{r_o + R_c} \frac{1}{1 + j\omega C_{in} r_b} \frac{1 + j\omega C_{out} r_c}{1 + j\omega C_{out} \dfrac{r_o R_c}{r_o + R_c}}, \tag{12.89}$$

It can be seen that the two poles exist in (12.89), they are

$$p_1 = -\frac{1}{C_{in} r_b} = -\frac{1}{r_b} \frac{1}{C_\pi + C_\mu \left(1 + g_m \dfrac{r_o}{r_o + R_c} R_c \right)}, \tag{12.90}$$

$$p_2 = -\frac{1}{C_{out} \dfrac{r_o R_c}{r_o + R_c}} = -\frac{1}{C_{cs} \dfrac{r_o R_c}{r_o + R_c} + C_\mu \left(\dfrac{r_o R_c}{r_o + R_c} + \dfrac{1}{g_m} \right)}. \tag{12.91}$$

By comparing (12.90) with (12.52), we see that the p_1 values are almost identical to each other. The difference is not too great. This implies that the Miller approximation is a good approach in circuit analysis.

By means of the equivalent model modified with the Miller effect as shown in Figure 12.11, the input and output impedances of a *CE* device can be modified from expressions (12.71) and (12.73) to the following:

$$Z_{in} = \frac{r_b + r_\pi + j\omega C_{in} r_\pi r_b}{1 + j\omega C_{in} r_\pi}, \tag{12.92}$$

$$Z_{out} = \frac{r_c + r_o + jC_{out}\omega r_c r_o}{R_c + r_c + r_o + jC_{out}\omega r_o(r_c + R_c)} R_c. \tag{12.93}$$

in which the capacitors C_π and C_{cs} in (12.71) and (12.73) are simply replaced by the capacitors C_{in} and C_{out}, respectively. In low-frequency cases the expressions (12.92) and (12.93) revert to the previous equations (12.74) and (12.75).

12.4.5 Emitter Degeneration

Figure 12.12 shows a bipolar device with a *CE* configuration and emitter degeneration. Its model is simplified by the additional negligence of C_μ.

Then, the equivalent model of the bipolar transistor with a CE configuration shown in Figure 12.12(b) is simplified to Figure 12.13.

KCL at node *E*

$$\frac{v_e}{R_e} + \frac{v_e + i_o(R_c // C_{cs})}{r_o} = g_m v_1 + i_b, \tag{12.94}$$

KCL at node *C*

$$i_o + \frac{v_e + i_o(R_c // C_{cs})}{r_o} = g_m v_1, \tag{12.95}$$

KVL at input loop

$$i_b = \frac{v_i - v_e}{r_\pi // C_\pi}, \tag{12.96}$$

$$v_1 = (r_\pi // C_\pi) i_b, \tag{12.97}$$

Solving (12.94) for i_o, then substituting into (12.95) and applying (12.97) for v_1, we have

$$v_e = \frac{R_e R_c}{R_c + R_e(1 + j\omega C_{cs} R_c) + r_o(1 + j\omega C_{cs} R_c)} \left[1 + \left(\frac{\beta_o}{1 + j\omega C_\pi r_\pi} + 1\right)(1 + j\omega C_{cs} R_c)\frac{r_o}{R_c}\right] i_b, \tag{12.98}$$

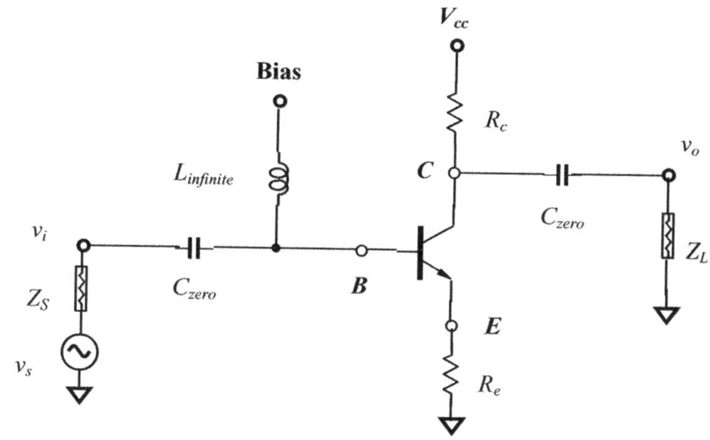

(a) Schematic of a bipolar transistor with *CE* configuration and emitter degeneration

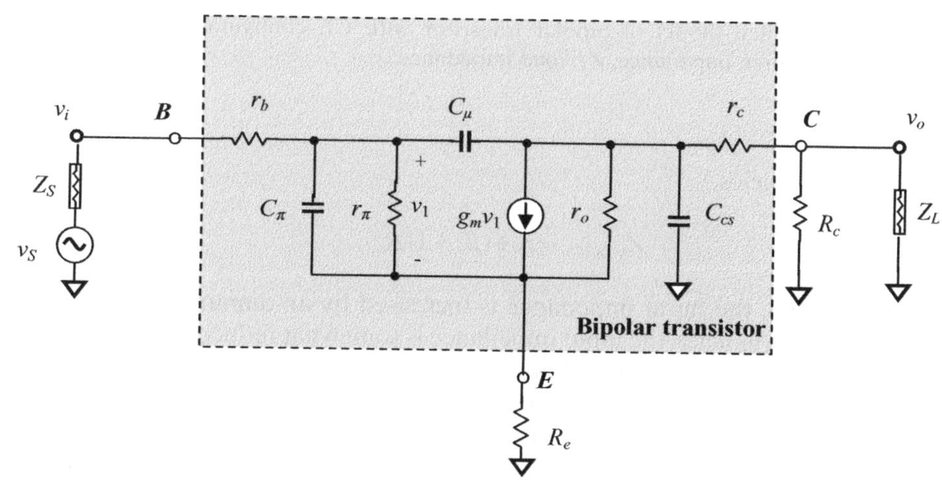

(b) Model of a bipolar transistor with *CE* configuration and emitter degeneration

Figure 12.12 A bipolar transistor with *CE* configuration and emitter degeneration. Z_S: source impedance, Z_L: load impedance, C_{zero}: "zero" capacitor, $L_{infinite}$: "infinite" inductor.

$$Z_{in} = \frac{v_i}{i_b} = \frac{1}{1 + j\omega C_\pi r_\pi}\left(r_\pi + \frac{r_o(1 + j\omega C_{cs}R_c)(\beta_o + 1 + j\omega C_\pi r_\pi) + R_c(1 + j\omega C_\pi r_\pi)}{R_c + (r_o + R_e)(1 + j\omega C_{cs}R_c)} R_e\right), \tag{12.99}$$

If

$$r_o >> R_c, \tag{12.100}$$

$$r_o >> R_e, \tag{12.101}$$

$$Z_{in} = \frac{v_i}{i_b} = \frac{1}{1 + j\omega C_\pi r_\pi}[r_\pi + (\beta_o + 1 + j\omega C_\pi r_\pi) R_e]. \tag{12.102}$$

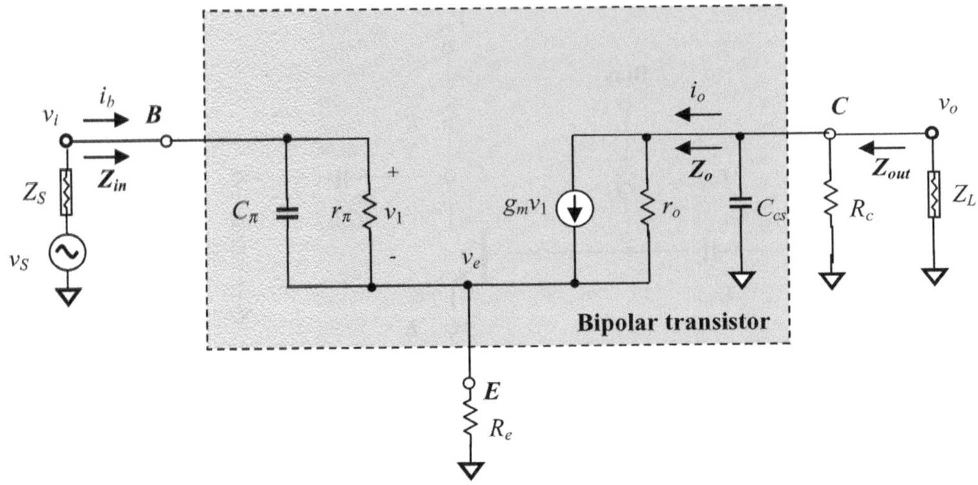

Figure 12.13 Simplified model of bipolar transistor with *CE* configuration and emitter degeneration. Z_S: source impedance, Z_L: load impedance.

And at low frequencies,

$$Z_{in}|_{\omega\to 0} \approx r_\pi + (\beta_o + 1) R_e. \tag{12.103}$$

At low frequencies, the input impedance is increased by an amount of $(\beta_o + 1)R_e$ from r_π. At high frequencies, the input impedance is somewhat reduced from (12.102) to (12.99) due to the existence of a finite r_o.

Let's calculate the trans-conductance $G_m = i_o/v_i$, with output short-circuited:

Substituting (12.98) into (12.94) with $R_c = 0$ and applying (12.97) for v_1, we have

$$v_e = \frac{\left(\dfrac{\beta_o}{1 + j\omega C_\pi r_\pi} + 1\right)}{\left(\dfrac{1}{R_e} + \dfrac{1}{r_o}\right) r_\pi \, // \, C_\pi + \left(\dfrac{\beta_o}{1 + j\omega C_\pi r_\pi} + 1\right)} v_i. \tag{12.104}$$

Substituting (12.98) and (12.104) into (12.97) with $R_c = 0$ and applying (12.97) for v_1, we have

$$G_m = \frac{i_o}{v_i} = g_m \frac{1 - \dfrac{R_e(1 + j\omega C_\pi r_\pi)}{\beta_o r_o}}{1 + g_m R_e \left[1 + \dfrac{(1 + j\omega C_\pi r_\pi)}{\beta_o} + \dfrac{1}{g_m r_o}\right]}. \tag{12.105}$$

Usually

$$\beta_o >> 1, \tag{12.106}$$

$$r_o >> R_e, \tag{12.107}$$

$$g_m r_o >> 1, \tag{12.108}$$

then

$$G_m \approx g_m \frac{1 - \dfrac{R_e j\omega C_\pi r_\pi}{\beta_o r_o}}{1 + g_m R_e \left[1 + \dfrac{j\omega C_\pi r_\pi}{\beta_o}\right]}. \tag{12.109}$$

At low frequencies,

$$G_m|_{\omega \to 0} \approx \frac{g_m}{1 + g_m R_e}. \tag{12.110}$$

From (12.105), (12.109), and (12.110) it can be seen that G_m drops from g_m due to the emitter resistor R_e. At low frequencies, G_m is dropped to half of g_m if $R_e = 1/g_m$.

The output impedance is calculated by the use of Figure 12.13 when input is short-circuited, that is

$$v_i = 0, \tag{12.111}$$

then

$$v_1 = -i_o(r_\pi // C_\pi // R_e). \tag{12.112}$$

The current flowing through r_o is

$$i_o|_{r_o} = i_o - g_m v_1 = i_o[1 + g_m(r_\pi // C_\pi // R_e)], \tag{12.113}$$

$$v_o = -v_1 + i_o|_{r_o} r_o. \tag{12.114}$$

Substituting (12.112) and (12.113) into (12.114), we have

$$v_o = i_o(r_\pi // C_\pi // R_e) + i_o[1 + g_m(r_\pi // C_\pi // R_e)]r_o, \tag{12.115}$$

$$Z_o = \frac{v_o}{i_o} = (r_\pi // C_\pi // R_e) + [1 + g_m(r_\pi // C_\pi // R_e)]r_o = r_o + (1 + g_m r_o)(r_\pi // C_\pi // R_e), \tag{12.116}$$

$$Z_{out} = Z_o // C_{cs} // R_c = [r_o + (1 + g_m r_o)(r_\pi // C_\pi // R_e)] // C_{cs} // R_c, \tag{12.117}$$

$$Z_{out} = r_o \frac{1 + \left(\dfrac{1}{r_o} + g_m\right)\dfrac{r_\pi R_e}{r_\pi + R_e(1 + j\omega C_\pi r_\pi)}}{1 + \dfrac{r_o}{R_c}(1 + j\omega C_{cs} R_c)\left[1 + \left(\dfrac{1}{r_o} + g_m\right)\dfrac{r_\pi R_e}{r_\pi + R_e(1 + j\omega C_\pi r_\pi)}\right]}. \tag{12.118}$$

Usually,

$$r_o >> 1, \tag{12.119}$$

then

$$Z_{out} \approx r_o \frac{1+\dfrac{\beta_o R_e}{r_\pi + R_e(1+j\omega C_\pi r_\pi)}}{1+\dfrac{r_o}{R_c}(1+j\omega C_{cs} R_c)\left[1+\dfrac{\beta_o R_e}{r_\pi + R_e(1+j\omega C_\pi r_\pi)}\right]}. \tag{12.120}$$

In the cases of infinite R_c, that is

$$R_c \to 0, \tag{12.121}$$

$$Z_{out}|_{R_c\to\infty} \approx r_o \frac{1+\dfrac{\beta_o R_e}{r_\pi + R_e(1+j\omega C_\pi r_\pi)}}{1+j\omega C_{cs} r_o\left[1+\dfrac{\beta_o R_e}{r_\pi + R_e(1+j\omega C_\pi r_\pi)}\right]}. \tag{12.122}$$

At low frequencies,

$$Z_{out}|_{\omega\to 0} \approx r_o \frac{1+\dfrac{\beta_o R_e}{r_\pi + R_e}}{1+\dfrac{r_o}{R_c}\left[1+\dfrac{\beta_o R_e}{r_\pi + R_e}\right]}. \tag{12.123}$$

In the cases of infinite R_c and low frequencies,

$$Z_{out}|_{\omega\to 0, R_c\to\infty} \approx r_o \frac{1+(\beta_o+1)\dfrac{R_e}{r_\pi}}{1+\dfrac{R_e}{r_\pi}}. \tag{12.124}$$

And usually if

$$r_\pi >> R_e, \tag{12.125}$$

$$Z_{out}|_{\omega\to 0, R_c\to\infty, r_\pi >> R_e} \approx r_o(1+g_m R_e). \tag{12.126}$$

12.5 BIPOLAR TRANSISTOR WITH *CB* (COMMON BASE) CONFIGURATION

A bipolar transistor with a *CB* configuration can be drawn as Figure 12.14. In Figure 12.14(a), two "zero" capacitors at the input and output, which approach zero impedance at the operating frequency, function as *DC* blocking capacitors and are

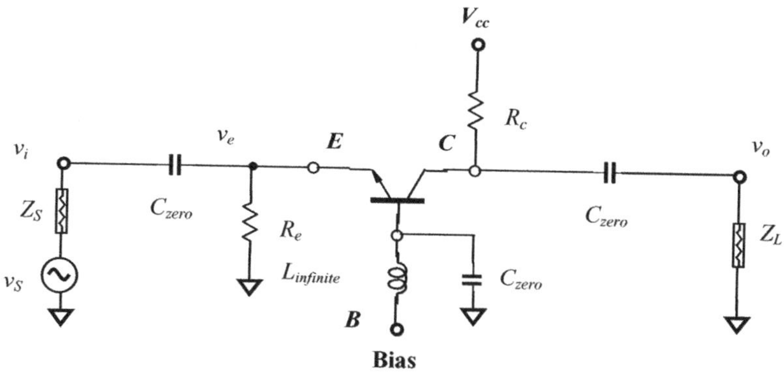

(a) Schematic of a bipolar transistor with *CB* configuration. C_{zero}: "zero" capacitor approaches z capacitance at operating frequencies, $L_{infinite}$: "infinite" inductor approaches infinite inductance at frequencies

CE configuration

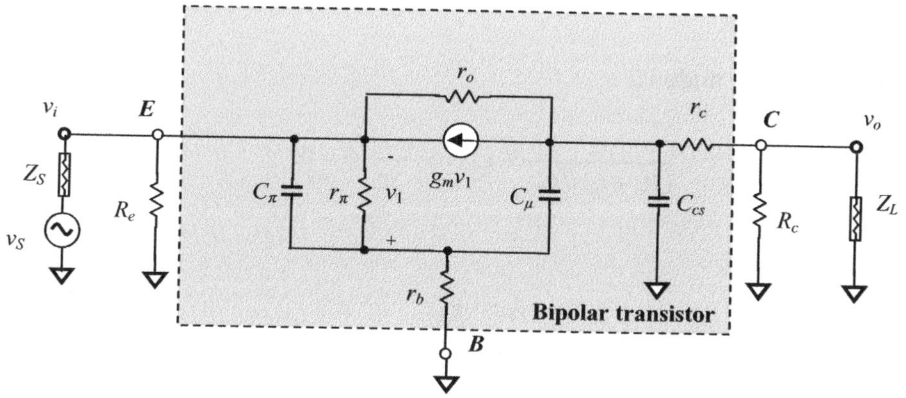

(b) Model of a bipolar transistor with *CB* configuration

Figure 12.14 Equivalent circuit of bipolar transistor with *CB* configuration. Z_S: source impedance, Z_L: load impedance.

symbolized as C_{zero}. Another "zero" capacitor, C_{zero}, is connected between the base and the ground so that the base is *AC* grounded. The "infinite" inductor functions as an *RF* choke, which approaches infinite impedance at the operating frequencies and is symbolized as $L_{infinite}$. The collector resistor R_c leads the *DC* power to the collector. The emitter resistor R_e leads the *DC* power to the ground.

12.5.1 Open-circuited Voltage Gain $A_{v,CB}$ of a *CB* Device

In the calculation of open-circuit voltage of a *CB* transistor, Figure 12.14 is simplified to Figure 12.15 by an additional approximation: the capacitor C_μ is neglected because

$$C_\mu << C_{cs}, C_\pi. \qquad (12.127)$$

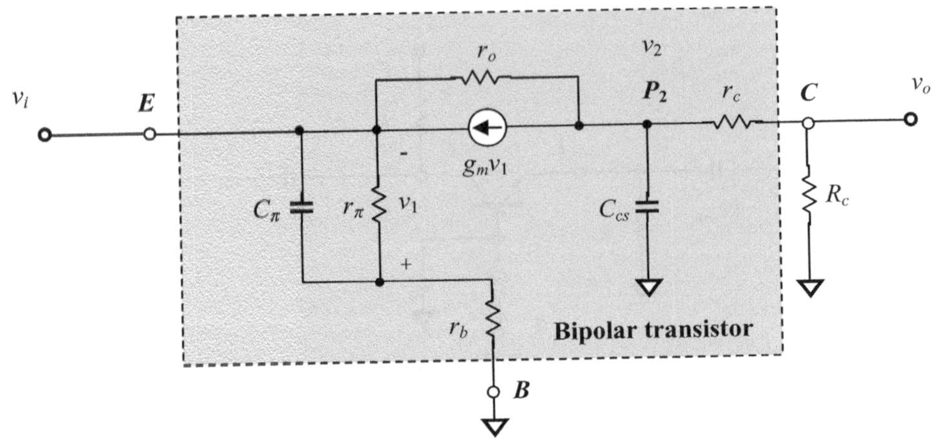

Figure 12.15 Equivalent circuit for calculation of open-circuited voltage gain of a *CB* device. Z_S: source impedance, Z_L: load impedance.

By applying *KCL* at node v_o,

$$\frac{v_2}{(R_c + r_c)//C_{cs}} + g_m v_1 + \frac{v_o - v_i}{r_o} = 0. \tag{12.128}$$

Note that

$$v_i = -v_1, \tag{12.129}$$

$$v_2 = \frac{R_c + r_c}{R_c} v_o. \tag{12.130}$$

By replacing v_1 by v_i in (12.128), from (12.129), and v_2 by v_o, from (12.130), the expression (12.128) becomes

$$\frac{\frac{R_c + r_c}{R_c} v_o}{\frac{R_c + r_c}{1 + j\omega C_{cs}(R_c + r_c)}} - g_m v_i + \frac{v_o - v_i}{r_o} = 0, \tag{12.131}$$

$$\left(\frac{1 + j\omega C_{cs}(R_c + r_c)}{R_c} + \frac{1}{r_o}\right) v_o = \left(g_m + \frac{1}{r_o}\right) v_i, \tag{12.132}$$

$$A_{v,CB} = \frac{v_o}{v_i} = \frac{g_m R_c \left(1 + \frac{1}{g_m r_o}\right)}{1 + j\omega C_{cs}(R_c + r_c) + \frac{R_c}{r_o}}. \tag{12.133}$$

Finally,

$$A_{v,CB} = \frac{r_o}{R_c + r_o} \frac{g_m R_c \left(1 + \frac{1}{g_m r_o}\right)}{1 + j\omega C_{cs} \frac{(R_c + r_c)}{(R_c + r_o)} r_o}. \tag{12.134}$$

At low frequencies,

$$A_{v,CB}\big|_{\omega \to 0} = g_m \left(1 + \frac{1}{g_m r_o}\right) \frac{R_c r_o}{R_c + r_o}. \tag{12.135}$$

If

$$r_o >> R_c, \tag{12.136}$$

$$g_m r_o >> 1, \tag{12.137}$$

then

$$A_{v,CB}\big|_{\omega \to 0} \approx g_m R_c. \tag{12.138}$$

12.5.2 Short-circuited Current Gain β_{CB} and Frequency Response of a *CB* Device

Figure 12.16 is drawn for the calculation of the short-circuited current gain and frequency response of a bipolar transistor with a *CB* configuration, in which the following additional approximations have been made: the resistors, r_b, r_c, are neglected because their values are much lower than the value of r_π, that is,

$$r_b, r_c << r_\pi. \tag{12.139}$$

At node *C*,

$$i_o = -g_m v_1. \tag{12.140}$$

At node *E*,

$$i_i + g_m v_1 + \frac{v_1}{C_\pi // r_\pi} = 0, \tag{12.141}$$

$$i_i = -\left(g_m + \frac{1}{r_\pi} + j\omega C_\pi\right) v_1. \tag{12.142}$$

From (12.140) and (12.142), we have

$$\beta_{CB}(j\omega) = \frac{i_o}{i_i} = \frac{g_m r_\pi}{g_m r_\pi + 1} \frac{1}{1 + j\omega C_\pi \frac{r_\pi}{g_m r_\pi + 1}}. \tag{12.143}$$

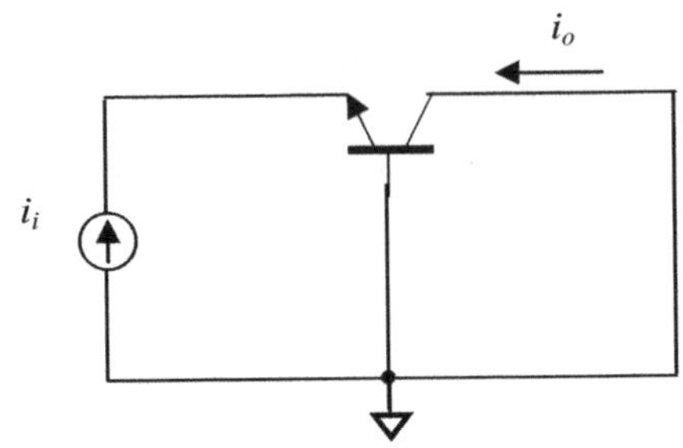

(a) Output is short-circuited for calculation of current gain β of a CB device

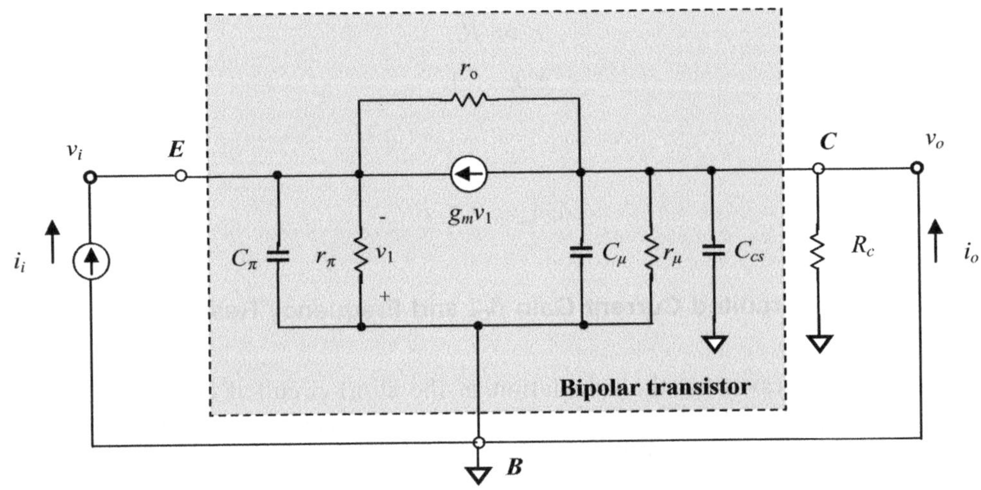

(b) Calculation of current gain β and frequency response for a CB device

Figure 12.16 Output short-circuited for calculation of current gain β and frequency response of a bipolar with CB configuration.

Note that in a CE device in low frequencies as shown in (12.69),

$$\beta_{CE}(j\omega)|_{\omega\to 0} = \beta_o = g_m r_\pi = \frac{\alpha_o}{1-\alpha_o}, \tag{12.144}$$

where

α_o = base transport factor,
β_o = small signal current gain of a CE device in low frequencies.

then

$$\beta_{CB}(j\omega) = \frac{\beta_o}{\beta_o+1}\frac{1}{1+j\omega\dfrac{C_\pi}{g_m}\dfrac{\beta_o}{\beta_o+1}} = \frac{\alpha_o}{1+j\omega\dfrac{C_\pi}{g_m}\alpha_o}. \tag{12.145}$$

It can be seen that the current gain of a *CB* device is less than 1, that is

$$|\beta_{CB}(j\omega)| < 1. \quad (12.146)$$

At low frequencies,

$$\beta_{CB}(j\omega)|_{\omega\to 0} \approx \alpha_o. \quad (12.147)$$

The short-circuited current gain of a *CB* device in low frequencies is close to but less than 1.

Rewriting (12.145) in terms of (12.147), when $\beta_{CB}(j\omega)$ is $3\,dB$ down from the low-frequency value $\beta_{CB}(j\omega)|_{\omega\to 0}$ that is,

$$|\beta_{CB}(j\omega)| = \frac{\beta_{CB}(j\omega)|_{\omega\to 0}}{\sqrt{2}} = \frac{\alpha_o}{\sqrt{2}}, \quad (12.148)$$

$$\omega = \omega_{\beta,CB} = \frac{g_m}{\alpha_o C_\pi}, \quad (12.149)$$

or,

$$f = f_{\beta,CB} = \frac{1}{2\pi}\frac{g_m}{\alpha C_\pi}. \quad (12.150)$$

In expressions (12.149) or (12.150), ω_β or f_β is determined by C_π only. This means that **$\omega_{\beta,CB}$ or $f_{\beta,CB}$ in a *CB* device is much higher than that in a *CE* device** as shown in expression (12.67) or (12.68), which is reproduced below:

$$\omega_{\beta,CE} = \frac{1}{\beta_o}\frac{g_m}{C_\pi + C_\mu} = \frac{\omega_T}{\beta_o}, \quad (12.67)$$

or,

$$f_{\beta,CE} = \frac{1}{2\pi\beta_o}\frac{g_m}{C_\pi + C_\mu} = \frac{f_T}{\beta_o}. \quad (12.68)$$

It must be noted that it is impossible to define a transition frequency $\omega_{T,CB}$ like that in expressions (12.64) to (12.65) for the case of a *CE* device, because the current gain of a *CB* device is always less than 1 as shown in (12.146). Reluctantly, we may have a special alternate definition of $\omega_{T,CB}$ for the transition frequency of a *CB* device, such as

$$|\beta_{CB}(j\omega)| = 0.5. \quad (12.151)$$

for which an international agreement must be reached before it could be applied to engineering design.

Figure 12.17 shows an example of the frequency response of a *CB* device.

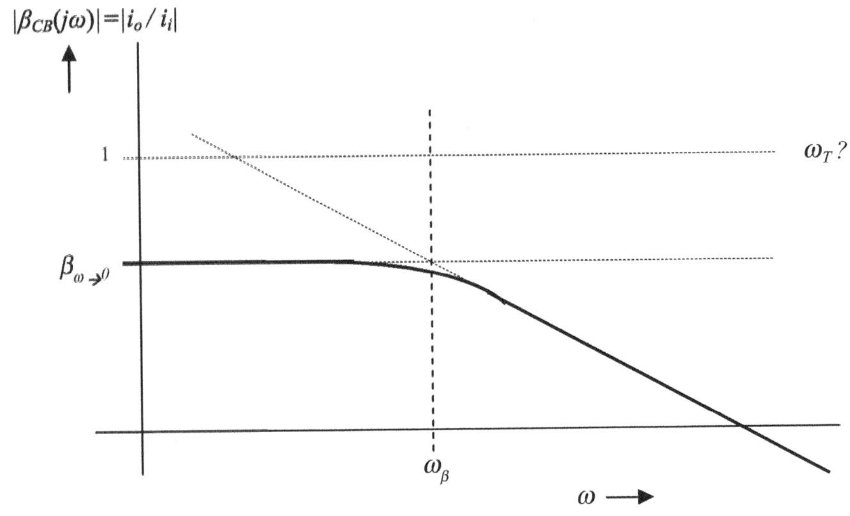

Figure 12.17 Frequency response of short-circuited current gain in a bipolar transistor with *CB* configuration.

12.5.3 Input and Output Impedances of a *CB* Device

Figure 12.14(b) is an equivalent circuit of a *CB* device in terms of its small signal hybrid-π model, in which all the "zero" capacitors and the "infinite" inductor are ignored. As a matter of fact, it is a copy of Figure 12.7(b), but with the positions of the emitter and the base exchanged, in which the emitter is at the input position which usually is drawn in the left side of the figure while the output is drawn in the right side of the figure.

From 12.14(b) it is found that the circuit analysis would be somewhat reluctant because the dependent current source is located between the collector and the emitter, which can be replaced by two identical current sources, one from the collector to the base and then another from the base to the emitter as shown in Figure 12.18. The reason for this is that the currents fed into and drawn out of the base are equal. The dependent current source from the base to the emitter is connected with the input resistor r_π and C_π in parallel and forms a special combined part, in which the current of the dependent current source depends on a voltage drop between two terminals of this special combined part.

Now we are going to apply Ohm's Law to this special combined part. By applying Ohm's Law, this dependent current source can be replaced by a resistor with a value of $1/g_m$, which is connected with either r_π or C_π in parallel. This resistor is called the emitter resistor R_e, that is,

$$r_e = r_\pi // \frac{1}{g_m} = \frac{1}{g_m + \dfrac{1}{r_\pi}} = \frac{\alpha_o}{g_m}, \tag{12.152}$$

$$\alpha_o = \frac{1}{1 + \dfrac{1}{\beta_o}} = \frac{\beta_o}{\beta_o + 1}, \tag{12.153}$$

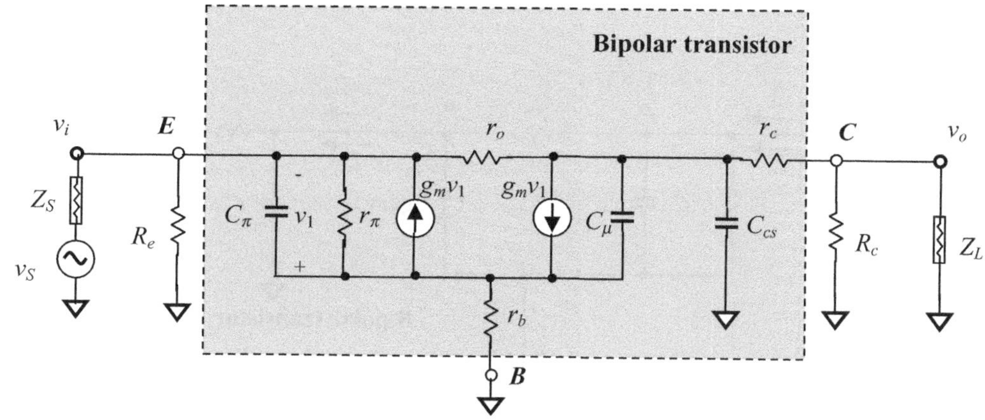

Figure 12.18 Equivalent circuit of bipolar transistor with *CB* configuration.

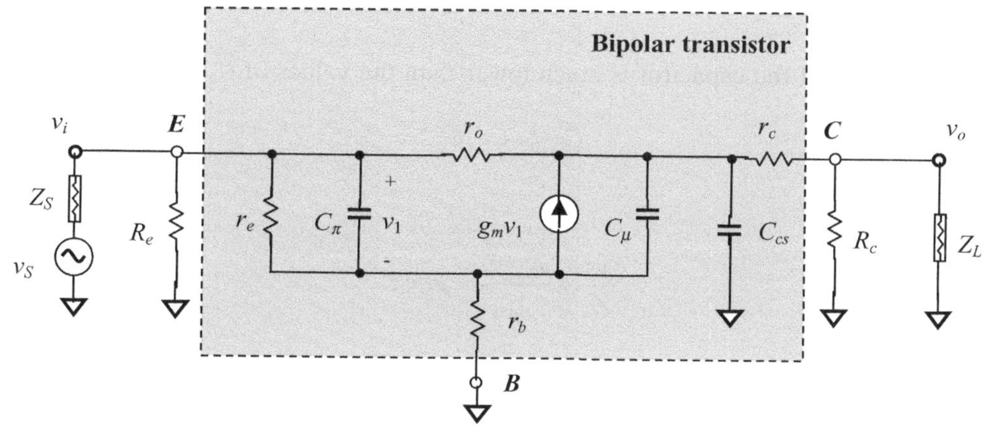

Figure 12.19 Equivalent circuit for calculation of input and output impedance of a *CB* device.

where

α_o = base transport factor,
β_o = small signal current gain in low frequencies.

Consequently, Figure 12.18 can be redrawn as Figure 12.19. It should be noted that the direction of the dependent current source in Figure 12.19 has been changed to upward from downward in Figure 12.15, while the direction of dependent voltage v_1 is correspondingly upside down. The following analysis of the *CB* device will be conducted on the basis of Figure 12.19.

In the calculation of input impedance of the *CB* transistor, Figure 12.19 is simplified to Figure 12.20 by the additional approximations: the C_μ and r_b are neglected because the value of resistor r_b is much lower than that of other resistors, that is,

$$r_b << r_\pi, r_o. \tag{12.154}$$

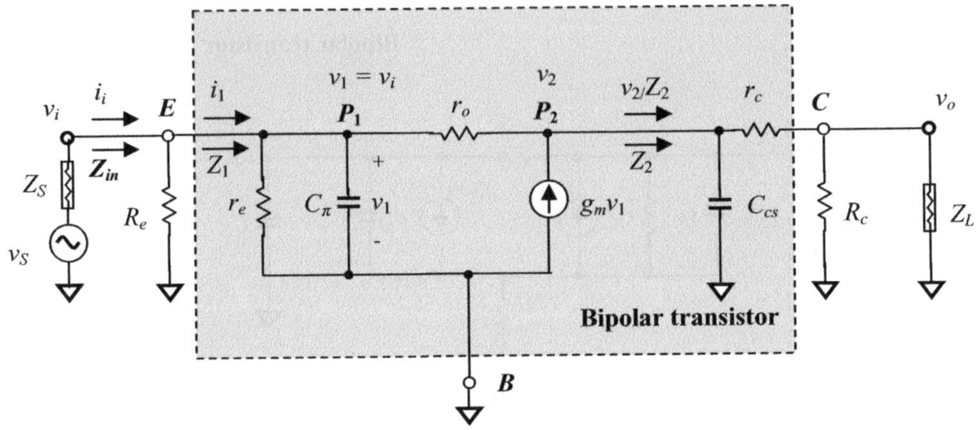

Figure 12.20 Equivalent circuit for calculation of input and impedance of a *CB* device.

and the value of the capacitor is much lower than the values of C_π and C_{cs}, that is,

$$C_\mu << C_{cs}, C_\pi. \tag{12.155}$$

KCL at node P_2

$$\frac{v_2}{Z_2} + \frac{v_2 - v_1}{r_o} = g_m v_1. \tag{12.156}$$

And *KCL* at node P_1

$$i_1 = \frac{v_1}{r_e // C_\pi} + \frac{v_1 - v_2}{r_o}. \tag{12.157}$$

Solving v_2 from (12.156) and substituting it into (12.157) gives

$$v_2 = \frac{\left(g_m + \frac{1}{r_o}\right)}{\left(\frac{1}{Z_2} + \frac{1}{r_o}\right)} v_1, \tag{12.158}$$

$$\frac{i_1}{v_1} = \frac{1}{r_e // C_\pi} + \frac{1}{r_o}\left(1 - \frac{g_m + \frac{1}{r_o}}{\frac{1}{Z_2} + \frac{1}{r_o}}\right), \tag{12.159}$$

$$Z_1 = \frac{v_1}{i_1} = \frac{r_o + Z_2}{1 - g_m Z_2 + \frac{r_o + Z_2}{r_e // C_\pi}} = \frac{(r_e // C_\pi)(r_o + Z_2)}{r_e // C_\pi + r_o + [1 - g_m (r_e // C_\pi)] Z_2}, \tag{12.160}$$

where

$$Z_2 = C_{cs} // (r_c + R_c) = \frac{r_c + R_c}{1 + j\omega C_{cs}(r_c + R_c)}, \tag{12.161}$$

$$v_1 = v_i, \tag{12.162}$$

$$v_2 = \frac{r_c + R_c}{R_c} v_o, \tag{12.163}$$

$$r_e = \frac{\alpha_o}{g_m}, \tag{12.164}$$

$$r_e // C_\pi = \frac{\alpha_o}{g_m} // C_\pi = \frac{\alpha_o}{g_m + j\omega C_\pi \alpha_o} = \frac{1}{\dfrac{\beta_o + 1}{r_\pi} + j\omega C_\pi}. \tag{12.165}$$

Substituting $r_e // C_\pi$ from (12.165) into (12.160), we have

$$Z_1 = \frac{\dfrac{1}{\dfrac{\beta_o + 1}{r_\pi} + j\omega C_\pi}(r_o + Z_2)}{\dfrac{1}{\dfrac{\beta_o + 1}{r_\pi} + j\omega C_\pi} + r_o + \left(1 - \dfrac{g_m}{\dfrac{\beta_o + 1}{r_\pi} + j\omega C_\pi}\right) Z_2}$$
$$= \frac{r_o + Z_2}{1 + \left(\dfrac{\beta_o + 1}{r_\pi} + j\omega C_\pi\right) r_o + \left(\dfrac{\beta_o + 1}{r_\pi} + j\omega C_\pi - g_m\right) Z_2}, \tag{12.166}$$

$$Z_1 = \frac{r_o + Z_2}{1 + (1 + \beta_o)\dfrac{r_o}{r_\pi} + j\omega C_\pi r_o + \left(\dfrac{1}{r_\pi} + j\omega C_\pi\right) Z_2}$$
$$= \frac{r_\pi}{1 + \beta_o} \frac{1 + \dfrac{Z_2}{r_o}}{1 + \dfrac{r_\pi}{1 + \beta_o}\left[\dfrac{1}{r_o} + j\omega C_\pi + \left(\dfrac{1}{r_\pi} + j\omega C_\pi\right)\dfrac{Z_2}{r_o}\right]}. \tag{12.167}$$

Substituting Z_2 from (12.161) into (12.167), we have

$$Z_1 = \frac{r_\pi}{1 + \beta_o} \frac{1 + \dfrac{1}{r_o} \dfrac{r_c + R_c}{1 + j\omega C_{cs}(r_c + R_c)}}{1 + \dfrac{r_\pi}{1 + \beta_o}\left[\dfrac{1}{r_o} + j\omega C_\pi + \dfrac{1}{r_o}\left(\dfrac{1}{r_\pi} + j\omega C_\pi\right)\dfrac{r_c + R_c}{1 + j\omega C_{cs}(r_c + R_c)}\right]}. \tag{12.168}$$

At low frequencies,

$$Z_1|_{\omega\to 0} = \frac{r_\pi}{1+\beta_o} \frac{1+\frac{r_c+R_c}{r_o}}{1+\frac{(r_\pi+r_c+R_c)}{r_o(1+\beta_o)}}, \tag{12.169}$$

$$Z_1|_{\omega\to 0} \approx \frac{r_\pi}{1+\beta_o}\left(1+\frac{r_c+R_c}{r_o}\right), \tag{12.170}$$

if

$$r_o(1+\beta_o) >> (r_\pi + r_c + R_c). \tag{12.171}$$

Finally, the input impedance of the *CB* device is

$$Z_{in} = \frac{v_i}{i_i} = R_e // Z_1 = R_e // \frac{r_\pi}{1+\beta_o} \frac{1+\frac{1}{r_o}\frac{r_c+R_c}{1+j\omega C_{cs}(r_c+R_c)}}{1+\frac{r_\pi}{1+\beta_o}\left[\frac{1}{r_o}+j\omega C_\pi+\frac{1}{r_o}\left(\frac{1}{r_\pi}+j\omega C_\pi\right)\frac{r_c+R_c}{1+j\omega C_{cs}(r_c+R_c)}\right]}, \tag{12.172}$$

It can be seen that the input impedance of a *CB* device is about β_o times lower than r_π. This is one of the special features of the *CB* device.

It should be noted that in Figures 12.18 to 12.20, the resistor R_e might not be needed in some applications, while the resistor R_c is an indispensable part because it leads the *DC* power supply to the device. For instance, in a *CE-CB* cascode amplifier, the implementation of the resistor R_e is unnecessary. In such a case, we have

$$Z_{in}|_{Re\to\infty} = Z_1 = \frac{r_\pi}{1+\beta_o} \frac{1+\frac{1}{r_o}\frac{r_c+R_c}{1+j\omega C_{cs}(r_c+R_c)}}{1+\frac{r_\pi}{1+\beta_o}\left[\frac{1}{r_o}+j\omega C_\pi+\frac{1}{r_o}\left(\frac{1}{r_\pi}+j\omega C_\pi\right)\frac{r_c+R_c}{1+j\omega C_{cs}(r_c+R_c)}\right]}, \tag{12.173}$$

which corresponds to the case of

$$R_e \to \infty, \tag{12.174}$$

In the cases of low frequencies,

$$Z_{in}|_{\omega\to 0, Re\to\infty} = r_\pi \frac{r_o+r_c+R_c}{(1+\beta_o)r_o+r_\pi+r_c+R_c} = r_\pi \frac{1}{1+(1+g_m r_o)\frac{r_\pi}{r_o+r_c+R_c}}. \tag{12.175}$$

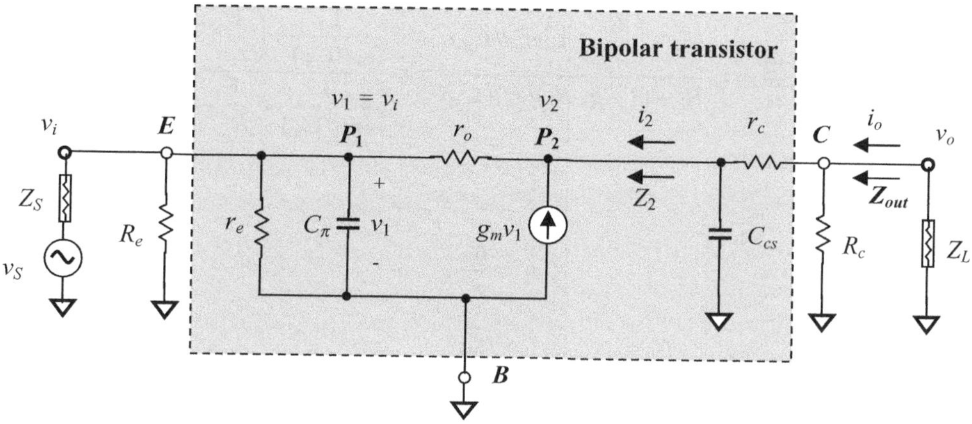

Figure 12.21 Equivalent circuit for calculation of output impedance of a *CB* device.

Now let's calculate the output impedance. Figure 12.21 shows the equivalent model for calculation of the output impedance of a *CB* device.

KCL at node P_1

$$\frac{v_1}{R_e} + \frac{v_1}{r_e // C_\pi} + \frac{v_1 - v_2}{r_o} = 0, \tag{12.176}$$

And *KCL* at node P_2

$$i_2 = -g_m v_1 + \frac{v_2 - v_1}{r_o} = \frac{v_2}{r_o} - \left(g_m + \frac{1}{r_o}\right) v_1. \tag{12.177}$$

Solving v_1 from (12.192) and substituting it into (12.193) gives

$$v_1 = \frac{\frac{1}{r_o}}{\left(\frac{1}{R_e} + \frac{1}{r_e // C_\pi} + \frac{1}{r_o}\right)} v_2 = \frac{R_e (r_e // C_\pi)}{R_e r_o + (R_e + r_o)(r_e // C_\pi)} v_2, \tag{12.178}$$

$$\frac{i_2}{v_2} = \frac{1}{r_o} - \frac{\left(g_m + \frac{1}{r_o}\right) R_e (r_e // C_\pi)}{R_e r_o + (R_e + r_o)(r_e // C_\pi)}, \tag{12.179}$$

$$\frac{i_2}{v_2} = \frac{1}{r_o}\left[1 - \frac{(1 + g_m r_o) R_e (r_e // C_\pi)}{R_e r_o + (R_e + r_o)(r_e // C_\pi)}\right] = \frac{1}{r_o} - \frac{1}{r_o} \frac{\left(\frac{1}{r_o} + g_m\right) R_e (r_e // C_\pi)}{R_e + \left(\frac{R_e}{r_o} + 1\right)(r_e // C_\pi)}, \tag{12.180}$$

$$\frac{i_2}{v_2} = \frac{1}{r_o} \frac{R_e + (1 - g_m R_e)(r_e // C_\pi)}{R_e + \left(\frac{R_e}{r_o} + 1\right)(r_e // C_\pi)}, \tag{12.181}$$

$$Z_2 = \frac{v_2}{i_2} = r_o \frac{R_e + \left(\frac{R_e}{r_o} + 1\right)(r_e // C_\pi)}{R_e + (1 - g_m R_e)(r_e // C_\pi)} = r_o \frac{\frac{1}{(r_e // C_\pi)} + \left(\frac{1}{r_o} + \frac{1}{R_e}\right)}{\frac{1}{(r_e // C_\pi)} + \frac{1}{R_e} - g_m}. \tag{12.182}$$

Note that

$$r_e = \frac{\alpha_o}{g_m}, \tag{12.183}$$

$$r_e // C_\pi = \frac{\alpha_o}{g_m} // C_\pi = \frac{\alpha_o}{g_m + j\omega C_\pi \alpha_o} = \frac{1}{\frac{\beta_o + 1}{r_\pi} + j\omega C_\pi}. \tag{12.184}$$

Substituting $r_e // C_\pi$ from (12.184) into (12.182), we have

$$Z_2 = r_o \frac{\frac{1}{\frac{\alpha_o}{g_m + j\omega C_\pi \alpha_o}} + \left(\frac{1}{r_o} + \frac{1}{R_e}\right)}{\frac{1}{\frac{\alpha_o}{g_m + j\omega C_\pi \alpha_o}} + \frac{1}{R_e} - g_m} = r_o \frac{1 + \left(\frac{1}{r_o} + \frac{1}{R_e}\right)\frac{\alpha_o}{g_m + j\omega C_\pi \alpha_o}}{1 + \left(\frac{1}{R_e} - g_m\right)\frac{\alpha_o}{g_m + j\omega C_\pi \alpha_o}}, \tag{12.185}$$

$$Z_2 = r_o r_\pi \frac{\left[1 + R_e\left(\frac{1}{r_o} + \frac{\beta_o + 1}{r_\pi} + j\omega C_\pi\right)\right]}{(r_\pi + R_e + j\omega C_\pi R_e r_\pi)}. \tag{12.186}$$

Finally, the output impedance is

$$Z_{out} = [Z_2 // C_{cs} // (r_c + R_c)] \frac{R_c}{r_c + R_c} = \frac{R_c}{r_c + R_c} \frac{(r_c + R_c)}{[1 + j\omega C_{cs}(r_c + R_c)] + \frac{(r_c + R_c)}{Z_2}}, \tag{12.187}$$

$$Z_{out} = R_c \frac{1}{1 + j\omega C_{cs}(r_c + R_c) + (r_c + R_c)\frac{(r_\pi + R_e + j\omega C_\pi R_e r_\pi)}{r_o r_\pi + R_e[r_\pi + r_o(\beta_o + 1) + j\omega C_\pi r_\pi r_o]}}. \tag{12.188}$$

In the low-frequency cases,

$$Z_{out}|_{\omega \to 0} = R_c \frac{1}{1 + (r_c + R_c)\frac{(r_\pi + R_e)}{r_o r_\pi + R_e[r_\pi + (\beta_o + 1) r_o]}}. \tag{12.189}$$

In cases without the resistor R_e, that is, $R_e \to \infty$,

$$Z_{out}|_{R_e \to \infty} = R_c \frac{1}{1 + j\omega C_{cs}(r_c + R_c) + \frac{(r_c + R_c)(1 + j\omega C_\pi r_\pi)}{r_\pi + (1 + \beta_o + j\omega C_\pi r_\pi) r_o}}. \tag{12.190}$$

In low-frequency cases,

$$Z_{out}|_{\omega\to 0, R_e\to\infty} = R_c \frac{1}{1+\frac{r_c+R_c}{r_\pi+(1+\beta_o)r_o}}. \tag{12.191}$$

12.6 BIPOLAR TRANSISTOR WITH *CC* (COMMON COLLECTOR) CONFIGURATION

Figure 12.22 shows the equivalent circuit of a bipolar transistor with a *CC* configuration, in which the collector is *AC* grounded, the input port is the base, and the output port is the emitter of the bipolar transistor. The notation "*CC*" originates in that the collector is a common ground terminal of the input (base) port and the output (emitter) port.

In Figure 12.22(a), two "zero" capacitors which approach zero impedance at operating frequencies function as *DC* blocking capacitors, and are symbolized as C_{zero}. The "infinite" inductor functions as an *RF* choke, which approaches infinite impedance at operating frequencies and is therefore is symbolized by $L_{infinite}$. The emitter resistor R_e is connected from the emitter to the ground.

Figure 12.22(b) is the equivalent circuit of a *CC* device in terms of its small signal hybrid-π model, in which all the "zero" capacitors and the "infinite" inductor are ignored.

12.6.1 Open-circuited Voltage Gain $A_{v,CC}$ of a *CC* Device

The voltage gain A_v of a *CC* device can be derived with the additional approximations: the resistor r_c is neglected because its value is much lower than the value of R_c, that is,

$$r_c << R_c. \tag{12.192}$$

And the value of the capacitor C_π is much higher than the values of C_μ and C_{cs}, that is,

$$C_\pi >> C_\mu, C_{cs}. \tag{12.193}$$

Then, the model of bipolar transistor with the *CC* configuration shown in Figure 12.22(b) can be re-drawn as Figure 12.23.

KCL at node P_2

$$\frac{v_i - v_o}{r_b + r_\pi // C_\pi} + g_m v_1 - \frac{v_o}{R_e // Z_L} - \frac{v_o}{r_o} = 0, \tag{12.194}$$

Note that the terminal V_{cc} is *AC* grounded and

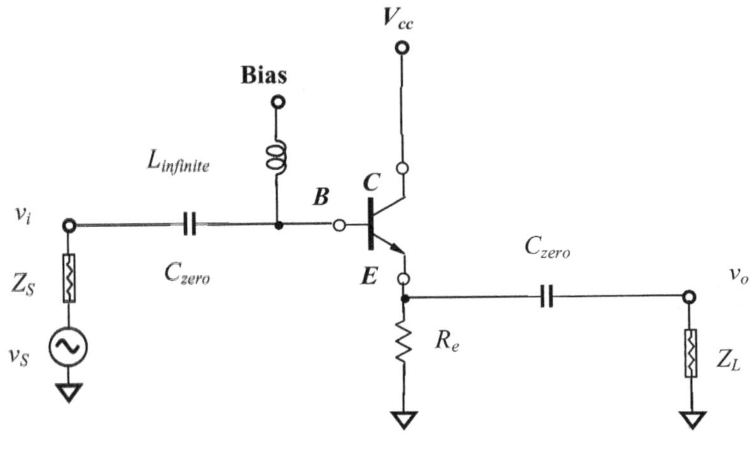

(a) Schematic of a bipolar transistor with *CC* configuration

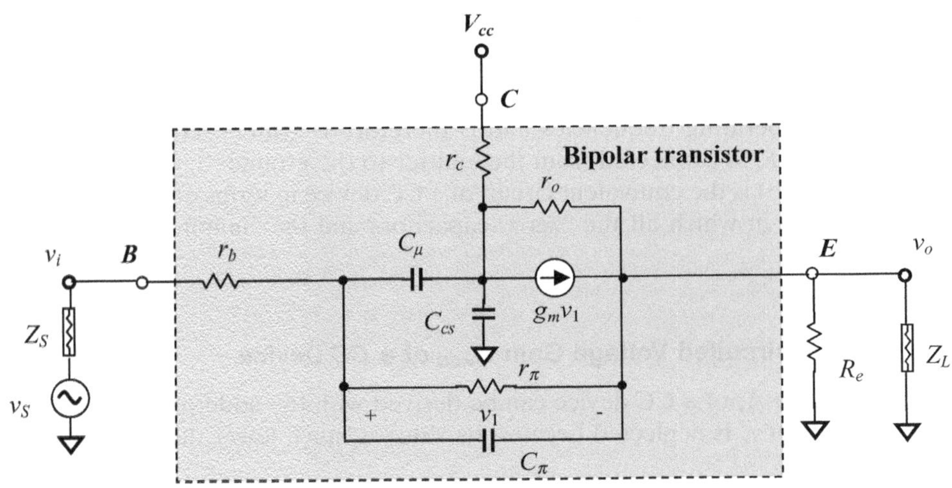

(b) Model of bipolar transistor with *CC* configuration

Figure 12.22 A bipolar transistor with *CC* configuration.
C_{zero}: "zero" capacitor Z_S: source impedance
$L_{infinite}$: "infinite" inductor Z_L: load impedance

$$g_m = \frac{\beta_o}{r_\pi}, \tag{12.195}$$

$$v_1 = \frac{r_\pi // C_\pi}{r_b + r_\pi // C_\pi}(v_i - v_o). \tag{12.196}$$

Substituting g_m in (12.195) and v_1 in (12.196) into (12.194), we have

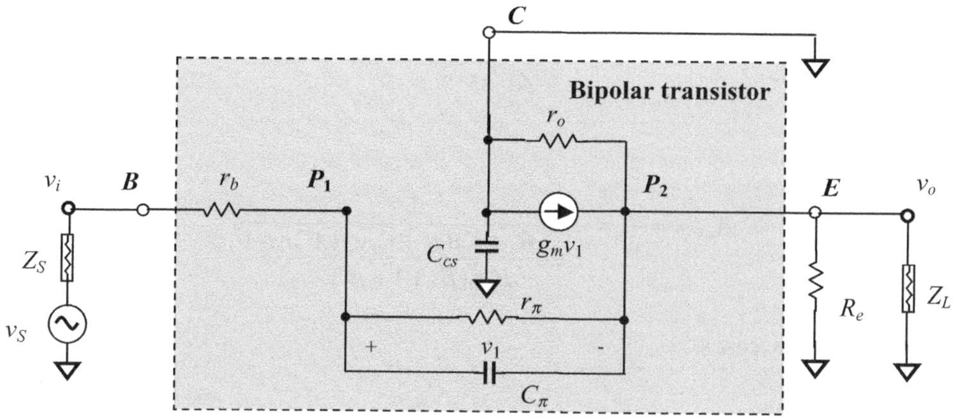

Figure 12.23 Equivalent model for calculation of open-circuited voltage gain of a *CC* device.

$$\frac{v_i - v_o}{r_b + r_\pi // C_\pi} + \frac{\beta_o}{r_\pi}\frac{r_\pi // C_\pi}{r_b + r_\pi // C_\pi}(v_i - v_o) - \frac{v_o}{R_e // Z_L} - \frac{v_o}{r_o} = 0. \tag{12.197}$$

Finally the voltage gain is

$$A_{v,CC} = \frac{v_o}{v_i} = \frac{\left(\frac{1}{r_b + r_\pi // C_\pi} + \frac{\beta_o}{r_\pi}\frac{r_\pi // C_\pi}{r_b + r_\pi // C_\pi}\right)}{\left(\frac{1}{r_b + r_\pi // C_\pi} + \frac{\beta_o}{r_\pi}\frac{r_\pi // C_\pi}{r_b + r_\pi // C_\pi}\right) + \left(\frac{1}{R_e // Z_L} + \frac{1}{r_o}\right)}. \tag{12.198}$$

Note that

$$\frac{1}{r_b + r_\pi // C_\pi} + \frac{\beta_o}{r_\pi}\frac{r_\pi // C_\pi}{r_b + r_\pi // C_\pi} = \frac{\beta_o + 1 + j\omega C_\pi r_b}{r_\pi\left[1 + \frac{r_b}{r_\pi}(1 + j\omega C_\pi r_b)\right]}, \tag{12.199}$$

$$\frac{1}{R_e // Z_L} + \frac{1}{r_o} = \frac{r_o(R_e + Z_L) + R_e Z_L}{r_o R_e Z_L}, \tag{12.200}$$

$$A_{v,CC} = \frac{v_o}{v_i} = \frac{\dfrac{\beta_o + 1 + j\omega C_\pi r_b}{r_\pi\left[1 + \frac{r_b}{r_\pi}(1 + j\omega C_\pi r_b)\right]}}{\dfrac{\beta_o + 1 + j\omega C_\pi r_b}{r_\pi\left[1 + \frac{r_b}{r_\pi}(1 + j\omega C_\pi r_b)\right]} + \dfrac{R_e Z_L + Z_L r_b + r R_e}{R_e Z_L r_b}}, \tag{12.201}$$

$$A_{v,CC} = \frac{v_o}{v_i} = \frac{1}{1 + \dfrac{(R_e Z_L + Z_L r_b + r R_e)(r_\pi + r_b + j\omega C_\pi r_b r_b)}{R_e Z_L r_b(\beta_o + 1 + j\omega C_\pi r_b)}}, \tag{12.202}$$

The open-circuited voltage gain $A_{v,CC}$ is in the case when

$$Z_L \to \infty, \tag{12.203}$$

then,

$$A_{v,CC} = \frac{v_o}{v_i} = \frac{1}{1+\dfrac{(R_e + r_b)(r_\pi + r_b + j\omega C_\pi r_b r_b)}{R_e r_b(\beta_o + 1 + j\omega C_\pi r_b)}}. \tag{12.204}$$

In low-frequency cases,

$$A_{v,CC}\big|_{\omega\to 0} = \frac{1}{1+\dfrac{(R_e + r_b)(r_\pi + r_b)}{R_e r_b(\beta_o + 1)}}. \tag{12.205}$$

Usually, we have

$$r_\pi >> r_b, \text{ and} \tag{12.206}$$

$$\beta_o >> 1, \tag{12.207}$$

then

$$A_{v,CC}\big|_{\omega\to 0} \approx \frac{1}{1+\dfrac{1}{g_m}\left(\dfrac{1}{R_e}+\dfrac{1}{r_o}\right)}. \tag{12.208}$$

12.6.2 Short-circuited Current Gain β_{CC} and Frequency Response of a Bipolar Transistor with *CC* Configuration

Figure 12.24 is drawn for the calculation of the short-circuited current gain and the frequency response of a bipolar transistor with a *CC* configuration, in which the resistor r_c and C_μ are neglected because

$$r_c << R_c. \tag{12.209}$$

$$C_\pi >> C_{cs}, C_\mu. \tag{12.210}$$

Also note that C_{cs} is shorted to the ground because in a *CC* device (emitter follower), the terminal *C* is grounded. Then,

KCL at node *E*

$$i_o = g_m v_1 + i_i. \tag{12.211}$$

Note that

$$v_1 = i_i C_\pi // r_\pi, \tag{12.212}$$

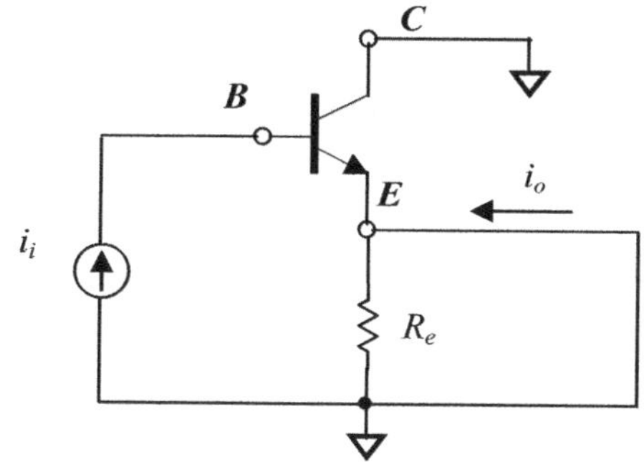

(a) Output is short-circuited for calculation of current gain β

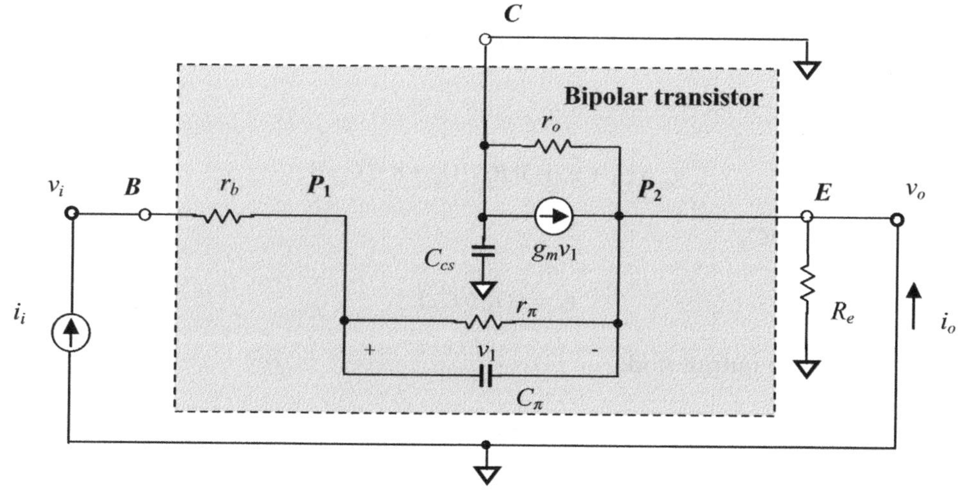

(b) Equivalent model for calculation of short-circuited current gain β_{CC}.

Figure 12.24 Schematic and equivalent for calculation of short-circuited current gain β_{CC} of a *CC* device.

then

$$i_o = (g_m C_\pi // r_\pi + 1) i_i = \beta_o \frac{\left(1 + \frac{1 + j\omega C_\pi r_\pi}{\beta_o}\right)}{(1 + j\omega C_\pi r_\pi)} i_i, \tag{12.213}$$

$$\beta_{CC}(j\omega) = \frac{i_o}{i_i} = \beta_o \frac{\left(1 + \frac{1 + j\omega C_\pi r_\pi}{\beta_o}\right)}{(1 + j\omega C_\pi r_\pi)} = \frac{1 + \beta_o + j\omega C_\pi r_\pi}{1 + j\omega C_\pi r_\pi}. \tag{12.214}$$

At low frequencies,

$$\beta_{CC}(j\omega)|_{\omega\to 0} = 1+\beta_o. \tag{12.215}$$

The short-circuited current gain of a CC device in low frequencies is $(1+\beta_o)$, which is in the same order as that in a CE device.

12.6.3 Input and Output Impedances of a *CC* Device

In the calculation of input impedance of a CC transistor, Figure 12.25 is simplified by the additional approximation: the capacitor C_μ is neglected because

$$C_\mu < C_{cs} << C_\pi. \tag{12.216}$$

Then, from the input to output node,

$$v_i = i_i(r_b + r_\pi // C_\pi) + v_o, \tag{12.217}$$

KCL at node P_2

$$v_o = (i_i + g_m v_1)[R_e // (r_o + r_c // C_{cs})]. \tag{12.218}$$

Ohm's Law at $r_\pi // C_\pi$,

$$v_1 = i_i(r_\pi // C_\pi), \tag{12.219}$$

From the input to output node,

$$v_i = i_i(r_b + r_\pi // C_\pi) + v_o, \tag{12.220}$$

KCL at node P_2

$$v_o = (i_i + g_m v_1)[R_e // (r_o + r_c // C_{cs})]. \tag{12.221}$$

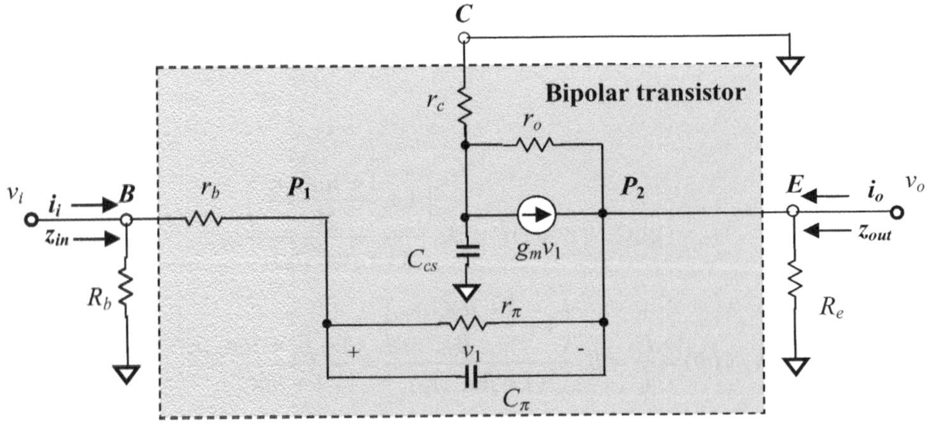

Figure 12.25 Equivalent circuit for calculation of input and output impedance of a CC device.

Ohm's Law at $r_\pi // C_\pi$,

$$v_1 = i_i(r_\pi // C_\pi), \tag{12.222}$$

Substituting (12.222) into (12.221),

$$v_o = i_i[1 + g_m(r_\pi // C_\pi)][R_e // (r_o + r_c // C_{cs})]. \tag{12.223}$$

Replacing v_o in (12.220) by (12.223),

$$v_i = i_i(r_b + r_\pi // C_\pi) + i_i[1 + g_m(r_\pi // C_\pi)][R_e // (r_o + r_c // C_{cs})]. \tag{12.224}$$

Then,

$$Z_{in} = \frac{v_i}{i_i} = (r_b + r_\pi // C_\pi) + [1 + g_m(r_\pi // C_\pi)][R_e // (r_o + r_c // C_{cs})]. \tag{12.225}$$

Note that

$$r_\pi // C_\pi = \frac{1}{1 + jC_\pi \omega r_\pi} r_\pi, \tag{12.226}$$

$$r_b + r_\pi // C_\pi = \frac{r_b + r_\pi + jC_\pi \omega r_b r_\pi}{1 + jC_\pi \omega r_\pi}, \tag{12.227}$$

$$r_o + r_c // C_{cs} = \frac{r_o + r_c + jC_{cs}\omega r_o r_c}{1 + jC_{cs}\omega r_c}, \tag{12.228}$$

$$R_e // (r_o + r_c // C_{cs}) = \frac{r_o + r_c + jC_{cs}\omega r_o r_c}{R_e + r_o + r_c + jC_{cs}\omega r_c(r_o + R_e)} R_e. \tag{12.229}$$

Then,

$$Z_{in} = \frac{r_\pi + r_b(1 + jC_\pi \omega r_\pi) + (1 + \beta_o + jC_\pi \omega r_\pi)\dfrac{r_o + r_c + jC_{cs}\omega r_o r_c}{R_e + r_o + r_c + jC_{cs}\omega r_c(r_o + R_e)} R_e}{1 + jC_\pi \omega r_\pi}. \tag{12.230}$$

In low-frequency cases,

$$Z_{in}|_{\omega \to 0} = r_\pi + r_b + (1 + \beta_o)\frac{r_o + r_c}{R_e + r_o + r_c} R_e. \tag{12.231}$$

A special feature of a *CC* device is that the input resistance is equal to r_π plus $(1 + \beta_o)$ times the incremental resistance connected from the emitter to the ground.

Now let's discuss the output impedance of a *CC* device.

The output current consists of three portions when an output voltage is applied to the terminal *E*. The first portion is the current flowing through r_o and r_c and C_{cs} to the *AC* ground due to the output voltage. The second portion is the current generated by the device, $g_m v_1$, in which v_1 is the voltage drop on r_π and C_π due to the output voltage. The third portion is the current flowing through r_π, r_b, and R_b to the ground due to the output voltage. We then have

$$i_o = \frac{v_o}{r_o + (r_c // C_{cs})} + g_m \frac{(r_\pi // C_\pi) v_o}{R_b + r_b + (r_\pi // C_\pi)} + \frac{v_o}{R_b + r_b + (r_\pi // C_\pi)}, \tag{12.232}$$

$$i_o = \frac{1 + j\omega C_{cs} r_c}{r_c + r_o(1 + j\omega C_{cs} r_c)} v_o + \frac{1 + \beta_o + j\omega C_\pi r_\pi}{r_\pi + (R_b + r_b)(1 + j\omega C_\pi r_\pi)} v_o, \tag{12.233}$$

$$\frac{1}{Z_{out}} = \frac{i_o}{v_o} = \frac{1 + j\omega C_{cs} r_c}{r_c + r_o(1 + j\omega C_{cs} r_c)} + \frac{1 + \beta_o + j\omega C_\pi r_\pi}{r_\pi + (R_b + r_b)(1 + j\omega C_\pi r_\pi)}, \tag{12.234}$$

$$Z_{out} = \frac{r_\pi + (R_b + r_b)(1 + j\omega C_\pi r_\pi)}{(1 + \beta_o + j\omega C_\pi r_\pi) + (1 + j\omega C_{cs} r_c) \dfrac{r_\pi + (R_b + r_b)(1 + j\omega C_\pi r_\pi)}{r_c + r_o(1 + j\omega C_{cs} r_c)}}. \tag{12.235}$$

At a low-frequency range, the capacitors C_π and C_{cs} could be neglected; thus, we have

$$Z_{in}|_{\omega \to 0} \approx r_\pi + r_b + (1 + \beta_o) \frac{1}{1 + \dfrac{R_e}{r_o + r_c}} R_e \approx r_\pi + r_b + \beta_o R_e. \tag{12.236}$$

$$Z_{out}|_{\omega \to 0} \approx \frac{(r_\pi + R_b + r_b)}{(1 + \beta_o) + \dfrac{(r_\pi + R_b + r_b)}{(r_c + r_o)}} \approx \frac{1}{g_m} + \frac{R_b + r_b}{\beta_o}. \tag{12.237}$$

It can be seen that the input impedance Z_{in} is dependent on the output resistors and load impedance R_e. Similarly, the output impedance Z_{out} is dependent on the input resistors and source impedance R_b.

A further approximation is that if

$$r_o >> z_L, \tag{12.238}$$

then

$$z_{in} \approx r_\pi + r_b + (1 + \beta) z_L, \tag{12.239}$$

and if

$$r_\pi >> z_s, \tag{12.240}$$

then

$$z_{out} \approx \frac{1}{1 + \beta + \dfrac{r_\pi}{r_c + r_o}} r_\pi, \tag{12.241}$$

furthermore if

$$r_o >> r_\pi, \tag{12.242}$$

then

$$z_{out} \approx \frac{1}{g_m}. \tag{12.243}$$

12.7 SMALL SIGNAL MODEL OF A *MOSFET* TRANSISTOR

Figures 12.26 and 12.27 show the *DC* characteristics of a *MOSFET*.

The *DC* characteristics of a *MOSFET* can be expressed by

At triode region 1:

$$I_d = \mu_n C_{ox} \frac{W}{L}(V_{gs} - V_{th})V_{ds}, \tag{12.244}$$

where

I_d = drain current,
g_m = trans-conductance of *MOSFET* transistor,
W = width of *MOSFET* transistor,

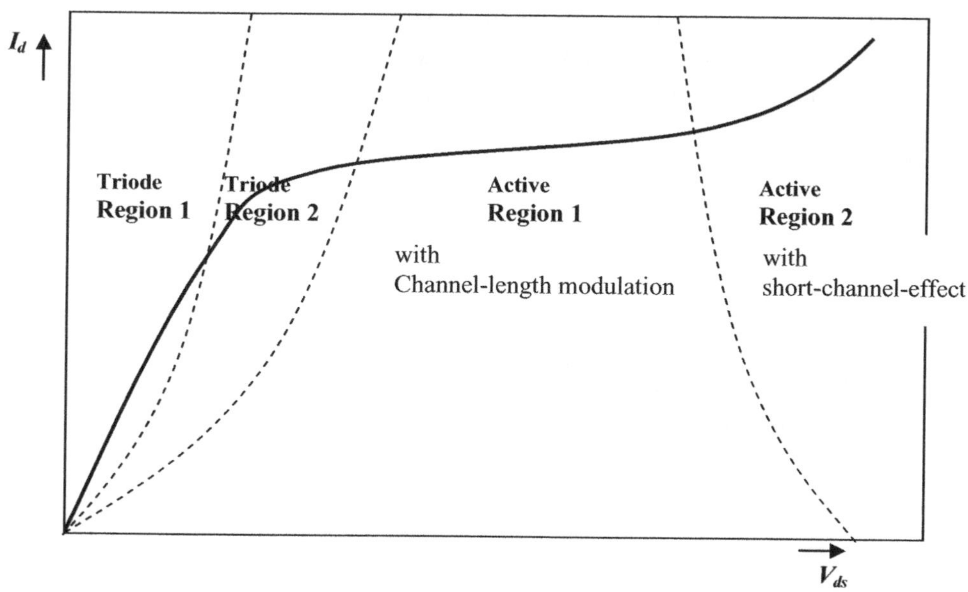

Figure 12.26 *DC* characteristics of a *MOSFET* transistor (I_d vs. V_{ds}) with short-channel effect.

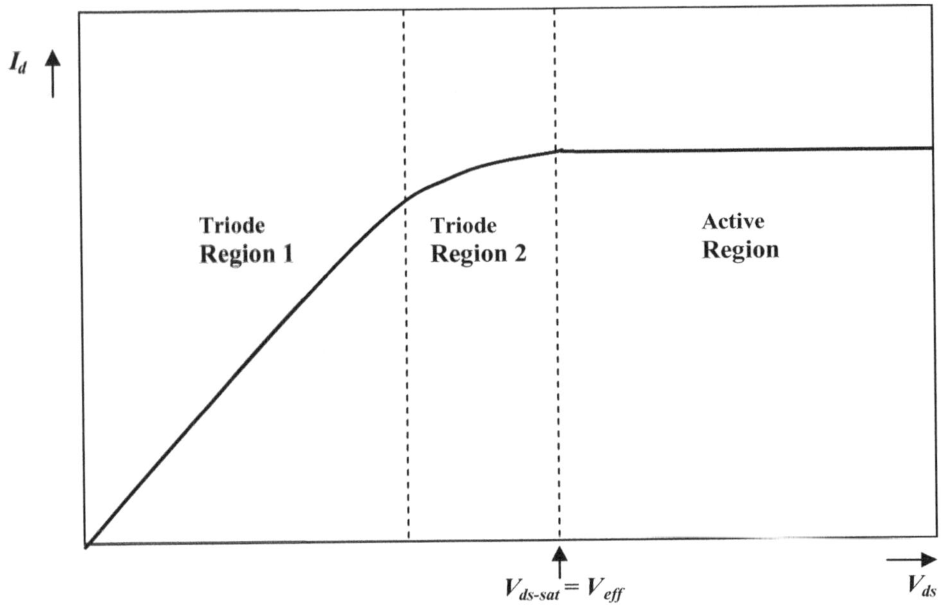

Figure 12.27 *DC* characteristics of a *MOSFET* transistor (I_d vs. V_{ds}) without short-channel effect.

L = length of *MOSFET* transistor,
V_{gs} = gate-source voltage for n channel *MOSFET*,
V_{th} = threshold voltage for n channel *MOSFET*, the minimum gate-to channel voltage needed for n carriers in the channel to exist,
V_{ds} = drain-source voltage for n channel *MOSFET*.
μ_n = channel mobility, typically $700\,cm^2/V\text{-}sec$,
C_{ox} = capacitance per unit area of the gate oxide,

and

$$C_{ox} = \frac{\varepsilon_{ox}}{t_{ox}}, \tag{12.245}$$

where

t_{ox} = thickness of the gate oxide,
ε_{ox} = permittivity of oxide.

For example,
if

$$t_{ox} = 400\,\text{A}^\circ = 0.4 \times 10^{-5}\,cm, \tag{12.246}$$

$$\mu_n = 700\,cm^2/V\text{-}sec, \tag{12.247}$$

$$\varepsilon_{ox} = 3.45 \times 10^{-13}\,F/cm, \tag{12.248}$$

then

$$C_{ox} = 0.86\, fF/\mu^2, \tag{12.249}$$

$$\mu_n C_{ox} = 60.2\,\mu A/V^2. \tag{12.250}$$

At triode region 2:

$$I_d = \mu_n C_{ox} \frac{W}{L}\left[(V_{gs} - V_{th})V_{ds} - \frac{V_{ds}^2}{2}\right], \tag{12.251}$$

As V_{ds} increases, I_d increases until the drain end of the channel becomes pinched off and then levels off. This pinch-off occurs for

$$V_{dg} = -V_{th}, \tag{12.252}$$

or approximately,

$$V_{ds} = V_{eff} = V_{gs} - V_{th}, \tag{12.253}$$

where V_{eff} = effective gate-source voltage.

Therefore, the active region equation can be obtained by substituting (12.253) into (12.251), that is,

At the active region:

$$I_d = \frac{\mu_n C_{ox}}{2}\frac{W}{L}(V_{gs} - V_{th})^2 = I_{d-sat}. \tag{12.254}$$

For $V_{ds} > V_{eff}$, the current stays constant at the value given by equation (12.283) and is denoted by $I_{d\text{-}sat}$. This implies that the drain current is independent of the drain-source voltage, V_{ds}.

In the active region,

$$g_m = \frac{\partial I_d}{\partial V_{gs}} = \mu_n C_{ox}\frac{W}{L}(V_{gs} - V_{th}) = \mu_n C_{ox}\frac{W}{L}V_{eff}, \tag{12.255}$$

Replacing V_{eff} by I_d, from (12.282) and (12.283) we have

$$V_{eff} = \sqrt{\frac{2I_d}{\mu_n C_{ox}(W/L)}}, \tag{12.256}$$

Consequently,

$$g_m = \sqrt{2\mu_n C_{ox}\frac{W}{L}I_d} = \frac{2I_d}{V_{eff}} = \frac{2I_d}{V_{gs} - V_{th}}. \tag{12.257}$$

It must be noted that the independence between I_d and V_{ds}, expressed by (12.254), is only true to a first order approximation. As a matter of fact, the channel shrinks as V_{ds} is increased. A pinched-off region with very little charge exists between the drain and the channel. The voltage at the end of the channel closest to the drain is fixed at $V_{gs} - V_{th} = V_{eff}$. The voltage difference between the drain and the near end of the channel lies across a short depletion region often called the pinch-off region. As V_{ds} becomes larger than V_{eff}, this depletion region surrounding the drain junction increases its width in a square-root relationship with respect to V_{ds}. This increase in the width of the depletion region surrounding the drain junction decreases the effective channel length. In turn, this decrease in effective channel length increases the drain current, resulting in what is commonly referred to as ***channel length modulation***. Then, expression (12.254) must be modified:

$$I_d = \frac{\mu_n C_{ox}}{2}\frac{W}{L}(V_{gs} - V_{th})^2 [1 + \lambda (V_{ds} - V_{eff})], \tag{12.258}$$

$$\lambda = \frac{k_{ds}}{2L\sqrt{V_{ds} - V_{eff} + \Phi_o}}. \tag{12.259}$$

$$k_{ds} = \sqrt{\frac{2K_S \varepsilon_o}{qN_A}}. \tag{12.260}$$

$$\Phi_o = V_T \ln \frac{N_A N_D}{n_i^2}. \tag{12.261}$$

where

λ = output impedance constant, (in units of V^{-1});
Φ_o = built-in voltage of an open-circuit PN junction (in units of V);
$V_T = kT/q$ = thermal voltage (at room temperature approximately $26 mV$),
K_S = relative permittivity of silicon (typically, it is 11.8),
ε_o = permittivity of free space equals to 8.854×10^{-12} F/m,
q = charge of an electron equals to 1.602×10^{-19} *Coulumb*,
N_A = acceptor concentration (in units of holes/m^3),
N_D = donor concentration (in units of electrons/m^3),
n_i = carrier concentration in intrinsic silicon (in unit of carriers/m^3).

Furthermore, equation (12.258) is accurate only in the cases without velocity saturation. Velocity saturation occurs when V_{ds} is large enough that the carrier speed no longer increases as the electric field increases. This second-order effect is often called the short-channel effect, in which I_d increases more than that calculated from (12.258) as V_{ds} is increased. Figure 12.26 shows the regions affected by the short channel effect.

In the actual device construction, the voltage of the source is sometimes different from that of the substrate (also known as the bulk). In other word, the voltage between the source of the device and the substrate, V_{sb}, exists, or in other words, is non-zero. The threshold voltage of an n-channel transistor is now given by

$$V_{tn} = V_{tno} + \gamma\left(\sqrt{V_{sb} + 2|\Phi_F|} - \sqrt{2|\Phi_F|}\right), \tag{12.262}$$

$$\gamma = \sqrt{\frac{2qN_A K_S \varepsilon_o}{C_{ox}}}, \tag{12.263}$$

$$\Phi_F = -\frac{kT}{q}\ln\left(\frac{N_A}{n_i}\right), \tag{12.264}$$

where

V_{tno} = threshold voltage with zero V_{sb};
Φ_F = built-in voltage of an open-circuit *PN* junction (in units of *V*).

It causes the body effect. For a small signal in the active region, this corresponds to having another current source, v_S, and its corresponding trans-conductance, g_s.

$$g_s = \frac{\gamma g_m}{2\sqrt{V_{sb} + 2|\Phi_F|}}, \tag{12.265}$$

$$V_s = V_{sb}, \tag{12.266}$$

$$\frac{1}{r_{ds}} = \lambda I_d. \tag{12.267}$$

Figure 12.28 shows the small signal model of a *MOSFET* when the body effect is considered. This is most commonly used hybrid-π model,
where

S = source of the *MOSFET* transistor,
G = gate of the *MOSFET* transistor,
D = drain of the *MOSFET* transistor,
C_{gs} = gate-source capacitance,

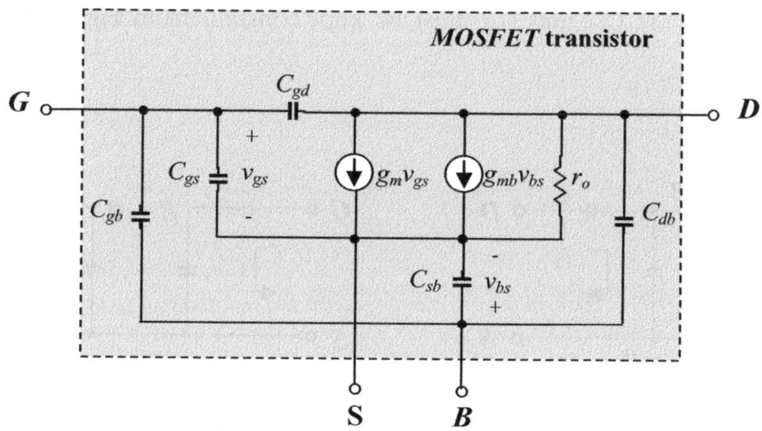

Figure 12.28 Small signal model of a *MOSFET* transistor.

C_{gd} = gate-drain capacitance,
C_{db} = drain-body depletion capacitance,
C_{gb} = gate-body capacitance,
C_{sb} = source-body depletion capacitance,
r_o = output resistance,
g_m = top-gate transconductance,
g_{mb} = body-effect transconductance.

12.8 SIMILARITY BETWEEN BIPOLAR AND *MOSFET* TRANSISTORS

A similarity exists between the small signal device models of bipolar and *MOSFET* transistors. Because of this similarity, the conclusions or rules acquired from the bipolar device may be popularized over the *MOSFET* device so as to create a shortcut to the derivation of the theory or formula. However, a discussion of the *MOSFET* model will be conducted before we compare *MOSFET* transistor models with bipolar transistor models.

12.8.1 Simplified Model of a *CS* Device

Looking at the *MOSFET* small signal model as shown in Figure 12.28, there are three small capacitors, C_{gb}, C_{sb}, and C_{db}, which are connected between G (gate), S (source), D (drain), and B (substrate) with a T configuration as sketched in Figure 12.29(a). In order to make the *MOSFET* model similar to the bipolar model, it is best to divide the capacitor C_{sb} into two capacitors, $C_{sb,g}$ and $C_{sb,d}$, so that they form two piggy-back Γ configurations as shown in Figure 12.29(b). Then, one Γ branch on the left-hand side of Figure 12.29(b) is combined with C_{gs} in parallel to form a new capacitor C_{gs}, while the other Γ branch on right-hand side of Figure 12.29(b) forms another new capacitor C_{ds}.

The question is: How will we divide the capacitor C_{sb}, and what are the values of $C_{sb,g}$ and $C_{sb,d}$?

In order to maintain the equivalence between Figure 12.29(a) and (b), the voltages v_G, v_B, v_D, i_1, i_2, C_{gb}, and C_{db} must be kept constant from Figure 12.29(a) to 12.29(b).

In Figure 12.29(a),

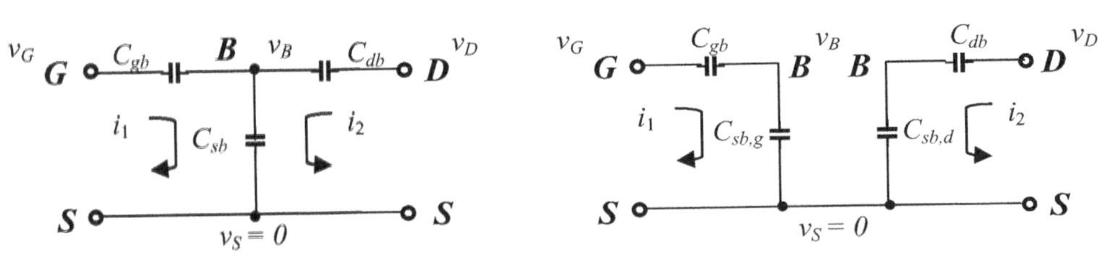

(a) T network (b) Γ network

Figure 12.29 T network and its equivalent two Γ networks in *CS* device's model, $v_S = 0$.

$$i_1 = \frac{v_G - v_B}{\dfrac{1}{j\omega C_{gb}}} = j\omega C_{gb}(v_G - v_B). \tag{12.268}$$

$$i_2 = \frac{v_D - v_B}{\dfrac{1}{j\omega C_{db}}} = j\omega C_{db}(v_D - v_B). \tag{12.269}$$

$$v_B = \frac{i_1 + i_2}{j\omega C_{sb}} = \frac{j\omega C_{gb}(v_G - v_B) + j\omega C_{db}(v_D - v_B)}{j\omega C_{sb}} = \frac{C_{gb}(v_G - v_B) + C_{db}(v_D - v_B)}{C_{sb}}. \tag{12.270}$$

In Figure 12.29(b),

$$v_B = \frac{i_1}{j\omega C_{sb,g}} = \frac{i_2}{j\omega C_{sb,d}}, \tag{12.271}$$

$$v_B = \frac{C_{gb}(v_G - v_B)}{C_{sb,g}} = \frac{C_{db}(v_D - v_B)}{C_{sb,d}}. \tag{12.272}$$

From (12.270) and the first equation of (12.272), we have

$$\frac{C_{gb}(v_G - v_B)}{C_{sb,g}} = \frac{C_{gb}(v_G - v_B) + C_{db}(v_D - v_B)}{C_{sb}}, \tag{12.273}$$

$$\left(\frac{1}{C_{sb,g}} - \frac{1}{C_{sb}}\right) C_{gb}(v_G - v_B) = \frac{C_{db}(v_D - v_B)}{C_{sb}}, \tag{12.274}$$

$$\frac{C_{sb} - C_{sb,g}}{C_{sb,g} C_{sb}} C_{gb}(v_G - v_B) = \frac{C_{db}(v_D - v_B)}{C_{sb}}, \tag{12.275}$$

$$\frac{C_{gb}(v_G - v_B)}{C_{sb,g}} = \frac{C_{db}(v_D - v_B)}{(C_{sb} - C_{sb,g})}. \tag{12.276}$$

Comparing (12.272) and (12.276), we have

$$C_{sb} = C_{sb,g} + C_{sb,d}. \tag{12.277}$$

From (12.272)

$$C_{sb,d} = C_{sb,g} \frac{C_{db}(v_D - v_B)}{C_{gb}(v_G - v_B)}. \tag{12.278}$$

Combining (12.277) and (12.278),

$$C_{sb,g} = \frac{1}{\left[1 + \dfrac{C_{db}(v_D - v_B)}{C_{gb}(v_G - v_B)}\right]} C_{sb} = \frac{C_{gb}(v_G - v_B)}{C_{gb}(v_G - v_B) + C_{db}(v_D - v_B)} C_{sb}. \tag{12.279}$$

From (12.277) and (12.279), we have

$$C_{sb,d} = C_{sb} - C_{sb,g} = \frac{C_{db}(v_D - v_B)}{C_{gb}(v_G - v_B) + C_{db}(v_D - v_B)} C_{sb}. \tag{12.280}$$

As an approximation, the values of C_{gb} and C_{db} are considered to be in the same order, that is,

$$C_{gb} \Leftrightarrow C_{db}, \tag{12.281}$$

and usually in a *CS* device,

$$v_B << v_D, \quad \text{and} \quad v_B << v_G \tag{12.282}$$

placing these two approximations into (12.279) and (12.280) we have

$$C_{sb,g} \approx \frac{C_{gb}}{A_{v,CS} C_{db}} C_{sb}, \tag{12.283}$$

$$C_{sb,d} \approx \frac{1}{1 + \dfrac{C_{gb}}{A_{v,CS} C_{db}}} C_{sb}. \tag{12.284}$$

where $A_{v,CS}$ is the open-circuited voltage gain of a *CS* device, that is,

$$A_{v,CS} = \frac{v_D}{v_G}. \tag{12.285}$$

Now, in terms of expressions (12.283) and (12.284), the small signal model as shown in Figure 12.28(b) can be modified to a new one as shown in Figure 12.30, in which the capacitor C_{sb} is split into two parts, $C_{sb,g}$ and $C_{sb,d}$. The value of $C_{sb,g}$ is much lower than C_{sb} while the value of $C_{sb,d}$ is lower than but almost equal to that of C_{sb}.

A further simplification of the modified model in Figure 12.30 is to combine C_{gb}, $C_{sb,g}$ and C_{gs} as one capacitor C'_{gs}, and combine C_{db} and $C_{sb,d}$ as another capacitor C'_{ds}, that is,

$$C'_{gs} = C_{gs} + \frac{1}{1 + A_{v,CS} \dfrac{C_{db}}{C_{sb}}} C_{gb}, \tag{12.286}$$

$$C'_{ds} = \frac{1}{1 + \dfrac{C_{db}}{C_{sb}} + \dfrac{C_{gb}}{A_{v,CS} C_{sb}}} C_{db}. \tag{12.287}$$

Expressions (12.286) and (12.287) show that the three small capacitors, C_{gb}, C_{sb}, and C_{db}, increase the capacitance of C_{gs} up to that of C'_{gs} with an amount of C_{gb} and $C_{sb,g}$ in series and forms a new capacitor C'_{ds} by C_{db} and $C_{sb,d}$ in series.

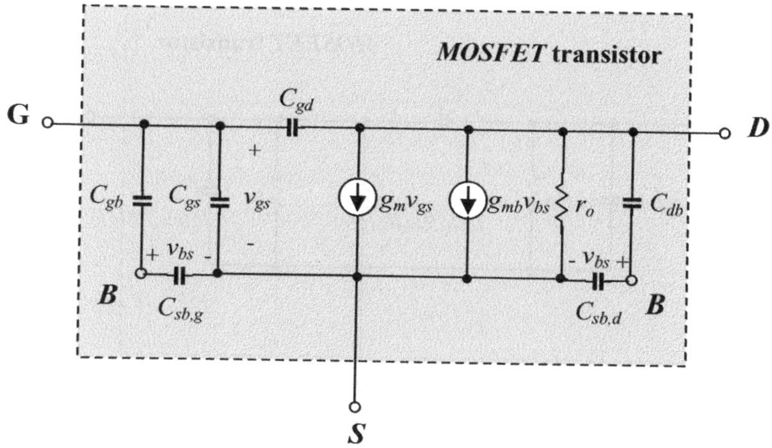

In *CS* Device: $C_{sb,g} \approx \dfrac{C_{gb}}{A_{v,CS} C_{db}} C_{sb}$, $C_{sb,d} \approx \dfrac{1}{1+\dfrac{C_{gb}}{A_{v,CS} C_{db}}} C_{sb}$

Figure 12.30 Modified model of a *CS* device when the capacitor C_{sb} is split into two capacitors $C_{sb,g}$ and $C_{sb,d}$.

Figure 12.31 shows the final simplified model of a *CS* device with the combined capacitors C'_{gs} and C'_{ds}. It looks very similar to the model of the *CE* device shown in Figure 12.6.

It must be noted that the discussion in this sub-section might not be important to the *RF* circuit design if the designed block is operated in a relatively low-frequency range because the three capacitors, C_{gb}, C_{sb}, and C_{db}, have much lower values than C_{gs}. A simple workaround is to just forget or erase them from the schematic. However, in a high-frequency range, say, in *GHz* or above, the difference between C'_{gs} and C_{gs}, and the value of the new capacitor C'_{ds} is not negligible, but an appreciable capacitance.

12.8.2 Simplified Model of a *CG* Device

Figure 12.32 is a copy of Figure 12.28 with the positions of *S* and *G* exchanged.

There are three small capacitors, C_{gb}, C_{sb}, and C_{db}, which are connected between *G* (gate), *S* (source), *D* (drain), and *B* (substrate) with a *T* configuration as sketched in Figure 12.33(a). In order to make the *MOSFET* model similar to the bipolar one, it is preferable to divide the capacitor C_{gb} into two capacitors $C_{gb,s}$ and $C_{gb,d}$, forming two piggy-back Γ configurations as shown in Figure 12.33(b). Then, one Γ branch on the left-hand side of Figure 12.33(b) is combined with C_{gs} in parallel to form a new capacitor C'_{gs}, while another Γ branch on the right-hand side of Figure 12.33(b) is combined with C_{gd} in parallel to form a new capacitor C'_{gd}.

The mathematical derivation for the values of $C_{gb,s}$ and $C_{gb,d}$ is similar to that for the capacitors $C_{sb,g}$ and $C_{sb,d}$ in Section 12.8.1. By comparing Figure 12.33 with Figure

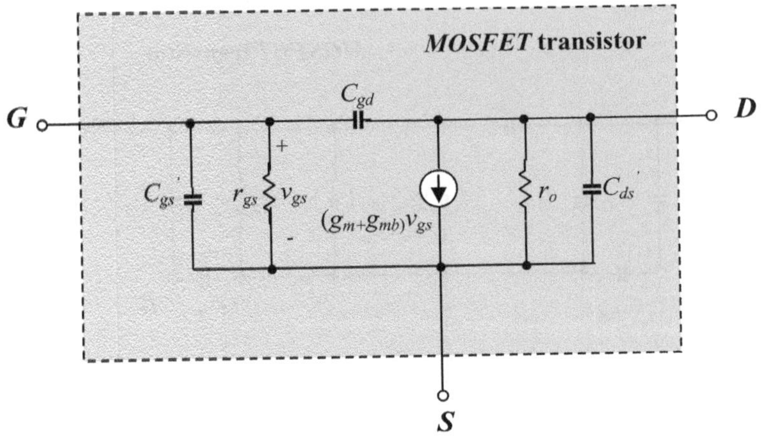

In *CS* Device: $C_{gs}' = C_{gs} + \dfrac{1}{1 + A_{v,CS}\dfrac{C_{db}}{C_{sb}}} C_{gb}$, $C_{ds}' = \dfrac{1}{1 + \dfrac{C_{db}}{C_{sb}} + \dfrac{C_{gb}}{A_{v,CS} C_{sb}}} C_{db}$

Figure 12.31 Simplified model of a *CS* device with combined capacitors C'_{gs} and C'_{ds}.

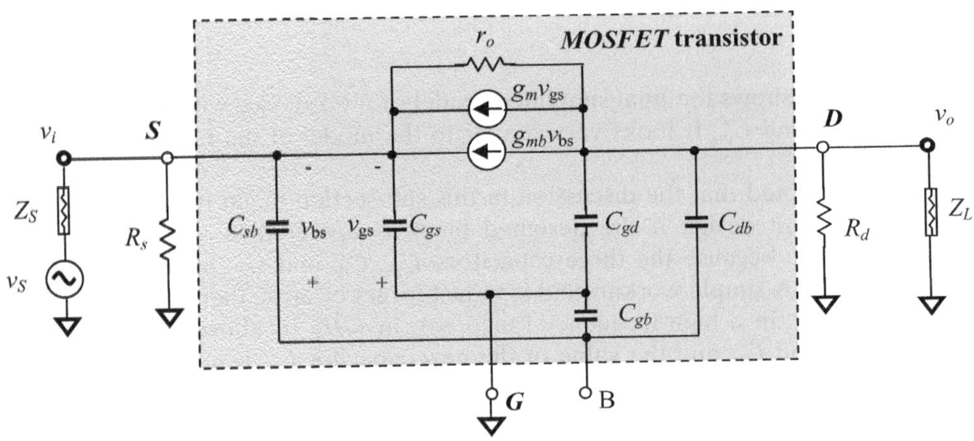

Figure 12.32 Model of a *MOSFET* transistor with *CG* configuration.

12.29, one finds that the Figure 12.33 becomes Figure 12.29 if the subscripts "*s*" or "*S*" and the symbol "*S*" are exchanged with the subscripts "*g*" or "*G*" and the symbol "*G*," respectively.

Therefore, it is not necessary to repeat the mathematical derivation. We need only copy expressions (12.283) to (12.285) with the subscripts and symbol exchanged as described; we then have

$$C_{gb,s} \approx \frac{C_{sb}}{A_{v,CG} C_{db}} C_{gb}, \qquad (12.288)$$

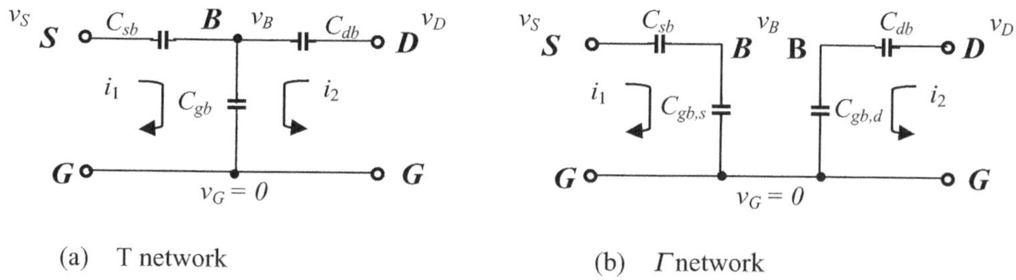

Figure 12.33 T network and its equivalent two Γ networks in CG device's model, $v_G = 0$.

$$C_{gb,d} \approx \frac{1}{1+\dfrac{C_{sb}}{A_{v,CG}C_{db}}} C_{gb}. \tag{12.289}$$

where $A_{v,CG}$ is open-circuited voltage gain of CG device, that is,

$$A_{v,CG} = \frac{v_D}{v_S}. \tag{12.290}$$

By replacing one T network by two Γ networks as shown in Figures 12.33 and 12.32, the equivalent model of CG device becomes Figure 12.34. To further simplify the model, we can combine the capacitors C_{sb}, $C_{gb,s}$, and C_{gs} as one capacitor C'_{gs}, and the capacitors C_{db}, $C_{gb,d}$, and C_{gd} as one capacitor C'_{gd}, in which the node B is "swallowed" or "submerged" inside C'_{gs} and C'_{gd}, which are

$$C'_{gs} = C_{gs} + \frac{C_{sb}C_{gb,s}}{C_{sb}+C_{gb,s}} = C_{gs} + \frac{C_{sb}}{1+A_{v,CG}\dfrac{C_{db}}{C_{gb}}}, \tag{12.291}$$

$$C'_{gd} = C_{gd} + \frac{C_{db}C_{gb,d}}{C_{db}+C_{gb,d}} = C_{gd} + \frac{C_{db}}{1+\dfrac{C_{db}}{C_{gb}}+\dfrac{1}{A_{v,CG}}\dfrac{C_{sb}}{C_{gb}}}. \tag{12.292}$$

Meanwhile, the two current sources are combined as one current source based on the same reasons mentioned in the discussion of the CS device. In addition, a resistor r_{gs} is added to increase the similarity though its value approaches infinity.

The simplified model of a $MOSFET$ transistor with a CG configuration with two combined capacitors C'_{gs} and C'_{gd} and a combined current source is shown in Figure 12.35, which is quite similar to the CB transistor model shown in Figure 12.36.

12.8.3 Simplified Model of a *CD* Device

Figure 12.36 is a copy of Figure 12.28 with a 90° counter-clockwise turning of the positions of D and S. The three small capacitors C_{gb}, C_{sb}, and C_{db} are connected between G (gate), S (source), D (drain), and B (substrate) with a T configuration

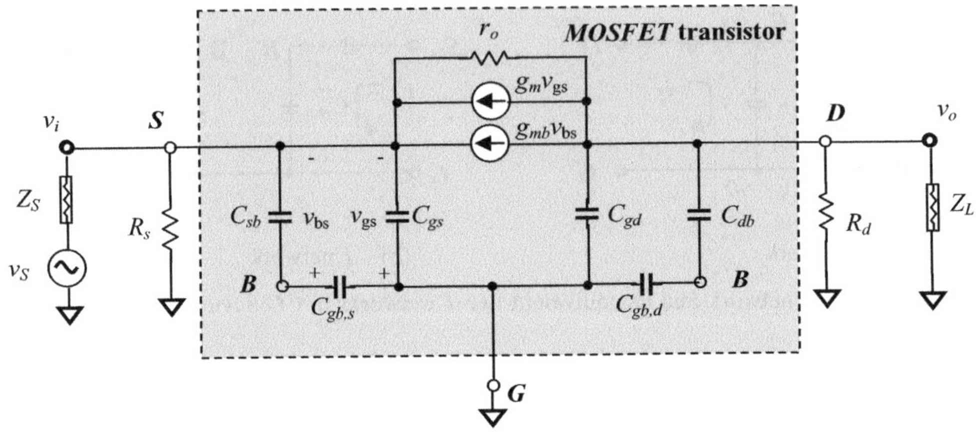

In CG Device: $C_{gb,s} = \dfrac{C_{sb}}{A_{v,CG} C_{db}} C_{gb}, \quad C_{gb,d} = \dfrac{1}{1 + \dfrac{C_{sb}}{A_{v,CG} C_{db}}} C_{gb}$

Figure 12.34 Modified model of a *CS* device when the capacitor C_{gb} is split into two capacitors $C_{gb,s}$ and $C_{gb,d}$.

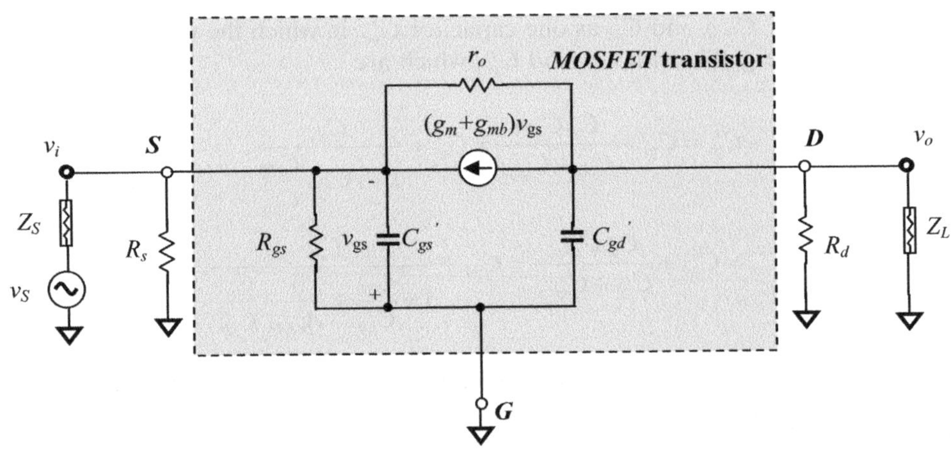

In CG Device: $C_{gs}' = C_{gs} + \dfrac{C_{sb}}{1 + A_{v,CG} \dfrac{C_{db}}{C_{gb}}}, \quad C_{gd}' = C_{gd} + \dfrac{C_{db}}{1 + \dfrac{C_{db}}{C_{gb}} + \dfrac{C_{sb}}{A_{v,CG} C_{gb}}}$

Figure 12.35 Simplified model of a *CS* device with two combined capacitors C_{gs}' and C_{gd}' and a combined current source.

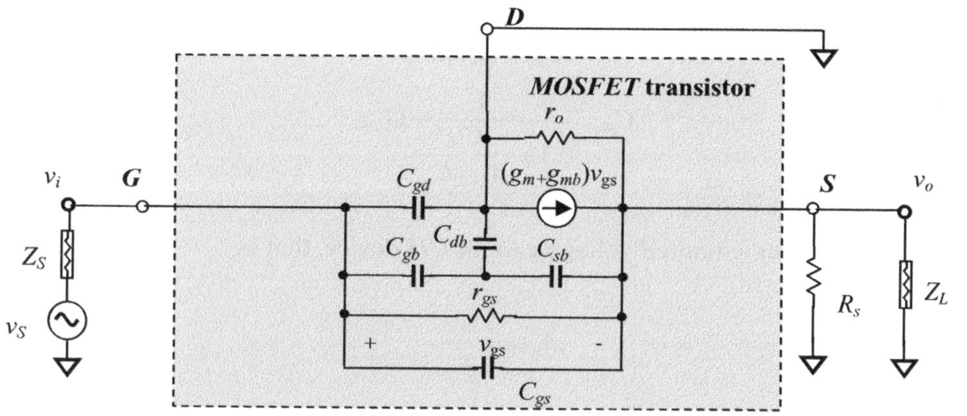

Figure 12.36 A *MOSFET* transistor with *CD* configuration, $r_{gs} \to \infty$. Z_S: source impedance, Z_L: load impedance.

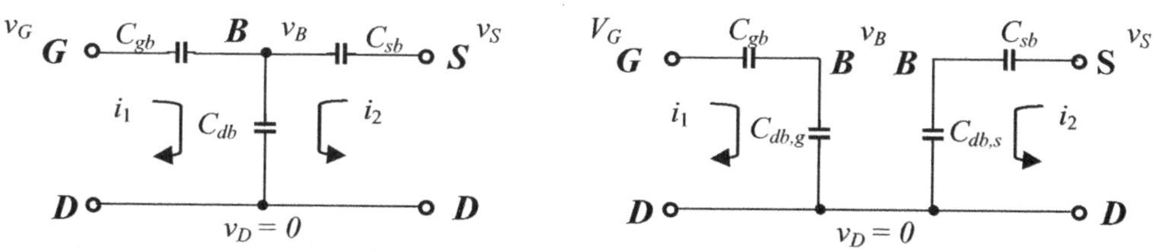

(a) *T* network (b) *Γ* network

Figure 12.37 *T* network and its equivalent two *Γ* networks in *CD* device's model, $v_D = 0$.

as sketched in Figure 12.37(a). In order to make the MOSFET model similar to the bipolar one, we divide the capacitor C_{db} into two capacitors $C_{db,g}$ and $C_{db,s}$, so that they form two piggy-back *Γ* configurations as shown in Figure 12.37(b). Then, one *Γ* branch on the left-hand side of Figure 12.37(b) is combined with C_{gd} in parallel to form a new capacitor C'_{gd}, while the other *Γ* branch othe right-hand side of Figure 12.37(b) forms a new capacitor C'_{ds}.

The mathematical derivation for the values of $C_{db,g}$ and $C_{db,s}$ is similar to that for the capacitors $C_{gb,g}$ and $C_{gb,d}$ in Section 12.8.2. By comparing Figure 12.37 with Figure 12.33, one can find that Figure 12.33 becomes Figure 12.37 if

- The subscripts or symbol "*s*" and "*S*," are changed to the subscripts or symbol, "*g*" and "*G*," respectively.
- The subscripts or symbol "*g*" and "*G*," are changed to the subscripts or symbol, "*d*" and "*D*," respectively.
- The subscripts or symbol "*d*" and "*D*," are changed to the subscripts or symbol, "*s*" and "*S*," respectively.

Therefore, it is not necessary to repeat the mathematical derivation. We need only copy expressions (12.288) to (12.290) with the subscripts and symbols exchanged as described; we then have

$$C_{db,g} \approx \frac{C_{gb}}{C_{gb} + A_{v,CD} C_{sb}} C_{db}, \tag{12.293}$$

$$C_{db,s} \approx \frac{1}{1 + \dfrac{C_{gb}}{A_{v,CD} C_{sb}}} C_{db}, \tag{12.294}$$

where $A_{v,CD}$ is open-circuited voltage gain of CD device, that is,

$$A_{v,CD} = \frac{v_S}{v_G}. \tag{12.295}$$

It should be noted that expression (12.293) is not exactly translated from expression (12.288) or (12.283). Instead, a capacitor C_{gb} is added in the denominator of expression (12.293). This term is recovered from the negligence in the derivation of expression (12.283) where $A_{v,CS} >> 1$. In the case of the CD device, $A_{v,CD}$ is not much greater than 1 but approximately equal to 1, that is,

$$A_v = \frac{v_S}{v_G} \approx 1. \tag{12.296}$$

Therefore, the neglected term due to the high open-circuited voltage gain must be recovered. On the other hand, from (12.293), (12.294), and (12.296), we have

$$C_{db,s} \approx C_{db,g} \approx \frac{1}{2} C_{db}. \tag{12.297}$$

By replacing one T network by two Γ networks as shown in Figure 12.37(b) into Figure 12.36, the equivalent model of the CG device becomes Figure 12.38. A further simplification of the model is to combine the capacitors C_{gb}, $C_{db,g}$, and C_{gd} as one capacitor, C'_{gd}, and the capacitors $C_{db,s}$ and C_{sb} as one new capacitor, C'_{ds}, in which the node B is "swallowed" or "submerged" inside the C'_{gd} and C'_{ds}, which are

$$C'_{gd} = C_{gd} + \frac{C_{gb} C_{db,s}}{C_{gb} + C_{db,s}} = C_{gd} + \frac{C_{db}}{1 + 2\dfrac{C_{gb}}{C_{db}}} C_{gb}, \tag{12.298}$$

$$C'_{ds} = \frac{C_{sb} C_{db,s}}{C_{sb} + C_{db,s}} = \frac{1}{1 + 2\dfrac{C_{sb}}{C_{db}}} C_{sb}. \tag{12.299}$$

The simplified model of a $MOSFET$ transistor with a CD configuration when two combined capacitors C'_{gd} and C'_{ds} and a combined current source is shown in Figure

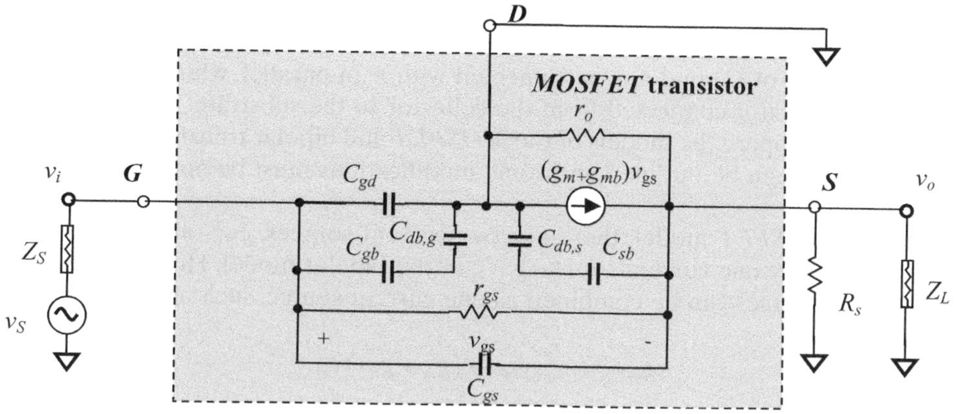

In *CD* Device: $C_{db,g} \approx \dfrac{1}{1+A_{v,CD}\dfrac{C_{sb}}{C_{gb}}}C_{db}$, $C_{db,s} \approx \dfrac{1}{1+\dfrac{C_{gb}}{A_{v,CD}C_{sb}}}C_{db}$

Figure 12.38 Modified model of a *CD* device when the capacitor C_{db} is split into two capacitors $C_{db,g}$ and $C_{db,s}$.

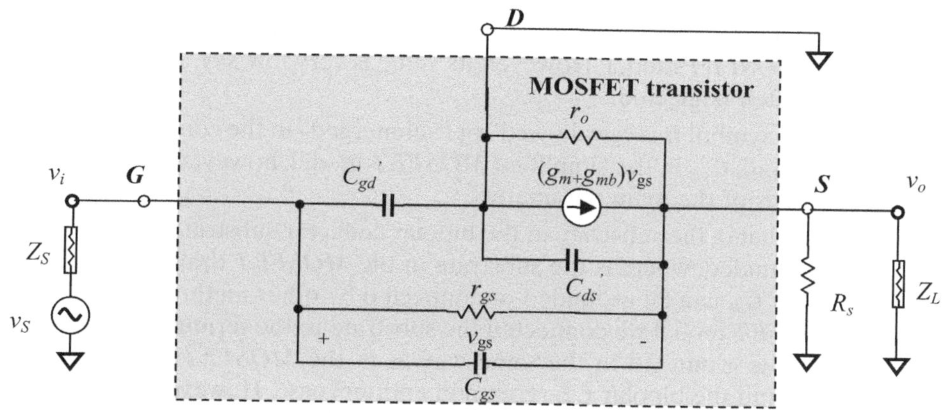

In *CD* Device: $C_{gd}' = C_{gd} + \dfrac{1}{1+2\dfrac{C_{gb}}{C_{db}}}C_{gb}$ · $C_{ds}' = \dfrac{1}{1+2\dfrac{C_{sb}}{C_{db}}}C_{sb}$

Figure 12.39 Simplified model of a CD device with two combined capacitors C'_{gd} and C'_{ds} and a combined current source.

12.39, which is quite similar to the *CB* transistor model in Figure 12.22(b), reproduced above. The apparent difference is that the capacitor C'_{ds} in the *CS* model, due to the existence of C_{sb} and C_{db}, is connected with r_o in parallel, while C_{cs} in the *CE* model is a capacitor connected from the collector to the substrate.

Now let's compare the models of the *MOSFET* and bipolar transistor. The apparent differences can be outlined and some modifications must be made as follows:

- In the *MOSFET* model, there are two current sources, $g_m v_1$ and $g_{mb} v_{bs}$, while there is only one current source, $g_m v_1$, in the bipolar model. However, the two current sources can be combined as one current source, such as

$$g_m v_{gs} + g_{mb} v_{bs} \rightarrow (g_m + g_{mb}) v_{gs}, \tag{12.300}$$

  because $v_{bs} = v_{gs}$ if the substrate or body connection is assumed to operate at *AC* ground. In most cases the value of g_{mb} is much lower than the value of g_m and is neglected.
- In order to easily compare the *MOSFET* and bipolar transistor model, a resistor r_{gs} is added to the *MOSFET* model, which is a resistor with a very high value; therefore its impact on the model is negligible.
- On the other hand, there is a resistor r_μ in the bipolar transistor model while there is nothing in the corresponding position in the *MOSFET* model. This difference does not impact their similarity because the resistor r_μ in the bipolar transistor model is usually neglected or is a resistor with a very high value.
- There is an emitter resistor r_x in the bipolar transistor model, which does not exist in the *MOSFET* model. However, its value is very low, say, a couple ohms, and is very often neglected.
- The substrate symbol *B* is "swallowed" or "submerged" in the combined capacitors C'_{gs}, C'_{ds}, and, C'_{gd} in the simplified *MOSFET* model; however, this does not impact analysis of the main parameters.
- On the other hand, the substrate in the bipolar collector-substrate capacitor C_{cs} is usually grounded, whereas the substrate in the *MOSFET* drain-body depletion capacitor C_{db} can be grounded or connected by other methods. In our discussed *MOSFET* model we connected the substrate to the terminal *S*. Actually, the substrate is grounded in the same way as in the *MOSFET CS* (common source) case and the bipolar *CE* (common emitter) case. However, the topologies of C_{cs} and C_{db} are slightly different because in the *MOSFET CG* (common gate) and *CD* (common drain) cases the substrate is connected to terminal *S*, while in the bipolar *CB* (common base) and *CC* (common collector) cases the substrate is grounded. This slight difference may somewhat impact the calculation of the input and output impedances, but could simply be taken as a tolerance or an allowable approximation.

The similarity between the bipolar and *MOSFET* models thus provides a short-cut to translate most of the formulas, such as the open-circuited voltage gain, the short-circuited current gain, and the input and output impedances, from the bipolar transistor to the *MOSFET*, as long as we perform the replacements shown in Table 12.1.

TABLE 12.1 The corresponding relationship of parameters between bipolar and *MOSFET* devices

From Bipolar	To *MOSFET* model		
CE*/*CB*/*CD	***CS***	***CG***	***CD***
r_x	0	0	0
r_b	0	0	0
r_c	0	0	0
r_π	r_{gs}	r_{gs}	r_{gs}
r_o	r_o	r_o	$r_o // C'_{ds}$
C_μ	C_{gd}		C'_{gd}
$C_\mu + C_{cs}$		$C'_{gd}(CG)$	
C_{cs}	C'_{ds}		
C_π	C'_{gs}	C'_{gs}	C_{gs}
$C_{cs}//R_c$			R_d
R_c	R_d	R_d	
g_m	$(g_m + g_{mb})$	$(g_m + g_{mb})$	$(g_m + g_{mb})$
β_o	$(g_m + g_{mb})r_{gs}$	$(g_m + g_{mb})r_{gs}$	$(g_m + g_{mb})r_{gs}$

Notes: $r_{gs} \to \infty$. Capacitors in *MOSFET* transistor models. $A_{v,CS}$ and $A_{v,CG}$ are open-circuited voltage gains of *CS* and *CG* devices, respectively.

12.9 *MOSFET* TRANSISTOR WITH *CS* (COMMON SOURCE) CONFIGURATION

A *MOSFET* transistor with a *CS* configuration can be drawn as Figure 12.40. Figure 12.40(a) plots its schematic, which contains two "zero" capacitors, C_{zero}, and one "infinite" $L_{infinite}$. The drain resistor R_d leads the *DC* power to the drain. Figure 12.40(b) is the simplified model of the *CS* device with the combined capacitors, C'_{gs} and C'_{ds}, in which all the "zero" capacitors and the "infinite" inductor are ignored.

12.9.1 Open-circuited Voltage Gain $A_{v,CS}$ of a *CS* Device

The simplified model of the *CS* device with the combined capacitors, C'_{gs} and C'_{ds} shown in Figure 12.40(b) is quite similar to the model of the *CE* device shown in Figure 12.7(b). Based on their similarity, the open-circuited voltage gain of the *CS* device can be obtained by translating from expression (12.49) in terms of the similar relationships shown in Table 12.1. Table 12.2 lists those relationships which must be particularly taken care of in the translation from the *CE* to the *CS* configuration.

$$A_{v,CS} = -(g_m + g_{mb} - j\omega C_{gd})\frac{r_o R_d}{r_o + R_d}\frac{1}{1 + j\omega(C_{gd} + C'_{ds})\dfrac{r_o R_d}{r_o + R_d}}, \quad (12.301)$$

It can be seen that this expression is simpler than (12.52) because there is no gate resistor like the base resistor r_b in *CE* device. In addition, instead of two poles in the expression (12.52), there is only one pole in the expression (12.301):

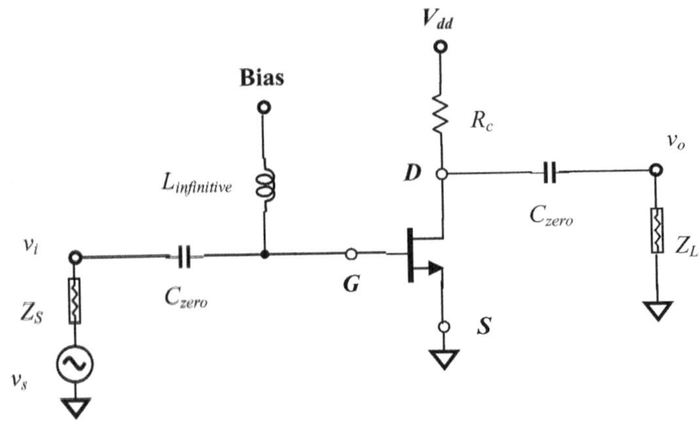

(a) Schematic of a *MOSFET* transistor with *CS* configuration C_{zero}: "zero" capacitor approaches zero capacitance at operating frequencies, $L_{infinite}$: "infinite" inductor approaches infinite inductance at operating frequencies.

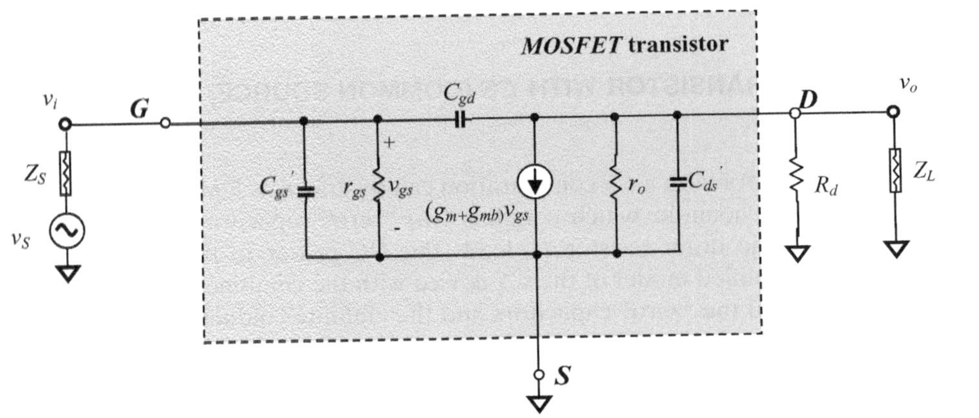

In CS Device: $C_{gs}' = C_{gs} + \dfrac{1}{1 + A_{v,CS}\dfrac{C_{db}}{C_{sb}}} C_{gb}$, $C_{ds}' = \dfrac{1}{1 + \dfrac{C_{db}}{C_{sb}} + \dfrac{C_{gb}}{A_{v,CS} C_{sb}}} C_{db}$

(b) Modified model of a *MOSFET* transistor with combined capacitors C_{gs}' and C_{ds}'.

Figure 12.40 A *MOSFET* transistor with *CS* configuration. Z_S: source impedance, Z_L: load impedance.

TABLE 12.2 Special relationships to be taken care of in the translation from *CE* to *CS* configuration

From Bipolar model *CE*	To *MOSFET* model *CS*
r_π	r_{gs}
r_o	r_o
C_μ	C_{gd}
C_{cs}	C'_{ds}
C_π	C'_{gs}
R_c	R_d
g_m	$(g_m + g_{mb})$
β_o	$(g_m + g_{mb})r_{gs}$

$$\omega_{-3dB} = |p_1| = \frac{1}{(C_{gd} + C'_{ds})\dfrac{r_o R_d}{r_o + R_d}}. \tag{12.302}$$

In low-frequency cases, the terms containing capacitors in expression (12.301) can be omitted, and then the open-circuited voltage gain becomes

$$A_{v,CS}|_{\omega\to 0} = -(g_m + g_{mb})\frac{r_o R_d}{r_o + R_d}, \tag{12.303}$$

which is similar to expression (12.55).

Should the Miller effect be considered, the voltage gain can be translated from expression (12.89) as

$$A_{v,CS} = \frac{v_o}{v_i} = -(g_m + g_{mb})\frac{r_o R_d}{r_o + R_d}\frac{1}{1 + j\omega C_{out}\dfrac{r_o R_d}{r_o + R_d}}. \tag{12.304}$$

It can again be seen that, instead of two poles as in the *CE* device, there is only one pole existing in (12.304) because there is no gate resistor like the base resistor r_b in the *CE* device, and

$$\omega_{-3dB} = |p_1| = -\frac{1}{C_{in} r_b} = \frac{1}{C_{out}\dfrac{r_o R_d}{r_o + R_d}}. \tag{12.305}$$

In low-frequency cases, the terms containing capacitors in expression (12.304) could be removed; then the open-circuited voltage gain becomes

$$A_{v,CS}|_{\omega\to 0} = -(g_m + g_{mb})\frac{r_o R_d}{r_o + R_d}, \tag{12.306}$$

Expressions (12.304) to (12.306) are quite similar to expressions (12.301) to (12.303) correspondingly.

12.9.2 Short-circuited Current Gain β_{CS} and Frequency Response of a *CS* Device

Figure 12.41 is drawn for the calculation of the short-circuited current gain and frequency response of a *MOSFET* transistor with a *CS* configuration.

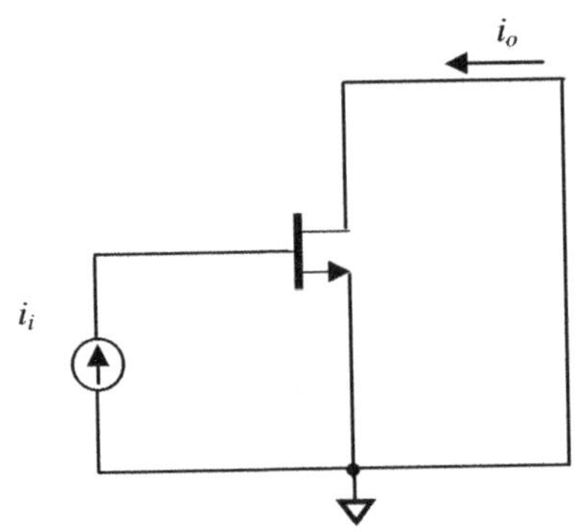

(a) Output is short-circuited for calculation of current gain β_{CS} and frequency response

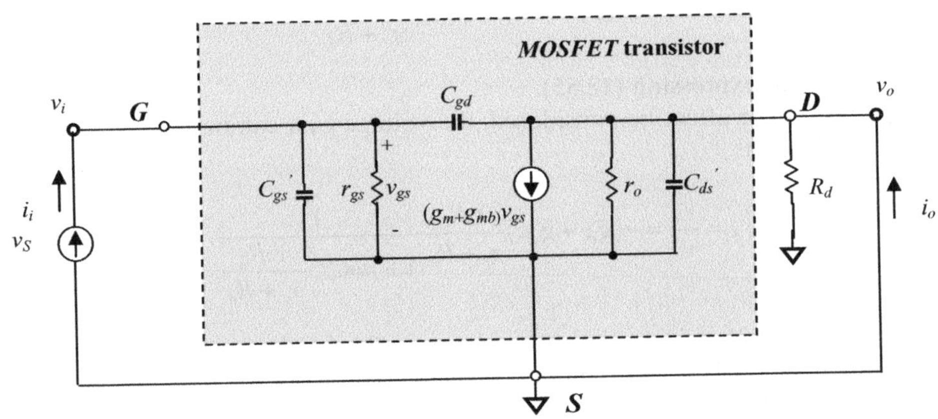

In *CS* Device: $C_{gs}' = C_{gs} + \dfrac{1}{1 + A_{v,CS}\dfrac{C_{db}}{C_{sb}}} C_{gb}, \quad C_{ds}' = \dfrac{1}{1 + \dfrac{C_{db}}{C_{sb}} + \dfrac{C_{gb}}{A_{v,CS} C_{sb}}} C_{db}$

(b) Modified model of a *MOSFET* transistor with combined capacitors C_{gs}' and C_{ds}'.

Figure 12.41 Output short-circuited for calculation of current gain β_{CS} and frequency response of a *MOSFET* transistor with *CS* configuration.

Similarly to the mathematical derivation in the corresponding section about the *CE* device, if the current going through the capacitor C_{gd} is neglected, then translated from (12.60), we have

$$\beta_{CS}(j\omega) = \frac{i_o}{i_i} = \frac{(g_m + g_{mb})r_{gs}}{1 + j(g_m + g_{mb})r_{gs}\dfrac{(C'_{gs} + C_{gd})}{g_m + g_{mb}}\omega} = \frac{1}{\dfrac{1}{(g_m + g_{mb})r_{gs}} + j\omega\dfrac{(C'_{gs} + C_{gd})}{g_m + g_{mb}}}. \tag{12.307}$$

In low-frequency cases, where $\omega \to 0$,

$$\beta_{CS}(j\omega)|_{\omega \to 0} \to \infty, \tag{12.308}$$

because $r_{gs} \to \infty$.

And in high-frequency cases,

$$\beta_{CS}(j\omega) \approx \frac{(g_m + g_{mb})}{j\omega(C'_{gs} + C_{gd})}. \tag{12.309}$$

When the input current is equal to the output current in magnitude, that is,

$$|\beta_{CS}(j\omega)| = 1, \tag{12.310}$$

then from (12.309) it can be found that the corresponding frequency $\omega_{T,CS}$ is

$$\omega_{T,CS} = \frac{g_m + g_{mb}}{C'_{gs} + C_{gd}}, \tag{12.311}$$

or,

$$f_{T,CS} = \frac{1}{2\pi}\frac{g_m + g_{mb}}{C'_{gs} + C_{gd}}. \tag{12.312}$$

$\omega_{T,CS}$ is called the transition frequency of the *CS* device.

On the other hand, the corresponding definition of $\omega_{\beta,CS}$ appears to be meaningless because $\beta_{CS}(j\omega)$ at low frequencies approaches infinity as shown in (12.309).

12.9.3 Input and Output Impedances of a *CS* Device

As we did for the *CE* device, in the calculation of input and output impedances for the *CE* device, the feedback capacitor C_{gd} will be omitted, and the input capacitor C'_{gs} and output capacitor C'_{ds} will be replaced by $C_{i,miller}$ and $C_{o,miller}$, which can be translated from the expressions (12.77) and (12.78) respectively as follows:

$$C_{i,miller} = C_{gd}\left[1 + (g_m + g_{mb})\frac{r_o}{r_o + R_d}R_d\right], \tag{12.313}$$

and

$$C_{o,miller} = C_{gd}\left[1+\frac{r_o + R_d}{(g_m + g_{mb})r_o R_d}\right]. \tag{12.314}$$

The total input capacitance is therefore

$$C_{in} = C'_{gs} + C_{i,miller} = C'_{gs} + C_{gd}\left[1+(g_m + g_{mb})\frac{r_o}{r_o + R_d}R_d\right]. \tag{12.315}$$

And the total output capacitance is

$$C_{out} = C'_{ds} + C_{o,miller} = C'_{ds} + C_{gd}\left[1+\frac{r_o + R_d}{(g_m + g_{mb})r_o R_d}\right]. \tag{12.316}$$

Figure 12.42 shows the modified model of a *CS* device. The input and output impedances are:

$$Z_{in} = \frac{1}{jC_{in}\omega} = \frac{1}{j\omega\left\{C'_{gs} + C_{gd}\left[1+(g_m + g_{mb})\frac{r_o}{r_o + R_d}R_d\right]\right\}}. \tag{12.317}$$

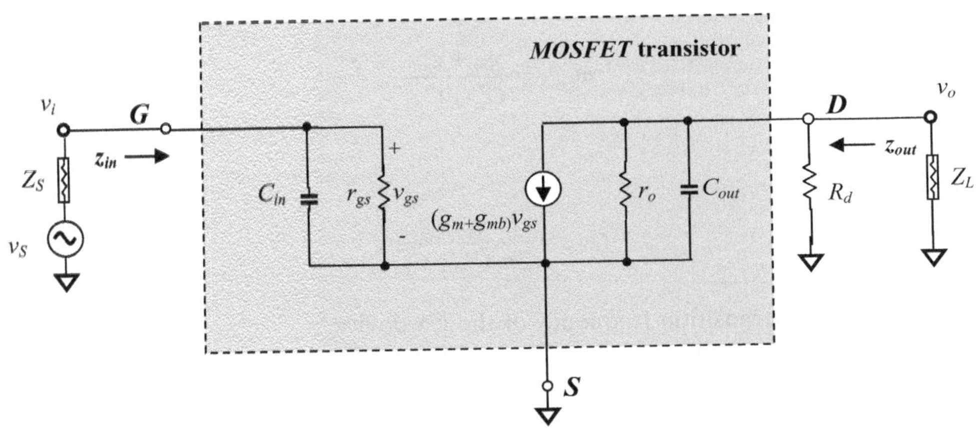

$$C_{in} = C_{gs}' + C_{gd}\left[1+(g_m + gmb)\frac{r_o}{r_o + R_d}R_d\right],$$

$$C_{out} = C_{ds}' + C_{gd}\left[1+\frac{r_o + R_d}{(g_m + g_{mb})r_o R_d}\right]$$

Figure 12.42 Modified model of a *CS* device with combined capacitors C_{in} and C_{out} for calculation of input and output impedance.

where

$$C'_{gs} = C_{gs} + \frac{1}{1 + A_{v,CS}\dfrac{C_{db}}{C_{sb}}} C_{gb}, \quad (12.286)$$

$$Z_{out} = \frac{r_o}{R_d + r_o + jC_{out}\omega r_o R_d} R_d. \quad (12.318)$$

In a low-frequency range, all the capacitors could be neglected. Thus we have

$$z_{in}|_{\omega \to 0} \to \infty, \quad (12.319)$$

$$z_{out\ \omega \to 0} = r_o \mathbin{//} R_d = \frac{r_o R_d}{r_o + R_d}. \quad (12.320)$$

12.9.4 Source Degeneration

Figure 12.43 shows a *MOSFET* device with a *CS* configuration and source degeneration. Its model would be simplified with the additional approximations:

- The capacitor C_{gd} is neglected because its value is much lower than that of C'_{gs}, that is,

$$C_{gd} << C'_{gs}. \quad (12.321)$$

Then, the equivalent model of the *CS* device with source degeneration shown in Figure 12.43(b) is simplified as shown in Figure 12.44.

Comparing Figure 12.44 with 12.13, we see that the capacitor C'_{ds} in Figure 12.44 does not correspond to the capacitor C_{cs} in Figure 12.13. Those relationships to be specially taken care of in the translation from the *CE* device with emitter degeneration to the *CS* device with source degeneration are listed in Table 12.3.

Consequently, from (12.99), we have

$$Z_{in} \approx \frac{1}{j\omega C'_{gs}} \left\{ 1 + \frac{r_o[(g_m + g_{mb}) + j\omega C'_{gs}] + j\omega C'_{gs} R_d (1 + j\omega C'_{ds} r_o)}{r_o + (R_d + R_s)(1 + j\omega C'_{ds} r_o)} R_s \right\}, \quad (12.322)$$

under the condition of

$$r_{gs} \to \infty. \quad (12.323)$$

If

$$r_o >> R_d, \quad (12.324)$$

$$r_o >> R_s, \quad (12.325)$$

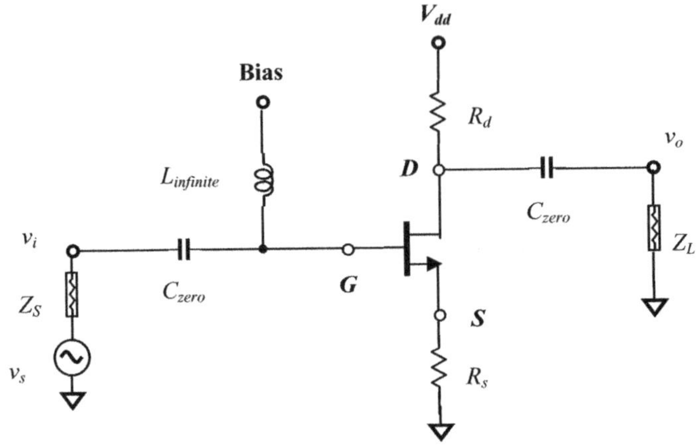

(a) Schematic of a *MOSFET* transistor with *CS* configuration and emitter degeneration C_{zero}: "zero" capacitor approaches zero capacitance at operating frequencies, $L_{infinite}$: "infinite" inductor approaches infinite inductance at operating frequencies.

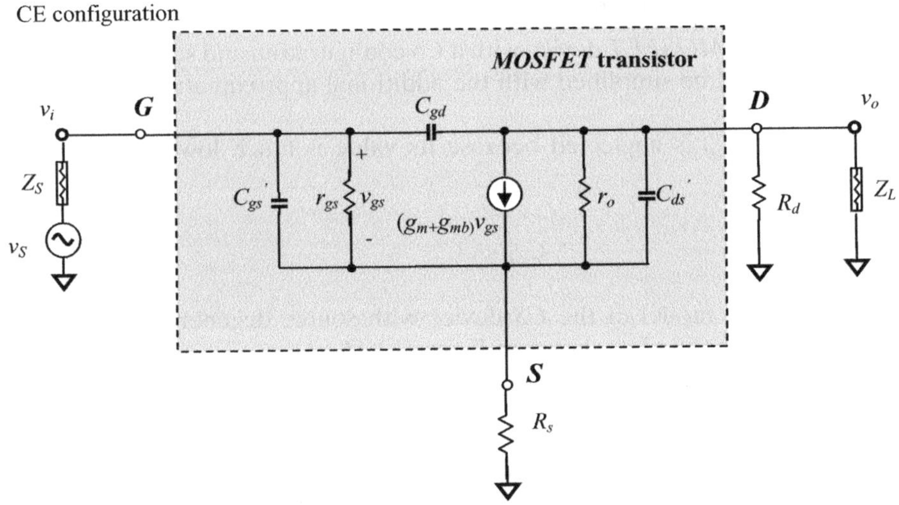

In *CS* Device: $C_{gs}' = C_{gs} + \dfrac{1}{1+A_v\dfrac{C_{db}}{C_{sb}}}C_{gb}$, $\quad C_{ds}' = \dfrac{1}{1+\dfrac{C_{db}}{C_{sb}}+\dfrac{C_{gb}}{A_v C_{sb}}}C_{db}$

(b) Modified model of a *MOSFET* transistor with *CS* configuration and source degeneration and with combined capacitors C_{gs}' and C_{ds}'

Figure 12.43 A bipolar transistor with *CE* configuration and emitter degeneration. Z_S: source impedance, Z_L: load impedance.

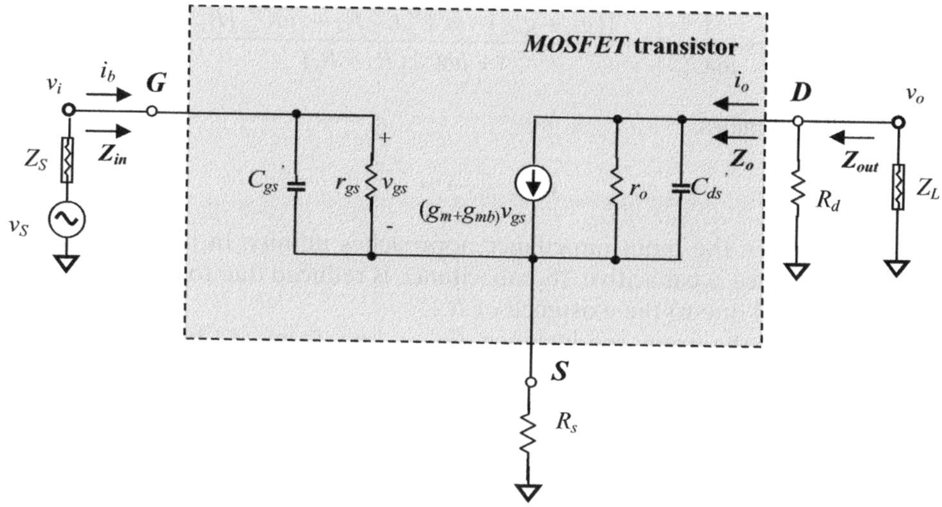

In *CS* Device: $C_{gs}' = C_{gs} + \dfrac{1}{1 + A_{v,CS}\dfrac{C_{db}}{C_{sb}}} C_{gb}, \quad C_{ds}' = \dfrac{1}{1 + \dfrac{C_{db}}{C_{sb}} + \dfrac{C_{gb}}{A_{v,CS} C_{sb}}} C_{db}$

Figure 12.44 Modified model of a *CS* device with source degeneration and with combined capacitors C_{gs}' and C_{ds}'.

TABLE 12.3 Special relationships to be taken care of in the translation of equations from *CE* with emitter degeneration to *CS* with source degeneration

From Bipolar model *CE*	To *MOSFET* model *CS*
r_π	r_{gs}
r_o	$r_o // C_{ds}'$
C_μ	C_{gd}
C_{cs}	C_{ds}'
C_π	C_{gs}'
$C_{cs} // R_c$	R_d
g_m	$(g_m + g_{mb})$
β_o	$(g_m + g_{mb}) r_{gs}$

$$Z_{in} \approx \frac{1}{j\omega C'_{gs}} \left\{ 1 + \frac{[(g_m + g_{mb}) - \omega^2 C'_{gs} C'_{ds} R_d + j\omega C'_{gs}] R_s}{1 + j\omega C'_{ds}(R_d + R_s)} \right\}. \tag{12.326}$$

And at low frequencies,

$$Z_{in}|_{\omega \to 0} \to \infty. \tag{12.327}$$

At low frequencies, the input impedance approaches infinity. In high frequencies, the input impedance is capacitive. Its capacitance is reduced due to the existence of R_s but is increased due to the existence of R_d.

Let's translate the trans-conductance $G_m = i_o/v_i$, from (12.105) with output short-circuited,

$R_d = 0.$

$$G_m = \frac{i_o}{v_i} = (g_m + g_{mb}) \frac{1 - \dfrac{R_s(1 + j\omega C'_{gs} r_{gs})}{(g_m + g_{mb}) r_{gs} \dfrac{r_o}{1 + j\omega C'_{ds} r_o}}}{1 + (g_m + g_{mb}) R_s \left[1 + \dfrac{(1 + j\omega C'_{gs} r_{gs})}{(g_m + g_{mb}) r_{gs}} + \dfrac{1 + j\omega C'_{ds} r_o}{(g_m + g_{mb}) r_o} \right]}. \tag{12.328}$$

Note that

$$r_{gs} \to \infty,$$

$$G_m \approx (g_m + g_{mb}) \frac{1 - \dfrac{j\omega C'_{gs} R_s (1 + j\omega C'_{ds} r_o)}{(g_m + g_{mb}) r_o}}{1 + R_s \left[(g_m + g_{mb}) + j\omega (C'_{gs} + C'_{ds}) + \dfrac{1}{r_o} \right]}, \tag{12.329}$$

and usually

$$r_o >> R_s, \tag{12.330}$$

$$(g_m + g_{mb}) r_o >> 1, \tag{12.331}$$

then

$$G_m \approx (g_m + g_{mb}) \frac{1 + \dfrac{\omega^2 C'_{gs} C'_{ds} R_s}{(g_m + g_{mb})}}{1 + R_s[(g_m + g_{mb}) + j\omega (C'_{gs} + C'_{ds})]}. \tag{12.332}$$

At low frequencies,

$$G_m|_{\omega \to 0} \approx \frac{(g_m + g_{mb})}{1 + (g_m + g_{mb}) R_s}. \tag{12.333}$$

From (12.328), (12.329), (12.332), and (12.333) it can be seen that G_m drops from $(g_m + g_{mb})$ due to the source resistor R_s. At low frequencies, G_m will drop to half of $(g_m + g_{mb})$ if $R_e = 1/(g_m + g_{mb})$.

The output impedance can be translated from (12.118), that is

$$Z_{out} = \frac{r_o}{1+j\omega C'_{ds}r_o} \frac{1+\left(\frac{1+j\omega C'_{ds}r_o}{r_o}+g_m+g_{mb}\right)\frac{r_{gs}R_s}{r_{gs}+R_s(1+j\omega C'_{gs}r_{gs})}}{1+\frac{r_o}{R_d}\frac{1}{(1+j\omega C'_{ds}r_o)}\left[1+\left(\frac{1+j\omega C'_{ds}r_o}{r_o}+g_m+g_{mb}\right)\frac{r_{gs}R_s}{r_{gs}+R_s(1+j\omega C'_{gs}r_{gs})}\right]}. \tag{12.334}$$

Note that

$$r_{gs} \to \infty,$$

$$Z_{out} = \frac{r_o}{1+j\omega C'_{ds}r_o} \frac{1+\left(\frac{1+j\omega C'_{ds}r_o}{r_o}+g_m+g_{mb}\right)\frac{R_s}{1+j\omega C'_{gs}R_s}}{1+\frac{r_o}{R_d}\frac{1}{(1+j\omega C'_{ds}r_o)}\left[1+\left(\frac{1+j\omega C'_{ds}r_o}{r_o}+g_m+g_{mb}\right)\frac{R_s}{1+j\omega C'_{gs}R_s}\right]}. \tag{12.335}$$

Usually,

$$r_o >> 1, \tag{12.336}$$

then

$$Z_{out} = \frac{1}{j\omega C'_{ds}} \frac{1+(j\omega C'_{ds}+g_m+g_{mb})\frac{R_s}{1+j\omega C'_{gs}R_s}}{1+\frac{r_o}{R_d}\frac{1}{(1+j\omega C'_{ds}r_o)}\left[1+(j\omega C'_{ds}+g_m+g_{mb})\frac{R_s}{1+j\omega C'_{gs}R_s}\right]}. \tag{12.337}$$

In cases of infinite R_d, that is,

$$R_d \to 0, \tag{12.338}$$

$$Z_{out}|_{R_d\to\infty} = \frac{1}{j\omega C'_{ds}}\left[\frac{1+j\omega C'_{gs}R_s+(j\omega C'_{ds}+g_m+g_{mb})R_s}{1+j\omega C'_{gs}R_s}\right]. \tag{12.339}$$

At low frequencies,

$$Z_{out}|_{\omega\to 0} \to \infty. \tag{12.340}$$

12.10 *MOSFET* TRANSISTOR WITH *CG* (COMMON GATE) CONFIGURATION

A *MOSFET* transistor with *CG* configuration is shown in Figure 12.45.

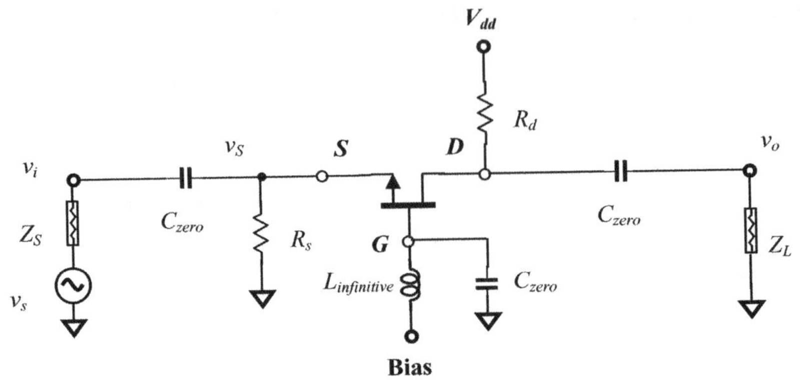

(a) Schematic of a *MOSFET* transistor with *CG* configuration. C_{zero}: "zero" capacitor approaches zero capacitance at operating frequencies, $L_{infinite}$: "infinitive" inductor approaches infinitive inductance at operating frequencies.

CE configuration

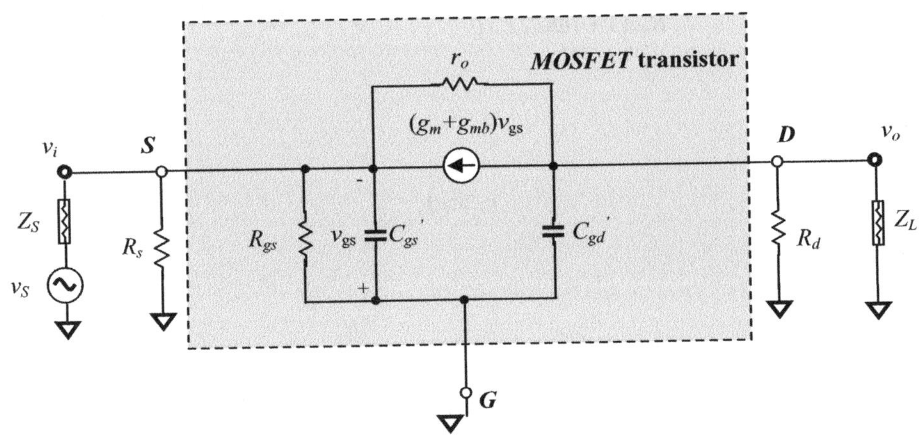

$$\textbf{In } \textit{\textbf{CG}} \textbf{ Device: } C_{gs}' = C_{gs} + \frac{1}{1 + A_v \dfrac{C_{db}}{C_{gb}}} C_{sb}, \quad C_{gd}' = C_{gd} + \frac{1}{1 + \dfrac{C_{db}}{C_{gb}} + \dfrac{C_{sb}}{A_v C_{gb}}} C_{db}$$

(b) Simplified model of a *MOSFET* transistor with *CG* configuration with two combined capacitors C_{gs}' and C_{gd}' and a combined current source

Figure 12.45 A *MOSFET* transistor with *CG* configuration. Z_S: source impedance, Z_L: load impedance.

12.10.1 Open-circuited Voltage Gain $A_{v,CG}$ of a *CG* Device

The modified model of a *CG* device with the combined capacitors, C'_{gs} and C'_{gd} shown in Figure 12.44(b) is quite similar to the model of the *CB* device shown in Figure 12.14. Based on the similarity between these two models, the open-circuited voltage gain of the *CG* device can be obtained by translation from the corresponding expressions in Section 12.5.1 in terms of the similar relationships listed in Table 12.1. Table 12.4 lists those relationships which must be particularly taken care of in the translation from a *CB* to a *CG* configuration.

The open-circuited voltage gain of a *CG* device can be translated from expression (12.133)

$$A_{v,CG} = \frac{r_o}{R_d + r_o} \frac{(g_m + g_{mb}) R_d \left[1 + \dfrac{1}{(g_m + g_{mb}) r_o}\right]}{1 + j\omega C'_{gd} \dfrac{(R_d + r_c)}{(R_d + r_o)} r_o}. \tag{12.341}$$

At low frequencies,

$$A_{v,CG}\big|_{\omega \to 0} = \frac{R_d r_o}{R_d + r_o} (g_m + g_{mb}) \left[1 + \frac{1}{(g_m + g_{mb}) r_o}\right]. \tag{12.342}$$

If

$$r_o >> R_d, \tag{12.343}$$

$$(g_m + g_{mb}) r_o >> 1, \tag{12.344}$$

then

$$A_{v,CG}\big|_{\omega \to 0} \approx (g_m + g_{mb}) R_d. \tag{12.345}$$

TABLE 12.4 Special relationships to be taken care of in the translation from *CB* to *CG* configuration

From Bipolar model *CB*	To *MOSFET* model *CG*
r_π	r_{gs}
r_o	r_o
$C_\mu + C_{cs}$	C'_{gd}
C_π	C'_{gs}
R_c	R_d
g_m	$(g_m + g_{mb})$
β_o	$(g_m + g_{mb}) r_{gs}$

12.10.2 Short-circuited Current Gain β_{CG} and Frequency Response of a *CG* Device

Figure 12.46 is drawn for the calculation of the short-circuited current gain and frequency response of a *MOSFET* transistor with a *CG* configuration. It is similar to Figure 12.14, which shows a bipolar transistor with a *CB* configuration. Therefore, its short-circuited current gain and frequency response can be translated from equation (12.143), as

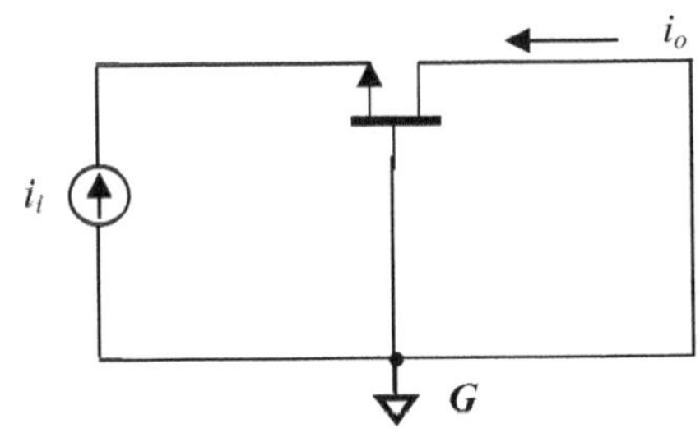

(a) Output is short-circuited for calculation of current gain β of a *CG* device

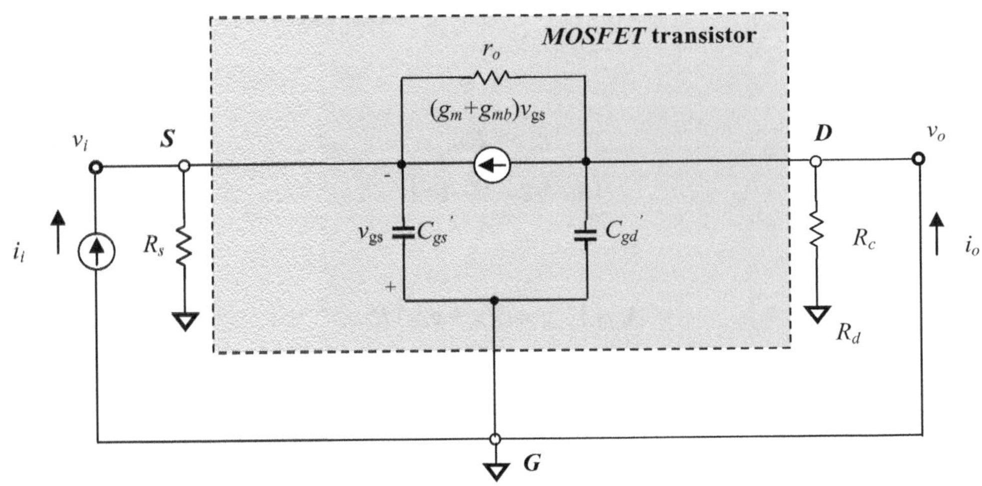

$$C_{gs}' = C_{gs} + \frac{C_{sb}}{1 + A_v \dfrac{C_{db}}{C_{gb}}}, \quad C_{gd}' = C_{gd} + \frac{C_{db}}{1 + \dfrac{C_{db}}{C_{gb}} + \dfrac{C_{sb}}{A_v C_{gb}}}$$

In *CG* Device

(b) Model of a *MOSFET* transistor with *CG* configuration with two combined capacitors C_{gs}' and C_{gd}' and a combined current source

Figure 12.46 Output short-circuited for calculating of current gain β and frequency response of a *MOSFET* with *CG* configuration.

$$\beta_{CG}(j\omega)=\frac{i_o}{i_i}=\frac{(g_m+g_{mb})r_{gs}}{(g_m+g_{mb})r_{gs}+1}\frac{1}{1+j\omega C'_{gs}\dfrac{r_{gs}}{(g_m+g_{mb})r_{gs}+1}}\approx\frac{1}{1+j\omega\dfrac{C'_{gs}}{(g_m+g_{mb})}},\tag{12.346}$$

because

$$r_{gs}\to\infty.\tag{12.347}$$

It can be seen that the current gain of the CG device is less than 1, that is,

$$|\beta_{CG}(j\omega)|<1.\tag{12.348}$$

At low frequencies,

$$\beta_{CG}(j\omega)|_{\omega\to 0}\approx 1.\tag{12.349}$$

The current gain of the CG device at low frequencies is approximately equal to 1.

Rewriting (12.346) in terms of (12.349), when $\beta_{CG}(j\omega)$ is $3\,dB$ lower from the low-frequency value β_{CG}, that is,

$$|\beta_{CG}(j\omega)|_{-3dB}=\frac{\beta_{CG}(j\omega)|_{\omega\to 0}}{\sqrt{2}}=\frac{1}{\sqrt{2}},\tag{12.350}$$

$$\omega=\omega_{\beta,CG}=\frac{g_m+g_{mb}}{C'_{gs}},\tag{12.351}$$

or,

$$f=f_{\beta,CG}=\frac{1}{2\pi}\frac{g_m+g_{mb}}{C'_{gs}}.\tag{12.352}$$

Figure 12.47 shows an example of the frequency response of a CG device.

12.10.3 Input and Output Impedances of a *CG* Device

Similarly to the discussion for the CB device, from Figure 12.46(b) we discuss the input and output impedances somewhat reluctantly because the dependent current source is located between the drain and the source. By the same scheme as in the discussion for a CB device, the equivalent model of the CG device can be modified to Figure 12.48.

Figure 12.48 is similar to Figure 12.20; the input impedances can be translated from the expressions (12.172) to (12.175), such that

$$Z_{in}=R_s\,//\frac{1}{(g_m+g_{mb})}\frac{1+\dfrac{1}{r_o}\dfrac{R_d}{1+j\omega C'_{gd}R_d}}{1+\dfrac{1}{(g_m+g_{mb})}\left[\dfrac{1}{r_o}+j\omega C'_{gs}+\dfrac{1}{r_o}j\omega C'_{gs}\dfrac{R_d}{1+j\omega C'_{gd}R_d}\right]},\tag{12.353}$$

It can be seen that the input impedance of a CG device is low, about $1/(g_m+g_{mb})$.

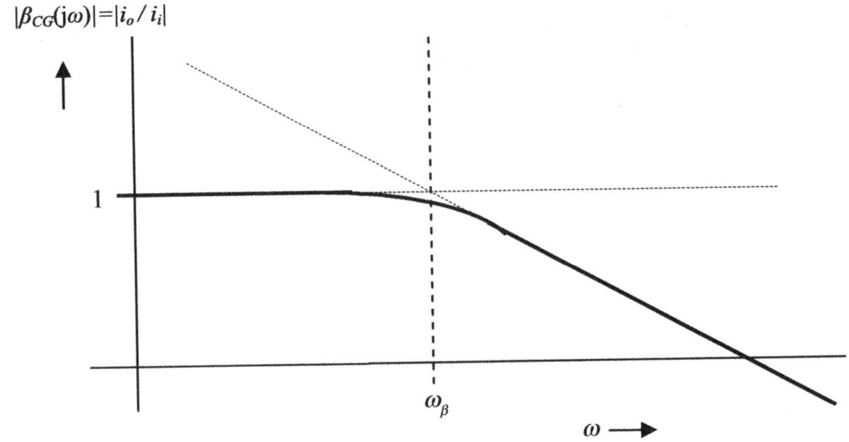

Figure 12.47 Frequency response of short-circuited current gain in a *CG* device.

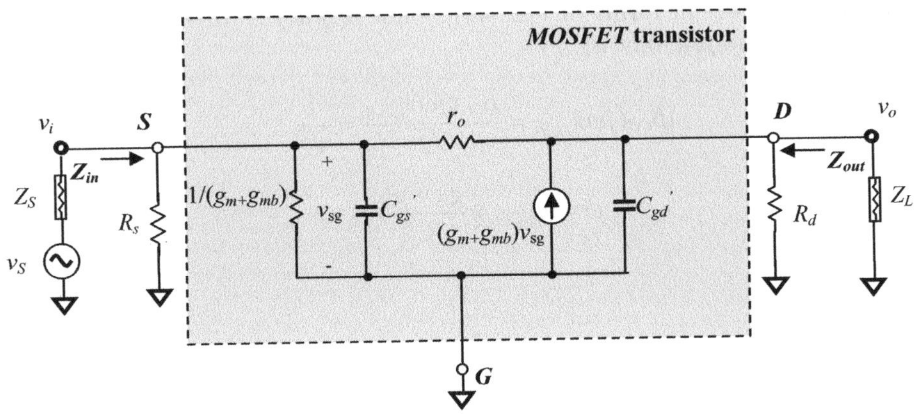

$$\text{In } CG \text{ Device: } C_{gs}' = C_{gs} + \frac{1}{1+A_v\dfrac{C_{db}}{C_{gb}}}C_{sb}, \quad C_{gd}' = C_{gd} + \frac{1}{1+\dfrac{C_{db}}{C_{gb}}+\dfrac{C_{sb}}{A_v C_{gb}}}C_{db}$$

Figure 12.48 Equivalent model of *MOSFET* transistor with *CG* configuration for calculation of input and output impedance.

It should be noted that the resistor R_s may be unnecessary in some applications, while the resistor R_d is an indispensable part because it leads the *DC* power supply to the device. For instance, in the *CS-CG* cascode amplifier, implementation of the resistor R_s is not necessary, so that in such a case, we have

$$Z_{in}\big|_{Rs\to\infty} = \frac{1}{(g_m+g_{mb})}\frac{1+\dfrac{1}{r_o}\dfrac{R_d}{1+j\omega C_{gd}'R_d}}{1+\dfrac{1}{(g_m+g_{mb})}\left[\dfrac{1}{r_o}+j\omega C_{gs}'+\dfrac{1}{r_o}j\omega C_{gs}'\dfrac{R_d}{1+j\omega C_{gd}'R_d}\right]}, \tag{12.354}$$

which corresponds to the case of

$$R_s \to \infty. \tag{12.355}$$

In low-frequency cases,

$$Z_{in}|_{\omega\to 0,\, R_s\to\infty} = \frac{1+\dfrac{R_d}{r_o}}{(g_m+g_{mb})+\dfrac{1}{r_o}}. \tag{12.356}$$

The output impedance is

$$Z_{out} = R_d \frac{1}{1+j\omega C'_{dg}R_d + R_d \dfrac{(1+j\omega C'_{gs}R_s)}{r_o + R_s[1+(g_m+g_{mb}+j\omega C'_{gs})r_o]}}. \tag{12.357}$$

In low-frequency cases,

$$Z_{out}|_{\omega\to 0} = R_d \frac{1}{1+R_d \dfrac{1}{r_o + R_s[1+(g_m+g_{mb})r_o]}}. \tag{12.358}$$

In cases without the resistor R_s, that is, $R_s \to \infty$,

$$Z_{out}|_{R_e\to\infty} = R_d \frac{1}{1+j\omega C'_{dg}R_d + \dfrac{j\omega C'_{gs}R_d}{1+(g_m+g_{mb}+j\omega C'_{gs})r_o}}. \tag{12.359}$$

In low-frequency cases,

$$Z_{out}|_{\omega\to 0,\, R_s\to\infty} = R_d. \tag{12.360}$$

12.11 *MOSFET* TRANSISTOR WITH *CD* (COMMON DRAIN) CONFIGURATION

Figure 12.49 shows the equivalent circuit of a *MOSFET* transistor with *CD* configuration, in which the drain is AC grounded, the input port is the base and the output port is the emitter of the bipolar transistor. The notation "*CD*" stands for the drain that is a common ground terminal of input (gate) port and output (source) port.

The modified model of the *CD* device with the combined capacitors C'_{gd} and C'_{ds} shown in Figure 12.49 is quite similar to the model of the *CC* device presented in Figure 12.22. Based on the similarity between these two models, most of the *CD* device parameters can be translated from the corresponding expressions of *CC* parameters in terms of the similar relationships listed in Table 12.1. Comparing

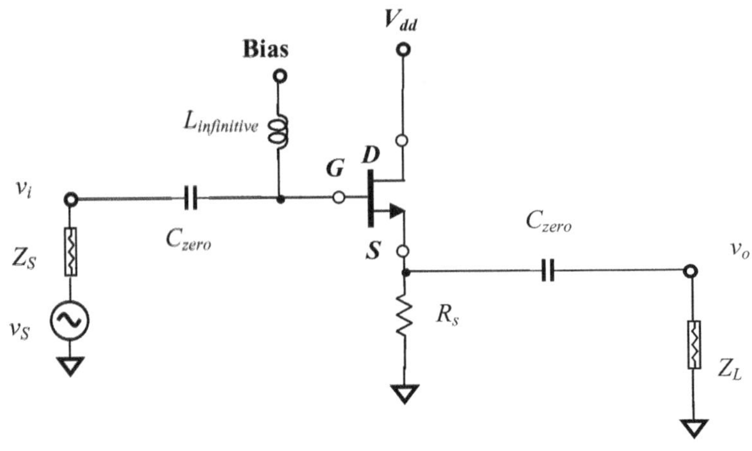

(a) Schematic of a *MOSFET* transistor with *CD* configuration. C_{zero}: "zero" capacitor, $L_{infinite}$: "infinite" inductor

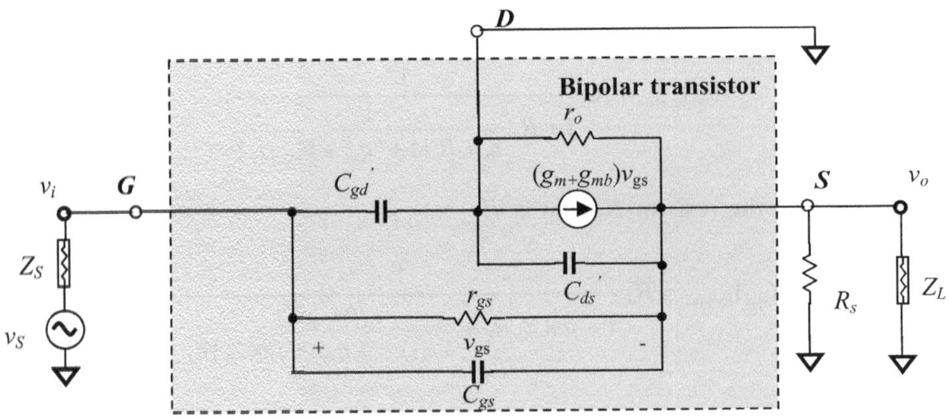

In *CD* Device: $C_{gd}' = C_{gd} + \dfrac{1}{1+2\dfrac{C_{gb}}{C_{db}}} C_{gb}$, $\quad C_{ds}' = \dfrac{1}{1+2\dfrac{C_{sb}}{C_{db}}} C_{sb}$

(b) Model of bipolar transistor with *CC* configuration

Figure 12.49 A *MOSFET* transistor with *CD* configuration, $r_{gs} \to \infty$. Z_S: source impedance, Z_L: load impedance.

Figure 12.49 with Figure 12.22, the capacitor C'_{ds} in the former does not correspond to the capacitor C_{es} in the latter. Therefore, in the translation of equations from the bipolar *CC* device with emitter degeneration to the *MOSFET CS* device with source degeneration, the special relationships listed in Table 12.5 must be taken care of.

TABLE 12.5 Special relationships to be taken care of in the translation from *CC* to *CD* configuration

From Bipolar model *CC*	To *MOSFET* model *CD*
r_π	r_{gs}
r_o	$r_o // C'_{ds}$
C_π	C'_{gs}
C_μ	C'_{gd}
$C_{cs}//R_c$	R_d
g_m	$(g_m + g_{mb})$
β_o	$(g_m + g_{mb})r_{gs}$

12.11.1 Open-circuited Voltage Gain $A_{v,CD}$ of a *CD* Device

The voltage gain A_v of a *CD* device is derived with the additional approximation: the capacitor C'_{gd} in Figure 12.49(b) is neglected.

The open-circuited voltage gain of a *CD* device can be translated from expression (12.204), that is,

$$A_{v,CD} = \frac{1}{1+\dfrac{1}{R_s(g_m+g_{mb})}+\dfrac{(1+j\omega C'_{ds}r_o)}{r_b(g_m+g_{mb})}}. \qquad (12.361)$$

In low-frequency cases,

$$A_{v,CD}\big|_{\omega\to 0} = \frac{1}{1+\dfrac{1}{(g_m+g_{mb})}\dfrac{r_b+R_s}{r_bR_s}}. \qquad (12.362)$$

12.11.2 Short-circuit Current Gain β_{CD} and Frequency Response of a *CD* Device

Figure 12.50 is drawn for the calculation of the short-circuited current gain and frequency response of a *MOSFET* transistor with a *CD* configuration, in which the additional approximation, $C'_{gd} \to 0$, is applied. It is similar to that of the *CC* device presented in Figure 12.24. Therefore, the short-circuited current gain and the frequency response can be translated from expression (12.214), that is,

$$\beta_{CD}(j\omega) = \frac{i_o}{i_i} = \frac{1+(g_m+g_{mb})r_{gs}+j\omega C_{gs}r_{gs}}{1+j\omega C_{gs}r_{gs}} \approx 1+\frac{(g_m+g_{mb})}{j\omega C_{gs}}. \qquad (12.363)$$

In low-frequency cases,

$$\beta_{CD}(j\omega)\big|_{\omega\to 0} \to \infty. \qquad (12.364)$$

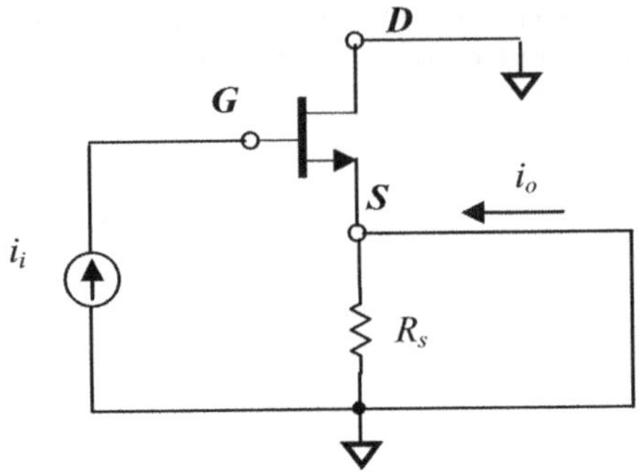

(a) Output is short-circuited for calculation of current gain β

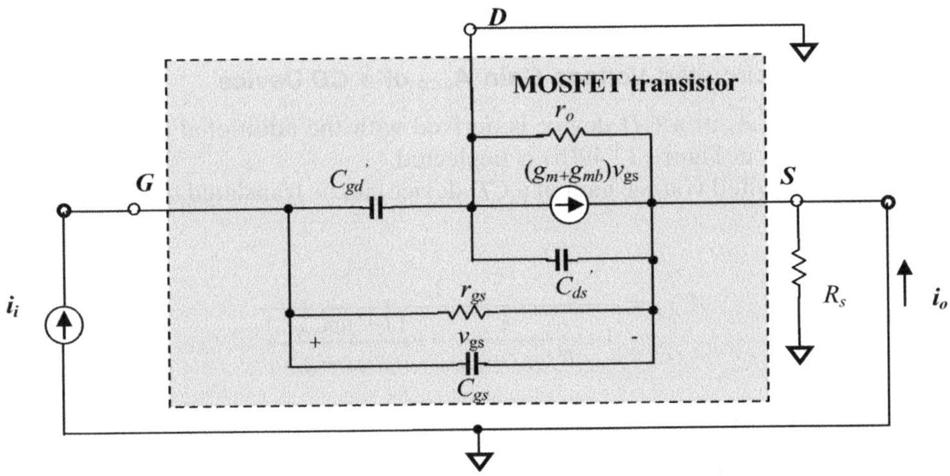

In *CD* Device: $$C_{gd}' = C_{gd} + \frac{1}{1+2\dfrac{C_{gb}}{C_{db}}}C_{gb}, \quad C_{ds}' = \frac{1}{1+2\dfrac{C_{sb}}{C_{db}}}C_{sb}$$

(b) Modified model of a *MOSFET* transistor with *CD* configuration, $r_{gs} \to \infty$

Figure 12.50 Schematic and equivalent for calculation of short-circuited current gain β of a *CD* device.

12.11.3 Input and Output Impedances of a *CD* Device

Figure 12.51 shows the schematic and its equivalent model of a *MOSFET* transistor with a *CD* configuration for the calculation of the input and output impedances. It is similar to that for the *CC* device as shown in Figure 12.25. Therefore, the input and output impedances can be translated from equation (12.230), that is

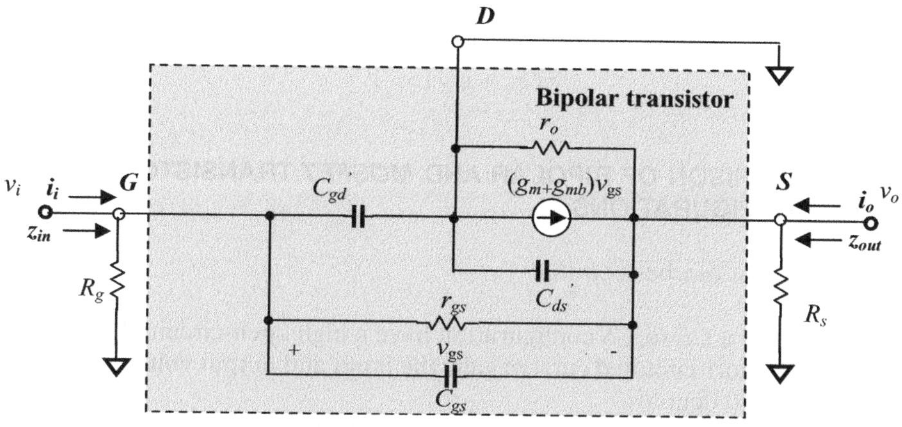

In *CD* Device: $C_{gd}' = C_{gd} + \dfrac{1}{1+2\dfrac{C_{gb}}{C_{db}}} C_{gb}, \quad C_{ds}' = \dfrac{1}{1+2\dfrac{C_{sb}}{C_{db}}} C_{sb}$

Figure 12.51 Equivalent circuit for calculation of input and output impedance of a *CD* device.

$$Z_{in} = \frac{1 + \dfrac{R_s r_o}{R_s(1+j\omega C_{ds}') + r_o}[(g_m + g_{mb}) + j\omega C_{gs}]}{jC_{gs}\omega}. \tag{12.365}$$

In low-frequency cases,

$$Z_{in}|_{\omega \to 0} \to \infty. \tag{12.366}$$

The output impedance of the *CD* device can be translated from (12.235), that is,

$$Z_{out} = \frac{1}{1 + \left[\dfrac{(g_m + g_{mb} + j\omega C_{gs})}{(1 + j\omega C_{gs} R_g)} + j\omega C_{ds}'\right] r_o} r_o, \tag{12.367}$$

In low-frequency cases,

$$Z_{out}|_{\omega \to 0} = \frac{1}{1 + r_o(g_m + g_{mb})} r_o, \tag{12.368}$$

$$Z_{out}|_{\omega \to 0} \approx \frac{1}{g_m + g_{mb}}, \tag{12.369}$$

if

$$r_o(g_m + g_{mb}) >> 1. \tag{12.370}$$

12.12 COMPARISON OF BIPOLAR AND *MOSFET* TRANSISTORS IN VARIOUS CONFIGURATIONS

From Table 12.6 it can be seen that

- Devices with a *CE* or *CS* configuration have a high open-circuited voltage gain and a high short-circuited current gain; the input and output voltages are phase-shifted by 180 degrees.
- Devices with a *CB* or *CG* configuration have the highest open-circuited voltage gain and lowest short-circuited current gain; their current gain is less than but close to 1;
- Devices with a *CC* or *CD* configuration have the lowest open-circuited voltage gain and highest short-circuited current gain.

From Table 12.7 it can be seen that

- Devices with a *CE* or *CS* configuration have a high input impedance and a high output impedance;
- Devices with a *CB* or *CG* configuration have the lowest input impedance and highest output impedance. These are the very special features of a *CB* or *CG* device;

TABLE 12.6 List of open-circuited voltages and short-circuited current gains of device with different configurations in the very low-frequency cases, $\omega \to 0$

Configuration	Voltage gain, A_v	Current gain, β or A_i
CE	$-g_m \dfrac{R_c r_o}{R_c + r_o}$	β_o
CB	$g_m\left(1+\dfrac{1}{g_m r_o}\right)\dfrac{R_c r_o}{R_c + r_o}$	α_o
CC	$\dfrac{1}{1+\dfrac{1}{g_m}\left(\dfrac{1}{R_e}+\dfrac{1}{r_o}\right)}$	$1 + \beta_o$
CS	$-(g_m + g_{mb})\dfrac{R_d r_o}{R_d + r_o}$	$\to\infty$
CG	$(g_m + g_{mb})\left[1+\dfrac{1}{(g_m + g_{mb})r_o}\right]\dfrac{R_d r_o}{R_d + r_o}$	$\to 1$
CD	$\dfrac{1}{1+\dfrac{1}{(g_m + g_{mb})}\left(\dfrac{1}{R_s}+\dfrac{1}{r_b}\right)}$	$\to\infty$

TABLE 12.7 Input and output impedances of device with different configuration in the very low-frequency cases, $\omega \to 0$

Configuration	*Input impedance*	*Output impedance*
CE	$r_\pi + r_b$	$(r_c + r_o)//R_c$
CB	$r_\pi \dfrac{1}{1+(1+g_m r_o)\dfrac{r_\pi}{r_o + r_c + R_c}}$	$R_c \dfrac{1}{1+\dfrac{r_c + R_c}{r_\pi + (1+\beta_o) r_o}}$
CC	$r_\pi + r_b + \beta_o R_e$	$\dfrac{1}{g_m} + \dfrac{R_b + r_b}{\beta_o}$
CS	$\to\infty$	$(r_c + r_o)//R_c \dfrac{r_o R_d}{r_o + R_d}$
CG	$\dfrac{1+\dfrac{R_d}{r_o}}{(g_m + g_{mb}) + \dfrac{1}{r_o}}$	R_d
CD	$\to\infty$	$\dfrac{1}{(g_m + g_{mb}) + \dfrac{1}{r_o}}$

TABLE 12.8 Transition frequency, general expression of input and output impedance

CE

$$\omega_{T,CE} = \frac{g_m}{C_\pi + C_\mu}, \tag{12.64}$$

$$\omega_{\beta,CE} = \frac{1}{\beta_o}\frac{g_m}{C_\pi + C_\mu} = \frac{\omega_{T,CE}}{\beta_o}, \tag{12.67}$$

$$Z_{in} = \frac{r_b + r_\pi + j\omega C_\pi r_\pi r_b}{1 + j\omega C_\pi r_\pi}, \tag{12.70}$$

$$Z_{out} = \frac{r_c + r_o + jC_{cs}\omega r_c r_o}{R_c + r_c + r_o + jC_{cs}\omega r_o(r_c + R_c)} R_c. \tag{12.73}$$

***CE* with emitter degeneration**

$$Z_{in} = \frac{v_i}{i_b} = \frac{1}{1 + j\omega C_\pi r_\pi}\left(r_\pi + \frac{r_o(1 + j\omega C_{cs} R_c)(\beta_o + 1 + j\omega C_\pi r_\pi) + R_c(1 + j\omega C_\pi r_\pi)}{R_c + (r_o + R_e)(1 + j\omega C_{cs} R_c)} R_e\right), \tag{12.99}$$

$$Z_{out} = r_o \frac{1 + \left(\dfrac{1}{r_o} + g_m\right)\dfrac{r_\pi R_e}{r_\pi + R_e(1 + j\omega C_\pi r_\pi)}}{1 + \dfrac{r_o}{R_c}(1 + j\omega C_{cs} R_c)\left[1 + \left(\dfrac{1}{r_o} + g_m\right)\dfrac{r_\pi R_e}{r_\pi + R_e(1 + j\omega C_\pi r_\pi)}\right]}, \tag{12.118}$$

TABLE 12.8 Continued

CB

$$\omega = \omega_{\beta,CB} = \frac{g_m}{\alpha_o C_\pi}, \tag{12.149}$$

$$Z_{in} = \frac{v_i}{i_i} = R_e // \frac{r_\pi}{1+\beta_o} \frac{1 + \frac{1}{r_o}\frac{r_c + R_c}{1 + j\omega C_{cs}(r_c + R_c)}}{1 + \frac{r_\pi}{1+\beta_o}\left[\frac{1}{r_o} + j\omega C_\pi + \frac{1}{r_o}\left(\frac{1}{r_\pi} + j\omega C_\pi\right)\frac{r_c + R_c}{1 + j\omega C_{cs}(r_c + R_c)}\right]}, \tag{12.172}$$

$$Z_{out} = R_c \frac{1}{1 + j\omega C_{cs}(r_c + R_c) + (r_c + R_c)\frac{(r_\pi + R_e + j\omega C_\pi R_e r_\pi)}{r_o r_\pi + R_e[r_\pi + r_o(\beta_o + 1) + j\omega C_\pi r_\pi r_o]}}, \tag{12.188}$$

CC

$$Z_{in} = \frac{r_\pi + r_b(1 + jC_\pi \omega r_\pi) + (1 + \beta_o + jC_\pi \omega r_\pi)\frac{r_o + r_c + jC_{cs}\omega r_o r_c}{R_e + r_o + r_c + jC_{cs}\omega r_c(r_o + R_e)}R_e}{1 + jC_\pi \omega r_\pi}. \tag{12.230}$$

$$Z_{out} = \frac{r_\pi + (R_b + r_b)(1 + j\omega C_\pi r_\pi)}{(1 + \beta_o + j\omega C_\pi r_\pi) + (1 + j\omega C_{cs} r_c)\frac{r_\pi + (R_b + r_b)(1 + j\omega C_\pi r_\pi)}{r_c + r_o(1 + j\omega C_{cs} r_c)}} \tag{12.235}$$

CS

$$\omega_{T,CS} = \frac{g_m + g_{mb}}{C'_{gs} + C_{gd}}, \tag{12.311}$$

$$Z_{in} = \frac{1}{jC_{in}\omega} = \frac{1}{j\omega\left\{C'_{gs} + C_{gd}\left[1 + (g_m + g_{mb})\frac{r_o}{r_o + R_d}R_d\right]\right\}}, \tag{12.317}$$

where

$$C'_{gs} = C_{gs} + \frac{1}{1 + A_{v,CS}\frac{C_{db}}{C_{sb}}}C_{gb}. \tag{12.286}$$

$$Z_{out} = \frac{r_o}{R_d + r_o + jC_{out}\omega r_o R_d}R_d. \tag{12.318}$$

***CS* with emitter degeneration**

$$Z_{in} \approx \frac{1}{j\omega C'_{gs}}\left\{1 + \frac{r_o[(g_m + g_{mb}) + j\omega C'_{gs}] + j\omega C'_{gs}R_d(1 + j\omega C'_{ds}r_o)}{r_o + (R_d + R_s)(1 + j\omega C'_{ds}r_o)}R_s\right\}, \tag{12.322}$$

$$Z_{out} = \frac{r_o}{1 + j\omega C'_{ds}r_o}\frac{1 + \left(\frac{1 + j\omega C'_{ds}r_o}{r_o} + g_m + g_{mb}\right)\frac{R_s}{1 + j\omega C'_{gs}R_s}}{1 + \frac{r_o}{R_d}\frac{1}{(1 + j\omega C'_{ds}r_o)}\left[1 + \left(\frac{1 + j\omega C'_{ds}r_o}{r_o} + g_m + g_{mb}\right)\frac{R_s}{1 + j\omega C'_{gs}R_s}\right]}, \tag{12.335}$$

TABLE 12.8 Continued

CG		
	$$\omega = \omega_{\beta,CG} = \frac{g_m + g_{mb}}{C'_{gs}},$$	(12.351)
	$$Z_{in} = R_s // \frac{1}{(g_m + g_{mb})} \frac{1 + \frac{1}{r_o} \frac{R_d}{1 + j\omega C'_{gd} R_d}}{1 + \frac{1}{(g_m + g_{mb})} \left[\frac{1}{r_o} + j\omega C'_{gs} + \frac{1}{r_o} j\omega C'_{gs} \frac{R_d}{1 + j\omega C'_{gd} R_d} \right]},$$	(12.353)
	$$Z_{out} = R_d \frac{1}{1 + j\omega C'_{dg} R_d + R_d \frac{(1 + j\omega C'_{gs} R_s)}{r_o + R_s[1 + (g_m + g_{mb} + j\omega C'_{gs}) r_o]}},$$	(12.357)
CD		
	$$Z_{in} = \frac{1 + \frac{R_s r_o}{R_s(1 + j\omega C'_{ds}) + r_o} [(g_m + g_{mb}) + j\omega C_{gs}]}{jC_{gs}\omega},$$	(12.365)
	$$Z_{out} = \frac{1}{1 + \left[\frac{(g_m + g_{mb} + j\omega C_{gs})}{(1 + j\omega C_{gs} R_g)} + j\omega C'_{ds} \right] r_o} r_o.$$	(12.367)

- Devices with a *CC* or *CD* configuration have the highest input impedance and lowest output impedance.

For convenience, the formula of the transition frequency and the general expressions of input and output impedances are listed in Table 12.8.

REFERENCES

[1] Jack Smith, *Modern Communication Circuits*, McGraw-Hill, 1986.

[2] Paul R. Gray, Paul J. Hurst, Stephen H. Lewis, and Robert G. Meyer, *Analysis and Design of Analog Integrated Circuits*, 4th ed., John Wiley & Sons, Inc., 2001.

[3] Richard Chi-Hsi Li, *Key Issues in RF/RFIC Circuit Design*, Higher Education Press, Beijing, 2005.

CHAPTER 13

IMPEDANCE MEASUREMENT

13.1 INTRODUCTION

As mentioned in Chapter 1, the task of a digital circuit block in a communication system is status transportation or manipulation. The so-called "status" of a digital signal is characterized by "0" or "1." Direct observation of the signal waveform, including voltage at a node or current flowing through a part, is the best way to judge the performance of the circuit block, including its sequence of digits, repetition frequency, voltage level, rise and fall time, jitter and stability, and so on. Therefore, the main test equipment is the oscilloscope since the testing of a digital circuit block is in the time domain.

On the other hand, the task of an *RF* circuit block in a communication system is power transportation or manipulation. In order to ensure effective transport and manipulation of power, the input and output impedances of the *RF* block must be matched with the output impedance of the previous block and the input impedance of the next block, respectively. On the other hand, the frequency bandwidth of the block must also be taken care of. The observation of the impedance matching state, power intensification, and attenuation over the entire frequency bandwidth is the best way to judge the performance of the circuit block, including spurious products, noise, non-linearity, and so on. Therefore, the main test equipment to be used is the network analyzer, since the testing of the *RF* circuit block is in the frequency domain.

Impedance measurement takes first priority in *RF* circuit testing. It is the starting point for the implementation of an impedance matching network. Also, it is a powerful means of examining the performance of an *RF* block. The performance of power transportation or manipulation of an *RF* block is directly related to how well its impedance matching has been taken care of.

In the case of a small signal, the impedance is usually tested by the network analyzer. In the case of a large signal, the impedance is usually tested with the assistance of the circulator. The impedance of discrete parts can be tested by the network analyzer as well as an impedance meter.

The *RF* circuit is always tested through power measurement by a variety of test equipment, such as the network analyzer, spectrum analyzer, power meter, and so on. Voltage measurement by an oscilloscope is basically not helpful in characterizing the performance of *RF* circuitry. Almost all the parameters which specify the performance, such as the power gain, noise figure, and intercept points, are computed through the power measurements. However, in *RF* circuit design, voltage measurement is not neglected. There are two kinds of voltage measurement. One is the measurement of scale voltage by an oscilloscope as mentioned above. A scale voltage is the resultant voltage of incident and reflective voltages at one node. The other is measurement of the vector voltage by a vector voltmeter. *RF* designers are not interested in the scale voltage, but in vector voltage. The vector voltmeter can distinguish the incident and reflected voltages at one node instead of just the resultant voltage. The operating principle of a network analyzer is based on the measurement of vector voltage, in which the incident and reflective voltages are measured simultaneously.

13.2 SCALE AND VECTOR VOLTAGE MEASUREMENT

13.2.1 Voltage Measurement by Oscilloscope

An oscilloscope can test or measure the voltage at any node in a circuitry and display its waveform on its screen. This intuitive feeling of display is very nice to everyone. As electronic technology progresses, the frequency response of an oscilloscope can approach up to tens of *GHz* and more than 10 waveforms can be displayed on the same screen simultaneously. It is a powerful tool in digital circuit design.

This powerful tool raises several questions: Why is there not an oscilloscope to be found in any advanced *RF* circuit test lab? Why do *RF* circuit designers never use the oscilloscope in *RF* circuit testing?

The answer can be explained by Figure 13.1. For simplicity, the actual layout on the *PCB* is replaced by the simple schematic. There are three circuitry branches coming together at node *P*: one with a *MOSFET* transistor *M*, one with an inductor *L*, and one with a capacitor *C*, while the rest of the circuitry is connected between the capacitor *C* and the output *SMA* connector in series. The impedance of a probe is usually quite high so that it will not disturb the performance of the circuitry when it touches the test node *P*. When it senses the voltage *V* at node *P*, the oscilloscope magnifies and displays the voltage on its screen as a waveform in the time domain. The waveform at node *P* is a resultant voltage contributed by the three circuit branches. For a digital circuit block, this does indeed describe the "status" at node *P*, either that of high voltage, corresponding to "1," or that of low voltage, corresponding to "0."

If the input impedance of the oscilloscope is not high enough, a buffer with high input and output impedances can be inserted between the oscilloscope and the *DTU* as shown in Figure 13.2.

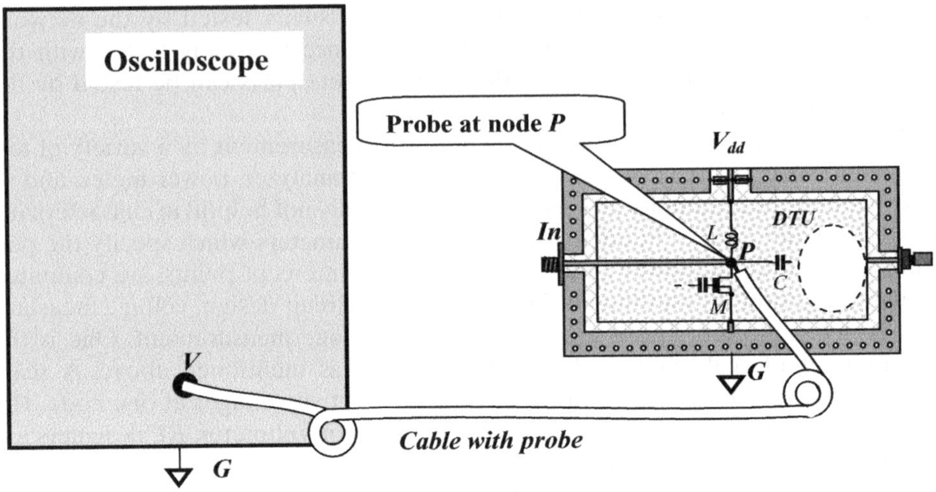

Figure 13.1 Voltage measured at node P by an oscilloscope or a regular voltmeter.
Top metallic area;
Bottom metallic area;
o Conductive via from top to bottom;
"Zero" capacitor
50 Ω terminator
DTU (Desired Test Unit)
Runner, $Z_o = 50\,\Omega$
The rest of circuitry
SMA connector
50 Ω cable

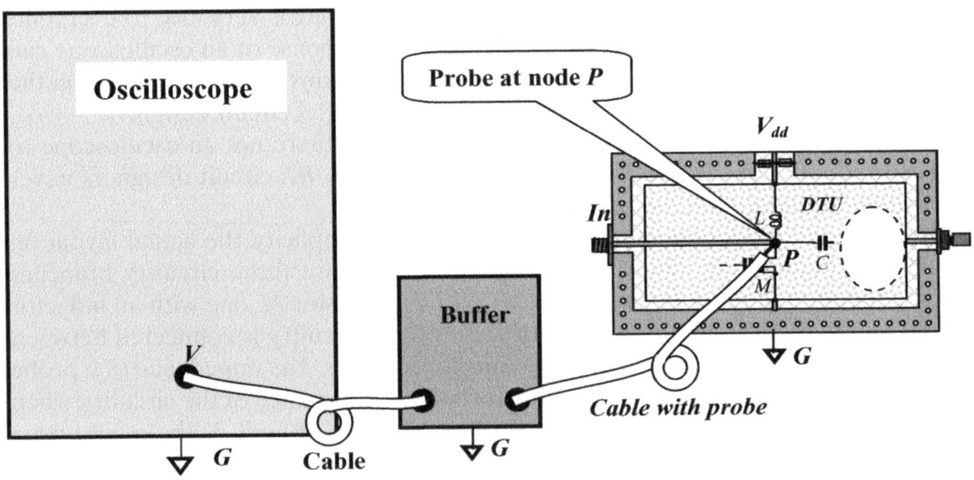

Figure 13.2 Voltage measured at node P by an oscilloscope with assistance of a buffer.
Top metallic area
Bottom metallic area
o Conductive via from top to bottom;
"Zero" capacitor
50 Ω terminator
DTU (Desired Test Unit)
Runner, $Z_o = 50\,\Omega$
The rest of circuitry
SMA connector
50 Ω cable

The oscilloscope can only read the resultant voltage; it cannot distinguish the incident and reflective voltages in the cable with the probe. Such a test or measurement is non-directive.

In *RF* circuit design, such a test is insufficient. The voltage measurement for an *RF* block must distinguish the incident and reflected voltages so that the impedance can be computed. Therefore, voltage measurement by oscilloscope is not what an *RF* circuit designer asks for and in most cases is of limited usefulness in *RF* circuit design.

13.2.2 Voltage Measurement by Vector Voltmeter

Voltage can also be tested or measured by a vector voltmeter. It can distinguish the incident and the reflective voltages at one node so that the test or measurement is directive.

Figure 13.3 shows the test setup. In addition to a vector voltmeter, a signal generator and a bidirectional coupler are needed. The signal generator provides a voltage signal to the input of the bidirectional coupler. The output of the bidirectional coupler feeds the voltage signal to the input *SMA* connector of the *PCB* with *DTU*.

An oscilloscope simply magnifies and displays the resultant voltage that is sensed by a cable with a probe at node *P*. It is a "passive test or measurement." In the voltage test or measurement by a vector voltmeter, a voltage signal is provided by the signal generator and is fed to the test node *P*. It is an "active test or measurement."

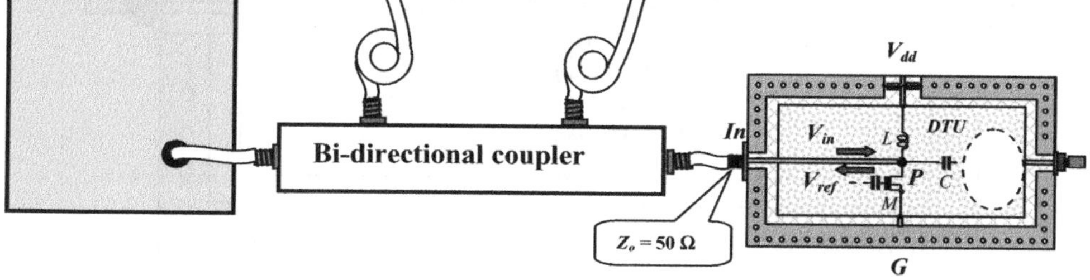

Figure 13.3 An example of test *PCB* and test setup.
Top metallic area
Bottom metallic area
o Conductive via from top to bottom
"Zero" capacitor
SMA connector
DTU
The rest of circuitry
50 Ω terminator
50 Ω cable

Another remarkable aspect is that in the voltage test by vector voltmeter, a micro strip line with 50 Ω of characteristic impedance must be connected from the input *SMA* connector to the tested node *P*, so that the vector voltmeter can sense the incident voltage from the signal generator at port A and the reflective voltage returned from the tested node *P* at port B without conceivable additional attenuation. In the oscilloscope testing or measuring, the micro strip line with 50 Ω is not needed. Its high-impedance probe touches the tested node *P* directly.

Now let's take a look at the various voltages around the test node *P* in a little more detail. Figure 13.4 depicts only the *DTU* with its various incident and reflective voltages.

There are two resultant voltages on the 50 Ω micro strip line: the resultant incident voltage V_{in} and the resultant reflective voltage V_{ref}, that is,

$$V_{in} = V_{in1} + V_{in2} + V_{in3}, \tag{13.1}$$

$$V_{ref} = V_{ref1} + V_{ref2} + V_{ref3}, \tag{13.2}$$

where

V_{in} = incident voltage at node *P*,
V_{ref} = reflected voltage at node *P*,
$V_{in,L}$ = incident voltage from branch *L* at node *P*,

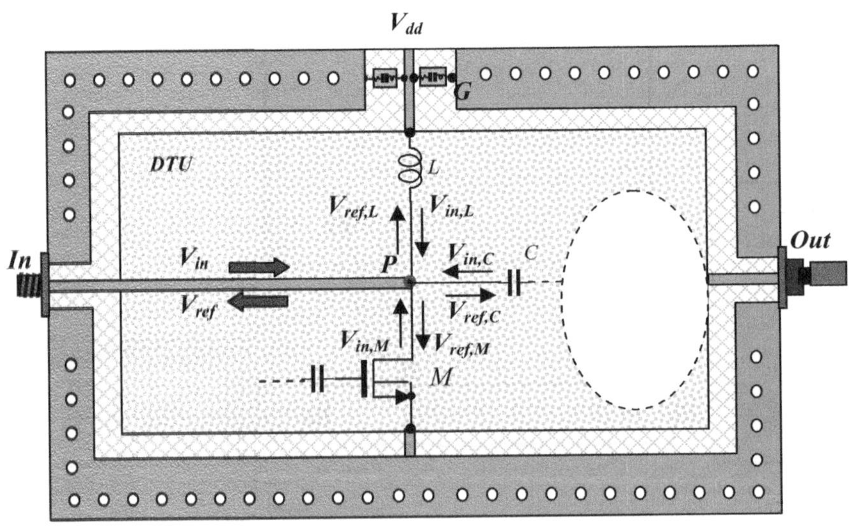

Figure 13.4 The various incident and reflective voltages at node ***P***.
Top metallic area
Bottom metallic area
o Conductive via from top to bottom
"Zero" capacitor
50 Ω terminator
DTU (Desired Test Unit)
Runner, Z_o = 50 Ω
The rest of circuitry
SMA connector

$V_{ref,L}$ = reflected voltage to branch L at node P,
$V_{in,C}$ = incident voltage from branch C at node P,
$V_{ref,C}$ = reflected voltage to branch C at node P,
$V_{in,M}$ = incident voltage from branch M at node P,
$V_{ref,M}$ = reflected voltage to branch M at node P.

The vector voltmeter can distinguish and read the resultant incident and reflective voltages, V_{in} and V_{ref}, but not the individual incident and reflective voltages, $V_{in,L}$, $V_{ref,L}$, $V_{in,C}$, $V_{ref,C}$, $V_{in,M}$, $V_{ref,M}$. However, this is enough to calculate the impedance at node P, such as

$$\Gamma = \frac{V_{ref}}{V_{in}}, \tag{13.3}$$

$$Z = \frac{1+\Gamma}{1-\Gamma}, \tag{13.4}$$

where

Γ = voltage reflection coefficient at node P;
z = normalized impedance looking into node P;

13.3 DIRECT IMPEDANCE MEASUREMENT BY NETWORK ANALYZER

13.3.1 Direction of Impedance Measurement

As mentioned above, the operation of the network analyzer is based on the operating principle of the vector voltmeter. The impedance is calculated from the directive voltage measurement. The direction of voltage is referenced to the node, whether the voltage is incident into or reflected from the node.

The direction of impedance is based on another reference. The direction of impedance is referenced to the input or output of a circuit stage, instead of a node. Figures 13.5 to 13.7 show three impedance measurements at node P.

Figure 13.5 shows an input impedance measurement. The impedance $\mathbf{Z}_{\text{in}}$ is looking into the node P, which is connected only with the input sub-circuit branch.

Figure 13.6 shows an output impedance measurement. The impedance $\mathbf{Z}_{\text{out}}$ is looking into the node P, which is connected with two output sub-circuit branches. One is the inductor L between the drain of the *MOSFET* transistor and the *DC* power supply V_{dd}, the other is the *MOSFET* transistor M. They are connected in parallel.

Figure 13.7 shows a non-directive impedance measurement. The impedance $\boldsymbol{Z}$ is looking into the node P, which is connected with three sub-circuit branches. The first is the inductor L between the drain of *MOSFET* transistor and the *DC* power supply V_{dd}; the second is the *MOSFET* transistor M; and the third is the branch with the capacitor C at output. They are connected in parallel.

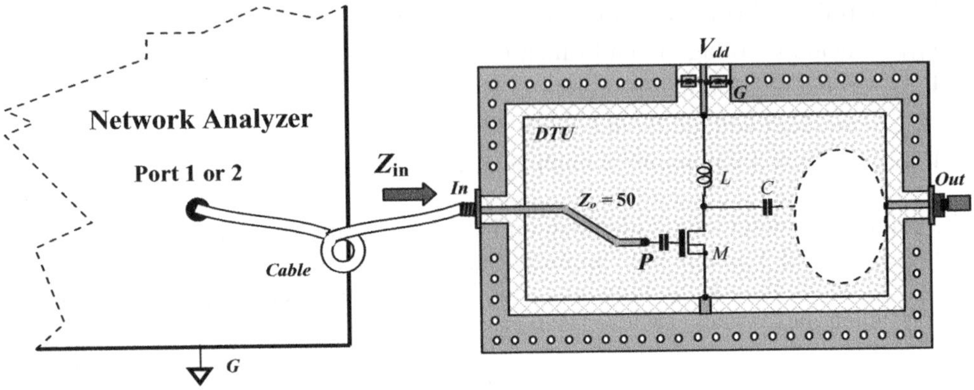

Figure 13.5 Directive impedance measurement for input impedance $\mathbf{Z}_{in}$. At node P, only input sub-circuit is connected.

- Top metallic area
- Bottom metallic area
- o Conductive via from top to bottom
- "Zero" capacitor
- *SMA* connector
- ***DTU***
- The rest of circuitry
- 50 Ω terminator
- 50 Ω cable

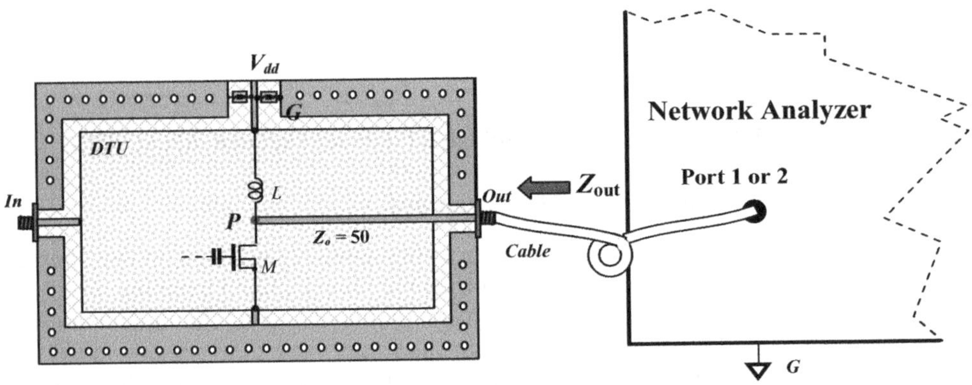

Figure 13.6 Output impedance measurement for output impedance $\mathbf{Z}_{out}$. At node P, 2 sub-circuit branches are connected in parallel.

- Top metallic area
- Bottom metallic area
- o Conductive via from top to bottom
- "Zero" capacitor
- SMA connector
- ***DTU***
- The rest of circuitry
- 50Ω terminator
- 50Ω cable

13.3.2 Advantage of Measuring *S* Parameters

In digital circuit design, or in the early stages of *RF* circuit design, there are many sets of parameters to describe a two-port network, including impedance $\mathbf{Z}$ parameters, admittance $\boldsymbol{Y}$ parameters, and hybrid $\boldsymbol{h}$ parameters. They are outlined in Figure 13.8 with both matrix and block diagrams.

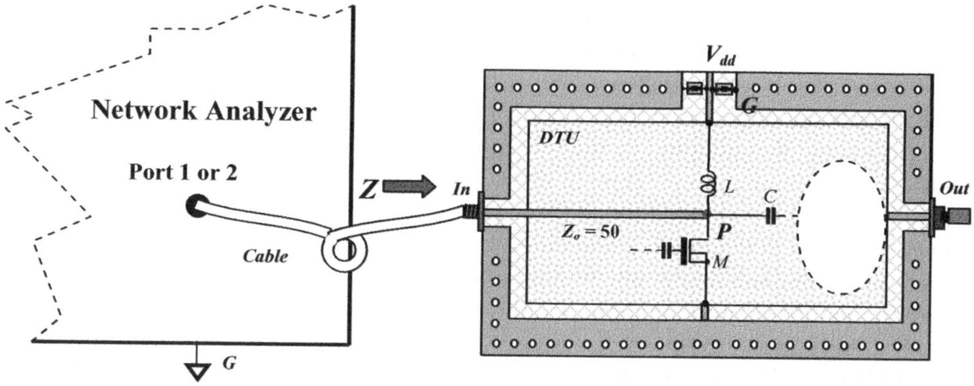

Figure 13.7 Non-directive impedance measurement for resultant impedance Z. At node P, 3 sub-circuit branches are connected in parallel.

Top metallic area
Bottom metallic area
o Conductive via from top to bottom
"Zero" capacitor
SMA connector
DTU
The rest of circuitry
50 Ω terminator
50 Ω cable

Parameters	Matrix	Coefficients	Block diagram
Z	$v_1 = z_{11}i_1 + z_{12}i_2$ $v_2 = z_{21}i_I + z_{22}i_2$	$z_{11} = v_1/i_1 \|_{i_2=0}$ ...	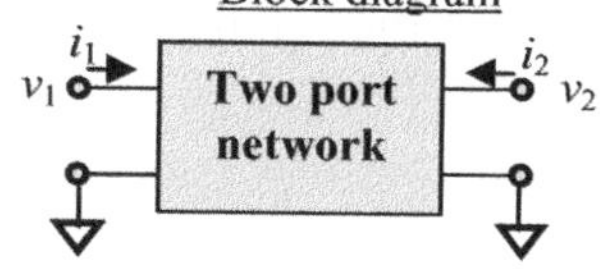
Y	$i_1 = y_{11}v_1 + y_{12}v_2$ $i_2 = y_{21}v_1 + y_{22}v_2$	$y_{11} = i_1/v_1 \|_{v_2=0}$ ...	
h	$v_1 = h_{11}i_1 + h_{12}v_2$ $i_2 = h_{21}i_1 + h_{22}v_2$	$h_{11} = v_1/i_1 \|_{v_2=0}$ $h_{12} = v_1/v_2 \|_{i_1=0}$ ...	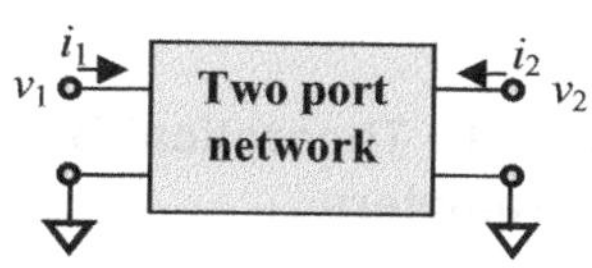
S	$b_1 = s_{11}a_1 + s_{12}a_2$ $b_2 = s_{21}i_1 + s_{22}b_2$	$s_{11} = b_1/a_1 \|_{a_2=0}$ ...	

Figure 13.8 The various parameters which characterize a two-port network.

The first three parameter sets in Figure 13.8 are based on the measurements of voltage and currents at the input or output terminals. In all of these three sets of parameters, inaccuracy occurs due to the measurement of their coefficients as shown in the third column of Figure 13.8. These coefficients must be measured or tested under the condition of either open-circuiting or short-circuiting, which is never absolutely true in reality. In the *RF* frequency range, the voltage of a short-circuited terminal is not absolutely equal to zero due to the isolation problem between the terminal and the ground. For the same reason, the current of an open-circuited terminal is not absolutely equal to zero.

The fourth set of parameters in Figure 13.8, called "scatter parameters" or simply "*S* parameters," was adapted in the second half of the twentieth century. Instead of resultant voltage or current, this set distinguishes the incident or reflective voltage or current at the input or output terminals in a two-port network. This idea, in fact, comes from the concept of a travel wave. The incident and reflected voltage or current are fully dependent on the impedances of the source and the load, that is, the source and load impedances,. In other words, the voltage, current, and impedance at a node or terminal are the directional parameters.

By means of the *S* parameters, the conditions of the open-circuiting and short-circuiting in the determination of their coefficients of the matrix are not required and therefore, the inaccuracy introduced in those three sets of parameters mentioned above is eliminated. Today, the *S* parameters are widely applied in *RF* circuit design; other parameter sets are still applied in the digital circuit design. The *S* parameters of a two-port network are measured by the network analyzer; hence the network analyzer becomes the most important test equipment in an *RF* circuit design lab.

Very often, *RF* designers read or calculate the input or output impedances of a two-port network directly from a conversion of *S* parameters, either S_{11} or S_{22}, through the relations

$$Z_{in} = \frac{1+S_{11}}{1-S_{11}}, \tag{13.5}$$

$$Z_{out} = \frac{1+S_{22}}{1-S_{22}}. \tag{13.6}$$

It should be noted that equations (13.5) and (13.6) are approximate expressions, valid under certain conditions which will be discussed in the following sections.

13.3.3 Theoretical Background of Impedance Measurement by *S* Parameters

In *RF* laboratories, the impedance of a basic part, such as a capacitor, inductor, or resistor, or the impedance of a block, sub-system, or an entire system, can be measured by either an impedance meter or a network analyzer. In the simulation phase, the impedance measurement for a *DTU* (Desired Test Unit) is usually executed by a network analyzer. In the early stages of development, the network analyzer is used for low power measurements, based on the principle of small signal measurement. Today, the network analyzer is able to test or measure a two-port network with high or large signals. This is outside the scope of this book. In the following discussion, the impedance measured directly by the network analyzer through *S* parameters is based on the principle of small signal measurement.

The network analyzer measures the *S* parameter at one or two ports, and then the impedance is calculated from the *S* parameters or directly read from the Smith chart. Figure 13.9 shows the *S* parameters and voltage reflection coefficients of a

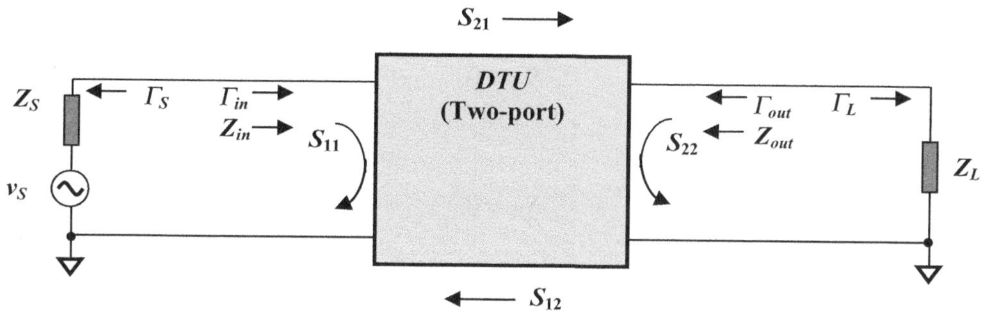

Figure 13.9 *S* parameters and voltage reflection coefficients of a two-port network.

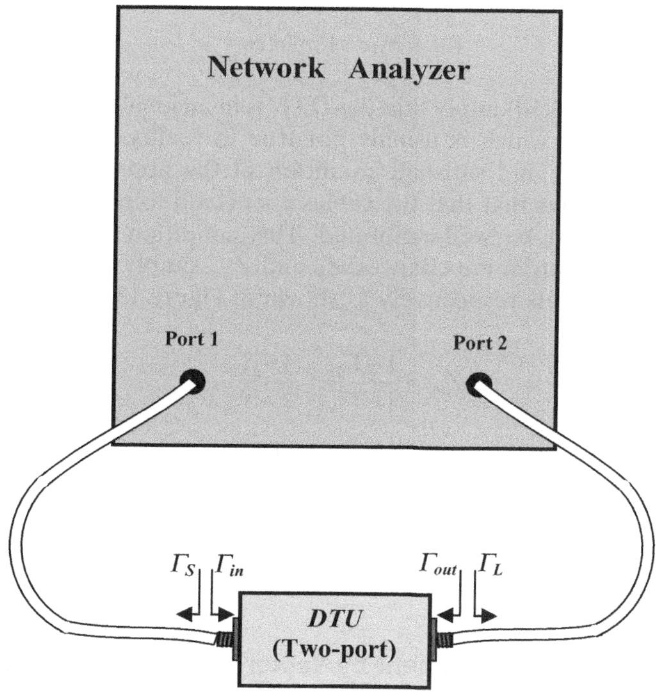

Figure 13.10 Impedance measured by network analyzer.

two-port network and Figure 13.10 shows the impedance measured by network analyzer.

Theoretically, the relations between the input and output voltage reflection coefficients, Γ_{in} and Γ_{out}, and input and output *S* parameters, S_{11} and S_{22}, are

$$\Gamma_{in} = S_{11} + \frac{S_{12}S_{21}\Gamma_L}{1 - S_{22}\Gamma_L}, \tag{13.7}$$

$$\Gamma_{out} = S_{22} + \frac{S_{12}S_{21}\Gamma_S}{1 - S_{11}\Gamma_S}. \tag{13.8}$$

If

$$S_{12} = 0, \tag{13.9}$$

or

$$S_{21} = 0, \tag{13.10}$$

or

$$\Gamma_S = \Gamma_L = 0, \tag{13.11}$$

then Γ_{in}, Γ_{out} would be equal to S_{11}, S_{22}, respectively, that is,

$$\Gamma_{in} = S_{11}, \quad \Gamma_{out} = S_{22}, \tag{13.12}$$

Conditions (13.9) or (13.10) imply that the *DTU* is in an ideal state of either forward or backward isolation, which is usually not true in reality. Relation (13.11) then becomes the necessary and satisfied condition of the approximations in (13.12). Condition (13.11) means that that the cables connected to port 1 and port 2 of the network analyzer must be well-calibrated. This condition (13.11) enables us to obtain the input and output impedances, Z_{in} and Z_{out}, simply by converted from the S_{11} and S_{22} measurements respectively as shown in Figure 13.10:

$$Z_{in} = \frac{1+\Gamma_{in}}{1-\Gamma_{in}} = \frac{1+S_{11}}{1-S_{11}}, \tag{13.13}$$

$$Z_{out} = \frac{1+\Gamma_{out}}{1-\Gamma_{out}} = \frac{1+S_{22}}{1-S_{22}}. \tag{13.14}$$

These values can be directly read off of the Smith chart displayed on the screen of the network analyzer.

13.3.4 *S* Parameter Measurement by Vector Voltmeter

The test fixture shown in Figure 13.11 consists of three portions. The upper portion, X_1-Y_1, tests the *S* parameters for a transistor. A bipolar transistor is soldered on the pads marked with *B, E*, and *C* which denote the base, emitter, and collector respectively. A *MOSFET* transistor can be also tested with this fixture if its *G, S*, and *D* (the gate, source, and drain), are instead soldered at the *B, E*, and *C* pads on the test fixture respectively. The intermediate portion, X_2, is for short-circuited calibration, while the bottom portion, X_3-Y_3, is for double length line calibration. It should be noted that the characteristic impedance of all the runners connected to the *SMA* connector must be 50 Ω.

The bias T_1 combines the *DC* voltage, V_1, from the *DC* power supply 1 and the *RF* signal, v_S, from the signal generator, and then delivers them to the base of the transistor through the bidirectional coupler 1. The bias T_2 passes the *DC* voltage, V_2, from the *DC* power supply 2 and then delivers it onto the collector of the tran-

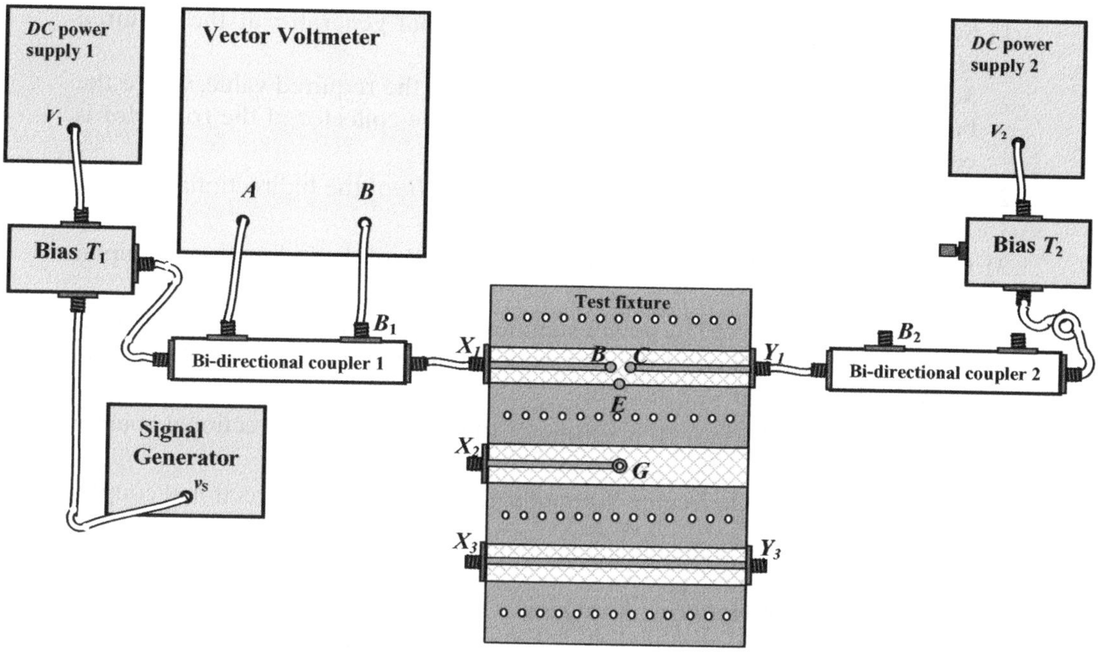

Figure 13.11 Setup of S parameter test for a transistor by vector voltmeter.
Top metallic area
Bottom metallic area
Conductive via from top to bottom
"Zero" capacitor
SMA connector
B,E,C of a bipolar device
The rest of circuitry
$50\,\Omega$ terminator
$50\,\Omega$ cable

sistor through the bidirectional coupler 2. A $50\,\Omega$ terminator must be connected to the bias T_2 output terminal.

The Vector Voltmeter is calibrated as follows:

1) Short-circuited calibration:
Port B of the Vector Voltmeter is connected to B_1 of the bidirectional coupler 1 and the output terminal of the bidirectional coupler 1, X, is connected to X_2, the input terminal of the intermediate portion on the test fixture. Then, $180°$ must be indicated in the Vector Voltmeter after an appropriate adjustment.
2) Double length line calibration:
Port B of the Vector Voltmeter is connected to the B_2 of the bidirectional coupler 2. The output terminal of the bidirectional coupler 1, X, is connected to X_3, the input terminal in the bottom portion on the test fixture. The output terminal of the bottom portion on the test fixture is connected to the input terminal of the bidirectional coupler 2. Then, $0°$ must be indicated in the Vector Voltmeter after an appropriate adjustment.

After the calibrations above have been done, the test is conducted step by step with the following procedures:

1) Set the reference RF input power of the signal generator at the required level.
   Set V_1 and V_2 of DC power supplies 1 and 2 to the required value, where the base of the transistor is connected to X_1 and the collector of the transistor is connected to Y_1.
2) Connect port B of the Vector Voltmeter to the B_1 of the bidirectional coupler 1, read S_{11}.
3) Connect port B of the Vector Voltmeter to the B_2 of the bidirectional coupler 2, read S_{21}.
4) Exchange V_1 and V_2 of DC power supplies 1 and 2, and reverse the test fixture, that is, connect the base of the transistor i to Y_1 and the collector of the transistor to X_1.
5) Connect port B of the Vector Voltmeter to the B_1 of the bidirectional coupler 1, read S_{22}.
6) Connect port B of the Vector Voltmeter to the B_2 of the bidirectional coupler 2, read S_{12}.

13.3.5 Calibration of Network Analyzer

The network analyzer is a powerful tool in the measurement of an *RF* circuit block or system; calibration is an essential step before the test or measurement is conducted.

Figure 13.12 depicts the layout for a *DTU* on a *PCB*. Its input terminal is point *A* and its output terminal is point *B*. For simplicity, the detailed layout of the *DTU* is replaced by a blank rectangular block with its *DC* power supply terminal V_{dd} and its "zero" capacitors. At the input and output ports, a micro strip line leads from the input and output *SMA* connector to the circuit block. The characterized impedance of the micro strip line is 50 Ω.

One can carry out calibration by means of the standard calibration kit, which is provided by the manufacturer. There are four basic calibration procedures: "open",

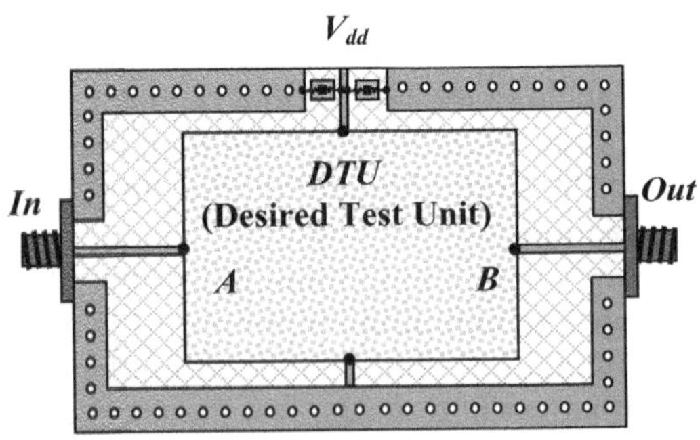

Figure 13.12 Layout of test *PCB*.
- Top metallic area
- Bottom metallic area
- o Conductive via from top to bottom
- "Zero" capacitor
- *DTU* (Desired Test Unit)
- Runner, $Z_o = 50\,\Omega$
- SMA connector

"short", "50 Ω", and "through." It should be noted that the calibration is "on the air" without considering the effect of the test environment and the tested *PCB*. The calibration accuracy may decline in the high *RF* frequency range.

Instead of using the standard calibration kit provided by the manufacturer, the *RF* design engineer may develop a custom calibration kit for more precise measurement. Figure 13.13 shows a set of self-supporting calibration kits: with the same *PCB* adapted for the *DTU* as shown in Figure 13.12, there are four sub-calibration kits for "open," "short," "50 Ω," and "through" calibration purposes. Each sub-calibration kit has the same grounding pattern and size as it does for the *DTU*. As shown in Figures 13.12 and 13.13, the input terminal is calibrated at point *A*, the output terminal at point *B*. By this method the calibrations represent the actual environment of the *DTU* more accurately than the standard calibration kit.

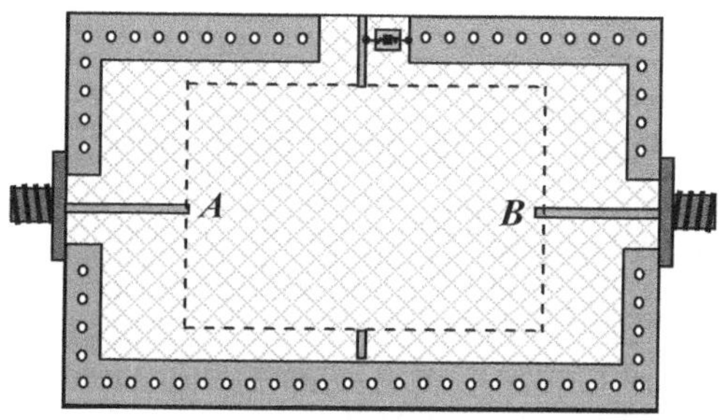

(a) Calibration kit: **Open**

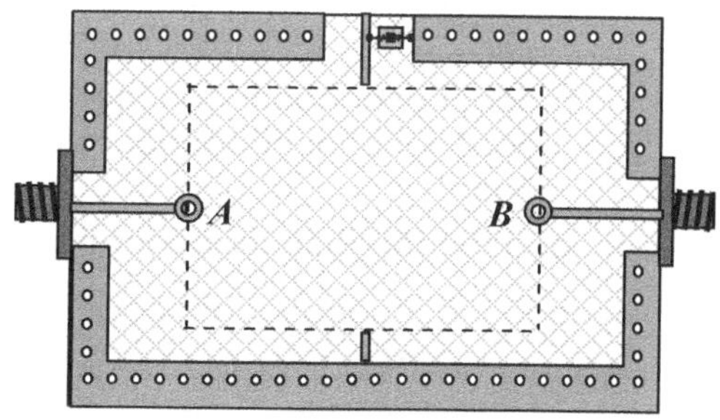

(b) Calibration kit: **Short**

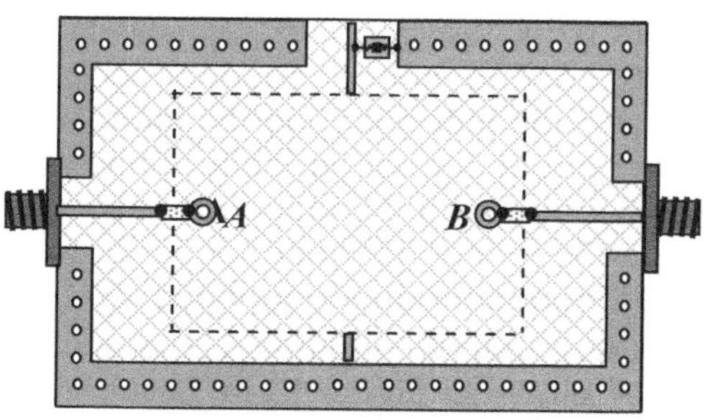

(c) Calibration kit: **50 Ω**

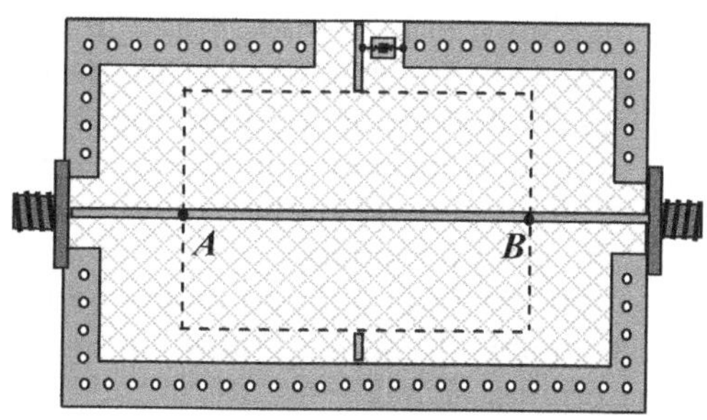

(d) Calibration kit: **Through**

Figure 13.13 Layout of a self-supporting calibration kit.
Top metallic area
Bottom metallic area
Conductive via from top to bottom
"Zero" capacitor
SMA connector
50 Ω resistor
Runner, $Z_o = 50\,\Omega$.

Experiments indicates that the values of measured S_{11} and S_{22} could be couple dB difference at the frequency range of GHz between the calibrations made by the standard calibration kit and that made by the favorite calibration kit.

It should be noted that in order to ensure the constancy of the $50\,\Omega$ resistor within a desired frequency bandwidth, a combination of resistors may be necessary. For instance, a $50\,\Omega$ resistor could be substituted by two $50\,\Omega$ resistors connected together in series and then connected with another $100\,\Omega$ resistor in parallel. This combination of multiple resistors tends to cancel the frequency variation of the resistance in the resulting resistor. Empirically, the value of $50\,\Omega$ can be kept unchanged by such a combination from DC up to $6\,GHz$.

There are four basic types of *PCB*:

1) *COB* (Chip On Board)
   Cut the block or system from a wafer as an *IC* die first, and then bond the *IC* die onto a test *PCB* for testing.
2) *POB* (Package On Board)
   Cut the block or system from a wafer as an *IC* die first, package it, and finally place it as a part on the test *PCB* for testing.
3) *MOB* (Module as one part On Board).
   Module is built by discrete parts and packaged as one part on the test board.
4) *DOB* (Module directly built by Discrete parts On Board).

Another method is to test the *RFIC* on the wafer directly in an *IC* probe station. The *IC* die could be an individual block, multiple blocks, or a *SOC* (System On Chip) system; the test *PCB*, of course, is not necessary when testing is conducted directly on the wafer.

Figure 13.14 illustrates the four basic types of *PCB*, in which one sub-figure contains only one type of *PCB* built by either *RFIC* or *RF* discrete parts. The test *PCB*s as shown in Figure 13.14(a) and 13.14(b) contain an *RFIC* die and *RFIC* package, respectively, while the test *PCB*s as shown in Figure 13.14(c) and 13.14(d) contain an *RF* module or *RF* discrete parts. For simplicity, only the main part and the main terminals, including the input, output, and *DC* power supply, are drawn, while other additional parts are neglected since we are only concerned with the main part and the main terminals.

If the *PCB* is tested by the network analyzer, calibration must be conducted up to points *A* and *B* as shown in Figure 13.14, but not just up to the input and output terminal. Unlike the *MOB* and *DOB* types of *PCB* shown in Figure 13.14, in the *COB* and *POB* types of *PCB*, the calibration points *A* and *B* are quite far from the input and output terminals. Using the calibration kit provided by the manufacturer, calibration may be reluctantly acceptable for the *MOB* and *DOB* types of *PCB*, where the calibration points are close to the input and output terminals, but this is definitely not acceptable for the *COB* and *POB* boards. Instead of the standard calibration kit, a self-supporting calibration kit must be designed and applied.

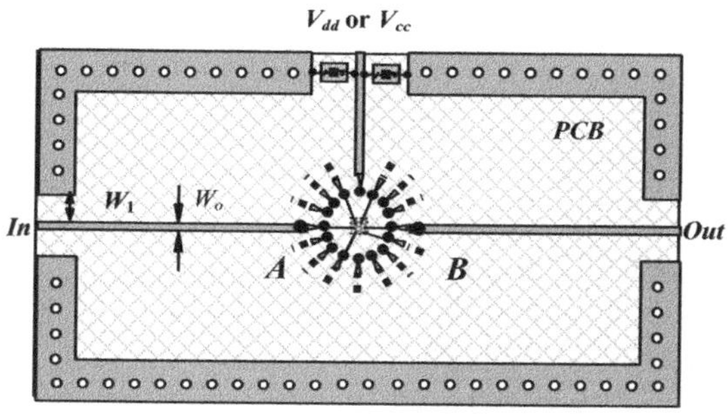

(a) *COB* (Chip On Board)

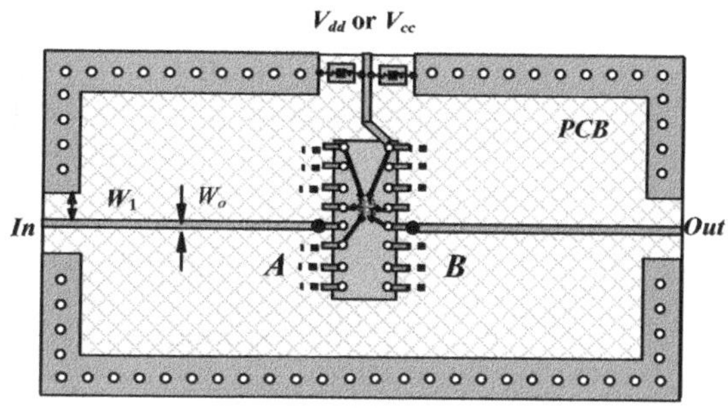

(b) *POB* (IC Package On Board)

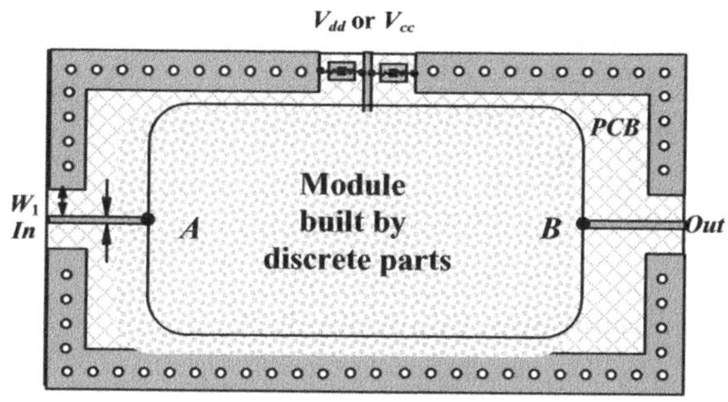

(c) *MOB* (Module On Board)

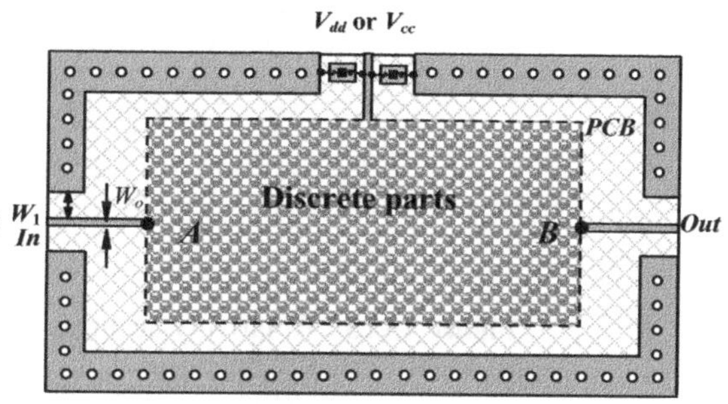

(d) *DOB* (Discrete parts On Board)

Figure 13.14 Basic types of test *PCB*.

- —— Bond wire
- • Pad
- ▪ *IC* die
- ▪ Top metallic area
- ▫ Bottom metallic area
- ▫ Module built by discrete parts
- ▫ Discrete parts
- ▭ 50 Ω runner
- ◅ End of a runner with beveled shape
- o Conductive via from top to bottom
- ▭ "Zero" capacitor
- —— Runner of *IC* package
- ▭ Pins of *IC* package

13.4 ALTERNATIVE IMPEDANCE MEASUREMENT BY NETWORK ANALYZER

13.4.1 Accuracy of a Smith Chart

It should be noted that the accuracy of a reading from a Smith chart depends on the location of the reading. Figure 13.15 shows the relatively inaccurate and accurate areas on a Smith chart. In areas where the impedance is not too low and not too high, the accuracy of an impedance reading is acceptable and reliable. However, in areas where impedance is very low or very high, the accuracy of the impedance reading may be questionable.

Evidence of low accuracy in the low impedance area can be found from the measurement of the Q value of an inductor. One can test and read the impedance of an inductor by a single-port test as shown in Figure 13.16; its display on the Smith chart is shown in Figure 13.17. The trace on the Smith chart is the measured result

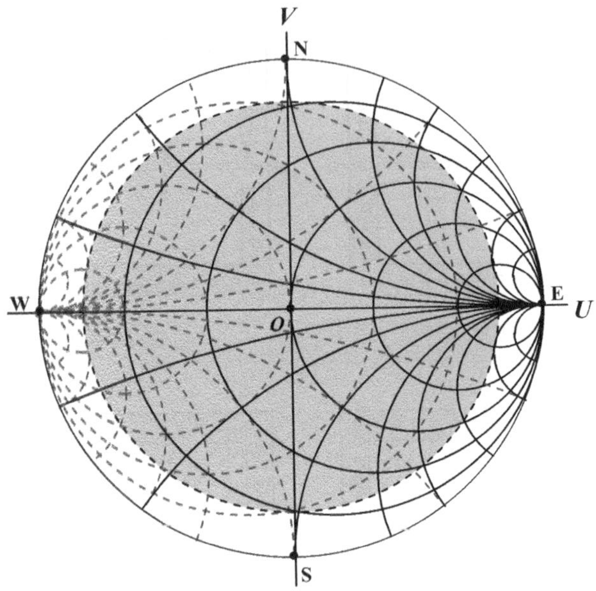

Figure 13.15 Relatively accurate and inaccurate areas of a Smith chart.

Relatively accurate area
Relatively inaccurate area

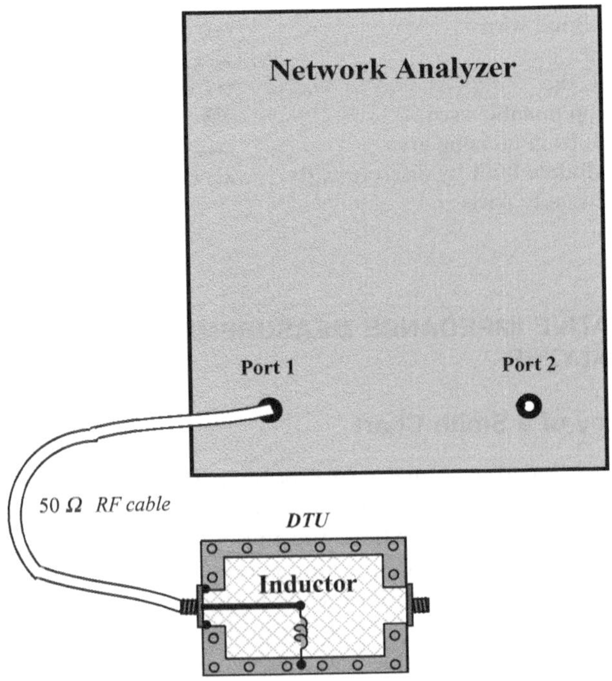

Figure 13.16 Single-port impedance test for an inductor.

Top metallic area
Bottom metallic area
SMA connector
Conductive via from top to bottom

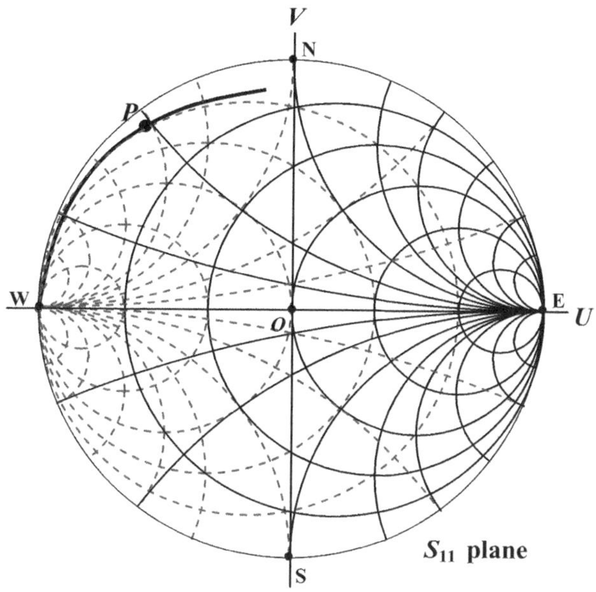

Figure 13.17 One-port test for the impedance of an inductor ($f = 0$ to $10\,GHz$).

of S_{11} or impedance Z in the frequency range from DC to $10\,GHz$. When $f = 0$, the impedance or its resistance and reactance are zero because the behavior of an inductor in the DC operation is short-circuited. In general, its inductance increases as the operating frequency increases, although it is not necessarily proportional. On the other hand, its resistance also increases as the operating frequency increases. On the trace, each point P corresponds to an operating frequency and from each point P, the reactance x_S and the resistance r_S can be directly read from the Smith chart. Finally, from the reactance the inductance of the inductor L can be calculated.

The single-port impedance measured directly by the network analyzer is simple and easy. Unfortunately, if the resistance or reactance of the inductor has a very low or very high value as shown on the Smith chart in Figures 13.15 or 13.17, then the readings of the reactance x_S or the resistance r_S are quite inaccurate, and hence the calculated L values can deviate from the actual values by around 10 to 50%. The actual values, of course, can be obtained from other test means with greater accuracy.

The second evidence can be found in the impedance measurement of a short whip antenna, which is directly connected to port 1 of the network analyzer as shown in Figure 13.18.

The antenna operates at around $27\,MHz$, which is a low RF frequency. Its impedance can be directly displayed on the screen of a network analyzer via the Smith chart, such as

- Real part: -1000 to $+1500\,\Omega$;
- Imaginary part: $3.0\,pF$.

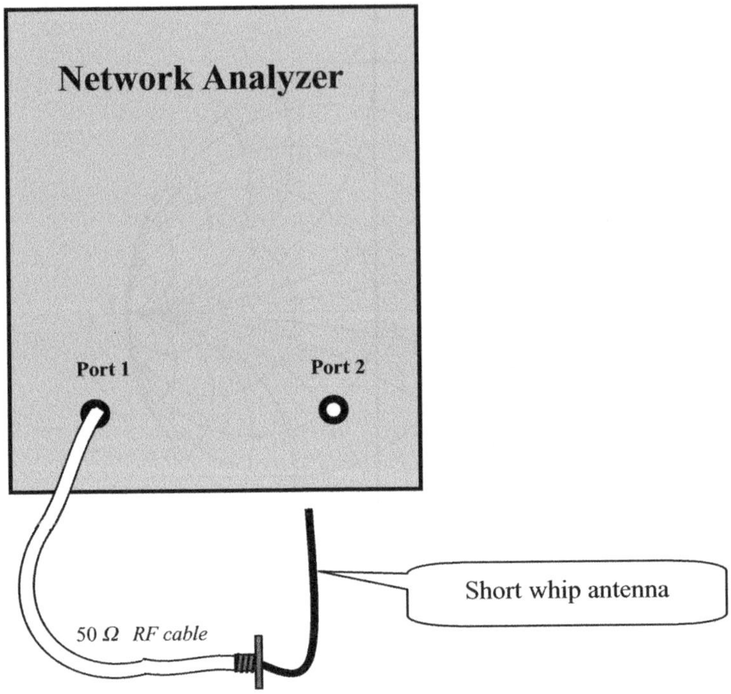

Figure 13.18 Single-port impedance test for a short whip antenna.
SMA connector

The real part of an impedance is unstable and rapidly flashes between −1000 Ω to +1500 Ω; therefore it cannot be determined to a definite value. The reactance is kept at about $3pF$.

These two evidences or examples prove the truth of the assertion about the measurement inaccuracy of the impedance on the Smith chart as shown in Figure 13.15. Consequently, other alternative measuring or testing ways must be sought.

13.4.2 Low and High Impedance Measurement

As a matter of fact, the developed *RF* impedance meter can be applied to measure or test parts which have very low or very high values of resistance or reactance. However, the developed meter brings about other inaccuracies and inconveniences. In addition, it is not a cost-effective piece of equipment.

One of the alternative ways is to keep using the network analyzer for testing or measuring impedance. However, the method of testing or measuring must be changed.

Instead of the single-port test through S_{11} measurement, it is possible to test or measure the impedance by a two-port test through S_{21} measurement. Figure 13.19(a) and (b) shows how to test or measure the impedance when the impedance is very low and very high. In Figure 13.19(a), the desired test part is soldered on the test *PCB* in parallel. Should the impedance of the desired test part be very low, the power delivered from port 1 to port 2 will be considerably attenuated and therefore,

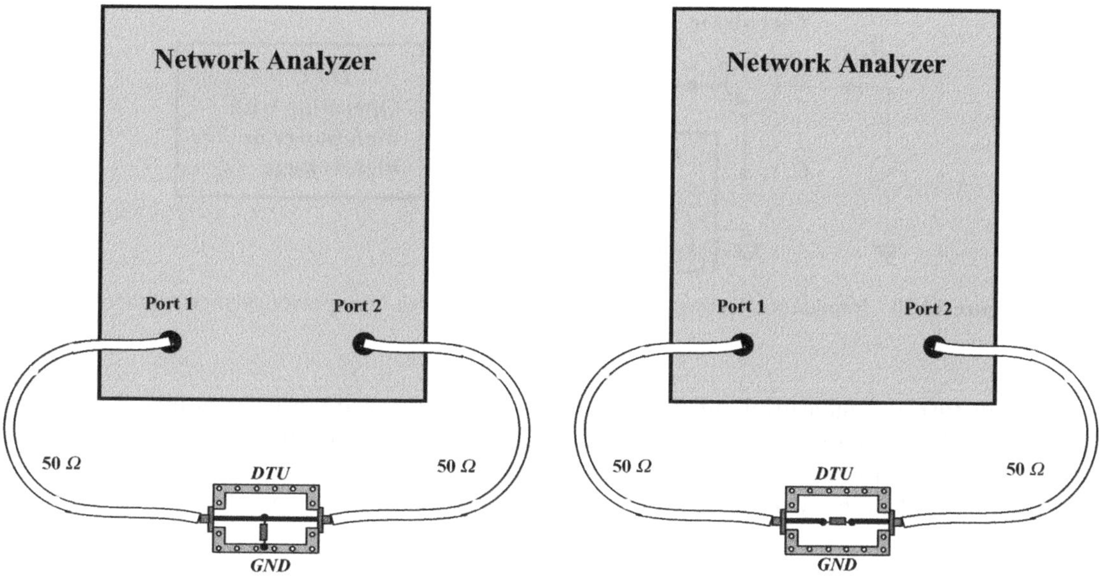

(a) Testing for the parts with very low impedance (b) Testing for the parts with very high impedance

Figure 13.19 Test setup for parts with very low and very high impedance *DTU* (Desired Test Unit).

its resistance can be calculated from the insertion loss S_{21}, while its reactance can be calculated from its *SRF* (Self-Resonance Frequency) in parallel. In Figure 13.19(b), the desired test part is soldered on the test *PCB* in series. Should the impedance of the desired test part be high, the power delivered from port 1 to port 2 will be considerably attenuated and therefore, its resistance can also be calculated from the insertion loss S_{21} while its reactance can be calculated from its *SRF* in series. We continue the discussion of this subject in more detail in Appendix 13.A.1.

13.5 IMPEDANCE MEASUREMENT WITH THE ASSISTANCE OF A CIRCULATOR

So far, we have discussed impedance measurement by means of *S* parameter testing. It is well-known that the *S* parameters are accurate only for a linear part, in which the testing signal is small. For the impedance measurement of a desired test unit which operates under high power or high voltage, *S* parameter testing is no longer suitable. The impedance measurement in this case can be conducted by means of a circulator and vector voltmeter as shown in Figure 13.20.

The special feature of a circulator is that the input power or voltage can be transported in only one direction, either clockwise or counter-clockwise. As shown in Figure 13.20, the input voltage or power at point *A* can reach point *B* as the incident voltage or power for the *DTU*. The reflected voltage or power from the *DTU* can only be transported to point *C* and cannot be returned to point *A*, so it will

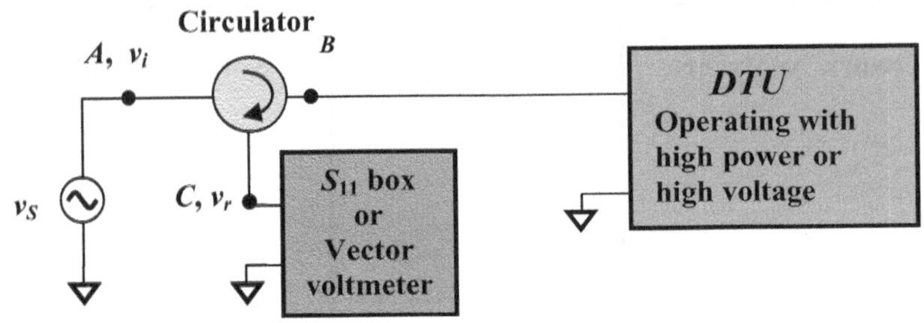

Figure 13.20 Impedance testing of a desired test unit with high power or high voltage.

not disturb the measurement at the input, point A. Consequently, at point C, the reflected voltage or power from the DTU can be correctly measured by the S_{11} box or Vector Voltmeter. The impedance can be calculated from the incident voltage measured at point A, v_i, and the reflected voltage measured at point C, v_r, by the relationship,

$$z = \frac{1+\Gamma}{1-\Gamma}, \tag{13.15}$$

where

$$\Gamma = \frac{v_r}{v_i}. \tag{13.16}$$

In a practical power amplifier design, the power amplifier operates with a high power input and output. In a practical mixer design, its RF input and IF output are usually treated as low power ports, whereas the LO injection must be treated as a high power port.

APPENDICES

13.A.1 Relationship between the Impedance in Series and in Parallel

The impedance reading from a test includes both real and imaginary parts, which are usually expressed in series as shown in equation (13.A.1). Sometimes the impedance with its real and imaginary parts in series must be converted into the impedance with its real and imaginary parts in parallel. Figure 13.A.1 sketches the real and imaginary parts of the impedance expressed in series (a) or in parallel (b). Their relations are

$$R_S + jX_S = R_P \,//\, jX_P = \frac{X_P^2 R_P + jX_P R_P^2}{R_P^2 + X_P^2}, \tag{13.A.1}$$

$$Q = \frac{|X_S|}{R_S} = \frac{R_P}{|X_P|}, \tag{13.A.2}$$

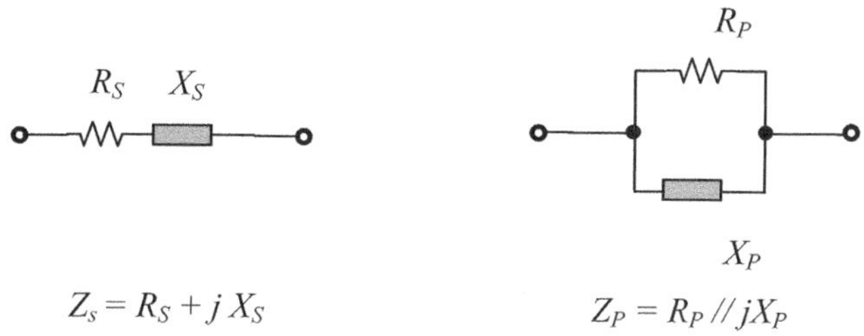

(a) Impedance in series (b) Impedance in parallel

Figure 13.A.1 Real and imaginary parts of impedance expressed in series (a) or in parallel (b).

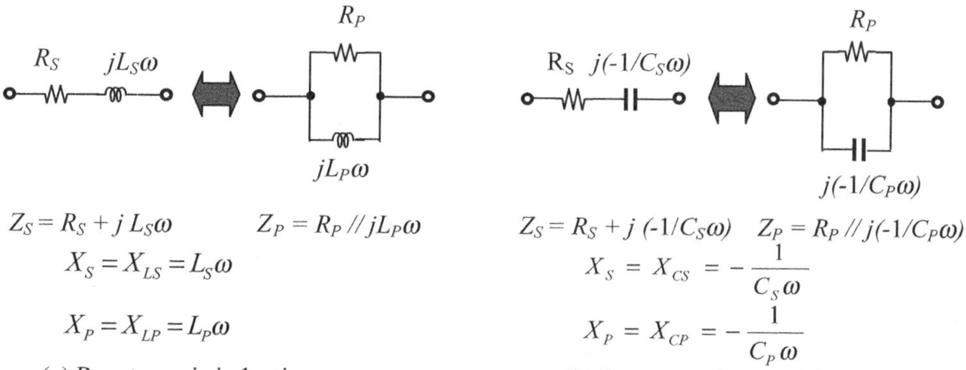

(a) Reactance is inductive (b) Reactance is capacitive

Figure 13.A.2 Conversion of impedance between in series and in parallel.

$$R_p = R_S(Q^2+1) \approx \frac{X_S^2}{R_S}, \quad \text{if } Q \gg 1, \tag{13.A.3}$$

$$X_p = X_S\frac{Q^2+1}{Q^2} \approx X_S, \quad \text{if } Q \gg 1, \tag{13.A.4}$$

where Q = quality factor.

When the reactance is inductive as shown in Figure 13.A.2(a), we have

$$R_S + jL_S\omega = R_P \,//\, jL_P\omega = \frac{L_P^2\omega^2 R_P + jL_P\omega R_P^2}{R_P^2 + L_P^2\omega^2}, \tag{13.A.5}$$

from (13.A.5), Q factor is

$$Q = \frac{L_S\omega}{R_S} = \frac{R_P}{L_P\omega}, \tag{13.A.6}$$

$$R_p = R_S(Q^2+1), \tag{13.A.7}$$

$$L_p = L_S \frac{Q^2+1}{Q^2}. \tag{13.A.8}$$

When the reactance is capacitive as shown in Figure 13.A.2(b), we have

$$R_S + j\left(-\frac{1}{C_S\omega}\right) = R_P \,//\, j\left(-\frac{1}{C_p\omega}\right) = \frac{R_P - jC_P\omega R_P^2}{1+R_P^2C_P^2\omega^2}, \tag{13.A.9}$$

from (13.A.9), Q factor is

$$Q = \frac{1}{R_S C_S \omega} = R_P C_P \omega, \tag{13.A.10}$$

$$R_p = R_S(Q^2+1), \tag{13.A.11}$$

$$C_p = C_S \frac{Q^2}{Q^2+1}. \tag{13.A.12}$$

REFERENCES

[1] G. J. Laughlin, "A New Impedance-Matched Wide-Band Balun and Magic Tee," *Microwave Theory and Techniques, IEEE Transactions*, Vol. 24, No. 3, March 1976, pp. 135–141.

[2] M. Lapinoja and T. Rahkonen, "An Active Tuning and Impedance Matching Element," *Circuits and Systems*, 1998. ISCAS '98. *Proceedings of the 1998 IEEE International Symposium*, Vol. 1, May 31–June 3, 1998, pp. 559–562.

[3] B. C. Wadell, "Smith Charts Are Easy. I," *Instrumentation & Measurement Magazine, IEEE*, Vol. 2, No. 1, March 1999, pp. 37–40.

[4] B. C. Wadell, "Smith Charts Are Easy. II," *Instrumentation & Measurement Magazine, IEEE*, Vol. 2, No.2, June 1999, pp. 45–47.

[5] B. C. Wadell, "Smith Charts Are Easy. III," *Instrumentation and Measurement Magazine, IEEE*, Vol. 2, No. 3, September 1999, pp. 38–42.

CHAPTER 14

GROUNDING

14.1 IMPLICATIONS OF GROUNDING

Grounding is a common concern in all circuit design for circuit blocks and systems, including *RF*, digital, and analog circuit designs. As long as the circuit operates in *AC* mode, grounding must be done properly no matter how low or high the operating frequencies are. In the *RF* frequency range, grounding becomes a more serious matter because *RF* frequencies are usually higher than other *AC* frequencies. Many problems or "bugs" arise from inappropriate *RF/AC* grounding. As a matter of fact, *RF/AC* grounding is one of the core technologies and an indispensable skill in *RF* circuit design. In high-speed digital circuit designs, grounding is as important as in *RF* circuit design.

In a circuit block or a system, there are many ground points or nodes marked in a schematic. Figure 14.1 shows a typical way to draw a schematic of circuitry, in which the positive terminal of the *DC* power supply point is marked with V_{cc} and the negative terminal is marked with *GND* or by an upside-down symbol "Δ." The positive terminal of the *DC* power supply point, V_{cc}, is connected to many sub-points of *DC* power supply, V_{cc1} *to* V_{cc8}, and the negative terminal of the *DC* power supply point, *GND*, is connected to many sub-points of ground points, G_1 *to* G_8.

First, a question is raised: which ground point is the reference ground point in a circuit block or system if these ground points are not equipotential? The answer is: the reference ground point is the terminal or ground point of the *DC* power supply. This is an irrefutable and unanimous assertion because all the currents flowing through all the branches of circuitry must eventually return to the terminal or ground point of the *DC* power supply. A real ground point or node in a schematic must be equipotential to the ground point of the *DC* power supply, *GND*.

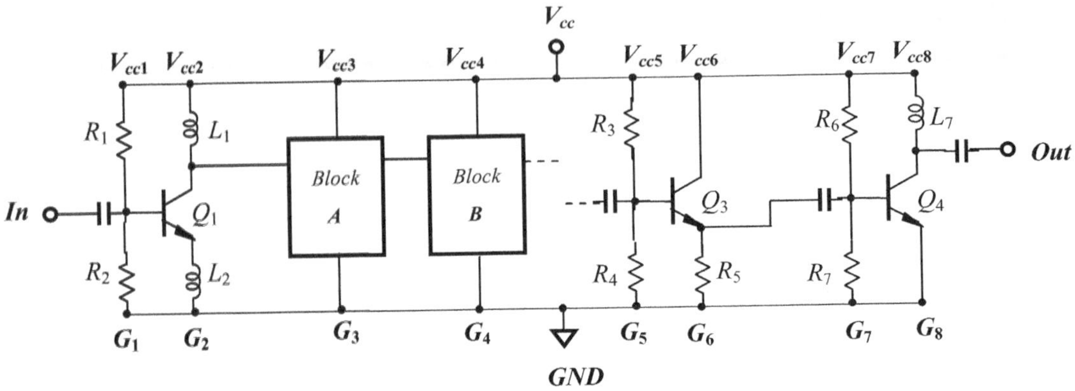

Figure 14.1 A typical way to draw a schematic of circuitry.

Second, the ground points or nodes, G_1 *to* G_8, are connected together by a straight line segment as shown in the bottom of Figure 14.1. At these ground points or nodes, both the *DC* and *AC* grounding are expected simultaneously. That is, in respect to a reference ground point or node, *GND*, the following conditions of the *AC* and *DC* components of voltage and impedance must be satisfied.

$$V_{AC} = 0, \quad V_{DC} = 0, \tag{14.1}$$

And

$$Z_{AC} = 0, \quad Z_{DC} = 0, \tag{14.2}$$

where the subscripts "*AC*" and "*DC*" denote the alternative and directive components of the parameters respectively.

In other words, these ground points are expected to be equipotential with or short-circuited to the reference ground point, *GND*. Consequently, these ground points, G_1 *to* G_8, are named as fully ground points, in which both the *AC* and *DC* components of voltage and impedance are equal to zero in respect to the reference ground point, *GND*.

In *RF* circuit design with discrete parts, these kinds of ground points are usually embedded or "swallowed" by a ground surface. In the *RFIC* circuit design, the *P+* guard ring in an individual *RF* block functions as a ground surface.

Third, the points or nodes V_{cc1} to V_{cc8} are *DC* power supply or *DC* bias terminals. They are connected by a straight line segment as shown on the top of Figure 14.1. At these ground points or nodes, *AC* but not *DC* grounding is expected. That is, in respect to a reference ground point or node, *GND*, the following conditions for the *AC* and *DC* components of voltage and impedance must be satisfied.

$$V_{AC} = 0, \quad V_{DC} \neq 0. \tag{14.3}$$

and

$$Z_{AC} = 0, \quad Z_{DC} \neq 0. \tag{14.4}$$

In other words, at all terminals of *DC* power supply or *DC* bias, both the *AC* impedance and *AC* voltage are expected to be equipotential with or short-circuited to the reference ground point, *GND,* while both the *DC* impedance and *DC* voltages must be kept at an expected non-zero value. These ground points, V_{cc1} to V_{cc8}, are therefore named as "half ground points."

Obviously, as distinguished from grounding for fully ground points (G_1 *to* G_8), grounding for these half ground points, V_{cc1} to V_{cc8}, must be done by means of special parts, which will be introduced and discussed later.

14.2 POSSIBLE GROUNDING PROBLEMS HIDDEN IN A SCHEMATIC

In the textbook and in the classroom, the schematic of circuitry is always drawn like Figure 14.1, in which many ambiguous grounding problems can hide in dark corners.

Figure 14.1 contains many full and half ground points but does not address the following questions:

- How to ensure that all the fully ground points *AC* and *DC* are equipotential with the reference ground point or terminal?
- How to ensure that all the half ground points are *AC* but not *DC* short-circuited to the reference ground point or terminal?
- Furthermore, if the conditions of equipotentiality, (14.1) to (14.4), are not satisfied, the problem of forward current and return current coupling will appear.

If a half ground point is not equipotential with the reference ground point or terminal, its *AC* forward current would flow from the *DC* power supply V_{cc} to the half ground point. (Usually, the internal resistance of a *DC* power supply is about a tenth of ohms, so that the positive and negative terminal is equipotential to the *AC* component of either voltage or current). The forward currents will be coupled as they flow from *DC* power supply, V_{cc}, to the half ground points.

Similarly, if a full ground point is not equipotential with the reference ground point or terminal, its *AC* return current will flow from the full ground point to the reference ground point, *GND*. The return currents will couple in the same way as when they flow from the full ground point to the reference ground point, *GND*.

Figure 14.2 shows the forward and return currents which are ignored in Figure 14.1, as is usual in a schematic drawing.

In the simulation design stage, the above questions are never raised. On the contrary, it is assumed that

- The fully ground points are equipotential with reference ground point, *GND*, and conditions (14.1) and (14.2) are satisfied.
- The half ground points are equipotential with the reference ground point, *GND*, for the *RF/AC* signal but not for the *DC* voltage, and conditions (14.3) and (14.4) are satisfied.
- The forward and return current couplings are negligible.

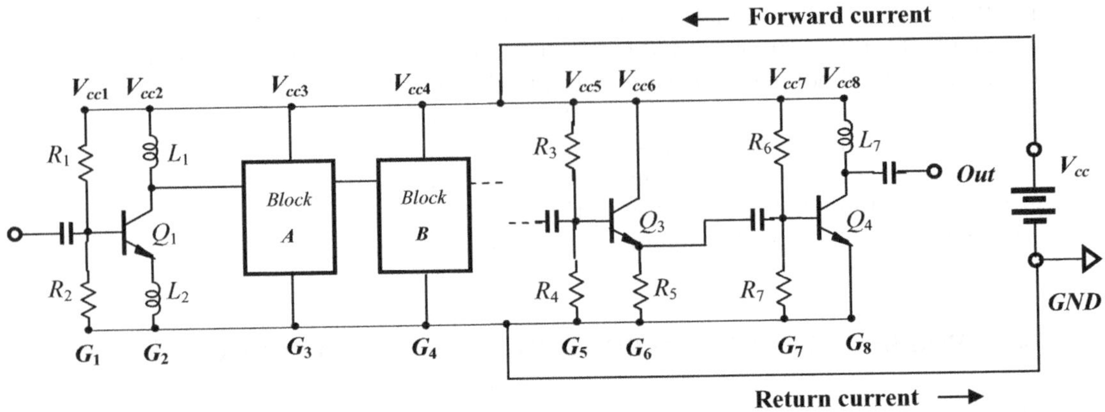

Figure 14.2 Forward and return currents added to the schematic shown in Figure 14.1.

This is, of course, far from reality! Should these hidden problems be erased, it is equal to saying that "there are no grounding problems in the circuit design."

14.3 IMPERFECT OR INAPPROPRIATE GROUNDING EXAMPLES

Based on the practical statistics of *RF* circuit design, most design problems do not exist in the simulation stage, but quite a high percentage of problems are due to imperfect or inappropriate *RF/AC* grounding. This grounding leads to performance with unexpected erroneous functions. At the extreme end, it can put an *RF* circuit block or system out of work.

The following examples expose some typical problems in *RF/AC* grounding.

14.3.1 Inappropriate Selection of Bypass Capacitor

14.3.1.1 Underestimation of or Ignoring the Bypass Capacitor Essentially, the *AC* bypass capacitor and the *DC* blocking capacitor are the same part in an *RF* circuit block. Their impedance approaches zero at the operating frequency or within the operating frequency bandwidth. They are commonly called "zero" capacitors, which will be discussed in Section 14.4. In this section, only the bypass capacitor is discussed because it is more often ignored than the *DC* blocking capacitor. The *DC* blocking capacitor is one of the parts in the circuit schematic and is taken care of in the simulation, whereas the bypass capacitor is very often not drawn in the circuit schematic and neglected in the simulation. Computers always assume that the *DC* power supply or *DC* bias is an ideal unit with short-circuited *AC*. It seems unnecessary to have a bypass capacitor connected with it in parallel. *RF* circuit designers do not need to worry about the bypass capacitor until they are going to test the circuit board. This might be one of reasons that the bypass capacitor is easily ignored.

Some think that the selection of the bypass capacitor is a tiny thing and not worth mentioning.. In electrical engineering texts and courses, it is hard to find any material dealing with the bypass capacitor.

Is the bypass capacitor very easy to deal with, and therefore, not worth mentioning? Statistics indicates that a very high percentage of failure rates in the testing of an *RF* circuit block is due to inappropriate selection of the bypass capacitor. Obviously, the task of selecting a bypass capacitor or "zero" capacitor is underestimated. In the testing stage, the bypass capacitor is an indispensable part not only for *RF* but also for digital circuit blocks. An improper bypass capacitor always leads to ridiculous test results and sometimes, the tested circuit block is simply out of work even though the circuit block has demonstrated excellent performance in the simulation.

14.3.1.2 Blind Selection of Bypass Capacitor Here is a true story: In a famous institute engaging in *R&D* for a wireless electronic product, an electrical engineer was about to test an *RF* block circuitry. He was aware that he had to attach a bypass capacitor at the *DC* power supply port. He asked his supervisor: "Which value of capacitor must be selected from the engineering stock room?" Without any hesitation, his boss answered his question with brimming confidence: "$0.01\,\mu F$"! From then on, the majority of the electrical engineers would apply a $0.01\,\mu F$ capacitor as a bypass capacitor in whatever circuit they designed.

Some people had different opinions: "The bypass capacitor must be different for various circuits. You can choose the bypass capacitor when you are going to adjust your circuit block into an optimum state by testing. For example, if you are designing a *LNA*, you may swap out bypass capacitors one by one until you get a maximum of gain or a minimum of noise figure, or until you get both simultaneously."

This sounds a pretty good idea, but in reality, it starts to look like little more than a dream. The performance of the *LNA* might be unchanged even with many different values of bypass capacitors.

An easy but lazy way to solve the selection problem for a bypass capacitor is to put a huge crowd of capacitors on the *DC* power supply port, by which the *AC* signal should be short-circuited over a very wide frequency range. This method is time-effective and universal, dealing with all the half ground points in the circuit, regardless of the operating frequencies in the circuit block.

For example, Figure 14.3 shows a circuit block in which the main operating frequency in the circuit block is $2.4\,GHz$. A huge crowd of capacitors with values ranging from very low, $10\,pF$, to very high, $100\,\mu F$, are connected between the *DC* power supply port and ground in parallel. The capacitors from C_1 to C_6 are chip capacitors, while the capacitors from C_7 to C_{12} are electrolytic. The designer thought that, with such a setup, there should be no problem in reaching the goal of *AC* short-circuiting and *DC* open-circuiting at *DC* power supply port.

Unfortunately, the huge crowd of capacitors is useless because the *AC* signal appearing on the *DC* power supply port is maintained at the same level whether they are connected to or removed from the *DC* power supply port. This inconceivable event frustrated the designer for a long time.

The answer to this problem can be found in Table 14.A.1, where the "zero" capacitor is discussed: The *SRF* (Self-Resonant Frequency) of any capacitor in the huge crowd of capacitors is below the operating frequency of $2.4\,GHz$. From Table 14.A.1 it can be found that the correct value of the bypass capacitor must be $5.1\,pF$.

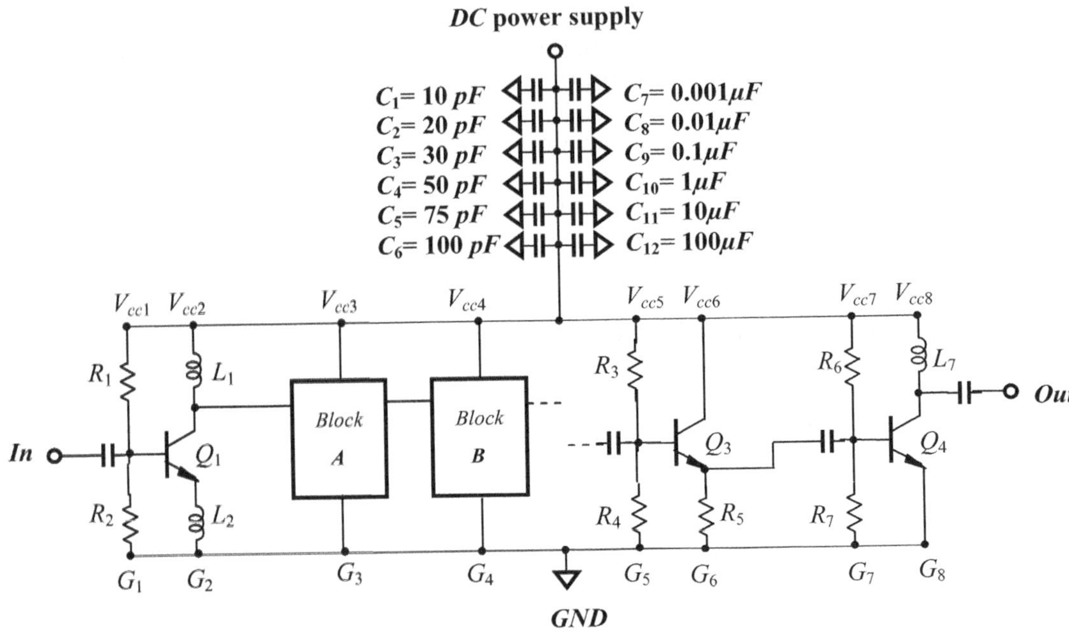

Figure 14.3 A huge crowd of capacitors are supposed to function as bypass capacitors.

14.3.2 Imperfect Grounding

14.3.2.1 The "Long March" Figure 14.4 shows a *PCB* (Printed Circuit Board) test for an 800 *MHz* transceiver, in which the transceiver *IC* die has been packaged with 20 pins and soldered onto the *PCB*. The purpose of this test setup is to check the receiver sensitivity, since the *IC* die is bought from an outside company. The input is connected to a generator that provides a small modulated signal so as to simulate a received signal from the antenna. The output is connected to an audio distortion meter so that the sensitivity of the receiver can be measured by 12 *dB SINAD* or 20 *dB* Quiet. For simplicity, only the parts and runners related to grounding are shown in Figure 14.4.

The tested sensitivity is found to be much lower than what the manufacturer specified. The problem is due to an inappropriate *PCB* layout. One can see its extremely poor *RF/AC* grounding as shown in Figure 14.4. Let's take a look at how flawed the *RF/AC* grounding of the input *SMA* connector is. Step by step, the ground port of the input *SMA* connector is eventually connected to the reference ground point of the *DC* power supply, *GND*, through a "Long March" of sorts, that is,

- Ground of *SMA* connector at input
- Runner 1 on *PCB*
- Pin 1 of *IC* package
- Pad 1 inside *IC* package
- Runner 2 inside *IC* package
- Bonding pad 2 inside *IC* package

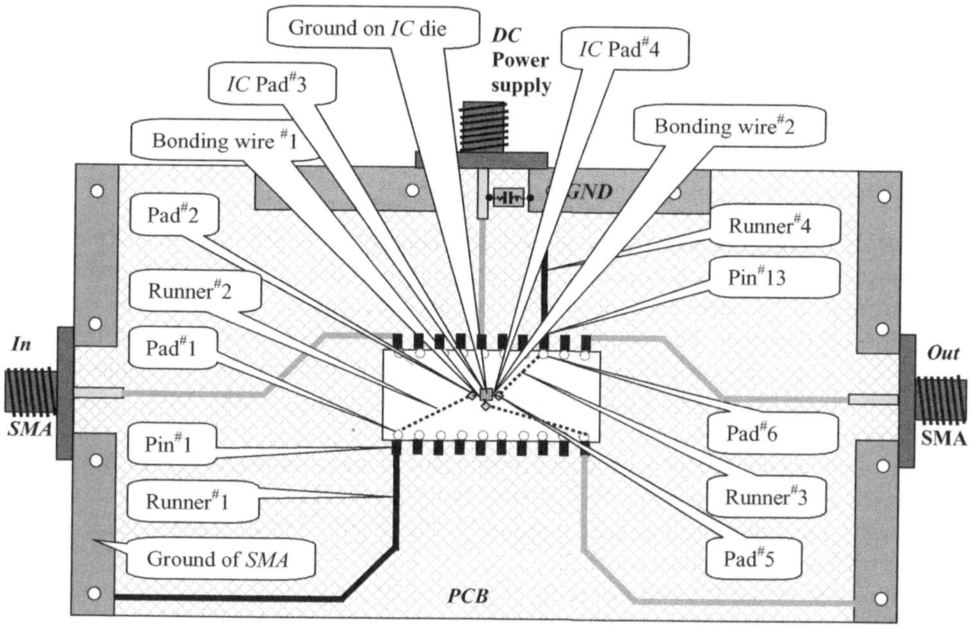

Figure 14.4 "Long march" from ground at *SMA* input to the *GND* at *DC* power supply.

.... Runner on *IC* package
Runner on *PCB*
Ground runner on *PCB*
Bottom ground area
Top ground area
"Zero" Capacitor
○ Pad inside package
▮ Pin
O Conductive via from top to bottom
IC die
◇ *IC* bonding pad
/ *IC* bonding wire

- Bonding wire 1
- *IC* pad 3 on *IC* die
- Ground on *IC* die
- *IC* pad 4 on *IC* die
- Bonding wire 2
- Bonding pad 5 inside *IC* package
- Runner 3 inside *IC* package
- Pad 6 inside package
- Pin 13 of *IC* package
- Runner 4 on *PCB*
- Destination: reference ground point, *GND*.

Should the thinness and tiny size of the bonding wires and ground wire inside an *IC* die come to mind, one would definitely feel sorry for such a grounding path, which mostly runs through micro strip lines with high characteristic impedance. Through such a grounding path, the ground of the input *SMA* connector is definitely not equipotential with the reference ground point of the *DC* power supply, *GND*, but is better compared to a "balloon" floating in the air. Consequently, the testing of the sensitivity of the receiver is of course incorrect and is usually greatly degraded.

It has been proven that the higher sensitivity of the receiver can be restored, in which it is enhanced by $6\,dB$ after the test *PCB* is re-designed with good *RF/AC* grounding.

14.3.2.2 Unequipotentiality on a Ground Surface

A good ground *RF* circuit block implies that equipotentiality is maintained over the entire ground surface. In other words, the entire ground surface is an electrically equipotential surface. Unequipotentiality on the ground surface is one of the imperfect grounding problems. The tested results are questionable if the ground surface is not an equipotential surface, because the tested parameter is not referenced to the same reference ground point, *GND*.

If the ground surface is not in an equipotential state, the current will flow over the surface and result in so-called current coupling, including both return current and forward current coupling, which might be the problem in the dark corner which puts the circuit block out of work!

It should be noted that equipotentiality is not directly related to high conductivity. For instance, a copper layer with gold plating has very high conductivity. However, its conductivity does not guarantee that its surface is electrically equipotential. On the contrary, the voltage is generally different from one point to another. If the shape of this copper plane is long and narrow on a substrate or *PCB*, it is a micro strip line. The voltage on a micro strip line varies from point to point or from line to line. Strictly speaking, equipotentiality on a metallic surface is never exactly true; however, it can be well approached if its size or dimension is greatly less than the quarter wavelength, that is,

$$L_d << \lambda/4, \tag{14.5}$$

where L_d is the maximum dimension of the metallic surface.

In order to ensure the equipotentiality of a ground surface, it is better to design the *PCB* or *IC* die with a dimension much less than the quarter wavelength. For complicated or large systems, in which the size of the *PCB* or *IC* die may be comparable with or larger than the quarter wavelength, a special means or scheme must be worked out to guarantee equipotentiality on the whole ground surface. This is discussed in Chapter 15 along with the current coupling problem.

14.3.3 Improper Connection

The following are two examples of improper connection about the ground points.

14.3.3.1 Connection from Tested PCB to DC Power Supply

Figure 14.5 shows a lazy approach to connecting the test *PCB* with the *DC* power supply. A short copper wire stub is erected and soldered at the terminal V_{dd} or V_{cc} of the test *PCB*. This stub is connected to the output of the *DC* power supply by a plastic-covered copper wire. The ground terminal of the *DC* power supply, *GND*, is connected to the test *PCB* by an enamel-covered copper wire. One end of the

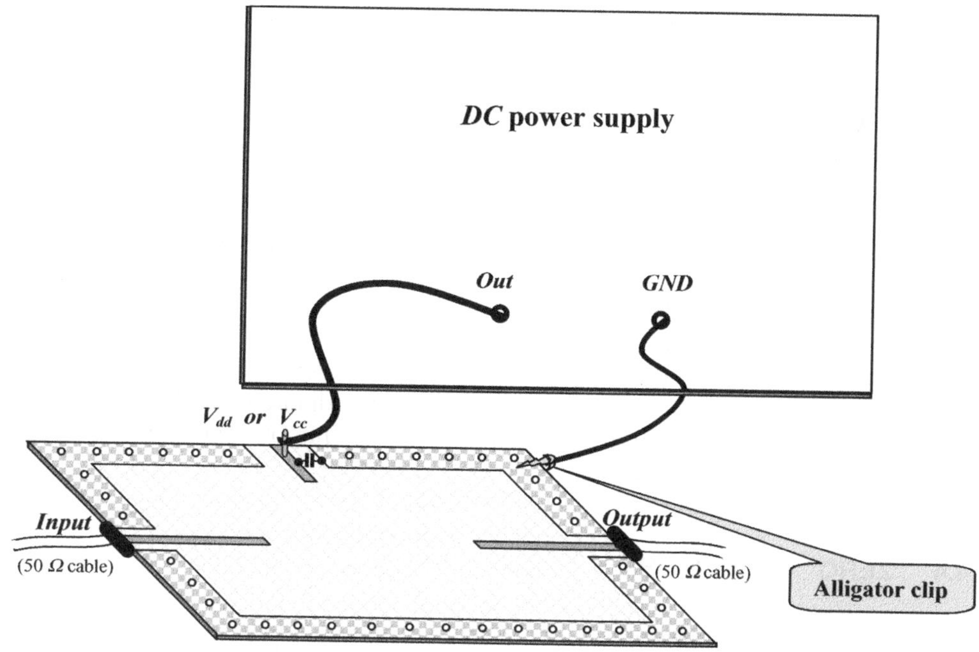

Figure 14.5 A last way to connect tested *PCB* with *DC* power supply.
Copper wire stub
Plastic-covered copper wire
Enamel-covered copper wire
Conductive via
Top ground metallic frame
Top metallic runner
Bottom metallic area
"Zero" capacitor

enamel-covered wire is connected to the ground terminal of the *DC* power supply. Another end of the enamel-covered wire is put on an alligator clip, which is placed between the top ground metallic frame of the test *PCB* in a convenient way.

Using an alligator clip, enamel-covered copper wire, and plastic-covered copper wire is really convenient. However, it brings about a lot of trouble:

- The short wire copper stub may function as a small antenna or an impedance compensator.
- The long plastic-covered copper wire may function as a big antenna with a heavy load, the output impedance of the *DC* power supply.
- The length of the enamel-covered copper wire may be comparable to or longer than the quarter wavelength corresponding to the operating frequency. Two ground terminals in the *DC* power supply and the tested *PCB* may be in *AC/RF* disconnected or open-circuited state.
- The "antennas" absorb whatever interference signals and noise they sense in the sky. The "zero" capacitor can *AC* short-circuit those frequency spectrums around the operating frequency, but does nothing for those interferences whose spectrums are far from the operating frequency; especially, it does nothing to alleviate white noise.
- Noise not only comes from the sky but also from the *DC* power supply.

Without hesitation I affirm that such a lax method must be strictly prohibited in *RF* circuit testing.

14.3.3.2 Connection from Ground to Ground Very often there is more than one piece of ground surface in an *RF* system or block. The connection between ground surfaces may bring about trouble. Figures 14.6 and 14.7 show two types of connections between ground surfaces.

In the circuit board of an *RF* module or *RF* system, two pieces of ground surfaces are very often connected by a runner as shown in Figure 14.6. If the length of this runner is comparable to the quarter wavelength, the two pieces of the ground surface are not equipotential. Consequently, grounding problems arise.

In an *RFIC* chip, two pieces of ground surfaces are very often connected by a bunch of conductive vias as shown in Figure 14.7. If the diameter of the conductive via is too small, the two ground surfaces may not be equipotential. Consequently, grounding problems may arise.

14.4 "ZERO" CAPACITOR

In this section, our discussion about the "zero" capacitor applied for *RF/AC* grounding is not for integrated circuits, but for circuits built with discrete parts. However,

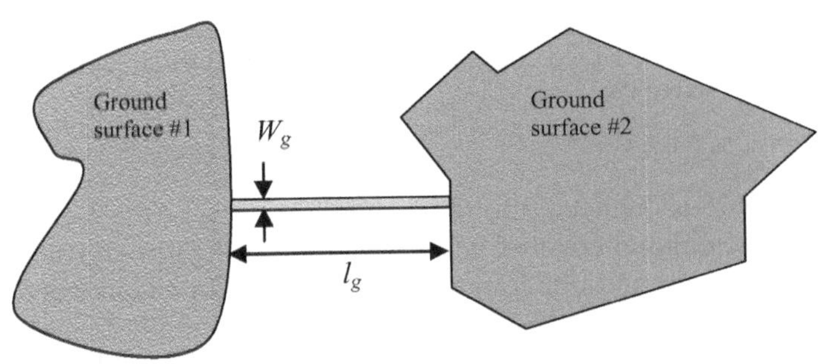

Figure 14.6 Two pieces of ground surface are connected by a runner.

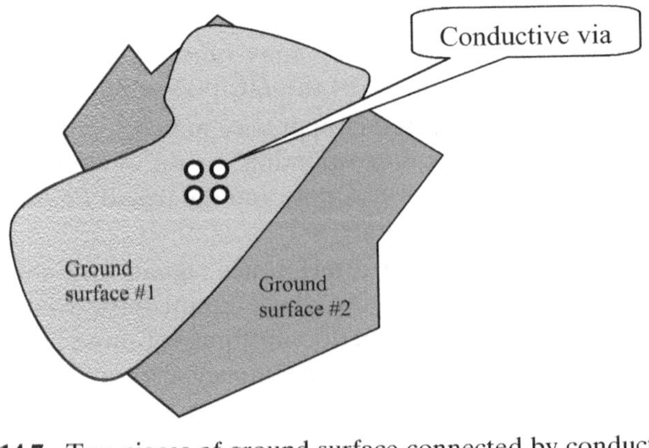

Figure 14.7 Two pieces of ground surface connected by conductive vias.

the discussion about the quarter wavelength for *RF/AC* grounding is suitable to both integrated circuits and circuits built by discrete parts. Among the discrete parts applied for *RF/AC* grounding, "zero" capacitors, and micro strip lines are the two main parts widely applied in RF blocks and systems.

14.4.1 What Is a "Zero" Capacitor?

The term "zero capacitor" is "professional" slang. It indicates that at the specified operating frequency the impedance of said capacitor approaches zero.

A "zero" capacitor can therefore function as a bypass capacitor if it is connected between the port to be *RF/AC* ground and the reference ground point, *GND*, in parallel. At the port to be ground, the *DC* voltage is kept unchanged, but the *RF/AC* signal is short-circuited to the ground. On the other hand, a "zero" capacitor can function as a *DC* blocking capacitor if it is connected with the port to be *RF/AC* blocked in series, in which the *DC* voltage is blocked but the *RF/AC* signal crosses over without any attenuation. Both applications, as *AC* bypass capacitor and as *DC* blocking capacitor, are shown in Figure 14.8. At the nodes of V_{cc} and Bias, the "zero" capacitors are connected in parallel and thus function as bypass capacitors; at the nodes of input and output, the "zero" capacitors are connected in series and function as *DC* blocking capacitors.

In short, either the bypass capacitor or the *DC* blocking capacitor is the "zero" capacitor.

14.4.2 Selection of the "Zero" Capacitor

It is well-known that the impedance for an ideal capacitor, Z_c, is

$$Z_c = \frac{1}{jC\omega}, \tag{14.6}$$

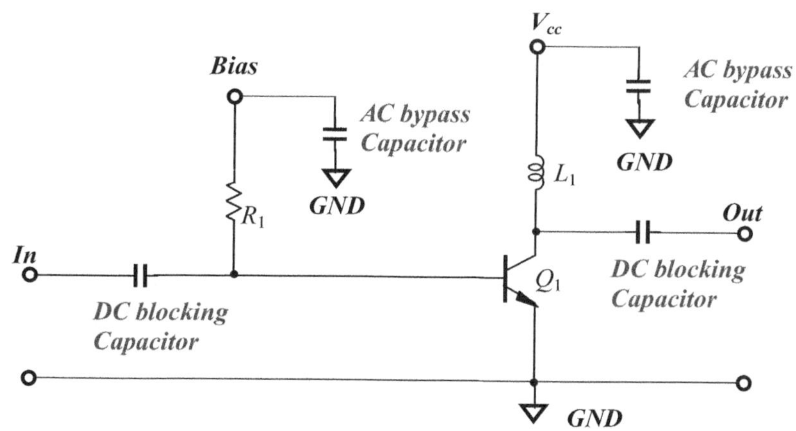

Figure 14.8 "Zero" capacitor functions as a bypass or a blocking capacitor.

where

C = capacitance of the capacitor,
ω = operating angular frequency.
The subscript "c" denotes capacitor's parameter.

It can be seen that the impedance of a capacitor, Z_c, is infinite to the *DC* current or voltage when $\omega = 0$ and that the impedance of a capacitor is decreased as the value of the capacitor increases when $\omega \neq 0$.

Theoretically, at the *RF* frequency range, the impedance of an ideal capacitor could approach zero by infinitely increasing the capacitor's value C when $\omega \neq 0$. In other words, the "zero" capacitor is an ideal capacitor with infinite capacitance when $\omega \neq 0$. *RF/AC* grounding could be accomplished by connecting an ideal capacitor with infinite capacitance between the desired *RF* ground terminal and a real ground point.

In reality, an ideal capacitor does not exist and applying infinite capacitance is not realistic or practical. In the *RF* frequency range, an actual capacitor is never close to being an ideal capacitor, so the ideal formula for the capacitor impedance is even less viable, although in the low-frequency range, an actual capacitor is pretty close to being ideal and can be described by expression (14.6).

There are many kinds of capacitors available in the market. In today's *RF* circuit design, implemented either by discrete parts or by integrated circuits, most "zero" capacitors are chip capacitors due to their advantage of small size, low cost, and reliable performance.

In the *RF* frequency range, a chip capacitor has additional inductance and additional resistance and can be modeled as shown in Figure 14.9, in which a chip capacitor is represented by an ideal capacitor C, an ideal inductor L_S, and an ideal resistor R_S in series. The additional inductance L_S brings about additional phase shift of the *RF/AC* signal and the additional resistance R_S represents additional attenuation of the *RF/AC* signal or the Q value of the actual capacitor.

Instead of infinitely increasing of the capacitor's value C, the impedance of an actual capacitor, Z_c, approaches R_S when its additional inductor is resonant with the capacitor itself at its *SRF* (Self-Resonant Frequency). The *SRF* of an actual chip capacitor can be expressed as

$$SRF_c = \frac{1}{2\pi\sqrt{L_S C}}, \tag{14.7}$$

where the subscript "c" denotes the capacitor's parameter.

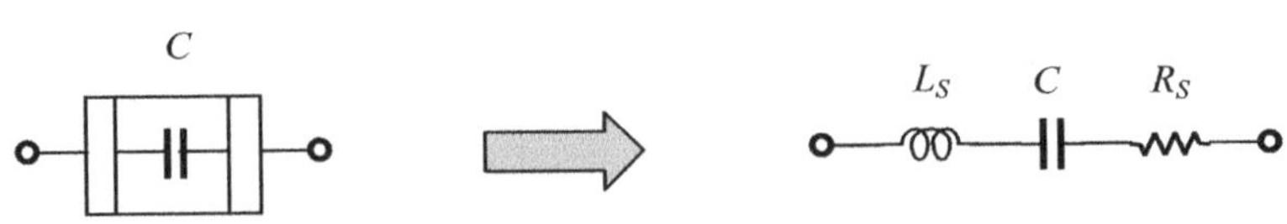

(a) An actual chip capacitor (b) Equivalent of an actual capacitor

Figure 14.9 An actual chip capacitor and its equivalent.

Should the *SRF* be identical to the operating frequency, the capacitor is then a "zero" capacitor since the value of the resistor R_S shown in Figure 14.9 is a tiny value or, approaches zero at operating frequency. In other words, the "zero" capacitor is re-recognized as an *AC* bypass or *DC* blocking capacitor with "zero impedance" when its operating frequency is equal to its *SRF*.

The *SRF* of a chip capacitor depends on the material, art, and technology adapted by the manufacturer. For the chip capacitor manufactured by MuRata, the *SRF* can be formularized as

$$SRF_c = \frac{5400}{\sqrt{C_{specicied}}}, \tag{14.8}$$

where

SRF_c = self-resonant frequency in MHz,
$C_{specified}$ = specified capacitance by the manufacturer in pF.

Formula (14.8) is available within the range of specified capacitance

$$C_{specified} = 0.5\,pF \text{ to } 18000\,pF,$$

for both sizes

$$W \times L = 50 \times 80 \; mils \text{ and } 30 \times 40 \; mils$$

where W and L are the width and length of the chip capacitor, respectively.

In terms of formula (14.8), one can easily find out the "zero" capacitor for a specific operating frequency. In Appendix 14.A.1, we discuss how to find the additional inductance and additional resistance from the actual measurement of a chip capacitor, and finally how to obtain formula (14.8).

If the "zero" capacitor is not manufactured by MuRata, the formula may be slightly different. By means of the same testing procedures, an appropriate formula can be found, in which the coefficient "5400" in formula (14.8) would be different and replaced with some other value.

Figure 14.10 draws a formal *PCB*, which is discussed in Chapter 15. The parts on the *PCB* could be discrete elements, *RFIC* chips, or both. For simplicity, the parts on the *PCB* are not drawn, since our concern is only focused on how to apply the "zero" capacitor for *RF/AC* grounding.

The terminals of the *DC* power supply and *DC* bias are half ground points, in which *RF/AC* signals must be ground but the *DC* voltage must not be short-circuited. A "zero" capacitor is therefore connected between the *DC* power supply terminal V_{dd} or V_{cc} and the point *GND*, which is the reference ground point on the *PCB* ground surface. Similarly, the "zero" capacitor can be applied to other half ground terminals, such as the *DC* bias, which are not shown in Figure 14.10.

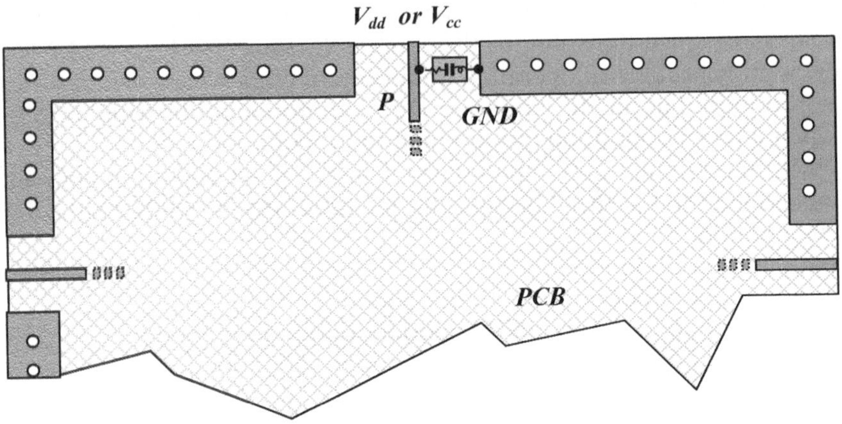

Figure 14.10 *RF* or *AC* ground by a "zero" capacitor on a *PCB* (Printed Circuit Board).
Top metallic area
Bottom metallic area
Conductive via from top to bottom
"Zero" capacitor
P: Point to be ground
G: Well-ground point

14.4.3 Bandwidth of the "Zero" Capacitor

The *SRF* of a chip capacitor can be found by means of a S_{21} measurement.

Figure 14.11 shows a S_{21} testing setup for a chip capacitor. As a *DTU* the tested chip capacitor is soldered between the $50\,\Omega$ micro strip line and ground on a small *PCB*. The small *PCB* is then connected to ports 1 and 2 of the network analyzer by $50\,\Omega$ *RF* cables. The tested chip capacitor is connected with port 1 and port 2 of the network analyzer in parallel accordingly.

By referring to the equivalent model of chip capacitor as shown in Figure 14.9 and the test setup as shown in Figure 14.11, the capacitor C is resonant with L_S in series at the self-resonant frequency, so that the S_{21} is significantly lowered due to its connection with port 1 and port 2 of network analyzer in parallel. Figure 14.12 shows a typical S_{21} frequency response of a tested chip capacitor, $C = 15pF$. It is found that

$$S_{21} = -45.1\,dB,$$

at

$$SRF = 1.394\,GHz.$$

An actual "zero" capacitor is self-resonant at its *SRF* and approaches zero impedance within a frequency range around the *SRF*. In other words, the "zero" capacitor can force a node to approach zero impedance within a certain bandwidth as long as it is connected with this port to the reference ground in parallel. The bandwidth depends on which level of S_{21} is assigned as the critical value of "approaching to zero impedance." From Figure 14.12, the bandwidth of the "zero" capacitor can be found and is listed in Table 14.1.

The lower the assigned critical value of S_{21}, the better the approach to zero impedance. However, as shown in Figure 14.12 and Table 14.1, the lower the assigned critical value, the narrower the bandwidth to be covered.

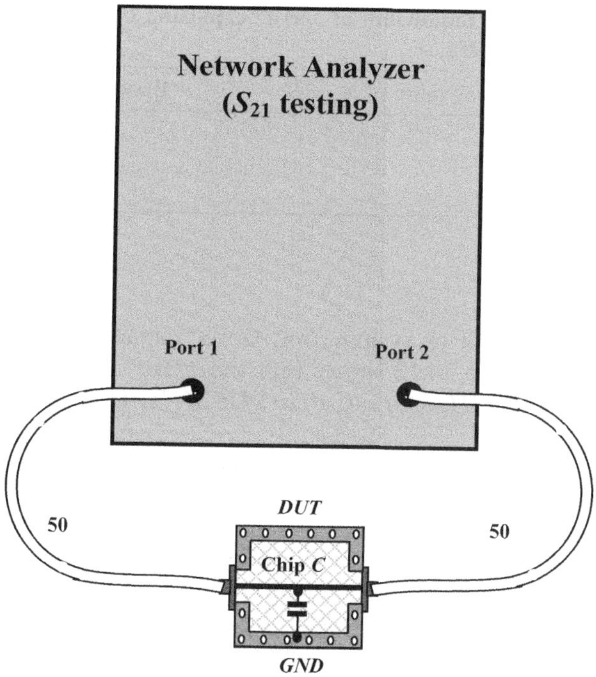

Figure 14.11 S_{21} testing for a chip capacitor.

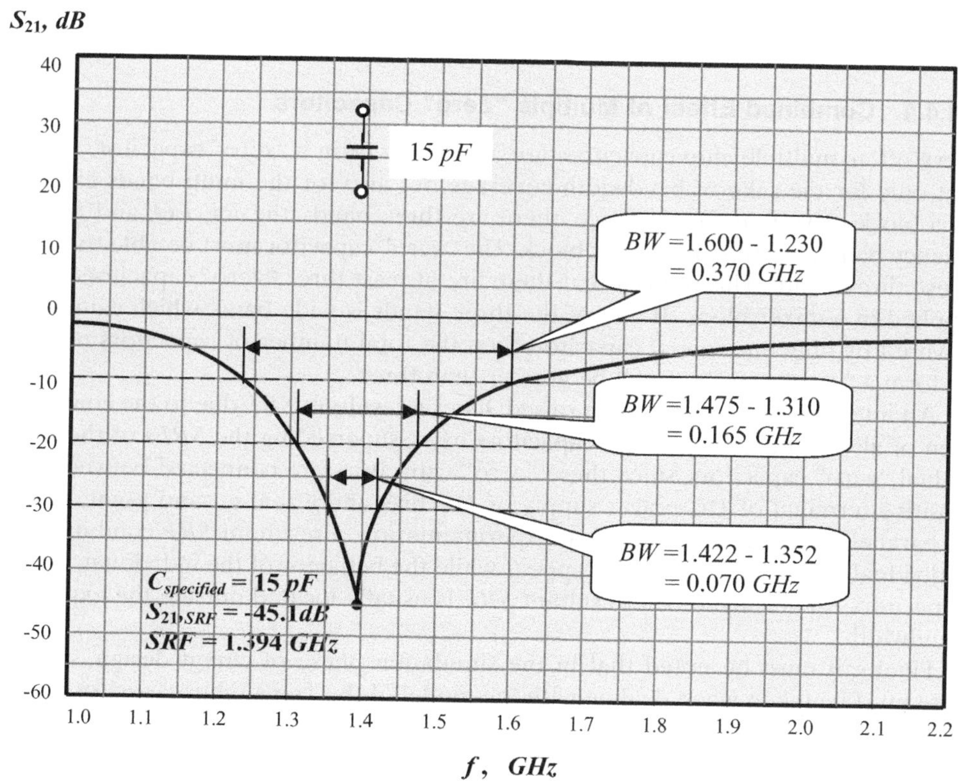

Figure 14.12 S_{21} frequency response of a chip capacitor tested with setup as shown in Figure 14.11.

TABLE 14.1 Bandwidth of "zero" capacitor, $C = 15 pF$, $SRF = 1.394 GHz$

Assigned critical value of S_{21}, dB	Bandwidth, MHz
–10	370
–20	165
–30	70

For a narrow-band block or system, one "zero" capacitor is usually enough to cover its bandwidth. Figure 14.12 shows that the "zero" capacitor, $C = 15 pF$, can cover a bandwidth of 165 MHz, or $f = 1310$ to 1475 MHz, if the assigned critical value of S_{21} is $-20 dB$.

However, for a wide-band block or system, the "zero" capacitor can and must be formed by a combination of several capacitors in parallel, in which the SRF of the capacitors is shifted in a way so that "zero impedance" can be approached as the S_{21} is lower than the assigned value over the expected bandwidth. For example, if the bandwidth to be covered is 250 MHz from $f = 1.310$ to 1.560 GHz and the critical value of S_{21} is assigned as $-20 dB$, the single "zero" capacitor, $C = 15 pF$, is not enough to cover the bandwidth. However, the "zero" capacitor can be formed by a combination of two capacitors, $C = 15 pF$ and $C = 13 pF$, connected together in parallel. The combined "zero" capacitor can cover a bandwidth even wider than the expected 250 MHz. Figure 14.13 shows the coverage of the expected bandwidth.

14.4.4 Combined Effect of Multiple "Zero" Capacitors

Very often, multiple chip capacitors are applied to form a "zero" capacitor. This is not only for the sake of bandwidth coverage but also for the multi-bands existing in a block or system. For instance, there are three bands, the *RF, LO,* and *IF* frequency bands, existing in a mixer block. The "zero" capacitor must be able to cover these three bands. This implies that there are at least three "zero" capacitors to be applied in a mixer block. If any of the three bands is wide band, which cannot be covered by only one "zero" capacitor, then the total number of capacitors needed to form a "zero" capacitor will be greater than three.

An interesting question is then raised: Does a resultant SRF due to the combination of all the individual "zero" capacitors exist, superseding the SRFs of the individual "zero" capacitors, since these "zero" capacitors are connected between the positive terminal of DC power supply or DC bias and a real ground point, GND, in parallel? According to empirical experimentation, a resultant SRF combined by individual "zero" capacitors may appear while the behavior of the individual "zero" capacitors is unchanged. The resultant SRF is usually located outside the expected bandwidth.

Finally, it must be noted that in the simulation phase of circuit design, special care must be taken when dealing with the model of the "zero" chip capacitor, especially when the operating frequency is higher than GHz. In addition to the equivalent model as shown in Figure 14.9, the soldering pad and runners must be attached to the equivalent model as shown in Figure 14.14.

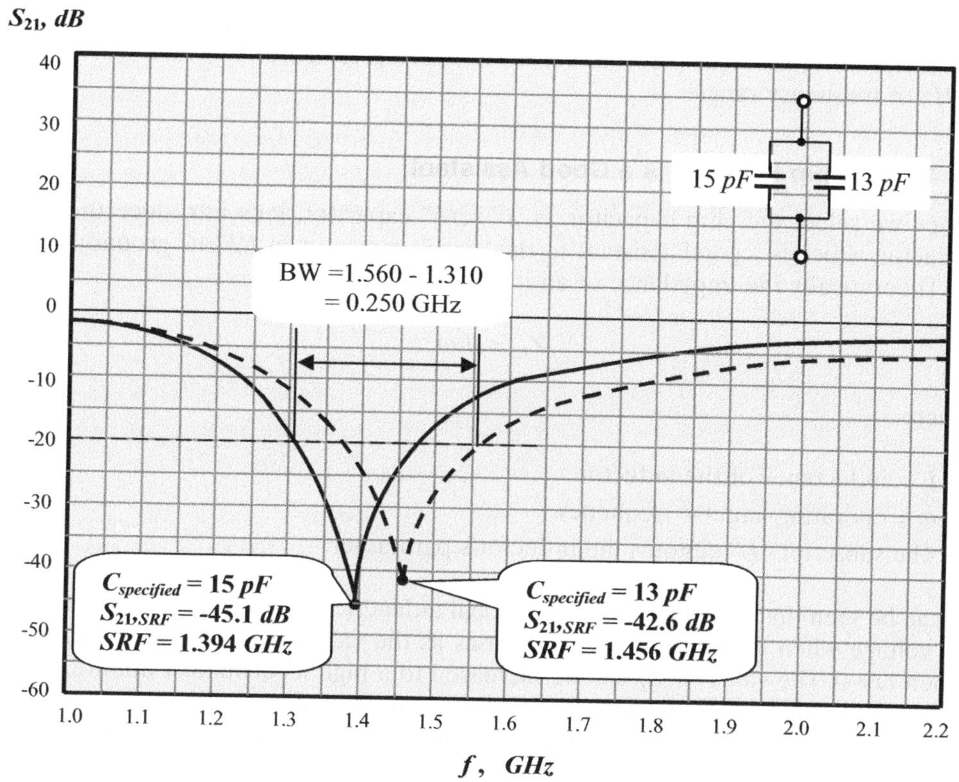

Figure 14.13 Frequency bandwidth $250 MHz = 1.560 - 1.310 GHz$ is covered by a "zero" capacitor combined by two capacitors $C = 15pF$ and $C = 13pF$ connected together in parallel.

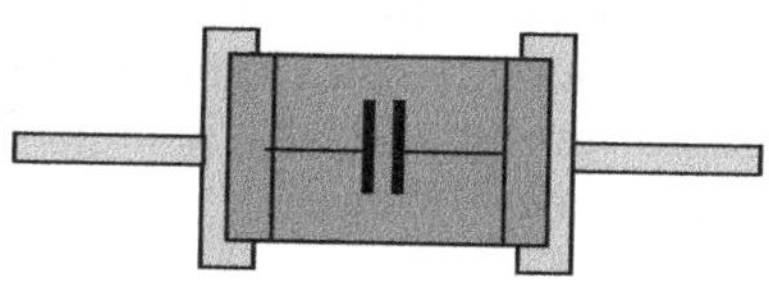

Figure 14.14 Modified equivalent model of "zero" chip capacitor by adding of soldering pads and runners for *RF* modules.
▭ Runner on *IC* die ▣ "Zero" chip capacitor
▯ Soldering pad on *PCB*

As shown in Figure 14.14, the modified equivalent model consists of the following five parts:

- A runner on the *PCB*,
- A soldering pad on the *PCB*,
- A "zero" chip capacitor,
- A soldering pad on the *PCB*,
- A runner on the *PCB*.

By design experience, the difference of *SRF* between the cases with and without the modification of the equivalent model of "zero" chip capacitor is around 5% in the *GHz* of frequency range.

14.4.5 A Chip Inductor Is a Good Assistant

Since we select the chip capacitor as a "zero" capacitor, let's introduce the chip inductor, which is a good assistant to the "zero" capacitor in *RF/AC* grounding.

Theoretically the impedance of an ideal inductor, Z_L, is

$$Z_L = jL\omega, \tag{14.9}$$

where

L = inductance of the inductor,
ω = operating angular frequency.
The subscript "L" denotes the inductor's parameter.

It can be seen that the impedance of an ideal inductor, Z_L, is zero to the *DC* current or voltage when $\omega = 0$ and that it increases as the value of the inductor increases when $\omega \neq 0$. The value of Z_L can be increased to a high level, almost infinite, if the value of the inductor, L, is high enough for the *RF* signal.

Just like the ideal capacitor, however, ideal inductors are never available, though a real inductor in the low-frequency range is pretty close to ideal and can be described by expression (14.9). In the *RF* range, this ideal formula for the inductor impedance is not viable. Instead, a practical chip inductor in the *RF* range has additional capacitance and additional resistance in parallel and can be modeled as shown in Figure 14.15.

The impedance of an actual chip inductor approaches R_P when its additional capacitor is resonant with the inductance itself in parallel at an operating frequency. It is called an "infinite" inductor because ideally, its impedance approaches infinity at its *SRF* (Self- Resonant Frequency), that is,

$$SRF_L = \frac{1}{2\pi\sqrt{LC_P}}, \tag{14.10}$$

where the subscript "L" denotes the inductor's parameter.

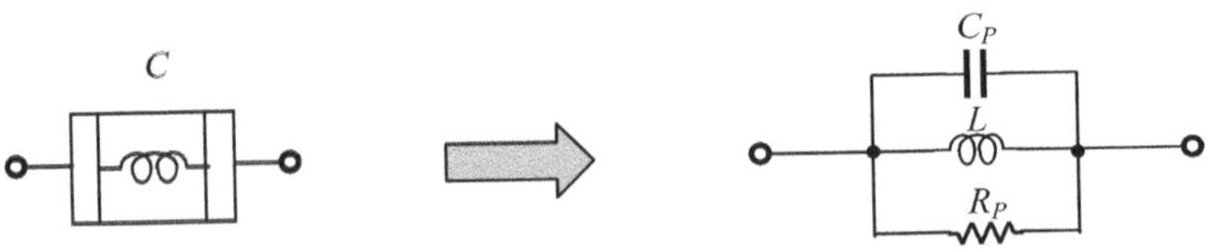

(a) An actual inductor (b) Equivalent of an actual inductor

Figure 14.15 An actual inductor and its equivalent.

Should the *SRF* be identical to the operating frequency, the inductor is then an inductor with "infinite" impedance, since the value of the resistor R_P shown in Figure 14.14 is a high value, or approaches infinity at operating frequency. In other words, the "infinite" inductor is re-recognized as an *AC* blocking inductor with "infinite impedance" when the operating frequency is equal to its *SRF*. As a matter of fact, the alias of the "infinite" inductor is the *RF* choke which most people are familiar with.

The SRF_L expression (14.10) looks the same with the SRF_c expression (14.8). However, one must bear in mind that the SRF_L is a resonant frequency when L and C_P are connected in parallel and its impedance approaches infinity at SRF_L, while SRF_c is a resonant frequency when C and L_S are connected in series and its impedance approaches zero at SRF_C,

The *SRF* of a chip inductor depends on the material, art, and technology adapted by the manufacturer. For the chip inductor manufactured by MuRata, the *SRF* can be formularized as

$$SRF_L = \frac{8920}{\sqrt{L_{specified}}}, \tag{14.11}$$

where

SRF_L = self-resonant frequency in MHz,
$L_{specified}$ = specified capacitance by the manufacturer in nH.

Formula (14.11) is available within the range of specified inductance

$$L_{specified} = 22\,nH \text{ to } 1800\,nH.$$

If the "infinite" inductor is not selected from the chip capacitor manufacturer by MuRata, the formula may be mostly different. By means of the same testing procedures, an appropriate formula can be found, in which the coefficient "8920" in formula (14.11) is different and is replaced with another value.

In terms of formula (14.11), one can easily find out the "infinite" inductor for a specific operating frequency.

In Appendix 14.A.1 we discuss how to find the additional capacitance and additional resistance from the actual measurement of a chip inductor, and finally how to obtain the formula (14.11).

Figure 14.16 shows how an "infinite" inductor functions as an assistant of "zero" capacitor in the *RF/AC* grounding.

For simplicity, Figure 14.16 duplicates everything from Figure 14.10 except the portion containing the point to be *RF/AC* ground, *P*. In Figure 14.16, an "infinite" inductor is inserted between point P_o and point P, where P_o is the outside adjacent point on the same runner before the "infinite" inductor is inserted. This "infinite" inductor blocks the *RF* signal from the outside point P_o to point P. At point P, the voltage or power of the *RF* signal from point P_o is reduced to below a significant level, but the *DC* voltage at point P is kept the same as at point P_o, since the impedance of the "infinite" inductor approaches infinity. Point P is in fact very close to being *RF/AC* ground even if the "zero" capacitor is imperfect, as long as the point P_o is the unique *RF* source.

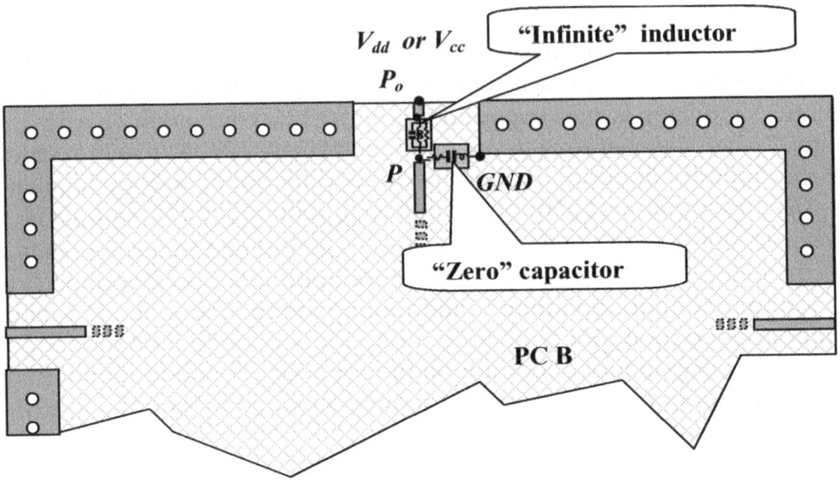

Figure 14.16 *RF* or *AC* ground by a "zero" capacitor with assistance of an "infinite" inductor on a *PCB*.
▩ Top metallic area
☐ Bottom metallic area
○ Conductive via from top to bottom
"Zero" capacitor
"Infinite" inductor
P: Point to be ground
GND: Reference ground point

14.4.6 "Zero" Capacitor in *RFIC* Design

Today, chip capacitors are widely applied in *RF/AC* grounding as "zero" capacitors for a test *PCB* or *RF* module. However, there does not seem to be a way that they can be directly applied in an *RFIC* die. The designer cannot but treat the "zero" capacitor as an off-chip discrete part. Even so, the chip capacitor is still a good part of *RF/AC* grounding for *RFIC*.

Figure 14.17 illustrates how the chip capacitor works together with other parts between the ground ring in the *IC* die and a ground frame on the *PCB*. It is assumed that the ground frame on *PCB* is equipotential with the reference ground point at the *DC* power supply terminal. The ground ring of an individual *RF* block in the *IC* die is a *P*+ guard ring. It must be forced to be equipotential with the reference ground point at the *DC* power supply terminal. As shown in Figure 14.17, it is connected to the ground frame on the *PCB* through the following nine parts, which form the modified equivalent model of a "zero" capacitor for an *RFIC* and is redrawn in Figure 14.18.

1. A short runner in the *IC* die from the ground ring to a tiny pad,
2. A tiny bond pad in the *IC* die,
3. A bond wire in the air between the tiny pad in the *IC* die and the bonding pad on the *PCB*,
4. A bonding pad on the *PCB*,
5. A runner on the *PCB* from the bonding pad on the *PCB* to the soldering pad on the *PCB*,

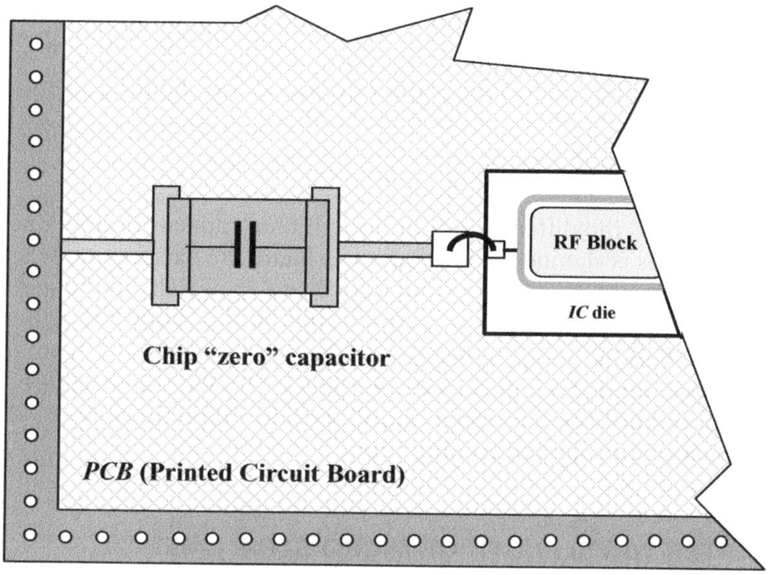

Figure 14.17 "Zero" capacitor is connected between the *P*+ guard ring in *IC* die and the ground frame on *PCB*.

Bond wire
Pad
Top metallic area
Bottom metallic area
IC die
— Runner in *IC* die
Runner on *IC* die
Soldering pad on *PCB*
O Conductive via from top to bottom
"Zero" Chip Capacitor
Ground ring of *RF* block

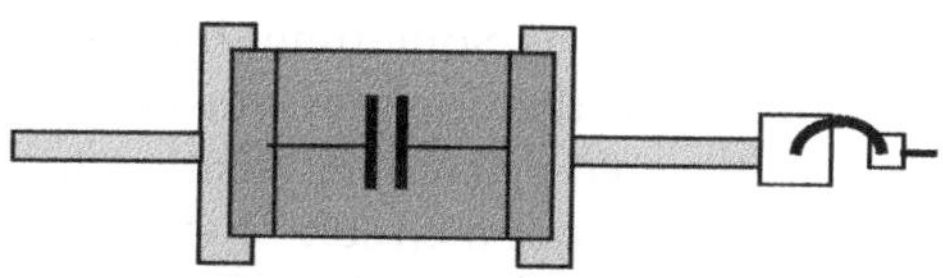

Figure 14.18 Modified equivalent model of "zero" chip capacitor by adding of soldering pads and runners for *RFIC*.

Runner on *IC* die
Soldering pad on *PCB*
"zero" chip capacitor

6. A soldering pad on the *PCB*,
7. A "zero" chip capacitor,
8. A soldering pad on the *PCB*,
9. A runner on the *PCB* from the "zero" chip capacitor to the ground frame on the *PCB*.

The equivalent model of the "zero" capacitor must be correspondingly modified from Figure 14.14 to Figure 14.18.

There are four differences in the modified equivalent model of the "zero" capacitor shown in Figures 14.14 and Figure 14.18:

1. A short runner in the *IC* die from the ground ring to a tiny pad
2. A tiny bond pad in the *IC* die,
3. A bond wire in the air between the tiny pad in the *IC* die and the bonding pad on the *PCB*,
4. A bonding pad on the *PCB*.

By design experience, the difference of the "zero" chip capacitor's *SRF* due to these four additional items is significant. On the other hand, the existing models for these four additional parts are complicated and inaccurate. Especially in the model of the bonding wire between pads, the inaccuracy of the model arises not only from the inconsistent length of the bonding wire in production lines, but also from the variation of its inclined angle in respect to the vertical jump. We are therefore aware of the importance and need to develop a "zero" capacitor directly in the *IC* die, which may be a remarkable project in coming years.

14.5 QUARTER WAVELENGTH OF MICRO STRIP LINE

14.5.1 A Runner Is a Part in *RF* Circuitry

A runner between parts in a circuitry can be a copper line with gold plating on the *PCB* or a gold line segment on an *IC* chip. Very often, the circuit designer just ignores its existence, only focusing on the lump parts, such as the inductor, capacitor, resistor, transistor, and so on. Very often, the test results of performance are quite far from agreement with the simulation results of performance. This may be partially due to the inaccuracies of the lump parts applied in the simulation, but also partially due to the disregarding of the existence of runners. In the *RF* frequency range, a runner is a part just like other lump parts. Sometimes it may even be more important than a lump part. All the runners in the circuitry must be taken care of in a good simulation. Otherwise, a considerable discrepancy between simulation and testing is inevitable, and sometimes this can put a circuit block out of work.

Essentially, a runner is a micro strip line in the *RF* frequency range. Its existence is the equivalent of an additional impedance existing between the two parts connected by the runner.

Based on transmission line theory, the impedance of a micro strip line can be depicted as Figure 14.19 and can be expressed as

$$Z_P = Z_o \frac{Z_L \cosh \gamma_t l + Z_o \sinh \gamma_t l}{Z_o \cosh \gamma_t l + Z_L \sinh \gamma_t l}, \tag{14.12}$$

where

Z_P = impedance at point P, looking into load with a distance l,
Z_o = characteristic impedance of the micro strip line,
Z_L = load impedance of the micro strip line,
l = length of the micro strip line,
γ_t = transmission coefficient.

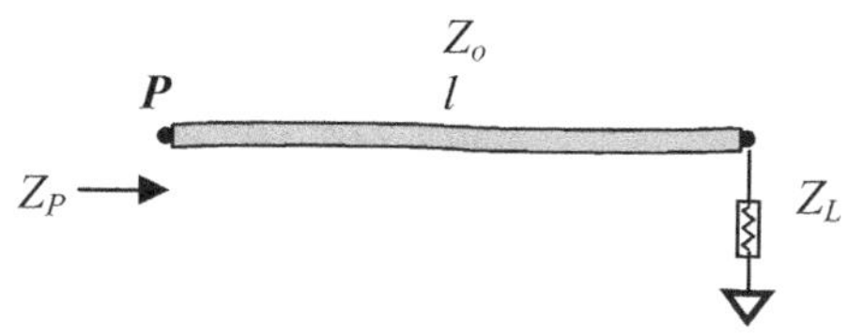

(a) A micro strip line with its characteristic and load impedances, Z_o, Z_L

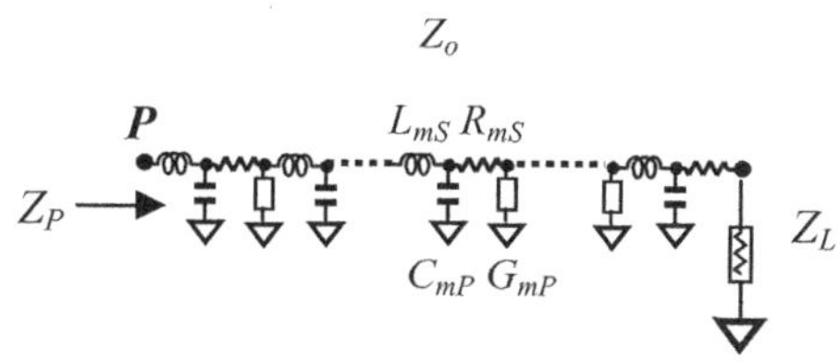

(b) A micro strip line with its distribution parameters, R_{mS}, L_{mS}, G_{mP}, C_{mP}

Figure 14.19 A micro strip line with its characteristic and load impedances, Z_o, Z_L, and distribution parameters, R_{mS}, L_{mS}, G_{mP}, C_{mP}.

The characteristic impedance Z_o and the transmission coefficient γ_t can be expressed as

$$Z_o = \sqrt{\frac{R_{mS} + j\omega L_{mS}}{G_{mP} + j\omega C_{mP}}} \approx \sqrt{\frac{L_{mS}}{C_{mP}}}, \tag{14.13}$$

$$\gamma_t = \sqrt{(R_{mS} + jL_{mS}\omega)(G_{mP} + jC_{mP}\omega)} \approx \alpha + j\beta, \tag{14.14}$$

if

$$R_{mS} << \omega L_{mS}, \tag{14.15}$$

$$G_{mP} << \omega C_{mP}, \tag{14.16}$$

where

R_{mS} = resistance per unit length of the micro strip line,
L_{mS} = inductance per unit length of the micro strip line,
G_{mP} = conductance per unit length of the micro strip line,
C_{mP} = capacitance per unit length of the micro strip line,
α = attenuation per unit length of the micro strip line,
β = phase shift per unit length of the micro strip line.

A closer approximation for α and β can be obtained by rearranging expression (14.14) for γ_t and using the binominal expression. Thus

$$\alpha \approx \frac{1}{2}\left(\frac{R_{mS}}{\sqrt{L_{mS}/C_{mP}}} + G\sqrt{L_{mS}/C_{mP}}\right), \tag{14.17}$$

$$\beta \approx \omega\sqrt{L_{mS}C_{mP}}. \tag{14.18}$$

Figure 14.19 shows a micro strip line with its characteristic load impedances, Z_o and Z_L, and its distribution parameters, R_{mS}, L_{mS}, G_{mP}, and C_{mP}.

In the cases where the attenuation is negligible, that is,

$$\alpha \to 0, \tag{14.19}$$

and conditions (14.15) and (14.16) are satisfied, expression (14.12) becomes

$$Z_P = Z_o \frac{Z_L \cos\beta l + jZ_o \sin\beta l}{Z_o \cos\beta l + jZ_L \sin\beta l}, \tag{14.20}$$

or

$$Z_P = Z_o \frac{Z_L \cos\frac{2\pi l}{\lambda} + jZ_o \sin\frac{2\pi l}{\lambda}}{Z_o \cos\frac{2\pi l}{\lambda} + jZ_L \sin\frac{2\pi l}{\lambda}}, \tag{14.21}$$

where λ = electrical wavelength,
and

$$\beta = \frac{2\pi}{\lambda}, \tag{14.22}$$

If the runner is taken as a part in the circuitry, its special feature is that its impedance depends not only on the self-parameters, the length of runner l and the characteristic impedance Z_o, but also on the load impedance, Z_L.

Now, let's examine two special cases:

1) When the load is short-circuited, that is,

$$Z_L = 0, \tag{14.23}$$

then

$$Z_P|_{Z_L \to o} = jZ_o \tan\beta l = jZ_o \tan\frac{2\pi}{\lambda} l. \tag{14.24}$$

2) When the load is open-circuited, that is,

$$Z_L \to \infty, \tag{14.25}$$

then

$$Z_P|_{Z_L\to\infty} = \frac{1}{j}Z_o \cot\beta l = -jZ_o \cot\frac{2\pi}{\lambda}l. \tag{14.26}$$

Figure 14.20 plots the curves of $Z_P|_{Z_L\to o}$ and $Z_P|_{Z_L\to\infty}$ vs. l. The figure indicates that

1) If the length of the runner is varied from 0 to $\lambda/4$, or from $\lambda/2$ to $3\lambda/4$,
   - The impedance of a runner Z_P is changed from 0 to ∞ when the load is short-circuited, that is, $Z_L = 0$,
   - The impedance of a runner ZP is changed from $-\infty$ to 0 when the load is open-circuited, that is, $Z_L = \infty$,
2) If the length of the runner is varied from $\lambda/4$ to $\lambda/2$, or from $3\lambda/4$ to λ,
   - The impedance of a runner Z_P is changed from $-\infty$ to 0 when the load is short-circuited, that is, $Z_L = 0$,
   - The impedance of a runner Z_P is changed from 0 to ∞ when the load is open-circuited, that is, $Z_L = \infty$.

Alternatively, in spite of the short-circuited load, $Z_L = 0$, or open-circuited load, $Z_L \to \infty$, it is concluded that

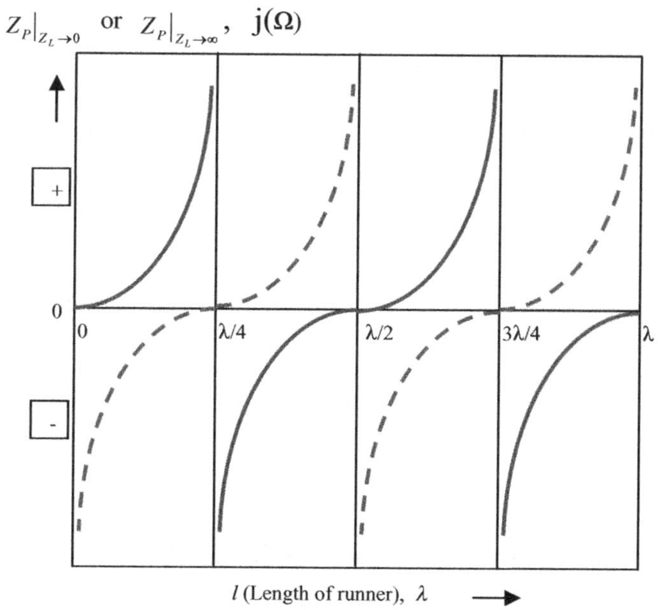

Figure 14.20 Impedance of runners when $Z_L = 0$ and $Z_L = \infty$.
——— $Z_P|_{Z_L\to o}$ - - - - $Z_P|_{Z_L\to\infty}$

1) The impedance of the runner Z_P can be ignored if

$$l << \lambda/4, \tag{14.27}$$

or

$$l \approx n\lambda/2 = 2n\lambda/4, \quad n = 1, 2, 3, 4 \ldots \tag{14.28}$$

because under conditions (14.27) or (14.28) expressions (14.24) and (14.26) become

$$Z_P|_{Z_L \to o} \to 0, \tag{14.29}$$

$$Z_P|_{Z_L \to \infty} \to \infty, \tag{14.30}$$

which implies that

$$Z_P \approx Z_L. \tag{14.31}$$

This means that the insertion of the runner does not introduce any conceivable change of impedance.

2) The impedance of the runner Z_P must be seriously taken care of if

$$l \approx (2n+1)\lambda/4, \tag{14.32}$$

because under this condition,

$$Z_P|_{Z_L \to o} \to \pm\infty, \tag{14.33}$$

$$Z_P|_{Z_L \to \infty} \to 0, \tag{14.34}$$

which implies that Z_P and Z_L are at extreme opposites respectively, that is,

$$Z_P \to \pm\infty, \quad \text{if } Z_L \to 0, \tag{14.35}$$

$$Z_P \to 0, \quad \text{if } Z_L \to \infty, \tag{14.36}$$

This indicates that the insertion of the runner brings about an earth-shaking change of impedance.

3) The impedance of runner Z_P must be taken care of if the length is neither in the cases of (14.27) and (14.28), nor in the case of (14.32), because in such a case, the impedance of the runner Z_P is usually comparable with the impedance of other parts.

14.5.2 Why Is the Quarter Wavelength So Important?

It should be noted that condition (14.28), $l \approx$ n $\lambda/2$, is intentionally rewritten as $l \approx$ n $\lambda/2$ = 2n $\lambda/4$. This is due to the recognition that all three conditions (14.27), (14.28), and (14.32) are related to the quarter wavelength $\lambda/4$.

Let us furthermore examine conditions (14.27), (14.28), and (14.32). In the layout for an *RF* circuit block or an *RF* system, it is always sound to shorten the length of runner between two parts as much as possible. The theoretical background is condition (14.27). It states that the length of the runner must be much shorter than the quarter wavelength. If so, the insertion of the runner does not introduce a conceivable change of impedance. In other words, the additional impedance due to the existence of the runner can be neglected or ignored.

If the length of a runner is about an even multiple of the quarter wavelength as shown in the condition (14.28), a case similar to (14.27) exists. If so, the insertion of a runner does not introduce a conceivable change of impedance. In other words, the additional impedance due to the existence of the runner can be neglected or ignored.

However, if the length of a runner is about one quarter wavelength or an odd multiple of the quarter wavelength as shown in condition (14.32), the insertion of a runner introduces an earth-shaking change of impedance. The additional impedance due to the existence of a runner approaches infinity. The function of the runner in the circuitry dominates that of the other parts and consequently, will put the circuitry out of work.

The three cases of runner's length, (14.27), (14.28), and (14.32), are the criteria used to judge the importance of a runner. These criteria become useful only if the quarter wavelength of the runner is well understood. This is why it is so important to understand the quarter wavelength.

No matter if the circuit is built by discrete parts as a module or built on an *IC* chip, the first thing or first step in designing the layout of the circuit is to know how long the quarter wavelength will be in the actual layout *PCB* or on the actual *IC* substrate. Otherwise, it is impossible to judge whether a runner is too short or too long.

The quarter wavelength can be calculated by formulas, but it is better obtained by actual testing.

14.5.3 The Magic of the Open-circuited Quarter Wavelength Micro Strip Line

If two parts are connected by a runner and the length of the runner is about a one quarter wavelength or an odd multiple of the quarter wavelength as shown in the condition (14.32), it equates to inserting a special part with infinite impedance between the two parts which leaves the connection between these two parts broken or open-circuited. This is, of course, an unexpected worst case.

Like many things in the universe, the worst and best cases switch positions under certain circumstances. As a runner to make a connection between two parts, the one or odd multiple of quarter wavelength runner is terrible. It is, however, a magic part in the task of *RF/AC* grounding or isolation if one end of the runner is kept in the open-circuited state.

Figure 14.21 shows a runner or a micro strip line a quarter wavelength long, with an open-circuited load. From (14.25), (14.26), and (14.32), we have

$$Z_P|_{Z_L\to\infty} = jZ_o \cot\frac{2\pi}{\lambda}l\bigg|_{l=(2n+1)\lambda/4} = 0, \tag{14.37}$$

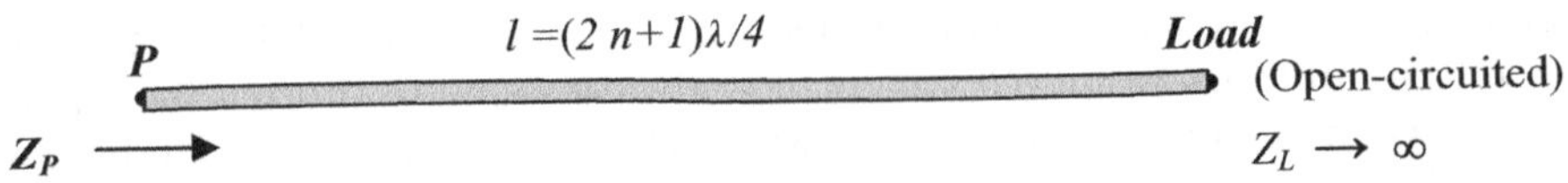

Figure 14.21 Point P is ground by micro strip line if $l = (2n + 1)\lambda/4$ with an open-circuited load, that is, $Z_L \rightarrow \infty$.

Expression (14.37) indicates that Point P of the runner is ground, since the values of Z_L and Z_P appearing in Figure 14.21 are the impedances in respect to the reference ground point.

It is therefore concluded that in order to ground point P, a simple connection by a conductive metallic runner will work well if one end of the runner is connected to point P, another end is open-circuited, and the length of the runner is an odd multiple of the quarter wavelength as shown in equation (14.32). This technology is called "compulsory," "enforced," or "coercing" grounding.

However, it should be noted that at the open-circuited load, the infinite impedance is just a theoretical approximation. Instead, some spray capacitance always exists, especially in a high RF frequency range. Consequently, in the practical product, the actual length of a quarter wavelength micro strip line for coercing grounding purposes is a little bit shorter than the quarter wavelength.

From expression (14.32) it can be seen that the shortest length of a micro strip line from point P to the open-circuited load is

$$l = \frac{\lambda}{4}, \tag{14.38}$$

when

$$n = 0. \tag{14.39}$$

A quarter wavelength micro strip line can be applied to improve isolation. The following example shows how the quarter wavelength micro strip line improves isolation in a mixer.

The device applied in the mixer shown in Figure 14.22 is a dual-gate *MOSFET*. One of the gates is injected with the *LO* injection and another gate is the input of *RF* signal. Due to the application of the dual-gate *MOSFET*, the isolation between the input and output is poor. The *LO* injection at the *LO* port could leak to the *RF* or *IF* ports. Similarly, *RF* power from the *RF* port could leak to the *LO* or *IF* ports. The *LO* power leakage into the *IF* portion usually dominates over *RF* power leakage to the *IF* port because the *LO* injection has the highest power among these three ports.

By means of a quarter-wavelength micro strip line, a mixer can perform with excellent isolation between the *LO*, *RF*, and *IF* ports. As shown in Figure 14.22, a quarter-wavelength micro strip line corresponding to the *LO* frequency is attached to the drain of the dual-gate *MOSFET* at one end. Another end of this quarter wavelength micro strip line is in an opened-circuited state. Consequently, at the

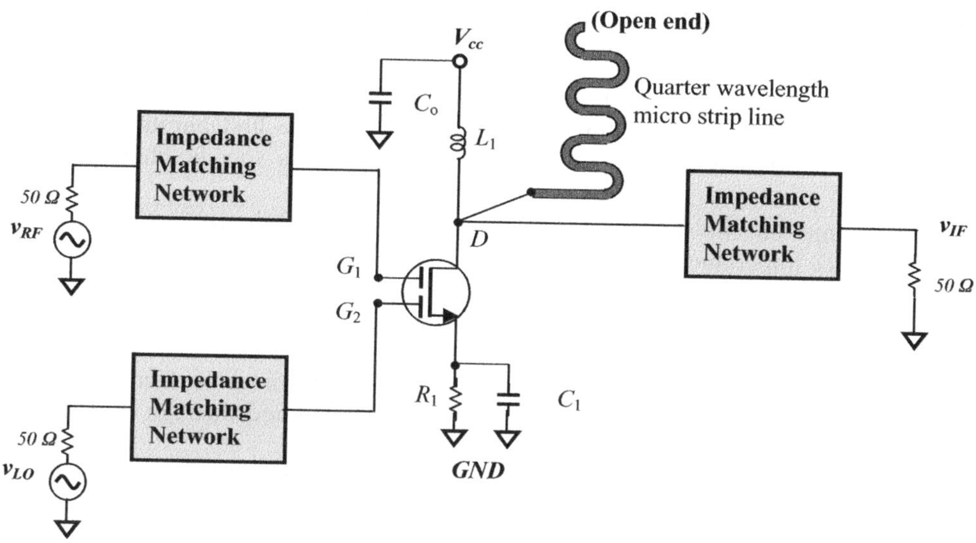

Figure 14.22 Isolation scheme by applying a quarter-wavelength micro strip lines in a mixer.

connected point at the drain of the *MOSFET*, the possible leakage site of the *LO* injection from gate 1 to point *D* is "absorbed" or "rejected" by the quarter wavelength micro strip line because at point *D*, any *LO* leaked injection is forced to be zero by the quarter wavelength micro strip line. In other words, any voltage at the *LO* frequency is impossible to be established at point *D*, the drain of the *MOSFET*.

On the other hand, the *RF* frequency is usually close to the *LO* frequency. The micro strip line with a *LO* quarter-wavelength can be very helpful to the isolation between the *RF* and *IF* ports at the same time, because the bandwidth of the attached quarter wavelength micro strip line usually covers the *RF* frequency as well so that the leaked *RF* signal from gate 2 to point *D* can likewise be "absorbed" or "rejected."

In general, isolation can be enhanced by more than 20*dB* with this technology.

Alternatively, the quarter wavelength micro strip line corresponding to the *LO* frequency in Figure 14.22 can be replaced by a quarter wavelength cable corresponding to the *LO* frequency. However, cables are too clumsy and thus are seldom applied in practical engineering.

A question is then raised: Does the attached quarter wavelength micro strip line radiate the power at the *LO* frequency since it looks like a small antenna? Yes, indeed! The primary purpose of the attached quarter wavelength micro strip line is to suck out the leakage of *LO* injection at point *D*, the drain of the dual-gate *MOSFET*. The absorbed *LO* leakage power therefore radiates to the space around the circuitry. However, the radiated power should be very low, negligible to the circuitry, and should not be a concern.

14.5.4 Testing for the Width of a Micro Strip Line with a Specific Characteristic Impedance

In terms of the network analyzer, the *RF* circuit designer should be able to characterize a runner or micro strip line on a specific *PCB*. The first parameter to be studied is the width of a micro strip line or runner with a specific characteristic impedance.

Let's assume that the specific characteristic impedance of the runner is the standard reference impedance, that is,

$$Z_o = 50\,\Omega. \tag{14.40}$$

It is well known that the characteristic impedance Z_o of a runner or micro strip line is related to the width of runner W, which can be determined through impedance testing by a network analyzer.

If the load is set with the specific characteristic impedance, that is,

$$Z_L = Z_o = 50\,\Omega, \tag{14.41}$$

then from equation (14.21), we have

$$Z_P = Z_o, \tag{14.42}$$

regardless of how long the micro strip line is. This is the theoretical background of our test.

A *PCB* to be tested is shown on the right hand side of Figure 14.23, in which many micro strip line or runners with different width W_i but the same length are printed. Terminal A_i is connected to port 1 of the network analyzer and Terminal B_i is supposed to be connected with a $50\,\Omega$ terminator for all of the micro strip lines. Testing can be started by connecting the $50\,\Omega$ *RF* cable to terminal A_1, and switching it to, A_2, A_3, A_4, and so on. (Due to space restrictions, in Figure 14.23 only eight runners are shown. The number of runners could be greater for higher accuracy of testing). When the trace of S_{11} on the Smith chart is moved to the center 50Ω as a terminal A_i and is connected, the micro strip line with terminal A_i is a micro strip line with characteristic impedance $50\,\Omega$.

In *RFIC* design, the test *PCB* would be replaced by the test *IC* die on the wafer.

14.5.5 Testing for the Quarter Wavelength

The quarter wavelength can be obtained from the same setup as shown in Figure 14.23. Let us assume that the fourth runner with terminal A_4 and B_4 was selected by the testing described in the previous section. Its width W_4 corresponds to the specific characteristic impedance $Z_o = 50\,\Omega$. We will focus our attention on this runner and ignore the others.

By simply replacing the $50\,\Omega$ terminator at terminal B_4 with either a "Short" or "Open" terminator, the quarter wavelength can be calculated from the readings of the impedance trace on the Smith chart.

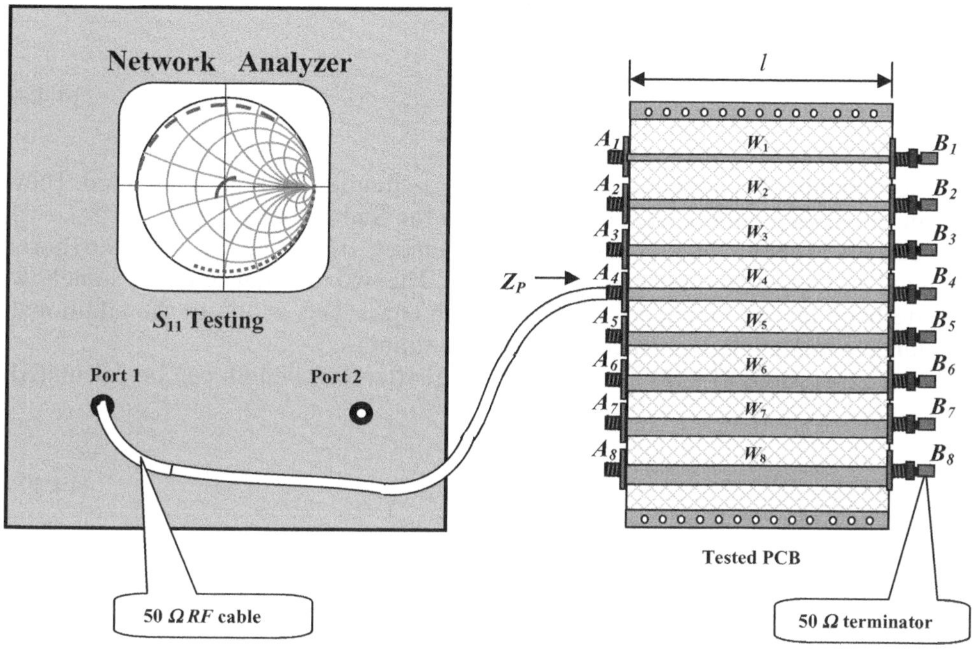

Figure 14.23 Impedance testing for determination of runner's width corresponding to 50 Ω characteristic impedance.

Top metallic area — Bottom metallic area — ○ Conductive via from top to bottom — 50 Ω terminator — *SMA* connector — 50 Ω cable

From equations (14.24) and (14.26) it can be seen that Z_P becomes a pure reactance in both cases when the 50 Ω terminator is replaced by either a "Short" or "Open" terminator, that is,

$$Z_P|_{Z_L \to o} = jX_P|_{Z_L \to o} = jZ_o \tan\frac{2\pi}{\lambda}l, \tag{14.43}$$

$$Z_P|_{Z_L \to \infty} = jX_P|_{Z_L \to \infty} = -jZ_o \cot\frac{2\pi}{\lambda}l, \tag{14.44}$$

Expressions (14.43) and (14.44) can be rewritten as

$$x|_{Z_L \to o} = \tan\frac{2\pi}{\lambda}l, \tag{14.45}$$

$$x|_{Z_L \to \infty} = -\cot\frac{2\pi}{\lambda}l, \tag{14.46}$$

where

$$x|_{Z_L \to o} = \frac{Z_P|_{Z_L \to o}}{Z_o}, \tag{14.47}$$

and

$$x|_{Z_L\to\infty} = \frac{Z_P|_{Z_L\to\infty}}{Z_o}, \tag{14.48}$$

are the normalized reactances when Z_L is short-circuited and open-circuited. They are the readings from the impedance trace on the Smith chart.

Figure 14.24 shows the two normalized reactances, $x|_{Z_L\to o}$ and $x|_{Z_L\to\infty}$, in the short-circuited and open-circuited cases respectively. The two impedance traces somehow deviate from the big circle, $r = 0$, of the Smith chart. This is due to the additional resistance which exists in a micro strip line or runner.

From expressions (14.45) and (14.46), the quarter wavelength can be calculated as

$$\frac{\lambda}{4} = \frac{\pi l}{2\tan^{-1}(x|_{Z_L\to o})}, \tag{14.49}$$

or

$$\frac{\lambda}{4} = -\frac{\pi l}{2\cot^{-1}(x|_{Z_L\to\infty})}. \tag{14.50}$$

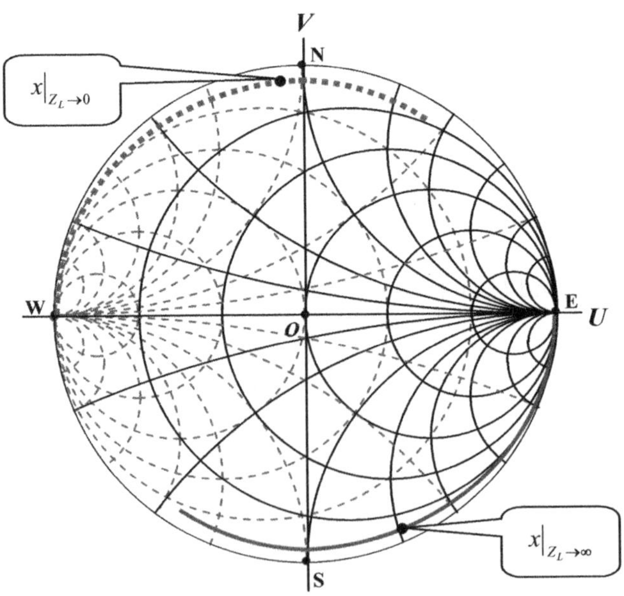

Figure 14.24 Two normalized reactances, $x|_{Z_L\to o}$ and $x|_{Z_L\to\infty}$, displayed on Smith chart.

APPENDICES

14.A.1 Characterizing a Chip Capacitor and Chip Inductor by Means of S_{21} Testing

As *IC* technology has developed in the past decades, discrete parts have been replaced by *IC* chips on a large scale. However, the chip capacitor, chip inductor, and chip resistor are still indispensable parts in the implementation of a circuit module or system today. The reasons are:

- They are the parts with advantages of small size, cost-efficiency, and reliability in the implementation of the circuit module or system,
- Even in a module or system replaced by *IC* chips, the "zero" chip capacitor must be applied in the testing.
- The chip inductor is very often applied as an off-chip part because the Q value of an on-chip inductor is too low.

This appendix is introduced to characterize the chip capacitor and chip inductor, and is mainly abstracted from author's design work in 1991.

The impedance on the Smith chart is a complex number, and its real and imaginary parts can be changed from zero to infinity. Theoretically the impedance of a chip part can be measured by single port testing, either S_{11} or S_{22}, using a network analyzer. Unfortunately, as pointed out in Section 13.4.1, at the very low- or high-impedance regions on the Smith chart, the readings are quite inaccurate. On the other hand, as shown in Figure 14.9 the value of additional spray resistance R_S in a chip capacitor is very low and, as shown in Figure 14.15, the value of additional spray resistance R_P in a chip inductor is very high. Consequently, it is inappropriate to characterize the chip capacitor and chip inductor by S_{11} or S_{22} testing.

Instead, the chip capacitor and chip inductor are characterized by S_{21} testing.

Figures 14.A.1 and 14.A.2 show the test setups for the characterization of the chip capacitor and the chip inductor, respectively.

Figure 14.A.1 shows the S_{21} testing setup for a chip capacitor. The tested chip capacitor is soldered between the $50\,\Omega$ micro strip line and ground in parallel on a small *PCB* as a *DTU*. Then the small *PCB* is connected with ports 1 and 2 of the network analyzer by $50\,\Omega$ *RF* cables. As a result, the tested chip capacitor is connected to ports 1 and 2 of the network analyzer in parallel accordingly.

Figure 14.A.2 shows the S_{21} testing setup for a chip inductor. The tested chip inductor is inserted and soldered between two short $50\,\Omega$ micro strip lines in series on a small *PCB* as a *DTU*. Then the small *PCB* is connected with ports 1 and 2 of the network analyzer by $50\,\Omega$ *RF* cables. As a result, the tested chip inductor is connected to ports 1 and 2 of the network analyzer in series accordingly.

By Carson's derivation (1975), S_{21} can be expressed as a function of the input and output parameters. Figure 14.A.3 shows the various parameters at the source and load. The relationship between S_{21} and the various parameters can be expressed as

$$S_{21} = 2\sqrt{\frac{R_{01}}{R_{02}}}\frac{v_2}{E_1}, \qquad (14.A.1)$$

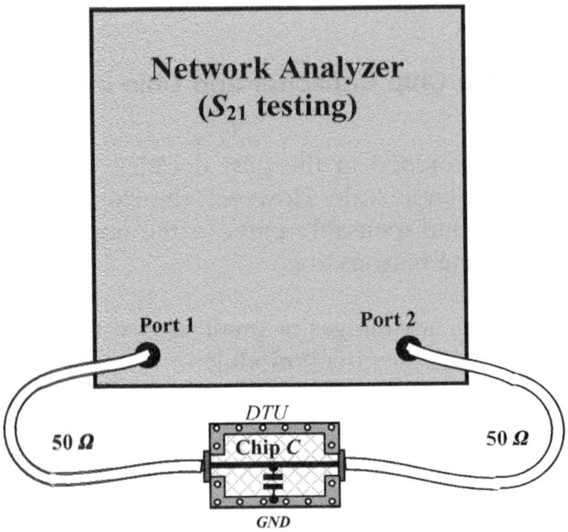

Figure 14.A.1 S_{21} testing for a chip capacitor.

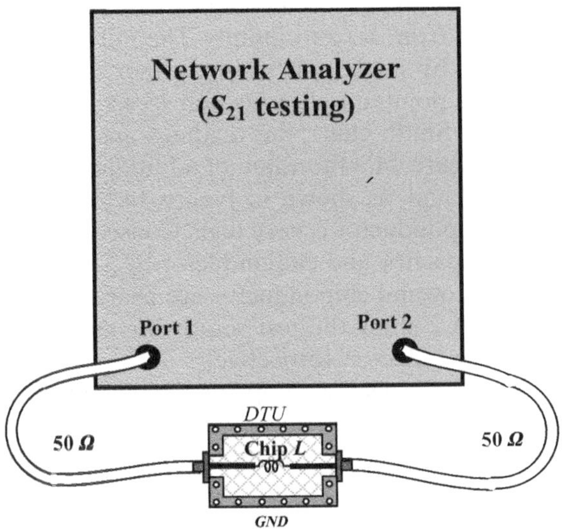

Figure 14.A.2 S_{21} testing for a chip. inductor.

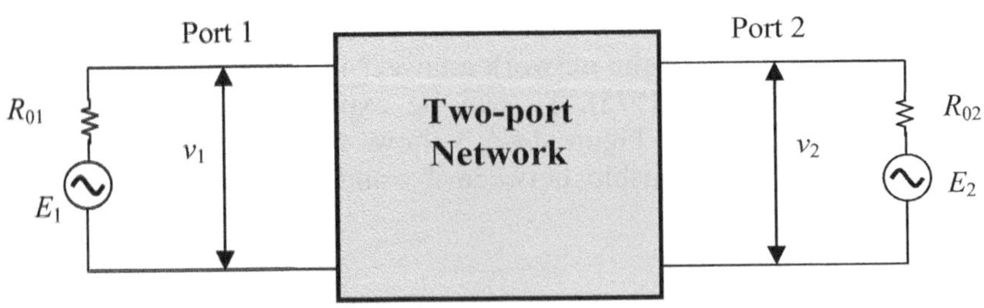

Figure 14.A.3 Source and load parameters in a two-port network.

if

$$E_2 = 0, \tag{14.A.2}$$

where

v_1 = input voltage,
v_2 = output voltage,
E_1 = voltage at source,
E_2 = voltage at load,
R_{o1} = source resistance,
R_{o2} = load resistance.

And if,

$$R_{01} = R_{02} = 50\,\Omega, \tag{14.A.3}$$

Then the equation (14.A.1) becomes

$$S_{21} = 2\frac{v_2}{E_1}. \tag{14.A.4}$$

The ratio v_2/E_1 is determined by the *DTU* (Desired Test Unit). It can be expressed by the parts of the desired test unit.

Let's replace the "two-port network" in Figure 14.A.3 by the equivalent model of the chip capacitor as shown in Figure 14.9; then, Figure 14.A.3 becomes 14.A.4.

The ratio v_2/E_1 in equation (14.A.4) can be found through a simple but tedious mathematic derivation

$$\frac{v_2}{E_1} = \frac{\sqrt{\left[R_S(50+R_S)+\left(L_S\omega-\dfrac{1}{C\omega}\right)^2\right]^2+\left[(50+R_S)\left(L_S\omega-\dfrac{1}{C\omega}\right)-R_S\left(L_S\omega-\dfrac{1}{C\omega}\right)\right]^2}}{(50+R_S)^2+\left(L_S\omega-\dfrac{1}{C\omega}\right)^2}, \tag{14.A.5}$$

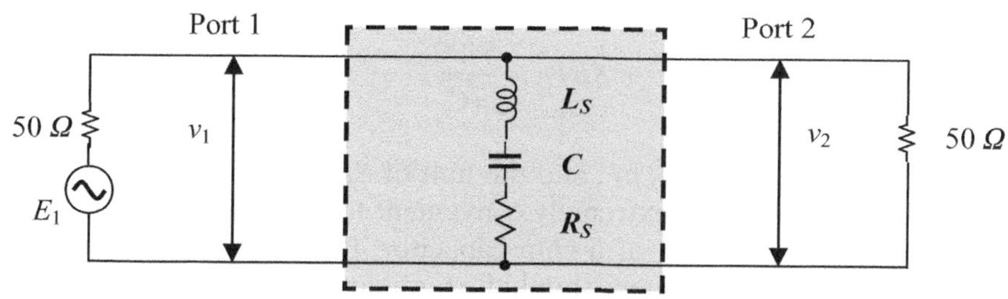

Figure 14.A.4 Equivalent model replacing "two-port network" in Figure 14.A.3 by chip capacitors.

At the self-resonant frequency,

$$\omega = \omega_{SRF} = \frac{1}{\sqrt{L_S C}}, \tag{14.A.6}$$

$$\frac{v_2}{E_1} = \frac{R_S}{50 + R_S}, \tag{14.A.7}$$

then

$$S_{21,SRF} = 20\log\left(2\frac{v_2}{E_1}\right) = 20\log\left(2\frac{R_S}{50 + R_S}\right), \tag{14.A.8}$$

or,

$$R_S = \frac{50}{2 \cdot 10^{\frac{S_{21,SRF}}{20}} - 1}, \tag{14.A.9}$$

$$L_S = \frac{1}{\omega_{SRF}^2 C}, \tag{14.A.10}$$

$$C = C_{specified}. \tag{14.A.11}$$

where

$S_{21,SRF}$ = S_{21} at the self-resonant frequency,
ω_{SRF} = angular self-resonant frequency,
$C_{specified}$ = specified value of capacitance by manufacturer.

From expressions (14.A.9), (14.A.10), and (14.A.11), it can be seen that the values of R_S and L_S can be obtained from the reading of $S_{21,SRF}$ at the self-resonant frequency, ω_{SRF}.

Figure 14.A.5 shows the S_{21} frequency response of chip capacitor $C = C_{specified} = 15pF$ with the test setup as shown in Figure 14.A.1.

Focusing on MuRata chip capacitors, the tested range of capacitance is from $1.8pF$ to $18{,}000pF$ with two different sizes: 30·60 $mils^2$ and 50·80 $mils^2$. It is found that their *SRF*s can be formulized as

$$SRF_C = \frac{5400}{\sqrt{C}}, \tag{14.A.12}$$

where the unit of $C = C_{specified}$ is pF, and the unit of SRF_C is MHz.

Equation (14.A.12) proves extremely convenient for designers in the calculation of the self-resonant frequency of a chip capacitor. However, instead of equation (14.A.12), an experienced engineer should be familiar with the values of the special chip capacitors listed in Table 14.A.1.

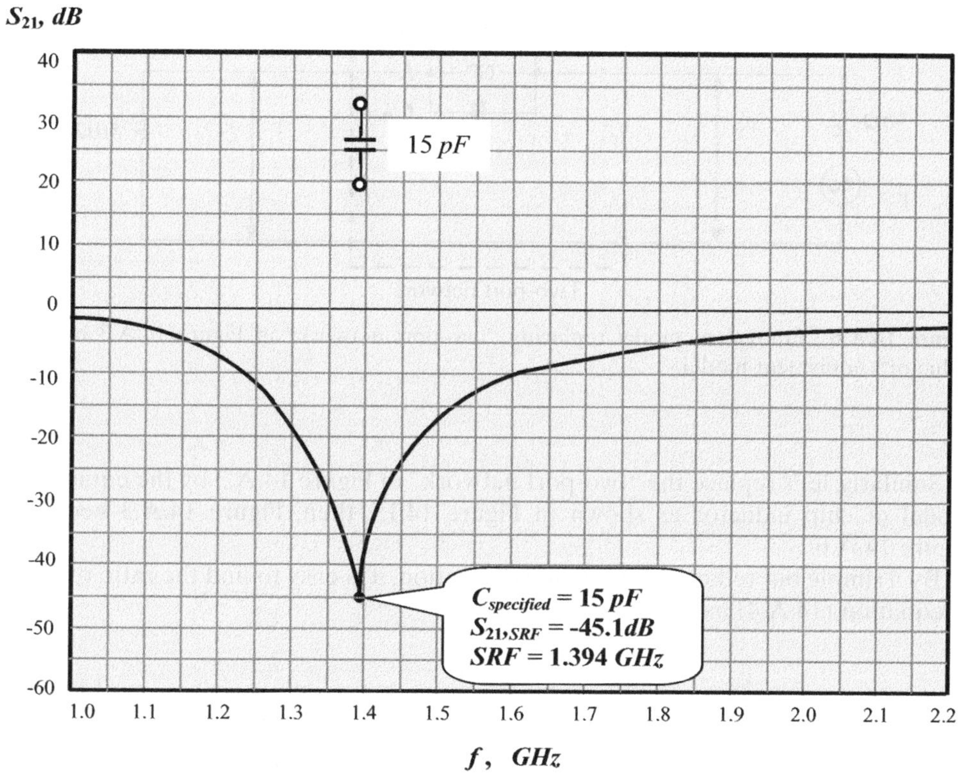

Figure 14.A.5 A.5 S_{21} frequency response of chip capacitor $C = C_{specified} = 15pF$ with test setup as shown in Figure 14.A.1.

TABLE 14.A.1 Some important SRF_C (Self-Resonant Frequency) values of chip capacitors

SRF_C (MHz)	Value of chip capacitor (pF)
40	18,000
150	1,296
450	144
500	117
800	46
900	36
1,000	29
1,500	13
2,400	5.1
5,400	1.0

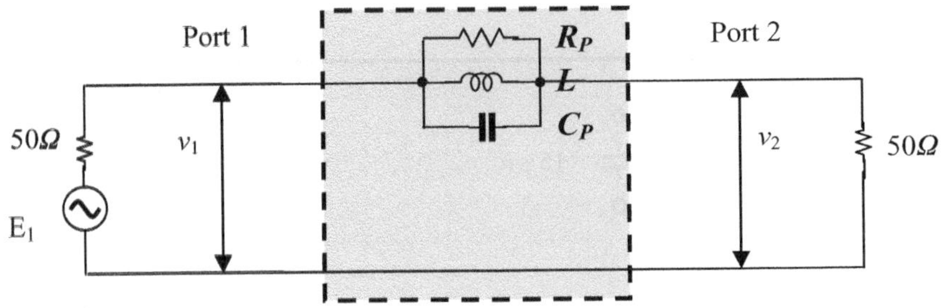

Figure 14.A.6 Equivalent model replacing "two-port network" in Figure 14.A.3 by chip inductor's equivalent model.

Similarly, let's replace the "two-port network" in Figure 14.A.3 by the equivalent model of chip inductor as shown in Figure 14.15: then, Figure 14.A.3 becomes Figure 14.A.6.

By a simple but tedious mathematic derivation, it is easy to find the ratio of v_2/E_1 in equation (14.A.4) as

$$\frac{v_2}{E_1} = \frac{50}{100 + R_p\left[1 + (R_P C_P \omega)^2\left(1 - \frac{1}{LC_P\omega^2}\right)^2\right]^{\frac{1}{2}}}, \tag{14.A.13}$$

we have

$$S_{21} = 20\log\left(2\frac{v_2}{E_1}\right) = 20\log\frac{100}{100 + R_p\left[1 + (R_P C_P \omega)^2\left(1 - \frac{1}{LC_P\omega^2}\right)^2\right]^{\frac{1}{2}}}. \tag{14.A.14}$$

At the self-resonant frequency,

$$\omega = \omega_{SRF} = \frac{1}{(LC_P)^{\frac{1}{2}}}, \tag{14.A.15}$$

$$S_{21,SRF} = 20\log\frac{100}{100 + R_P}, \tag{14.A.16}$$

or

$$R_P = 100\left[10^{-\frac{S_{21,SRF}}{20}} - 1\right]. \tag{14.A.17}$$

$$C_P = \frac{1}{\omega_{SRF}^2 L}. \tag{14.A.18}$$

$$L = L_{specified}. \tag{14.A.19}$$

where

$S_{21,SRF} = S_{21}$ at the self-resonant frequency,
ω_{REF} = angular self-resonant frequency,
$L_{specified}$ = specified value of inductance by manufacturer.

From equations (14.A.17) to (14.A.19), it can be seen that the values of R_P and C_P can be obtained from the reading of $S_{21,REF}$ at self-resonant frequency, ω_{SRF}.

Figure 14.A.7 shows the S_{21} frequency response of chip inductor $L = L_{specified} = 62\,nH$ with test setup as shown in Figure 14.A.2.

Focusing on MuRata chip inductors, the tested range of inductance is from $22\,nH$ to $18{,}00\,nH$ with one size: 80·120 $mils^2$. It is found that their *SRF*s can be formulized as

$$SRF_L = \frac{8920}{\sqrt{L}}, \tag{14.A.20}$$

where the unit of $L = L_{specified}$ is nH, and the unit of SRF_L is MHz.

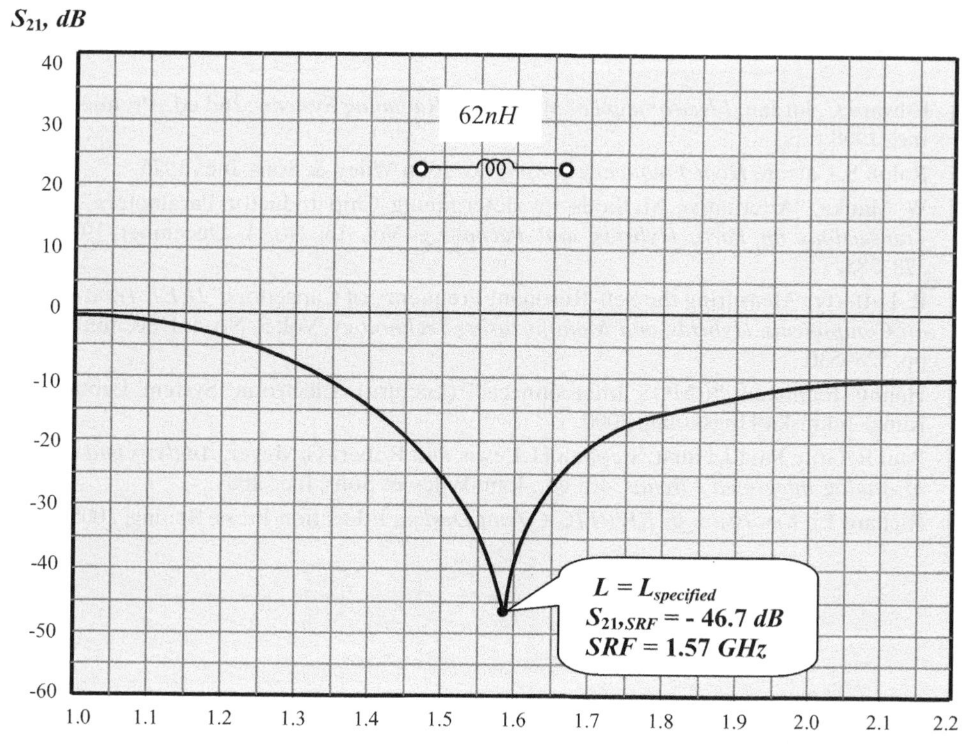

Figure 14.A.7 S_{21} frequency response of chip capacitor $L = 62\,nH$ with test setup as shown in Figure 14.A.2.

TABLE 14.A.2 Some SRF_L (Self-Resonant Frequency) values of chip

SRF_L (MHz)	*Value of chip inductor* (nH)
210	1,800
300	884
450	393
500	318
800	124
900	98
1,000	79.6
1,500	35.4
2,400	13.8
5,400	2.7

Equation (14.A.20) proves extremely convenient to designers in the calculation of the self-resonant frequency of a chip inductor. However, instead of equation (14. A.20), an experienced engineer should be familiar with the values of the special chip inductors listed in Table 14.A.2.

REFERENCES

[1] Edward C. Jordan, *Electromagnetic Waves and Radiating Systems*, 2nd ed., Prentice-Hall, Inc., 1968.

[2] Ralph S. Carson, *High-Frequency Amplifiers*, John Wiley & Sons, Inc., 1975.

[3] W. Manka, "Alternative Methods for determining Chip Inductor Parameters," *IEEE Transactions on Parts, Hybrids, and Packaging*, Vol. 13, No. 4, December 1977, pp. 378–385.

[4] R. Lafferty, "Measuring the Self-Resonant Frequency of Capacitors," *IEEE Transactions on Components, Hybrids, and Manufacturing Technology*, Vol. 5, No. 4, December 1982, pp. 528–530.

[5] Hannu Tenhunen, "CMOS Interconnects" (Lecture), Electronic System Laboratory, Kungl Tekniska Hogskolan, 2000.

[6] Paul R. Gray, Paul J. Hurst, Stephen H. Lewis, and Robert G. Meyer, *Analysis and Design of Analog Integrated Circuits*, 4th ed., John Wiley & Sons, Inc., 2001.

[7] Richard Li, *Key Issues in RF/RFIC Circuit Design*, Education Press, Beijing, 2005.

CHAPTER 15

EQUIPOTENTIALITY AND CURRENT COUPLING ON THE GROUND SURFACE

15.1 EQUIPOTENTIALITY ON THE GROUND SURFACE

15.1.1 Equipotentiality on the Ground Surfaces of an *RF* Cable

Let's start our discussion of the equipotentiality of ground surfaces with a regular *RF* cable. Simply speaking, a regular *RF* cable is a conductive wire encircled by a conductive cylinder (Figure 15.1). The outer cylinder at the two ends is usually connected as a ground terminal.

In the *RF* frequency range, the following question must be answered:

- Are the ground points on the outside cylinder, G_a and G_b, equipotential?

$$v_{G_a} = v_{G_b} \ ? \tag{15.1}$$

- Are the ground points of an *RF* cable at two ends, G_1 and G_2, equipotential?

$$v_{G_1} = v_{G_2} \ ? \tag{15.2}$$

In the *DC* or low-frequency range, these points are, approximately, equipotential:

$$v_{G_a} = v_{G_b} = v_{G_1} = v_{G_2}. \tag{15.3}$$

In the *RF* frequency range, they are in most cases not equipotential, that is,

$$v_{G_a} \neq v_{G_b}, \tag{15.4}$$

RF Circuit Design, by Richard Chi-Hsi Li

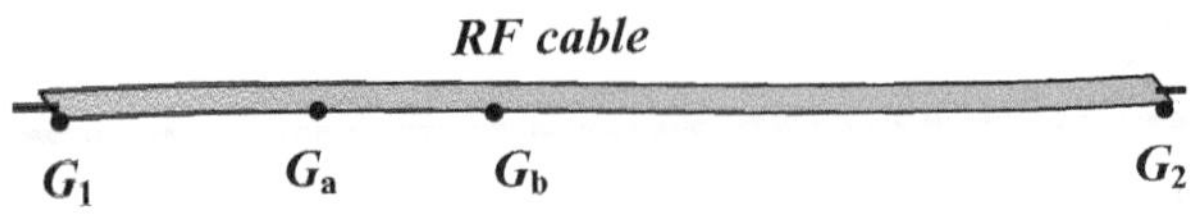

Figure 15.1 A regular *RF* cable.

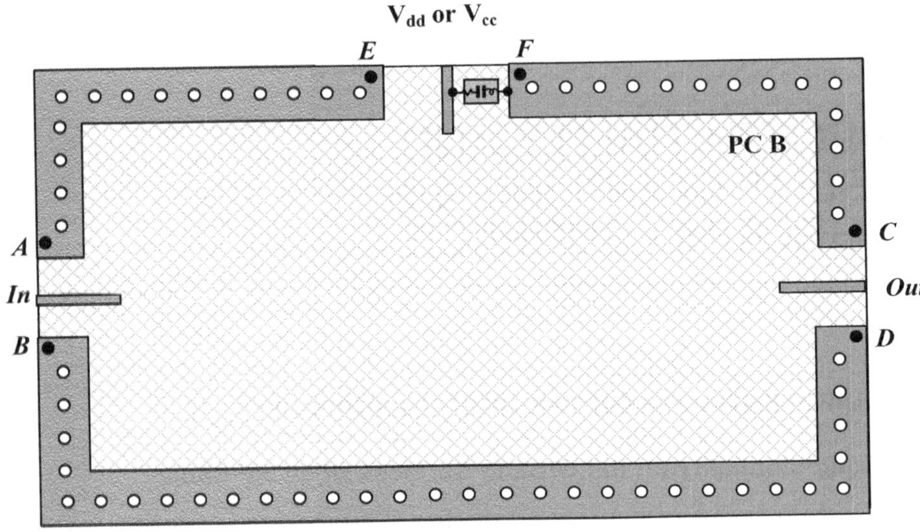

Figure 15.2 Discussion of equipotentiality between points, *A, B, C, D, E, F*. (The printed circuit on the PCB is neglected.)

Top metallic area
Bottom metallic area
50 Ω runner
Conductive via from top to bottom
"Zero" capacitor

$$v_{G_1} \neq v_{G_2}, \tag{15.5}$$

unless the distances between *a* and *b*, or 1 and 2, are equal to a multiple of the half wavelength.

An *RF* cable can be simply considered as a wave-guide tube. The transportation of the *RF* power or energy is conducted and propagated along the central conductive wire. In most cases, not only on the outside cylinder but also on the central conductive wire, two points on the cable are not equipotential unless the distance between these two points is equal to the multiple of the half wavelength.

15.1.2 Equipotentiality on the Ground Surface of a *PCB*

The same question is asked of the ground surface of a *PCB*: Are the ground points on the ground surface of a *PCB* equipotential? That is, as shown in Figure 15.2,

$$v_A = v_B = v_C = v_D = v_E = v_F\ ? \tag{15.6}$$

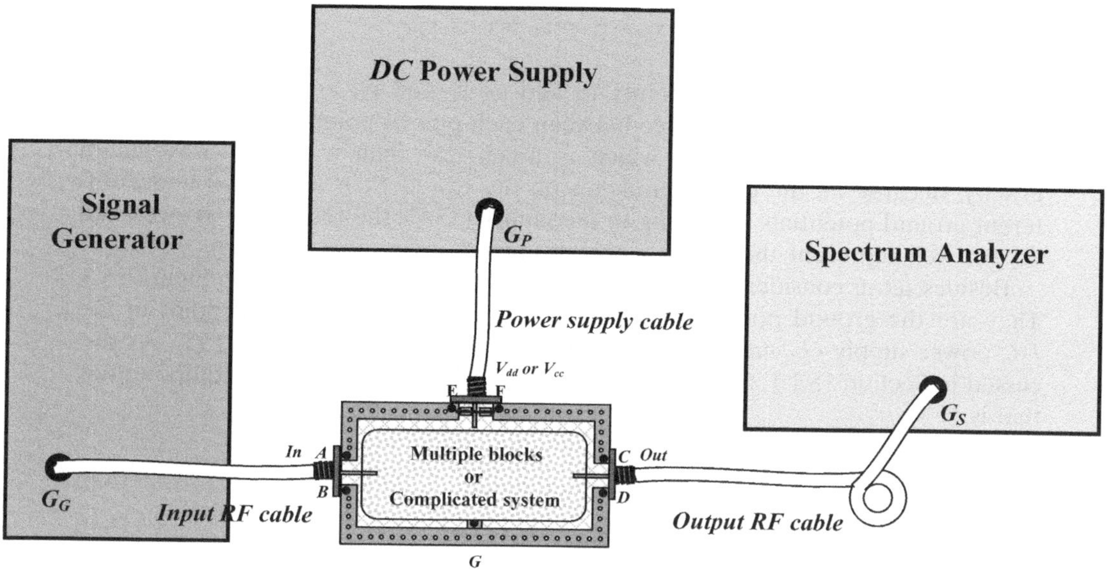

Figure 15.3 An example test setup with a large *PCB*.

- top metallic area
- bottom metallic area
- conductive via from top to bottom
- "zero" capacitor
- *SMA* connector

G_G: ground of signal generator
G_S: ground of spectrum analyzer
G_P: ground of DC power supply
A, B, C, D, E, F, G: expected ground points on *PCB*.

The answer is

- Yes, if the size of *PCB* is small, that is, the maximum dimension of *PCB* is much shorter than the quarter wavelength of the operating frequency.
- No, if the size of *PCB* is large, that is, the maximum dimension of *PCB* is comparable or longer than the quarter wavelength of the operating frequency.

15.1.3 Possible Problems of a Large Test *PCB*

Now let's examine a test setup when the test *PCB* is large. Figure 15.3 shows a large test *PCB* connected with a signal generator, *DC* power supply, and spectrum analyzer.

In the large *PCB*, points *A* and *B* are ground points of the input, points *C* and *D* are ground points of the output, and points *E* and *F* are ground points of the *DC* power supply. All these ground points are expected to be equipotential so that test results really represent the circuit's performance.

The ground surface on a small *PCB* is equipotential and so equation (15.6) is correct. Consequently, testing is carried out under the same reference ground point so that test results do represent the circuit performance.

On the contrary, on a large *PCB*, the ground points as shown in Figure 15.3 are not equipotential, that is,

$$v_A = v_B \neq v_C = v_D \neq v_E = v_F, \tag{15.7}$$

although the three pairs of points, v_A and v_B, v_C and v_D, v_E, and v_F, might each be equipotential since the distance between each pair of points is not greater than the size of a *SMA* connector, which is much less than a quarter wavelength usually. Because all the circuit branches on the tested *PCB* are tested under different ground potentials according to inequality (15.7), the results are skewed and do not well represent the circuit's performance.

Besides, let us consider the ground points connected by *RF* cables in Figure 15.3. They are the ground point of the signal generator G_G, the ground point of the *DC* power supply G_P, and the ground point of the spectrum analyzer G_S. As discussed in Section 15.1.1, the two ends of an *RF* cable are usually not equipotential, that is,

$$v_{GG} \neq v_A\,(\text{or } v_B), \tag{15.8}$$

$$v_{GS} \neq v_C\,(\text{or } v_D), \tag{15.9}$$

$$v_{GP} \neq v_E\,(\text{or } v_F). \tag{15.10}$$

Also, usually

$$v_{GG} \neq v_{GS} \neq v_{GD}, \tag{15.11}$$

because v_{GG}, v_{GS}, and v_{GD} are the ground points of individual equipments so that they are not necessarily equal. Consequently, the expected ground points, *A*, *B*, *C*, *D*, *E*, *F*, *G*, on the *PCB* are usually not equipotential through the connections between the *PCB* and the equipment by means of three cables.

On the other hand, it can be seen that the expected ground points, *A*, *B*, *C*, *D*, *E*, *F*, *G*, on the *PCB* are connected together by the rectangular metallic frame. Does it ensure that these ground points are equipotential? In most cases it doesn't because this is a large *PCB*. Two points on the ground surface of a large *PCB* are usually not equipotential.

Consequently, if expressions (15.7) and (15.11) hold true, the unequipotentiality will make the testing of the circuitry on the *PCB* meaningless.

In order to make testing meaningful **the design task of a test *PCB* is to force the expected ground points on *PCB* to be equipotential, that is,**

$$v_A = v_C = v_E. \tag{15.12}$$

The equipotentiality condition (15.12) ensures that testing is conducted under "common grounding" conditions for the input, output, and other terminals so that correct test results can be expected, although equations (15.8) to (15.11) are still true. In *RF* circuit testing, conditions (15.8) to (15.11) are allowed as long as the power can be well-transported by a cable from one end to the other.

15.1.4 Coercing Grounding

There are many methods to force the expected ground points on the *PCB* to achieve equipotentiality. The following ways to force the ground surface to equipotentiality are recommended in *AC/RF* grounding for a large *PCB*:

1) **To force the ground surface equipotential by "zero" chip capacitors**
When the dimensions of the test *PCB* are comparable or larger than the quarter wavelength at the operating frequency, a long runner or a large ground surface is usually in an unequipotentiality status.
Figure 15.4(a) shows a long runner on a *PCB*. Usually the *AC/RF* voltage on points P_1, P_2, P_3, and P_4 are unequipotential, that is,

$$v_{P_1} \neq v_{P_2} \neq v_{P_3} \neq v_{P_4}. \tag{15.13}$$

Now, as shown in Figure 15.4(b), if this long runner is cut into four or more small pieces so that the length of each small line segment is much less than quarter wavelength of the operating frequency and then they are connected by the "zero" capacitors, the *AC/RF* voltage on points P_1, P_2, P_3, and P_4 become approximately equipotential, that is,

$$v_{P_1} \approx v_{P_2} \approx v_{P_3} \approx v_{P_4}. \tag{15.14}$$

Similarly, Figure 15.5(a) shows a large ground surface on a *PCB*. Usually the *AC/RF* voltage on points P_1, P_2, P_3, and P_4 are unequipotential, that is,

$$v_{P_1} \neq v_{P_2} \neq v_{P_3} \neq v_{P_4}. \tag{15.15}$$

Now, as shown in Figure 15.5(b), if this large ground surface is cut into four or more small pieces so that the maximum dimension of each small ground area is much less than quarter wavelength of the operating frequency and then they are connected by the "zero" capacitors, the *AC/RF* voltage at points P_1, P_2, P_3, and P_4 become approximately equipotential, that is,

$$v_{P_1} \approx v_{P_2} \approx v_{P_3} \approx v_{P_4}. \tag{15.16}$$

It should be noted that, if the operating frequency covers a wide band or consists of more than one frequency band, a single "zero" capacitor between two small pieces of metal in Figures 15.4 and 15.5 is insufficient and must be replaced by multiple "zero" capacitors.
This technology is easier and more direct than the other technologies to be introduced below; however, many "zero" capacitors must be applied.

2) **To force the ground surface equipotential by a half wavelength of micro strip line**
Instead of applying "zero" capacitors to force the ground surface into an equipotential state as shown in Figures 15.4 and 15.5, half-wavelength runners as shown in Figures 15.6 can be used, because the voltage is the same at both ends of a half-wavelength runner.

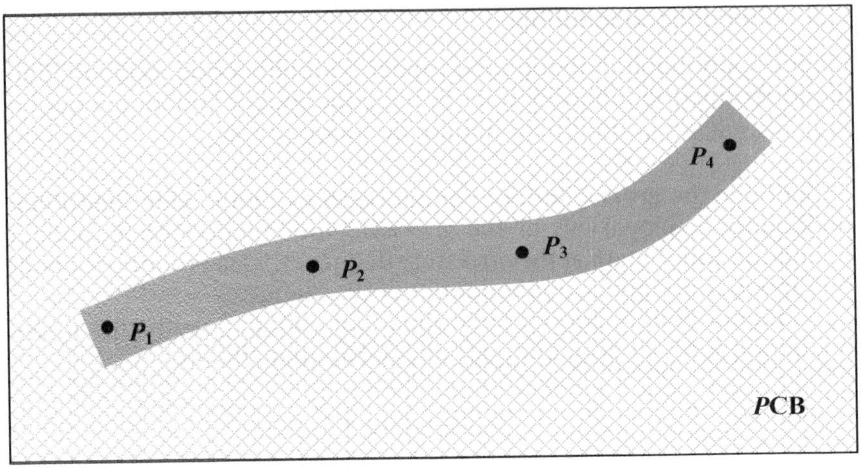

(a) Unequipotentiality on a long runner, $v_{P_1} \neq v_{P_2} \neq v_{P_3} \neq v_{P_4}$

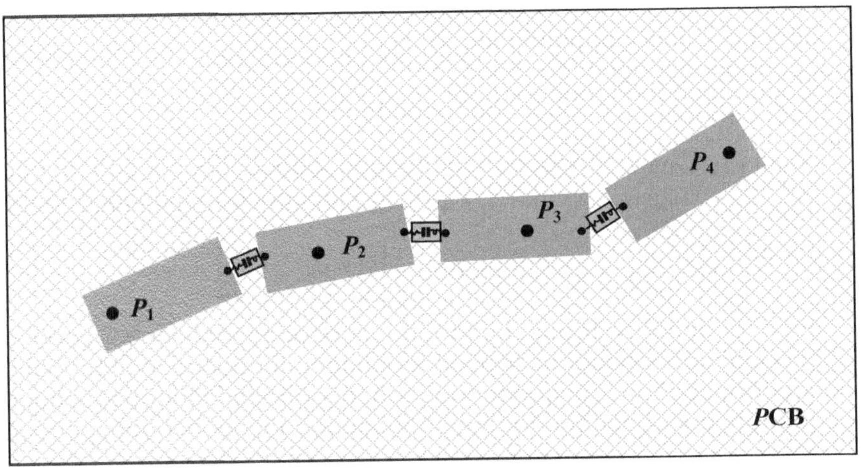

(b) Equipotentiality on a long runner resumed by "zero" capacitors, $v_{P_1} = v_{P_2} = v_{P_3} = v_{P_4}$

Figure 15.4 Unequipotentiality of a long runner changed to equipotentiality by means of "zero" chip capacitors.
"Zero" chip capacitors
Top metallic area
Bottom metallic area

Figure 15.6(a) shows a long runner on a *PCB*. Usually the *AC/RF* voltage at points P_1, P_2, P_3, and P_4 are unequipotential, that is,

$$v_{P_1} \neq v_{P_2} \approx v_{P_3} \neq v_{P_4}, \tag{15.17}$$

where v_{P_2} is approximately close to v_{P_3} because the points P_2 and P_3 are quite closed from each other.

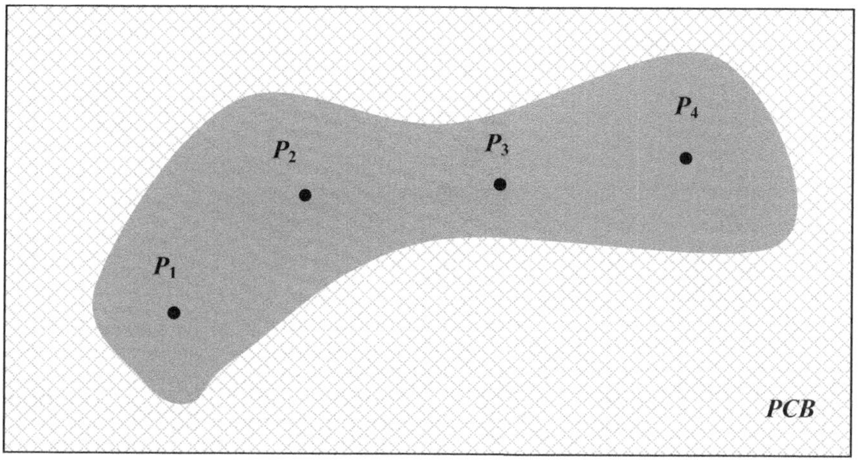

(a) Unequipotentiality on large ground surface, $v_{P_1} \neq v_{P_2} \neq v_{P_3} \neq v_{P_4}$

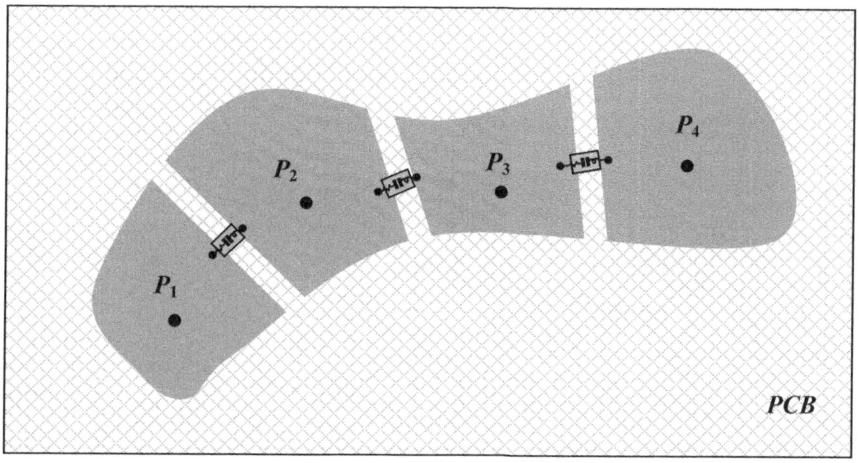

(b) Equipotentiality on a large ground surface resumed by "zero" capacitors, $v_{P_1} \approx v_{P_2} \approx v_{P_3} \approx v_{P_4}$

Figure 15.5 Unequipotentiality of a large ground surface changed to equipotentiality by means of "zero" chip capacitors.
"Zero" chip capacitors — Top metallic area — Bottom metallic area

Now, as shown in Figure 15.6(b), if two micro strip lines with half wavelength are connected between P_1 and P_2, P_3, and P_4 on this long runner, the *AC/RF* voltage at points P_1, P_2, P_3, and P_4 become approximately equipotential, that is,

$$v_{P_1} = v_{P_2} \approx v_{P_3} = v_{P_4}. \tag{15.18}$$

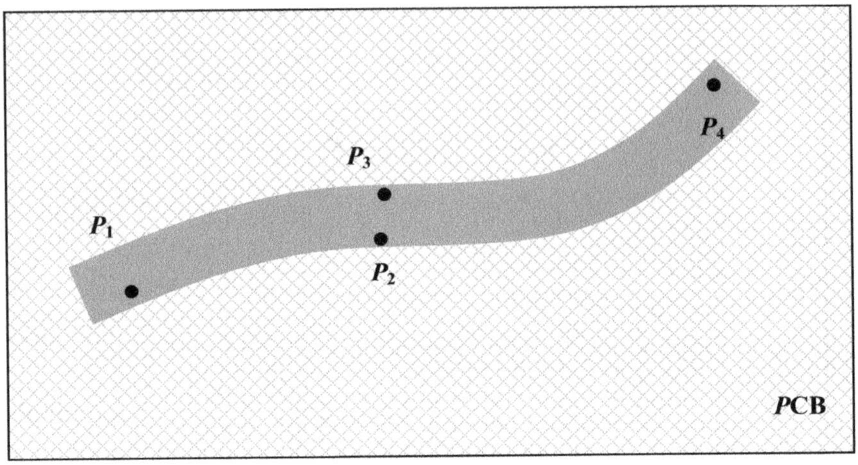

(a) Unequipotentiality on a long runner, $v_{P_1} \neq v_{P_2} \approx v_{P_3} \neq v_4$

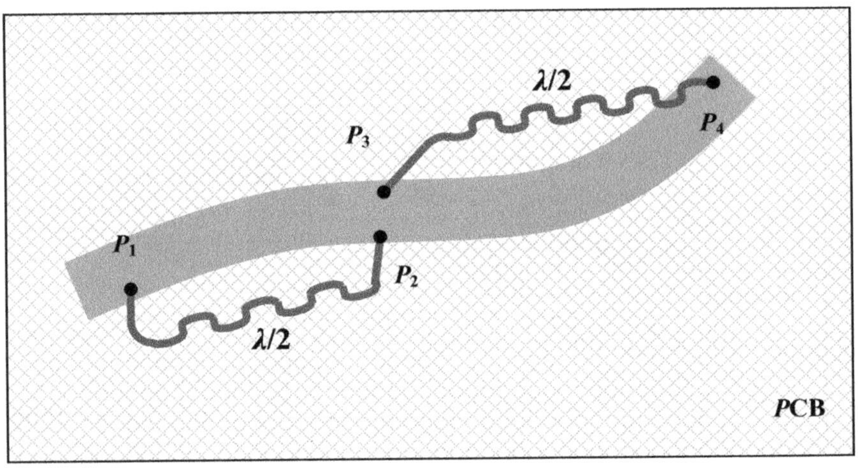

(b) Equipotentiality on a long runner resumed by λ/2 micro strip line, $v_{P_1} = v_{P_2} \approx v_{P_3} = v_{P_4}$

Figure 15.6 Unequipotentiality of a large ground surface changed to equipotentiality by means of λ/2 micro strip line.

Micro strip line — Top metallic area — Bottom metallic areaa

Figure 15.7(a) shows a large ground surface on a *PCB*. Usually the *AC/RF* voltage at points P_1, P_2, P_3, and P_4 are unequipotential, that is,

$$v_{P_1} \neq v_{P_2} \approx v_{P_3} \neq v_{P_4}, \tag{15.19}$$

where v_{P_2} is approximately close to v_{P_2}, because the points P_2 and P_3 are quite closed from each other.

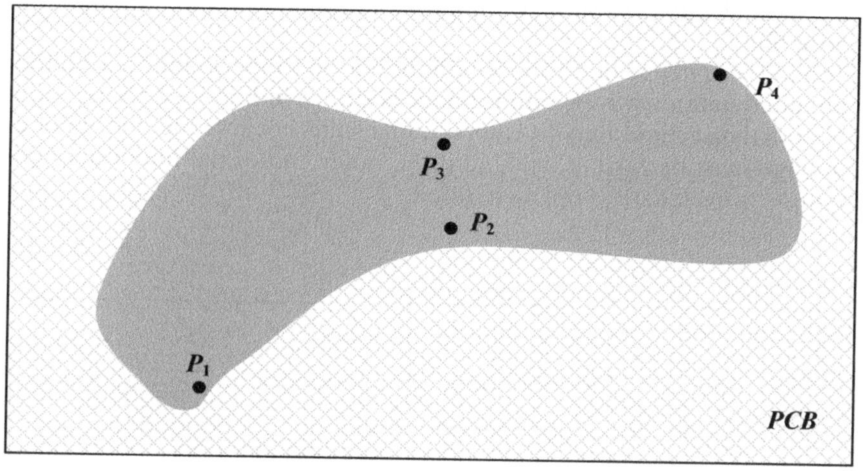

(a) Unequipotentiality on large ground surface, $v_{P_1} \neq v_{P_2} \approx v_{P_3} \neq v_{P_4}$

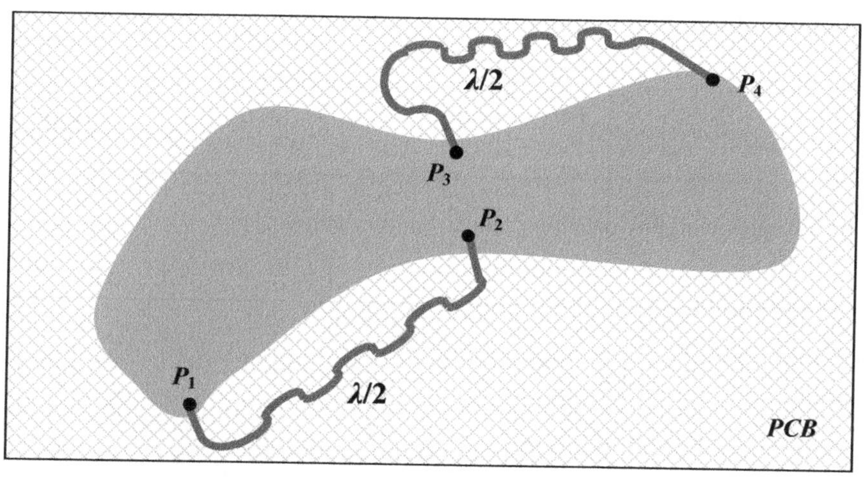

(b) Equipotentiality on a large ground surface resumed by λ/2 micro strip line, $v_{P_1} = v_{P_2} \approx v_{P_3} = v_{P_4}$

Figure 15.7 Un-equipotentiality of a large ground surface changed to equipotentiality by means of λ/2 micro strip line.

Micro strip line

Top metallic area

Bottom metallic area

Now, as shown in Figure 15.7(b), if two micro strip lines with half wavelength are connected between P_1 and P_2, P_3, and P_4, on this large ground surface, then the *AC/RF* voltage at points P_1, P_2, P_3, and P_4 becomes approximately equipotential, that is,

$$v_{P_1} = v_{P_2} \approx v_{P_3} = v_{P_4}. \tag{15.20}$$

3) **To force the ground surface equipotential by a quarter wavelength of micro strip line**
Instead of applying half-wavelength runners as shown in Figure 15.7, the open-circuited quarter-wavelength runners as shown in Figures 15.8 can be used to force a long runner into an equipotential state, because the voltage or electric voltage must be zero at one end if the other end of the micro strip line with quarter wavelength is open-circuited.

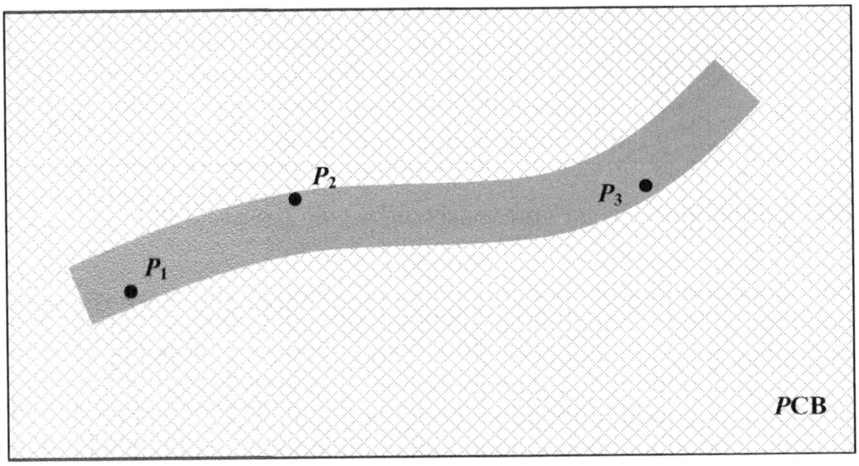

(a) Unequipotentiality on a long runner, $v_{P_1} \neq v_{P_2} \neq v_{P_3}$

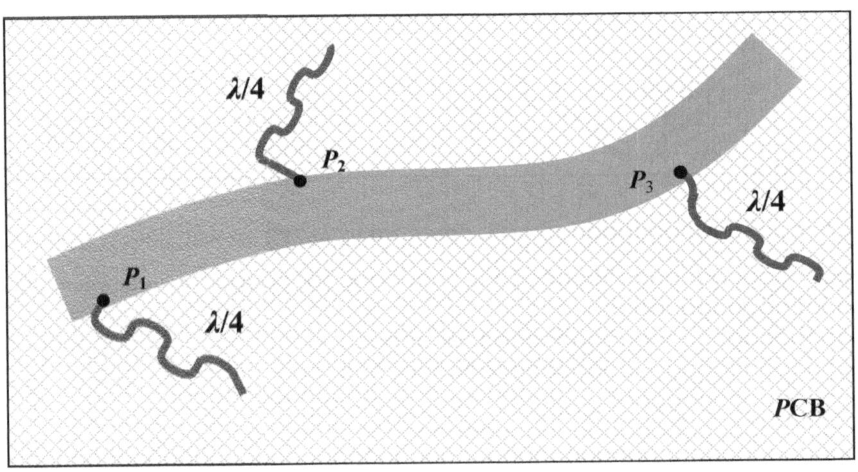

(b) Equipotentiality on a long runner resumed by λ/4 micro strip line, $v_{P_1} = v_{P_2} = v_{P_3}$

Figure 15.8 Unequipotentiality of a large ground surface changed to equipotentiality by means of λ/4 micro strip line.
Micro strip line
Top metallic area
Bottom metallic area

Figure 15.8(a) shows a long runner on a *PCB*. Usually the *AC/RF* voltage at points P_1, P_2, and P_3 are unequipotential, that is,

$$v_{P_1} \neq v_{P_2} \neq v_{P_3}, \tag{15.21}$$

Now, as shown in Figure 15.8(b), if at points P_1, P_2, and P_3 a micro strip line with open-circuited quarter wavelength is connected as shown in Figure 15.8(b), then the *AC/RF* voltage at points P_1, P_2, and P_3 becomes equipotential, that is,

$$v_{P_1} = v_{P_2} = v_{P_3} = 0. \tag{15.22}$$

Figure 15.9(a) shows a large ground surface on a *PCB*. Usually the *AC/RF* voltage at points P_1, P_2, P_3, and P_4 are unequipotential, that is,

$$v_{P_1} \neq v_{P_2} \neq v_{P_3} \neq v_{P_4}, \tag{15.23}$$

Now, as shown in Figure 15.9(b), if two micro strip lines with open-circuited quarter wavelength are connected between P_1 and P_2, P_3, and P_4 on this large ground surface, then the *AC/RF* voltage at points P_1, P_2, P_3, and P_4 becomes approximately equipotential, that is,

$$v_{P_1} = v_{P_2} = v_{P_3} = v_{P_4} = 0. \tag{15.24}$$

A half-wavelength runner is somewhat lengthy. The length of a quarter-wavelength micro strip line is 50% reduced from that of a half wavelength, so that it is a very effective part in the *RF* circuit grounding.

Besides micro strip lines, the half/quarter-wavelength cables could be applied in the same manner as half/quarter-wavelength runners; however, the cables are too clumsy and thus are seldom applied in practical engineering.

15.1.5 Testing for Equipotentiality

We have talked a lot about equipotentiality on the ground surface. It is desirable to know how to test the equipotentiality on a large test *PCB*; however, there is no formal equipment available for this task to date.

Figure 15.10 shows a tentative setup for the testing of equipotentiality of a large test *PCB*. The basic idea of the setup is to discover unequipotentiality on the ground surface through the testing of S_{11} or S_{22} by a network analyzer.

First, three *RF* cables must be prepared for the calibration. The first cable is termed the calibrated cable as shown in Figure 15.10. One end of this cable is connected to port 1 or port 2 on the network analyzer. Another end is denoted by the symbol *C* and must be calibrated in a "pick-tail" manner.

Calibration by the "pick-tail" method is a special calibration in which the standard calibration kits' "short," "open," "50 Ω," and "through" are replaced by a set of cables (called "tails") with one end in "short," "open," "50 Ω," and "through" and another end connected to the network analyzer. The length of these cables is equal

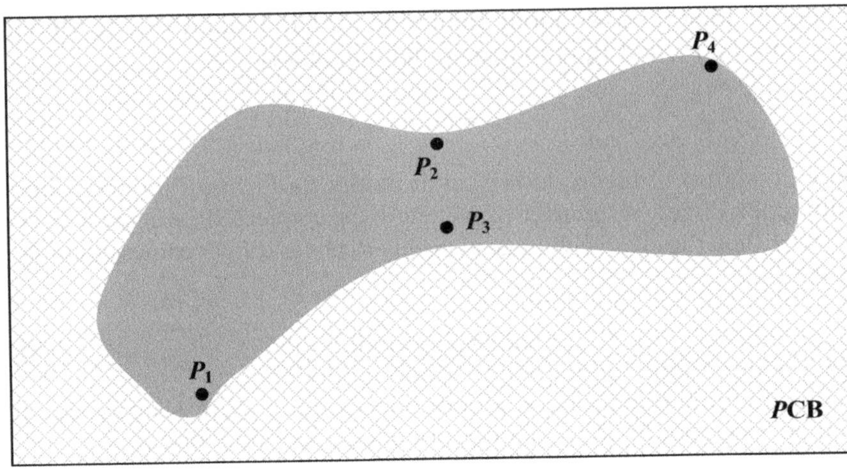

(a) Unequipotentiality on large ground surface, $v_{P_1} \neq v_{P_2} \neq v_{P_3} \neq v_{P_4}$

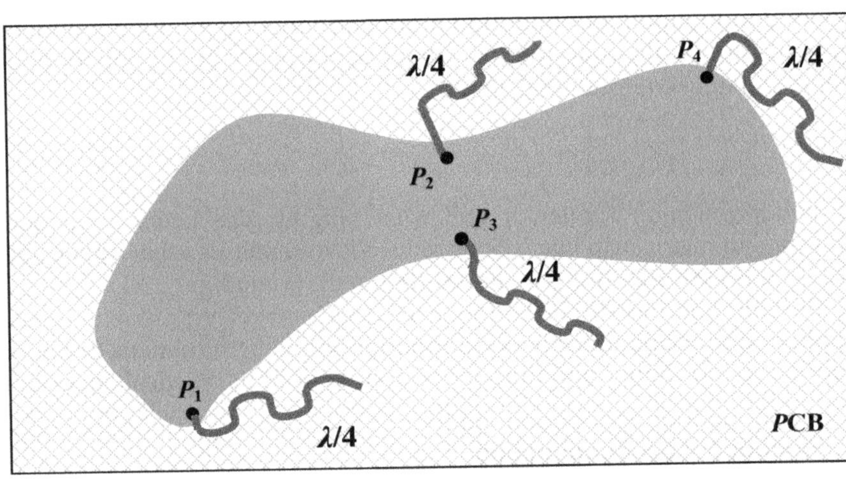

(b) Equipotentiality on a large ground surface resumed by $\lambda/4$ micro strip line, $v_{P_1} = v_{P_2} = v_{P_3} = v_{P_4}$

Figure 15.9 Unequipotentiality of a large ground surface changed to equipotentiality by means of $\lambda/4$ micro strip line.
Micro strip line ▇ Top metallic area □ Bottom metallic area

to the desired length of the first cable, the calibrated cable. A set of calibrated cables is shown in Figure 15.11.

The second cable is termed the extended cable, as shown in Figure 15.10. Its length must be a multiple of the half wavelength of the operating frequency, which can be determined through careful measurement by a network analyzer. The length of this cable must be long enough so that the probe at its end can reach anywhere on the *PCB*. The expected minimum length is the half wavelength, of course.

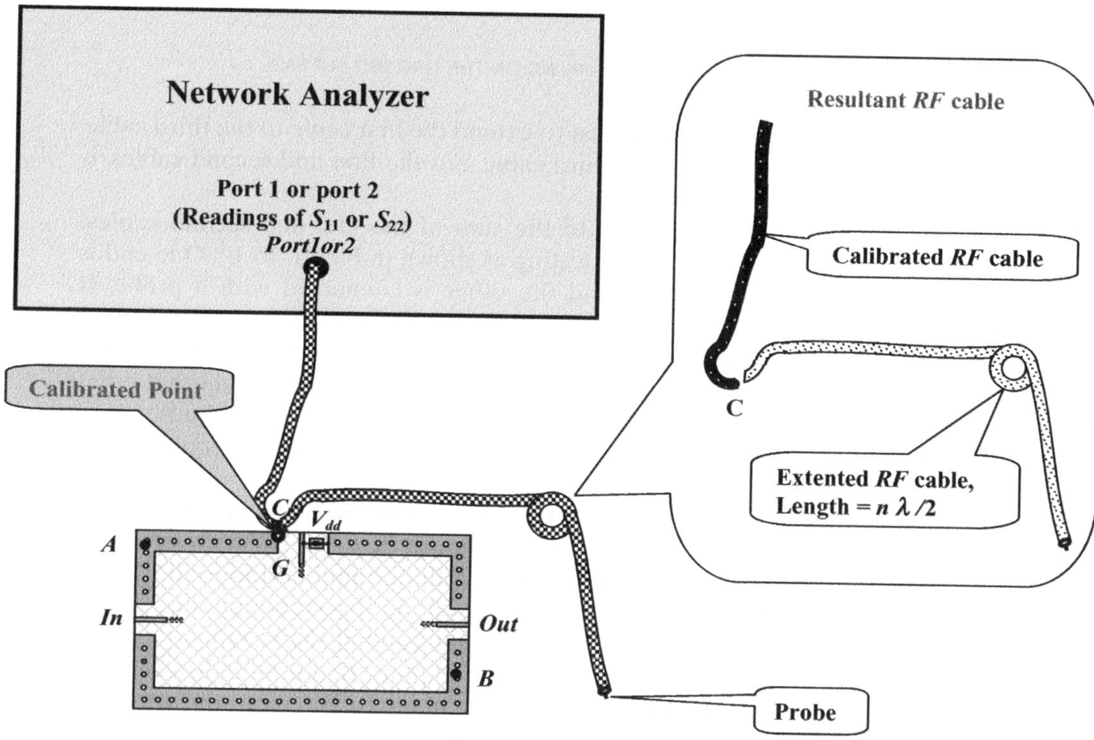

Figure 15.10 Testing of equipotentiality of a *PCB*.

- Top metallic area
- Bottom metallic area
- ○ Conductive via from top to bottom
- ● Reference ground point, ***G***
- Calibrated cable
- Extended cable
- "Zero" capacitor

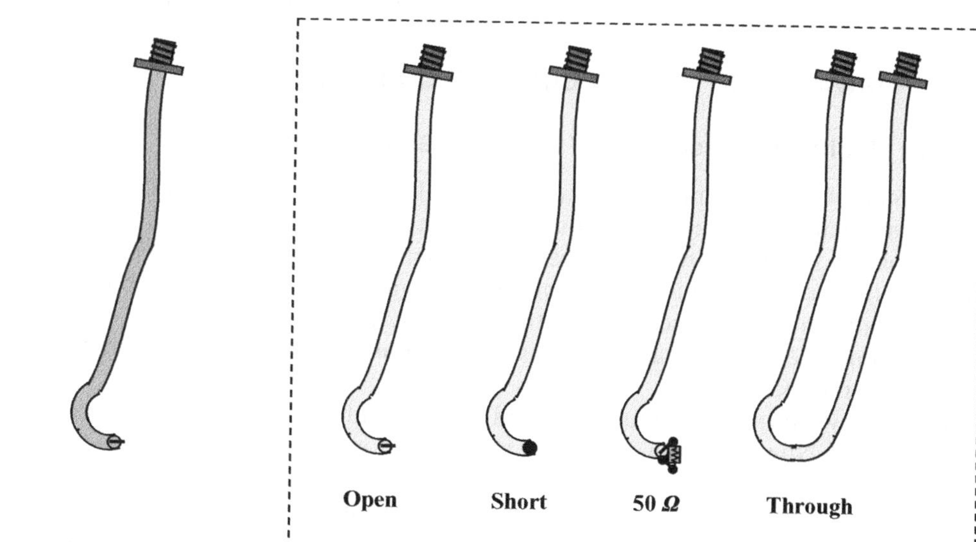

(a) Calibrated cable

(b) Calibration kit

Figure 15.11 Calibrated cable and calibration kits for testing of equipotentiality of a *PCB*.

- Calibrated cable
- Cable for calibration
- Resistor 50 Ω
- SMA connector

The purpose of the second cable is just to extend the first cable to the third cable. In the testing of equipotentiality, the third cable, not the first and second cables, is actually connected in the test setup.

The third cable has a length equal to the sum of the first and second cables' length. It is the actual cable applied to testing as shown in Figure 15.10. One end is connected to the network analyzer and the other is connected with a probe. It intermediate point *C* is soldered to the reference ground point at the *DC* power supply terminal. The length between point *C* and the end connected to the network analyzer must be equal to the length of the first cable. Owing to the fact that point *C* has been calibrated and that the length between point *C* and the probe is a multiple of the half wavelength, the potential that the probe senses is equal to the potential at point *C*.

After calibration is complete and the third cable is fixed as shown in Figure 15.10, testing can begin. By moving the probe around on the *PCB*, the S_{11} or S_{22} at different points on the *PCB* can be compared. Should the ground surface and ground points on *PCB* be equipotential, the readings of S_{11} or S_{22} should remain unchanged as the probe moves around the *PCB*. On the other hand, if the readings of S_{11} or S_{22} fluctuate when the probe moves from point *A* to point *B*, it is an indication that that the ground point *A* is not equipotential with point *B*. The more the variation of the readings, the more unequipotentiality exists.

It should be noted that the testing is sensitive to the human body when the third cable is held by hand. It is suggested that a protective isolation glove be put on the cable where it is held by hand; it is even better to have a static protective metal ring put on the wrist or arm of the tester.

This testing scheme has been developed by the author in recent years. It is only tentative and not yet satisfactory, because it is qualitative, not quantitative.

15.2 FORWARD AND RETURN CURRENT COUPLING

15.2.1 "Indifferent Assumption" and the "Great Ignore"

In Chapter 14 it was pointed out that the typical schematic of a circuitry is always drawn with many "indifferent assumptions" which can cause possible *AC/RF* grounding problems. People allow these oversights because their attention is focused on the circuitry itself, or because they think that these assumptions are just temporarily ignored and can be easily realized by the circuit designer.

Experienced engineers cannot but say: "These are not indifferent assumptions, but great ignorances!"

The first two problems about fully- and half-ground points have been described in Chapter 14. Now let's discuss another problem about forward current and return current coupling.

When the half ground points in a circuit block are not equipotential due to imperfect or inappropriate *RF/AC* grounding, the forward currents will be coupled magnetically on the path from the positive terminal of *DC* power supply to every branch of the circuitry in a cable or a common runner. Should each *DC* power supply node for each branch be equipotential by means of "zero" capacitors, the forward current coupling would disappear. Similarly, when the full ground points in a circuit block are not equipotential due to imperfect or inappropriate *RF/AC* grounding, the return currents will be coupled magnetically on the path from every branch to the negative terminal of *DC* power supply. Usually the return current has complicated paths because it flows over the entire ground surface in a circuit block. Should the entire ground surface be equipotential, the return current coupling would disappear.

In short, there will be no current coupling problems if the ground surface is perfectly equipotential. In reality, un-equipotentiality exists more or less on the ground surface. Therefore, reduction or removal of forward and return current coupling becomes an important subject in *RF* circuit design.

Figure 15.12 creates images of the forward and return currents which are more semblant of actual performance. The forward current flows from the positive pole of the *DC* power supply to the circuitry through a cable or a wire. The return current flows from the circuitry toward the negative pole or ground of the *DC* power supply spread over the entire ground surface.

From Figure 15.12, it can be seen that the four forward sub-currents flowing into the four branches of circuitry are coupled together in a cable or a wire, while the four return sub-currents returning to the *DC* power supply are spread and coupled over the entire ground surface. In reality, the return currents flowing on the *PCB* form a complicated pattern, depending on the placement of parts and the arrangement of the ground areas.

The current coupling between adjacent branches is a kind of magnetic cross-talk. Signals from different branches coupling with each other in either the forward

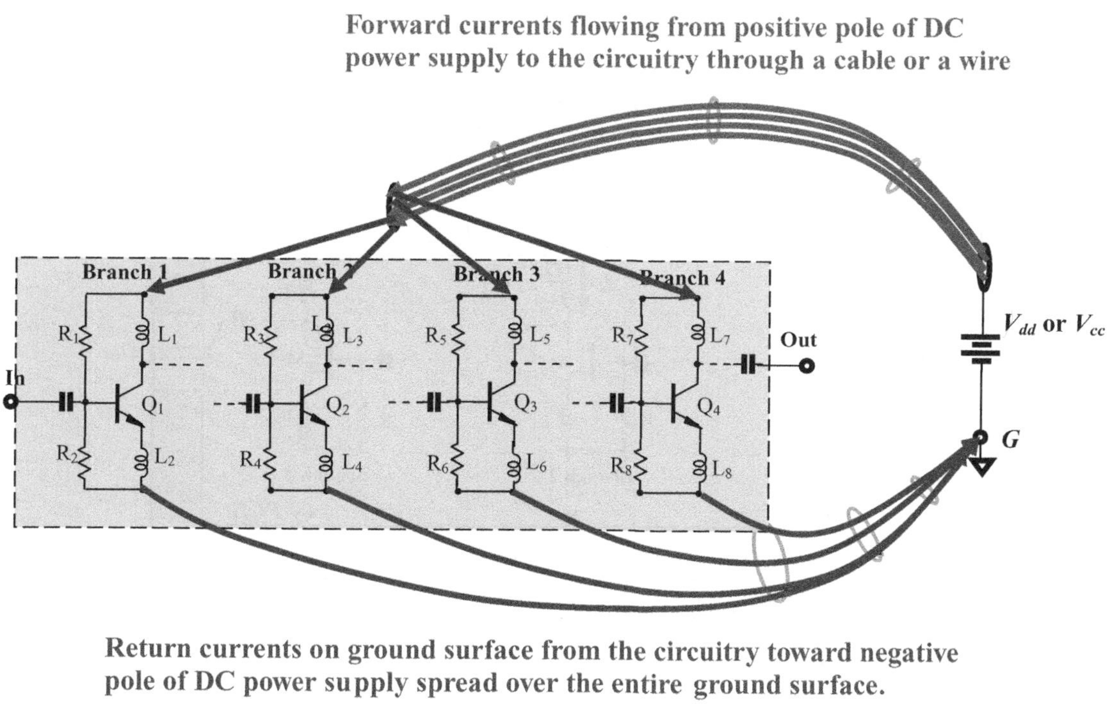

Figure 15.12 Images of forward and return currents between circuitry and *DC* power supply.

current flow path or the return current flow path are equivalent to the addition of a transformer into the circuitry. The transformer has complicated inputs and outputs from the four branches. It is particularly hard to figure out the return current coupling pattern on the ground surface. As a result, the circuit performance could become ridiculously wrong if the ground surface is unequipotential. This is the problem of forward and return current coupling.

15.2.2 Reduction of Current Coupling on a *PCB*

Figure 15.13 shows an example of forward current coupling on a *PCB*. For simplicity, the actual layout is replaced by the corresponding schematic in this section. On the *PCB*, it can be seen that there are three branches or three stages of this *RF* block. A *DC* power supply V_{dd} is provided to point *P*. Obviously, the currents for three branches I_1, I_2, and I_3 are actually magnetically coupled along the runner *PQ* even though there are two "zero" capacitors connected from point *P* to *Q*. This is called forward current coupling. The equivalent of this current coupling is a common transformer between these three branches. The equivalent transformer mainly depends on the length of the runner *PQ*. The length of *PQ* would be extended to include the length of the cable which connects the *DC* power supply to point *P* if there are not two "zero" capacitors connected from point *P* to the ground.

In the low-frequency range, the equivalent transformer plays a tiny role in the circuitry and is negligible. At *RF* frequencies, however, the equivalent transformer becomes important and is sometimes the main source of trouble screwing up circuit

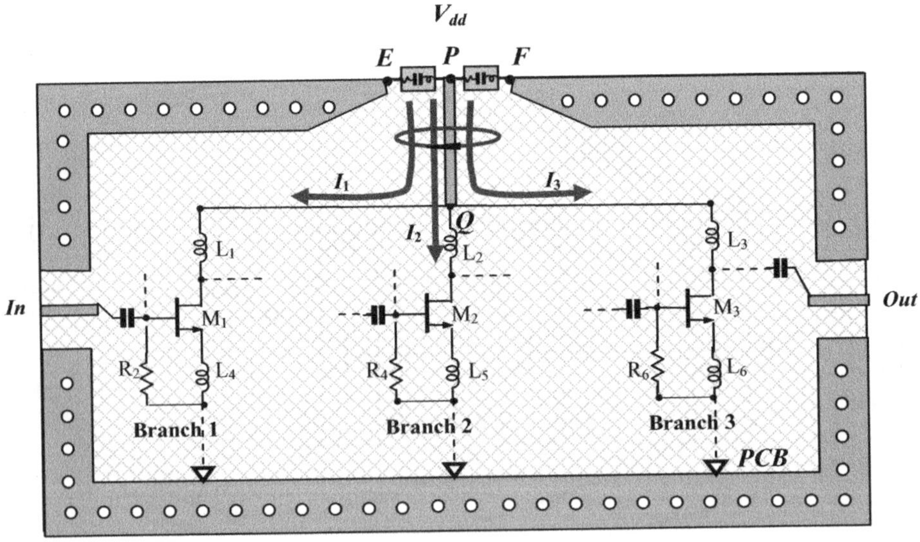

Figure 15.13 Reduction of forward current coupling by means of "zero" capacitor.

- Top metallic area
- Bottom metallic area
- "Zero" capacitor
- o Conductive via from top to bottom
- Magnetic flux

P: Point to be ground
E,F: Well-ground points
Forward current
Return current

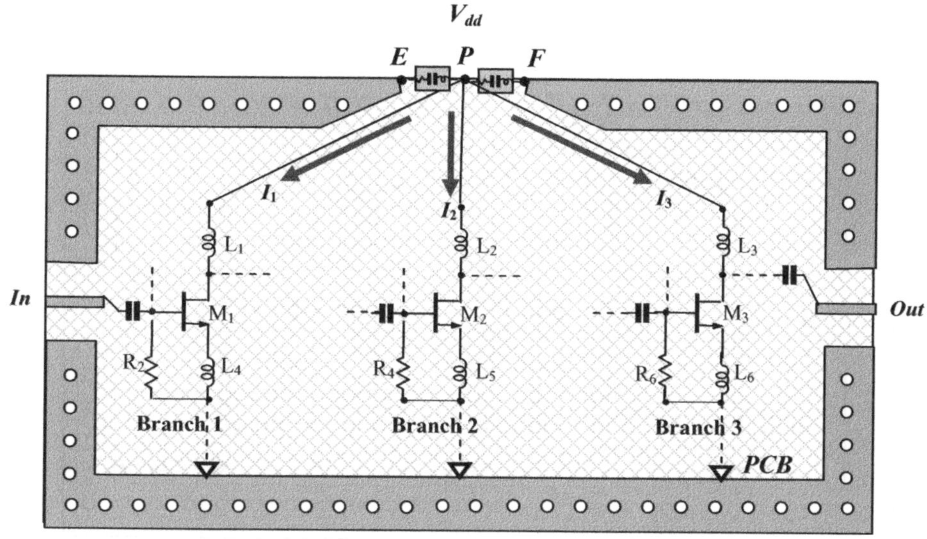

Figure 15.14 Reduction of forward current coupling by means of "zero" capacitor.

Top metallic area

Bottom metallic area

"Zero" capacitor

Conductive via from top to bottom;

Magnetic flux

P: Point to be ground

E,F: Well-ground points

Forward current

Return current

function. Unfortunately, to new designers, this source of trouble is hidden in a dark corner.

A very straightforward method of reducing forward current coupling is to provide a *DC* power supply to each circuit branch separately. This method is also called "to provide a *DC* power supply to each circuit branch or block in parallel." Figure 15.14 illustrates such a scheme. The *DC* currents for each branch, I_1, I_2, and I_3, are provided separately from point *P* to each branch without a common runner. Magnetic coupling still exists between the three paths for I_1, I_2, and I_3, but the coupling is greatly reduced to a minuscule level.

Now, let's consider the return current coupling. Unlike the forward current flowing along a runner, the return current actually flows over the entire ground surface and then from the area around point *E* or *F*, returning to the reference ground point of the battery or the *DC* power supply.

Figure 15.15 roughly depicts both forward and return currents. The solid bold arrow line represents the direction of the forward current and the dash bold arrow line represents the direction of the return current. It can be seen that the return currents are strongly coupled in the area between two branches: this is equal to having a transformer between the two branches in the circuitry. All of the equivalent transformers would seriously disturb the performance of the circuitry. If a designer ignores return current coupling, he might never know what's going on when his *RF* block does not work well or simply goes out of work altogether.

In order to reduce return current coupling, as shown in Figure 15.16, two slots are cut on the bottom metal plane so that the ground areas of adjacent branches of

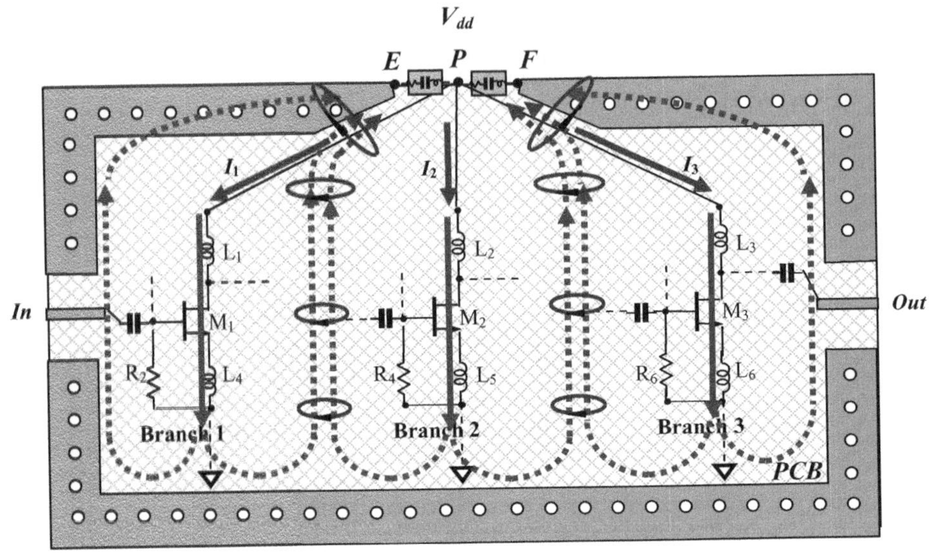

Figure 15.15 Reduction of forward current coupling by means of "zero" capacitors.

- Top metallic area
- Bottom metallic area
- "Zero" capacitor
- Conductive via from top to bottom
- Magnetic flux

P: Point to be ground
E,F: Well-ground points
Forward current
Return current

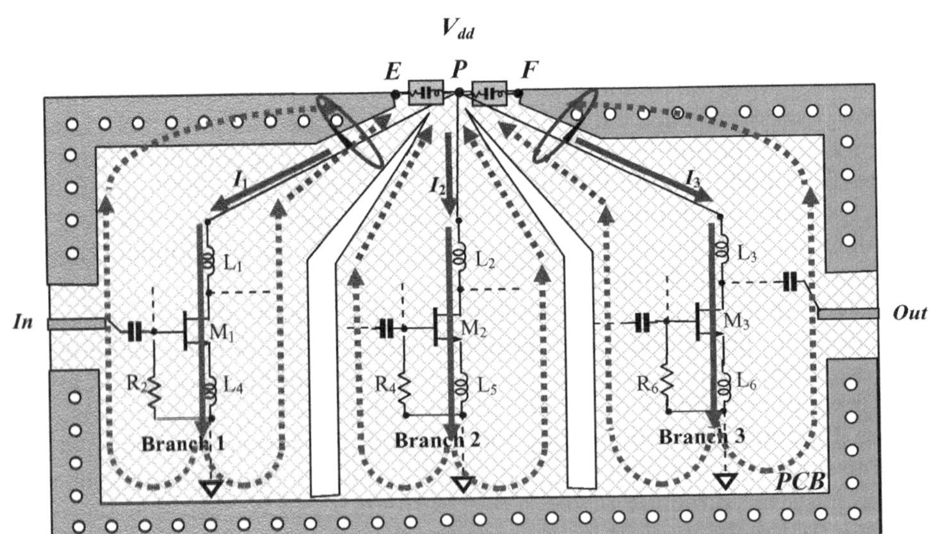

Figure 15.16 Reduction of forward current coupling by means of "zero" capacitor.

- Top metallic area
- Bottom metallic area
- "Zero" capacitor
- Conductive via from top to bottom
- Magnetic flux

P: Point to be ground
E,F: Well-ground points
Forward current
Return current

circuitry are isolated from each other. Each return current from the individual branches returns along its own ground area to points *E* or *F*. Consequently, the return current coupling is significantly reduced, though it is still exists around points *E*, or *F*, or *P*.

15.2.3 Reduction of Current Coupling in an *IC* Die

If a single *RF* block is built in an *IC* die, the scheme discussed above can be adapted; however, some modification is necessary. Figure 15.17 shows a single *RF* block built in an *IC* die for the reduction of forward and return current coupling.

In order to reduce forward current coupling in an *RFIC* die, the same scheme shown in Figure 15.16 can be applied to the *RFIC* die containing an *RF* block. The three branches are provided separate *DC* power supplies to greatly reduce forward current coupling. However, three separate "zero" capacitors are chip capacitors and must be allocated outside the *IC* die as shown in Figure 15.17, so that three bonding wires and six pads are needed for the chip capacitors outside the *IC* die. For a simple *RFIC* block, this is not a problem; however, if an *IC* die contains multiple *RF* blocks, the number of bonding wires and pads quickly increases up to an unaffordable amount. We are therefore aware of the importance and need to develop a "zero" capacitor directly in the *IC* die in the coming years.

In order to reduce return current coupling in an *RFIC* die, the ground metal in the *IC* die is divided into three pieces so that the three branches of circuitry have their individual ground areas. The individual ground areas are connected to the *P*+ guard ring, which is the ground ring encircling the *RF* block. The *P*+ guard ring then is connected to a pad and extruded out of the die by a bonding wire so as to connect to the metal frame on the *PCB*, which is equipotential with the reference ground point. The return currents from the 3 branches of circuitry are therefore isolated from each other, and the return current coupling is significantly reduced.

In Figure 15.17, the metal layer in the *IC* die is etched in a pattern with three oval areas. This beautiful and neat pattern is drawn just for illustration of the principle. In reality, the areas can take on any shape as long as the return current in the three branches of circuitry can be isolated. Instead of an oval shape, a pattern with etched long slots in the *IC* die appears more often in engineering design.

Finally, it should be noted that when circuitry is discussed, much attention is paid to the forward current which is drawn from the *DC* power supply. Very often, the return currents, which originate from all the ground points of the circuit to the ground point of the *DC* power supply, are ignored. In an actual *RF* layout or a test board, return current coupling could completely screw up the performance of a circuit due to the coupling between branches through the return current. This problem is especially a dark corner for a new engineer who has little experience in layout work.

15.2.4 Reduction of Current Coupling between Multiple *RF* Blocks

In a single *RF* block, the basic principles for reducing forward and return current coupling can be outlined as

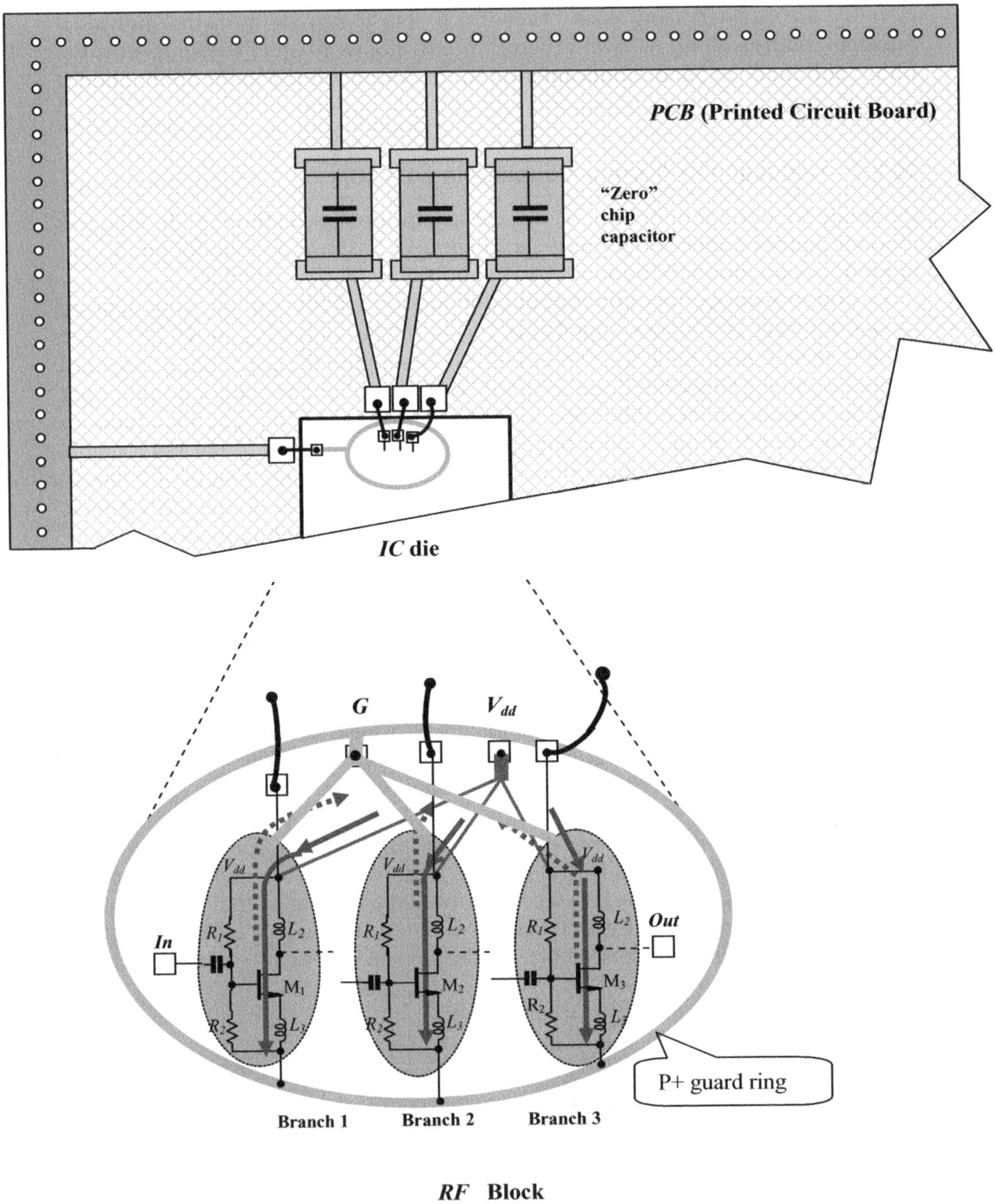

Figure 15.17 Reduction of forward and return current coupling simultaneously.

- Top metallic area
- Bottom metallic area
- Conductive via from top to bottom
- "Zero" capacitor
- Bonding pad
- *P*+ guard ring
- *P*: Point to be ground
- *G*: Reference ground point
- Forward current
- Return current
- Bonding wire
- Metal layer in *IC* die

- Providing *DC* power supply to branches of circuitry separately,
- Inserting long slots in the ground area so that each branch has its own ground area. All individual ground areas then are connected in parallel to the common ground surface, which is equipotential with the reference ground point at the *DC* power supply terminal.

The same principles can be applied to reduce forward and return current between multiple *RF* blocks. In other words, the above principles can be simply copied, replacing the words "branch" with "block," and "block" with "multiple blocks."

That is, in cases with multiple *RF* blocks, the basic principles for reducing forward and return current coupling can be outlined as

- Providing *DC* power supply to blocks of circuitry separately,
- Inserting long slots in the ground area so that each block has its individual ground area. All individual ground areas then are connected in parallel to the common ground surface, which is equipotential with the reference ground point at the power supply terminal.

In practical engineering designs, flexibility always exists for the separation of branches or blocks. In other word, a few branches could be combined as a big branch or a few blocks combined as a big block in the execution of the design principles mentioned above. These choices depend on the complexity of the circuitry, the current drain, and so on. As long as forward and return current coupling do not seriously impact performance, the designer may try whatever might work better to reach the goals.

15.2.5 A Plausible System Assembly

The "multi-closets" assembly type is a popular type of system assembly appearing in electronic products. Figure 15.18 shows its configuration, which can be described as follows.

- It is basically a metallic box with many closets. Each closet contains one block.
- There are 12 closets shown in Figure 15.18. The raw material of the metallic box is usually aluminum; in order to enhance its shielding function, the entire surface of the metallic box is gold-plated. As shown in Figure 15.18, a gold-plated aluminum cover is screw connected to the main body of the metallic box. (In Figure 15.18, the screw holes are marked by dark circles.) Consequently, the isolation between blocks and closets is, expectedly, very good.
- A block in a closet consists of many electronic parts, including resistors, capacitors, inductors, *IC* dies or packages, and so on. All the electronic parts are assembled on a *PCB* or some other type of board.
- On the assembled *PCB*, a terminal marked with a bold black circle is connected to the *DC* power supply V_{dd}. The *DC* power supply, denoted by V_{dd} is provided and goes from the side-connector to the terminals in block 1, block 2, block 3 ... block 12 in numerical order. In Figure 15.18, the path of the *DC* power

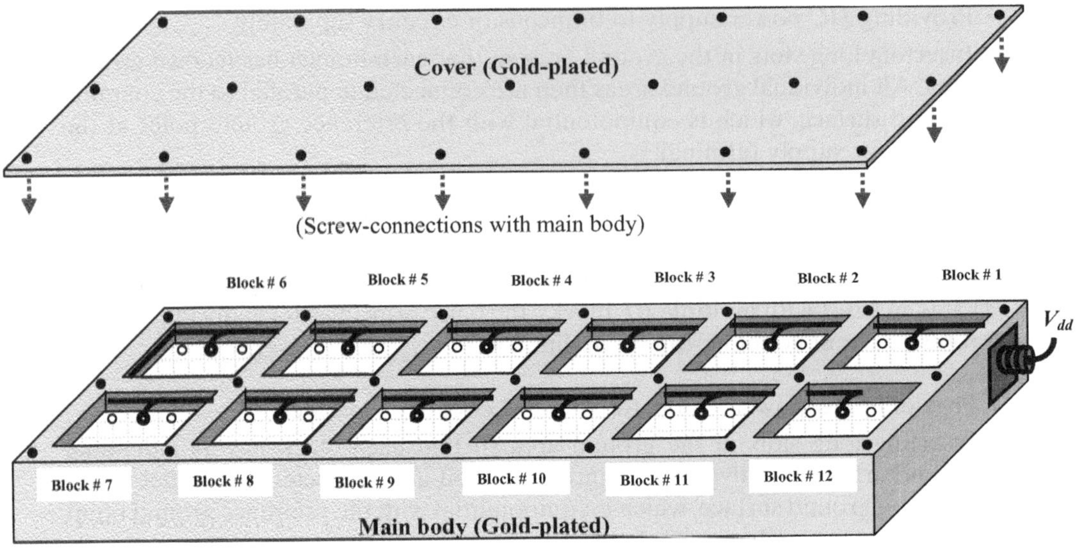

Figure 15.18 Configuration of a "multi-closet" type system assembly.

supply is shown with solid bold runners. Each individual block is fed with the *DC* power supply from the common connector, V_{dd}, in series.

- The ground surface on each *PCB* contains two holes marked with two white circles, which are electrically connected to the bottom side of the closet by conductive screws. Additional ground points on each block's board can be added and electrically connected to the main body of the box wherever it is short and convenient.

The "multi-closets" system assembly type looks nice and neat. There is no doubt that the shielding of the entire system assembly against outside interference is very good; however, its cost is high because the casting and gold plating of the metallic box is expensive. Thus, the question is whether it is worth the cost to construct such an assembly or not, or alternatively, how valuable its performance would be.

The performance of this system assembly, as a matter of fact, can be characterized by the following three factors, in which problems are found correspondingly.

- Equipotentiality over the entire gold-plated surface of the box
  The ground potential may be different from block to block if the gold-plated surface of the box is not equipotential, especially when the size of the box is in the same order of the quarter wavelength or is greater than quarter wavelength. This is important, though **the additional ground point on each block board can be added and electrically connected to the main body of the box wherever it is short and convenient.**
- Forward current magnetic coupling
  If the *DC* power supply nodes on each branch are not equipotential, forward current magnetic coupling could greatly degrade the performance of the circuitry. Obviously, this is a problem in the "multi-closets" type of system

assembly as shown in Figure 15.18 **because each individual block is fed with *DC* power supply from the common connector, V_{dd}, in series.**

- Return current magnetic coupling
  If the voltage over all the gold-plated surface is not equipotential, return current magnetic coupling could also greatly degrade the performance of circuitry, and is likewise an obvious problem in the "multi-closets" type of system assembly as shown in Figure 15.18 because the return currents from all individual blocks will flow over all the gold-plated surface with a complicated pattern and couple to each other.

15.2.5.1 An Improved Path of DC Power Supply In Figure 15.18, the *DC* power supply from the common connector, V_{dd}, to the individual blocks is connected by a path in series. This inevitably results in magnetic coupling in the forward current path. Figure 15.19 shows an alternative *DC* power supply path by which magnetic coupling of the forward current could be greatly reduced.

Distinguished from the path shown in Figure 15.18,

- The terminal connector of the *DC* power supply is allocated at the center of the top side of the box, instead of the lateral side of the box.
- Ideally, the *DC* power supply must be provided from this *DC* terminal separately. This implies that 12 wires must go to their individual blocks respectively from the common *DC* power supply at the center of the box. In order to alleviate the crowding caused by 12 wires, the *DC* power supply can be divided into three groups of four wires, in which groups *A* and *B* are connected to the *DC* terminal along with group *C*, as shown in Figure 15.19.

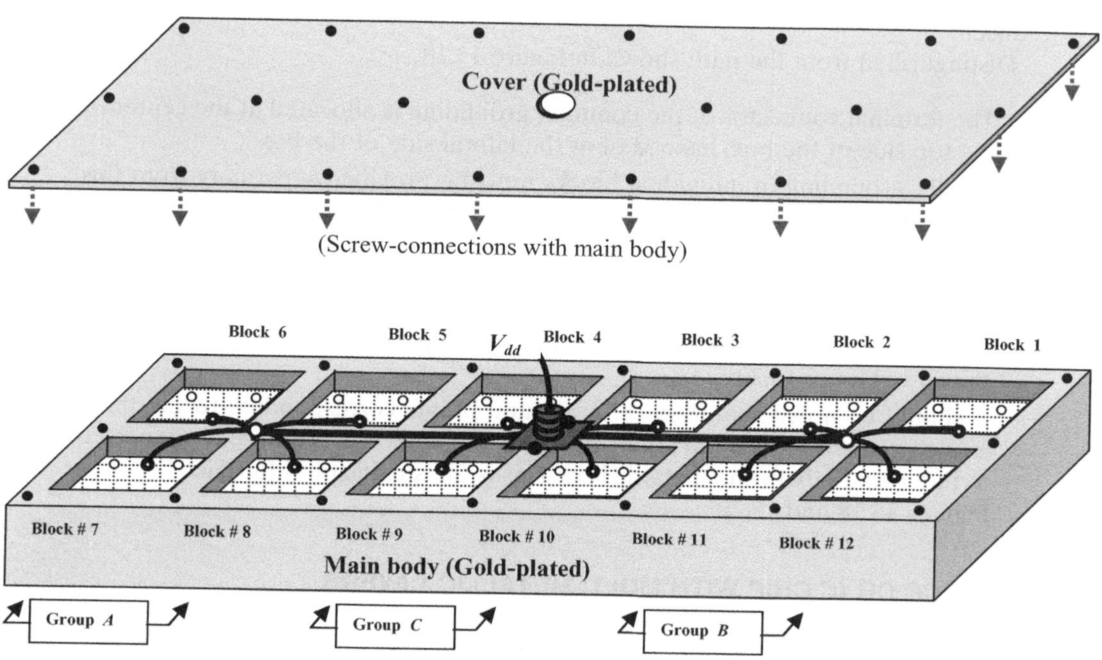

Figure 15.19 An improved connection path for power supply.

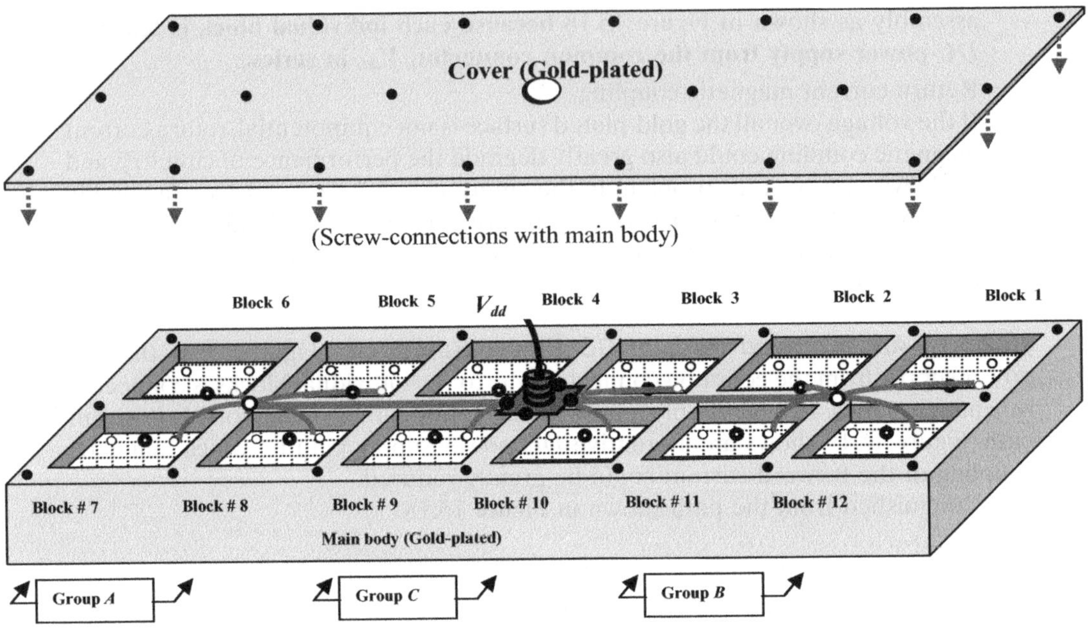

Figure 15.20 An improved connection path for grounding.

15.2.5.2 An Improved Path of Grounding In Figure 15.18, the common ground point to the individual blocks is provided by the path in series. This inevitably results magnetic coupling in the return current path. Figure 15.20 shows an improved ground path, by which the magnetic coupling of the return current can be greatly reduced.

Distinguished from the path shown in Figure 15.18,

- The terminal connector of the common grounding is allocated at the center of the top side of the box, instead of at the lateral side of the box.
- Ideally, grounding in individual blocks must be provided separately from this common grounding terminal. This implies that 12 wires must go to their individual blocks respectively from the common ground terminal at the center of the top side of box. In order to alleviate the crowding caused by 12 wires, the ground points can be divided into three groups of four wires, in which groups A and B are connected to the center of the top side along with group C, as shown in Figure 15.20.
- It should be noted that only three ground points corresponding to the three groups, and not the ground points from individual blocks, are directly connected to the main body of the metallic box. This is the significant difference between Figures 15.18 and 15.20.

15.3 *PCB* OR *IC* CHIP WITH MULTI-METALLIC LAYERS

A multi-layered *PCB* is often adapted in a complicated system. In parallel with multi-layer technology, there are multi-metallic layers available in the *IC* circuit design today. Undoubtedly, multi-layer technology is an important progression in the field of electrical engineering. It is very helpful to layout a complicated system.

For instance, if the *PCB* of *IC* chip contains only one metallic layer, some runners must be replaced by jumping wires or "bridges" in the air where two runners must cross over each other. If the *PCB* contains a top and bottom metallic layer or if an *IC* chip contains two metallic layers, a runner can be moved up and down between the two metallic layers so that the jumping wires or "bridges" can be removed.

A *PCB* with more than three metallic layers not only allows the runners laid on the *PCB* to be free of jumping wires or "bridges," but also provides more flexibility to the runners' layout. For instance,

- In order to get better isolation between the *RF* and digital signals, one intermediate metallic layer can be assigned to function as a standard ground surface while the rest of the layers are assigned to contain all the runners. For example, all the runners involved with *RF* signals are collected in the top layers above the ground layer, while all the runners involved with digital signals are collected in the bottom layers below the ground layer.
- In order to get better isolation between *RF* and low-frequency signals, those runners involved with *RF* frequencies are collected in one metallic layer while those runners involved with low frequencies are collected in another metallic layer.
- In order to get better isolation of a special part, this special part is "sandwiched" between two big ground metallic portions in the metallic layers.
- And so on.

These all seem like good ideas on the surface. However, for some reason, the actual performance is often far from expectations. The expected good isolation seldom comes true although the expense for multiple metallic layers has been paid.

Therefore, it seems worthwhile to study and then answer the following questions:

- First, is the entire metallic layer an equipotential ground surface? If it is not equipotential, then the parts in the circuit connected to this metallic layer are referenced at different ground potential levels, and the performance, of course, would then be greatly different from that if the parts were referenced to an equipotential ground surface. The designer must verify the equipotentiality of the entire metallic layer before it is applied to the layout of the circuit. A demarcation line for equipotentiality is to see whether the maximum dimension of the metallic layer is greatly lower than the quarter wavelength of the highest operating frequency or not.
- If the entire metallic layer is verified to be an equipotential ground surface, then the second question is, is it equipotential with the main ground reference point, the negative pole of the *DC* power supply? If not, then forward and return currents will happen. Especially, the return current will flow over the entire metallic layer with a very complicated pattern, making the case worse than when the entire metallic layer did not exist. In addition, if the metallic

layer is not equipotential with the main ground reference point, not only *RF* return currents but also digital return currents would flow over the entire metallic layer and couple with the *RF* return currents. It would result a big mess!

- Third, is the entire metallic layer equipotential with those ground points not on the entire metallic layer? By what method are they connected together?
- Now, let's consider another potential big trouble source, the conductive vias from one layer to another layer. Conductive vias are the main tool to connect runners jumping from one layer to another. One should aware that a conductive via is by no means a zero impedance part. Instead, it is a part with parasitic inductance, capacitance, and resistance even though its entire surface is gold-plated. In either *PCB* or *IC* design, the equivalent model of a via should be developed and put into the circuit simulation. The parasitic inductance and resistance can screw up normal performance in a very bad way. For example, a conductive via with 10 mils of diameter on a *FR4* plastic *PCB* with *25* mils of thickness may attenuate a 10 *GHz* signal by 5–10 *dB*.
- As for the sandwich configuration, one must worry about the additional capacitance between the top and bottom surface of sandwich, which could be considerable, or in extreme cases, form a resonant cavity.
- In practical design for *PCB*s or *IC* chips with multiple metallic layers, many more problems and questions can be found. For instance, forward current coupling may be due to providing an inappropriate *DC* power supply mode to an individual RF or digital block. Similarly, return current coupling may be due to an inappropriate grounding mode from the reference ground point to the individual ground point of an RF or digital block.

Summarily, a good design for a *PCB* or *IC* chip with multi-metallic layers must have these conditions:

- The ground surface is equipotential with the reference ground point, the negative pole of *DC* power supply.
- Forward and return current coupling are reduced to an insignificant level.
- The *RF* signal with the highest operating frequency, which flows through the conductive vias, does not have considerable attenuation.
- The additional capacitance, inductance, and resistance due to the multi-metallic configuration is insignificant.

Additional interference between digital and digital, *RF* and *RF*, and *RF* and digital areas due to the multi-metallic configuration are negligible.

APPENDICES

15.A.1 Primary Considerations of a *PCB*

In *RF* circuit design, *PCB*s can be roughly categorized into two categories: (1) test *PCB*s, which are designed for testing, and (2) circuit *PCB*s, which are designed as the circuit board of the final product. However, there is no clear demarcation between the circuit *PCB* and the test *PCB*. It often happens that a circuit *PCB* is directly tested in the laboratory or on the production line, so that it is a circuit *PCB* as well as a test *PCB*. No matter for a circuit *PCB* or for a test *PCB*, the design of a good *PCB* is a "must" skill for an *RF* circuit designer.

The questions that need to be answered are:

- How should a *PCB* be selected?
- What should the layout of a *PCB* be?
- What should the size of a *PCB* be?
- How should a large *PCB* be handled?
- How should the equipotentiality of the desired ground surface and desired ground points on a large *PCB* be measured?
- Why does current coupling appear on the ground surface?
- How should a multi-layer *PCB* be organized?

We will answer these questions one by one.

15.A.1.1 Selection

- **Type**

  There are many types of *PCB,* such as plastic, ceramic, glass, and so on. The most popular type of board is the plastic *PCB*, but sometimes the ceramic *PCB* is adapted for its special features.

  Table 15.A.1 compares the main features of the plastic *PCB* and ceramic *PCB*. Processing for a plastic *PCB* is easier. It takes only a couple hours to process a *PCB* with two metal layers, the top and bottom layers. But, the plastic *PCB* cannot be directly heated over an electric stove. The discrete parts, modules, or *IC* die must be soldered by an electric soldering iron or a special soldering machine.

  On the contrary, processing a ceramic *PCB* is somewhat more difficult. Processing a ceramic board takes at least a few days, which is much longer than the time needed to process a plastic *PCB*. However, the ceramic *PCB* has a remarkable advantage: it can be heated over an electric stove, and therefore, the discrete parts, modules, or *IC* die can be directly soldered onto the board

TABLE 15.A.1 Comparison of features between plastic and ceramic *PCB*

	Plastic *PCB*	Ceramic *PCB*
Processing time	A couple of hours for two metal layers	A couple of days for top and bottom layer
Soldering of parts	A couple of hours by soldering iron or machine	A couple of minutes by heating over an electric stove
Size	Large	Small
Mechanical property	Robust and sturdy	Fragile, easily broken
Cost	Low	High

in minutes. In addition, the dielectric constant ε of a ceramic *PCB* is higher than that of a plastic one. The quarter wavelength for a given frequency on a ceramic *PCB* should be shorter that on a plastic *PCB* and therefore, the size of a ceramic *PCB* should be smaller than that of a plastic one.

From the viewpoint of cost-saving and mechanics, people prefer the plastic *PCB* to the ceramic *PCB*.

- **Upper frequency limit**
  On the other hand, the most important feature of a *PCB* is attenuation. The signal transported and manipulated on the *PCB* inevitably suffers attenuation, since the PCB is not a super-conductor. The higher the attenuation factor of the *PCB* material, the worse the attenuation. On the other hand, the higher the frequency, the higher the attenuation. In the low-frequency range, then, attenuation is negligible and does not impact the performance of the tested block or system. The additional attention of the signal on *PCB* increases as the operating frequency increases. A frequency is reached when the performance of the tested circuit block or system is degraded by a considerable level due to the attenuation of the *PCB*. This frequency is called the upper frequency limit of the *PCB*. For example, the upper frequency limit of a *FR4*-type plastic *PCB* is about 2 *GHz*. This indicates that this *PCB* does not have an unacceptable amount of attenuation for a given operating frequency up to 1 to 2 *GHz*. However, it becomes a very lossy medium if the operating frequency gets any higher than about 2 *GHz*. In general, the upper frequency limit of a *PCB* must be guaranteed to be at least 1.5 times higher than the highest operating frequency. Otherwise, unacceptable attenuation will occur.
- **Number of metallic layers**
  For simplicity, the test *PCB* contains only two metal layers: the top and bottom metal layers. It is not encouraged to apply a test *PCB* with more than two metallic layers, as such a board takes a much longer processing time and increases the cost a lot. In addition, it is difficult to do *RF/AC* grounding well in a *PCB* with multiple metal layers, especially in the high end of the *RF* frequency range. If a *PCB* is selected not for the testing purposes but for the building of a formal product, a *PCB* with multi-metal layers may be reasonable, in which, however, careful work must be done on *RF/AC* grounding.
- **Electromagnetic parameters**
  The performance of a *PCB* is determined by its electromagnetic parameters and other physical properties, such as the dielectric constant, ε, attenuation factor, coefficient of temperature variation, and so on. A *PCB* with higher values of ε has the advantage of smaller size if it is necessary for the micro strip line or printed inductor or capacitor to be printed directly on the *PCB*. However, it might bring about higher tolerance of a runner's impedance and other distributed parameters. The value ε of a ceramic *PCB* is higher than that of a plastic *PCB*. In addition, it is possible to solder parts onto the board by heating on a ceramic *PCB*, but not on a plastic *PCB*.

15.A.1.2 Layout A preferred layout of the test *PCB* is recommended and depicted in Figure 15.A.1, in which only the main ports, including input, output, and

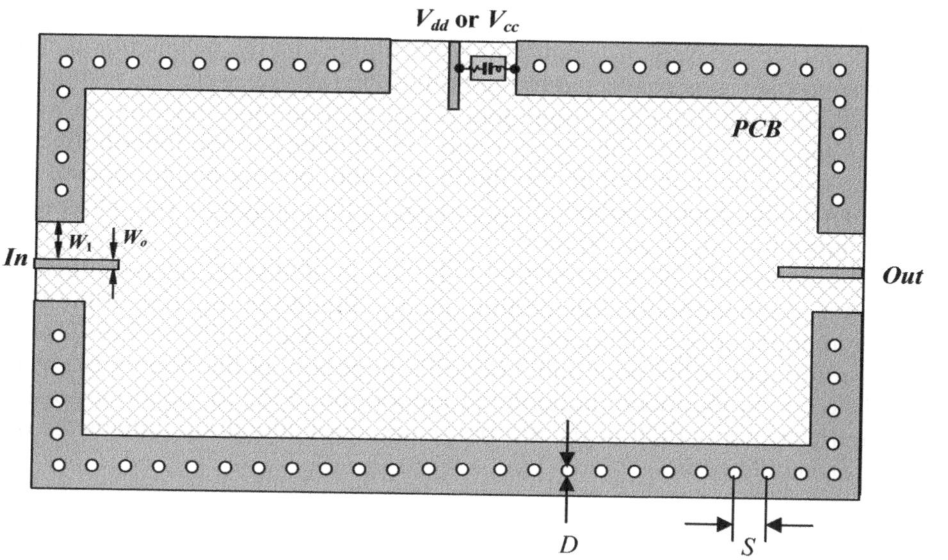

PCB. (The printed

Figure 15.A.1 A preferred layout of a test *PCB*. (The printed circuit on the *PCB* is neglected.)

▩ Top metallic area
□ Bottom metallic area
▭ 50 Ω runner
○ Conductive via from top to bottom
"Zero" capacitor

DC power supply, are shown, while the detailed circuit items are neglected, since our discussion is confined to *AC/RF* grounding.

The main features of the recommended layout of a test *PCB* are:

- **Two metal layers are preferred**
  As mentioned above, a test *PCB* with two metal layers, top and bottom layer, is preferred for simplicity. Usually the bottom metal layer is fully unetched and kept as the main ground surface. Another portion of the ground surface is a rectangular metallic frame on the top as shown in Figure 15.A.1, which is connected to the bottom metal plane by a number of conductive via.
- **Conductive via**
  A via is a cylindrical conductive hole. The cylindrical surface inside the hole is gold-plated, or plated with some other high conductivity material. The existence of vias is important. They connect the rectangular metallic frame on the top to the bottom metal layer together so that the top and bottom ground surfaces are expected to be equipotential. According to experimental testing, the performance of a circuit block is quite different between two cases with and without conductive holes. For example, if a filter is built on the *PCB*, the insertion loss of a filter might have a couple *dB* difference between the two cases. Consequently, the design of vias becomes one of important issues in the layout of a *PCB*.

- **Diameter of vias**
  In the *DC* or low-frequency range, it is indubitable that the rectangular metallic frame on the top is connected to the bottom metal layer well. The rectangular metallic frame on the top is equipotential with the bottom metal layer. However, in *RF* frequency range, they may or may not be equipotential to each other. In other words, the rectangular metallic frame may or may not be ground well though it is connected to the bottom ground by the conductive vias. This mainly depends on the diameter of the via.

  The equivalent of a conductive hole is a combination of an inductor, a resistor, and a capacitor. Their values mainly depend on the size of the holes and other parameters related to the *PCB*'s material and configuration. Usually, the equivalent inductance and resistance increases as the size of the via is reduced. It is therefore desirable to enlarge the hole as much as possible. Empirically, for a *PCB* in the *RF* frequency range, the desired diameter of a via, *D*, should be

$$D > 10\ mils. \tag{15.A.1}$$

- **Spacing between vias**
  On the other hand, in order to connect the top and bottom ground surfaces better, it is desirable to punch more vias in the board. Empirically, the space between vias can be taken as

$$S \approx 4D \text{ to } 10D. \tag{15.A.2}$$

  depending on the operating frequency of the *PCB*. The space $S \approx 4D$ is for higher *RF* operating frequencies while $S \approx 10D$ is for lower *RF* operating frequencies.

- **Special function of rectangular metallic frame on top**
  The rectangular metallic frame on the top is an important configuration of the test *PCB*. It functions not only as RF grounding but also partially as shielding. Figure 15.A.2 shows its shielding performance.

  In Figure 15.A.2(a), it is assumed that the *AC/RF* grounding is perfect so that the rectangular metallic ground frame is equipotential with the reference ground point, *GND*. Most of the electric lines produced and then radiated from the circuit inside the *PCB* would be terminated on the rectangular ground frame. In addition, most of the electric lines from any interference source outside the *PCB* would be stopped by the rectangular ground frame as well. We say "most of the electric lines" but not "all of the electric lines" because some of them would cross over the barrier of the rectangular ground frame since the environment of the test *PCB* is somewhat complicated, rather than a free space; in addition, some of the electric lines could cross over the frame through the plastic medium of a plastic *PCB*. Nevertheless, the isolation between the circuits inside and outside the rectangular ground frame is greatly improved due to the existence of the rectangular ground frame. This is a very simple but powerful idea, by which the development of the *RFIC* becomes realistic.

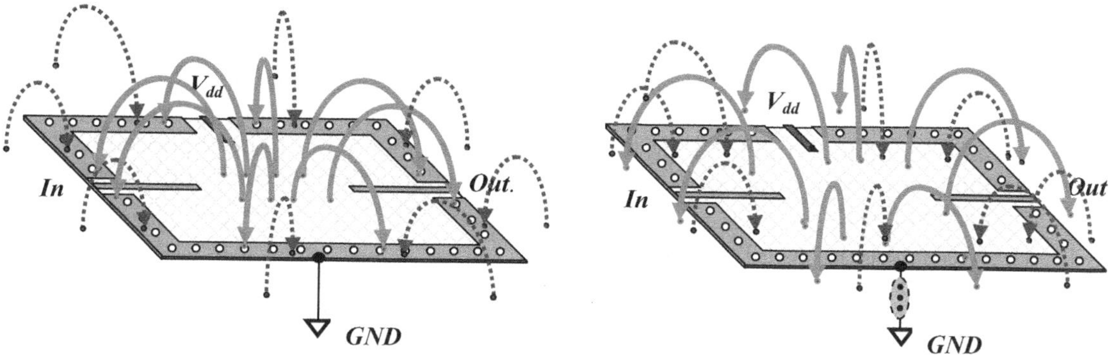

(a) Perfect *RF/AC* grounding (b) Imperfect *RF/AC* grounding

Figure 15.A.2 Electric lines radiated either from the test circuit inside the *PCB* or from any interference source outside the *PCB* would be terminated on the rectangular ground frame.

- Electric line from inside
- Electric line from outside
- Imperfect grounding symbol
- *DC* power supply runner
- Top metallic area or runner
- Bottom metallic area
- o Conductive via from top to bottom side

On the contrary, in Figure 15.A.2(b), it is assumed that the *AC/RF* grounding is imperfect so that the rectangular metallic ground frame is not equipotential with the reference ground point, *GND*. Most of the electric lines produced and then radiated from the circuit inside the *PCB* would not be terminated on the rectangular ground frame, but instead jump out and disturb the adjacent circuit blocks. In addition, most of the electric lines from any interference source outside the *PCB* would not be stopped on the rectangular ground frame either, but would arrive to disturb the circuit inside the rectangular ground frame.

- **Spacing between the input or output runner and the adjacent ground edge**
  At the input, output, and *DC* power supply ports, one might like to paste more ground area beside the runners. However, the spacing between the input or output runners and the adjacent ground edge, W_1, must be wide enough so that the edge spray capacitance between the input or output runners and the ground area in both sides can be neglected. As shown in Figure 15.A.1, the width of the input or output runners, W_o, must correspond to the 50 Ω characteristic impedance of the micro strip line because it is supposed to be connected to the input signal generator or output analyzer. In order to keep its characteristic impedance around 50 Ω without being affected by the rectangular ground frame, empirically the following condition must be satisfied:

$$W_1 > 3\,W_o. \tag{15.A.3}$$

Of course, if the runner for the input and output is designed on the basis of a coplanar waveguide, condition (15.A.3) is not necessary. As a result, more metallic area can be added to the ground surface. Under the coplanar waveguide design, the width of input or output runners is usually narrower than that corresponding to a micro strip line with 50 Ω of characteristic impedance, W_o. Also, the spacing between the runner and the adjacent ground edge is much narrower than $3W_o$. However, the width of the runner needs to be expanded in the non-coplanar-waveguide area so that it can keep the characteristic impedance at 50 Ω.

- **Addition of a "zero" capacitors**
  With the exception of the above effort for *RF* grounding, "zero" capacitors must be added between the V_{dd} or V_{cc} terminals and the adjacent ground area because the V_{dd} or V_{cc} terminal must be *AC/RF* ground as well. The ground potential is supposed to be the same for all operating frequencies. Should the circuit be involved in more than one operating frequency, such as the *RF, LO*, and *IF* frequencies, more than one "zero" capacitor must be added accordingly. When testing the *RFIC* die, the "zero" capacitors should be put as near the *IC* die as possible.
- **Other considerations**
  In addition to *RF* grounding, there are some other considerations for the improvement of a test *PCB*, such as having gold plating for the terminal runners, or adding resistive material on both top and bottom sides so as to protect the *PCB* from damage. However, these are beyond our topics of this section.

15.A.1.3 Size In the design of a test *PCB*, the first objective is to make sure what size the test *PCB* must be, because designing a large *PCB* is much more difficult than designing a small *PCB*. Let us define the small and large sizes of a *PCB*.

If the dimension of a *PCB*, either the length or width, is much less than the quarter wavelength of a runner with 50Ω of characterized impedance, that is,

$$L \ll \lambda/4, \tag{15.A.4}$$

where

L = dimension of the *PCB*, and
λ = electric wavelength of a runner on the *PCB*, when the characteristic impedance of the runner, Z_o, is 50 Ω

then we define it as a small *PCB*. Equation (15.A.4) is the definition of a small-sized *PCB* electrically.

On the contrary, if the dimension, either length or width, of the test *PCB* is comparable to or greater than the quarter wavelength of a runner, that is,

$$L \sim \text{or} > \lambda/4, \tag{15.A.5}$$

then we define it as a large *PCB*. Equation (15.A.5) is the definition of a large-sized *PCB* electrically.

It should be noted that the electric wavelength of a runner on the *PCB*, λ, depends on the width of the runner or its related characteristic impedance, Z_o, since a runner on the *PCB* is a segment of a micro strip line. It is quite reasonable to take 50Ω as a standard characteristic impedance of the runner because in the actual testing the input and output is usually connected to the equipment with 50*Ω* of impedance. By taking 50Ω as a standard reference, the electric wavelength of a runner on the *PCB*, λ, can be calculated on the basis of the electric medium constant or the permittivity, the permeability, and the thickness of the *PCB*.

When testing an individual block, such as the *LNA*, Mixer, *VCO* and so on, condition (15.A.4) is usually satisfied in both *RF* module and *RFIC* circuit design.

In cases where the *PCB* is of a small size, the ground surface of the *PCB* mentioned above is unanimously equipotential. This is a remarkable advantage and therefore, it is encouraged to apply small *PCB*s for testing or packaging as much as possible.

In cases where the *PCB* is of a large size as shown in inequality (15.A.5), the ground surface is not equipotential in general. Then, special treatment must be conducted.

REFERENCES

[1] Edward C. Jordan, *Electromagnetic Waves and Radiating Systems*, 2nd ed., Prentice-Hall Inc., 1968.

[2] Hannu Tenhunen, "CMOS Interconnects" (Lecture), Electronic System Laboratory, Kungl Tekniska Hogskolan, 2000.

[3] H. W. Shim, T. M. Zeeff, and T. H. Hubing, "Decoupling Strategies for Printed Circuit Boards without Power Planes," *2002 IEEE International Symposium on Electromagnetic Compatibility*, Vol. 1, August 19–23, 2002, pp. 258–261.

[4] Richard Chi-Hsi Li, *Key Issues in RF/RFIC Circuit Design*, Higher Education Press, Beijing, 2005.

CHAPTER 16

RFIC (RADIO FREQUENCY INTEGRATED CIRCUIT) AND *SOC* (SYSTEM ON CHIP)

16.1 INTERFERENCE AND ISOLATION

16.1.1 Existence of Interference in Circuitry

An electrical circuit consists of basic parts, such as resistors, capacitors, inductors, and transistors. Each of these has its specific function in the circuitry. One should be aware of the fact that the parts' performance is different between *DC* and *AC* operation cases. When a circuit operates with *AC* or *RF* signals, every part appears as a small antenna since there is *AC* current flowing through it or *AC* voltage across it. This alternative electromagnetic field radiates from the part to its whole surroundings. All these alternative electromagnetic fields can mix to form an interference, which would disturb the desired signal at any node in the circuitry.

According to the theory of electromagnetic radiation, the interference thus produced in the low-frequency range is negligible and usually can be ignored. However, in the high-frequency range, such as the *RF* frequency range, the interference due to the radiation is considerable and it does interfere with the circuit at every node in the circuitry.

It must be noted that these electromagnetic fields disturb the circuitry along two paths. One path is that of the electromagnetic field radiated to the surrounding space and then returned to the circuitry. This can be categorized as an "air path." The second path is that of the electromagnetic field moving along the *PCB* in the *RF* module or along the *IC* substrate in the *IC* chip. It can be categorized as an "underground" path.

In general, as long as the circuit block is in *AC* or *RF* operation, the desired signal at any node in the circuitry is more or less disturbed by the interference from all the parts in the circuitry. Besides, interference may come from external sources outside the block. The interference might be especially strong if the adjacent block is a power amplifier or if powerful oscillators without any means of shielding are adapted.

16.1.2 Definition and Measurement of Isolation

Interference inevitably exists when a circuitry is operated in *AC* or *RF* frequency. It is expected to be attenuated as much as possible. For instance, if the interference can be attenuated 100 *dB* from its origin to the observed node in an *RF* module or *RFIC* die, then it becomes negligible and can usually be ignored. If the interference can be attenuated infinitely from its origin to the observed node, then the node is perfectly isolated from the origin. Therefore, isolation can be simply defined as the attenuation of the interference in magnitude, but with the opposite sign, that is,

$$\text{isolation} = -\text{attenuation of interference}. \tag{16.1}$$

Quantitatively, isolation between points *A* and *B* can be measured by the interference attenuation from node *A* to *B* as shown in Figure 16.1.

Assuming that the power of the interference at node *A* is P_A and the power at node *B* is P_B, and both of them are expressed by *dB*, the isolation between *A* and *B* is

$$Attenuation = 10\log\frac{P_B}{P_A},\ dB \tag{16.2}$$

$$Isolation = 10\log\frac{P_A}{P_B},\ dB \tag{16.3}$$

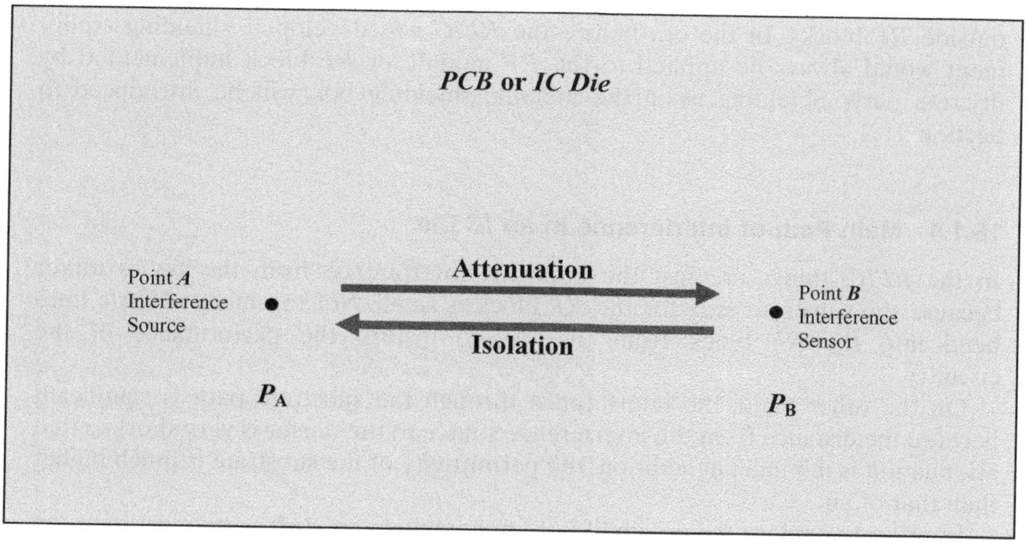

Figure 16.1 Definition of isolation between nodes *A* and *B*.

For example, if an interference is $-10\,dB$ at node A and $-30\,dB$ at node B, then the attenuation of interference is $-20\,dB$ from node A to node B, and the isolation is $20\,dB$ at node B with respect to node A.

A perfect isolation occurs when

$$P_B = 0, \tag{16.4}$$

then,

$$Isolation \to \infty. \tag{16.5}$$

16.1.3 Main Path of Interference in an *RF* Module

In an *RF* module, the interference moving along the substrate path is insignificant because the plastic or ceramic *PCB* is usually a good insulator in the *RF* frequency range below couple *GHz*.

The main path of interference in an *RF* module is from the sky because the dimension of the *PCB* is greater than the dimension of a part, or is sometimes comparable with the quarter wavelength of the operating frequency. The electric lines bending into the *RF* module from the sky could seriously disturb the performance of the circuitry. These electric lines consist of two portions. One is made up of the electric lines radiated from local parts of the *RF* block. The other is made up of the electric lines radiated from outside blocks.

Nothing can be done about the electric lines radiated from the local *RF* block; however, electric lines radiated from outside blocks can be shielded against. In most cases, the electric lines radiating from outside blocks are the main interference source rather than those radiating from the local block. This implies that a good shielding can usually ensure good isolation of an *RF* block from outside blocks and thus ensure the performance without interference from outside *RF* blocks. In the era before the *RFIC* was developed, shielding equipment would always be applied to the *RF* module or *RF* block implemented by discrete parts. Shielding using the metallic shielding box will be introduced in Section 16.2.

16.1.4 Main Path of Interference in an *IC* Die

In the *RFIC*, things are just the opposite. Interference from the sky is minor because the substrate area for the *RF* block is small. Not too many electric lines bend into the *RF* block from the sky to disturb the performance of the circuitry.

On the other hand, the interference through the substrate path is significant because the distance from the interference source to the sensor is very short so that attenuation is minimal; in addition, the permittivity of the substrate is much higher than that of air.

In order to explore the possibility of implementing an *RFIC*, one must focus on the study of the interference moving along the path of the substrate.

16.2 SHIELDING FOR AN *RF* MODULE BY A METALLIC SHIELDING BOX

At high or radio frequencies, the conceivable electromagnetic field radiates from each part in the circuit, including runners. Each part starts to look like a small antenna. The electromagnetic waves radiate from the surrounding sky and could mingle together as a feedback signal returning to the circuitry so as to cause disorder in the circuit performance.

This problem was recognized as the key barrier in the early stages of *RF* circuit design, which were fabricated with discrete parts before the 1970s.

In order to avoid the electromagnetic wave radiation from the *RF* block to the sky, each individual *RF* block is shielded by a small metallic box. Figure 16.2 shows a typical shielding scheme. The top portion is a metallic shielding box. The hole on the top is for receiving the short stub on the *PCB* so that the *PCB* can be fixed after it is slid into the metallic shielding box. The *PCB* of the *RF* block is depicted as the middle portion in Figure 16.2, in which the parts and runners are neglected. The *PCB* is slid into the metallic shielding box through the slot in the box and its position is fixed by the slot and by the short stub inserted into the hole on the top of

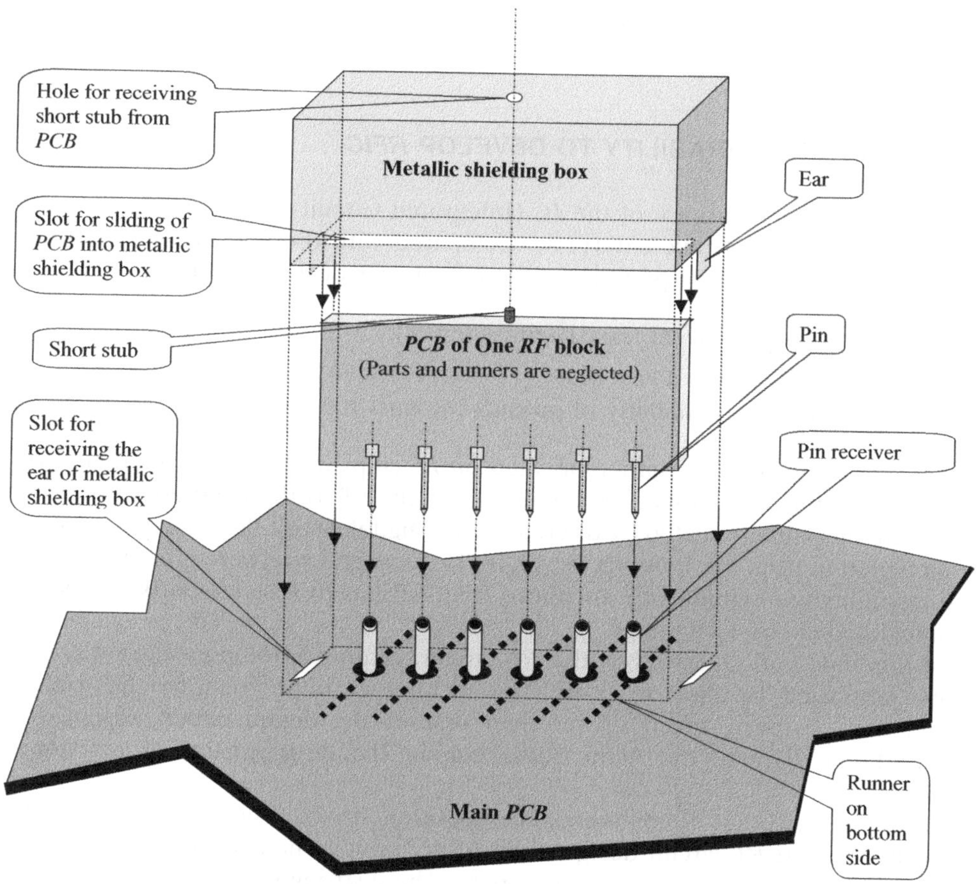

Figure 16.2 Shielding for a *RF* module.

the metallic shielding box. The input, output, bias, and *DC* power supply are assigned to six pins; these are inserted into the pin receivers on the main *PCB* as the two ears on the metallic shielding box are inserted into two slots on the main *PCB* simultaneously. After that, the two ears are bent on the bottom side of the main *PCB* so that the *PCB* of the *RF* block is fixed on the main *PCB* along with the metallic shielding box. The runners on the main *PCB* are on the bottom side of *PCB*, so that they are not short-circuited by the metallic shielding box when it touches the main *PCB*.

The addition of the metallic shielding box is not an easy task.

First, the cost for the addition of the metallic shielding box is usually much higher than the direct material cost of the *RF* block itself.

Second, the insertion configuration of the pin and pin receivers is designed with a really high technical level. They required not only gold plating but also elastic contact with both compression and elongation. The most outstanding feature of the pin and pin receivers is that their characteristic impedances are kept at 50 over all the operating frequency ranges. Much R&D effort has been put into this sophisticated mechanical-electrical design.

In the 1990s, a portable radio had approximately 10 to 15 *RF* modules with metallic shielding boxes. More than one hundred pins must have been in the main *PCB*. The reliability of all these pins is another subject of concern on the production line.

16.3 STRONG DESIRABILITY TO DEVELOP *RFIC*

In the 1960s, the appearance of the *IC* (Integrated Circuit) brought about a great revolution in the electronic enterprise. Compared with those electronic products implemented with discrete parts, the great advantages of the *IC* were:

- Greatly reduced cost (at least 10 times lower),
- Greatly reduced size (more than 1000 times smaller),
- Greatly enhanced reliability of product (at least 100 times more reliable).

The digital circuit was partially fabricated on *IC* chips in the 1960s; the digital *IC* became quite popular in digital circuit designs in the 1970s, though there were a few digital circuits still fabricated by discrete parts. Digital circuit design experienced a real revolution from the 1960s to the 1970s. The cost and size were greatly reduced and reliability was significantly enhanced from designs by discrete parts to designs by integrated circuit technology.

In this time period, however, the *RF* circuit in an electronic product was still 100% fabricated by discrete parts. The technological progress for *RF* circuit design was obviously much slower than that of the digital circuit design. *RF* circuit design became the main bottleneck in the development of electronic products.

From the viewpoint of commercial competition, it would be desirable to apply *IC* technology to *RF* circuit design since *IC* technology had so many advantages compared to the fabrication by discrete parts. These strong incentives existed in all parts of the electronics industry, because it was well known that the way to become a product winner in a serious commercially competitive environment was to lower cost, reduce size, increase reliability, and enhance performance.

On the other hand, from the engineering design viewpoint, it seemed hopeless to apply *IC* technology to *RF* circuit design because electromagnetic radiation might still exist even if very careful shielding was applied to *RF* blocks. It was absolutely impossible to put metallic shielding boxes as shown in Figure 16.1 into an *IC* die. Consequently, at that time, very few people thought that it would be possible to fabricate an *RF* circuit block on an *IC* chip.

Was it really impossible or hopeless to apply *IC* technology into *RF* circuit design? In other words, was it impossible to have *RFIC* appearing in the world?

16.4 INTERFERENCE GOING ALONG AN *IC* SUBSTRATE PATH

As mentioned above, the main interference in an *RF* module is from the sky. In the *RFIC*, cases are just the opposite. The interference from the sky is minor but interference through the substrate is significant. In order to explore the possibility of implementing an *RFIC*, we must focus on the study of interference through the substrate.

16.4.1 Experimentation

Special experiments have been conducted to study the interference path along the *IC* substrate. Figure 16.3 shows the vertical profile of an *IC* die sample in such an experiment.The purpose of this experiment is to provide an interference source

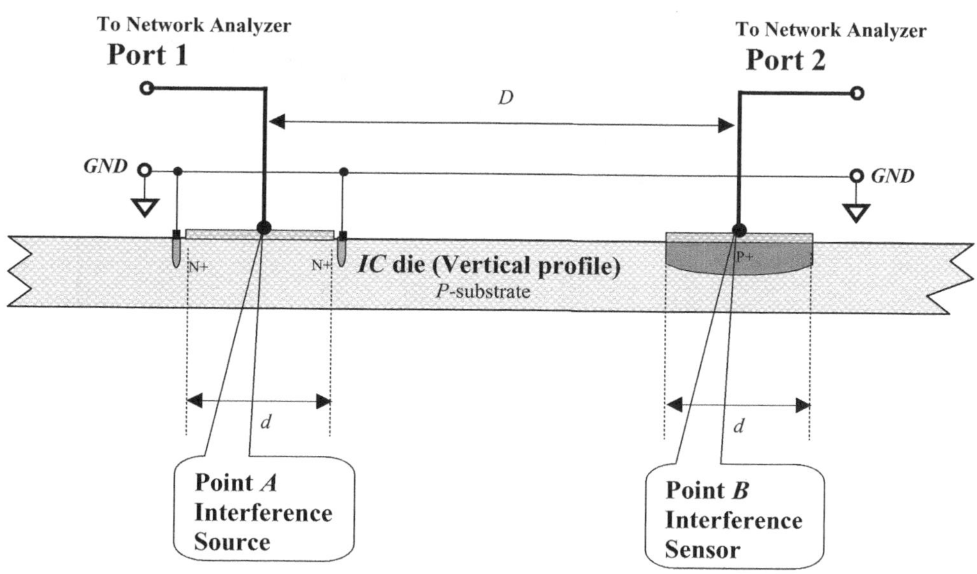

Figure 16.3 *IC* die sample for experiment of interference from substrate.

N+ P substrate
P+ Contact

to a specific *IC* substrate and to study its attenuation as it moves along the substrate path.

Point *A* is an interference source and point *B* is the sensor for receiving the interference signal moving along the substrate path from *A* to *B*.

The source at point *A* looks like an *n*-channel *MOSFET* on the left hand side, which will be connected to the network analyzer port 1. The network analyzer at port 1 provides an *RF* signal to this *IC* die sample as an interference source. The sensor at point *B* is a *P*+ buried pad with a metallic contact on the right hand side, which will be connected to port 2 of the network analyzer and will function as an interference sensor. If the substrate is not a perfect insulator in the interference frequency, the interference signal from port 1 of the network analyzer will move along the substrate path from point *A* to *B* and be received by the sensor. The attenuation of the interference signal from *A* to *B* can be found from the reading of S_{21} in the network analyzer.

If the substrate were an ideal insulator, the attenuation would be infinite and the interference would completely disappear. Usually, the substrate is not an ideal insulator and the interference will appear with a certain amount of attenuation. As a matter of fact, the attenuation from source *A* to sensor *B* is a measure of isolation between the sensor and the source.

Quantitatively, the value of S_{21} depends on the distance *D* between the sensor and source, the size of the source contact area and the sensor contact area *d*, and other parameters of the *IC* die, such as the thickness of the substrate, the characteristics of the substrate, and so on. As an example, Figure 16.4 shows a measured

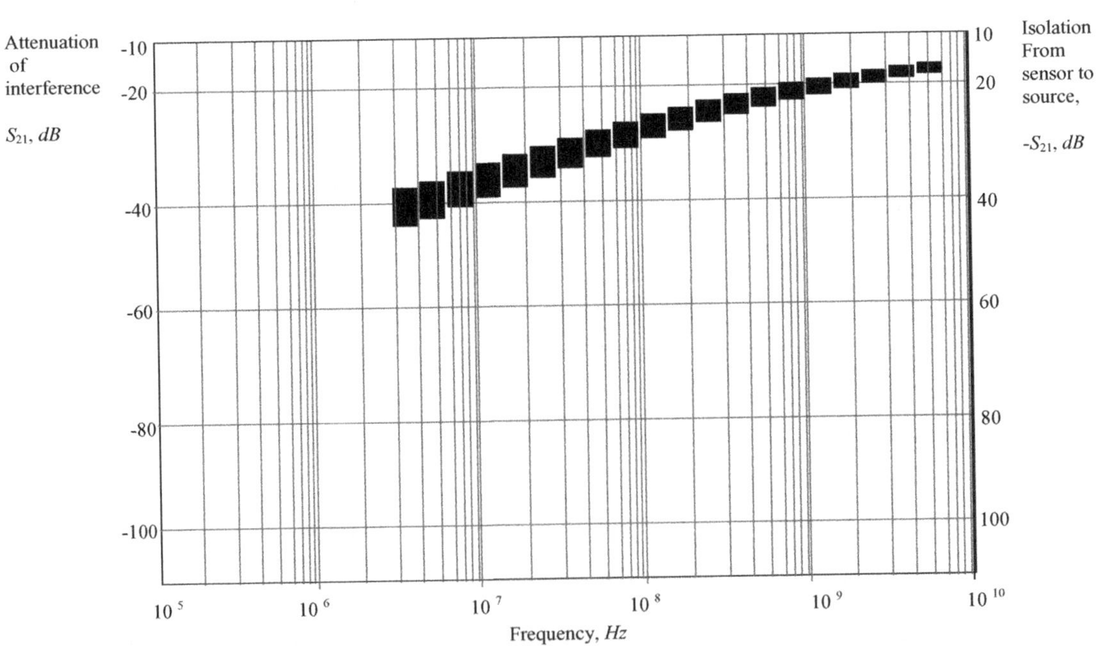

Figure 16.4 Measured attenuation of interference from source *A* to sensor *B* or isolation from sensor *B* to source *A* when $D \approx 150\,\mu m$, $d \approx 50\,\mu m$.

TABLE 16.1 Interference attenuation or isolation when interference signal goes along *IC* substrate path

$S_{21} \approx -40\,\text{dB}$	when $f = 10\,MHz$
$S_{21} \approx -30\,\text{dB}$	when $f = 100\,MHz$
$S_{21} \approx -20\,\text{dB}$	when $f = 1000\,MHz$

attenuation of the interference signal from source A to sensor B. Its negative value represents the value of the corresponding isolation. Both ordinates of attenuation and isolation are shown in Figure 16.4. In this measurement, the distance between the source and sensor is $150\,\mu m$ and the size of the source or sensor contact area is $50\,\mu m$. The experimental results are shown in Figure 16.4 and their approximate values are listed in Table 16.1.

Obviously, the interference going along the substrate path is significant in all of the *RF* frequency range. At the high *RF* frequency range, say, $f > 1000\,MHz$, attenuation from source to sensor, or isolation from sensor to source, is only $20\,dB$! The interference going along the substrate path will considerably disturb the performance of the *RF* circuit. We conclude that the *RF* circuitry implemented by *IC* technology is impossible if no special *IC* scheme is developed. Even in the low *RF* frequency range, say, $f \approx 10\,MHz$, the attenuation from source A to sensor B, or the isolation from sensor B to source A, is only $40\,dB$. This is better than that of the high *RF* frequency range, but not good enough to implement the *RF* circuit by *IC* technology.

16.4.2 Trench

The experimental results as shown in Figure 16.4 and Table 16.1 are very disappointing. However, after carefully re-examining the problem, one may find the solution by some other means.

The simplest solution to attenuate or remove the interference moving along the substrate path is to remove part of the substrate so as to cut off the path by which the interference moves along from the source to the sensor. This is called trenching of an *RF* block.

As a concrete example, let's do a trench for an *RF* block. That is, we dig a deep ditch circling around the *RF* block on *IC* substrate as shown in Figure 16.5. For simplicity, the area of the *RF* block and the deep ditch are simply represented by an ellipse. In a practical *IC* layout, the shape of *RF* block area and the ditch are irregular and, in addition, the ditch is broken into several segments because the *RF* block must be connected to outside blocks.

As shown in Figure 16.5, owing to the existence of the encompassing ditch, the external interference source is considerably blocked by the deep ditch in the path that the interference crosses over the ditch inward. Similarly, any internal interference source is considerably blocked by the deep ditch in the path that the interference must crosses over.

Experiments indicate that the trenching for an *RF* block is an effective means to suppress interference moving along the *IC* substrate path.

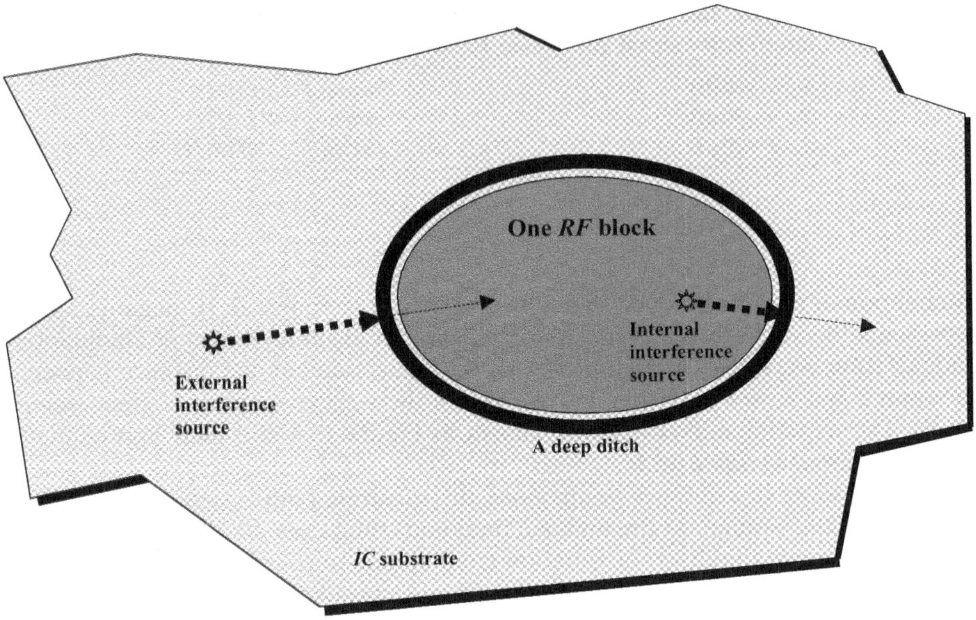

Figure 16.5 Trenching an *RF* block by digging a deep ditch surrounding the *RF* block.

Of course, the attenuation of the interference depends on the width and depth of the ditch. The wider and deeper the ditch, the more attenuated the interference. In practical layouts, the area of the *IC* die and its cost increase as the width of the ditch increases. Too wide a ditch and the cost may be pumped up to an unaffordable level. Additionally, the mechanical rigidity of the *IC* die is affected by the depth of the ditch. Too deep a ditch and the *IC* die may be easily damaged. Besides, this trenching increases the complexity of *IC* processing.

Nevertheless, in the early stages of *RFIC* development, much effort was placed on researching the topic of ditch width and depth.

16.4.3 Guard Ring

By means of trenching around an *RF* block, the fabrication of the *RFIC* becomes possible. However, considering the many drawbacks and complexities of trenching, we cannot but ask: Is it possible to replace trenching by other technology or scheme? The answer is yes.

Based on the experimental setup as shown in Figure 16.3, Figure 16.6 shows another experimental setup, in which a *P*+ guard ring is added which encompasses the interference source at point *A*. The purpose of this guard ring is to block the interference going along the substrate path from the interference source at point *A* to the interference sensor at point *B*. Hopefully, this guard ring will function similarly to a trenched ditch as mentioned above.

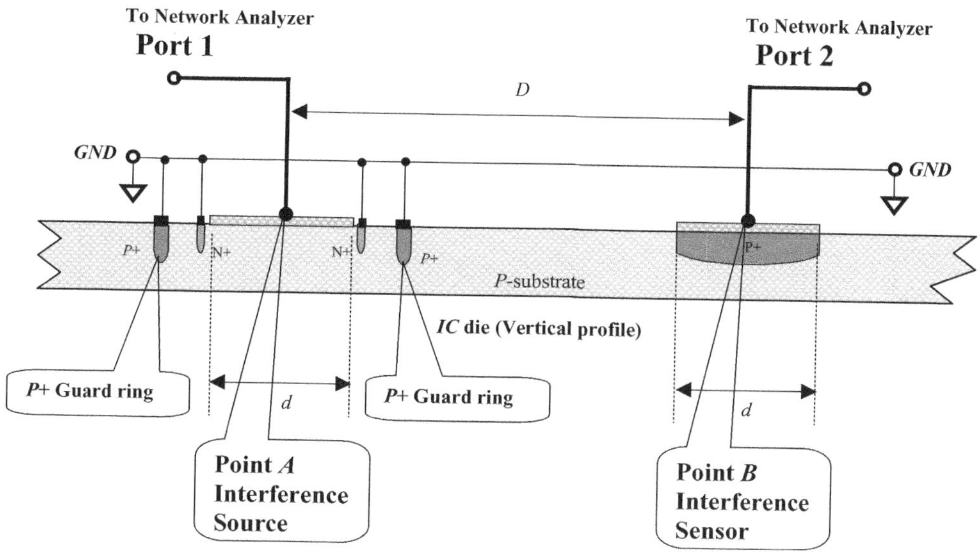

Figure 16.6 *IC* die sample for experiment of interference from substrate. Interference source is circled by a *P*+ guard ring.

N+   P substrate
P+   Contact

The experimental results with the *P*+ guard ring shown in Figure 16.7, are positive and encouraging. For comparison, the experimental results without the *P*+ guard ring shown in Figure 16.4 are also plotted in Figure 16.7.

It can be seen that the attenuation of the interference or isolation is significantly improved from the case without the *P*+ guard ring to the case with the *P*+ guard ring. Table 16.2 lists the approximate values of S_{21} at the frequencies, 10, 100, and 1000 *MHz*. The improvement of S_{21} is 30 to 40 *dB* at 10 *MHz*, 25 to 30 *dB* at 100 *MHz*, and 20 *dB* at 1000 *MHz*.

In the past decade, a lot of effort has been put into research on the guard ring, including its type, shape, width, spacing, multiple rings, and distance between multiple rings. An *IC* die with a guard ring has a better mechanical rigidity than an *IC* die with a trench. At the same time, the width and depth of a guard ring is easier to control than that of a trenching ditch in the *IC* processing.

At present, most trenching processing has been replaced by a *P*+ guard rings and a deep *N*–well. Figure 16.8(a) and (b) shows the *RF* block encircled by a trenching ditch and (*P*+ and deep *N*–well) guard rings, respectively. Experiments indicate that the attenuation of the interference, or isolation, is increased if the width of the guard ring or the spacing between the guard rings is increased. Circuit designers can select the appropriate width and spacing for guard rings based on the specific requirements of the attenuation of the interference or the isolation.

In order to keep the size of the *IC* die small so as to keep the cost low, the width and spacing of guard rings are confined to less than 15 μm. A set of typical values

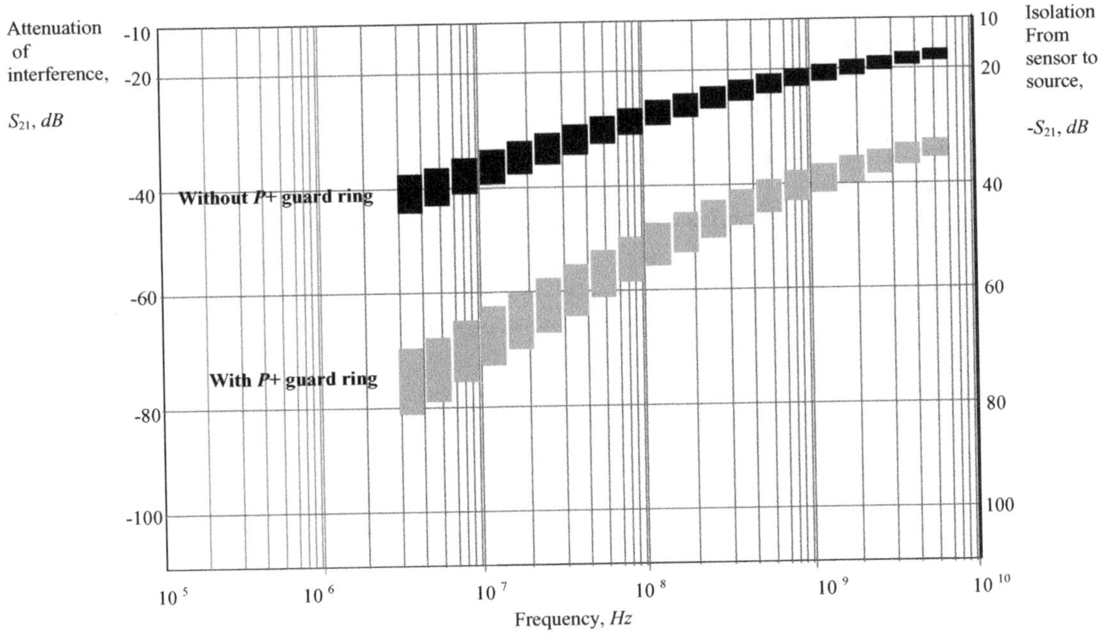

Figure 16.7 Measured attenuation of interference from source A to sensor B or isolation from sensor B to source A when $D \approx 150\,\mu m$, $d \approx 50\,\mu m$.

TABLE 16.2 Comparison of interference attenuation or isolation between the cases with and without *P*+ guard ring

Without P+ guard ring	With P+ guard ring	Frequency
$S_{21} \approx -40\,dB$	≈ -80 to $-70\,dB$	$10\,MHz$
$S_{21} \approx -30\,dB$	≈ -60 to $-55\,dB$	$100\,MHz$
$S_{21} \approx -20\,dB$	$\approx -40\,dB$	$1000\,MHz$

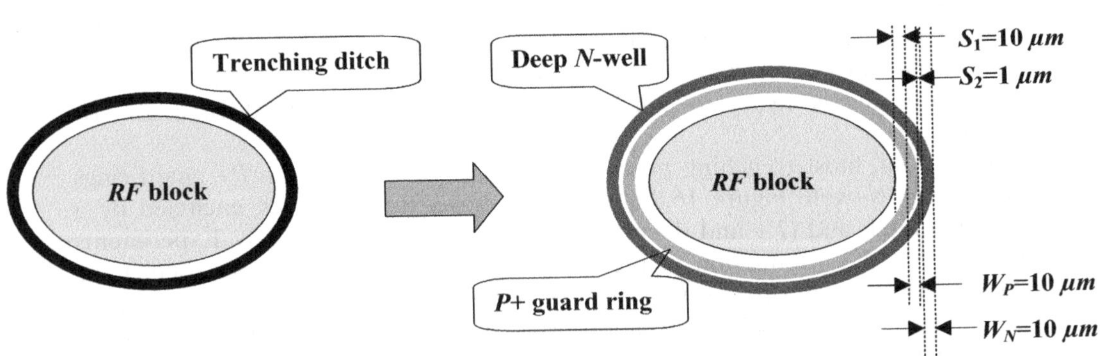

(a) Trenching ditch (b) *P*+ and deep *N*–well

Figure 16.8 Trenching ditch is replaced by *P*+ and deep *N*–well.

TABLE 16.3 Typical width of guard ring and spacing between guard rings in a *RFIC* layout

Item	Value
Spacing between *RF* block and *P*+ guard ring	$S_1 = 10\,\mu m$
Spacing between *P*+ guard ring and deep *N*–well	$S_2 = 1\,\mu m$
Width of *P*+ guard ring	$W_P = 10\,\mu m$
Width of deep *N*–well guard ring	$W_N = 10\,\mu m$

is marked in Figure 16.8 and listed in Table 16.3, which is applied in *RFIC* layout currently.

16.5 SOLUTION FOR INTERFERENCE COMING FROM THE SKY

As mentioned above, the interference from the sky in the *RFIC* die is minor because the substrate area for the *RF* block is small. Not too many electric lines bend into the *RF* block from the sky which could disturb the performance of the circuitry. Nevertheless, it should still be reduced or removed as much as possible.

In the discussion about *PCB* in Appendix 15.A.1, it was pointed out that the rectangular metallic grounded frame functions not only as a grounded surface but also partially plays a role as shielding. Recall that I said that most of electric lines produced and then radiated from the circuit inside the *PCB* would be terminated on the rectangular grounded frame. On the other hand, most of electric lines from any interference source outside the *PCB* would be stopped on the rectangular grounded frame as well. We say "most of electric lines" but not "total electric lines" because some of them would cross over the barrier of the rectangular grounded frame since the environment of the test *PCB* is somewhat complicated rather than a free space. Nevertheless, the isolation between the circuits inside and outside the rectangular grounded frame is greatly improved due to the existence of the rectangular grounded frame. This is a very simple but very powerful idea, by which the development of the *RFIC* becomes realistic. We now apply this idea to the *RFIC* fabrication. Connecting the *P*+ guard ring to the ground as shown in Figure 16.9, helps to reduce interference from the sky. Most of the electric lines either radiated from the *RFIC* die internally or radiated to the *RFIC* die externally would terminate on the grounded *P*+ guard ring.

Based on the same principle, the deep *N*–well guard ring is connected to V_{dd} or V_{cc}. The voltage drop between the *P*+ and *N*–well guard rings further improves the attenuation of the interference, or isolation. Figure 16.10 illustrates these two guard rings and marks their widths and spacings.

In terms of *P*+ and *N*–well guard rings, and by connecting them to the ground and the V_{dd} or V_{cc}, respectively, the problem of interference or isolation between *RF* blocks is basically solved. Thus, *RFIC* fabrication becomes realistic. This milestone opened a new era in the history of the electronic enterprise.

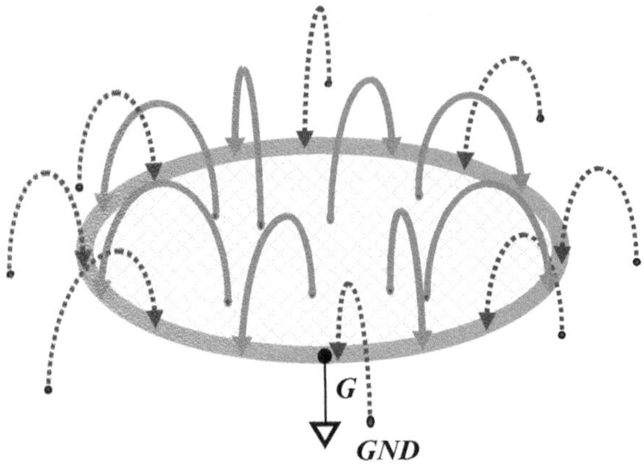

Figure 16.9 Electric lines either radiated from *RFIC* die internally or radiated to *RFIC* die externally will be terminated on the grounded guard ring.

Electric line from inside | *P*+ guard ring
Electric line from outside | *RFIC* — One *RF* block

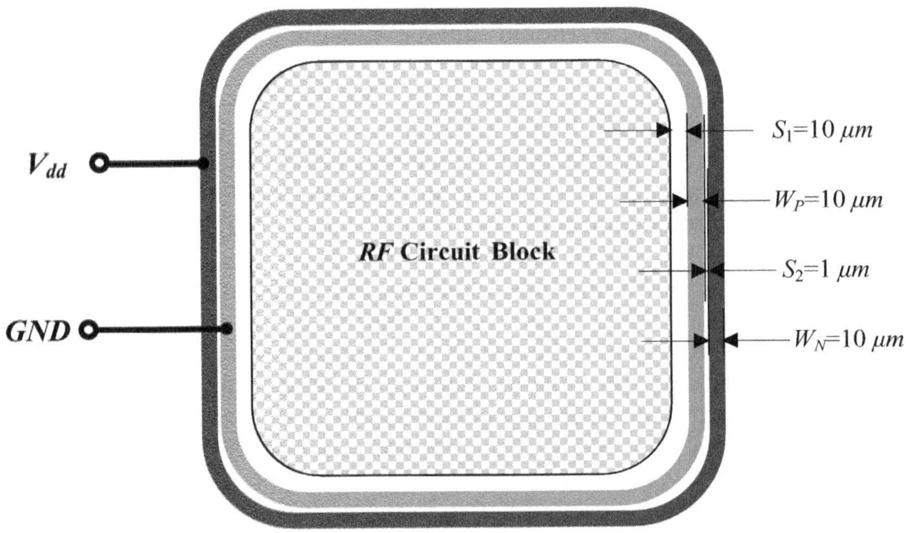

Figure 16.10 Typical width, spacing, and connections of *P*+ and *N*–well guard ring.

P+ guard ring | *N*–well guard ring

16.6 COMMON GROUNDING RULES FOR AN *RF* MODULE AND *RFIC* DESIGN

In addition to solving the problem of interference or isolation between *RF* blocks, some common grounding rules must be followed for both *RF* modules and *RFICs*.

16.6.1 Grounding of Circuit Branches or Blocks in Parallel

In Chapter 15, grounding of a *PCB* was discussed. In order to reduce or avoid return current coupling between multiple circuit branches or blocks, some long slots were inserted between the circuit branches or blocks so that the grounding surfaces of individual circuit branches or individual blocks were separated from one another.

In *RFIC* design, the grounding of circuit branches or blocks must also be separated so as to reduce or avoid return current coupling between multiple circuit branches or blocks.

As matter of fact, "grounding separately" is equivalent to "grounding in parallel." Its opposite would be "grounding stacked together" or "grounding in series." In an *RFIC* chip or a *PCB* containing multiple circuit branches or multiple blocks, grounding of circuit-branches or blocks must be connected separately but not stacked together, that is, in parallel but not in series. Figure 16.11 shows correct and incorrect grounding methods.

16.6.2 *DC* Power Supply to Circuit Branches or Blocks in Parallel

As mentioned in Chapter 14, the *DC* power supply point is a half ground point. Similarly, the *DC* power supply to a *PCB* was discussed in Chapter 15. In order to reduce or avoid forward current coupling between multiple circuit branches or blocks, the *DC* power supply must be provided separately or in parallel.

In *RFIC* design, the *DC* power supply to circuit branches or blocks must also be separated or provided in parallel so as to reduce or avoid forward current coupling between multiple circuit branches or blocks.

As matter of fact, the statement "*DC* power supply provided separately" is equivalent to "*DC* power supply provided in parallel." Its opposite is "*DC* power supply provided stacked together" or "*DC* power supply provided in series." In an *RFIC* chip or a *PCB* containing multiple circuit branches or multiple blocks, the *DC* power supply to the circuit branches or blocks must be connected separately but not stacked together, that is, in parallel but not in series. Figure 16.12 shows correct and incorrect methods of providing the *DC* power supply.

16.7 BOTTLENECKS IN *RFIC* DESIGN

16.7.1 Low *Q* Inductor and a Possible Solution

In *RFIC* design, the inductor is built on the *IC* substrate. On a two-dimensional plane, it cannot but take on a spiral configuration. Figure 16.13 shows a typical spiral inductor on the *IC* substrate and its equivalent model is shown in Figure 16.14.

There are three problems affecting a spiral inductor:

1) Extremely low Q value: In respect to L_S, the value of R_S is usually high, so that the Q value of the spiral inductor is low.
   In the *RF* frequency range, the Q value of the spiral inductor is about 10 or so at present in most foundries. In an *RF* circuit design with discrete parts, the Q value of inductors is generally over 100. For example, the Q value of a chip inductor is around 120. This is the biggest bottleneck in *RFIC* design.

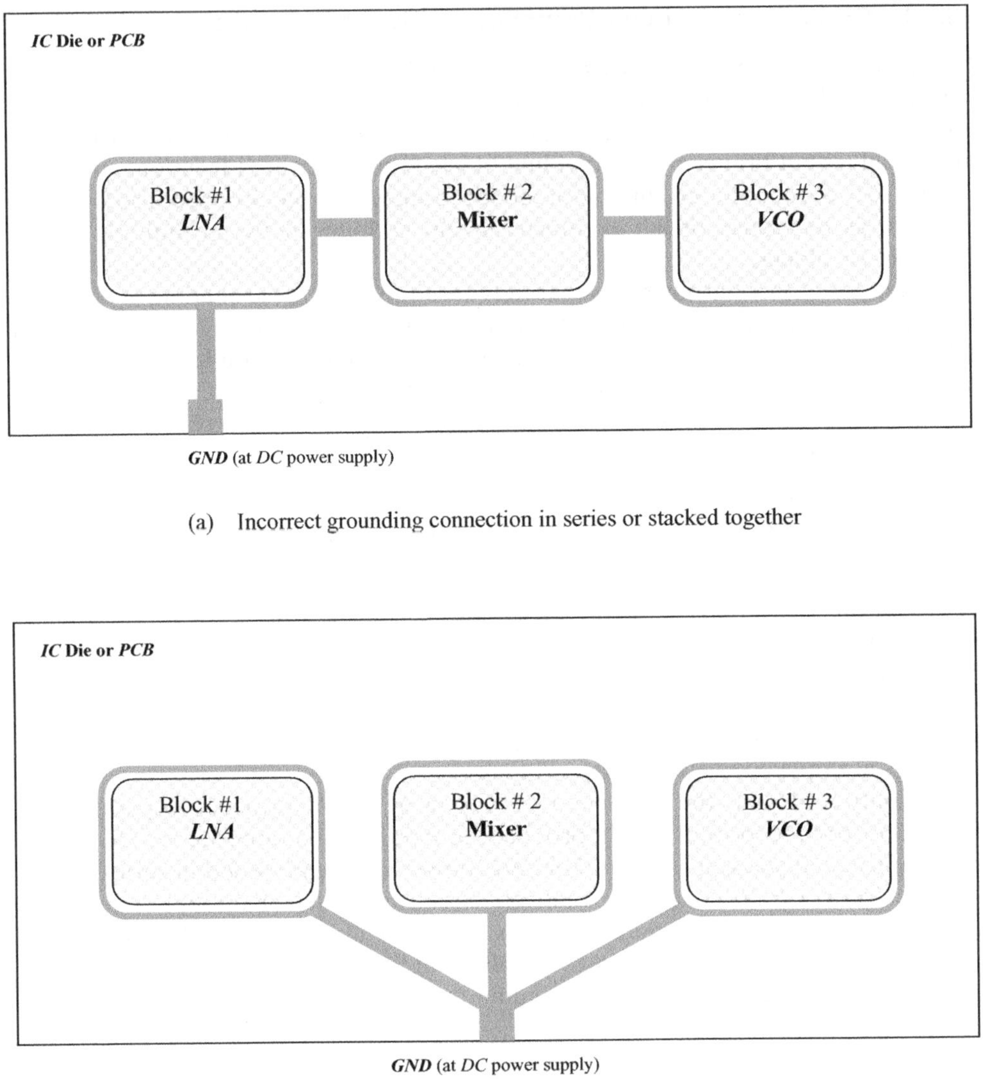

(a) Incorrect grounding connection in series or stacked together

(b) Correct grounding connection in parallel or separate

Figure 16.11 Grounding connection for multiple circuit branches or blocks. ▢ Ground ring

2) Large area: The spiral inductor's size is usually greater than 100 μm^2. This area is more than ten or hundred times that of a resistor. The large area impacts the cost of the *IC* chip directly since the cost of *IC* chip is directly proportional to its die area.
3) It is easily interfered with by other interference sources and, additionally, is an interference source to disturb other parts or blocks, because its size is large and its magnetic flux or electric lines are open wide to the sky.

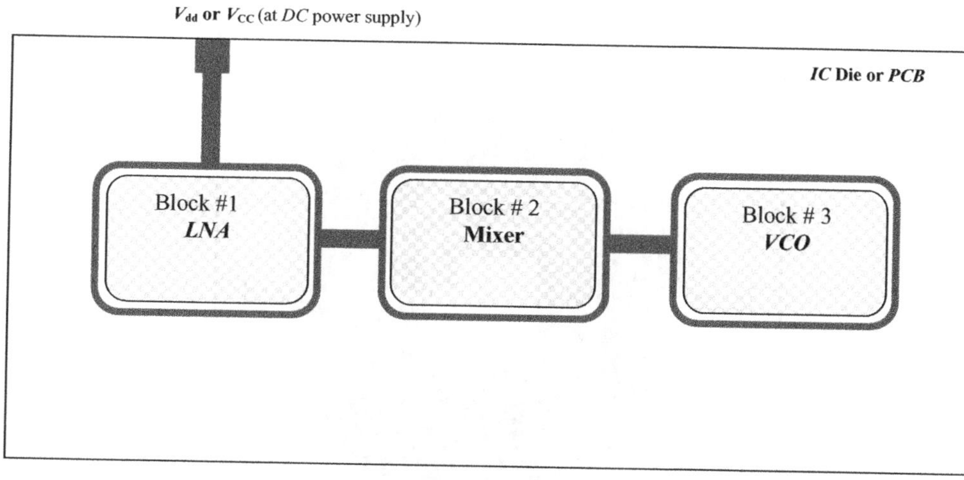

(a) Incorrect *DC* power supply connection in series or stacked together

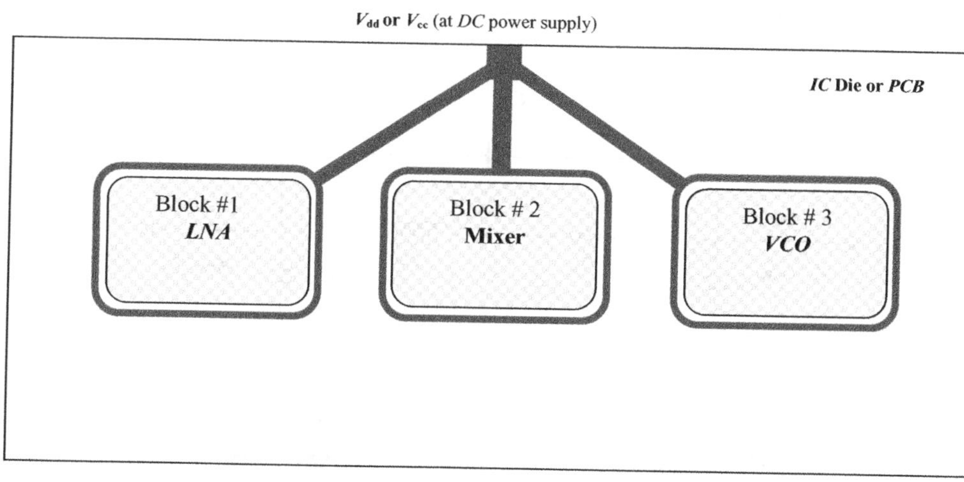

(b) Correct *DC* power supply connection in parallel or separate

Figure 16.12 *DC* power supply connection for multiple circuit branches or blocks.
□ *DC* power supply ring (*N*–well guard ring in *IC* die)

It seems that we cannot do too much about the last two problems, except that the interference could be somewhat alleviated by a reasonable layout. We will focus on the low *Q* problem, which leads many problems in *RFIC* design such as:

- It is impossible to build a filter directly on an *IC* chip. In the fabrication of a filter, the *Q* value of the inductor must be around 100 or higher. Otherwise, the insertion loss of the filter would attain an unacceptable level.
- Besides the filter, some other *RF* circuit blocks cannot be built on *RFIC* chip due to the low *Q* value of inductor. For example, the input impedance matching

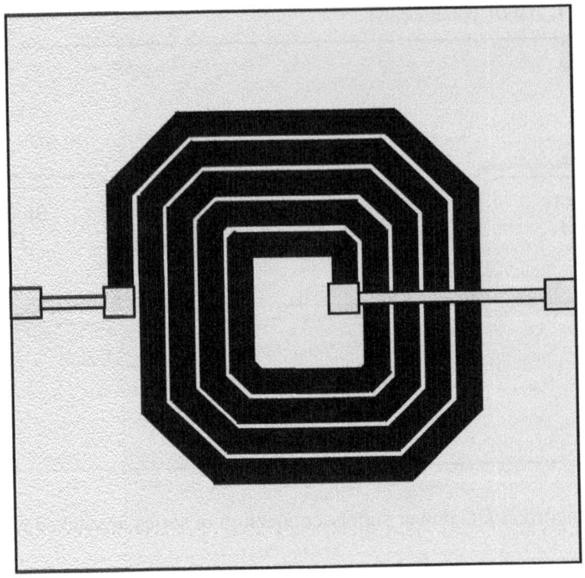

Figure 16.13 Spiral configuration of an inductor in *IC* chip.

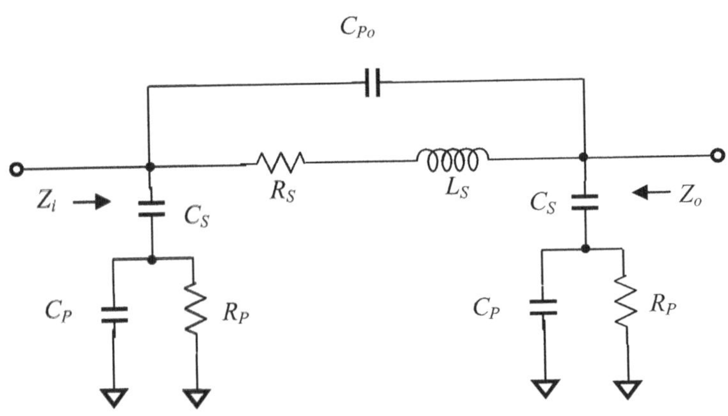

Figure 16.14 Model of a spiral inductor.

network of the *LNA* may not be built on the *IC* chip because the low *Q* value of the inductor could change a *LNA* (Low Noise Amplifier) to a *HNA* (High Noise Amplifier).

- An off-chip inductor must be applied and connected to the *RFIC* chip if a high *Q* value of inductor is required in the circuitry. A pair of input and output bonding wires, pads and pins for an off-chip inductor must be provided. The bonding wires and pads bring about serious uncertainty and inconsistency in the production line. An *IC* package cannot afford too many additional bonding wires and pins. For instance, if 10 off-chip inductors were needed, the additional number of bonding wires and pins would be greater than 20.

In the last decade, some scientists and engineers have attempted to avoid in the use of inductors in certain *RF* circuit designs such as the filter design. For instance, the log-domain filtering scheme is one of the technologies in the constructing of a filter by only devices and capacitors. Unfortunately, this is only realistic when the operating frequency is low, say, below 1 *GHz*. Second, this filter is an active circuit so that the cost of *DC* current or power consumption must be paid. Similar to the log-domain filtering technology, much effort has been put on the development of the equivalent inductor circuit, the gyrator. The restrictions or jeopardizes of the gyrator are the same as those of the log-domain filtering technology.

The inductor is one of the indispensable parts of an *RF* circuit design because it is a phase-shift part. To avoid the use of inductors in an *RFIC* is almost impossible. People have been putting a lot of effort into enhancing the Q value of the spiral inductor. Unfortunately, it looks like not too much exciting progress has been made up to date.

Why is the Q value of a spiral inductor so low? Many assertions have been presented and many experiments have been conducted. They can be outlined as follows:

1) Skin effect

Some people believe that the low Q value of the spiral inductor is due to the skin effect on the spiral metallic loop. The depth of the skin effect for copper can be evaluated and is found approximately,

$$\delta \approx 0.66\,\mu m, \qquad \text{when frequency} = 10\,GHz, \tag{16.6}$$

and

$$\delta \approx 6.6\,\mu m. \qquad \text{when frequency} = 100\,MHz. \tag{16.7}$$

On the other hand, the thickness of the metal layer in *IC* is in the order of

$$T \sim 0.1\,\mu m. \tag{16.8}$$

One might conclude that the low Q value of the spiral inductor on an *IC* chip is due to the fact that the metal layer is too thin when the operating frequency is below 10 *GHz*.

On the basis of such a judgment, many experiments in which the wire thickness of the spiral inductor is increased have been conducted in order to enhance the Q value of *IC* inductors. Unfortunately, the Q value of *IC* inductors is not increased conceivably even though the thickness of the metallic wire is piled up to several μm.

This indicates that the thinness or thickness of the metal layer is not the main reason that brings about the low Q value of the *IC* spiral inductor.

2) Attenuation due to the existence of substrate

The metallic winding of an *IC* spiral inductor is laid on a substrate. The eddy current may appear on the substrate, which leads to the attenuation of the *RF* signal. On the other hand, the electric permittivity of the substrate, ε_s, is generally much higher than the permittivity of the free space, so that the electromagnetic field as well as its power in the substrate is much stronger than in the free space. The electromagnetic field can be stored, propagated, and consequently attenuated in the substrate.

Therefore, the substrate may be the main factor causing the low value of the *IC* spiral inductor.

An outstanding experiment was conducted at UCLA, in which the substrate beneath the spiral inductor was dug away. This is a difficult and highly technical task. Professors and students, scientists and engineers had been working hard and finally completed this special treatment successfully. The Q value of this spiral inductor did indeed increase, but, unfortunately, it was still not high enough to satisfy most *RFIC* designs.

In recent years, the micro-machined *RF* inductor was developed and reported in *RFIC Circuit Design*. The idea is basically the same as the UCLA one. In addition to the highly technological work involved in the *IC* processing, the high cost is another problem.

3) Flux leakage

Flux can leak from the gap between windings. The flux leak implies energy loss or attenuation of signal power. If the gap between windings is reduced or narrowed, the flux leak would be reduced and the Q value of the spiral inductor should be increased. Unfortunately, the increase of the Q value through this method is very small.

Flux can be emitted from the gap between the windings to the sky. The flux escape also implies energy loss or attenuation of signal power. A "sandwich" experiment was conducted, in which the spiral inductor was enclosed by a magnetic-shielding material. The purpose of this experiment was to enhance the Q value of the spiral inductor by stopping the flux escaping to the sky. Unfortunately, the Q value of the spiral inductor did not show a large difference before and after the spiral inductor was "sandwiched."

Therefore, we again concluded that the low Q value of the spiral inductor was not mainly due to the flux leakage or escape from the gap between the windings to the sky.

4) Flux cancellation

Owing to the inherent drawback of the spiral configuration, the cancellation of flux between the windings is significant. As shown in Figure 16.15, in the spacing between two windings, the flux produced by the inner winding is cancelled by the flux produced by the outer winding. The flux cancellation seems to be more important than the flux leakage to the reduction of the Q value.

So far, many theories have been presented and many experiments have been conducted. There may be multiple reasons for the low Q value of spiral inductors. However, flux cancellation seems to be the most important.

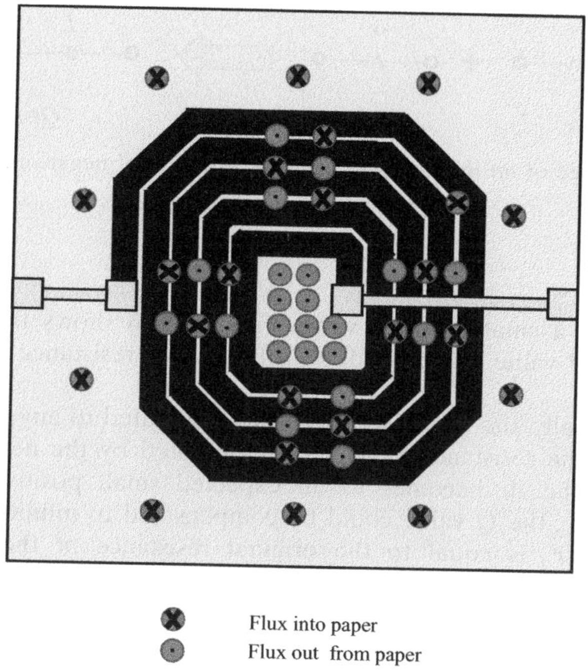

Figure 16.15 Cancellation of flux between two windings.

If so, then the low Q problem inherently exists due to the spiral configuration.

If the spiral configuration of the inductor is kept unchanged, a possible solution for the low Q problem is to compensate for its Q value by adding negative resistance into the spiral inductor. It is well known that the Q value of an inductor L can be expressed by its reactance $L\omega$ divided by an equivalent resistance r in series, that is,

$$Q = \frac{L\omega}{r}. \tag{16.9}$$

If a resistor with negative resistance r' is connected to this inductor L in series, the Q value will be changed to

$$Q = \frac{L\omega}{r - r'} = \frac{L\omega}{\Delta r}, \tag{16.10}$$

where

$$\Delta r = r - r'. \tag{16.11}$$

Theoretically, Q can be enhanced to any value as needed, as long as the resistance r can be compensated for by the negative resistance r' so that Δr

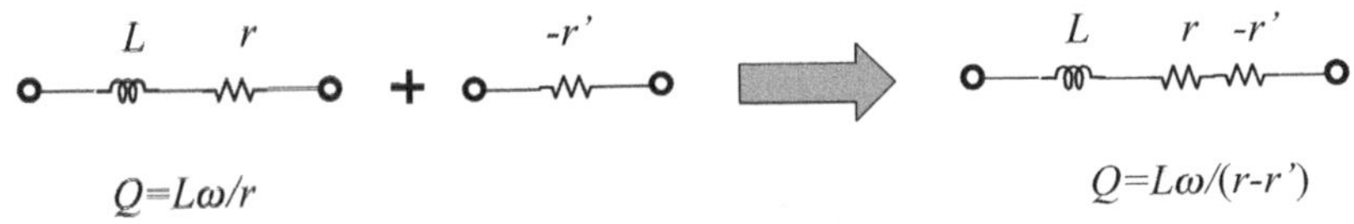

Figure 16.16 Change of an inductor's Q value by addition of negative resistance into an inductor in series.

stays below a small positive value. Figure 16.16 shows the change of an inductor's Q value by the addition of negative resistance into inductor in series.

Theoretically, the Q value could be compensated to any value as needed, as long as the resistance r can be compensated by the negative resistance r' so that the Δr becomes to an expected small positive value. In the extreme case, the Q value could be compensated to infinity if the negative resistance, $-r'$, is equal to the original resistance of the inductor, r, in magnitude.

However, it is not so simple in actual engineering design. The difficult points are

- Generating a negative resistance,
- Ensuring that there is not negative resistance outside the expected bandwidth,
- Keeping the remaining negative resistance inside the bandwidth below a small positive value,
- Reducing current consumption of generating negative resistance, which is usually done by an active device,
- Handling the noise generated due to the existence of the active device.

16.7.2 "Zero" Capacitors

The second bottleneck in *RFIC* design is the development of a "zero" capacitor directly on the *IC* chip. In the *RF* module design with *PCB* testing, "zero" capacitors can be selected from chip capacitors based on their self-resonant principle. However, the chip "zero" capacitor cannot be directly applied to the *IC* chip due to its large size.

As mentioned in Section 14.4.6, when a "zero" capacitor is needed in an *RF* block fabricated by *RFIC*, the circuit on the *IC* chip must jump out of the *IC* chip to connect to the "zero" capacitor by bonding wires. This brings about much inaccuracy due to the addition of the bonding wires and the additional runners. Second, if many off-chip "zero" capacitors are needed and the *IC* chip is packaged with multiple pins, the number of pins required may be unacceptably high. Finally, if many off-chip "zero" capacitors are needed, the cost will be raised considerably.

Therefore, it is strongly desirable to develop a "zero" capacitor directly on the *IC* chip. Hopefully, progress in such a development will appear in the near future.

16.7.3 Bonding Wires

The bonding wire may create other problems for the *RFIC* designer. Two ends of a bonding wire are connected to the pads, which are the points of impedance discontinuity, or to the terminals where the impedance is unmatched. In order to ensure the impedance matching or smooth transition, an accurate model of the bonding wire is much in demand. Unfortunately, the current models of bonding wire provided by foundries are very inaccurate.

The bonding wire is a very unstable part. The inaccuracy of the model is due to many factors: for example, bonding wires with the same length could have different heights or different tilted angles in the air. When it is soldered on the pad, the uncertainty of the soldered point may cause the inaccuracy of the model.

16.8 PROSPECT OF *SOC*

The development of the *RFIC* was a milestone in the history of the electronic enterprise. The next milestone will be the *SOC* (System On Chip). On a *SOC* chip, not only are all *RF* circuit blocks integrated on the same chip substrate, but so are all the digital and analog circuit blocks. In short, all the circuit blocks of a system are integrated together as an *IC* chip.

The *SOC* is the highest goal in circuit design. We face a lot of challenging tasks today in trying to reach the *SOC* design goal, though many positive achievements have been attained. The challenges are as follows.

16.8.1 Remove All Bottlenecks in *RFIC* Design

The main bottlenecks in *RFIC* design are:

- Enhancing the low Q value of the spiral inductor,
- Developing a "zero" capacitor directly on the *RFIC* chip,
- Modeling the bonding wire with higher accuracy.

16.8.2 Continue to Study Isolation

- Study isolation between *RF* blocks.
  By means of double guard rings, $P+$ and N–well, the isolation between *RF* blocks can reach 70 to $90\,dB$ usually, depending on the width and spacing between rings. This is basically satisfactory in the fabrication of *RFIC*s for general purposes, such as in cellular phones or portable radios. However, this is not satisfactory for the fabrication of those *RFIC*s needed for the most advanced electronic products, such as 128 *QAM* or military communication systems.

- Study isolation between digital blocks.
  In contrast with the *RF* signal, the digital signal contains many abundant harmonics. Isolation between digital blocks with high data rate is important because the frequency of their harmonics can be much higher over the *RF* frequency range.
- Study isolation between *RF* and digital blocks.
  On average, the power of the *RF* block is much higher than that of the digital block since the current in the *RF* block is in the order of *mA* while the current in the digital block is μA. From the power viewpoint, a digital block more easily disturbed than an *RF* block. From the frequency viewpoint, an *RF* block is more easily disturbed than a digital block because a digital signal is basically a pulse which contains many harmonics. A narrow pulse is a wideband signal. It contains high frequencies, including *RF* components, even if its pulse repetition rate is lower than the *RF* frequency. Therefore, isolation between *RF* and digital blocks is somewhat more complicated than other isolation cases.

In the wireless communication system, the ideal isolation level should be around $130\,dB$, which is $10\,dB$ higher than the total gain of the useful signal from the antenna to the data output terminal. Assuming that an interference with $0\,dB_m$ power appears somewhere in the system, the maximum power of interference at at the antenna is less than $-130\,dB_m$ because the isolation in the system is $130\,dB$. The interference with $-130\,dB_m$ power at the antenna is lower than the sensitivity, say, $-120\,dB_m$, of a communication system. Consequently, the communication system never perceives the existence of the interference. Of course, if the power of the interference is much greater than $10\,dB_m$, then $130\,dB$ of isolation is not enough for the system; however, cases like this seldom occur.

16.9 WHAT IS NEXT?

Digital and analog *IC*s have been developing since the 1960s and the *RFIC* has been developing worldwide since 1995. For the reduction of cost and size and the enhancement of reliability, our goal is the development of the *SOC IC*, which gained common-recognition or common-agreement over the world in the 1990s. A *SOC IC* chip is supposed to contain three kinds of integrated circuits together, the digital *IC*, analog *IC*, and *RFIC*.

Today, we are just on the way to approaching the goal of the *SOC*, which is expected to be reached in the near future if all the tasks listed in Section 16.8 are fulfilled.

As a matter of fact, the concept of the *SOC* concerns only hardware but not software. What will be the relationship between hardware and software in the future? This is a very important and interesting question in the circle of electronic enterprises. Let's take a communication system as an example and try to figure it out in Figure 16.17.

As shown in Figure 16.17, a *SOC* communication system consists of 10 big blocks. They are:

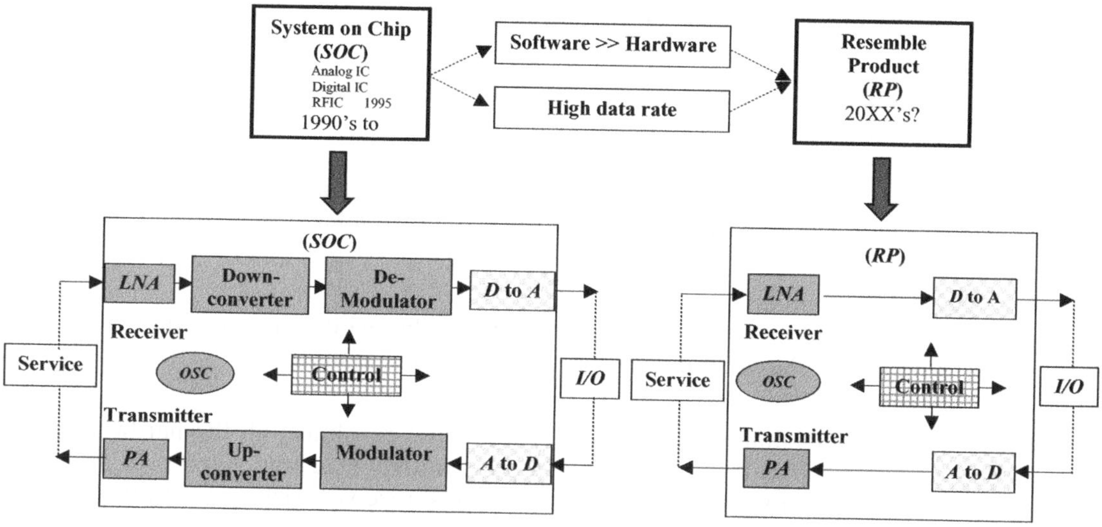

Figure 16.17 Prospect of a communication system.
Hardware
Hardware-Software
Software

Receiver:

- *LNA* (hardware)
- down-converter, (hardware)
- demodulator, (hardware)
- *D* to *A* converter; (software-hardware)

Synchronization and Control

- oscillator, (hardware)
- μP control; (software)

Transmitter

- *A* to *D* converter, (hardware-software)
- modulator, (hardware)
- up-converter, (hardware)
- *PA* (hardware)

In recent years, replacement of hardware by software is a general and irresistible tendency. The reasons are very simple. The software is much superior to the hardware in many aspects, such as

- Lower cost
  The physical cost of software is merely that of memory chips, while the cost of hardware is that of a bunch of parts, devices, *PCB*, and so on. The big difference of cost between software and hardware is obvious.
- Smaller physical size
  The size of software is only the size of the memory chip it resides in, while the size of hardware is that of a bunch of parts, devices, *PCB*, and

so on. The big difference in size between software and hardware is also obvious.

- Higher reliability
  The software statement is executed by two symbols, "1" and "0," which can be distinguished with quite a high reliability, while hardware operates with voltage and current, which are very dependent on the impedance, type and size of device, and many environmental factors, so that its reliability is much lower than that of software.
- Power consumption
  Power consumption for software is almost negligible, while power consumption for hardware is the main power consumption of the system.

In terms of *DSP* (Digital Signaling Processing) software, almost all the hardware for gate circuits, such as the *AND* gate, *OR* gate, *NAND* gate, and *NOR* gate, are replaced by software. In terms of *DSP* software, most receiver back ends and transmitter front ends have been replaced by software.

Traditionally, the *QPSK* modulator and demodulator were built by hardware. Nowadays, software *QPSK* modulators and demodulators are available for application.

The transition of circuit blocks from hardware to software is restricted by the data rate or clock frequency. Should the data rate or clock frequency be high enough, so that within one cycle the micro-processor can sample an *RF* signal many times and thus can manipulate the *RF* signal effectively, more *RF* circuit blocks built by hardware can be replaced by software. Going along with this general tendency of the circuit block built by hardware being replaced by software, not only modulators and demodulators, but also filters, down-converters, up-converters, and other circuit blocks may be "swallowed up" by software sooner or later.

The "software radio" has sounded loudly since the 1990s. Yes, it represents the general tendency that circuit blocks built by hardware will be "swallowed" by software. However, is the "software radio" 100% implemented by software without any hardware? The answer is no!

No matter how sophisticated the software is, a very weak signal sensed by the antenna of the receiver is impossible to be intensified by software. The *LNA* cannot be replaced by software, but must be fabricated by hardware. Similarly, the *PA* cannot be replaced by software but must be fabricated by hardware. Furthermore, let's examine the total 10 big blocks of the *SOC* communication system as shown in Figure 16.17. Two converters, *D* to *A* and *A* to *D*, are transformers between external analog parameters and internal digital sequences. It is pointless to talk about replacement of hardware by software here because the blocks themselves function as the transition of software to hardware. The unique block, the oscillator, must be fabricated by hardware and is by no means built by software. Finally, the entire system must be controlled by the micro-processor controller, which is indeed a software block.

Consequently, the number of circuit blocks in a *SOC* communication system is revised from the existing 10 big blocks to the six big blocks, which are:

Receiver:

- *LNA*, (hardware)
- *D* to *A* converter (software-hardware)

Synchronization and Control

- Oscillator, (hardware)
- μP control, (software)

Transmitter

- *A* to *D* converter, (hardware-software)
- *PA* (hardware)

In Figure 16.17, the *SOC* system becomes the *RP* (Resemble Product). The *RP* integrated circuit consists of six big blocks as shown above. It can be produced as different products. The essential difference between its products is the software block, the μP control, while all the other hardware blocks basically "resemble" each other.

APPENDICES

16.A.1 Notes about RFIC Layout

16.A.1.1 Runner

- Length: A runner usually is a micro strip line. The existence of a runner does not impact the performance of the circuitry if the length of the runner is much shorter than a quarter wavelength of the operating frequency, that is,

$$l << \frac{\lambda}{4}, \tag{16.A.1}$$

In other words, if condition (16.A.1) is satisfied, the runner can be neglected in the simulation stage for the circuitry.

On the contrary, if the length of the runner as short is much shorter than a quarter wavelength of the operating frequency, that is,

$$l \sim \frac{\lambda}{4} \quad \text{or} \quad > \frac{\lambda}{4}, \tag{16.A.2}$$

the runner must be treated as a micro strip line and taken care of in the simulation stage for the circuitry.

It is therefore encouraged to keep the runner as short as possible so that the condition (16.A.1) is satisfied,

- Width: The width of a runner is directly related to its characteristic impedance. It is better to select the width of the runner so that its characteristic impedance is $50\,\Omega$ if the runner is not designed for special purpose. In input or output of block, the characteristic impedance must be $50\,\Omega$ so as to be able to match the impedance of the test equipment, which is usually $50\,\Omega$.

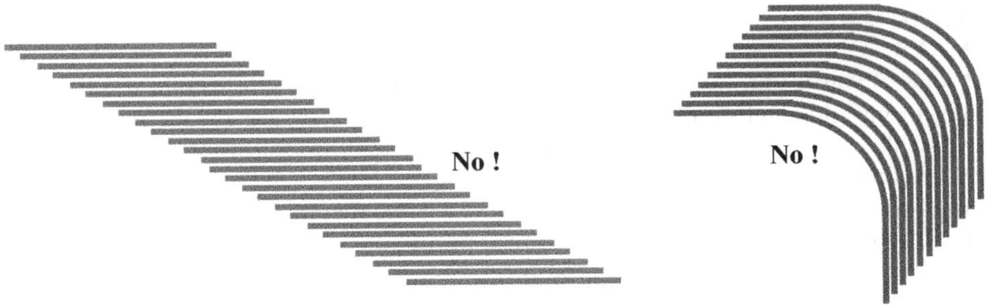

(a) Multiple runners in parallel! (b) Multi curves in parallel!

Figure 16.A.1 Multiple runners or multiple curves in parallel are not welcome.

- Multiple runners or curves in parallel: In order to reduce the area of *IC* die, it is quite popular on the digital *IC* layout with multiple runners or curves in parallel as shown in Figure 16.A.1.

However, this style of layout is absolutely not allowed to appear in the *RF* circuit layout, because it brings about the co-planar capacitance, mutual inductance, and so on, between runners, and hence brings about a strong coupling between runners and eventually, leads the malfunction of circuitry accordingly.

- Style of runner: In the layout for the digital circuitry, people like a layout with "nice looking" style. The runner is therefore aligned with "East-West, South-North" orientation, which looks like "Street, Boulevard" in a city traffic map. This is not the expected style in the layout for *RF* circuitry. Instead, the encouraged style of the runner is "as short as possible."
- Smoothness of the runner: the characteristic impedance of a runner is directly related to its width. Therefore, the width of a runner must not be changed suddenly so as to avoid the suddenly change of the characteristic impedance. A runner with its width being suddenly changed is equivalent to an impedance transformer. It is encouraged that the width of a runner must be changed gradually if the variation of a runner's width is necessary.
- Corner of the runner: A strong electromagnetic field appears in the adjacent area of a runner with a sharp angle where the *AC/RF* signal is going through the runner. The radiation of the strong electromagnetic field consumes the power of the signal as well as disturbs the signals in the circuitry. Therefore, one of the basic layout principles is to make the corner of the runner as smooth as possible.
- Placement of runners: If the runners are going along together in parallel, as mentioned above, it brings about co-planar capacitance, mutual inductance, and so on between the runners. Therefore, it is encouraged to lay out the runners not in parallel but in perpendicular if a *PCB* or an *IC* die has more than two metallic layers.
- Runners in parallel: The runners must be in parallel for a special purpose. The spacing between two runners in parallel, as shown in Figure 16.A.2, must be three times wider than the runner's width so that the electromagnetic interfer-

ence between two adjacent runners can be reduced below a conceivable level, that is,

$$S_{rr} > 3W. \quad (16.A.3)$$

- Similarly, as shown in Figure 16.A.3, the spacing between a runner and its adjacent ground surface in parallel must be greater than 3 times of the runner's width, that is,

$$S_{rg} > 3W. \quad (16.A.4)$$

- As shown in Figure 16.A.4, the spacing between a runner and its two adjacent ground surfaces in parallel must be greater than 3 times of the runner's width, that is,

$$S_{rgg} > 3W. \quad (16.A.5)$$

16.A.1.2 Parts

- Device: A *MOSFET* device may contain many fingers. All the fingers aligned together may become a long "dragon." The delay appears when an *AC/RF* signal goes along the "dragon" from one end to the other. Consequently, the

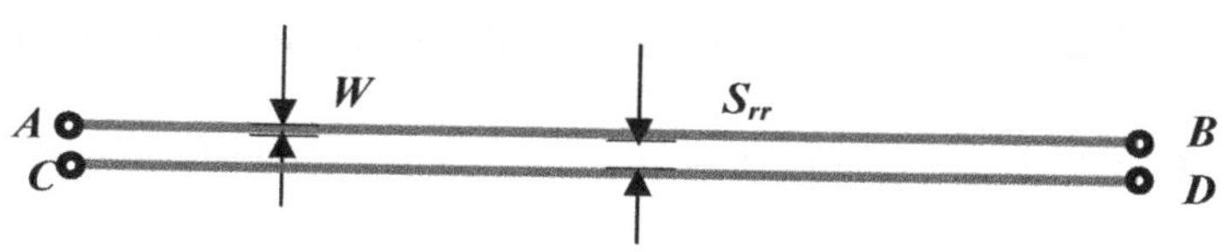

Figure 16.A.2 Spacing between two runners in parallel.

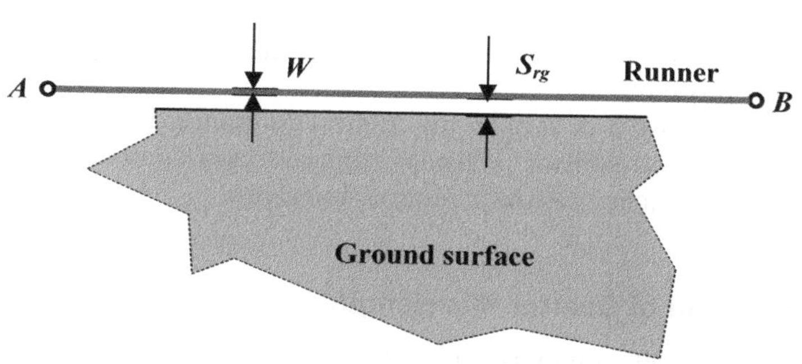

Figure 16.A.3 Spacing between runner and ground surface in parallel.

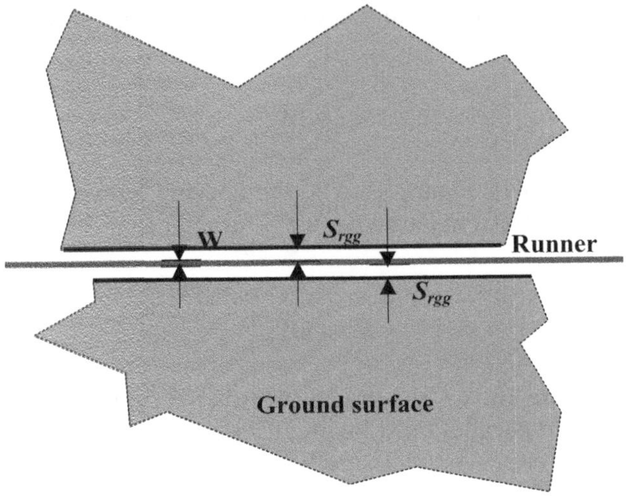

Figure 16.A.4 Spacing between runner and ground surface in parallel.

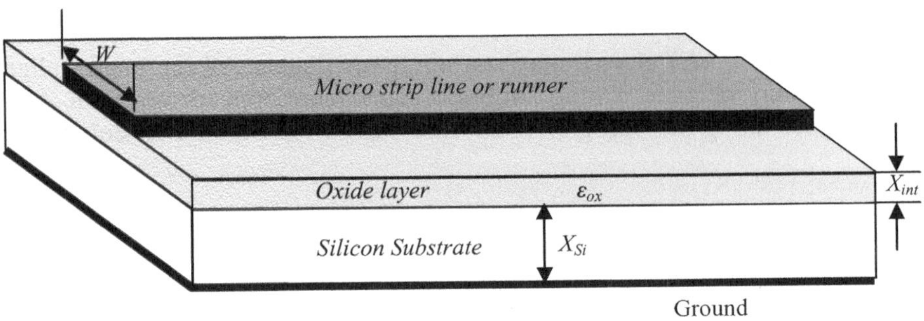

Figure 16.A.5 Various parameters of a micro strip line on silicon substrate.

phase shift appears from finger to finger so that it leads to the malfunction of the device. It is therefore encouraged to arrange all the fingers of the device in a "square" shape instead of a "dragon"!

- Inductor: Inductors in an *IC* die would occupy the largest area. Unfortunately more area must be provided for a circuit block that contains more than two inductors. The distance between inductors must be equal to at least two diameters' width of the spiral inductor. Otherwise, the strong coupling between the inductors will occur.
- Resistor: The resistor in *IC* has the highest tolerance. In order to reduce its tolerance, a desired resistor is always replaced by a combination of resistors with higher resistance but lower relative tolerance.

16.A.2 Calculation of Quarter Wavelength

Hannu Tenhunen (2000) summarized a set of equations for calculation of quarter wavelength in *CMOS* processing. Figure 16.A.5 shows the various parameters of a

TABLE 16A.1 Example of calculated quarter wavelength in *CMOS*

W μm	C pF/cm	L nH/cm	Zo Ω	f GHz	$\lambda/4$ mm	$\lambda/360$ μm	f GHz	$\lambda/4$ mm	$\lambda/360$ μm
1	1.45	16.59	106.8	1.0	16.09	178.83	2.4	6.71	74.51
2	2.24	15.20	82.4	1.0	13.55	150.58	2.4	5.65	62.74
3	3.00	14.39	69.3	1.0	12.04	133.79	2.4	5.02	55.74
4	3.73	13.82	60.9	1.0	11.01	122.34	2.4	4.59	50.98
5	4.45	13.37	54.8	1.0	10.25	113.84	2.4	4.27	47.43
6	5.17	13.01	50.2	1.0	9.64	107.15	2.4	4.02	44.65
7	5.88	12.70	46.5	1.0	9.15	101.70	2.4	3.81	42.37
8	6.58	12.43	43.5	1.0	8.74	97.13	2.4	3.64	40.47
9	7.28	12.20	40.9	1.0	8.39	93.22	2.4	3.50	38.84
10	7.98	11.98	38.8	1.0	8.08	89.82	2.4	3.37	37.42
11	8.68	11.79	36.9	1.0	7.81	86.82	2.4	3.26	36.18
12	9.38	11.62	35.2	1.0	7.57	84.16	2.4	3.16	35.06
13	10.07	11.46	33.7	1.0	7.36	81.76	2.4	3.07	34.07
14	10.77	11.31	32.4	1.0	7.16	79.59	2.4	2.98	33.16
15	11.46	11.17	31.2	1.0	6.99	77.62	2.4	2.91	32.34
16	12.16	11.04	30.1	1.0	6.82	75.81	2.4	2.84	31.59
17	12.85	10.92	29.2	1.0	6.67	74.14	2.4	2.78	30.89
18	13.54	10.81	28.3	1.0	6.53	72.60	2.4	2.72	30.25
19	14.23	10.70	27.4	1.0	6.41	71.17	2.4	2.67	29.65
20	14.93	10.60	26.6	1.0	6.29	69.84	2.4	2.62	29.10
25	18.39	10.15	23.5	1.0	5.79	64.29	2.4	2.41	26.79
30	21.84	16.79	21.2	1.0	5.41	60.07	2.4	2.25	25.03
35	25.30	16.48	19.4	1.0	5.10	56.72	2.4	2.13	23.63
40	28.75	16.21	17.9	1.0	4.86	53.97	2.4	2.02	22.49
45	32.21	8.98	16.7	1.0	4.65	51.66	2.4	1.94	21.53
50	35.66	8.77	15.7	1.0	4.47	49.68	2.4	1.86	20.70
55	39.11	8.58	14.8	1.0	4.32	47.96	2.4	1.80	19.98
60	42.56	8.40	14.1	1.0	4.18	46.45	2.4	1.74	19.35
65	46.01	8.24	13.4	1.0	4.06	45.11	2.4	1.69	18.79
70	49.46	8.09	12.8	1.0	3.95	43.90	2.4	1.65	18.29
75	52.92	7.96	12.3	1.0	3.85	42.81	2.4	1.61	17.84
80	56.37	7.83	11.8	1.0	3.76	41.82	2.4	1.57	17.42
85	59.82	7.71	11.4	1.0	3.68	40.91	2.4	1.53	17.05
90	63.27	7.59	11.0	1.0	3.61	40.08	2.4	1.50	16.70
95	66.72	7.48	10.6	1.0	3.54	39.31	2.4	1.47	16.38
100	70.17	7.38	10.3	1.0	3.47	38.59	2.4	1.45	16.08
200	139.17	6.00	6.6	1.0	2.74	30.39	2.4	1.14	12.66
500	346.18	4.22	3.5	1.0	2.07	22.98	2.4	0.86	9.57

Note: $X_{int} = 0.5\,\mu m$, $X_{si} = 500\,\mu m$, and $\varepsilon_{ox} = 3.45 \times 10^{-13}\,F/cm$.

micro strip line fabricated on silicon substrate in *CMOS* processing. It can be described by the distribution parameters, the capacitance per unit length in respect to the substrate, C_{msl}, and the self-inductance per unit length along the runner, L_{msl}.

$$C_{msl} = \varepsilon_{ox}\left[2.42 + \frac{W}{X_{\text{int}}} - 0.44\frac{X_{\text{int}}}{W} + \left(1 - \frac{X_{\text{int}}}{W}\right)^{6}\right],\ F/cm \qquad (16.A.6)$$

TABLE 16A.2 Example of calculated quarter wavelength in *CMOS* processing (continued)

w *μm*	C *pF/cm*	L *nH/cm*	Zo *Ω*	*f* *GHz*	λ/4 *mm*	λ/360 *μm*	*f* *GHz*	λ/4 *mm*	λ/360 *μm*
1	1.45	16.59	106.8	5.8	2.77	30.83	10.0	1.61	17.88
2	2.24	15.20	82.4	5.8	2.34	25.96	10.0	1.36	15.06
3	3.00	14.39	69.3	5.8	2.08	23.07	10.0	1.20	13.38
4	3.73	13.82	60.9	5.8	1.90	21.09	10.0	1.10	12.23
5	4.45	13.37	54.8	5.8	1.77	19.63	10.0	1.02	11.38
6	5.17	13.01	50.2	5.8	1.66	18.47	10.0	0.96	10.72
7	5.88	12.70	46.5	5.8	1.58	17.53	10.0	0.92	10.17
8	6.58	12.43	43.5	5.8	1.51	16.75	10.0	0.87	9.71
9	7.28	12.20	40.9	5.8	1.45	16.07	10.0	0.84	9.32
10	7.98	11.98	38.8	5.8	1.39	15.49	10.0	0.81	8.98
11	8.68	11.79	36.9	5.8	1.35	14.97	10.0	0.78	8.68
12	9.38	11.62	35.2	5.8	1.31	14.51	10.0	0.76	8.42
13	10.07	11.46	33.7	5.8	1.27	14.10	10.0	0.74	8.18
14	10.77	11.31	32.4	5.8	1.24	13.72	10.0	0.72	7.96
15	11.46	11.17	31.2	5.8	1.20	13.38	10.0	0.70	7.76
16	12.16	11.04	30.1	5.8	1.18	13.07	10.0	0.68	7.58
17	12.85	10.92	29.2	5.8	1.15	12.78	10.0	0.67	7.41
18	13.54	10.81	28.3	5.8	1.13	12.52	10.0	0.65	7.26
19	14.23	10.70	27.4	5.8	1.10	12.27	10.0	0.64	7.12
20	14.93	10.60	26.6	5.8	1.08	12.04	10.0	0.63	6.98
25	18.39	10.15	23.5	5.8	1.00	11.08	10.0	0.58	6.43
30	21.84	9.79	21.2	5.8	0.93	10.36	10.0	0.54	6.01
35	25.30	9.48	19.4	5.8	0.88	9.78	10.0	0.51	5.67
40	28.75	9.21	17.9	5.8	0.84	9.31	10.0	0.49	5.40
45	32.21	8.98	16.7	5.8	0.80	8.91	10.0	0.46	5.17
50	35.66	8.77	15.7	5.8	0.77	8.57	10.0	0.45	4.97
55	39.11	8.58	14.8	5.8	0.74	8.27	10.0	0.43	4.80
60	42.56	8.40	14.1	5.8	0.72	8.01	10.0	0.42	4.65
65	46.01	8.24	13.4	5.8	0.70	7.78	10.0	0.41	4.51
70	49.46	8.09	12.8	5.8	0.68	7.57	10.0	0.40	4.39
75	52.92	7.96	12.3	5.8	0.66	7.38	10.0	0.39	4.28
80	56.37	7.83	11.8	5.8	0.65	7.21	10.0	0.38	4.18
85	59.82	7.71	11.4	5.8	0.63	7.05	10.0	0.37	4.09
90	63.27	7.59	11.0	5.8	0.62	6.91	10.0	0.36	4.01
95	66.72	7.48	10.6	5.8	0.61	6.78	10.0	0.35	3.93
100	70.17	7.38	10.3	5.8	0.60	6.65	10.0	0.35	3.86
200	139.17	6.00	6.6	5.8	0.47	5.24	10.0	0.27	3.04
500	346.18	4.22	3.5	5.8	0.36	3.96	10.0	0.21	2.30

$$L_{msl} = 2\ln\left[\frac{8X}{W} + \frac{W}{4X}\right] nH/cm, \quad (16.A.7)$$

$$\varepsilon_{ox} = 3.45 \cdot 10^{-13}\ F/cm, \quad (16.A.8)$$

$$X = X_{\text{int}} + X_{si}, \quad (16.A.9)$$

where

C_{msl} = capacitance per unit length in respect to the substrate,
W = width if micro strip line
X_{int} = thickness of oxide layer,
ε_{ox} = electric permittivity of the silicon-oxide layer, and
L_{msl} = self-inductance per unit length along the runner,
X_{si} = thickness of silicon substrate.

The characteristic impedance is

$$Z_o = \sqrt{\frac{L_{msl}}{C_{msl}}}. \tag{16.A.10}$$

The quarter wavelength is

$$\frac{\lambda}{4} = \frac{2\pi}{4}\sqrt{C_{msl}L_{msl}}. \tag{16.A.11}$$

As an example, assuming that $X_{int} = 0.5\,\mu m$, $X_{si} = 500\,\mu m$, and $\varepsilon_{ox} = 3.45 \times 10^{-13}$ F/cm, the quarter wavelength with different frequencies and the width of the micro strip line are listed in Tables 16.A.1 and 16.A.2.

In the practical *RFIC* design, the width of the runner is typically taken in the range of

$$100\,\mu m \geq W \geq 5\,\mu m. \tag{16.A.12}$$

REFERENCES

[1] K. B. Ashby, I. A. Koullias, W. C. Finley, J. J. Bastek, and S. Moinian, "High Q Inductors for Wireless Applications in a Complementary Silicon Bipolar Process," *IEEE Journal of Solid-State Circuits*, Vol. 31, No. 1, January 1996, pp. 4–16.

[2] C. Nelson, R. Sigliano, and C. Makihara, "Built-in Passive Components in Multilayer Ceramics for Wireless Applications," *Proceedings of the 3rd International Symposium on Advanced Packaging Materials, March 9–12, 1997*, pp. 75–80.

[3] N. Klemmer and J. Hartung, "High Q Inductors for MCM-Si Technology," 1997 *IEEE Multi-Chip Module Conference*, 1997, *February 4–5, 1997*, pp. 33–37.

[4] Hirad Samavati et al., "Fractal Capacitors," *IEEE Journal of Solid-State Circuits*, Vol. 33, No. 12, December 1998, pp. 2035–2041.

[5] U. Erben, P. Abele, D. Behammer, H. Ergraber, and H. Schmacher, "High-Q Inductors on Silicon Using a quasi Thin Film Microstrip Technique," *Topical Meeting on Silicon Monolithic Integrated Circuits in RF Systems, Digest of Papers*, September 17–18, 1998, pp. 155–159.

[6] R. Groves, J. Malinowski, R. Volant, and D. Jadus, "High Q Inductors in a SiGe BiMOS Process Utilizing a Thick Metal Process Add-on Module," *Proceedings of the 1999 Bipolar/BiCMOS Circuits and Technology Meeting,* September 26–28, 1999, pp. 149–152.

[7] K. Kamogawa, K. Nishikawa, T. Tokumitsu, and M. Tanaka, "A Novel High-Q Inductor Based on Si 3D MMIC Technology and Its Application," *IEEE Radio Frequency Integrated Circuits (RFIC) Symposium*, June 13–15, 1999, pp. 185–188.

[8] Sung-Jin Kim, Yong-Goo Lee, Sang-Ki Yun, and Hai-Young Lee, "Realization of High-Q Inductors Using Wirebonding Technology," *First IEEE Asia Pacific Conference on ASICs,* August 23–5, 1999, pp. 13–16.

[9] Hannu Tenhunen, "CMOS Interconnects" (Lecture), Electronic System Laboratory, Kungl Tekniska Hogskolan, 2000.

[10] U. Yodprasit and J. Ngarmnil, "Q-enhancing Technique for RF CMOS Active Inductor," ISCAS Geneva Switzerland, *IEEE International Symposium on Circuits and Systems*, May 28–31, 2000.

[11] J.-B. David, F.-X. Musalem, and P. Albert, "RF High-Q Spiral Inductor Design," *Proceedings of the First ISA/IEEE Conference, Sensor for Industry,* November 5–7, 2001, pp. 78–82.

[12] Behzad Razavi, *Design of Analog CMOS Integrated Circuits*, McGraw-Hill, 2001.

[13] S. Jenei, S. Decoutere, S. Van Huylenbroeck, G. Vanhorebeek, and B. Nauwelaers, "High Q Inductors and Capacitors on Si Substrate," *2001 Topical Meeting on Silicon Monolithic Integrated Circuits in RF Systems, Digest of Papers*, September 12–14, 2001, pp. 64–70.

[14] Theerachet Soorapanth and S. Simon Wong, "A 0-dB IL 2140+/–30 MHz Bandpass Filter Utilizing Q-enhanced Spiral Inductors in Standard CMOS," *IEEE Journal of Solid-State Circuits*, Vol. 37, No. 5, May 2002, pp. 579–586.

[15] K. Van Schuylenbergh, C. L. Chua, D. K. Fork, Jeng-Ping Lu, and B. Griffiths, "On-chip Out-of-plane High-Q Inductors," *Proceedings of IEEE Lester Eastman Conference on High Performance Devices,* August 6–8, 2002, pp. 364–373.

[16] H. Lakdawala, X. Zhu, H. Luo, S. Santhanam, L. R. Carley, and G. K. Fedder, "Micromachined High-Q Inductors in a 0.18-μm Copper Interconnect Low-k Dielectric CMOS Process," *IEEE Journal of Solid-State Circuits*, Vol. 37, No. 3, March 2002, pp. 394–403.

[17] F. Rotella, D. Howard, M. Racanelli, and P. Zampardi, "Characterizing and Optimizing High Q Inductors for RFIC Design in Silicon Processes," *2003 IEEE Radio Frequency Integrated Circuits (RFIC) Symposium*, June 8–10, 2003, pp. 339–342.

[18] Jr-Wei Lin, C. C. Chen, J. K. Huang, and Y. T. Cheng, "An Optimum Design of the Micromachined RF Inductor," *IEEE Radio Frequency Integrated Circuits Symposium*, 2004, pp. 639–642.

[19] Madjid Hafizi, Shen Feng, Taoling Fu, Kim Schulze, Robert Ruth, Richard Schwab, Per Karlsen, David Simmonds, and Qizheng Gu, "RF Front-End of Direct Conversion Receiver RFIC for CDMA-2000," *IEEE Journal of Solid-State Circuits*, Vol. 39, No. 10, October 2004, pp. 1622–1632.

[20] Jr-Wei Lin, C. C. Chen, and Yu-Ting Cheng, "A Robust High-Q Micromachined RF Inductor for RFIC Applications," *IEEE Transactions on Electron Devices*, Vol. 52, No. 7, July 2005, pp. 1489–1496.

[21] Chih-Yuan Lee, Tung-Sheng Chen, Joseph Der-Son Deng, and Chin-Hsing Kao, "A Simple Systematic Spiral Inductor Design with Perfected Q Improvement for CMOS RFIC Application," *IEEE Transactions on Microwave Theory and Techniques*, Vol. 53, No. 2, February 2005, pp. 523–528.

[22] Choon Beng Sia, Beng Hwee Ong, Kwok Wai Chan, Kiat Seng Yeo, Jian-Guo Ma, and Manh An Do, "Physical Layout Design Optimization of Integrated Spiral Inductors for Silicon-Based RFIC Applications," *IEEE Transactions on Electron Devices*, Vol. 52, No. 12, December 2005, pp. 2559–2567.

[23] David G. Rahn, Mark S. Cavin, Fa Foster Dai, Neric H. W. Fong, Richard Griffith, José Macedo, A. David Moore, John W. M. Rogers, and Mike Toner, "A Fully Integrated Multiband MIMO WLAN Transceiver RFIC," *IEEE Journal of Solid State Circuits*, Vol. 40, No. 8, August 2005, pp. 1629–1641.

[24] Daniel Kaczman, Charles Dozier, Nihal Godambe, Manish Shah, Homero Guimaraes, Mohammed Rachedine, Mohammed Alam, Lu Han, Wayne Shepherd, David Cashen, Jeff Ganger, Karl Couglar, Bill Getka, Eddie Brotkowski, Derek Wong, and Don Hayes, "A Tri-band (2100/1900/800 MHz) Single-Chip Cellular Transceiver for WCDMA/HSDPA" (Invited Paper), *2005 IEEE Radio Frequency Integrated Circuits Symposium*, pp. 281–284.

[25] Fu-Yi Han, Jian-Ming Wu, Tzyy-Sheng Horng, and Cheng-Chia Tu, "A Rigorous Study of Package and PCB Effects on W-CDMA Upconverter RFICs," *IEEE Transactions on Microwave Theory and Techniques*, Vol. 54, No. 10, October 2006, pp. 3793–3803.

[26] Tao Wang, Yo-Sheng Lin, and Shey-Shi Li, "An Ultralow-Loss and Broadband Micromachined RF Inductor for RFIC Input-Matching Applications," *IEEE Transactions on Electron Device*, Vol. 53, No. 3, March 2006, pp. 568–570.

[27] Yo-Sheng Lin, Hsiao-Bin Liang, Chi-Chen Chen, Tao Wang, and Shey-Shi Lu, "A High Quality Factor and Low Power Loss Micromachined RF Bifilar Transformer for UWB RFIC Applications," *IEEE Electron Device Letters*, Vol. 27, No. 8, August 2006, pp. 684–688.

CHAPTER 17

MANUFACTURABILITY OF PRODUCT DESIGN

17.1 INTRODUCTION

The development process of a product can probably be divided into two stages: R&D and production. Let's take a look what happens when a product is developed in this way:

- In the R&D stage, the main design steps, including simulation, layout, and implementation of module or *IC* tape-out, are completed based on the product specifications.
- Some samples based on the prototype design are fabricated by hand.
- Testing of these samples is carried out.
- Patents are awarded and papers published.
- Product is put on the production line for mass production.
- However, the yield rate of the product is very poor, say, less than 40%.
- The product uses more than 60% of the funds on the direct and indirect material costs, including manpower, the manufacturing facility, and time.

  People from R&D and manufacturing scream, shout, and blame each other for the product failure.
- Result: the company goes bankrupt.

What is going on?

The failure of this product is due to the lack of tolerance analysis prior to being put on the production line. To ensure the manufacturability of a product, tolerance analysis in the simulation stage must be carried out. Obviously, this is an important step guaranteeing the success of a new product.

The unique criterion to measure success or failure of product development is its manufacturability. What is the manufacturability of a product? Manufacturability of a product means that the product can be manufactured as a mass production line and retain the features of

1) High yield rate, including
   - Satisfication of specifications,
   - Good reputation or identity,
   - High reliability.
2) Low cost, including the cost of
   - Material and parts,
   - Manpower,
   - Factory maintenance and equipment.

Therefore, the answer is simple: the yield rate of the product on the production line is higher than the expected goal. The definition of the expected goal, of course, is somewhat artificial but by summing up all the features of manufacturability there is a bottom line: the product must be profitable.

In the disastrous example above, not only is the tolerance analysis missing, so are some other important segments of the design process, such as *ALT* (Accelerate Life Testing), *TOP* (Technical Operating Production), and pilot production. The typical design procedures for a new product are shown in Figure 17.1.

Instead of the two stages in the example, Figure 17.1 shows that three stages required for a new product design to reach an acceptable manufacturability, which are:

First stage:	R&D,
Second stage:	Pre-production,
Third stage:	Production.

In order to ensure the manufacturability of a new product, the designer must be a qualified 6σ engineer. Tolerance analysis is the key component in the 6σ design.

At present, in the simulation stage, tolerance analysis is usually conducted through the so-called "corner analysis," in which tolerance analysis is done only for the devices, but not for other parts such as resistors, capacitors, inductors, transformers, and so on. This may be fine in digital circuit design because in digital circuits the devices may play the decisive role in the contribution of tolerance. However, in *RF* circuit design, the devices and the other parts contribute to the tolerance with the same weight. Sometimes the tolerance contributed by other parts may even dominate devices. Conducting tolerance analysis for devices only is not enough in a 6σ design.

One of main reasons that the yield rate in the production line is poor is that a high percentage of *RF* circuit designs do not employ tolerance analysis, or they confine tolerance analysis to devices only. Often, tolerance analysis is ignored until the low yield rate shows up. In electrical engineering courses, tolerance analysis is often set aside or ignored so that new engineers fresh from school are somewhat green on it.

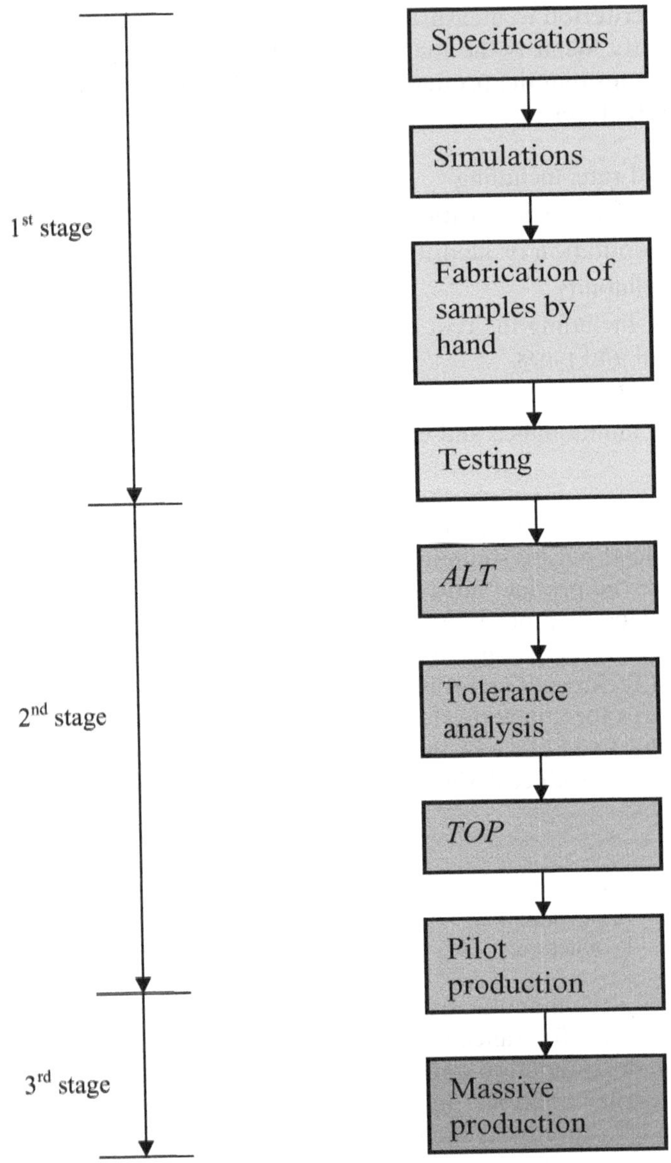

Figure 17.1 Typical design procedures for a new product with acceptable manufacturability.

17.2 IMPLICATION OF 6σ DESIGN

17.2.1 6σ and Yield Rate

Appendix 17.A.1 introduces random variables and the fundamentals of random process. For a random variable C, the ratio of relative deviation from the average m to the standard deviation σ, is defined as

$$z = \frac{C - m}{\sigma}, \tag{17.1}$$

where

z = ratio of relative deviation from the average m to the standard deviation σ,
C = random variable,
m = average value of random variable,
σ = variance or standard deviation, or average square root of deviation,

Figure 17.2 plots a random variable with Gaussian distribution. The ordinate coordination is probability density $\rho(z)$ while the abscissas is z, the ratio of relative deviation from the average m to the standard deviation σ.

The design tolerance Tol_{design} is defined as

$$Tol_{design} = USL - LSL, \tag{17.2}$$

where

USL = upper specification limit, σ,
LSL = lower specification limit, σ.

The magnitudes of USL and LSL are the same but the LSL is negative while the USL is positive because the Gaussian distribution is symmetrical to its average value m. The design tolerance Tol_{design} represented the design capability of a circuit designer.

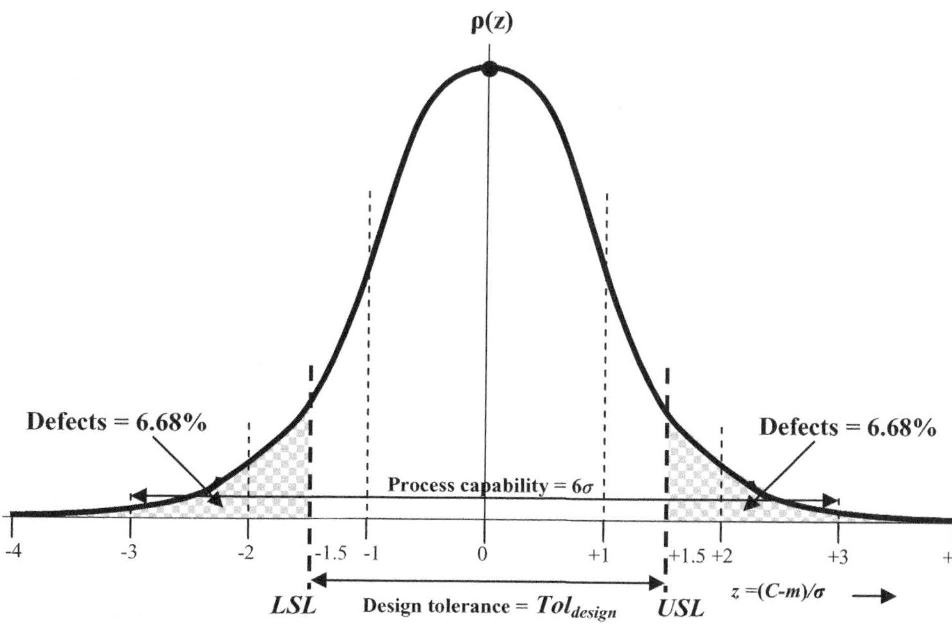

Figure 17.2 Definition of design tolerance Tol_{design} and process capability 6σ.

In Figure 17.2,

$$USL = 1.5\sigma, \tag{17.3}$$

and

$$LSL = -1.5\sigma, \tag{17.4}$$

so that

$$Tol_{design} = 3\sigma. \tag{17.5}$$

The upper and lower specification limits, USL and LSL, and hence the design tolerance Tol_{design}, are specified by the designer. The shadowed area in Figure 17.2 represents the defects of the product. In the case as shown in Figure 17.2,

$$Tol_{design} = 3\sigma, \tag{17.6}$$

$$Defects = 1 - f(z = USL) - f(z = LSL) = 1 - 43.32\% - 43.32\% = 13.36\%, \tag{17.7}$$

because from Figure 17.A.4,

$$f(z = USL) = f(z = 1.5) = 19.15\% + 14.98\% + 9.19\% = 43.32\%, \tag{17.8}$$

$$f(z = LSL) = f(z = -1.5) = 19.15\% + 14.98\% + 9.19\% = 43.32\%. \tag{17.9}$$

Obviously, if the design tolerance Tol_{design} is high, the defects will be low. Therefore, the design tolerance Tol_{design} is a measure of the design capability of a designer. The expected high value or goal of Tol_{design} is called the process capability, which is

$$Process\ capability = 6\sigma. \tag{17.10}$$

Based on Figure 17.A.4, Figure 17.3 plots the relationship between the yield rate $f(z)$ and the design tolerance Tol_{design}, by which the yield rate can be directly found from the value of design tolerance Tol_{design}. The design tolerance Tol_{design} is usually measured in units of σ. This is why people talk about product with "how many σ design." As a matter of fact, the relation between the yield rate $f(z)$ and the design tolerance Tol_{design} in Figure 17.3 can also be listed as in Table 17.1.

Figure 17.3 and Table 17.1 show the increase of the yield rate as the design tolerance Tol_{design} is increased. The yield rate is quite sensitive to the design tolerance Tol_{design} when design tolerance Tol_{design} is much less than 6σ. A higher yield rate can be obtained if higher design tolerance Tol_{design} is allowed. The yield rate reaches 99.74% if the design tolerance Tol_{design} is increased to 6σ. In such a case only 0.26% of products are defects. However, if further effort is put on the design so that the design tolerance Tol_{design} is continuously increased from 6σ up to 7σ as seen in Figure 17.3 the increase of yield rate is very little. This is why the 6σ of the design tolerance Tol_{design} is ruled as an expected high design goal but not 7σ, 8σ, 9σ, or higher.

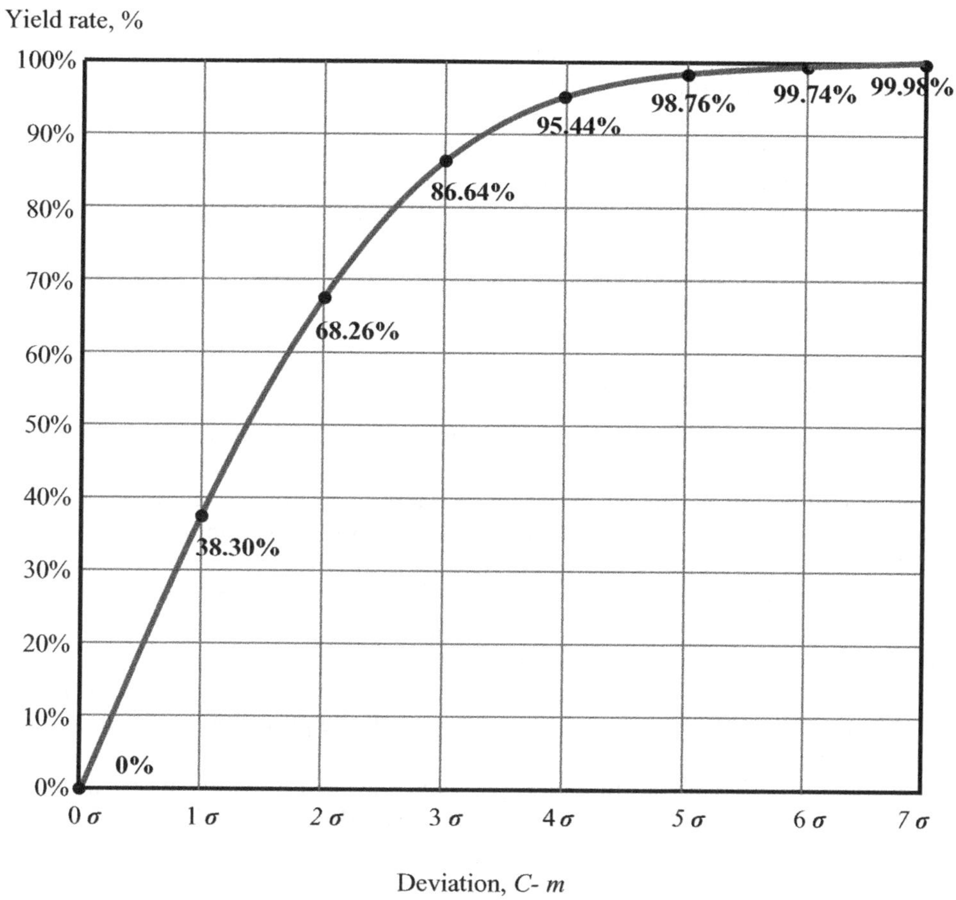

Figure 17.3 Relationship between yield rate and deviation.

TABLE 17.1 Relationship between the yield rate *f*(*z*) and the design tolerance Tol_{design}

Design tolerance Tol_{design}	Yield rate *f*(*z*)
0 σ	0%
1 σ	38.30%
2 σ	68.26%
3 σ	38.30%
4 σ	86.64%
5 σ	98.76%
6 σ	**99.74%**
7 σ	99.98%

Also, it can be seen that the standard deviation of product σ is a key parameter in the 6σ circuit design.

For example, a hypothetical customer is going to buy a capacitor specified as $1\,pF \pm 0.1\,pF$. Two designers are in charge of the design works in the factory. Designer *A* designs the $1\,pF$ capacitor with $\sigma = 0.1\,pF$, while designer *B* designs the $1\,pF$ capacitor with $\sigma = 0.2pF$. The specification from the customer indicates that he wants to buy 2σ products designed by designer A and 1σ products designed by designer B. Consequently, the manufacturer can provide the customer with 68.26% of the mass-produced products designed by designer A and only 38.30% of the products designed by designer B. Obviously, designer A makes more profit for the factory.

Who is the 6σ designer in respect to this customer's specification? A designer who can design the $1pF$ capacitor with a $\sigma = 0.033pF$ is the 6σ designer, because his 6σ range $-3\sigma = -0.1pF < z\sigma < +3\sigma = +0.1pF$ is what the customer asks for. The manufacturer can provided the customer with 99.74% of the mass-produced products designed by this designer.

17.2.2 6σ Design for a Circuit Block

In the previous section, the discussion was confined to a single part. Now, we expand the discussion to a circuit block design.

A 6σ circuit design is a design such that the yield rate of this circuit block in mass production reaches 99.74% when 6σ tolerance of all the parts applied in the circuit block is allowed. Actually, the expected yield rate is 100%.,

Similarly, referring to Figure 17.3, a 5σ circuit design is a design in which the yield rate of the circuit block in mass-production reaches 98.76% when 5σ tolerance of **all the parts** applied in the circuit block is allowed.

A 4σ circuit design is one in which the yield rate of the circuit block in mass-production reaches 95.44% when 4σ tolerance of **all the parts** applied in the circuit block is allowed.

A 3σ circuit design is one in which the yield rate of the circuit block in mass-production reaches 86.64% when 3σ tolerance of **all the parts** applied in the circuit block is allowed.

A 2σ circuit design is a design in which the yield rate of the circuit block in mass production reaches 68.26% when 2σ tolerance of **all the parts** applied in the circuit block is allowed.

A 1σ circuit design is a design in which the yield rate of the circuit block in mass production reaches 38.30% when 1σ tolerance of **all the parts** applied in the circuit block is allowed.

The value of σ in the 6σ design for a circuit block is contributed by the individual σ values of all the parts in the circuit block. It depends not only on the individual σ values but also on the weight of individual parts in the circuit block. Usually it is hard to calculate this by hand. Monte Carlo analysis in most circuit simulation programs provides a powerful tool to obtain the value of σ and other parameters for a circuit block, which will be introduced below.

17.3 APPROACHING 6σ DESIGN

17.3.1 By Changing Parts' 6σ Value

The 6σ design is a tough task to a circuit design engineer because the design

- Must ensure that the circuit block is in normal performance or that all the goals of performance are satisfied.

- Must allow a 6σ deviation for all the parts of the circuit block, such as resistors, capacitors, inductors, transistors, and so on.

It is possible to approach a 6σ design by changing the component's σ value. Figure 17.4 illustrates such an approach.

The original design is depicted in the upper portion of Figure 17.4. By applying the values of $USL = 1.5\sigma$ and $LSL = -1.5\sigma$ as shown in the upper portion of Figure 17.4 into expression (17.2), the design tolerance Tol_{design} of 3σ is found so that the original design is a 3σ design. The defects of the product are $2 \cdot 6.68\% = 13.36\%$, and the yield rate is 86.64%.

With the same specified values of USL and LSL, the original 3σ design in the upper portion of Figure 17.4 can be improved and converted to 6σ design as shown in the bottom portion of Figure 17.4. Such a conversion can be realized by replacing the original parts by new parts if the average value m of the part is kept unchanged, and if the standard deviation of the new part is lower than that of the original part, that is,

$$\sigma' = \frac{\sigma}{2}, \tag{17.11}$$

where

σ' = new standard deviation of part,
σ = old standard deviation of part.

Consequently, the original defect rate of 13.36% would be reduced to 0.26%, and the yield rate increased from 86.64% to 99.74%.

The cost of this increased yield rate is price, because parts with low standard deviation are more expensive than parts with high standard deviation. Expression (17.11) indicates that the standard deviation of the new parts is lower than that of the original parts.

As a matter of fact, the relation (17.11) is specific to the conversion from 3σ to 6σ circuit design or for the conversion from the yield rate 86.64% to 99.74%. In general, by changing the standard deviation of parts, a circuit design can be converted from any yield rate to another. If the conversion is made from a low to a high yield rate, the cost of the higher yield rate is price.

17.3.2 By Replacing a Single Part with Multiple Parts

In the real parts, such as the resistor, capacitor, and inductor, the σ value decreases as the value of parts increases if they are produced with the same processing. By utilizing this attribute, we have another way to approach the 6σ design either for the discrete *RF* module or the *RFIC* circuit. That is, we replace the original parts with multiple parts which together equal the higher part's value.

Let's illustrate this idea by an example. Assume that a circuit block consists of three parts. The original parts are listed in the second column of Table 17.2.

In order to convert the circuit design from a low-sigma circuit design to a high-sigma circuit design, or from a low yield rate to a high yield rate, without changing

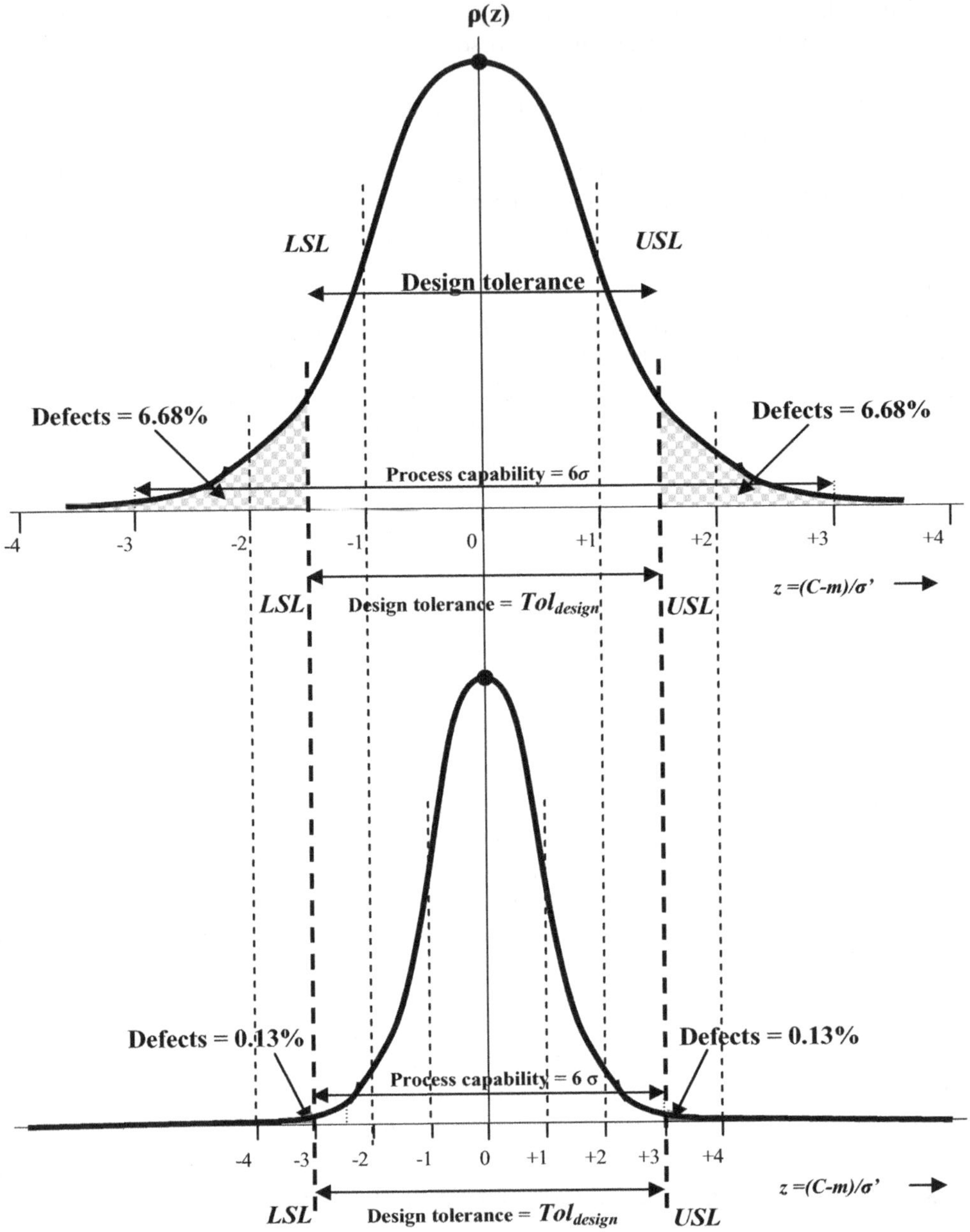

Figure 17.4 Conversion of 3σ design to 6σ design by halving standard deviation from σ to σ', that is, $\sigma' = \sigma/2$.

the parts' standard deviation as discussed in Section 17.3.1, the original parts are replaced by multiple parts which together have the higher parts' value. The resultant group of multiple parts' value is kept the same as the original part's value through a reasonable combined configuration either in parallel or in series as shown in Figure 17.5. The resultant standard deviation of the multiple parts is lower than that

TABLE 17.2 Replacement of original one resistor, one capacitor, and one inductor by three resistors, four capacitors, and two inductors

	Original part		Multi-parts	
	m	σ	m	σ'
Resistor	$2k\Omega$ (1 resistor)	**200 Ω**	$6k\Omega/3$ (3 resistors in parallel)	**120 Ω**
Capacitor	$4pF$ (1 capacitor)	**0.5*pF***	16 pF/4 (4 capacitors in series)	**0.2*pF***
Inductor	$7nH$ (1 inductor)	**0.2*nH***	$14nH/2$ (2 inductors in parallel)	**0.15*nH***

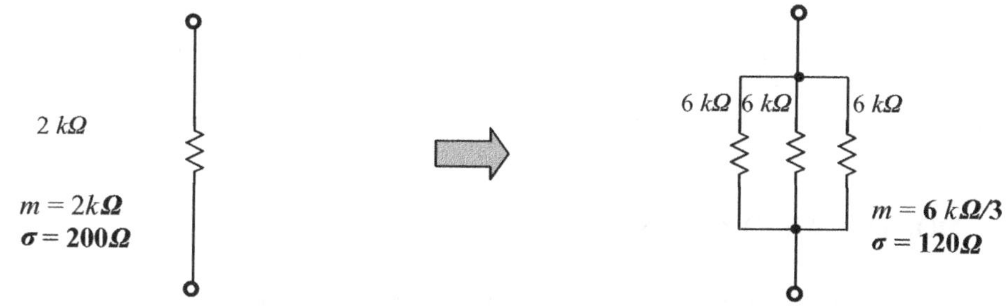

(a) One resistor of 2 $k\Omega$ is replaced by three 6 $k\Omega$ resistors in parallel

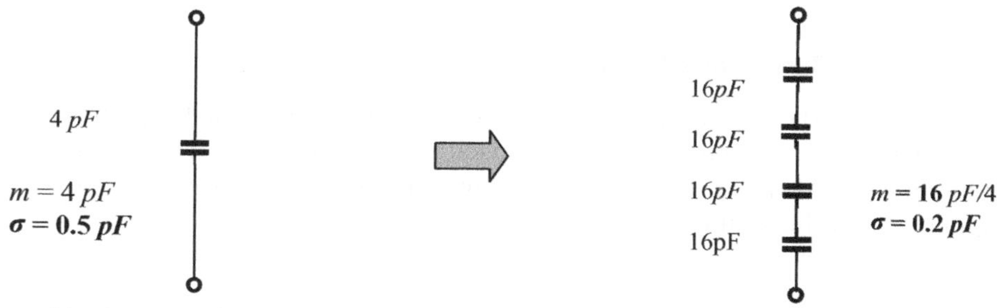

(b) One capacitor of 5 pF is replaced by four 16 pF capacitors in series

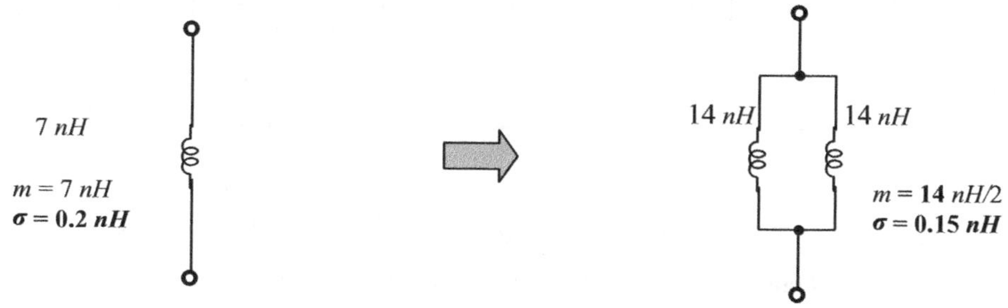

(c) One inductor of 7 nH is replaced by two 14 nH inductors in parallel.

Figure 17.5 One part is replaced by multiple parts.

of the original part because the multiple parts have the higher part's value. The multiple parts are listed in the third column of Table 17.2.

In Table 17.2,

- The original $2k\Omega$ resistor with $\sigma = 200\Omega$ is replaced by three $6k\Omega$ resistors with $\sigma' = 120\Omega$ in parallel.
- The original $4pF$ capacitor with $\sigma = 0.5pF$ is replaced by four $16pF$ capacitors with $\sigma' = 0.2$pF in series.
- The original $7nH$ inductor with $\sigma = 0.2nH$ is replaced by two $14nH$ inductors with $\sigma' = 0.15nH$ in parallel.

The yield rate in either the circuit block built by original parts or by multiple parts depends on the topology of the circuit block and all the parts in the circuit block. Unfortunately the data listed in Table 17.2 is not enough to calculate the yield rate in either the circuit block built by original parts or by multiple parts since it is a block built by multiple parts. However, the yield rate in the circuit block built by multi-parts is undoubtedly higher than that of the circuit block built by original parts; that is, the circuit block built by multi-parts is a higher sigma circuit design than the circuit block built by original parts.

17.4 MONTE CARLO ANALYSIS

The tolerance evaluation of a circuit block can be conducted by a Monte Carlo analysis. Let us take a band pass filter as an example to illustrate the key features of this analysis.

17.4.1 A *BPF* (Band Pass Filter)

Figure 17.6 shows the schematic of a practical band pass filter in the frequency range of 403 to 470 MHz. The parts, including capacitors and inductors, are the chip parts manufactured by MuRata. The corresponding models for capacitors and inductors are applied in all the simulations.

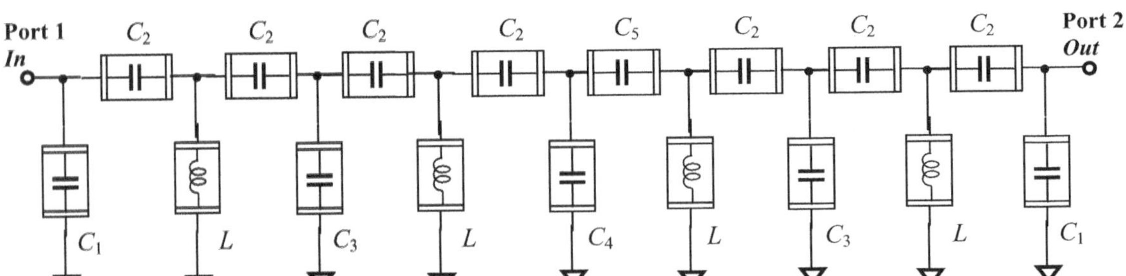

Figure 17.6 *BPF* (Band Pass Filter), 403–470 MHz.
$L = 20nH$ $C_1 = 3.9pF$
$C_2 = 4.3pF$ $C_3 = 7.5pF$
$C_4 = 9.1pF$ $C_5 = 3.9pF$

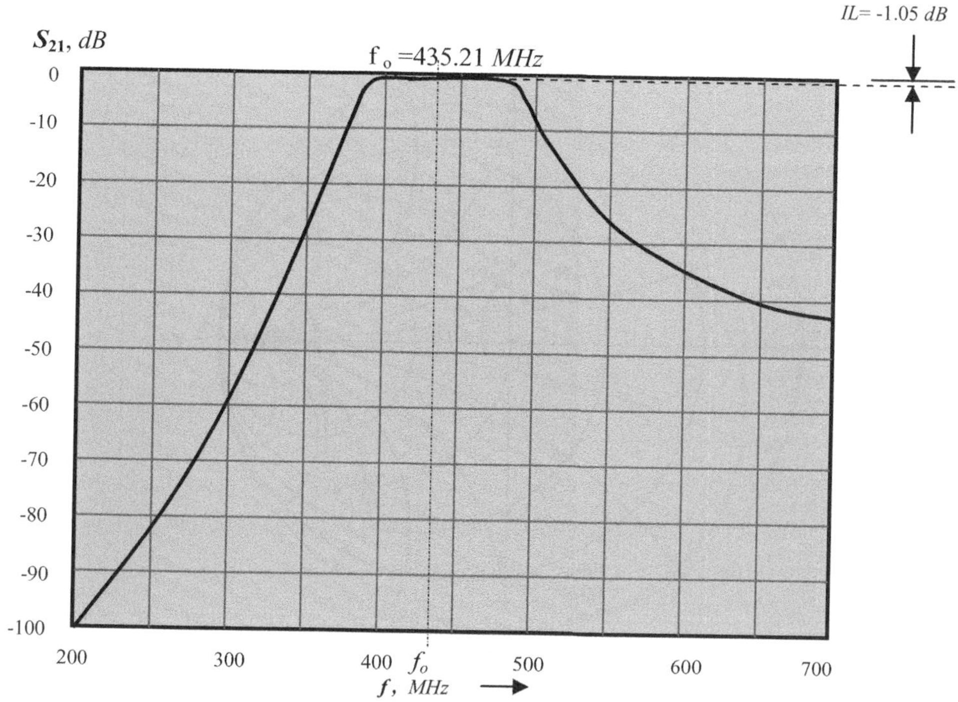

Figure 17.7 Frequency response of band pass filter, 403–470 MHz.

Figure 17.7 shows the frequency response of the band pass filter in the frequency range from 200 to 700 MHz. Its insertion loss, IL, is

$$IL = -1.05\,dB, \tag{17.12}$$

when

$$403 < f < 407\,MHz. \tag{17.13}$$

And,

$$f_o = \sqrt{403 \cdot 407} = 435.21\,MHz, \tag{17.14}$$

Figure 17.8 shows the return loss, S_{11}, on the Smith chart, which indicates that the input impedance of the band pass filter is well matched to $50\,\Omega$ in the frequency range from 403 to 470 MHz.

17.4.2 Simulation with Monte Carlo Analysis

Figure 17.9 shows a simulation page for Monte Carlo analysis. A schematic of the BPF as shown in Figure 17.6 is represented by the circuit block labeled "BPF (Band Pass Filter), 403–407 MHz."

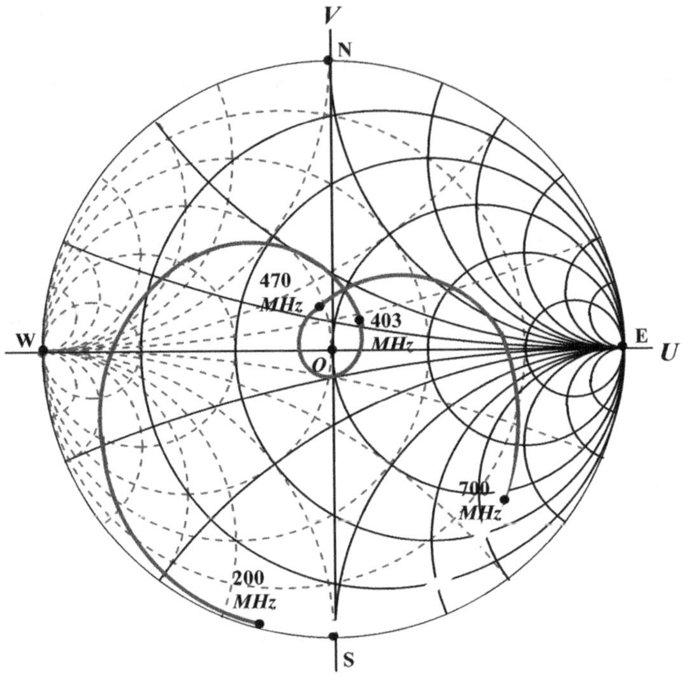

Figure 17.8 S_{11} of band pass filter, 403–470 MHz.

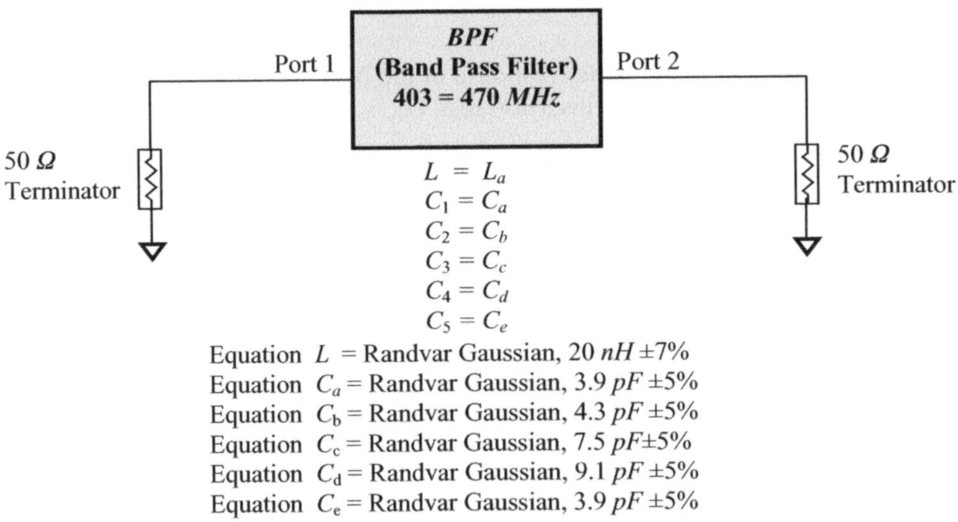

Figure 17.9 Simulation page for Monte Carlo analysis. Number of iterations = 50.

The purpose of the Monte Carlo analysis is to find the effect of the parts' tolerance on the performance of the tunable filter. In Figure 17.6, there are six different parts: one type of inductor, and five kinds of capacitors; they are all involved in the tolerance analysis.

The first objective in the simulation with Monte Carlo analysis is to specify that

- All the values of different parts are: Random variables,
- The distribution type of the variables is: Gaussian distribution,
- Their nominal value or mean is: m,
- The tolerance of each part is: σ (by percentage).

The nominal value or mean, m, and tolerance by percentage, σ, of parts are provided by the manufacturers. In the Monte Carlo analysis, the second objective is to ask for the iteration number in sampling the random variables.

The iteration number must be set high enough so that the simulated results can reflect all the possible variations due to the parts' tolerance. However, too many iterations may increase simulation time to an unacceptable level. In a practical simulation, the number of iterations is usually set to

$$Iterations \Rightarrow 50-100. \tag{17.15}$$

Figures 17.10 and 17.11 show the frequency response and the input return loss of the band pass filter. For clarity, Figure 17.10 contains only 10 iterations, that is, it contains only 10 curves of frequency response.

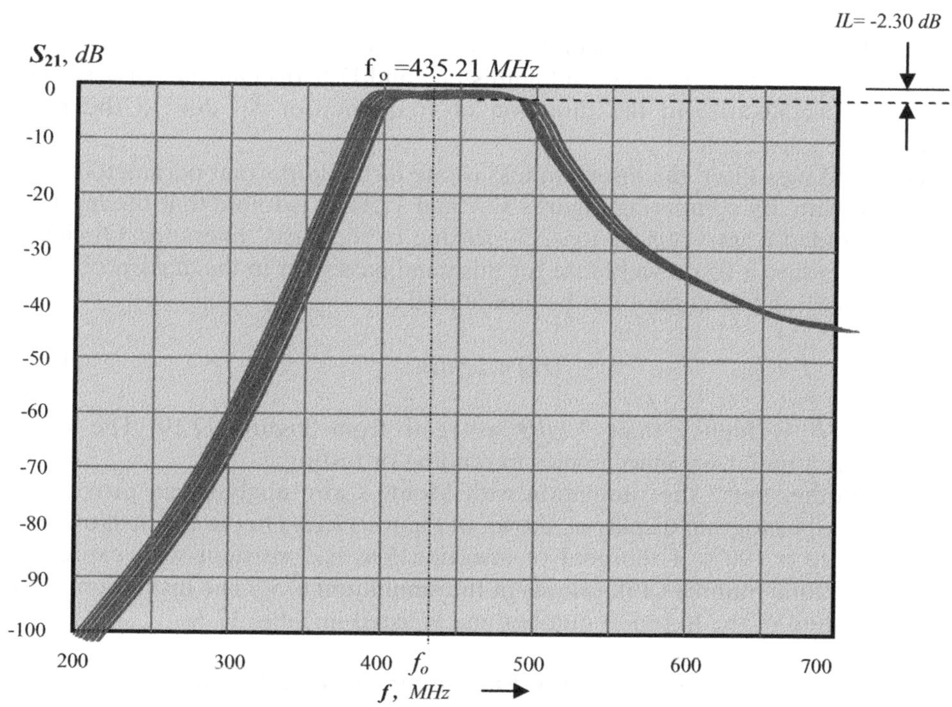

Figure 17.10 Frequency response of band pass filter, 403–470 MHz. Number of iterations = 50.

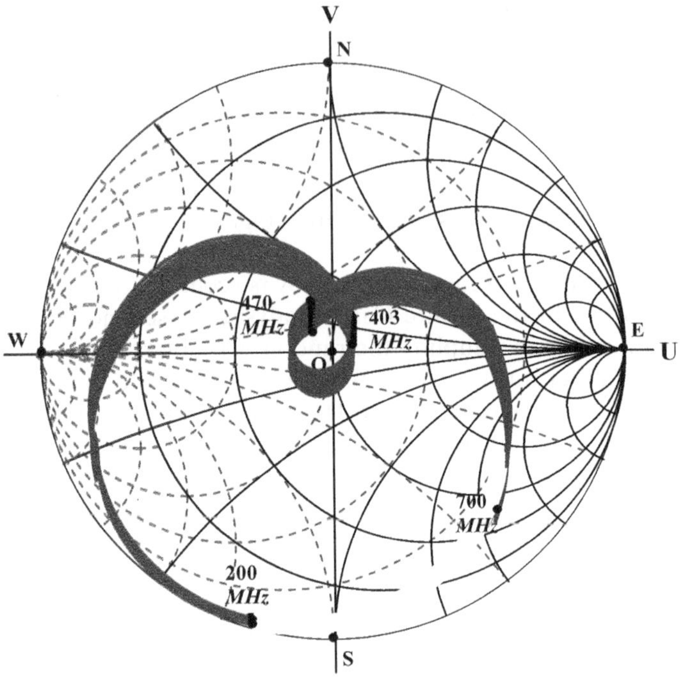

Figure 17.11 S_{11} of band pass filter, 403–470 MHz. Number of iterations = 50.

From Figure 17.10 it can be seen that the curve of the frequency response is no longer a clear-cut curve as shown in Figures 17.7 and 17.8. Instead, the curve spreads out over a small area in the direction of frequency or S_{21} due to the parts' tolerance.

As a band pass filter, the insertion loss is one of the important parameters to be concerned with. By comparing Figures 17.7 and 17.11, it is found that the insertion loss is degraded from $-1.05\,dB$ to $-2.30\,dB$ due to the parts' tolerance. Obviously, in order to ensure a 100% yield rate for this band pass filter in the mass-production line, the insertion loss cannot but be announced as

$$|IL| < 2.5\,dB, \tag{17.16}$$

where $2.5\,dB$ is higher than $2.3\,dB$ as read from Figure 17.10. The $0.2\,dB$ difference is a buffer for inaccuracies in reading or testing.

On the other hand, the simulation with Monte Carlo analysis can provide the yield rate and histogram display as shown in Figure 17.12. On the top, it shows that the yield rate is 100% if the goal of insertion loss is consistent with expression (17.16). The total number of iterations in the simulation is 50. The histogram shows the distribution of the iteration number and is listed in Table 17.3.

17.4.3 Sensitivity of Parts on the Parameter of Performance

Another special function of the Monte Carlo analysis is to examine the sensitivity of each part on the parameters of performance. Figure 17.13(a) to (f) shows the

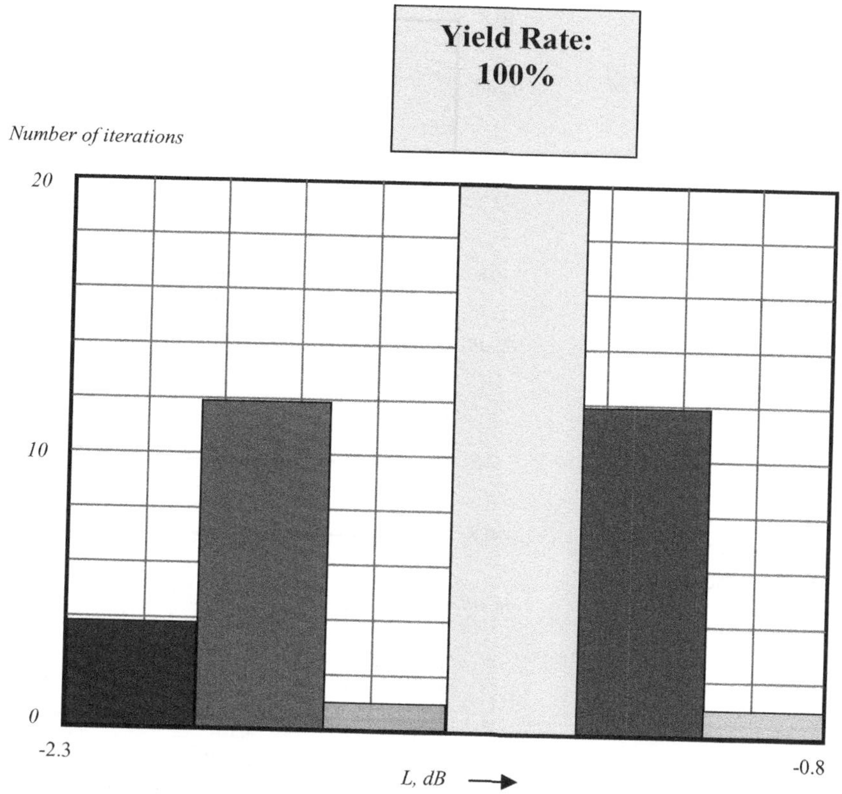

Figure 17.12 Display of insertion loss histogram and yield rate for $|IL| < 2.5\,dB$. Number of iterations = 50.

TABLE 17.3 Distribution of iteration number vs. insertion loss

Insertion loss	−2.30 to −2.05	−2.05 to −1.80	−1.80 to −1.55	−1.55 to −1.30	−1.30 to −1.05	−1.05 to −0.8 dB
Number in iteration	4	12	1	20	12	1

sensitivity of the six individual parts, L, C_1, C_2, C_3, C_4, and C_5, on an insertion loss respectively. The insertion loss is less than $2.5\,dB$. Fifty iterations are conducted in the analysis so that there are 50 points on each plot. Each point therefore represents one random value of the part in each iteration.

In the plots of Figure 17.13(a) to (d), and (f) for IL vs. L, C_1, C_2, C_3, and C_5, the tolerance of these five parts looks quite adequate. All the points are scattered within the range of insertion loss in an approximately homogeneous pattern. However, in the plot of Figure 17.13(e) for IL vs. C_4, the tolerance of C_4 seems somewhat smaller than what is needed because all the points are crowded in the middle area within

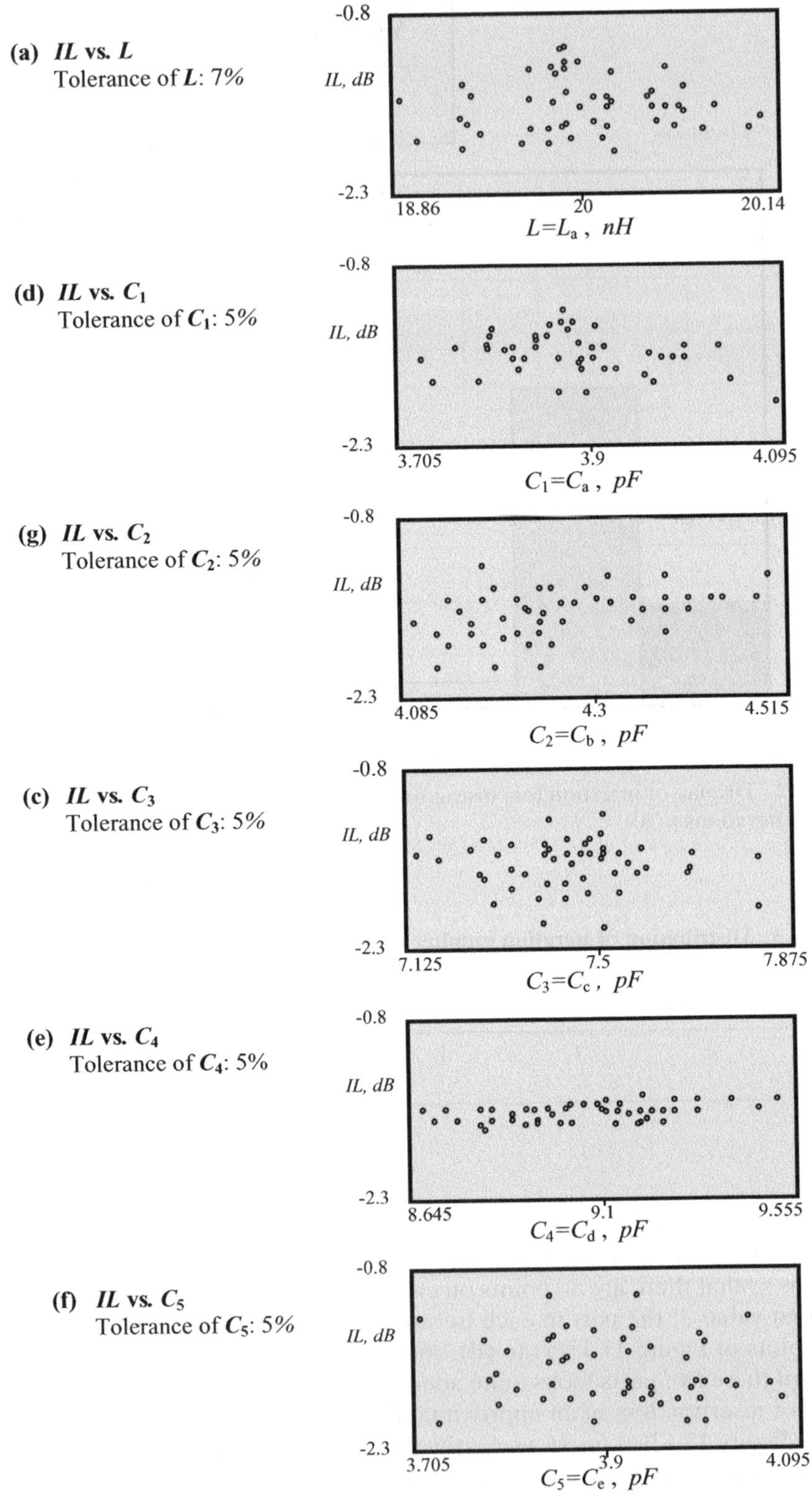

Figure 17.13 Sensitivity of individual part's value on insertion loss.

(b) ***IL*** **vs.** C_4
Tolerance of C_4: 7%

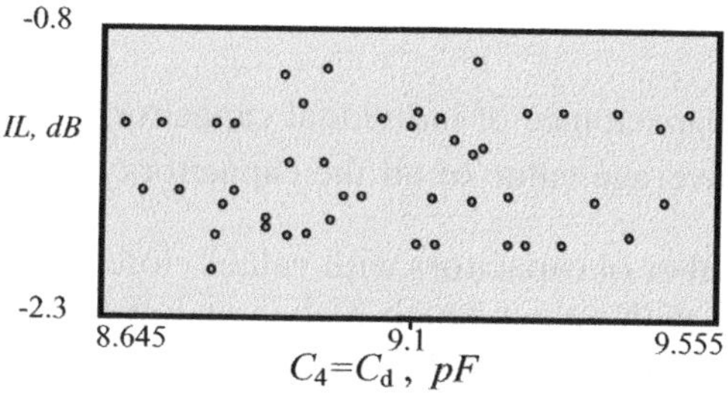

Figure 17.14 Sensitivity of C_4 on insertion loss after its tolerance of 5% is replaced by 7%.

the *IL* range. In other words, the tolerance of C_4 can be relaxed somewhat, say, from the original 5% as shown in Figure 17.13 to 7%. After a re-simulation with Monte Carlo analysis, replacing the 5% tolerance by 7% for C_4, Figure 17.13(e) is changed as shown in Figure 17.14,

As seen in Figure 17.14, all the points are now scattered within the range of the insertion loss in an approximately homogeneous pattern. This implies that the replacement of 5% tolerance by 7% tolerance is adequate. This replacement is valuable because the C_4 capacitors with 7% tolerance are much cheaper than C_4 capacitors with 5% tolerance. The cost therefore can be reduced by examination of the sensitivity of parts.

Similarly, in addition to the insertion loss, the Monte Carlo analysis can be conducted for other parameters so as to examine all the parts' sensitivity and finally to ensure the 100% yield rate. Monte Carlo analysis is a powerful tool to realize the 6σ circuit design for the mass production of the product.

The *BPF* example introduced above is a simple example. Based on the same principle, readers can expand the Monte Carlo analysis to more complicated circuit blocks, including transistors, micro strip lines, and so on.

APPENDICES

17.A.1 Fundamentals of Random Process

Both regular processes and random processes exist in the world. They also exist in circuit design. The fabrication of the prototype samples in the first design stage is a regular process, while the massive production in the third design stage is a random process. For example, if the designed product is a $1 pF$ capacitor, then in the first design stage, after the simulation, a couple of capacitor samples are built and tested. The capacitance of each capacitor sample is under control and non-random. However, once this product is put on the production line for large-scale production, the value of individual capacitors becomes a random variable. Some of the values are greater than $1 pF$, while some are$F$. Their average value, of course, is $1 pF$, that is,

$$m = \overline{\sum_i C_i} = 1\, pF, \tag{17.A.1}$$

where

C_i = capacitance of individual capacitor,
m = average value of all the capacitors.

The number of capacitors with values close to the average value is greater than that of those with values much higher or lower than the average value. However, the sum of all the individual deviations from the average value would be zero because the chance of the capacitance of individual capacitor being "higher or lower than the average value" is the same, that is,

$$D_i = \Delta C_i = m - C_i, \tag{17.A.2}$$

$$\sum_i D_i = \sum_i \Delta C_i = \sum_i (m - C_i) = 0, \tag{17.A.3}$$

where D_i = deviation of individual capacitor's value from the average value.

In order to characterize the deviation status of the random variable, a new parameter called variance or standard deviation is defined as the average value of the square root of deviation, that is

$$\sigma = \sqrt{\sum_i \Delta C_i^2}\, pF, \tag{17.A.4}$$

where σ = variance or standard deviation of the capacitor.

The standard deviation is always a positive value because every individual square root of deviation is positive. The tolerance is the ratio of the standard deviation to the average value, that is,

$$Tol = \frac{\sqrt{\sum_i \Delta C_i^2}}{m} = \frac{\sigma}{m}, \tag{17.A.5}$$

where Tol = relative tolerance of the capacitance.

If all the capacitors are tested for their values, which are grouped with a small constant interval of capacitance, a plot of the values vs. the number of capacitors can be shown as Figure 17.A.1. This is a histogram of the relative number of capacitors *vs.* the value of the capacitor. It is symmetrical with respect to the point m.

Based on statistics theory, the histogram as shown in Figure 17.A.1 approaches a normal probability distribution with a probability density, $\rho(z)$,

$$\rho(z) = \frac{e^{-\frac{z^2}{2\sigma^2}}}{\sqrt{2\pi}}, \tag{17.A.6}$$

where z = ratio of deviation from the average to the standard deviation, say, a dimensionless capacitance, which is relative to the average value of the capacitor and is normalized by the variance σ.

$$z = \frac{C - m}{\sigma}, \tag{17.A.7}$$

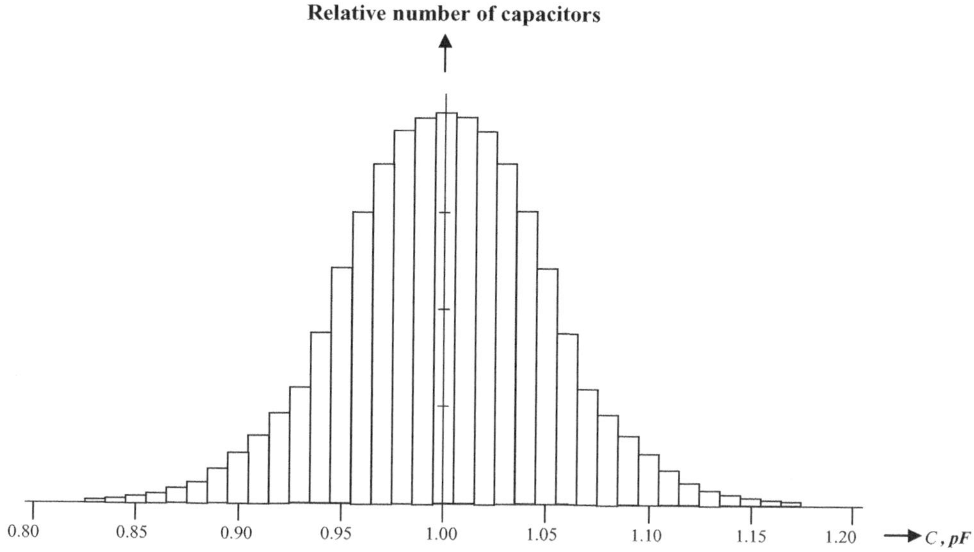

Figure 17.A.1 Histogram of relative number of capacitors vs. the value of capacitor.

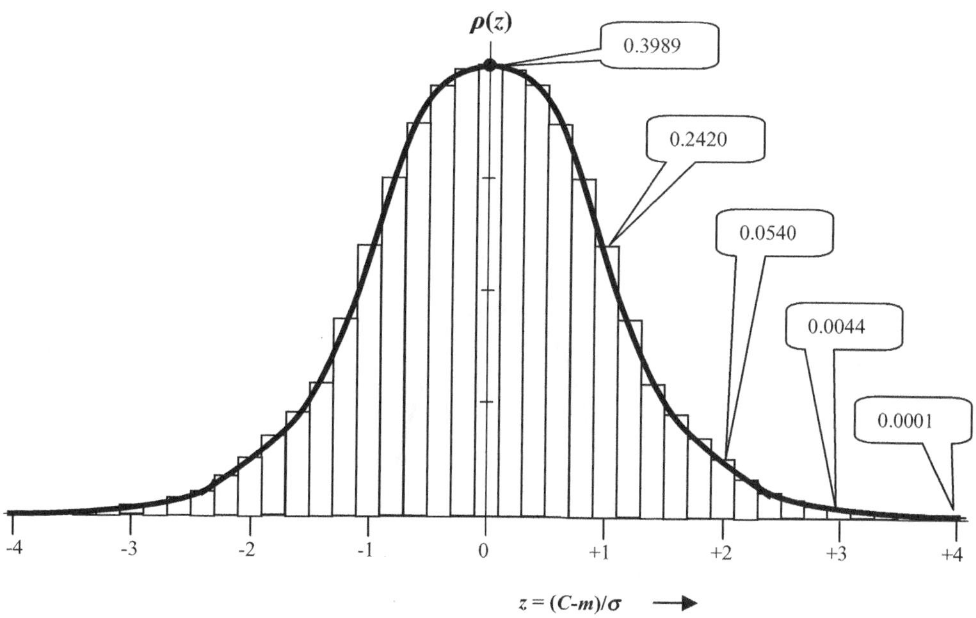

Figure 17.A.2 Distribution of the random variable, C, is a normal probability function.

Figure 17.A.2 shows the normal probability density superimposed on the histogram of the capacitor as shown in Figure 17.A.1; the two are well-matched. This verifies the assertion that the random variable C is a Gaussian or normal probability function. The abscissa scale, z, is a dimensionless and normalized scale, which corresponds to its normal Gaussian distribution.

A Gaussian distribution is a function symmetrical to $z = 0$, and it becomes a normal distribution when

$$m = 0, \tag{17.A.8}$$

and

$$\sigma = 1. \tag{17.A.9}$$

It is interesting to see how many capacitors there would be when z is varied from 0 to z, or C is varied from m to $m + z\sigma$. This is an integral of the function $\rho(z)$, integrated from $z = 0$ to $z = z$, or from $C = m$ to $C = m + z\sigma$, that is,

$$f(z) = \int_0^z \rho(x)\,dx = \int_0^z \frac{e^{-\frac{x^2}{2\sigma^2}}}{\sqrt{2\pi}}\,dx, \tag{17.A.10}$$

The integral, $f(z)$, is the shadowed area as shown in Figure 17.A.3.

The normal probability function is related to error function, *erf(x)*, such that

$$f(z) = \int_0^z \frac{e^{-\frac{x^2}{2}}}{\sqrt{2\pi}}\,dx = \frac{1}{2}\,erf\left(\frac{z}{\sqrt{2}}\right), \tag{17.A.11}$$

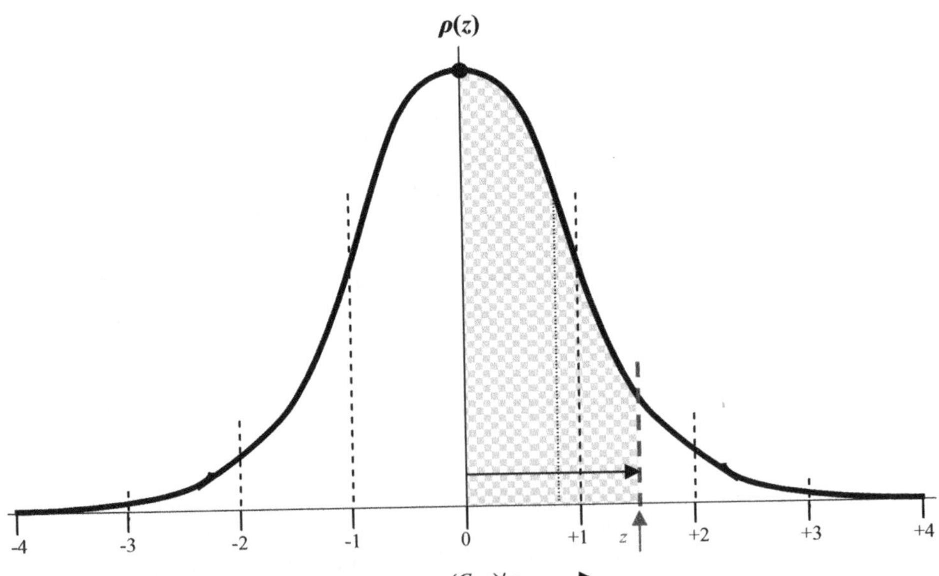

Figure 17.A.3 Shadowed area, $f(z)$, is an integral of $\rho(z)$ from 0 to z.

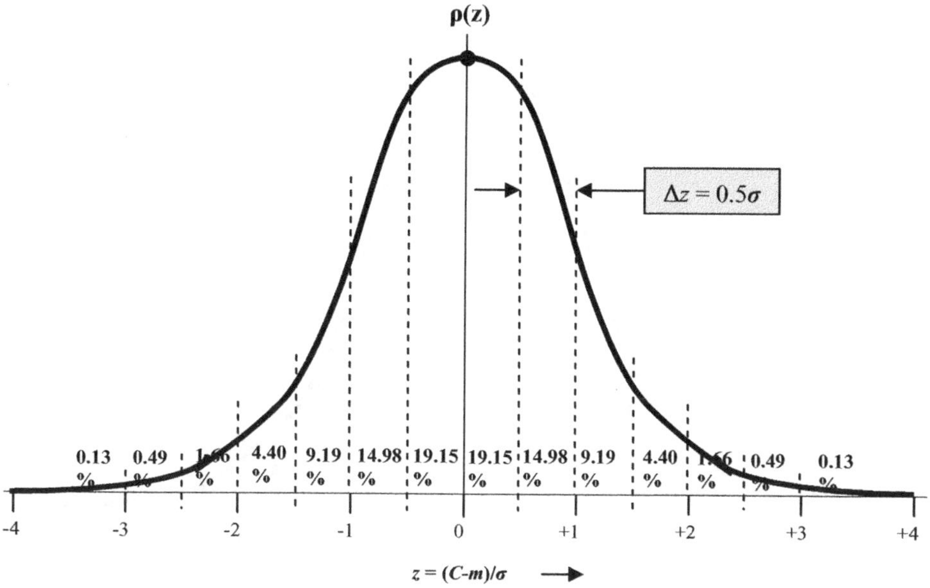

Figure 17.A.4 Yield rate $f(z)$, the percentage of a random variable with a normal distribution at each interval $\Delta z = 0.5$.

and

$$erf(x) = \frac{2}{\sqrt{\pi}} \int_0^x e^{-y^2} dy. \qquad (17.A.12)$$

The value of $f(z)$ or the shadowed area in Figure 17.A.3, depends on the integrated interval along the abscissa. Figure 17.A.4 shows the values of $f(z)$ in each interval of σ.

It can be seen that

when the interval is	$-0.0\sigma < z\sigma = (C\text{-}m) < +0.0\sigma$,	then the area is $f(z) = 0.00\%$,
when the interval is	$-0.5\sigma < z\sigma = (C\text{-}m) < +0.5\sigma$,	then the area is $f(z) = 38.30\%$,
when the interval is	$-1\sigma < z\sigma = (C\text{-}m) < +1\sigma$,	then the area is $f(z) = 68.26\%$,
when the interval is	$-1.5\sigma < z\sigma = (C\text{-}m) < +1.5\sigma$,	then the area is $f(z) = 86.64\%$,
when the interval is	$-2\sigma < z\sigma = (C\text{-}m) < +2\sigma$,	then the area is $f(z) = 95.44\%$,
when the interval is	$-2.5\sigma < z\sigma = (C\text{-}m) < +2.5\sigma$,	then the area is $f(z) = 98.75\%$,
when the interval is	$-3\sigma < z\sigma = (C\text{-}m) < +3\sigma$,	then the area is $f(z) = 99.74\%$,
when the interval is	$-3.5\sigma < z\sigma = (C\text{-}m) < +3.5\sigma$,	then the area is $f(z) = 99.98\%$,
when the interval is	$z\sigma < -3\sigma$, and $z\sigma > +3\sigma$,	then the area is $f(z) = 0.26\%$,

Figure 17.A.5 shows the various terminologies of a process with normal distribution.

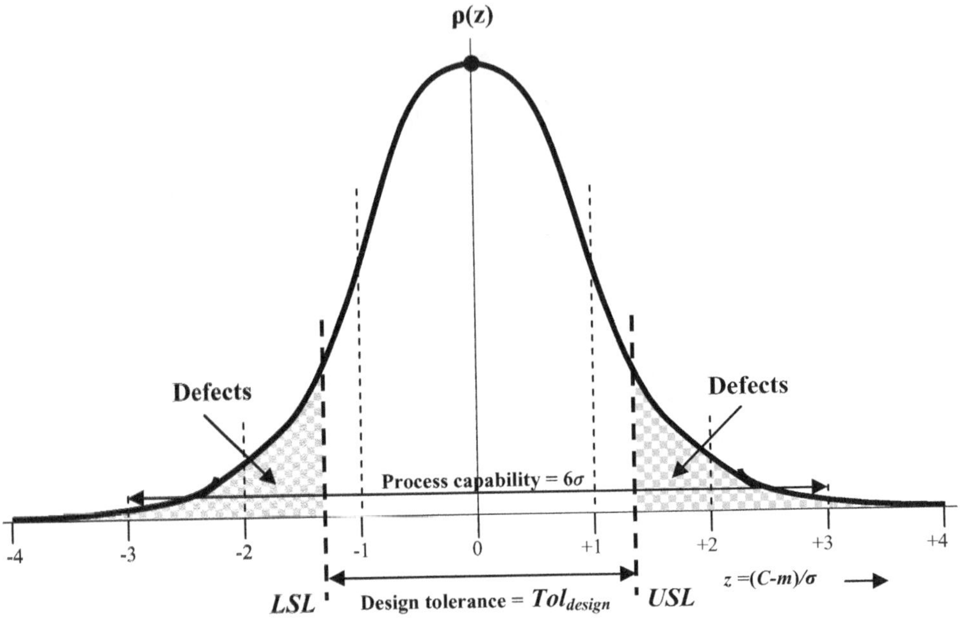

Figure 17.A.5 Definition of design tolerance Tol_{design} and process capability 6σ.

17.A.2 Indices C_p and C_{pk} Applied in 6σ Design

- **The Capability Index C_p**

  In order to quantitatively describe the distance of the design tolerance from the expected high process capability 6σ, the capability index is defined as

$$C_p = \frac{design\_tolerance}{process\_capability} \approx \frac{USL - LSL}{6\sigma}, \tag{17.A.13}$$

where

USL = upper specification limit, σ,
LSL = lower specification limit, σ.

- **Adjusted Capability Index, C_{pk}**

  In order to quantitatively describe the deviation of the actual process mean from the nominal mean, another index, the **Adjusted Capability Index**, C_{pk}, is defined as

$$C_{pk} = C_p(1 - K), \tag{17.A.14}$$

where

$$K = \frac{m - m'}{(USL - LSL)/2}, \tag{17.A.15}$$

where

m = nominal process mean, σ,
m' = Actual process mean, σ.

TABLE 17.1 Table of the normal distribution

Z	0	1	2	3	4	5	6	7	8	9
0.0	.0000	.0040	.0080	.0120	.0160	.0199	.0239	.0279	.0319	.0359
0.1	.0398	.0438	.0478	.0517	.0557	.0596	.0636	.0675	.0714	.0754
0.2	.0793	.0832	.0871	.0910	.0948	.0987	.1026	.1064	.1103	.1141
0.3	.1179	.1217	.1255	.1293	.1331	.1368	.1406	.1443	.1480	.1517
0.4	.1554	.1591	.1628	.1664	.1700	.1736	.1772	.1808	.1844	.1879
0.5	.1915	.1950	.1985	.2019	.2054	.2088	.2123	.2157	.2190	.2224
0.6	.2258	.2291	.2324	.2357	.2389	.2422	.2454	.2486	.2518	.2549
0.7	.2580	.2612	.2642	.2673	.2704	.2734	.2764	.2794	.2823	.2852
0.8	.2881	.2910	.2939	.2967	.2996	.3023	.3051	.3078	.3016	.3133
0.9	.3135	.3186	.3212	.3238	.3264	.3289	.3315	.3340	.3365	.3389
1.0	.3413	.3438	.3461	.3485	.3508	.3531	.3554	.3577	.3599	.3621
1.1	.3643	.3665	.3686	.3708	.3729	.3749	.3770	.3790	.3810	.3830
1.2	.3849	.3869	.3888	.3907	.3925	.3944	.3962	.3980	.3997	.4015
1.3	.4032	.4049	.4066	.4082	.4099	.4115	.4131	.4147	.4162	.4177
1.4	.4192	.4207	.4222	.4236	.4251	.4265	.4279	.4292	.4306	.4319
1.5	.4332	.4345	.4357	.4370	.4382	.4394	.4406	.4418	.4429	.4441
1.6	.4452	.4463	.4474	.4484	.4495	.4505	.4515	.4525	.4535	.4545
1.7	.4554	.4564	.4573	.4582	.4591	.4599	.4608	.4616	.4625	.4633
1.8	.4641	.4649	.4656	.4664	.4671	.4678	.4686	.4693	.4699	.4706
1.9	.4713	.4719	.4726	.4732	.4738	.4744	.4750	.4756	.4761	.4767
2.0	.4772	.4778	.4783	.4778	.4793	.4798	.4803	.4808	.4812	.4817
2.1	.4821	.4826	.4830	.4834	.4838	.4842	.4846	.4850	.4854	.4857
2.2	.4861	.4864	.4868	.4871	.4875	.4878	.4881	.4884	.4887	.4890
2.3	.4893	.4896	.4898	.4901	.4904	.4906	.4909	.4911	.4913	.4916
2.4	.4918	.4920	.4922	.4925	.4925	.4929	.4931	.4932	4934	.4936
2.5	.4938	.4940	.4941	.4943	.4945	.4946	.4948	.4949	.4951	.4952
2.6	.4953	.4955	.4956	.4957	.4959	.4960	.4961	.4962	.4963	.4964
2.7	.4965	.4966	.4967	.4968	.4969	.4970	.4971	.4972	.4973	.4974
2.8	.4974	.4975	.4976	.4977	.4977	.4978	.4979	.4979	.4980	.4981
2.9	.4981	.4982	.4982	.4983	.4984	.4984	.4985	.4985	.4986	.4986
3.0	.4987	.4987	.4987	.4988	.4988	.4989	.4989	.4989	.4990	.4990
3.1	.4990	.4991	.4991	.4991	.4992	.4992	.4992	.4992	.4993	.4993
3.2	.4993	.4993	.4994	.4994	.4994	.4994	.4994	.4995	.4995	.4995
3.3	.4995	.4995	.4995	.4996	.4996	.4996	.4996	.4996	.4996	.4997
3.4	.4997	.4997	.4997	.4997	.4997	.4997	.4997	.4997	.4997	.4998
3.5	.4998	.4998	.4998	.4998	.4998	.4998	.4998	.4998	.4998	.4998
3.6	.4998	.4998	.4999	.4999	.4999	.4999	.4999	.4999	.4999	.4999
3.7	.4999	.4999	.4999	.4999	.4999	.4999	.4999	.4999	.4999	.4999
3.8	.4999	.4999	.4999	.4999	.4999	.4999	.4999	.4999	.4999	.4999
3.9	.5000	.5000	.5000	.5000	.5000	.5000	.5000	.5000	.5000	.5000

17.A.3 Table of the Normal Distribution

Table 17.A.1 shows the table of the normal distribution expressed by the equation

$$f(z)=\int_0^z \frac{e^{-\frac{x^2}{2}}}{\sqrt{2\pi}}dx=\frac{1}{2}erf\left(\frac{z}{\sqrt{2}}\right) \tag{17.A.16}$$

REFERENCES

[1] G. Taguchi, *Experimental Designs*, 3rd ed., Vol. 1 (in Japanese), Maruzen Publishing Company, Tokyo, Japan, 1976.

[2] G. Taguchi, *Experimental Designs*, 3rd ed., Vol. 2, (In Japanese), Maruzen Publishing Company, Tokyo, Japan, 1977.

[3] G. Taguchi, "Off-Line and On-Line Quality Control System," Proceedings of International Conference on Quality Control, Tokyo, Japan, 1978.

[4] J. M. Juran, *Quality Control Handbook*, 3rd ed., McGraw-Hill, New York, 1979.

[5] R. K. Brayton, S. W. Director, and G. D. Hachtel, "Yield Maximization and Worst-Case Design with Arbitrary Statistical Distributions," *IEEE Transactions on Circuits and Systems*, CAS-27, 1908, pp. 756–764.

[6] Raghu N. Kackar, "Off-Line Quality Control, Parameter Design, and the Taguchi Method," *Journal of Quality Technology*, Vol. 17, No. 4, October 1985, pp. 176–188.

[7] Victor E. Kane, "Process Capability Indices," *Journal of Quality Technology*, Vol. 18, No. 1, January 1986, pp. 41–52.

[8] Jonghae Kim, Jean-Olivier Plouchart, and Noah Zamdmer, "Design and Manufacturability Aspect of SO1 CMOS RFICs," IEEE 2004 Custom Integrated Circuits Conference, 2004, pp. 541–548.

[9] Richard C. Li, *Key Issues in RF/RFIC Circuit Design*, Higher Education Press, Beijing, 2005.

PART III

RF SYSTEM ANALYSIS

CHAPTER 18

MAIN PARAMETERS AND SYSTEM ANALYSIS IN *RF* CIRCUIT DESIGN

18.1 INTRODUCTION

With the exceptions of the oscillator or the *VCO*, the main task of *RF* circuit design is to ensure good power transportation or power manipulation in most electronic products or systems, such as the wireless communication system, navigation system, electronic control system, and so on. Consequently, the main parameters in *RF* circuit design to be seriously considered are:

1) Power gain

   In *RF* circuit design, it is very often necessary to intensify a weak signal to an appropriate power amount when the signal is passed over an *RF* block. In other words, the designed *RF* block must have enough power gain. For instance, in the receiver of a communication system, the signal sensed by the antenna is a very weak carrier modulated with desired digital data or other signals. The demodulator is able to detect the desired signal from the carrier only if the power of the weak carrier is intensified to reach or exceed a "conceivable" level. The conceivable level depends on the incoming noise power, type of demodulator, and so on. In a transmitter, the power of the modulated signal must be intensified to become a strong signal so that it can be transmitted by the antenna and then propagated to the receiver, which is located a long distance from the transmitter.

   It should be noted that power gain is not equal to voltage gain. It is well known that power is equal to the product of voltage and current, that is,

$$P = v \cdot i = \frac{v^2}{z} = i^2 z, \tag{18.1}$$

RF Circuit Design, by Richard Chi-Hsi Li

where

P = power of signal,
v = voltage of signal,
i = current of signal,
z = impedance, which the signal voltage is dropped over or the current of signal is flowing through.

Power gain is emphasized in *RF* circuit design while voltage gain is emphasized in digital circuit design. A *RF* block with a high voltage gain may have very low power gain if its input impedance is very low and its output impedance is very high, or vice versa.

2) Noise
Whether the power of an *RF* signal is strong or weak is relative to the power of noise "attached" to the *RF* signal. If the noise power is higher than the power of the signal, the signal will be "swallowed" by the noise and become undetectable. For instance, the ratio of carrier to noise ratio, *CNR*, at the demodulator input must higher than a threshold value, say, $10\,dB$. Otherwise, the signal cannot be demodulated and separated from the carrier.

Therefore, noise is another important parameter in the *RF* circuit design.

3) Linearity
The third important parameter in *RF* circuit design concerns the distortion of the weak signal. In order to ensure the fidelity of power transportation and manipulation, distortion of the desired signal must be avoided or reduced as much as possible.

The source of the distortion is mainly caused by the non-linearity of the device and all the interferences in the circuit block. A practical device has linear and non-linear portions. Its non-linear terms produce many harmonics and spurious products which contribute to the distortion of the signal.

Just like noise, the harmonics, spurious products, and interferences in the circuit block are not welcome in *RF* circuit design. However, they are essentially different from noise. Noise is a random object. The frequency spectrum of a "white" noise spreads from minus infinity to plus infinity so that it cannot be effectively filtered out entirely by a practical filter. On the contrary, the spurious products and interferences are definite disturbing signals with definite frequencies, so that they can be reduced or removed by means of special circuit design schemes.

The effects of noise and spurious products on a receiver are different. Noise is directly related to the sensitivity of the receiver. The sensitivity of a receiver is high if its noise figure is low and vice versa. The spurious products do cause distortion of the signal, but they are not directly related to the sensitivity of a receiver. For instance, a receiver with low noise has high sensitivity. However, it could have either high or low distortion of the voice. In other words, a receiver can sense the very weak signal at the antenna, but the purity of the signal after demodulation mainly depends on the distortion. In cases where distortion is high, a girl's voice could have a boy's sound if the sensitivity of the receiver is high. In cases when distortion is low, the fidelity of the voice could be kept even though the sensitivity of the receiver is low.

In addition to the three main parameters mentioned above, other important parameters commonly concerned in *RF* circuit design are

4) Stability,
5) *DC* power supply,
6) Current drain,
7) Part count,
8) Special parameters in some special *RF* circuit blocks, such as
   - Load pulling effect to a *VCO*,
   - Phase noise of a *VCO*,
   - *PAE* of a power amplifier,
   - Sensitivity of a receiver,
   - Output power of a transmitter.

18.2 POWER GAIN

Generally, there are three well-known gains in an electronic circuit block: the power gain, voltage gain, and current gain. The meaning of voltage and current gain is quite straight forward. It is simply the ratio between input and output. However, there are different power gains with different meanings.

Power gains can be explained by the signal flow graph as introduced in Appendix 18.A.3. Instead of the signal flow graph, we are going to explain power gains through a new concept, reflection power gain. The derivation of reflection power gain seems more intuitive than the derivation by means of the signal flow graph.

18.2.1 Basic Conception of Reflection Power Gain

It is easy to understand that amplification power gain can be obtained from the input to the output power of an amplifier due to amplification. However, it may be a little hard to understand that reflection power gain can be obtained due to power reflection at a terminal or at a load. Let's introduce the concept of reflection power gain through the concept of amplification power gain.

For an amplifier, power gain due to amplification is defined as the ratio of output power and input power. Usually, this definition is related to two nodes, the input and output nodes of the amplifier, when the amplifier exists. However, this definition of power gain due to amplification can be replaced by a new concept, in which the definition is not related to both input and output nodes, but to the output node only. Now, the power gain due to amplification is defined as a ratio of the output power when the amplifier exists to the output power when the amplifier does not exist and the output node is directly connected to the input node. Figure 18.1 illustrates this new definition.

The expression of power gain is kept the same before and after the definition of power gain is replaced by such a new concept, that is,

$$G_{amp} = \frac{P_o}{P_i}, \tag{18.1}$$

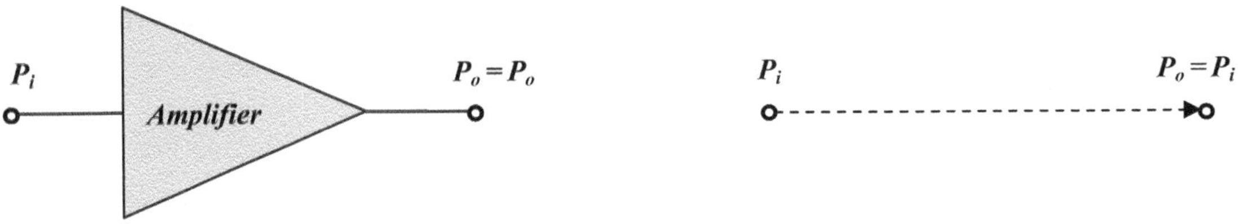

(a) At output of amplifier, $P_o = P_o$ when amplifier exists

(b) At output of amplifier, $P_o = P_i$ when amplifier does not exist

Figure 18.1 Definition of power gain is based on the readings only at output node in two cases when amplifier does or does not exist, $G_{amp} = P_o/P_i$.

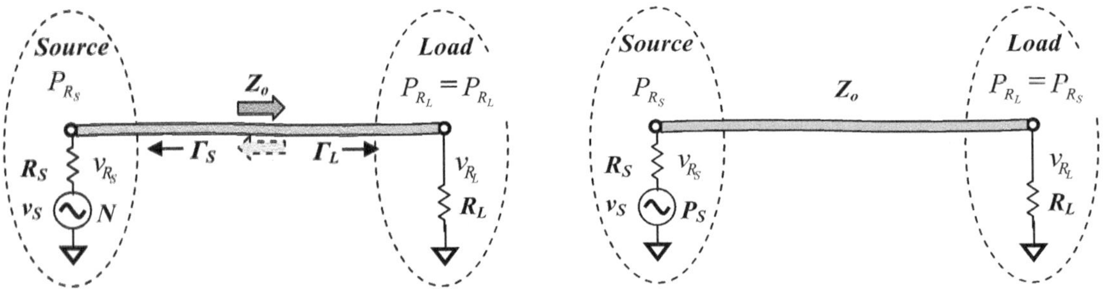

(a) At load, $P_{R_L} = P_{R_L}$ when reflection is existed

(b) At load, $P_{R_L} = P_{R_S}$ when reflection is not existed

Figure 18.2 Definition of power gain is based on the readings at the load in two cases when reflections do or do not exist.

or,

$$G_{amp,dB} = P_o - P_i \ dB, \tag{18.2}$$

where

G_{amp} = power gain due to amplification, measured by *Watt* or *mW*,
$G_{amp,dB}$ = power gain due to amplification, measured by *dB*.

In terms of this new concept, power gain due to reflection existing in either the source or load can be well defined. Figure 18.2 illustrates the definition of power gain due to reflection based on the readings at load only.

Similar to the definition of amplification power gain in expression (18.1), the reflection power gain is defined as a ratio of the delivered power from source to load when reflections Γ_S and Γ_L exist and the delivered power from source to load when reflections Γ_S and Γ_L do not exist, that is,

TABLE 18.1 Corresponding relations between power gain due to amplification and power gain due to reflection

Power gain due to amplification	Power gain due to reflection
Amplification	Reflection
Input	Source
Output	load
P_i	P_{R_S}
P_o	P_{R_L}

$$G_{ref} = \frac{P_{R_L}\big|_{with-reflection}}{P_{R_L}\big|_{without-reflection}}, \tag{18.3}$$

where

G_{ref} = power gain due to reflection.

$P_{R_L}\big|_{with-\mathrm{Re}\,flections}$ = delivered power from source to load when reflections exist

$P_{R_L}\big|_{without-\mathrm{Re}\,flections}$ = delivered power from source to load when reflections do not exist

By comparing power gain due to amplification as shown in Figure 18.1 and power gain due to reflection as shown in Figure 18.2, the corresponding relations of these two power gains are listed in Table 18.1.

Now let's derive the expressions for $P_{R_L}\big|_{with-\mathrm{Re}\,flections}$ and $P_{R_L}\big|_{without-\mathrm{Re}\,flections}$ so that the expression of power gain due to reflection can be described as a function of the reflection coefficients Γ_S and Γ_L.

In cases where reflections exist, recalling expressions (9.7), (9.9), (9.11), and (9.14),

$$v_S = v_{So}e^{j\omega t}, \tag{9.7}$$

$$v_{R_L}\big|_{t=T_d} = v_{So}\frac{R_L}{Z_S + Z_L}(1-\Gamma_L)\sum_{n=0}^{\infty} e^{j\omega[t-(2n+1)T_d]}(\Gamma_S\Gamma_L)^n. \tag{9.9}$$

If

$$T_d \to 0, \tag{9.11}$$

Then

$$v_{R_L}\big|_{t=T_d} = v_{So}e^{j\omega t}\frac{R_L}{Z_S + Z_L}\frac{1-\Gamma_L}{1-\Gamma_S\Gamma_L}, \tag{9.14}$$

where T_d = delay time while the delivered signal travels from source to load.

Without loss of generality, for simplicity let's assume that both the impedances of source and load are pure resistance. Then, Figure 9.1 can be re-drawn as Figure 18.2(a) and expression (9.14) becomes

$$v_{R_L}\big|_{t=T_d} = v_{So}e^{j\omega t}\frac{R_L}{R_S+R_L}\frac{1}{1-\Gamma_S\Gamma_L} - v_{So}e^{j\omega t}\frac{R_L}{R_S+R_L}\frac{\Gamma_L}{1-\Gamma_S\Gamma_L}, \quad (18.4)$$

It should be noted that the delivered voltage from source to load v_{R_L} in expression (18.4) contains two terms. The first term is the incident voltage and the second is the resultant voltage reduction due to reflection. Their corresponding incident power and the resultant power reduction due to reflection are

$$P_{R_L}\big|_{with-\mathrm{Re}flection} = v_{So}^2 e^{j2\omega t}\frac{R_L}{(R_S+R_L)^2}\frac{1}{|1-\Gamma_S\Gamma_L|^2} - v_{So}^2 e^{j2\omega t}\frac{R_L}{(R_S+R_L)^2}\frac{|\Gamma_L|^2}{|1-\Gamma_S\Gamma_L|^2}, \quad (18.5)$$

They can be combined together as

$$P_{R_L}\big|_{with-reflection} = v_{So}^2 e^{j2\omega t}\frac{R_L}{(R_S+R_L)^2}\frac{1-|\Gamma_L|^2}{|1-\Gamma_S\Gamma_L|^2}, \quad (18.6)$$

Obviously, for the cases when the reflections do not exist, that is, $\Gamma_S = \Gamma_L = 0$, then

$$P_{R_L}\big|_{without-reflection} = v_{So}^2 e^{j2\omega t}\frac{R_L}{(R_S+R_L)^2}, \quad (18.7)$$

Substituting (18.6) and (18.7) into (18.3), we have

$$G_{ref} = \frac{1-|\Gamma_L|^2}{|1-\Gamma_S\Gamma_L|^2}. \quad (18.8)$$

From expression (18.8) it can be seen that the reflection power gain can be positive if both Γ_L and Γ_S are less than 1 but greater than zero. This is not surprising because the power on the load in the case with reflection can be higher than that in the case without reflection at the load but with reflection at the source.

18.2.2 Transducer Power Gain

Figure 18.3 shows the various powers, power gains, voltage reflection coefficients, and S parameters in a two-port block.

In Figure 18.3 there are four powers denoted by

1) P_{IB} = power input to the block,
2) P_{AVS} = power available from the source,
3) P_L = power delivered to the load,
4) P_{AVB} = power available from the block, P_{AVB}.

The transducer power gain is a power gain from source to load in cases where neither input nor output impedances are matched. As shown in Figure 18.3, the

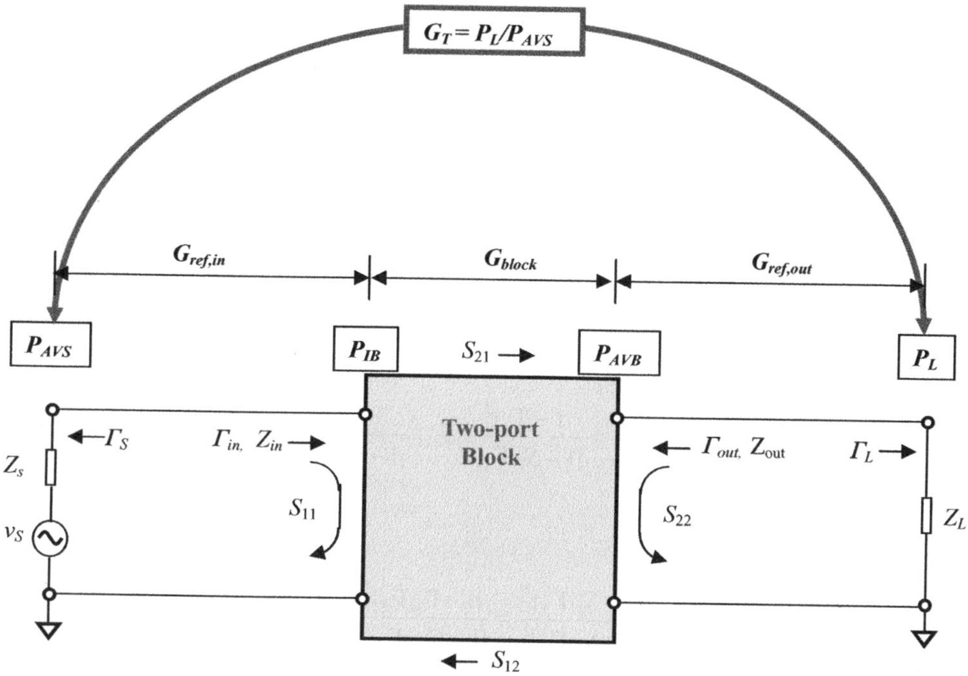

Figure 18.3 The various powers, power gains, voltage reflection coefficients, and S parameters in a two-port block.

transducer power gain is defined as a ratio of power delivered to the load P_L and power available from the source P_{AVS}. It consists of three portions, that is,

$$G_T = \frac{P_L}{P_{AVS}} = G_{ref,in} \cdot G_{block} \cdot G_{ref,out}. \tag{18.9}$$

According to the discussion in Section 18.2.1, the reflection power gains at the input and output portions are

$$G_{ref,in} = \frac{1-|\Gamma_{in}|^2}{|1-\Gamma_S\Gamma_{in}|^2}, \tag{18.10}$$

and

$$G_{ref,out} = \frac{1-|\Gamma_L|^2}{|1-S_{22}\Gamma_L|^2}, \tag{18.11}$$

respectively.

Instead Γ_{out}, S_{22} in expression (18.11) represents the reflection at output of the block when the incident power is coming from the load. S_{22} in expression (18.11) looks just like Γ_S in expression (18.10).

The power gain G_{block} is obviously

$$G_{block} = |S_{21}|^2. \tag{18.12}$$

Substituting expressions (18.10) to (18.12) into (18.9), we have

$$G_T = \frac{P_L}{P_{AVS}} = \frac{1-|\Gamma_s|^2}{|1-\Gamma_{in}\Gamma_s|^2}|S_{21}|^2\frac{1-|\Gamma_L|^2}{|1-S_{22}\Gamma_L|^2}, \tag{18.13}$$

or

$$G_T = \frac{P_L}{P_{AVS}} = \frac{1-|\Gamma_s|^2}{|1-S_{11}\Gamma_s|^2}|S_{21}|^2\frac{1-|\Gamma_L|^2}{|1-\Gamma_{out}\Gamma_L|^2}, \tag{18.14}$$

because

$$\frac{|1-S_{11}\Gamma_S|^2}{|1-S_{22}\Gamma_L|^2} = \frac{|1-\Gamma_{in}\Gamma_S|^2}{|1-\Gamma_{out}\Gamma_L|^2}. \tag{18.15}$$

In addition to the transducer power gain, G_T, the operating power gain G_P is defined as a ratio of the power delivered to the load P_L to the power input to the block P_{IB}, that is,

$$G_P = \frac{P_L}{P_{IB}} = G_T|_{\Gamma_{in}=\Gamma_S^*} = \frac{1}{1-|\Gamma_{in}|^2}|S_{21}|^2\frac{1-|\Gamma_L|^2}{|1-S_{22}\Gamma_L|^2}, \tag{18.16}$$

because

$$P_{IB} = P_{AVS}|_{\Gamma_S^*=\Gamma_{in}^*}, \tag{18.17}$$

when

$$\Gamma_S^* = \Gamma_{in}. \tag{18.18}$$

The third power gain, the available power gain G_A, is defined as the ratio of the power available from the block, P_{AVB}, to the power available from the source, P_{AVS}, that is,

$$G_A = \frac{P_{AVB}}{P_{AVS}} = G_T|_{\Gamma_{out}=\Gamma_L^*} = \frac{1-|\Gamma_s|^2}{|1-S_{11}\Gamma_s|^2}|S_{21}|^2\frac{1}{1-|\Gamma_{out}|^2}, \tag{18.19}$$

because

$$P_{AVB} = P_L|_{\Gamma_{out}=\Gamma_L^*}, \tag{18.20}$$

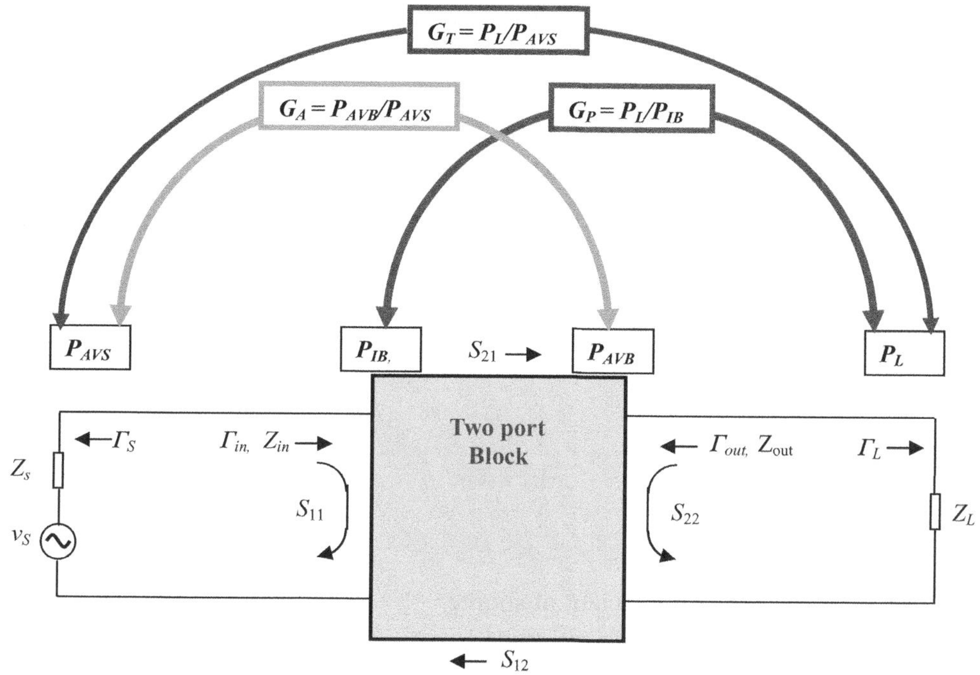

Figure 18.4 Transducer power gain, operating power gain, and available power gain in a two-port block.

when

$$\Gamma_{out} = \Gamma_L^*. \tag{18.21}$$

Figure 18.4 shows the relationships between the four powers, $\boldsymbol{P}_{IB}$, $\boldsymbol{P}_{\text{AVS}}$, $\boldsymbol{P}_L$, $\boldsymbol{P}_{\text{AVB}}$, and the three power gains G_T, G_P, $\boldsymbol{G}_A$.

The transducer power gain G_T describes the power gain of a 2 port block in a general form. The operating power gain G_P is a special case of the transducer power gain G_T when the input impedance of the block is well matched. The available power gain $\boldsymbol{G}_A$ is a special case of the transducer power gain G_T when the output impedance of the block is well matched. In general, the transducer power gain G_T is lower than G_P or $\boldsymbol{G}_A$.

18.2.3 Power Gain in a Unilateral Case

In general cases, the relations of the various voltage reflection coefficients, Γ_S, Γ_L, Γ_{in}, Γ_{out}, in a two-port block are

$$\Gamma_{in} = S_{11} + \frac{S_{12}S_{21}\Gamma_L}{1 - S_{22}\Gamma_L}, \tag{18.22}$$

$$\Gamma_{out} = S_{22} + \frac{S_{12}S_{21}\Gamma_s}{1 - S_{11}\Gamma_s}. \tag{18.23}$$

When the voltage reflections at the source and load of a two-port block exist, that is

$$\Gamma_s \neq 0, \tag{18.24}$$

and

$$\Gamma_L \neq 0, \tag{18.25}$$

then

$$\Gamma_{in} \neq S_{11}, \tag{18.26}$$

and

$$\Gamma_L \neq S_{22}, \tag{18.27}$$

where

Γ_s = voltage reflection coefficient at source,
Γ_L = voltage reflection coefficient at load,
Γ_{in} = voltage reflection coefficient at input of two-port block,
Γ_{out} = voltage reflection coefficient at output of two-port block.

In the special case when the block becomes **unilateral**, that is, when

$$S_{12} = 0, \tag{18.28}$$

then,

$$\Gamma_{in} = S_{11}, \tag{18.29}$$

$$\Gamma_{out} = S_{22}, \tag{18.30}$$

and then expressions (18.13) and (18.14) become

$$G_T = G_{T,S_{12}=0} = G_S G_o G_L = \frac{1-|\Gamma_s|^2}{|1-S_{11}\Gamma_s|^2}|S_{21}|^2\frac{1-|\Gamma_L|^2}{|1-S_{22}\Gamma_L|^2}, \tag{18.31}$$

where

$$G_S = \frac{1-|\Gamma_s|^2}{|1-S_{11}\Gamma_s|^2}, \tag{18.32}$$

$$G_o = |S_{21}|^2, \tag{18.33}$$

$$G_L = \frac{1-|\Gamma_L|^2}{|1-S_{22}\Gamma_L|^2}. \tag{18.34}$$

It can be seen that the transducer power gain G_T is composed of three different independent power gains when $S_{12} = 0$.

In the unilateral case, the operating power gain and available power gain become

$$G_P = G_{P,S12=0} = \frac{1}{1-|S_{11}|^2}|S_{21}|^2\frac{1-|\Gamma_L|^2}{|1-S_{22}\Gamma_L|^2}, \quad (18.35)$$

and

$$G_A = G_{A,S12=0} = \frac{1-|\Gamma_s|^2}{|1-S_{11}\Gamma_s|^2}|S_{21}|^2\frac{1}{1-|S_{22}|^2}. \quad (18.36)$$

respectively.

18.2.4 Power Gain in a Unilateral and Impedance Matched Case

Let's examine an ideal case, in which the block is not only unilateral but also reflection-free at both the source and the load, that is

$$S_{12} = 0, \quad (18.37)$$

and

$$\Gamma_s = \Gamma_L = 0, \quad (18.38)$$

which are the conditions of impedance matching.

In the unilateral and impedance matched case,

$$G_S = G_{S,S12=0,\Gamma_S=\Gamma_L=0} = 1, \quad (18.39)$$

$$G_o = G_{o,S12=0,\Gamma_S=\Gamma_L=0} = |S_{21}|^2, \quad (18.40)$$

$$G_L = G_{L,S12=0,\Gamma_S=\Gamma_L=0} = 1, \quad (18.41)$$

$$G_P = G_{P,S12=0,\Gamma_S=\Gamma_L=0} = \frac{1}{1-|S_{11}|^2}|S_{21}|^2, \quad (18.42)$$

$$G_A = G_{A,S12=0,\Gamma_S=\Gamma_L=0} = \frac{1}{1-|S_{22}|^2}|S_{21}|^2, \quad (18.43)$$

$$G_T = G_{T,S12=0,\Gamma_S=\Gamma_L=0} = |S_{21}|^2. \quad (18.44)$$

Design engineers usually present their simulation results of $|S_{21}|^2$ as power gain. This implies that it is a transducer power gain G_o, not only in the unilateral case, $S_{12} = 0$, but also in cases without reflection at both the source and the load, $\Gamma_S = \Gamma_L = 0$.

18.2.5 Power Gain and Voltage Gain

In a digital circuit design, the main task is to realize the digital status transportation. The engineers always try to reduce power consumption as much as possible. Power gain is not important in digital circuit design. Instead, the voltage gain is the important parameter, by which the status transportation is executed.

On the other hand, in an *RF*/*RFIC* circuit design, the main task is to realize the power transportation. Consequently, power gain is a perpetual concern. The voltage gain is meaningless until the corresponding impedance is specified.

However, as long as the impedance is given, one can transfer the voltage gain to power gain through the well-known relation:

$$G_P = \frac{\dfrac{V_{out}^2}{Z_{out}}}{\dfrac{V_{in}^2}{Z_{in}}} = G_v^2 \frac{Z_{in}}{Z_{out}}. \tag{18.45}$$

where G_p and G_v are the power and voltage gains respectively.

18.2.6 Cascaded Equations of Power Gain

In the system analysis for multiple *RF* blocks or an *RF* subsystem, the system power gain is contributed by the gain of individual *RF* blocks, which can be calculated in terms of so-called "cascaded equation of power gain.".

Let us derive the cascaded equation for power gain of a system containing only two blocks as shown in Figure 18.5.

$$G_k = G_{T,k} = \frac{P_{L,k}}{P_{S,k}}, \tag{18.46}$$

$$G_{k+1} = G_{T,k+1} = \frac{P_{L,k+1}}{P_{S,k+1}}, \tag{18.47}$$

where transducer power gain is the assigned power to be discussed.

Assuming that the input and output impedances between blocks are well matched. then the system power gain, G_{SYS}, should be

$$G_{SYS} = \frac{P_{L,k+1}}{P_{S,k}} = \frac{P_{L,k}}{P_{S,k}} \frac{P_{L,k+1}}{P_{L,k}} = G_k G_{k+1}, \tag{18.48}$$

where

$P_{L,k}$ = power delivered to the load of the kth block,
$P_{L,k+1}$ = power delivered to the load of the $(k + 1)^{th}$ block,
$P_{S,k}$ = power available from the source of the kth block,
$P_{S,k+1}$ = power available from the source of the $(k + 1)^{th}$ block,
G_k = power gain of the kth block,
$G_{T,k}$ = transducer power gain of the kth block,

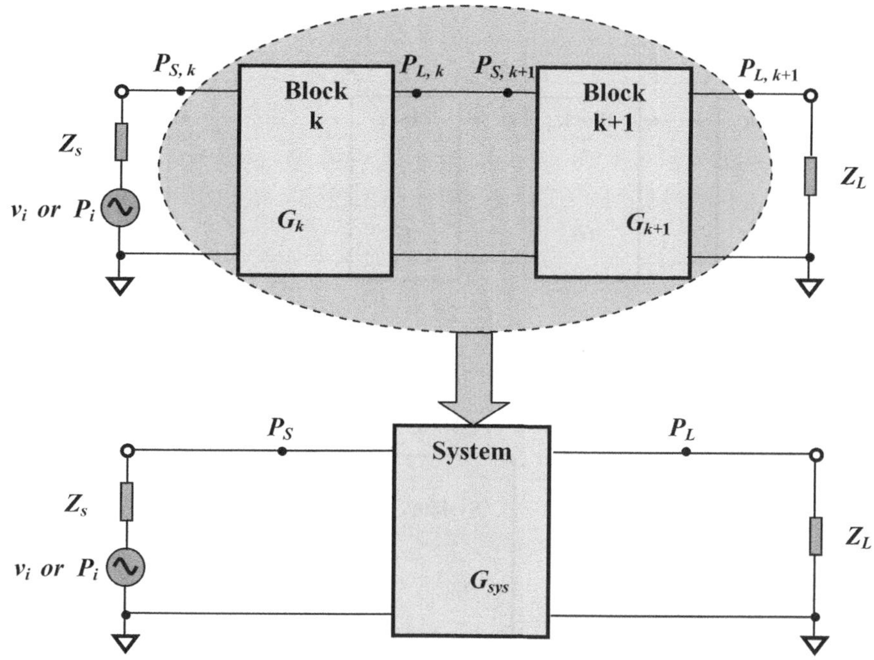

Figure 18.5 Gain of a system cascaded by two blocks k and $k + 1$.

G_{k+1} = power gain of the $(k + 1)^{th}$ block,
$G_{T,k+1}$ = transducer power gain of the $(k + 1)^{th}$ block,
G_{SYS} = power gain of the system in the numeric scale.

and, in the logarithic form or by the unit of dB, equation (18.48) becomes

$$G_{SYS,dB} = G_k + G_{k+1}. \tag{18.49}$$

The system gain, G_{SYS}, is a simple product of the individual gains, G_k and G_{k+1}, in the numeric scale, or in other words, $G_{SYS,dB}$ is a simple sum of the individual gains, G_k and G_{k+1}, in the logarithic scale.

Figure 18.6 shows a general case of a system containing n blocks; equations (18.48) and (18.49) can in this case be extended to

$$G_{SYS} = G_1 G_2 G_3 \ldots . G_n, \tag{18.50}$$

$$G_{SYS,dB} = G_1 + G_2 + G_3 + \ldots + G_n, \tag{18.51}$$

or,

$$G_{SYS} = \prod_1^n G_k, \tag{18.52}$$

$$G_{SYS,dB} = \sum_1^n G_k. \tag{18.53}$$

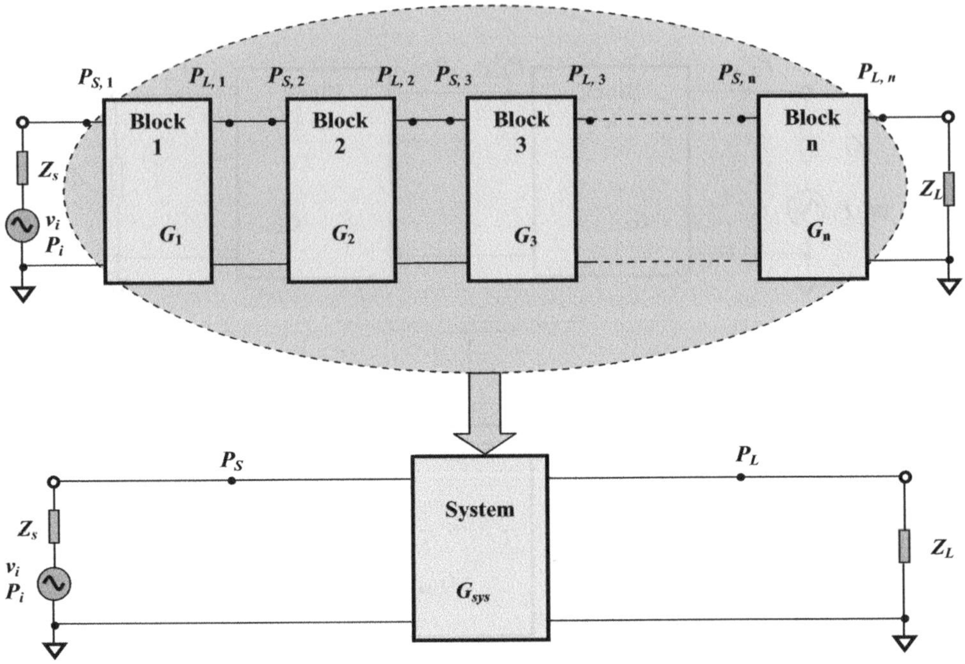

Figure 18.6 Gain of a system cascaded by *n* blocks.

All the equations from (18.48) to (18.53) are called cascaded equations of power gain.

18.3 NOISE

Noise inherently exists in either the active or passive parts of an *RF* circuit. In the micro electrical world, the charge amount of the main electric carriers, electrons, and the positive-charged holes is not a continuous quantity, and the motion of the carriers is a chaotic or random process in the time domain. The macro parameter, which describes the random process, always contains two portions. One represents its average value, while the other represents its random fluctuation from its average value. The random fluctuation can be described by its root-square-mean value or variance. For example, the electric current always contains two portions. One is the current with its average value that we are familiar with. The other is the noise current that is always neglected in fields where such a tiny fluctuation of current is not important. In the electronic circuit design field, the random fluctuation of the charge carriers is one of the main noise sources and should never be neglected. Noise current, noise voltage, and noise power are some of the most important parameters in the performance of a block or a system.

In a circuit or a system, another noise source is its outside environment. When a signal is transported or operated in a circuit block or system, it is inevitably interfered with or imposed on by the noise which radiates from other blocks or systems. This topic is outside limited topics of this book and will not be discussed here;

Figure 18.7 Various parameters related to the definition of noise figure.

readers who are interested in this subject can refer to other specific books or papers.

Noise degrades the performance of the signal. One of the main tasks in circuit design is to reduce or suppress the noise produced by the circuit itself or from the environment outside so as to ensure a good performance of the signal.

18.3.1 Significance of the Noise Figure

Figure 18.7 shows the various parameters in the input and output of a practical *RF* circuit block.

In reality, a noise-free circuit block does not exist. All practical blocks are noisy blocks: a practical circuit block always produces an additional part of noise in addition to the signal. Therefore, its total output noise power is greater than the output noise power due to input noise, that is

$$N_o = GN_i + \Delta N, \tag{18.54}$$

where

ΔN = additional noise power produced by the circuit block,
N_i, N_o = noise power at the input and output respectively.

The additional noise means that the increase of noise is positive:

$$\Delta N > 0. \tag{18.55}$$

On the other hand, the input signal is intensified by the *RF* circuit block to the output signal by the power gain factor G, that is,

$$S_o = GS_i, \tag{18.56}$$

where

S_i, S_o = signal power at input and output respectively,
G = power gain of the circuit block.

From expressions (18.54) and (18.56), the ratio of output signal to output noise is

$$\frac{S_o}{N_o} = \frac{GS_i}{GN_i + \Delta N} = \frac{S_i}{N_i + \frac{\Delta N}{G}} < \frac{S_i}{N_i}, \tag{18.57}$$

It can be seen that the additional noise therefore reduces the ratio of the signal to noise from the input to output.

At any point of a circuit block, the signal can be detected only if the power of the signal is greater than the power of the noise by a certain amount. The power ratio of the signal to noise is a measure of the possibility for detecting a signal in a practical circuit.

Instead of ΔN, a new parameter, the noise figure, NF, can be defined as

$$NF = \frac{\frac{S_i}{N_i}}{\frac{S_o}{N_o}}, \tag{18.58}$$

By the substitution of relation (18.56) into (18.58), we have

$$NF = \frac{N_o}{GN_i} = \frac{N_o}{N_{o,i}}, \tag{18.59}$$

where $N_{o,i} = GN_i$ = Output noise power portion due to the input noise power.

Or, from expressions (18.54) and (18.59),

$$NF = \frac{GN_i + \Delta N}{GN_i}. \tag{18.60}$$

In other words,

$$NF = \frac{\text{Total output noise power}}{\text{Output noise power due to the input noise}}. \tag{18.61}$$

The expressions from (18.58) to (18.61) have the same meaning:

The noise figure of an *RF* circuit block represents additional noise existing in the *RF* circuit block.

In an ideal case when the circuit block is noise free, that is, the additional noise is zero, then from the definition of the noise figure from (18.58) to (18.61),

$$\frac{S_o}{N_o} = \frac{S_i}{N_i}, \tag{18.62}$$

$$N_o = GN_i = N_{o,i}, \tag{18.63}$$

Or,

$$NF = 1\,(\text{or } 0\,dB). \tag{18.64}$$

A noise figure NF of 1 indicates that the input ratio of the signal to noise is equal to the output ratio of signal to noise; in other words, the total output noise is equal to only the output noise caused by the input noise power. In short, there is no additional noise existing in the circuit block if the noise figure of the circuit block is 1, or $0\,dB$.

An ideal or noise-free circuit block never exists in practical RF circuit blocks. Therefore, in a practical RF circuit block, the ratio of signal to noise, S/N, the noise, and the noise figure are always

$$\frac{S_o}{N_o} < \frac{S_i}{N_i}, \tag{18.65}$$

$$N_o > GN_i = N_{o,i}, \tag{18.66}$$

or,

$$NF > 1\,(\text{or } 0\,dB). \tag{18.67}$$

18.3.2 Noise Figure in a Noisy Two-port *RF* Block

As mentioned in Chapter 1, based on Haus's theory (1960), the noise figure of a noisy block can be expressed by

$$NF = NF_{min} + \frac{R_n}{G_s}\left[(G_S - G_{S,opt})^2 + (B_S - B_{S,opt})^2\right]. \tag{18.68}$$

where

$$Y_S = G_S + jB_S, \tag{18.69}$$

$$Y_{S,opt} = G_{S,opt} + jB_{S,opt}, \tag{18.70}$$

and

NF = noise figure of the noisy block,
NF_{min} = minimum of noise figure of the noisy block,
R_n = equivalent noise resistance,
Y_S = admittance of input source,
G_S = conductance of input source,
B_S = subceptance of input source,
$Y_{S,opt}$ = optimum admittance of input source,
$G_{S,opt}$ = optimum conductance of input source,
$B_{S,opt}$ = optimum subceptance of input source.

The profound significance of this equation is that there is an optimum admittance of input source existing in a noisy RF block, when

$$Y_S = Y_{S,opt}, \tag{18.71}$$

or, when

$$G_S = G_{S,opt}, \tag{18.72}$$

$$B_S = B_{S,opt}, \tag{18.73}$$

the noisy two-port block reaches a minimum of noise figure,

$$NF = NF_{\min}. \tag{18.74}$$

The expressions of optimum admittance, $Y_{S,\text{opt}}$, or $G_{S,\text{opt}}$, $B_{S,\text{opt}}$, and the minimum of noise figure, $NF_{\min}$, have been derived from and can be found in Paul R. Gray et al., *Analysis and Design of Analog Integrated Circuits* (2001), and Thomas Lee, *The Design of CMOS Radio-Frequency Integrated Circuits* (1998).

18.3.3 Notes for Noise Figure Testing

In the low *RF* frequency range, say, below the *UHF* frequency range, the noise figure of an *RF* circuit block is directly measured by the noise figure meter as shown in Figure 18.8. Special attention must be given to the setup adjustments, such as

- Testing must be conducted in a shielding room to shield from the *RF* signal;
- All the electric lights and *DC* power supplies, except for the *DC* power supplies of the *DTU* and the noise figure meter, must be turned OFF, as they produce a lot of noise in low frequencies as well as in high frequencies.
- All the cables must be placed in appropriate positions so that the reading of the noise figure in the noise figure meter is at a minimum.

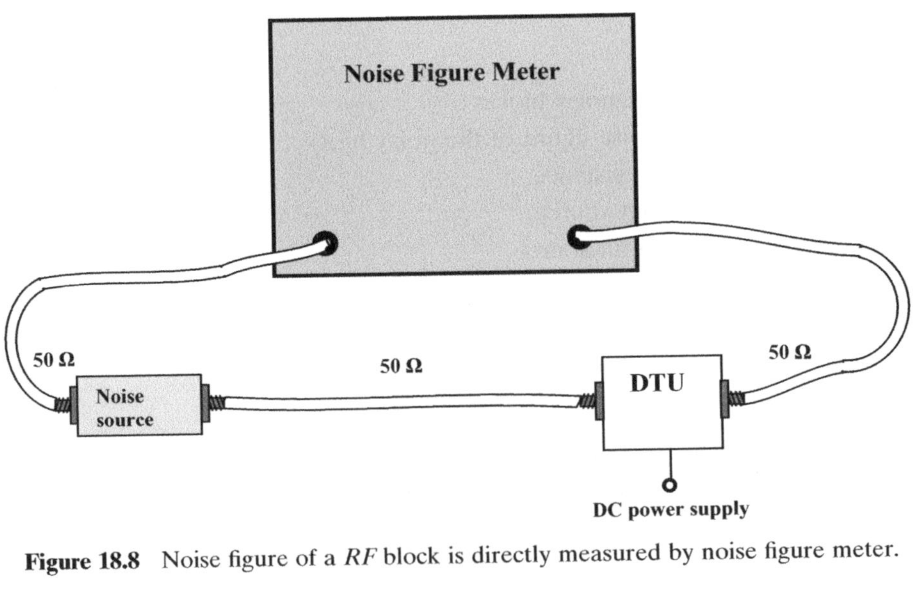

Figure 18.8 Noise figure of a *RF* block is directly measured by noise figure meter.

In the high *RF* frequency range, the noise figure meter is not available. An alternative method is to measure the noise figure by the network analyzer. This is not as simple as measurement by the noise figure meter, but the test is not impacted seriously by the environment as long as the network analyzer is well-calibrated and the noise floor is taken care of.

18.3.4 An Experimental Method to Obtain Noise Parameters

By means of the experimental method one can obtain the noise parameters of an *RF* block. Let us recall expression (18.68),

$$NF = NF_{\min} + \frac{R_n}{G_s}\left[(G_S - G_{S,opt})^2 + (B_S - B_{S,opt})^2\right]. \tag{18.68}$$

There is a remarkable significance in this expression of the noise figure. It indicates that for any noisy two-port circuit block, a minimum of the noise figure can be obtained as long as the source admittance, $Y_s = G_s + jB_s$, is adjusted to its optimum values, $Y_{s,opt} = G_{s,opt} + jB_{s,opt}$.

Mathematically, this equation can be considered as the noise figure *NF* being a function of the variables G_s and B_s, with four noise parameters, NF_{min}, R_n, $G_{s,opt}$, and $B_{s,opt}$. By testing the noise figure *NF* four times with four different source admittances, G_s and B_s, four equations can be established on the basis of expression (18.68) and the four unknown parameters, NF_{min}, R_n, $G_{s,opt}$, and $B_{s,opt}$ can be solved. This is called the four-point experimental method of noise figure testing.

$$NF_1 = NF_{\min} + \frac{R_n}{G_{s1}}\left[(G_{s1} - G_{s,opt})^2 + (B_{s1} - B_{s,opt})^2\right], \tag{18.75}$$

$$NF_2 = NF_{\min} + \frac{R_n}{G_{s2}}\left[(G_{s2} - G_{s,opt})^2 + (B_{s2} - B_{s,opt})^2\right], \tag{18.76}$$

$$NF_3 = NF_{\min} + \frac{R_n}{G_{s3}}\left[(G_{s3} - G_{s,opt})^2 + (B_{s3} - B_{s,opt})^2\right], \tag{18.77}$$

$$NF_4 = NF_{\min} + \frac{R_n}{G_{s4}}\left[(G_{s4} - G_{s,opt})^2 + (B_{s4} - B_{s,opt})^2\right], \tag{18.78}$$

where the subscripts 1, 2, 3, 4 denote four sets of tested *NF* and variables G_s and B_s in each test, respectively.

The test setup is shown in Figure 18.9. The output impedance of the noise source is converted to $Z_S = R_S + jX_S$ via an impedance converter. In order to calculate the four noise parameters via equations (18.75) to (18.78), the impedance Z_S must be converted to admittance Y_S through the following relations:

$$Y_s = G_s + jB_s = \frac{1}{Z_s} = \frac{1}{R_s + jX_s} = \frac{R_s - jX_s}{R_s^2 + X_s^2}, \tag{18.79}$$

$$G_s = \frac{R_s}{R_s^2 + X_s^2}, \tag{18.80}$$

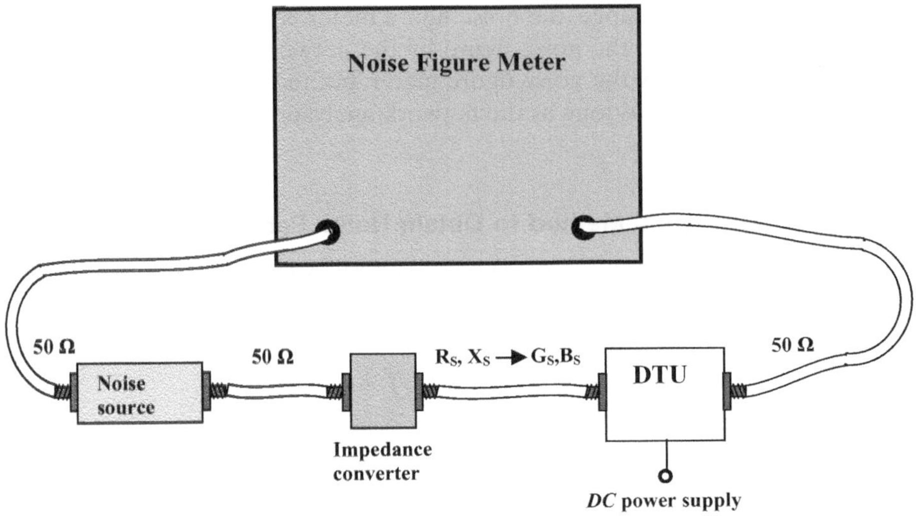

Figure 18.9 Setup for four-point experimental method of noise figure testing.

$$B_s = -\frac{X_s}{R_s^2 + X_s^2}. \tag{18.81}$$

Once these four parameters, NF_{min}, R_n, G_o, and B_o, are known, we can calculate the noise figure for the case with any source admittance.

18.3.5 Cascaded Equations of Noise Figure

Let's derive the cascaded equation for power gain of a system containing only two blocks as shown in Figure 18.10, where G_k and G_{k+1} are the power gains of blocks k and $k + 1$ respectively, and NF_k and NF_{k+1} are the noise figures of blocks k and $k + 1$ respectively.

By the definition of the noise figure, the resulting noise figure is a ratio of the total output power to the output power due to the input noise power, that is,

$$NF_k = \frac{G_k N_i + \Delta N_k}{G_k N_i} = 1 + \frac{\Delta N_k}{G_k N_i}, \tag{18.82}$$

$$NF_{k+1} = \frac{G_{k+1} N_i + \Delta N_{k+1}}{G_{k+1} N_i} = 1 + \frac{\Delta N_{k+1}}{G_{k+1} N_i}. \tag{18.83}$$

Then,

$$NF_{SYS} = \frac{G_{k+1}(G_k N_i + \Delta N_k) + \Delta N_{k+1}}{G_k G_{k+1} N_i} = 1 + \frac{G_{k+1}\Delta N_k + \Delta N_{k+1}}{G_k G_{k+1} N_i}, \tag{18.84}$$

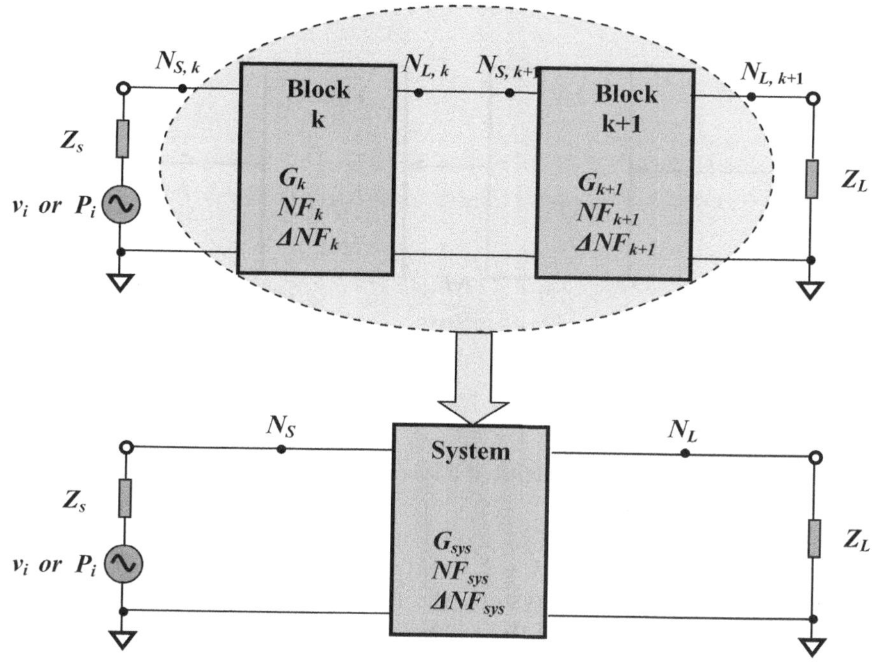

Figure 18.10 Noise figure of a system cascaded by two blocks k and $k + 1$.

where

N_i = noise source power delivered to k^{th} block,
ΔN_k = additional noise power of kth block,
ΔN_{k+1} = additional noise power of $(k + 1)^{th}$ block,
ΔN_{sys} = additional noise power of system,
G_k = power gain of kth block,
G_{k+1} = power gain of $(k + 1)^{th}$ block,
NF_k = noise figure of kth block,
NF_{k+1} = noise figure of $(k + 1)^{th}$ block,
NF_{SYS} = noise figure of system in the numeric scale.

Substituting (18.82) and (18.83) into (18.84), we have

$$NF_{SYS} = 1 + \frac{G_{k+1}(NF_k - 1)G_k N_i + (NF_{k+1} - 1)G_{k+1}N_i}{G_k G_{k+1} N_i}, \tag{18.85}$$

$$NF_{SYS} = NF_k + \frac{(NF_{k+1} - 1)}{G_k}. \tag{18.86}$$

An important feature of the noise figure can be found from equation (18.86). The system noise figure, or resulting noise figure, NF_{sys}, consists of two terms. The first term NF_k is contributed by the first block k. The second term $(NF_{k+1} - 1)/G_k$, is

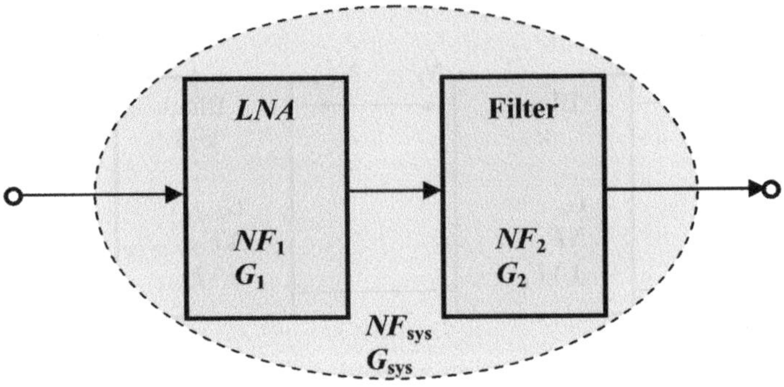

$$NF_{SYS} = NF_1 + \frac{(NF_2 - 1)}{G_1}$$

Figure 18.11 System noise figure if a system is cascaded by blocks 1 and 2.

contributed by the second block $k + 1$, but it is reduced by a factor G_k. This means that, from the viewpoint of the system, the noise figure of the first block plays a more important role than that of the second, while the noise figure of the second block can be reduced by the gain of the first block if the gain of the first block is positive and appreciable. In a receiver, in order to minimize the noise figure of the system or to enhance the sensitivity of a receiver, the first objective is to design the first block *LNA* with a low noise figure but high gain. This is the main goal for an *LNA* design in a receiver because the *LNA* is always located in the most front end of the receiver.

For example, there are two *RF* blocks, *LNA* and a filter, cascaded together as shown in Figure 18.11. Table 18.2 lists design goals in two examples. The noise figure of the first block *LNA* is the same 2.5 *dB* while the insertion loss is different in both examples. The 7 *dB* insertion loss of filter in the second example is higher than 3.5 *dB* insertion loss of filter in first example. The system noise figure is the same 2.67 *dB* in both examples. This is not surprising because the gain of *LNA* block in the second example 17.4 *dB* is 4.9 *dB* higher than that in the first example 12.5 *dB*.

In general cases, in a system containing n blocks as shown in Figure 18.12, equation (18.86) can be extended to

$$NF_{SYS,watt} = NF_1 + \frac{(NF_2 - 1)}{G_1} + \frac{(NF_3 - 1)}{G_1 G_2} + \frac{(NF_4 - 1)}{G_1 G_2 G_3} + \ldots + \frac{(NF_n - 1)}{G_1 G_2 \ldots G_{n-1}}, \tag{18.87}$$

or,

$$NF_{SYS,watt} = NF_1 + \sum_{2}^{n} \frac{(NF_k - 1)}{\prod_{1}^{k-1} G_j}. \tag{18.88}$$

Expressions (18.87) or (18.88) are the cascaded equations of the noise figure.

TABLE 18.2 System noise figure contributed from two blocks with different settlements of noise figure and gain

Example 1	Block 1 *LNA*	Block 2 *FLT*	System
NF_{sys}, *dB*	2.5	3.5	2.67
G_{sys}, *dB*	12.5	–3.5	9.00

Example 2	Block 1 *LNA*	Block 2 *FLT*	System
NF_{sys}, *dB*	2.5	7	2.67
G_{sys}, *dB*	17.4	–7	10.40

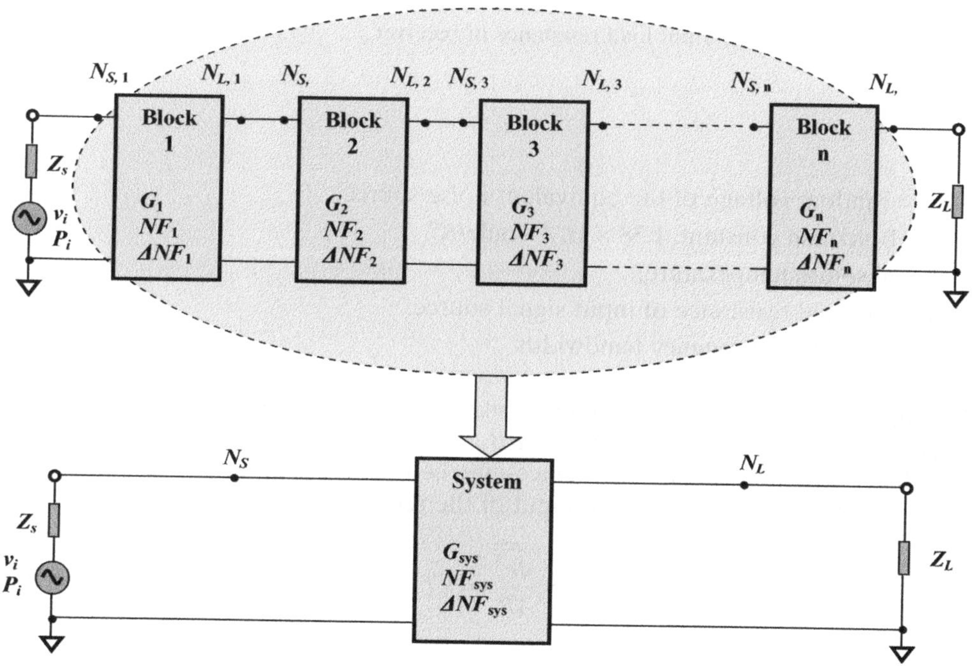

Figure 18.12 Noise figure of a system cascaded by n blocks.

18.3.6 Sensitivity of a Receiver

Figure 18.13 shows the basic parameters applied in the sensitivity testing of a receiver.

There is an equivalent noise source $\sqrt{\overline{e_{nt}^2}}$ and a signal source E_i in Figure 18.13. The input equivalent noise source is the thermal noise of the resistor R_i. Referring to Appendix 18.A.4, the average square root of its voltage is

$$\sqrt{\overline{e_{nt}^2}} = 4kTR_i\Delta f, \tag{18.89}$$

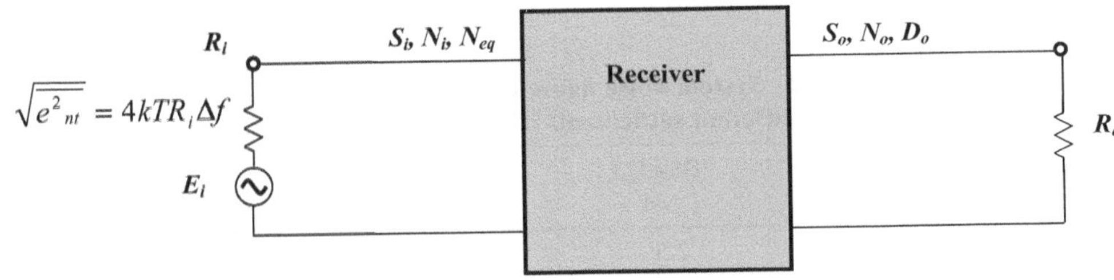

Figure 18.13 The basic parameters applied in the sensitivity testing for a receiver.
S_i = input signal power,
N_i = input noise power,
N_{eq} = equivalent input noise power,
S_o = output signal power,
N_o = output noise power,
D_o = output distortion power,
E_i = input signal source, μV or dBm,
R_i = input source resistance of receiver,
R_o = output load resistance of receiver.

where

e_{nt} = random voltage of the equivalent noise source,
k = Boltzman constant, 1.38×10^{-23} Joule/K°,
T = absolute temperature,
R_i = internal resistance of input signal source,
Δf = operating frequency bandwidth.

The input signal source E_i is usually expressed by the voltage with units of μV, or sometimes by the power with units of dB_m.

If these two sources are impedance-matched with the input impedance of the receiver, both their powers at the input of the receiver are

$$N_i = \frac{\sqrt{e^2_{nt}}}{4R_i} = kT\Delta f, \tag{18.90}$$

$$S_i = \frac{E_i^2}{4R_i}, \tag{18.91}$$

respectively.

The new parameter N_{eq} is defined as an equivalent input power, which becomes the output noise power after it is intensified by G times, that is,

$$N_{eq} = \frac{N_o}{G}. \tag{18.92}$$

Substituting N_o from (18.59) into (18.92), we have

$$N_{eq} = NF \cdot N_i. \tag{18.93}$$

The sensitivity of a receiver is defined as the minimum of an input signal level which is detectable at the output of the receiver.

At the output of the receiver, test equipment such as the audio distortion meter are able to separate the signal power from the composed power of the distortion and the noise. It is therefore an appropriate parameter defined as *SINAD*, which is a ratio of the output signal power to the output composed power of noise and distortion, that is,

$$SINAD = \frac{S_o}{N_o + D_o}. \tag{18.94}$$

In the input portion, another ratio, called *RISE*, is the ratio of the sum of the input signal power and the input equivalent noise power, $(S_i + N_{eq})$, to the input equivalent noise power, N_{eq}, that is,

$$RISE = \frac{S_i + N_{eq}}{N_{eq}} = \frac{S_i}{N_{eq}} + 1. \tag{18.95}$$

It indicates the strength of the input signal power over the input equivalent noise power.

If

$$S_i = N_{eq}, \tag{18.96}$$

then,

$$RISE = 2. \tag{18.97}$$

Condition (18.96) implies that the input signal power is about to exceed the input equivalent noise power at the input of the receiver. In other words, it is a critical point which can be considered as the minimum input signal detectable in the output of receiver. This point is termed the receiver's sensitivity.

From (18.95), we have

$$\frac{S_i}{N_{eq}} = RISE - 1. \tag{18.98}$$

Then, from (18.91) and (18.98), we have

$$E_i = 2\sqrt{S_i R_i} = 2\sqrt{N_{eq}(RISE - 1) R_i}, \tag{18.99}$$

By substituting (18.90) and (18.93) into (18.99), we have

$$E_i = 2\sqrt{NF(RISE - 1) R_i kT\Delta f}, \tag{18.100}$$

or,

$$S_i = \frac{E_i^2}{4R_i} = NF(RISE - 1)kT\Delta f, \tag{18.101}$$

where

S_i, S_o = signal power at input and output respectively, dB_w or dB_m;
N_i, N_o = noise power at input and output, respectively, dB;
D_o = distortion power at output, dB;
R_i = input resistance of generator, Ω;
N_{eq} = equivalent input noise power of output noise;
E_i = sensitivity looking from input, μV.

If condition (18.96) or (18.97) is satisfied, then the expressions for the sensitivity of the receiver (18.100) and (18.101) become

$$E_i = 2\sqrt{NF \cdot R_i kT\Delta f}, \tag{18.102}$$

or,

$$S_i = \frac{E_i^2}{4R_i} = NF \cdot kT\Delta f. \tag{18.103}$$

Obviously, the sensitivity of a receiver consists of three portions:

1) Noise figure, NF, of the entire receiver,
2) Noise floor (noise spectrum density), kT, in the environment of receiver,
3) Frequency bandwidth, Δf, of receiver.

Then,

$$S_{i,dB} = 10Log_{10}S_i = 10Log_{10}NF + 10Log_{10}(kT) + 10Log_{10}(\Delta f). \tag{18.104}$$

If

$$T = 290K^o, \tag{18.105}$$

the noise floor (noise spectrum density), kT, is

$$(kT)_{Watt/Hz} = 1.38 \cdot 10^{-23} \cdot 290 = 4.002 \cdot 10^{-21}\ Watt/Hz. \tag{18.106}$$

$$10Log_{10}(kT)_{dBW/Hz} = 10Log_{10}(4.002 \cdot 10^{-21}) = -204\,dBW/Hz. \tag{18.107}$$

$$10Log_{10}(kT)_{dBm/Hz} = 10Log_{10}(4.002 \cdot 10^{-21}/10^{-3}) = -174\,dBm/Hz. \tag{18.108}$$

For example,
If

$$NF = 6\,dB, \tag{18.109}$$

$$\Delta f = 1\,MHz, \tag{18.110}$$

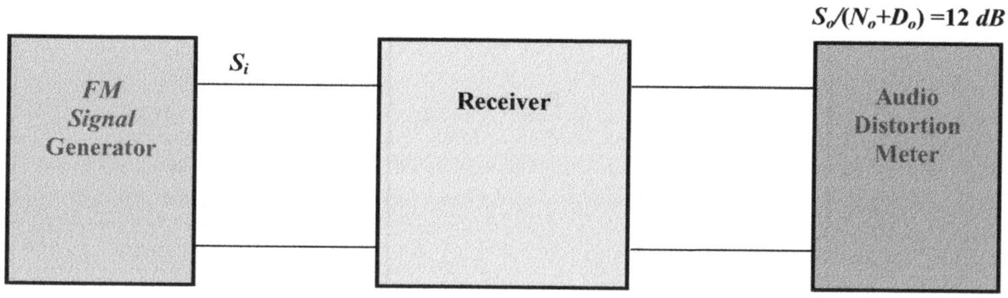

(a) Setup for testing of 12 *dB SINAD* sensitivity of receiver

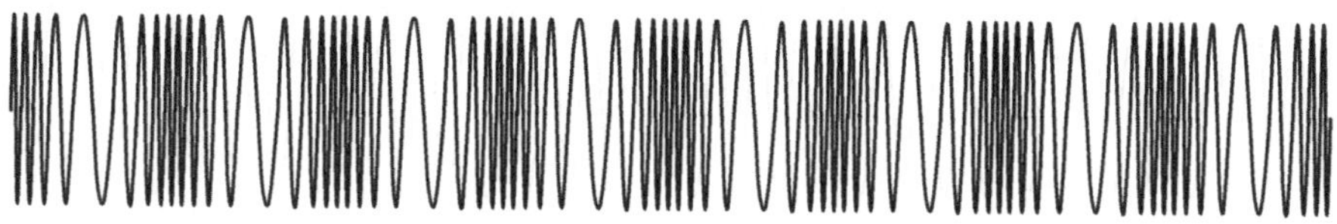

(b) S_i = Carrier f_o modulated with f_D=1 *kHz* audio signal

Figure 18.14 12 dB *SINAD* sensitivity testing for receiver.

Then

$$10Log_{10}(\Delta f) = 60\,dB, \tag{18.111}$$

$$S_{i,dB} = 6 - 174 + 60 = -108\,dBm, \tag{18.112}$$

If

$$R_i = 50\,\Omega, \tag{18.113}$$

$$E_i = 10^{[-108-30+10\log(4Ri)]/20} = 10^{-5.75} = 1.78\,\mu V, \tag{18.114}$$

There are typically two ways to test the sensitivity of the receiver:

1) 12 *dB SINAD*
Figure 18.14 shows the setup and the input signal from the generator for the testing of 12 *dB SINAD* sensitivity of receiver.

The input signal power S_i is provided by an *RF* signal generator. It is a *FM* modulated signal, in which the carrier of the desired frequency in the operating frequency range is modulated with a 1 *kHz* sinusoidal signal. At the beginning, the output power of the generator is turned off. It is then increased by very tiny increments at a slow speed. The *SINAD* value at the audio distortion meter will increase accordingly. Finally, the *SINAD* value is increased up to 12 *dB* as the input signal power from generator is increased. The corresponding value of the input signal power S_i when the *SINAD* value reaches 12 *dB* is the sensitivity S_{sen12}, that is,

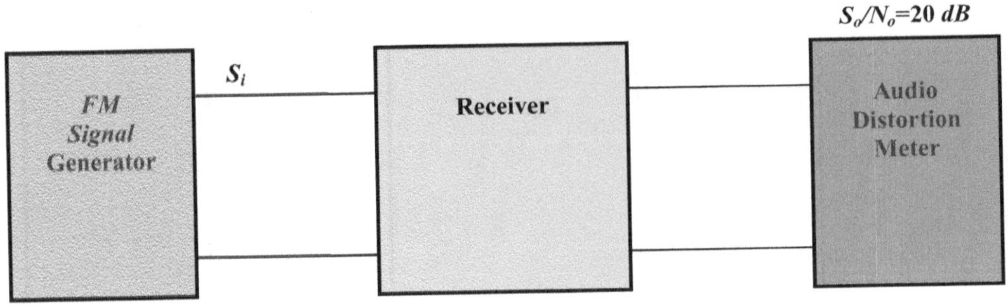

(a) Setup for testing of 20 dB quieting sensitivity of receiver

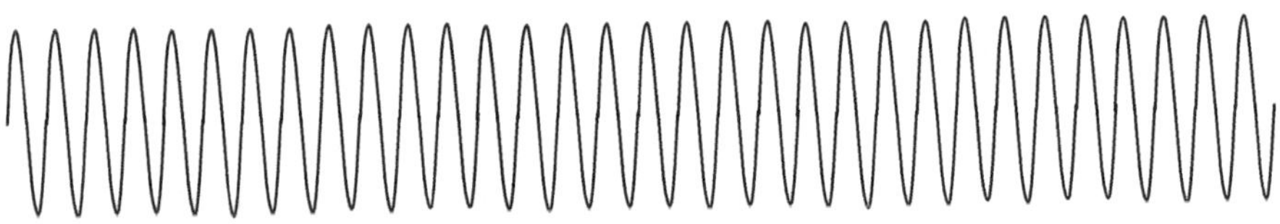

(b) S_i = Carrier signal with frequency f_o

Figure 18.15 20 dB quieting sensitivity testing of receiver.

$$S_{i,dB}|_{SINAD=12dB} = S_{sen12}, \tag{18.115}$$

where $S_{i,sen12}$ = 12 dB $SINAD$ sensitivity.

2) 20 dB Quieting
Figure 18.15 shows the setup and the input signal from the generator for testing the 20 dB quieting sensitivity of receiver.

The input signal power S_i is provided by an RF signal generator. Is is a pure carrier signal with 1 kHz frequency shift from the desired carrier frequency. At the beginning, the output power of the generator is turned off. The only noise can be found at the audio distortion meter. The output power of the generator then is increased by tiny increments at a slow speed. The noise at the audio distortion meter gradually quiets down and the ratio of signal to noise S_o/N_o increases accordingly as the input signal power is increased. Finally, the ratio S_o/N_o is increased by up to 20 dB as the input signal power from the generator is increased. The corresponding value of the input signal power S_i when the ratio S_o/N_o reaches 20 dB, is the sensitivity S_{sen20}, that is,

$$S_{i,dB}|_{S_o/N_o=20dB} = S_{sen20}, \tag{18.116}$$

where S_{sen20} = 20 dB quieting sensitivity.

As a matter of fact, experiments indicate that either 12 dB $SINAD$ or 20 dB quieting is close to the condition of (18.96), though their corresponding sensitivities, $S_{i,sen12}$ and S_{sen20}, are not exactly the same due to different input signals.

18.4 NON-LINEARITY

18.4.1 Non-linearity of Devices

Non-linearity of a system mainly arises from the non-linearity of the device. Non-linearity exists more or less in a practical device. In reality, an ideal linear device or system does not exist, and any "linear" device or system is just a reasonable approximation.

Non-linearity of the devices or system produces spurious products. All the spurious products are sources of interference to the signal and eventually cause distortion of the signal. There are two kinds of spurious products: harmonics and complicated spurious products.

1) Harmonics when Input Is a Signal with Only One Frequency

Assuming that the input voltage of the signal is a pure sinusoidal wave as shown in Figure 18.16, that is,

$$v_{in} = v_{io} \cos \omega t, \tag{18.117}$$

The non-linearity of a device can be expressed in a general form, that is

$$v_{out} = a_o + a_1 v_{in} + a_2 v_{in}^2 + a_3 v_{in}^3 + a_4 v_{in}^4 + \ldots, \tag{18.118}$$

where

$a_i = i^{th}$ *order* non-linear coefficients,
$i = 0, 1, 2, 3, 4 \ldots$

Substituting (18.102) into (18.103), we have

$$v_{out} = V_{dc} + \sum_{1}^{\infty} A_n \cos n\omega t, \tag{18.119}$$

where

$V_{dc} = DC$ component of output voltage,
A_n = amplitude of nth harmonic; they are

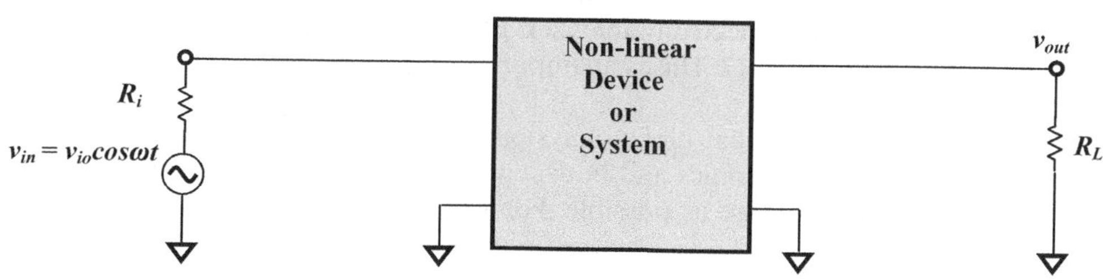

Figure 18.16 Harmonics are produced from a non-linear device or system even if the input is a signal with only one frequency.

$$V_{dc} = a_o + \frac{1}{2}a_2 v_{io}^2 + \frac{3}{8}a_4 v_{io}^4 + \ldots \tag{18.120}$$

$$A_1 = a_1 v_{io} + \frac{4}{3}a_3 v_{io}^3 + \ldots \tag{18.121}$$

$$A_2 = \frac{1}{2}a_2 v_{io}^2 + \frac{1}{2}a_4 v_{io}^4 + \ldots \tag{18.122}$$

$$A_3 = \frac{1}{4}a_3 v_{io}^3 + \ldots \tag{18.123}$$

$$A_4 = \frac{1}{8}a_4 v_{io}^4 + \ldots \tag{18.124}$$

$$\ldots$$

The output voltage is a complicated entity with infinite harmonics.

These coefficients A_n may be dependent or independent of the current flowing through the device or system, or of the voltage drop across the device or system. As a general trend, these coefficients are reduced as the order increases. However, sometimes there are extraordinary cases in which the non-linearity coefficients of higher orders have higher values than those of lower orders. The infinite number of harmonics, the terms containing the factors $cos\omega t$, $cos2\omega t$, $cos3\omega t$, $cos4\omega t$, etc., are produced due to the non-linearity of the device or system. The new resulting coefficient of each harmonic A_n is another infinite series containing the original non-linearity coefficients a_i. These new coefficients look very clumsy at first glance. In a practical engineering design, it is sufficient to keep only the first two or three terms containing the non-linear coefficients a_i with low orders.

A remarkable feature of these new coefficients can be discovered from equations (18.120) to (18.124), that is, the new coefficients with odd harmonics A_{2n+1} contain only odd coefficients with odd orders a_{2i+1} while the new coefficients with even harmonics A_{2n} contain only even coefficients with even orders a_{2i}. This is important to engineers who are tasked with the cancellation of spurious products. In *RF* or *RFIC* circuits where a high linearity is required, such as the *LNA* or *PA*, devices with higher a_1 but lower other coefficients should be chosen. In the circuit design of the mixer, which operates on the basis of the second order of non-linearity, devices with higher a_2 but lower other coefficients should be chosen.

By means of the spectrum analyzer, the harmonics can be seen on-screen as shown in Figure 18.17. This is an important test to measure the non-linearity of a device.

For an expected linear block or system, the power differences between the first and second harmonics and between the second and third harmonics are expected to be as large as possible. For instance, in a *LNA* design, they are expected to be

$$\Delta_1 = P_1 - P_2 > 20\,dB, \tag{18.125}$$

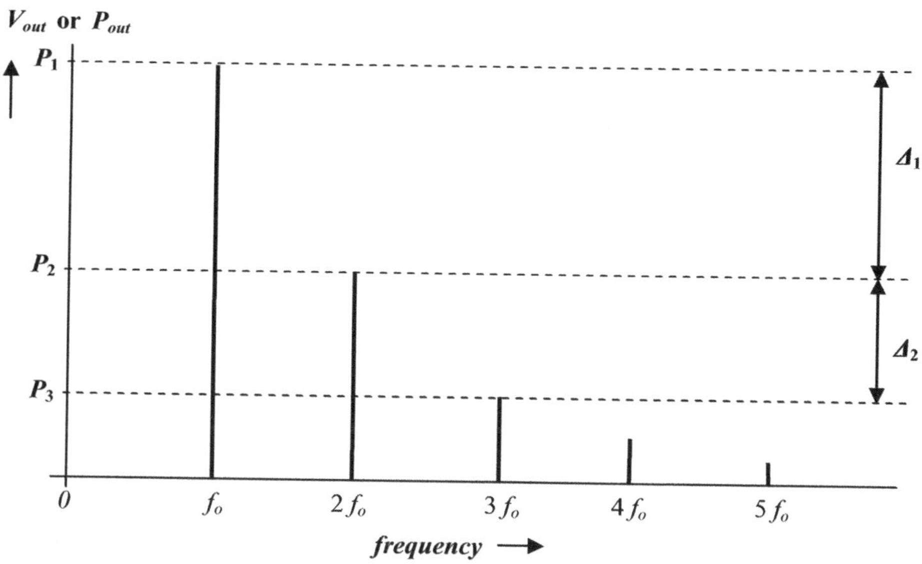

Figure 18.17 The frequency spectrum of a device when the input is a signal with only one frequency.

and

$$\Delta_2 = P_2 - P_3 > 10\,dB. \tag{18.126}$$

2) Spurious Products When Input Is a Signal with Two Frequencies

If the input voltage signal is a signal with two frequencies as shown in Figure 18.18, that is,

$$v_{in} = v_{io} \cos \omega_1 t + v_{io} \cos \omega_2 t, \tag{18.127}$$

Substituting (18.127) into (18.118), we have

$$v_{out} = \sum_{m=0,n=0}^{\infty} A_{mn} \cos(m\omega_1 \pm n\omega_2)t, \tag{18.128}$$

where A_{mn} = amplitude of the harmonic with angular frequency = $m\omega_1 \pm n\omega_2$; they are

$$A_{00} = a_o + a_2 v_{io}^2 + \frac{4}{9} a_4 v_{io}^4 + \ldots \tag{18.129}$$

$$A_{10} = A_{01} = a_1 v_{io} + \frac{4}{9} a_3 v_{io}^3 + \ldots \tag{18.130}$$

$$A_{20} = A_{02} = \frac{1}{2} a_2 v_{io}^2 + 2a_4 v_{io}^4 \ldots \tag{18.131}$$

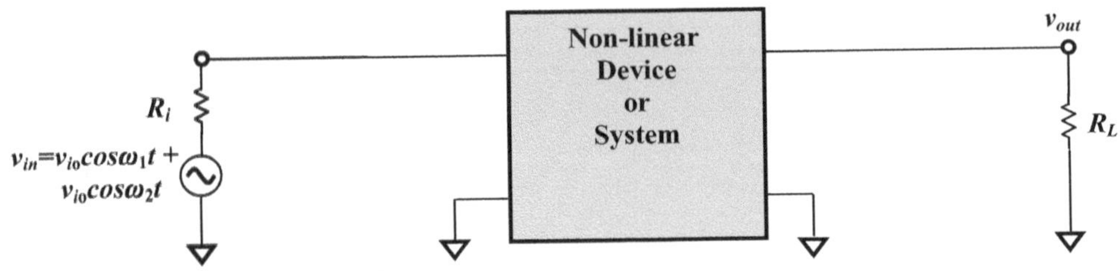

Figure 18.18 Spurious products are produced from a non-linear device or system if the input is a signal with two frequencies.

$$A_{11} = a_2 v_{io}^2 + 3a_4 v_{io}^4 + \dots \tag{18.132}$$

$$A_{30} = A_{03} = a_3 v_{io} 3 + \dots \tag{18.133}$$

$$A_{21} = A_{12} = \frac{3}{4} a_4 v_{io}^4 + \dots \tag{18.134}$$

$$A_{40} = A_{04} = \frac{1}{8} a_4 v_{io}^4 + \dots \tag{18.135}$$

$$A_{31} = A_{13} = \frac{1}{2} a_4 v_{io}^4 + \dots \tag{18.136}$$

$$A_{22} = \frac{3}{4} a_4 v_{io}^4 + \dots \tag{18.137}$$

$$\dots$$

It should be noted that A_{00} is the DC component of output voltage.

If the output voltage is a complicated entity with infinite harmonics. then the output voltage is a complicated transfer function with many harmonics and composed frequencies. It can be derived as follows:

A remarkable feature of these new coefficients is that the coefficients of the odd spurious products or harmonics contain only odd coefficients with odd orders while the coefficients of the even spurious products or harmonics contain only even coefficients with even orders.

From expression (18.128), it can be seen that the spectrum becomes quite complicated. The frequencies of the spurious products can be summarized as

$$\omega_{spur} = |m\,\omega_1 \pm n\,\omega_2|. \tag{18.138}$$

It is a DC component of output voltage when both m and n are equal to zero. When either m or n is zero, the other is non-zero, the output contains the desired signal and spurious product with harmonics. In cases when both m and n are non-zero positive integers, the output contains spurious products with composed frequencies. For example, in an LNA design, the third order inter-modulation is due to the spurious product with

$$m = 2, \quad \text{and} \quad n = 1, \tag{18.139}$$

or,

$$m = 1, \quad \text{and} \quad n = 2, \tag{18.140}$$

and, in the mixer design, the second order inter-modulation is due to the spurious product with

$$m = 2, \quad \text{and} \quad n = 2, \tag{18.141}$$

which are called half *IF* spurs or Able-Baker spurs.

The resulting output contains not only the harmonics of the two frequencies, but also the other combined spurious products. The order of the spurious products is the number of frequencies involved in the terms, which are listed in Table 18.3.

As examples, Table 18.4 lists three sets of spurious products when the input signals with two frequencies are

$$1)\quad \omega_1 = 850\,MHz, \omega_2 = 828.6\,MHz, IF = \omega_1 - \omega_2 = 21.4\,MHz \tag{18.142}$$

$$2)\quad \omega_1 = 850\,MHz, \omega_2 = 805.0\,MHz, IF = \omega_1 - \omega_2 = 45.0\,MHz \tag{18.143}$$

$$3)\quad \omega_1 = 850\,MHz, \omega_2 = 776.65\,MHz, IF = \omega_1 - \omega_2 = 73.35\,MHz \tag{18.144}$$

It should be noted that one of the two frequencies is kept the same at 850 MHz, that is

$$\omega_1 = 850\,MHz. \tag{18.145}$$

And, the difference of the two frequencies is denoted by IF, that is,

$$IF = \omega_1 - \omega_2, \tag{18.146}$$

The three sets of spurious products listed in Table 18.4 are re-listed, sorted from low to high frequency, in Table 18.5.

It is very interesting to compare these three sets of spurious products. Figures 18.19(a), (b), and (c) plot the first nine rows for the three sets of spurious products listed in Table 18.4 respectively, which are the low-frequency spurious products below 1000 MHz. The order of the spurious products is marked beside the spectrum accordingly.

By comparing Figures 18.19(a), (b) and (c), it can be found that

- All of them have low-frequency spurious products of *IF*, $2 \cdot IF$, and $3 \cdot IF$ with second, fourth, and sixth orders of non-linearity in the low-frequency end. These three frequency components are squeezed toward the low-frequency end more as the *IF* is decreased.
- All of them have some spurious products around ω_1. They scatter farther from ω_1 as the *IF* is increased.

TABLE 18.3 Order of spurious products and their frequencies if the input is a signal with two frequencies

Order of the spurious products	Frequency
1	ω_1
1	ω_2
2	$2\omega_1$
2	$2\omega_2$
2	$\omega_1 - \omega_2$
2	$\omega_1 + \omega_2$
3	$3\omega_1$
3	$3\omega_2$
3	$2\omega_1 - \omega_2$
3	$2\omega_1 + \omega_2$
3	$2\omega_2 - \omega_1$
3	$2\omega_2 + \omega_1$
4	$4\omega_1$
4	$4\omega_2$
4	$3\omega_2 - \omega_1$
4	$3\omega_2 + \omega_1$
4	$3\omega_1 - \omega_2$
4	$3\omega_1 + \omega_2$
4	$2(\omega_1 - \omega_2)$
4	$2(\omega_1 + \omega_2)$
5	$5\omega_1$
5	$5\omega_2$
5	$4\omega_2 - \omega_1$
5	$4\omega_2 + \omega_1$
5	$4\omega_1 - \omega_2$
5	$4\omega_1 + \omega_2$
5	$3\omega_1 - 2\omega_2$
5	$3\omega_1 + 2\omega_2$
5	$3\omega_2 - 2\omega_1$
5	$3\omega_2 + 2\omega_1$
6	$6\omega_1$
6	$6\omega_2$
6	$5\omega_2 - \omega_1$
6	$5\omega_2 + \omega_1$
6	$5\omega_1 - \omega_2$
6	$5\omega_1 + \omega_2$
6	$4\omega_1 - 2\omega_2$
6	$4\omega_1 + 2\omega_2$
6	$4\omega_2 - 2\omega_1$
6	$4\omega_2 + 2\omega_1$
6	$3(\omega_1 - \omega_2)$
6	$3(\omega_1 + \omega_2)$

TABLE 18.4 Examples of spurious products and their frequencies if the input is a signal with two frequencies

Order	*Frequency*	(1) Spurious, *MHz*	(2) Spurious, *MHz*	(3) Spurious, *MHz*
1	ω_1	**850.0**	**850**	**850.00**
1	ω_2	**828.6**	**805**	**776.65**
2	$2\omega_1$	1700.0	1700	1700.00
2	$2\omega_2$	1657.2	1610	1553.30
2	$\omega_1 - \omega_2$	**21.4**	**45**	**73.35**
2	$\omega_1 + \omega_2$	1678.6	1655	1626.65
3	$3\omega_1$	2550.0	2550	2550.00
3	$3\omega_2$	2485.8	2415	2329.95
3	$2\omega_1 - \omega_2$	871.4	895	923.35
3	$2\omega_1 + \omega_2$	2528.6	2505	2476.65
3	$2\omega_2 - \omega_1$	807.2	760	703.30
3	$2\omega_2 + \omega_1$	2507.2	2460	2403.30
4	$4\omega_1$	3400.0	3400	3400.00
4	$4\omega_2$	3314.4	3220	3106.60
4	$3\omega_1 - \omega_2$	1721.4	1745	1773.35
4	$3\omega_1 + \omega_2$	3378.6	3355	3326.65
4	$3\omega_2 - \omega_1$	1635.8	1565	1479.95
4	$3\omega_2 + \omega_1$	3335.8	3265	3179.95
4	$2(\omega_1 - \omega_2)$	42.8	90	146.70
4	$2(\omega_1 + \omega_2)$	3357.2	3310	3253.30
5	$5\omega_1$	4250.0	4250	4250.00
5	$5\omega_2$	4143.0	4025	3883.25
5	$4\omega_1 - \omega_2$	2571.4	2595	2623.35
5	$4\omega_1 + \omega_2$	4228.6	4205	4176.65
5	$4\omega_2 - \omega_1$	2464.4	2370	2256.60
5	$4\omega_2 + \omega_1$	4164.4	4070	3956.60
5	$3\omega_1 - 2\omega_2$	892.8	940	996.70
5	$3\omega_1 + 2\omega_2$	4207.2	4160	4103.30
5	$3\omega_2 - 2\omega_1$	785.8	715	629.95
5	$3\omega_2 + 2\omega_1$	4185.8	4115	4029.95
6	$6\omega_1$	5100.0	5100	5100.00
6	$6\omega_2$	4971.6	4830	4659.90
6	$5\omega_1 - \omega_2$	3421.4	3445	3473.35
6	$5\omega_1 + \omega_2$	5078.6	5055	5026.65
6	$5\omega_2 - \omega_1$	3293.0	3175	3033.25
6	$5\omega_2 + \omega_1$	4993.0	4875	4733.25
6	$4\omega_1 - 2\omega_2$	1742.8	1790	1846.70
6	$4\omega_1 + 2\omega_2$	5057.2	5010	4953.30
6	$4\omega_2 - 2\omega_1$	1614.4	1520	1406.60
6	$4\omega_2 + 2\omega_1$	5014.4	4920	4806.60
6	$3(\omega_1 - \omega_2)$	64.2	135	220.05
6	$3(\omega_1 + \omega_2)$	5035.8	4965	4879.95

TABLE 18.5 Examples of the spurious products and their sorted frequencies if the input is a signal with two frequencies

Order	Frequency	(Sorted) (1) Spurious, *MHz*	(Sorted) (2) Spurious, *MHz*	(Sorted) (3) Spurious, *MHz*
2	$\omega_1 - \omega_2$	21.4	45	73.35
4	$2(\omega_1 - \omega_2)$	42.8	90	146.70
6	$3(\omega_1 - \omega_2)$	64.2	135	220.05
5	$3\omega_2 - 2\omega_1$	785.8	715	629.95
3	$2\omega_2 - \omega_1$	807.2	760	703.30
1	ω_2	828.6	805	776.65
1	ω_1	850.0	850	850.00
3	$2\omega_1 - \omega_2$	871.4	895	923.35
5	$3\omega_1 - 2\omega_2$	892.8	940	996.70
6	$4\omega_2 - 2\omega_1$	1614.4	1520	1406.60
4	$3\omega_2 - \omega_1$	1635.8	1565	1479.95
2	$2\omega_2$	1657.2	1610	1553.30
2	$\omega_1 + \omega_2$	1678.6	1655	1626.65
2	$2\omega_1$	1700.0	1700	1700.00
4	$3\omega_1 - \omega_2$	1721.4	1745	1773.35
6	$4\omega_1 - 2\omega_2$	1742.8	1790	1846.70
5	$4\omega_2 - \omega_1$	2464.4	2370	2256.60
3	$3\omega_2$	2485.8	2415	2329.95
3	$2\omega_2 + \omega_1$	2507.2	2460	2403.30
3	$2\omega_1 + \omega_2$	2528.6	2505	2476.65
3	$3\omega_1$	2550.0	2550	2550.00
5	$4\omega_1 - \omega_2$	2571.4	2595	2623.35
6	$5\omega_2 - \omega_1$	3293.0	3175	3033.25
4	$4\omega_2$	3314.4	3220	3106.60
4	$3\omega_2 + \omega_1$	3335.8	3265	3179.95
4	$2(\omega_1 + \omega_2)$	3357.2	3310	3253.30
4	$3\omega_1 + \omega_2$	3378.6	3355	3326.65
4	$4\omega_1$	3400.0	3400	3400.00
6	$5\omega_1 - \omega_2$	3421.4	3445	3473.35
5	$5\omega_2$	4143.0	4025	3883.25
5	$4\omega_2 + \omega_1$	4164.4	4070	3956.60
5	$3\omega_2 + 2\omega_1$	4185.8	4115	4029.95
5	$3\omega_1 + 2\omega_2$	4207.2	4160	4103.30
5	$4\omega_1 + \omega_2$	4228.6	4205	4176.65
5	$5\omega_1$	4250.0	4250	4250.00
6	$6\omega_2$	4971.6	4830	4659.90
6	$5\omega_2 + \omega_1$	4993.0	4875	4733.25
6	$4\omega_2 + 2\omega_1$	5014.4	4920	4806.60
6	$3(\omega_1 + \omega_2)$	5035.8	4965	4879.95
6	$4\omega_1 + 2\omega_2$	5057.2	5010	4953.30
6	$5\omega_1 + \omega_2$	5078.6	5055	5026.65
6	$6\omega_1$	5100.0	5100	5100.00

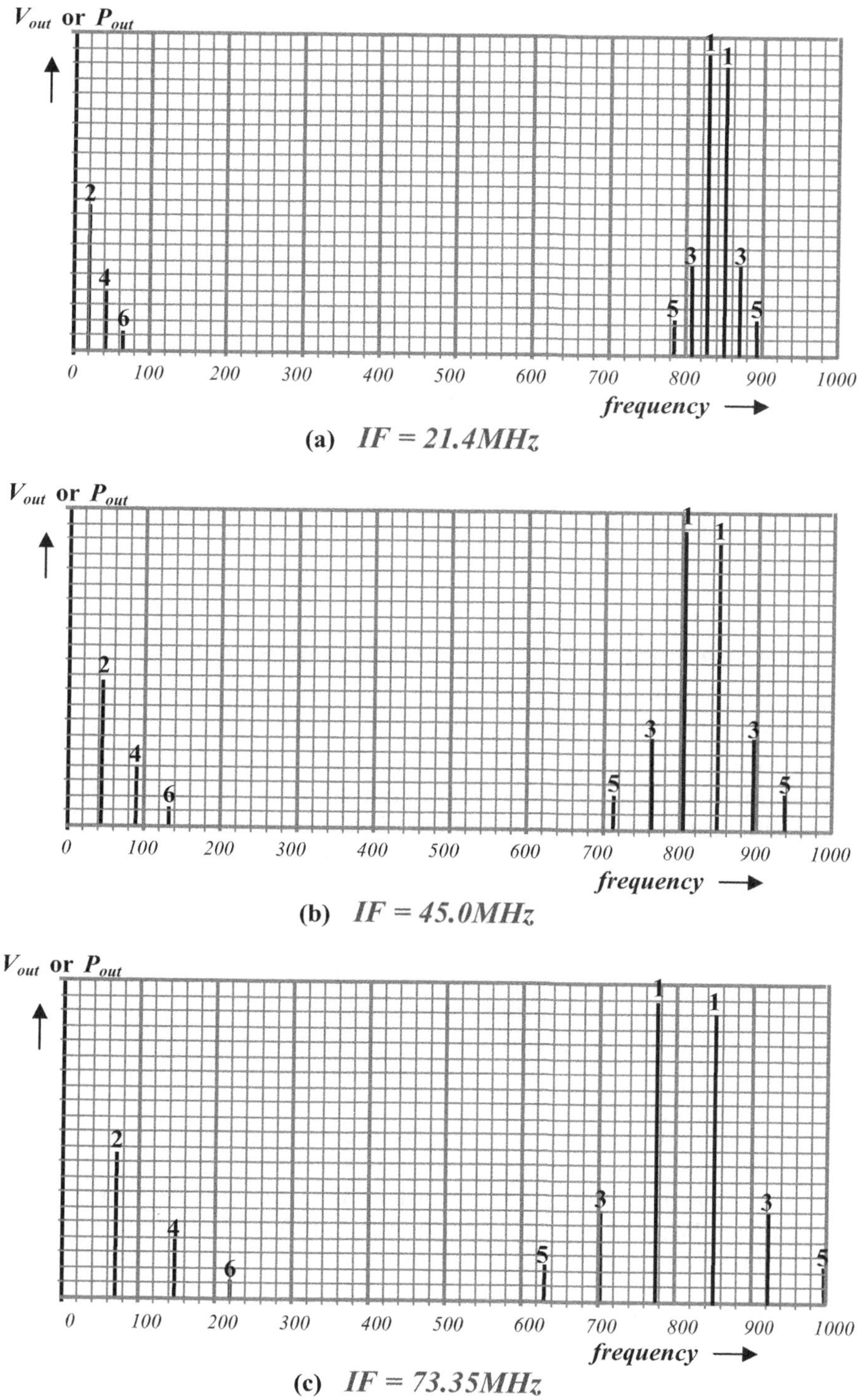

Figure 18.19 Frequency spectrum of spurious products when the input is a signal with two frequencies.

Consequently, the first set of spurious products, $IF = 21.4\,MHz$, may have the least opportunity to interfere with other circuit blocks or systems because there are no spurious products in a wide frequency range from about from 65 to $780\,MHz$. In other words, if there are other circuit blocks or systems operating in the frequency range from 65 to $780\,MHz$, the first set of spurious products is no longer an interference source.

This feature is applied in communication system designs, where the selection of the *LO* frequency, and hence the *IF* frequency for a given *RF* frequency, is important. If $\omega_1 = 850\,MHz$, as in our example above, is selected as the *RF* frequency in the mixer design, then the selection of $\omega_2 = 828.6\,MHz$ or $IF = 21.4\,MHz$ is better than the selections of $\omega_2 = 805\,MHz$ or $IF = 45\,MHz$ and $\omega_2 = 776.65\,MHz$ or $IF = 73.35\,MHz$.

As a matter of fact, in the selection of the *LO* and hence the *IF* frequency for a communication system, not only low-frequency but also high-frequency spurious products must be analyzed. In other words, the analysis for only the first nine rows listed in Table 18.4 must be extended to cover higher orders of non-linearity of device, and thc plot of Figure 18.19 must be extended to higher frequencies. This is tedious but important work in system design.

18.4.2 *IP* (Intercept Point) and *IMR* (Inter-Modulation Rejection)

Inter-modulation of an *RF* block mainly comes from the non-linearity of the device or the block. It therefore depends on the device characteristic itself. On the other hand, it also depends on the circuit topology and other parts' characteristics. This implies that the inter-modulation would be different for different devices, different topologies, or different parts applied in the circuit block. Conversely, different *RF* blocks have different levels of capability to reject inter-modulation. *IMR* (Inter-Modulation Rejection) therefore becomes an important specification to measure the non-linearity or linearity of an *RF* block.

The intercept point is just an intermediate parameter by which the inter-modulation rejection can be evaluated. Intercept points are categorized on the basis of their order of non-linearity. For instance, the *m*th order intercept point is defined as the intercept point of the input/output power curves of first order signal and the *m*th order *IM* (Inter-Modulation) product for a given frequency. Figure 18.20 shows the mth order intercept point IP_{m}.

The straight line with a slope of 1 depicts the relation between input and output signal power, where G is the power gain of the device or system. Another line with a slope of m represents the *m*th order *IM* (Inter-Modulation) product or spurious product. When the input signal power P_i is low, the output power of the *m*th order *IM* product P_u is still much lower than the output power of the first order output signal P_o. Let Δ denote the output power difference between the first order and the *m*th order when input signal power is P_i. The *m*th order *IM* product output increases much faster than the first order output signal power as the input signal power is increased from P_i, and eventually the two lines intercept as shown in Figure 18.20. This point is called the intercept point of the first order signal and the *m*th order *IM* product. The corresponding abscissa and ordinate coordinates are called the *m*th order output intercept point OIP_m and the *m*th order input intercept point IIP_m, respectively.

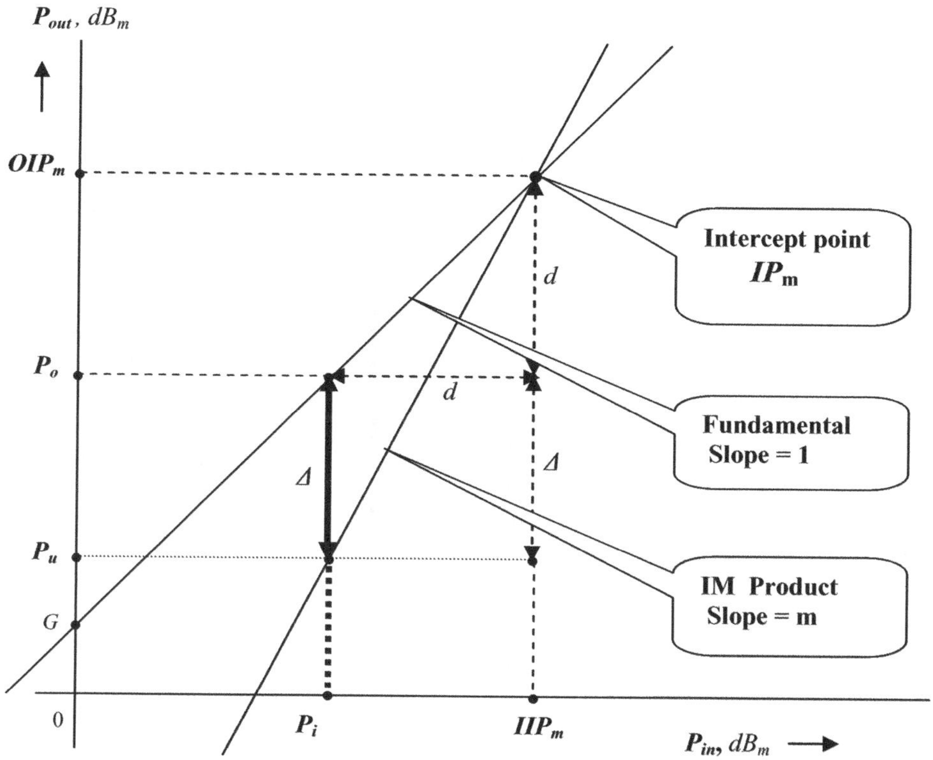

Figure 18.20 Plot of first-order signal and mth order IM product on the input-output power plane.
P_i = input power of signal in dB,
P_o = output power of signal in dB,
P_u = undesired output power due to mth non-linearity in dB,
G = power gain of the device, or block, or system in dB,
OIP_m = mth order output intercept point in dB,
IIP_m = mth order input intercept point in dB,
d = power difference between OIP_m and P_o in dB,
Δ = output power difference between the first order and the mth order in dB.

Let's discuss how to find out the mth order IMR from the mth order intercept point IP_m since the intercept point is just an intermediate parameter.

First, let's find out the relationship between the input intercept point IIP_m and the input signal power P_i. By simple geometric derivation,

$$\Delta = m(IIP_m - P_i) - 1(IIP_m - P_i) = (m-1)(IIP_m - P_i), \tag{18.147}$$

$$IIP_m = \frac{\Delta}{(m-1)} + P_i, \tag{18.148}$$

$$OIP_m = G + IIP_m. \tag{18.149}$$

By means of equations (18.148) and (18.149), either the input intercept point IIP_m or the output intercept point OIP_m can be obtained by the measurement of Δ at an appropriate value of P_i.

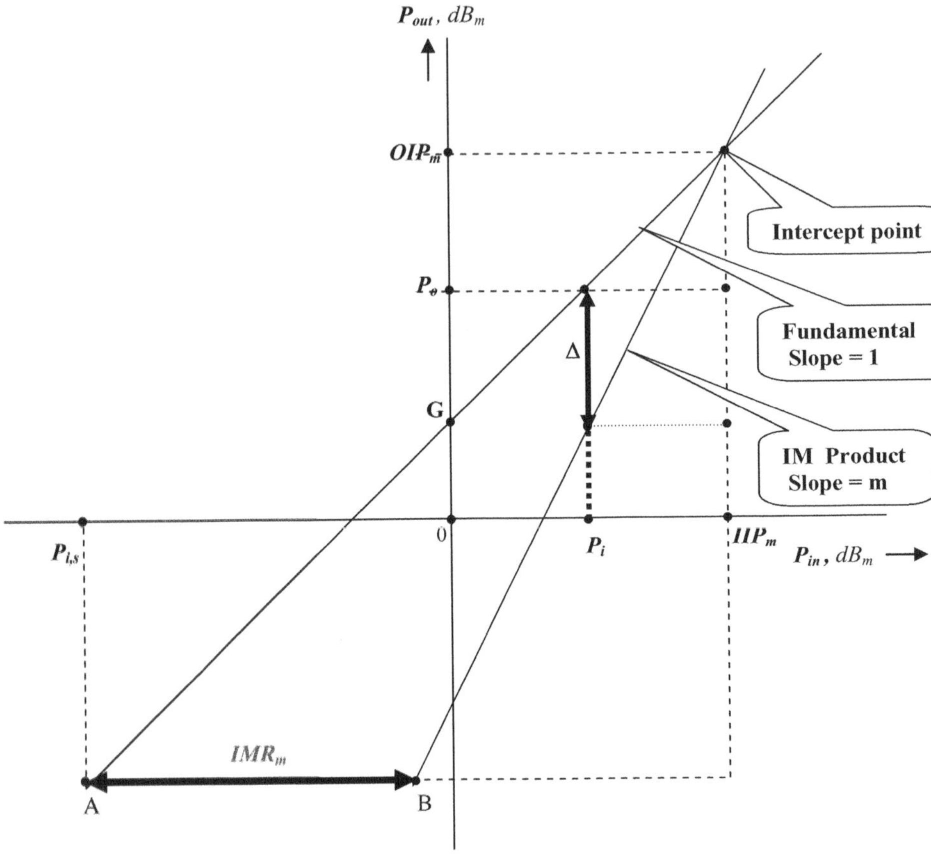

Figure 18.21 Calculation of IMR_m from intecept point, IIP_m.

Now, let's calculate the mth order inter-modulation rejection, IMR_m, from the mth order input intercept point, IIP_m. By simple geometry as shown in Figure 18.21, we have

$$IMR_m = (IIP_m - P_{i,s}) - \frac{(IIP_m - P_{i,s})}{m}, \quad (18.150)$$

$$IMR_m = \alpha(IIP_m - P_{i,s}), \quad (18.151)$$

where

$$\alpha = \frac{(m-1)}{m}. \quad (18.152)$$

Figure 18.21 illustrates the meaning of the mth order inter-modulation rejection, IMR_m, which is expressed by the length of AB. In Figure 18.21, $P_{i,s}$ denotes the input power at the sensitivity point, the minimum of a detectable input signal level.

- Point A represents the input signal power at the sensitivity point and its corresponding output signal power.
- Point B is on the mth order IM product line with the same output power as point A.

The input power at point B is (IMR_m) dB_m higher than point A. In other words, when the input signal is at the sensitivity point, the mth order IM product in the input is (IMR_m) dB_m stronger than the input signal power. As both input powers are increased toward IIP_m, they have the same output at the intercept point. Therefore, it implies that this system has the capability to suppress or to reject the mth order IM product down (IMR_m) dB.

18.4.2.1 IP_2 As mentioned above, the power or amplitude of an IM or spurious product is relatively reduced as its order is increased. Therefore, more attention is paid to the IM or spurious products with lower orders. The IM or spurious products with $m = 2$ or 3 have been more focused on than the others since the IM or spurious products with $m = 1$ is the desired signal.

In this section let's discuss the second order spurious product. When

$$m = 2, \tag{18.153}$$

the expressions (18.148), (18.149), (18.150) and (18.151) become

$$IIP_2 = \Delta + P_i, \tag{18.154}$$

$$OIP_2 = G + IIP_2, \tag{18.155}$$

$$\alpha = \frac{1}{2}, \tag{18.156}$$

$$IMR_2 = \alpha(IIP_2 - P_{i,s}) = \frac{1}{2}(IIP_2 - P_{i,s}). \tag{18.157}$$

By direct measurement of Δ in the test laboratory, one can calculate IIP_2, OIP_2, and IMR_2 by equations (18.154), (18.155), and (18.157).

The setup and display of IP_2 testing is shown in Figure 18.22. Figure 18.22(a) shows its setup. There are two generators generating the same power level, v_{io} or P_{in}, but with different frequencies, f_d and f_u. Two input signals are toggled by a switch. The power output P_{out} from the desired test block or system is displayed on the screen of the spectrum analyzer. There are three experimental steps:

The first step is to test for the response from the input to the output when the switch is thrown on position 1. The input signal is a signal with the desired frequency f_d and input voltage v_{io} or input power P_{io}. The output power P_{o1} *is* read at the desired frequency f_d from the screen on the spectrum analyzer. The display is shown in Figure 18.22(b).

The second step is to test for the response from the input to output when the switch is thrown on position 2. The input signal is a signal with the undesired frequency $f_u = f_d/2$ and the same input voltage v_{io} or input power P_{io} as in step 1.

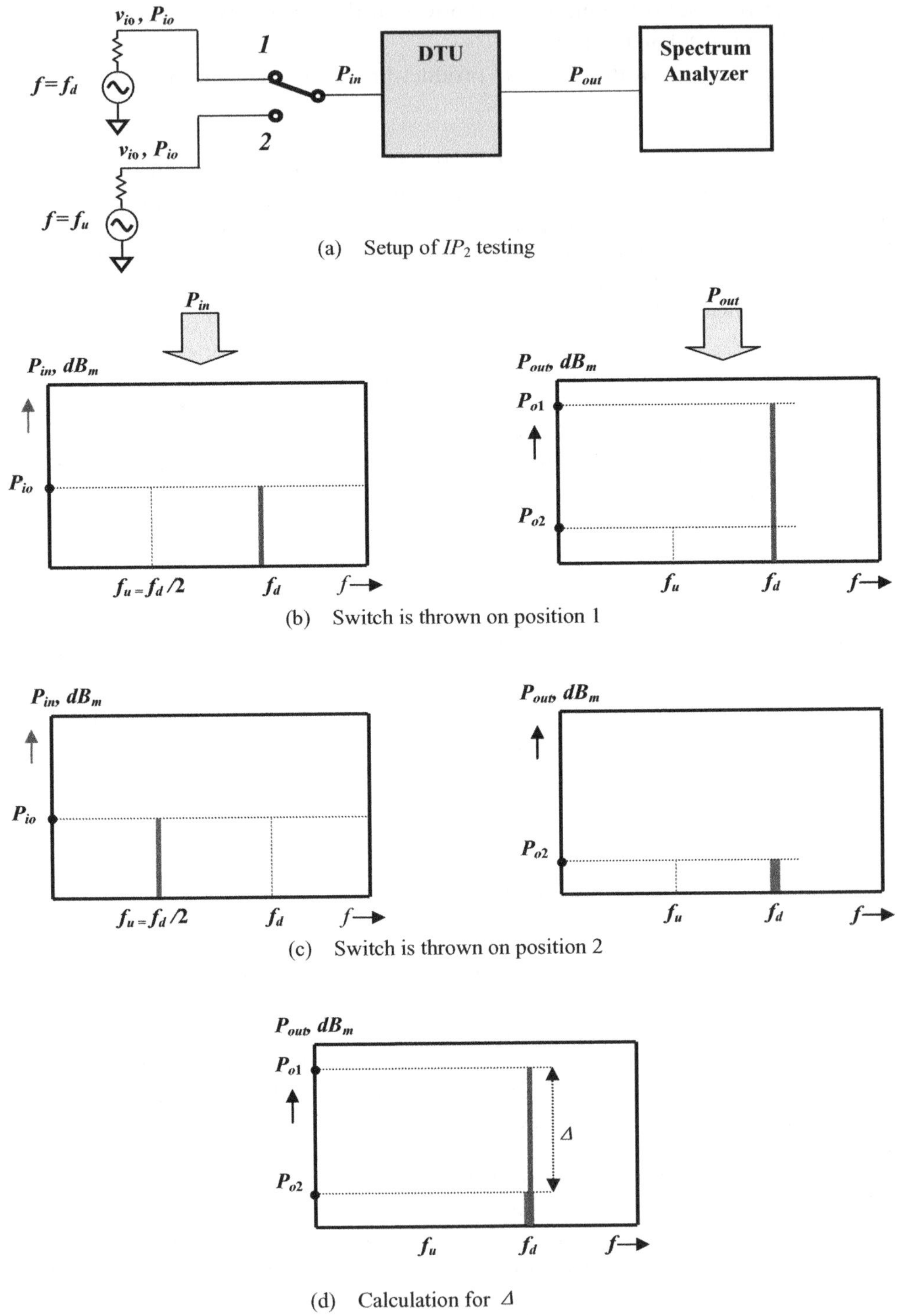

Figure 18.22 Setup and display of IP_2 testing.

The output power P_{o2} is read at the desired frequency f_d from the screen on the spectrum analyzer. The display is shown in Figure 18.22(c).

As shown in Figure 18.22(d), the third step is to calculate the difference of output powers at the desired frequency f_d based on the readings of output power in two previous test steps, P_{o1} and P_{o2}, that is,

$$\Delta = P_{o1} - P_{o2}, \tag{18.158}$$

And note that

$$P_i = P_{io}, \tag{18.159}$$

By substituting (18.158) and (18.159) into (18.154), (18.155), and (18.157), the values of IIP_2, OIP_2, and IMR_2 can be obtained, respectively.

This is one way to obtain the value of IMR_2 by means of testing with the setup as shown in Figure 18.22(a).

Another way to obtain the value IMR_2 is to calculate it from the non-linear coefficients of the non-linear device or system if the transfer function as shown in expression (18.104) is well known.

Recalling the expressions from (18.119) to (18.124), the transfer function (18.119) can be rewritten as

$$\begin{aligned} v_{out} = & \left[a_o + (1/2)a_2 v_{io}^2 + (3/8)a_4 v_{io}^4 + \ldots\right] + \left[a_1 v_{io} + (3/4)a_3 v_{io}^3 + \ldots\right]\cos\omega t + \\ & \left[(1/2)a_2 v_{io}^2 + (1/2)a_4 v_{io}^4 + \ldots\right]\cos 2\omega t + \left[(1/4)a_3 v_{io}^3 + \ldots\right]\cos 3\omega t + \\ & \left[(1/8)a_4 v_{io}^4 + \ldots\right]\cos 4\omega t + \ldots \end{aligned} \tag{18.160}$$

Assuming that the ω in equation (18.119) is the angular frequency of an undesired signal ω_u, which is half *of* ω_d, the angular frequency of desired signal, that is

$$\omega_u = \frac{\omega_d}{2}, \quad \text{or} \quad f_u = \frac{f_d}{2}, \tag{18.161}$$

where

ω_u or f_u = undesired angular frequency or frequency,
ω_d or f_d = desired angular frequency or frequency.

The *IM* product of the undesired signal with frequencies ω_u in transfer function (18.160) containing the term with frequency $2\omega_u = \omega_d$ is

$$v_{out,u} = \ldots + \left[(1/2)a_2 v_{iu}^2 + (1/2)a_4 v_{iu}^4 + \ldots\right]\cos 2\omega_u t + \ldots \tag{18.162}$$

where v_{iu} = voltage amplitude of undesired signal.

According to the measurement of *IM* rejection, we raise this un-desired signal's level $v_{out,u}$ up to a reference level of desired signal $v_{out,d}$, which is

$$v_{out,d} = \ldots + [a_1 v_{io} + \ldots]\cos\omega_d t \tag{18.163}$$

where v_{io} = voltage amplitude of the desired signal.

This implies that the term with $cos2\omega_u t$ in expression (18.162) for the transfer function of the undesired signal corresponds to the term with $cos\omega_d t$ in expression (18.163) for the transfer function of the desired signal, if the terms higher than fourth order in expression (18.162) and the terms higher than second order in expression (18.163) are neglected, that is,

$$a_1 v_{io} \cos \omega_d t \approx \frac{1}{2} a_2 v_{iu}^2 \cos 2\omega_u t, \tag{18.164}$$

then,

$$v_{iu} \approx \left(2 \frac{a_1}{a_2} v_{io}\right)^{1/2}. \tag{18.165}$$

Consequently,

$$IMR_{2,v} = \frac{v_{iu}}{v_{io}} \approx \left(2 \frac{a_1}{a_2}\right)^{1/2} v_{io}^{-1/2}, \tag{18.166}$$

$$IMR_{2,dB} = 20 \log \frac{v_{iu}}{v_{io}} \approx 20 \log \left[\left(2 \frac{a_1}{a_2}\right)^{1/2} v_{io}^{-1/2}\right]. \tag{18.167}$$

The second order inter-modulation rejection, IMR_2, is dependent not only on the ratio of the non-linearity coefficients, a_1 and a_2, but is also sensitive to the voltage amplitude of the input signal, v_{io}. The higher the voltage amplitude of the input signal v_{io}, the lower the IMR_2.

18.4.2.2 IP_3

When

$$m = 3, \tag{18.168}$$

then expressions (18.148) to (18.151) become

$$IIP_3 = \frac{\Delta}{2} + P_i, \tag{18.169}$$

$$OIP_3 = G + IIP_3, \tag{18.170}$$

$$\alpha = \frac{2}{3}, \tag{18.171}$$

$$IMR_3 = \alpha(IIP_3 - P_{i,s}) = \frac{2}{3}(IIP_3 - P_{i,s}). \tag{18.172}$$

By direct measurement of Δ in the test laboratory, one can calculate IIP_3 and then IMR_3 by equations (18.169) and (18.172).

Figure 18.23(a) shows the setup of IP_3 testing. There are two generators generating two undesired signals with the same power level, v_{io} or P_{io} but with two different

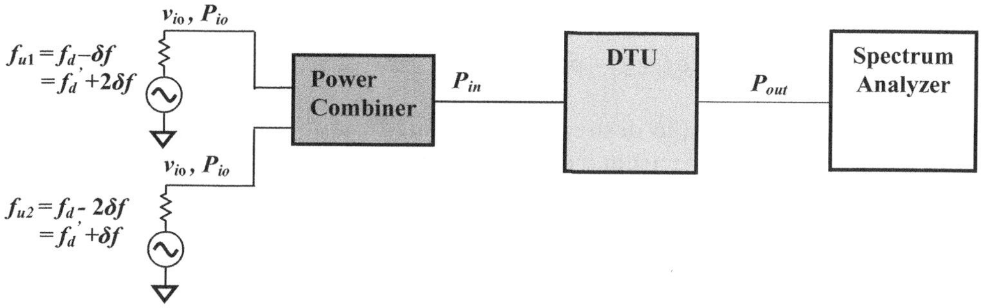

(a) Setup of IP_3 testing

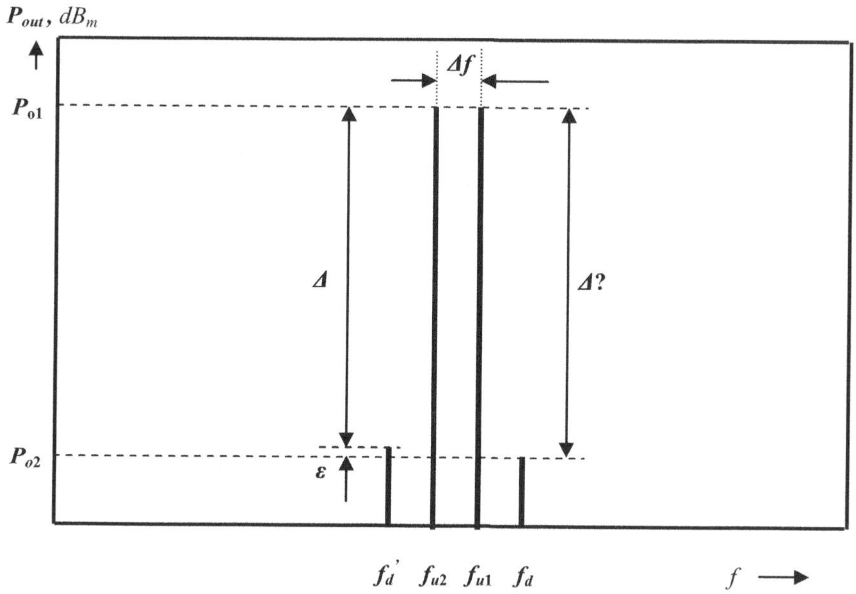

(b) Display of IP_3 testing

Figure 18.23 Setup and display of IP_3 testing.

undesired frequencies, f_{u1} and f_{u2}. These two input signals are combined by a power combiner and then delivered to the DTU (Desired Tested Unit). The power output from the DTU, P_{out}, is displayed on the screen of the spectrum analyzer.

Figure 18.23(a) shows the display of IP_3 testing on the screen of the spectrum analyzer. In addition to the two undesired signals at the frequencies f_{u1} and f_{u2}, two third order spurious products with frequency f_d and f_d' appear on the screen of spectrum analyzer.

$$f_d = 2f_{u1} - f_{u2}, \quad \text{or} \quad \omega_d = 2\omega_{u1} - \omega_{u2}, \tag{18.173}$$

$$f_d' = 2f_{u2} - f_{u1}, \quad \text{or} \quad \omega_d' = 2\omega_{u2} - \omega_{u1}, \tag{18.174}$$

This is due to the two input undesired frequencies, which are

$$f_{u1} = f_d - \delta f = f_d' + 2\delta f, \quad \text{or} \quad \omega_{u1} = \omega_d - \delta\omega = \omega_d' + 2\delta\omega, \tag{18.175}$$

$$f_{u2} = f_d - 2\delta f = f_d' + \delta f, \quad \text{or} \quad \omega_{u2} = \omega_d - 2\delta\omega = \omega_d' + \delta\omega. \tag{18.176}$$

The difference between the desired and undesired output powers, Δ, can be read from the screen of the spectrum analyzer, that is,

$$\Delta = P_{o1} - P_{o2}, \tag{18.177}$$

And note that

$$P_i = P_{io}, \tag{18.178}$$

By substituting (18.177) and (18.178) into (18.169), (18.170), and (18.172), the values of IIP_3, OIP_3, and IMR_3 can be obtained, respectively.

This is one way to obtain the value of IMR_3 by means of testing with the setup as shown in Figure 18.23(a).

There is a minor problem that happens quite often in the actual testing. Sometimes, the two values of Δ at f_d and f_d' appearing on the display screen are not the same, but have a difference ε. This is due to the large number of computation iterations in the computer system and results from the truncation of decimal numbers. The worst case resides with the lesser Δ, and should be taken into account in the calculation of IP_3 or IMR_3.

Another way to obtain the value of IMR_3 is to calculate it from the non-linear coefficients of the non-linear device or system if the transfer function as shown in expression (18.128) is well known.

Recalling the expressions from (18.128) to (18.137), the transfer function (18.128) can be rewritten as

$$\begin{aligned}
V_{out} = {} & [a_o + a_2 V_o^2 + (9/4) a_4 V_o^4 + \ldots] + \\
& [a_1 v_{io} + (9/4) a_3 v_{io}^3 + \ldots] \cos\omega_1 t + [a_1 v_{io} + (9/4) a_3 v_{io}^3 + \ldots] \cos\omega_2 t + \\
& [(1/2) a_2 v_{io}^2 + 2a_4 v_{io}^4 + \ldots] \cos 2\omega_1 t + [(1/2) a_2 v_{io}^2 + 2a_4 v_{io}^4 + \ldots] \cos 2\omega_2 t + \\
& [a_2 v_{io}^2 + 3a_4 v_{io}^4 + \ldots] \cos(\omega_1 - \omega_2) t + [a_2 v_{io}^2 + 3a_4 v_{io}^4 + \ldots] \cos(\omega_1 + \omega_2) t + \\
& [a_3 v_{io}^3 + \ldots] \cos 3\omega_1 t + [a_3 v_{io}^3 + \ldots] \cos 3\omega_2 t + \\
& [(3/4) a_3 v_{io}^3 + \ldots] \cos(2\omega_2 - \omega_1) t + [(3/4) a_3 v_{io}^3 + \ldots] \cos(2\omega_2 + \omega_1) t + \\
& [(3/4) a_3 v_{io}^3 + \ldots] \cos(2\omega_1 - \omega_2) t + [(3/4) a_3 v_{io}^3 + \ldots] \cos(2\omega_1 + \omega_2) t + \\
& [(1/8) a_4 v_{io}^4 + \ldots] \cos 4\omega_1 t + [(1/8) a_4 v_{io}^4 + \ldots] \cos 4\omega_2 t + \\
& [(1/2) a_4 v_{io}^4 + \ldots] \cos(3\omega_2 - \omega_1) t + [(1/2) a_4 v_{io}^4 + \ldots] \cos(3\omega_2 + \omega_1) t + \\
& [(1/2) a_4 v_{io}^4 + \ldots] \cos(3\omega_1 - \omega_2) t + [(1/2) a_4 v_{io}^4 + \ldots] \cos(3\omega_1 + \omega_2) t + \\
& [(3/4) a_4 v_{io}^4 + \ldots] \cos 2(\omega_1 - \omega_2) t + [(3/4) a_4 v_{io}^4 + \ldots] \cos 2(\omega_1 + \omega_2) t + \ldots
\end{aligned} \tag{18.179}$$

Let's focus on the term containing the frequency $(2\omega_2 - \omega_1)$ in the transfer function (18.179), that is,

$$v_{out} = \ldots + [(3/4) a_3 v_{io}^3 + \ldots] \cos(2\omega_1 - \omega_2) t + \ldots \tag{18.180}$$

Now if

$$f_1 \to f_{u1}, \quad \text{or} \quad \omega_1 \to \omega_{u1}, \tag{18.181}$$

$$f_2 \to f_{u2}, \quad \text{or} \quad \omega_2 \to \omega_{u2}. \tag{18.182}$$

Then from (18.173), we have

$$2\omega_1 - \omega_2 = 2\omega_{u1} - \omega_{u2} = \omega_d, \tag{18.183}$$

Equation (18.180) can be rewritten as

$$v_{out,u} = \ldots + [(3/4)a_3 v_{iu}^3 + \ldots]\cos\omega_d t + \ldots \tag{18.184}$$

where

v_{iu} = voltage amplitude of undesired signal,
$v_{out,u}$ = undesired signals' level.

This is the *IM* product of the two undesired signals with frequencies ω_{u1} or ω_{u2} and with the same amplitude v_{io}. The measurement of *IM* rejection raises this undesired signals' level $v_{out,u}$ up to a reference level of desired signal $v_{out,d}$, which is

$$v_{out,d} = \ldots + [a_1 v_{io} + \ldots]\cos\omega_d t + \ldots \tag{18.185}$$

where v_{io} = voltage of the desired signal.

This implies that the term with $cos 2\omega_d t$ in expression (18.184) for the transfer function of the undesired signal corresponds to the term with $cos\omega_d t$ in expression (18.185) for the transfer function of the desired signal, if the terms higher than fourth order in expression (18.184) and the terms higher than second order in expression (18.185) are neglected, that is,

$$a_1 v_{io}\cos\omega_d t \approx \frac{3}{4}a_3 v_{iu}^3 \cos(2\omega_{u1} - \omega_{u2})t, \tag{18.186}$$

then,

$$v_{iu} \approx \left(\frac{4}{3}\frac{a_1}{a_3}v_{io}\right)^{\frac{1}{3}}, \tag{18.187}$$

Consequently,

$$IMR_{3,v} = \frac{v_{iu}}{v_{io}} \approx \left(\frac{4}{3}\frac{a_1}{a_3}\right)^{\frac{1}{3}} v_{io}^{-\frac{2}{3}}, \tag{18.188}$$

$$IMR_{3,dB} = 20\log\frac{v_{iu}}{v_{io}} \approx 20\log\left[\left(\frac{4}{3}\frac{a_1}{a_3}\right)^{\frac{1}{3}} v_{io}^{-\frac{2}{3}}\right]. \tag{18.189}$$

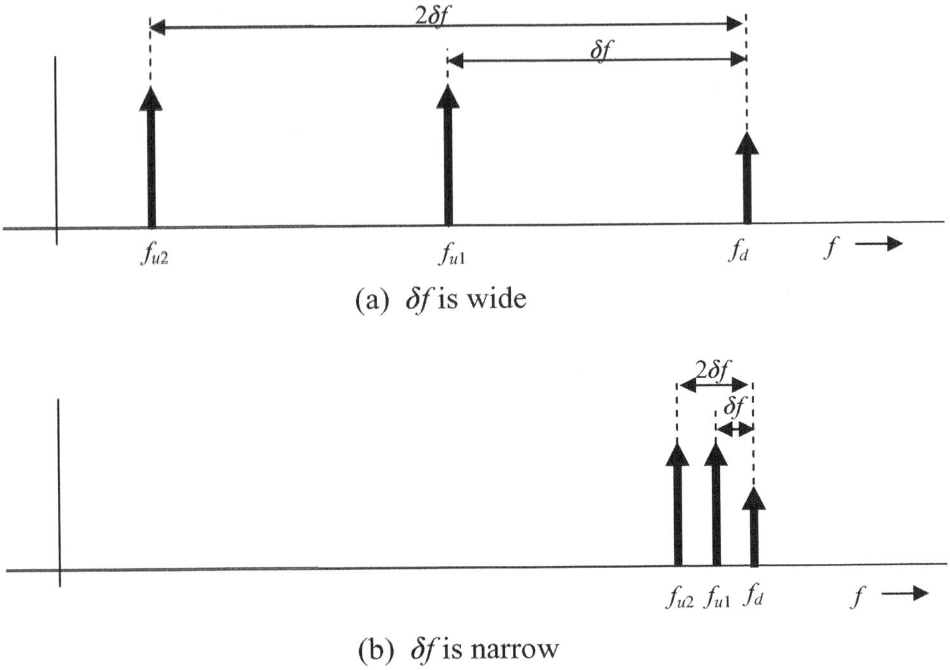

Figure 18.24 Undesired signal frequencies, f_{u1} and f_{u2}, and desired signal frequency, f_d.

The third order inter-modulation rejection, IMR_3, is not only dependent on the ratio of the non-linearity coefficients a_1 and a_3, but is also sensitive to the input signal voltage amplitude, v_{io}. The higher the signal voltage amplitude, the less the IMR_3.

Finally, it must be pointed out that there is a unique feature existing in the third order spurious product: the difference of the two un-desired frequencies δf could be any value! There is no restriction on δf either in the IP_3 testing setup as shown in Figure 18.23(a) or in expressions (18.175) and (18.176). The range of δf could be very wide, say, several thousand MHz, or very narrow, say, a few Hz, as shown in Figure 18.24.

Consequently, in practical circuit and system designs, testing for the third order spurious product must be carefully conducted., The two undesired frequencies δf must cover a wide frequency range from a few kHz to couple hundred MHz.

Theoretically, the non-linearity of a device or a system could be described by any order of IM or spurious product. However, by means of the unique feature of the third order spurious product, the third order intercept point, IP_3, has been chosen by the U.S. government as a criterion to judge the non-linearity of a device or a system. Let's demonstrate this with the example of two cellular phones produced by Motorola and Nokia respectively.

Assume that both phones are operating in the same operating frequency range and have the same channel spacing or bandwidth, say, 12.5 kHz. In order to compare their non-linearity performance without prejudice, the FCC (Federal Communications Commission) could choose a value of δf less than 6.25 KHz. Due to this choice, in the examination by the FCC, the two undesired signals are entirely located within

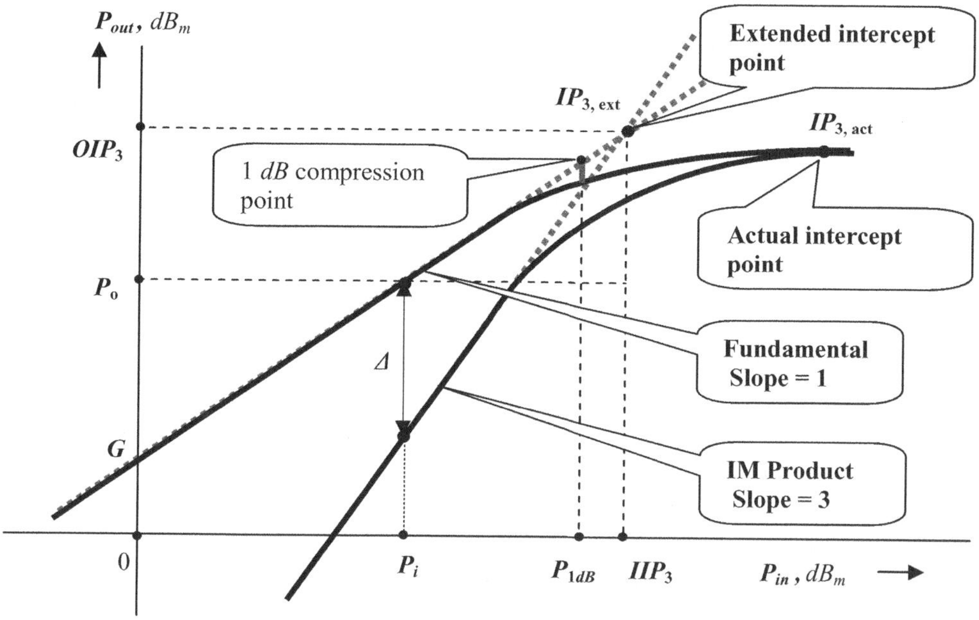

Figure 18.25 1 dB compression point and IP_3.

the channel bandwidth and can never be filtered away by either Motorola or Nokia's phone because a filter design with a few kHz of bandwidth in the RF frequency range is almost impossible. The examination by the FCC would do justice to both Motorola and Nokia phones without any prejudice. This is why the third-order spurious product, but not the other orders of spurious products, is the criterion of the non-linearity of a communication system.

***18.4.2.3 1 dB Compression Point and* IP$_3$** The intercept point as shown in Figures 18.20 and 18.21 is only a linear approximation of an actual measurement. For instance, the actual testing result for IP_3 can be represented by the Figure 18.25. There are two curves shown on the plot. They are straight segments with slopes of $m = 1$ or $m = 3$ when the input power P_{in} is low, but start to bend downward when the input power P_{in} is increased to a certain level. As the input power continuously increases, the plots then deviate more and more from straight lines. The actual intercept point is marked as $IP_{3,\text{act}}$ in Figure 18.25. The two straight segments can be extended in the increasing direction of input power. The extended intercept point $IP_{3,\text{ext}}$ of these two straight segments with slopes of $m = 1$ and $m = 3$ is the third-order intercept point that we are looking for. The corresponding input and output intercept points are higher IIP_3 and OIP_3.

Testing for the third intercept point is somewhat more complicated than testing only for the curve containing the line with a slope of $m = 1$. It would be easy to get stuck on testing only the curve containing the line with a slope of $m = 1$ if the third-order intercept point $IP_{3,\text{ext}}$ can be evaluated from the bending point because the bending point is correlated with the third-order intercept point and is not too far from the extended intercept point $IP_{3,\text{ext}}$.

Instead of the bending point, the $1\,dB$ compression point is utilized to evaluate the third-order intercept point $IP_{3,\text{ext}}$. As shown in Figure 18.25, the $1\,dB$ compression point is a special input power, P_{1dB}, at which the output power deviates $1\,dB$ from the extended straight line with a slope of $m = 1$. The relationship between $1\,dB$, P_{1dB}, and the 3rd order input intercept point, IIP_3, can be evaluated as

$$IIP_3 \to P_1 dB + (3 \text{ to } 10)\,dB, \tag{18.190}$$

The uncertainty factor, (3 to 10) dB, depends on the type of device, circuit topology, current drain, DC voltage supply, and other parameters of the device or system.

In order to cover enough of the linear portion, the input power in the third-order Intercept point testing should be started from

$$P_i < -30\,dB_m, \tag{18.191}$$

to ensure that the plot can be extended from the linear portion to get the third-order intercept point $IP_{3,\text{ext}}$.

18.4.3 Cascaded Equations of Intercept Point

Let's derive the cascaded equation for the intercept point of a system containing only two blocks as shown in Figure 18.26, where G_k and G_{k+1} are the power gains of blocks k and $k + 1$, respectively, and IPm_k and IPm_{k+1} are the power gains of blocks k and $k + 1$, respectively.

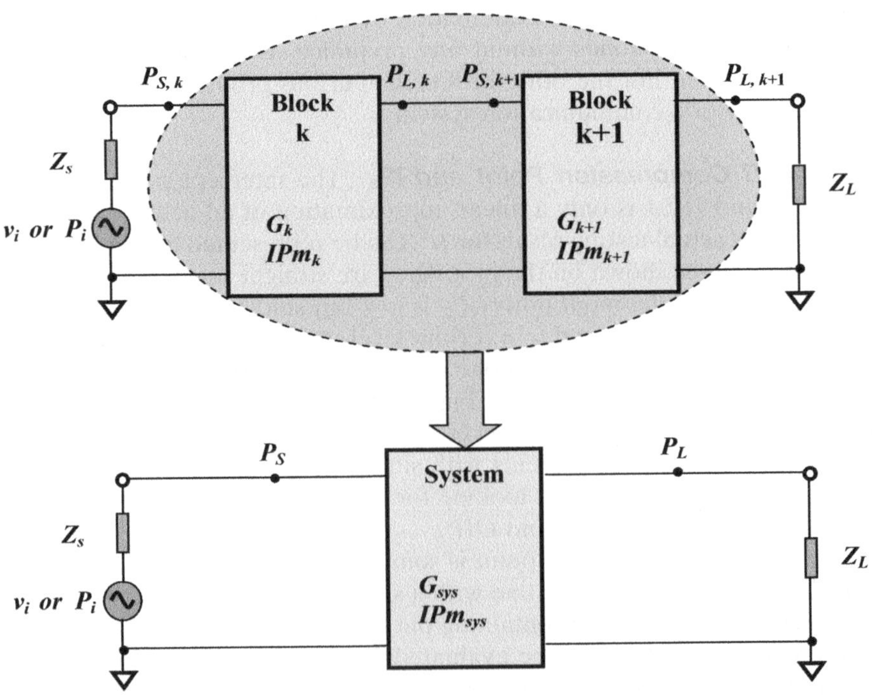

Figure 18.26 Intercept point of a system cascaded by two blocks k and $k + 1$.

Simple geometrical relations are shown in Figure 18.20:

$$\Delta = d(m-1), \tag{18.192}$$

$$\Delta = P_o - P_u, \tag{18.193}$$

$$d = OIP_m - P_o. \tag{18.194}$$

We have

$$P_u = P_o - \Delta = P_o - d(m-1), \tag{18.195}$$

$$P_u = P_o - (OIP_m - P_o)(m-1), \tag{18.196}$$

$$P_u = mP_o - (m-1)OIP_m, \tag{18.197}$$

In the numeric scale or in units of *watts*, equation (18.197) becomes

$$P_u = \frac{P_o^m}{OIP_m^{(m-1)}}, \tag{18.198}$$

Referring to the individual blocks k and $k+1$ in Figure 18.26, equation (18.198) can be rewritten as

$$P_{u,k} = \frac{P_{o,k}^m}{OIP_{m,k}^{(m-1)}}, \tag{18.199}$$

$$P_{u,k+1} = \frac{P_{o,k+1}^m}{OIP_{m,k+1}^{(m-1)}}, \tag{18.200}$$

where

$P_{o,k}$ = desired output power from the kth block,
$P_{o,k+1}$ = desired output power from the $(k+1)$th block,
$P_{u,k}$ = undesired output power from the kth block,
$P_{u,k+1}$ = undesired output power from the $(k+1)$th block,
$OIP_{m,k}$ = output intercept point from the kth block,
$OIP_{m,k+1}$ = output intercept point from the $(k+1)$th block,

Note that

$$P_{o,k} = \frac{P_{o,k+1}}{G_{k+1}}, \tag{18.201}$$

Then

$$P_{o,k}^m = \frac{P_{o,k+1}^m}{G_{k+1}^m}. \tag{18.202}$$

The undesired output power at Z_L from block k is

$$P_{u,k}G_{k+1} = \frac{P_{o,k}^{m}G_{k+1}}{OIP_{m,k}^{(m-1)}} = \frac{P_{o,k+1}^{m}G_{k+1}}{G_{k+1}^{m}OIP_{m,k}^{(m-1)}} = \frac{P_{o,k+1}^{m}}{G_{k+1}^{(m-1)}OIP_{m,k}^{(m-1)}}. \tag{18.203}$$

The undesired voltage at Z_L from block k is

$$V_{u,k} = \left[\frac{P_{o,k+1}^{m}Z_L}{G_{k+1}^{(m-1)}OIP_{m,k}^{(m-1)}}\right]^{\frac{1}{2}}. \tag{18.204}$$

From equation (18.200), the undesired voltage at Z_L from block $k+1$ is

$$V_{u,k+1} = \left[\frac{P_{o,k+1}^{m}Z_L}{OIP_{m,k+1}^{(m-1)}}\right]^{\frac{1}{2}}. \tag{18.205}$$

Then, the total undesired voltage or the system undesired voltage across Z_L is

$$V_{u,sys} = V_{u,k} + V_{u,k+1} = \sqrt{P_{o,k+1}^{m}Z_L}\left[\left(\frac{1}{G_{k+1}OIP_{m,k}}\right)^{\frac{(m-1)}{2}} + \left(\frac{1}{OIP_{m,k+1}}\right)^{\frac{(m-1)}{2}}\right]. \tag{18.206}$$

The total undesired output power or the system undesired output power across Z_L is

$$P_{u,sys} = \frac{V_{u,sys}^2}{Z_L} = P_{o,k+1}^{m}\left[\left(\frac{1}{G_{k+1}OIP_{m,k}}\right)^{\frac{(m-1)}{2}} + \left(\frac{1}{OIP_{m,k+1}}\right)^{\frac{(m-1)}{2}}\right]^2. \tag{18.207}$$

$$\frac{P_{u,sys}}{P_{o,k+1}} = P_{o,k+1}^{(m-1)}\left[\left(\frac{1}{G_{k+1}OIP_{m,k}}\right)^{\frac{(m-1)}{2}} + \left(\frac{1}{OIP_{m,k+1}}\right)^{\frac{(m-1)}{2}}\right]^2. \tag{18.208}$$

At the system intercept point,

$$P_{u,sys} = P_{o,k+1}, \tag{18.209}$$

$$P_{o,k+1} = OIP_{m,sys}, \tag{18.210}$$

then, equation (18.208) becomes

$$\left(\frac{1}{OIP_{m,sys}}\right)^{\frac{(m-1)}{2}} = \left(\frac{1}{G_{k+1}OIP_{m,k}}\right)^{\frac{(m-1)}{2}} + \left(\frac{1}{OIP_{m,k+1}}\right)^{\frac{(m-1)}{2}}. \tag{18.211}$$

Mathematically, equation (18.211) looks like a very nice form of the relationship between the total output intercept point, $OIP_{m,sys}$, and the individual output intercept point, $OIP_{m,k}$ and $OIP_{m,k+1}$. The total output intercept point, $OIP_{m,sys}$, is not only a function of the individual output intercept points $OIP_{m,k}$ and $OIP_{m,k+1}$, but also is related to the gain of the second individual block, G_{k+1}.

Now let's transfer equation (18.211) from the output intercept point, OIP_m, to the input intercept point, IIP_m. Note that

$$IIP_{m,sys} = \frac{OIP_{m,sys}}{G_T} = \frac{OIP_{m,sys}}{G_k G_{k+1}}, \tag{18.212}$$

also,

$$IIP_{m,k} = \frac{OIP_{m,k}}{G_k}, \tag{18.213}$$

and,

$$IIP_{m,k+1} = \frac{OIP_{m,k+1}}{G_{k+1}}, \tag{18.214}$$

then, we have

$$\left(\frac{1}{IIP_{m,sys}}\right)^{\frac{(m-1)}{2}} = \left(\frac{1}{IIP_{m,k}}\right)^{\frac{(m-1)}{2}} + \left(\frac{G_k}{IIP_{m,k+1}}\right)^{\frac{(m-1)}{2}}. \tag{18.215}$$

This can be rewritten as

$$\left(\frac{1}{IIP_{m,sys}}\right)^{\frac{(m-1)}{2}} = \left(\frac{1}{IIP_{m,k}}\right)^{\frac{(m-1)}{2}} + \left(\frac{1}{\frac{IIP_{m,k+1}}{G_k}}\right)^{\frac{(m-1)}{2}}. \tag{18.216}$$

A special feature of the intercept point can be found from equations (18.215) and (18.216). The system intercept point or resulting intercept point OIP_{sys} *or* IIP_{sys} looks like the formula of resistance resulted by two resistors connected together in parallel. From the cascaded equation (18.216) it can be seen that the intercept point of second block IIP_{k+1} contributes more to the system intercept point IIP_{sys} than the intercept point of first block IIP_k because IIP_k has the same weight as IIP_{k+1}/G_k in the equation (18.216). In other words, the gain in the first block G_k promotes the function of the intercept point in the second block. This is the reverse feature of that in the cascaded equation of noise figure described in Section 18.3.5. High power gain in the first block is harmful to the system intercept point but beneficial to the system noise figure.

TABLE 18.6 Different system intercept points contributed from two blocks with different settlements of intercept point and gain

Example 1	Block *k* *LNA*	Block *k* + 1 Mixer	System
IIP_3, dB_n	3	100	3.0
Gain, dB	10	5	15.0

Example 2	Block *k* *LNA*	Block *k* + 1 Mixer	System
IIP_3, dB_n	100	5	0.0
Gain, dB	5	5	10.0

Example 3	Block *k* *LNA*	Block *k* + 1 Mixer	System
IIP_3, dB_n	3	5	–5.6
Gain, dB	10	5	15.0

Example 4	Block *k* *LNA*	Block *k* + 1 Mixer	System
IIP_3, dB_n	3	5	–18.2
Gain, dB	16	5	21.0

In order for the reader to become more familiar with this feature of cascaded intercept points, Table 18.6 lists different system intercept points contributed from two blocks with different settlements of intercept point and gain.

In example 1, the system $IIP_{3,sys}$ is equal to the IIP_3 of block k because the $IIP_{3,k+1}$ of block $k + 1$ is $100\,dB_m$ which is extremely high.

In example 2, the system $IIP_{3,sys}$ is almost contributed by only the $IIP_{3,k+1}$ of block $k + 1$ because the $IIP_{3,k}$ of block k is $100\,dB_m$ which is also extremely high. However, the system $IIP_{3,sys}$ is not the same as $IIP_{3,k+1}$ of block $k + 1$ because the *gain* G_k of block k diminishes the contibution of $IIP_{3,k+1}$ to the system $IIP_{3,sys}$.

In example 3, the system $IIP_{3,sys}$ is lower than that in example 1 because the $IIP_{3,k+1}$ of block $k + 1$ is much lower than that in example 3 than in example 1.

In example 4, the system $IIP_{3,sys}$ is lower than that in example 3 because the gain $G_k = 16\,\text{dB}$ of block k is higher than that in example 3.

The opposite feature of cascaded equations between noise figure and intercept point leads to different considerations in the front end design of a receiver. Figure 18.27(a) is one kind of front end receiver design which is beneficial to the system noise figure while Figure 18.27(b) is a kind of front end receiver design which is beneficial to the system intercept point. It should be noted that the sensitivity of a receiver is directly related to the system noise figure and that the linearity of a system is directly related to the system intercept point.

Now let's extend our discussion to a general case: a system consists of n blocks as shown in Figure 18.28. Equations (18.211) and (18.216) can be extended as

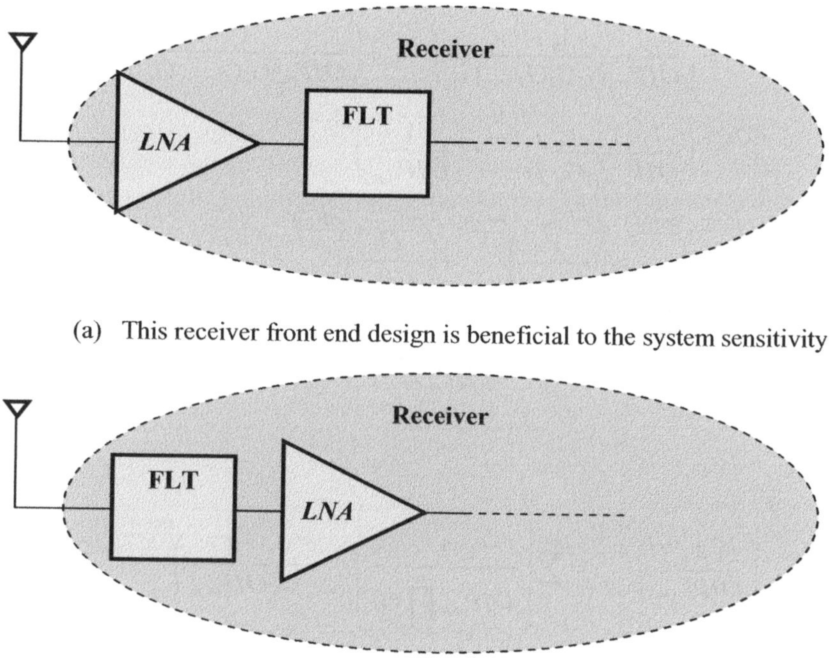

(a) This receiver front end design is beneficial to the system sensitivity

(b) This receiver front end design is beneficial to the system linearity

Figure 18.27 Two kinds of receiver front end design.

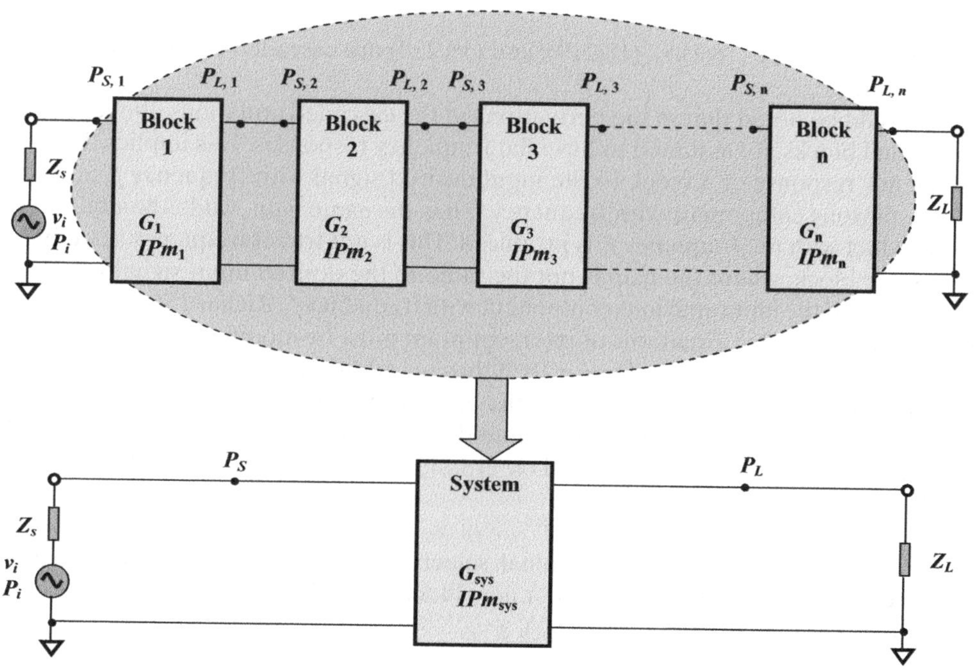

Figure 18.28 Intercept point of a system cascaded by *n* blocks.

$$\left(\frac{1}{OIP_{m,sys}}\right)^{\frac{(m-1)}{2}} = \left(\frac{1}{OIP_{m,1}G_2G_3G_4\ldots G_n}\right)^{\frac{(m-1)}{2}} + \left(\frac{1}{OIP_{m,2}G_3G_4\ldots G_n}\right)^{\frac{(m-1)}{2}} + \ldots + \left(\frac{1}{OIP_{m,n-1}G_n}\right)^{\frac{(m-1)}{2}} + \left(\frac{1}{OIP_{m,n}}\right)^{\frac{(m-1)}{2}}. \quad (18.217)$$

$$\left(\frac{1}{IIP_{m,sys}}\right)^{\frac{(m-1)}{2}} = \left(\frac{1}{IIP_{m,1}}\right)^{\frac{(m-1)}{2}} + \left(\frac{G_1}{IIP_{m,2}}\right)^{\frac{(m-1)}{2}} + \left(\frac{G_1G_2}{IIP_{m,3}}\right)^{\frac{(m-1)}{2}} + \ldots + \left(\frac{G_1G_2G_2\ldots G_{n-1}}{IIP_{m,n}}\right)^{\frac{(m-1)}{2}}. \quad (18.218)$$

Or

$$\left(\frac{1}{OIP_{m,sys}}\right)^{\frac{(m-1)}{2}} = \sum_{1}^{n-1}\left(\frac{1}{OIP_{m,k}\prod_{k+1}^{n} G_j}\right)^{\frac{(m-1)}{2}} + \left(\frac{1}{OIP_{m,n}}\right)^{\frac{(m-1)}{2}}, \quad (18.219)$$

$$\left(\frac{1}{IIP_{m,sys}}\right)^{\frac{(m-1)}{2}} = \left(\frac{1}{IIP_{m,1}}\right)^{\frac{(m-1)}{2}} + \sum_{2}^{n}\left(\frac{\prod_{1}^{k-1} G_j}{IIP_{m,k}}\right)^{\frac{(m-1)}{2}}. \quad (18.220)$$

Equations (18.211), (18.215), (18.219), and (18.220) are cascaded equations of intercept point.

It should be noted that in the derivations of the cascaded equations above, all the individual blocks are assumed to have flat frequency responses. This implies that the frequency response of a block to the input desired signal with frequency f_d and the input spurious component with frequency f_s has the same gain, while the undesired *IM* product with the frequency f_u is produced. This is not true if frequency selectivity exists in a block, where the gain is not the same to the desired input signal with frequency f_d and the input spurious component with frequency f_s. Richard Sagers pointed out that the cascaded equations of intercept point must be modified as follows.

Assuming that block k in Figure 18.29 has an additional selectivity SEL_k dB at the input spurious frequency f_s, where the gain at the input spurious frequency f_s is SEL_k dB is lower than the gain at the input desired signal frequency f_d, the cascaded equations of intercept points (18.211), (18.215), (18.219), and (18.220), must be modified using three steps:

1) The block k having an additional selectivity should be combined with the sequential block $k + 1$ to form a new block k' as shown in Figure 18.30.
2) The power gain of the new block k' is

$$G'_k = G_k + G_{k+1}, \quad (18.221)$$

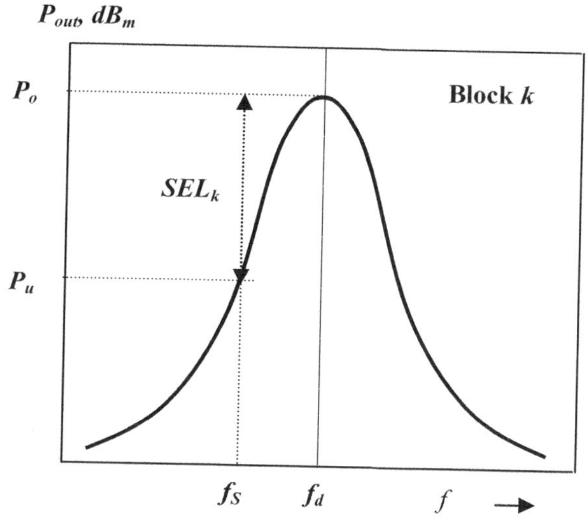

Figure 18.29 Selectivity at the input spurious frequency f_s in block k.

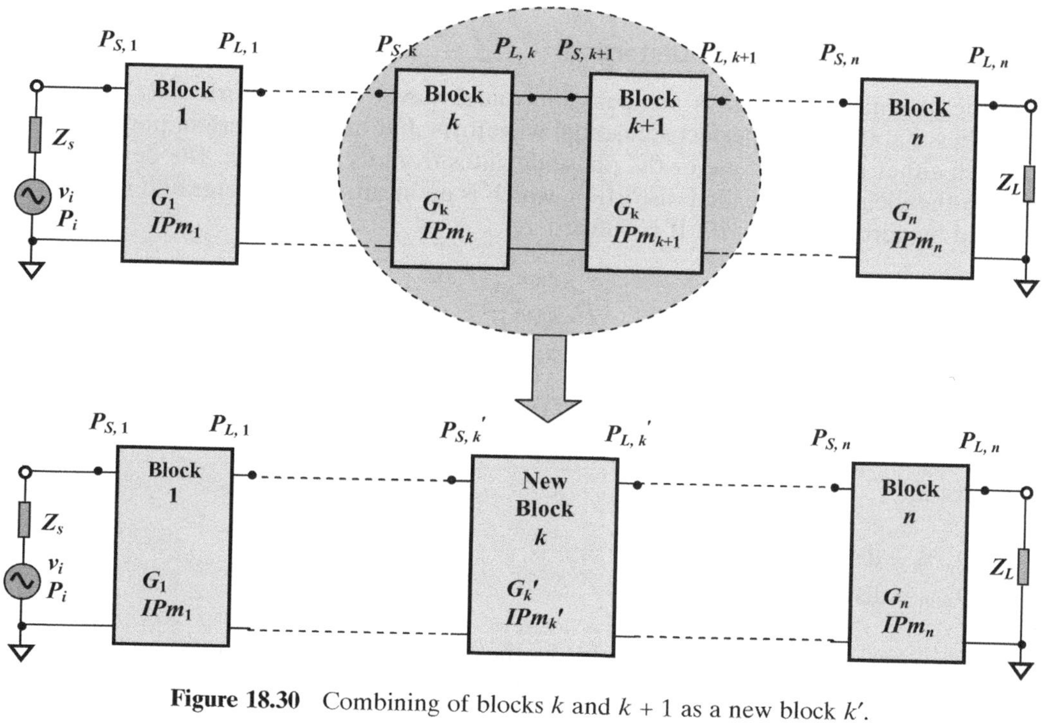

Figure 18.30 Combining of blocks k and $k+1$ as a new block k'.

where

G_k' = combined gain to the desired input signal in the new block k',
G_k = gain to the desired input signal in block k,
G_{k+1} = gain to the desired input signal in block $k + 1$.

3) The input intercept point, *IIPm* of the new block k' is

$$IIPm'_k = IIPm_{k+1}(SEL_k)^{\frac{m}{m-1}}, \tag{18.222}$$

where

$IIPm'_k$ = combined *IIPm* to the desired input signal in the new block k',
$IIPm_k$ = IIP_m to the desired input signal in block k,
IIP_{mk+1} = IIP_m to the desired input signal in block $k + 1$,
SEL_k = selectivity of block k at spurious frequency f_s.

4) The output intercept point, *OIPm* of the new block k' is,

$$OIPm'_k = G'_k IIPm'_k = G'_k IIP_{k+1} SEL^{\frac{m}{m-1}}. \tag{18.223}$$

where $OIPm'_k$ = The combined *OIPm* to the desired input signal in the new block k'.

18.4.4 Non-linearity and Distortion

If the original *RF* signal is an ideal sinusoidal waveform, a distorted *RF* signal implies that it is not a perfect sinusoidal waveform, but has an imperfect portion in which either the amplitude or the phase deviates from its ideal state. The deviation from the ideal state is called distortion, which is an intuitive parameter and is measured by percentage, or *dB*. It is defined as

$$D_{v,\%} = \frac{\Delta V}{V}, \tag{18.224}$$

$$D_{P,dB} = 20\log\frac{\Delta V}{V} \approx 20\log(D_{v,\%}). \tag{18.225}$$

where

$D_v,\%$ = distortion of voltage in %,
$D_{P,dB}$ = distortion in *dB*.

In terms of relation (18.225), distortion by the unit of % and by *dB* can be exchanged with one another. For example, 5% of distortion is equal to about –26.02 *dB*:

$$(-26.02)\,dB = 20\log\left(\frac{5}{100}\right). \tag{18.226}$$

It should be noted that the mechanism of the distortion and the noise of an *RF* signal are essentially different. The noise spreads or swallows its waveform trace but does not bring about distortion, while all the harmonics and all the *IM* or spurious products contribute to distortion but not to noise. Their impacts on a communication system are essentially different. For instance, a receiver with low noise has a high sensitivity in which the antenna can sense a very weak input signal. However, a considerable distortion could be present in its voice. A girl's voice may end up sounding like a boy's voice. On the other hand, a receiver with high noise has low sensitivity in which the antenna can sense an input signal only when the input signal power exceeds a certain level. However, its distortion may be low so that the voice can be easily recognized. In short, both spurious products and noise are not welcome in a communication system, but their impacts on the system are different. Spurious products cause distortion while noise reduces the sensitivity in the communication system.

18.5 OTHER PARAMETERS

18.5.1 Power Supply Voltage and Current Drain

DC power supply voltage applied into a circuit block or system becomes lower and lower. The reasons are simple: to reduce the cost and shrink the size of the block or system.

For instance, in a wireless communication system, the *DC* power supply and current drain is not considered as an important subject in the design of basic stations or repeaters, but it is quite an important subject in the design of the hand-set. More battery cells must be provided, and thus higher costs must be paid, if a high *DC* power supply voltage is required.

At present, the *DC* power supply voltage in most handsets of cellular phones is $3V$. The competition between cellular phone manufacturers is centered on current drain. High current drain in the circuits means high current consumption from the battery. The current consumption of a handset is the product of the total current drain and the total operating hours per day. The units of current consumption are ($mA \cdot$ hour) or ($A \cdot$ hour).

Total current consumption is important to customers. It is acceptable if a handset can be normally operated all day and requires battery charging at night. However, the customer will be upset if a handset can only be operated in the morning and must be charged in the afternoon.

The total current drain i_{tot} or current consumption $i_{tot} \cdot H_{tot}$ of a handset can be evaluated as follows.

First, the total current drain $i_{\rm tot}$ can be categorized into three types of sub-current drain:

1. Standby current drain, $i_{\rm standby}$: This is the current drain when the handset is operating in "waiting" status, that is, the handset is operating neither in transmit mode nor in receive mode, but is merely waiting for incoming signals.
2. Receive mode current drain, $i_{\rm receive}$: This is the current drain when the handset is operating in receive mode, listening to the incoming signals.
3. Transmit mode current drain, $i_{\rm transmit}$: This is the current drain when the handset is operating in transmit mode to deliver outbound signals.

Second, the corresponding operating periods of how much time the handset operates in its respective modes, that is,

- Operating period when the handset is in standby mode, $T_{standby}$.
- Operating period when the handset is in receive mode, $T_{receive}$.
- Operating period when the handset is in transmit mode, $T_{transmit}$.

The ratio of these three operating periods differs between customers. For example, for a policeman, the period ratio of these three current drains might be approximately

$$T_{standby}:T_{receive}:T_{transmit} = 1:2:2.$$

For a student, the period ratio of these three current drains might be

$$T_{standby}:I_{receive}:T_{transmit} = 5:1:1,$$

and for someone caring for a house and children, the period ratio of these three current drains could be

$$T_{standby}:I_{receive}:T_{transmit} = 10:1:1.$$

Third, another parameter is the total operating hours per day, H_{tot}.

Consequently, the current drain of a handset i_{tot} can be calculated by the following formula:

$$i_{tot} = i_{standby}\frac{T_{standby}}{T_{tot}} + i_{receive}\frac{T_{receive}}{T_{tot}} + i_{transmit}\frac{T_{transmit}}{T_{tot}}. \tag{18.227}$$

And the total current consumption can be calculated by the following formula:

$$i_{tot}H_{tot} = \left(i_{standby}\frac{T_{standby}}{T_{tot}} + i_{receive}\frac{T_{receive}}{T_{tot}} + i_{transmit}\frac{T_{transmit}}{T_{tot}}\right)H_{tot}. \tag{18.228}$$

18.5.2 Part Count

Part count is the total number of parts in a circuit block or a system. It is a simple statistical number, but it is also an important parameter in a circuit block design or a system design. This is because the reliability of a product, either a circuit block or a system, depends significantly on its part count. In the early stages of developing cell phones, the part count of the entire handset was about 200 to 300; the reliability was quite low and its price was high. At present, the part count has been significantly reduced from a few hundreds down to a few tens, resulting in increased reliability; therefore its price correspondingly rolls down. Should the expected *SOC* (System On Chip) goal be realized, the part count will approach one, theoretically increasing reliability to infinity.

18.6 EXAMPLE OF *RF* SYSTEM ANALYSIS

Usually, an *RF* system consists of a number of blocks. The system parameters must be calculated from the corresponding parameters of all the individual blocks. The basic skill for this task is calculating the system parameters from the corresponding parameters of two individual blocks. As long as we master such a skill for two blocks, it can be extended to a system constructed with more than two blocks.

18.6.1 Application of Cascaded Equation in System Analysis

Today, computer simulation is a convenient and sophisticated tool not only in circuit design, but also in system performance analysis. However, applying cascoded equations on a Microsoft Excel worksheet enables engineers to do a system analysis in a fast and handy way, although such an analysis is confined to the primary stage. Table 18.7 shows such an example.

This is a system analysis table for the front end of a receiver. It consists of eight blocks:

1. Harmonic Filter/Balun,
2. *LNA*
3. Mixer
4. *IF VGA*
5. *IF LPF*
6. *IF VGA*
7. *IF LPF*
8. *SP* amplifier
9. Back end (base band portion).

TABLE 18.7 System analysis of a receiver front end by the cascaded equations

System goals

Lower Frequency : 2412 *MHz* | Temperature : 300 K^o

Upper Frequency : 2902.5 *MHz* | *RISE@12dBSINAD* : 6.0 *dB*

System BW : 13.5 *kHz* | *RISE@20dBQuiet* : 7.6 *dB*

Worst case Factor : 0.1

Performance										
(Column)		**1**	**2**	**3**	**4**	**5**	**6**	**7**	**8**	**9**
		HF/Balun	*LNA*	Mixer.	*IF VGA*	*IF LPF*	*IF VGA*	*IF LPF*	*SP* Amp	Bk.End
Gain	*dB*	-2.0	16	1	25	-2	25	-2	0	
NF	*dB*	2	3	12	5	2	5	2	0	7
IP3	dB_m	20	0	12	0	18	0	18	100	-13
IP2	dB_m	100	80	40	100	100	100	100	100	1000
SEL@Δf	*dB*	0	0	0	10	0	10	0	0	0
SEL@2Δf	*dB*	0	0	0	15	0	15	0	0	0
SEL@1/2 IF	*dB*	0	10	0	0	0	0	0	0	0
Calculations			*H*	*G*	*F*	*E*	*D*	*C*	*B*	*A*
Gain (worst)	*dB*	**-2.2**	**14.4**	**0.9**	**22.5**	**-2.2**	**22.5**	**-2.2**	**-2.2**	0
NF (worst)	*dB*	**2.2**	**3.3**	**13.2**	**5.5**	**2.2**	**5.5**	**2.2**	**2.2**	**8.4**
NFsys	*dB*	**5.8**	**3.8**	**12.4**	**5.0**	**7.0**	**5.0**	**9.0**	**7.0**	**7.0**
12dBSINAD	*uV*	**6.87**	**5.46**	**14.7**	**6.26**	**7.89**	**6.27**	**9.90**	**7.86**	**7.86**
12dBSINAD	dB_m	**-90.3**	**-92.3**	**-83.6**	**-91.1**	**-89.0**	**-91.0**	**-87.1**	**-89.1**	**-89.1**
20dBQuiet	*uV*	**8.67**	**6.89**	**18.6**	**7.91**	**9.97**	**7.92**	**12.5**	**9.93**	**9.93**
20dBQuiet	dB_m	**-88.2**	**-90.2**	**-81.6**	**-89.0**	**-87.0**	**-89.0**	**-85.0**	**-87.0**	**-87.0**
IIP3sys	dB_m	**-39.3**	**-41.3**	**-25.3**	**-24.3**	**-16.8**	**-18.8**	**-18.2**	**-13.2**	**-13.2**
IMR3	*dB*	**31.6**	**31.6**	**36.5**	**42.1**	**45.8**	**45.8**	**48.2**	**48.2**	**48.2**

The ninth block in the last column is the back end of the receiver and is excluded and constructed in the base-band portion. The values of performance in the ninth block are provided by the base band designer.

There are three portions in Table 18.7:

1) System goals: General goals for all the blocks which are not involved in the calculation.
2) Performance: The expected goals of performance for each individual block are listed, such as gain, *NF*, IP_3 ...
3) Calculation: Calculations are only conducted in this portion on the basis of the expected goals in the performance portion. In the process of calculation, the calculated results are registered in the corresponding cells in this portion.

Instead of simultaneously doing the calculations for all the blocks, the calculations are confined to only two blocks at a time. In other words, only cascaded equations for two blocks are applied in the calculations in Table 18.7.

In order to show the calculation path, let's take the calculation of the noise figure as an example. In Table 18.7, the *NF* values of every block in the Performance portion and the NF_{sys} values of every block in the Calculation portion are circled with an ellipse. The *NF* values of each block in the Performance portion are the respective expected goals of each individual block and the NF_{sys} values of each block in the Calculation portion is a systematic value of noise figure looking from the block toward all the following blocks in the right hand side. These values are grouped and named group *A*, *B*, *C*, *D*, *E*, *F*, *G*, and H. The calculation path for noise figure is also shown in Table 18.7.

Now calculation is started from group *A* in the last two blocks. The original noise figure of the back end block, NF_{sys}, shown in the calculation portion is $7\,dB$, which is provided by the base band circuit designer. The noise figure of *SP* amplifier, *NF*, is $0\,dB$. The new NF_{sys} is the original noise figure of the back end block, NF_{sys}, cascaded with the noise figure of *SP* amplifier, *NF*. In terms of the cascaded equation for two blocks, the new or resultant noise figure NF_{sys} can be calculated and is obtained as $7\,dB$, which is registered into the NF_{sys} cell of the *SP* amplifier located in the calculation portion.

Following a similar calculation path, the calculation moves from group *A* to *B*, and then from *B* to *C*, from *C* to *D*, from *D* to *E*, and so on. At last, the final value of NF_{sys} appears in group *H*. This value is calculated and is obtained as $5.8\,dB$. This is the value of the system *NF* for the entire receiver.

By a similar calculation path as shown for the noise figure above, other parameters, such as gain, intercept point, etc. can be calculated starting from the last block and working toward the first block. All the values located in the column of the first block in Table 18.7, the *HF*/Balun, are the entire system values for the receiver. In Table 18.7, not only the gain, noise figure, and intercept point, but also other parameters, such as the $12\,dB$ *SINAD*, $20\,dB$ quieting, and *IMR* are presented.

It should be noted that in Table 18.7 there are two rows which present the worst-case gain and noise figures, in which the worst case factor is $0.1 = 10\%$. The purpose of system analysis using the worst case is to retain the necessary "room" for the high performance reliability of the system.

The calculations performed in Table 18.7 are quite flexible. If using a Microsoft Excel spread sheet, the calculated results are automatically corrected if any expected goal in the performance portion in Table 18.7 is varied. This is very helpful to the system engineer in the adjustment of the design goals between the individual blocks. Table 18.7 shows a successful design since the final results would produce excellent performance:

- $12\,dB,\ SINAD = -90.3\,dB_m,$
- $20\,dBQuiet = -88.2\,dB_m,$
- $IMR_3 = 31.6\,dB,$

The first two items represent the sensitivity of the receiver and the last one item is related to the distortion of the receiver.

APPENDICES

18.A.1 Conversion between Watts, Volts, and dB_m in a System with 50Ω of Input and Output Impedance

The reader is referred to Tables 18.A.1A and 18.A.1B.

Equations

Voltage:

$$V = \sqrt{50 * 10^{\frac{P_{dBm}-30}{10}}}, \tag{18.A.1}$$

Power:

$$P = \frac{V^2}{50}, \tag{18.A.2}$$

Power in dB:

$$P_{dBm} = 10 Log\left(\frac{V^2}{50}\right) + 30. \tag{18.A.3}$$

18.A.2 *VSWR* and Other Reflection and Transmission Coefficients (R_o = 50Ω, Standard Characteristic Resistance)

The reader is referred to Table 18.A.2.

Equations

Voltage reflection coefficient:

$$\Gamma = \frac{R - R_o}{R + R_o}, \tag{18.A.4}$$

TABLE 18.A.1A Conversion between watts, Volts, and dB_m in a System with 50 Ω of input/output impedance

P, dBm	Voltage, *mV*	Power, *mW*	*P, dBm*	Voltage, *μV*	Power, *nW*
0	223.61	1.0000	−50	707.11	10.0000
−1	199.29	0.794328	−51	630.21	7.943282
−2	177.62	0.630957	−52	561.67	6.309573
−3	158.30	0.501187	−53	500.59	5.011872
−4	141.09	0.398107	−54	446.15	3.981072
−5	125.74	0.316228	−55	397.64	3.162278
−6	112.07	0.251189	−56	354.39	2.511886
−7	99.88	0.199526	−57	315.85	1.995262
−8	89.02	0.158489	−58	281.50	1.584893
−9	79.34	0.125893	−59	250.89	1.258925
−10	70.71	0.1000	−60	223.61	1.0000
−11	63.02	0.079433	−61	199.29	0.794328
−12	56.17	0.063096	−62	177.62	0.630957
−13	50.06	0.050119	−63	158.30	0.501187
−14	44.62	0.039811	−64	141.09	0.398107
−15	39.76	0.031623	−65	125.74	0.316228
−16	35.44	0.025119	−66	112.07	0.251189
−17	31.59	0.019953	−67	99.88	0.199526
−18	28.15	0.015849	−68	89.02	0.158489
−19	25.09	0.012589	−69	79.34	0.125893
−20	22.36	0.0100	−70	70.71	0.1000
−21	19.93	0.007943	−71	63.02	0.079433
−22	17.76	0.006310	−72	56.17	0.063096
−23	15.83	0.005012	−73	50.06	0.050119
−24	14.11	0.003981	−74	44.62	0.039811
−25	12.57	0.003162	−75	39.76	0.031623
−26	18.21	0.002512	−76	35.44	0.025119
−27	9.99	0.001995	−77	31.59	0.019953
−28	8.90	0.001585	−78	28.15	0.015849
−29	7.93	0.001259	−79	25.09	0.012589
−30	7.07	0.0010	−80	22.36	0.0100
−31	6.30	0.000794	−81	19.93	0.007943
−32	5.62	0.000631	−82	17.76	0.006310
−33	5.01	0.000501	−83	15.83	0.005012
−34	4.46	0.000398	−84	14.11	0.003981
−35	3.98	0.000316	−85	12.57	0.003162
−36	3.54	0.000251	−86	18.21	0.002512
−37	3.16	0.000200	−87	9.99	0.001995
−38	2.82	0.000158	−88	8.90	0.001585
−39	2.51	0.000126	−89	7.93	0.001259
−40	2.24	0.0001	−90	7.07	0.0010
−41	1.99	0.000079	−91	6.30	0.000794
−42	1.78	0.000063	−92	5.62	0.000631
−43	1.58	0.000050	−93	5.01	0.000501
−44	1.41	0.000040	−94	4.46	0.000398
−45	1.26	0.000032	−95	3.98	0.000316
−46	1.12	0.000025	−96	3.54	0.000251
−47	1.00	0.000020	−97	3.16	0.000200
−48	0.89	0.000016	−98	2.82	0.000158
−49	0.79	0.000013	−99	2.51	0.000126

TABLE 18.A.1B Conversion between watts, Volts, and dB_m in a System with 50 Ω of input/output impedance

P, dBm	Voltage, μV	Power, *nW*	*P, dBm*	Voltage, *nV*	Power, *fW*
–50	707.11	10.0000	–100	2236.07	100.0000
–51	630.21	7.943282	–101	1992.90	79.432823
–52	561.67	6.309573	–102	1776.17	63.095734
–53	500.59	5.011872	–103	1583.01	50.118723
–54	446.15	3.981072	–104	1410.86	39.810717
–55	397.64	3.162278	–105	1257.43	31.622777
–56	354.39	2.511886	–106	1120.69	25.118864
–57	315.85	1.995262	–107	998.81	19.952623
–58	281.50	1.584893	–108	890.19	15.848932
–59	250.89	1.258925	–109	793.39	12.589254
–60	223.61	1.0000	–110	707.11	10.0000
–61	199.29	0.794328	–111	630.21	7.943282
–62	177.62	0.630957	–112	561.67	6.309573
–63	158.30	0.501187	–113	500.59	5.011872
–64	141.09	0.398107	–114	446.15	3.981072
–65	125.74	0.316228	–115	397.64	3.162278
–66	112.07	0.251189	–116	354.39	2.511886
–67	99.88	0.199526	–117	315.85	1.995262
–68	89.02	0.158489	–118	281.50	1.584893
–69	79.34	0.125893	–119	250.89	1.258925
–70	70.71	0.1000	–120	223.61	1.0000
–71	63.02	0.079433	–121	199.29	0.794328
–72	56.17	0.063096	–122	177.62	0.630957
–73	50.06	0.050119	–123	158.30	0.501187
–74	44.62	0.039811	–124	141.09	0.398107
–75	39.76	0.031623	–125	125.74	0.316228
–76	35.44	0.025119	–126	112.07	0.251189
–77	31.59	0.019953	–127	99.88	0.199526
–78	28.15	0.015849	–128	89.02	0.158489
–79	25.09	0.012589	–129	79.34	0.125893
–80	22.36	0.0100	–130	70.71	0.1000
–81	19.93	0.007943	–131	63.02	0.079433
–82	17.76	0.006310	–132	56.17	0.063096
–83	15.83	0.005012	–133	50.06	0.050119
–84	14.11	0.003981	–134	44.62	0.039811
–85	12.57	0.003162	–135	39.76	0.031623
–86	11.21	0.002512	–136	35.44	0.025119
–87	9.99	0.001995	–137	31.59	0.019953
–88	8.90	0.001585	–138	28.15	0.015849
–89	7.93	0.001259	–139	25.09	0.012589
–90	7.07	0.0010	–140	22.36	0.0100
–91	6.30	0.000794	–141	19.93	0.007943
–92	5.62	0.000631	–142	17.76	0.006310
–93	5.01	0.000501	–143	15.83	0.005012
–94	4.46	0.000398	–144	14.11	0.003981
–95	3.98	0.000316	–145	12.57	0.003162
–96	3.54	0.000251	–146	18.21	0.002512
–97	3.16	0.000200	–147	9.99	0.001995
–98	2.82	0.000158	–148	8.90	0.001585
–99	2.51	0.000126	–149	7.93	0.001259
			–150	7.07	0.0010

TABLE 18.A.2 Conversion between *VSWR* and other reflection and transmission coefficients (R_o = 50 , standard characteristic resistance)

R_o	R	Γ	$VSWR$	$VSWR_{dB}$	RL_{dB}	TL_{dB}	Γ_P	$P_{T,\%}$	$P_{R,\%}$
50.00	2500.00	0.96	50.00	33.98	−0.35	7.69	0.92	7.69	92.31
50.00	1250.00	0.92	25.00	27.96	−0.70	14.79	0.85	14.79	85.21
50.00	1000.00	0.90	20.00	26.02	−0.87	18.14	0.82	18.14	81.86
50.00	750.00	0.88	15.00	23.52	−1.16	23.44	0.77	23.44	76.56
50.00	500.00	0.82	10.00	20.00	−1.74	33.06	0.67	33.06	66.94
50.00	400.00	0.78	8.00	18.06	−2.18	39.51	0.60	39.51	60.49
50.00	300.00	0.71	6.00	15.56	−2.92	48.98	0.51	48.98	51.02
50.00	200.00	0.60	4.00	12.04	−4.44	64.00	0.36	64.00	36.00
50.00	100.00	0.33	2.00	6.02	−9.54	88.89	0.11	88.89	11.11
50.00	90.00	0.29	1.80	5.11	−10.88	91.84	0.08	91.84	8.16
50.00	80.00	0.23	1.60	4.08	−12.74	94.67	0.05	94.67	5.33
50.00	70.00	0.17	1.40	2.92	−15.56	97.22	0.03	97.22	2.78
50.00	60.00	0.09	1.20	1.58	−20.83	99.17	0.01	99.17	0.83
50.00	50.00	0.00	1.00	0.00	Infinite	100.00	0.00	100.00	0.00
50.00	41.50	−0.09	1.20	1.62	−20.64	99.14	0.01	99.14	0.86
50.00	35.60	−0.17	1.40	2.95	−15.48	97.17	0.03	97.17	2.83
50.00	31.20	−0.23	1.60	4.10	−12.71	94.64	0.05	94.64	5.36
50.00	27.80	−0.29	1.80	5.10	−10.89	91.86	0.08	91.86	8.14
50.00	25.00	−0.33	2.00	6.02	−9.54	88.89	0.11	88.89	11.11
50.00	12.50	−0.60	4.00	12.04	−4.44	64.00	0.36	64.00	36.00
50.00	8.33	−0.71	6.00	15.57	−2.92	48.97	0.51	48.97	51.03
50.00	6.25	−0.78	8.00	18.06	−2.18	39.51	0.60	39.51	60.49
50.00	5.00	−0.82	10.00	20.00	−1.74	33.06	0.67	33.06	66.94
50.00	3.33	−0.87	15.00	23.52	−1.16	23.44	0.77	23.44	76.56
50.00	2.50	−0.90	20.00	26.02	−0.87	18.14	0.82	18.14	81.86
50.00	2.00	−0.92	25.00	27.96	−0.70	14.79	0.85	14.79	85.21
50.00	1.00	−0.96	50.00	33.98	−0.35	7.69	0.92	7.69	92.31
50.00	0.00	−1.00	Infinite	Infinite	0.00	0.00	1.00	0.00	100.00

Voltage standing wave ratio:

$$VSWR = \frac{1+\Gamma}{1-\Gamma}, \tag{18.A.5}$$

Voltage standing wave ratio in *dB*:

$$VSWR_{dB} = 20Log(VSWR) \tag{18.A.6}$$

Return loss in *dB*:

$$RL_{dB} = S_{11,dB} = 20Log\Gamma, \tag{18.A.7}$$

Transmission loss in *dB*:

$$TL_{dB} = 10Log(1-\Gamma^2), \tag{18.A.8}$$

Power reflection coefficient:

$$\Gamma_P = \gamma = \Gamma^2, \tag{18.A.9}$$

Transmitted power in %:

$$P_{T,\%} = 100(1 - \Gamma^2), \tag{18.A.10}$$

Reflected power in %:

$$P_{R,\%} = 100\Gamma^2. \tag{18.A.11}$$

18.A.3 Definition of Powers in a Two-Port Block by Signal Flow Graph

Figure 18.A.1 shows the various powers, voltage reflection coefficients, and S parameters in a two-port block and its signal flow graph. There are four powers in a two-port block. According to the signal flow graph as shown in Figure 18.A.1, they are:

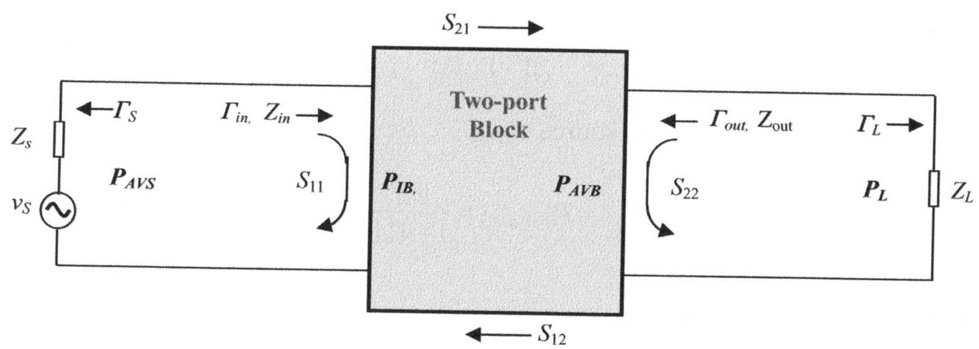

(a) The various powers, voltage reflection coefficients, and S parameters in a two-port block

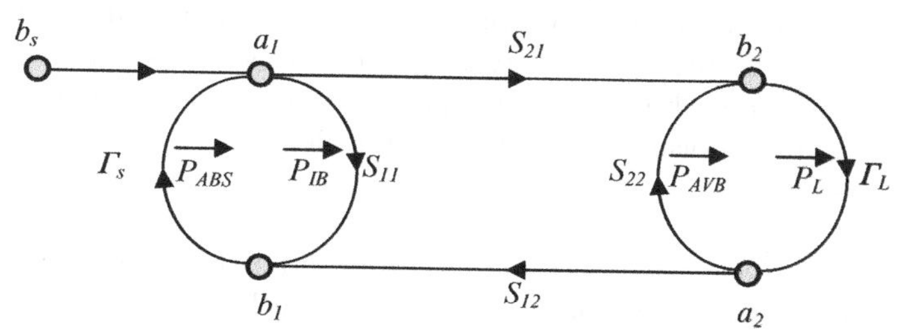

(b) Signal flow graph of a two-port block

Figure 18.A.1 The various powers, voltage reflection coefficients, S parameters, and signal flow graph in a two-port block.

1) Power delivered to the load P_L

$$P_L = \frac{|b_s|^2}{|1-\Gamma_{in}\Gamma_S|^2}|S_{21}|^2\frac{1-|\Gamma_L|^2}{|1-S_{22}\Gamma_L|^2}, \tag{18.A.12}$$

or

$$P_L = \frac{|b_s|^2}{|1-S_{11}\Gamma_S|^2}|S_{21}|^2\frac{1-|\Gamma_L|^2}{|1-\Gamma_{out}\Gamma_L|^2}, \tag{18.A.13}$$

Because

$$\frac{|1-S_{11}\Gamma_S|^2}{|1-S_{22}\Gamma_L|^2} = \frac{|1-\Gamma_{in}\Gamma_S|^2}{|1-\Gamma_{out}\Gamma_L|^2}. \tag{18.A.14}$$

2) Power available from the block, P_{AVB}

$$P_{AVB} = P_L|_{\Gamma_L=\Gamma_{out}^*} = \frac{|b_s|^2}{|1-S_{11}\Gamma_S|^2}|S_{21}|^2\frac{1}{1-|\Gamma_{out}|^2}, \tag{18.A.15}$$

3) Power input to the block P_{IB}

$$P_{IB} = \frac{|b_s|^2}{|1-\Gamma_{in}\Gamma_s|^2}, \tag{18.A.16}$$

4) Power available from the source, P_{AVS}

$$P_{AVS} = P_{IB}|_{\Gamma_{in}=\Gamma_S^*} = \frac{|b_s|^2}{1-|\Gamma_s|^2}, \tag{18.A.17}$$

18.A.4 Main Noise Sources

Any random process of current or voltage in a part or a circuit block results in noise. In other words, noise inevitably exists in a part or a circuit block because in the basic parts, such as the resistor, capacitor, inductor, transistor, and so on, there are many random processes in the motions of electrons and other charge carriers. Consequently, it is found that there are many kinds of noise sources, such as shot noise, thermal noise, flicker noise, burst noise (popcorn noise), and avalanche noise.

There are three main noise sources in a circuit block:

1) Shot noise
A current in a device, part, or runner is composed of a large number of moving carriers, either positive charges or electrons with their random velocities. The product of the electric charge and velocity of a charge or an electron forms a current element. Owing to the different velocities, these current elements are different from each other. The sum of all the current elements at a junction

of semiconductor or a cross-section of a part looks like a large number of current pulses. The average value of these random current pulses is called current and its fluctuation around the average value is called shot noise. According to statistics, the mean-square value of its fluctuation is

$$\overline{i_n^2} = 2qI\Delta f, \tag{18.A.18}$$

where

i_n = current fluctuation, a random variable,
I = average value of current,
q = charge of an electron,
Δf = bandwidth in which noise source is acting.

The special feature of thermal noise is that it is directly related to the DC current I. The higher the DC current flowing through the part, the higher the shot noise that the part has.

2) Thermal noise

Essentially, thermal noise is due to the fluctuation of resistance in a device, a part, or a runner. It is well known that resistance is directly associated with the collision between electrons and between electrons and other charge carriers or particles. Fluctuation of the resistance in parts or devices implies a fluctuation of collision, which is a random process.

The random thermal motion and collision of electrons produce thermal noise, whereas the random drift velocity of the moving charge-carriers or current fluctuation produces shot noise. The velocity of the electron thermal motion is much higher than the drift velocity of the electron and other charge-carriers. Therefore, thermal noise is produced by a completely different mechanism than shot noise. Thermal noise still exists even when shot noise is gone due to a zero current.

According to statistics, the thermal noise of a resistor can be represented by either voltage or current noise source, that is,

$$\overline{e_n^2} = 4kTR\Delta f, \tag{18.A.19}$$

or,

$$\overline{i_n^2} = 4kT\frac{1}{R}\Delta f, \tag{18.A.20}$$

where

e_n = voltage fluctuation,
i_n = current fluctuation,
k = Boltzmann constant,

T = room Kelvin temperature,

R = resistance of resistor,

Δf = bandwidth in which noise source is acting.

The special feature of thermal noise is that its noise power spectrum density is independent of frequency, that is,

$$\frac{\overline{e_n^2}}{4R\Delta f} = kT, \tag{18.A.21}$$

or,

$$\frac{\overline{i_n^2}R}{4\Delta f} = kT. \tag{18.A.22}$$

3) Flicker noise

Flicker noise is due to contamination and crystal defects. The traps capture and release carriers in a random fashion associated with a noise with energy concentrated at low frequencies. Just like the shot noise, the existence of flicker noise is always associated with a direct current. Therefore, a resistor that is not carrying a direct current does not have flicker noise. According to statistics, flicker noise can be represented by a noise current source as follows:

$$\overline{i_n^2} = kI^a \frac{\Delta f}{f}, \tag{18.A.23}$$

where

i_n = current fluctuation,

k = constant for a particular device,

I = direct current,

a = 0.5 to 2,

f = operating frequency,

Δf = operating bandwidth.

REFERENCES

[1] Ralph S. Carson, *High-Frequency Amplifiers*, John Wiley & Sons, Inc., 1975.

[2] R. E. Ziemer and W. H. Tranter, *Principles of Communications*, Houghton Mifflin Company, 1976.

[3] Richard C. Sagers, "Intercept Point and Undesired Responses," *Transactions of IEEE Vehicular Technology*, Vol. VT–32, No. 1, February 1983, pp. 121–133.

[4] B. P. Lathi, *The Design of CMOS Radio-Frequency Integrated Circuits*, Cambridge University Press, 1983.

[5] Robert M. Gagliardi, *Satellite Communications*, Lifetime Learning Publications, 1984.

[6] Guillermo Gonzalez, *Microwave Transistor Amplifiers, Analysis and Design*, Prentice-Hall, Inc., 1984.

[7] Jack Smith, *Modern Communication Circuits*, McGraw-Hill Publishing Company, 1986.

[8] Bernard Sklar, *Digital Communications*, Prentice Hall, 1988.

[9] Lawrence E. Larson, *Modern Digital and Analog Communication Systems*, 2nd ed., Holt, Rinehart and Winston, Inc., 1989.

[10] George D. Vendelin, Anthony M. Pavio, and Ulrich L. Rohde, *Microwave Circuit Design Using Linear and Nonlinear Techniques*, John Wiley & Sons, Inc., 1990.

[11] T. S. Chu, "Intermodulation in CDMA," *Personal, Indoor and Mobile Radio Communications, 1994, Wireless Networks: Catching the Mobile Future, 5th IEEE International Symposium*, Vol. 2, September 18–23, 1994, pp. 595–600.

[12] Thomas H. Lee, *The Design of CMOS Radio-Frequency Integrated* Circuits, Cambridge University Press, 1998.

[13] Keng Leong Fong and R. G. Meyer, "High-frequency Nonlinearity Analysis of Common-emitter and Differential-pair Transconductance Stages," *Solid-State Circuits, IEEE Journal*, Vol. 33, No. 4, April 1998, pp. 548–555.

[14] Sung-Mo (Steve) Kang and Yusuf Leblebici, *CMOS Digital Integrated Circuits*, WCB/McGraw-Hill, Inc., 1999.

[15] E. E. Bautista, B. Bastani, and J. Heck, "A High IIP2 Downconversion Mixer Using Dynamic Matching," *Solid-State Circuits, IEEE Journal*, Vol. 35, No. 12, December 2000, pp. 1934–1941.

[16] K. Kivekas, A. Parssinen, and K. A. I. Halonen, "Characterization of IIP2 and DC-offsets in Transconductance Mixers," *Circuits and Systems II: Analog and Digital Signal Processing, IEEE Transactions*, Vol. 48 , No. 11, November 2001, pp. 1028–1038.

[17] C. R. Iversen and T. E. Kolding, "Noise and Intercept Point Calculation for Modern Radio Receiver Planning," *Communications, IEE Proceedings*, Vol. 148, No. 4, August 2001, pp. 25–259.

[18] A. Geens and Y. Rolain, "Noise Figure Measurements on Nonlinear Devices," *Instrumentation and Measurement, IEEE Transactions*, Vol. 50, No. 4, August 2001, pp. 971–975.

[19] Paul R. Gray, Paul J. Hurst, Stephen H. Lewis, and Robert G. Meyer, *Analysis and Design of Analog Integrated Circuits*, 4th ed., John Wiley & Sons, Inc., 2001.

[20] E. Allen Phillip and Douglas R. Holberg, *CMOS Analog Circuit Design*, 2nd ed., Oxford University Press, 2002.

[21] Min Lin, Yongming Li, and Hongyi Chen, "A Novel IP3 Boosting Technique Using Feedforward Distortion Cancellation Method for 5 GHz CMOS LNA," *Microwave Symposium Digest, 2003 IEEE MTT-S International*, Vol. 1, June 8–13, 2003, pp. A185–A188.

[22] A. R. Petrov, "System Approach for Low 1/f Noise, High IP2 Dynamic Range CMOS Mixer Design," *University/Government/Industry Microelectronics Symposium, Proceedings of the 15th Biennial*, June 30–July 2, 2003, pp. 74–77.

[23] A. A. Abidi, "General Relations between IP2, IP3, and Offsets in Differential Circuits and the Effects of Feedback," *Microwave Theory and Techniques, IEEE Transactions*, Vol. 51, No. 5, May 2003, pp. 1610–1612.

[24] D. Manstretta, M. Brandolini, and F. Svelto, "Second-order Intermodulation Mechanisms in CMOS Downconverters," *Solid-State Circuits, IEEE Journal*, Vol. 38, No. 3, March 2003, pp. 394–406.

[30] H. A. Haus et al., "Representation of Noise in Linear Twoports," *Proceedings of the IRE*, Vol. 48, January 1960, pp. 69–74.

INDEX

Printed and bound by CPI Group (UK) Ltd, Croydon, CR0 4YY
16/09/2026
14963463-0007